Mechanisms of Photophysical Processes and Photochemical Reactions in Polymers

'The essence of knowledge is,
having it, to apply it.'

Confucius
(551–479 BC)

Mechanisms of Photophysical Processes and Photochemical Reactions in Polymers

Theory and Applications

Jan F. Rabek

Department of Polymer Technology
The Royal Institute of Technology
Stockholm, Sweden

JOHN WILEY & SONS

Chichester · New York · Brisbane · Toronto · Singapore

CHEMISTRY

Copyright © 1987 by John Wiley & Sons Ltd.

Library of Congress Cataloging-in-Publication Data:

Rabek, J. F.
  Mechanisms of photophysical processes and photochemical reactions in polymers.

  Bibliography: p.
  Includes index.
  1. Polymers and polymerization.   2. Photochemistry.
  I. Title. II. Title: Photophysical and photochemical reactions in polymers.
QD381.9.P56R33   1987   547.7'0455   86–15693
ISBN 0 471 91180 1

British Library Cataloguing in Publication Data:

Rabek, J. F.
  Mechanisms of photophysical processes and photochemical reactions in polymers: theory and applications.
  1. Polymers and polymerization
  I. Title
  547.7'0455   QD381

  ISBN 0 471 91180 1

Type set by Macmillan India Ltd, Bangalore 25
Printed and bound in Great Britain

Dedicated to my wife Ewelina
the best friend in my life

Preface

This book has been written in an attempt to cover the remarkable progress made in past decade in the field of polymer photophysics and photochemistry.

The book has a twofold purpose. Essentially it presents, in organized form, most of all available knowledge on the photophysical and photochemical reactions in polymers. In addition, however, it attempts to gather some of the important practical and industrial applications of these reactions.

From my own 25 years experience in experimental work in the photochemistry of polymers, I have learned that the most interesting and productive areas of science are at the interfaces of the disciplines of physics and chemistry. An understanding of the mechanisms of photophysical and photochemical reactions in polymers provides an important background for all polymer chemists involved in the study of photopolymerization, photocrosslinking, photodegradation, photostabilization, solar energy conversion processes, and also for polymer photophysicists, polymer photobiochemists, and people working with photocuring, photoresists, and photostabilizers. This book should be a useful aid to all of these people, but also to graduate or advanced undergraduate students.

As a monograph this book lists 4000 references including older fundamental studies and those of recent years, up to January 1985. The space limitations and the tremendous number of publications in the past decade have made a detailed presentation of all important results difficult. I apologize to those whose work has not been quoted or widely presented in this book. The chemical nomenclature of organic compounds and polymers in this book is the same as was used by authors in their original publications. In spite of the recommendation given by the IUPAC Commission on the Nomenclature of Organic Chemistry, I did not make any change in order to avoid some difficulties in using and reading of original papers. My task was limited to using the same nomenclature throughout the book.

Since the mechanisms of photophysical and photochemical reactions in polymers are initiated by absorption of ultraviolet or visible radiation, a brief introduction to the electronic states and transitions in molecules was necessary. The text is further designed to introduce the number of reactions involved in the initiation of processes which occur as a result of absorbed light. Most of the photochemical reactions are followed by radical (or radical-ion) reactions. My attempt was to gather these important mechanisms into certain arbitrary but hopefully systematic groups.

Photophysical reactions which include the formation and deactivation of electronically excited states and electronic energy transfer processes are fundamental for the further course of photochemical reactions. Practical aspects of the photophysical processes have been described in Chapter 4, Application of Luminescence Spectroscopy in Polymers. Of course an understanding of the various energy states available to polymer molecules, or to systems of reacting molecules, is the province of spectroscopists rather than polymer photochemists.

The is an effort to integrate this vitally needed information on mechanisms of photophysical and photochemical reactions in polymers in a single book written as a monograph of the subject.

In conclusion I should like to express my gratitude for the patience of members of my family who, during the time in which this book was being prepared, have had to forgo my company during innumerable evenings, weekends and holidays.

J. F. Rabek

Acknowledgments

Grateful acknowledgments are due to the following for permission to use published data. The first number in each of the following sets indicates the chapter. References to the authors concerned are made in the text.

Academic Press: Figs. 16.1, 16.2, 16.8, 16.13.

Akademische Verlagsgesellschaft: Fig. 3.38.

American Chemical Society: Figs. 3.6, 5.2, 6.3, 10.1, 12.7, 14.2, 14.3, 14.4, 14.5, 15.4, 15.5, 15.6, 15.7; Tables 5.5, 5.7, 5.8, 5.9, 7.14, 12.6, 14.2, 15.3.

American Institute of Physics: Fig. 3.30; Tables 15.10, 15.11.

Butterworth & Co. Ltd: Figs. 3.9, 3.11, 3.12, 3.16, 3.29, 3.34, 7.11, 10.4

J. G. Calvert and J. N. Pitts: Figs. 7.3, 7.6, 7.12; Table 7.12.

Marcel Dekker Inc.: Tables 8.1, 8.2, 8.3, 8.6.

Elsevier: Figs. 3.17, 4.11, 4.12, 4.13, 4.14, 13.1, 14.9, 14.10, 14.11, 14.12, 14.20, 14.21; Tables 15.1, 15.8.

IUPAC: Figs. 3.5, 3.39, 9.4, 15.8; Tables 3.7, 3.8, 12.3, 12.7, 12.8, 12.9.

Journal of Oil and Colour Chemists' Association: Figs. 7.12, 15.3.

Makromolekulare Chemie: Figs. 3.32, 3.33, 14.7, 14.8, 16.7; Table 7.22.

New York Academy of Sciences: Figs. 2.7, 3.15, 3.31, 3.36.

North-Holland Physics Publishing: Fig. 3.25.

Open University Press: Figs. 16.4, 16.5, 16.6.

Optical Society of America: Figs. 3.2, 16.18.

Pergamon Press: Figs. 3.13, 3.14, 3.37, 4.15, 4.17, 4.18, 6.17, 7.10, 14.1, 14.6, 14.14, 14.16, 14.17, 15.10, 16.10; Tables 15.4, 15.6, 16.2, 16.16.

Royal Society of Chemistry: Table 4.6.

Society of Photographic Scientists and Engineers: Figs. 12.5, 12.6, 15.2, Tables 7.1, 7.5, 7.6, 7.7, 7.8, 12.4.

Society of Plastics Engineers: Fig. 12.8.

Springer-Verlag: Figs. 9.3, 10.3, 12.9, 12.10, 16.15.

Technology Marketing Corporation: Figs. 7.14, 14.23; Tables 7.20, 13.1, 13.2, 13.4, 13.5, 15.2.

VCH Verlagsgesellschaft: Fig. 16.11.

Contents

1. Electronically excited states

1.1. NATURE AND PROPERTIES OF LIGHT

From the definition the word *light* should be used for optical radiation perceived and evaluated by the human eye. Light is psychophysical, being neither purely physical nor purely psychological.

OPTICAL RADIATION is a part of electromagnetic radiation (Fig. 1.1), and can be divided into different ranges and groups (Table 1.1). The ultraviolet (u.v.) and infrared (i.r.) ranges are additionally divided into subgroups A, B, and C. The visible light is divided into the relevant colours.[1601,3023]

ELECTROMAGNETIC RADIATION is a form of energy which travels through space unaccompanied by any matter. The behaviour of electromagnetic

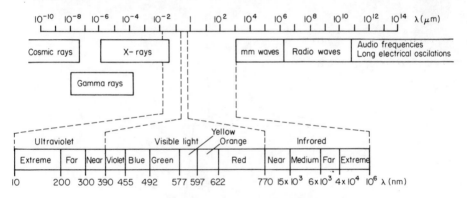

Fig. 1.1   Electromagnetic radiation spectrum

Table 1.1   Subdivision of the optical radiation spectrum

Wavelength range	Designation of radiation
100– 280 nm	UV-C
280– 315 nm	UV-B
315– 380 nm	UV-A
380– 440 nm	Light—violet
440– 495 nm	Light—blue
495– 558 nm	Light—green
580– 640 nm	Light—yellow
640– 750 nm	Light—red
750–1400 nm	IR-A
1.4– 3 μm	IR-B
3 –1000 μm	IR-C

1

radiation can be attributed either to its wave-like character or to its corpuscular character. Figure 1.2 shows a plane-polarized electromagnetic wave of a single frequency, i.e. a MONOCHROMATIC BEAM. If an electromagnetic wave is PLANE-POLARIZED the electric vector (**E**) vibrates in a single plane and the magnetic field vector (**H**) vibrates in another plane perpendicular to the electric field. In practice most of electromagnetic radiation is UNPOLARIZED, i.e. has electric and magnetic vectors at all orientations perpendicular to the direction of propagation. Different kinds of electromagnetic radiation are usually characterized by either the wavelength (λ) or the frequency (v).[3023]

WAVELENGTH (λ) is defined as the length of the cycle or the distance between successive maxima or minima (Fig. 1.2). The following are the units of wavelength commonly used:

$$1 \ \mu m = 10^{-6} \ m = 10^{-4} \ cm = 10{,}000 \ \text{Å} \quad (\mu m = \text{micrometre})$$

$$1 \ nm = 10^{-9} \ m = 10^{-7} \ cm = 10 \ \text{Å} \qquad (nm = \text{nanometre})$$

$$1 \ \text{Å} \ = 10^{-10} \ m = 10^{-8} \ cm \qquad\qquad (\text{Å} = \text{angstrom})$$

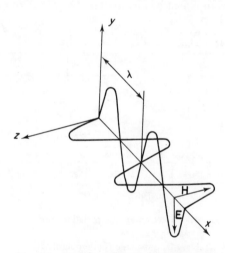

Fig. 1.2 A plane-polarized electromagnetic wave of a single frequency. **E** is the electric vector and **H** is the magnetic vector

FREQUENCY (v) is the number of cycles per unit time (in cycles per second or hertz, where $1 \ \text{Hz} = 1 \ \text{cycle s}^{-1}$):

$$v = \frac{1}{t} \tag{1.1}$$

where t is time it takes for one complete wave to pass.

The wavelength and the frequency are related by:

$$\lambda = \frac{v}{v} = \frac{c}{v} \tag{1.2}$$

where v is the velocity of propagation. All electromagnetic radiation (light) travels through a vacuum with the same velocity (c), which has the value 2.9979×10^8 m s^{-1}. The frequency (v) is the only true characteristic of a particular radiation; both the velocity (v) and the wavelength (λ) depend on the nature of the medium through which the electromagnetic wave travels.

The WAVENUMBER ($\bar{v}$) is the number of waves per unit length (numbers of waves per centimetre):

$$\bar{v} = \frac{1}{\lambda} = \frac{v}{c} \tag{1.3}$$

The units of wavenumber and wavelength are related as follows:

$$cm^{-1} = \frac{1}{\mu m} \times 10^4 \tag{1.4}$$

Typical values for the optical spectrum such as wavelength (λ), frequency (v) and wavenumber ($\bar{v}$) are given in Table 1.2.

Table 1.2 Typical values of the optical spectrum

	X-rays	Vacuum ultra-violet	Near-ultra-violet	Visible	Near-infrared		Far-infrared
λ, Å	1	2000	4000	7000	10,000		
nm	0.1	200	400	700	1000		
μm			0.4	0.7	1	50	500
v, Hz		1.5×10^{15}	7.5×10^{14}		3×10^{14}	6×10^{12}	6×10^{11}
$\bar{v}$, cm^{-1}		50,000	25,000	14,300	10,000	200	20

The PHOTON or QUANTUM of RADIANT ENERGY is a quantized form of electromagnetic wave. The energy of a photon is given by the PLANCK EQUATION:

$$E = hv = \frac{ch}{\lambda} = ch\bar{v} \tag{1.5}$$

where:

h = Planck's constant, which has the value = 6.626×10^{-34} Js,

v = frequency of a photon (s^{-1}),

λ = wavenumber of a photon (m^{-1}),

c = electromagnetic radiation velocity (m s^{-1}).

The energy of a photon can be calculated in various energy units using the following equations in which λ is expressed in Å:

$$E = \frac{2.865 \times 10^5}{\lambda} \text{ (kcal mol}^{-1}) = \frac{1.24 \times 10^4}{\lambda} \text{ (eV)} = \frac{1.986 \times 10^{-15}}{\lambda} \text{ (J)} \tag{1.6}$$

The amount of energy equal to that of 1 mol of photons (6.02×10^{23} photons) is called an EINSTEIN. Thus an einstein of 3000 Å radiation is given by:

$$E = \frac{2.865 \times 10^5}{3000} = 95.5 \text{ kcal} \tag{1.7}$$

1.2. FORMATION OF THE ELECTRONICALLY EXCITED STATES

The PHOTOCHEMICAL REACTION occurs by the activation of a molecule provided by absorption of a photon of light by the system. A photochemical reaction can be divided into three stages:[497, 591, 767, 2196, 2600, 2664, 3678, 3680, 3815]

(i) the ABSORPTION step, which produces an electronically excited state;
(ii) the PRIMARY PHOTOCHEMICAL PROCESSES, which involve electronically excited states;
(iii) the SECONDARY or DARK (thermal) REACTIONS of the various chemical species produced by the primary processes.

Important information on these reactions in the field of organic and polymer chemistry can be find in many reviews and books.[18, 314, 438, 573, 769, 1156, 1429, 1434, 1435, 1441, 2073, 2273, 2469, 2583, 2850, 2935, 2969, 3060, 3212, 3216, 3228, 3429, 3678, 3680, 3798, 3815, 3889]

The nature of photochemical activation differentiates it from thermal activation. Absorption of a photon (quantum) of light can specifically excite (activate) a particular bond or group in a given molecule. Use of the proper frequency of exciting light allows activation of a solute in the presence of a large excess of a transparent solvent. Thermal activation of the same molecule or a particular bond can only be achieved by an increase in the over-all molecular energy of the environment.

The PRIMARY PROCESSES of a photochemical reaction include:

(i) the initial ABSORPTION STEP;
(ii) all of the PRIMARY PHOTOCHEMICAL PROCESSES, which involve an electronically excited state of the absorbing molecule.

When a molecule absorbs electromagnetic radiation (light), its energy increases by an amount equal to the energy of the absorbed photon (E):

$$E = E_2 - E_1 = h\nu \tag{1.8}$$

where:

E_2 and E_1 = energies of a single molecule in the final (excited) and initial (ground) states, respectively;
h = Planck's constant;
ν = frequency of radiation.

All photochemical reactions involve electronically excited states (SINGLET and/or TRIPLET excited states). Each excited state has:

(i) definite energy,

(ii) lifetime,

(iii) structure.

In addition, the excited states are different chemical entities from the ground singlet electronic state and are expected to behave differently.

1.3. ABSORPTION OF RADIATION

The absorption process can be written as an elementary reaction:

$$M + h\nu \rightarrow M^* \tag{1.9}$$

where (M^*) means a molecule in an excited state. The rate of absorption (I_a) of the amount of photons per unit volume is given by equation:

$$I_a = -\frac{d[M]}{dt} = \frac{d[M^*]}{dt} \tag{1.10}$$

I_a has dimension (amount of photon) (volume)$^{-1}$ (time)$^{-1}$.

When a parallel beam of monochromatic light of intensity (I_0) passes through a homogeneous absorbing substance, the intensity of the transmitted beam (I_t) is given by the BEER–LAMBERT LAW:

$$I_t = I_0 \, 10^{-\varepsilon cl} \tag{1.11}$$

where:

ε = molar absorptivity (molar extinction coefficient),

c = concentration of the absorbing species,

l = path length traversed by the beam.

The absorbed intensity (I_a) is given by equation:

$$I_a = I_0 - I_t = I_0 \, (1 - 10^{-\varepsilon cl}) \tag{1.12}$$

In photochemical studies I_a, I_0 and I_t, which have dimensions (energy)(time^{-1}), are usually measured in units of amount of photons per second. Thus the rate of absorption expressed by equation 1.9 can be calculated from the absorbed intensity I_a (equation 1.12) and a knowledge of the volume irradiated.

The ABSORBANCE (referred also as OPTICAL DENSITY (D) or EXTINCTION (E)) is given by:

$$A = \varepsilon cl \tag{1.13}$$

Instead of absorbance TRANSMITTANCE is often measured ($T(\%)$).

$$A = -\log_{10} T = \log_{10}(100/T) \tag{1.14}$$

Figure 1.3 presents the initial distribution of excited molecules (M^*) in a reaction vessel for various values of the absorbance (A), which is far from homogeneous:

6

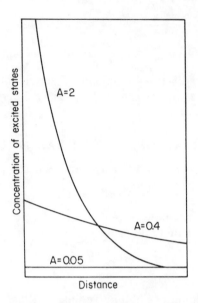

Fig. 1.3 Concentration of excited states
(M*) as a function of distance within an
absorbing medium for various values of the
absorbance A

(i) The concentration of excited molecules is homogeneous to within 5 % for values of $\varepsilon cl < 0.02$.

(ii) If $\varepsilon cl \ll 1$, the I_a can be expressed as $I_a \approx 2.303\ \varepsilon cl I_0$.

The MOLAR ABSORPTIVITY (ε), which is an experimental measure of the probability of absorption at a particular frequency, is constant for a particular compound at a given wavelength. When more than one absorbing substance is present in a homogeneous mixture:

$$I_t = I_0\ 10^{-\varepsilon_i c_i l} \tag{1.15}$$

and the fraction absorbed by the jth species is given by

$$\frac{I_{a(j)}}{I_0} = \frac{\varepsilon_j c_j}{\sum_i \varepsilon_i c_i}\ (1 - 10^{\varepsilon_i c_i l}) \tag{1.16}$$

The ABSORBANCE is a cumulative property for a mixture of two or more absorbing substances:

$$A = 1(\varepsilon_1 c_1 + \varepsilon_2 c_2 + \ldots + \varepsilon_n c_n) \tag{1.17}$$

The concentration of two absorbing substances may be determined when four values of the molar absorptivity (ε) are known, and when the measurements are made for two wavelengths:

$$A' = 1(\varepsilon_1' c_1 + \varepsilon_2' c_2) \quad \text{for } \lambda' \tag{1.18}$$

$$A'' = 1(\varepsilon_1'' c_1 + \varepsilon_2'' c_2) \quad \text{for } \lambda'' \tag{1.19}$$

The wavelengths λ' and λ'' are chosen such that one component absorbs these wavelengths strongly, whereas the other component absorbs them much less (Fig. 1.4).

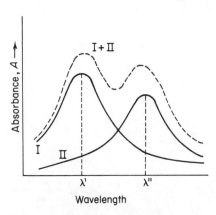

Fig. 1.4 Overlapping absorption spectra of two components I and II and the spectrum of a mixture of the two components I + II

Electronic absorption spectra (Fig. 1.4) arise from the absorption of radiation in the ultraviolet (u.v.) or visible regions of the spectrum (Fig. 1.1) which causes transition between electronic states. The absorption of radiation occurs only if the difference between two energy levels is exactly to the energy of a quantum (cf. equation 1.8). For the absorption of radiation CHROMOPHORES are most responsive, i.e. functional groups which contain electrons originating from π and n orbitals (Table 1.3) (cf. Section 4.9.4.2.1).

Table 1.3 Typical chromophores and their characteristics

Chromophore	Wavelengh, λ_{max} (nm)	Molar absorptivity, ε_{max}
C=C	175	14,000
	185	8000
C≡C	175	10,000
	195	2000
	223	150
C=O	160	18,000
	185	5000
	280	15
C=C–C = C	217	20,000
	184	60,000
	200	4400
	255	204

8

There are three main types of orbital involved in an electronic transition (Fig. 1.5):[591, 3060, 3131, 3678, 3680]

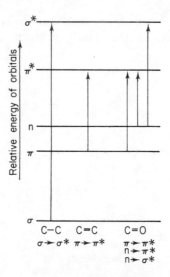

Fig. 1.5 Relative energies of molecular orbitals and different types of electronic transitions involved in electronic spectroscopy

(i) π Orbital. The electron originates from a bonding π orbital and is promoted to a π anti-bonding orbital (π*) of higher energy. The transition is noted as (π, π*). Removal of an electron from the bonding orbital in the ground state will weaken the bond to an extent depending on whether the π orbital is localized.

(ii) σ Orbital. The electron originates on a bonding σ orbital, being promoted into a σ* orbital. As σ orbitals are generally of lower energy than π orbitals, this will require absorption of a quantum of a higher energy than the (π, π*) transition. Removal of the electron from the bonding σ orbital may result in a considerable weakening of the bond and its dissociation into free radicals. Transitions of this type occur in saturated organic molecules and macromolecules, and usually occur under irradiation below 200 nm.

(iii) n Orbital. The electron from an n (non-bonding) orbital may be promoted to a π* orbital (n, π*) or σ* orbital (n, σ*). The removal of this electron from the non-bonding orbital has less effect on the bonding. Since non-bonding orbitals have relatively high energies, lower-energy quanta are required for this type of transition.

Some of the most important transitions are n → π*, π → π*, σ → π*, n → σ*, σ → σ*, etc., given here roughly in order of increasing energy. The excited states formed by these promotions which have the separated electrons with spins paired will be singlet states and they can be labelled ^{1}(n, π*) and ^{1}(π, π*) states, whereas those which have the separated electrons with parallel spins will be triplet state and are labelled as $^3(n, \pi*)$ and $^3(\pi, \pi*)$ states.

The polarity of the liquid solvent (or solid polymer matrix) plays a very important role in the SOLVATOCHROMIC EFFECTS. The polarity of solvent determines to a large extent the relative energies of electronic states of dipolar solute molecules.[579] This effect can in some cases be so large that inversion of electronic states occurs in different solvents with related changes in the photochemical properties of the solute.

1.4. ENERGY LEVEL DIAGRAMS

Absorption and emission processes can be easily described by Jablonsky or Franck–Condon energy level diagrams:[573, 591, 3678, 3680]

(i) The ENERGY LEVEL JABLONSKY-TYPE DIAGRAM of a single electron within an atom is shown in Fig. 1.6. The GROUND STATE OF THE ATOM (GROUND SINGLET ELECTRONIC STATE) is indicated by S_0 and its successive EXCITED ELECTRONIC STATES (SINGLETS (S) or TRIPLETS (T)) by E_1* (S_1 or T_1), E_2* (S_2 or T_2). Higher excited states of the electron differ successively, as shown, by progressively smaller increments of energy.

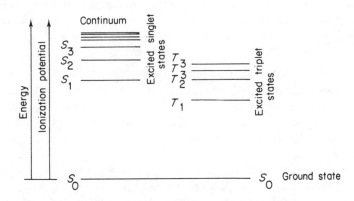

Fig. 1.6 Jablonsky diagram of energy levels of an atomic electron

The UPPER LIMIT OF THE ENERGY LEVELS of the molecule is determined by the energy required for IONIZATION (the electron escapes altogether from the atom and may then exist at a continuum of energy levels) or for rupture (DISSOCIATION) of a chemical bond between nuclei.

In the case of a regular array of molecules—that is, of crystalline or paracrystalline structures, certain additional types of energy level exist. Electrons may acquire energies that are sufficient for their dissociation from specific molecules and yet remain in association with the array as a whole. There exists a narrow continuum of energies which constitute the EXCITON BAND of an array (Fig. 1.7). A somewhat more extensive continuum of energies constitutes the CONDUCTION BAND (Fig. 1.7). Electrons with energies in the conduction band may conduct electric current within a crystal.

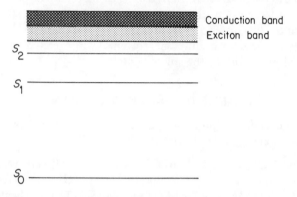

Fig. 1.7   Jablonsky diagram of energy levels in the case of regular array of some molecules

Each electronic state is split into a series of vibrational levels, and each vibrational level is itself split into a series of rotational levels (Fig. 1.8). Overlapping of vibrational energy levels of different excited electronic states is a common feature. It occurs whenever the ground vibrational levels of the respective electronic states are of comparable energy.

(ii) The FRANCK-CONDON DIAGRAM (Fig. 1.9) emphasizes more the physical significance of the vibrational energy levels, and is a plot of the energy content of a diatomic molecule as a function of the separation of the two nuclei. Thus, while the molecule is at any of the quantized vibrational levels, the energy content is constant and the interatomic separation (r) oscillates between two extreme values.

In the Franck–Condon diagram, 1 and 1', as shown in the lowest vibrational level, are extreme interatomic separations corresponding to the energy content E_1. Higher vibrational levels correspond to larger total energy content and to more extensive oscillations of the internuclear distance. The spacing of the levels (2–2', 3–3', 4–4') is very close. The envelope of all the quantized vibrational levels is the continuous curve AOA'. The left side of the curve tends to become a vertical line at high energy levels. The maximum separation of nuclei becomes increasingly great at high energy levels. Ultimately the two nuclei become separated, i.e. the chemical bond between the nuclei is broken. The right side of the curve thus becomes a horizontal line at high energy levels. The lower point (O) of the curves represents the limiting energy of the molecule in the absence of any vibration, a condition which, hypothetically, would occur at 0° absolute temperature. Vibrational level 1–1' is, however, the vibrational ground level of the molecule at the temperature to which the diagram applies.

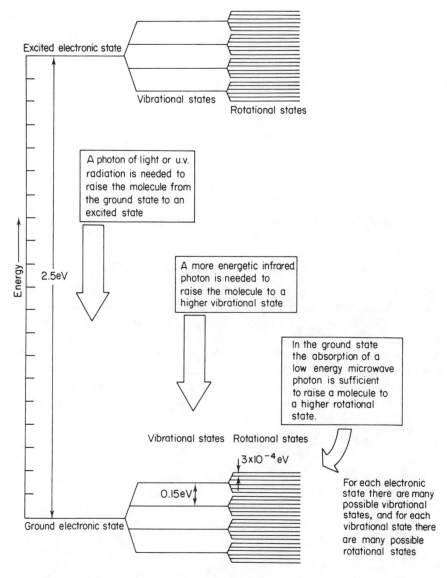

Fig. 1.8 Diagram showing that a molecule can occupy only discrete energy levels

The Franck–Condon diagram (Fig. 1.9) applies strictly to diatomic molecules only. The vibrational levels of a triatomic molecule could be represented by an analogous three-dimensional diagram, but polyatomic molecules in which many internuclear distances must be considered cannot be graphically represented thus. In general, a molecule containing n atoms would require a hypothetical n-dimensional diagram.

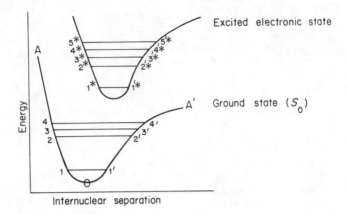

Fig. 1.9 Franck–Condon diagram for two electronic states of a molecule

1.5. EXCITED SINGLET AND TRIPLET STATES

Most of the molecules exists in the GROUND SINGLET ELECTRONIC STATE (S_0), a state in which orbitals contain two spin paired electrons (Fig. 1.10).

EXCITED SINGLET STATES ($S_1, S_2, S_3, \ldots, S_i$) are formed after absorption of the photon (Fig. 1.11). In this process one of electrons is shifted into a higher orbital, but spins off, electrons are still paired (Fig. 1.10). The ABSORPTION SPECTRUM of a molecule provides information concerning the lifetimes, energies, and electronic configurations of the excited singlet states. For reactions in condensed phases, the LOWEST EXCITED SINGLET STATE

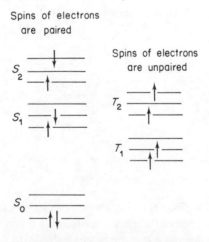

Fig. 1.10 Representation of spins of electrons in the ground (S_0), excited singlet (S_1 and S_2), and excited triplet (T_1 and T_2) states

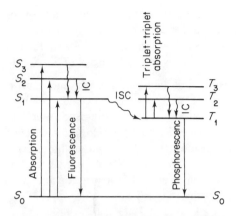

Fig. 1.11 Jablonsky diagram representing the most important photophysical processes. Radiative and radiationless transitions are indicated with straight (→) and wavy (⤳) arrows, respectively. IC is an internal conversion process; ISC is an intersystem crossing process

(S_1) is a state from which photochemical reactions occur. The very fast rate of internal conversion (cf. Section 1.7) from upper singlet states $(S_2, S_3, \ldots)$ to the lowest excited singlet state (S_1) makes photochemical reaction unlikely from the upper states.[573, 591, 3678, 3680]

The LOWEST EXCITED TRIPLET STATE (T_1) is formed mainly by radiationless transition called INTERSYSTEM CROSSING (cf. Section 1.7) from the lowest excited singlet state (S_1) (Fig. 1.11). The formation of a triplet state by direct absorption of a photon by a molecule in its ground singlet electronic state (S_0) is a spin-forbidden transition. The spins of electrons are unpaired in the triplet states (Fig. 1.10).

The HIGHER TRIPLET STATES $(T_2, T_3, \ldots, T_i)$ may be formed only when a molecule in its lowest triplet state (T_1) absorbs a new photon (TRIPLET–TRIPLET ABSORPTION) (Fig. 1.11).

The excitation energy of a molecule in its excited state may be dissipated by the following processes:

(i) radiative processes: luminescence (fluorescence and phosphorescence),
(ii) radiationless processes,
(iii) bimolecular deactivation processes (energy transfer processes),
(iv) dissociation processes.

1.6. RADIATIVE TRANSITIONS

An electronically excited molecule can lose its excitation energy by emission of radiation, which is known as LUMINESCENCE. There are two main kinds of luminescence:[314, 573, 591, 1137, 2109, 2196, 2896, 3228, 3678, 3680, 3889]

(i) FLUORESCENCE, which is a spin-allowed radiative transition between two states of the same multiplicity ($S_1 \rightarrow S_0$) (Fig. 1.11).

(ii) PHOSPHORESCENCE, which is a spin-forbidden radiative transition between two states of different multiplicity ($T_1 \rightarrow S_0$) (Fig. 1.11).

These radiative transitions occur between electronic states of different energy.

The wavelength of fluorescence emission extends to longer values than those absorbed. In many compounds some overlap can be observed between the shorter wavelength of fluorescence emission and the longer wavelengths absorbed by the same molecule. Relationships among absorption, fluorescence, and phosphorescence spectra are shown in Fig. 1.12. When spectra are plotted on a frequency scale (rather than a wavelength scale) the absorption and fluorescence spectra are mirror-images of each other.

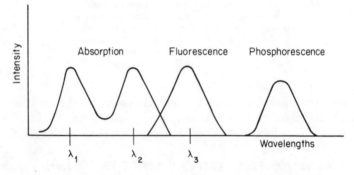

1.12   Relationship between absorption and emission spectra

Figure 1.12 shows an absorption spectrum of a given compound which has two absorption maxima at λ_1 and λ_2. These maxima correspond, respectively, to excitation to two different excited singlet states, S_2 and S_1. Employing the Franck–Condon diagram it is easy to show relations between absorption and fluorescence processes (Fig. 1.13). This diagram represents the absorption energy at the longer of the two wavelengths (λ_2) shown in Fig. 1.12, and of the subsequent fluorescent decay of the excited singlet state of the molecule. As shown, photon absorption raises the molecule at the most probable internuclear radius to an excited vibrational level (in this case, to the level 4*–4'* of the excited singlet state (S_1)). After absorption the molecule undergoes many cycles of vibration during its excited lifetime. Vibrational relaxation by quantum losses of vibrational energy thus take place. Fluorescent decay to the ground singlet electronic state (S_0) then occurs after a total excited lifetime which is of the order of 10^{-8} s. At the time of decay the molecule has reached the lowest vibrational level of the excited singlet state (S_1), in which, as shown, the internuclear separation oscillates between extreme values of r' and r''. The wavelength of the MAXIMUM FLUORESCENCE INTENSITY is λ_3, corresponding to decay from the most probable internuclear separation in the excited states.

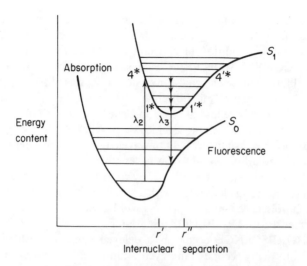

Fig. 1.13 Franck–Condon diagram representing absorption
and fluorescence

Figure 1.14 shows the absorption of light at λ_1, the shorter of the two wave-lengths of maximum absorption marked in Fig. 1.12. The molecule is excited, by this absorption, to the excited singlet state (S_2). Decay of vibrational energy proceeds from the level v^{**}–v'^{**} to which the molecule is excited. A range of values of the internuclear separation (between r_1 and r_2) is common to both the

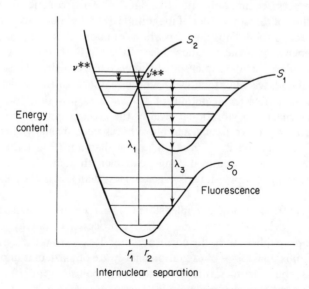

Fig. 1.14 Franck–Condon diagram representing fluorescence by
a molecule following excitation to a higher excited state (S_2)

excited singlet states (S_1 and S_2). Molecules decay to an excited electronic state of lower energy (S_1) during the process of vibrational relaxation. A change in electronic state at a condition of constant internuclear separation (called the INTERNAL CONVERSION PROCESS) is achieved without the emission of a photon. The lifetime of the internal conversion process is of the order of 10^{-13} s, i.e. shorter than the period of molecular vibration. The subsequent decay of vibrational energy and the loss of electronic energy by emission of λ_3 and other wavelengths are identical with those which occur when S_1 is directly excited. The fluorescence spectrum excited by light of wavelength λ_1 is thus identical with that emitted upon irradiation by light of wavelength λ_2.

If the excited singlet state is populated by way of a strongly allowed transition (compound has a strong absorption with a molar absorptivity (ε_{max}) of 10^5) then the reverse fluorescence emission process will be very short-lived. Such fluorescence is termed PROMPT FLUORESCENCE.

PHOSPHORESCENCE occurs as a delayed emission with lifetime from about 10^{-3} s to many seconds (and even minutes). Phosphorescence occurs at longer wavelengths than fluorescence (Fig. 1.12).

Direct excitation of a molecule to the excited triplet state by absorption of a photon is improbable. The energy levels of excited singlet and triplet states are often very close, however, so that singlet and triplet states may share certain values of the internuclear separation. Processes of internal conversion then occur with a reasonably high probability. The process of internal conversion between states of different multiplicity is termed INTERSYSTEM CROSSING.

Whereas INTERNAL CONVERSION between states of identical multiplicity occurs during a time of about 10^{-13} s, INTERSYSTEM CROSSING, which is a forbidden process, occurs only after 10^{-7}–10^{-8} s. The frequency of intersystem crossing is thus of the same order of magnitude as that of fluorescent decay from the singlet excited state. Thus the proportions of fluorescent and phosphorescent emission depend upon the specific values of the life-times in any given system. Once a molecule enters the triplet state by intersystem crossing, it loses energy by vibrational decay, reaching the lowest vibrational level of the triplet state (T_1). From there it may, after a lifetime of 10^{-4} s, decay by radiative transitions (phosphorescence) to a vibrational level of the ground state. The processes of absorption, intersystem crossing, and phosphorescence are shown in a Jablonski diagram (Fig. 1.11) and in a Franck–Condon diagram (Fig. 1.15). The phosphorescence emission occurs at longer wavelength than does fluorescence, because the energy of triplet states is in general less than that of the corresponding singlets.

While decay from the lowest triplet state (T_1) may occur by phosphorescence, this process is delayed, so that non-radiative processes of energy loss become much more probable than in the corresponding decays from the excited singlet state (S_1). Collisional losses of excitation energy are of particular importance. In solutions at room temperature intermolecular collisions are frequent, and normally account almost completely for energy losses from triplet states. Thus phosphorescence is rarely observed at room temperature, although it may be

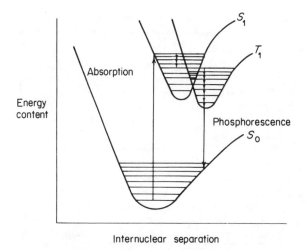

Fig. 1.15 Franck–Condon diagram representing absorption,
intersystem crossing, (ISC) and phosphorescence

observed from many compounds at low temperatures. Phosphorescent emission
is usually studied at liquid nitrogen temperature.

1.7. RADIATIONLESS TRANSITIONS

RADIATIONLESS TRANSITIONS occur between different electronic states
and are induced by molecular or crystal vibrations. There are two types of
radiationless transitions (Fig. 1.11):[591, 3678, 3680]

(i) INTERNAL CONVERSION (IC) is the spin-allowed radiationless transition between two states of the same multiplicity: $(S_i \to S_1$ and $T_i \to T_1)$.
(ii) INTERSYSTEM CROSSING (ISC) is the spin-forbidden radiationless
transition between two states of the different multiplicity: $(S_1 \to T_1)$.

The phosphorescence emission depends to a large extent on the population of the
triplet state from the excited singlet state (S_1) which occurs by the intersystem
crossing process.

1.8. LIFETIME OF PHOTOPHYSICAL PROCESSES
INVOLVED IN ELECTRONICALLY
EXCITED STATES

Data on lifetime, or reciprocal rates, in solution of photophysical processes
involving electronically excited states (Fig. 1.11) are collected in Table
1.4.[591, 2196, 3679, 3680] The lifetime of excited singlet state (S_1) and triplet state
(T_1) depends on the competition of different photophysical processes, which are
collected in Table 1.5.

Table 1.4 The lifetime, or reciprocal rates, in solution of photophysical processes involving electronically excited states (cf. Fig. 1.11)

Step	Process	Lifetime (s)
1. Excitation	$S_0 + hv \rightarrow S_1$	10^{-15}
2. Internal conversion (IC)	$S_1 \rightarrow S_1 + \Delta$	10^{-11}–10^{-14}
3. Fluorescent (F) emission	$S_1 \rightarrow S_0 + hv_F$	10^{-6}–10^{-11}
4. Intersystem crossing (ISC)	$S_1 \rightarrow T + \Delta$	10^{-8}–10^{-11}
5. Internal conversion (IC)	$T_1 \rightarrow T_1 + \Delta$	10^{-11}–10^{-14}
6. Phosphorescent (P) emission	$T_1 \rightarrow S_0 + hv_P$	10^{2}–10^{-3}

Table 1.5 Photophysical processes involving electronically excited states (S_1) and (T_1)

Step	Process	Rate
1. Excitation	$S_0 + hv \rightarrow S_1$	$I_A = k_{S_1}[S_0]$
2. Fluorescence (F) emission	$S_1 \rightarrow S_0 + hv_F$	$k_F[S_1]$
3. Internal conversion (IC)	$S_1 \rightarrow S_0 + \Delta$	$k_{IC}[S_1]$
4. Intersystem crossing (ISC)	$S_1 \rightarrow T_1 + \Delta$	$k_{ISC(S)}[S_1] = k_{T_1}[S_1]$
5. Phosphorescence (P) emission	$T_1 \rightarrow S_0 + hv_P$	$k_P[T_1]$
6. Intersystem crossing (ISC)	$T_1 \rightarrow S_0 + \Delta$	$k_{ISC(T)}[T_1]$

The population of the excited singlet state (S_1) and triplet state (T_1) at any given time are presented by following equations:

$$-\frac{d[S_1]}{dt} = k_F[S_1] + k_{IC}[S_1] + k_{ISC(S)}[S_1] \tag{1.20}$$

$$-\frac{d[T_1]}{dt} = k_P[T_1] + k_{ISC(T)}[T_1] \tag{1.21}$$

where k_F, k_{IC}, $I_{ISC(S)}$, k_P and $k_{ISC(T)}$ are defined in Table 1.5. Solutions of these equations are:

$$\ln \frac{[S_1]}{[S_1]_0} = -(k_F + k_{IC} + k_{ISC(S)})t \tag{1.22}$$

$$\ln \frac{[T_1]}{[T_1]_0} = -(k_P + k_{ISC(T)})t \tag{1.23}$$

There are two important definitions of the lifetime of emission:

(i) UNIMOLECULAR LIFETIME (τ)

This is the time (in seconds) required for the concentration of molecules in an excited state to decay to $1/e$ of the initial value:
for excited singlet state (S_1)

$$\ln \frac{1}{e} = -(k_F + k_{IC} + k_{ISC(S)})\tau_F \qquad (1.24)$$

for excited triplet state (T_1)

$$\ln \frac{1}{e} = -(k_P + k_{ISC(T)})\tau_P \qquad (1.25)$$

or

$$\tau_F = \frac{1}{k_F + k_{IC} + k_{ISC(S)}} \qquad (1.26)$$

$$\tau_P = \frac{1}{k_P + k_{ISC(T)}} \qquad (1.27)$$

An increase in the solution molar concentration causes collisional losses of excitation energy (concentration quenching) and eqs (1.26) and (1.27) should be revised as:

$$\tau_F = \frac{1}{k_F + k_{IC} + k_{ISC(S)} + k_{Q(S)}} \qquad (1.28)$$

$$\tau_P = \frac{1}{k_P + k_{ISC(T)} + k_{Q(T)}} \qquad (1.29)$$

In general

$$\tau = \sum_{ik_i} \frac{1}{} \qquad (1.30)$$

(ii) INHERENT RADIATIVE LIFETIME (TRUE RADIATIVE LIFETIME) (τ^0)

This is the time (in seconds) required to deactivate the molecule provided that no radiationless processes occur from S_1 or T_1 states:

$$\varphi_F \tau_F^0 = \tau_F \qquad (1.31)$$

$$\tau_P \left(\frac{1 - \varphi_F}{\varphi_P} \right) = \tau_P^0 \qquad (1.32)$$

where φ_F and φ_P are quantum yields of fluorescence (F) and phosphorescence (P), respectively.

The INHERENT RADIATIVE LIFETIME OF FLUORESCENCE (τ_F^0) can be calculated by the equation

$$\tau_F^0 = \frac{3.5 \times 10^8}{\bar{v}_m^2 \int \varepsilon \, dv} \text{ (s)} \tag{1.33}$$

where:

$\bar{v}_m$ = mean frequency of the absorption in wavenumbers,

$\varepsilon \, dv$ = the experimental molar absorptivity integrated over the width of the absorption band.

It is possible to make certain approximations reducing the equation to

$$\tau_F^0 = \frac{10^{-4}}{\varepsilon_{max}} \text{ (s)} \tag{1.34}$$

and from this rough values of the fluorescent lifetime can be obtained. Thus a strongly absorbing compound with an ε_{max} of 10^5 would have a lifetime of 10^{-9} s.

1.9. QUANTUM YIELDS OF PHOTOPHYSICAL PROCESSES

The QUANTUM YIELD (QUANTUM EFFICIENCY) of any photo-physical process is defined as a ratio of the fraction of the excited molecules in a given excited state which decay by that process to the total number of excited molecules in a given state.[591, 2196, 3678, 3680]

The QUANTUM YIELD OF LUMINESCENCE (Φ_L) is defined as:

$$\Phi_L = \frac{\text{Number of einsteins emitted}}{\text{Number of einsteins absorbed}} \tag{1.35}$$

An EINSTEIN is the amount of energy equal to that of 1 mol of photons.

The QUANTUM YIELD OF FLUORESCENCE (Φ_F) and PHOS-PHORESCENCE (Φ_P) can be also defined as:

$$\Phi_F = \frac{\text{Rate of fluorescent emission}}{\text{Rate of excitation}} = \frac{k_F[S_1]}{I_a} \tag{1.36}$$

$$\Phi_P = \frac{\text{Rate of phosphorescent emission}}{\text{Rate of excitation}} = \frac{k_P[T_1]}{I_a} \tag{1.37}$$

where I_a = intensity of radiation (light) absorbed.

1.10. FLUORESCENCE TIME-RESOLVED SPECTRA

LUMINESCENCE (fluorescence or phoshorescence) LIFETIME (or LUMINESCENCE DECAY) measurements of the order of microseconds or milliseconds do not lead to serious difficulties, and can be measured by kinetic spectroscopic methods. Most common fluorescent substances have lifetimes ranging from 10^{-6} to 10^{-10} s and experimental measurement of fluorescent decay in this region is difficult.[3023] Fluorescence spectra may change with time after excitation for the following reasons;

  (i) vibrational relaxation (a picosecond phenomenon in condensed media);
 (ii) solvent relaxation;
(iii) interconversion of electronic states;
 (iv) electronic energy transfer to non-identical species;
  (v) complex formation (excimer, exciplex);
 (vi) formation and decay of secondary excited species (Fig. 1.16);
(vii) presence of several excited species with different lifetimes and emission characteristics.

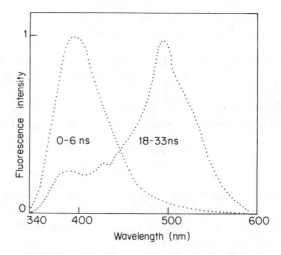

Fig. 1.16   Typical time-resolved fluorescence spectra

Decay times coupled with fluorescence time-resolved spectra frequently provide sufficient data to characterize a substance.

1.11. DELAYED FLUORESCENCE

DELAYED FLUORESCENCE is the emission of wavelengths characteristic of the fluorescence spectrum of a molecule after delays characteristic of phosphorescence.[591, 2196, 3678, 3680] Delayed fluorescence can exist in two forms:

(i) E-TYPE DELAYED FLUORESCENCE (eosin type because it occurs in eosin and other dye molecules). It arises from the process:

$$T_1 \rightarrow S_1 \rightarrow S_0 + h\nu' \text{ (delayed fluorescence)} \tag{1.38}$$

The thermal activation of molecules in the lowest triplet state (T_1) repopulates the fluorescent singlet state (S_1), resulting in the delayed fluorescence (Fig. 1.17).

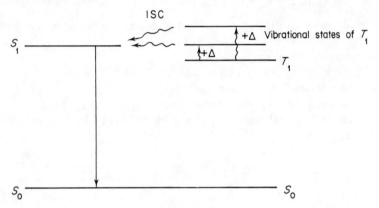

Fig. 1.17 Mechanism of the formation of the E-type delayed fluorescence (Δ = thermal activation)

(ii) P-TYPE DELAYED FLUORESCENCE (pyrene type because it occurs in pyrene and other aromatic hydrocarbons). It arises from the process:

$$T_1 + T_1 \rightarrow S_1 \text{ (or } S_i) + S_0 \qquad (1.39)$$

$$S_1 \rightarrow S_0 + h\nu' \text{ (delayed fluorescence)} \qquad (1.40)$$

Diffusional interaction between two molecules in the lowest triplet state (T_1) produces an excited singlet state $(S_1$ or $S_i)$ which emits delayed fluorescence.

SENSITIZED P-TYPE DELAYED FLUORESCENCE is produced in the diffusional interaction between the donor (sensitizer) (D) and acceptor (A) molecules in the lowest excited triplet states (T_1):

$$D(T_1) + A(T_1) \rightarrow A(S_1) + D(S_0) \qquad (1.41)$$

$$A(S_1) \rightarrow A(S_0) + h\nu' \text{ (delayed fluorescence)} \qquad (1.42)$$

Delayed fluorescence differs from 'normal fluorescence' because:

(i) It has the spectral properties of fluorescence, but with a much longer rise and decay than 'normal fluorescence'.
(ii) Its decay time is frequently of the same order of magnitude as the phosphorescent decay time.
(iii) It depends upon the square of the incident light intensity because the triplet state (T_1) arises only from an initial excitation via the excited singlet state (S_1), and two such events are required to obtain fluorescence of this type.
(iv) It is sensitive to oxygen, which quenches the triplet state (T_1) with great efficiency.

Delayed fluorescence is a typical emission observed in the triplet–triplet (T–T) annihilation process (cf. Section 2.15.2).

1.12. QUENCHING PROCESSES OF EXCITED STATES

Excited singlet and/or triplet states can be deactivated by interaction of the excited molecules with the components of a system, and can be considered as bimolecular processes.[591, 2196, 3678, 3680]

Quenching processes are often divided into:

(i) viscosity-dependent (dynamic) type,
(ii) viscosity-independent (static) type.

The following types of quenching processes can be distinguished:

  (i) collisional quenching (diffusion or non-diffusion controlled),
 (ii) concentration quenching,
(iii) oxygen quenching,
(iv) energy transfer quenching,
 (v) radiative migration (self-quenching).

In the simplified kinetic consideration, the following photophysical processes can be considered in a quenching process of an excited state of molecule (D*) by the addition of a quencher molecule (Q) (Table 1.6).

Table 1.6   Photophysical processes involved in quenching mechanism

Step	Process	Rate
1. Excitation (absorption)	$D_0 + h\nu \rightarrow D^*$	I_a
2. Emission	$D^* \rightarrow D_0 + h\nu'$	$k_L [D^*]$
3. Deactivation	$D^* \rightarrow D_0 + \Delta$	$k_D [D^*]$
4. Quenching	$D^* + Q \rightarrow D_0 + Q$	$k_Q [D^*][Q]$

where:

  I_a = rate of radiation (light) absorption,
  k_L = rate constant for emission from excited molecule D*,
  k_D = rate constant for deactivation (e.g. thermal deactivation) of molecule D*,
  k_Q = rate constant for quenching of molecule D* by a quencher molecule (Q).

The concentration of the excited molecule (under conditions of steady illumination and no irreversible photochemical reactions) is

$$\frac{d[D^*]}{dt} = I_a - (k_L + k_Q[Q] + k_D)[D^*] \qquad (1.43)$$

where I_a is the rate of light absorption equal to the formation of an excited state of molecule (D^*) (which may be singlet or triplet excited state).

In the steady-state condition:

$$I_a = (k_L + k_Q[Q] + k_D)[D^*] \tag{1.44}$$

The quantum yield for emission from an excited molecule (D^*) in the absence of quencher (Q) is given by

$$\Phi_0 = \frac{k_L[D^*]}{I_a} = \frac{k_L}{k_L + k_D} \tag{1.45}$$

The quantum yield for emission from an excited molecule (D^*) in the presence of quencher (Q) is given by

$$\Phi_0 = \frac{k_L[D^*]}{I_a} = \frac{k_L}{k_L + k_D + k_Q[Q]} \tag{1.46}$$

Dividing equation (1.45) by equation (1.46) gives the well-known STERN–VOLMER EQUATION:

$$\frac{\Phi_0}{\Phi_Q} = \frac{k_L + k_D + k_Q[Q]}{k_L + k_D} \tag{1.47}$$

or

$$\frac{\Phi_0}{\Phi_Q} = 1 + \frac{k_Q}{k_L + k_D}[Q] \tag{1.48}$$

or

$$\frac{\Phi_0}{\Phi_Q} = 1 + k_Q\tau[Q] \tag{1.49}$$

where

$$\tau = \frac{1}{k_L + k_D} \tag{1.50}$$

is the measured lifetime of an excited molecule (D^*) in the absence of a quencher molecule (Q). The plot of Φ/Φ_Q versus [Q] produces a straight line with a slope of $k_Q\tau$ (Fig. 1.18).

The concentration-quenching process is conventionally described in terms of the concentration $[Q]_{1/2}$ which reduces Φ_0 to one-half of its original value (50 %):

$$\frac{\Phi_0}{\Phi_Q} = \frac{1}{0.5} = 2 = 1 + k_Q\tau[Q]_{1/2} \tag{1.51}$$

or

$$k_Q\tau[Q]_{1/2} = 1 \tag{1.52}$$

Evaluation of the Stern–Volmer plot gives several important conclusions on quenching processes.

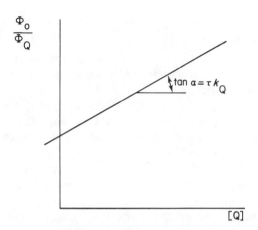

Fig. 1.18   The Stern–Volmer plot

Sometimes a few anomalies in the typical Stern–Volmer plot are observed:

(i) The plot of Φ_0/Φ_Q versus [Q] is parallel to the concentration axis (Fig. 1.19a). In this case there is no quenching, because the reaction is too fast to be quenched or the quenching is endothermic.

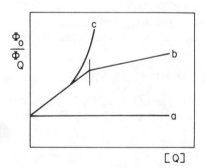

Fig. 1.19   Anomalies in the Stern–Volmer plot

(ii) The slope of plot Φ_0/Φ_Q versus [Q] changes with the concentration of the quencher [Q] (Fig. 1.19b). In this case two different excited states, the singlet and the triplet, can take part in the quenching reaction.

(iii) The plot of Φ_0/Φ_Q versus [Q] increases rapidly with increasing concentration of the quencher [Q] (Fig. 1.19c). In this case the statistical examination of the quenching reaction improves the results, and the Stern–Volmer equation acquires a new form:

$$\frac{\Phi_0}{\Phi_Q} = \frac{1 + \tau k_Q[Q]}{1 + \alpha u} \tag{1.53}$$

where:

α = the probability of energy transfer from triplet donor molecule to the quencher during the lifetime of solution encounter,

u = fraction of donor molecules which has at least one quencher molecule as a nearest neighbour.

When a triplet energy transfer is actually controlled by diffusion the $\alpha = 1$. When the quencher concentration is low or when the energy transfer is inefficient, the value of u is small and equation (1.53) is reduced to the original Stern–Volmer equation (1.51).

(iv) Other anomalies can follow when a part of incident radiation can be absorbed by the quencher at a high concentration, and when the acceptor has polar characteristics.

(v) The formation of a metastable complex, e.g. charge transfer complex (CT) between donor and quencher molecules, which is known as STATIC QUENCHING, represents another type of anomaly. The excitation of the CT complex by light or the formation of (CT)* complex in a collisional quenching process may result in the quenching of the singlet (S_1) or triplet (T_1) excited states in various ways:

   (a) By an external heavy atom or by paramagnetic effects.

   (b) By radiationless transition in donor–quencher excited complexes leading in some cases to a stable donor–quencher molecule, when the quencher is for example oxygen.

   (c) Radiationless or radiative transition in quencher.

   (d) Charge transfer leading to free radical formation.

1.13. QUENCHING BY MOLECULAR OXYGEN

Ground state molecular oxygen $O_2(^3\Sigma_g^-)$ is an efficient quencher of the excited singlet (S_1) and triplet (T_1) states.[438, 3678-3680] The quenching mechanism include:

(i) Physical quenching:

$$S_1 \text{ (or } T_1) + {}^3O_2 \rightarrow S_0 + {}^3O_2 + \text{heat} \tag{1.54}$$

$$S_1 \text{ (or } T_1) + {}^3O_2 \rightarrow S_0 + {}^1O_2 \text{ (singlet oxygen)} \tag{1.55}$$

$$S_1 \text{ (or } T_1 ?) + {}^3O_2 \rightarrow (S \cdots O_2)^* \text{ (oxciplex)} \tag{1.56}$$

SINGLET OXYGEN is an excited form of molecular oxygen (cf. Section 14.4).

OXCIPLEX is an exciplex formed between excited molecule and molecular oxygen.

(ii) Chemical quenching:

$$M^*(S_1 \text{ or } T_1) + {}^3O_2 \rightarrow MO_2 \tag{1.57}$$

$$M^*(S_1 \text{ or } T_1) + {}^3O_2 \rightarrow M^+ + O_2^- \tag{1.58}$$

With few exceptions, the rate constant for quenching is within an order of magnitude of the diffusional quenching constant.

1.14. PHOTOCHEMICAL PROCESSES

All photochemical processes obey four photochemical laws which can be applied generally in photochemistry:

(i) A photochemical reaction may occur only if light of sufficient energy is absorbed by the system.

(ii) Each photon or quantum absorbed activates only one molecule in the primary excitation step of a photochemical sequence.

(iii) Each photon or quantum absorbed by a molecule has a certain probability of populating either the excited singlet state (S_1) or triplet state (T_1).

(iv) The lowest excited singlet state (S_1) and triplet (T_1) states are the starting levels (in solution) of most organic photochemical processes. Since the lifetime of the lowest triplet state (T_1) is usually about 10^5 times longer than that of the lowest excited singlet state (S_1) it is often suggested that the majority of photochemical reactions involve triplet states. While this may generally be true, reactions involving excited singlet states can and do occur, and the greater energy available compared with the triplet state may accelerate these reactions. Therefore it is necessary to characterize the photoreactive state before a reaction mechanism can be postulated.

Photochemical processes usually occur in two stages:

(i) The PRIMARY PHOTOCHEMICAL REACTION is the reaction which is directly due to the absorbed photon or quantum involving electronically excited states. This process has been found to be independent of temperature.

(ii) The SECONDARY PHOTOCHEMICAL REACTIONS (also called DARK REACTIONS) are reactions of radicals, radical ions, ions and electrons which were produced by the primary photochemical reaction.

The quantitative relationship between the number of molecules which react or which are formed in a photochemical reaction and the number of photons absorbed in unit time is expressed by the QUANTUM YIELD (Φ):

$$\Phi = \frac{\text{Number of molecules reacting in a particular process}}{\text{Number of quanta absorbed by the system}} \qquad (1.59)$$

The number of molecules reacting in a particular process can be determined by any convenient method, and the number of quanta absorbed by the system per unit time can be measured by chemical or physical methods.[3023]

The knowledge of the value of quantum yield (Φ) is important for the understanding of the mechanism and course of a photochemical reaction:

When $\Phi = 1$, every absorbed quantum produces one photochemical reaction.
When $\Phi < 1$, other reactions compete with the main photochemical reaction.
When $\Phi > 1$, a chain reaction takes place.

The quantum yield defined above is wavelength-dependent. Most photochemical reactions are complex and occur in many stages; thus only in rare cases is the measured change that which was originally produced by the absorption of light.

The quantum yield of reaction should not be confused with the CHEMICAL YIELD OF REACTION. A low quantum yield of reaction can sometimes lead to high chemical yield if no other reactions take place, and if one irradiates long enough. However, the chemical yield is of little use in helping to establish the mechanism of a photochemical reaction.

1.15. PHOTOREACTIONS IN POLYMER MATRIX

Most photophysical and photochemical processes which occur or which are carried out in solid polymer matrices, such as poly(methyl methacrylate), poly(vinyl acetate) and polystyrene, differ considerably from similar reactions in solution on account of limitation of diffusion, and of translation and rotation possibilities.

Photophysical processes which occur in a polymer matrix depend on:

(i) Free volume, i.e. the space between macromolecules which is available for unhindered diffusion of low molecular quenching compounds, e.g. oxygen. In gels and polymeric films oxygen quenching of the excited triplet state (T_1) is markedly reduced because of the low diffusion rate through polymer matrix. For example, the oxygen quenching rate of the triplet state of 2-naphthol in gelatin gels is markedly reduced.[553]

(ii) Microscopic and macroscopic viscosity. In polymer solutions (coatings) microscopic viscosities can vary from site to site as a result of the change positioning of polymer chains during the solvent-coating process or extrusion of a given film or coating. Just as the local viscosity in a polymer film can vary from site to site, so can the microscopic polarity. Furthermore, the ability of a given chain to solvate either additional polymer chains or guest molecules varies from site to site.

Non-associative excited species cannot diffuse rapidly through the polymer network as though they were dissolved in a pure solvent. For example, the translational coefficient of anthracene is unaltered in solutions or in polyisobutylene gels, whereas it is reduced in polyisobutylene films.[1567]

(iii) Chemical structure of polymeric matrices. The translational coefficient of anthracene is reduced in poly(vinyl alcohol) or gelatin gels due to hydrogen bonding between polymer chains.[1567] The oxygen quenching rate of the triplet state (T_1) of 2-naphthol and erythrosine is decreased in poly(vinyl alcohol) or gelatin gels due to reduced migration of oxygen molecules between hydrogen bonds.[553]

(iv) Glass transition temperature (T_g). Below the T_g in rigid glassy polymer matrix limitations of translation and rotation possibilities will be most pronounced, whereas above the T_g in the rubbery state the polymer matrix behaves as a high-viscous medium and this affects only the rate of rotation. The photoisomerization process, which involves an appreciable change of conformation of the photosensitive groups, is strongly restricted in a polymer matrix below the T_g in the rigid glassy state. In photoisomerization, rotation possibilities are limited below T_g. Above T_g in the rubbery state deviations from solution behaviour will be much less pronounced (cf. Chapter 10).

The polymer matrix usually causes a considerable reduction in photolysis reaction velocity, and due to diffusion-controlled processes and cage-effects, reaction kinetics are greatly affected.[3386] Sometimes even the nature of the reaction products or their relative properties differ considerably.

When photodissociation produces reactive radicals which may recombine thermally or photochemically, the kinetics do not follow a bimolecular reaction mechanism, as in solution; they follow a stepwise process. As a first step, rapid 'CAGE RECOMBINATION' (cf. Section 5.1.7) occurs followed by a second very slow step corresponding to the diffusion-controlled recombination of isolated species (radicals or monomers). The stepwise reaction kinetics have been illustrated by three different dissociation–recombination equilibria:[3386]

(i) Thermal dissociation of benzopinacol and its derivatives, and radical recombination.[507, 3837]

(ii) Photochemical dissociation of hexaphenyl-1,2'-bisimidazole into triphenyl-imidazolyl radicals and their recombination.[341, 1549, 1550, 2365, 3870]

(iii) Photochemical dissociation of substituted anthracene- and benz(acridinium) dimers, and photochemical recombination of monomers.[2475, 3386, 3620]

Photolysis of low molecular compounds in polymer matrix such as poly(methyl methacrylate), poly(vinyl acetate), poly(vinyl chloride) and polystyrene may differ considerably from that in solution. For example in the photolysis of cyclooctyl nitrate, limited interconversion of ring conformation is responsible for a lower quantum yield of photolysis and for formation of cyclooctanone as the main reaction product in polymer matrix, instead of a nitroso compound as expected from the Barton reaction (photolysis of alkyl nitrates to alkoxy radicals, followed by intermolecular hydrogen abstraction by the latter and recombination of the resulting carbon radicals with nitrogen oxide to form nitroso derivatives and oximes).[1850]

Photochemical dissociation reactions of p-quinone bis-nitrones (1.1) which occurs in the two step photolysis mechanism is also influenced by polymer matrix, which stabilize the oxazirane reaction intermediate:[3387]

Step I. Formation of mononitrone (1.2) (which absorbs at longer wavelengths) and can dissociate to p-quinone nitrone (1.3):

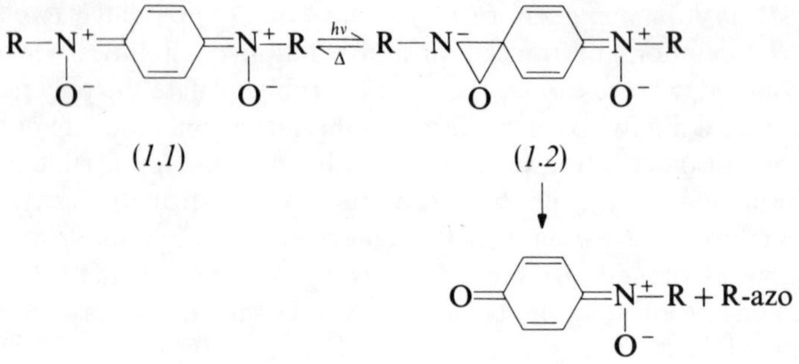

$$\text{R—N}^+ \underset{\text{O}^-}{=} \text{=N}^+_-\text{R} \quad \underset{\Delta}{\overset{h\nu}{\rightleftharpoons}} \quad \text{R—N}^-\underset{\text{O}}{\triangleleft} \text{=N}^+_-\text{R}$$

$(1.1) \qquad\qquad\qquad (1.2)$

(1.60)

$$\text{O}=\underset{\text{O}^-}{=}\text{=N}^+_-\text{R} + \text{R-azo}$$

(1.3)

Step II: p-Quinone nitrone (1.3) absorbs light and undergo to quinone-mononitrone (1.4), which dissociate to p-quinone (1.5) and R-azo:

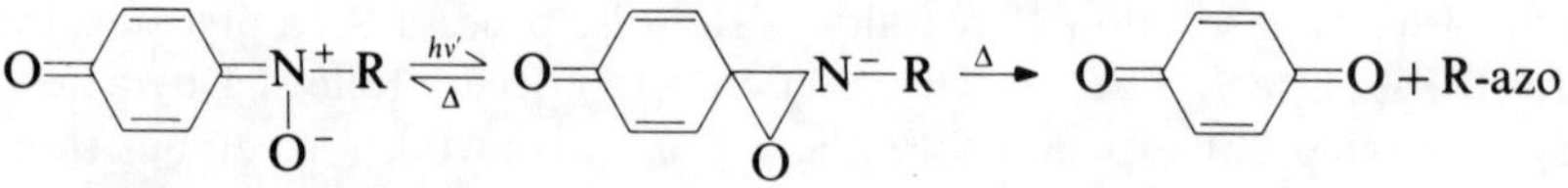

$$\text{O}=\underset{\text{O}^-}{=}\text{=N}^+_-\text{R} \quad \underset{\Delta}{\overset{h\nu'}{\rightleftharpoons}} \quad \text{O}=\underset{\text{O}}{\triangleleft}\text{—N}=\text{R} \quad \overset{\Delta}{\longrightarrow} \quad \text{O}=\text{=O} + \text{R-azo}$$

$(1.3) \qquad\qquad\qquad (1.4) \qquad\qquad\qquad (1.5)$

(1.61)

Reverse isomerization of nitrone into oxazirane involves rotation of a part of the molecule and therefore undergoes hindrance from the polymer matrix (when the glass transition temperature (T_g) is high).

Polymer matrix such as poly(methyl methacrylate) or polystyrene has remarkable influence on the thermally reversible photocyclization of α-p-dimethylaminophenyl-M-m-nitrophenyl nitrone (1.6) to corresponding oxaziridine (1.7) (Fig. 1.20):[3385]

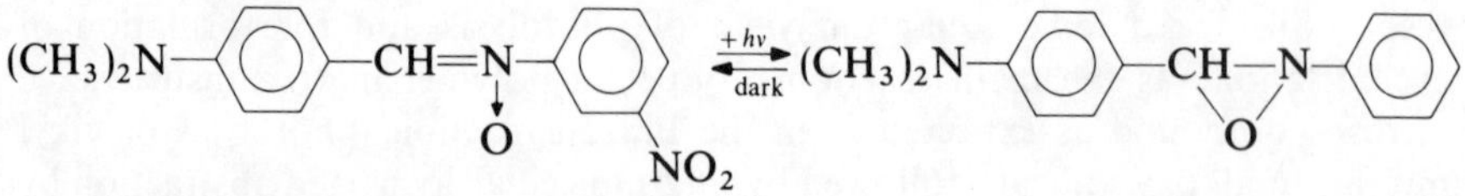

$$(\text{CH}_3)_2\text{N}\text{—}\underset{\text{O}}{\text{—CH}=\text{N}}\text{—}\underset{\text{NO}_2}{} \quad \underset{\text{dark}}{\overset{+h\nu}{\rightleftharpoons}} \quad (\text{CH}_3)_2\text{N}\text{—}\text{—CH}\underset{\text{O}}{\triangleleft}\text{N}\text{—}$$

$(1.6) \qquad\qquad\qquad\qquad (1.7)$

(1.62)

The kinetics of these reactions are strongly affected by the rigid matrix. The photocyclization is enhanced both in poly(methyl methacrylate) and poly(styrene) in comparison with that in solution. Since the quantum yield in solution is almost independent of the nature of the solvent, this enhancement

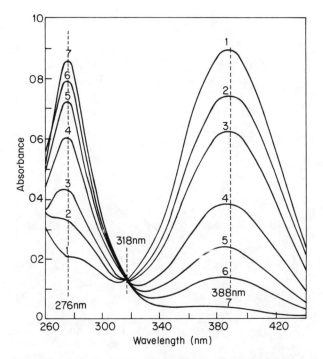

Fig. 1.20 Photochemical transformation of nitrone (*1.6*) into oxaziridine (*1.7*) with irradiation at 380 nm and various reaction times: (1)
0, (2) 15 s, (3) 30 s, (4) 60 s, (5) 90 s, (6) 120 s, (7) 180 s[3385]

must be due to the rigid nature of the polymer matrix. The stabilization of the
excited state by the rigid matrix gives the excited state a higher probability of
reacting chemically.

The rate of the thermal recovery of nitrone (*1.6*) to corresponding oxaziridine
(*1.7*) in solution is strongly dependent on the solvent polarity. A rigid polymer
matrix such as poly(methyl methacrylate) also enhances the thermal recovery
reaction (Fig. 1.21). At the temperature well below T_g, where the movement of the
polymer chain is practically frozen in, the oxaziridine lacks stabilization by
rearrangement of the solvation shell, whereas such stabilization occurs very
rapidly in solution. Consequently the rate of the thermal return within the
polymer matrix is greater than in the solution. Another possible explanation of
these effects could be the *cis–trans* isomerization of nitrone, which is inhibited in
the rigid matrix.

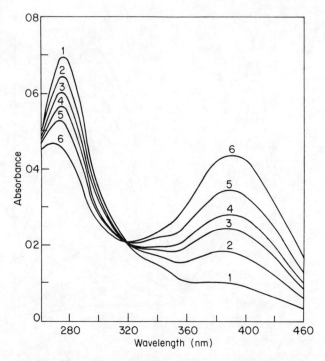

Fig. 1.21 Thermal conversion of oxazidirine (*1.7*) to the nitrone (*1.6*) in poly(methyl methacrylate) film at 25°C at various reaction times: (1) 0, (2) 10 s, (3) 20 s, (4) 45 s, (5) 60 s, (6) 80 s[3385]

2. Electronic energy transfer processes

ELECTRONIC ENERGY TRANSFER PROCESS is the one-step transfer of electronic excitation from an excited DONOR molecule (D*) to an ACCEPTOR molecule (A) in separate molecules (INTERMOLECULAR ENERGY TRANSFER) or in different parts of the same molecule (INTRAMOLECULAR ENERGY TRANSFER):

$$D + h\nu \to D^* \qquad \text{(Excitation of a donor)} \qquad (2.1)$$

$$D^* + A \to D + A^* \qquad \text{(Electronic energy transfer)} \qquad (2.2)$$

Electronic energy transfer process may occur by the following mechanisms:[162, 401, 573, 820, 970, 1423, 1989, 2205, 3678–3680, 3771, 3774, 3888, 3889, 4013]

(I) Radiative energy transfer (cf. Section 2.1).
(II) Non-radiative energy transfer:
    (i) Electron exchange energy transfer (cf. Section 2.2).
    (ii) Resonance excitation energy transfer (cf. Section 2.3).

Various factors affect the extent and rate of energy transfer between an excited donor (D*) and the acceptor (A):

  (i) distance between D* and A,
 (ii) relative orientation to each other,
(iii) spectroscopic properties of D and A,
(iv) optical properties of a medium,
 (v) effect of molecular collisions on the motion of the excited donor and an acceptor in the period during which the donor is excited.

In general:

 (i) the energy of the excited state (A*) must be lower than that of D* (the energy transfer process is efficient),
(ii) the sensitized excitation of A by D* must occur within the time (τ) that the molecule D remains in the excited state.

Energy transfer studies have potential application to:

  (i) polymer photodegradation,
 (ii) polymer photoconductivity,
(iii) use of polymer arrays as photo-harvesting 'antennas' for solar energy collection (cf. Section 1.6).

Energy transfer in polymeric systems has been a subject of several published reviews.[1269, 1436, 2353, 2354, 2733, 3294, 3679, 3680]

2.1. RADIATIVE ENERGY TRANSFER

RADIATIVE ENERGY TRANSFER (THE 'TRIVIAL' PROCESS) of electronic energy involves the possibility of reabsorption of donor emission. The process requires two steps with an intermediate photon:

$$D + h\nu \rightarrow D^* \qquad \text{(Excitation of a donor)} \qquad (2.3)$$

I step: $\qquad D^* \rightarrow D + h\nu' \qquad$ (Emission from a donor) $\qquad (2.4)$

II step: $\qquad A + h\nu' \rightarrow A^* \qquad$ (Excitation of an acceptor) $\qquad (2.5)$

No direct interaction of the donor with the acceptor is involved. Radiative energy transfer occurs only in that region where the emission spectrum of the donor overlaps with the absorption spectrum of the acceptor.

The EFFICIENCY OF THE RADIATIVE ENERGY TRANSFER depends on:

(i) Quantum yield of emission from the donor.
(ii) Absorption of this radiation by the acceptor (governed by the Beer–Lambert law).

The radiative energy transfer may occur over very long distances (relative to molecular diameters) and the probability that an acceptor molecule reabsorbs the light emitted by a donor at a distance R varies as R^{-2}.

The radiative energy transfer is characterized by:

(i) Invariance of the donor emission lifetime. If the donor and acceptor are identical molecules, there could be a lengthening of the donor emission lifetime if multiple reabsorption and re-emissions occur.
(ii) Change in the emission spectrum of the donor which can be accounted for on the basis of the acceptor absorption spectrum.
(iii) Lack of dependence of the transfer efficiency upon the viscosity of the medium.

Efficient energy transfer can take place if the donor and acceptor are different species and the spectral overlap is large. This can lead to the INTERNAL FILTER PHENOMENA, registered as a serious distortion of the emission spectrum of a donor.[2900]

2.2. ELECTRON EXCHANGE ENERGY TRANSFER

ELECTRON EXCHANGE ENERGY TRANSFER occurs when an excited donor molecule (D^*) and an acceptor molecule (A) are close enough (10–15 Å) that they may be considered to be in molecular contact, i.e. their centres are separated by the sum of their molecular radii. Their electron clouds may overlap each other and an electron on D* may also appear on A.

The RATE CONSTANT OF ELECTRON EXCHANGE ENERGY TRANSFER (k_{ET}) is given by the DEXTER EQUATION:[983]

$$k_{D^* \to A} = \frac{2\pi K^2}{\hbar} e^{-2R/L} \int_0^\infty F_D(v) \varepsilon_A(v) \, dv \qquad (2.6)$$

where:

K and L = constants not available from experimental data,

R = distance between the centres of the donor and the acceptor molecules,

$\hbar = h/2$ is read 'h bar', h = Planck constant,

$F_D(\bar{v})$ = spectral distribution of the donor emission (fluorescence intensity) in the infinitessimally small wavenumber range $\bar{v}$ to $(\bar{v} + dv)$ (in quanta) normalized so that

$$\int_0^\infty F_D(\bar{v}) \, d\bar{v} = 1 \qquad (2.7)$$

$\varepsilon_A(\bar{v})$ = molar absorptivity (extinction coefficient) of the acceptor absorption in the infinitessimally small wavenumber range $\bar{v}$ to $(\bar{v} + dv)$ ($1 \text{ mol}^{-1} \text{ cm}^{-1}$) normalized so that

$$\int_0^\infty \varepsilon_D(\bar{v}) \, d\bar{v} = 1 \qquad (2.8)$$

$\bar{v}$ = wavenumber (cm^{-1}).

The SPECTROSCOPIC OVERLAP INTEGRAL $\int_0^\infty F_D(\bar{v}) \varepsilon_A(\bar{v}) d\bar{v}$ is a measure of the overlap of the donor emission and acceptor absorption.

The following spin-allowed electron exchange energy transfer processes may occur:

$$^1D^* \text{ (singlet)} + A \qquad \to D + {}^1A^* \text{ (singlet)} \qquad (2.9)$$

$$^3D^* \text{ (triplet)} + A \qquad \to D + {}^1A^* \text{ (singlet)} \qquad (2.10)$$

$$^1D^* \text{ (singlet)} + {}^3A^* \text{ (triplet)} \quad \to D + {}^3A_x^* \text{ (triplet)} \qquad (2.11)$$

$$^3D^* \text{ (triplet)} + {}^3A^* \text{ (triplet)} \quad \to D + {}^1A^* \text{ (or } {}^3A^*) \text{ (triplet–triplet annihilation)} \qquad (2.12)$$

The triplet–triplet energy transfer, which is forbidden by the resonance–excitation energy transfer (cf. Section 2.3) is allowed by an electron exchange energy transfer:

$$^3D^* \text{ (triplet)} + A \to D + {}^3A^* \text{ (triplet)} \qquad (2.13)$$

The energy transfer by the following route is forbidden both by an electron exchange energy transfer and the resonance–excitation energy transfer:

$$^1D^* \text{ (singlet)} + A \xrightarrow{\hspace{0.3cm}\diagup\hspace{-0.55cm}\diagdown\hspace{0.3cm}} D + {}^3A^* \text{ (triplet)} \qquad (2.14)$$

The electron exchange energy transfer is less important than the resonance–excitation energy transfer (cf. Section 2.3).

2.3. RESONANCE–EXCITATION ENERGY TRANSFER

RESONANCE–EXCITATION (DIPOLE–DIPOLE) ENERGY TRANSFER occurs when an excited donor molecule (D^*) can transfer its excitation energy to an acceptor (A) molecule over distances much greater than collisional diameters (e.g. 50–100 Å). In this mechanism energy transfer occurs via dipole(donor)–dipole(acceptor) interaction (Coulombic interaction). When an acceptor (A) is in the vicinity of an excited donor (D^*) (an oscillating dipole), it causes electrostatic forces which can be exerted on the electronic systems of an acceptor.

The RATE CONSTANT OF THE RESONANCE–EXCITATION ENERGY TRANSFER (k_{ET}) is given by the FÖRSTER EQUATION:[1131–1137]

$$k_{ET} = \frac{9000\,(\ln 10)\,K^2\,\Phi_D}{128\,\pi^5 n^4 N \tau_D R^6} \int_0^\infty F_D(v)\varepsilon_A(v)\,\frac{dv}{v^4} \qquad (2.15)$$

where

K = orientation factor for dipole–dipole interaction (cf. Section 2.4),

Φ_D = quantum yield of the donor emission (fluorescence) in the absence of the acceptor,

n = refractive index of the medium (solvent) between the donor and the acceptor,

    Note: The refractive index of the medium between the donor and the acceptor is not easy to measure when these are different parts of the same molecule. In many biopolymers n ranges from 1.3 to 1.6.

N = Avogadro's number,

τ_D = actual mean donor (D^*) emission lifetime (s),

R = distance between the centres of the donor and the acceptor molecules (cm),

$F_D(\bar{v})$ = spectral distribution of the donor emission (fluorescence intensity), defined as in equation 2.7.

$\varepsilon_A(\bar{v})$ = molar absorptivity (extinction coefficient) of the acceptor absorption, defined as in equation 2.8.

Equation (2.15) applies strictly only to those cases where:

(i) Donor and acceptor are well separated (20 Å, at least) and are immobile (rigid system).

(ii) The donor and acceptor exhibit broadened relatively unstructured spectra.
(iii) The spectral overlap is significant.
(iv) There are no important medium or solvent interactions.
(v) The solvent-excited states lie much higher than those of the donor and acceptor.

The RATE CONSTANT OF ENERGY TRANSFER (k_{ET}) can also be presented as a function of distance between the excited donor (D^*) and the acceptor (A) molecule:[1135, 1137–1139]

$$k_{ET} = \frac{1}{\tau_D} \left(\frac{R_0}{R} \right)^6 \qquad (2.16)$$

where:

τ_D = actual mean lifetime of the excited donor (D^*),
R = separation between the centres of D^* and A,
R_0 = CRITICAL RADIUS, the separation of donor and acceptor for which energy transfer from D^* to A and emission from D^* are equally probable (in other words the distance of separation of the donor and acceptor at which the rate of intermolecular energy transfer is equal to the sum of the rates for all other donor de-excitation processes).

$$R_0^6 = \frac{9000 (\ln 10) K^2 \Phi_D}{128 \pi^5 n^4 N} \int_0^\infty F_D(v)\varepsilon_A(v) \frac{dv}{v^4} \qquad (2.17)$$

Note: It is generally rather difficult to obtain the value of the quantum yield of fluorescence of the donor (Φ_D) when donor and acceptor are different parts of the same molecule. One approach is to use the typical quantum yield for the donor in the model compounds. The R_0 value is not critically sensitive to uncertainties in Φ_D because it depends only on the sixth root of Φ_D.

The rate of energy transfer at $R_0 = R$ is equal to $1/\tau_D$ and if $R < R_0$, energy transfer dominates.

$$k_{ET}(\text{at } R_0) = 1/\tau_D \qquad (2.18)$$

The 'EXPERIMENTAL' CRITICAL RADIUS (R_0) can be calculated from equation:

$$R_0 = \sqrt[3]{\frac{3,000}{4\pi N[A]_{1/2}}} = \frac{7.35}{\sqrt[3]{[A]_{1/2}}} \qquad (2.19)$$

where:

N = Avogadro's number,
$[A]_{1/2}$ = CRITICAL CONCENTRATION OF ACCEPTOR, the concentration at which the energy transfer is 50 % efficient (emission from D^* is half-quenched) it can be found experimentally when the

deactivation of D* in the presence of acceptor (A) is equal to one-half of its value in the absence of A, i.e.:

$$A = [A]_{1/2} \qquad (\text{mol}\,l^{-1}) \tag{2.20}$$

The 'THEORETICAL' CRITICAL RADIUS (R_0) can be determined from an approximate equation (2.21), where the emission spectrum of the excited donor is expressed in terms of the absorption spectrum of the donor by using the assumed mirror-image symmetry of these spectra:

$$R_0^6 \approx \frac{k\tau_D}{\bar{v}_0^2} \int_0^\infty \varepsilon_A(v)\varepsilon_D(\bar{v})\,(2\bar{v}_0 - \bar{v})\,\mathrm{d}v \tag{2.21}$$

where

k = constant,
τ_D = actual mean donor (D*) emission life time (s),
$\varepsilon_A(\bar{v})$ = extinction coefficient of the acceptor absorption at the wavenumber $\bar{v}$ (1 mol^{-1} cm^{-1}),
$\varepsilon_D(\bar{v})$ = extinction coefficient of the donor absorption at the wavenumber $\bar{v}$ (1 mol^{-1} cm^{-1}),
$\bar{v}$ = wavelength (wavenumber) for the 0–0 band in the donor spectrum (0–0 donor transition).

The RESONANCE–EXCITATION (DIPOLE–DIPOLE) ENERGY TRANSFER PROBABILITY is independent of the wavelength of the exciting radiation and increases, for a given donor molecule, as the extinction coefficient of the acceptor and the overlap of the donor emission spectrum and acceptor absorption spectrum increases.

The following spin-allowed resonance–excitation energy transfer processes may occur:

$$^1D^* \text{ (singlet)} + A \rightarrow D + {}^1A^* \text{ (singlet)} \tag{2.22}$$

$$^3D^* \text{ (triplet)} + A \rightarrow D + {}^1A^* \text{ (singlet)} \tag{2.23}$$

The triplet–triplet energy transfer, which is allowed by the electron exchange energy transfer (cf. Section 2.2) is forbidden by the resonance–excitation energy transfer:

$$^3D^* \text{ (triplet)} + A \nrightarrow D + {}^3A^* \text{ (triplet)} \tag{2.24}$$

The singlet–triplet energy transfer is forbidden both by the resonance–excitation energy transfer and an electron-exchange energy transfer:

$$^1D^* \text{ (singlet)} + A \nrightarrow D + {}^3A^* \text{ (triplet)} \tag{2.25}$$

The resonance–excitation energy transfer is more important than the electron-exchange energy transfer (cf. Section 2.2).

2.4. ORIENTATION FACTOR

The resonance–excitation energy transfer (DIPOLE–DIPOLE ENERGY TRANSFER) has an angular dependence expressed by the ORIENTATION FACTOR FOR DIPOLE–DIPOLE INTERACTION (κ^2), which is given by:

$$\kappa = (\cos \alpha - 3 \cos \beta \cos \gamma)^2 = 0\text{--}4 \qquad (2.26)$$

where

α = angle between the donor and acceptor transition moments,
β = angle between the donor moment and the line joining the centres of the donor and acceptor,
γ = corresponding angle between the acceptor moment and the line joining the centres of the donor and acceptor.

In many polymers and biopolymers the donor and acceptor molecules undergo essentially free rotation during the lifetime of the donor. Then an average $\kappa^2 = 2/3$ (for the random distribution of the donor and acceptor molecules) should be used. In such a case R_0 should be quoted as $\overline{R}_0$.

2.5. EFFICIENCY OF ENERGY TRANSFER

The EFFICIENCY OF ENERGY TRANSFER (Φ_{ET}) is given by equations:

$$\Phi_{ET} = \frac{(R_0/R)^j}{1 + (R_0/R)^j} = \frac{(R_0/R)^6}{1 + (R_0/R)^6} \qquad (2.27)$$

where

R = distance between the centres of the donor (D) and the acceptor (A) molecules (cm),
R_0 = critical radius (cm) (cf. Section 2.3).

The values R_0 and j can be obtained from the experimental data by plotting log $(\Phi_{ET}^{-1} - 1)$ vs. log R.[3466] The slope of this line is j, while R_0 is given by the value of R at $\Phi_{ET} = 0.5$. In most cases $j = 6$.

The efficiency of energy transfer (Φ_{ET}) can be determined in the following ways:

(i) By measuring of the quenching of donor fluorescence as a function of acceptor concentration. The energy transfer efficiency (Φ_{ET}) is given by:[2676]

$$\Phi_{ET} = 1 - \frac{I_{F(D+A)}}{I_{F(D)}} \qquad (2.28)$$

where:

$I_{F(D+A)}$ and $I_{F(D)}$ = fluorescence intensities of the donor (D) in the presence and absence of acceptor (A), respectively;

or

by measuring the quenching of acceptor fluorescence as a function of donor concentration. The energy transfer efficiency (Φ_{ET}) is given by:

$$\Phi_{ET} = \frac{I_{F(A)}}{I_{F(A+D)}} \qquad (2.29)$$

where:

$I_{F(A)}$ and $I_{F(A+D)}$ = fluorescence intensities of the acceptor (A) in the absence and presence of donor (D), respectively. In the majority of applications these procedures are equivalent, and generally the energy transfer efficiency (Φ_{ET}) is estimated from donor fluorescence quenching data as no calibration of excitation sources intensity is required.

(ii) By measuring the decreased fluorescence lifetime of the excited donor in the presence of the acceptor (τ_{D+A}) compared with that without the acceptor (τ_D):

$$\Phi_{ET} = 1 - (\tau_{D+A}/\tau_D) \qquad (2.30)$$

(iii) From the excitation spectrum corresponding to fluorescence from the acceptor, provided it is sufficiently fluorescent:

$$e = [G(\bar{v}_2)/G(\bar{v}_1) - \varepsilon_D(\bar{v}_2)/\varepsilon_A(\bar{v}_1)] \times [\varepsilon_A(\bar{v}_1)/\varepsilon_D(\bar{v}_2)] \qquad (2.31)$$

where

$G(\bar{v})$ = magnitude of the corrected excitation spectrum of the acceptor at $\bar{v}$. $G(\bar{v})$ is measured at $\bar{v}_1$, where the donor has no absorption, and at v_2, where the absorption coefficient of the donor is large compared with that of the acceptor;

$\varepsilon_D(v)$ and $\varepsilon_A(v)$ = corresponding absorption coefficients of the donor and the acceptor, respectively.

The two first methods are commonly used, whereas the third method can be used even if the local environment of the donor is different in the presence of an acceptor, provided that the absorption spectra of both the donor and acceptor groups are known.

From measurements of efficiency of energy transfer (Φ_{ET}) it is possible to calculate the distance (R) between a donor and an acceptor molecules (cf. equation 2.27).

Measurements of energy transfer efficiencies from studies of quenching of donor luminescence and/or enhancement of acceptor emission yield valuable information regarding relatively long-range migration effects. Down-chain energy migration, which involves relatively few energy transfers, is amenable to study by luminescence depolarization techniques.[195, 2465, 3080, 3081]

2.6. SPECTROSCOPIC OVERLAP INTEGRAL

The SPECTROSCOPIC OVERLAP INTEGRAL $\int_0^\infty F_D(\bar{v})\varepsilon_A(\bar{v})(d\bar{v}/\bar{v}^4)$ is large if the emission spectrum of an excited donor (D*) overlaps strongly with the absorption spectrum of an acceptor (A) (Fig. 2.1). The spectroscopic overlap integral can be determined by graphical integration (Fig. 2.1). It can also be calculated from Gaussian distributions of the absorption and emission bands:

$$\varepsilon_A(\bar{\nu}) = \varepsilon_{A_{max}} \exp[-\{(\bar{\nu}-\bar{\nu}_A)/\sigma_A\}^2] \tag{2.32}$$

$$F_D(\bar{\nu}) = \varepsilon_{D_{max}} \exp[-\{(\bar{\nu}-\bar{\nu}_D)/\sigma_D\}^2] \tag{2.33}$$

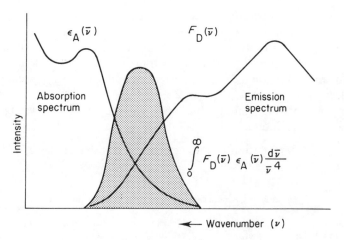

Fig. 2.1   The spectroscopic overlap integral (shaded area)

where

$\varepsilon_{D_{max}}$ and $\varepsilon_{A_{max}}$ = maximum molar absorption coefficients of the longest wavelength absorption bands of the donor and aceptor molecules, respectively;

ν_D and ν_A = wavenumber maxima of the donor emission and the acceptor absorption spectra;

σ_D and σ_A = standard deviations, which can be calculated by considering the overlapping sides of the absorption and emission bands (Fig. 2.1).

With the Gaussian approximations the spectroscopic overlap integral is given by:

$$J = \varepsilon_{A(max)}\varepsilon_{D(max)} \frac{\sqrt{\pi}}{\left(\dfrac{1}{\sigma_A^2} + \dfrac{1}{\sigma_D^2}\right)} \exp[-(\bar{\nu}_A-\bar{\nu}_D)^2/(\sigma_A^2+\sigma_D^2)] \tag{2.34}$$

Overall probability of the energy transfer is proportional to the spectroscopic overlap integral. In the case of energy transfer between equivalent molecules the spectroscopic overlap integral represents the extent of overlap of the emission and absorption of the donor molecule (D). In general this overlap is most favourable for chromophores that undergo minor geometric change in the excited state.

2.7. ELECTRONIC ENERGY TRANSFER PROCESSES IN SOLUTION

Electronic transfer processes in solution may occur in the following ways:

(i) Radiative energy transfer (cf. Section 2.1).
(ii) Collisional transfer via electron exchange energy transfer which requires close approach (on the order of collisional diameters) (cf. Section 2.2).
(iii) Long-range, single-step radiationless transfer via dipole–dipole interaction (resonance–excitation energy transfer) over distances large compared to molecular diameters (cf. Section 2.3).
(iv) Excitation migration among donor (solvent) molecules, followed by transfer to acceptor (solute).

Experimental differentiation of these electronic transfer processes is shown in Table 2.1.

Table 2.1   Characteristics of electronic transfer processes

Experimentally measured donor (D) characteristic	Radiative energy transfer	Electron exchange energy transfer	Resonance excitation energy transfer	Energy migration
Absorption spectrum	Unchanged	Unchanged	Unchanged	
Emission spectrum	Changed	Unchanged	Unchanged	Small change depending on magnitude of donor–donor interaction
Lifetime	Unchanged	Decreased	Decreased	Decreased
Dependency of rate on increasing viscosity	None	Decreased	Decreased	

2.7.1. Solvents Effects

The solvent may influence the rate or efficiency of an electronic energy transfer in the following ways:

(i) Viscosity effects (cf. Section 2.7.2).
(ii) Solvents effects on the energy levels of the donor and acceptor (absorption spectra and the spectra overlap integral).
(iii) Solvent effect on the excited-donor lifetime (e.g. a reversible photochemical reaction between excited donor and solvent, such as a proton transfer, which leads to quenching of the donor).

(iv) The solubility of the donor and acceptor. Good solvent provides a random distribution of donor and acceptor molecules, whereas a poor solvent might lead to a non-statistical distribution with clumping together of donor molecules, acceptor molecules, or donor and acceptor molecules. The non-statistical distribution could lead to anomalously high or low electronic energy transfer rates and efficiencies.

(v) Polarity effects. The field in a dielectric solvent has an effect on long-range resonance–excitation (dipole–dipole) energy transfer.

There is an exceptional mechanism in which electronic energy transfer from an excited donor molecule (D*) to an acceptor molecule (A) occurs via a series of transfers initiated by energy transfer from D* to a solvent molecule, mediated by hopping or migration of the excitation energy through the solvent and terminated by energy transfer from a solvent molecule to an acceptor molecule (A) (Fig. 2.2).[3679, 3680] In this case the solvent serves as an ELECTRONIC ENERGY CONDUCTOR.

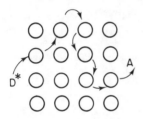

Fig. 2.2　Migration of electronic excitation via hops through the solvent

2.7.2. Viscosity Effects

MOLECULAR DIFFUSION in solution or fluid media can be depicted schematically as the relative motion of excited donor (D*) and acceptor (A) molecules through the empty space between solvent molecules (or macromolecules. In this case a free volume plays an important role) (Fig. 2.3).

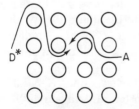

Fig. 2.3　Mutual molecular diffusion of an excited donor (D*) and an acceptor (A) molecule

To gain an appreciation of distance–time relations for molecular diffusion (or electronic energy transfer in solution) a plot of the distance (r), an excited donor molecule (D*) will diffuse in a time period (τ) is shown in Fig. 2.4.[3679, 3680]

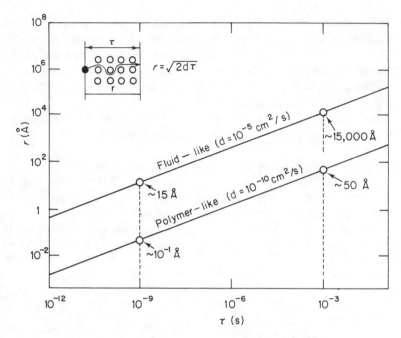

Fig. 2.4 Mean molecular displacement of a molecule in a non-viscous fluid $(d = 10^{-5}\ \text{cm}^2\ \text{s}^{-1})$ and a viscous fluid $(d = 10^{-10}\ \text{cm}^2\ \text{s}^{-1})$[3679]

The relationship between the distance (r) and time (τ) is given by the equation:

$$r = \sqrt{2d\tau} \qquad (2.35)$$

where d = diffusion coefficient.

In Fig. 2.4, equation 2.35 is plotted for $d = 10^{-5}\ \text{cm}^2\ \text{s}^{-1} = 10^{11}\ \text{Å}^2\ \text{s}^{-1}$, a value typical for a molecule diffusing in a fluid organic solvent, and $d = 10^{-10}\ \text{cm}^2\ \text{s}^{-1} = 10^6\ \text{Å}^2\ \text{s}^{-1}$, a value typical of very viscous, nearly rigid solvent (e.g. a polymer matrix).[3679] Considering 1 ns as typical of the lifetime of donor molecule (D*) in the excited singlet state (S_1), this molecule in the fluid solvent will diffuse roughly 15 Å during its life-time. In the more viscous environment, the same molecule will only diffuse about 10^{-1} Å. On other hand, a donor molecule (D*) in the triplet excited state (T_1), whose lifetime is 10^{-3} s may, during its lifetime, diffuse up to 15,000 Å in the fluid solvent or up to 50 Å in the viscous solvent.

If every collision in solution between excited donor (D*) and acceptor (A) molecules leads to energy transfer, the transfer rate will be a diffusion-controlled rate. That is, the rate constant of energy transfer (k_{ET}) is governed by the rate of diffusion of excited donor and acceptor molecules, i.e. $k_{ET} \approx k_{diff}$.

Approximate bimolecular rate constants for diffusion-controlled reactions (k_{diff}) can be obtained from:

(i) The simplified DEBAYE EQUATION:[496, 907, 1006]

$$k_{\text{diff}} = \frac{8R_g T}{3000\eta} \quad (\text{litre mol}^{-1}\,\text{s}^{-1}) \tag{2.36}$$

where:

R_g = universal gas constant,
T = thermodynamic temperature (in kelvin),
η = viscosity of solvent (poise).

For typical organic solvents at room temperature, diffusion-controlled rate constants (k_{diff}) are 10^9 to 10^{10} litre mol^{-1}s^{-1} (Table 2.2).

Table 2.2 Viscosities and diffusion-controlled rate constants ($k_{\text{diff.}}$) at $20°C$

Solvent	Viscosity $(0.001\,\eta)$	k_{diff} $(1\text{ mole}^{-1}/\text{s}^{-1})$
Hexane	3.26	2.0×10^{10}
Benzene	6.47	1.0×10^{10}
Cyclohexane	9.65	6.9×10^9
Water	10.05	6.5×10^9
Ethanol	11.94	5.4×10^9
Ethylene glycol	173.00	3.8×10^8
Glycerol	10690.00	6.0×10^6

(ii) The SMOLUCHOWSKI EQUATION:[3399, 3805, 3975]

$$k_{\text{diff}} = \frac{4\pi R_f dN}{1000} \tag{2.37}$$

where

R_f = distance between the centres of the donor (D) and the acceptor (A) (i.e. collisional radii),
d = sum of the donor (d_D) and acceptor (d_A) diffusion coefficients, i.e.

$$d = \frac{d_D + d_A}{2} \tag{2.38}$$

N = Avogadro's number,

$$R_f = \left(1 + \frac{R}{(d\tau_0)^{1/2}}\right) \tag{2.39}$$

where τ_0 = fluorescence lifetime of the excited donor in the absence of acceptor,
R_f can be taken from equation (2.39)

$$R_f \approx 0.676\,(\alpha/d)^{1/4} \tag{2.40}$$

$$\text{where } \alpha = R_0^6/\tau_0 \tag{2.41}$$

R_0 = critical radius (cf. Section 2.3).

In the case of intramolecular energy migration (cf. Section 2.12), the rate constant for diffusion-controlled reactions (k_{diff}) reflects a collision rate arising from both diffusion and energy migration:

$$k_{\text{diff}} = \frac{4\pi R(d + \Lambda)N}{1000} \left(1 + \frac{R}{[(d + \Lambda)\tau_0]^{1/2}} \right) \tag{2.42}$$

where Λ = energy migration constant ($\text{cm}^2\,\text{s}^{-1}$).

The EFFICIENCY OF ENERGY TRANSFER in solution (Φ_{ET}) can be obtained from the simplified equation:

$$\Phi_{\text{ET}} = \frac{k_{\text{ET}}}{k_{\text{diff}}} = \frac{k_Q}{k_{\text{diff}}} \tag{2.43}$$

where

k_{ET} = rate constant of energy transfer,
k_Q = rate constant of quenching.

The FREQUENCY OF ENERGY MIGRATION (ω) can be calculated from

$$\omega = \frac{6\Lambda}{l^2}\,(\text{s}^{-1}) \tag{2.44}$$

where

Λ = energy migration constant ($\text{cm}^2\,\text{s}^{-1}$),
l = separation between groups at which efficient energy transfer can occur.

Diffusion-controlled quenching of small molecules by polymers in solution, shows that quenching rate per quencher unit is smaller than that for a model compound of low molecular weight.[1055,2788,2988] The quenching rate constants decrease when the molecular weight of polymers used as quenchers increases. This result allows for the determination of QUENCHING OF MACRO-MOLECULAR VOLUME from quenching rate constants. This quenching volume can be related to the actual volume of the macromolecular coil and to the volume of the equivalent hydrodynamic sphere.

2.8. SPECTROSCOPIC METHODS FOR THE DETECTION OF ELECTRONIC ENERGY TRANSFER PROCESSES

When an electronic energy transfer process is determined by spectroscopic methods,[3023] the donor (D) is usually excited by radiation not absorbed by the acceptor (A), and then spectroscopic evidence for the formation of an excited acceptor (A*) is sought.

In order to establish that the resonance–excitation (dipole–dipole) energy transfer mechanism is operating, it must be shown that since resonance excitation does not require collisions, it:

(i) must be able to occur efficiently over distances considerably greater than molecular diameters,

(ii) must be insensitive to the viscosity of the solvents or medium.

Complex formation and the radiative energy transfer (the 'trivial' process) mechanism must also be carefully eliminated as the mechanism responsible for an electronic energy transfer process.

2.9. SINGLET–SINGLET ENERGY TRANSFER

Singlet–singlet energy transfer between a donor (D) and an acceptor (A)

$$^1D \text{ (singlet)} + A \rightarrow D + {}^1A \text{ (singlet)} \tag{2.45}$$

may occur in two ways, e.g.:

(i) Intermolecular energy transfer from donor to acceptor:

$$\boxed{^1D^*} + \boxed{A} \longrightarrow \boxed{D} + \boxed{^1A^*} \tag{2.46}$$

Some of examples of singlet–singlet energy transfer in polymeric systems where a polymer is a donor and low molecular compounds are acceptors is given in Table 2.3. Intermolecular singlet–singlet energy transfer can be indicated easily by using fluorescent probe compounds such as 4,4'-bis-(benzoxazyl)biphenyl or $\alpha,\alpha,\alpha,\alpha',\alpha',\alpha'$-hexachloro-$p$-xylene.[2059,2060,2509,3094] If a polymer film is doped with a fluorescent probe, a probe fluorescence is observed on irradiation with light which is only absorbed by the chromophore present in a given polymer (c.f. Section 4.1).

Table 2.3 Examples of singlet–singlet energy transfer in polymeric systems

Donor (polymer)	Acceptor (low molecular compound)	R_0 (A)	k_{ET}	References
Polystyrene	2,5-Diphenyloxazole			2943, 3120
	1,1,4,4-Tetraphenylbutadiene	20	3×10^9	324
Co(styrene/methyl methacrylate)	1,1,4,4-Tetraphenylbutadiene			851, 3344
Poly(vinylnaphthalenes)	Benzophenone	15	1×10^9	842, 845
	Anthracene			240
	Pyrene			1499, 2025
Co(isobutylene/methyl methacrylate) containing naphthalene groups	Anthracene	23	1.1×10^{10}	3918
Poly(acenaphthylene)	Benzophenone	15		845
Poly(9-phenanthryl methacrylate)	Anthracene			2685
Poly(methyl vinyl ketone)	Benzophenone	8	1.0×10^9	853

(ii) Intramolecular energy transfer from donor to acceptor:

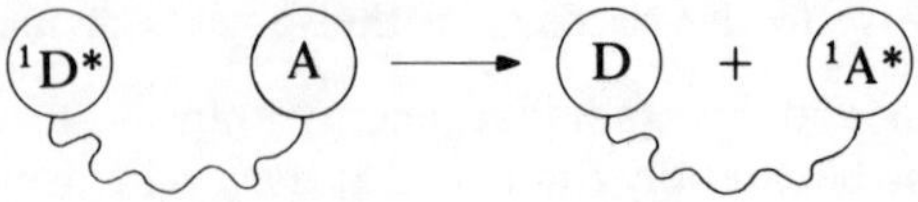

$$^1D^* \quad A \longrightarrow D + {}^1A^* \qquad (2.47)$$

and is allowed by:

(i) Electron exchange (Dexter) mechanism (cf. Section 2.2).
(ii) Resonance excitation (Förster) mechanism (cf. Section 2.3).

A typical example of intramolecular singlet–singlet energy transfer can be presented with following compound (2.1):[3219]

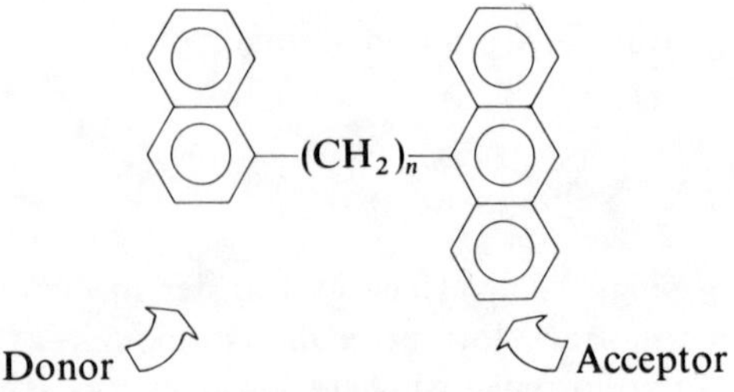

$$(CH_2)_n \qquad n = 1{-}3$$

Donor   Acceptor

(2.1)

A constant quantum yield of fluorescence ($\Phi_F = 0.3$) has been observed for all members of the series independent of whether the anthracene moiety absorbed and emitted the energy or the naphthalene moiety absorbed the energy and transferred it to the anthracene moiety. Energy transfer is here 100% efficient.

For another compound such as dansyl-(L-prolyl)-α-naphthyl (2.2), it has been

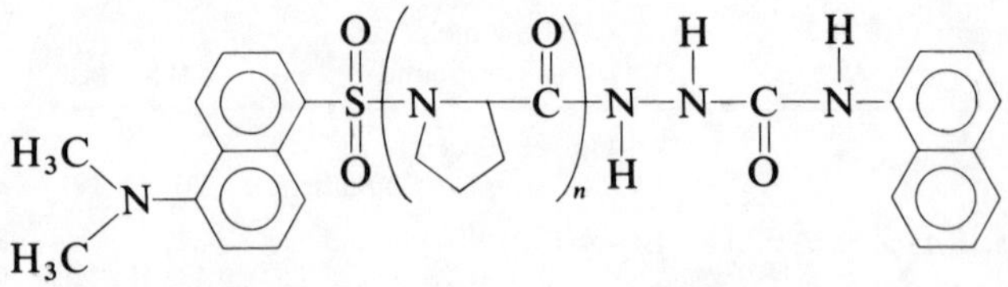

(2.2)

$$n = 1 - 12$$

found that energy transfer efficiency decreases from a value close to 100% for short oligomers to 16% for the $n = 12$ oligomer.[3466] The efficiency of energy transfer as a function of distance is shown in Fig. 2.5.

Similar results have been reported for the energy transfer in polysarcosines having a terminal β-naphthylamide groups (2.3).[3356]

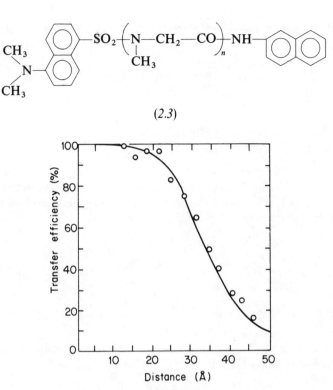

(2.3)

Fig. 2.5 Efficiency of energy transfer as a function
of distance[3466]

Intramolecular energy transfer in polymeric systems has special importance in copolymers (cf. Section 2.12.1).

Singlet–singlet energy transfer can be inferred from the studies of fluorescence or fluorescence depolarization.

Fluorescence quenching studies of a donor (D) in the presence of an acceptor (A) (i.e. quencher (Q)) can be carried out using the Stern–Volmer equation:

$$\frac{I_{F(D)}}{I_{F(D+Q)}} = 1 + k_Q \tau [Q] \tag{2.48}$$

where

$I_{F(D)}$ and $I_{F(D+Q)}$ = fluorescence intensities of the donor (D) in the absence and presence of the quencher (Q),

k_Q = bimolecular quenching rate constant,

τ = fluorescence lifetime of the donor (D) in the absence of quencher (Q),

[Q] = concentration of the quencher.

The plot of $I_{F(D)}/I_{F(D+A)}$ versus [Q] produces a straight line with a slope of $k_Q\tau$ (Fig. 1.18). The τ can be measured independently, and k_Q is easy to calculate.

The ratio $I_{F(D)}/I_{F(D+Q)}$ can be expressed according to the active-sphere model of Perrin as:[2926]

$$\frac{I_{F(D)}}{I_{F(D+Q)}} = \exp(VN[Q]) \tag{2.49}$$

where:

V = volume of active sphere for quenching,
N = Avogadro's number.

The SINGLET ENERGY MIGRATION CONSTANT (Λ_S) can be estimated by measuring the relative quenching efficiency of a contact quencher (typically CCl_4) for a polymer-bound chromophoric group and monomeric model compound:[5, 1502, 1806, 3821]

$$\frac{\Lambda_S}{d} = \frac{k_{Q(polymer)} - 1/2k_{Q(model)}}{1/2k_{Q(model)}} \tag{2.50}$$

where

$d = \dfrac{d_D + d_Q}{2}$ sum of the donor (D) and quencher (Q) diffusion coefficients.

The quenching rate constant k_Q can be calculated using the SMOLUCHOWSKI–EINSTEIN EQUATION:[3768]

$$k_Q = \frac{4\pi N(d_D + d_Q + \Lambda_S)R\rho}{1000} \tag{2.51}$$

where:

N = Avogadro's number,
R = collisional radius for donor (D) and quencher (Q),
ρ = quenching probability per collision.

with the assumption that the product R is equivalent for the polymer-bound chromophoric group and the monomeric model compound.

For the model compound it is assumed that $\Lambda_S = 0$, and for the polymer $d_D = 0$ (i.e. segmental diffusion can be ignored during the lifetime of the excited state).

Singlet energy migration constants (Λ_s) ($cm^2 s^{-1}$) determined by this method are collected in Table 2.4.

2.10. TRIPLET–TRIPLET ENERGY TRANSFER

Triplet–triplet energy transfer between a donor (D) and an acceptor (A)

$$^3D^* \text{ (triplet)} + A \rightarrow D + {}^3A^* \text{ (triplet)} \tag{2.52}$$

may occur in two ways as:

(i) Intermolecular energy transfer from donor to acceptor:

$$\boxed{^3D^*} \;+\; \boxed{A} \;\longrightarrow\; \boxed{D} \;+\; \boxed{^3A^*} \qquad\qquad (2.53)$$

Table 2.4 Examples of singlet energy migration constants (Λ_S)

Polymer	Λ_S (cm^2 s^{-1})	References
Poly(2-vinylnaphthalene)	0	3821
Poly(N-vinylcarbazole)	10^5	3821
Poly[(10-phenyl-9-anthryl) methyl methacrylate]	2×10^{-5}	1502
Poly (9-anthrylmethyl methacrylate)	0	1502
Poly(9-anthrylmethyl ethenyl ether)	0	1502

The triplet–triplet energy transfer in polymeric systems may occur in two cases:

(a) where the polymer is a donor and the low molecular compound is acceptor (Table 2.5),

Table 2.5 Examples of triplet–triplet energy transfer in polymeric systems

Donor (polymer)	Acceptor	R_0 (A)	k_{ET}	References
Co(ethylene/carbon monoxide)	1-cycloactadiene			
	Naphthalene			1581, 1582
Polystyrene	Piperylene			1163
Poly(methyl vinyl ketone)	1-cis, 3-cis-cyclo-octadiene			826
	Naphthalene	11		853
Poly(phenyl vinyl ketone)	Naphthalene	26		839, 843, 2021, 2050
	1-methylnaphthalene			3174
	Trans-stilbene			2401
Poly(vinylbenzophenone)	Naphthalene	36	1.0×10^5	840, 841
	1-Methylnaphthalene			3719
	Ferrocene			3176
Co(styrene/vinylbenzophenone)	Naphthalene	30		848
Poly(acrylophenone)	Naphthalene			2021
	1-methylnaphthalene			3174
Poly(1-benzoylethylene)	1-Methylnaphthalene			3174
Poly(vinylnaphthalenes)	Biacetyl			3820
	1,3-Pentadiene	15	1.0×10^2	845
	Piperylene	11		773, 2905
Poly(naphthyl methacrylate)	Piperylene			3408
	1,3-Cyclooctadiene			3408
Poly(acenaphthylene)	Biacetyl			3820
	Piperylene	15		845
Poly(ethylene terephthalate)	4,4'-bisphenyldicarboxylate			927
Poly(m-phenyleneiso-phtalamide)	2-(2'-methoxy-t'-t-butylphenyl)benzotriazole			3865

(b) where the low molecular compound is a donor and the polymer is an acceptor (this case is called TRIPLET SENSITIZATION OF POLYMERS and the donor is called a SENSITIZER) (Table 2.6).

Table 2.6 Examples of triplet sensitization of polymers

Donor (sensitizer)	Acceptor (polymer)	References
Biacetyl Benzyl Benzophenone	Polystyrenes Poly(naphthalenes) Poly(4-vinylbiphenyl)	2788
Benzophenone N-ethylphenone Triphenylene Phenanthrene 3-Acetylphenanthrene Chrysene Fluoroxanthene	Poly(4-vinylbiphenyl)	2990, 2991
Benzyl Benzophenone Fluorenone Anthracene Phenanthrene 3-Ethylphenanthrene Chrysene Fluoroanthrene Pyrene	Poly(2-vinylnaphthalene)	773, 2988

(ii) Intramolecular energy transfer from donor to acceptor:

$$^3D^* \quad A \longrightarrow D \quad {}^3A^* \tag{2.54}$$

and is allowed only by the electron exchange (Dexter) mechanism (cf. Section 2.2) but forbidden by the resonance excitation (Förster) mechanism (cf. Section 2.3).

When the triplet energy of the donor (E_T) is 3 kcal mol^{-1} or more higher than the acceptor triplet energy, the energy transfer rate (k_{ET}) is about the diffusion-controlled rate constant. When the triplet energy level of the donor is not 3 kcal mol^{-1} above the acceptor level, energy transfer becomes rather inefficient.[1568, 1569, 2974, 3454]

When a triplet donor has insufficient excitation energy to promote an acceptor to its triplet state, this deficiency can be supplied as an activation energy (ΔE):

$$\frac{\Delta (\log k)}{\Delta E} = -\frac{1}{2.303 \, R_g T} \tag{2.55}$$

where

R_g = universal gas constant,
T = thermodynamic temperature (in kelvin).

A typical example of intramolecular triplet–triplet energy transfer can be presented with the following compound (2.4):[2207, 2246]

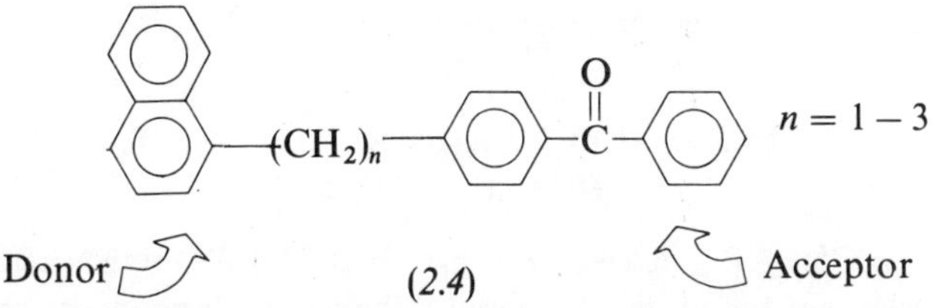

$$n = 1 - 3$$

Donor   (2.4)   Acceptor

Excitation of the naphthalene moiety with radiation at 313 nm produces emission of phosphorescence from a benzophenone moiety at 360 nm. Triplet–triplet energy transfer for $n = 1, 2, 3$ occurs with 100 % efficiency. For the above group of three compounds singlet–singlet energy transfers with high efficiency ($n = 1$, 98 %; $n = 2$, 80 %; and $n = 3$, 94 %) also occur.

Triplet–triplet energy transfer can be inferred from the studies of phosphorescence or delayed fluorescence.

2.11. INTERMOLECULAR ENERGY MIGRATION BETWEEN DONOR AND ACCEPTOR RANDOMLY DISTRIBUTED IN AN INERT POLYMER MATRIX

Low molecular solid donor (D) and acceptor (A) molecules can be randomly distributed in an inert polymer matrix (Table 2.7). Energy transfer from the excited donor (D*) to an acceptor molecule (A) occurs via interactions during which the polymer serves as an inert matrix which prevents molecular diffusion of D* during the lifetime of D* (Fig. 2.3).

Table 2.7 Examples of energy migration between donor and acceptor randomly distributed in an inert polymer matrix

Donor	Acceptor	Polymer matrix	Reference
Diphenylanthracene	Tetrabromo-o-quinone	Polystyrene	3100
N-isopropylcarbazole	Dimethylterephthalate Perylene	Polystyrene	1884
Pyrene	Sevron yellow	Poly(acrylonitryl)	402

If the intermolecular separation between donor molecules is so large that the probability for multi-step excitation becomes vanishingly small, the transfer of electronic excitation energy from an initially excited donor (D*) to the acceptor

(A) occurs in a single step via resonance–excitation energy transfer between the transition moments of the donor–acceptor pair (cf. Section 2.3).

Net molecular crystals or polymer films (matrices) are examples in which energy transfer is usually dominated by a process involving multi-step migration of the excitation among molecules of the host followed by transfer to a guest impurity, whereas in dilute rigid solutions energy transfer is a single-step process where the rate constant for transfer is time-dependent due to the random nature of the excited donor–acceptor separation.

2.12. INTERMOLECULAR ENERGY MIGRATION IN POLYMERS

INTRAMOLECULAR TRANSFER OF EXCITATION ENERGY (singlet and/or triplet energy migration) in vinyl polymers with pendant chromophoric groups (M) (Table 2.8) may occur:[1441, 2335, 2733, 3214, 3216, 3679]

Table 2.8   Examples of intramolecular energy migration in polymers

Polymer	Structure	Energy migration	Reference
Polystyrene	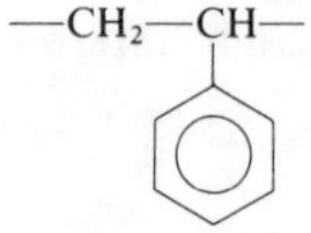 —CH_2—CH— ⬡	Singlet	845, 1558, 2354, 3726
Poly(vinyltoluene)	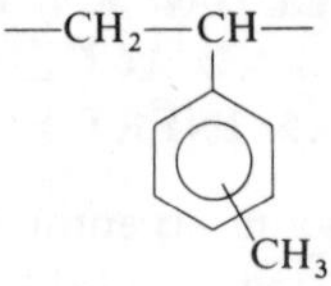 —CH_2—CH— ⬡—CH_3	Singlet	2983
Polystyrene/polyphenylene oxide blends		Singlet	2138, 2139
Poly(benzyl methacrylate)	CH_3 | —CH_2—CH—	Singlet	6
Poly(2-phenyl-ethyl methacrylate)	| C=O		
Poly(3-phenyl-propyl methacrylate)	| O | $(CH_2)_n$ ⬡		
Poly(methylvinyl ketone)	—CH_2—CH— | C=O | CH_3	Triplet	826, 3407

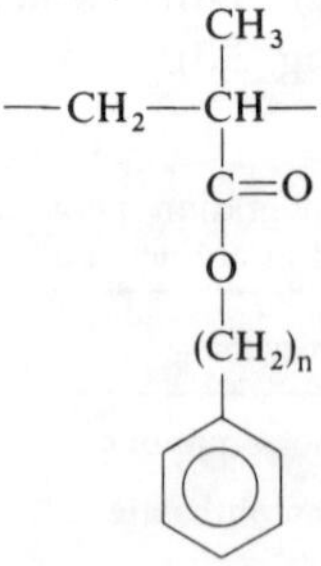

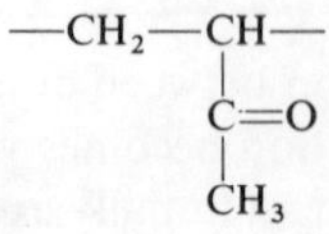

Table 2.8 (*Contd.*)

Polymer	Structure	Energy migration	Reference
Poly(phenylvinyl ketone)	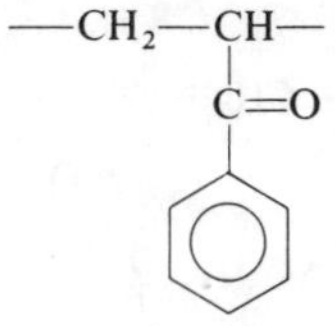 —CH₂—CH—	Triplet	843, 1047, 2021, 2050
Poly(*p*-methoxy-phenylvinyl ketone)	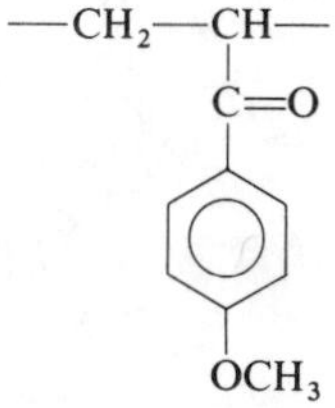 —CH₂—CH—	Triplet	3198, 3290
Poly(vinylbenzophenone)	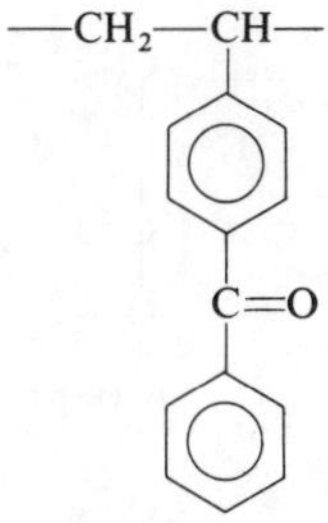 —CH₂—CH—	Triplet	841, 3719
Poly(1-vinylnaphthalene)	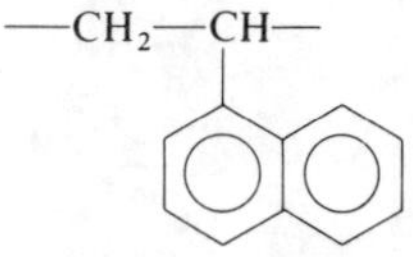 —CH₂—CH—	Singlet and triplet	773, 845, 2726
Poly(2-vinylnaphthalene)	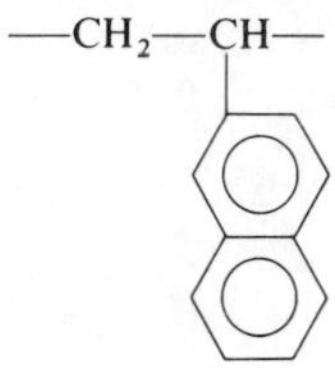 —CH₂—CH—	Singlet	2354, 2903, 3821
Poly(2-*tert*-butyl-6-vinyl-naphthalene)	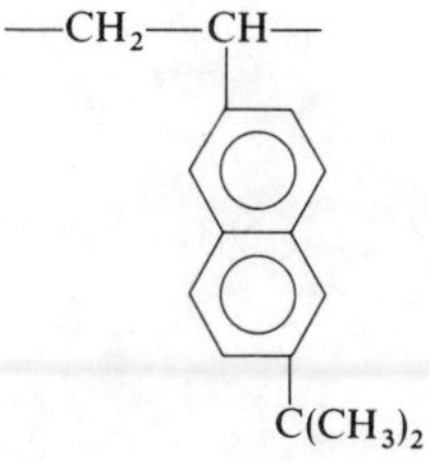 —CH₂—CH—	Singlet	2618

The structures show the following pendant groups:
- Poly(phenylvinyl ketone): $C=O$ attached to a phenyl ring
- Poly(*p*-methoxy-phenylvinyl ketone): $C=O$ attached to a phenyl ring with OCH_3 para substituent
- Poly(vinylbenzophenone): phenyl ring with $C=O$ linked to another phenyl ring
- Poly(1-vinylnaphthalene): naphthalene attached at 1-position
- Poly(2-vinylnaphthalene): naphthalene attached at 2-position
- Poly(2-*tert*-butyl-6-vinyl-naphthalene): naphthalene with $C(CH_3)_2$ group

Table 2.8 (*Contd.*)

Polymer	Structure	Energy migration	Reference
Poly(2-naphthyl methacrylate)	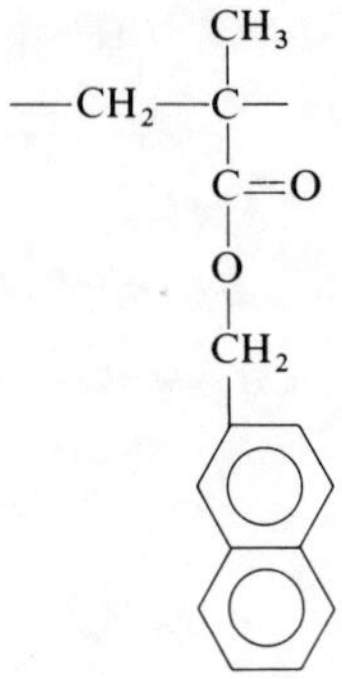	Singlet and triplet	1622, 2024, 3408
Poly(2-naphthylalkyl methacrylate)	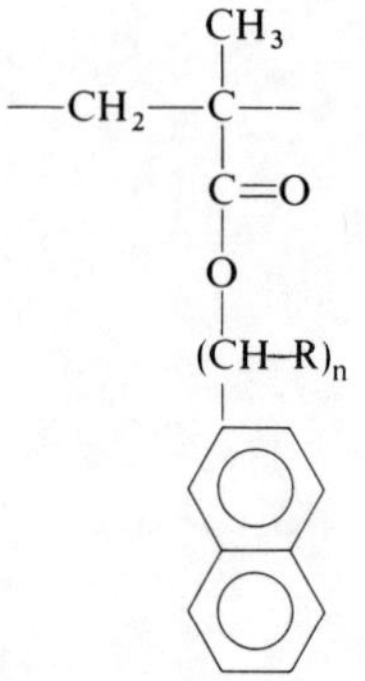	Singlet	2615
Co(1-vinylnaphthalene/styrene)		Triplet	1163
Co(1-vinylnaphthalene/methyl methacrylate)		Triplet	1163
Polypeptydes with 1-naphthyl group	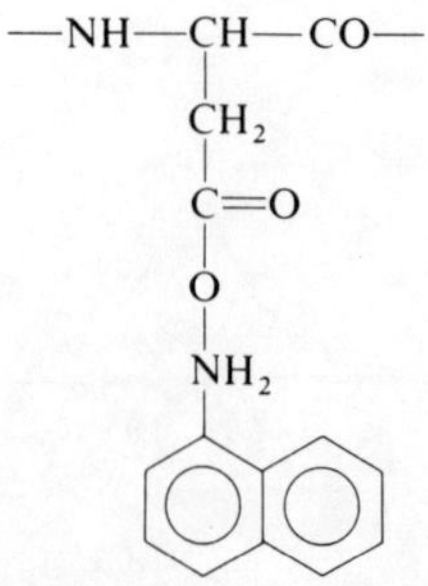	Singlet	3702, 3704

Table 2.8 (*Contd.*)

Polymer	Structure	Energy migration	Reference
Poly (γ-1-naphthyl methyl L and DL glutamates) Co (γ-1-naphthyl methyl-L-glutamate/γ-benzyl-L-glutamate)	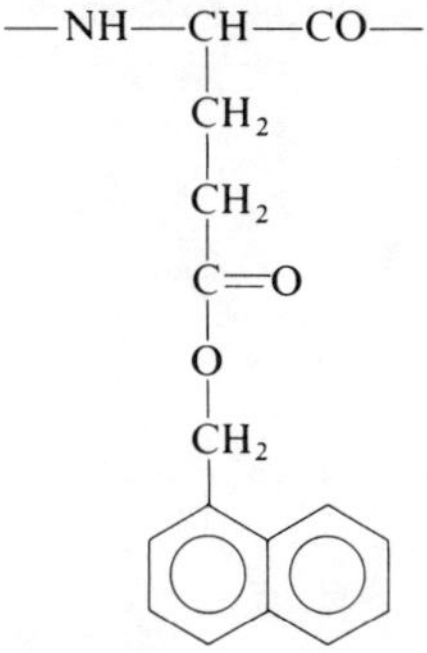 —NH—CH—CO— CH_2 CH_2 C=O O CH_2 (naphthyl)	Singlet	3703
Poly(acenaphthylene)	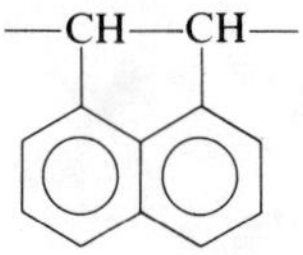 —CH—CH— (naphthalene)	Singlet and triplet	583, 845
Poly(methoxyphenyl-acetylene)	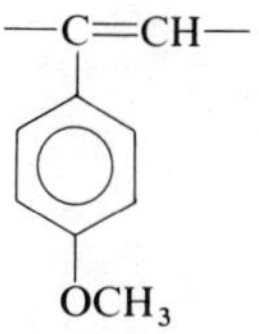 —C=CH— (phenyl) OCH_3	Singlet	2734
Poly(9, 10-diphenyl-anthracene)	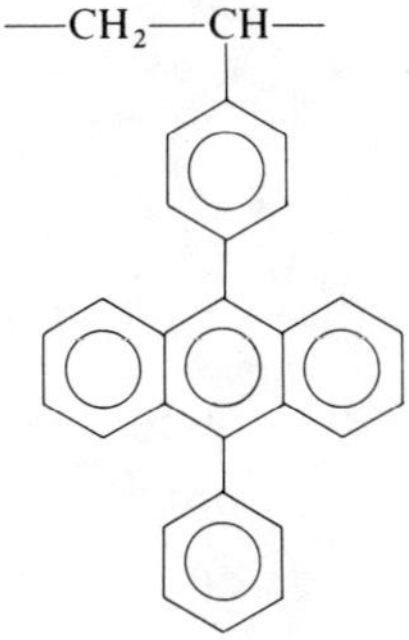 —CH_2—CH— (phenyl) (anthracene) (phenyl)	Singlet	1503
Poly(9-anthrylmethyl methacrylate)	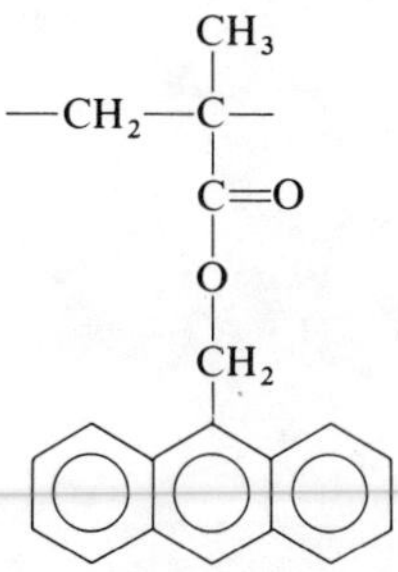 CH_3 —CH_2—C— C=O O CH_2 (anthracene)	Singlet	1502, 1515, 2684

Table 2.8 *(Contd.)*

Polymer	Structure	Energy migration	Reference
Poly(9-phenylanthryl methyl methacrylate)	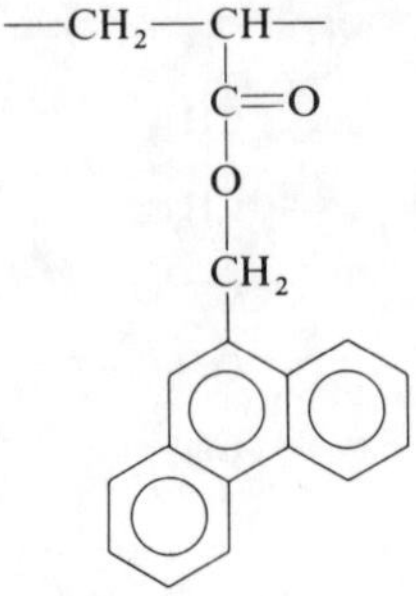	Singlet	2684
Poly[(10-phenyl-9-anthryl) methyl methacrylate]	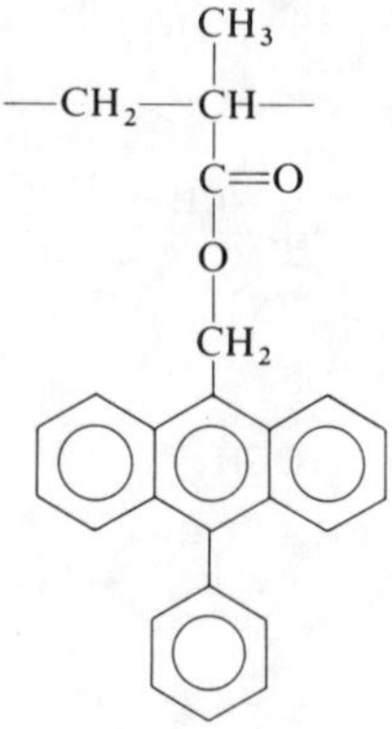	Singlet	1501, 1502
Poly(9-anthrylmethyl ethenyl ether)	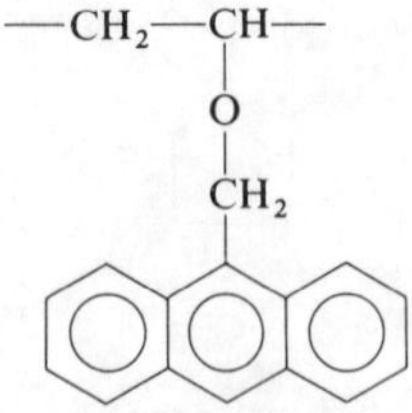	Singlet	1502
Poly(N-vinylcarbazole)	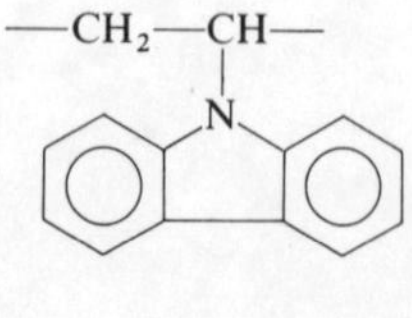	Singlet and triplet	569, 1815, 1887, 2058, 2059, 2062, 2071, 2072, 2078, 3114, 3222, 3821, 3985

Table 2.8   (*Contd.*)

Polymer	Structure	Energy migration	Reference
Poly(ethylene terephthalate)		Singlet	3521
Poly(vinylnaphthalimide)		Triplet	2212
Polynucleotides (DNA)		Singlet and triplet	1424

$$-\overset{O}{\underset{\|}{C}}-\bigcirc-\overset{O}{\underset{\|}{C}}-O-CH_2CH_2-O-$$

(i) from group to group along the chain (DOWN-CHAIN ENERGY MIGRATION),

(ii) from group to group in a coiled chain (INTRACOIL ENERGY MIGRATION) (Fig. 2.6).

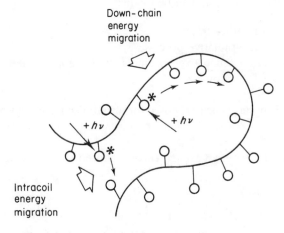

Fig. 2.6   Intramolecular energy migration in polymers

Intramolecular transfer of excitation energy depends on:

(i) chromophores orientation in different polymer conformations by a rotation of a chain segments (down-chain energy migration);

(ii) cross-chain segmental diffusion, which may occur within some 'critical distance' between segments. The probability of segments lying within this 'critical radius' of each other depends on coil density and hence molecular weight.[830, 2988, 2991]

Energy absorbed by an aromatic chromophore group attached to polymer chains can be transferred to TRAPS located in the polymer chains.[240, 248, 849–851, 1155, 1156, 1164, 1441, 1559, 1883, 2073, 2235, 2449, 2465, 2608, 2699, 2983, 3408, 3409, 3726, 3983, 3984]. There are the following types of TRAPS:

(i) Low-energy impurity sites[1165]:
    (a) Impurities that are not part of a polymer chain.
    (b) Chemical anomalies, such as oxidation sites, or those introduced in an initiation or termination step.
    (c) Physical defects in a relatively small number of polymer chains, such as those due to chain folding or strain points in a solid polymer.
(ii) Excimer forming sites (cf. Section 3.3).
(iii) Charge-transfer (CT) complex sites. For example a charge-transfer acceptor such as 1,2,4,5-tetracyanoethylene quenches fluorescence of an excited charge transfer donor, e.g. poly(N-carbazolyl methyl methacrylate),[362] whereas 1,2,4,5-tetracyanobenzene can quench fluorescence of poly(2-vinylnaphthalene).[2025]

Energy accumulated in a trap can be further:

(i) transferred to another traps,
(ii) dissipated by non-radiative processes,
(iii) emitted as fluorescence (e.g. excimers emit trapped energy as excimer fluorescence, c.f. Section 3.2).

Several mechanisms have been proposed for intramolecular energy migration in a polymer chain:

(i) Energy transfer from an excited chromophore (M*) to a trap (A)[1164, 1598]

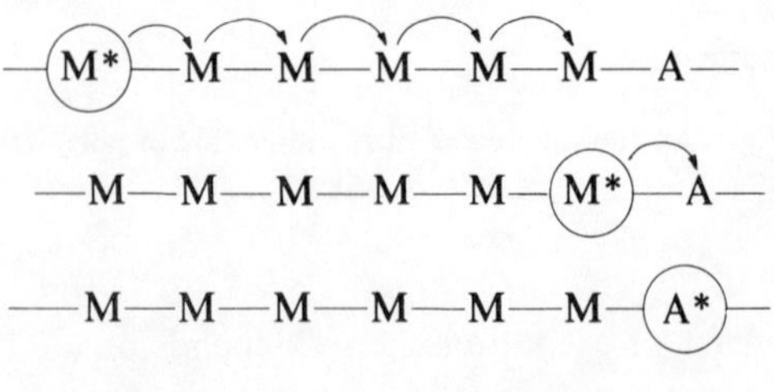

$$—(M^*)—M—M—M—M—M—A—$$
$$—M—M—M—M—M—(M^*)—A—$$
$$—M—M—M—M—M—M—(A^*)—$$

(2.56)

(ii) Successive migration from an excited chromophore (M*) to excimer forming site (M* ... M) and next to a trap (A)[1559]

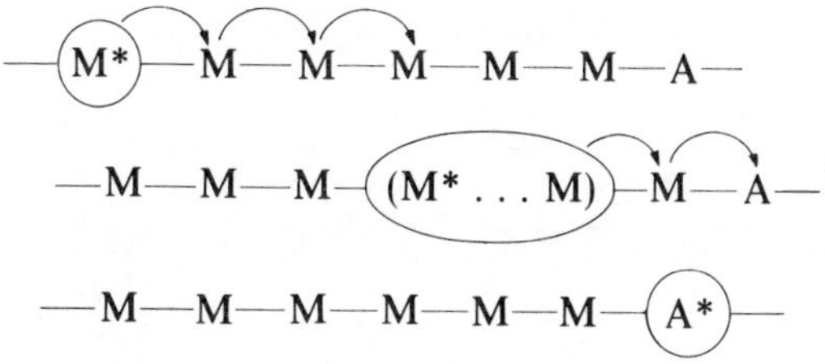

$$—(M^*)—M—M—M—M—M—A—$$

$$—M—M—M—\big((M^* \ldots M)\big)—M—A— \qquad (2.57)$$

$$—M—M—M—M—M—M—(A^*)—$$

(iii) Independent energy transfer from both excited chromophore (M*) and an excimer (M* . . . M) to a trap (A)[851, 2983]

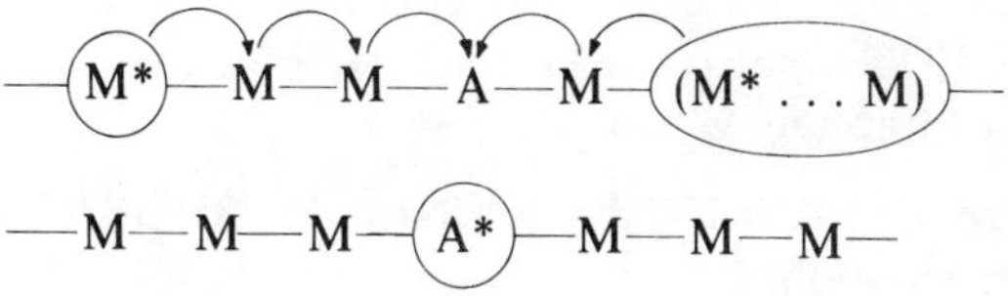

$$—(M^*)—M—M—A—M—\big((M^* \ldots M)\big)— \qquad (2.58)$$

$$—M—M—M—(A^*)—M—M—M—$$

(iv) Energy transfer via excimer (M* . . . M) which dissociate to excited chromophore (M*) which then transfers energy to a trap (A)[2943]

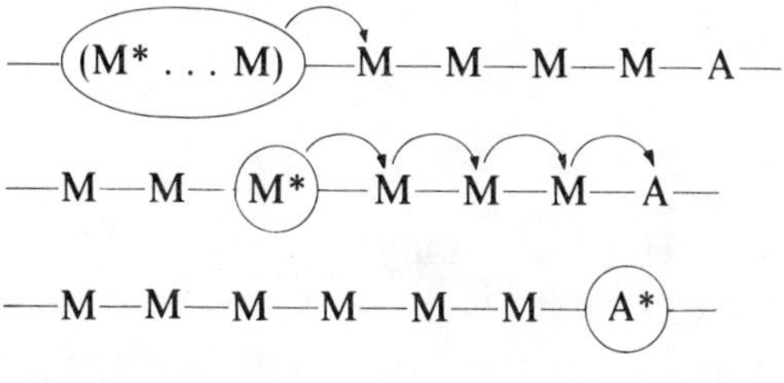

$$—\big((M^* \ldots M)\big)—M—M—M—M—A—$$

$$—M—M—(M^*)—M—M—M—A— \qquad (2.59)$$

$$—M—M—M—M—M—M—(A^*)—$$

Time-resolved fluorescence spectroscopy shows that in the case of polystyrene labelled with phenyl oxazole (trap), energy transfer is more efficient from the excited monomer (M*) than from excimer (M* . . . M). One possible explanation is that the concentration of excimer sites in the polymer is low compared to the monomeric chromophores. In this case the high yield of excimer to monomer fluorescence observed for polystyrene must imply singlet energy migration in the polymer so that any exciton (cf. Section 2.13) may have a reasonable probability of encountering a potential excimer forming site (cf. Section 3.3).

There are several polymers or copolymers where the absorbing chromophore group (M) is separated from the trap (A) by inert spacers (C) which do not participate in energy transfer:

$$—M—C—C—C—C—C—A— \qquad (2.60)$$

In such polymers only intracoil energy migration can be considered

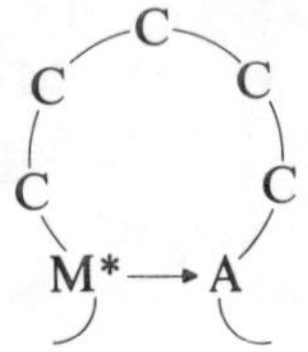

$$
\begin{array}{c}
\text{—C—} \\
\text{C} \qquad \text{C} \\
\text{C} \qquad \text{C} \\
\text{M}^* \longrightarrow \text{A}
\end{array}
\qquad (2.61)
$$

The fraction of the polymer coil over which the energy migration occurs depends on:

(i) The average 'HOPPING TIME' (τ_h) between chromophores relative to the lifetime of the excited state of chromopheres (τ_M^*). The APPROXIMATE NUMBER OF TRANSFER 'STEPS' is τ_M^*/τ_h.

(ii) The density of trapping groups (A).

The QUANTUM EFFICIENCY OF ENERGY MIGRATION (Φ) can be determined from equation:

$$
\Phi = \frac{\text{Number of excited state of traps (A*)}}{\text{Number of photons absorbed by chromophore (M)}} \qquad (2.62)
$$

The ENERGY MIGRATION CONSTANT (Λ) is given by:

$$
\Lambda = \frac{l^2}{2\tau_h} \; (\text{cm}^2 \, \text{s}^{-1}) \qquad (2.63)
$$

where:

l = average chromophore separation,

τ_h = average hopping time for the nearest neighbour transfer.

The MEAN FREE DISTANCE OF ENERGY MIGRATION (L) is given by

$$
L = (2\Lambda\tau_M^*)^{1/2} = l(\tau_M^*/\tau_h)^{1/2} \qquad (2.64)
$$

The AVERAGE NUMBER OF ENERGY TRANSFER STEPS (S) is given by

$$
S = (\tau_M^*/\tau_h)^{1/2} \qquad (2.65)
$$

The RATE OF ENERGY MIGRATION between polymer-bound chromophores depends on:

(i) Polymer-bound chromophore average separation.

(ii) Average mutual chromophore separations.

(iii) Distribution of chromophore separations. Breaks in a sequence of chromophore can impose a barrier to energy migration.

(vi) Spatial relationship that exist between adjacent or non-neighbouring chromophores.

Energy migration in polymers depends on macromolecular structures:

(i) Type of polymer backbone.

(ii) Type of bonding sequence by which the chromophore is attached to the backbone.

(iii) Tacticity of the polymer.

(iv) Molecular weight of the polymer. The molecular weight of a polymer affects both the coil density for a given solvent and the number of chromophores per coil.

An increased molecular weight increases the probability of multiple photon absorption on the same coil. Intracoil state annihilation (cf. Section 2.15) requires at least two excitations per coil. The increasing of the molecular weight will tend to enhance annihilation processes.[2075, 2076, 2903, 2992]

(v) Coil density, which affects the extent of intracoil energy migration ('cross-chain energy migration') between non-adjacent chromophores that are on different segments of the polymer chain. However, it is not possible to clearly distinguish intracoil energy migration from down-chain energy migration (i.e. between adjacent chromophores).

Energy migration in polymers in solution is affected in two ways:

(i) Excited state of chromophore groups quenching by:
  (a) External heavy atoms. Iodo- or bromo-solvents and even CCl_4 quench effectively excited states. Dichloromethane (CH_2Cl_2) does not quench excited states.
  (b) Charge-transfer (CT) complex formation. Solvents with cyano, amino or aromatic ester groups may form exciplexes (cf. Section 3.5) with aromatic chromophores, which may give rise to an exciplex fluorescence or simply quench.
  (c) Energy transfer to solvent molecules. In such a case it is very difficult to excite the polymer-bound chromophore in such solvents.

(ii) Coil density. The thermodynamic quality of the solvent affects the coil density, and in turn affects the average chromophore separation.[3022] The poorer solvents enhance the relative intensity of excimer fluorescence (cf. Section 3.4.7). Poor solvents increase efficiency of intracoil energy migration.

2.12.1. Intramolecular Energy Migration in Copolymers

In copolymers one chromophore can act as an absorbing chromophore (donor) and another chromophore as an energy trap (acceptor) (Table 2.9). The degree of randomness in the copolymer is critical to the energy transfer properties.[194, 195, 3417] Alternating copolymers have photophysical properties considerably different from normal homopolymers or random copolymers.[1162, 3985]

2.13. EXCITONS

The migrating quantum of electronic energy is termed an EXCITON. This term, carried over from crystal theory,[1368, 2073, 3680] has been applied to energy migration in polymers.[1883, 2011, 2073, 2947, 3409, 3740] This movement of an exciton wave down a polymer chain implies a highly ordered state with translational and orientational equivalence of each dipole and sufficiently small interchromophore distance that there is a weak interaction between an excited chromophore and its ground-state neighbours. Considering a polymer as an ideal 'one-dimensional crystal' model the existence of exciton can be proved.[2194] There is as yet no evident experimental proof for exciton migration in polymers. However, it is useful to use singlet (or triplet) exciton formulation in the consideration of energy migration.

2.14. ANTENNA EFFECT

The LIGHT-HARVESTING ANTENNA EFFECT is a well known factor in photobiology, and plays an important role in photosynthesis.[309, 2969, 2970, 4013] In green plants the primary photochemical step occurs by a one-electron transfer from the excited singlet state (S_1) of chlorophyll (2.5) to an electron acceptor which contains of quinone moiety such as ubiquinone (2.6):

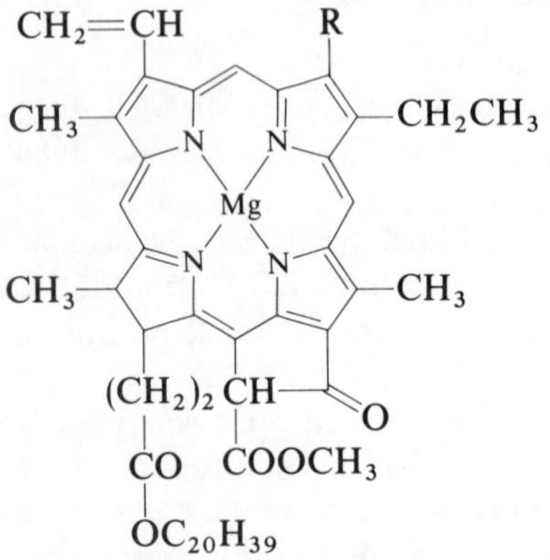

$$CH_3O \quad \left[CH_2CH=C{\overset{\displaystyle CH_3}{|}}CH_2 \right]_n H \qquad (2.6)$$

R = CH$_3$: chlorophyll *a*
R = CHO: chlorophyll *b*

(2.5)

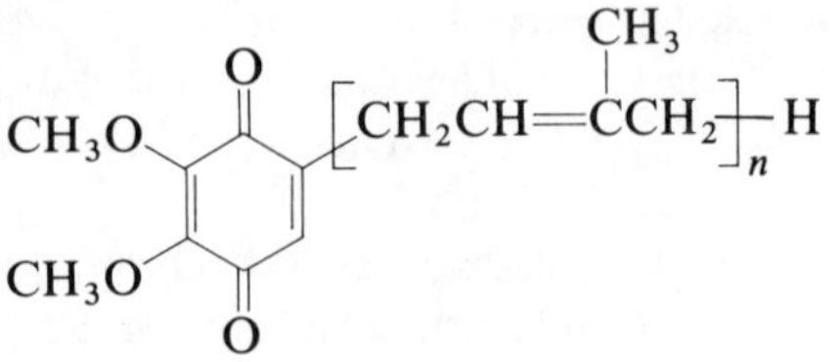

Chlorophyll in the chloroplast is photochemically inert, but acts as an ANTENNA, absorbing and transferring light through non-radiative interactions to the reactive centres. This reaction takes place within a reaction centre protein that spans the thylacoid membrane of the chloroplast organelle of green leaves and algae.

Chlorophyll absorbs at longer wavelengths than any of the other pigments[1372] and will therefore always be the ultimate acceptor in the antenna. Chlorophyll (*a*) is present in all photosynthetic systems capable of evolving oxygen (i.e. all except bacteria) and is usually the major pigment. In green plants and algae the principal pigments are chlorophyll (*b*) and carotene. In red and blue-green algae the

Table 2.9   Examples of intramolecular energy transfer in copolymers

Name	References
co(styrene/methyl methacrylate)	1162
co(styrene/1-vinylnaphthalene)	2465, 3079, 3080
co(styrene/2-vinylnaphthalene)	1162, 3079, 3080
co(styrene/2-naphthylmethacrylate)	1162
co(styrene/bis(1-naphthylmethyl fumarate))	1617
co(styrene/10-phenyl-9-anthryl-methyl methacrylate)	1501
co(styrene/phenylacetylene)	1444, 2361
co(styrene/2-phenyl-5(p-vinyl)phenyl-oxazole)	2465
co(styrene/methacrylophenone)	3198
co(styrene/phenyl vinyl ketone)	2049
co(1-vinylnaphthalene/methyl methacrylate)	194
co(2-vinylnaphthalene/methyl methacrylate)	1162, 3079, 3080
co(2-vinylnaphthalene/phenylvinyl ketone)	830, 1620
co(2-vinylnaphthalene/3-vinylpyrene)	1499
co(9-anthryl methyl methacrylate/9-phenylanthryl methyl methacrylate)	2684
co(phenylvinyl ketone/methyl methacrylate)	830, 2049
co(phenylvinyl ketone/vinyl acetate)	2049
co(phenylvinyl ketone/2-naphthyl methacrylate)	1622
co(methacrylophenone/methyl methacrylate)	2022
co(vinylbenzophenone/vinyl ferrocene)	3176

principal pigments are phycoerithrin and phycocyanin. These pigments extend the region of the sun's spectrum that is absorbed in the direction of shorter wavelengths.

2.14.1. Antenna Effect in Polymeric Systems

By analogy with the biological process of photosynthesis, the antenna term has been transferred to macromolecules containing aromatic chromophores[1436, 1441] which may also display electronic energy transfer to low-energy traps[240, 1616, 1620, 1621, 2617, 2620, 2684, 2685] (cf. Section 2.12).

The function of the connecting macromolecular chains and the plant thylacoid membrane is similar, in that both serve as anchors, supporting high local concentrations of chromophoric groups.

The most important process in photo-harvesting systems is singlet–singlet energy transfer (cf. Section 2.9). The photon collecting efficiency is not entirely due to energy migration among the chromophoric groups making up the antenna but to a combination of energy migration (cf. Section 2.12) and direct energy transfer (cf. Section 2.9) to the trap.[240, 1622, 1679] Fluorescence from these traps provides information on the mechanism and time scale of singlet–singlet energy transfer from the donor chromophores to the acceptor trap. On the basis of the fluorescence decay curves of the acceptor (trap) following pulsed excitation of the donor it was proposed that energy transfer involves two processes (Fig. 2.7).[1620, 1621]

66

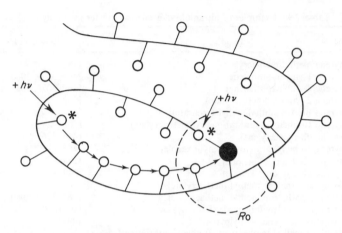

Fig. 2.7 Mechanism of the antenna effect in naphthalene (○) substituted
polymer, containing an anthracene (●) trap[1621]

(i) One-step resonance excitation (dipole–dipole) energy transfer (cf. Section 2.3) from the donor to the acceptor. This transfer occurs when the photoinitiated donor lies within a sphere of radius R_0 characteristic of the strength of the dipole–dipole interaction between two chromophores. But in this process a rate of energy transfer is angular-dependent.[820]

(ii) A process which involves singlet–singlet energy migration between donor chromophores,[2353] followed by the dipole–dipole energy transfer to the acceptor once the excitation resides on a chromophore group a distance of the order of R_0 from the trap. It has been proposed that this energy migration can be described by a random walk of discrete hops of the excitation,[1472, 1884, 3858] and that the chromophore group separations over which donor–donor energy transfer occurs are consistent with those predicted by the dipole–dipole energy transfer mechanism.[1343, 1617]

Extensive studies have been carried out of a variety of polymers containing naphthalene and phenanthrene donors with traps containing the anthracene moiety,[239, 240, 1439, 1616, 1617, 1619, 1620, 1679, 2613, 2616, 2619, 2684–2686, 3599] and copolymers of poly(1- and 2-naphthylmethyl methacrylate) with trap sites including methyl and phenyl ketones.[1620, 1622, 2490]

The detection and quantification of the role of energy migration in antenna polymer systems requires very careful experimentation and analysis. The antenna effect can be employed to

(i) stabilize synthetic polymers against photodegradation,
(ii) create polymers with enhanced photodegradation rate,
(iii) create polymeric photocatalysts that can be effective at extremely small concentration,
(iv) extend solar energy conversion processes. Shifting the absorption of a

polymer 'antenna' into the visible region is of significance from the point of view of possible solar energy applications.

2.15. EXCITED STATE ANNIHILATION PROCESSES

There are two types of excited state annihilation processes:

(i) HOMOGENEOUS ANNIHILATION

$$A^* + A^* \rightarrow A^{**} + A \tag{2.66}$$

(ii) HETEROGENEOUS ANNIHILATION

$$A^* + B^* \begin{cases} \nearrow A^{**} + B \\ \searrow A + B^{**} \end{cases} \tag{2.67}$$

where A^*, B^* are molecules (or groups) in excited singlet or triplet states, and A^{**} or B^{**} are molecules (or groups) in highly excited states. In these processes the highly excited state (A^{**} or B^{**}) may rapidly degrade to the lowest excited state of that multiplicity, or may undergo a chemical reaction such as ionization or dissociation.

The annihilation mechanism requires an overlap of molecular orbitals, and hence a close approach by the annihilating pair of excited state molecules.[3822]

Excited state annihilation can occur if the excited state molecules diffuse together in solution, or if the excitations can themselves migrate within a crystal lattice or along a polymer chain.

Singlet–singlet (S–S) annihilation and triplet–triplet (T–T) annihilation are both homogeneous annihilation processes.

S–S and T–T intracoil annihilation processes in solution are more important in polymer photopysics than is usually the case with small molecules in solution.[403, 2992, 3822]

2.15.1 Singlet–singlet Annihilation Process

Two photons are absorbed by two similar chromophore groups on the same chain, forming excited singlet states (S^*). The singlet excitation energy migrates as singlet exciton (either as a delocalized wave or by series of hopping events, cf. Section 2.12) along the polymer chain until two singlet excitons come close enough to undergo annihilation (S–S exciton annihilation) with formation of a ground state (S_0) and highly excited singlet state (S^{**}):

$$S_0 + S_0 + 2h\nu \rightarrow S^* + S^* \tag{2.68}$$

$$S^* + S^* \xrightarrow{\text{singlet energy migration}} S^{**} + S_0 \tag{2.69}$$

The highly excited S^{**} can be deactivated by emission of fluorescence or dissociate into free radicals

$$S^{**} \rightarrow S^* \rightarrow S_0 + h\nu \tag{2.70}$$

$$S^{**} \rightarrow R\cdot + \cdot R \tag{2.71}$$

Singlet–singlet (S–S) annihilation is characterized by a strong non-linear intensity dependence of S* fluorescence, transient absorptions, photoproducts formation and yield of intersystem crossing (ISC) to the triplet state.

S–S annihilation has been found in several polymers (Table 2.10). A recent discovery has been enhanced photoionization in poly(vinyl-aromatics) such as poly(2-vinylnaphthalene) and poly(4-vinylbiphenyl) solutions in CH_2Cl_2 at higher laser intensities.[2992] It has been suggested that this photoionization is the result of the S–S annihilation reaction occurring within the polymer chain. This annihilation results in a highly excited singlet state (S**) which donates an electron to the CH_2Cl_2 solvent. After capturing an electron the alkyl halide solvent undergoes a dissociative attachment reaction to form a neutral radical and a chloride ion that eventually protects the radical cation from the recombination reaction with the electron.

Table 2.10 Examples of polymers which exhibit singlet–singlet (S–S) annihilation process

Polymer	Structure	References
Poly(2-vinylnaphthalene)	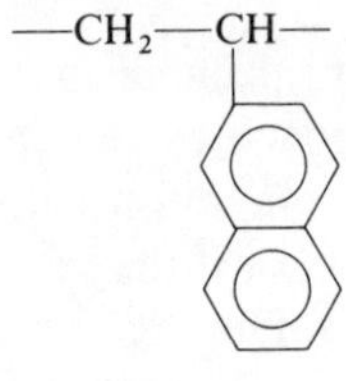	403, 2992
Poly(4-vinylbiphenyl)	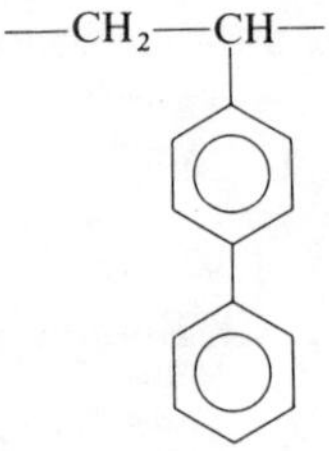	2992
Poly[(10-phenyl)-9-anthryl methyl methacrylate]	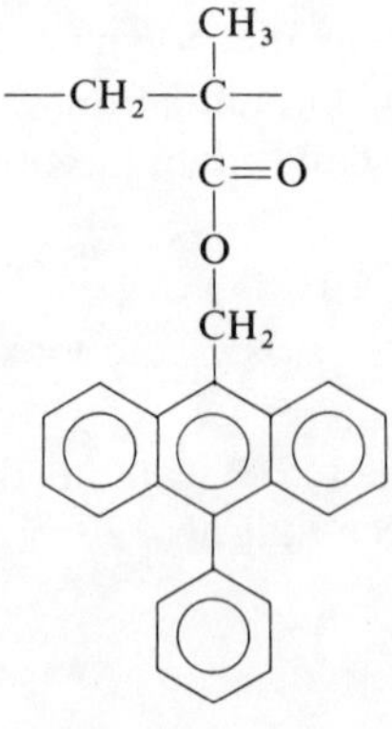	1502

Table 2.10 (*Contd.*)

Polymer	Structure	References
Polyesters with pendant 1-pyrenyl group	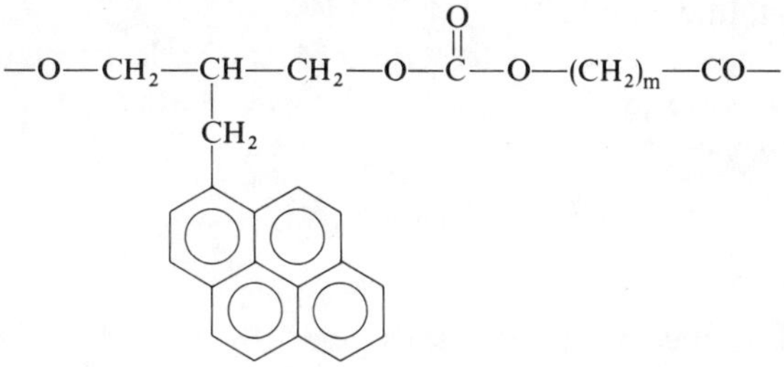	2407

$$-O-CH_2-CH-CH_2-O-\overset{\overset{\displaystyle O}{\|}}{C}-O-(CH_2)_m-CO-$$

Polymer	Structure	References
Poly(*N*-vinylcarbazole)	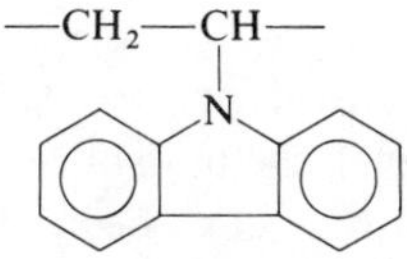 $-CH_2-CH-$	2406, 2410

Polymer	Structure	References
Polymethacrylates, polyacrylates, and polyurethanes with pendant 1,2-*trans*-dicarbazolylcyclobutane	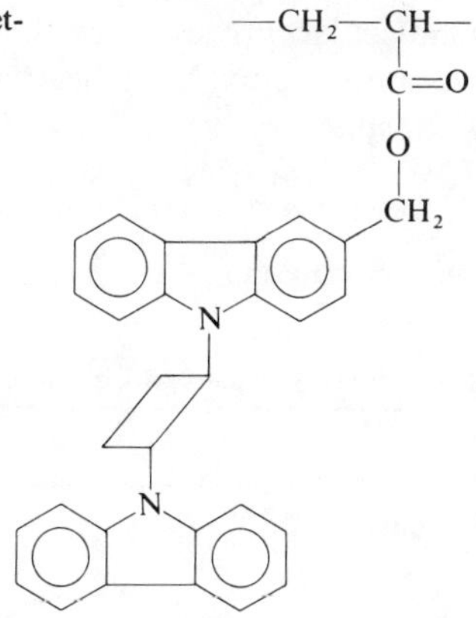 $-CH_2-CH-$ $C=O$	2405, 2408

2.15.2. Triplet–triplet Annihilation Process

Two photons are absorbed by two similar chromophore groups on the same chain, forming excited singlet states (S^*) which intersystem cross (ISC) to the triplet excited states (T^*). The triplet excitation energy migrates as triplet exciton (as a delocalized wave or by series of hopping events) along the polymer chain until two triplets exciton come close enough to undergo annihilation ($T–T$ exciton annihilation) with the formation of a ground state (S_0) and a highly excited triplet state (T^{**}):

$$S_0 + S_0 + 2h\nu \rightarrow S^* + S^* \qquad (2.72)$$

$$2S^* \xrightarrow{\text{ISC}} 2T^* \qquad (2.73)$$

$$T^* + T^* \xrightarrow{\text{triplet energy migration}} T^{**} + S_0 \qquad (2.74)$$

There is another T–T annihilation mechanism which relates to exciton-trapped annihilation. The physical nature of various types of traps (deep or shallow) is not yet well understood.[845,1162,2024,3217,3408,3820]

The highly excited triplet state (T^{**}) can be deactivated with the formation of a ground state (S_0) and an excited singlet state (S^*) which then emits delayed fluorescence (hv')

$$T^{**} \rightarrow S_0 + S^* \qquad (2.75)$$

$$S^* \rightarrow S_0 + hv' \quad \text{(delayed fluorescence)} \qquad (2.76)$$

or dissociates into free radicals

$$T^{**} \rightarrow R\cdot + \cdot R \qquad (2.77)$$

The T–T annihilation process is a typical intramolecular process, whereas quenching of triplet excited states by quenchers is a intermolecular process.

T–T annihilation has been found in several polymers (Table 2.11). Measurement of T–T annihilation rates is exceptionally difficult. Separation of intra- from bimolecular processes in these measurements is even more difficult.[1406] For example, poly(N-vinylcarbazole) and monomeric carbazole derivatives show phosphorescence in the wavelength region 400 and 500 nm, and delayed fluorescence which is caused by triplet–triplet (T–T) annihilation process.[563, 565, 568–571, 2072, 2074, 2075, 3113, 3686] The latter depends on the molecular weight of the polymer.[2076]

Table 2.11 Examples of polymers which exhibit triplet–triplet (T–T) annihilation process

Polymer	Structure	References
Poly(phenylvinyl ketone)	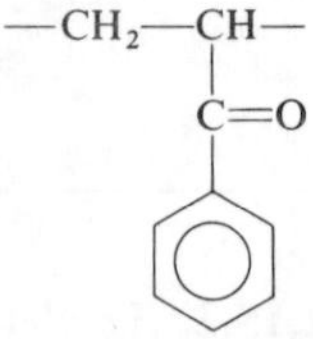 —CH₂—CH— C=O phenyl	2050
Co(methyl methacrylate/phenylvinyl ketone) Co(vinyl acetate/phenylvinyl ketone) Co(styrene/phenylvinyl ketone)		2049
Poly(p-methoxy-phenylvinyl ketone)	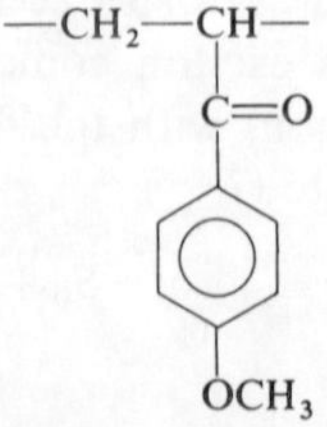 —CH₂—CH— C=O phenyl OCH₃	3198, 3290

Table 2.11 (*Contd.*)

Polymer	Structure	References
Poly(vinylbenzophenone)	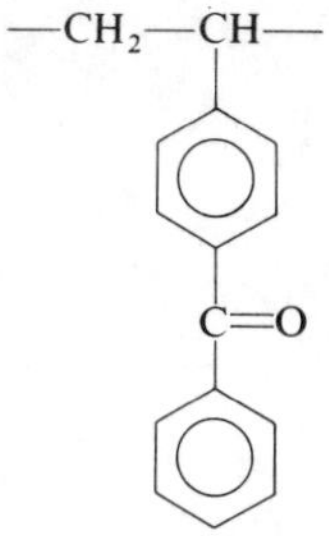	3214
Poly(1-vinylnaphthalene)	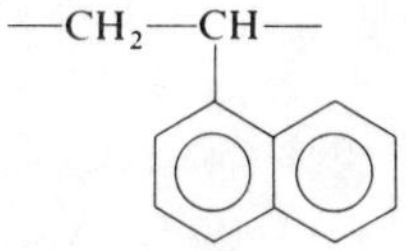	773, 843, 2024
Poly(2-vinylnaphthalene)	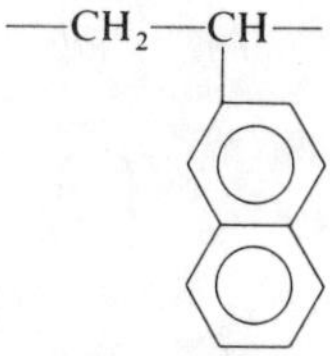	2024, 2025, 2903, 2992
Poly(2-isopropenyl-6-vinylnaphthalene)	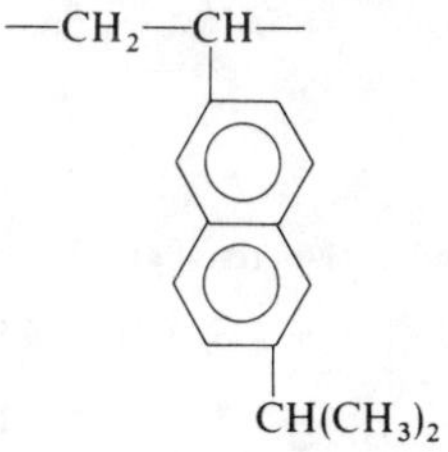	1618
Poly(2-naphthyl methacrylate)	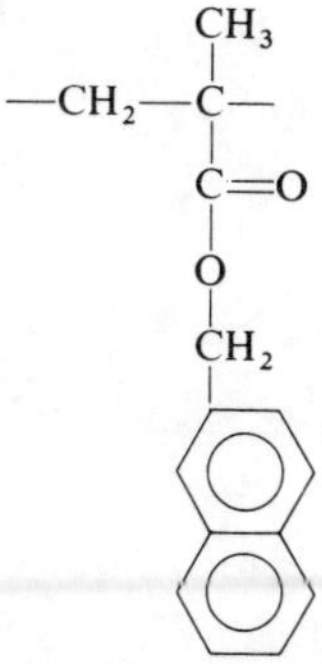	2904, 2905, 3408

Table 2.11   (*Contd.*)

Polymer	Structure	References
Poly(4-vinylbiphenyl)	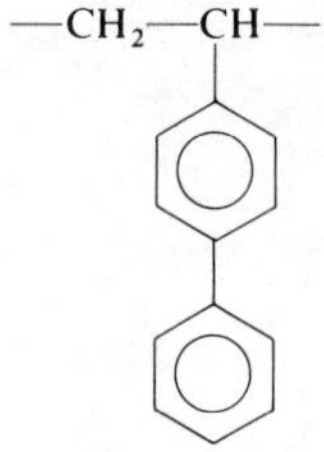 —CH$_2$—CH—	2991, 2992
α-Anthryl-polystyrene	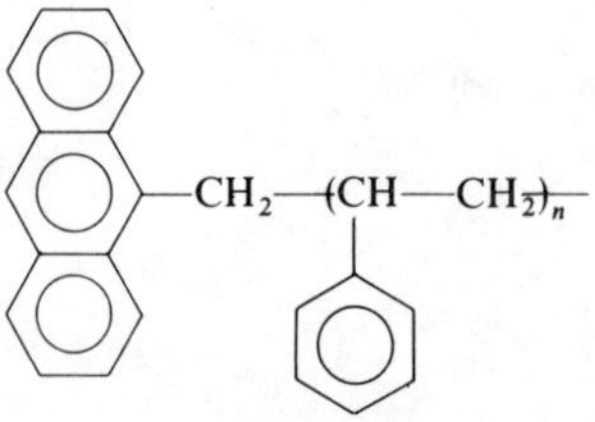 —CH$_2$—(CH—CH$_2$)$_n$—	1667, 3720, 3722
Di(α-anthryl)-polystyrene	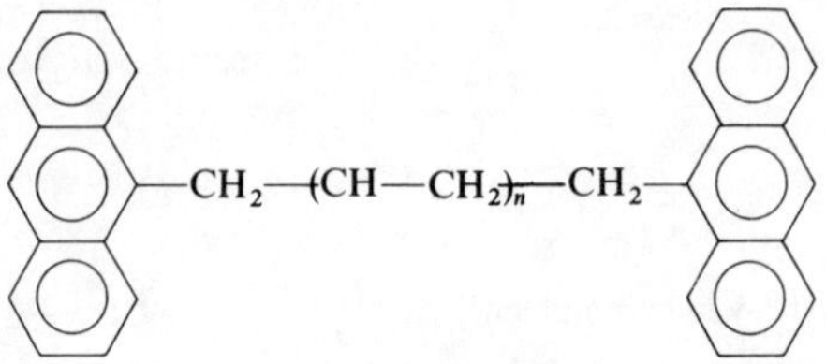 —CH$_2$—(CH—CH$_2$)$_n$—CH$_2$—	1667, 3722
Polyesters with pendant 1-pyrenyl group	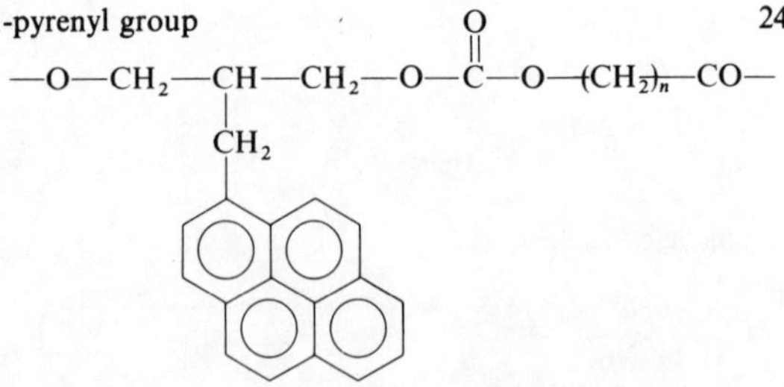 —O—CH$_2$—CH—CH$_2$—O—C—O—(CH$_2$)$_n$—CO—	2407
Poly(*N*-vinylcarbazole)	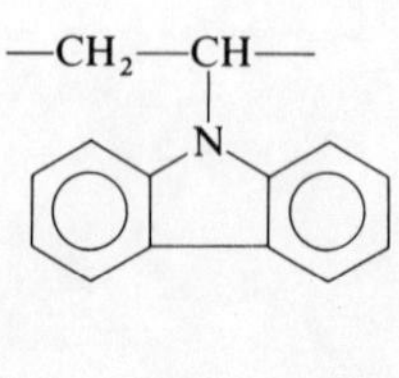 —CH$_2$—CH—	2075, 2076, 2989, 3985

Table 2.11 *(Contd.)*

Polymer	Structure	References
Polymethacrylates, polyacrylates, and polyurethanes with pendant 1,2-*trans*-dicarbazolyl-cyclobutane	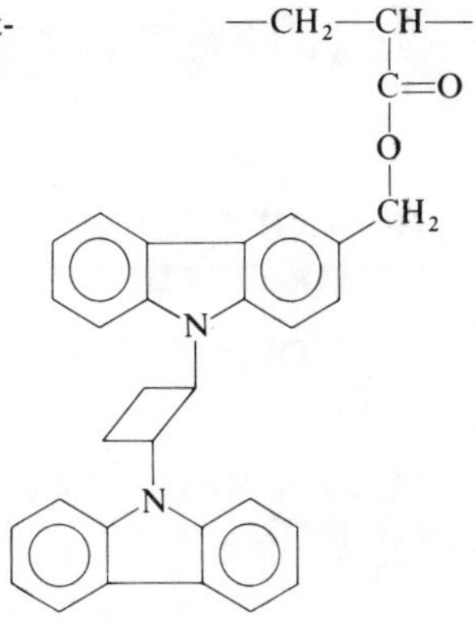	2408

There are several interesting dependencies:

(i) Delayed fluorescence of several polymers such as poly(2-vinylnaphthalene), poly(2-naphthyl methacrylate), poly(4-vinylbiphenyl) and poly(*N*-vinylcarbazol) is dependent on molecular weight of a polymer.[2024, 2903–2905, 3985] The intensity of delayed fluorescence increases with molecular weight, accompanied by a slight decrease in phosphorescence intensity. At higher molecular weight these intensities tend to become constant (i.e. the molecular weight effect 'saturates').

(ii) The phosphorescence at long time intervals is exponential, but the lifetime depends on molecular weight.

(iii) The rate of decay of delayed fluorescence for short intervals (ca. 2.4 ms) increases with molecular weight.

(iv) At longer time intervals the rate of decay of delayed fluorescence and phosphorescence is much less molecular weight-dependent.

These dependencies arise from the increased probability of multiple excitation occurring in polymers with a larger number of chromophores.

It has been suggested that the $T–T$ annihilation process is a cause for the non-exponential decay of several chromophores in poly(methyl methacrylates).[2358] Similarly a mechanism has been proposed which consists of energy transfer from excited triplet chromophore to polymer, triplet energy migration, and $T–T$ annihilation process between polymer triplet state and chromophore triplet.[1854] Recent measurements of the intensity dependency of the profile decay of benzophenone in poly(methyl methacrylate) support importance of $T–T$ annihilation.[1177] On other hand similar measurements support the diffusion-controlled dynamic quenching mechanism rather than the $T–T$ annihilation process.[1664]

3. Excimers and exciplexes

3.1. INTRAMOLECULAR DIMERS FORMED FROM A PAIR OF IDENTICAL MOLECULES

INTRAMOLECULAR DIMERS formed from a pair of identical molecules can be classified as follows:

(i) DIMER COMPLEX, which may be formed by interaction between two unexcited molecules (M):

$$M + M \rightarrow (MM) \text{ or } M_2 \qquad (3.1)$$

A dimer complex can absorb light and be excited to singlet and/or triplet states:

$$(MM) + h\nu \rightarrow {}^1(MM)^* \qquad (3.2)$$
$${}^1(MM)^* \rightarrow {}^3(MM)^*$$

A dimer complex absorption, fluorescence, and phosphorescence spectra are more diffuse than those of the parent molecule, and they are usually shifted to the region of longer wavelengths.

(ii) PHOTODIMER, which may be formed by the chemical interaction of an excited molecule in the lowest singlet state ($^1M^*$) and an unexcited molecule (M):

$$^1M^* + M \rightarrow MM \text{ or } M_2 \qquad (3.3)$$

M_2 is stable in the ground state (S_0), but it is usually dissociated in the excited state $^1M_2{}^*$. The M_2 absorption lies at much shorter wavelengths that of M, and it is unrelated to it. An example of such a type of photodimer is the anthracene photodimer (Fig. 3.1).

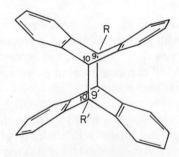

Fig. 3.1 Structural diagram of the anthracene photodimer[3620]

(iii) SANDWICH DIMER (M ‖ M), which may be formed by the photolysis of a photodimer (M_2) in a rigid matrix:

$$M_2 + h\nu \to {}^1M_2{}^* \to {}^1(M \| M)^* \to (M \| M) \tag{3.4}$$

and yielding a pair of adjacent molecules in a sandwich (face-to-face) configuration, e.g. anthracene sandwich dimer (Fig. 3.2).

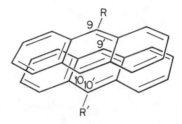

Fig. 3.2 Structural diagram of the anthracene sandwich dimer[3620]

(iv) NON-SANDWICH DIMER (M × M), corresponds to a pair of identical molecules which are not in a sandwich or displaced sandwich configuration, but which are sufficiently close and suitably oriented to interact in their ground and/or excited state.

(v) EXCIMER.

3.2. EXCIMERS

An EXCIMER (EXCITED DIMER) is a molecular aggregate formed between an excited molecule in the lowest excited singlet state $({}^1M^*)$ and a molecule in the singlet ground state (M):[2066, 3294]

$$^1M^* + M \to ({}^1M \ldots M)^* \tag{3.5}$$

Excimers (considered also as electronically excited complexes) are formed in solutions, liquids, or in the solid state if the crystal structures or chain conformations allow a close overlap of the molecular planes of paired molecules. Excitation of one member of this paired molecule by direct absorption of light or by energy transfer from a nearby excited molecule may lead to formation of an excimer. Excimer formation is a common phenomenon observed in aromatic molecules[188, 439, 441, 736, 1173, 1597, 1883, 3800, 3867] and in polymers (Table 3.1) and copolymers (Table 3.2) having aromatic groups.[354, 960, 1441, 2704, 3409]

A critical requirement for the formation of excimers is the formation of a coplanar sandwich-like orientation of at least two aromatic groups (Fig. 3.3) affording the maximum π-orbital overlap and in interchromophore separation in the range 3–3.7 Å.[439,1173,1597,1883,3800]

The excimer formation time of 1,3-diphenylpropane[1597,2065,2066] and of toluene[440,2587] (the simplest compounds for intramolecular interaction) are 9.1 $\times 10^{-10}$ s and 1.6×10^{-12} s (in concentrated solutions), respectively. Excimers

Table 3.1 Polymers with aromatic groups which forms excimers

Name	Structure	References
Polystyrene	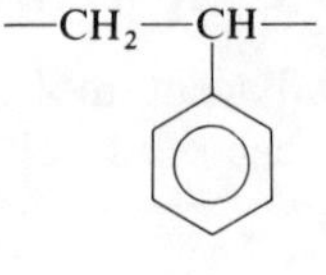 $-CH_2-CH-$ (phenyl)	7, 164, 322, 331, 461, 557, 854, 1000, 1173, 1211, 1212, 1245, 1246, 1290, 1294, 1463, 1558, 1559, 1597, 1598, 1805–1808, 2747, 2303, 2314, 2315, 2355, 2356, 2361, 2362, 2699, 2729, 2747, 2936, 2943, 3136, 3137, 3417, 3449, 3516, 3629, 3630, 3726, 3936, 3958.
Poly(α-methylstyrene)	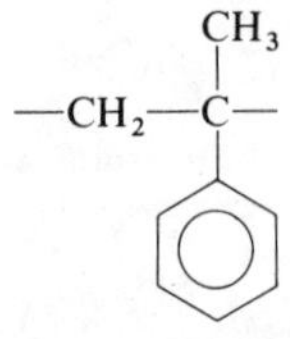 $-CH_2-\underset{\text{(phenyl)}}{\overset{CH_3}{C}}-$	461, 1290, 1294, 2356, 2752, 3516, 3801
Poly(o,m,p-methylstyrene)	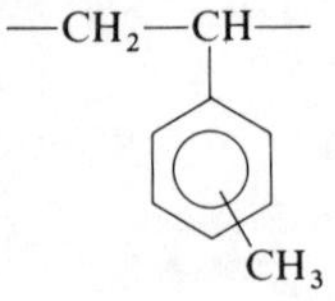 $-CH_2-CH-$ (phenyl)$-CH_3$	1804–1807, 2983, 3838, 3983
Poly(styrene sulfonic acid)	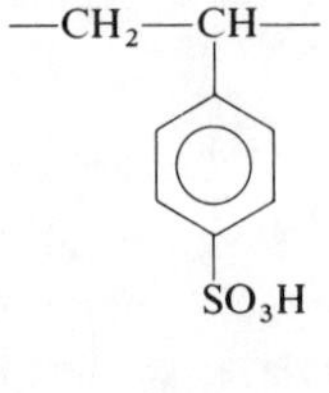 $-CH_2-CH-$ (phenyl)$-SO_3H$	190
Poly(1-vinylnaphthalene)	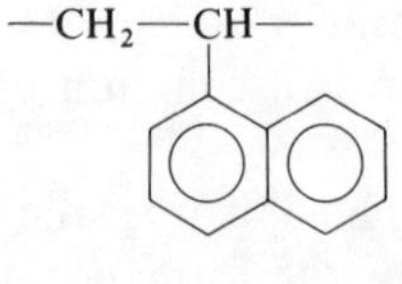 $-CH_2-CH-$ (naphthalene)	845, 849, 1164, 1294, 1457, 1684, 2705, 2937, 2941, 3123, 3726.
Poly(2-vinylnaphthalene)	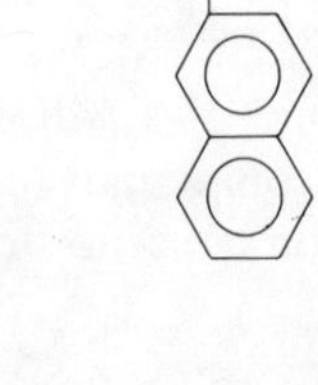 $-CH_2-CH-$ (naphthalene)	947, 967, 1112–1114, 1164, 1167, 1173, 1508, 2941, 2987, 3292, 3821

Table 3.1 (*Contd.*)

Name	Structure	References
Poly(4-methoxy-1-vinyl-naphthalene)	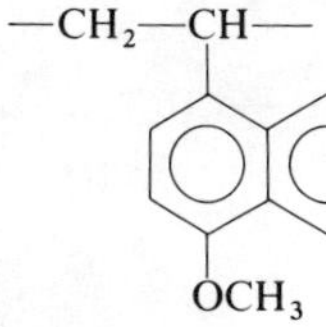	1809, 1814
Poly(2-isopropenyl-6-vinyl-naphthalene)	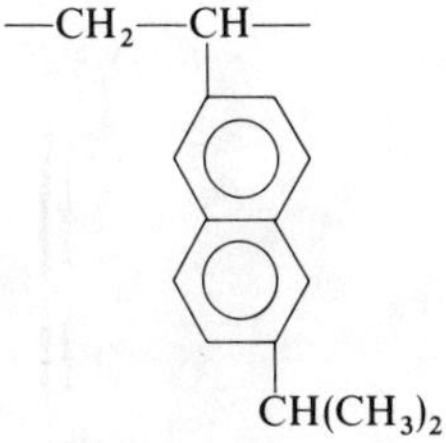	1618
Poly(2-*tert*-butyl-6-vinyl-naphthalene)	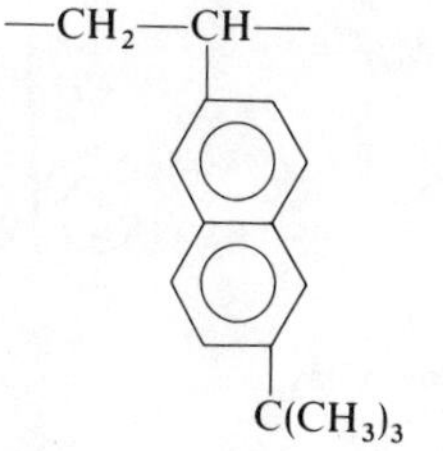	2614, 2618
Poly(1-naphthyl methacrylate)	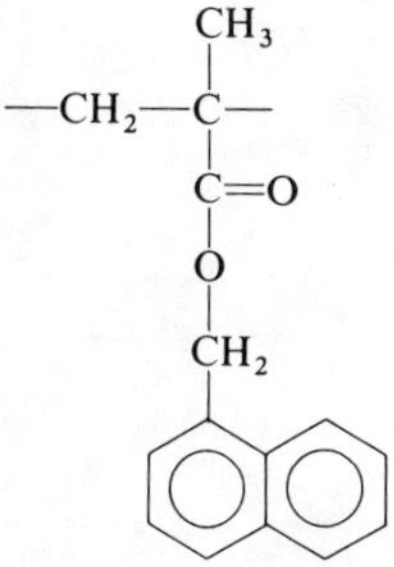	5,239, 1622, 1623, 1679, 2490, 2941, 3408, 3918
and Poly(2-naphthyl methacrylate)	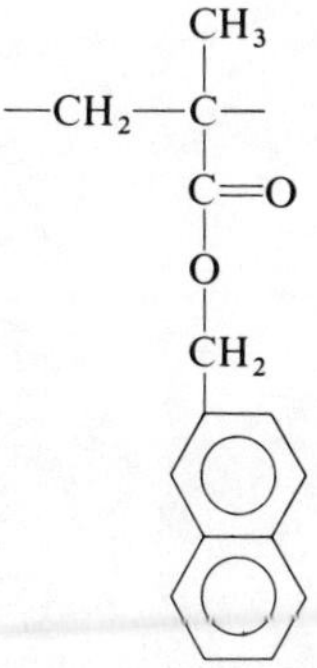	

Poly(4-methoxy-1-vinylnaphthalene):

—CH$_2$—CH—
with naphthalene ring, OCH$_3$

Poly(2-isopropenyl-6-vinylnaphthalene):

—CH$_2$—CH—
with naphthalene ring, CH(CH$_3$)$_2$

Poly(2-*tert*-butyl-6-vinylnaphthalene):

—CH$_2$—CH—
with naphthalene ring, C(CH$_3$)$_3$

Poly(1-naphthyl methacrylate):

CH$_3$
—CH$_2$—C—
C=O
O
CH$_2$
naphthalene ring

Poly(2-naphthyl methacrylate):

CH$_3$
—CH$_2$—C—
C=O
O
CH$_2$
naphthalene ring

Table 3.1   (*Contd.*)

Name	Structure	References
Poly(1-naphthyl-ethyl methacrylate)	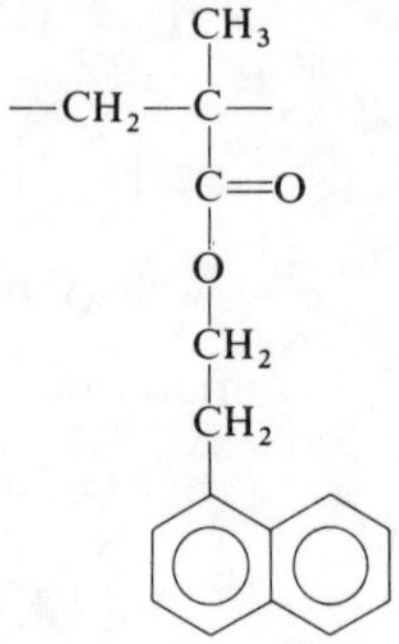	1623, 2490
Polypeptides with 1-naphthyl group	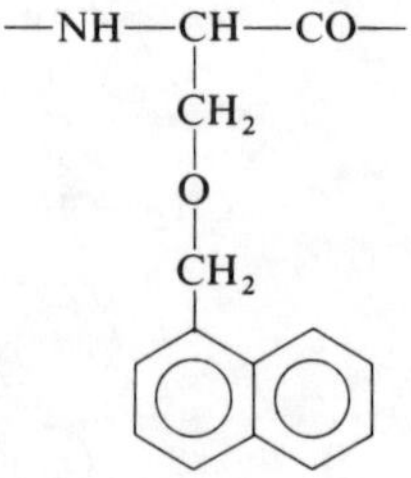	3703, 3704
Polyamides with widely separated naphthyl groups	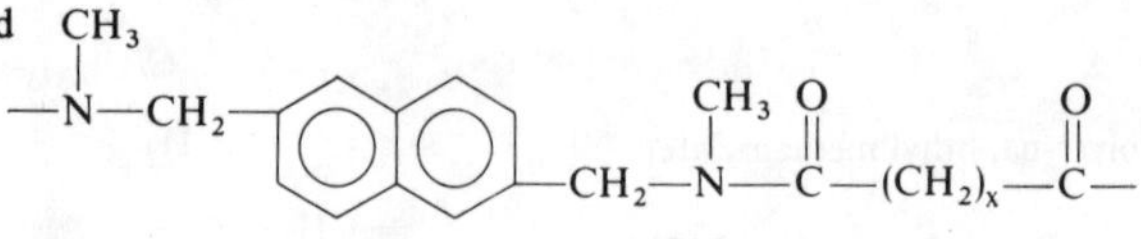	1735
Poly(benzyl methacrylate)	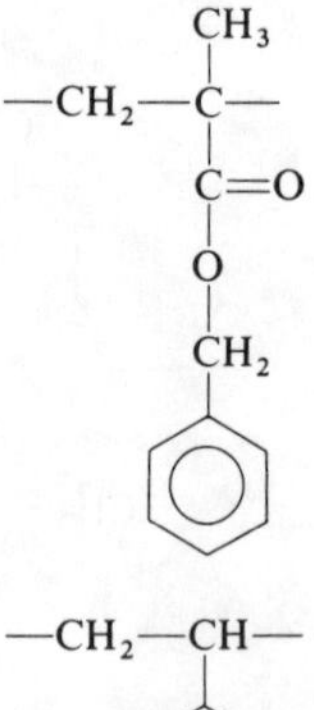	1842
Poly(4-vinylbiphenyl)	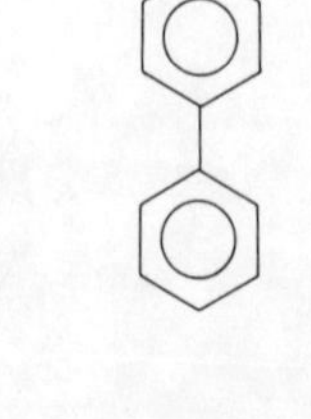	1166

Table 3.1 (*Contd.*)

Name	Structure	References
Poly(4-acryloxy-benzophenone)	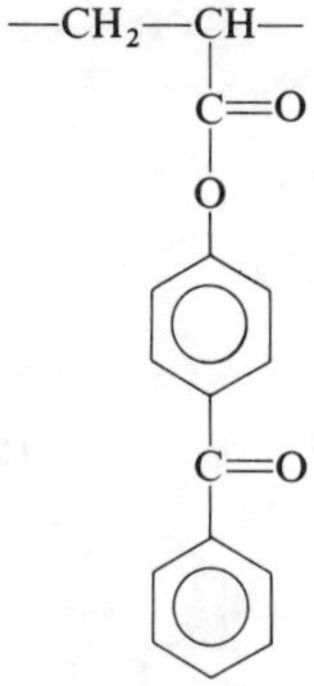 —CH₂—CH—	1117
Poly(1-vinylpyrene)	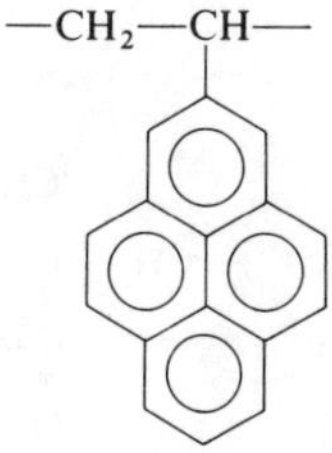 —CH₂—CH—	1600, 2449, 3984
Poly(vinyl acetate) bearing pyrene groups	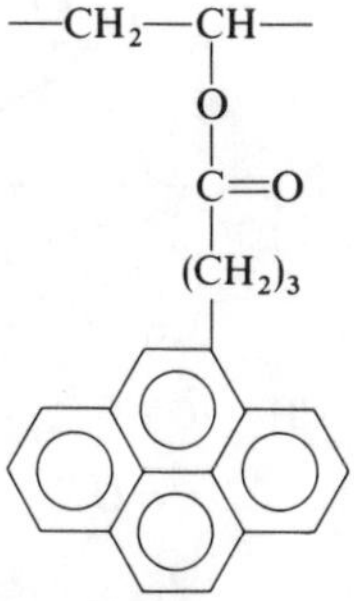 —CH₂—CH—	891
Poly(acenaphthylene)	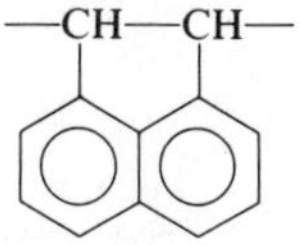 —CH—CH—	836, 845, 849, 1164, 2938, 3123, 3221
Poly(N-vinylcarbazole)	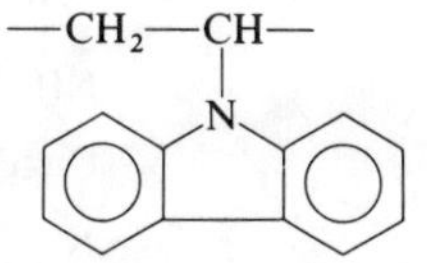 —CH₂—CH—	569, 704, 850, 971, 1288, 1290, 1292–1294, 1684, 1815, 1816, 1883, 1884, 1886, 1887, 2058, 2059, 2061, 2063, 2072, 2074, 2192, 2683, 2984, 3114, 3122 3125, 3222, 3359, 3517, 3740, 3821, 3983

Table 3.1 (*Contd.*)

Name	Structure	References
Poly(3,6-dibromo-*N*-vinylcarbazole)		3982
Poly(*N*-epoxycarbazole)		1695
Poly(*N*-vinyl-5H-benzo[b]carbazole)		1817, 1818
Poly(*N*-carbazolylmethyl methacrylate)		362, 2011
Poly[2-(9-ethyl)carbazolylmethyl methacrylate)]		362, 2011, 2012, 2241
Poly[*N*-(9-carbazolylcarbonyl)-L-lysine]		432, 433, 3126

Table 3.1 (*Contd.*)

Name	Structure	References
Poly(2-vinylfluorenone)	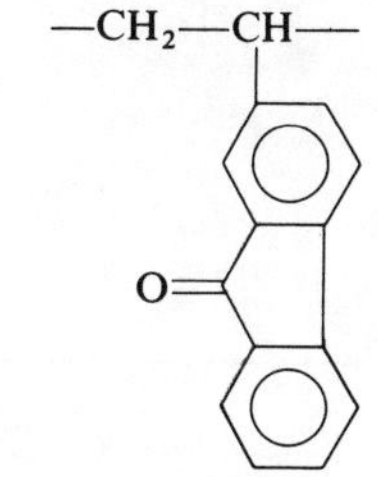 —CH$_2$—CH—	2391
Poly(vinyl alcohol)ester of 1,2-diphenyl-cyclopropene-3-carboxylic acid	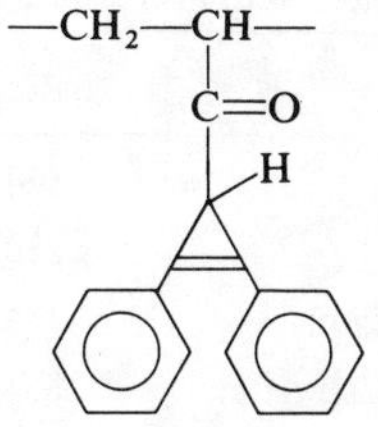 —CH$_2$—CH—	2509
Polyiononenes bearing pendant (9-anthryl) methyl groups	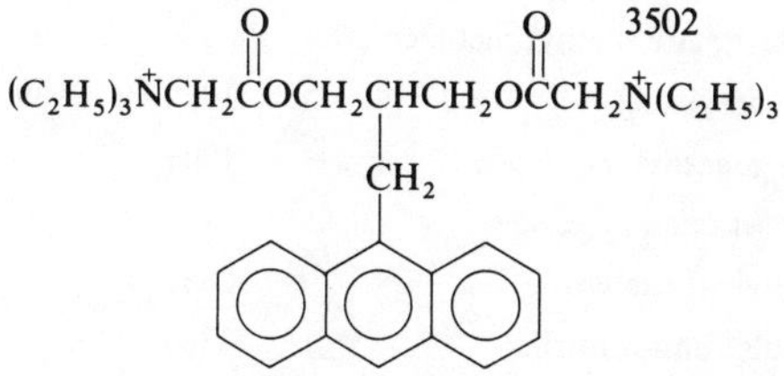 $(C_2H_5)_3\overset{+}{N}CH_2COCH_2CHCH_2OCCH_2\overset{+}{N}(C_2H_5)_3$	3502
Poly(phenylsiloxanes)	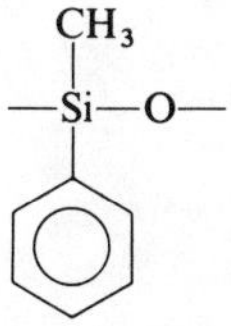 CH$_3$ —Si—O—	3944
Poly(ethylene terephthalate)	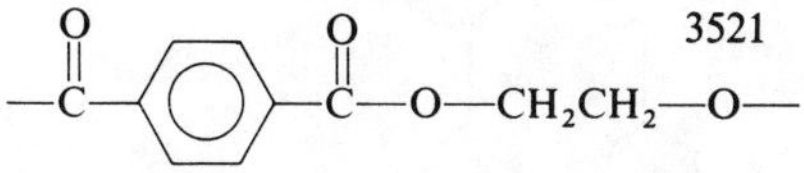 —C—$\bigcirc$—C—O—CH$_2$CH$_2$—O—	3521
Poly(2-phenyl-5(p-vinyl) phenyl oxazole)		2466
Bisphenol A diglycidil ether resin* *Probably triplet excimer		59,61
Polyamides bearing phenyl groups in the main chain		2562

Table 3.1 (*Contd.*)

Name	Structure	Reference
Polyesters bearing the following groups:		
p-phenylene diacrylate		1033, 1374, 1375
1-naphthylmethyl		3564
anthryl		3565, 3567, 3587
pyrenylmethyl		3579
ω-carbazolylalkyl		3574, 3576

Table 3.2 Copolymers with aromatic groups which forms excimers

Name	References
co(styrene/methyl methacrylate)	864, 1162, 2356, 2362, 3080, 3149, 3417, 3801
co(styrene/butadiene)	2303, 2936
co(1-vinylnaphthalene/methyl acrylate)	194, 2940, 2942
co(1-vinylnaphthalene/methyl methacrylate)	834, 852, 856, 1162, 2937, 3079, 3080, 3123
co(2-vinylnaphthalene/methyl methacrylate)	1162, 3080
co(1-naphthylmethacrylate/butyl methacrylate)	2721
co(2-naphthylmethacrylate/methyl methacrylate)	483
co(acenaphthylene/acrylates)	195, 836, 2939, 3081
co(acenaphthylene/menthyl acrylates)	1203
co(acenaphthylene/maleic anhydride)	2938
co(acenaphthylene/vinyl carbazole)	583
co(*N*-vinyl carbazole/fumaronitrile)	3360, 3983
co(*N*-vinyl carbazole/diethylmaleate)	3983
co(stilbene/maleic anhydride)	3801
co(phenyloxazole/methyl methacrylate)	193

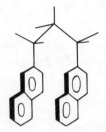

Fig. 3.3 Typical sandwich structure of the poly(2-vinyl-naphthalene) excimer

are usually unstable in their ground state but are stable under electronic excitation. A singlet excimer may undergo intersystem crossing (ISC) to a triplet state, which is usually dissociative.

During the decomposition of the singlet excimer an EXCIMER FLUORESCENCE is produced, which differs from 'normal fluorescence' (Fig. 3.4):

$$(^1M \ldots M)^* \rightarrow M + M + h\nu \text{ (excimer fluorescence)} \qquad (3.6)$$

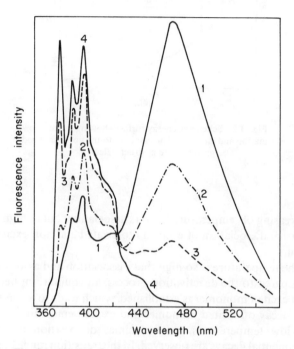

Fig. 3.4 Normal fluorescence from pyrene and excimer in ethanol: (1) 3×10^{-3} M, (2) 10^{-3} M, (3) 3×10^{-4} M, (4) 2×10^{-6} M[2897] (*Reproduced by permission of the Royal Society of Chemistry*)

Excimer fluorescence differs from 'normal fluorescence' because:

(i) It lies in the region of longer wavelengths than 'normal fluorescence' because the excited state of the singlet excimer ($^1M \cdots M$)* lies below the excited singlet state of M ($^1M^*$) (Fig. 3.5).

(ii) It shows a broad characteristic spectrum without vibrational structure (Fig. 3.4) because the ground state of the excimer is repulsive (unstable).

(iii) It depends on molecular structure. The number and nature of bond rotations necessary to create the excimer from the ground-state conformation, and the extent of excimer formation, are functions of structure.

(iv) Formation of excimers in solution is diffusion-controlled and viscosity of the solvent plays an important role.

(v) It is dependent on temperature:

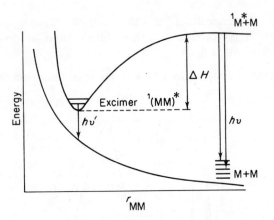

Fig. 3.5 Schematic energy surface showing excimer formation and emission. The emission to the non-quantized ground state is structureless[2235]

(a) Increasing of temperature favours association, but at high temperatures thermal dissociation of excimers to excited and non-excited monomers occurs.

(b) If the temperature is so high that dissociation and association are rapid compared to the deactivation process, an equilibrium between excimer and excited monomer is established, which leads to a common expotential decay of excited monomer and excited emission.

(c) At low temperatures, when excimer dissociation is negligible, two exponential decays are observed. In the transition range the decay is non-exponential.

(vi) It is dependent on pressure which increases excimer formation in the dissociation equilibrium range.[513, 514, 1112, 1139, 1187–1890, 3137, 3287, 3418]

(vii) Excimer fluorescence may occur from various $^1(M \parallel M)^*$ and $^1(M \times M)^*$ dimer configurations, depending on conditions. The excimer fluorescence of aromatic polymers is due to intramolecular and intermolecular $^1(M \times M)^*$ and $1(M \parallel M)^*$ dimers, suitably oriented and with sufficient freedom of motion to form excimers.

Absence of excimer fluorescence at room temperature does not imply the absence of excimer formation, because a small enthalpy of formation may lead to rapid dissociation of an excimer.

The failure to detect excimers in poly(vinylbenzophenone) (3.1) can be attributed to only slight flexibility of the pendant benzophenonic group compared with the poly(4-acryloxybenzophenone) (3.2) where a two-member chain connects the benzophenone chromophore to the vinylic backbone:[1117]

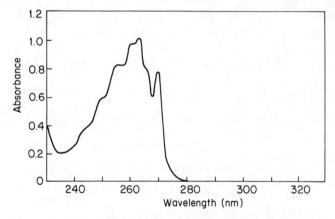

$$-CH_2-CH-$$

(3.1)

$$-CH_2-CH-$$

(3.2)

If the lifetime of the excited monomer is short, fairly high concentrations are needed in order to observe the formation of excimer. The concentration quenching of fluorescence is often due to excimer formation, followed in some cases by a photochemical reaction.

3.2.1. Polystyrene Excimer

The polystyrene absorption spectrum in dilute solution (Fig. 3.6) is essentially independent of molecular weight.[3630] The dependence of the absorbance at 260 nm on polystyrene concentration is a linear function (Fig. 3.7).[981, 2243, 3630]

Fig. 3.6 Absorption spectrum of 100,000 MW polystyrene in 1,2-dichloroethane (concentration is 0.488 g l^{-1}, cell path 1 cm)[3630]

The typical fluorescence spectrum of polystyrene in 1,2-dichloroethane solution (Fig. 3.8) shows two emission bands, which were assigned to:[1597, 1806, 3630, 3726, 3958]

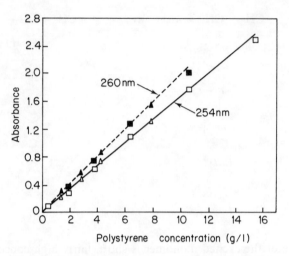

Fig. 3.7 Concentration dependence of polystyrene absorbance at 260 and 254 nm in 1,2-dichloroethane (cell path 0.1 cm): (□) and (○) 100,000 MW, (△) and (▲) 900,000 MW[3630]

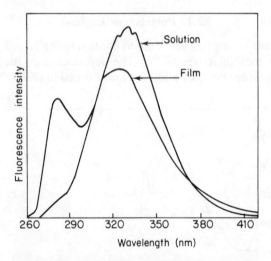

Fig. 3.8 Fluorescence spectra of polystyrene in methylene dichloride solution and of film cast from CH_2Cl_2 solution ($\lambda_{ex} = 250$ nm)[1463]

(i) monomer emission (bands at 283 nm (or 286 nm)),

(ii) excimer emission (band at 335 nm).

The ratio of excimer to monomer fluorescence intensity ($I_{F(E)}/I_{F(M)}$) changes with:

(i) Residual styrene monomer,[322]

(ii) Type of solvent used,[7, 2699, 3629, 3630] (polystyrene excimer is readily quenched by carbon tetrachloride[1558, 1807]).

(iii) Concentration of polystyrene in solution.[3136, 3629, 3726]

(iv) Molecular weight of polystyrene sample.[7, 1173, 1806, 3136, 3629]

(v) Configuration (isotactic polystyrene is more favourable for excimer formation than syndiotactic sample[459, 461], and has a considerably greater $I_{F(E)}/I_{F(M)}$ than atactic polystyrene (c.f. Fig. 3.34).[461, 1805, 1807, 1808, 2314]

(iv) Structural isomers (head-to-head).[2182, 2303, 3629] Head-to-head polystyrene, in which the phenyl rings are separated either by two or four carbon atoms (3.3), displays predominantly fluorescence from monomeric units.[2182, 2303] No excimer emission is observable even when the polymer is dissolved in increasingly thermodynamically poorer solvents. Head-to-tail polystyrene samples (3.4) may form excimer between nearest-neighbouring phenyl ring chromophores.

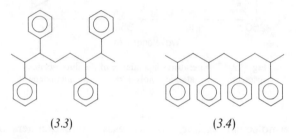

(3.3)   (3.4)

(vii) Orientation in the bulk phase.[854]

(viii) Copolymer structure.[833, 1582, 2303, 2936]

(ix) Pressure.[3137]

(x) Viscosity.[2699]

(xi) Oxygen (excimer fluorescence is quenched by oxygen).[2092, 2747]

The excimer intensity (Fig. 3.9) and lifetime of polystyrene excimer are approximately independent of temperature for $-40°$ to $+80°C$.[461, 1558] A similar result has been observed for 2,4-diphenylpentanes, the polystyrene model compounds.[460]

Depolarization of excimer fluorescence has been observed only in solid state,[1211, 1463, 2936] but two other reports suggest that polystyrene excimer emission is polarized both in film and solution.[2355, 2361]

Time-resolved emission measurements (Fig. 3.10) show that the early gated spectra are dominated by monomer fluorescence and the relative proportion of excimer to monomer ($I_{F(E)}/I_{F(M)}$) increases.[3417] In the late gated spectra, a small contribution from monomer is still observable. The ratio $I_{F(E)}/I_{F(M)}$ tends to a constant value at long times, indicating that the monomeremission observed in these spectra results from dissociation of excimer.

Fluorescence from solid polystyrene (film) is mainly from the excimer (Fig. 3.8).[1558, 1598, 3726] In solid polystyrene a clear-cut distinction between

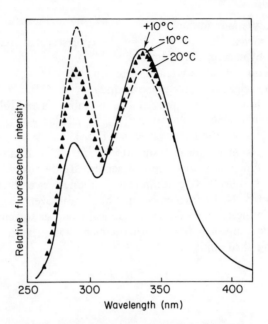

Fig. 3.9 Temperature-dependence of the fluorescence spectrum of atactic polystyrene in methylene dichloride[461]

intra- and intermolecular excimers is not easy. Here excimers of a similar symmetry may be formed either within the same polymer chain or between different chains.

The excimer formation time and the reorientation correlation time for polystyrene (with $\overline{M}_n > 1.0 \times 10^4$) have the same order of magnitude and are 10^{-10} s.[164, 331, 557, 1463, 2729, 3449] This result indicates that the excimer formation of high molecular weight polystyrene is related to local relaxation. On the other hand, for polystyrene with $\overline{M}_n < 1.0 \times 10^4$, the rate constant for excimer formation does not correlate to the reorientational correlation time. This discrepancy is characteristic of intramolecular excimer formation.[1806]

The excimer formation of polystyrenes can be attained more effectively at trapping sites such as crystalline or structural defects rather than intermolecular interaction. The excimer trapping sites in a polymer chain play an important role in the excimer formation of polystyrene.[1173, 1806]

The excimer of polystyrene cannot dissociate quickly to the excited monomer state, since phenyl groups in a polymer chain cannot move as quickly as in an aromatic liquid. This restriction of molecular motion induces the long resident time of the pair of excimers in polymer systems, during which they dissipate energy through non-radiative deactivation.

The photophysical behaviour of polystyrene is different from that of other polymers such as poly(vinylnaphthalene), poly(N-vinylcarbazole) and copolymers, mainly in that:[3417]

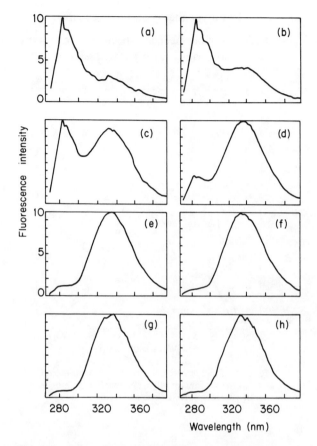

Fig. 3.10 Time-resolved fluorescence spectra of polystyrene in degassed methylene dichloride recorded at delays of (a) 0, (b) 3.8, (c) 7.7, (d) 11.5, (e) 15.4, (f) 19.3, (g) 28.8, and (h) 38.4 ns following excitation[3417]

(i) emission of monomer and excimer fluorescence decays can be described by the kinetic Model I and biexponential functions (cf. Section 3.4);

(ii) population of excited-state monomer by dissociation of excimer does occur in polystyrene;

(iii) the dissociation of excimer appears to be of lesser importance in polystyrene than in the polymers bearing naphthyl chromophores;

(iv) there is no detectable influence from 'isolated' chromophores (M_2^*) (cf. Section 3.4);

(v) in intramolecular energy transfer to guest traps in polystyrene the excimers play insignificant roles as donor relative to that from unassociated (or monomeric) excited-state donors[2943] (cf. Section 2.12).

3.2.2. Poly(α-methylstyrene) Excimer

The poly(α-methylstyrene) fluorescence spectrum in methylene chloride (Fig. 3.11) show two emission bands, which were assigned to[461]

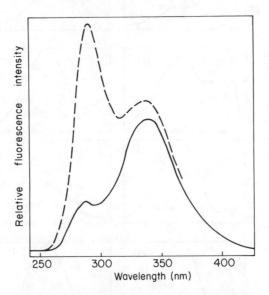

Fig. 3.11 Fluorescence spectra of: (–––) cationically and (——) anionically polymerized poly(α-methylstyrene) in methylene dichloride solution[461]

(i) monomer emission (band at 288 nm),
(ii) excimer emission (band at 335 nm).

The large differences in intensity of both bands are probably due to the difference in tacticity of the two samples. Poly(α-methylstyrene) exhibits a higher efficiency of excimer formation than polystyrene. Activation energy for excimer formation in the case of a syndiotactic diad is lower in poly(α-methylstyrene) than in polystyrene.

For poly(α-methylstyrene), as the temperature of the solution is increased from low temperatures, the intensity of the monomer is observed to decrease while the intensity of the excimer band increases (Fig. 3.12). But at temperatures exceeding − 60°C for poly(α-methylstyrenes) cationically polymerized and − 30°C for poly(α-methylstyrenes) anionically polymerized, the excimer band decreases greatly with increasing temperature. This behaviour is quite different from that of polystyrene, for which both excimer intensity and lifetime are approximately independent of temperature for − 40°C to + 80°C.[1558] A similar result has been obtained for 2,4-diphenylpentanes, the polystyrene model compounds.[460]

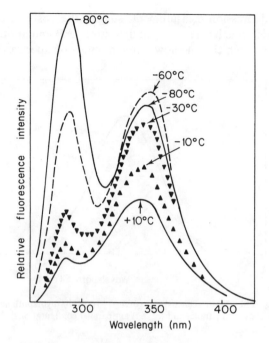

Fig. 3.12 Temperature-dependence of poly(α-methylstyrene) in methylene dichloride[461]

3.2.3. Poly(1-vinylnaphthalene)

The poly(1-vinylnaphthalene) (*3.7*) fluorescence spectrum in 1,2-dichloroethane is shown in Fig. 3.13. Emissions of monomer and excimer fluorescence decays are described by two kinetic models (Model 1 or Model 2, cf. Section 3.4).

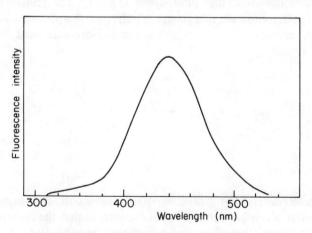

Fig. 3.13 Fluorescence spectrum of poly(1-vinyl-naphthalene) in methylene dichloride[1457]

Time-resolved emission measurements suggest the existence of two types of excimers, of which one has parallel configuration (*3.5*) and another antiparallel configuration (*3.6*), which can show different emission spectra (Fig. 3.14).[1457]

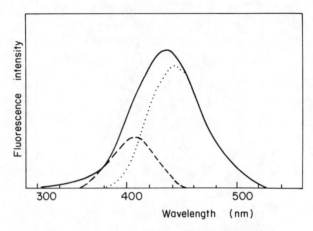

Fig. 3.14   Fluorescence spectra of: (——) poly(1-vinylnaphthalene) with (---) parallel and (. . . . .) antiparallel configurations[1457]

(*3.5*)   (*3.6*)

It has also been reported that poly(1-vinylnaphthalene) in cyclohexane solution forms intramolecular photodimer (*3.8*).[1927] The photocycloaddition reaction (*3.7*) of naphthalene side groups in poly(1-vinylnaphthalene) occurs with high conversion in cyclohexane (70%), while in dichloromethane it occurs at very low conversion (20%).

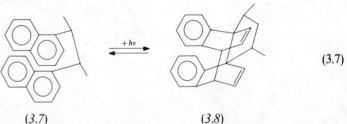

(*3.7*)   (*3.8*)

Phosphorescence from solid films of poly(1-vinylnaphthalene) originates from trapped triplets which have characters of excimers in that the emission is broad and red-shifted from that of isolated naphthalene groups. The delayed fluorescence arises from triplet–triplet annihilation[571] (cf. Section 2.15.2).

3.2.4. Poly(N-vinylcarbazole)

The poly(N-vinylcarbazole) fluorescence spectrum (Fig. 3.15) does not show emission from a monomer. This broad fluorescence emission spectrum has been attributed to the formation of two spectrally distinct excimers:

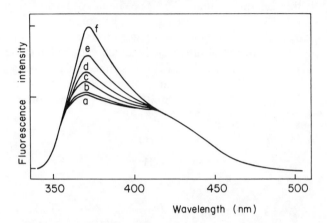

Fig. 3.15 Fluorescence spectra of poly(N-vinylcarbazole) as a function of temperature: (a) 295K; (b) 275K; (c) 260K; (d) 240K; (e) 220K and (f) 200K[3123]

(i) Low-energy excimer with emission maximum at 420 nm, with a rise of several nanoseconds, assigned to the formation of the normal sandwich-type excimer between neighbouring carbazole groups on the polymer chain in a totally eclipsed conformation.[1684, 1883]

(ii) High-energy excimer with emission at 380 nm, which is formed immediately after light pulse in the range of picoseconds, assigned to the formation of a dimeric structure with considerable deviation from coplanarity of the two carbazole rings.[1292, 1293, 1684, 3517]

The formation mechanism of these excimers is still unsettled at the present stage of investigation. The following mechanism of formation of these excimers has been proposed:[2683]

$$M + h\nu \rightarrow M^* \underset{\searrow\ \nwarrow}{\overset{E_1}{\underset{\rightleftarrows}{\ \ }}} E_2 \tag{3.8}$$

where:

M   = monomer carbazole chromophore in the ground state (S_0),
M* = excited uncomplexed monomer carbazole,
E_1 = high-energy excimer emitting at 380 nm,
E_2 = low-energy excimer emitting at 420 nm.

Interconversion between these two excimers (there is no isoemissive point in the steady state fluorescence spectra) has been demonstrated using time-resolved fluorescence spectroscopy.[1290, 1292–1294, 2683, 3122] This mechanism explains well the observed fluorescence from poly(1-vinylcarbazole) but is not sufficient to support an explanation that the main route populating the low-energy excimer (E_2) is directly through the monomer.[3124] The kinetics of excimer formation and decay in the case of poly(N-vinylcarbazole) is very complex (cf. Section 3.4).

When the carbazolyl chromophore is attached to the backbone by a spacing group, as in series of carbazolyl substituted methacrylate polymers (e.g. poly[2-(9-ethyl)carbazolyl-methyl methacrylate] both excimer and exciton migration are reduced.[362, 2011, 2012, 2241] It is very interesting that poly(9-carbazolyl-methyl methacrylate) (*3.9*) does not exhibit excimer emission or exciton migration.[362, 2011]

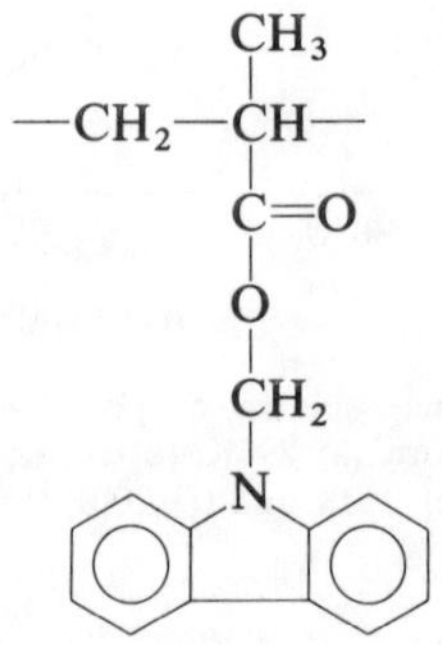

(*3.9*)

3.2.5. High-energy Excimers

The high-energy excimers have been observed in several polymers such as

(a) poly(N-vinylcarbazole) (cf. Section 3.2.4),[971, 1292, 1293, 1883, 2683, 3122–3124, 3517, 3740, 3983]

(b) poly(N-vinyl-5H-benzo[b]carbazole,[1817, 1818]

(c) poly(N-carbazolyl-methyl acrylate),[2011]

(d) poly[2-(9-ethyl)carbazolyl-methyl methacrylate)],[2011, 2012]

(e) poly(2-tert-butyl-6-vinylnaphthalene),[2614]

(f) poly(4-acryloxybenzophenone).[1117]

The high-energy excimers (traps) can probably be formed between chromophores when

(i) they are separated by steric or other structural causes or by extensive host–guest polymer interpretation, by distances longer than that for excimer-forming interaction;

(ii) they can be fixed in place by decreased chain mobility, as in a film or in poor solvent.

The stabilization due to charge resonance interaction, both in the excited state and in the ground state, is rather small compared with that of excitation resonance in a low-energy excimer.[1883]

3.2.6. Model Compounds for the Study of Polymeric Excimers

The formation of excimers in polymers having isotactic or syndiotactic sequences can be studied by analogy with formation of excimers in distereoisomeric model compounds (Table 3.3) with different geometrical configurations.

Table 3.3 Low molecular model compounds for the study of polymeric excimers

Polymer	Model compound	References
Polystyrene	1,3-diphenylpropane	1597
	2,4-diphenylpentanes 2,4,6-triphenylpentanes 2,4,6,8-tetraphenylpentanes	460, 969, 1364
Poly(1-vinylnaphthalene)	α,α-dinaphthylpropane α,β-dinaphthylpropane β,β-dinaphthylpropane	1457, 2944
	bis[1-(1-naphthyl)methyl]ether bis[1-(1-naphthyl)ethyl]ether	965–967, 3616
	1,3-bis[1-(4-methoxynaphthyl)]propane 1,3-bis[1-(4-hydroxynaphthyl)]propane	1813
Poly(2-vinylnaphthalene)	bis[1-(2-naphthyl)ethyl]ether	967
Poly(N-vinylcarbazol)	2,4-di(N-carbazolyl)pentane	971, 2409, 2412
	N-ethylcarbazole 1,2-trans-di-N-carbazolyl-cyclobutane 1,3-di-N-carbazolylpropane	2409
Poly(1-vinylpyrene)	1,3-di(1-pyrenyl)propane bis[1-(1-pyrenyl)ethyl]ether	1254, 3995 751, 2411

Formation of excimers for a low molecular compound such as 1,3-diphenyl-propane (3.10)[1597] and stereoisomers of 2,4-diphenylpentane (3.11)[459-461, 969, 1364] have been studied as models for excimer formation in polystyrene.

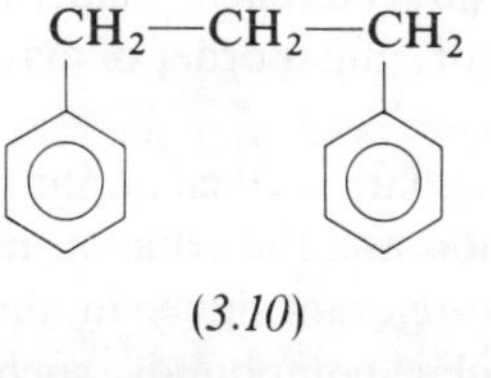

$$(3.10) \qquad\qquad (3.11)$$

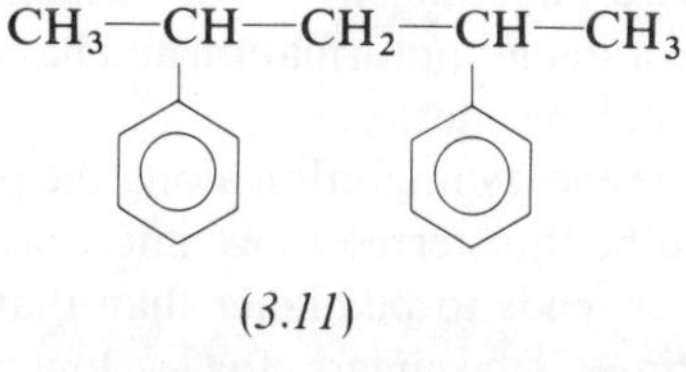

Figure 3.16 shows the emission spectra of *meso* and *dl* 2,4-diphenylpentane (3.11), a polystyrene model molecule which can be considered as the first step in the isotactic and syndiotactic chains.[461] The spectral patterns of the two

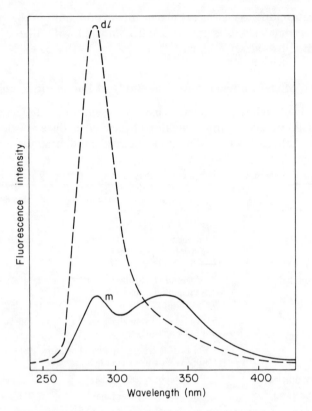

Fig. 3.16 Fluorescence spectra of (– – –) dl isomer and (——)
meso isomer of 2,4-diphenylpentanes[461]

compounds differ significantly, the *meso* isomer exhibits a greater ratio of $I_{F(E)}/I_{F(M)}$ intensity than the *dl* isomer. These results suggest that excimer formation is extremely sensitive to conformational changes due to a change of tacticity.[461] Conformational analysis for 2,4-diphenylpentanes shows that, in the case of a syndiotactic diad, the rotational process between the ground state conformation (tt or $g^- g^-$) and the higher energy excimer conformation $g^+ t$ (or tg^+) can barely be reached.[1364] In particular for the syndiotactic diad, the excimer state $g^+ t$ is a stable conformation and has energy of similar order of magnitude to the other stable states.

When the energy migration along the polymer chain is efficient, the excitation energy will be transferred to excimer-forming sites and the excimer intensity of the polymer tends to be higher than that of the relevant dimer model.

Formation of excimers for a low molecular compounds such as α,α-dinaphthylpropane   *(3.12)*,   α,β-dinaphthylpropane   *(3.13)*   and   β,β-dinaphthylpropane *(3.14)* have been studied as models for excimer formation in poly(vinylnapthalenes) and their copolymers.[1457, 2944] Fluorescence spectra of

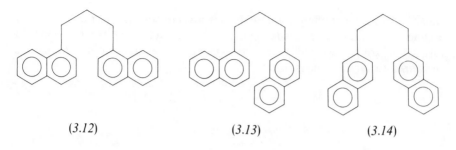

(3.12)   (3.13)   (3.14)

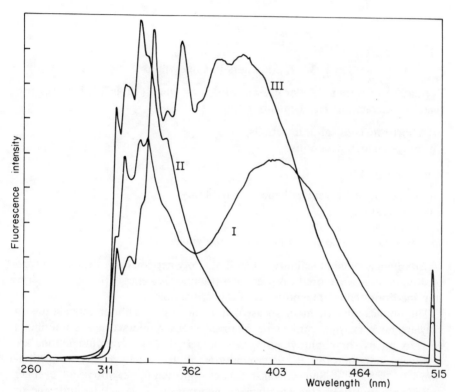

Fig. 3.17 Fluorescence spectra of: (I) α,α-dinaphthylpropane (3.12), (II) α,β-dinaphthylpropane (3.13), and (III) β,β-dinaphthylpropane (3.14)[2944]

these three dinaphthylpropanes (Fig. 3.17) show that the excimer formation and decay in these simple model compounds is very complex.

The *meso* isomer of bis[1-(naphthyl)ethyl]ether has two spectrally different excimers, due to strong steric interaction between hydrogen and methyl-group.[966, 967] This interaction is also evident in bis[1-(9-anthryl)ethyl]ether[365] and 1,3-di(9-anthryl)-1,1,3,3-tetramethyldisiloxane.[1103]

The *meso* and racemic isomers of 2,4-di(N-carbazolyl)pentane (3.15) model compounds for poly(N-vinylcarbazole) give the low- and high-energy excimers.[971] From a configuration analysis it was concluded that the low-energy excimer and high-energy excimer have full and partial overlap between two carbazolyl groups, respectively. The *meso* isomer has a larger ratio of $I_{F(E)}/I_{F(M)}$, compared to the racemic isomer.[971]

$$CH_3-CH---CH_2---CH-CH_3$$

(3.15)

3.2.7. Excimers in Copolymers

Formation of intramolecular excimers in copolymers (Table 3.2) involves two types of interactions between chromophores

(i) 'short-range' (local) interactions,
(ii) 'long-range' interactions,

which depend on

(i) variations of solvent/polymer compatibility,
(ii) temperature,
(iii) comonomer identity,
(iv) structure (regular, random, block copolymers),

which influence chain flexibility and coiling. Polymer coil dimension can have a profound influence upon the process of intramolecular energy transfer to energy trap incorporated as integral units of a polymer chain.

The applicability of models based upon nearest neighbour interactions in vinylaromatic polymers involving relatively weakly interacting chromophore, such as phenyl or naphthyl groups, implies that long-range interactions are relatively unimportant in determination of the excimer site concentration.[38, 194, 3079, 3080] Such conclusions are supported by the absence of excimer emission from alternating copolymers incorporating styrene, naphthalene or carbazole groups,[1162, 3983] and from the regular α-methylstyrene copolymers.[847, 2303]

Time-resolved spectra show that several copolymers reveal two types of spectrally distinct fluorescent sites which can be described by the kinetic Models 2 and 3 (cf. Section 3.4). The triexponential functions are necessary to fit adequately the empirical decays of fluorescence intensity in the regions of both monomer and excimer emission. The existence has been postulated of a conventional monomer–excimer interaction, and of 'isolated' monomer species which cannot populate excimer sites through a 'down-chain' energy migration scheme. The isolated monomer species can form excimer through two alternative mechanisms:[2939]

(i) 'Long-range' interactions of chromophores distant upon the polymer chain, which are brought into the correct configuration for excimer interaction through segmental diffusion of the macromolecular coil.

(ii) Migrative, long-range, energy transfer into a sequence of chromophores followed by conventional 'down-chain hopping' and excimer site sampling.

3.3. MECHANISMS OF EXCIMER FORMATION IN POLYMERS

The mechanism for excimer formation in the solid vinyl polymers involves initial excitation of an aromatic chromophore followed by singlet energy (exciton) migration along the polymer chain until the excitation is competitively trapped at a chain conformation which is geometrically suitable for excimer formation.[1155, 2059, 2070, 2071, 2073] Such a chain conformation is termed an EXCIMER-FORMING SITE.

There are two types of excimer-forming mechanism that have been considered in detail:

(i) Energy transfer to 'excimer-forming sites'

$$M^* + M_2 \rightarrow M + {}^1(M \ldots M)^* \tag{3.9}$$

where M_2 represents a pair of chromophores in a configuration that is favourable for excimer formation.

(ii) Dynamic formation of an excimer. In this mechanism, excimer formation requires some interchain rotations.

These two types of excimer-forming mechanisms cannot be uniquely distinguished in general:

(a) Excimer fluorescence observed in films at low temperatures supports mechanism (i).

(b) Many polymers in the solution phase do not retain excimer fluorescence at lower temperatures, especially in a case where the solution is frozen. These support the existence of mechanism (ii).

In aromatic vinyl polymers there are three different ways in which an EXCIMER SITE can be formed (Fig. 3.18).

(i) INTERMOLECULAR, by association between aromatic rings from two different polymer chains (INTERMOLECULAR EXCIMER). The number of these sites is directly proportional to the local concentration of aromatic rings.

(ii) INTRAMOLECULAR (ADJACENT), by association between aromatic rings on adjacent repeating units in the chain backbone (INTRAMOLECULAR EXCIMERS). The number of these sites is independent of the concentration of chain molecules, and only dependent upon the steric effects associated with backbone bonds.

Intermolecular

Intramolecular (adjacent)

Intramolecular (non–adjacent)

Fig. 3.18 Types of excimer-
forming sites

(iii) INTRAMOLECULAR (NON-ADJACENT), by association between aromatic rings on non-adjacent chain segments (INTRAMOLECULAR
EXCIMERS). The number of these sites is also independent of chain
concentration. However, such an excimer site depends upon the polymer
chain bending back upon itself, which is of low probability for the rather stiff
aromatic vinyl polymers.

In solid polymers a clear-cut distinction between inter- and intramolecular
excimers is sometimes difficult.

In INTRAMOLECULAR EXCIMERS two types of molecular structures can
be considered:

(i) Molecules in which at least two chromophores are connected by a flexible
atomic chain (Fig. 3.19a). Since, in the ground state, other conformations are
energetically preferred, the sandwich structure is reached during the lifetime
of the excited state. This type of excimer forms only in fluid solutions.

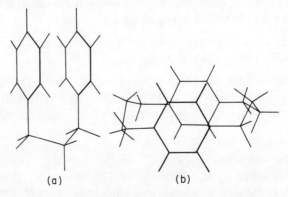

(a)   (b)

Fig. 3.19 Probable excimer structure of: (a) 1,3-diphenyl
propane and (b) (4,4)-paracyclophane

(ii) Molecules in which the sandwich structure is fixed by more or less rigid links (Fig. 3.19b) in both the excited and the ground states. Formation of excimer should be independent of the rigidity of the solvent.

In addition to these geometrical requirements, excimer formation is controlled by the dynamic processes of chromophores, including the surrounding viscosity and the interchain dynamic.[674, 1508, 1173, 2065]

Intramolecular excimer formation for polymer is controlled by both configuration and conformation in a polymer chain.[1173, 1294, 1804, 1805] In addition, the intramolecular dynamic of macromolecules is important for the transient phenomena of excimer formation in the excited state.

The intramolecular dynamics of macromolecules are controlled by the relaxation processes, which consist of two components:

(i) A molecular weight-dependent component which can be associated with the overall tumbling of the whole molecule.
(ii) A molecular weight-independent component which can be associated with local relaxation.

The observable relaxation time is considered to be the sum of these two separate mechanisms.[3449]

In fluid solutions of low viscosity, interconversion of chain conformations proceeds rapidly with the lifetime of a particular conformation limited by collision with solvent molecules:[1114, 1173, 1508]

(i) If the residence time of the random hopping singlet exciton on a particular chromophore at a non-excimer forming site is long enough, solvent collisions with the polymer will cause a rotational sampling of chain conformations leading to possible excimer formation.
(ii) If the residence time is too short and/or rotational sampling rate too slow (such as obtains in high viscosity or at low temperature) exciplex sampling must occur by virtue of competitive trapping at a suitable excimer site formed prior to the arrival of the exciton. The reduced rotational freedom in the solid state precludes extensive rotational conformational sampling.
(iii) Adjacent chromophores can also be in a marginal excimer-forming site which is not geometrically suitable to appreciably stabilize excimer formation.

Analysis of the photophysics of excimer formation in solution is complicated because sampling of suitable excimer-forming chain conformations may result both from exciton migration and from rotational transformation due to solvent collision.[1112]

The excimer state has been theoretically described as configurational mixing of exciton–resonance states and charge–resonance states:[436, 437, 1173, 3242]:

(a) The exciton–resonance interaction is polarized and has a strong inverse distance dependence.

(b) The charge–resonance interaction is fairly isotropic and has a moderate inverse distance dependence.

Both interactions favour the sandwich structure of the excimer.

Considering exciton migration in solid polymers, there are two ways in which excimer can be formed:[1173]

(i) An activated exciton migrates to preformed excimer-forming sites (temperature-dependent process) (EXCITON MIGRATION MODEL).
(ii) An unactivated exciton migrates to a marginal excimer site at which a geometrical barrier exists (temperature-independent process) (SITE BARRIER MODEL).

Excimer-forming sites are of greatest photopysical importance in aromatic vinyl polymers, even at low temperatures, because the electronic excitation energy in the form of excitons can migrate from chromophore to chromophore. This migration takes place non-radiatively by resonance transfer between an electronically excited chromophore and a ground-state chromophore. A theory for the incoherent transport and trapping of electronic excitations among chromophores attached to polymeric chains has been discussed in detail in several publications.[1178-1183]

3.4. KINETIC ANALYSIS OF EXCIMER FORMATION AND DISSOCIATION

Kinetics of excimer formation and dissociation differs for solution and solid-state conditions.[239, 442, 1000, 1346, 1614, 1623, 2065, 2077, 2294, 3121, 3294] Several kinetic models have been proposed.

Model 1 (Fig. 3.20)
This is the simplest model which describes only small molecules in dilute solution, where excimer formation is diffusion-controlled and can be presented by the photophysical processes shown in Table 3.4 and Fig. 3.21.

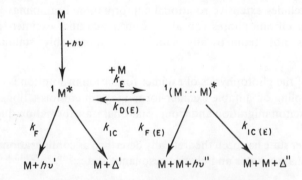

Fig. 3.20 Scheme of photophysical processes involved in excimer formation and dissociation (Model 1)

Table 3.4   Photophysical processes involved in the formation and dissociation of an excimer (E)

Step	Process	Rate	Rate constant
1. Excitation of molecule (energy absorption)	$M + h\nu \rightarrow {}^1M^*$	I_a	
2. Monomer fluorescence	${}^1M^* \rightarrow M \rightarrow h\nu'$	$k_{F(M)} [{}^1M^*]$	$k_{F(M)}$
3. Monomer internal conversion (IC)	${}^1M^* \rightarrow M + \Delta'$	$k_{IC(M)} [{}^1M^*]$	$k_{IC(M)}$
4. Monomer intersystem crossing (ISC)	${}^1M^* \rightarrow {}^3M^*$	$k_{ISC(M)} [{}^1M^*]$	$k_{ISC(M)}$
5. Excimer formation	${}^1M^* + M \rightarrow E$ (excimer)	$k_E [{}^1M^*] [M]$	k_E
6. Excimer dissociation	$E \rightarrow {}^1M^* + M$	$k_{D(E)} [E]$	$k_{D(E)}$
7. Excimer fluorescence	$E \rightarrow M + M + h\nu''$	$k_{F(E)} [E]$	$k_{F(E)}$
8. Excimer internal conversion	$E \rightarrow M + M + \Delta''$	$k_{IC(E)} [E]$	$k_{IC(E)}$
9. Excimer intersystem crossing	$E \rightarrow {}^3M^* + M$	$k_{ISC(E)} [E]$	$k_{ISC(E)}$

Note: Steps involving deactivation of ${}^3M^*$ have been omitted from the scheme as they are irrelevant to the kinetic analysis

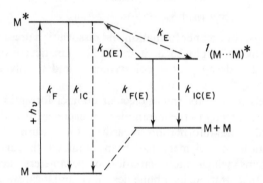

Fig. 3.21   Photophysical processes in excimer formation and dissociation

In the photostationary state the formation of an excited molecule (${}^1M^*$) and an excimer ($E({}^1(M \ldots M)^*)$) are given by the following equations:[849, 3079-3081]

$$\frac{d[M^*]}{dt} = I_a - (k_{F(M)} + k_{IC(M)} + k_{ISC(M)} + k_E[M])[{}^1M^*] + k_{D(E)}[E] = 0$$

(3.10)

$$\frac{d[E]}{dt} = k_E[{}^1M^*][M] - (k_{D(E)} + k_{F(E)} + k_{IC(E)} + k_{ISC(E)})[E] = 0 \qquad (3.11)$$

Solution of equations (3.10) and (3.11) yields the excimer ($I_{F(E)}$)-to-monomer ($I_{F(M)}$) fluorescence intensity ratio:

$$\frac{I_{F(E)}}{I_{F(M)}} = \frac{k_{F(E)}[E]}{k_{F(M)}[{}^1M^*]} - \frac{k_{F(E)}k_E[{}^1M]}{k_{F(M)}(k_{IC(M)} + k_{ISC(M)} + k_{F(E)} + k_{D(E)})} \qquad (3.12)$$

at constant temperature equation (3.12) may be written as:

$$\frac{I_{F(E)}}{I_{F(M)}} = kk_E[^1M] \tag{3.13}$$

Fluorescence intensity decays for the monomer ($I_{F(M)(t)}$) and excimer ($I_{F(E)(t)}$) are (for the kinetic scheme shown in Table 3.4) described by the following biexponential functions:

$$I_{F(M)(t)} = A_1 \exp(-t/\tau_{F(M)}) + A_2 \exp(-t/\tau_{F(E)}) \tag{3.14}$$

$$I_{F(E)(t)} = A_3 \exp(-t/\tau_{F(M)}) - \exp(-t/\tau_{F(E)}) \tag{3.15}$$

where:

A_1, A_2 and A_3 = pre-exponential factors;

$\tau_{F(M)}$ and $\tau_{F(E)}$ = decay times of monomer and excimer fluorescence, respectively;

$$\tau_{F(M)}{}^{-1} + \tau_{F(E)}{}^{-1} = (k_{F(M)} + k_{IC(M)} + k_{ISC(M)} + k_E[M])$$
$$+ (k_{F(E)} + k_{IC(E)} + k_{ISC(E)} + k_{D(E)}) \tag{3.16}$$

The kinetic analysis of excimer formation and dissociation presented in Table 3.4 and Fig. 3.20 can be employed as a model for kinetic consideration of the formation and decay of polystyrene and poly(α-methylstyrene) excimers.[1290, 1294, 3417]

Kinetic models to describe intermolecular excimer formation in fluid media cannot be applied directly to the intramolecular excimer formation in polymeric systems. Considerable information regarding the nature of intramolecular excimer formation in polymers has been obtained through the study of copolymer systems, which allow variation of parameters such as intrachain chromophore concentration, backbone flexibility, and local steric constrains (cf. Section 3.4.10).

Polymers and copolymers having naphthyl[239,947,967,1294,1457,1623,1809, 1811,1812,1814,2614,2937,2939−2941,3123,3616], anthryl[1551,1824] and carbazolyl[2683, 3122, 3123, 3125, 3517] show existence of a 'second excimer' which partially overlaps the sandwich structure of two aromatic rings and shows its emission peak in the middle of the monomer and the normal excimer peak. The second excimer can be experimentally confirmed by time-resolved spectra, e.g. poly(vinylnaphthalene/methyl methacrylate) copolymer (Fig. 3.22). For these polymers and copolymers fluorescence intensity decays for the monomer and excimer cannot be described by biexponential functions presented by equations (3.10–3.16). The three exponential kinetic equations are required on the basis of the two different kinetic models:

Model 2 (Fig. 3.23)
A second excimer can be formed by exciton diffusion mechanism. It has been considered that polymers contain isolated chromophores (M_2) which after excitation (M_2^*) can transfer energy into other excited chromophores (M_1^*) with

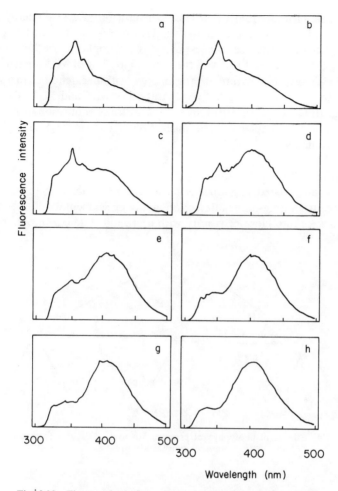

Fig. 3.22 Time-resolved fluorescence spectra of co(vinylnaphthalene/methyl methacrylate) recorded at delays of (a) 0, (b) 3.2, (c) 6.4, (d) 12.8, (e) 19.2, (f) 25.6 and (g) 32.0 ns following excitation[2937]

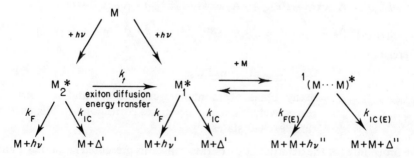

Fig. 3.23 Scheme of photophysical processes involved in excimer formation and dissociation (Model 2) considering the exciton diffusion energy transfer process

a transfer rate characterized by the rate constant (k_t). Reverse energy transfer from M_1^* to M_2^* is considered unimportant since exciton diffusion is expected to be very efficient within sequences of M chromophores within the chain comprising the M_1^* sites. In view of the reduced lifetime of M_1^* relative to M_2^*, and of the delocalized nature of the energy within extended chromophore sequences this increases the effective separation of M_1^* and M_2^*. The M_1^* to M_2^* energy transfer by electron exchange energy transfer or resonance excitation energy transfer (cf. Chapter 2) is diminished relative to the M_2^* to M_1^* exciton diffusion mechanism.

Model 3 (Fig. 3.24)
The M_2^* is an isolated naphthalene chromophore which can form a second excimer through a long-range interaction involving segmental diffusion to bring the M_2^* species into close proximity with a ground-state chromophore (M).

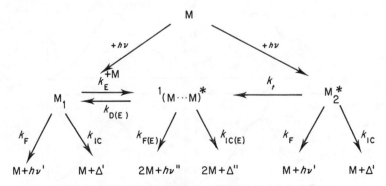

Fig. 3.24 Scheme of photophysical processes involved in excimer formation and dissociation (Model 3)

Fluorescence intensity decays for the monomer $(I_{F(M)(t)})$ and excimer $(I_{F(E)(t)})$ are described by the following three exponential functions:

$$I_{F(M)(t)} = A_1 \exp(-t/\tau_{F(M)}) + A_2 \exp(-t/\tau_{F(E_1)}) + A_3 \exp(-t/\tau_{F(E_2)}) \quad (3.17)$$

$$I_{F(E)(t)} = A_4 \exp(-t/\tau_{F(M)}) + A_5 \exp(-t/\tau_{F(E_1)}) + A_6 \exp(-t/\tau_{F(E_2)}) \quad (3.18)$$

where:

$A_1 \cdots A_6$ = pre-exponential factors,

$\tau_{F(M)}, \tau_{F(E_1)}, \tau_{F(E_2)}$ = decay times of monomer, excimer (1) and excimer (2), respectively. The existence times correspond to existence of three excited states.

The procedure for the determination of rate constants presented in models 2 and 3 are given in original papers.[194, 239, 1423, 1821, 1822, 2937, 2939, 2940, 3079–3081, 3408, 3415, 3417]

In the case of poly(N-vinylcarbazol) (cf. Section 3.2.4) there is no monomer emission (Fig. 3.15) but two spectrally distinct excimer sites (low-energy excimer (E_2) with emission at 420 nm and high-energy excimer (E_1) with emission at 380 nm) are distinguishable.[3122, 3517]

In the solution time-resolved fluorescence spectra from poly(1-vinylcarbazole) (Fig. 3.25) show emission from three sites:[3122, 3125, 3517]

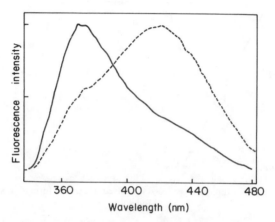

Fig. 3.25 Time-resolved fluorescence spectra for a dilute solution of poly(N-vinylcarbazole) in tetrahydrofurane: (——) early-gated and (– – –) late-gated spectrum[3122]

(i) Low-energy excimer at 420 nm.
(ii) High-energy excimer at 380 nm which inconverts with the low-energy excimer giving fluorescence at 420 nm.
(iii) Third species (E_3) which emits at 370 nm but not at 420 nm, and which does not convert to low-energy excimer.

The existence of more than one emitting excimer sites of poly(N-vinylcarbazole) dissolved in a variety of polymer films under high pressure (40 kbar) has been reported.[704] Changes of the ratio of respective intensities with pressure were interpreted according to the hopping of singlet exciton diffusion model with the two sites acting as competing traps. There are two distinct mechanisms; one predominant at low and the other at high pressures.

A fourth decay component was proposed in the poly(N-vinylcarbazole), which has been generally confirmed to have a second excimer. The emitting component was described as a 'relaxed monomer' in pulse radiolysis studies (time resolution 10 ps),[3517] or a third excimer in laser photolysis studies.[3122]

The kinetics of excimer formation and decay in the case of poly(N-vinylcarbazole) is very complex and can be presented by Model 4 (Fig. 3.26).[2683, 3122, 3123, 3125, 3517] Completely another explanation considers the relaxation process of the non-equilibrium state, because the time scale of fluorescence

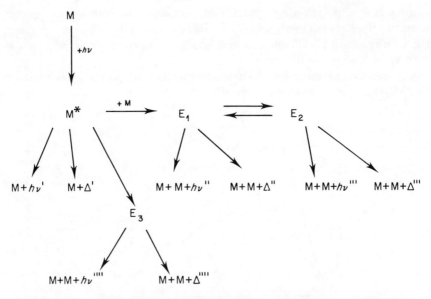

Fig. 3.26 Scheme of photophysical processes involved in excimer formation and dissociation (Model 4)

measurements (nano- and picosecond) is supposed to be comparable with the relaxation time of internal rotation.[1810]

Any compound at thermal equilibrium consists of molecules having an equilibrium distribution of conformations before excitation. If one of the molecules with a conformation ready to change to the excimer conformation is excited, it turns immediately to excimer. If another molecule with a conformation far from the excimer conformation is excited, the excited side-chain ring changes its conformation during the lifetime of the excited state and finally turns to an excimer or emits the monomer fluorescence before forming excimer.

3.4.1. Kinetic Analysis of Excimer Quenching

Kinetics of excimer formation, dissociation and quenching in the presence of a quencher (Q) (by Model 1) can be presented by the photophysical processes shown in Table 3.5.[1559, 1806, 2066, 2077, 3726, 3772]

The Stern–Volmer equations for monomer (M) and excimer (E) fluorescence emission intensity in the absence of quencher ($I_{F(M)}$, $I_{F(E)}$) and in the presence of quencher ($I_{F(M)(Q)}$, $I_{F(E)(Q)}$) under steady-state conditions are:[441, 442]

$$\frac{I_{F(M)}}{I_{F(M)(Q)}} = 1 + \tau_{F(M)} + k_{Q(M)}[Q] \tag{3.19}$$

$$\frac{I_{F(E)}}{I_{F(E)(Q)}} = 1 + \tau_{F(E)}k_{Q(E)}[Q] \tag{3.20}$$

Table 3.5   Photophysical processes involved in the formation and dissociation of an excimer (E) in the presence of a quencher (Q)

Step	Process	Rate	Rate constant
1. Excitation of molecule (energy absorption)	$M + hv \rightarrow {}^1M^*$	I_a	
2. Monomer fluorescence	${}^1M^* \rightarrow M + hv'$	$k_{F(M)} [{}^1M^*]$	$k_{F(M)}$
3. Monomer internal conversion (IC)	${}^1M^* \rightarrow M + \Delta'$	$k_{IC(M)} [{}^1M^*]$	$k_{IC(M)}$
4. Monomer quenching	${}^1M^* + Q \rightarrow M + Q^*$	$k_{Q(M)} [{}^1M^*][Q]$	$k_{Q(M)}$
5. Excimer formation	${}^1M^* + M \rightarrow E$ (excimer)	$k_E [{}^1M^*] [M]$	k_E
6. Excimer dissociation	$E \rightarrow {}^1M^* + M$	$k_{D(E)} [E]$	$k_{D(E)}$
7. Excimer fluorescence	$E \rightarrow M + M + hv''$	$k_{F(E)} [E]$	$k_{F(E)}$
8. Excimer internal conversion (IC(E))	$E \rightarrow M + M + \Delta''$	$k_{IC(E)} [E]$	$k_{IC(E)}$
9. Excimer quenching	$E + Q \rightarrow M + M + Q^*$	$k_{Q(E)} [E] [Q]$	$k_{Q(E)}$

Note: Steps involving monomer intersystem crossing (ISC) and excimer intersystem crossing (ISC(E)) have been omitted from the scheme for simplicity

where:

$\tau_{F(M)}$ and $\tau_{F(E)}$ = decay times of monomer and excimer fluorescence, respectively, in the absence of quencher (Q)

$$\frac{1}{\tau_{F(M)(Q)}} = \frac{1}{\tau_{F(M)}} + k_{Q(M)}[Q] \qquad (3.21)$$

$$\frac{1}{\tau_{F(E)(Q)}} = \frac{1}{\tau_{F(E)}} + k_{Q(E)}[Q] \qquad (3.22)$$

where:

$\tau_{F(M)(Q)}$ and $\tau_{F(E)(Q)}$ = decay times of monomer and excimer fluorescence in the presence of quencher (Q).

The plot $\tau_{F(E)(Q)}^{-1}$ versus [Q] gives the linear relation in Fig. 3.27. The slope yield value $\tau_{F(E)(Q)} k_{Q(E)}$ and excimer quenching rate constant ($k_{Q(E)}$) can be evaluated by using equation (3.22).

3.4.2. Activation Energy of Intramolecular Excimer Formation

The ACTIVATION ENERGY OF INTRAMOLECULAR EXCIMER FORMATION (E_a) can be obtained from the relation:[1173]

$$\ln(I_{F(E)}/I_{F(M)}) = -E_a/R_g T \qquad (3.23)$$

where:

$I_{F(E)}$ and $I_{F(M)}$ = the respective excimer and monomer fluorescence intensities in solution,

R_g = universal gas constant,

T = thermodynamic temperature (Kelvin).

110

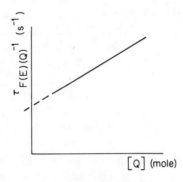

Fig. 3.27 The Stern–Volmer plot
for an excimer quenching

The activation energy (E_a) from solution measurements represents the height of the barrier between the ground state and excimer conformation for a dynamic sampling process, while the activation energy (E_a) measured in solid state reflects the energy difference between conformational minima in an equilibrium distribution. Thus, the solid state (E_a) must be less than or equal to the solution value of E_a.[1173]

3.4.3. Effect of Temperature on Excimer Formation

Excimer formation is observed in polymer films from very low temperatures (liquid helium temperatures)[1173] up to a temperature of 100°C.[1558] The temperature dependence of excimer formation differs for different polymers and has been reported in a number of publications.[461, 674, 849, 1173, 1508, 1558, 1821, 1883, 2466, 2706, 2709, 3123, 3359]

The temperature dependence of excimers can be assessed using a standard equation for $I_{F(E)}/I_{F(M)}$ (equation 3.23). Typical results are shown in Fig. 3.28. It has been found that the rate constants $k_{F(M)}$ and $k_{F(E)}$ are independent of temperature. More detailed discussion of temperature dependence of different rate constant of temperature is given in the paper quoted in Ref. 849.

3.4.4. Fraction of Excimer-forming Sites

The FRACTION OF EXCIMER-FORMING SITES (f_E) can be calculated from the following equation:[1173]

$$f_E = \frac{W_E \exp(-E_a/R_g T)}{W_G + W_E \exp(-E_a/R_g T)} \tag{3.24}$$

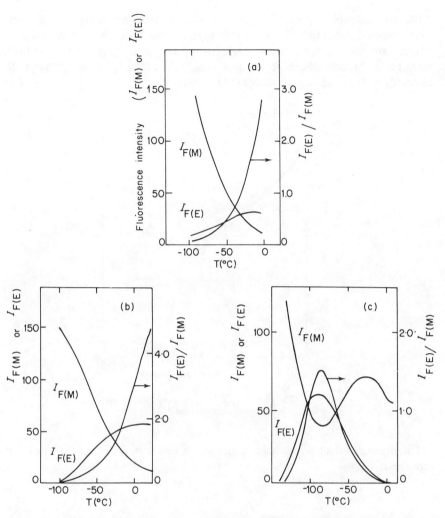

Fig. 3.28 Temperature dependence of monomer $(I_{F(M)})$ and excimer $(I_{F(E)})$ fluorescence emissions, and $I_{F(E)}/I_{F(M)}$ plot for: (a) poly(2-vinylnaphthalene), (b) 1,3-di-2-naphthylpropane, and (c) 2-ethylnaphthalene in α-methyl-tetrahydrofurane[1821,2706]

where:

W_E and W_G = degeneracies of the excimer (E) and ground (G) states, respectively.

E_a = energy of the excimer site conformation with respect to the ground state.

3.4.5. Effect of Pressure on Excimer Formation

Effect of pressure on excimer formation has been reported in a number of publications.[704, 1003, 3137]

Excimer emission of polymers in solutions under high pressure is dependent on their increased viscosity.[3137] Hydrostatic pressure influences the viscosity of both solvents and the thermodynamic properties of the theta system, which at high pressures approach phase separation. The ratio $I_{F(E)}/I_{F(M)}$ for polystyrene decreases with increasing pressures (Fig. 3.29).[3137]

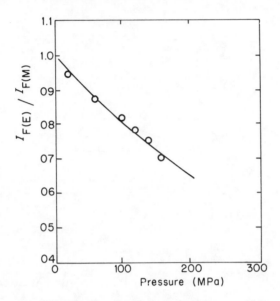

Fig. 3.29   The plot of $I_{F(E)}/I_{F(M)}$ dependence on high
pressures for polystyrene[3137]

The pressure effect on excimer formation in dilute polymer solutions can be expressed:

$$\frac{I_{F(E)}}{I_{F(M)}} = A\eta^{-\beta} \qquad (3.25)$$

where A is a constant and β is an exponent, both parameters characteristic of the solute–solvent pair system, whereas η is the solution viscosity.

Poly(vinylcarbazole) dissolved in a variety of polymers such as poly(methyl methacrylate), polystyrene, and poly(isobutylene) under very high pressures > 10 kbar show changes in emission characteristics.[704, 1003] An increase in pressures produces an increase in intensity of low-energy excimer at the expense of high-energy excimer intensity (cf. Section 3.2.4) (Fig. 3.30). Two different mechanism of excimer interaction are involved. One is evident at high pressures, and the other is present only at low pressures. The pressure at which the transition from one to the other occurs is polymer matrix-dependent. Very high pressures change main chain conformation and side group motion, which are important for excimer formation during the lifetime of excitation.

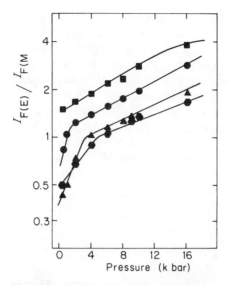

Fig. 3.30 The plot of $I_{F(E)}/I_{F(M)}$ dependence on low pressures for: ($\bullet$) poly(N-vinylcarbazole) film, and for poly(N-vinylcarbazole) dissolved in: ($\blacktriangle$) polystyrene, (O) poly(methyl methacrylate) and ($\blacksquare$) poly(isobutylene)[704]

3.4.6. Effect of Concentration on Excimer Formation

Excimer formation in polymer–solvent systems may result from both intra- and intermolecular interactions dependent on the concentration of the polymer solution. In very dilute solutions excimer formation is governed almost entirely by intrachain interactions between chromophores on the same chain. This is shown by the fact that the ratio of the excimer to monomer fluorescence intensities is independent of polymer concentration in this limited regime. At sufficiently high concentrations, intermolecularly formed excimers, i.e. excimers formed between chromophores on adjacent polymer chains, will probably contribute significantly to the observed excimer fluorescence. This is likely to be reflected in an increase in the ratio of excimer to monomer emission intensities.

In homopolymers where the local chromophore concentration is high, excimer emission predominates.[842, 849, 854] In general polymer matrix can be considered as a poor solvent. In such a matrix, polymer-forming excimer should have an increased $I_{F(E)}/I_{F(M)}$ ratio.[847, 1169, 1623]

3.4.7. Effect of Solvent on Excimer Formation

Formation of excimers of low molecular compounds is diffusion-controlled and viscosity of the solvent plays an important role.

Formation of excimers of polymers varies with solvents. In good solvents where expansion of macromolecular coils occurs, the average interchromophore distances increase and formation of excimers decreases, lowering the ratio $I_{F(E)}/I_{F(M)}$. Polymers whose $I_{F(E)}/I_{F(M)}$ varies with solvent are generally assumed to form excimers in part across loops in individual macromolecules.[239, 1169, 3408]

The extent to which excimer emission is observed is a function of the nature of the solvent in polymers containing vinylnaphthalene derivatives (Fig. 3.31) and in poly(naphthyl methacrylate)s. A minimum of $I_{F(E)}/I_{F(M)}$ is observed as the solubility parameter of the solvent matches that of the polymer, or that of the segment governing excimer formation.[3415] Variations of $I_{F(E)}/I_{F(M)}$ with solubility parameter have been used to investigate polymer compatibility in solid solutions (cf. Section 4.2.1).

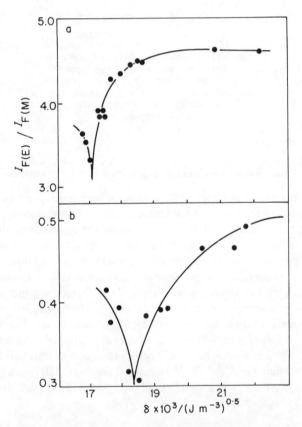

Fig. 3.31 The plot of $I_{F(E)}/I_{F(M)}$ dependence as a function of solubility parameter δ for the mixed solvent systems toluene/methanol and toluene/cyclohexane: (a) poly(1-vinylnaphthalene) and (b) co(1-vinylnaphthalene/methyl methacrylate)[3415]

3.4.8. Effect of Molecular Weight on Excimer Formation

Polymer molecular weight may affect the ratio $I_{F(E)}/I_{F(M)}$.[239, 1805, 1821] For polystyrene this effect is most pronounced at low molecular weights and becomes essentially constant at higher molecular weights (Fig. 3.32).[7, 1173, 1806, 3136, 3629] There is no reasonable explanation of this molecular weight dependence. It has been postulated that the end-group of macromolecules affects the ratio $I_{F(E)}/I_{F(M)}$. Formation of excimer sites across a chain loop may also increase with molecular weight.

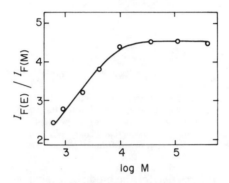

Fig. 3.32 Variation of the ratio $I_{F(E)}/I_{F(M)}$ fluorescence intensity with number-average molecular weight ($\overline{M}_n$) for anionic polystyrene in 1,2-dichloroethane[1805]

3.4.9. Effect of Polymer Structure on Excimer Formation

Polymer structure affects the $I_{F(E)}/I_{F(M)}$ ratio considerably. The $I_{F(E)}/I_{F(M)}$ decreases, which increases distance between the chromophore and the main chain[1883, 2011, 2616] as illustrated by the homologous series of poly[2-(1-naphthyl)ethyl methacrylate] (3.16) in Fig. 3.33.

$$x = 1, 2, 3$$

(3.16)

116

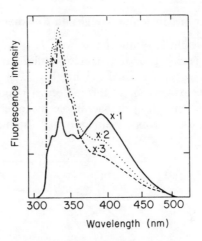

Fig. 3.33 Effect of side-chain length on extent of excimer formation for the homologous series of poly[2-(1-naphthyl)ethyl methacrylate)] (*3.16*) in tetrahydrofurane[2616]

The $I_{F(E)}/I_{F(M)}$ is very sensitive for the mode of attachment of the group which exhibit excimer fluorescence (e.g. carbazole ring) to the chain backbone.[2011]

In polymers in which chromophoric groups are apart from the main chain "dead-end" conformations may exist, especially in film below the glass transition. For example in poly[2-(2-naphthyl)ethyl methacrylate] (*3.17*) many bond rotations are possible and many configurations which do not allow for excimer formation within the excited state lifetime.[1623] A chromophore in such a polymer may be sufficiently far from its neighbours that it cannot transfer energy to an excimer site.

$$CH_3$$
$$-(C-CH_2)_n-$$
$$C=O$$
$$O$$
$$(CH_2)_2$$

(*3.17*)

The effect of chain tacticity on the ratio $I_{F(E)}/I_{F(M)}$ has been reported in a number of publications.[854, 1173, 1294, 1673, 1804, 1805] The $I_{F(E)}/I_{F(M)}$ for isotactic polystyrene is higher than for atactic or syndiotactic polymers of comparable molecular weight (Fig. 3.34).[7, 459, 461, 1173, 1806, 3136, 3629] Studies of model

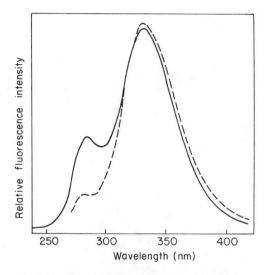

Fig. 3.34   Fluorescence spectra of: (——) atactic acid
(− −) isotactic polystyrene[461]

compounds (cf. Section 3.2.6) provide useful information on the role of chain tacticity in excimer formation.[249, 459, 674, 1597, 1882, 2077, 2315, 2708, 3081, 3802, 3994]

Chain flexibility determines the extent to which chromophore can form excimer within the excited state lifetime. Poly(1- and 2-naphthyl acrylates) show higher values of $I_{F(E)}/I_{F(M)}$ than the less flexible poly(naphthyl methacrylates).[2490]

3.4.10. Effect of Copolymer Structure on Excimer Formation

Copolymer composition also affects the $I_{F(E)}/I_{F(M)}$ ratio. The $I_{F(E)}/I_{F(M)}$ increases with increasing monomer-1 in random copolymer of fluorescent monomer-1 with a non-fluorescent monomer-2 (Fig. 3.35).[38, 846, 2708, 3079, 3080]

Excimers can also be formed in alternating copolymers, where non-nearest excimer formation may occur, for example excimers are formed in stilbene/maleic anhydride and acenaphthylene/maleic anhydride alternating copolymers[3801], whereas they are not formed in alternating copolymers of vinylcarbazole/fumaronitrile or diethylfumarate.[3983]

In styrene–butadiene–styrene–butadiene (SBSB) block copolymers the $I_{F(E)}/I_{F(M)}$ is a function of the number of bonds between styrene monomers in styrene block.[3415] A smooth curve (Fig. 3.36) is obtained for all polymers provided the styrene sequences is the SBSB block copolymers are regarded as isolated blocks.

The independence of $I_{F(E)}/I_{F(M)}$ upon the existence of a second styrene sequence in the SBSB block copolymers suggests that non-nearest neighbour interactions are of minor importance in excimer formation in the vinyl-aromatic polymers.

Excimer formation and decay in copolymers (Table 3.2) gives valuable information on macromolecular photoprocesses:[2939]

118

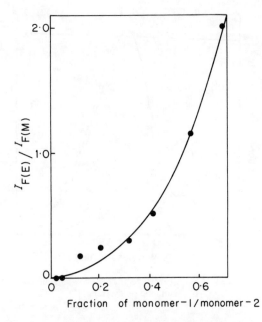

Fig. 3.35 The plot of $I_{F(E)}/I_{F(M)}$ dependence as a function of the fraction of links between naphthalene derivatives in co(1-vinylnaphthalene/methyl methacrylate)[3079]

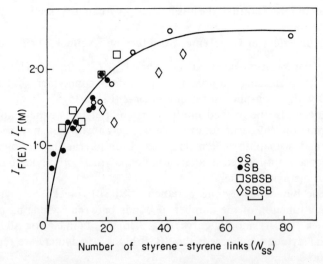

Fig. 3.36 The plot $I_{F(E)}/I_{F(M)}$ as a function of the number of styrene–styrene links (N_{SS}): (○) polystyrenes, (●) block-co(styrene/butadiene), (□) block-co(styrene/butadiene) sequences as independent block and (◇) block-co(styrene/butadiene) counting all styrene species[3415]

(i) Variation of copolymer structure allows study of the influence of steric constraints and overall chain flexibility on the process of excimer formation.[38,194,846,1162,2937,2939,2940,3079-3081,3801,3983]

(ii) Studies of the dependence of excimer emission upon copolymer microcomposition can yield information regarding the nature of inter-chromophore interactions which dominate the creation of excited state dimers.[38, 194, 846, 2937, 2940, 3079-3081]

(iii) Transient decay studies of copolymer systems using pulsed laser sources can be applied to the determination of individual rate coefficients in the excimer photophysical reactions models.[2937, 2939, 2940]

Energy transfer and excimer formation in copolymers requires generation of statistical functions descriptive of the pentad sequence distribution.[195, 3081] The probabilities (P_{ii}) and (P_{ij}) of a monomer (i) being placed adjacent to a like monomer or a dissimilar monomer, respectively, can be given by:[1515]

$$P_{ii} = \frac{r_i[M_i]}{r_i[M_i]+[M_j]} = \frac{f_i - R/200}{f_i} \tag{3.26}$$

$$P_{ij} = 1 - P_{ii} \tag{3.27}$$

where:

r_i = reactivity ratio,
$[M_i]$ = feed concentration,
f_i = mole fraction in the copolymer of monomer i,
R = run number.

The pentad fractions of the copolymer are generated from the pair probabilities. For example, the fraction of i units centred in $jiiij$ type pentads (f_{jiiij}) is given by

$$f_{jiiij} = P_{ii}^2 P_{ij}^2 \tag{3.28}$$

Two summation terms, Σ_1 and Σ_2, are invoked to characterize the concentration of excimer sites in the copolymers in which it is assumed that excimer formation results from interactions between next-to-nearest neighbour chromophores in the polymer chain. The excimer concentration terms can be calculated as follows:

$$\Sigma_1 = 2f_{AMAMA} + 2(f_{AMAAA} + f_{MMAMA} + f_{MAAMA}) \tag{3.29}$$

$$\Sigma_2 = 2f_{AMAMA} + 3xf_{AAAAA} + xf_{MAAAM} + 2[(1+x)f_{AMAAA}$$
$$+ 2xf_{MAAAA} + xf_{AAAMM} + f_{MMAMA} + f_{MAAMA})] \tag{3.30}$$

where A and M represent comonomer repeat units, and x = factor which quantifies the efficiency of excimer sites centred on comonomer A units relative to triads containing centrally located species (AAA sites).

The excimer site concentration may be described in terms of a function Σ_2, which assumes that excimer formation occurs as a consequence of interactions between next-to-nearest neighbour chromophores. Construction of Σ_2 requires knowledge of the relative efficiencies of excimer formation in sites of the form AMA and AAA respectively. Such a procedure could be accomplished through interactive fitting of x in equation (3.30) to obtain optimum description of $I_{F(E)}/I_{F(M)}$ as

$$\frac{I_{F(E)}}{I_{F(M)}} = kk_E f_a \Sigma_2 \qquad (3.31)$$

where:

k = for explanation cf. equations (3.10–3.13),
k_E = rate constant of excimer formation,
f_a = mole fraction of aromatic chromophore.

Figure 3.37 shows a plot of $I_{F(E)}/I_{F(M)}$ versus the function $f_a \Sigma_2$ generated according equation (3.31).

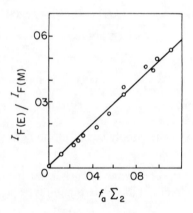

Fig. 3.37 The plot $I_{F(E)}/I_{F(M)}$ as a function of the term $f_a \Sigma_2$ of co(acenaphthylene/methyl acrylate) in dichloromethane: (●) homopolymer[195]

3.5. EXCIPLEXES

An EXCIPLEX (EXCITED CHARGE TRANSFER COMPLEX) is a well-defined complex which exists in electronically excited states.[1083, 1140, 1277, 1363, 1823, 1825–1827, 1836, 2235, 2413, 3568, 3575, 3581, 3582, 3588, 3590, 3591, 3859] An exciplex is formed between an excited donor (D*) and acceptor (A) or an excited acceptor (A*) and donor (D) molecules:

$$D^* + A \rightarrow (D^-A)^* \leftrightarrow (D^+A^-)^* \rightarrow D^*A \leftrightarrow DA^* = (DA)^* \qquad (3.32)$$

$$A^* + D \rightarrow \underbrace{(A^-D^+)^* \leftrightarrow (A^+D^-)^*}_{\text{Excited charge-transfer complex}} \rightarrow A^*D \leftrightarrow AD^* = (DA)^* \qquad (3.33)$$

The difference between the donor ionization energy and the acceptor electron affinity is a decisive criterion for the exciplex formation. If this difference is too small, ground-state charge-transfer (CT) complexes are formed.

Exciplexes (DA)* have important features:[778, 779, 2774]

(i) Exciplex formation is a thermodynamically reversible process in competition with subsequent formation of useful chemical intermediates:

$$D* + A \leftrightarrow (DA)* \rightarrow \text{reacting intermediates} \qquad (3.34)$$

$$A* + D \leftrightarrow (DA)* \rightarrow \text{reacting intermediates} \qquad (3.35)$$

(ii) Exciplexes may undergo substitution with a quenching compound (Q):

$$(DA)* + Q \leftrightarrow (DQ)* + A \qquad (3.36)$$

$$(DA)* + Q \leftrightarrow (AQ)* + D \qquad (3.37)$$

(iii) Exciplexes may be quenched with a quenching compound (Q):

$$(DA)* + Q \rightarrow D + A + Q \text{ (or products)} \qquad (3.38)$$

$$(DA)* + Q \rightarrow D + A^{\dot-} + Q^{\dot+} \qquad (3.39)$$

$$(DA)* + Q \rightarrow D^{\dot+} + A + Q^{\dot-} \qquad (3.40)$$

The last three reactions, (3.38)–(3.40), are very important in initiation of vinyl polymerization, where donor (D) is amine, acceptor (A) ketone, and quenching compound (Q) vinyl monomer (cf. Section 7.9.4.2.1.2).

Typical exciplexes are formed by pairs of such compounds as anthracene/N,N-dialkylaniline,[2392, 3859] N-isopropylcarbazole/dimethylterephthalate,[2191, 2192, 2776, 3859, 3957] 1,3-dinaphthylpropane/1,4-cyano-benzene[2525] and many other polyaromatic hydrocarbons with amines bounded to polymer chains (Table 3.6).

Figure 3.38 shows an example of experimental observation of exciplex formation between low molecular compounds such as biphenyl and diethylaniline. It is important to mention here that not all exciplexes emit fluorescence.

The fluorescence from an exciplex is dependent on the solvent polarity.[3574, 3576] With increasing solvent polarity the fluorescent yield decreases. Dipole

Table 3.6 Polymers with bounded groups forming exciplexes

Exciplex components		Reference

—(OCH$_2$—CH—CH$_2$—OCO)—
   |
   CH$_2$
 (anthracene)   +
	—(OCH$_2$—CH—CH$_2$—OCO)—
	   |
	   CH$_2$
	 (phenyl—N(CH$_3$)(CH$_3$))
		3580

122

Table 3.6   (Contd.)

Exciplex components	Reference
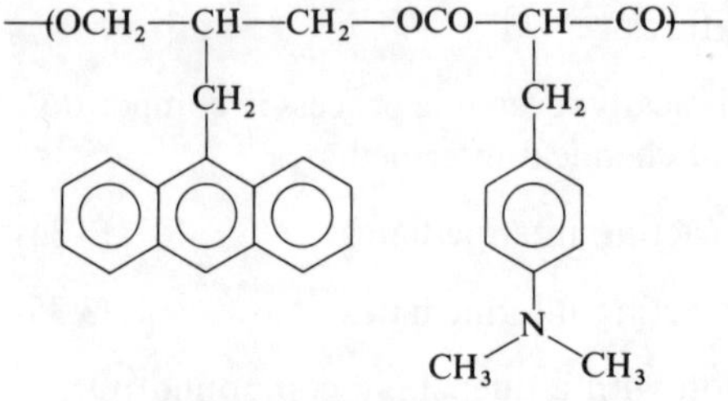 $-(OCH_2-CH-CH_2-OCO-CH-CO)-$ (with CH_2 substituents bearing anthracene and p-dimethylaminophenyl groups)	3581
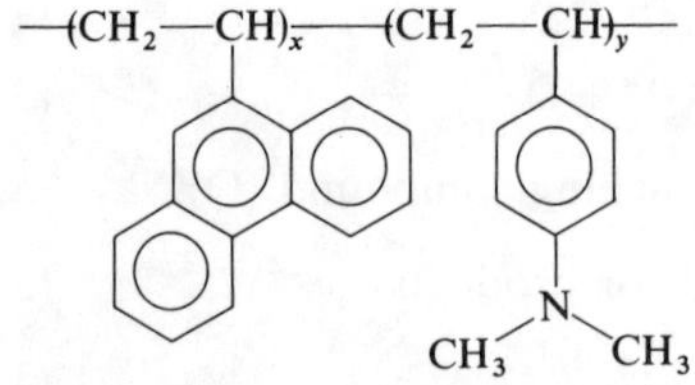 $-(CH_2-CH)_x-(CH_2-CH)_y-$ (bearing phenanthrene and p-dimethylaminophenyl groups)	1833
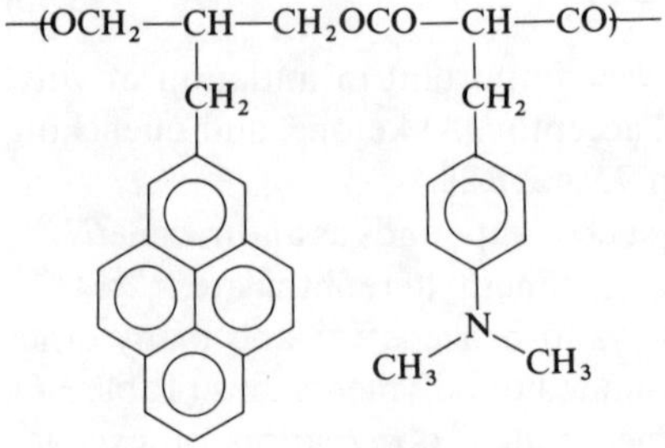 $-(OCH_2-CH-CH_2OCO-CH-CO)-$ (with CH_2 substituents bearing pyrene and p-dimethylaminophenyl groups)	3588, 3589, 3591, 3592
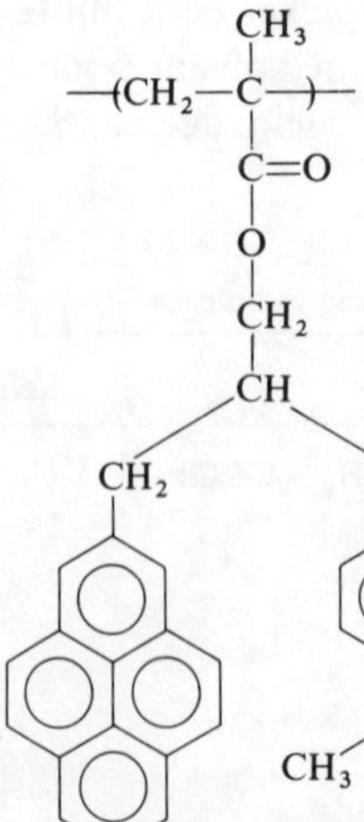 $-(CH_2-C(CH_3)-)-$ with $C=O$, O, CH_2, CH, CH_2 bearing pyrene and CH_2 bearing p-dimethylaminophenyl groups	1836 3569

Table 3.6 (Contd.)

Exciplex components	Reference

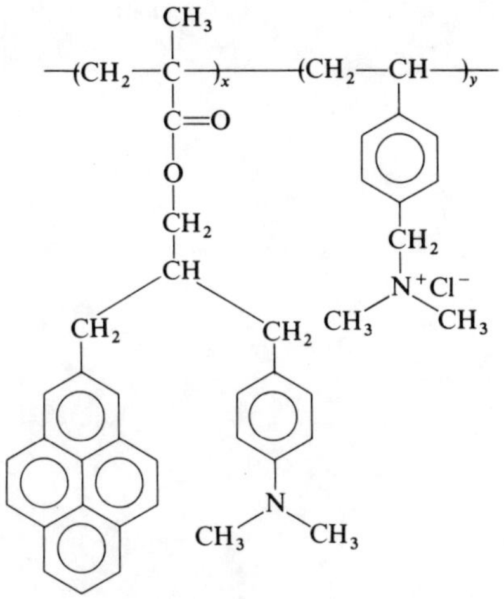

$x = 6.9, y = 93.1$
$x = 0.96, y = 99.04$

3569

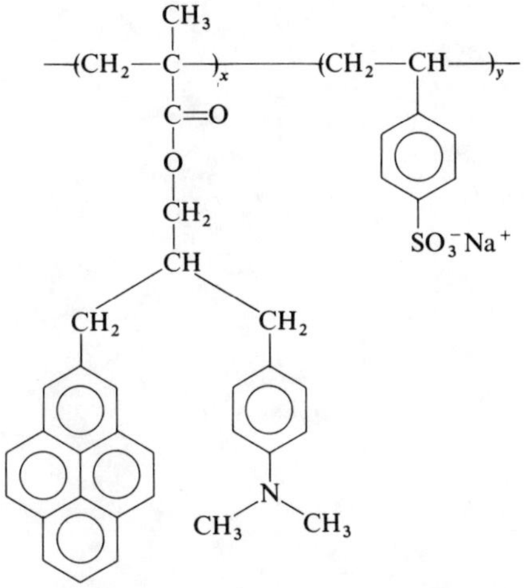

$x = 6.5, \; y = 83.5$

3569

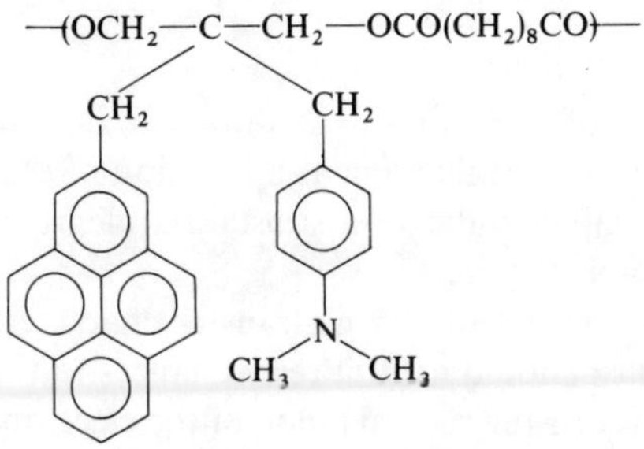

3990

Table 3.6 (Contd.)

Exciplex components	Reference
$-(OCH_2-C-CH_2-OCONH(CH_2)_6NHCO)-$ with pendant CH_2 groups bearing a pyrene unit and a p-dimethylaminophenyl unit (CH_3—N—CH_3)	3991
$-(CH_2-C(CH_3)-)_x-CH_2-C(CH_3)-)_y$ with $C=O$, O, $(CH_2)_2$, carbazole N; and $C=O$, O, $(CH_2)_2$, $C=O$, O, phenyl, $COOCH_3$	2010, 2011, 2235
$-(OCH_2-C(CH_3)(CH_2)_4)-CH_2-O-C(=O)-\text{(phenyl)}-C(=O)-O-)$ with carbazole N	2434, 3574, 3576

moments of the fluorescent states of molecules in solution can be estimated by measuring the energy of their fluorescent maxima as a function of the dielectric constants of the solvent. These values indicate a substantial degree of charge separation in the fluorescent exciplex state.

If the exciplex state has a high degree of charge-transfer character, then the absorption spectrum of an exciplex should exhibit resemblances to those of the radical cation of the electron donor or the radical anion of the electron acceptor.

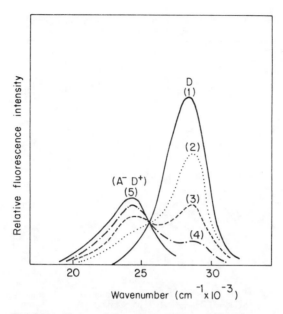

Fig. 3.38   Change of fluorescence spectra during the exciplex formation between low molecular biphenyl (acceptor (A)) and diethylaniline (donor (D)): (1) 10^{-2} M (D), (2) 10^{-2} M (D) + 0.03 M (A), (3) 10^{-2} M (D) + 0.1 M (A), (4) 10^{-2} M (D) + 0.3 M (A), (5) 10^{-2} M (D) + high concentration of (A)[2081] (*Reproduced by permission of Akademische Verlagsgesellschaft Geest & Portig K.-G.*)

Singlet exciplexes are very short-lived species; hence the absorption spectrum of an exciplex can be obtained only by using a laser which emits light pulses in the nanosecond or picosecond time domain. Fluorescence lifetimes of singlet exciplexes can be measured only by employing a nanosecond or picosecond spectroscopical method.

Triplet exciplexes can be identified in several cases by measuring phosphorescence and by triplet–triplet absorption.

Exciplexes are formed mainly in solutions, but in polymer matrix they are less stable. The decreased stability of the exciplex in a rigid polymer could be attributed to two factors:[2393, 3591]

(i) Improper orientation (alignment) of the exciplex constituents.
(ii) Poor solvation of the rather polar exciplex by the polymer.

Wavelength maxima of the exciplexes formed in polymer matrix are shifted towards shorter wavelengths (Table 3.7).[1084, 2392, 2393] Maxima of the exciplex emission are temperature-dependent.[1823] It has been suggested that the change in wavelength maxima of the exciplex with temperature in a fluid medium or in polymers above the glass transition temperature (T_g) may be due to a change in the energy profile of the exciplex.[2392] The shift to shorter wavelength with

Table 3.7 Wavelength maxima of the exciplex anthracene/dimethyltoluidinine in fluid and polymeric media[1084]

Acceptor (fluorescing compound)	Electron-donor	Medium	λ_{max} of exciplex emission (nm)
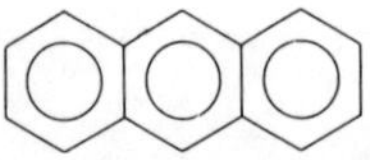	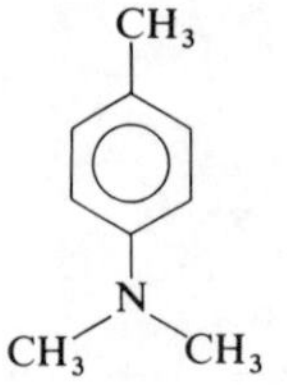	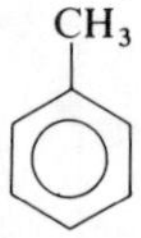	515
		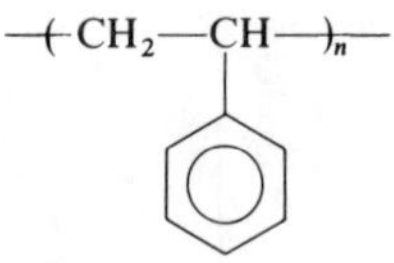 $-(CH_2-CH-)_n-$	490
	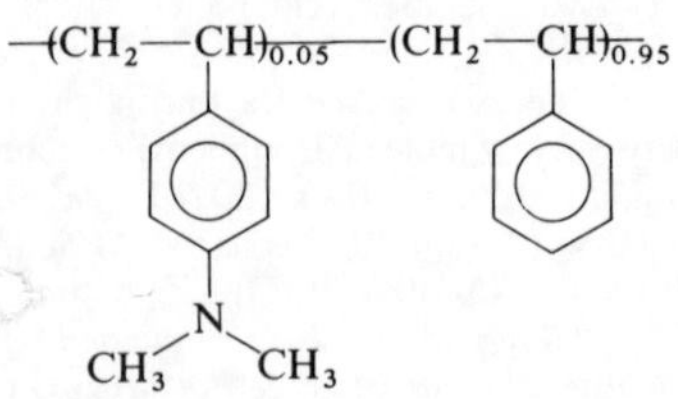 $-(CH_2-CH)_{0.05}-(CH_2-CH)_{0.95}$		480

decreasing temperature in polymeric matrices at temperatures below T_g can be due to improper orientation and/or larger separation of the exciplex components as compared with those in fluid media.

Attachment of either component of the exciplex to the polymer does not further reduce the degree of interaction with dissolved component (Table 3.8).

In the case of copolymers containing N-vinylcarbazole and terephthalate exciplex components it has been shown that a such copolymers in dilute solutions exhibit fluorescence emission both from carbazole units and from exciplex, whereas a cast film emits only exciplex fluorescence (Fig. 3.39).[2010, 2011, 2235] These results indicate for the inter-polymeric exciplex interaction with efficient energy migration between carbazole groups prior to population of exciplex sites. Varying compositions of the copolymer have the effect of changing the ratio of carbazole to exciplex fluorescence in dilute solutions, but in all cases emission from films arises almost exclusively from the exciplex.

Exciplex emission can be used to study polymer miscibility.[2177] Exciplex quenching can be employed to investigate photochemical reaction mechanisms.[2177] The product of a photochemical reaction is generally formed in one of the following three ways:

Table 3.8 Wavelength maxima of exciplexes from derivatives of pyrene and dimethylaniline[1084]

Exciplex components	Medium	λ_{max} of exciplex emission (nm)
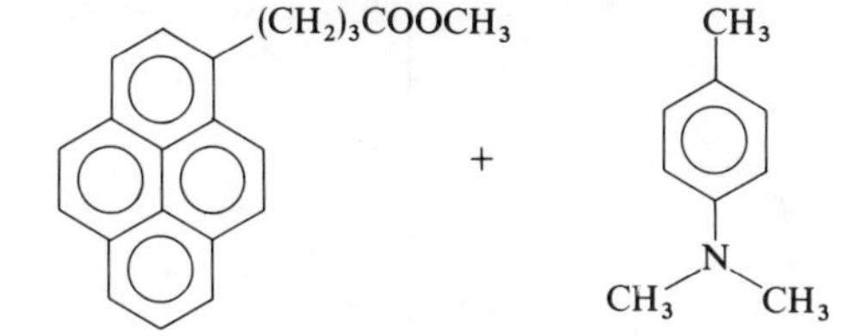	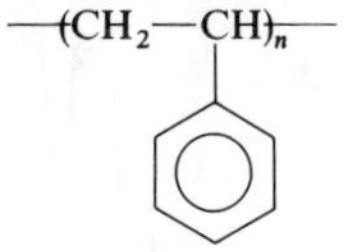	463
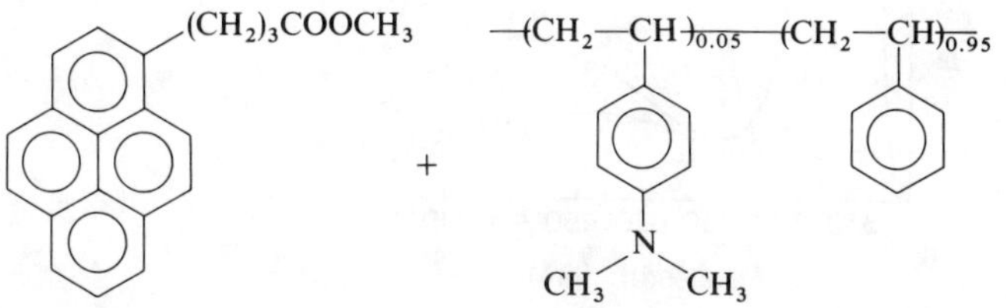		465
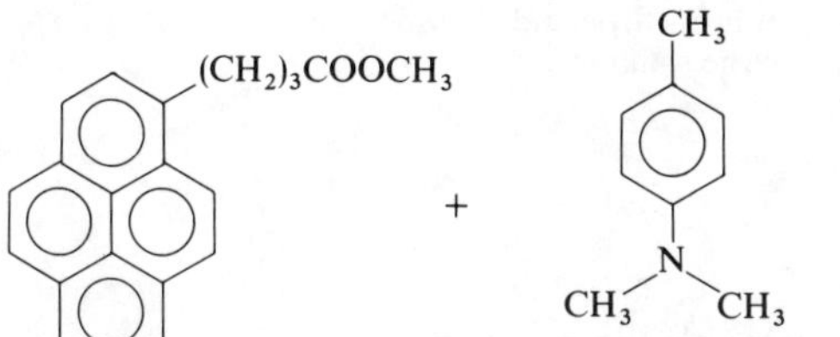	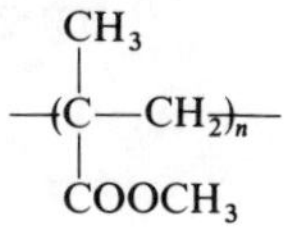	463
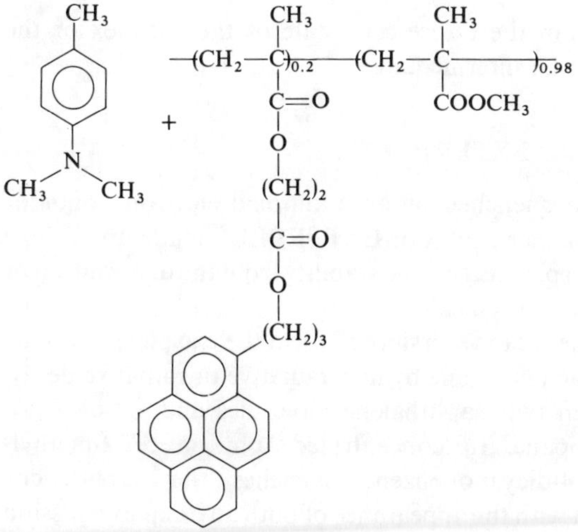		460

128

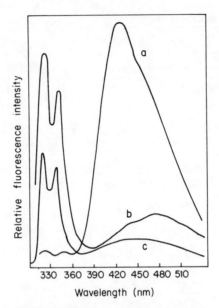

Fig. 3.39 Exciplex fluorescence emission spectra of co(N-vinylcarbazole/terephthalate) (21:79): (a) film, (b) in methylene dichloride, and (c) benzene solutions[2235]

(i) The product is formed in parallel with the exciplex.
(ii) The product and the exciplex are both formed from a common non-relaxed charge-transfer state.
(iii) The product is formed from the fluorescent state of the exciplex or the reaction via the exciplex as an intermediate.

3.6. EXTERPLEX

Highly polar exciplexes can be quenched by an additional electron donor or acceptor molecule to give an excimer triplex or EXTERPLEX made up of three molecules.[586, 777, 2774] The exterplex receives its stability from the distribution or separation of charge over three molecules.

The exterplex, once it is formed, can reversibly re-form the exciplex, or it may return to the ground state of each molecule by non-radiative or radiative decay.

The exterplex forms between two naphthalene molecules and 1,4-dicyanobenzene or 1,3-dinaphthylpropane. In concentrated solution of 2-methylnaphthalene, addition of 1,4-dicyanobenzene quenches the naphthalene monomer and excimer emission with the appearance of both an exciplex emission ($\lambda_{max} = 420$ nm) and the exterplex emission ($\lambda_{max} = 490$ nm).

The exterplex formation in polymeric systems has been found for:

(a) poly(N-vinylcarbazole) (two molecules) and dimethyl terephthalate (one molecule).[1680, 1681]
(b) Poly(1-vinylnaphthalene) (two molecules) and dicyanobenzene (one molecule).[1681]

3.7. PHOTOINDUCED ELECTRON TRANSFER AT THE POLYMER–SOLUTION INTERFACES

Electron transfer from leuco crystal violet in solution to the singlet excited state (S_1) of pyrene bonded to polymer film has been observed.[3584, 3585] This process has special importance at mimicking photosynthesis, where separation of oxidized and reduced products is a special requirement. If photooxidation on one side of a polymer film can be coupled with photo-reduction on the other, the film is electron transporting.

4. *Applications of luminescence spectroscopy in polymers*

Polymers which exhibit luminescence (fluorescence and/or phosphorescence) can be classified into two main types:[1614, 3409]

(i) Those which absorb and emit light through isolated chromophore groups (X) situated as in-chain or end-groups (termed here as type A):

In-chain chromophore

Side-chain chromophore

End-chain chromophore

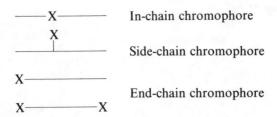

In type A polymers the chromophore density is low, so that deactivation processes are characteristic of the isolated emitting species.

(ii) Those which absorb and emit light through chromophore groups (X) present in the repeat unit (or units) that form the backbone structure of the polymer (termed here as type B):

—X—X—X—X—X—

or

Regular (isotactic)

Regular (syndiotactic)

Irregular (atactic)

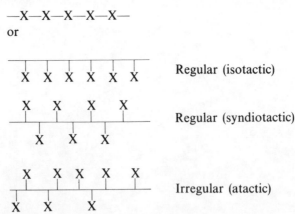

In type B polymers the chromophore density is higher and interactions between chromophores are considerable.

Luminescence spectroscopy (fluorescence and/or phosphorescence) can be applied to the study physicochemical properties of polymers.[162, 205, 1291, 1614, 2196, 2357, 2562-2564, 2733, 2876] Examples of such application are given in the following sections.

4.1. FLUORESCENT PROBES

Fluorescent emission of low molecular compounds (FLUORESCENT PROBES), doped in bulk polymers or polymer solutions, can yield valuable information on physical properties of macromolecules (Table 4.1).

The choice of fluorescent probe depends on:

(i) chemical structure;
(ii) place of location in polymer matrix, e.g. some fluorescent compounds may be located only in the amorphous regions, others only in crystalline regions;
(iii) photostability of probe;
(iv) high fluorescence efficiency;
(v) interaction with macromolecules (the fluorescent molecule must not perturb the polymer nature).

Compounds which satisfy such criteria are shown in Table 4.1.

4.1.1. Studies of Singlet–singlet Energy Transfer in Polymerizing Media

Measurements of fluorescence yield of low molecular compounds such as anthracene,[2129] 9-methylanthracene,[2676] 2,5-diphenyloxazole,[1417, 1436] bis[1, 4(2-(5-phenyloxazolyl)benzene)][2676] and 1,1',4,4'-tetraphenylbutadiene[2676] (acceptors) as a function of polymerization time (i.e. percentage conversion) have been recorded for a variety of styrene-based systems.

In a complex system such as a polymerizing donor, many effects will combine to determine the observed fluorescence yield:[2676]

(i) The chemical nature of the donor (polymer) changes continuously during the chain growth process.
(ii) Acceptor fluorescence depends on a variety of microenvironments which influence the energy transfer process. The quantum yield of fluorescence may also be affected by the variations in the microviscosity which it experiences. The viscosity of the medium will dictate the operative mechanism of energy transfer. At low viscosities collisional exchanges of energy characterized by Stern–Volmer kinetics will predominate. At high viscosity long-range energy migration (transfer) will predominate.
(iii) Fluorescence from acceptor can be overlapped by the excimer fluorescence of the donor. The relative importance of the excimer emission in the polymer spectrum is dependent upon the viscosity of the medium (which influences backbone rotational sampling of intramolecular excimer sites), and the polymer concentration (upon which depends the extent to which inter-molecular excimers may form) (cf. Section 3.3).

Table 4.1 Examples of fluorescent probes which have been used for the study of different polymer properties

Probe	Structure	Doped polymer	Problem	References
Pyrene	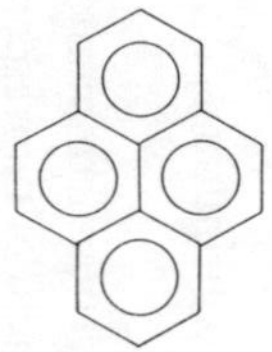	Polystyrene	Study of excimer formation in polymer matrix	1885
		Poly(methyl methacrylate)	Conformation state of polymer	691
		Water-soluble polymers i.e. poly(N-vinylpyrrolidone) and poly(ethylene oxide)	Interaction between sodium dodecyl sulphate and soluble polymers (the solubilization mechanism) 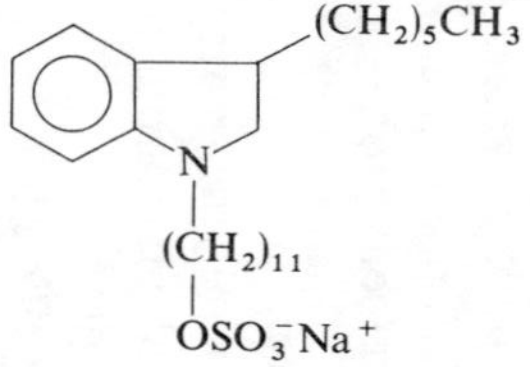	3682
Pyrene carboxaldehyde	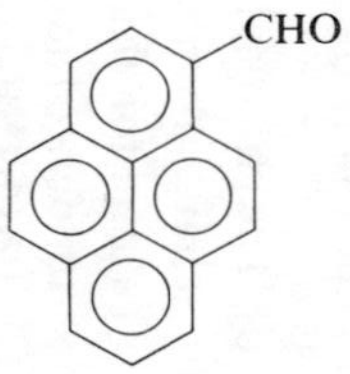 CHO	Block co(ethylene oxide/propylene oxide)	Study of 'low', 'medium', and 'high' concentration domains in water-soluble block copolymers	3688

$(CH_2)_5CH_3$

$(CH_2)_{11}$

$OSO_3^-Na^+$

N

1,3-di(α-Naphthyl)propane

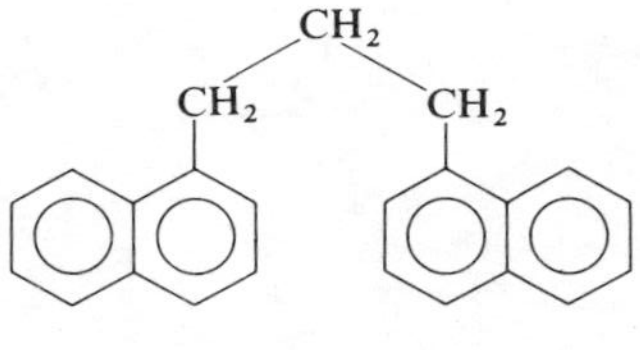

CH_2
CH_2
CH_2

Poly(styrene sulphonate) macro-ions

Study of microscopic environments of poly(styrene sulphonate) macro-ions with [11-(3-hexyl-1-indonyl) undecyl] trimethyl ammonium bromide

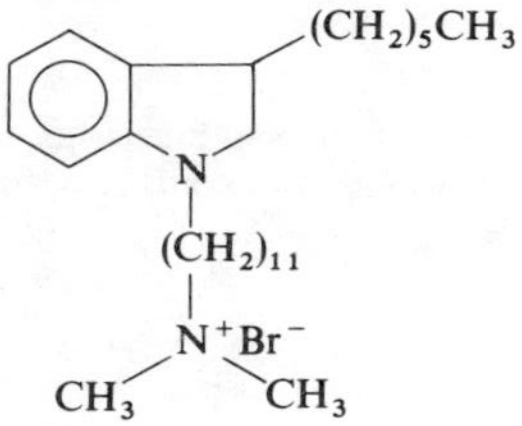

$(CH_2)_5CH_3$

N—$(CH_2)_{11}$—N^+Br^-—CH_3
CH_3
CH_3

3693

1,3-di(α-Naphthyl)propane

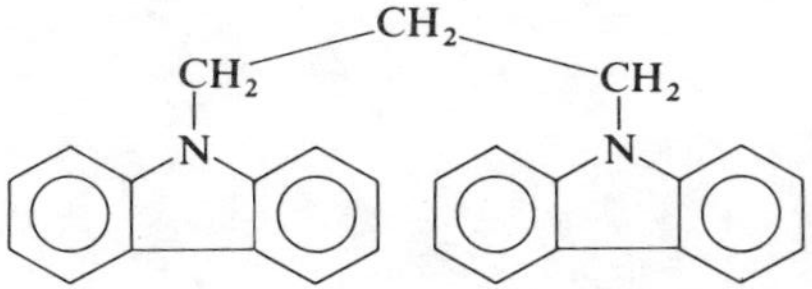

CH_2
CH_2—N
CH_2
N

Benzyl ammonium chloride

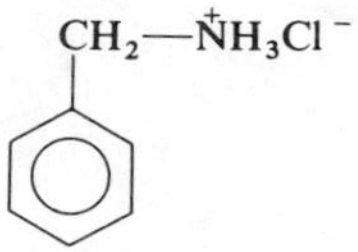

CH_2—$\overset{+}{N}H_3Cl^-$

Anionic polyelectrolytes, i.e. sodium (ethylene sulphonate), sodium poly(styrene sulphonate), sodium (2-acryloamido-2-methyl-1-propane sulphonate)

Electrostatic interactions of anionic polyelectrolytes with neutral polymer poly(ethylene oxide)

3694–3696

Dibenzyl ammonium chloride

CH_2—NH_2—CH_2
$+Cl^-$

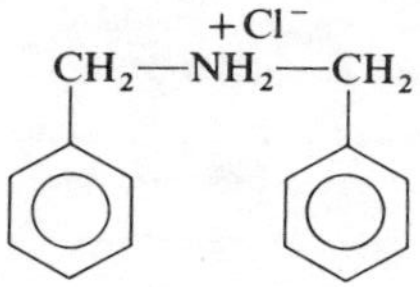

Table 4.1 (*Contd.*)

Probe	Structure	Doped polymer	Problem	References
(α-Naphthylmethyl) ammonium chloride	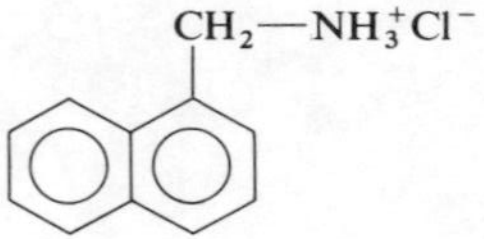 CH$_2$—NH$_3^+$Cl$^-$	Anionic polyelectrolytes, i.e. sodium (ethylene sulphonate), sodium poly(styrene sulphonate), sodium (2-acryloamido-2-methyl-1-propane sulphonate)	Electrostatic interactions of anionic polyelectrolytes with neutral polymer poly(ethylene oxide)	3694–3696
Bis(α-naphthyl-methyl)ammonium chloride	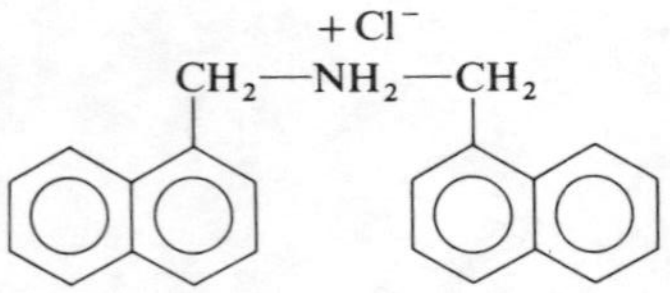			
5-(Dimethylamino)-1-[p(m)-vinyl-benzylamino)]-sulphonyl]naphthalene	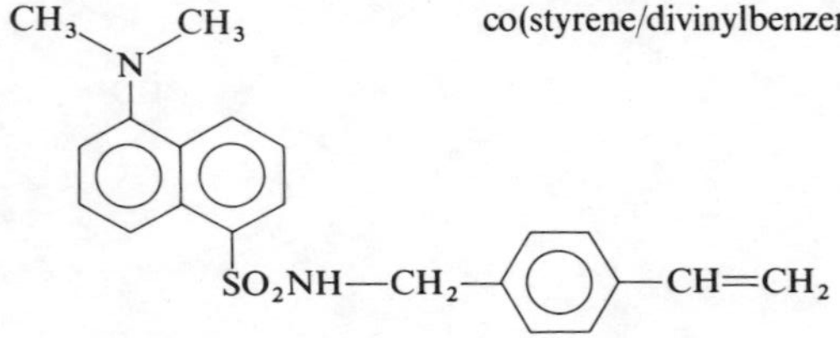	co(styrene/divinylbenzene)	Study of solvation	3303
2,5-Diphenyloxazole	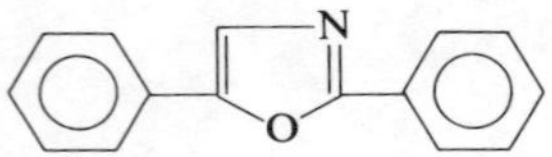	Polymerization of vinyltoluene	Study of microviscosity of polymerizing media	2695
Bis-1,4-[2-5-phenyl)oxazolyl)] benzene	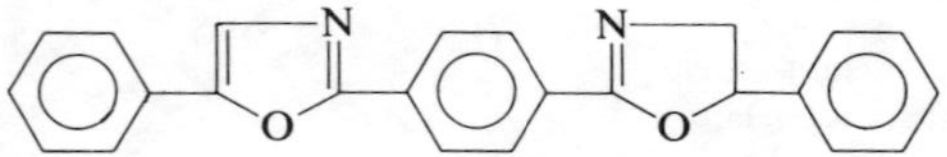			

1,1',4,4'-Tetraphenylbutadiene	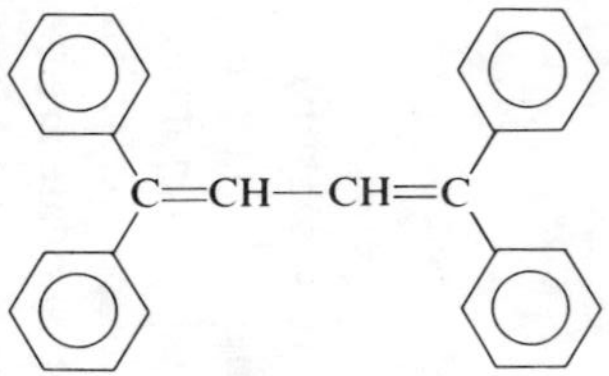	Polymerization of vinyltoluene	Study of microviscosity of polymerizing media	2695
4,4'-(dibenzoxazolyl)stilbene	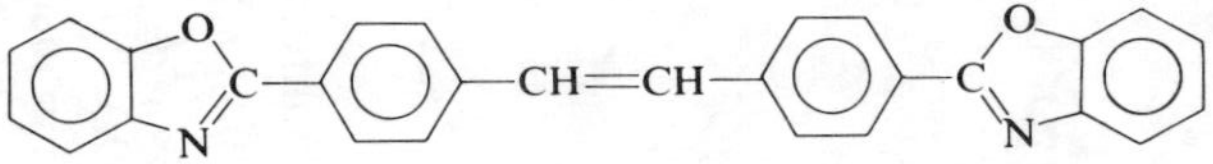	Poly(ethylene terephthalate)	Study of orientation of uniaxially drawn polymer	2728
Auramina		Poly(methyl methacrylate)	Relaxation phenomena	2009
Rare earth metal ions	Eu^{3+}, Tb^{3+}, Co^{2+}, and UO_2^{2+} salts of polymers containing acid and sulphonic acid ligands	Poly(acrylic acid), co(styrene/acrylic acid), co(styrene/maleic acid), co(styrene/methacrylic acid), co(methyl methacrylate/methacrylic acid)	Characterization of ion containing polymer structures	308, 2607, 2609, 2777, 2778

4.1.2. Donor–acceptor Probes

Specially interesting group of fluorescent probes are DONOR–ACCEPTOR PROBES such as benzylidenemalonitrile (*4.1* and *4.2*) or juloidinemalonitrile (*4.3*) probes.[2216, 2217, 2316, 2317, 2319] The absorption and fluorescence emission spectra of these probes are shown in Fig. 4.1 and Fig. 4.2, respectively.

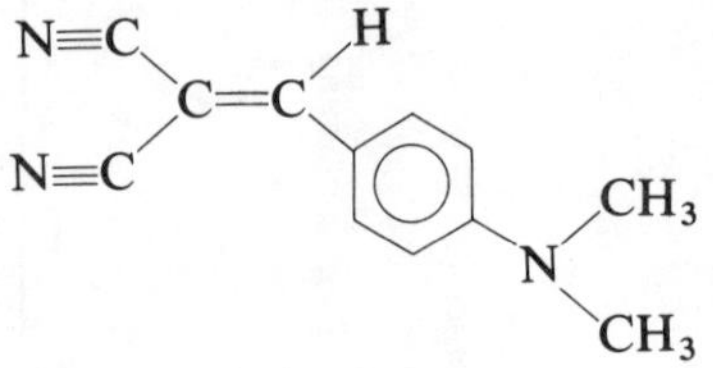

$$N{\equiv}C \diagdown \atop N{\equiv}C \diagup C{=}C \diagup{H} \diagdown \text{(aryl)}{-}N(CH_3)_2$$

(*4.1*)

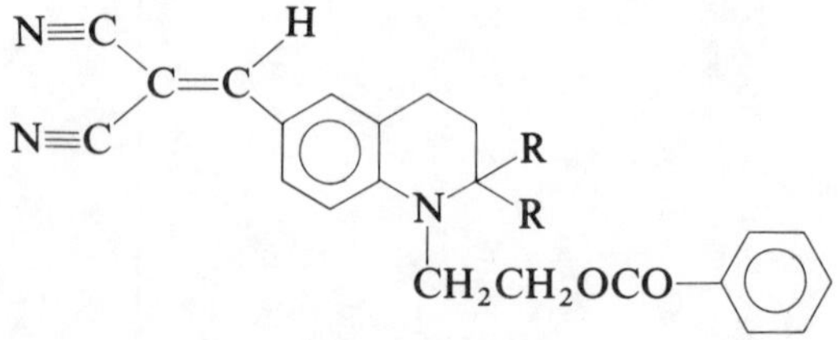

(*4.2*)

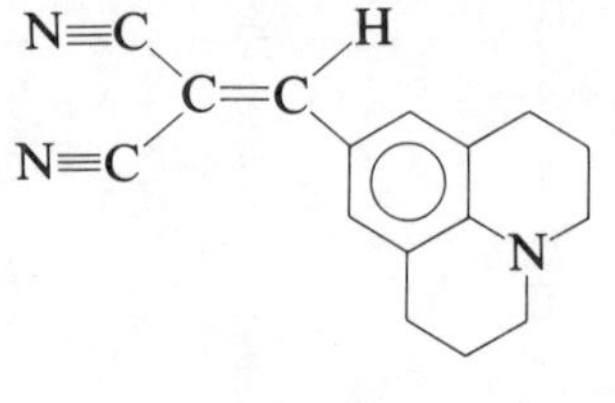

(*4.3*)

All of these donor–acceptor probes have quantum yield very sensitive to molecular rigidity, viscosity, temperature, and matrix properties.[2316] The non-radiative decay rates of excited molecular probes (*4.1*), (*4.2*), and (*4.3*) in methyl methacrylate are 2.5×10^{11}, 1.33×10^{11}, and $9.3 \times 10^{10} \, s^{-1}$ respectively. The singlet excited state (S_1) lifetime of these probes in methyl methacrylate are 4.6, 7.5, and 10.7 ps respectively. The extremely short lifetime of the singlet excited state (S_1) of these molecules and many intramolecular charge transfer complexes can be attributed to fast intrinsic internal relaxation.[638, 1373, 2318] Among the possible internal relaxation mechanism are rotations of the R_2N group (e.g. in the case of (*4.1*)) and/or twisting around the donor–acceptor as shown in Fig. 4.3. While probe (*4.1*) has both routes available to dissipate the excitation energy, the probes (*4.2*) and (*4.3*), due to structural restriction, have only the donor–acceptor route to channel the excess excitation energy. The internal rotation process with the fastest rate will control the non-radiative decay rate and consequently

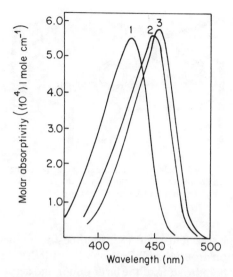

Fig. 4.1 Visible absorption spectrum of (1),(2) benzylidenemalonitriles (*4.1*) and (*4.2*) respectively, and (3) juloidinemalonitrile (*4.3*) in methyl methacrylate[2316]

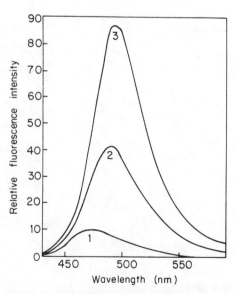

Fig. 4.2 Fluorescence spectra of (1),(2) benzylidenemalonitriles (*4.11*) and (*4.2*) respectively, and (3) juloidinemalonitrile (*4.3*) in methyl methacrylate[2316]

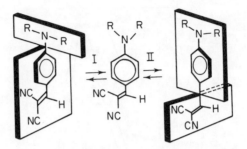

Fig. 4.3 Schematic diagram showing the internal
molecular rotation of the R_2N group (route I) and
of the donor and acceptor moiety (route II)[2316]

determine the singlet excited state (S_1) lifetime in this class of compounds.
International rotation around the R_2N group (2×10^{11} s^{-1}) in the case of probe
(*4.1*) must be faster than twisting of the large donor and acceptor moiety, leading
to the very short singlet lifetime (5 ps in methyl methacrylate). The quantum yield
of fluorescence of probes (*4.1*), (*4.2*), and (*4.3*) increases by a factor of 10, 20, and
40, respectively, when the probes are embedded in poly(methyl methacrylate)
matrix.

The fluorescence intensity of these probes changes drastically during polymerization of methyl methacrylate (Fig. 4.4). The fluorescence intensity changes

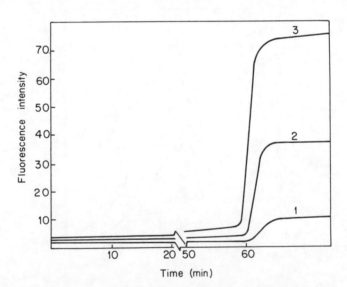

Fig. 4.4 Dependence of the fluorescence intensity of (1),(2) benzylidenemalonitriles (*4.1*) and (*4.2*) respectively and juloidinemalonitrile (*4.3*) in
methyl methacrylate polymerization time (polymerization carried out at
70°C)[2316]

only slightly for a period of time until a critical moment is reached when a sharp rise in conversion is reached. The fluorescence of the probe increased gradually as polymer conversion increased. Finally the fluorescence intensities levelled off as the polymer limiting conversion was reached. The polymerization region in which fluorescence intensity increases sharply corresponds to the increase of the rigidity of the medium at the glass transition. There is a correlation between the limiting quantum yield of fluorescence of a probe and the polymer glass transition (T_g) and expansion of coefficient. This method of fluorescence probe measurements has been applied for the study of bulk polymerization of different methacrylates and styrene.[2316, 2317]

A juloidinemalonitrile (*4.3*) probe has been used as fluorescence molecular rotor probe in a series of stereoregular (isotactic, syndiotactic, and atactic) poly(methyl methacrylates) (Fig. 4.5).[2319] Absorption and fluorescence emission

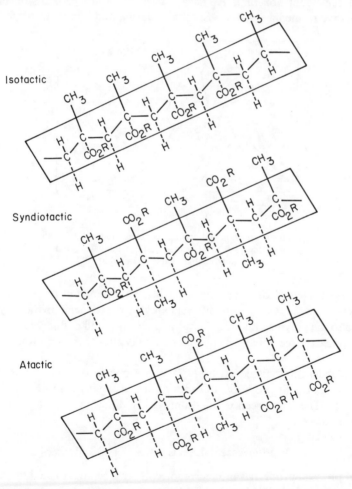

Fig. 4.5　Chemical structure of stereoregular poly(methyl methacrylate)[2319]

spectra of juloidinemalonitrile differ in different stereoregular poly(methyl methacrylates) (Fig. 4.6). The extremely fast deactivation rate of juloidinemalonitrile excited singlet state (S_1) is attributed to fast intrinsic internal rotation of the donor–acceptor moieties, a process involving movement of one part of the molecule with respect to the other (MOLECULAR ROTOR). Environmental factors restricting the internal rotation of the probe lead to decrease in the non-radiative decay rate and consequently increases fluorescence. The quantum yield for fluorescence of juloidinemalonitrile is markedly enhanced in isotactic poly(methyl methacrylate) as compared to the syndiotactic and atactic polymers. The enhancement of fluorescence of juloidinemalonitrile is caused by inhibition of the internal molecular rotation via rigidization of the probe by local environment. These results show that the flexibility of isotactic chains is lower than that of syndiotactic or atactic chains. A similar situation exists in dyes with rotation-dependent non-radiative decay, where fluorescence emission becomes very sensitive to media rigidity, viscosity, polarity, and temperature.[2316, 2318]

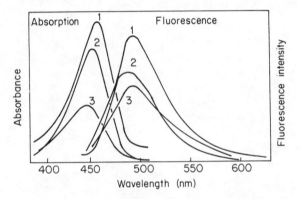

Fig. 4.6   Absorption and fluorescence emission bands of juloli-dinemalonitrile (4.3) in (1) isotactic, (2) syndiotactic, and (3) atactic poly(methyl methacrylate) film[2319]

Fluorescence emission properties of a benzylidenemalonitrile probe (4.2) in poly(methyl methacrylates) with different alkyl groups depend on the alkyl group attached to the polymer, and the change in quantum yield of fluorescence reflects the difference in free volume or polymer chain flexibility in various locations of these polymers.[2217]

A benzylidenemalonitrile probe (4.2) allows one to determine the effect of solvent vapour on the properties of vapour-swollen co(vinylchloride/vinylacetate) (87/17).[2216] The fluorescence quantum yields differ in the various swollen polymer matrices.

4.1.3. Fluorescent Labels Attached to Polymers

Some of physical properties of macromolecules can be investigated by the study of fluorescence of labelled (markered) chains with chromophore groups (labels or markers) in the following way:

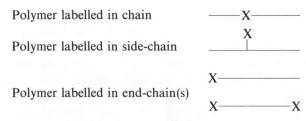

Polymer labelled in chain

Polymer labelled in side-chain

Polymer labelled in end-chain(s)

Fluorescent label (marker) should[3465]

(i) specifically label one region of the macromolecules,
(ii) not perturb the system under investigation (in the case of biopolymers the physiological function of the system must also remain unchanged[3598]),
(iii) have well-quantified spectral properties suitable for the measurements required.

Usually, fluorescent marker content in the polymer does not exceed 0.1 M% (i.e. 1 mole per 1000 monomeric units of polymer). Various methods of synthesis permit the preparation of polymers with different structures and different contents, and the required location of fluorescent marker in macromolecules. Luminescent markers can be introduced into macro-molecules in all the stages of formation (initiation, propagation, or termination) as well as in the reactions involving functional groups of preformed or natural polymers. The choice of method for the synthesis of a labelled polymer depends on its chemical structure and the required arrangement of luminescent marker in the polymer chain.

Polymers with luminescence markers at their chain ends can be prepared by the synthesis of macromolecules (in initiation or termination reactions) or by reactions involving terminal functional groups of macromolecules, e.g. anthracene compounds react with growing polymer radicals of some monomers such as vinyl acetate or styrene to form structures (*4.4*) (9,10-dihydroanthracene structures) or structures (*4.5*) (terminal anthryl group):[205,206]

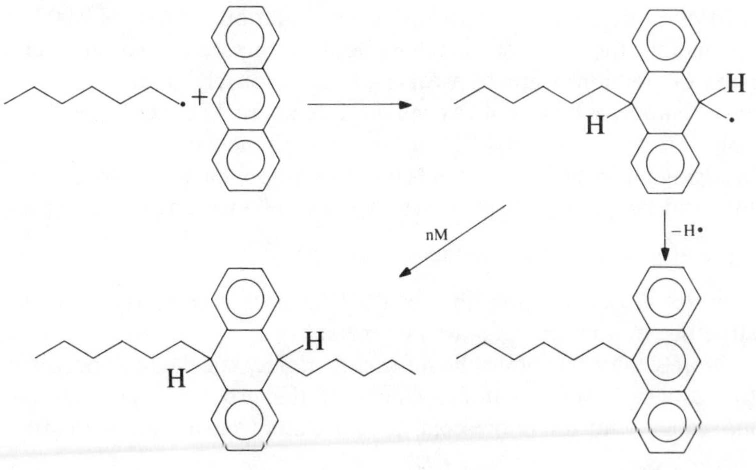

(4.1)

(4.2)

(*4.4*)

(*4.5*)

However, anthracene does not interfere with the propagating methyl methacrylate radicals, at least in the absence of u.v. radiation.[255] The light-initiated polymerization of methyl methacrylate at $40°C$ in the presence of benzyl peroxide is also inhibited by anthracene.[3389] In order to obtain anthracene-labelled polymethyl methacrylates, monomers with attached anthracene groups have been synthesized, which are further polymerized by free radical mechanism.[3451]

Polymers with luminescence markers in side groups of macromolecules can be obtained by the following methods:[205]

(i) copolymerization of the main monomer with the monomers bearing the luminescent groups,

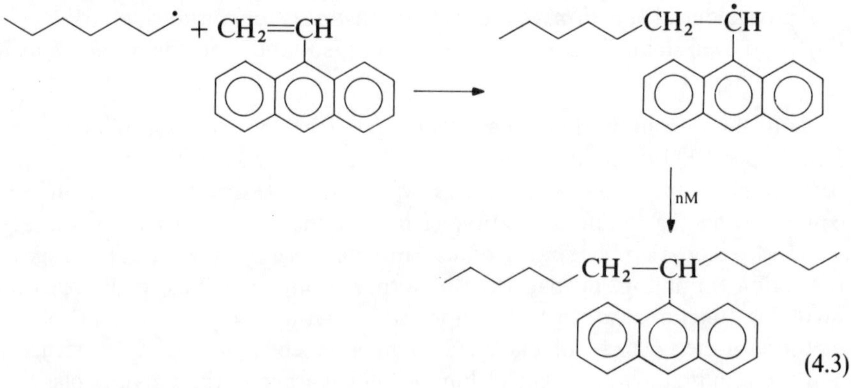

(4.3)

(ii) reactions involving macromolecules and reagents containing luminescence groups.

Study of fluorescence of labelled polymers has several advantages such as possibilities of:[205]

(i) Carrying out experiments in any solvents (polar, non-polar, water) and at any temperature.
(ii) Carrying out experiments in very dilute solutions as low as 0.001%. This means that the effect of interchain contacts superimposed on macromolecular interaction is greatly weakened or completely absent.
(iii) Investigating polymers of any chemical structure (linear or branched soluble polymers and even cross-linked insoluble polymers).
(iv) Studying individual components in multicomponent polymer system (structure and properties of block copolymers, polymer–polymer complexes),

These are also several disadvantages, such as:[2735]

(i) It is not always possible to introduce luminescence markers of the same structure into all investigated polymers.
(ii) A fluorescent marker must be sufficiently rigidly attached to a polymer chain by covalent bonds, that movement of the group truly monitors chain movement, and the fluorescent marker cannot move independently of the chain.

(iii) It cannot be employed with polymers which deactivate luminescence of the attached luminescent markers.

(iv) Theoretical problems in the interpretation of measurement results, when they were made under continuous irradiation rather than using a time-resolved pulsed irradiation.

4.1.4. Applications of Fluorescence Labelling Technique in Polymers

The fluorescence labelling technique can be employed for the study of:

(i) Polymer compatibility.[187,1478,2510,2511,2563,2564] If two different polymers are labelled with donor (D) and acceptor (A), e.g. by copolymerization of the labelled monomers (Table 4.2) the emission spectrum of their blends (Fig. 4.7 and Fig. 4.8) will depend on whether the two polymers are compatible or segregate into separated phases (incompatible). The energy transfer form donor molecules to the acceptor molecules occurs over distances of the order 2–5 nm (cf. Section 2.12). The donor fluorescence intensity ($I_{F(D)}$) will be

Table 4.2 Examples of label group employed for the study of polymer compatibility[187, 2565]

Donor label group	Acceptor label group
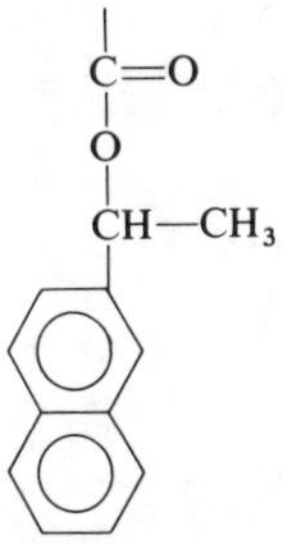 (2-naphthyl)-	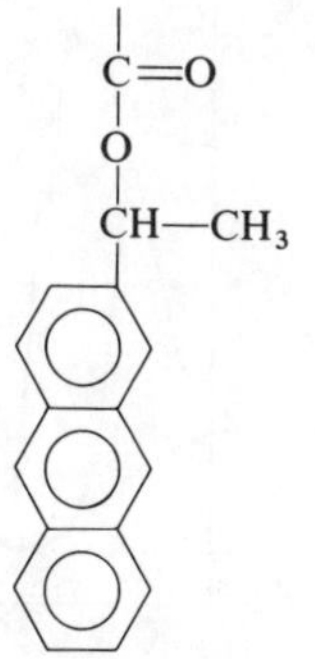 (2-anthryl)-
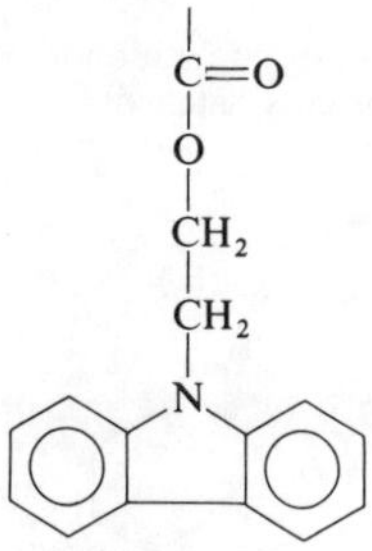 (N-carbazolyl)-	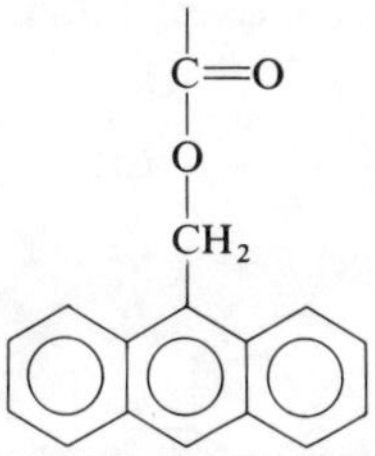 (9-anthryl)-

144

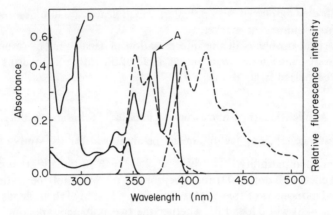

Fig. 4.7 Absorption (——) and emission (---) spectra of donor (D)–acceptor (A) pairs. The overlap integral is crosshatched[187]

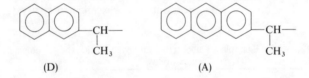

(D)   (A)

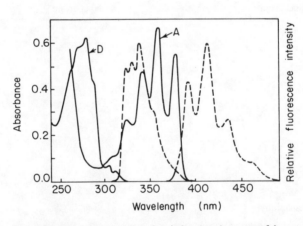

Fig. 4.8 Absorption (——) and emission (---) spectra of donor (D)–acceptor (A) pairs. The overlap integral is crosshatched[187]

(D)   (A)

diminished and the acceptor fluorescence intensity $(I_{F(A)})$ will be enhanced. The ratio $I_{F(D)}/I_{F(A)}$ is therefore decreased to the extent that the two polymers are intertwined, allowing the donor and acceptor fluorescence labels to approach one another within a critical distance.

Phase separation in incompatible polymer blends of poly(methyl methacrylate) and poly(benzyl methacrylate) may be followed by the measuring of the fluorescence emission of the latter.[2564]

Carbazole (donor) labelled co(styrene/acrylonitrile) and anthracene (acceptor) labelled poly(methyl methacrylate) emission fluorescence studies indicate that compatibility is optimized when styrene copolymer contains 30 to 40 mol % acrylonitrile (Fig. 4.9).[187,2564]

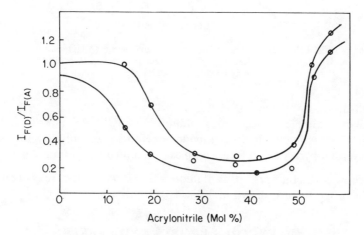

Fig. 4.9 Ratio of donor (D) and acceptor (A) fluorescence intensity $(I_{F(D)}/I_{F(A)})$ in films containing equal weights of 1.45 wt % carbazole-labelled co(styrene/acrylonitrile) (○) and 1 wt % anthracene-labelled poly(methyl methacrylate) (●)[187]

(ii) Extent of chain interpenetration in solution with different concentrations.[1842,2564] Rapid freeze-drying of polymers, e.g. poly(ethyl methacrylate) labelled with N-carbazole and 9-anthryl groups and subsequent sublimation of the solvent preserves the extent of chain interpenetration which exists in solution. The energy transfer between labelled polymers yields information on the extent of such penetration. If the solution contained two polymers labelled with donor and acceptor labels, respectively, the two labels will be segregated in two micro-phases if the original solution was highly diluted, but will come to lie increasingly close to one another as the solution concentration is increased beyond the point leading to chain entaglements. As a consequence, reflectance fluorescence spectra of pellets pressed from the freeze-dried solutions will exhibit increasing energy transfer with an increasing concentration of the original solution (Fig. 4.10).

146

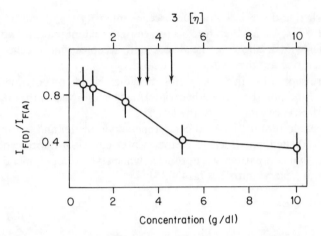

Fig. 4.10 Ratio of carbazole (D) and anthracene (A) intensity $(I_{F(D)}/I_{F(A)})$ in freeze-dried mixtures of carbazole-labelled poly(ethyl methacrylate)[1842]

(iii) Self-diffusion[3311] and interdiffusion.[2564]

(iv) Polymer complexes in aqueous solutions.[370,689,690] Poly(acrylic acid) with polymers acting as hydrogen bond acceptors, e.g. poly(oxyethylene) or poly(vinyl pyrrolidone) form complexes in aqueous solution.

More examples on the application of fluorescence labels for the study of different polymer properties are shown in Table 4.3.

4.2. EXCIMER PROBES AND LABELS

Low molecular compounds or chromophoric groups which exhibit excimer and monomer fluorescence can be used as probes or labels, respectively, to study the microscopic physical properties of the medium or polymer matrix. Characteristics of excimer as a probe are as follows:

(i) Polarization or ionic effects can be neglected.

(ii) Thermodynamic measurements will provide information on the environment of the excimer-forming site.

(iii) Since a fluorescence measurement technique is used, the sensitivity of measurements is very high.

(iv) Excimer-forming chromophores are, in general, stable aromatic hydrocarbons and easy to handle.

(v) Employing polymers of well-defined structure bearing excimer-forming chromophores, it is possible to obtain information on the interaction between fluorescent groups as a function of polymer structure.

(vi) Comparison of excimer-forming polymers with appropriate dimer (e.g. 1,3-diphenylpropane or 2,4-diphenylpentane) or oligomer model compounds, specificity in molecular interaction in polymer can be clarified.

Table 4.3 Examples of fluorescence labels which have been used for the study of different polymer properties

Label	Structure	Labelled polymer	Problem	References
Naphthalene	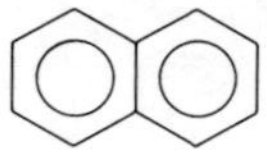	Poly(methyl methacrylate)	Anisotropy measurements	1000
			Rotation diffusion and rotation relaxation time	2008, 2735
		Co(2(6-tert-butyl-2-naphthoxy)ethyl methacrylate/N-vinyl pyrrolidone)	Role of the hydrophobic tert-butyl-naphthoxy chromophores in absorption of low molecular substances	2621
Anthracene	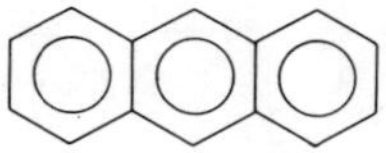	Poly(ethylene)	Orientation measurement in polymer	3972
		Co(ethylene/propylene/1,4-hexadiene) (EPDM)	Energy transfer between oxygen, and triplet and singlet states	1150
		Poly(methyl methacrylate)	Non-aqueous dispersion of polymer colloids	3914
		Polystyrene	Intermolecular end-to-end photodimerization	3721, 3723
			Molecular dynamics of polymer solutions	2549
			Rates of diffusion-controlled intermolecular reactions between chain end-groups of polymer	1661, 1662
			Orientation measurements during stretching	3554
9,10-dimethylanthracene	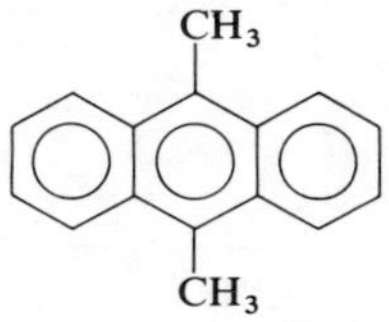	Polyisoprene	Glass rubber relaxation and orientation measurements	1851, 1852

Table 4.3 (*Contd.*)

Label	Structure	Labelled polymer	Problem	References
Pyrene	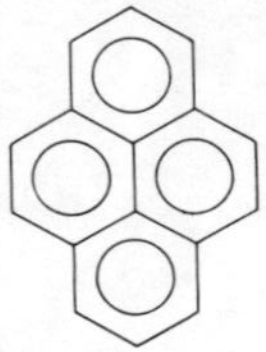	Polyethylene oligomers	Dynamic properties of hydrocarbon chains	3994
		Poly(ethylene oxide)	Solubility parameters	810
		Poly(methyl methacrylic acid)	Nature of chromatographic environments on the polymer role of pH and polymer architecture	706
		Polystyrene	Dynamics end-to-end cyclization	3076, 3915, 3919, 3920
			Solubility parameters	2289
		Polysiloxanes	Influence of polymer concentration on internal motion	3506, 3921
1,3,5-triphenyl-2-pyrazoline	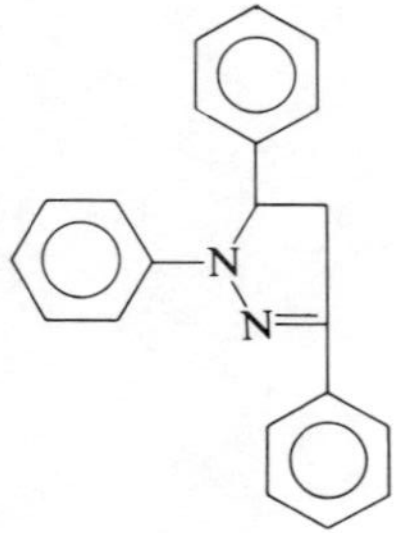	Poly(methacrylates)	Absorption and fluorescence characterization	1749
4-diethanolamino-7-nitrobenzofuran	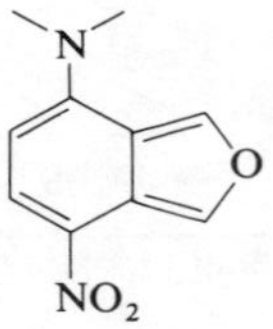	Poly(propylene oxide)	Polymer diffusion in molten polymer	3391
Isothiocyanate bridge		Amylose and dextran	Microviscosity in polymer solutions and gels	1044

Some examples of application of excimer probes for the solution of some problems in polymer chemistry are given below:

(a) conformational mobility of polymer chains,[960]
(b) probing molecular motions in bulk polymers,[462,463,2873,2874]
(c) the molecular-weight dependence of end-to-end cyclization in polystyrene,[3920]
(d) mechanical deformation and fracture in quenched as well as annealed polymers,[1463]
(e) solution behaviour of polymers,[1375]
(f) polymer chain interpretation in solutions,[2564]
(g) hindered rotation (crankshaft-like motions) in dilute solutions of flexible-chain motions,[688,2293]
(h) polymer colloids (mainly to the study of small-molecule penetrations into dispersions of colloidal polymer particles),[2918–2920, 3913, 3914]
(i) spinoidal decomposition in polymer fluids,[705]
(j) polymerization on the basis of the solvent-viscosity dependence of the excimer emission,[3797]

4.2.1. Application of Excimer Fluorescence to the Study of Polymer Structures

Excimer fluorescence yields information on the molecular structure of macro-molecules. This method is sensitive to molecular motion in solution and even in the solid state below glass transition temperature.[1167, 1171, 1173, 3294]

Excimer-forming site (cf. Section 3.3) is sensitive to different aspects of chain structure:

(i) Intermolecular excimer sites yield information on aggregation of several polymer chains.
(ii) Intramolecular excimer sites provide information on the conformation of a single chain.

The most important excimer site for the study of polymer blend compatibility is the intermolecular site. This yields information on the clustering of GUEST aromatic vinyl polymers physically dispersed in HOST polymer matrix (Table 4.4).[1168, 1171] The identification of intermolecular excimer formation may be obtained from measurements of excimer emission intensity versus concentration.

If the two polymers are incompatible, phase separation will occur, leading to micro- or macrophase clustering of the guest. This will cause a local increase in the concentration of aromatic rings to which the increase in the number of intermolecular excimer-forming sites should be directly proportional.

If the two polymers are compatible, the mixing will be on the molecular level so that each guest-polymer segment will be in a matrix of host-polymer segments. This will cause a local decrease in the concentration of aromatic rings to which the decrease in the number of intermolecular excimer-forming sites should also be directly proportional.

Table 4.4   Examples of applications of excimer fluorescence to the study of polymer blends (guest and host)

Guest	Host	Reference
Polystyrene	Poly(vinylmethyl ether)	1245–1248
Poly(2-vinylnaphthalene)	Poly(alkyl methacrylates)	1168–1172, 1229, 3291, 3293, 3295
Poly(N-vinylcarbazole)	Polystyrene and poly(methyl methacrylate)	1886
Poly(acenaphthylene)	Poly(alkyl methacrylates)	1168, 1172
Poly(4-vinylbiphenyl)	Poly(alkyl methacrylates)	1168, 1172
Co(1-naphthylmethyl methacrylate/9-anthryl-methyl methacrylate)	Poly(methyl methacrylate)	3599
Co(2(1-naphthyl)ethyl methacrylate/9-anthryl-methyl methacrylate)		

The measure of excimer fluorescence of poly(2-vinylnaphthalene) (Fig. 4.11), poly(acenaphthylene) (Fig. 4.12) and poly(4-vinylbiphenyl) (Fig. 4.13) (guest polymers) in a series of poly(alkyl methacrylates) (host polymers) show dependence of $I_{F(E)}/I_{F(M)}$ on solubility parameters of the guest polymer in a host polymer.[1168, 1172, 1229, 3292]

In order to distinguish between the intramolecular (non-adjacent) and the adjacent intermolecular excimer sites, a concentration study has to be performed for several host matrices.[1171, 1172] The effect due to intramolecular (non-adjacent) excimer formation should be independent of guest polymer concentration, while intermolecular interaction leading to excimer formation should depend directly upon concentration. The effect of concentration of poly(2-vinylnaphthalene) (guest polymer) in different poly(alkyl methacrylates) (host polymers) is shown in Fig. 4.14). The major feature of these results is that $I_{F(E)}/I_{F(M)}$ increases linearly with concentration. This is direct proof for the formation of intermolecular excimer sites.

Excimer fluorescence provides a good method for monitoring polymer-chain aggregation resulting from phase separation. The disadvantages of this method are:

(i) host polymers must contain an aromatic group which can produce excimer emission,
(ii) guest polymers must be transparent for radiation required for the excitation of host polymers.

End-to-end cyclization can be studied by using polymers with chromophores attached at opposite ends of a chain (Table 4.5).

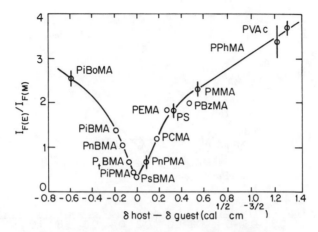

Fig. 4.11   Dependence of $I_{F(E)}/I_{F(M)}$ on host–solubility parameter for poly(2-vinylnaphthalene) guest[1168,1172]. Acronyms used in Figs 4.11–4.14 are shown in the table below:

Poly(alkylmethacrylates)		*Other vinyl polymers*	
Alkyl group	Acronym	Polymer	Acronym
methyl	PMMA	polystyrene	PS
ethyl	PEMA	polyvinylacetate	PVAc
n-propyl	PnPMA		
iso-propyl	PiPMA		
n-butyl	PnBMA		
iso-butyl	PiBMA		
sec-butyl	PsBMA		
tert-butyl	PtBMA		
isobornyl	PiBoMA		
cyclohexyl	PCMA		
phenyl	PPhMA		
benzyl	PBzMA		

The fraction of chain conformations which can cyclize end-to-end within the excited state lifetime decreases with increasing molecular weight.[810,3356] The effects are illustrated for poly(ethylene oxide) terminated with pyrene groups (Fig. 4.15). On the other hand, the fraction of chain-end to chain-middle conformations increases with molecular weight in the small molecule model systems.[3912]

4.3. FLUORESCENCE ANISOTROPY

The polarization of fluorescence can be characterized by an optical coordinate system (x, y, z) (Fig. 4.16). The total intensity of fluorescence emitted in all directions (I_F) is:

$$I_F = I_{F(x)} + I_{F(y)} + I_{F(z)} \qquad (4.4)$$

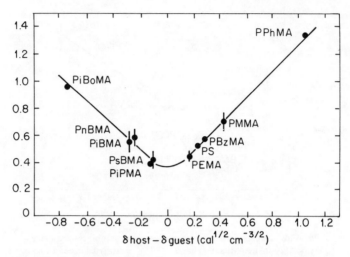

Fig. 4.12 Dependence of $I_{F(E)}/I_{F(M)}$ on host–solubility parameter for poly(acenaphthylene) guest.[1168,1172] Acronyms used are shown in Fig. 4.11

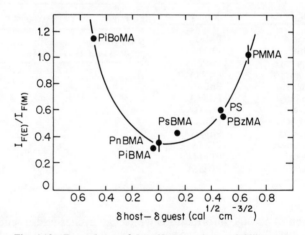

Fig. 4.13 Dependence of $I_{F(E)}/I_{F(M)}$ on host–solubility parameter for poly(4-vinylbiphenyl) guest.[1168,1172] Acronyms used are shown in Fig. 4.11

The incident light (exciting beam) employed in experiments consists of unpolarized (natural) light or linearly polarized light whose direction of polarization is along the z or x axis:

(i) If the incident light is unpolarized (its electric vector may have any orientation with the xy plane), then:

$$I_{F(x)} = I_{F(z)} \qquad (4.5)$$

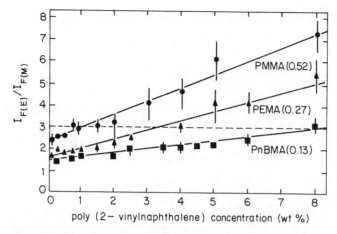

Fig. 4.14   Dependence of $I_{F(E)}/I_{F(M)}$ on poly(2-vinylnaphthalene) guest concentration.[1168,1172] Acronyms used are shown in Fig. 4.11

Table 4.5   Examples of application of excimer fluorescence to the study end-to-end cyclization

Polymer	End-groups	References
Polythene wax $(CH_2)_{n=1-32}$	Pyrene	3994
Poly(ethylene oxide)	Pyrene	810
Polystyrene	Pyrene	3920
Polysiloxane-$(CH_2)_5$-$Si(CH_3)_2$-O-$Si(CH_3)_2$-$(CH_2)_5$-	Pyrene	3506, 3921
Polypeptide:		
N^5-(2-hydroxyethyl)-L-glutamine	Naphthalene	1473, 1474

and the total intensity of fluorescence (I_F) is given by:

$$I_F = 2I_{F(z)} + I_{F(y)} \qquad (4.6)$$

(ii) If the incident light is polarized along the z axis, then:

$$I_{F(x)} = I_{F(y)} \qquad (4.7)$$

and

$$I_F = I_{F(z)} + 2I_{F(y)} \qquad (4.8)$$

(iii) If the incident light is polarized along the x axis, then

$$I_{F(y)} = I_{F(z)} \qquad (4.9)$$

and

$$I_F = 2I_{F(z)} + I_{F(x)} \qquad (4.10)$$

Fluorescence anisotropy measurements are usually made with incident light which is polarized in the z direction or, less commonly, with unpolarized light. The quantities observed are the components of fluorescence intensity polarized in the z and y directions.

The FLUORESCENCE ANISOTROPY (A_F) is defined by

$$A_F = \frac{I_{F(\parallel)} - I_{F(\perp)}}{I_F} \tag{4.11}$$

where:

$I_{F(\parallel)}$ = fluorescence intensity measured when the polarizer (P_1) and analyser (P_2) are set parallel ($P_1 \parallel P_2$), then $I_{F(\parallel)} = I_{F(z)}$,

$I_{F(\perp)}$ = fluorescence intensity measured when the polarizer (P_1) and analyser (P_2) are set perpendicular ($P_1 \perp P_2$), then $I_{F(\perp)} = I_{F(y)}$.

The fluorescence anisotropy for:

(i) Unpolarized light ($A_{F(u)}$)

$$A_{F(u)} = \frac{I_{F(\parallel)} - I_{F(\perp)}}{2I_{F(\parallel)} + I_{F(\perp)}} \tag{4.12}$$

(ii) Vertically polarized incident light ($A_{F(v)}$)

$$A_{F(v)} = \frac{I_{F(\parallel)} - I_{F(\perp)}}{I_{F(\parallel)} + 2I_{F(\perp)}} \tag{4.13}$$

The fluorescence anisotropies for the two cases are related by

$$A_{F(v)} = 2A_{F(u)} \tag{4.14}$$

(iii) Horizontally polarized incident light:

$$I_{F(\parallel)} = I_{F(\perp)} \tag{4.15}$$

$$\text{and} \quad A_F = 0 \tag{4.16}$$

If more than one fluorescent chromophore is present, the observed emission anisotropy is given by

$$A_F = A_{F(i)} f_i \tag{4.17}$$

where:

$A_{F(i)}$ = fluorescence anisotropy of fluorescent species i,

f_i = fractional contributions of fluorescent species i.

The DEGREE OF FLUORESCENCE POLARIZATION (P_F) is defined as follows:

$$P_F = \frac{I_{F(\parallel)} - I_{F(\perp)}}{I_{F(\parallel)} + I_{F(\perp)}} \tag{4.18}$$

The fluorescence intensity (I_F) (or fluorescence anisotropy (A_F) decays with time according to a simple exponential law:

$$I_F = I_{F(0)} \exp(-t/\tau) \tag{4.19}$$

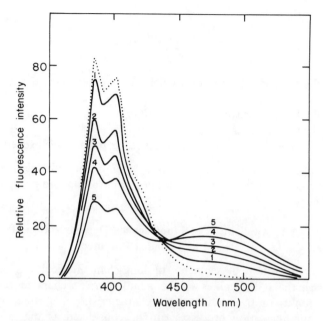

Fig. 4.15 Effect of molecular weight (M) of polyethylene oxide on the extent of intramolecular excimer formation between terminal pyrene groups in tetrahydrofuran at 20°C. M = (1) 195,000, (2) 8900, (3) 6500, (4) 3800, (5) 1650. The dotted line corresponds to a monomeric model compound[810]

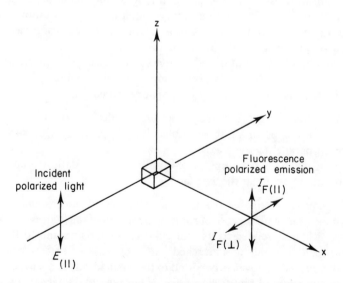

Fig. 4.16 Optical coordinate system for measurements of the degree of fluorescence polarization

or

$$A_F = {}_{A_{F(0)}}\exp(-t/\tau) \qquad (4.20)$$

where:

$I_{F(0)}$ = fluorescence at zero time (which corresponds to the time of excitation light pulse),

$A_{F(0)}$ = fluorescent anisotropy at zero time,

t = time after excitation,

τ = average decay time of fluorescence, or the average excited lifetime.

The observation of fluorescent anisotropy depends upon the selective and non-random excitation of fluorescent molecules or chromophore groups.

4.3.1. Application of Fluorescence Depolarization of Doped or Labelled Polymers

The rotational motion of segments of long-chain molecules is of technical importance in that it determines whether a bulk sample is characterized by a high (glassy) modulus or by a lower value characteristic of a rubber. The glass transition and relaxation processes can be considered as the movement of molecular segments of polymer molecules.

More detailed information on the rotational behaviour of luminescent species can be obtained from observation of the polarization of luminescence stimulated by absorption of polarized radiation. In such a technique only those molecules with their absorption vector correctly aligned with respect to the polarization of incident light will absorb energy, and so initially all the excited state species are in the same orientation. This orientation is randomized by rotation during the excited state lifetime, and then the depolarization of the emission measures how 'far' rotation has progressed during the lifetime.

The probability of excitation of a fluorescent molecule or group will depend upon the angle (θ) between its absorption transition moment and the electric vector of the incident light, being proportional to $\cos^2 \theta$:

(i) if the incident light (exciting beam) is vertically polarized, preferential excitation will occur of those molecules whose absorption moments are oriented parallel to the z axis;

(ii) if the incident light is unpolarized, preferential excitation occurs of those whose moments lie in the xz plane.

The information obtainable from the time profile of fluorescence anisotropy decay is concerned with the rate of rotational motion of the fluorescent molecules or groups, which is related to its molecular characteristics and to those of the macromolecule to which it is attached.

The label should be covalently bound to the chain ideally in such a manner that motion independent of that to be studied is not possible. Alternatively, the label may be bounded in such a manner that motion independent of the chain does not result in depolarization of the emission. The label should not be so bulky as to impede the relaxation interest. Several labels such as xanthydrol, dichlorofluorescein, rhodamin B, naphthalene, and anthracene have been used in monitoring the backbone relaxations of flexible macromolecules.[2008, 2735, 3817]

Figure 4.17 shows the orientation of the label with respect to chain backbone for the samples with different substituted groups:[1001, 2008, 3416]

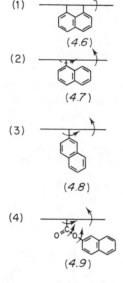

(1)

(4.6)

(2)

(4.7)

(3)

(4.8)

(4)

(4.9)

Fig. 4.17 Relative backbone/label geometries for (1) poly(acenaphthylene), (2) poly(1-vinylnaphthalene), (3) poly(2-vinylnaphthalene), and (4) poly(2-naphthylmethacrylate). Unbroken arrows represent motions resulting in depolarization of emitted radiation. Broken arrows represent motions which do not result in fluorescence depolarization[2008]

(i) Motion of the label independent of the macromolecule is impossible for poly(acenaphthylene) *(4.6)*.

(ii) Motion independent of polymer about the bond of attachment is possible in poly(1-vinylnaphthalene) *(4.7)* but it does not produce depolarization additional to that experienced due to chain motions.

(iii) Motion in poly(2-vinylnaphthalene) *(4.8)* and poly(2-naphthyl methacrylate *(4.9)* independent of the chain produce depolarization, because the transition vector corresponding to absorption and emission does not lie parallel to the bond of attachment to the chain.

If the fluorescent molecule is rigidly (covalently or non-covalently) attached to the macromolecule, with a fixed orientation of its transition moments with respect to the coordinate axes, the time decay of anisotropy depends not only upon the characteristics of the macromolecule, but also upon the respective orientations of the transition moments of absorption and emission.

In practice, the case of a fluorescent label which is rigidly oriented with respect to the coordinate axes of a rigid macromolecule is often not realized. For many fluorescent conjugates of polymers, internal rotation is present in some degree. The following forms of internal rotation can be considered:

(i) rotation of the fluorescent group about the bond linking it to the macromolecule,

(ii) rotational wobble of that portion of the macromolecule adjacent to the chromophore,

(iii) rotation of molecular domain as a unit about a flexible hinge point.

If the rotation of the fluorescent group is not hindered by interaction with the structural features of the macromolecule, or by other factors, then the rotation of the label group is governed by its own rotational diffusion coefficient.

The rotational diffusion of the macromolecule as a whole is much slower than local conformational transitions, and the depolarization of fluorescence may be interpreted in terms of the conformational mobility of the polymer chain backbone.

Because of the occurrence of multiple orientation of fluorescent labels with respect to the coordinates of the macromolecules to which they are attached, it is difficult to observe the effects of molecular asymmetry for fluorescent polymer conjugates.

Steady-state measurements of segmental motion in macromolecules can be determined using the PERRIN EQUATION:[2008, 2735, 2927]

$$\left(\frac{1}{P_F} - \frac{1}{3}\right) = \left(\frac{1}{P_{F_0}} - \frac{1}{3}\right)\left(1 + \frac{3\tau}{\rho}\right) \tag{4.21}$$

where:

P_F = degree of fluorescence polarization (of labelled group);

P_{F_0} = intrinsic polarization, i.e. that observed in the absence of all depolarizing effects, related to the angle (θ) between the transition moments of absorption and emission by equation:[1841]

$$P_{F_0} = \frac{3\cos^2\theta - 1}{\cos^2\theta + 3} \tag{4.22}$$

for vertically polarized excitation $0.5 \geqslant P_{F_0} \geqslant -0.33$;

τ = fluorescence lifetime of a labelled group (ns);

ρ = rotational relaxation time (ns). Examples are given in Table 4.6.

The intrinsic polarization (P_{F_0}) can be evaluated by two extrapolation procedures:[2008]

(i) The P_{F_0} may be estimated as the value of P_F in the limit as the rotational relaxation time (ρ) tends to infinity at constant fluorescence lifetime (τ). Application of such extrapolation is achieved through approximation of the rotational relaxation time via a STOKES–EINSTEIN relationship to yield the PERRIN EQUATION in the form:

$$\left(\frac{1}{P_F} - \frac{1}{3}\right) = \left(\frac{1}{P_{F_0}} - \frac{1}{3}\right)\left(1 + \frac{R_g T}{\eta V}\right) \tag{4.23}$$

Table 4.6 Rotational relaxation times (ρ) of different fluorescent species labelled to polymers[2735]

Polymer	Fluorescent species	Solvent	Rotational relaxation time (ρ) (ns)
Poly(methyl acrylate)	9-Vinylanthracene copolymer	Toluene	1.2
	Bis(methylbenzoxazolyl)ethylene end group	Toluene	1.2
	Dichlorofluorescein end-group	DMF	3.4
	Anthracene end-group	Toluene	<0.3
Poly(vinyl acetate)	9-Vinylanthracene copolymer	Toluene	3.4
	Dichlorofluorescein end-group	DMF	1.4
	9-Methylanthracene end-group	Toluene	0.7
Poly(methyl methacrylate)	9-Vinylanthracene copolymer	Toluene	3.9
	Dichlorofluorescein end-group	DMF	5.4
Poly(n-butyl methacrylate)	9-Vinylanthracene copolymer	Toluene	4.0
	Bis(methylbenzoxazolyl)ethylene end group	Toluene	2.2
	Dichlorofluorescein end-group	DMF	6.5
	9-Methylanthracene end-group	Toluene	0.8
Polystyrene	9-Vinylanthracene copolymer	Toluene	6.0
	Bis(methylbenzoxazolyl)ethylene (high temp. preparation)	Toluene	5.1
	Bis(methylbenzoxazolyl)ethylene (low temp. preparation)	Toluene	5.9
	Anthracene end-group	Toluene	1.7
	9-Methylanthracene end-group	Toluene	0.8
	2-Vinylnaphthalene copolymer	Toluene	5.5
Poly(p-chlorostyrene)	9-Vinylnthracene-copolymer (average of two samples	Toluene	6.6
	9-Methylanthracene copolymer	Toluene	1.4
Poly(m-chlorostyrene)	9-Vinylanthracene copolymer	Toluene	6.5
Poly(p-methoxystyrene)	9-Vinylanthracene copolymer	Toluene	5.0
Poly(p-bromostyrene)	9-Vinylanthracene copolymer	Toluene	5.5
Poly(N-vinylcarbazole)			
(M_w, 91,900)	9-Vinylanthracene copolymer	Toluene	26.2
(M_n, 3,700)	Dichlorofluorescein end-group	DMF	3.8
(M_w, 6,000	Rhodamine B end group	DMF	31.4
(M_w, 53,500)	Carbazole residues of homopolymer	Toluene	52.6

DMF = dimethylfuran

where:
R_g = universal gas constant,
T = thermodynamic temperature (in kelvin),
V = molar volume of fluorescent label,
η = viscosity of the medium.

The intrinsic polarization (P_{F_o}) is obtained as the intercept of a plot of P_F^{-1} against T/η.

This technique suffers due to the general inability of the Stokes–Einstein relationship to describe adequately the temperature and viscosity behaviour of the rotating group. Thus results may depend upon whether viscosity variation has been achieved by solvent variation at constant temperature or through temperature variation. Changes in solvent are particularly un-desirable in the case of macromolecular samples due to the resultant changes in polymeric chain geometries which may be encountered in addition to the fluorescence lifetime variations. Hence, accurate estimation of P_{F_0} is ham-pered by the requirement for extrapolation over T/η ranges and by the occurrence of non-linear Perrin plots.[434, 2735]

(ii) The P_{F_0} may be estimated by extrapolation of P_F^{-1} to zero while the rotational relaxation time (ρ) is held constant. The addition of small amounts of a species capable of dynamic quenching of the excited state responsible for emission allows reduction in fluorescence lifetime of a labelled group (τ) over a considerable range without essentially altering the nature of the solvent (or hence ρ). This technique offers several advantages over the use of a T/η plot:

   (a) the form of autocorrelation function may be checked,[408, 3835]
   (b) estimation of ρ_0 involves relatively short extrapolations,
   (c) the use of quenching species effectively produces (from a single fluor) a continuous range of molecular probes of varying and constant molar volume,
   (d) it is possible to use as fluorescent label groups species whose small molar dimensions will produce minimal perturbation of the relaxation under study. Such species generally have relatively long-lived excited states which, without quenching, are incapable of monitoring the rapid relaxation processes of flexible macromolecules.

The bimolecular rate constant for a diffusion-controlled reaction (k_{diff}) involving large spherical molecules may be estimated from the simplified Debaye (2.36)[496] or the Smoluchowski (2.37)[3399] equations.

Fluorescence depolarization technique has been also employed for the study of down-chain singlet energy migration which involves few energy transfers.[194, 195, 834, 855, 2465, 3079–3081, 3220, 3818, 3983]

The extent to which randomization of the photoselected species occurs during the excited state lifetime may be estimated from the degree of fluorescence polarization (P_F).[2927, 3817] The extent to which depolarization occurs as a result of an average energy transfers $(\bar{n})$ is given by:[3818]

$$\left(\frac{1}{P_F} - \frac{1}{3}\right) = \left(\frac{1}{P_{F_0}} - \frac{1}{3}\right)(1 + C\bar{n}) \tag{4.24}$$

where:

P_{F_0} = intrinsic polarization, i.e. that observed in the absence of all depolarizing effects,

C = constant, the value of which is dependent upon the mechanism of energy transfer.

Employment of fluorescence depolarization technique in the study of energy migration requires elimination of the depolarizing effects of micro-Brownian molecular rotations. This can be eliminated by use of dilute solutions of polymer in glass matrices (e.g. 2-methyltetrahydrofurane glass) at 77 K or in a polymer matrix. The degree of fluorescence polarization (P_F^{-1}) presented as a function of mol fraction of fluorescence chromophore group (f_a) gives a straight line (Fig. 4.18).

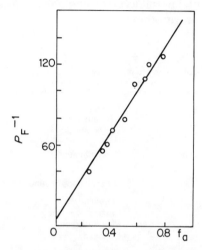

Fig. 4.18 Fluorescence polarization (P_F^{-1}) as a function of mol fraction (f_a) of acenaphthylene-derived chromophores in co(acenaphthylene/methyl acrylate) in methyltetrahydrofuran solvent at 77K[195]

Fluorescence depolarization is thus a convenient technique for observing rotational processes occurring on a nanosecond time scale, and these usually involve polymer chain (segments, or substituted groups) either in bulk or in solution[313, 408, 434, 534, 2008, 2701, 2707, 2735, 3416, 3593, 3594, 3727, 3782, 3835] and measure of rotational relaxation times of biopolymers.[3429, 3817, 3974]

Fluorescence depolarization has also been applied to the study of segment density in a particular case of cross-linked swollen polymers,[1188] the dynamic of biopolymers[3084, 3464] and membranes.[3318]

The application of fluorescence depolarization measurements to macromolecules has a number of limitations:[2735]

(i) The brief existence of singlet excited states precludes the use of polarized fluorescence in the investigations of segmental processes in gels, rubber, and glasses. This technique is limited in application to relatively fast relaxations such as those of polymers in solutions.

(ii) The use of steady-state techniques is inapplicable in cases of anisotropic rotation.[2703, 3819]

(iii) Experimental difficulties in the nanosecond domain.[3023]

4.4. POLYMERIC MATRICES FOR THE STUDY OF PHOSPHORESCENCE DECAY OF LOW MOLECULAR COMPOUNDS

Several polymers (Table 4.7) have been used as matrices for the study of the excited triplet states of low molecular compounds, instead of conventional alcohol or 2-methyltetrahydrofuran.

Table 4.7 Examples of polymeric matrices for the study phosphorescence decay of low molecular compounds

Polymer matrix	Doped low molecular compoud	References
Poly(methyl methacrylate)	Benzophenone	1177, 1663, 1664, 1854, 2358, 3169
	Dibenzothiophenone	435
	Triphenylene Carbazole Phenanthrene Coronene	435,1854
Polystyrene	Benzophenone	3169
	Anthracene	564,572
	1,2-Benzanthracene	564,566,567
	Triphenylene Carbazole Phenanthrene Coronene	1854
Co(styrene/methyl methacrylate)	Triphenylene Coronene	1855

It has been observed that the phosphorescence of low molecular compounds in polymeric matrices decays non-exponentially at room temperature.[1177, 1661, 1663, 1664, 1854, 2358] This process can be explained as eventual intermolecular triplet–triplet energy transfer from doped low molecular compound to polymer matrix (cf. Section 2.10).

Phosphorescence emission of low molecular compounds in solutions markedly decreases when temperature increases, whereas in polymer matrices emission at room temperature is only slightly reduced from that at low temperature. This difference indicates that the modes of radiationless decay are decreased in the polymer system since the mobility in the polymer film of the high microscopic viscosity is limited.

When a polymer–chromophore-bearing guest molecule combination is cast into polymer film by evaporation of a solvent, the solute or guest can find itself in an environment which can differ greatly in terms of the local viscosity or free volume unless thermal relaxation is permitted.[1896]

The triplet lifetime of benzene and related systems increases over a period of 5–6 hours in different glasses at low temperature, 77 K (Fig. 4.19).[2399]

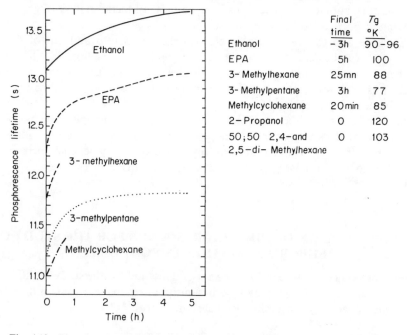

	Final time	T_g °K
Ethanol	− 3h	90 − 96
EPA	5h	100
3- Methylhexane	25mn	88
3- Methylpentane	3h	77
Methylcyclohexane	20 min	85
2 − Propanol	0	120
50;50 2,4−and 2,5−di− Methylhexane	0	103

Fig. 4.19   Phosphorescence lifetimes of benzene − d_6 in different organic glasses at 77K[2399]

4.4.1. Phosphorescent Probes and Labels

Phosphorescent low molecular compounds can be employed for the study of some physical processes in polymers. For example phosphorescence of benzophenone in poly(methyl methacrylate)[1663] and other acrylic and methacrylic polymers[1665] yields information on molecular motion, glass transition, and secondary transition processes.

Phosphorescent probes such as 10-(4-bromo-1-naphthoyl)dodecyltrimethyl-ammonium bromide (*4.10*) has been used successfully for the study of polyelectrolyte effects on fast interionic reactions.[466, 2784, 3692] Phosphorescence quenching of this probe by hexa-amminecobalt(III) chloride ([Co(NH₃)₆]Cl) yields information of the micelle-probe dynamics or kinetic analysis of gegenion binding in polyelectrolyte solutions.[2784]

Entrance and exit rates of aromatic hydrocarbons into and out of micelles[174, 3681] can be determined by the measurements of phosphorescent lifetimes in the presence of soluble quenchers. With excess quencher the phosphorescence decay was determined by a sum of the exit rate of the probe and the decay rate in the absence of quencher.[174]

Study of phosphorescence of labelled polymers with benzophenone has been employed for the study of end-to-end cyclization and intramolecular hydrogen abstraction,[2373, 3194] with naphthalene for the study of polymer chain behaviour in diluted solutions,[3919] or nonaqueous dispersions in polymer colloids.[2918-2920, 3913]

4.5. APPLICATION OF PHOSPHORESCENCE FOR THE STUDY OF SUBGROUP MOTION IN POLYMERS

Phosphorescence can be employed as a useful method for the study of different types of relaxations which may take in amorphous polymers containing small amounts of carbonyl or naphthalene chromophores.[3406]

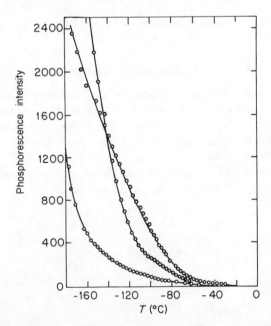

Fig. 4.20 Phosphorescence intensity (I_P) as a function of temperature in: (Δ) co(styrene/phenylvinyl ketone), (○) co(styrene/naphthyl methacrylate), and (□) co(styrene/methylvinyl ketone)[3406]

The phosphorescence emission intensity (I_P) is directly proportional to the intensity of light absorbed (I_a) and quantum yield of phosphorescence

$$I_P \sim I_a \Phi_P \qquad (4.25)$$

where

$$\Phi_P = \Phi_{ISC} \frac{k_P}{k_P + k_N + k_R} \qquad (4.26)$$

where:

Φ_{ISC} = quantum yield of intersystem crossing (cf. Section 1.7),
k_P = rate constant of phosphorescence,
k_N = rate sum of all other non-radiative processes from the triplet state (T_1),
k_R = rate of chemical reaction.

The rate constant of phosphorescence emission (k_P) is generally regarded as being independent of temperature. The quantum yield of intersystem crossing (Φ_{ISC}) for the n, π^* state in ketones is near unity and may be considered to be independent of temperature, whereas Φ_{ISC} for the π, π^* state in naphthalene show a small temperature dependence.

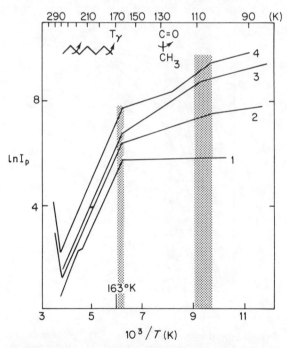

Fig. 4.21 Arrhenius curves for: (1) polyethylene, (2) co(ethylene/carbon monoxide), (3) co(ethylene/methyl vinyl ketone), (4) co(ethylene/methyl isopropenyl ketone) phosphorescence[3406]

166

It has been observed that the phosphorescence intensity changes drastically as the polymers are warmed to room temperature (Fig. 4.20).[3406] The decrease in phosphorescence with increasing temperature must therefore be attributed to the discontinuous increase of some non-radiative processes included in the k_N term. The term (k_N) may be divided into a temperature-independent part (k_0) and a temperature-dependent term (k_t)

$$k_N = k_0 + k_t = k_0 + A e^{-E_a/R_g T} \qquad (4.27)$$

where:

A = pre-exponential term which is related to the probability of the reaction occurring at any given temperature (T),
E_a = the Arrhenius activation energy,
R_g = universal gas constant,
T = thermodynamic temperature (in kelvin).

Using this basic assumption[494, 3153, 3406]

$$1/I_{P(T)} - 1/I_{P(0)} = A e^{-E_a/R_g T} \qquad (4.28)$$

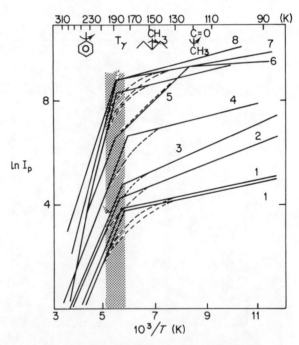

Fig. 4.22 Arrhenius curves for: (1) polystyrene, (2) co(styrene/methyl isopropenyl ketone), (3) co(styrene/methyl vinyl ketone), (4) co(styrene/5-hexene-2-one), (5) co(styrene/benzophenone), (6) co(styrene/naphthyl methacrylate), (7) co(styrene/phenyl vinyl ketone), (8) co(styrene/methyl isopropenyl ketone) phosphorescence[3406]

where:

$I_{P(T)}$ = phosphorescence emission intensity at a given temperature T,
$I_{P(0)}$ = phosphorescence emission at the absolute zero.

Extrapolation of $I_{P(T)}$ to low temperatures shows that ignoring the term $1/I_{P(0)}$

$$\ln I_{P(T)} = A' e^{-E_a/R_g T} \qquad (4.29)$$

the Arrhenius plot of $\ln I_{P(T)}$ versus $1/T$ gives a straight line of slope E_a/R_g. Arrhenius plots for different polymers' phosphorescence are shown in Figs. 4.21–4.24.

These results show the discontinuous behaviour of phosphorescence as a function of temperature in a wide variety of polymer films which contain phosphorescent chromophores. The effects on phosphorescence are directly related in some way to the local relaxations in the amorphous polymers such as:[3406]

(i) Primary main-chain motions involving segments containing up to 20–40 monomer units in hindered dynamic fluctuation, alternately creating and filling free-volume spaces in the polymer matrix. In the narrow temperature

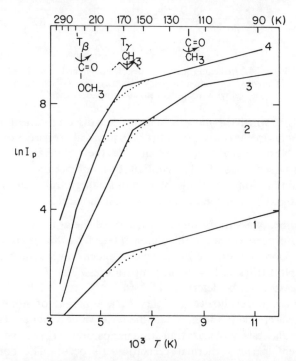

Fig. 4.23   Arrhenius curves for: (1) poly(methyl methacrylate), (2) co(methyl methacrylate/naphthyl methacrylate), (3) co(methyl methacrylate/methyl vinyl ketone), (4) co(methyl methacrylate/naphthyl methacrylate/phenylvinyl ketone) phosphorescence[3406]

168

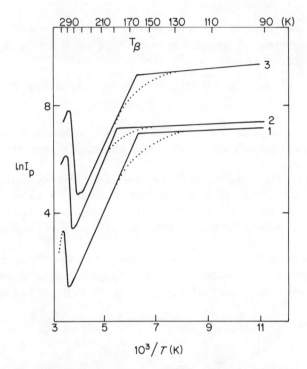

Fig. 4.24 Arrhenius curves for: (1) poly(vinyl chloride), (2) co(vinyl chloride/methylvinyl ketone), (3) co(vinyl chloride/methyl isopropenyl ketone) phosphorescence[3406]

range (T_g) in which this motion set in, marked discontinuities are observed in the temperature derivatives of many bulk physical properties of the polymer.

(ii) Secondary main-chain motions involve the cooperative but restricted motions of segments of a few contiguous monomer units.

(iii) Side-chain motions involving rotation of small groups such as alkyl or aryl groups analogous to the behavior in small molecules.

The main rubber–glass transition at T_g is termed the α-process. The secondary main chain β-process occurs in the region 200–300 K, whereas other processes, designated $\gamma,\delta,\varepsilon$-processes, are observable at much lower temperatures. From Arrheniums plots (Figs 4.21–4.24) the temperatures $T_\alpha, T_\beta, T_\gamma, \ldots$ for different α, β, γ, ... processes can be determined.[494, 2824, 3406]

If the glasses were allowed to relax over a period of hours before the measurements were made, the terminal time-series values resulted. This effect was found only in glasses whose glass transition temperature (T_g) was within 30–70 K. This treatment permits thermal relaxation to occur. The configurational relaxation processes in the glass were directly related to the phosphorescing guest's triplet lifetimes.[2399]

Phosphorescence depolarization has been employed for the study of relaxation processes in bulk polymeric systems,[2703, 2825, 3153] and for probing molecular motion in the millisecond time scale.[2520]

5. Photoinitiated polymerizations

PHOTOINITIATED (PHOTOINDUCED) POLYMERIZATIONS can be divided into four major classes:[162, 452, 1043, 1144, 1146, 1147, 1555, 1729, 1871, 2029, 2186, 2231, 2232, 2239, 2455, 2816, 2829, 2879, 2892, 3016, 3131, 3654]

(i) Photoinitiated radical polymerization (cf. Section 5.1).
(ii) Photoinitiated charge-transfer polymerization (cf. Section 5.3).
(iii) Photoinitiated cation–radical polymerization (cf. Section 5.4).
(iv) Photoinitiated simultaneous radical and cation–radical polymerization (cf. Section 5.4.6).

5.1. PHOTOINITIATED RADICAL POLYMERIZATION

PHOTOINITIATED RADICAL POLYMERIZATION is initiated by PHOTOINITIATORS (I) which can be divided into three groups:

(i) Photoinitiators which intramolecularly cleave into radicals ($R_1 \cdot$, $R_2 \cdot$), which initiate free radical polymerization:

$$I + h\nu \rightarrow I^* \tag{5.1}$$

$$I^* \rightarrow R_1 \cdot + R_2 \cdot \tag{5.2}$$

where I^* is an excited singlet (S_1) or triplet (T_1) state of an initiator.

(ii) Photoinitiators which abstract hydrogen intermolecularly from hydrogen-donor molecules (RH):

$$I^* + RH \rightarrow \cdot IH + R \cdot \tag{5.3}$$

(iii) Photoinitiators which form, with a coinitiator, a charge-transfer complex (CT complex) and further dissociate into radicals, which initiate free radical polymerization:

$$I^* + AH \rightarrow [I \ldots A]^* \rightarrow \cdot IH + A \cdot \tag{5.4}$$
$$\text{CT COMPLEX}$$

The terminology employed to describe a PHOTOINITIATOR or PHOTOSENSITIZER is unfortunately wrongly used by several authors. The term PHOTOINITIATOR should be used to describe a chemical compound or a chemical system which absorbs light and dissociates into free radicals. The photochemistry of photoinitiators is presented in Chapter 7. The term COINITIATOR (ACTIVATOR) should be used to describe a chemical compound or a chemical system which does not absorb light, but is nevertheless indirectly involved in the production of initiator radicals in a photochemical reaction. The term PHOTOSENSITIZER should be reserved to describe a chemical compound or a chemical system which SENSITIZES (PHOTOSENSITIZES) photoreaction by an ENERGY TRANSFER MECHANISM (cf. Chapter 2).

Photoinduced radical polymerization is important in the following areas:

(i) photoactive relief printing plates,
(ii) printing circuits,
(iii) photochemically cured surface coatings,
(iv) photochemically curable printing inks.

In all of these processes the final polymeric product is usually a crosslinked resin produced by photoinitiated polymerization of mixtures of suitable prepolymers with mono- and polyfunctional olefins. (c.f. Chapters 12 and 13).

5.1.1. Initiator Radical Formation

INITIATOR RADICALS are formed by one of the following processes:

(i) INTRAMOLECULAR PHOTOCLEAVAGE (reaction 5.2),
(ii) INTERMOLECULAR HYDROGEN ABSTRACTION from a hydrogen-donor molecule (reaction 5.3),
(iii) INTERMOLECULAR CHARGE-TRANSFER COMPLEXES (reaction 5.4).

Radical formation from the singlet excited state (S_1) competes with the following reactions:

(i) Deactivation of excited singlet state (S_1) to the singlet ground state (S_0) by radiative process (fluorescence) and/or radiationless processes (cf. Section 1.6).
(ii) Intersystem crossing (ISC) to the triplet state (T_1) (radiationless process) (cf. Section 1.7).
(iii) Energy transfer process (if it does occur?) (cf. Chapter 2).
(iv) Deactivation by monomer.

Radical formation from the triplet state (T_1) competes with the following reactions:

(i) Deactivation of triplet state (T_1) to the singlet ground state (S_0) by radiative process (phosphorescence) and/or radiationless processes (cf. Section 1.6).
(ii) Deactivation by oxygen.
(iii) Deactivation by bimolecular quenching processes.
(iv) Deactivation by self-quenching.
(v) Energy transfer process (if it does occur?) (cf. Chapter 2).
(vi) Deactivation by monomer.

All of these processes depend on the lifetime of a given excited state (singlet or triplet), e.g. the longer lifetime of the triplet state (T_1) makes it more susceptible than the excited singlet state (S_1) to deactivation by bimolecular quenching processes.

INTRAMOLECULAR PHOTOCLEAVAGE may occur only if the singlet and/or triplet states possess sufficient energy to allow efficient bond dissociation into free radicals. In many cases the dissociation energy lies much higher than the energy of excited singlet and/or triplet states (cf. Section 1.4).

5.1.2. Photoinitiator Efficiency

PHOTOINITIATOR EFFICIENCY depends on several factors:

(i) Efficient population of the reactive excited singlet and/or triplet state requires desirable absorptivity characteristics, and a high efficiency of intersystem crossing (ISC) process from the excited singlet state to the excited triplet state.
(ii) Efficient initiator radical formation by;
   (a) intramolecular photocleavage,
   (b) intermolecular hydrogen abstraction.
(iii) Initiator radicals must be highly reactive with monomers and oligomers.[2997]
(iv) The initiator molecule, or any of its photolysis products, should not function as a chain transfer or termination agent in polymerization (cf. Section 5.1.6).

5.1.3. Efficiency of Photoinitiation

The OVERALL EFFICIENCY OF PHOTOINITIATION depends on:

(i) The fraction of incident light which is absorbed by the photoinitiator.[2309]
(ii) The fraction of excited singlet state (S_1) molecules which yield initiator radicals, if the photocleavage process may occur in the singlet state (S_1).
(iii) The fraction of excited singlet state (S_1) molecules which are transformed by intersystem crossing into excited triplet state (T_1) molecules.
(iv) The fraction of excited triplet state (T_1) molecules which yield initiator radicals.
(v) The fraction of initiator radicals which initiate the polymerization process. The major competing processes here are:
   (a) initiator radical recombination;
   (b) initiator radical termination by a growing macroradical,
   (c) initiator radical reactions with oxygen,
   (d) initiator radical reactions with solvent.
(vi) Reactivity of free radicals formed from photocleavage of initiator with monomer.

There is an OPTIMAL CONCENTRATION OF PHOTOINITIATOR which is determined by the overall efficiency of photoinitiator and also by self-quenching (concentration quenching) and light screening properties of the photoinitiator.

5.1.4. Requirements for an Ideal Photoinitiator

An ideal photoinitiator should be:

(i) Completely stable by itself and not initiate spontaneous polymerization when dissolved in reactive monomers.
(ii) When irradiated it should undergo photolysis with a high quantum efficiency and without liberation of by-products that inhibit polymerization or degrade the quality of the final product.
(iii) The synthesis of the prospective photoinitiator should be reasonably straightforward and inexpensive.
(iv) The photoinitiator should have as low toxicity as possible.

5.1.5. Kinetics of Photoinitiation of Radical Polymerization

The PHOTOINITIATION OF FREE RADICAL POLYMERIZATION occurs in two steps:[3213]

(i) Formation of free radicals (R·) from photoinitiator (I)

$$I + hv \xrightarrow{\ k_d\ } R\cdot + R\cdot \tag{5.5}$$

The RATE OF THE INITIATOR DECOMPOSITION (R_d) is given by:

$$R_d = k_d[I] \tag{5.6}$$

where k_d = radical formation rate constant ($1 \, \mathrm{mol^{-1} \, s^{-1}}$).
(ii) Addition of the first monomer molecule (M) to the initiator radical (R·):

$$R\cdot + M \xrightarrow{\ k_i\ } R\text{--}M\cdot \tag{5.7}$$

The RATE OF THE CHAIN INITIATION (R_i) is given by

$$R_i = k_i[R\cdot] \tag{5.8}$$

where k_i = chain initiation rate constant ($1 \, \mathrm{mol^{-1} \, s^{-1}}$). Because the number of moles of free radicals (R·) formed per second by decomposition of photoinitiator (I) (reaction 5.6) is twice as large as the number of moles of photoinitiator disappearing per second, the EFFECTIVE RATE OF THE CHAIN INITIATION (R_i) is given by:

$$R_i = 2k_i[I] \tag{5.9}$$

The reaction (5.7) has no influence on the kinetics of initiation of free radical polymerization, because it occurs much more rapidly than reaction (5.6) ($k_i \gg k_d$).

For the simplest case in which all radicals formed by reaction (5.5) can react further by reaction (5.7) with monomer molecules:

$$R_i = R_d \tag{5.10}$$

In practice the kinetics of photoinitiation of radical polymerization is much more complicated because:

(i) Several monomers are capable of quenching (deactivating) the excited state (singlet and/or triplet state) of photoinitiator molecule, decreasing the amount of free radicals (R·) formed:

$$I^* + M \xrightarrow{k_Q} \begin{cases} \text{non-radical products} & (5.11) \\ \text{radical products not capable of starting} \\ \text{initiation of polymerization} & (5.12) \end{cases}$$

where k_Q = quenching rate constant $(1 \text{ mol}^{-1} \text{ s}^{-1})$.

(ii) Radicals of different type formed from photocleavage have different reactivities with monomers.

(iii) If the monomer reacts with the free radicals (R·) only very slowly, the radical recombination reaction may occur:

$$R \cdot + R \cdot \xrightarrow{k_r} R\text{-}R \qquad (5.13)$$

where k_r = initiator radical recombination rate constant. $(1 \text{ mol}^{-1} \text{ s}^{-1})$.

(iv) If the initiator used has a high reactivity for a chain transfer to the initiator, or a part of the radical produced by the photolysis of the initiator easily undergoes primary radical termination, a polymer with two initiator fragments can be obtained:[2840]

$$R\text{-}R' + nM \rightarrow R(M)_n R' \qquad (5.14)$$

This type of initiators are called INIFERTER (initiator–transfer agent–terminator). A number of photoinitiators such as phenylazotriphenylmethane, phenylphenylazosulphide, 1,3-diphenyltriazene, alkyl perbenzoates and thiuramdisulphides (cf. Section 5.5) are effective iniferters.

The RATE OF THE INITIATOR DECOMPOSITION (i.e. THE RATE OF RADICAL PRODUCTION) in a stirred solution is often approximated by:[1903, 2232, 2829]

$$R_d = I_0 \Phi \varepsilon [I] l = \Phi I_a \qquad (5.15)$$

where:

I_0 = intensity of incident light,
Φ = quantum yield for production of radicals,
ε = molar absorptivity of initiator at the wavelength employed,
$[I]$ = initiator concentration,
l = path length,
I_a = absorbed intensity (cf. Section 1.3).

For most of photoinitiators the light absorption is not small, and I_a will vary within the reaction vessel even though the concentration of initiator is kept uniform by efficient stirring. This is quite different from thermal initiation, where the rate of radical production is uniform throughout the reaction vessel. Assuming the validity of Beer–Lambert's law:

$$R_d = \Phi I_0 (1 - e^{-\varepsilon[I]l}) \tag{5.16}$$

The change in light intensity with depth has a marked effect on the kinetics of photoinitiated polymerization and should not be ignored in real systems.[3341] The kinetics of photoinitiated polymerization in a well-stirred system[1903, 1904] differs significantly from that in an unstirred system,[3635] especially when the major portion of incident light is absorbed in the system. In a well-stirred system at high optical density the initial rate of polymerization varies inversely with the square root of the system thickness, and is independent of the initiator concentration.[3341] On the other hand, in an unstirred system at high optical density the initial rate of polymerization varies inversely with both the system thickness and the square root of initiator concentration. When components of the polymerization system other than the photoinitiator absorb light, but only the light absorbed by the photoinitiator produces radicals, the apparent order of the polymerization reaction with respect to initiator, as determined by conventional experiments, becomes greater than the normally expected value of 0.5.[3341] Studies of photopolymerization kinetics in patterned space-intermittent systems have also been examined.[877, 3170, 3390]

The rate of radical production (the rate of initiator decomposition) can be easily measured by nitroxide termination using ESR spectroscopy.[1255]

The PROBABILITY (α) for the formation of radicals from a photoinitiator capable of initiating the radical polymerization of a monomer (M) is for:[2152, 3213]

(i) Intramolecular photocleavage:

$$I^* \xrightarrow{k_1} R\cdot + R\cdot \tag{5.17}$$

$$I^* + M \xrightarrow{k_Q} I + M \text{ (non-radical products)} \tag{5.18}$$

$$\alpha = \frac{k_1}{k_1 + k_Q[M]} \tag{5.19}$$

(ii) Intermolecular hydrogen abstraction from hydrogen-donor molecule (RH):

$$I^* + RH \xrightarrow{k_2} \cdot IH + R\cdot \tag{5.20}$$

$$I^* + M \xrightarrow{k_Q} I + M \text{ (non-radical products)} \tag{5.21}$$

$$\alpha = \frac{k_2[RH]}{k_2[RH] + k_Q[M]} \tag{5.22}$$

5.1.6. Kinetics of Free Radical Polymerization

CHAIN ADDITION POLYMERIZATION of vinyl monomers by free radical mechanism, involving a series of very rapid repetitive steps, is shown in the following scheme:[902, 2585, 3762]

(i) INITIATION:

$$R\cdot + CH_2 = CHX \xrightarrow{k_i} R-CH_2\dot{C}HX \qquad (5.23)$$

(ii) PROPAGATION:

$$R-CH_2\dot{C}HX + x(CH_2 = CHX) \xrightarrow{k_p} R(CH_2CHX)_x CH_2\dot{C}HX$$

$$(5.24)$$

(iii) TERMINATION:
via combination reaction:

$$R(-CH_2CHX)_x CH_2\dot{C}HX + R(-CH_2CHX)_y CH_2\dot{C}HX \xrightarrow{k_{tc}}$$

$$R(-CH_2CHX)_{\overline{x+y}} R \qquad (5.25)$$

via disproportionation reaction:

$$R(-CH_2CHX)_x CH_2 \dot{C}HX + R(-CH_2CHX)_y \dot{C}HX \xrightarrow{k_{td}}$$
$$(5.26)$$
$$R(-CH_2CHX)_x CH_2CH_2X + R(-CH_2CHX)_y CH = CHX$$

where:

k_i = initiation rate constant (1 mol^{-1} s^{-1}),

k_p = propagation rate constant (1 mol^{-1} s^{-1}),

k_{tc} = termination rate constant (via combination reaction) (1 mol^{-1} s^{-1}),

k_{td} = termination rate constant (via disproportionation reaction) (1 mol^{-1} s^{-1}).

Chain addition polymerization is a typical chain reaction, in which the polymer chain grows very rapidly by successive additions of monomer units during the short interval between the initiation and termination of the chain.

In the absence of any side reaction, the kinetic chain length is given by the value x (or y), which also represents the number of monomer units in the chain molecule.

The actual lifetime of a growing chain radical is of the order of a few seconds, but this interval is sufficient for the successive addition of thousands of monomer units. Thus the free radical is a short-lived but very reactive species, leading to the formation of long chains almost instantaneously. As usual for chain reactions, the values of the individual rate constants are not readily accessible. However, the overall characteristics of the reaction can be deduced by means of the very useful 'steady-state' assumption, i.e. the rate of initiation becomes equal to the rate of termination very soon after the reaction starts. This leads to relatively simple rate equations, which can be verified experimentally:

INITIATION RATE:

$$R_i = d[M\cdot]/dt = 2k_i[I] \tag{5.27}$$

PROPAGATION RATE:

$$R_p = -d[M]/dt = k_p[M\cdot][M] \tag{5.28}$$

TERMINATION RATE:

$$R_t = -d[M\cdot]/dt = 2k_t[M\cdot]^2 \tag{5.29}$$

At steady state condition:

$$R_i = R_t \text{ or } [M\cdot] = (k_i/k_t)^{1/2}[I]^{1/2} \tag{5.30}$$

POLYMERIZATION RATE:

$$R_p = k_p(k_i/k_t)^{1/2}[M][I]^{1/2} = k_p\left(\frac{I_0\,\varepsilon\,1\,\Phi\,[I]^{1/2}}{k_t}\right)[M] \tag{5.31}$$

where:

[M] = monomer concentration,

[M·] = concentration of propagating radical of any size,

[R·] = concentration of initiator radicals,

k_i = initiation rate constant (1 mol^{-1} s^{-1}),

k_p = propagation rate constant (1 mol^{-1} s^{-1}),

k_t = termination rate constant (1 mol^{-1} s^{-1}),

I_0 = intensity of incident light,

ε = molar absorptivity of initiation of the wavelength employed,

Φ = quantum yield for production of radicals,

l = path length,

[I] = initiator concentration.

It can be easily seen here that the polymerization rate (R_p) should be proportional to the probability (α) for the formation of radicals from a photoinitiator if the rate of initiation (k_i) is proportional to α:

$$R_p \approx \alpha^{1/2} \approx k_i^{1/2} \tag{5.32}$$

At a constant concentration of photoinitiator [I], light intensity (I_0) and path length (l), a linear plot of $\log R_p$ vs $\log[M]$ should be obtained (Fig. 5.1).

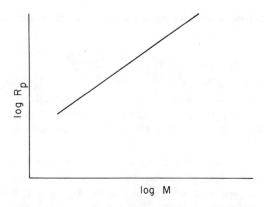

Fig. 5.1   Plot of polymerization rate (R_p) versus log of
monomer (M) concentration

The KINETIC CHAIN LENGTH, i.e. the number of monomer units consumed per free radicals is given by:

$$\text{KINETIC CHAIN LENGTH} = \frac{R_p}{R_i} = \frac{k_p[M]}{2(k_i k_t)^{1/2}[I]^{1/2}} \qquad (5.33)$$

In the absence of any side reactions, the kinetic chain length also represents the ACTUAL NUMBER OF UNITS PER CHAIN (x_n). In practice the side reactions, as transfer reactions, made the actual chain size smaller than the kinetic chain length.

The following TRANSFER REACTIONS should be considered:[901, 2781, 3274]

(i) TRANSFER TO MONOMER:

$$M_j\cdot + M \xrightarrow{\;k_{trm}\;} M_j + M\cdot \qquad (5.34)$$

(ii) TRANSFER TO SOLVENT:

$$M_j\cdot + S \xrightarrow{\;k_{trs}\;} M_j + S\cdot \qquad (5.35)$$

Transfer reactions can lead to the initiation of a new chain. Taking into account the possibility of such transfer reactions, it is possible to derive an expression for the ACTUAL CHAIN SIZE (x_n):

$$x_n = \frac{\text{Rate of propagation}}{\text{Rate of termination and transfer}} =$$

$$= \frac{k_p[M\cdot][M]}{k_{tc} + 2k_{td}[M\cdot]^2 + k_{trm}[M\cdot][M] + k_{trs}[M\cdot][S]} \qquad (5.36)$$

This equation can be simplified by elimination of the troublesome parameter [M·]:

$$1/x_n = \frac{R_p(k_{tc} + 2k_{td})}{k_p^2[M]^2} + \frac{k_{trm}}{k_p} + \frac{k_{trs}[S]}{k_p[M]}$$ (5.37)

This equation can be used to determinate experimentally such parameters as k_{trs}/k_p and k_{trm}/k_p, i.e. the relative rates of transfer to propagation, usually referred to as the TRANSFER CONSTANTS for solvent and monomer, respectively.

The monomer transfer constant (k_{trm}/k_p), which is usually very small, can be obtained by carrying out polymerizations in undiluted monomer at different initiator concentrations, and noting the relation between x_n and R_p.

The solvent transfer constant (k_{trs}/k_p) is usually more important, and can be obtained from uncatalysed polymerizations at different initial monomer concentrations. Under such circumstances it is usually found that

$$R_p \propto [M]^2$$ (5.38)

so that the first two terms on the right-hand side of equation (5.37) are constant, and independent of monomer concentration, thus representing the value of $1/x_n$ for undiluted monomer. Equation (5.37) can than be written as:

$$\frac{1}{x_n} = \frac{1}{x_{n0}} + \frac{k_{trs}[S]}{k_p[M]}$$ (5.39)

where x_{n0} = number of units per chain for polymerization of undiluted monomer. The value of the solvent transfer constant (k_{trs}/k_p) is obtained from a plot of $1/x_n$ against $[S]/[M]$.

Actually such transfer constants are really a measure of the reactivity of various monomers toward a given initiation free radical. Measurements of chain lengths (x_n) in polymerizations can yield information about the reactivity of monomers in free radical reactions.

Chain lengths (x_n) can only be measured for the truly soluble polymers, whereas several photoinduced free radical polymerizations yield insoluble cross-linked resins as product.

Neglecting chain transfer and assuming termination by bimolecular polymer chain combination only (at lower concentrations of initiator), the reciprocal DEGREE OF POLYMERIZATION ($1/\overline{P}$), can be calculated from equation:

$$\frac{1}{\overline{P}} = \frac{1}{k_p^2/k_t} \frac{R_p}{[M]^2}$$ (5.40)

From the slope of the linear portion (low concentration of initiator) on the plot of $1/\overline{P}$ vs R_p, k_p^2/k_t can be determined.

The EFFICIENCY OF THE RADICALS TO INITIATE POLYMERIZATION (i.e. fraction of initiator radicals starting kinetic chains)

(f), in the absence of any appreciate amount of chain transfer, can be defined as:

$$f = \frac{\text{Moles of polymer formed}}{\text{Moles of initiator decomposed}}$$ (5.41)

assuming bimolecular termination by the combination of two growing chains only

$$f = \frac{R_p}{P} k_1 \Phi[I]$$ (5.42)

where:

k_1 = rate of photodecomposition of photoinitiator in the absence of monomer (mol l^{-1} s^{-1}),
Φ = the quantum yield of radical production from the photoinitiator,
$[I]$ = initiator concentration (mol l^{-1}).

In the case of some monomers the presented kinetics can be complicated by side-effects. In the photopolymerization of diacrylates, double-bond conversion causes a volume shrinkage.[2056, 2057] This leads to the trapping of free radicals in partly polymerized samples. This trapping is responsible for a considerable 'dark reaction' observed upon heating after partial polymerization at room temperature. If partially polymerized samples are allowed to shrink, their capability of further photopolymerization at room temperatures is greatly reduced.

In order to evaluate kinetic data of photoinduced radical polymerization the following sets of experiments must be carried out:

(i) Measurements of the decay of a photoexcited initiator molecule in the absence and presence of monomer.
(ii) Measurements of the decay initiator radicals in the absence and presence of monomer.
(iii) Polymerization of the monomers in the absence and presence of photoinitiator in different solvents.
(iv) Polymerization of the monomers as in (iii) in the absence and presence of oxygen.

In many polymerization kinetics deviation from normal photopolymerization kinetic orders has been observed which can be attributed to primary radical or photoproduct termination interactions with growing polymer radicals.[3274, 3275]

The low kinetic order in initiator is also consistent with primary initiator radical competition for monomer and other deactivation pathways:

$$R\cdot + M \rightarrow M\cdot$$ (5.43)

$$M\cdot + M \rightarrow P\cdot$$ (5.44)

$$R\cdot + R\cdot \rightarrow \text{absorbing photoproducts}$$ (5.45)

$$R\cdot + \text{photoproducts} \rightarrow \text{termination}$$ (5.46)

$$R\cdot + P\cdot \rightarrow \text{termination}$$ (5.47)

$$P\cdot + \text{photoproducts} \rightarrow \text{termination}$$ (5.48)

Variations in light intensity orders are consistent with screen effects due to photoproduct formation. The photoproducts and primary radical fragments absorb light in the same area as the photoinitiator, and could lead to internal quenching of the reaction rate.[163, 2462]

In order to explain observed kinetic orders for the photopolymerization, it is always necessary to understand the photochemistry associated with the photoinitiator.

The most common monomer used as a model system for the study of photopolymerization kinetics is methyl methacrylate.[345, 602, 1307, 1323, 1326, 2238, 2307, 2462, 2463, 2883]

5.1.7. Cage Effect

When a molecule (or polymer molecule) is dissociated (photodissociated) in the liquid phase into two radicals they can undergo the following reactions:[3060, 3131]

(i) PRIMARY RECOMBINATION within the SOLVENT CAGE, which occurs at less than a molecular diameter, during the time of the order of a vibration ($\sim 10^{-13}$ s) and less than the time between diffusive displacements ($\sim 10^{-11}$ s). The solvent is able to remove the excess kinetic energy by collision and thermalize the formed radicals within a few molecular diameters.

(ii) SECONDARY DIFFUSIVE RECOMBINATION within the SOLVENT CAGE, which occurs at about one molecular diameter and is rate-controlled by diffusion of two radicals ($\sim 10^{-9}$ s).

(iii) OTHER RECOMBINATIONS, which occur outside of the SOLVENT CAGE. Radicals which escaped off the solvent cage may recombine with radicals from another dissociation event or undergo reactions with other species (e.g. double-bonds).

The CAGE EFFECT, which includes the combination of the first two recombination processes, is very much dependent on solvent viscosity, strength of hydrogen bonding, and mass and geometry of radicals formed. For dissociation products lighter than the solvent molecules (e.g. $HO\cdot$ or $CH_3\cdot$ radicals) the cage effect is very high, whereas if the dissociation products are heavier than the solvent, molecules will escape through the walls of the solvent cage.

There are many examples demonstrating the cage effect in the photochemistry of photoinitiators and macromolecules, some of which will be discussed.

5.1.8. Encounter Phenomena

The interaction of two molecules (e.g. small molecule with a polymer) can be treated as the encounter phenomena which consist of three consecutive processes:[3922]

(i) Diffusion together of the two reactants from some random position in solution so that they are not longer separated by a solvent molecule.

(ii) Rearrangement or rotation of the macromolecular encounter pair so that the reactive portions of the molecule collide, and so that these groups are correctly oriented for chemical reaction to take place.

(iii) Chemical reaction of the active groups.

These three stages can be represented in a kinetic scheme as:

$$A + B \underset{k_{-1}}{\overset{k_1}{\rightleftharpoons}} (AB) \tag{5.49}$$

$$(AB) \underset{k_{-2}}{\overset{k_2}{\rightleftharpoons}} (A \ldots B) \tag{5.50}$$

$$(A \ldots B) \overset{k_3}{\longrightarrow} \text{reaction} \tag{5.51}$$

where:

(AB) = encounter pairs with the reactive groups in positions which are unfavourable for chemical reaction (unreactive complex),

$(A \ldots B)$ = encounter pairs with the reactive groups in positions which are favourably oriented for chemical reaction (reactive complex).

The equilibrium constant for formation of these two kinds of intermediate pairs is given by:

$$k_{obs} = \left(\frac{k_1}{k_{-1}}\right)\left(\frac{k_2}{k_{-2}}\right)k_3 \tag{5.52}$$

5.1.9. Magnetic Field and Magnetic Isotope Effects on Photoinduced Emulsion Polymerization

The photochemically induced emulsion polymerization of styrene, methyl methacrylate, or acrylic acids is photoinitiated by oil-soluble ketone initiators, e.g. dibenzyl ketone. The efficiency of polymerization and the average molecular weight of the polymers formed are significantly increased by the application of magnetic field when polymerization is photoinitiated by ketones (Fig. 5.2).[3685, 3687] No magnetic field effects are observed for photoinitiated polymerization of styrene and methyl methacrylate in toluene solution. This phenomenon has been explained by the fact that the external magnetic field decreases the efficiency of triplet to singlet radical-pair intersystem crossing within micelles, and accordingly increases the fraction of radical pairs that escape without terminating polymerization chains.

5.2. PHOTOPOLYMERIZATION OF PURE MONOMERS

A number of vinyl monomers, such as methyl methacrylate, styrene, and acrylonitrile, may undergo very slow free radical polymerization when exposed to u.v. irradiation. An initiating radical can be formed by direct photolysis of a

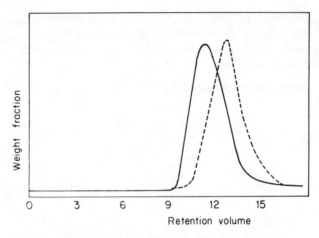

Fig. 5.2 Typical gel permeation chromatograms (GPC) of polystyrene polymerized in: (– – –) earth field and (———) magnetic field of 5000 G[3687]

monomer, or from impurities, or charge-transfer (CT) complexes between monomer and oxygen (cf. Section 5.3.2). With the exception of some monomers (Table 5.1), most vinyl monomers require photoinitiators (cf. Chapter 7) to achieve a reasonable rate of polymerization.

Table 5.1 A list of photosensitive monomers which undergo efficient u.v.-induced polymerization and copolymerization

Allyl methacrylate	2-Hydroxypropyl acrylate
Barium acrylate	N,N′-Methylenebisacrylamide
Cinnamyl methacrylate	Pentaerythritol tetramethacrylate
Diallyl phthalate	Tetraethylene glycol dimethacrylate
Diallyl isophthalate	N-Vinylcarbazole
Diallyl terephthalate	Vinyl cinnamate
2-Ethylhexyl acrylate	Vinyl 2-fuorate
2-Hydroxyethyl methacrylate	Vinyl 2-furylacrylate

Some monomers can be polymerized in a gas phase, e.g. ethylene at 130–225°C under 2500 bar pressure,[546] ethyl acrylate,[993] or benzaldehyde.[412, 941, 943]

A number of monomers, such as tetrafluoroethylene and hexachlorobutadiene, have been directly photopolymerized at polymer surfaces.[3939] There is a group of monomers which can be photopolymerized in a solid state (cf. Section 9.9).

5.2.1. Photopolymerization of Tetrafluoroethylene

Tetrafluoroethylene undergoes polymerization in the gas phase when exposed to u.v. radiation of 185–300 nm.[3890, 3891, 3939] With monomer pressure less than 10 Torr, the polymer forms deposits on the walls of the container as an adherent polymeric film. With monomer pressure ranging from 10 to 760 Torr a white powdery polymer ('floc') is formed. The i.r. spectra of the polymer indicate the polymer structure to be different from that of poly(tetrafluoroethylene) in that it contains CF_3 groups and perhaps some cyclic fluorocarbon groups. The polymer, unlike poly(tetrafluoroethylene), can be fused to transparent films at 300°C.

Several photoinitiators such as mercury bromide,[244] phosgene,[2390] nitrous oxide,[3193] trifluoroiodomethane,[1528] tribromofluoroethane,[3370] and 1,2-dibromotetra fluoroethane[2594] have been used to accelerate polymerization of tetrafluoroethylene in the gas phase.

5.2.2. Photopolymerization of Styrene

During irradiation (313 nm) of styrene (5.1) main products are low molecular compounds such as trimers (5.2), oligomers (5.3) and small amount of polystyrene (5.4).[1976–1978] Trimers and oligomers are formed by successive photocycloaddition reactions: (see 5.1–5.4). Higher molecular weight oligomers may also be formed in analogous processes.

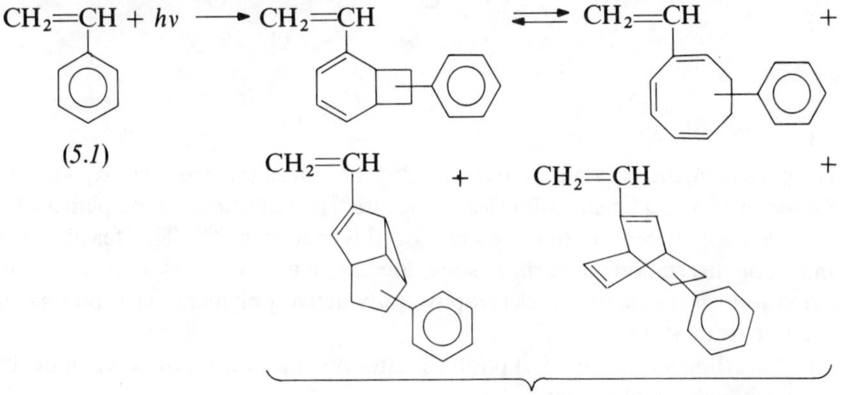

(5.1)

$\sim 80\%$
(5.2)

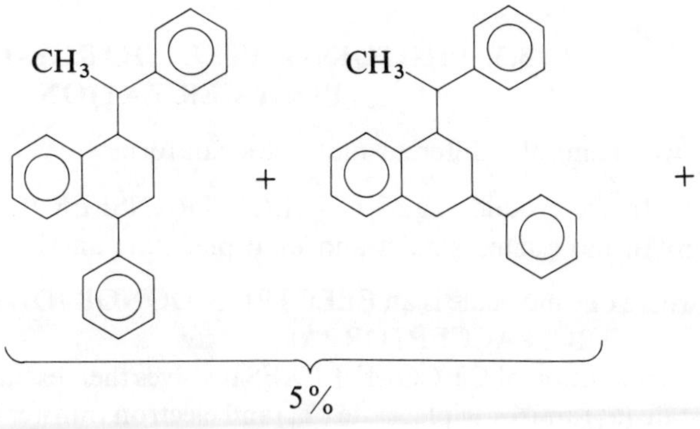

5%
(5.3)

$$\text{other oligomers} \ + \underset{\underset{\displaystyle \bigcirc}{\big|}}{\left(CH_2-CH\right)_n} \tag{5.53}$$

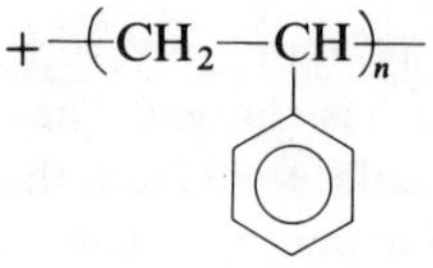

$$(5.4)$$

5.2.3. Photopolymerization of Carbon Suboxide

Carbon suboxide (5.5) photopolymerize[3396] to a ladder polymer having the poly-α-pyrone ladder structure (5.6):[3962]

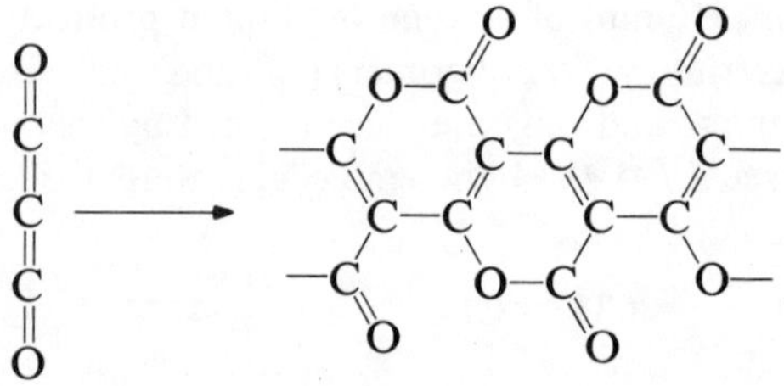

$$(5.54)$$

$$(5.5) \qquad\qquad (5.6)$$

The paramagnetic polymer (5.6) is a reddish brown called 'red carbon', while the monomer (5.5) is a linear, colourless molecule. The red colour of the planet Mars has been attributed to the presence of this polymer.[2965] The feasibility of photopolymerization of carbon suboxide can lend to the system, i.e. image formation on magneto- or electrophysically active polymers via topologically selective irradiation.

Poly(carbon suboxide) (5.6) photoinitiates polymerization of acrylamide, but the mechanism is not given.[3962]

5.3. PHOTOINITIATED CHARGE-TRANSFER POLYMERIZATION

Two chemically different molecules can form:

(i) In the ground state a CHARGE TRANSFER (CT) COMPLEX.
(ii) In the excited singlet and/or triplet state an EXCIPLEX (cf. Section 3.5),

where one molecule is an ELECTRON DONOR (D) and another molecule is an ELECTRON ACCEPTOR (A).

Irradiation of CT COMPLEXES involves their excitation to the excited singlet state (S_1) and/or triplet state (T_1) and electron transfer from the donor (D) to the acceptor (A) molecules resulting in the formation of radical cations (D^+) and radical anions (A^-):[1543, 1547, 1548, 1782, 1785–1787, 3158, 3559, 3563]

$$D + A \rightleftarrows (D\text{–}A)(CT\ COMPLEX) \tag{5.55}$$

$$(D\text{–}A) + h\nu \rightarrow (D\text{–}A)^* (S_1\ or\ T_1) \tag{5.56}$$

or

$$A + h\nu \rightarrow A^* (S_1\ or\ T_1) \tag{5.57}$$

$$A^* + D \rightarrow (D\text{–}A)^* (S_1\ or\ T_1) \tag{5.58}$$

$$(D\text{–}A)^* \xrightarrow{\text{solvent (s)}} (D_s^+ \dots A_s^-) \rightarrow D_s^+ + A_s^- \tag{5.59}$$

Polymerization of the radical cation (D^+) formed from electron-donating monomers proceeds from either the solvent-separated $(D_s^+ \dots A_s^-)$ or contaction pairs $(D_s^+ + A_s^-)$:

$$(D_s^+ \dots A_s^-) \rightarrow polymerization \tag{5.60}$$

$$D_s^+ + A_s^- \rightarrow polymerization \tag{5.61}$$

Cationic polymerization occurs in rigorously dried monomers only. This mechanism has been confirmed by ESR, absorption/emission spectroscopy, laser photolysis, and photoconductivity.[1546, 1776, 1783, 1785] Several examples of photoinitiated charge-transfer cationic polymerization are shown in Table 5.2.

Photoinitiated charge-transfer cationic polymerization of N-vinylcarbazole (5.7) in the presence of an electron acceptor, e.g. chloranil (5.8) (or phthalic anhydride) depends on solvent basicity.[3511–3513, 3563] The strength of the solvent basicity exerts a striking effect on the reaction course. In basic solvents such as methanol, acetone, or acetonitrile, the cyclodimerization of N-vinylcarbazole occurs exclusively according to the mechanism:

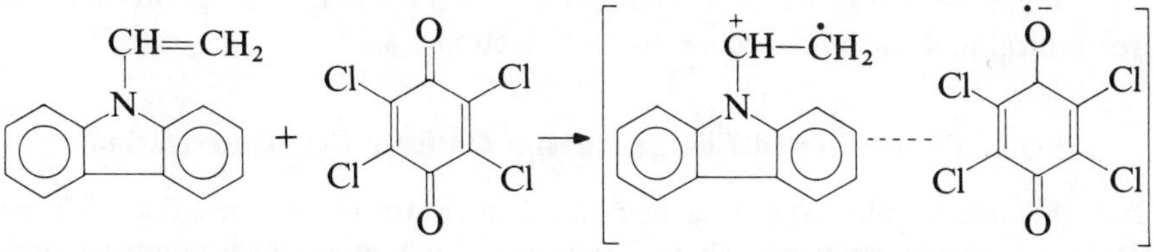

(5.7)   (5.8)   CT COMPLEX

$$\tag{5.62}$$

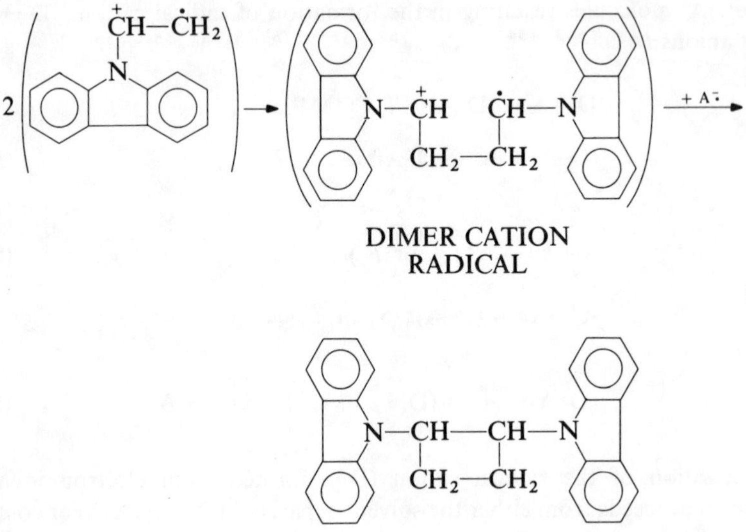

DIMER CATION
RADICAL

(5.63)

PHOTOCYCLODIMER

In stronger basic solvents such as N-methyl-2-pyrrolidone, N,N-dimethyl-formamide or dimethyl sulphoxide simultaneous radical cationic polymerization and cyclodimerization occur:

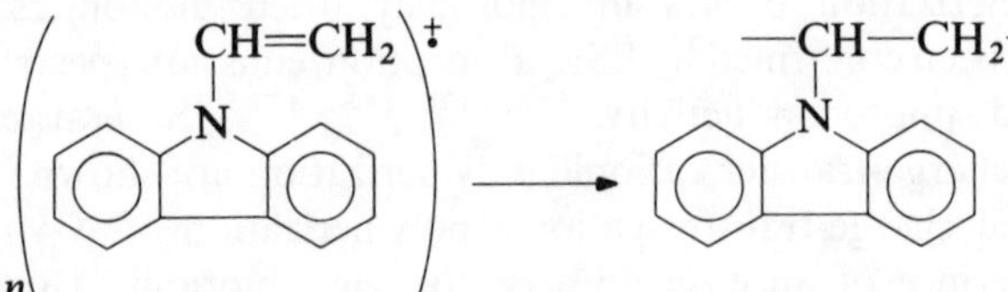

(5.64)

In strongest basic solvent, such as hexamethylphosphoric triamide, no more cyclodimerization of N-vinylcarbazole occurs and only radical cationic poly-merization take place. In non-polar dichloromethane, cationic polymerization occurs irrespective on the atmosphere.[33, 1898, 3512]

N-vinylcarbazole can also be simultaneously polymerized by photoinitiated free radical and cationic mechanisms (cf. Section 5.4.6.1).

5.3.1. Photoinitiated Charge-transfer Cationic Copolymerization

Two chemically different monomers, one is an electron donor and the other an electron acceptor, can form a charge-transfer (CT) complex which, when exposed to irradiation, forms radical cations and radical anions. For example styrene (5.9) (and its derivatives such as p-methylstyrene or p-tertbutylstyrene) can form, with maleic anhydride (5.10) in the ground state, a CT complex, which under further irradiation initiates alternating copolymerization:[3158]

Table 5.2 Examples of photoinitiated charge-transfer cationic polymerization of electron donating monomers

Electron donor monomer (D)	Electron acceptor (A)	References
Styrene	Pyromellitic dianhydride	3498
α-Methylstyrene	Pyromellitic dianhydride	1776, 1783, 1785
	Tetracyanobenzene	1543, 1544, 1548, 1769, 1782, 1784–1786
Isobutyl vinyl ether	Maleic anhydride	3523
2-Vinylnaphthalene	Diethyl fumarate	3903
	Fumaronitrile	3903
Cyclohexene	N-ethylmaleimide	3927
Cyclohexene oxide	Pyromellitic dianhydride	1779, 1785, 1786
	Tetracyanobenzene	1779, 1785
Trioxane, 3,3-Bis(chloromethyl) oxetane Oxepane	Pyromellitic dianhydride Tetracyanobenzene	3522
N-vinylcarbazole	Carbon tetrabromide	1979, 2787
	Thiobenzophenone	2773
	p-Chloranil	1042, 2660, 3328, 3511, 3513,
	Nitrobenzene	3562
	Nitrobenzene sensitized by sodium chloroaurate (NaAuCl$_4$ · 2H$_2$O). Au(III) enhances spin density of N-vinylcarbazole–Au(III) system	231, 3562
	Phthalic anhydride	3511
	Fumaronitrile Diethyl fumarate	3329, 3510

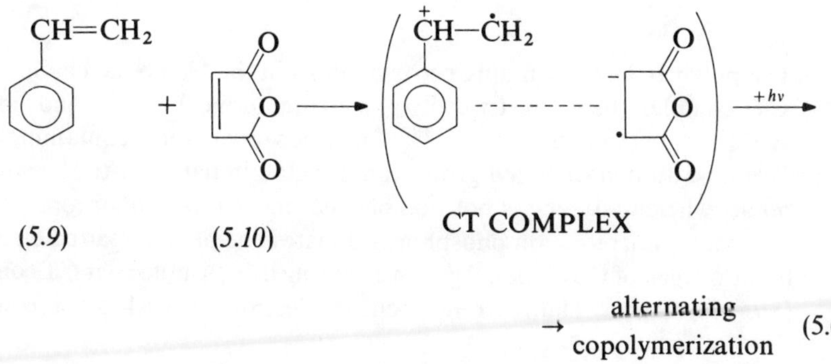

(5.9) (5.10) CT COMPLEX

$$\rightarrow \text{alternating copolymerization} \quad (5.65)$$

Trans-stilbene and maleic anhydride,[1480] styrene, or methyl methacrylate and alkylvinylimidazoles[2597], acrylic acid and substituted imidazoles,[876] styrene and diethyl fumarate,[875] can be photocopolymerized by an analogous CT complex mechanism.

5.3.2. Different Types of Charge-transfer Complex Photoinitiated Polymerizations

Different types of CT complexes formed between two chemically different molecules may photoinitiate polymerization of different monomers:

(i) CT complexes between oxygen and monomers. It is well known that oxygen can form CT complexes with several organic compounds such as alcohols,[2595, 3637] ethers,[2595, 3432] hydrocarbons,[698, 1793, 2595, 3637] amines,[1792, 3637] unsaturated compounds such as ethylene, butadiene, hexatriene,[1070] and vinyl monomers such as styrene, α-methylstyrene, isoprene, vinyl acetate and methyl methacrylate.[2091, 2092] In the last case the CT complex between oxygen and monomer (5.11) may photoinitiate the polymerization of the latter. For styrene the following mechanism has been suggested:[2091, 2092]

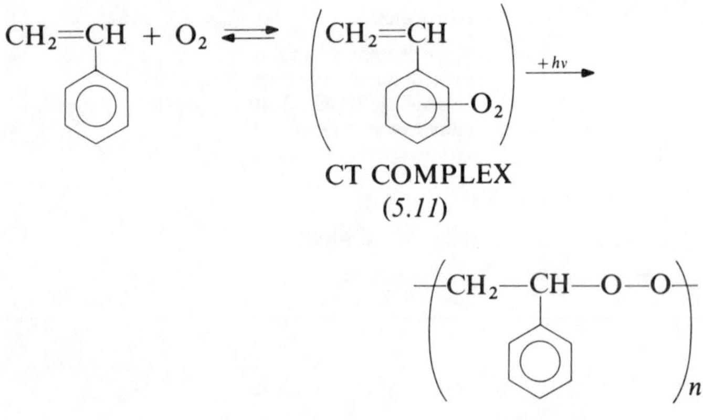

CT COMPLEX
(5.11)

(5.66)

(5.12)

The polymer (5.12) contains peroxide links in the backbone chain.

(ii) CT complex between triphenylphosphine (donor) (5.13) and methyl methacrylate (acceptor) (5.14).[2380] Only those monomers containing an α- or β-unsaturated carbonyl group can be photoinitiated by triphenylphosphine, whereas styrene is not polimerized under these conditions.

Under u.v. irradiation phosphorus donates one of its unpaired electrons to the oxygen of the carbonyl group, resulting in intermolecular CT complex (5.15) formation. The rearrangement of electrons in such a CT complex produces a methyl methacrylate radical (*5.16*), which can initiate polymerization of a monomer:

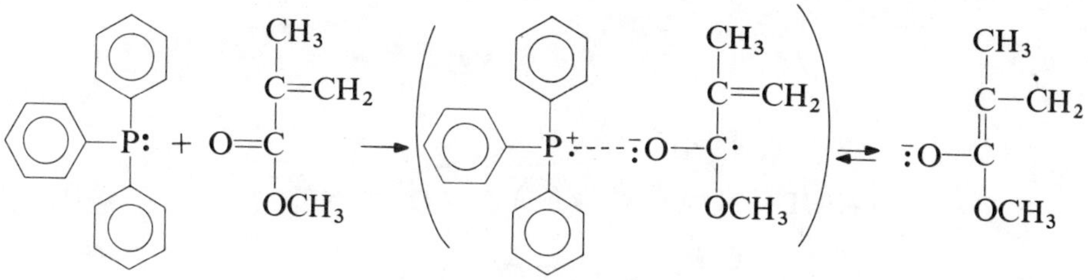

(*5.13*) (*5.14*) CT COMPLEX
(*5.15*) (*5.16*)

(5.67)

(iii) CT complex between dimethylamine (donor) (*5.17*) and nitrosobenzene (acceptor) (*5.18*):[1321, 1322]

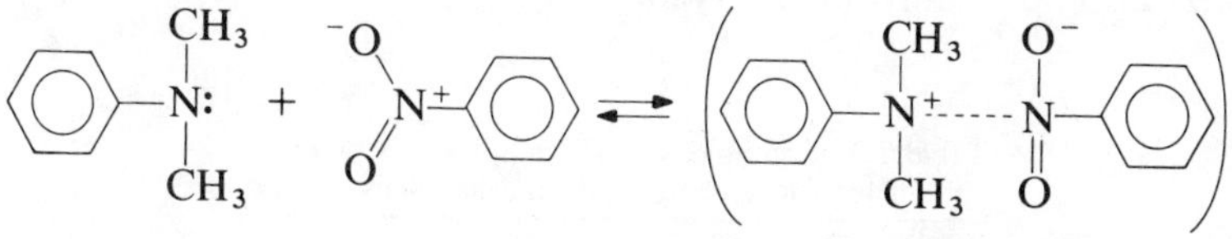

(*5.17*) (*5.18*) CT COMPLEX
(*5.19*)

(5.68)

Such a CT complex (*5.19*), which is formed in the ground state, absorbs light and is photolysed to *N*-methylanilinomethyl radical (*5.20*), hydrogen radical (H·) and nitrosobenzene (*5.21*):

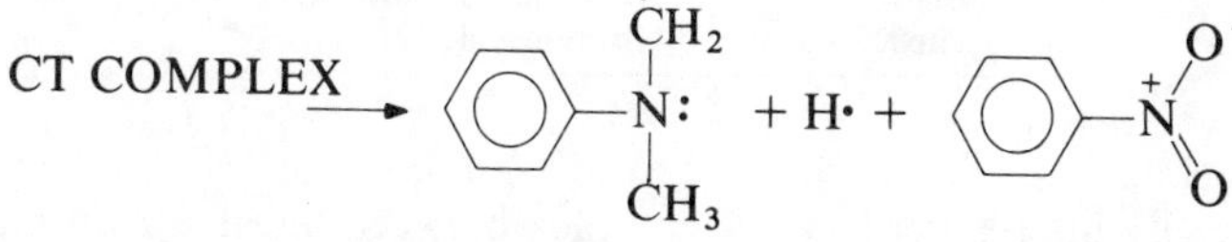

CT COMPLEX ⟶ ... + H· + ...

(*5.20*) (*5.21*)

(5.69)

N-methylanilinomethyl radical may initiate further polymerization of methyl methacrylate.

(iv) CT complex between triethylamine (donor) (*5.22*) and benzoyl peroxide (acceptor) (*5.23*):[1320]

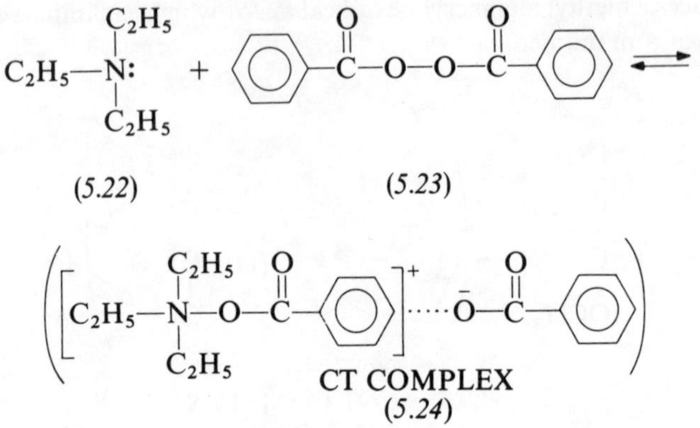

$$\underset{\overset{|}{C_2H_5}}{\overset{\overset{\displaystyle C_2H_5}{|}}{C_2H_5\!-\!N\!:}} \;+\; \langle\bigcirc\rangle\!-\!\overset{\overset{\displaystyle O}{\|}}{C}\!-\!O\!-\!O\!-\!\overset{\overset{\displaystyle O}{\|}}{C}\!-\!\langle\bigcirc\rangle \;\rightleftharpoons$$

(5.22)   (5.23)

(5.70)

$$\left(\left[\underset{\overset{|}{C_2H_5}}{\overset{\overset{\displaystyle C_2H_5}{|}}{C_2H_5\!-\!N\!-\!O\!-\!\overset{\overset{\displaystyle O}{\|}}{C}\!-\!\langle\bigcirc\rangle}} \right]^{+} \!\cdots\! \overset{\overset{\displaystyle O}{\|}}{\underset{}{O}}\!-\!\overset{}{C}\!-\!\langle\bigcirc\rangle \right)$$

CT COMPLEX
(5.24)

Such a CT complex (5.24), which is formed in the ground state, absorbs light and is photolysed to triethylamine radical which may initiate polymerization of methyl methacrylate.

(v) CT complexes between heterocyclic compounds (donors) and chlorine or bromine (acceptors) (Table 5.3). These CT complexes photoinitiate polymerization of methyl methacrylate.

Table 5.3 Examples of CT complexes between heterocyclic compounds and halogens

Donor	Acceptor	References
Tetrahydrofurane	Bromine	1299
Pyridine	Bromine	1317
Quinoline	Bromine	1316, 1318
Lutidine	Bromine	2662
Quinaldine	Bromine	2257, 2662
Isoquinaldine	Chlorine	2259
Isoquinaldine	Bromine	2261
γ-Picoline	Chlorine	2258
γ-Picoline	Bromine	2263
Lepidine	Bromine	2256
Acridone	Bromine	1296
Poly(vinylcarbazol)	Bromine	1312

Specially interesting is a CT complex between tetrahydrofuran (donor) and bromine (Br_2) (acceptor):[2162]

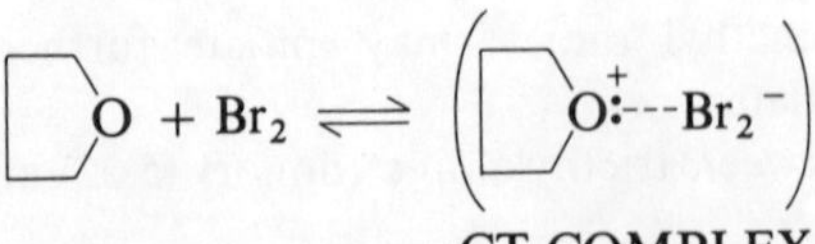

$$\Box O + Br_2 \;\rightleftharpoons\; \left(\Box O\!\overset{+}{:}\!\cdots\!Br_2^{-} \right)$$

CT COMPLEX
(5.25)

(5.71)

This deep reddish-brown CT complex (*5.25*) may photoinitiate polymerization of methyl methacrylate.[1299] As polymerization proceeds the initial reddish-brown colour of the solution disappears.

(vi) CT complexes between tetrahydrofuran and sulphur dioxide (SO_2),[1314] benzophenone–SO_2,[1315] and pyridine–SO_2 (in the presence of carbon tetrachloride)[1304, 1305] may photoinitiate polymerization of methyl methacrylate.

(vii) CT complexes between isobutylene (donor) (*5.26*) and Lewis acids ($TiCl_4$, VCl_4, $SnCl_4$, $TiBr_4$, or TiI_4) (acceptors) are responsible for the light-induced cationic polymerization of isobutylene:[2383–2385, 3617, 3618]

$$
\underset{\underset{CH_3}{|}}{\overset{\overset{CH_3}{|}}{CH_2{=}C}} \; + \; TiCl_4 \; \rightleftharpoons \; \left(\overset{\overset{CH_3}{|}}{\underset{\underset{CH_3}{|}}{\dot{C}H_2{-}\overset{+}{C}\cdots TiCl_4^{\bar{\cdot}}}} \right) \overset{+h\nu}{\longrightarrow}
$$

$$
\begin{array}{c} \text{CT COMPLEX} \\ (5.26) \end{array} \hspace{3cm} (5.72)
$$

$$
\underset{\underset{CH_3}{|}}{\overset{\overset{CH_3}{|}}{\dot{C}H_2{-}C}} \;\; (TiCl_4^{\bar{\cdot}}) \; + \text{monomer} \rightarrow \text{polymerization}
$$

This reaction is completely inhibited by the presence of oxygen.

In the presence of VCl_4 photoinitiated copolymerization of isobutylene with butadiene[2949] and isoprene[3619] occurs.

(viii) CT complex between methylacrylolyl-L-vaniline and maleic anhydride photoinitiates copolymerization of these two monomers.[2698]

5.3.3. Different Types of Exciplex Photoinitiated Polymerizations

Different types of exciplexes formed between two chemically different molecules may photoinitiate polymerization:

(i) Exciplexes between amines: triethylamine,[3980,3981] aniline and its derivatives,[1791,2150,3981] and monomers such as methyl methacrylate and styrene may photoinitiate polymerization. It has been reported that triethylamine has no ability to photoinitiate polymerization of acrylonitrile.[2148]

(ii) Exciplexes between amines and ketones (cf. Section 7.9.4.2.1.2), amines, and polycyclic hydrocarbons (cf. Section 7.11) effectively photoinitiate polymerization of many vinyl monomers.[2148,2149]

(iii) Exciplex which is formed between two monomer molecules of *p*-dimethylaminobenzyl methacrylate (*5.27*) is responsible for the two radicals (*5.28*) and (*5.29*), which may both initiate polymerization:[3943]

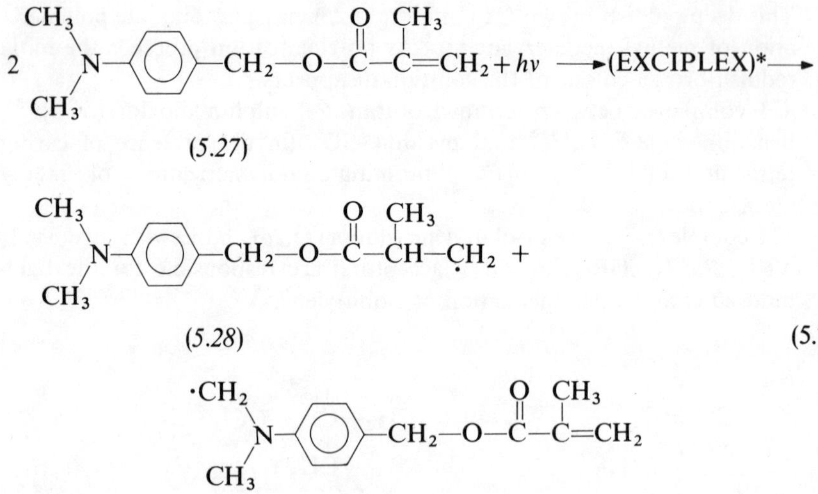

2 (5.27) $+ h\nu \longrightarrow$ (EXCIPLEX)* $\longrightarrow$

(5.28)

(5.29)

(5.73)

(iv) Exciplex which is formed between acetone and methyl methacrylate is probably responsible for the photoinitiation of polymerization of methyl methacrylate in acetone solution.[318] Addition of $ZnCl_2$ retards polymerization remarkably, by quenching the exciplex formation.

(v) Exciplex (5.32) which is formed between the excited triplet state (T_1) of poly(vinyl benzophenone) (5.30) and indole-3-ylacetic acid (5.31) can photoinitiate polymerization of methyl methacrylate:[2238]

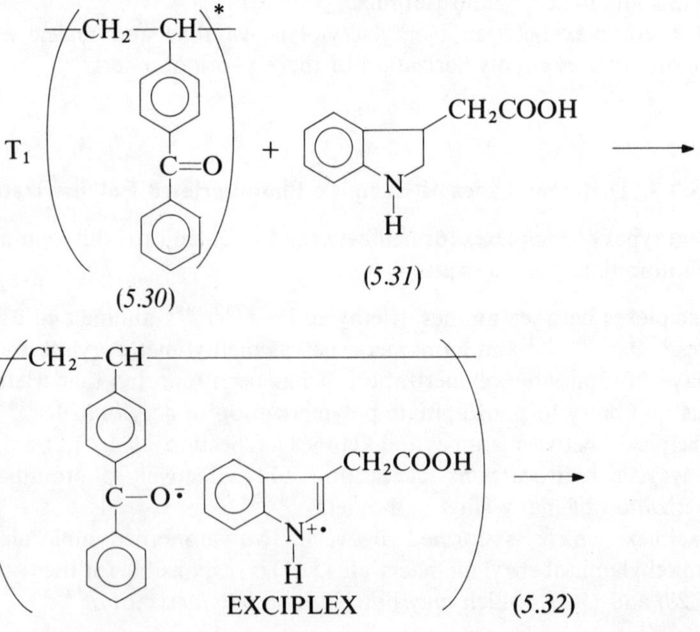

T_1 (5.30) + (5.31) $\longrightarrow$

EXCIPLEX (5.32)

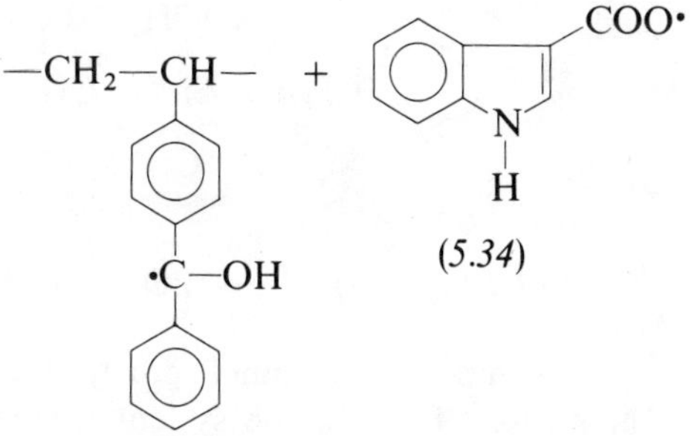

(5.34) (5.74)

(5.33)

Indole-3-ylacetic radical (5.34) initiates polymerization of methyl methacrylate, whereas polymeric ketyl radical (5.33) recombines and forms crosslinked structures. It can also be grafted to a growing poly(methyl methacrylate) chain.

(vi) Photoreducible dyes (cf. Section 7.25) and suitable reducing agents or activators under irradiation form exciplexes which decompose further into free radical ions, which may initiate polymerization:[686,2820]

$$D \text{ (dye)} + h\nu \rightarrow D^* \text{ (singlet and/or triplet)} \tag{5.75}$$

$$D^* + A \text{ (activator)} \rightarrow E^* \text{ (EXCIPLEX)} \tag{5.76}$$

$$E^* \rightarrow D^{\bar{\cdot}} + A^{\ddagger}. \tag{5.77}$$

$$D^{\bar{\cdot}} \rightarrow DH^- \tag{5.78}$$

$$A^{\ddagger} \rightarrow R\cdot + X^+ \text{ (radical formation)} \tag{5.79}$$

$$R\cdot + M \rightarrow \text{initiation of polymerization} \tag{5.80}$$

For example methylene blue or rose bengal (D) in the presence of reducing agent (activator) such as trimethylbenzylstannane (5.35) under visible irradiation produces benzyl radical (5.36) which may initiate polymerization of vinyl monomers (M):[1016]

$$D + h\nu \rightarrow {}^1D \xrightarrow{\text{ISC}} {}^3D \tag{5.81}$$

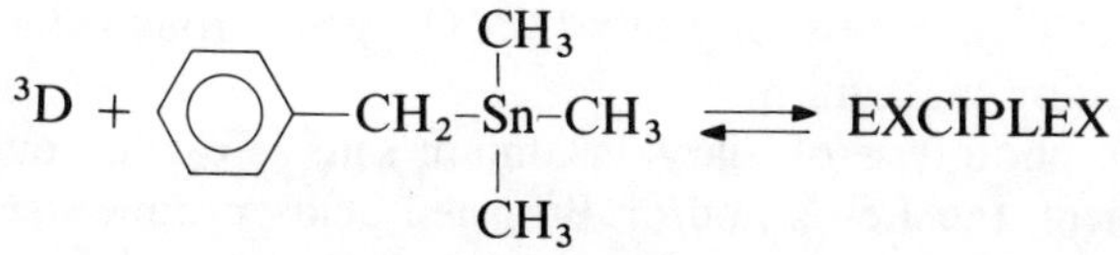

(5.35)

$$\text{EXCIPLEX} \rightarrow D^{\bar{\cdot}} + A^{\cdot+} \tag{5.83}$$

$$A^{\ddot{+}} \longrightarrow \langle\bigcirc\rangle{-}CH_2\cdot + {}^+Sn(CH_3)_3 \quad (5.84)$$

(with CH_3 groups on Sn)

$$(5.36)$$

$$\langle\bigcirc\rangle{-}CH_2\cdot + M \longrightarrow POLYMER \quad (5.85)$$

This type of polymerization has found practical application in the PHOTOIMAGING PROCESS. In the system leuco dyes–diimidazoles (activator), colour formation is prevented in non-image areas by exposure to visible radiation (panchromatic sensitization) which results in formation of free-radical traps that interfere with the imaging process.[648,1016] Polymerization can give positive or negative image formation, depending on the sequence of exposures. An intense exposure causes simultaneous colour formation and polymerization, whereas low-intensity exposure results only in polymerization without colour formation.

5.4. PHOTOINITIATED CATION–RADICAL POLYMERIZATION

5.4.1. Arlyldiazonium Salts

Aryldiazonium salts (5.37) are easily photolysed to aryl halides, nitrogen gas and Lewis acids (5.38), e.g.:[771,2933]

$$Ar-N_2{}^+BF_4^- \rightarrow Ar-F + N_2 + BF_3. \quad (5.86)$$

$$(5.37) \qquad\qquad\qquad (5.38)$$

Similarly, diazonium salt possessing the PF_6^-, AsF_6^-, SbF_6^-, $FeCl_4^-$ and $SbCl_6^-$ anions on photolysis generate PF_5, AsF_5, SbF_5, $FeCl_3$, and $SbCl_5$, respectively. In the presence of traces of water Brönsted acids are produced; e.g.:

$$BF_4 + H_2O \rightarrow H^+ + BF_4OH^- \quad (5.87)$$

The elimination of nitrogen requires that an electron is donated to the N_2^+ group. This may come from the aryl ring, in which case an aryl cation is formed, or from another molecule, in which case an aryl radical is formed.[1878]

Quantum yields of diaryldiazonium salts photolysis in the absence of oxygen are greatly in excess of unity.[3379] Oxygen strongly inhibits the chain decomposition mechanism.

If the photolysis of diaryldiazonium salts is carried out in the presence of monomers, the Lewis and/or Brönsted acids produced may initiate cationic polymerization, which is not sensitive to the presence of oxygen.

By modification of the structure of aryldiazonium salts it is possible to obtain photoinitiators sensitive in the blue region of visible light (300–400 nm).[2604,3205,3206,3484] Aryldiazonium salts are specially suitable for the u.v. curing of epoxy resins, but there are several disadvantages with application of this type of photoinitiators:[782,801]

(i) The diazonium salt–epoxy resin systems gel on standing in the dark and must be stabilized by the addition of Lewis and complexing agents. Maximum stability (pot-life) in the presence of stabilizers is only about 2 weeks.

(ii) Diazonium salts undergo low-temperature thermolysis as well as photolysis, and thus may present a serious safety hazard due to spontaneous highly exothermic bulk polymerization.

(iii) Diazonium salts are very sensitive to u.v./visible radiation and may gel in a normally lit room.

(iv) The photodecomposition of diazonium salts involves the evolution of nitrogen gas which causes bubbles and pinholes in u.v.-cured films thicker than a few tenths of a mil. For that reason they cannot be used for container coatings and photoresists.

(v) Polymers crosslinked using diazonium salts are often highly coloured and thus may be objectionable for certain applications.

Aryldiazonium salts were also used as photoinitiators of cationic polymerization of 1,2-epoxypropane and 1,2-epoxybutane,[264] crosslinking of glycydyl methacrylate polymers,[3379] and poly(vinyl alcohol).[2245,3663,3664]

5.4.2. Quaternary Ammonium Salts

Different types of quaternary ammonium salts (Table 5.4) may photoinitiate radical and/or cation–radical polymerization of different vinyl monomers such as methyl methacrylate, acrylonitrile, and styrene.[2083,2084]

Depending on their structure the quaternary ammonium salts (5.39) can be photolysed into cation–radicals (5.40) and benzyl radicals (5.41) according to the mechanism:[2083,2090,2313,2865,2866]

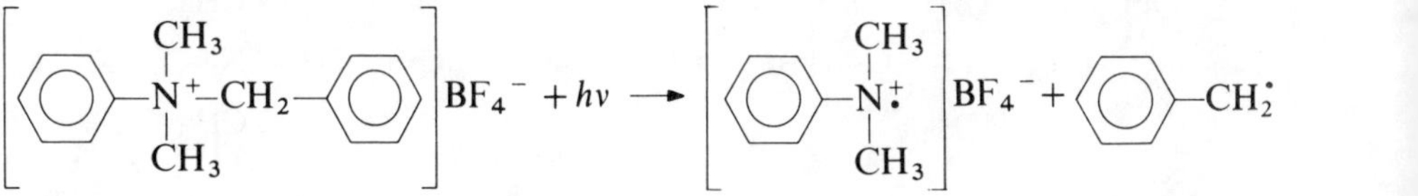

$$(5.39) \qquad (5.40) \qquad (5.41)$$

$$(5.88)$$

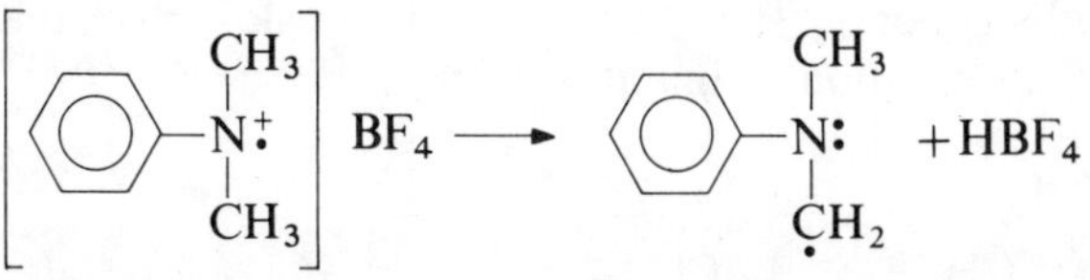

$$(5.42)$$

$$(5.89)$$

The N-methylanilinomethyl radical (5.42) has been suggested to be the initiating polymerization radical.

Table 5.4 Quaternary ammonium salts[2083]

Structure	X
$\left[\begin{array}{c} CH_3 \\ \mid \\ C_2H_5{-}N^+{-}CH_2C_6H_5 \\ \mid \\ CH_3 \end{array}\right] X^-$	Cl
	Br
	BF$_4$
$\left[\begin{array}{c} CH_3 \\ \mid \\ C_2H_5{-}N^+{-}C_2H_5 \\ \mid \\ CH_3 \end{array}\right] X^-$	Br
	BF$_4$
$\left[\begin{array}{c} C_2H_5 \\ \mid \\ C_6H_5CH_2{-}N^+{-}C_2H_5 \\ \mid \\ C_2H_5 \end{array}\right] X^-$	BF$_4$
	Br
$\left[\begin{array}{c} C_2H_5 \\ \mid \\ C_2H_5{-}N^+{-}C_2H_5 \\ \mid \\ C_2H_5 \end{array}\right] X^-$	BF$_4$
$\left[\begin{array}{c} CH_3 \\ \mid \\ C_2H_5{-}N^+{-}CH_2COC_6H_5 \\ \mid \\ CH_3 \end{array}\right] X^-$	BF$_4$
$\left[\begin{array}{c} C_2H_5 \\ \mid \\ C_2H_5{-}N^+{-}CH_2COC_6H_5 \\ \mid \\ C_2H_5 \end{array}\right] X^-$	BF$_4$

During photolysis of another quaternary ammonium salt (5.43), a benzoyl radical (5.44) has been considered to be the initiating polymerization radical:

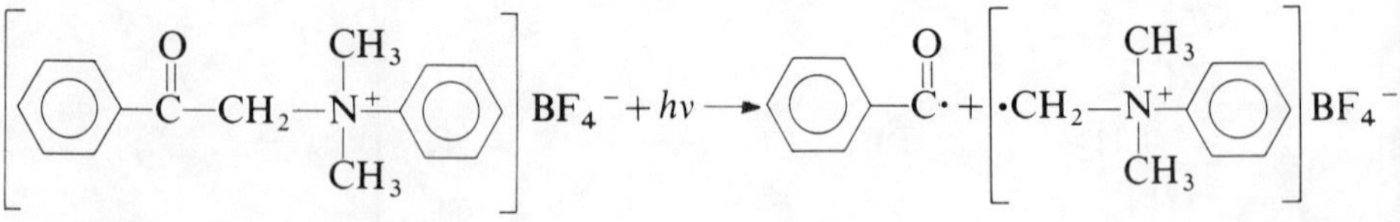

$$(5.43) \qquad\qquad (5.44) \qquad\qquad (5.90)$$

It has been observed that the halide salts show higher photoinitiating activity than the tetrafluoroborate salts. The following redox mechanism has been proposed for the formation of initiating radicals:[2083,2865,2866]

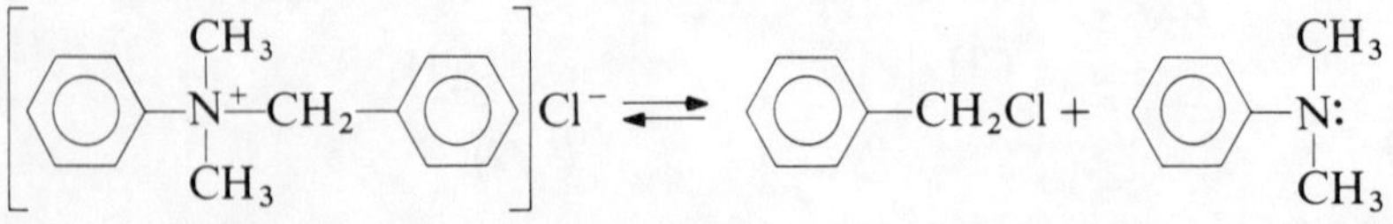

$$(5.45) \qquad\qquad\qquad\qquad (5.91)$$

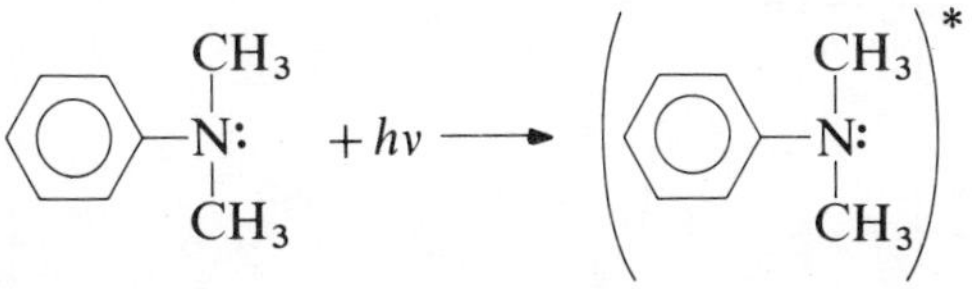

$$(5.92)$$

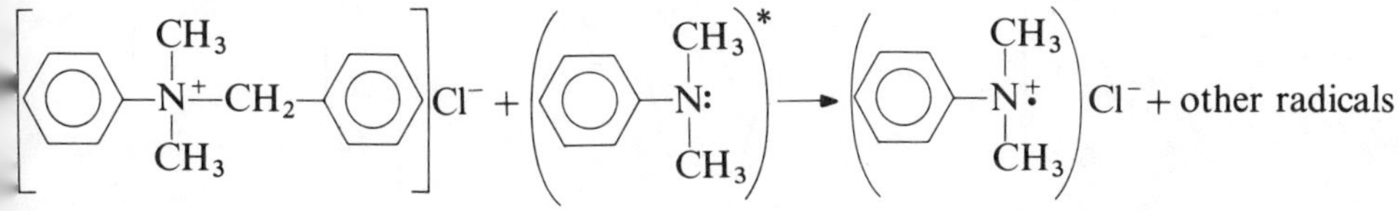

$$(5.93)$$

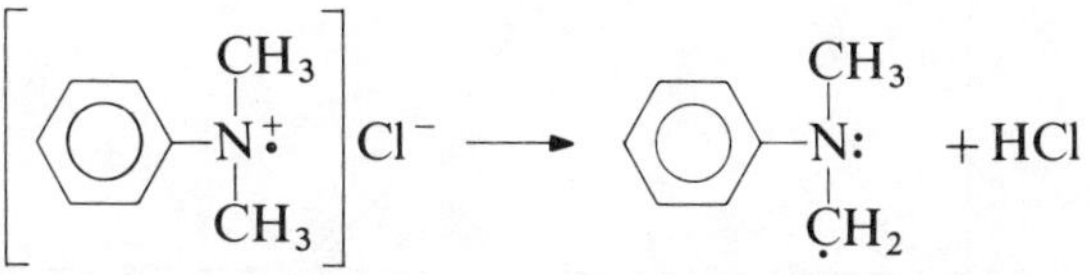

$$(5.94)$$

Another possibility is a photochemical reduction of the anilinium cation radical with a halide ion in the excited quaternary ammonium halide.[2080]

5.4.3. Onium Salts Photoinitiators

Diaryliodonium $(Ar_2I^+X^-)$,[782,784,788,789,801,860], triarylsulphonium $(Ar_3S^+X^-)$,[782-784,786,791,792,794,798-800,803,805,860] triarylselenonium $(Ar_3Se^+X^-)$,[782,793] tetraarylphosphonium $(Ar_4P^+X^-)$,[229,2106] and oxosulphonium[2107] salts (where X^- is a non-nuclephilic anion such as BF_4^-, PF_6^-, SbF_6^-, and AsF_6^-) are highly efficient thermally stable photoinitiators for cation–radical polymerization.

Onium salts do not absorb light at visible range (Fig. 5.3) and for that reason they cannot be used directly to initiate polymerization in this range. The photolysis products have absorption spectra almost identical to those of the parent compounds, and therefore compete for the radiation absorption. In order to overcome this problem it is possible to use:

(i) Complex salts with a larger chromophore (5.46 or 5.47) instead of simple triarylsulphonium salts.[789,799,3814]

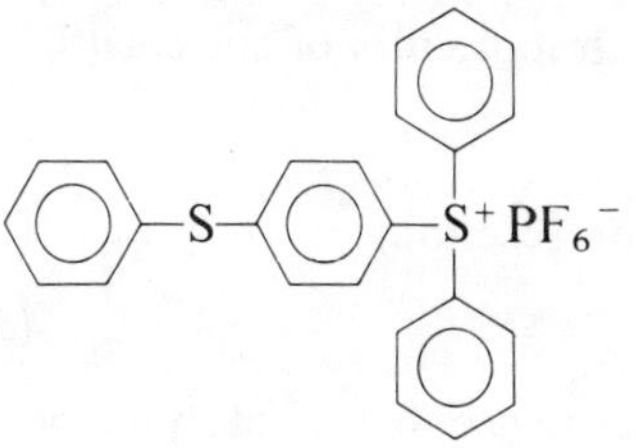

(5.46)

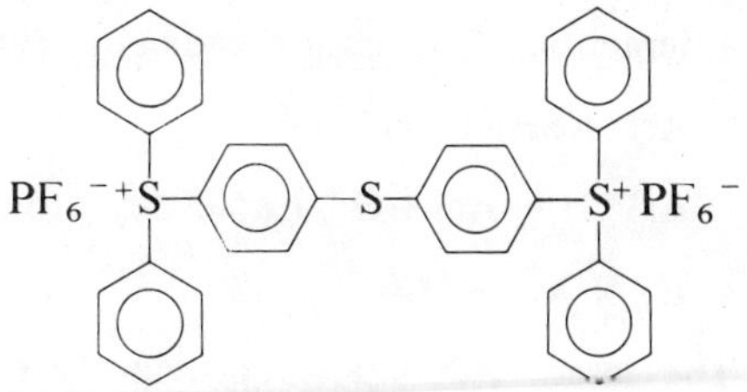

(5.47)

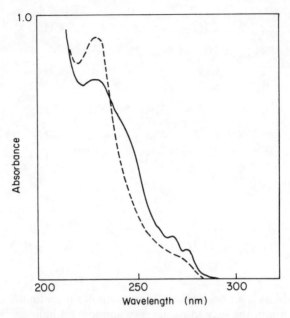

Fig. 5.3 Ultraviolet absorption spectra of: (---) diaryliodonium $(Ar_2I^+X^-)$ and (——) triarylsulphonium
$(Ar_3S^+X^-)$ salts[789, 792]

(ii) Photosensitizers which will absorb at longer wavelength.

The photolysis mechanism of diaryliodonium[781,782,788,789,801,2079,2881,2893]
triarylsulphonium[782,783,791,792,800,2080,2893] and triarylselenium[782,793,2881]
salts proceeds by the same mechanism, which is presented here as an example of
diaryliodonium salt (5.48) and includes two pathways:

(i) The major process (95%) involves the photoinduced homolytic cleavage of
the diaryliodonium salt into radical, radical–cationic and anionic
fragments:[782,788,789]

$$Ar_2I^+X^- + hv \rightarrow (Ar_2I^+X^-)^* \rightarrow Ar^{\cdot+} + Ar\cdot + X^- \qquad (5.95)$$

$$(5.48) \qquad\qquad\qquad (5.49)$$

Interaction of the cation–radical (5.49) with molecules of solvent (SH):

$$ArI^{\cdot+} + SH \rightarrow ArI^+H + S\cdot \qquad (5.96)$$

$$ArI^+H \rightarrow ArI + H^+ \text{ (rapid deprotonation reaction)} \qquad (5.97)$$

$$H^+ + X^- \rightarrow HX \qquad (5.98)$$

The laser flash studies provide direct evidence for photolysis of diaryliodonium salts by the above mechanism.[2894]

(ii) The minor process (less than 5 %) involves a nucleophilic attack of the solvent on the photoexcited diaryliodonium salt:

$$(ArI_2^+X^-)^* + SH \rightarrow (ArSH)^+ + ArI + X^- \tag{5.99}$$

$$(ArSH)^+ \rightarrow ArS + H^+ \tag{5.100}$$

The photolysis of onium salts is independent of the presence of oxygen and singlet and/or triplet quenchers.

The quantum yield of photolysis of onium salts which have the same cation structure is independent of the different anion structures (Table 5.5). This means that during photolysis of a given onium salt group, identical amounts of Brönsted acid per unit time are generated.

Table 5.5 Quantum yields of diaryliodonium salts[789]

Diaryliodonium salt	$\Phi_{313.0\,nm}$	$\Phi_{365.0\,nm}$
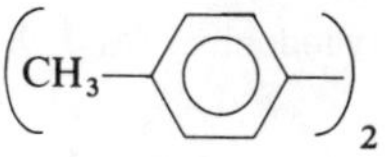 $\left(CH_3-\!\!\!\bigcirc\!\!\!-\right)_2 I^+AsF_6^-$	0.21	
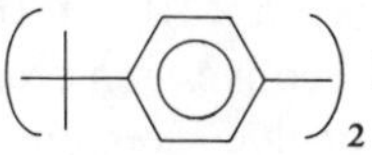 $\left(+\!\!\!\bigcirc\!\!\!-\right)_2 I^+AsF_6^-$	0.20	0.19
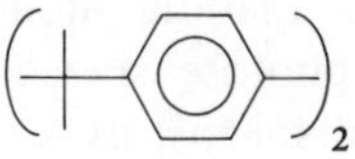 $\left(+\!\!\!\bigcirc\!\!\!-\right)_2 I^+PF_6^-$	0.22	
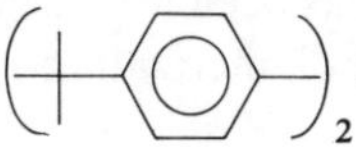 $\left(+\!\!\!\bigcirc\!\!\!-\right)_2 I^+SbF_6^-$	0.22	

Onium salts with common substituents on the aromatic rings absorb strongly in the region of 230–260 nm, with very little absorption above 300 nm. In contrast to the situation with diazonium salts, simple substituents on the aromatic rings of onium salts have very little effect on their absorption (Fig. 5.4).

The photolysis sensitivity of diaryliodonium salts, for example, can be increased by using different types of dyes (Table 5.6) as photosensitizers.[790] Figure 5.5 shows the acridine orange and benzoflavin-sensitized photolysis of 4,4-dimethyldiphenyliodonium hexafluoroarsenate with the photolysis of the same salt in the absence of a dye.

Only small amounts (0.07 M) of the dyes are necessary to sensitize diaryliodonium salt photolysis. In the case of 4,4-dimethyldiphenyliodonium

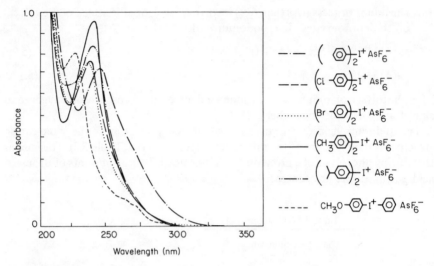

Fig. 5.4   Ultraviolet absorption spectra of substituted diaryliodonium ($Ar_2I^+X^-$) salts[789]

hexafluoroarsenate photolysis sensitized by benzoflavin, the conversion curve exhibits a maximum at the concentration of the dye of 3.5×10^{-3} M (Fig. 5.6). As the concentration of dye is increased, the percentage of conversion decreases. This effect can be explained as dye–dye quenching and/or dye screening effects.

Triarylsulphonium salts ($Ar_3S^+X^-$), like diaryliodonium salts ($Ar_2I^+X^-$) do not absorb light in the visible range. It has been found that diaryliodonium salts (with more favourable reduction potentials) can be photosensitized by a larger range of compounds than can triarylsulphonium salts. Most dyes which show pronounced sensitizing activity with diarylsulphonium salts are completely inactive when applied to triarylsulphonium salts. It has been found that perylene can photosensitize triarylsulphonium salt cationic polymerization in the range 390–475 nm.[783, 794]

Free radical photoinitiators are able to act as photosensitizers (P) for onium salts because the radicals formed on photolysis are powerful electron donors:[2, 783, 2233, 2881, 2888, 2893]

$$P + h\nu \rightarrow P^* \tag{5.101}$$

$$P^* + Ar_3S^+X^- \rightarrow [P \cdots Ar_3S^+X^-]^* \tag{5.102}$$

$$\text{EXCIPLEX}$$

Exciplex formed can further decompose by one of the following mechanisms:

(i) bond cleavage:

$$[P \cdots Ar_3S^+X^-]^* \rightarrow Ar_2S^{+\cdot} X^- + Ar\cdot + P \tag{5.103}$$

$$Ar_2S^{+\cdot} X^- + M \text{ (monomer)} \rightarrow H^+X^- \xrightarrow{+M} \text{POLYMER} \tag{5.104}$$

Table 5.6 Sensitizing dyes for the photodecomposition of diaryliodonium salts[790]

Dye	Structure	Max absorption (nm)
Acridine orange	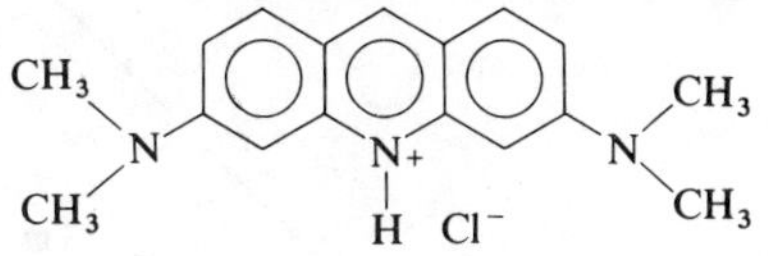	539
Acridine yellow	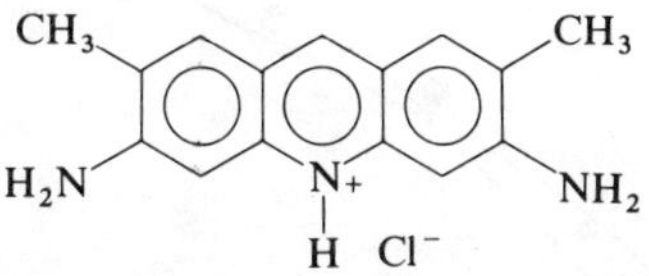	411
Phosphine R	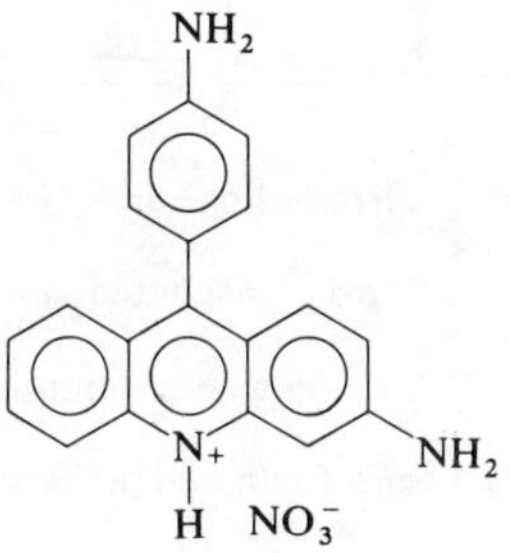	492
Benzoflavin	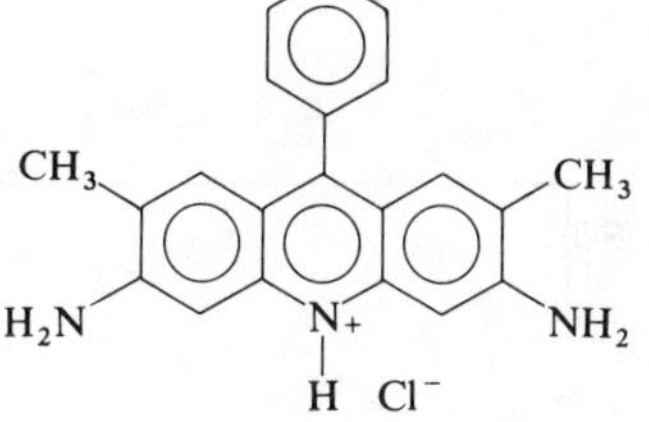	460
Setoflavin T	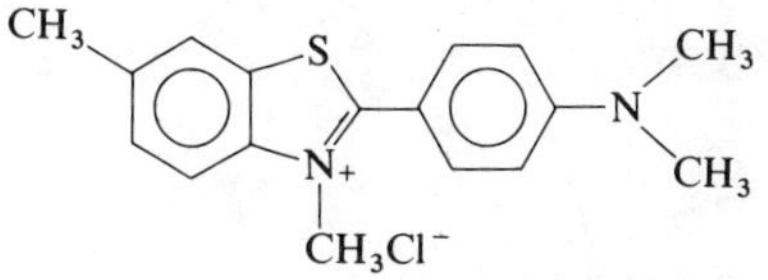	380

(ii) electron transfer:

$$[P \cdots Ar_3S^+X^-]^* \rightarrow Ar_3S\cdot + P^{\cdot+}X^- \tag{5.105}$$

$$P^{\cdot+}X^- + M \rightarrow H^+X^- \xrightarrow{+M} POLYMER \tag{5.106}$$

$$Ar_3S\cdot \rightarrow Ar_2S + Ar\cdot \tag{5.107}$$

202

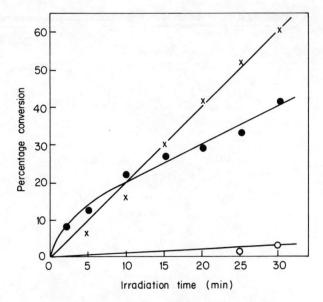

Fig. 5.5 Dye-sensitized photodecomposition of $(CH_3-\langle\bigcirc\rangle)_2\ I^+AsF_6^-$ salt in acetonitrile solution in the presence of: ($\times$) benzoflavin and acridine orange ($\bullet$), ($\circ$) without dye[790]

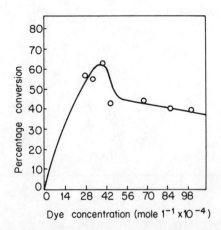

Fig. 5.6 Dye-sensitized photodecomposition of $(CH_3-\langle\bigcirc\rangle)_2\ I^-AsF_6^-$ salt in tetrahydrofuran with various concentrations of benzoflavin[790]

For electron transfer to be energetically favourable, the excitation energy of the photosensitizer (P*) must be greater than the net energy required to oxidize the photosensitizer, and reduce the onium salt.[2889]

The free energy of this photosensitized electron transfer (ΔG) can be calculated from the oxidation (E_{ox}) and excitation (E_p) energies of the photosensitizer and the reduction energy of the onium salt (E_{red}):[225, 1230, 2888, 2893]

$$\Delta G = E_{ox} - E_p - E_{red} \qquad (5.108)$$

The rate of electron transfer approaches diffusion control when ΔG is more negative than $-10\,\mathrm{kcal\,mol^{-1}}$. The sensitizer systems with ΔG larger than $-10\,\mathrm{kcal\,mol^{-1}}$ exhibit relatively large quantum yields of photolysis and high relative polymerization rates (Tables 5.7 and 5.8).

Table 5.7   Sensitized photolysis of diaryliodonium salt ($Ar_2I^+X^-$) to iodobenzene and H^{+}[1230]

Photosensitizer	Quantum yield of formation of iodobenzene	Quantum yield of formation of H^+	Relative polymerization rates (to perylene)	Free energy of photosensitized electron transfer ΔG(kcal mol^{-1})*
Anthracene	0.56	0.6	1.1	-45
Perylene	0.28	0.3	1.0	-41
Phenothiazine	0.42	1.0	0.7	-39
Xanthone	0.41	0.5	1.0	-27
Thioxanthone	0.32	0.7	2.3	-22
Acetophenone	0.05	0.1	0.5	-7
Benzophenone	0.01	0.2	0.8	-3
Benzophenone + tetrahydrofurane	3.3	8.0	3.7	
Benzophenone + isopropanol	2.4	5.0		

* For conversion energy units, use 1 kcal mol^{-1} = 4.1868 × 10^3 J

Chain reactions can increase the efficiency of diaryliodonium salt-initiated cationic polymerization.[1230, 2234, 2888] Chain reactions require initiation via the hydrogen abstraction process (i.e. most aromatic carbonyl photoinitiators (cf. Section 7.9.4.2)) and a substrate which will form electron-rich (oxidizable) free radicals. The reduction potentials of triphenyl-sulphonium salts are too large to enter into a chain process.

The addition of tetrahydrofurane to the benzophenone photosensitized polymerization of epoxy resin dramatically increases the relative polymerization rate in the presence of diaryliodonium salts but not in the presence of triarylphenylsulphonium salts. The following indirect electron transfer mechanism for this reaction has been proposed:[1230]

Table 5.8 Sensitized photolysis of triarylsulphonium $(Ar_3S^+X^-)$ to benzylsulphide and H^+[1230]

Photosensitizer	Quantum yield of formation benzyl-sulphide	Quantum yield of formation of H^+	Relative polymerization rates (to perylene)	Free energy of photosensitized electron transfer ΔG(kcal mol^{-1})*
Anthracene	0.59	1.0	1.0	-22
Perylene	0.20	0.3	1.0	-18
Phenothiazine	0.69	1.0	1.1	-16
Xanthone	<0.02	0.1	<0.1	-4
Thioxanthone	<0.02	0.3	<0.1	$+1$
Acetophenone	<0.02	0.1	<0.1	$+16$
Benzophenone	<0.02	0.3	<0.1	$+20$
Benzophenone + tetrahydrofurane			0.14	

* For conversion energy units, use 1 kcal mol^{-1} = 4.1868 × 10^3 J

(i) The triplet state of benzophenone (5.50) abstracts hydrogen from tetrahydrofurane (5.51) producing benzophenone ketyl radical (5.52) and tetrahydrofurane radical (5.53):

$$\phi-\overset{O}{\overset{\|}{C}}-\phi + h\nu \longrightarrow {}^1(\phi-\overset{O}{\overset{\|}{C}}-\phi) \longrightarrow {}^3(\phi-\overset{\overset{\cdot}{O}}{\overset{\|}{C}}-\phi) \qquad (5.109)$$

$$(5.50)$$

$$\overset{3}{(\phi-\overset{O}{\overset{\|}{\underset{\cdot}{C}}}-\phi)} + \underset{O}{\square} \xrightarrow{\text{hydrogen abstraction}} \phi-\overset{OH}{\underset{\cdot}{\overset{|}{C}}}-\phi + \underset{O}{\square}\cdot \qquad (5.110)$$

$$(5.51) \qquad\qquad (5.52) \qquad (5.53)$$

(ii) Both radicals (5.52 and 5.53) are easily oxidized by diaryliodonium salts, resulting in the formation of possible initiating species (5.54 and 5.55) for cationic polymerization:

$$\phi-\overset{OH}{\underset{\cdot}{\overset{|}{C}}}-\phi + Ar_2IAsF_6 \xrightarrow{\text{electron transfer}} \phi-\overset{OH}{\underset{+}{\overset{|}{C}}}-\phi + Ar-I + Ar\cdot + AsF_6^- \quad (5.111)$$

$$\phi-\overset{OH}{\underset{+}{\overset{|}{C}}}-\phi \longrightarrow \phi-\overset{O}{\overset{\|}{C}}-\phi + H^+ \qquad (5.112)$$

$$(5.54)$$

$$H^+ + \text{monomer} \longrightarrow \text{polymerization} \qquad (5.113)$$

$$\underset{O}{\square}\cdot + Ar_2IAsF_6 \xrightarrow{\text{electron transfer}} \underset{O}{\square}{}_+ + Ar-I + Ar\cdot + AsF_6^- \qquad (5.114)$$

$$(5.55)$$

$$\underset{O}{\square}_+ + \text{monomer} \longrightarrow \text{polymerization} \tag{5.115}$$

(iii) Hydrogen abstraction by aryl radicals ($Ar\cdot$) propagates chain:

$$Ar\cdot + \underset{O}{\square} \xrightarrow{\underset{\text{abstraction}}{\text{hydrogen}}} Ar{-}H + \underset{O}{\square}\cdot \tag{5.116}$$

In another photosensitizing system, benzophenone–propanol, the following mechanism for indirect electron transfer has been proposed:[2080, 2881, 2888, 2893, 2894]

$$^3(\phi{-}\overset{\overset{\bullet}{O}}{\underset{\bullet}{C}}{-}\phi) + (CH_3)_2CHOH \xrightarrow{\underset{\text{abstraction}}{\text{hydrogen}}} \phi{-}\overset{OH}{\underset{\bullet}{C}}{-}\phi + (CH_3)_2\overset{\bullet}{C}OH \tag{5.117}$$

$$\phi{-}\overset{OH}{\underset{\bullet}{C}}{-}\phi + Ar_2IAsF_6 \longrightarrow \phi{-}\overset{OH}{\underset{+}{C}}{-}\phi + Ar_2I\cdot + AsF_6^- \tag{5.118}$$

$$(CH_3)_2\overset{\bullet}{C}OH + Ar_2IAsF_6 \longrightarrow (CH_3)_2\overset{+}{C}OH + ArI\cdot + AsF_6^- \tag{5.119}$$

$$Ar_2I\cdot \longrightarrow Ar{-}I + Ar\cdot \tag{5.120}$$

$$Ar\cdot + (CH_3)_2CHOH \longrightarrow Ar{-}H + (CH_3)_2\overset{\bullet}{C}OH \tag{5.121}$$

$$\phi{-}\overset{OH}{\underset{+}{C}}{-}\phi \longrightarrow \phi{-}\overset{O}{\overset{\|}{C}}{-}\phi + H^+ \tag{5.122}$$

$$(CH_3)_2\overset{+}{C}OH \longrightarrow (CH_3)_2CO + H^+ \tag{5.123}$$

The features of the indirect electron transfer mechanism are:

(i) generation of radicals by hydrogen abstraction,
(ii) oxidation of the radicals by the iodonium salt to produce protons,
(iii) regeneration of radicals in a chain process involving hydrogen abstraction by aryl radicals from dissociation of diaryliodine radical.

The direct electron transfer process constitutes an amplification of photons in the formation of protons (or other reactive cations), which are further amplified by initiating polymerization of reactive monomers.

Using the diaryliodonium $(Ar_2I^+X^-)$[781,782,784,789], triarylsulphonium $(Ar_3S^+X^-)$[782,784,792,801,805,2925] and triarylselenium $(Ar_3Se^+X^-)$[782,793] salt photoinitiators the polymerization of any cationically polymerizable monomer can be carried out, such as olefins, dienes, epoxides, cyclic ethers, sulphides, acetals, and lactones (Table 5.9).

Table 5.9   Examples of photoinitiated cation polymerizations using diaryliodonium salt photoinitiators[781]

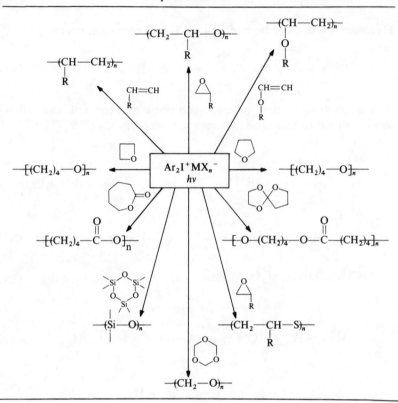

Onium salts have also been used to design novel photoimaging systems.[785,1819] (cf. Section 12.8.2).

Polymerization proceeds for some time after u.v. irradiation, and this process depends on the type of monomer and photoinitiator used, and the temperature.

The mechanism of cation–radical polymerization initiated by products of photolysis of onium salts is not well established yet. Several suggestions have been proposed:

(i) Cationic polymerization can be initiated by strong protonic Brönsted acids of the type HBF_4, $HAsF_6$, HPF_6, $HSbF_6$, etc.:[782]

$$M \text{ (monomer)} + HX \rightarrow HM^+X^- \tag{5.124}$$

$$M + HM^+X^- \rightarrow HMM^+X^- \tag{5.125}$$

$$nM + HMM^+X^- \rightarrow H\text{-}M(M)_{n+1}^+ X^- \tag{5.126}$$

The observed polymerization rates depend only on the nature of the acid which is produced. The order of reactivity is thus:

$$SbF_6^- > AsF_6^- > PF_6^- > BF_4^-$$

Diaryliodonium and triarylsulphonium salts exhibit different activities in the presence of various photosensitizers[1230,1231,2889] and rather similar differences are observed in rates of decomposition of the same salts when promoted by reducing organic radicals.[2233,2234,2237] These results give little doubt that Brönsted acid is primarily responsible for the initiation of cationic polymerization.

(ii) Cationic polymerization can be initiated by the cation–radicals formed in a chain reaction from a solvent.

(iii) Radical polymerization can be initiated by initially formed cation–radical fragments ($Ar_2I^{+\cdot}$, $Ar_2S^{+\cdot}$, $Ar_2Se^{+\cdot}$), which may abstract hydrogen from a monomer (MH) and initiate free radical polymerization:[782,788,789,2894]

$$Ar_2I^{+\cdot} + MH \rightarrow Ar_2I^+H + M\cdot \qquad (5.127)$$

$$Ar_2S^{+\cdot} + MH \rightarrow Ar_2S^+H + M\cdot \qquad (5.128)$$

(iv) Onium salts may fragment on electron transfer reduction to given free radicals:[2, 2233]

$$Ar_2I^+ + e^- \rightarrow Ar_2I\cdot \rightarrow ArI + Ar\cdot \qquad (5.129)$$

$$Ar_2S^+ + e^- \rightarrow Ar_2S\cdot \rightarrow ArS + Ar\cdot \qquad (5.130)$$

It has been reported that triarylsulphonium iodide, chloride, and tetrafluoroborate[2108,2163] photoinitiate free radical polymerization of methyl methacrylate and styrene.

Several other onium salts have been proposed as useful photoinitiators for cation–radical polymerization, such as:

(i) Dialkylphenacylsulphonium salts (5.56)[783,795,804]

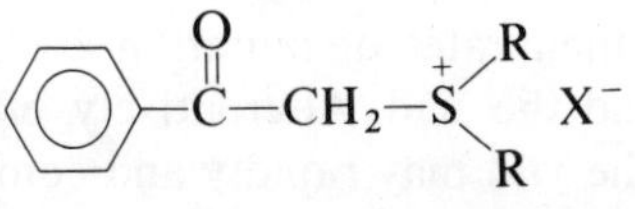

$$(5.56) \quad (\text{where } X^- = BF_4^-, AsF_6^-, SbF_6^-, PF_6^-)$$

for polymerization of difunctional epoxy monomers.

(ii) Hydroxyphenylsulphonium salts (5.57)[783, 797, 802]

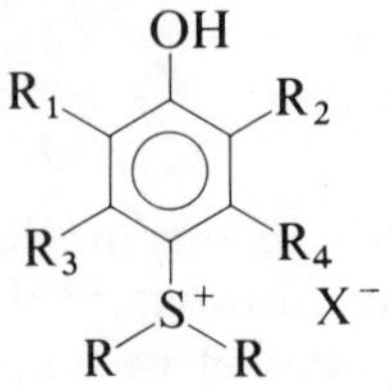

$$(5.57)$$

(where $X^- = BF_4^-$, AsF_6^-, SbF_6^-, ClO_4^-)

for polymerization of epoxides, cyclic acetals, and vinyl ethers. The photolysis of dialkyl-4-hydroxyarylsulphonium salt (5.58) occurs with formation of the ylid (5.59) and acid HBF_4:

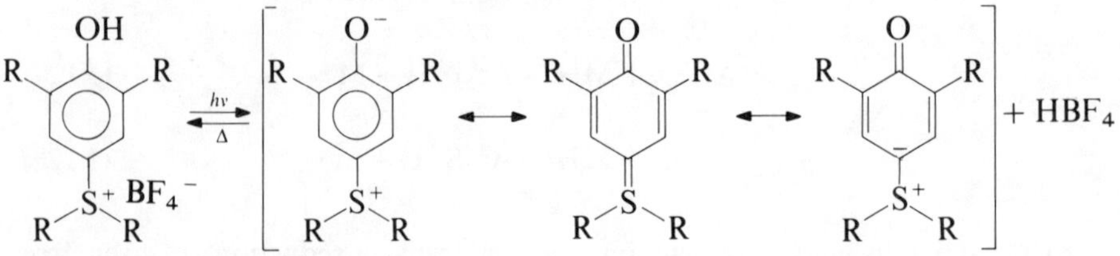

$$(5.131)$$

Polymerization can be initiated both by the ylid and/or solvated acid HBF_4. The appearance of a strong band at 272 nm in polycyclohexane oxide polymers produced using dialkyl-4-hydroxyphenylsulphonium salt photo-initiators is strongly indicative of the presence of phenoxy end groups derived from the photoinitiator in these materials. The presence of phenoxy end groups in polymers may be determined by their absorption at 273 nm.[3160]

The polymerization is typically inhibited by bases as tertiary amines and unaffected by radical inhibitors such as alkylated phenols and catechols, which indicates that polymerization occurs rather by a cationic than a radical mechanism.

Tetrahydrofuran, ε-caprolactam and α-methylstyrene did not undergo polymerization in the presence of dimethyl(3,5-dimethyl-4-hydroxyphenyl) sulphonium hexafluoroarsenate. In this case the monomers may fail to polymerize because their rates of initiation are slower than the recombination of the ylid and the acid. Alternatively, efficient termination of the growing chains by the ylid may rapidly and completely remove the active cationic species from the polymerization mixture.

Aromatic hydrocarbons (anthracene, 1,2-benzanthracene, perylene, etc.) photosensitize the photolysis of dialkylphenacylsulphonium salts, while aryl ketones (benzophenone, 2-chlorothioxanthone, 9-fluorenone, thioxanthone, etc.) photosensitize the photolysis of dialkyl(4-hydroxyphenyl)sulphonium salts.[802]

(iii) Thiopyrylium salts (5.60):[2007]

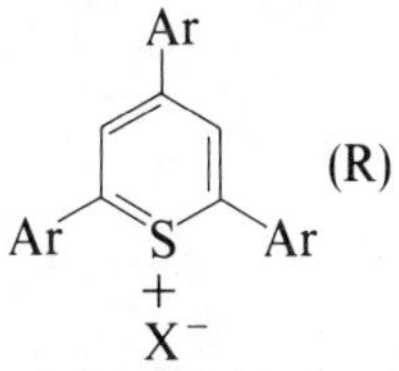

(5.60)

(where $X^- = BF_4^-$, PF_6^-)

for polymerization of epoxides.

The related thioxanthylium salts were also examined as potential photoinitiators, but were found too hydrolytically unstable to be useful.[1011]

(iv) Bis[(4-diphenylsulphonio)phenyl]sulphide-bis-hexafluorophosphate (5.61)[3814]

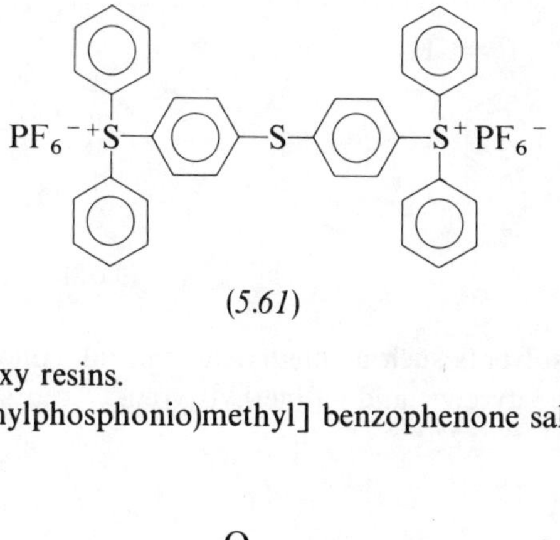

(5.61)

for curing expoxy resins.

(v) p,p-Bis[(triphenylphosphonio)methyl] benzophenone salt (5.62):[2667]

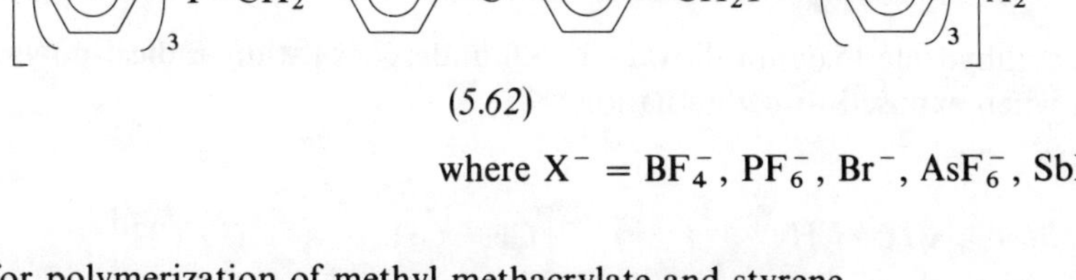

(5.62)

where $X^- = BF_4^-$, PF_6^-, Br^-, AsF_6^-, SbF_6^-

for polymerization of methyl methacrylate and styrene.

(vi) Diphenylsulphonium bis(methoxycarbonyl)methylide (5.63) which under u.v. irradiation is decomposed into phenyl radical and cation radical (5.64).[7105]

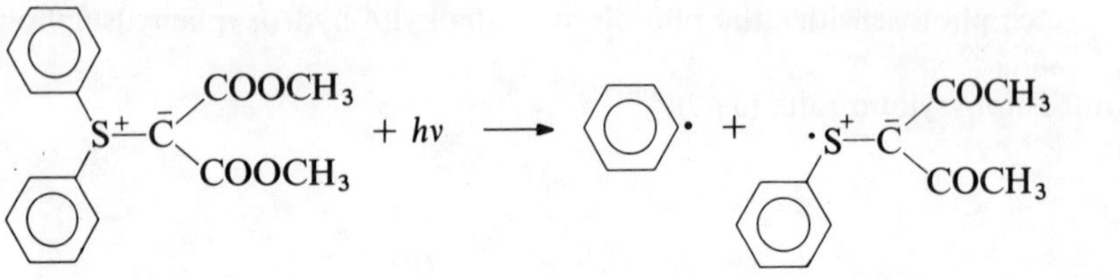

$$(5.132)$$

Both radicals may initiate polymerization of methyl methacrylate and styrene.

5.4.4. Photoinitiated Cationic Polymerization of Styrenes

During u.v. irradiation of styrene, α-methylstyrene or p-methylstyrene in polar solvents such as acetonitrile or dimethoxymethane photodimerization into 1-phenyl-1,2,3,4-tetrahydronaphthalene (tetralin) (5.65) occurs via cation radicals:[234-236,3956]

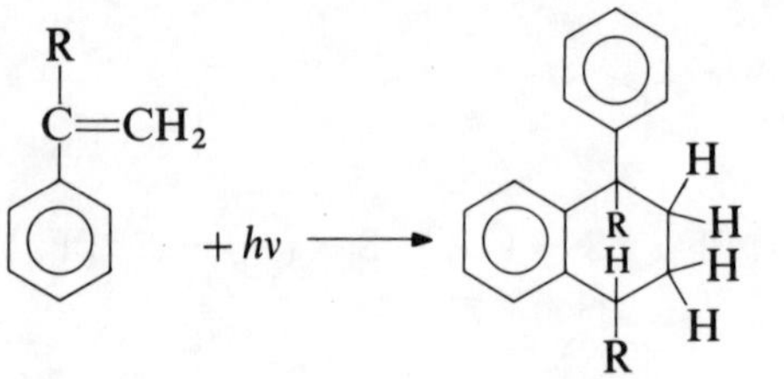

$$(5.133)$$

In halogenated solvents such as methylene chloride, photoinitiated cationic polymerization of styrene and α-methylstyrene occurs in parallel with photodimerization.[1370,1548,1782]

5.4.5. Photopolymerization of Maleic Anhydride in Dioxane

Maleic anhydride (5.66) in dioxane (5.67) undergoes cation–radical polymerization when exposed to u.v. radiation.[2606]

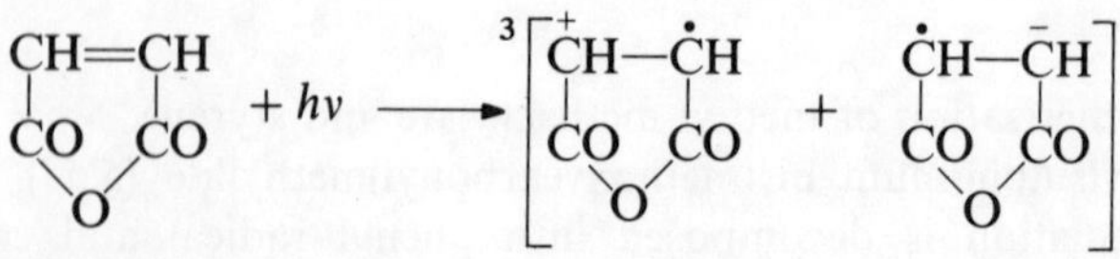

$$(5.134)$$

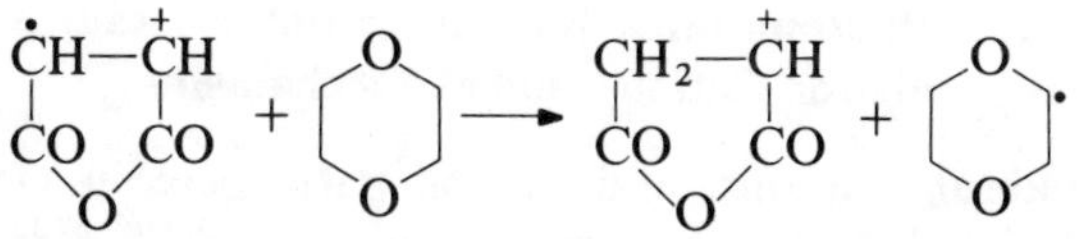

$$\underset{(5.67)}{\overset{\bullet}{C}H-\overset{+}{C}H} + \underset{}{\bigcirc} \longrightarrow \underset{(5.68)}{CH_2-\overset{+}{C}H} + \underset{}{\bigcirc}\!\!\bullet \tag{5.135}$$

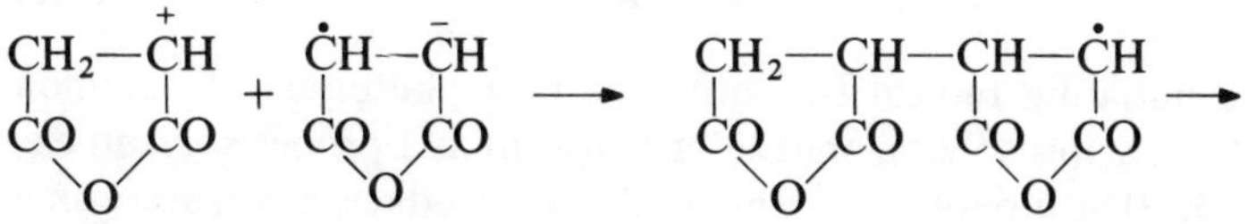

$$CH_2-\overset{+}{C}H + \overset{\bullet}{C}H-\overset{-}{C}H \longrightarrow CH_2-CH-CH-\overset{\bullet}{C}H \longrightarrow \text{propagation} \tag{5.136}$$

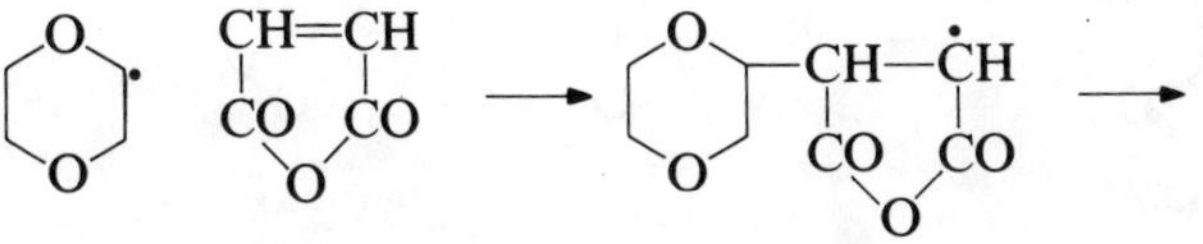

$$\bigcirc\!\!\bullet \quad CH{=}CH \longrightarrow \bigcirc\!-CH-\overset{\bullet}{C}H \longrightarrow \text{propagation} \tag{5.137}$$

The low molecular weight of the oligomer (5.69) suggests that most of the oligomers terminate by the chain transfer to dioxane:

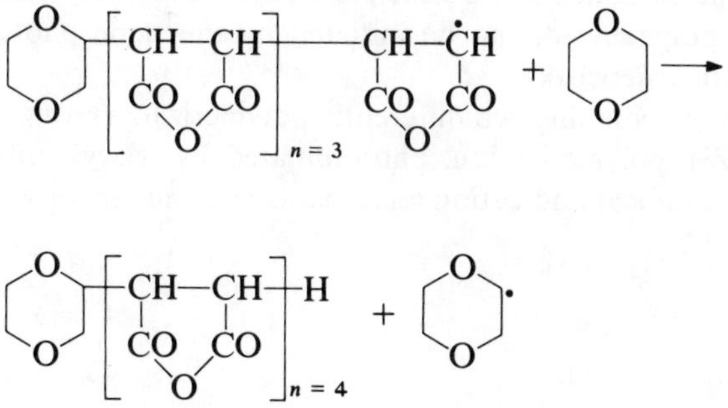

$$\bigcirc\!\!-\!\!\left[\underset{CO\ CO}{CH-CH}\right]_{n=3}\!\!\!-\underset{CO\ CO}{CH-\overset{\bullet}{C}H} + \bigcirc \longrightarrow \tag{5.138}$$

$$\bigcirc\!\!-\!\!\left[\underset{CO\ CO}{CH-CH}\right]_{n=4}\!\!\!-H + \bigcirc\!\!\bullet$$

(5.69)

and subsequent propagation will occur from dioxane radicals (5.68). An additional product of low molecular oligomers are cyclic dimers (5.70), which are produced according to the reaction:

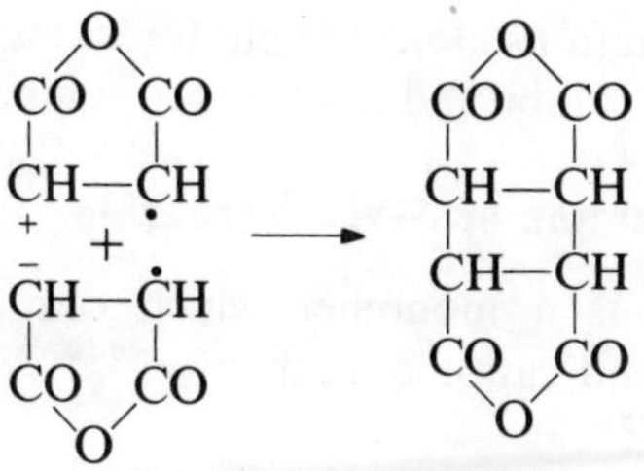

$$\begin{array}{c}\underset{CO\ CO}{\overset{O}{\frown}}\\ CH-\overset{\bullet}{C}H \\ \overset{-}{C}H-\overset{+}{\overset{\bullet}{C}H} \\ \underset{O}{CO\ CO}\end{array} \longrightarrow \begin{array}{c}\underset{CO\ CO}{\overset{O}{\frown}}\\ CH-CH \\ CH-CH \\ \underset{O}{CO\ CO}\end{array} \tag{5.139}$$

(5.70)

5.4.6. Photoinitiated Polymerization by Radical and/or Cation–radical Mechanisms

Organic free radicals, cations, and anions participate in electron transfer equilibria which may be represented as in equation:[273,500,2233]

$$R^+ \underset{-e}{\overset{+e}{\rightleftarrows}} R\cdot \underset{-e}{\overset{+e}{\rightleftarrows}} R^- \tag{5.140}$$

For example the initiating radical (R·) obtained in a photolysis of common photoinitiators (e.g. azobisisobutyronitrile or benzoin) in a proper solvent, e.g. tetrahydrofuran (5.71) or n-butyl vinyl ether (5.72) can produce cation according to reactions:[2]

$$R\cdot + \underset{(5.71)}{\square_O} \longrightarrow RH + \square_{O}\cdot \xrightarrow{-e} \square_{O^+} \tag{5.141}$$

$$R\cdot + CH_2{=}CHOR^1 \longrightarrow RCH_2\overset{\cdot}{C}HOR^1 \longrightarrow RCH_2\overset{+}{C}H{=}OR^1 \tag{5.142}$$
(5.72)

Two different monomers, one which is free radically polymerized and another cationically polymerized, can be initiated by the same photo-initiator giving interpenetrating networks.

If a monomer contains two differently polymerizable groups, such as glycidyl acrylate (5.73), polymerization, photoinitiated by triaryl sulfonium salt, may occur by free radical and cationic mechanisms simultaneously:[785,2925]

$$
\begin{array}{ccc}
CH_2{=}CH & & {-}CH_2{-}CH{-} \\
| & & | \\
C{=}O & & C{=}O \\
| & & | \\
O & \xrightarrow[Ar_3S^+x^-]{+h\nu} & O \\
| & & | \\
CH_2 & & CH_2 \\
| & & | \\
CH\diagdown_O & & {-}CH \\
| & & | \\
CH_2 & & CH_2{-}O{-}
\end{array}
\tag{5.143}
$$
(5.73)

Simultaneous radical and cationic polymerizations are superior to free radical photoinitiated polymerizations alone for photoresist applications, because they are extremely rapid and exhibit reduced oxygen inhibition effects.

5.4.6.1. Photopolymerization of N-vinylcarbazole

N-vinylcarbazole (5.74) is a monomer which can photopolymerize simultaneously by a radical and cationic mechanism.[519, 637, 1483, 2230, 3329, 3330, 3511–3513, 3577, 3588, 3595]

In oxygen-free solvents such as methanol, acetone, ethyl methyl ketone, tetrahydrofuran, acetonitrile, sulpholane, N,N-dimethylformamide, or dimethylsulphoxide both free radical and cationic mechanism occur. Addition of a free radical scavenger such as 1,1-diphenyl-2-picrylhydrazyl greatly reduces the polymerization rate:

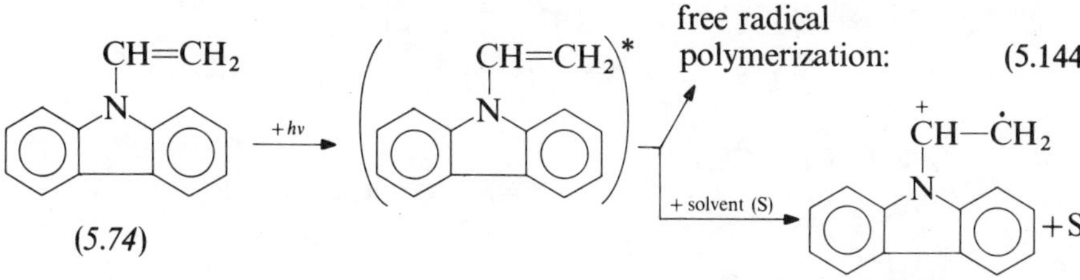

$$\text{(5.74)} \xrightarrow{+h\nu} \left(\text{...} \right)^* \begin{array}{l} \text{free radical} \\ \text{polymerization:} \quad (5.144) \end{array}$$

$$\xrightarrow{+\text{solvent (S)}} \quad (5.75) \quad +S^{\bar{\cdot}} \quad (5.145)$$

The major reaction of the cation–radical (5.75) is cyclodimerization (cf. reaction 5.63) accompanied by minor cationic polymerization.

In the presence of oxygen in polar solvents such as methanol, acetone, ethyl methyl ketone, or acetonitrile the main reaction is free radical polymerization.[3511,3512] Oxygen acts as an electron acceptor:

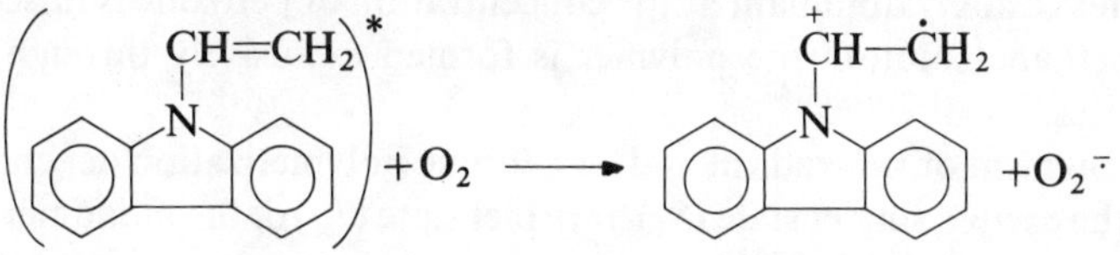

$$+O_2 \longrightarrow \quad +O_2^{\bar{\cdot}} \quad (5.146)$$

N-vinylcarbazole can be simultaneously polymerized by two different radical and cationic mechanisms initiated by the same photoinitiators, e.g. benzoyl peroxide (5.76)[33, 2173] or azobisisobutyronitrile (5.77)[1897, 1898, 3595] which can

(i) be photolysed into free radicals (which initiate free radical polymerization) (cf. Section 7.12),
(ii) form in the ground state an charge–transfer (CT) complex with monomer (cf. Section 5.3) (which after light excitation can further initiate cation–radical polymerization):

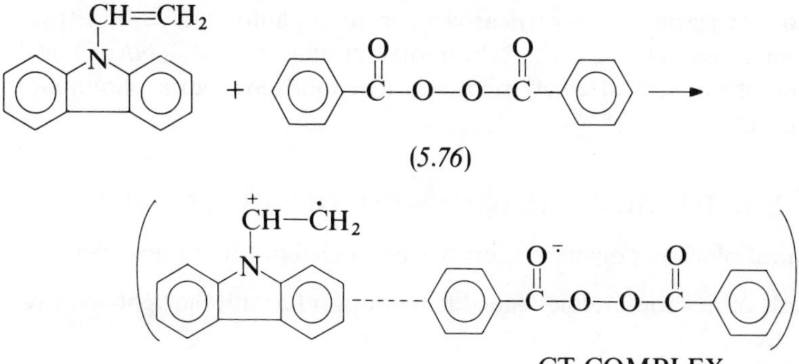

(5.76)

CT COMPLEX

$$\xrightarrow{+h\nu} \quad \text{cationic polymerization} \quad (5.147)$$

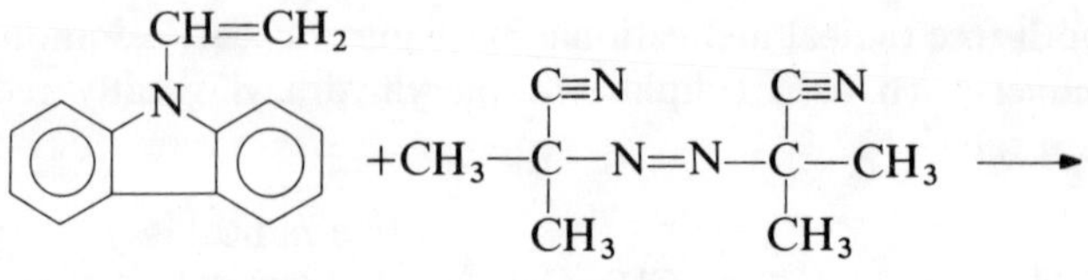

$$
\text{(5.77)}
$$

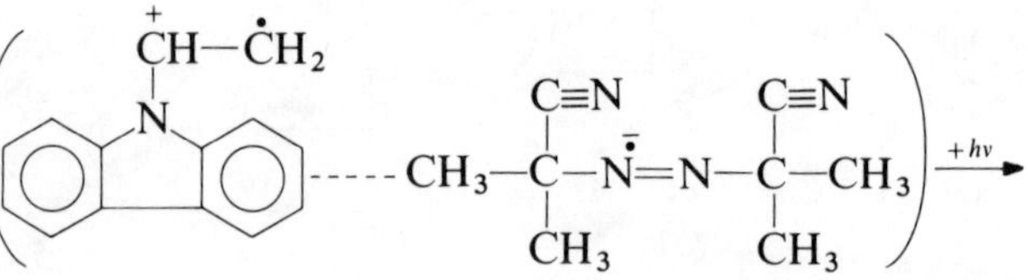

$$
\text{CT COMPLEX} \qquad \text{(5.148)}
$$

cationic polymerization

In benzene solution, polymer is formed mainly by a radical mechanism at low concentrations of benzoyl peroxide, but its formation by cationic mechanisms becomes increasingly dominant as the concentration of peroxide is raised.[2173] In dichloromethane solution the polymer is formed exclusively through cationic intermediates.

The mechanism of this radical and/or cationic polymerization depends on the nature of the excited singlet state (S_1) or triplet state (T_1) of the monomer, and can be summarized as follows:[1898]

(i) If the first excited triplet state of the monomer is dynamically quenched by the initiator, it later behaves as an electron acceptor, and cationic propagation results.

(ii) If the first excited singlet state of the monomer is dynamically quenched by the initiator then energy transfer followed by homolytic bond fission occurs, and free radical propagation ensues.

(iii) If the monomer and initiator form a ground state donor–acceptor complex, excitation of this system will always lead to cationic polymerization.

Cationic polymerization of N-vinylcarbazole can be photoinitiated be tetra-n-butylammonium tetrachloro (or tetrabromo)aurate(III) $((n\text{-}C_4H_9)_4\,NAuCl_4)$.[230] Gold(III) complexes are extremely photosensitive, and undergo a photoredox decomposition.[25, 232, 3118, 3930]

5.5. PHOTOCHEMICAL BLOCK COPOLYMERIZATION

Photochemical block copolymerization can be carried out by several methods:

(i) Synthesis of macromolecules ended from both sides with the light-sensitive groups (X)

$$
\fbox{X}\!-\!\fbox{POLYMER A}\!-\!\fbox{X}
$$

The photodissociation of groups X normally produces two free radicals, of which only one is a macroradical. Upon further u.v. irradiation in the presence of monomer B, the block copolymer BAB can be obtained:

$$\boxed{X}-\boxed{\text{POLYMER B}}-\boxed{\text{POLYMER A}}-\boxed{\text{POLYMER B}}-\boxed{X}$$

(ii) Synthesis of macromolecules in which the light-sensitive groups (X) are incorporated within the polymer chain (Table 5.10).

$$\boxed{\text{POLYMER A}}-\boxed{X}-\boxed{\text{POLYMER A}}$$

Upon u.v. irradiation the photoreactive group dissociates into free radicals, e.g. a ketooxime group:[2210]

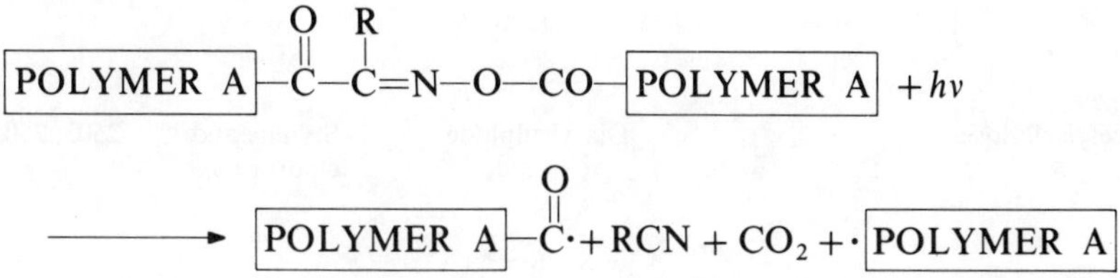

$$\boxed{\text{POLYMER A}}-\overset{\overset{\displaystyle O}{\|}}{C}-\overset{\overset{\displaystyle R}{|}}{C}=N-O-CO-\boxed{\text{POLYMER A}}+h\nu$$

$$\longrightarrow \boxed{\text{POLYMER A}}-\overset{\overset{\displaystyle O}{\|}}{C}\cdot+RCN+CO_2+\cdot\boxed{\text{POLYMER A}}$$

$$(5.78) \qquad\qquad\qquad (5.79)$$

$$(5.149)$$

Macroradicals (5.78) and (5.79) may further initiate copolymerization of a monomer B to produce the AB type block copolymer:

$$\boxed{\text{POLYMER A}}-\overset{\overset{\displaystyle O}{\|}}{C}\cdot+B \longrightarrow \boxed{\text{POLYMER A}}-\overset{\overset{\displaystyle O}{\|}}{C}-\boxed{\text{POLYMER B}}$$

$$(5.150)$$

For example thiuramdisulphides (5.80) dissociate upon u.v. irradiation to free radicals, which may initiate polymerization of methyl methacrylate and styrene.[2832,2834,2835,2837,2840,3190,3639] They act as INITIATOR–TRANSFER AGENT–TERMINATOR (INIFERTER):

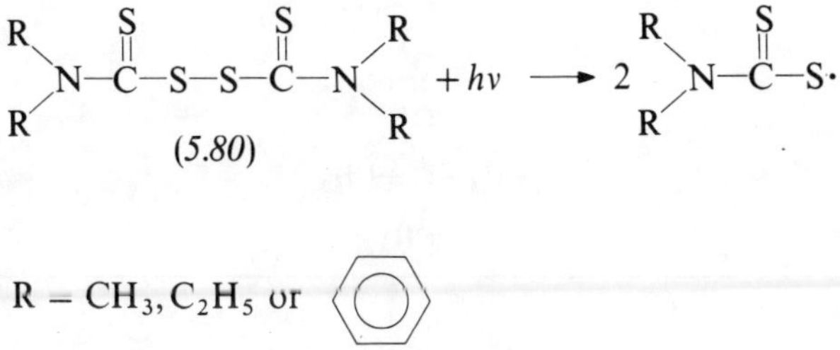

$$\overset{R}{\underset{R}{>}}N-\overset{\overset{\displaystyle S}{\|}}{C}-S-S-\overset{\overset{\displaystyle S}{\|}}{C}-N\overset{R}{\underset{R}{<}}+h\nu \longrightarrow 2\ \overset{R}{\underset{R}{>}}N-\overset{\overset{\displaystyle S}{\|}}{C}-S\cdot$$

$$(5.80)$$

$$R - CH_3, C_2H_5 \text{ or } \langle\bigcirc\rangle$$

$$(5.151)$$

Table 5.10   Examples of photochemical block copolymerization

Polymer A— X —Polymer A	Light-sensitive group	Monomer B	References
Poly(tetrachlorobisphenoladipate)	Ketooxime 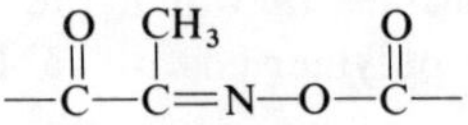	Styrene	2210
Polycarbonates	Trichloromethyl 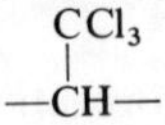	Acrylonitrile	1736
Polyamides or polyurethanes	Nitroso 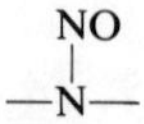	Vinyl monomers	775, 776
Triacetylcellulose	Diarylsulphide 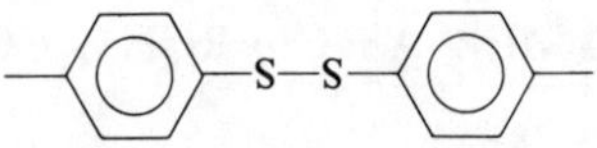	Styrene and chloroprene	2503, 2505
Polypeptides	Diarylsulphide 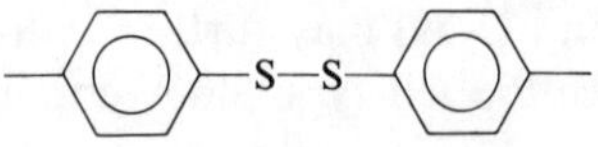	Vinyl monomers	3759
Co(styrene/p-vinylbenzopheonone-p-tert-butyl benzoate)	Benzophenone peresters 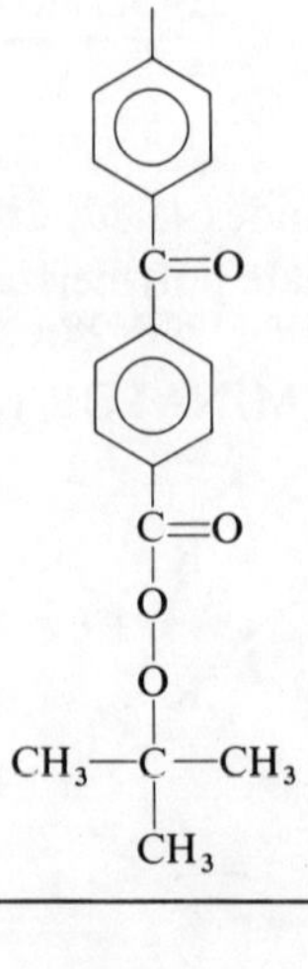	Methyl methacrylate	3, 1462, 1466, 2666

Ketooxime group:

$$-\overset{\overset{\displaystyle O}{\|}}{C}-\overset{\overset{\displaystyle CH_3}{|}}{C}=N-O-\overset{\overset{\displaystyle O}{\|}}{C}-$$

Trichloromethyl group:

$$-\overset{\overset{\displaystyle CCl_3}{|}}{CH}-$$

Nitroso group:

$$-\overset{\overset{\displaystyle NO}{|}}{N}-$$

Diarylsulphide group:

$$-\!\!\bigcirc\!\!-S-S-\!\!\bigcirc\!\!-$$

Benzophenone peresters group:

$$CH_3-\overset{\overset{\displaystyle CH_3}{|}}{\underset{\underset{\displaystyle CH_3}{|}}{C}}-CH_3$$

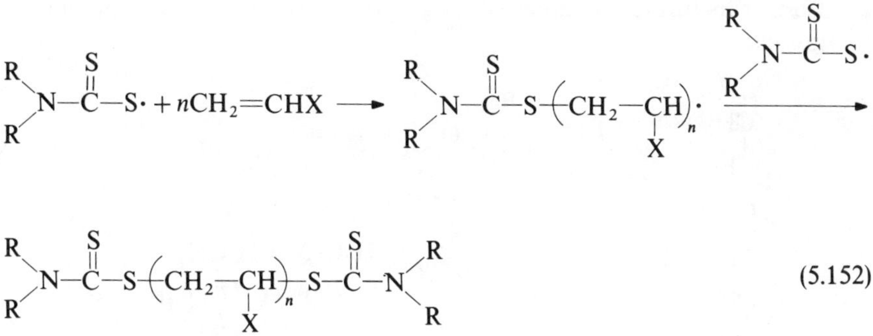

$$R \overset{R}{\underset{R}{\diagdown}} N - \overset{\overset{S}{\parallel}}{C} - S\cdot + n CH_2{=}CHX \longrightarrow \overset{R}{\underset{R}{\diagdown}} N - \overset{\overset{S}{\parallel}}{C} - S + CH_2 - \underset{X}{CH} \cdot_n \qquad \overset{R}{\underset{R}{\diagdown}} N - \overset{\overset{S}{\parallel}}{C} - S\cdot \longrightarrow$$

$$\overset{R}{\underset{R}{\diagdown}} N - \overset{\overset{S}{\parallel}}{C} - S + CH_2 - \underset{X}{CH}_n - S - \overset{\overset{S}{\parallel}}{C} - N \overset{R}{\underset{R}{\diagup}} \qquad (5.152)$$

POLYMER A

These α,ω-bifunctional polymers (POLYMER A) are useful telechelic polymers and potential chain-extending backbone polymers, applied to thermal and/or photochemical synthesis of increased molecular weight polymers and block copolymers. The polymer A, in the presence of another monomer B, upon u.v. irradiation can give the BAB type co-polymer (7.81):[2837,2839]

$$\overset{R}{\underset{R}{\diagdown}} N - \overset{\overset{S}{\parallel}}{C} - S - \boxed{\text{POLYMER B}} - \boxed{\text{POLYMER A}} - \boxed{\text{POLYMER B}} - S - \overset{\overset{S}{\parallel}}{C} - N \overset{R}{\underset{R}{\diagup}}$$

(7.81)

By this method prepolymer polystyrene has been transformed into block copolymer with poly(methyl methacrylate), poly(ethyl acrylate), poly(vinyl acetate)[2833,2838] and with poly(2-acryloamido-2-methylpropane sulphonic acid).[2110]

(iii) Synthesis of macromolecules with the help of initiators which can initiate polymerization photochemically and chemically. An interesting example of this type of reaction is photoinitiated polymerization of acrylamide by eosin which proceeds through the semiquinone of eosin to yield a growing chain of macromolecules terminated at one end by eosin (7.82):[3382]

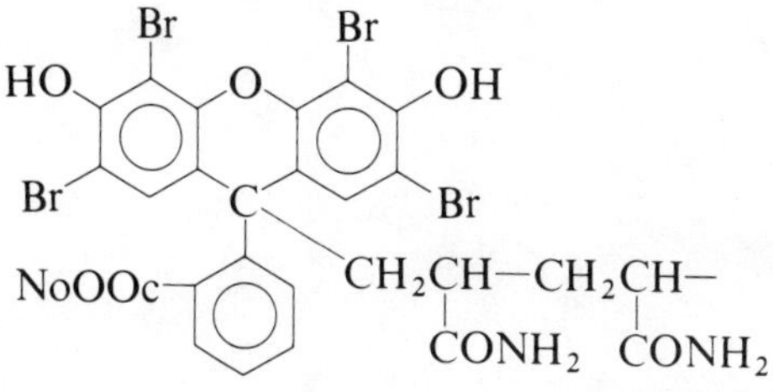

(7.82)

Eosin can further be attached to poly(vinyl amine) through the amino groups:

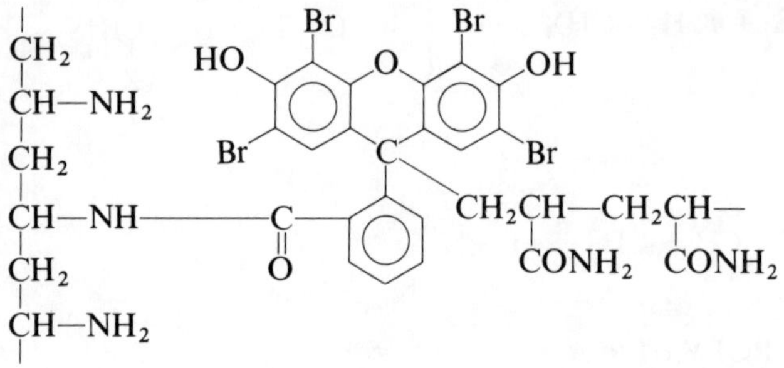

(7.83)

In this way block copolymer of poly(vinyl amine) with poly(vinyl amine) can be obtained (7.83).

(iv) Special photochemical methods, e.g. application of metal carbonyl/coinitiator system (cf. Section 7.21).

Photochemical block copolymerization methods have several advantages as:

 (i) Unlimited choice of different photosensitive groups, monomers, etc.
 (ii) Selectivity of initiation by use of a specific irradiation wavelength.
(iii) Carrying out of reactions in low temperatures, which avoids side and transfer chain reactions.
(iv) High yield of block copolymer.

The main disadvantages of these methods are:

 (i) Monomer sequences may have a molecular weight distribution characteristic for radical polymerization, and lack of homogeneity.
(ii) A high efficiency of block polymerization requires a high quantum yield of photolysis of the photosensitive group into free radicals.

6. Photochemistry of solvents

Many common solvents are usually considered to be inert in photochemical experiments. Only a few solvents, such as alcohols and paraffinic hydrocarbons, are unreactive when irradiated in the range 200–700 nm. Photochemistry of other solvents is discussed in detail in the following sections:

(a) organic halides—Section 6.2
(b) aromatic hydrocarbons (benzene)—Section 6.3
(c) tetrahydrofurane—Section 6.4, and dioxane—Section 6.5
(d) ketones—Section 7.9.

Many solvents form, with oxygen contact, charge-transfer (CT) complexes. Aromatic hydrocarbons, e.g. benzene[698, 1068, 1069, 1900, 1901, 3637], toluene,[3833] and even alkanes, e.g. cyclohexane,[1900, 1901, 2595] interact with oxygen and form CT complexes. These compounds, in the presence of oxygen, exhibit electronic absorption bands which are not characteristic of either the hydrocarbon or oxygen. The stabilization energy of these complexes is extremely low and they have therefore been termed CONTACT CHARGE-TRANSFER COMPLEXES.[1900, 1901, 2796] Excitation of the aromatic hydrocarbon–oxygen systems via the contact CT complexes absorption band gives rise to excited complexes.[1685] Oxygen quenches the excited singlet state (S_1) and/or triplet state (T_1), probably by the formation of CT complexes.[416, 498, 524, 1792, 1793, 2908, 2982, 3804, 3973] Oxygen quenches excimers of aromatic hydrocarbons.[416, 3973]

Oxygen quenching of aromatic hydrocarbon triplets states leads to the formation of the singlet oxygen[645, 769, 1121, 1351, 1352, 1987] (cf. Section 14.4).

The photochemical reactions of pure solvents in an oxygen-free atmosphere and in the presence of oxygen interfere in photochemical studies of polymers in solution. Solvent radicals can initiate radical polymerization instead of radicals formed from photodecomposition of initiator, and can completely change the mechanism and kinetics of observed reactions. Many of such examples are discussed in detail in Chapter 7. Knowledge of the photochemistry of solvents is necessary in order to understand all reactions which may be involved in the photoreactions of polymeric systems in solutions.

6.1. ALCOHOLS

Aliphatic alcohols have their longest absorption band at wavelengths below 200 nm $(\varepsilon = 10^2–10^3$ litre $g^{-1} cm^{-1})$, which is associated with an (n,π^*)

transition involving the promotion of a non-bonding electron of oxygen to an anti-bonding (π^*) group orbital.[591]

Photolysis of methanol, ethanol, isopropanol, and n-pentanol in the condensed phase becomes negligible under u.v. irradiation over 200 nm, whereas photolysis of methanol in the gas phase occurs by the following reactions:

$$CH_2O + H_2 \quad 20\% \tag{6.1}$$

$$CH_3O\cdot + H\cdot \tag{6.2}$$

$$CH_3OH \xrightarrow{+h\nu} \qquad 79\% $$

$$HOCH_2\cdot + H\cdot \tag{6.3}$$

$$CH_3\cdot + HO\cdot \quad 1\% \tag{6.4}$$

Radiolysis and photolysis of alcohols have been reviewed.[3412, 3413]

6.2. ORGANIC HALIDES

Organic halides having the structure CX_4 (e.g. CCl_4, CBr_4, CCl_3Br, etc.) are very easily photolysed to very reactive radicals such as $X\cdot$ and $CX_3\cdot$:[2, 2014–2018, 2325, 2481, 2934]

$$CCl_4 + h\nu \rightarrow Cl\cdot + CCl_3\cdot \tag{6.5}$$

In this reaction chlorine gas and hexachloroethane formed:

$$Cl\cdot + Cl\cdot \rightarrow Cl_2 \tag{6.6}$$

$$CCl_3\cdot + CCl_3\cdot \rightarrow C_2Cl_6 \tag{6.7}$$

$$Cl\cdot + CCl_4 \rightarrow Cl_2 + CCl_3\cdot \tag{6.8}$$

In the presence of methanol the following reaction occurs:[2934]

$$CCl_4 + CH_3OH \xrightarrow{+h\nu} CHCl_3 + CH_2O + HCl \tag{6.9}$$

A mixture of benzene–carbon tetrachloride shows the existence of charge-transfer addition (C_6H_6–CCl_4 and C_6H_6–$2CCl_4$).[192, 733, 2928, 3072]

In the presence of oxygen, formation of phosgene ($COCl_2$) may occur:[2325]

$$2CCl_4 + O_2 \xrightarrow{+h\nu} 2COCl_2 + Cl_2 \tag{6.10}$$

Both radicals—chlorine ($Cl\cdot$) and trichloromethyl ($CCl_3\cdot$)—are very reactive and may initiate, e.g., polymerization of styrene and vinyl acetate.[2015, 2017] More examples are given in other sections.

6.3. BENZENE

The absorption and emission spectroscopy of benzene has been studied in detail.[591, 808, 809] Absorption and fluorescence spectra are shown in Fig. 6.1, and the phosphorescence spectrum is given in Fig. 6.2.

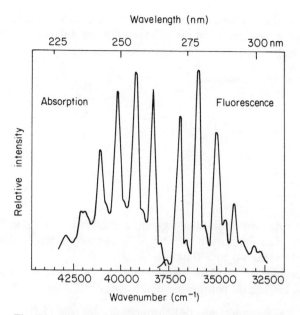

Fig. 6.1 Absorption and fluorescence spectra of benzene[809]

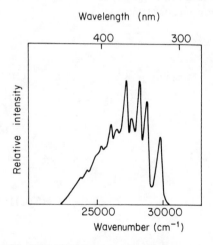

Fig. 6.2 Phosphorescence spectrum of benzene at 77K[809]

Benzene can form excimer[439, 809, 824, 1832] (cf. Section 3.2). Broad structureless emission on the long wavelength side of benzene monomer emission at room temperature is clearly observed at concentration of about 1 M, and further increasing concentration becomes the dominant component in the spectrum.

Benzene in an oxygen-free atmosphere under u.v. irradiation (254 nm) undergoes slow isomerization to yellow-coloured fulvene (6.1)[204,3803] and benzalvene (6.2)[3910]:

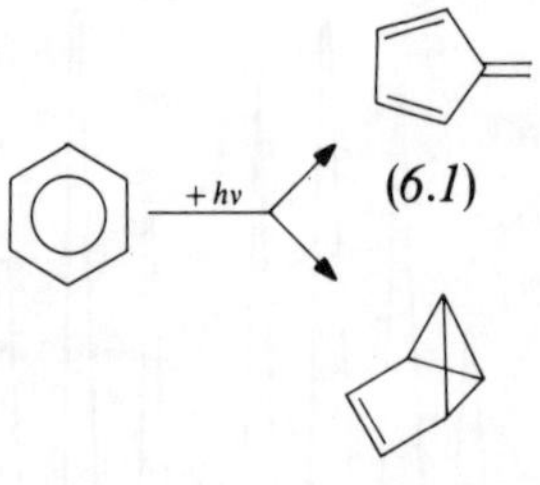

(6.1) (6.11)

           (6.12)

(6.2)

Benzene forms, with molecular oxygen, a contact CT complex.[698,1068,1069,1900,1901,3637] The formation of this CT complex extends the absorption spectrum of benzene up to 340 nm.

In the presence of oxygen, photooxidation of benzene occurs with the formation of long-chain conjugated dialdehydes:[3834]

(i) trans-trans-2,4-hexadiene-1,6-dial(mucondialdehyde) (6.3)
(ii) 2,4,6,8,10-dodecapentaene-1,12-dial (6.4)

The initial photochemical attack of oxygen on benzene leads to ring-opening as shown in the following mechanism:

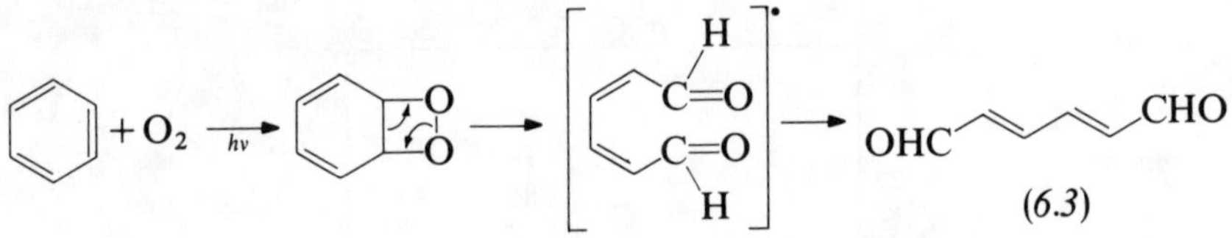

(6.3)

(6.13)

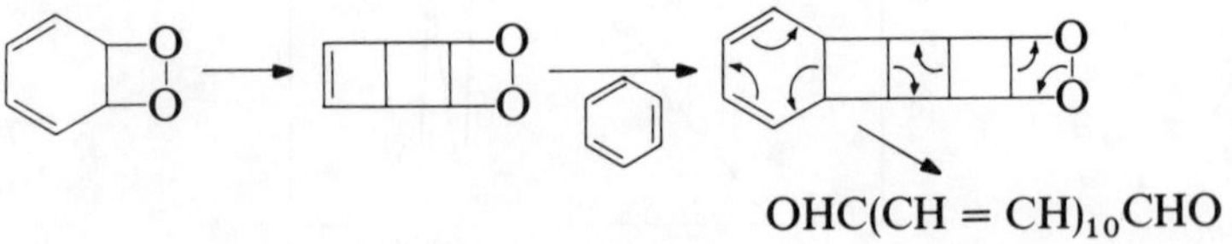

$OHC(CH = CH)_{10}CHO$

(6.4)

(6.14)

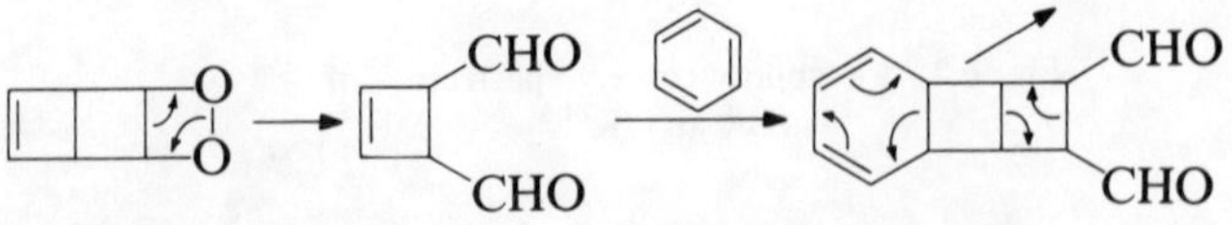

(6.15)

The long-chain conjugated dialdehydes (6.3) and (6.4) are strongly yellow-coloured and their spectra are shown in Fig. 6.3. Additional products of photooxidation of benzene are phenol and o-quinone.[1998] The detailed mechanism for the photooxidation of benzene is as yet unresolved.

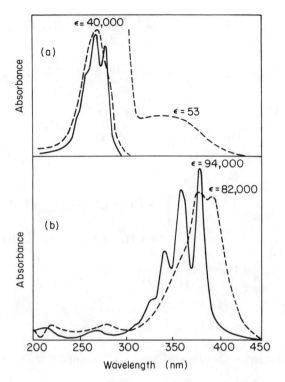

Fig. 6.3 Absorption spectra of long chain conjugated dialdehydes: (a) compound (6.3) (b) compound (6.4)[3834]

During u.v. irradiation of a benzene (gas)–oxygen mixture singlet oxygen can be formed.[858, 3400] Reaction occurs via energy transfer from the excited triplet state (T_1) of benzene to molecular oxygen:

$$\text{benzene } (T_1) + O_2 \rightarrow \text{benzene } (S_0) + {}^1O_2 \tag{6.16}$$

6.4. TETRAHYDROFURANE

Tetrahydrofurane (6.5) is a commonly used polymer solvent but it is very reactive photochemically.

Photolysis of tetrahydrofurane with a medium-pressure lamp led to several products:[1359,1985,3138]

$$\overset{\displaystyle \dot{C}H_2\ \dot{C}H_2}{\underset{\displaystyle CH_2}{}} + CH_2O \qquad (6.17)$$

$$\longrightarrow \overset{\displaystyle CH_2\text{---}CH_2}{\underset{\displaystyle CH_2}{}}, \ CH_2{=}CH\text{---}CH_3$$

$$\qquad\qquad\qquad (6.18)$$

$$\overset{\displaystyle \dot{C}H_2\ \dot{O}}{\underset{\displaystyle CH_2}{}} + CH_2{=}CH_2 \qquad (6.19)$$

$$\longrightarrow CH_3\text{---}CHO, \ \overset{\displaystyle CH_2\text{---}CH_2}{\underset{\displaystyle O}{\diagdown\diagup}}$$

$$\qquad\qquad\qquad (6.20)$$

$$\longrightarrow CH_2{=}CH\text{---}CH_2\text{---}CH_2OH \qquad (6.21)$$

$$\longrightarrow CH_3\text{---}CH_2\text{---}CH_2\text{---}CHO \qquad (6.22)$$

Central intermediate:

$$\overset{\displaystyle CH_2\text{---}CH_2}{\underset{\displaystyle \underset{O\,\cdot}{CH_2\ \dot{C}H_2}}{}}$$

Starting material (6.5):

$$\overset{\displaystyle CH_2\text{---}CH_2}{\underset{\displaystyle \underset{O}{CH_2\ CH_2}}{}} \xrightarrow{\ h\nu\ }$$

$$(6.5)$$

$$\overset{\displaystyle CH_2\text{---}CH_2}{\underset{\displaystyle \underset{O}{CH_2\ \underset{|}{C}\text{---}H}}{}} \ \overset{\displaystyle \cdot}{} + H^{\cdot} \qquad (6.23)$$

$$(6.6)$$

Tetrahydrofurane with oxygen forms a CT complex (6.7) according to the reaction:[698]

$$\square O + O_2 \longrightarrow \overset{\overset{\displaystyle O_2}{\vdots}}{\square O} \ \rightleftharpoons \ \overset{\overset{\displaystyle O_2^{\,-}}{\displaystyle O}}{\square \overset{+}{O}} \qquad (6.24)$$

$$(6.7)$$

During u.v. irradiation of tetrahydrofurane in the presence of oxygen the photooxidation reaction occurs. The following main products are formed: butyrolactone (6.7), α-hydroxytetrahydrofurane (6.8), and α-hydroperoxytetrahydrofurane (6.9) according to the mechanism:[2159,3433]

$$\square O{\cdots}O_2 \xrightarrow{\ h\nu\ } \square \overset{\pm}{O}\text{---}O_2^{\,-} \qquad (6.25)$$

$$\square \overset{\pm}{O}\text{---}O_2^{\,-} \ \rightleftharpoons \ \square O + {\cdot}O_2^{\,-} \qquad (6.26)$$

$$\square \overset{+}{O} + O_2^{\,-} \longrightarrow \square \overset{\cdot}{O} + {\cdot}O_2H \qquad (6.27)$$

$$(6.6)$$

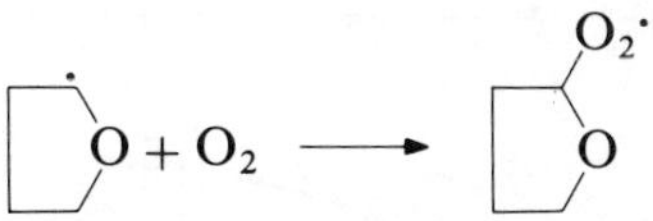

$$\overset{\cdot}{\square}O + O_2 \longrightarrow \square O_2^{\cdot}$$
(6.28)

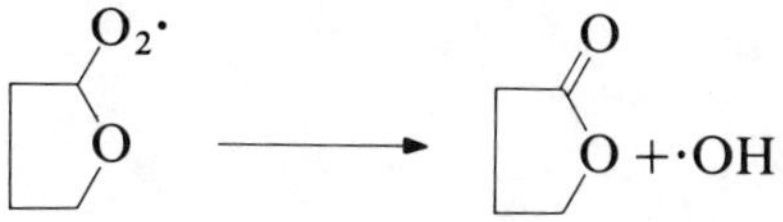

$$\square O_2^{\cdot} \longrightarrow \square O + \cdot OH$$
(6.29)

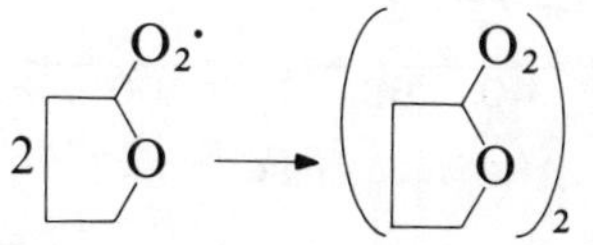

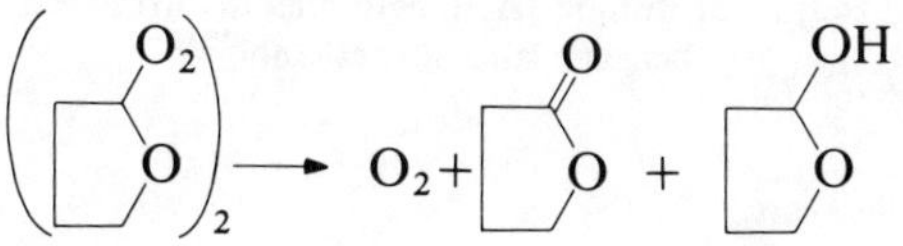

(6.7)

$$2\,\square O_2^{\cdot} \longrightarrow \left(\square O_2\right)_2$$
(6.30)

$$\left(\square O_2\right)_2 \longrightarrow O_2 + \square O + \square OH$$
(6.31)

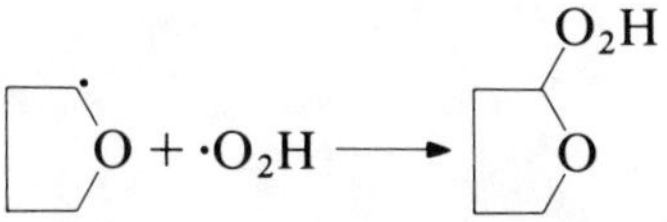

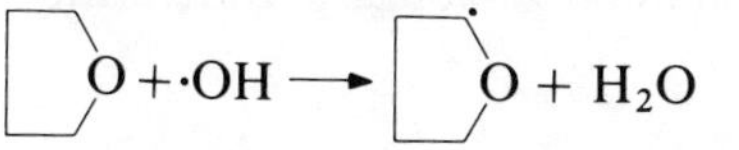

(6.8)

$$\overset{\cdot}{\square}O + \cdot O_2H \longrightarrow \square O_2H$$
(6.32)

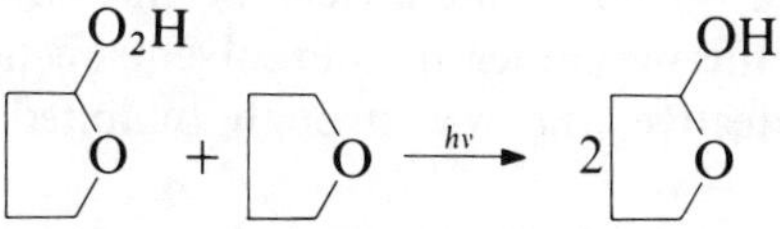

(6.9)

$$\square O_2H + \square O \xrightarrow{\ h\nu\ } 2\,\square OH$$
(6.33)

$$\square O + \cdot OH \longrightarrow \overset{\cdot}{\square}O + H_2O$$
(6.34)

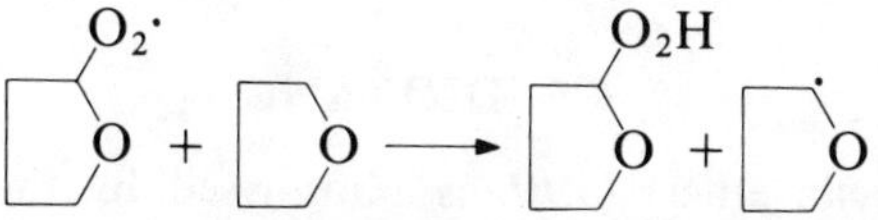

$$\square O_2^{\cdot} + \square O \longrightarrow \square O_2H + \overset{\cdot}{\square}O$$
(6.35)

It is well known that traces of hydroperoxides are present in commercial tetrahydrofurane. They are formed during the storage of the solvent under light. The best method for purifying tetrahydrofurane is redistillation over calcium hydride or sodium borohydride under nitrogen.

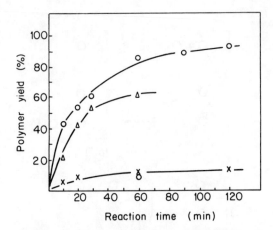

Fig. 6.4 Kinetics of photopolymerization of hydroquinone diacrylate in the presence of different solvents:
(O) tetrahydrofurane, (Δ) dimethyltetrahydrofurane, (×)
benzene, and (□) ethanol[2753]

It has been reported that tetrahydrofurane photoinitiates polymerization of hydroquinone diacrylate, giving a high yield of the polymer (Fig. 6.4).[2753] Photopolymerization of this monomer in other solvents such as ethanol, acetone, and benzene was not effective, and was strongly inhibited by the presence of oxygen.

The tetrahydrofurane radical (6.6) is formed easily in the presence of benzophenone and may initiate free radical polymerization of vinyl monomers (cf. Section 7.9.4.2.3.5).

6.5. DIOXANE

The p-dioxane (cyclic ether) (6.10) is photolysed by the ring-opening and formation of biradicals (6.11) and (6.12). Such radicals are often stable at room temperature, but may also begin to undergo a fragmentation reaction, depending on substituents:[1416, 3413, 3790]

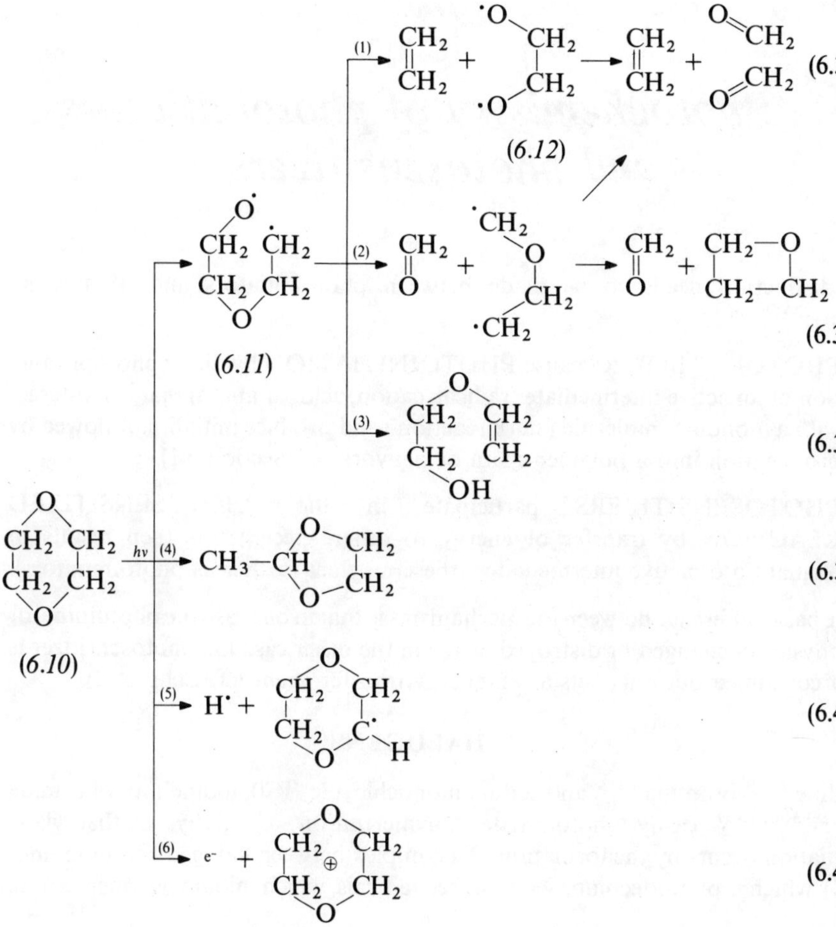

$$\text{(6.36)}$$

$$\text{(6.37)}$$

$$\text{(6.38)}$$

$$\text{(6.39)}$$

$$\text{(6.40)}$$

$$\text{(6.41)}$$

7. Photochemistry of photoinitiators and photosensitizers

A distinction needs to be made between photoinitiators and photosensitizers:[2458, 3019]

PHOTOINITIATORS cause PHOTOINITIATION by direct photoproduction of an active intermediate (radical, cation, acid, or anion) that can interact with a monomer molecule ('dark reaction') and produce initiation followed by propagation into a polymer chain or network (cf. Section 5.1).

PHOTOSENSITIZERS participate in the PHOTOSENSITIZED REACTION by transfer of energy to other species; this then results in formation of active intermediates (these may act further as photoinitiators).

The basic difference between the mechanisms is that in one case the photoinitiator is physically changed or distroyed, while in the other case the photosensitizer is not consumed but only acts as an energy transfer agent (cf. Chapter 2).

7.1. HALOGENS

Iodine,[1298] bromine,[1319] and iodine monochloride (ICl), iodine monobromide (IBr)[1300, 1301] easily photoinitiate polymerization of methyl methacrylate. Initiation occurs by the formation of a complex between halogen and monomer (M), which is photodecomposed into free radicals, which initiate polymerization:

$$I_2 + 2M \rightleftharpoons [M \ldots I - I \ldots M] \xrightarrow{+h\nu} RADICALS \qquad (7.1)$$
$$COMPLEX$$

It has also been reported that combination of Cl_2, Br_2, or I_2 with sulphur dioxide (SO_2) is effective in photoinitiating the system for polymerization of methyl methacrylate.[1313]

7.2. HALOGENATED ORGANIC COMPOUNDS

Photochemistry of halogenated organic compounds depends on their chemical structure. The fragmentation of α-halogenated compounds (Table 7.1) into free radicals occurs by the general mechanism:[2454, 2458, 3769]

$$R-CH_2-X + h\nu \rightarrow RCH_2\cdot + X\cdot \qquad (7.2)$$

$$R-SO_2-X + h\nu \rightarrow RSO_2\cdot + X\cdot \qquad (7.3)$$

where: R is the photosensitive light absorbing group and X = Cl, Br, or I.

Photolysis of photochemically labile α-halogenoketones, e.g. α-chloroacetone (*7.1*)[3453] or α-bromoacetophenone (*7.2*)[191, 3453] produces reactive chlorine or bromine radicals, respectively:

$$CH_3COCH_2Cl + h\nu \rightarrow CH_3COCH_2\cdot + Cl\cdot \quad (7.4)$$

(*7.1*)

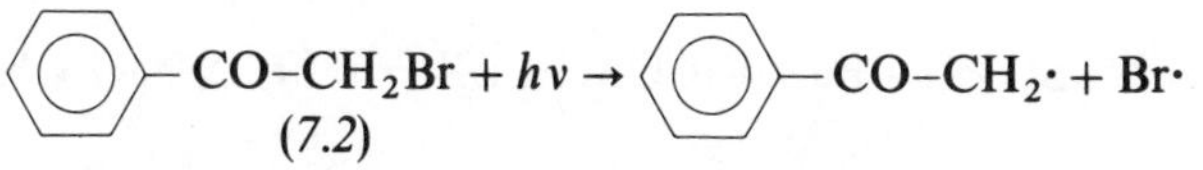

(*7.2*) (7.5)

These photoinitiators were used for polymerization of several vinyl monomers.[2446, 2836, 2863, 3483]

The ease of photocleavage of C–Cl and S–Cl bonds allows initiators with low triplet energies to fragment, and therefore fragmentation can result from long wavelength absorption. Sulphonyl chlorides have triplet energies which are about 10 kcal mol^{-1} lower than the chloromethyl derivatives. A number of halogenated organic compounds (Table 7.1) have been used as effective photoinitiators for

Table 7.1 Examples of halogenated compounds used as photoinitiators[2458]

Compound	Compound
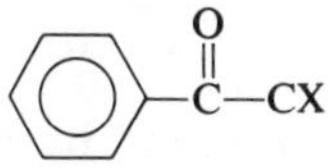	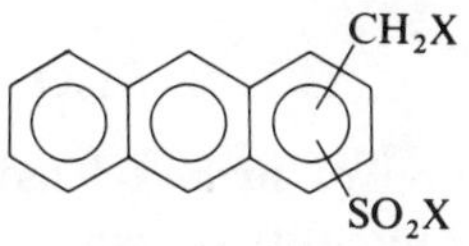
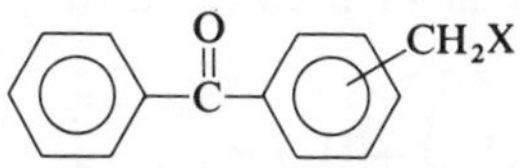	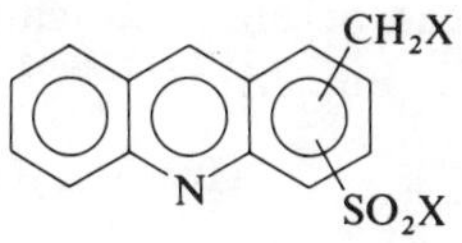
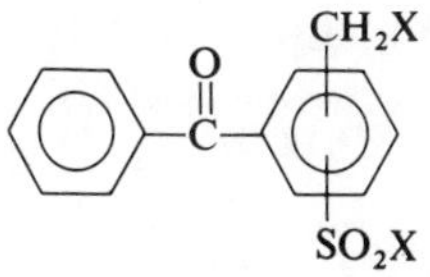	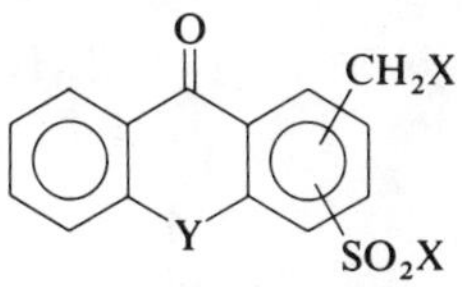 Y = O, CH_2, CHX S, C = O
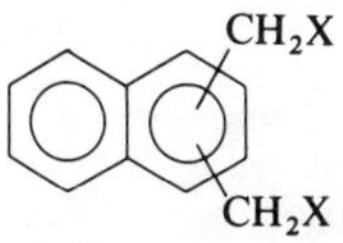	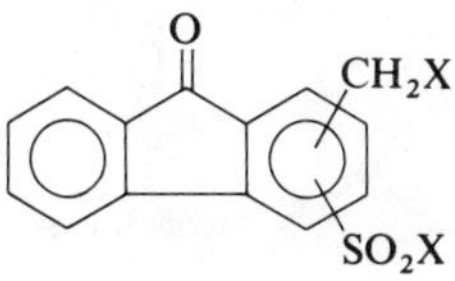

X = Cl, Br, I

curing pigmented coatings.[2454, 2458] A practical problem with halogenated photoinitiators is the possible corrosion which may be produced by their by-products, especially hydrogen chloride.

Halogenated compounds can also form photoactive complexes with ferrocene (7.3):[2464]

$$R-CH_2X \text{ (or } R-SO_2X) + Fe \longrightarrow (COMPLEX) \xrightarrow{+h\nu} Fe^+X^- + RCH_2 \cdot \text{ (or } RSO_2 \cdot)$$

$$(7.3) \hspace{6cm} (7.6)$$

It has been reported that phenyliododichloride ($C_6H_5ICl_2$) is an effective photoinitiator for polymerization of methyl methacrylate.[1302]

7.3. HYDROGEN PEROXIDE

Hydrogen peroxide (H_2O_2) absorbs in the ultraviolet region (200–300 nm) but it has no absorption peak.[3233] Under u.v. irradiation (254 nm) it gives almost completely hydroxyl radicals (HO·) with a quantum yield $\Phi = 1.0$:[348, 2758, 3447, 3823]

$$H_2O_2 + h\nu \rightarrow 2HO \cdot \tag{7.7}$$

The concentration of hydroxyl radicals can be controlled by the change of wavelength and intensity of light.

The hydroxyl radicals react further with H_2O_2 to give hydroperoxy radicals (HO$_2$·), which can be observed with ESR at high concentrations of hydrogen peroxide at 77K (Fig. 7.1):[2977, 3062, 3395]

$$HO \cdot + H_2O_2 \rightarrow H_2O + HO_2 \cdot \tag{7.8}$$

$$HO_2 \cdot + H_2O_2 \rightarrow H_2O + O_2 + HO \cdot \tag{7.9}$$

$$2 HO_2 \cdot \rightarrow H_2O_2 + O_2 \tag{7.10}$$

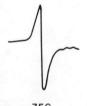

35G

Fig. 7.1 ESR spectrum of photolyzed hydrogen peroxide

The photodecomposition of hydrogen peroxide in alkaline media gives ozone via ozonide.[2209]

The photodecomposition of hydrogen peroxide is markedly accelerated by the presence of carbon monoxide (CO) which reduces hydroxyl radicals to hydrogen in a chain reaction:[581,1367,2476]

$$HO \cdot + CO \rightarrow CO_2 + H \cdot \qquad (7.11)$$

$$H \cdot + H \cdot \rightarrow H_2 \qquad (7.12)$$

$$H \cdot + H_2O_2 \rightarrow HO \cdot + H_2O \ (or \ H_2 + HO_2 \cdot) \qquad (7.13)$$

Very low concentrations of transition metals are capable of accelerating the photolysis of hydrogen peroxide.[2347]

Hydroxyl radicals (HO·) are very reactive. They can add to aromatic rings yielding phenols after elimination of hydrogen from adduct:

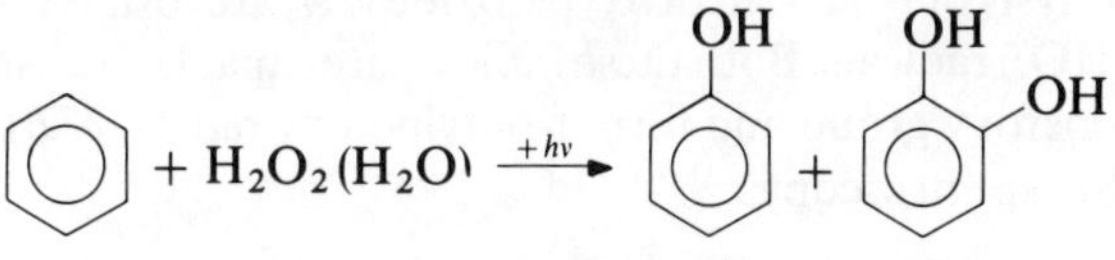

$$+ H_2O_2 (H_2O) \xrightarrow{+h\nu} \qquad (7.14)$$

Further oxidation reaction leads to ring-opening products:[991,1844]

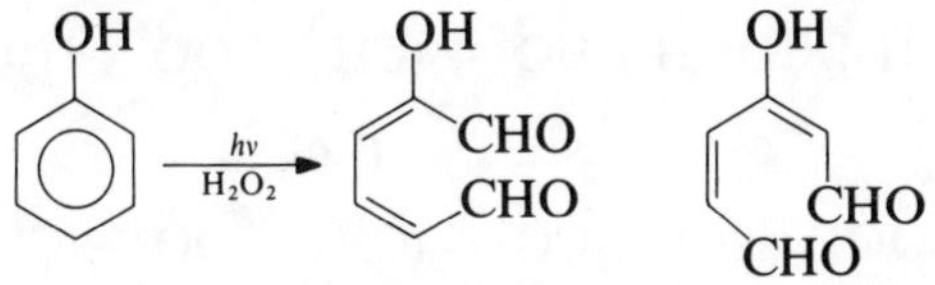

$$\xrightarrow[H_2O_2]{h\nu} \qquad (7.15)$$

In the oxidation of phenols the *ortho* and *para* positions are hydroxylated, because of the electrophilic nature of the hydroxyl radical, with the *ortho* position being preferred.[2428,2795]

A few papers report successful photoinitiation of methyl methacrylate[1326] and isoprene[3545] polymerization in different solvents using hydrogen peroxide.

Hydroxyl radicals (HO·) are capable of abstracting hydrogen from a polymer macromolecule, producing water and polymer alkyl radical.[2468, 3020, 3060]

7.4. ALKYL HYDROPEROXIDES

Alkyl hydroperoxides (R–OOH) absorb in the ultraviolet region (200–300 nm) but the absorption curves for most of them do not form a peak.[1076]

Alkyl hydroperoxides undergo two general types of photochemical reactions:

(i) Unimolecular photolysis into alkyloxy (RO·) and hydroxyl (HO·) radicals:[998,2198,2471,2472,2758,3309]

$$ROOH + h\nu \rightarrow RO \cdot + \cdot OH \qquad (7.16)$$

The quanta 280–480 nm have sufficient energy to cleave the RO–OH bond (bond energy 42–43 kcal mol^{-1})[404–406] The RO–OH bond energy should be relatively independent of R.[406,2005]

(ii) In concentrated hydroperoxide solutions a second-order process may also occur, in which alkylperoxy radicals $(RO_2 \cdot)$ are formed:

$$2 ROOH + h\nu \rightarrow RO \cdot + RO_2 \cdot + H_2O \qquad (7.17)$$

These two reactions are more complicated because the radicals thus formed may further react with alkyl hydroperoxide molecules (in dark reaction) to decompose them efficiently:[404,2198,2267]

$$ROOH + RO \cdot \rightarrow RO_2 \cdot + ROH \qquad (7.18)$$

$$ROOH + R \cdot \rightarrow RO_2 \cdot + RH \qquad (7.19)$$

$$2 RO_2 \cdot \rightarrow 2 RO \cdot + O_2 \qquad (7.20)$$

During photolysis of t-butyl hydroxyperoxide (7.4) are formed t-butyloxy (7.5) and hydroxyl $(HO \cdot)$ radicals. Both these radicals are capable to abstract hydrogen from the hydroperoxy group and form t-butylperoxy radical (7.6), which can be detected by ESR spectroscopy:[2369]

$$(CH_3)_3COOH + h\nu \rightarrow (CH_3)_3CO \cdot + HO \cdot \qquad (7.21)$$

$$(7.4) \qquad\qquad (7.5)$$

$$(CH_3)_3COOH + HO \cdot \rightarrow (CH_3)_3COO \cdot + H_2O \qquad (7.22)$$

$$(7.6)$$

$$(CH_3)_3COOH + (CH_3)_3CO \cdot \rightarrow (CH_3)_3COO \cdot + (CH_3)_3COH \qquad (7.23)$$

$$2(CH_3)_3COO \cdot \rightarrow (CH_3)_3COOC(CH_3)_3 + O_2 \qquad (7.24)$$

Alkoxy radicals abstract hydrogen atoms from the solvents, e.g. t-butyloxy radical (7.5) from isopropanol (7.7):[2548,2758,3402]

$$(CH_3)_3CO \cdot + (CH_3)_2CHOH \rightarrow (CH_3)_3COH + (CH_3)_2\dot{C}OH \qquad (7.25)$$

$$(7.5) \qquad\qquad (7.7)$$

$$(CH_3)_2\dot{C}OH \rightarrow CH_2{=}\underset{\underset{CH_3}{|}}{C}{-}OH + (CH_3)_2CO \qquad (7.26)$$

$$(CH_3)_3COO \cdot + (CH_3)_2\dot{C}OH \rightarrow (CH_3)_3COOC(CH_3)_2OH \qquad (7.27)$$

7.5. CUMENE HYDROPEROXIDE

Cumene hydroperoxide (7.8) dissolved in cumene (7.12) is very efficiently photolysed under u.v. irradiation (254 nm) with quantum yield $\Phi = 0.97$.[1276] The main products are 2-phenylpropanol-2 (7.10) $(\Phi = 0.93)$, acetophenone (7.11) $(\Phi = 0.03)$, hydroxyl, and methyl radicals which originate from the cumyloxy radical (7.9):[1276,4012]

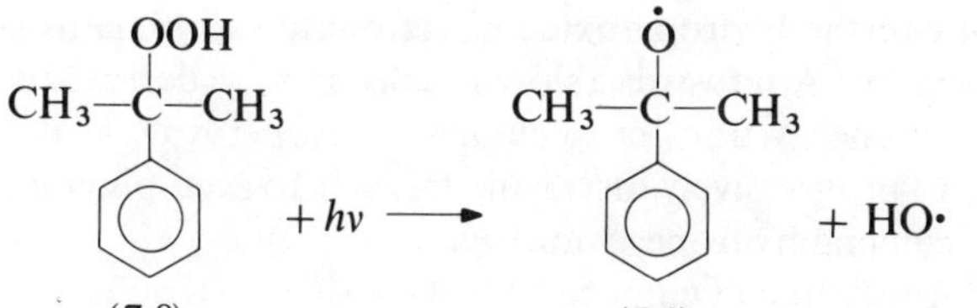

$$CH_3-\underset{\underset{(7.8)}{\displaystyle C_6H_5}}{\overset{\overset{\displaystyle OOH}{|}}{C}}-CH_3 \;+\; h\nu \;\longrightarrow\; CH_3-\underset{\underset{(7.9)}{\displaystyle C_6H_5}}{\overset{\overset{\displaystyle \dot{O}}{\|}}{C}}-CH_3 \;+\; HO\cdot \tag{7.28}$$

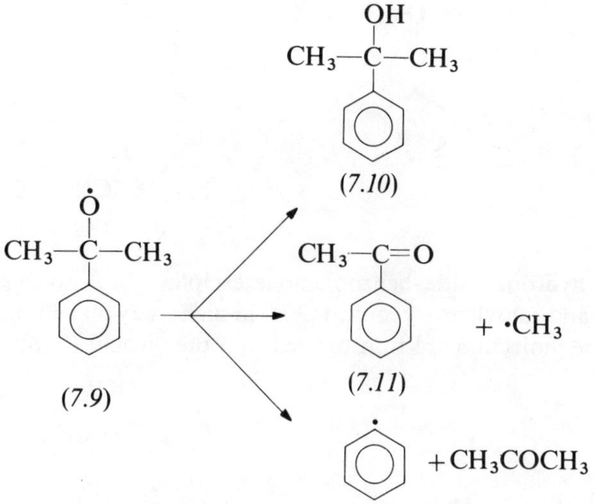

$$CH_3-\underset{\underset{(7.10)}{\displaystyle C_6H_5}}{\overset{\overset{\displaystyle OH}{|}}{C}}-CH_3 \tag{7.29}$$

$$CH_3-\underset{\underset{(7.9)}{\displaystyle C_6H_5}}{\overset{\overset{\displaystyle \dot{O}}{\|}}{C}}-CH_3$$

$$CH_3-\underset{\underset{(7.11)}{\displaystyle C_6H_5}}{C}=O \;+\; \cdot CH_3 \tag{7.30}$$

$$C_6H_5\cdot \;+\; CH_3COCH_3 \tag{7.31}$$

These results indicate that an efficient energy transfer process occurs from the excited phenyl chromophore of cumene (7.12) to cumene hydroperoxide:[1276]

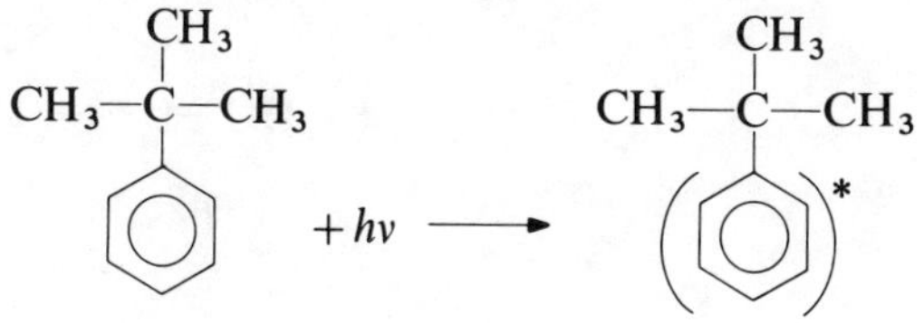

$$CH_3-\underset{\underset{(7.12)}{\displaystyle C_6H_5}}{\overset{\overset{\displaystyle CH_3}{|}}{C}}-CH_3 \;+\; h\nu \;\longrightarrow\; CH_3-\underset{\displaystyle (C_6H_5)^*}{\overset{\overset{\displaystyle CH_3}{|}}{C}}-CH_3 \tag{7.32}$$

$$CH_3-\underset{\displaystyle (C_6H_5)^*}{\overset{\overset{\displaystyle CH_3}{|}}{C}}-CH_3 \;+\; CH_3-\underset{\displaystyle C_6H_5}{\overset{\overset{\displaystyle OOH}{|}}{C}}-CH_3 \;\longrightarrow\; CH_3-\underset{\displaystyle C_6H_5}{\overset{\overset{\displaystyle CH_3}{|}}{C}}-CH_3 \;+\; CH_3-\underset{\displaystyle (C_6H_5)^*}{\overset{\overset{\displaystyle OOH}{|}}{C}}-CH_3 \tag{7.33}$$

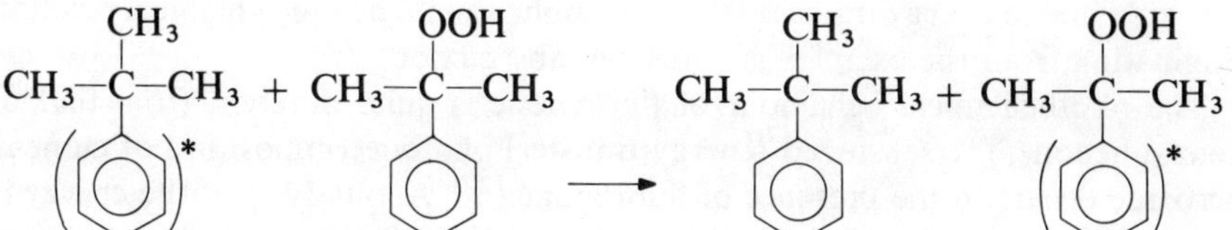

The rate of cumene hydroperoxide direct photolysis remains constant up to 40 % conversion.[3277] Afterwards a slow and progressive decrease is observed, due to a less efficient energy transfer to cumene hydroperoxide as its concentration decreases and to progressively increasing transfer to acetophenone formed from photolysis of cumene hydroperoxide.

During u.v. irradiation of cumene hydroperoxide–benzophenone formation of an exciplex has been observed (cf. Section 3.5) between the excited triplet state of benzophenone and cumene hydroperoxide:[1276]

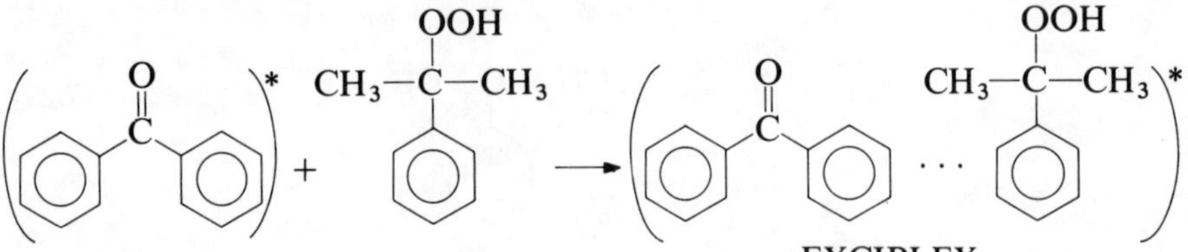

$$\text{EXCIPLEX}$$
$$(7.13) \qquad\qquad (7.34)$$

The cumene hydroperoxide–benzophenone exciplex (7.13) dissociates to ketyl radical (7.14) and cumylperoxy radical (7.15) in such a way that 80 % of the excited benzophenone molecules are transformed into the pinacol (7.16):

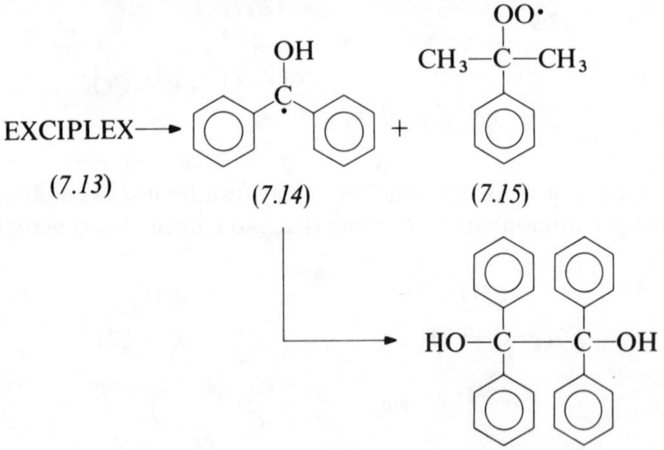

$$\text{EXCIPLEX} \longrightarrow \qquad + \qquad\qquad (7.35)$$
$$(7.13) \qquad\qquad (7.14) \qquad\qquad (7.15)$$

$$(7.16) \qquad\qquad (7.36)$$

There is no evidence for sensitized photodecomposition of the cumene hydroperoxide due to energy transfer from benzophenone. The only chemical reaction originating from the exciplex is hydrogen abstraction.

The photochemical behaviour of fluorenone is quite different from that of benzophenone; the sensitized (energy transfer) photodecomposition of cumene peroxide occurs in the presence of fluorenone.[1276] About 14 % of the energy is transferred from the excited singlet state (S_1) of fluorenone to cumene hydroperoxide.

7.6. PEROXIDES

Peroxides absorb in the ultraviolet region (200–300 nm), but the absorption curves for most of them do not form a peak.[3346,3632] Diaryl peroxides, however, have two absorption peaks at 230 and 273 nm.[517,3346].

The primary photochemical reaction is the O–O bond scission giving:

(i) alkoxy radicals (RO·) from dialkyl peroxides:

$$RO-OR_1 + hv \rightarrow RO\cdot + \cdot OR_1 \qquad (7.37)$$

(ii) Acyloxy (R–$\overset{\overset{\displaystyle O}{\|}}{C}$–O·) and alkyl (R·) radicals from diacyl peroxides (and also from peroxy esters):

$$RCOO-OOCR_1 + hv \rightarrow RCOO\cdot + \cdot OOCR_1 \rightarrow R\cdot + 2CO_2 + R_2\cdot \quad (7.38)$$

(iii) Alkyl and hydroxyl radicals from peroxy acids:

$$RCOO-OH + hv \rightarrow RCOO\cdot + \cdot OH \rightarrow R\cdot + CO_2 + \cdot OH \qquad (7.39)$$

(iv) Alkyl, alkoxy and acyloxy radicals from peroxy esters:

$$RCOO-OR_1 + hv \rightarrow RCOO\cdot + \cdot OR_1 \rightarrow R\cdot + CO_2 + \cdot OR_1 \qquad (7.40)$$

Alkyl radicals, formed from acyl peroxides and peroxy esters, can be identified by ESR spectroscopy at low temperatures.[2089] Relative stabilities of alkoxy radicals derived from unsymmetric dialkyl peroxides are in the following order:[372]

$$CH_3O\cdot > CH_3CH_2O\cdot > CH_3CH(CH_3)O\cdot$$
$$> (CH_3)_2CH-CH_2O\cdot \approx CH_3-C(CH_3)_2O\cdot$$

Some of produced radicals are excited ('hot radicals'),[3878–3880] e.g. diisopropyl peroxide (7.17) photolysis gives an excited isopropyloxy ('hot') radical (7.18), which readily decomposes to give mostly methyl radical and acetaldehyde before it reacts with other molecules:[2471]

$$(CH_3)_2CHOOCH(CH_3)_2 + hv \rightarrow ((CH_3)_2CHO\cdot)^* + ((CH_3)_2CHO\cdot)^* \quad (7.41)$$

$$(7.17) \qquad\qquad (7.18) \text{ HOT RADICALS}$$

$$((CH_3)_2CHO\cdot)^* \rightarrow CH_3\cdot + CH_3CHO \qquad (7.42)$$

An alternative path for photolysis is a considered multiple scission in the primary photoprocess:

$$(CH_3)_2CHOOCH(CH_3)_2 + hv \rightarrow (CH_3)_2CHO\cdot + CH_3CHO + CH_3\cdot \rightarrow$$
$$2CH_3\cdot + 2CH_3CHO \quad (7.43)$$

Ground-state (ordinary) alkoxy radicals formed by thermolysis of diisopropyl peroxide can be trapped by scavengers, e.g. nitric oxide[2471]

$$(CH_3)_2CHO\cdot + NO \rightarrow (CH_3)_2CHONO\cdot \rightarrow CH_3COCH_3 + HNO\cdot \quad (7.44)$$

whereas, excited alkoxy radicals (if they exist) are not trapped by nitric oxide.

Photolysis of dicumyl peroxide (*7.19*) in *n*-hexane with 313 nm radiation gives mainly (95–100 %) of α,α-dimethylbenzyl alcohol (*7.21*) by hydrogen abstraction of cumyloxy radical (*7.20*) from the solvent (RH):[2732]

$$
\underset{(7.19)}{\text{C}_6\text{H}_5\text{-C(CH}_3)_2\text{-OO-C(CH}_3)_2\text{-C}_6\text{H}_5} + h\nu \longrightarrow 2\,\underset{(7.20)}{\text{C}_6\text{H}_5\text{-C(CH}_3)_2\text{-O}\cdot} \qquad (7.45)
$$

$$
2\,\text{C}_6\text{H}_5\text{-C(CH}_3)_2\text{-O}\cdot + 2\text{RH} \longrightarrow 2\,\underset{(7.21)}{\text{C}_6\text{H}_5\text{-C(CH}_3)_2\text{-OH}} + 2\text{R}\cdot \qquad (7.46)
$$

The photolysis of dicumyl peroxide in carbon tetrachloride with the same 313 nm radiation gives mainly (95–100 %) acetophenone (*7.23*) and some acetone, which are products of the C–C cleavage of the 'hot' cumyloxy radical (*7.22*):[2732]

$$
2\left(\underset{(7.22)}{\text{C}_6\text{H}_5\text{-C(CH}_3)_2\text{-O}\cdot}\right)^{*} \longrightarrow \underset{(7.23)}{\text{C}_6\text{H}_5\text{-COCH}_3} + \text{CH}_3\text{COCH}_3 + \text{C}_6\text{H}_5\cdot + \text{CH}_3\cdot \qquad (7.47)
$$

A number of different sensitizers such as acetophenone, benzophenone, and aromatic hydrocarbons (e.g. anthracene) have been proposed for the photolysis of peroxides, peresters, and hydroperoxides with >305 nm radiation.[2652, 3199, 3792] Sensitizers absorb light energy of longer wavelength and transfer excitation energy to peroxides. Sensitization by ketones occurs via the excited triplet state (T_1) of the sensitizer, while the sensitization by hydrocarbons may involve the excited singlet state (S_1). The rate constant for the energy transfer from benzophenone to benzoyl peroxide is very large (3.2×10^6 M^{-1} s^{-1}), but only 25 % of the excited peroxides decompose because of deactivation or cage recombination.[3792]

7.7. BENZOYL PEROXIDE

Benzoyl peroxide (*7.24*) has two absorption peaks at 230 and 273 nm.[517,3346] Under u.v. irradiation it is photolysed to pairs of benzoyloxy radicals (*7.25*), some of which may further dissociate to phenyl radicals (*7.26*) and carbon dioxide:[3296]

$$
\underset{(7.24)}{\text{C}_6\text{H}_5\text{-CO-O-O-CO-C}_6\text{H}_5} + h\nu \longrightarrow 2\,\underset{(7.25)}{\text{C}_6\text{H}_5\text{-CO-O}\cdot} \longrightarrow 2\,\underset{(7.26)}{\text{C}_6\text{H}_5\cdot} + 2\text{CO}_2 \qquad (7.48)
$$

Phenyl radicals can be identified by ESR spectroscopy during irradiation of benzoyl peroxide crystal at 4.2K.[502] It is of interest to note that no benzoyloxy radicals were observed, which shows that the photoinduced benzoyloxy radicals are very unstable and rapidly eliminate CO_2, affording phenyl radicals.

Both types, benzoyloxy and phenyl radicals, may initiate polymerization of methyl methacrylate[345,746,1881] and styrene,[425,428] and copolymerization of styrene or methyl methacrylate with N-methylmaleimide.[3166]

7.8. NON-KETONIC PERESTERS

Tert-butyl peresters based on aromatic hydrocarbons:

4-[(1-pyrenyl)carbonyl]peroxybenzoic acid tert-butyl ester (7.27),

4-[((1-pyrenyl)carbonyl)oxy)methyl] peroxybenzoic acid tert-butyl ester (7.28),

4-[(((9H-fluorenon-9-on-4-yl)carbonyl)oxy)methyl]peroxybenzoic acid tert-butyl ester (7.29),

4-[(((9-anthryl)carbonyl)oxy)methyl)]]peroxybenzoic acid tert-butyl ester (7.30)

are efficient photoinitiators of the polymerization of methyl methacrylate and styrene.[4]

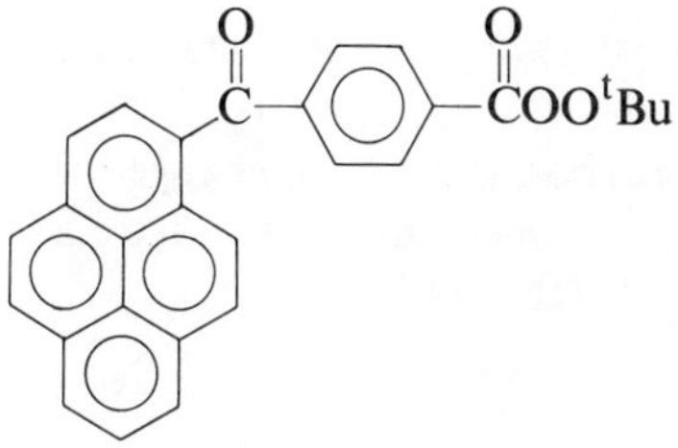

(7.27)

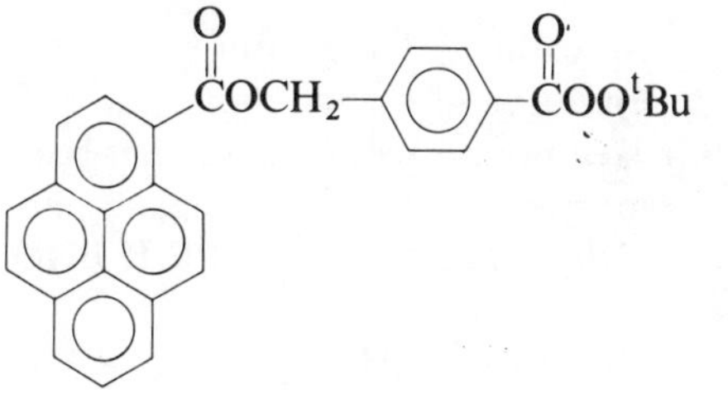

(7.28)

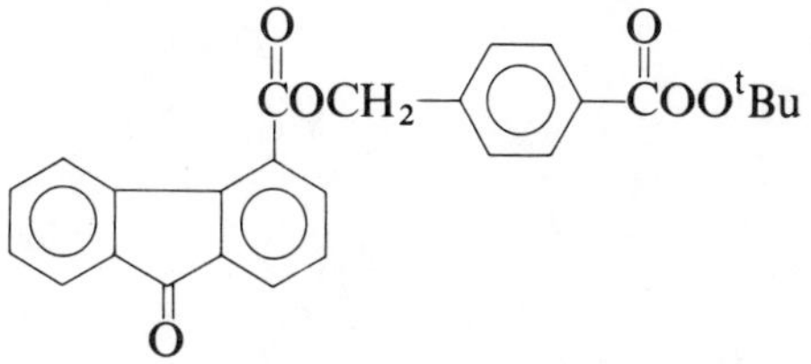

(7.29)

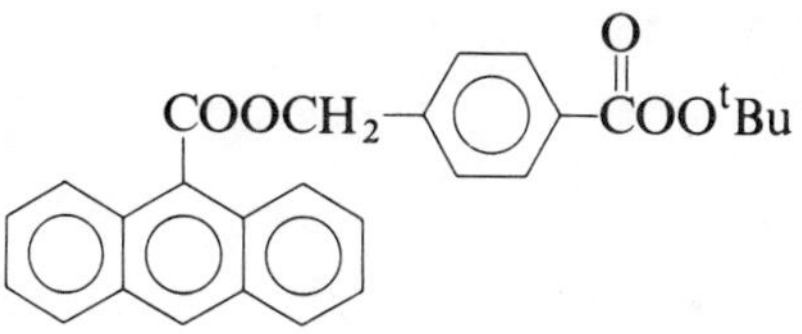

(7.30)

They are photolysed according to the mechanism (7.48) producing aryloxy (7.31) and t-butoxy (7.32) free radicals:

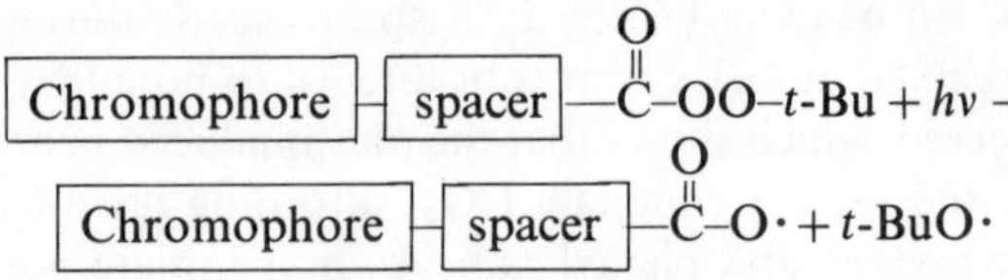

$$\boxed{\text{Chromophore}} - \boxed{\text{spacer}} - \overset{\overset{\text{O}}{\|}}{\text{C}} - \text{OO} - t\text{-Bu} + h\nu \rightarrow$$

$$\boxed{\text{Chromophore}} - \boxed{\text{spacer}} - \overset{\overset{\text{O}}{\|}}{\text{C}} - \text{O} \cdot + t\text{-BuO} \cdot \qquad (7.49)$$

$$(7.31) \qquad\qquad (7.32)$$

The rate of methyl methacrylate photoinitiation decreases in the following sequence of perester structure: $(7.29) > (7.30) > (7.31) > (7.32)$.

The rate of styrene photoinitiation decreases in the following order: $(7.27) > (7.28) > (7.30) > (7.31)$.

7.9. KETONES

Photochemistry of ketones (and aldehydes) has been the subject of several publications.[412, 591, 744, 822, 823, 1049, 1050, 1057, 1059, 1513, 1514, 2285, 2324, 2911, 2958, 3773, 3775]

In general excited ketones may participate in the following types of photochemical reactions:

(i) Photocleavage of a bond α to the carbonyl group (α-CLEAVAGE or NORRISH TYPE I CLEAVAGE). During this reaction acyl($R–CO\cdot$) and alkyl($R\cdot$) radicals are formed:

$$R\text{--CO--R} + h\nu \rightarrow R\text{--CO} \cdot + R \cdot \rightarrow 2R \cdot + CO \qquad (7.50)$$

At elevated temperatures the acyl radical can decarbonylate, with the evolution of carbon monoxide.

(ii) Ketones possessing hydrogen-bearing γ-carbon atoms can change via six-membered cyclic intermediate photorearrangement to methyl ketone and olefine (NORRISH TYPE II PHOTOELIMINATION):

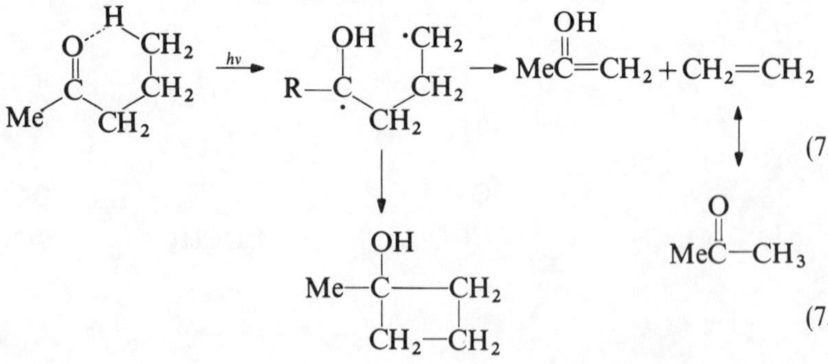

$$(7.51)$$

$$(7.52)$$

$$(7.33)$$

Few aliphatic ketones can photorearrange to cyclobutanol derivatives (7.33).[3960] But this reaction does not appear to be an important process observed during photolysis of aliphatic ketones. Both Norrish Type I and Type II reactions are important in polymer systems.[1513, 1514]

(iii) Addition of an excited triplet state of carbonyl compound to an olefin to form an oxetane (7.34):

$$(R-\overset{\overset{\displaystyle O}{\|}}{C}-R)^* + CH_2{=}CH\phi \longrightarrow R{\underset{\underset{\displaystyle R}{|}}{\overset{\overset{\displaystyle O-CH_2}{|}}{-}}}\underset{H}{\overset{|}{C}}{-}\phi \qquad (7.53)$$

(7.34)

(iv) Excited aromatic ketones can intermolecularly abstract hydrogen from hydrogen-donor molecules (RH) (cf. Section 7.9.4.2):

$$\left(\underset{}{\bigcirc}{-}\overset{\overset{\displaystyle O}{\|}}{C}{-}\bigcirc\right)^* + RH \longrightarrow \bigcirc{-}\underset{\cdot}{\overset{\overset{\displaystyle OH}{|}}{C}}{-}\bigcirc + R{\cdot} \qquad (7.54)$$

(v) Electron transfer.
(vi) Electronic energy transfer.

All of these processes may occur from both excited singlet (S_1) or triplet (T_1) states.[822,3773] The latter are more important as regards competition with conformational motion,[3775] because of the rapid intersystem crossing (ISC) rates in ketones.[2324]

7.9.1. Aliphatic Ketones

Diisopropyl ketone (7.35) undergoes only Norrish Type I photocleavage with a quantum yield of carbon monoxide production $\Phi_{CO} = 0.93$.[3872] Photolysis of diisopropyl ketone in the presence of oxygen is more complicated; the main products are acetone and isopropanol with traces of isobutyric acid (7.36):[1098]

$$(CH_3)_2CH{-}CO{-}CH(CH_3)_2 + h\nu \rightarrow$$
$$CH_3COCH_3 + (CH_3)_2CHOH + (CH_3)_2COOH \quad (7.55)$$

(7.35)   (7.36)

Isopropyl radical (7.37) and isopropylacyl radical (7.38) formed in the Norrish Type I photocleavage react further with oxygen by the following mechanism:

$$(CH_3)_2CH{-}CO{-}CH(CH_2)_2 + h\nu \rightarrow (CH_3)_2CH{\cdot} + {\cdot}COCH(CH_3)_2 \qquad (7.56)$$

(7.37)   (7.38)

$$(CH_3)_2CHCO{\cdot} \rightarrow (CH_3)_2CH{\cdot} + CO \qquad (7.57)$$

$$2(CH_3)_2CH{\cdot} + O_2 \rightarrow CH_3COCH_3 + (CH_3)_2CHOH \qquad (7.58)$$

$$(CH_3)_2CH{\cdot} + (CH_3)_2CHCO{\cdot} + O_2 \rightarrow CH_3COCH_3 + (CH_3)_2CHCOOH$$
$$(7.59)$$

$$(CH_3)_2CHCO{\cdot} + O_2 \rightarrow (CH_3)_2CHO{-}OO{\cdot} \qquad (7.60)$$

$$2(CH_3)_2CHO-OO\cdot \rightarrow [(CH_3)_2CHO-OO-OO-COCH(CH_3)_2]$$
$$\rightarrow 2(CH_3)_2C\cdot + 2CO_2 + O_2 \qquad (7.61)$$

During photoirradiation of aliphatic ketones (e.g. acetone, 4-heptanone, 8-pentadecanone, 12-tricosanone, and γ-ketopimellate) in carbon tetrachloride at 314 nm, energy transfer occurs from the $^1(n, \pi^*)$ state of ketone to CCl_4 leading to the photosensitized decomposition of the CCl_4 into $Cl\cdot$ and $\cdot CCl_3$ radicals:

$$^1(\text{ketone}^*) + CCl_4 \rightarrow \text{ketone} + Cl\cdot + \cdot CCl_4 \qquad (7.62)$$

which may further react accordingly to the mechanism:[2911]

$$2\cdot CCl_3 \rightarrow C_2Cl_6 \qquad (7.63)$$

$$RCH_2CO + \cdot Cl \rightarrow R\dot{C}HCO + HCl \qquad (7.64)$$

$$R\dot{C}HCO + CCl_4 \rightarrow \underset{\displaystyle |}{RCHCO} + \cdot CCl_3 \qquad (7.65)$$
$$Cl$$

$$R\dot{C}HCO + \cdot Cl \rightarrow \underset{\displaystyle |}{RCHCO} \qquad (7.66)$$
$$Cl$$

$$R\dot{C}HCO + \cdot CCl_3 \rightarrow \underset{\displaystyle |}{RCHCO} \qquad (7.67)$$
$$CCl_3$$

In the presence of air and moisture the yield of HCl increases as other reactions become important:

$$\cdot CCl_3 + \tfrac{1}{2}O_2 \rightarrow COCl_2 + \cdot Cl \qquad (7.68)$$

$$COCl_2 + 2H_2O \rightarrow CO(OH)_2 + 2HCl \qquad (7.69)$$

$$CO(OH)_2 \rightarrow CO_2 + H_2O \qquad (7.70)$$

A few examples of application of ketones as photoinitiators of polymerization of vinyl monomers are listed in Table 7.2.

7.9.2. Aliphatic Diketones

Aliphatic diketones such as biacetyl (7.39) have two absorption bands at 254 nm and 430 nm (Fig. 7.2).[591] Under u.v. irradiation biacetyl is photolysed by the following mechanism:[3144]

$$CH_3CO-COCH_3 + hv \underset{(7.39)}{\overset{\nearrow 2CH_3CO\cdot \rightarrow 2CH_3\cdot + 2CO \qquad (7.71)}{\underset{\searrow CH_3COCH_3 + CO \qquad (7.72)}{\big\langle \quad (7.40)}}}$$

Acyl radicals (7.40) are capable of initiating polymerization of several vinyl monomers.[746,2278,2819]

Table 7.2 Polymerization of vinyl monomers photoinitiated by ketones

Photoinitiator	Monomer	References
Acetone 3-Pentanone 2,4-Dimethyl-3-pentanone 2,2,4,4-Tetramethyl-3-pentanone 2-Methyl-pentanone 2-Heptanone 5-Methyl-2-hexanone 2,3-Butanedione 2,3-Pentanedione	Methyl methacrylate Styrene	1048, 1050–1053 1058, 2306, 2307
Methoxy-acetone 1,3-Dihydroxy-acetophenone 3-Hydroxy-3-methyl-2-butanone 3-Hydroxy-2-butanone 1,4-Dibromo-2,3-butenedione	Methyl methacrylate Styrene	1053, 1054, 1056
1-Phenyl-acetone 1,1-Diphenyl-acetone 1,3-Diphenyl-acetone 3-Phenyl-2,3-propanedione	Methyl methacrylate Styrene	1056

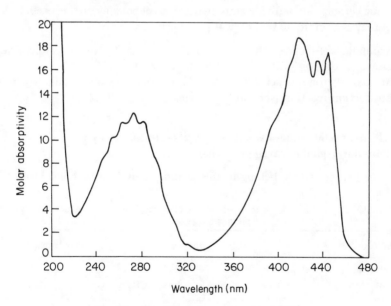

Fig. 7.2 Absorption spectrum for biacetyl[591]

7.9.3. Biacetyl Monooximes

Biacetyl monooximes (7.41), upon u.v. irradiation, are photolysed according to mechanism:[1659]

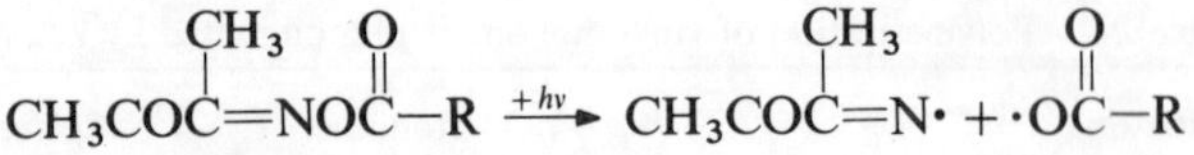

$$CH_3COC{=}NOC{-}R \xrightarrow{+h\nu} CH_3COC{=}N{\cdot} + {\cdot}OC{-}R \qquad (7.73)$$

with structures labeled:

$$\overset{CH_3}{\underset{(7.41)}{}} \overset{O}{} \qquad \overset{CH_3}{\underset{(7.42)}{}} \qquad \overset{O}{\underset{(7.43)}{}}$$

$$\overset{O}{R{-}CO{\cdot}} \longrightarrow R{\cdot} + CO_2 \qquad\qquad (7.74)$$

$$(7.44)$$

where R $= -CH_3,\ -OC_2H_5,\ -\langle\bigcirc\rangle,\ -NH{-}\langle\bigcirc\rangle$

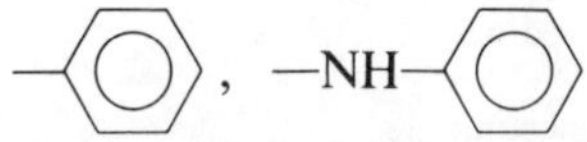

Radicals (7.43) and (7.44) can initiate radical polymerization, whereas the radical (7.42) is not capable of acting as initiating radical.

Other oxime derivatives, such as dimethylglyoximes or ketone oximes, are not able to initiate radical formation.

7.9.4. Aromatic Ketones

Aromatic ketones can be divided into two groups depending on the mechanism in which they photoinitiate free radical polymerization:

(i) Aromatic ketones which act as photoinitiators by the unimolecular α-cleavage reaction (Table 7.3) (cf Section 7.9.4.1).

(ii) Aromatic ketones which act as photoinitiators by the bimolecular hydrogen abstraction reaction (see Table 7.10) (cf. Section 7.9.4.2).

7.9.4.1. Aromatic ketones which act as photoinitiators by a unimolecular α-photocleavage reaction

Aromatic ketones which belong to this group are listed in Table 7.3.

Table 7.3 Aromatic ketones which act as photoinitiators by a unimolecular α-photocleavage reaction

Name	Structure	Trade mark
Acetophenone	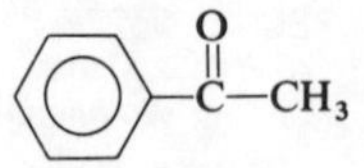 $\langle\bigcirc\rangle{-}\overset{O}{\overset{\|}{C}}{-}CH_3$	
Chlorinated acetophenone	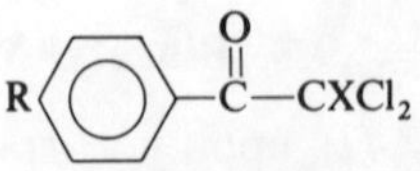 $R\langle\bigcirc\rangle{-}\overset{O}{\overset{\|}{C}}{-}CXCl_2$ X = H, Cl	

Table 7.3 (*Contd*)

Name	Structure	Trade mark
Dialkoxyacetophenones		Irgacure 651 (R = CH$_3$)
Dialkylhydroxyacetophenones		Darocure 1173 (R = CH$_3$)
Dialkylhydroxyacetophenone alkyl ethers		Darocure 1116 (R' = (CH$_3$)$_3$ C, R = CH$_3$, R'' = H)
		Trigonal P1 (R' = (CH$_3$)$_3$C, R = R'' = Cl)
1-Benzoylcyclohexanol-2		Irgacure 184
Benzoin		
Benzoin acetate		
Benzoin alkyl ethers		Trigonal 14 (R = butyl)
Dimethoxybenzoin		
Deoxybenzoin		
Dibenzyl ketone		

Table 7.3 (*Contd*)

Name	Structure	Trade mark
Acyloxime esters	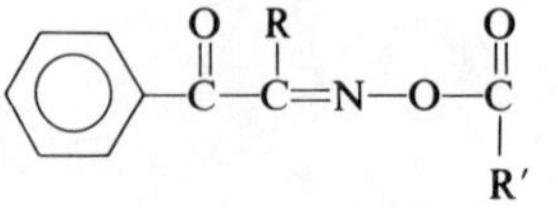	Quantaqure PDO ($R = CH_3$, $R' = C_2H_5$)
Acylphosphine oxides	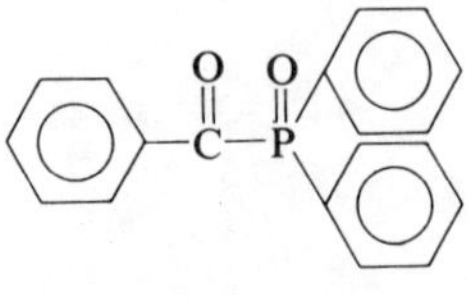	
Acylphosphonates	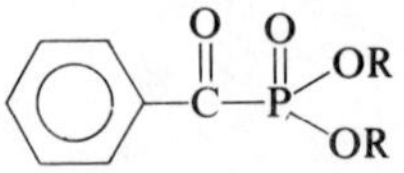	
Ketosulphides	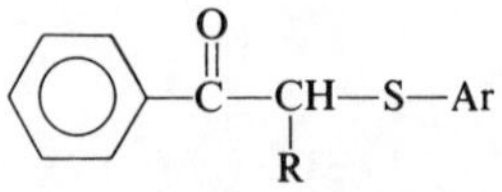	
Dibenzoyl disulphides	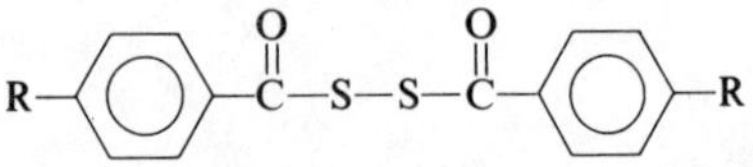	
Diphenyldithiocarbonate	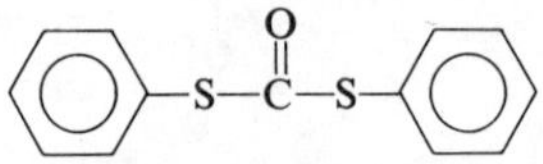	

Note: Darocur is a trade mark of E. Merck; Irgacure is a trade mark of Ciba Geigy; Trigonal is a trade mark of Akzo; and Quantaqure is a trade mark of Ward Blenkinsop.

7.9.4.1.1. Acetophenone

Acetophenone (*7.45*) has an absorption spectrum (Fig. 7.3) with maximum near 280 nm, which is related to (n, π^*) transition involving the non-bonding electrons of the oxygen atom. The absorption band at 330 nm is related to the (π, π^*) transition in aromatic ketones[591] (cf. Section 4.9.4.2.1).

Acetophenone undergoes α-photocleavage at elevated temperatures in the vapour phase:[1007]

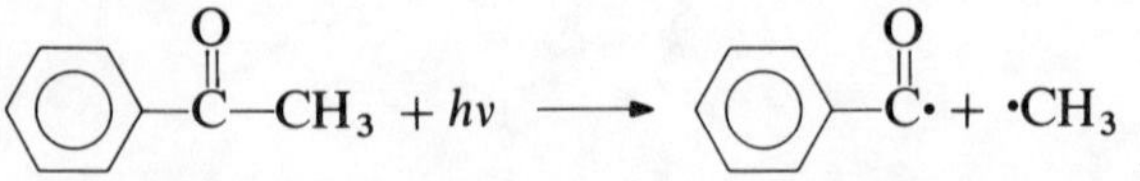

$$(7.75)$$

$$(7.45) \qquad (7.46)$$

Benzoyl radical (*7.46*) is a reactive radical which may initiate free radical polymerization of vinyl monomers.

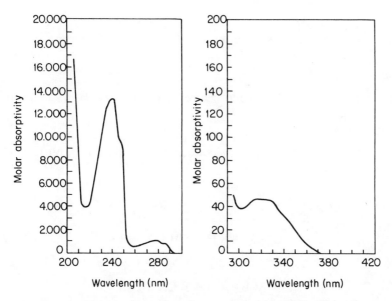

Fig. 7.3 Absorption spectra for acetophenone[591]

Acetophenone is also photoreduced by hydrogen-donor molecules (RH) such as benzene, 2-propanol, thiols, and disulphides:[744]

$$\underset{}{\bigcirc}\!-\!\overset{O}{\underset{}{C}}\!-\!CH_3 + h\nu \longrightarrow \left(\underset{}{\bigcirc}\!-\!\overset{O}{\underset{}{C}}\!-\!CH_3\right)^1 \longrightarrow \left(\underset{}{\bigcirc}\!-\!\overset{O}{\underset{}{C}}\!-\!CH_3\right)^3$$

(7.76)

$$\left(\underset{}{\bigcirc}\!-\!\overset{O}{\underset{}{C}}\!-\!CH_3\right)^3 + RH \longrightarrow \underset{}{\bigcirc}\!-\!\overset{OH}{\underset{\cdot}{C}}\!-\!CH_3 + R\cdot$$

(7.47)

(7.77)

Acetophenone ketyl radicals (7.47) react with each other and produce pinacol (2,2-diphenyl-2,3-butyl glycol) (7.48).

$$2\,\underset{}{\bigcirc}\!-\!\overset{OH}{\underset{\cdot}{C}}\!-\!CH_3 \longrightarrow \underset{}{\bigcirc}\!-\!\overset{OH}{\underset{CH_3}{C}}\!-\!\overset{OH}{\underset{CH_3}{C}}\!-\!\bigcirc$$

(7.78)

(7.48)

7.9.4.1.2. Chlorinated acetophenone derivatives

Substituted di- or tri-chloroacetophenone derivatives (7.49) undergo β-photocleavage in which an reactive chlorine radical is formed:[1555]

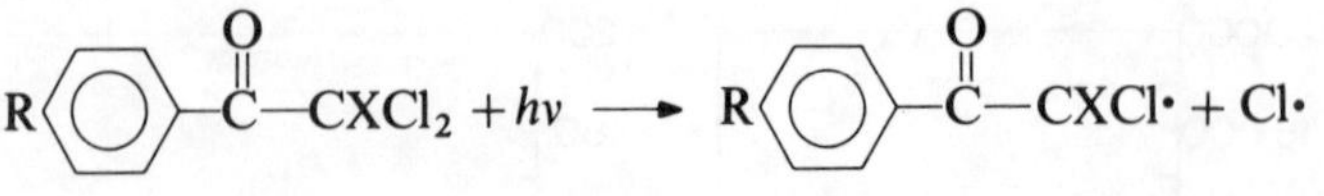

$$R\underset{}{\bigcirc}\overset{O}{\overset{\|}{C}}-CXCl_2 + h\nu \longrightarrow R\underset{}{\bigcirc}\overset{O}{\overset{\|}{C}}-CXCl\cdot + Cl\cdot \qquad (7.79)$$

(7.49)

Substituted di- and tri-chloroacetophenones may also undergo α-photocleavage:

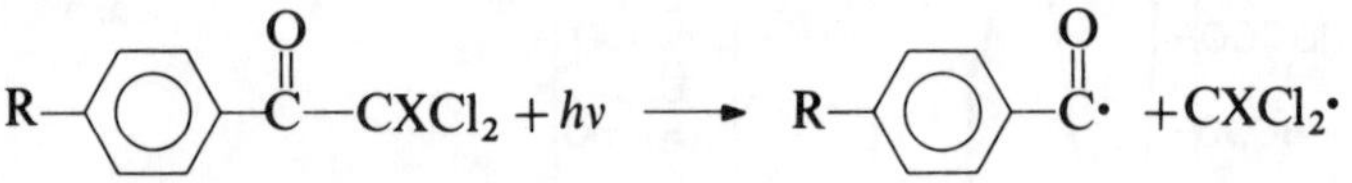

$$R-\underset{}{\bigcirc}-\overset{O}{\overset{\|}{C}}-CXCl_2 + h\nu \longrightarrow R-\underset{}{\bigcirc}-\overset{O}{\overset{\|}{C}}\cdot + CXCl_2\cdot \qquad (7.80)$$

$$R-\underset{}{\bigcirc}-\overset{O}{\overset{\|}{C}}\cdot + R'H \longrightarrow R-\underset{}{\bigcirc}-\overset{O}{\overset{\|}{C}}H + R'\cdot \qquad (7.81)$$

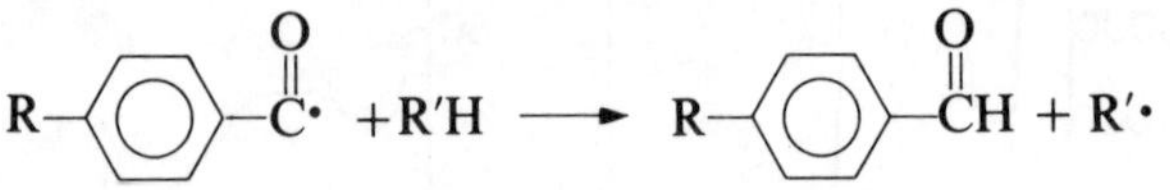

$$CXCl_2\cdot + R'H \rightarrow CHXCl_2 + R'\cdot \qquad (7.82)$$

$$CXCl_2\cdot \rightarrow CXCl\cdot + Cl\cdot \qquad (7.83)$$

$$CXCl\cdot + R'H \rightarrow R'CX\cdot + HCl \qquad (7.84)$$

The chlorine radical (Cl·) efficiently initiates radical polymerization. A strong HCl acid generated from chlorinated acetophenone photoinitiators can be used to cure acid curing resins such as urea formaldehyde and melamine types.[371] A disadvantage of this type of photoinitiators is that the HCl formed is highly corrosive.

7.9.4.1.3. Dialkoxyacetophenones

A number of dialkoxyacetophenones based on α-substitution of the acetophenone nucleus have been synthesized (Table 7.4). The photochemistry of one of them, i.e. diethoxyacetophenone (7.50), has been studied in detail.[473, 1484, 2816, 2817, 2881, 3178] This photoinitiator undergoes two types of reaction:

Table 7.4 Different types of dialkoxyacetophenones used as photoinitiators[703]

Name	Structure
Dimethoxyacetophenone	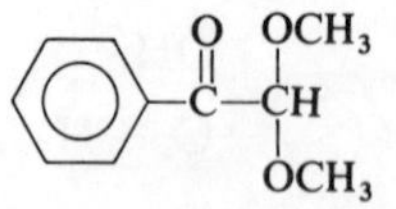 $\bigcirc-\overset{O}{\overset{\|}{C}}-\overset{OCH_3}{\underset{OCH_3}{CH}}$
Diethoxyacetophenone	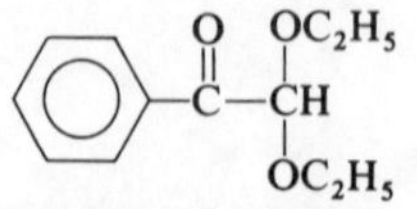 $\bigcirc-\overset{O}{\overset{\|}{C}}-\overset{OC_2H_5}{\underset{OC_2H_5}{CH}}$
Bis(2-chloroethoxy)acetophenone	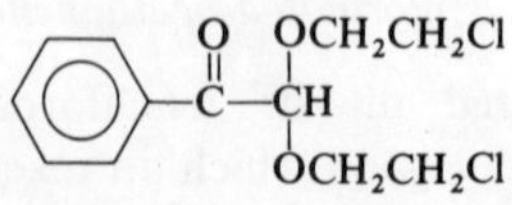 $\bigcirc-\overset{O}{\overset{\|}{C}}-\overset{OCH_2CH_2Cl}{\underset{OCH_2CH_2Cl}{CH}}$

Table 7.4 (*Contd.*)

Name	Structure
Bis(2-methoxyethoxy)acetophenone	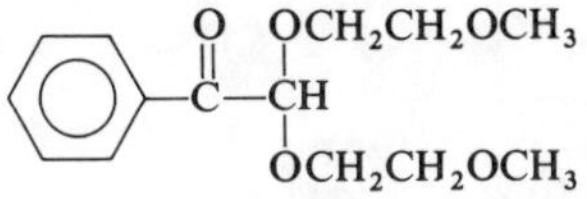
Bis(2-methylpropoxy)acetophenone	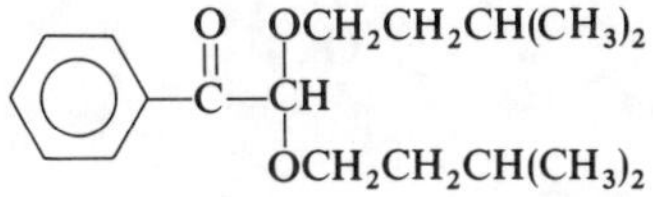
Bis(2-propoxyacetophenone)	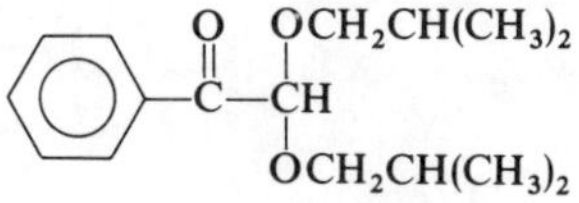
Ethoxybutoxyacetophenone	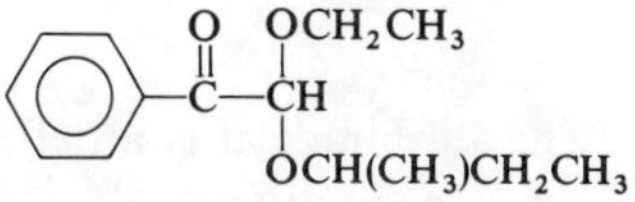
Bis(cyclohexyloxy)acetophenone	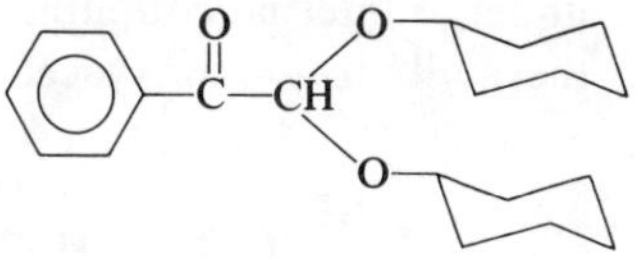
Bis(benzyloxy)acetophenone	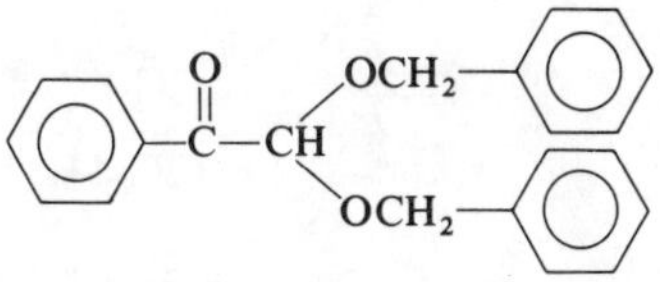

Note: The photochemistry of these dialkoxyacetophenones has been expected to be similar to that outlined for diethoxyacetophenone.

(i) α-Photocleavage (Norrish Type I cleavage) which gives benzoyl radical (7.51) and diethoxyalkyl radical (7.52):

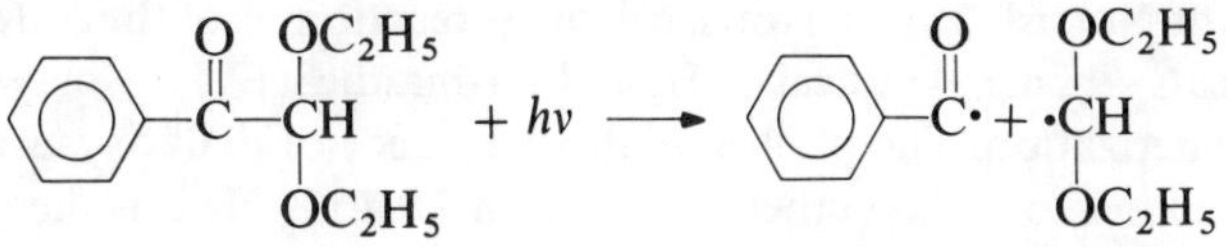

$$(7.85)$$

$$(7.50) \qquad (7.51) \quad (7.52)$$

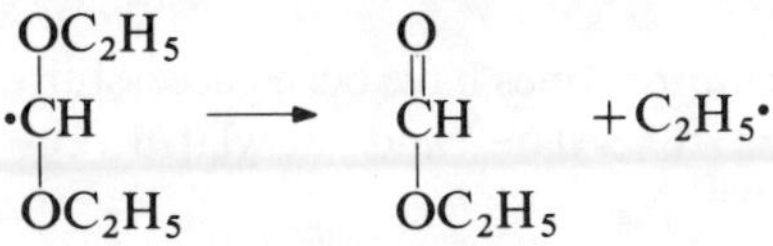

$$(7.86)$$

(ii) Intramolecular hydrogen abstraction (Norrish Type II cleavage) with formation of 1,4-biradical (7.53):[473,703,2816]

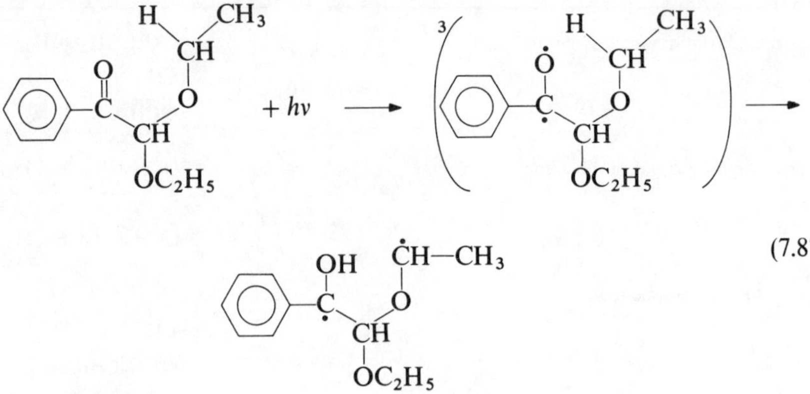

(7.87)

(7.53)

The 1,4-biradical is an effective initiator radical of the polymerization of acrylate monomers.[1484,2816] In the absence of monomers, the 1,4-biradicals undergo internal coupling to form oxetanol (7.54) which disproportionates thermally to ω-ethoxyacetophenone (7.55) and acetaldehyde:

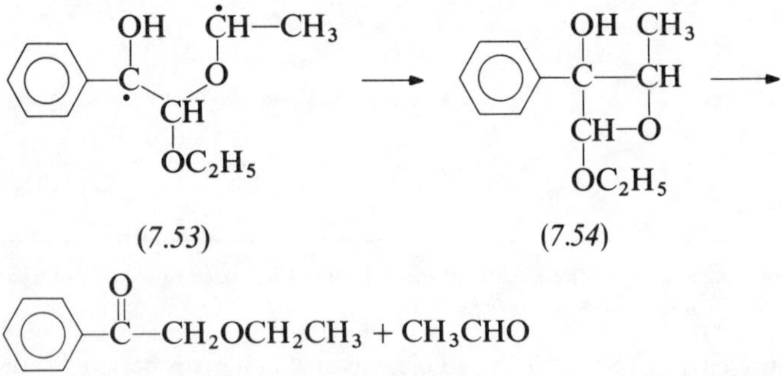

(7.53)   (7.54)

(7.88)

(7.55)

The ratio of Norrish Type I/Norrish Type II reactions is of the order 2:1. Both type of radicals—benzoyl radical (7.51) and 1,4-biradical (7.53)—may initiate free radical polymerization. The photoinitiator efficiency of dialkoxyacetophenones is greater than benzoin alkyl ethers (cf. Section 7.9.4.1.8). This is the result of the secondary fragmentation of the diethoxyalkyl radical (7.52) to alkyl radical ($C_2H_5\cdot$) which may participate in the initiation and radical combination reactions.

Dialkoxyacetophenones have been successfully employed for the photocuring of hexanediol diacrylate and acrylated epoxy, and acrylated urethane formulations.[703]

7.9.4.1.4. Dialkylhydroxyacetophenones and their alkyl ethers

The α,α-dimethyl-α-hydroxyacetophenone (7.56) undergoes α-photocleavage to benzoyl radical (7.57) and dimethylhydroxymethyl radical (7.58):[1034, 1035, 1579, 2881, 3168]

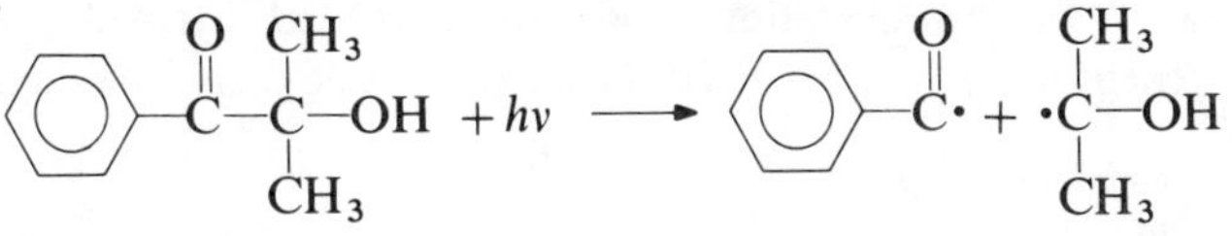

$$\text{(7.89)}$$

(7.56)   (7.57)   (7.58)

whereas α,α-dimethyl-α-hydroxyacetophenone alkyl ether (7.59) abstracts hydrogen from a hydrogen-donor molecules (RH) and produce ketyl radicals (7.60):

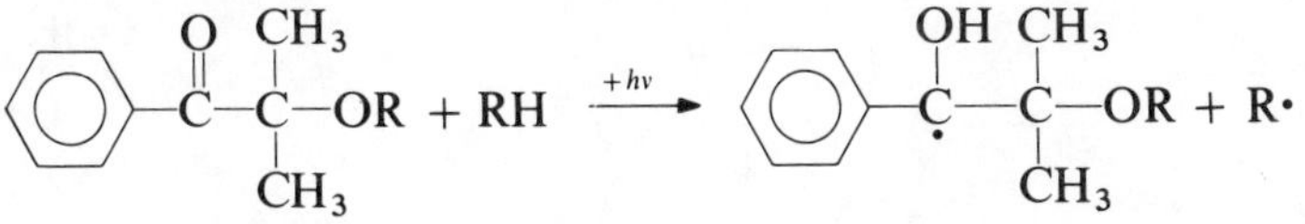

$$\text{(7.90)}$$

(7.59)   (7.60)

Dialkylhydroxyacetophenones are capable of photoinitiating curing of epoxyacrylates, whereas their alkyl ethers are not capable of photoinitiating curing without hydrogen-donor molecules e.g. N-methyldiethanolamine. Dialkylhydroxyacetophenones were also used for photo-curing of trimethylpropane triacrylates.[920]

The dimethylhydroxyacetophenone alkyl ether having a structure (7.61), employed at concentration of 0.1 wt%, enhances the surface hardness of films cured in air.[1706,1707]

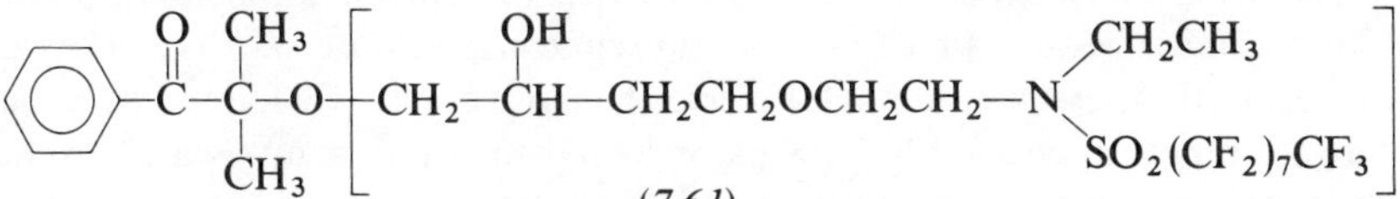

(7.61)

7.9.4.1.5. 1-Benzoylcyclohexanol-2

1-Benzoylcyclohexanol-2 (7.62) undergoes α-photocleavage to a benzoyl radical (7.63) and a hydroxycyclohexyl radical (7.64):

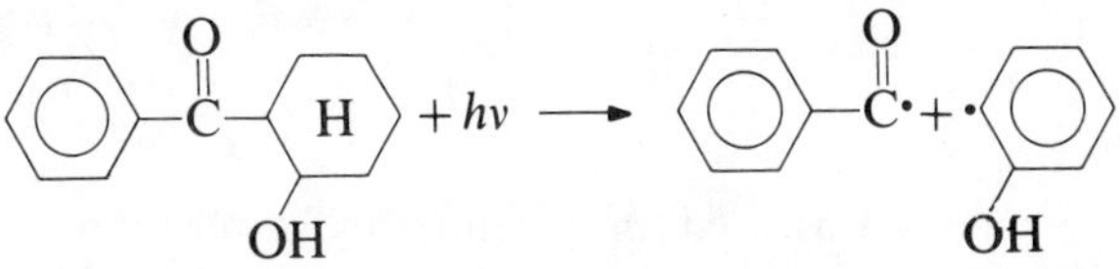

$$\text{(7.91)}$$

(7.62)   (7.63)   (7.64)

Both radicals produced may, e.g., initiate polymerization of methyl methacrylate (in sodium dodecyl sulphate micelles),[2320] curing of trimethylpropane triacetate,[920] or grafting epoxy acrylate coatings on poly(vinyl chloride).[910]

7.9.4.1.6. Benzoin

Benzoin (7.65) is one of the most investigated and used photoinitiators.[701,732, 734,2111,2231,2242,2276,2459,2884,3178]

The benzoin absorption spectrum is shown in Fig. 7.4. Benzoin undergoes α-photocleavage via dissociation of the excited triplet state (T_1) $(E_T = 73.0 \, kcal \, mol^{-1})$ to a benzoyl radical (7.66) and a hydroxybenzyl radical (7.67):

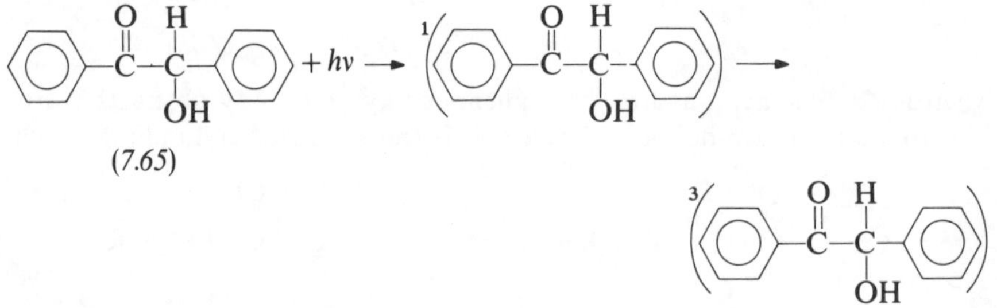

(7.65)

(7.92)

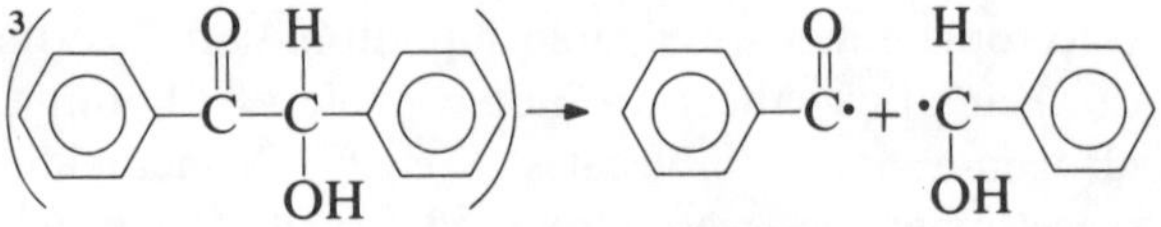

(7.93)

(7.66)   (7.67)

This mechanism has been proved by employing nitroso compounds as spin traps and further characterization of new radicals formed by ESR spectroscopy,[2242] and also by CIDNP experiments.[732] In the case of benzoin and benzoin acetate (7.70), α-photocleavage, which occurs via excited triplet state (T_1), is a relatively slow reaction, whereas α-photocleavage of benzoin ethers (7.74) appears to be a relatively rapid process (about 100 ps or less), because it occurs via an excited singlet state (S_1) or a very short-lived excited triplet state (T_1).[2155,2276,2884,3178]

The benzoin molecule in the excited triplet state (T_1) can abstract hydrogen from hydrogen-donor molecules (RH) and produce benzoin ketyl radicals (7.68):

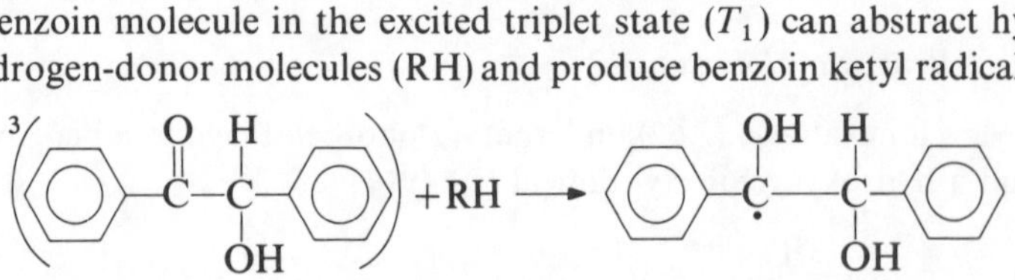

(7.68)   (7.94)

This process has been observed in solution by nanosecond laser spectroscopy,[488,1149] and can be concomitant with the generation of benzoyl radicals (7.66).

Benzoin itself is rather poor photoinitiator of free radical polymerization, and has a low solubility in resin–monomer systems, and therefore benzoin alkyl ethers have much more interest (cf. Section 7.9.4.1.8).

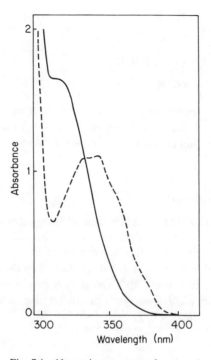

Fig. 7.4 Absorption spectrum for: (——)
benzoin and (– – –) benzoin methyl ether

A very effective photoinitiator such as sulphonic ester of α-hydroxymethyl benzoin (7.69) has been reported recently.[1234]

(7.69)

7.9.4.1.7. Benzoin acetate

Benzoin acetate (7.70) undergoes α-photocleavage via dissociation of the excited triplet state (T_1) to benzil (7.71) and pinacol acetate (7.72), along with 10–20% of cyclization product 2-phenylbenzofuran (7.73):[2155, 2276, 3213, 3304, 3305]

(7.70)   (7.71)

$$+ \quad \underset{\substack{CH_3OCO \quad OCOCH_3 \\ (7.72)}}{\overset{\overset{\displaystyle H \quad H}{\underset{|}{\underset{\displaystyle C}{—}}\underset{|}{\underset{\displaystyle C}{—}}}}{\bigcirc}} \quad + \quad \underset{(7.73)}{\bigcirc\hspace{-0.3em}\bigcirc} \hspace{4em} (7.95)$$

Triplet quenching experiments and CIDNP studies established that α-photocleavage occurs from a triplet excited state (T_1) for benzoin acetate.[732] Both α-photocleavage and cyclization processes are completely quenched by 1 M of dodecanethiol.[2276]

7.9.4.1.8. Benzoin alkyl ethers

Benzoin alkyl ethers (7.74) have similar absorption spectra in ethanol to benzoin (Fig. 7.4), whereas spectra are different when the solvent is methyl methacrylate.[1728,1729] In this less polar solvent there is intramolecular hydrogen bonding in the case of benzoin, which is disrupted when the solvent is changed to methanol or ethanol, and results in a red shift for the n,π* absorption.

Benzoin alkyl ethers (7.74) undergo α-photocleavage to benzoyl radical (7.76) and alkoxybenzyl radical (7.77):[417,602,1418,1476,1477,1552,1557,2155,2274,2276, 2277,2321,2817,2879,2881,2883,2885,3146,3178]

$$\underset{(7.74)}{\overset{\overset{\displaystyle O \quad H}{\underset{|}{\underset{\displaystyle C}{—}}\underset{|}{\underset{\displaystyle C}{—}}}}{\bigcirc—\underset{OR}{}—\bigcirc}} + h\nu \longrightarrow \left[\underset{(7.75)}{\overset{\overset{\displaystyle O \quad H}{\underset{\sigma^-}{C}\cdots\underset{\sigma^+ \atop OR}{C}}}{\bigcirc—\hspace{2em}—\bigcirc}} \right]$$

$$\longrightarrow \underset{(7.76)}{\overset{\overset{\displaystyle O}{\underset{|}{C}}}{\bigcirc—C\bullet}} + \underset{(7.77)}{\overset{\overset{\displaystyle H}{\underset{OR}{C}}}{\bullet C—\bigcirc}} \hspace{4em} (7.96)$$

The inability to quench photocleavage of benzoin alkyl ethers with triplet quenching agents suggested that reaction may occur via dissociation of the excited singlet states. This lack of quenching of benzoin alkyl ether photocleavage indicates that the rate constant for α-cleavage is of the order of 10^8–10^{10} s^{-1}.[1552,1557,2155,2276,2884] The differences in rate constants have been attributed to polar contributions in the activated complex (7.75) which is cleaved to benzoyl radical (7.76) and alkoxybenzyl radical (7.77). Electron-donating groups on the benzyl group and electron-attracting groups on the benzoyl group facilitates α-photocleavage.[1552,1557,2276,2884,3403] Hypothesis of the activated complex explains why benzoin alkyl ethers possess large rate constants as well as relatively high quantum yield of α-photocleavage, and why they are not equally effective as photoinitiators for polymerization (cf. Section 7.9.4.1.9).

Benzoyl radical (7.76) can dimerize to benzil (7.78) or can abstract hydrogen from a hydrogen-donor molecule (RH) and produce benzaldehyde (7.79) and alkyl radical (R·):

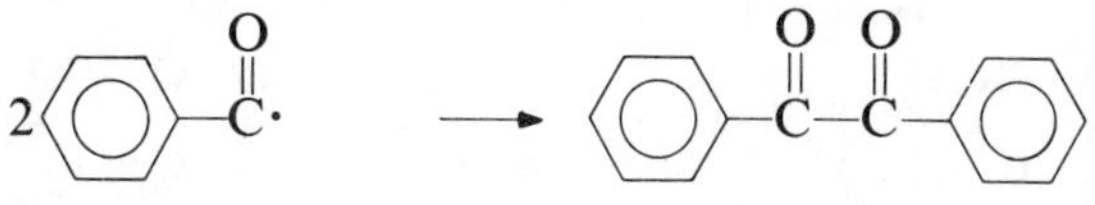

$$(7.97)$$

(7.76)   (7.78)

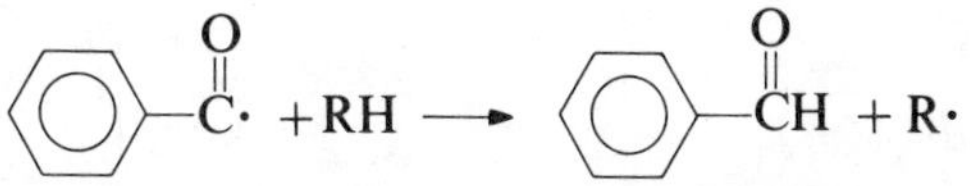

$$(7.98)$$

(7.79)

Addition of low concentration (0.01 M) of dodecanethiol which is an efficient scavenger of benzoyl radicals,[1553] results in an increased yield of benzaldehyde and complete suppression of benzyl formation.[2276]

Alkoxybenzyl radicals (7.77) can dimerize to pinacol ether (7.80) with a yield of 60–70% or be cleaved to benzaldehyde (7.79) and alkyl radical (R·):

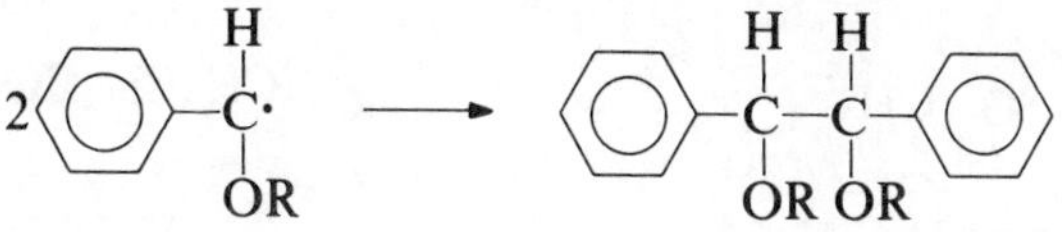

$$(7.99)$$

(7.77)   (7.80)

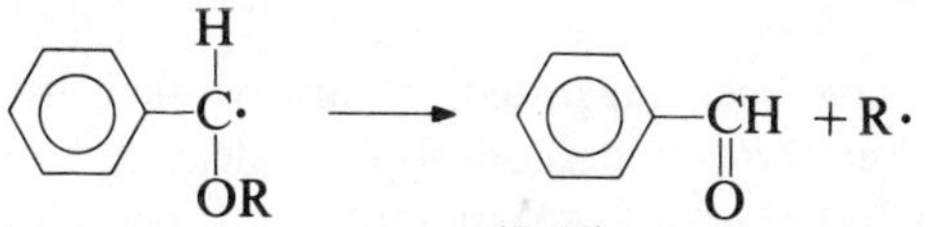

$$(7.100)$$

(7.79)

Benzoin alkyl ethers are easily oxidized, probably owing to the activated hydrogen in the α-position of the ether group.[1557] Benzoin ether hydroperoxides (7.81) are formed, which may cleave into reactive benzoyl radicals and hydroxyl radicals (HO·):[2757]

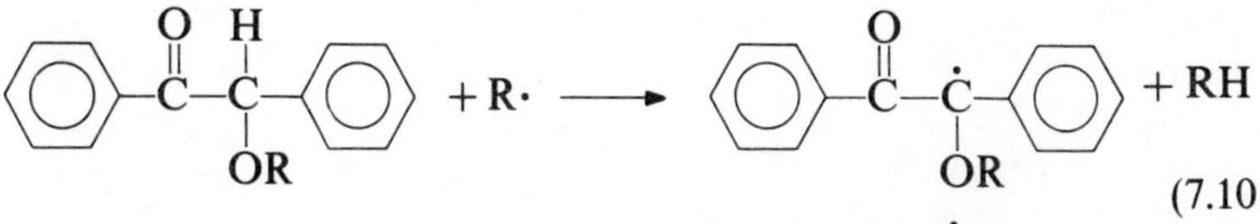

$$(7.101)$$

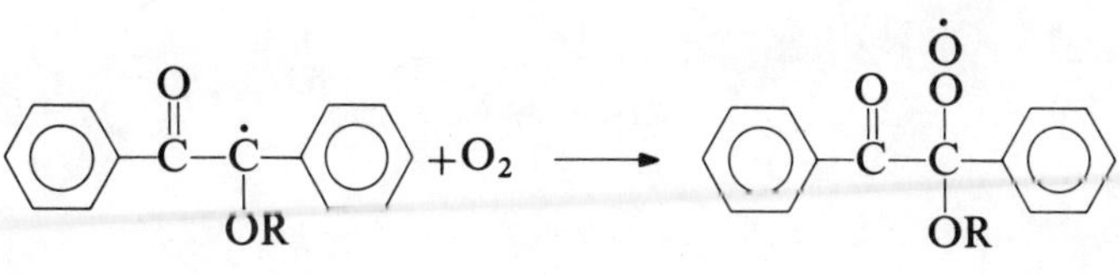

$$(7.102)$$

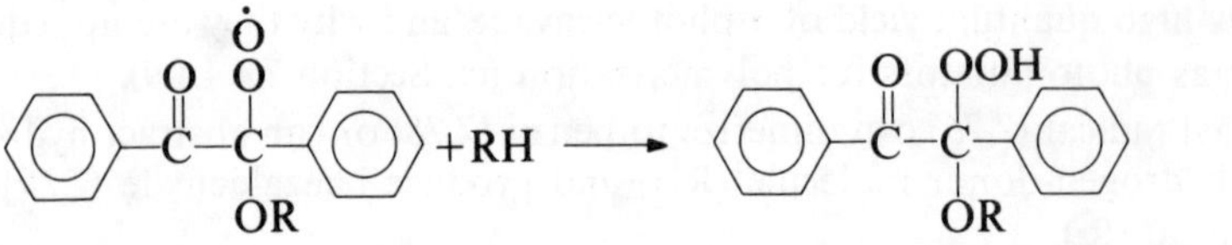

$$(7.103)$$

$$(7.81)$$

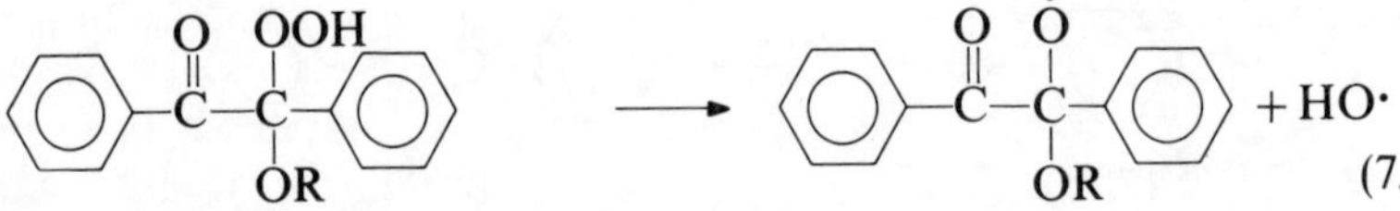

$$(7.104)$$

$$(7.105)$$

These reactions lead to poor pot stability (shelf-life).[1557] Benzoin ether hydroperoxides (7.81) are thermally unstable, particularly in the presence of transition metals. The structure of alkyl benzoin ethers affects the pot stability. Short-chain alkyl or non-branched ethers, such as benzoin methylether, are the worst. Substitution of the benzylic H-atom, as in the case of α-methyl benzoin methyl ether (7.82), or the ester derivative (7.83), enhance shelf-life.[1557]

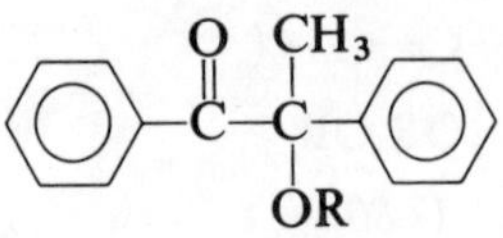

$$(7.82)$$

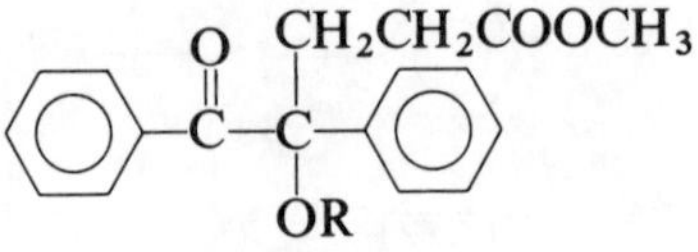

$$(7.83)$$

Various additives have been suggested to improve shelf-life stability.[2996]

Benzoin phenyl ether (7.84) undergoes β-photocleavage to small amounts of deoxybenzoin (7.85) and phenol (7.86) in addition to benzaldehyde, benzil and pinacol ethers:[2276]

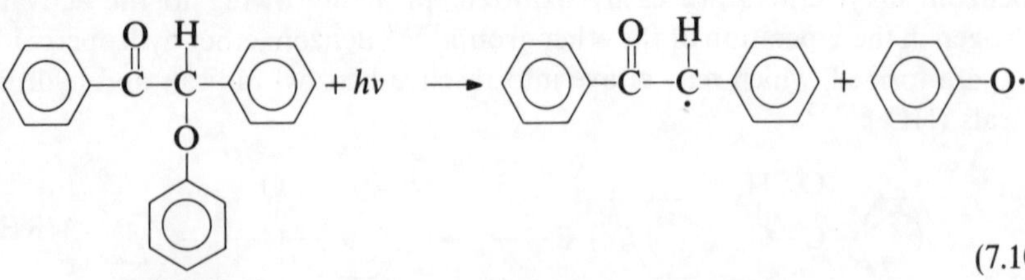

$$(7.84)$$

$$(7.106)$$

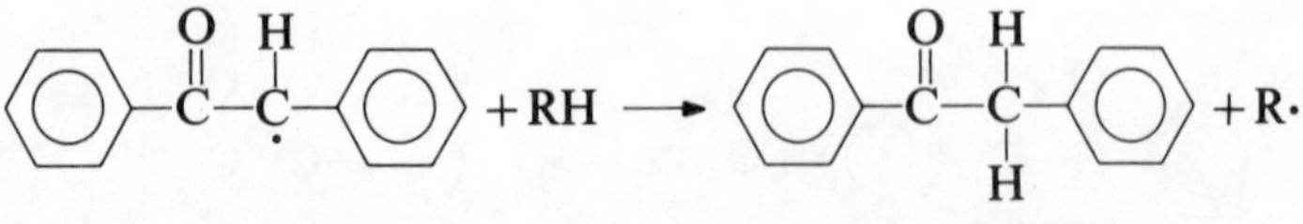

$$(7.85) \qquad (7.107)$$

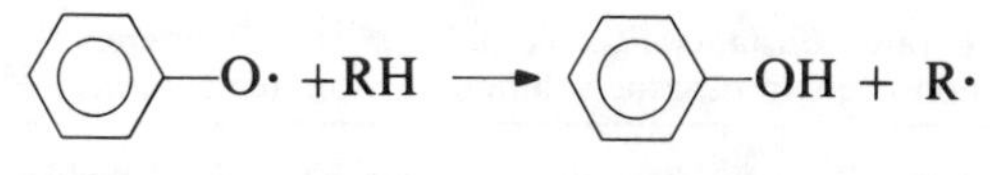

$$\text{C}_6\text{H}_5\text{–O}\cdot + \text{RH} \longrightarrow \text{C}_6\text{H}_5\text{–OH} + \text{R}\cdot$$

(7.86)

(7.108)

Benzoin phenyl ether is $> 10^3$ times more reactive than benzoin alkyl ethers.

7.9.4.1.9. Photoinitiation of polymerization by benzoin and its derivatives

Polymerization is initiated by radicals (R·) formed upon the photocleavage of initiator (I) (benzoin or its derivatives):

$$I + h\nu \rightarrow I^* \tag{7.109}$$

$$I^* \overset{k_d}{\rightarrow} R_1\cdot + R_2 + \ldots + R_n\cdot \tag{7.110}$$

where I* is the excited singlet and/or triplet state of photoinitiator. The rate constants of radical formation (rate constants of dissociation) (k_d) for different benzoin derivatives are listed in Table 7.5. In the case of benzoin methyl ether and benzoin isopropyl ether it is difficult to determine the k_d values, and they were estimated as being greater than 10^9 s^{-1}.

Table 7.5 Rate constants of the α-photocleavage (k_d) of benzoin derivatives[3213]

Photoinitiator	k_d (s^{-1})
Benzoin	1.1×10^6
Benzoin acetate	5.3×10^7
Deoxybenzoin	1.3×10^6
α-Methyldeoxybenzoin	1.1×10^7
α-Phenyldeoxybenzoin	3.7×10^6

Vinyl monomers (M) at high concentrations quench effectively the excited triplet state of benzoin and its derivatives:

$$I^* + M \overset{k_Q}{\rightarrow} I + M \tag{7.111}$$

The quenching rate constants (k_Q) for different benzoin derivatives are given in Table 7.6.

Initiation of monomer (M) polymerization by initiator radicals (R·) formed from photocleavage of initiator (I):

$$R\cdot + M \overset{k_i}{\rightarrow} R\text{–}M\cdot \tag{7.112}$$

Table 7.6 Quenching rate constants (k_Q) $(1\ mol^{-1}\ s^{-1})$ for different benzoin derivatives by monomers in benzene solution at room temperature[3213]

Monomer	Benzoin	Benzoin acetate	Deoxy-benzoin	α-Methyldeoxy benzoin	α-Phenyldeoxy-benzoin
Styrene	7.5×10^8* 8.1×10^9†	4.8×10^9	6.5×10^9	1.5×10^9	1.9×10^8
Methyl methacrylate	4.4×10^7* 4.8×10^8†	8.0×10^8	1.0×10^9	2.0×10^8	1.3×10^7
Acrylonitrile		1.3×10^9	7.0×10^8	3.8×10^8	1.8×10^8
Vinyl acetate		6.0×10^7	1.7×10^7	$< 10^7$	5.7×10^6

Note: *Calculated with $k_d = 1.1 \times 10^8\ s^{-1}$, †calculated with $k_d = 1.1 \times 10^9\ s^{-1}$

depends on the following factors:

(i) Structure of free radicals formed and their reactivity. Under conditions of relatively high radical and low monomer concentrations (styrene), the benzoyl radical (7.76) is primarily responsible for initiation of polymerization, whereas the alkoxybenzyl radical (7.77) participates, predominantly, as a chain terminating agent.[1477]

(ii) Secondary reactions of free radicals, e.g. dimerization (cf. reactions 7.97 and 7.99), hydrogen abstraction (cf. reaction 7.98), or secondary cleavages (cf. reaction 7.100). The last reaction is not important in the case of the α-methoxybenzyl radical,[602] and this has been additionally supported by ESR studies.[3428]

(iii) Structure of monomer used for polymerization. In the case of benzoin methyl ethers it has been shown that both benzoyl and α-methoxybenzyl radicals formed during photocleavage are equally effective as initiator radicals in the polymerization of methyl acrylate and methyl methacrylate.[602,701,1147,2545,2883,2885,2886] On the other hand, with styrene, the benzoyl radical was more effective.[1477,1557,2155,2276] With equal concentrations (0.1 M) of benzoin methyl ether and styrene, the major products were pinacol (7.87) and an adduct (7.88). Formation of the adduct with n = 2 has importance at higher styrene concentrations (1.0 M).

(7.87)   (7.88)

The initiation rate constants (k_i) for radicals formed from different benzoin derivatives are given in Table 7.7.

Table 7.7 Initiation rate constants (k_i) $(l \, mol^{-1} \, s^{-1})$ for radicals formed from different benzoin derivatives[3213]

Monomer	Benzoin	Benzoin acetate	Benzoin methyl ether	α-Phenoxy-benzoin
Methyl methacrylate			0.9×10^5	5×10^4
Vinyl acetate	1.2×10^5	1.2×10^5	1.5×10^5	
Styrene			1.6×10^5	
Acrylonitrile			0.2×10^5	1.5×10^3

The PROBABILITY (α_R) for the formation of radicals (R·) capable of initiating the radical polymerization of monomer (M) is given by equation:

$$\alpha_R = \frac{k_i[M]}{k_i[M] + k_Q[M]} \tag{7.113}$$

The values of α_R and quantum yields of initiation of radical chains (Φ) for methyl methacrylate and styrene polymerization are given in Table 7.8.

Table 7.8 The probability (α_R) values and quantum yields of initiation of radical chains (Φ) of monomers in benzene solution at room temperature[3213]

Photoinitiator	Monomer	α_R	Φ
Benzoin	Methyl methacrylate	3.3×10^{-1}	2.35×10^{-1}
Benzoin	Styrene	2.9×10^{-1}	5.0×10^{-1}
Benzoin acetate	Methyl methacrylate	1.3×10^{-2}	2.7×10^{-3}
Benzoin acetate	Styrene	2.2×10^{-3}	
Benzoin methyl ether	Methyl methacrylate	9.1×10^{-1}	4.8×10^{-1}
Benzoin methyl ether	Styrene	5.4×10^{-1}	2.5×10^{-1}

Addition of triethanolamine causes a slight deterioration in the polymerization of methyl acrylate with benzoin methyl ether.[3179]

Benzoin and its derivatives were widely used as photoinitiators for polymerization of vinyl monomers (Table 7.9). Benzoin has also been used as photoinitiator

Table 7.9 Vinyl monomers polymerized with benzoin and its derivatives as photoinitiators

Monomer	References
Methyl methacrylate	602, 1151, 1728, 2308, 2320, 2492, 2883, 2885, 2886, 3213
Butyl acrylate	728, 1151
Lauryl acrylate	1750
Vinyl acetate	3213
Styrene	1187, 1477, 1557, 2155, 3213
Acrylonitrile	3213
N-vinylimidazole	1905

of polymerization of dimethyl-acrylate-based dental filling materials to a depth of 1–2 mm,[754,3129] whereas benzoin methyl ether was employed for photoinitiation of polymerization of N-vinylsuccinimide crystal monomer in crystalline matrix photoimaging.[1578]

Benzoin and its derivatives are commercially important photoinitiators of free radical polymerization.[933,2321,2540] Industrial applications of these photoinitiators are:

(i) Curing, which involves the light-induced copolymerization of unsaturated polyesters with styrene.[1552,1553,1555,1556,3146]
(ii) Photocrosslinking of epoxy-acrylate resin.[911]
(iii) Surface coating polymerization.
(iv) Printing inks and printing plates.
(v) Photoimaging.

7.9.4.1.10. Dimethoxybenzoin

Dimethoxybenzoin (α,α-dimethoxy-α-phenylacetophenone) (7.89) has enhanced reactivity as compared with benzoin ethers due partly to an extra electron-donating group, and partly to the secondary fragmentation of the dimethoxybenzyl radical (7.91) following the initial α-photocleavage, yielding a highly reactive methyl radical:[417,2459,2881,2892,3178]

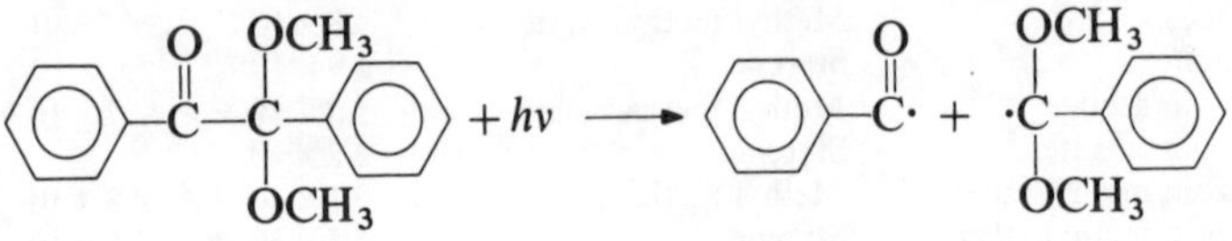

$$(7.89) \qquad (7.90) \qquad (7.91) \qquad (7.114)$$

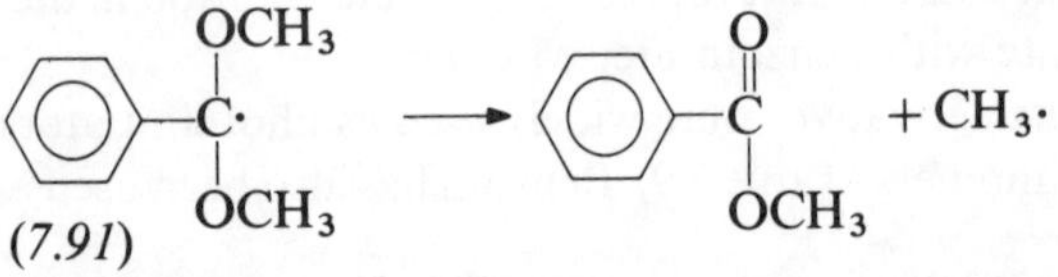

$$(7.91) \qquad \qquad (7.115)$$

The CIDNP studies show that a variety of radical recombination and hydrogen abstraction reactions are possible during the photolysis of dimethoxybenzoin.[473]

Addition of benzoyl radical (7.90) in the *ortho* or *para* position of dimethoxybenzyl radical (7.91) gives semiquinoid structures (7.92) which may cause yellowing during the polymerization process:[417,2881]

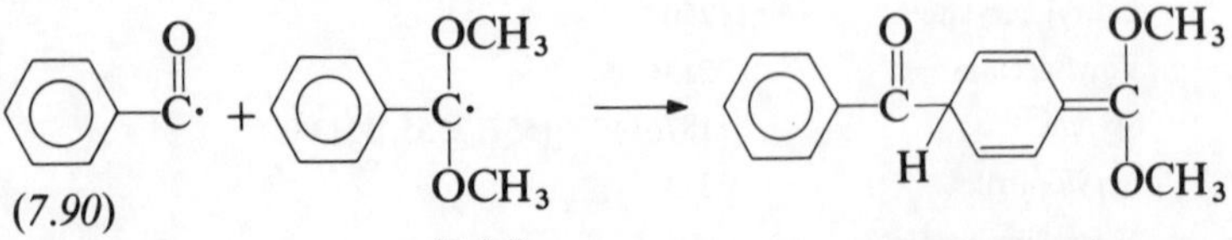

$$(7.90) \qquad (7.91) \qquad (7.92) \qquad (7.116)$$

The photoinitiator efficiency of dimethoxybenzoin is similar to that of benzoin alkyl ethers.[1557,2817] Dimethoxybenzoin has been used as photoinitiator of polymerization of acrylic monomers in bulk[729] or in micelle,[1148] methyl methacrylate,[2491] and trimethylpropane triacrylate.[920] Addition of aliphatic amines provides a marked improvement in surface curing photoinitiated by dimethoxybenzoin.[417]

7.9.4.1.11. Deoxybenzoin and analogues

Deoxybenzoin (α-phenylacetophenone) (7.93) and its analogues such as α-methyldeoxybenzoin (7.96) and α-phenyldeoxybenzoin (7.98) undergo α-photocleavage to produce benzoyl radical (7.94) and phenylmethyl radical (7.95), methylphenylmethyl radical (7.97), and diphenylmethyl radical (7.99), respectively:[184,2491,3146,3213]

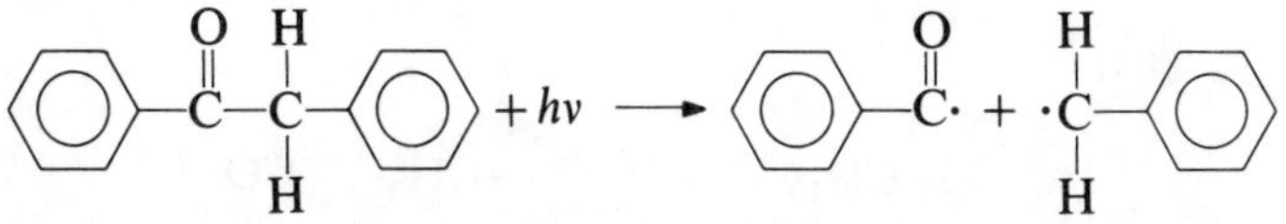

$$(7.117)$$

(7.93)   (7.94)   (7.95)

$$(7.118)$$

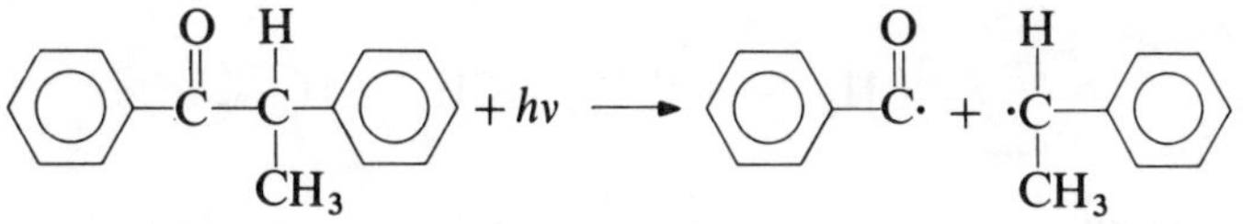

(7.96)   (7.94)   (7.97)

$$(7.119)$$

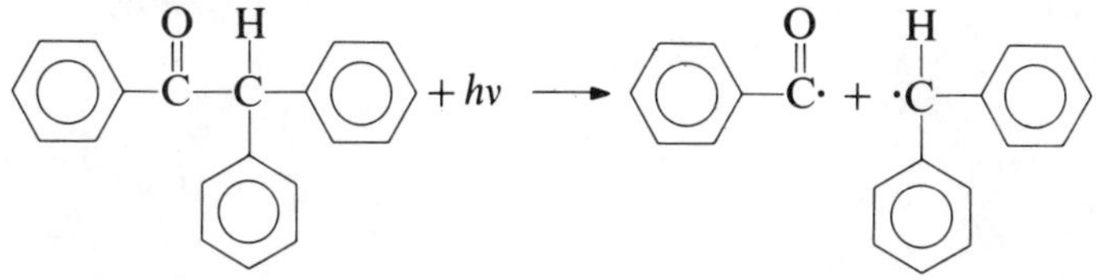

(7.98)   (7.94)   (7.99)

The deoxybenzoin compounds in the excited triplet state (T_1) can abstract hydrogen from hydrogen-donor molecules (RH) (e.g. from tetrahydrofurane (7.100)) and produce a deoxybenzoin ketyl radical (7.101):

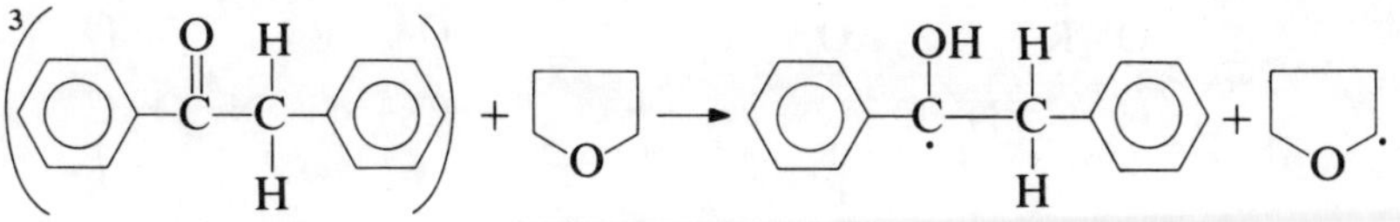

$$(7.120)$$

(7.100)   (7.101)

Deoxybenzoin itself is a rather poor photoinitiator of free radical polymerization because vinyl monomers such as methyl methacrylate, styrene, and acrylonitrile quench the triplet state (T_1) of the deoxybenzoin.

7.9.4.1.12. Dibenzyl ketone

Dibenzyl ketone (7.102) undergoes α-photocleavage via dissociation of the excited triplet state (T_1) to a benzyl radical (7.103) and a phenylacyl radical (7.104):[3683–3685,3691]

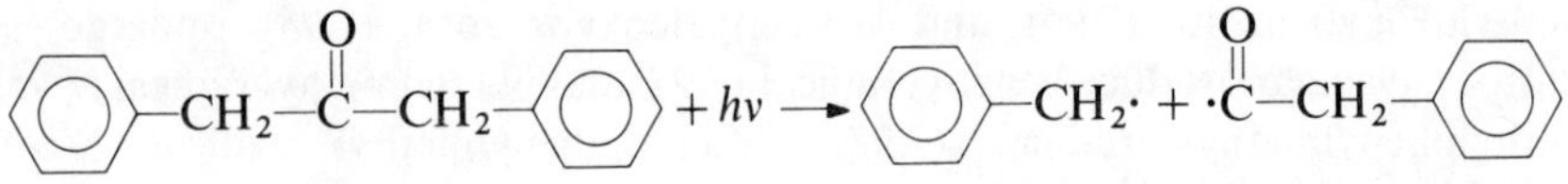

(7.102)　　　　　　　　　　(7.103)　　　　　(7.104)

(7.121)

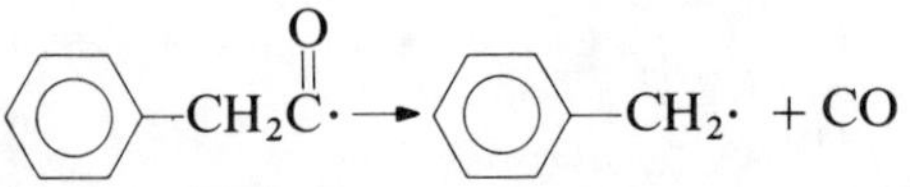

(7.122)

Benzyl radicals dimerize easily to 1,2-diphenylethane:

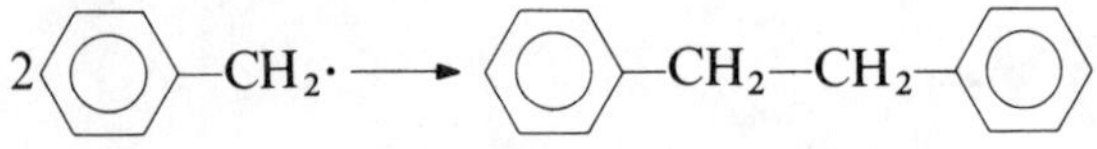

(7.123)

Photolysis of dibenzyl ketone is enhanced by the oxygen:[1062,3119]

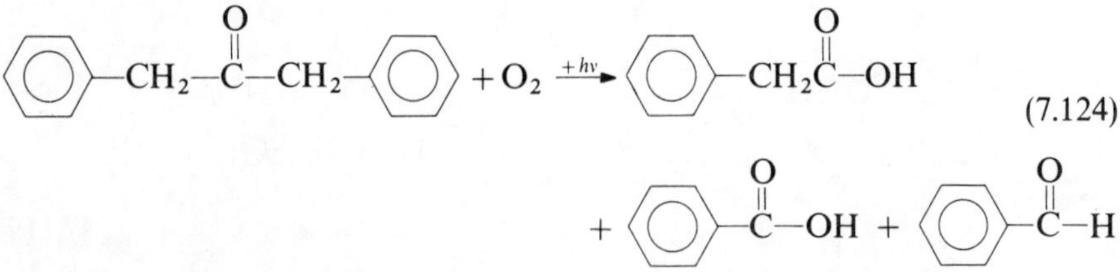

(7.124)

7.9.4.1.13. Acyloxime esters

α-Acyloxime esters (O-acetylated α-oximinoketones) (7.105) undergo α-photocleavage of the N–O bond following n, π* absorption by the $>C=N-$ group and/or n, π* absorption by the C=O group of the $-\underset{\underset{O}{\|}}{C}-C=N-$ fragment:[418,940,1557,2455,2456]

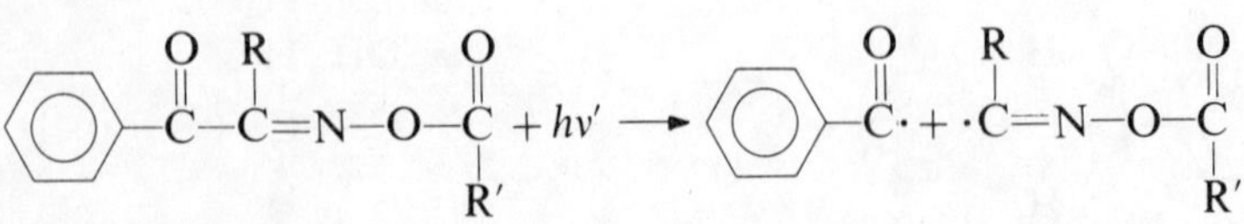

(7.105)　　　　　　　　　　　　　　　　　(7.125)

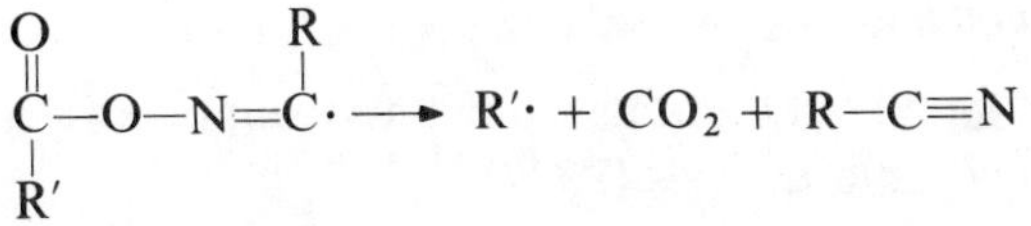

$$\underset{\underset{R'}{|}}{\overset{\overset{O}{\|}}{C}}-O-N=\overset{\overset{R}{|}}{C}\cdot \longrightarrow R'\cdot + CO_2 + R-C\equiv N \qquad (7.126)$$

A similar initial bond cleavage was proposed for the photolysis of camphore oxime[3191] and O-oxime acetates.[3746]

During the photolysis of 1-phenyl-1,2-propanedione-2-O-benzoyl oxime (7.105) in benzene or carbon tetrachloride several decomposition products such as acetonitrile, biphenyl, phenylbenzoate and benzoic acid are formed. Benzophenone and benzyl were found in minor quantities.[940] The higher quantum yield observed in CCl_4 can be attributed to an induced decomposition of the initiator by the solvent radicals.

α-Acyloxime esters photoinitiate polymerization of methyl methacrylate and acrylamide[940, 1659] and curing of unsaturated polyester resins with styrene.[2885] These photoinitiators are particularly effective in hybrid initiator system with benzophenone[2171] and dyes or amines for visible light curing of glass-reinforced polyesters.[35]

7.9.4.1.14. Acylphosphine oxides and acylphosphonates

Acylphosphine oxides (e.g. benzoyldiphenylphosphine oxide) (7.106) and acylphosphonates (e.g. diethylbenzoyl phosphonate) (7.109) undergo α-photocleavage to benzoyl radicals (7.107), diphenylphosphonyl radical (7.108), and diethylphosphonate radical (7.109) with quantum yields 0.3 and 1.0, respectively:[1845, 3490]

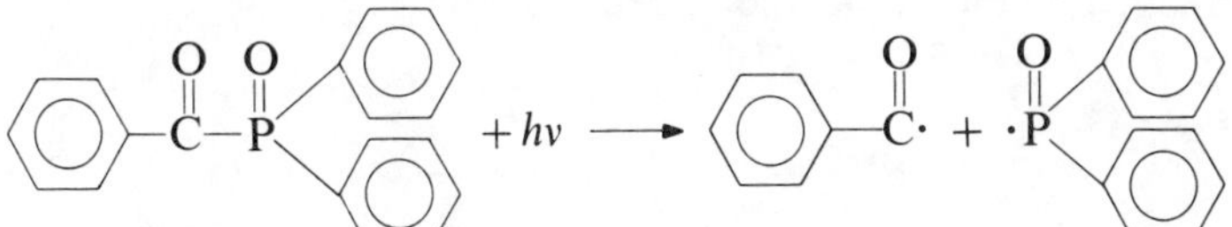

(7.106)   (7.107)   (7.108)   (7.127)

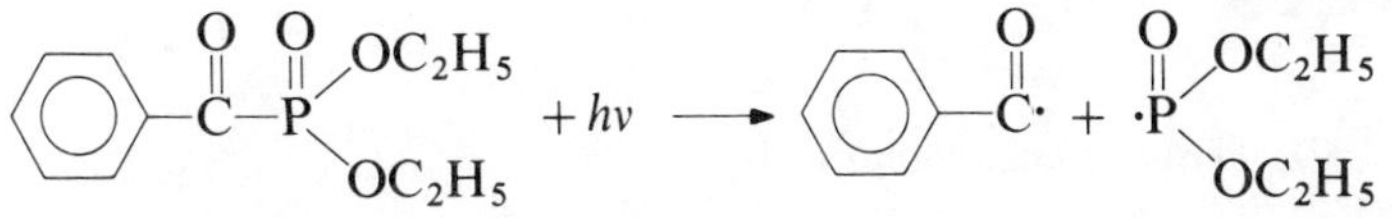

(7.109)   (7.107)   (7.109)   (7.128)

The major problem reported with these photoinitiators is that they are easily decomposed by water and alcohols. The most stable, and showing the greatest photoinitiating efficiency, is 2,4,6-trimethylbenzoyldiphenylphosphine oxide. These photoinitiators have found practical application in photocuring epoxy acrylates containing pigments (TiO_2)[1845] and in photocuring of thick-walled glass-fibre reinforced polyester laminates.[2694]

7.9.4.1.15. Ketosulphides

Phenyl phenacyl sulphide (*7.110*) and desyl aryl sulphides (*7.112*) undergo α-photocleavage to arylthio radicals (*7.111*):[3223, 3671, 3675]

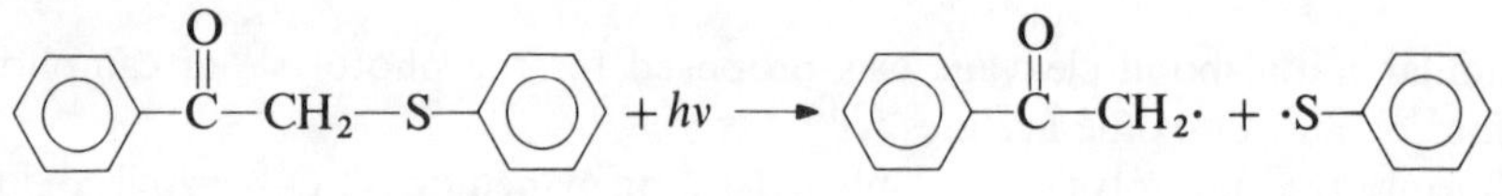

(*7.110*)   (*7.111*)

(7.129)

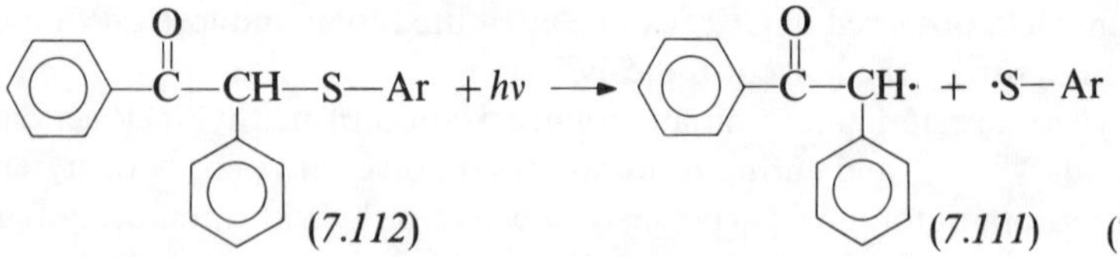

(*7.112*)   (*7.111*)   (7.130)

(Ar = phenyl, *o*-tolyl, *p*-tolyl, *β*-anisyl, *β*-naphthyl)

Arylthio radicals (*7.111*) can initiate free radical polymerization of methyl methacrylate, vinyl acetate, styrene and acrylonitrile,[3671] and tetra-ethylene glycol dimethacrylate.[2930]

7.9.4.1.16. Dibenzoyl disulphides

p,p'-Substituted dibenzoyl disulphides (*7.113*) undergo α-photocleavage to thiobenzoyl radicals (*7.114*), which can both initiate and terminate polymerization of methyl methacrylate and styrene:[3640]

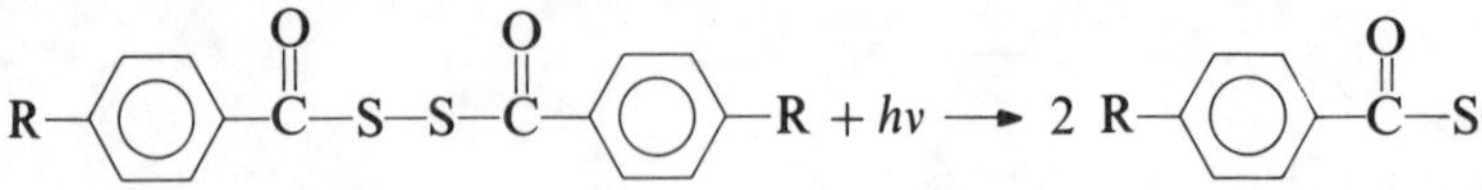

(*7.113*)   (*7.114*)   (7.131)

R = Cl, Br, CH_3, OCH_3, NO_2, CN

7.9.4.1.17. Diphenyldithiocarbonate

S,S'-Diphenyldithiocarbonate (*7.115*) undergoes α-photocleavage to phenylthio radicals (*7.116*), which can both initiate and terminate polymerization of methyl methacrylate and styrene:[3639]

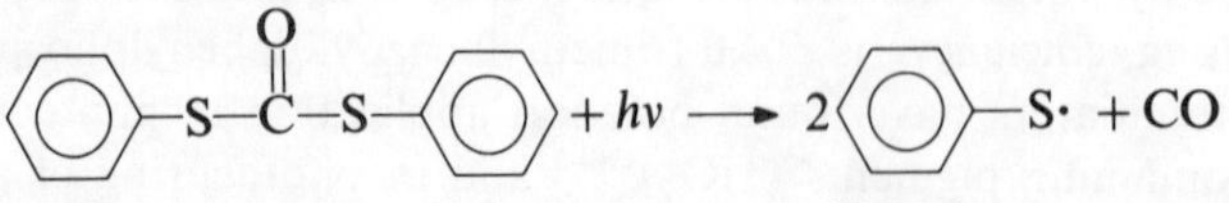

(*7.115*)   (*7.116*)

(7.132)

7.9.4.2. Aromatic ketones which act as photoinitiators by an intermolecular hydrogen abstraction from donor molecules

Aromatic ketones which belongs to this group are listed in Table 7.10.

Table 7.10 Aromatic ketones which act as photoinitiators by inter-molecular hydrogen abstraction from donor molecules

Name	Structure	E_T (kcal mol^{-1})*
Benzophenone	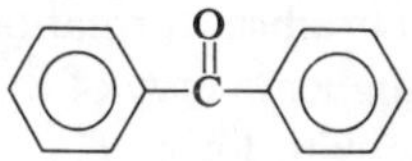	69.3
4,4'-bis(N,N-dialkyl-amino)benzophenone	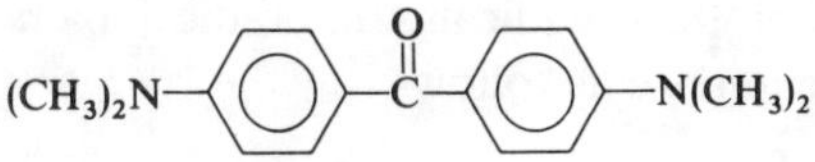	62
Fluorenone	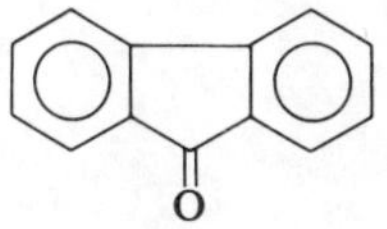	53
Thioxanthone	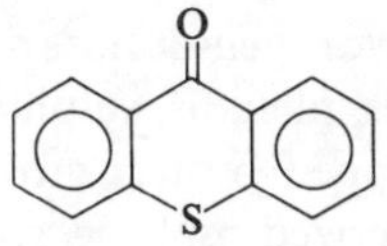	65.5
Benzil	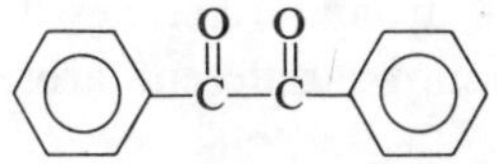	53

*For conversion energy units, use 1 kcal mol^{-1} = 4.1868 × 10^3 J

7.9.4.2.1. Photoreduction of aromatic ketones

PHOTOREDUCTION of aromatic ketones (aryl ketones) occurs by an INTERMOLECULAR HYDROGEN ABSTRACTION from hydrogen-donor molecules (RH) such as alcohols, ethers, amines, thiols, sulphides, phenols, etc., by the following mechanism:

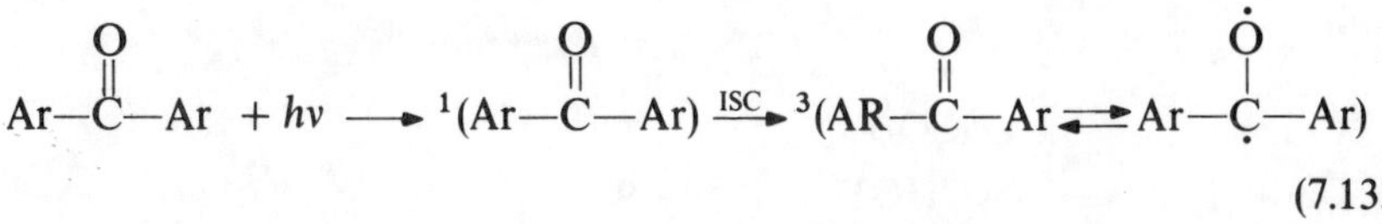

$$Ar-\overset{O}{\overset{\|}{C}}-Ar + h\nu \longrightarrow {}^1(Ar-\overset{O}{\overset{\|}{C}}-Ar) \xrightarrow{ISC} {}^3(AR-\overset{O}{\overset{\|}{C}}-Ar \rightleftharpoons Ar-\overset{\overset{\cdot}{O}}{\underset{\cdot}{C}}-Ar)$$

(7.133)

$$^3(Ar-\overset{\overset{\cdot}{O}}{\underset{\cdot}{C}}-Ar) + RH \longrightarrow Ar-\overset{OH}{\underset{|}{C}}-Ar + R\cdot$$

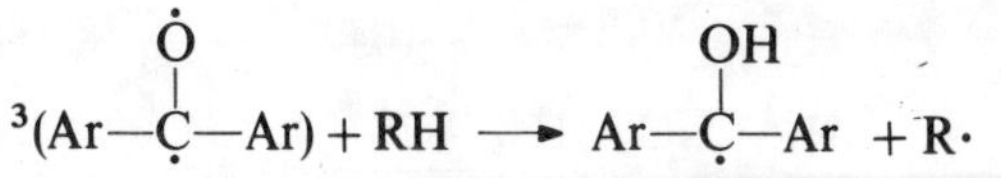

(7.134)

(7.117)   (7.118)

Ketyl radicals (7.118) recombine and form pinacol (7.119):

$$2\ Ar-\underset{\underset{\displaystyle \cdot}{|}}{\overset{\displaystyle OH}{\overset{|}{C}}}-Ar \longrightarrow \underset{Ar}{\overset{Ar}{>}}\underset{|}{\overset{OH}{\overset{|}{C}}}-\underset{Ar}{\overset{OH}{\overset{|}{C}}}<\overset{Ar}{} \qquad (7.135)$$

(7.118)   (7.119)

The excited triplet state of the carbonyl group can be considered as a 1,2-diradical (7.117).[244, 823, 831, 1560, 3780, 4005] Diradical formation can occur without spin inversion.[3778] Photoinitiators with higher energy of the excited triplet state (E_T) yield 1,2-diradicals easier than those with lower E_T.

The reactions of the diradicals and radical pairs formed by triplet state hydrogen abstraction are as follows:

(i) coupling (cyclization),
(ii) disproportionation,
(iii) cleavage (1,4-diradicals),
(iv) radical rearrangements,
(v) further hydrogen abstraction.

Radical intermediates formed from hydrogen-donors during the photoreduction of benzophenone by alcohols, amines, phenols, sulphides, thiols, ethers, hydrocarbons, and amides have been trapped by tert-nitrosobutane, and resulting nitroxide radicals have been characterized by ESR, spectroscopy.[2227]

In general, photoreduction of aromatic ketones depends on their chemical structure.[2664,3773] Table 7.11 shows different aromatic ketones which are photoreduced to pinacols.[2664]

The rate-limiting step in the photoreduction of aromatic ketones is the abstraction of a hydrogen from a hydrogen-donor molecules (RH) to produce a ketyl radical (7.118). This reaction is not dependent on the triplet energy (E_T) of the excited carbonyl groups. The triplet energies of some aromatic ketones are shown in Table 7.10, and energies of some C–H bonds are given in Table 7.12. It is easy to see that no one excited triplet state of carbonyl group has enough energy to abstract hydrogen to form the ketyl radical.

The energetics of the hydrogen abstraction reaction for benzophenone show that the overall reaction is exothermic by 26 kcal mol^{-1}.[3791]

$$\Delta H \text{ (kcal mol}^{-1})$$

$^3(\phi CO\phi)$	$\rightarrow$	$\phi CO\phi$	-69	(7.136)
$\phi CO\phi + H_2$	$\rightarrow$	$\phi_2 CHOH$	-9	(7.137)
$2H\cdot$	$\rightarrow$	H_2	-104	(7.138)
$\phi_2 CHOH$	$\rightarrow$	$H\cdot + \phi_2\dot{C}\text{–OH}$	$+78$	(7.139)
$^3(\phi CO\phi) + H\cdot$	$\rightarrow$	$\phi_2\dot{C}\text{–OH}$	-104	(7.140)

Table 7.11 Photoreduction characteristics of diaryl
ketones[2644]

Diaryl ketone	Pinacol produced	References
o-Chlorobenzophenone	Yes	
m-Chlorobenzophenone	Trace	
p-Chlorobenzophenone	No	
p-Methoxybenzophenone	Yes	
p-Methylbenzophenone	Yes	
p-p'-Dimethylbenzophenone	Yes	
p-Bromobenzophenone	Yes	
p-p'-Dichlorobenzophenone	Trace	
o,o',p,p'-Tetrachlorobenzophenone	Yes	
o-Chloro-p-methylbenzophenone	Yes	750
p-Chloro-p-methylbenzophenone	Yes	
m-Methylbenzophenone	Yes	
o-Methylbenzophenone	No	
p-Phenylbenzophenone	No	
Phenyl-1-naphthyl ketone	No	
Fluorenone	No	
Michler's ketone	No	
Anthrone	No	
Xanthone	No	3224
2-Acetonaphthone	No	
1-Naphthaldehyde	No	
1-Acetonaphthone	No	
Di-α-naphthyl ketone	No	
o-Phenylbenzophenone	No	
Desoxybenzoin	Yes	
Benzylacetophenone	No	
Benzylacetophenone	No	413,1486,
1,4-Diphenylbutanone	Trace	3348
α-Indanone	No	
Benzalacetophenone	No	
Benzalacetone	No	
p-Acetoxypropiophenone	No	
Phenalen-1-one	No	2104
Cyclopropyl phenyl ketone	No	2663
Cyclobutyl phenyl ketone	No	2644

$$RH \rightarrow R\cdot + H\cdot \qquad +78 \qquad (7.141)$$

$$^3(\phi CO \phi) + RH \rightarrow \phi_2 \overset{\cdot}{C}\text{–OH} + R\cdot \quad -26 \qquad (7.142)$$

whereas for acetophenone it is $-31\,\text{kcal mol}^{-1}$ and for fluorenone $-10\,\text{kcal mol}^{-1}$. It is also clearly shown that the energetics of the hydrogen abstraction reaction is not the reason why benzophenone and acetophenone photoreduce whereas fluorenone (7.189) does not. It is much more reasonable that hydrogen abstraction for fluorenone requires a relatively higher activation energy than for benzophenone and acetophenone.

Table 7.12 The C–H bond dissociation
energies

Compound	C–H (kacl mol^{-1})*
CH_3CH_2–H	98
$(CH_3)_2CH$–H	91.5
$(CH_3)_3C$–H	91
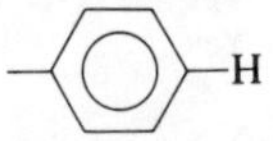 —H	103
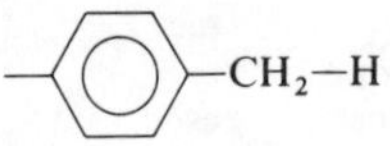 —CH_2—H	85
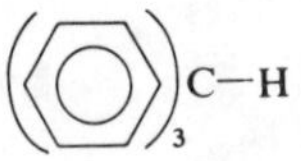 C—H	75
$HOCH_2$–H	93

*For conversion energy units, use
1 kcal mol^{-1} = 4.1868 × 10^3 J

The photoreduction of carbonyl compounds depends on the nature of the excited state.[1729] Aromatic ketones have two distinct chromophores:

(i) aromatic nucleus which produce excited triplet state

$T_1(\pi, \pi^*)$ ($\pi \to \pi^*$ transition)

(ii) carbonyl group which produce excited triplet state

$T_1(n, \pi^*)$ ($n \to \pi^*$ transition)

Aromatic ketones such as benzaldehyde, benzophenone, benzil, benzoin, benzoin ethers, α,α-dimethoxy-α-phenyl acetophenone, α,α-diethoxyacetophenone, and quinone have $T_1(n, \pi^*)$, whereas aromatic ketones such as naphthyl ketones, fluorenone, xanthone, and thioxanthone have $T_1(\pi, \pi^*)$.

$T_1(\pi, \pi^*)$ and $T_1(n, \pi^*)$ states have quite different physical and chemical properties. Differences in reactivity towards photoreduction result from differences in the character of the lowest excited triplet state. Aromatic ketones with a lowest $T_1(n, \pi^*)$ will be most reactive towards photoreduction. Those aromatic ketones having $T_1(\pi, \pi^*)$ are only poorly photoreduced, and those having lowest triplets of the charge-transfer (CT) type are least reactive towards photoreduction.

In an $n \to \pi^*$ transition an electron originally located on the oxygen atom is removed to a region between the carbon and the oxygen, leaving the latter with a partial positive charge.

In a $\pi \to \pi^*$ transition of an aromatic ketone the carbonyl group does not participate to any great extent, hence the normal polarization of the carbonyl group is preserved.

In a CT transition where the carbonyl group is the electron-acceptor a formal negative charge appears on the oxygen atom.

$$\underset{Ar}{\overset{Ar}{>}}\overset{-}{C} = \overset{+}{O} \qquad \underset{Ar}{\overset{Ar}{>}}\overset{\sigma^+}{C} = \overset{\sigma^-}{O} \qquad \underset{Ar}{\overset{Ar}{>}}\overset{+}{C} = \overset{-}{O}$$

$$n \rightarrow \pi^* \qquad\qquad \pi \rightarrow \pi^* \qquad\qquad CT$$

In order for photoreduction to be efficient the carbonyl oxygen must bear a positive charge (or at least a greater amount of positive charge than in the ground-state configuration) such as would result from an $n \rightarrow \pi^*$ transition. Electron-donating substituents stabilize the $T_1(\pi,\pi^*)$ and destabilize the $T_1(n,\pi^*)$ and solvents do the same.

In several aromatic ketones a $T_1(\pi,\pi^*)$ state has a degree of $F_1(n,\pi^*)$ state resulting from vibronic and spin-orbit coupling. It is this degree of T_1 (n,π^*) state which is responsible for the observation of photoreduction of a compound with a $T_1(\pi,\pi^*)$ state. Factors that reduce the energy gap between $T_2(\pi,\pi^*)$ and the $T_1(n,\pi^*)$ state will lower the photoreactivity. For aromatic ketones having $T_1(\pi,\pi^*)$ states, factors that reduce the energy gap between the $T_2(n,\pi^*)$ and the $T_1(\pi,\pi^*)$ will increase the reactivity towards photoreduction.

For aromatic ketones having T_1 (π,π^*) states, factors that increase vibronic coupling with higher $T_2(n,\pi^*)$ states and spin-orbit coupling with $n \rightarrow \pi^*$ singlets will increase reactivity through mixing of the states. Since vibronic and spin-orbit coupling depend inversely upon the energy separation between the states, closely adjacent $n \rightarrow \pi^*$ singlet and triplet states will couple more strongly with the $T_1(\pi,\pi^*)$ than more widely separated states.

Alcohols and ethers react efficiently only with n, π^* electronic transition state of carbonyl group, whereas amines[748] are more generally reactive with π, π^* than n, π^* transition state.

Michler's ketone,[3494] xanthone,[2985] and thioxanthone[821, 3976] show two different electronic transitions of the carbonyl group in the excited triplet state depending on solvent polarity: $T_2(n,\pi)$ in non-polar solvents (e.g. 3-methylpentane) and $T_1(\pi,\pi^*)$ in polar solvents (e.g. EPA (ethanol-isopentane-ether = 5:2:2)).

Intersystem crossing (ISC) in thioxanthone occurs predominantly from $S_1(\pi,\pi^*)$ to $T_2(n,\pi^*)$ in non-polar poor hydrogen bonding solvents, e.g. hexane (Fig. 7.5). As the solvent becomes more polar or a better hydrogen bonder, the energy of the state $T_2(n,\pi^*)$ will become greater than that of the $S_1(\pi,\pi^*)$ state, meaning that intersystem crossing (ISC) from $S_1(\pi,\pi^*)$ to $T_2(n,\pi^*)$ will require a larger activation energy and will become slower. The greater the solvent polarity or hydrogen bonding ability, the larger the energy difference between $T_2(n,\pi^*)$ and $S_1(\pi,\pi^*)$, the greater activation energy and the lower the rate constant for intersystem crossing (ISC) from $S_1(\pi,\pi^*)$ to $T_2(n,\pi^*)$. Furthermore, it is well known that increasing solvent polarity or hydrogen bonding ability stabilizes π, π^* states and destabilizes n, π^* states.[2896]

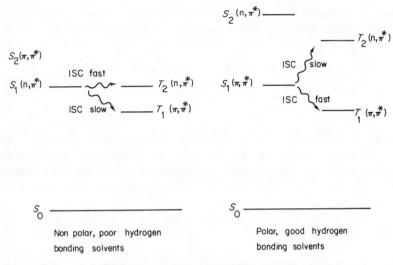

Fig. 7.5 Energy level diagram for thioxanthone in different solvents

Generally the photoinitiators with n, π^* electronic transition of the carbonyl group in the excited triplet state (T_1) show higher overall efficiency of photoinitiation than with π, π^* electronic transition.

The chemical structures of the hydrogen-donor molecules play also important role in hydrogen abstraction by the carbonyl groups in the excited triplet state (T_1). Particularly reactive hydrogen-donor molecules are those in which hydrogen atom is in α-position to an oxygen, as in

$$\text{alcohols } (R_2 \overset{\alpha}{\underset{\overset{|}{H}}{C}}\text{-OH}) \quad \text{and ethers} \quad (R_2 \overset{\alpha}{\underset{\overset{|}{H}}{C}}\text{H-OR)},$$

or to nitrogen in tertiary amines

$$(R_2 \overset{\alpha}{\underset{\overset{|}{H}}{C}}\text{-NR}_2)$$

and directly attached to sulphur, as in thiols (RS–H).

Rates of hydrogen abstraction are strongly dependent on C–H strength. For a given C–H bond there is very little difference in rate constant among the excited triplet state of benzophenone, benzil, or benzoin ethers, etc., all of which possess n, π^* lowest triplet. The activation energy for abstraction of an unactivated secondary hydrogen is of the order of 3–3.5 kcal/mole. The activation entropy varies strongly from system to system.[1335,2275]

7.9.4.2.1.1. Photoreduction of aromatic ketones with alcohols

The excited triplet states (T_1) of a carbonyl group, which may exist as a 1,2-diradical (7.120), abstract hydrogen on α carbon to the hydroxyl group in preference to more distant hydrogens:

$$^3(Ar-\overset{\cdot}{\underset{\cdot}{\overset{\overset{\cdot}{O}}{C}}}-Ar) + \overset{H}{\underset{R}{\overset{}{\underset{R}{C}}}}\overset{OH}{\diagup} \longrightarrow Ar-\overset{OH}{\underset{\cdot}{\overset{|}{C}}}-Ar + \overset{\cdot}{\underset{R}{\overset{}{\underset{R}{C}}}}\overset{OH}{\diagup} \qquad (7.143)$$

$$(7.120) \qquad\qquad\qquad\qquad (7.121) \qquad (7.122)$$

This reaction is affected by oxygen, because the deactivation of the triplet state (T_1) of a carbonyl group by oxygen is generally faster than the intermolecular hydrogen abstraction process.

Ketyl radicals (7.121) formed recombine to pinacol (7.119), whereas alkylhydroxyl (7.122) radicals give ketone and alcohol:

$$2 \overset{\cdot}{\underset{R}{\overset{}{\underset{R}{C}}}}\overset{OH}{\diagup} \longrightarrow RCOR + \overset{H}{\underset{R}{\overset{}{\underset{R}{C}}}}\overset{OH}{\diagup} \qquad (7.144)$$

Several aromatic ketones such as fluorenone, xanthone, aminobenzophenones, and α- and β-naphthyl ketones are not photoreduced by alcohols. Photoreduction of ketones by alcohols in aqueous media is complicated by transients.[738, 741]

The ketone–alcohol hybrid system is not an effective polymerization initiation system, because the hydrogen abstraction reaction (7.143) is slow in comparison with quenching of the triplet state (T_1) of the ketone group by oxygen or by a monomer.[3179]

If the monomer contains an ether group, as in diethylene glycol diacrylate, ketone- photoinitiated polymerization has been observed.[1555] Polyether groups in the oligomer have also been found to serve as hydrogen-donating sites.[3208]

7.9.4.2.1.2. Photoreduction of aromatic ketones with amines

The excited triplet state (T_1) of a carbonyl group forms, with an amine, an EXCIPLEX (7.123):[213,214,2238,2461,3179,3689]

$$^3\left(\underset{Ar}{\overset{O}{\underset{\diagdown}{\overset{\|}{C}}}}\overset{}{Ar}\right) + \underset{R}{\overset{}{:N-CHR_2}}\overset{}{R} \rightarrow {}^3\left[\underset{Ar}{\overset{\overset{\cdot}{O}\,\sigma^-}{\underset{\diagdown}{\overset{|}{C}}}}\overset{}{Ar} \cdots \cdots \cdots \underset{R}{\overset{\sigma^+}{\overset{\cdot}{N}-CHR_2}}\overset{}{R}\right] \qquad (7.145)$$

$$\text{EXCIPLEX}$$
$$(7.123)$$

The exciplex (cf. Section 3.5) is internally stabilized by a formation of a weak charge-transfer (CT) complex due to the low ionization potential of amine.

The rate of CT complex formation increases with:

(i) Decreasing ionization potential (oxidation potential) of the amine.
(ii) Increasing electron affinity (reduction potential) of ketone. It is comparable with the rate of the deactivation of the triplet state (T_1) of carbonyl group by oxygen.

The exciplex may participate in the following reactions:

(i) Efficient quenching of the excited triplet state (T_1) of carbonyl group in ketone:

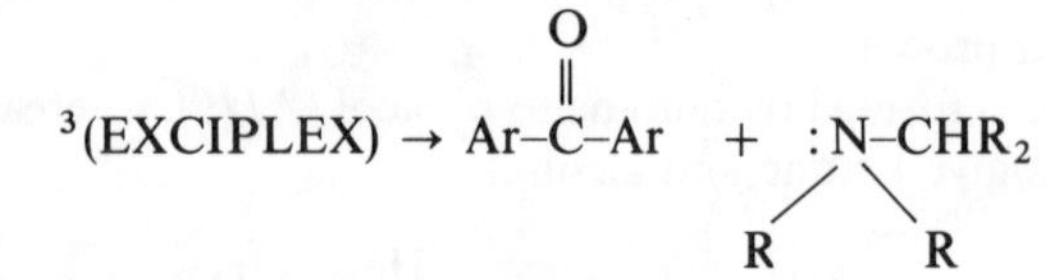

$$^3(EXCIPLEX) \rightarrow Ar-\overset{\overset{\displaystyle O}{\|}}{C}-Ar \;\; + \;\; :N-CHR_2 \qquad (7.146)$$

(ii) Electron-transfer from the amine to the ketone in which are produced radical ions:[213,214,317,2231]

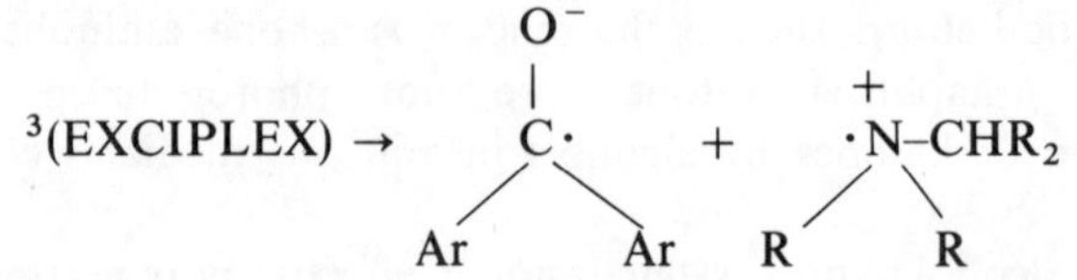

$$^3(EXCIPLEX) \rightarrow \overset{\displaystyle O^-}{\underset{}{C}}\cdot \;\; + \;\; \overset{\displaystyle +}{\cdot N}-CHR_2 \qquad (7.147)$$

(iii) Hydrogen-abstraction from amine to ketone in which are produced ketyl (7.124) and initiator amine radicals (7.125):[213, 2231]

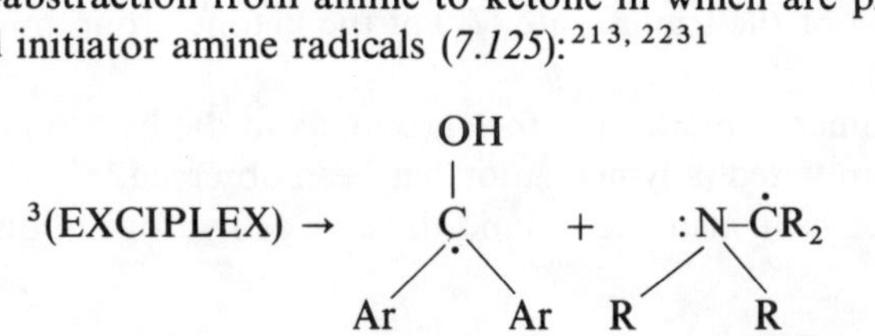

$$^3(EXCIPLEX) \rightarrow \overset{\displaystyle OH}{\underset{}{\overset{.}{C}}} \;\; + \;\; :N-\overset{.}{C}R_2 \qquad (7.148)$$

(7.124)   (7.125)

These three reactions may compete one with another; for example the quenching reaction with hydrogen abstraction, and the final result depends on:

(i) The rate of amine complexation, which is favoured by low ionization potential.
(ii) The partitioning of the charge-transfer complex between the quenching reaction, and hydrogen abstraction where quenching may also be favoured by the low ionization potential of the amine and hydrogen abstraction reaction.
(iii) The efficiency of initiation of the polymerization reaction by the amine initiator radicals.

(iv) Participation by a neighbouring groups (e.g. OH, COOH) in the hydrogen abstraction reaction.[870,872,2240]

The relative reactivity of amines (tertiary > secondary > primary) is proportional to the ease with which they are oxidized. Ketones with π,π^* lowest excited triplet states are photoreduced by tertiary amines, but quantum efficiency varies with solvent polarity. For fluorenone photoreduction quantum yields decrease with increasing solvent polarity,[742,743,889] whereas for naphthyl ketones the dependence is reversed.[3777] In both cases rate constants for quenching of the excited triplet state of ketone by amine are much higher in polar (both protic and aprotic) than in non-polar solvents, which suggests a high degree of charge transfer in the interaction.

In contrast, ketones with n,π^* lowest excited triplet state are quenched by tertiary amines no faster in polar than in non-polar solvents, and appreciably more slowly in protic solvents.[748,3779] Hydrogen bonding to the amine lone pairs can explain the retardation of charge-transfer (CT) quenching of n,π^* excited triplet states in protic solvents.

The identity of the radical intermediates produced in the reaction of ketones with tertiary amines has been examined by flash photolysis and ESR spectroscopy.[215,317,865,873]

During the photoreduction of ketones by amines formations cross-coupled products was observed (probably as result of cage recombination) and disproportionation of initially produced radicals.[859,866,1760]

Ketone–amine hybrid systems are efficient photoinitiators of free radical polymerization (cf. Section 7.9.4.2.3.6).

7.9.4.2.2. Intramolecular hydrogen abstraction

INTRAMOLECULAR HYDROGEN ABSTRACTION is dominated by conformational factors which limit the accessibility of most hydrogens in a given molecule, e.g. long-chain n-alkyl esters of benzophenone-4-carboxylic acid (7.126):

(7.149)

(7.126)

cyclized product

The rate constant of any intramolecular hydrogen abstraction of a flexible molecule includes rotational equilibrium constants:

unfavourable conformations $(\overset{*}{u}) \rightleftarrows$ favourable conformations $(\overset{*}{f}) \xrightarrow{k_H}$ H-abstraction (7.150)

$\uparrow hv$ $\uparrow hv$

unfavourable conformations $(u) \rightleftarrows$ favourable conformations (f) (7.151)

The rate constant for the remote intramolecular hydrogen abstraction from para-substituted benzophenones increases as the probability of a favourable conformation increases.[521-523,3923]

There are three boundary conditions which should be considered:

(i) Excited state conformational changes are faster than triplet decay.
(ii) Conformational changes are slower than triplet decay.
(iii) Conformational changes are rate-determining.

7.9.4.2.3. Benzophenone

Benzophenone (7.127) is a one of the most investigated photoinitiators and photosensitizers.[23,367,373,554,1729,2068,2152,2229,2447,2455,2560,2726,2975,2976,3237,3239,3349,3959]

The benzophenone absorption spectrum is shown in Fig. 7.6. Information obtained from absorption and emission spectra allows for the presentation of an energy level diagram of benzophenone, as is shown in Fig. 7.7.

Irradiation of benzophenone with radiation 340 nm or 254 nm excites the benzophenone molecule to an excited singlet state $S_1(n, \pi^*)$ or $S_2(\pi, \pi^*)$, respectively. The internal conversion (IC) from S_2 to S_1 state is very rapid $(10^{-12}$ s). The intersystem crossing (ISC) from $S_1(n, \pi^*)$ to the triplet state

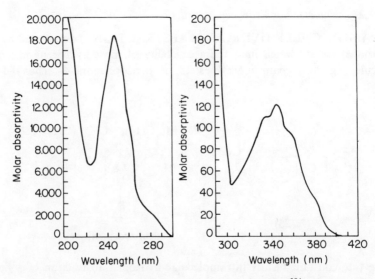

Fig. 7.6 Absorption spectra for benzophenone[591]

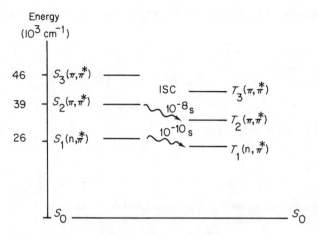

Fig. 7.7 Energy level diagram for benzophenone[367]

$T_1(n,\pi^*)$ occurs very efficiently ($\Phi_{ISC} \sim 1$) during 10^{-10} s, so that the chemically active state is actually the triplet state $T_1(n,\pi^*)(E_T = 69.3$ kcal mol^{-1}) but not the excited singlet state $S_1(n,\pi^*)$.[3101]

$$\text{(7.127)} \qquad + h\nu \longrightarrow {}^1\!\left(\qquad\right) \qquad (7.152)$$

or

$$(\phi - \overset{\overset{\displaystyle O}{\|}}{C} - \phi) + h\nu \longrightarrow {}^1(\phi - \overset{\overset{\displaystyle O}{\|}}{C} - \phi) \qquad (7.153)$$

$${}^1\!\left(\qquad\right) \xrightarrow{ISC} {}^3\!\left(\qquad = \qquad\right) \qquad (7.154)$$

$$(7.128)$$

or

$${}^1(\phi - \overset{\overset{\displaystyle O}{\|}}{C} - \phi) \longrightarrow {}^3(\phi - \overset{\overset{\displaystyle \dot{O}}{|}}{\underset{.}{C}} - \phi) \qquad (7.155)$$

During the excitation of carbonyl group, the π bond is broken and the n-electron on oxygen is very weakly correlated with the electron in the π-system. The triplet state T_1 of benzophenone can be considered as 1,2-diradical (7.128). Conjugation of the carbonyl group with benzene ring lowers the carbonyl π-bond energy since the resulting 1,2-diradical is resonance-stabilized.

The triplet state (T_1) of benzophenone can be deactivated by the following reactions:

(i) Concentration deactivation (self-quenching):

$$^3(\phi\text{-C-}\phi) + \phi\text{-C-}\phi \xrightarrow{k_D} \phi\text{-C-}\phi + \phi\text{-C-}\phi \tag{7.156}$$

(ii) Triplet-triplet annihilation:

$$^3(\phi\text{-C-}\phi) + {}^3(\phi\text{-C-}\phi) \xrightarrow{k_T} \phi\text{-C-}\phi + \phi\text{-C-}\phi \tag{7.157}$$

(iii) Energy transfer quenching (to quencher (Q) molecule):

$$^3(\phi\text{-C-}\phi) + Q \xrightarrow{k_Q} \phi\text{-C-}\phi + Q^* \tag{7.158}$$

(iv) Intermolecular hydrogen abstraction from a hydrogen-donor molecule (RH):

$$^3(\phi\text{-}\overset{\cdot}{\underset{\cdot}{C}}\text{-}\phi) + RH \xrightarrow{k_H} \phi\text{-}\overset{OH}{\underset{}{C}}\text{-}\phi + R\cdot \tag{7.159}$$

where:

k_D = rate constant for concentration deactivation,
k_T = rate constant for triplet–triplet annihilation
   ($k_T = 10^6 \text{ mol}^{-1}\text{s}^{-1}$),[3349,3959]
k_Q = rate constant for energy transfer to quencher,
k_H = rate constant for hydrogen abstraction.

Note: A step involving deactivation of triplet (T_1) by emission of phosphorescence has been omitted from the above scheme as it is irrelevant to the modes of deactivation presented. The phosphorescence of benzophenone (at 450–470 nm) can be observed when a degassed benzene solution of ketone is irradiated at 366 nm.

The rate constants k_Q and k_H can be evaluated from the following equations:[373,1485,2560]

$$\frac{1}{\tau} = \frac{1}{\tau_0} + k_Q[Q] \tag{7.160}$$

$$\frac{1}{\tau} = \frac{1}{\tau_0} + k_H[RH] \tag{7.161}$$

where:

τ_0 and τ = benzophenone triplet (T_1) lifetimes in the absence and in the presence of Q or RH, respectively.

The lifetime of the triplet state of benzophenone (T_1) is shorter in benzene (10^{-5}–10^{-6} s) than in other inert solvents (10^{-3} s) such as carbon tetrachloride or perfluoroalkanes.[373, 554, 726, 1335, 2088, 2152, 2447, 2898, 2899, 2975, 2976, 3172, 3238, 3239]

7.9.4.2.3.1. Benzophenone ketyl radical

The 1,2-diradical (7.129) formed in the triplet state (T_1) of benzophenone can abstract hydrogen from hydrogen-donor molecules (RH) such as hydrocarbons, alcohols, ethers, amines, thiols, sulphides, phenols, and even from macromolecules and produce a BENZOPHENONE KETYL RADICAL (also called a DIPHENYLHYDROXYMETHYL RADICAL or SEMIPINACOL RADICAL (7.130)):[367, 507, 554, 924, 2497, 2956, 2971, 2975, 3171, 3236, 3621, 3791]

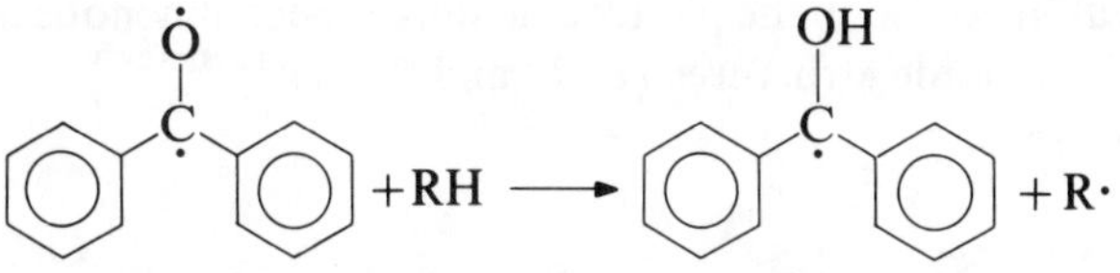

$$(7.129) \qquad\qquad (7.130) \tag{7.162}$$

The molar absorptivity for the ketyl radical has been determined as $\varepsilon = 5.1 \times 10^3$ (mol^{-1} cm^{-1}) (at 543 nm).[367]

The decay of ketyl radicals (10^{-1} to 10^{-4} M solutions of benzophenone in isopropanol) is 80 ms in ketyl radical concentration and independent of benzophenone concentration. For that reason it is difficult to determine their ESR spectra.[3062] Application of t-nitrosobutane (7.131) as a spin trap allowed for trapping different radicals (R ·) formed from hydrogen donor molecules (RH) but not ketyl radicals:[2227]

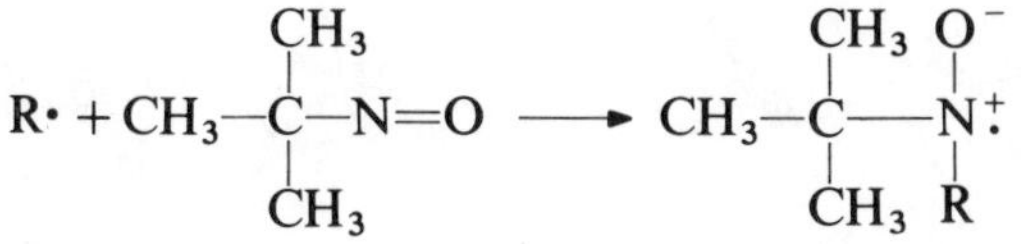

$$(7.131) \tag{7.163}$$

Ketyl radicals react with t-nitrosobutane by the mechanism:

$$\tag{7.164}$$

and for that reason cannot be detected.

Two ketyl radicals formed react with each other to form benzopinacol (1,1,2,2-tetraphenyl-1,2-ethanediol) (7.132):

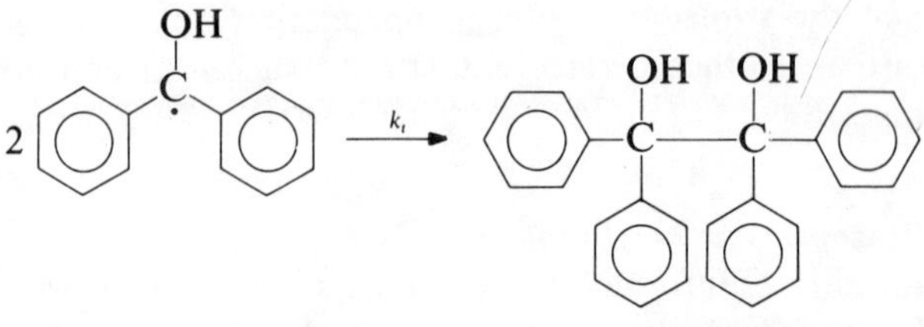

(7.165)

(7.132)

The rate constant of this reaction, $k_t = 5.9 \times 10^7 \text{ mol}^{-1}\text{s}^{-1}$. The amount of benzopinacol formed is proportional to the amount of ketyl radicals produced.[367]

Ketyl radicals involved in the photoreduction of benzophenone may also form products with quinoide structures (7.133 and 7.134):[1090,3203]

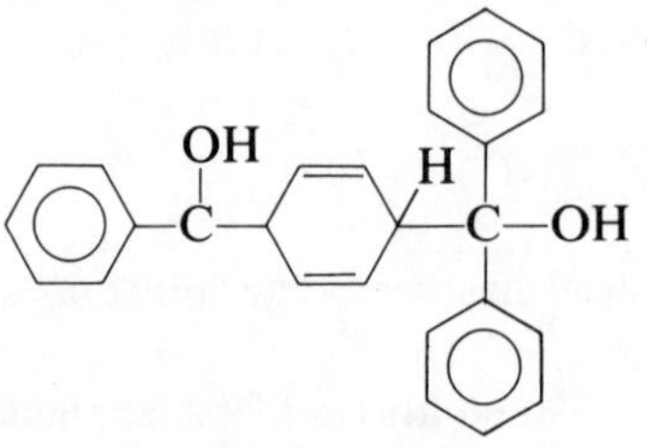

(7.133) 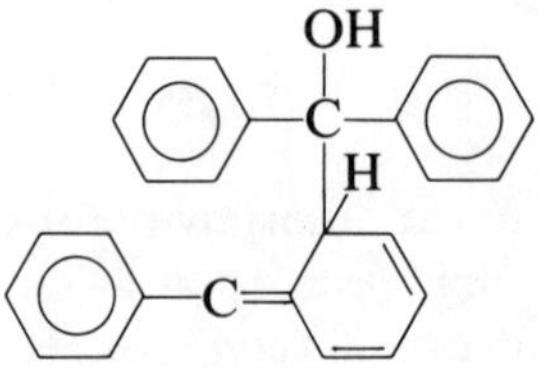   (7.134)

Ketyl radicals are efficiently scavenged by camphorquinone (7.135) by hydrogen transfer mechanism:[2555,2556]

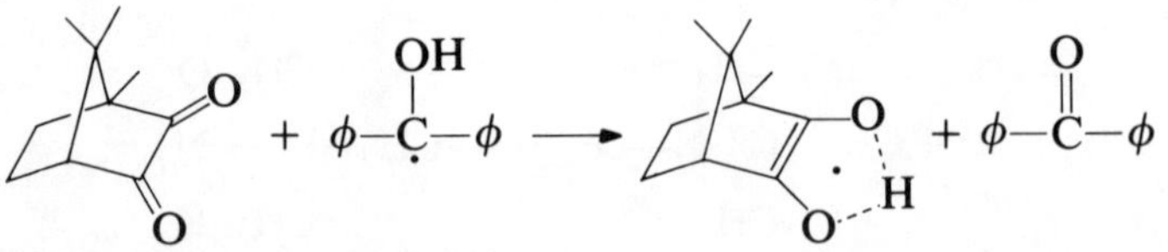

(7.135)   (7.166)

Ketyl radicals in the reaction with $\cdot CCl_3$ radicals dimerize selectively to benzopinacol:[3791]

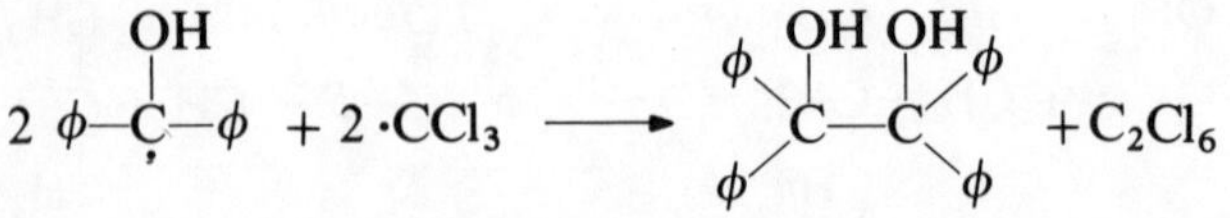

$$2 \; \phi-\underset{\phi}{\underset{|}{C}}-\phi \; + 2 \cdot CCl_3 \longrightarrow \underset{\phi}{\overset{\phi}{}}\underset{\phi}{\overset{OH}{\underset{|}{C}}}-\underset{\phi}{\overset{OH}{\underset{|}{C}}}\overset{\phi}{} \; + C_2Cl_6 \qquad (7.167)$$

In the presence of oxygen, ketyl radicals can react with oxygen to regenerate benzophenone molecules and form hydroperoxide radicals ($HO_2\cdot$) and hydroperoxide:[2956]

$$\phi\text{-}\underset{|}{\overset{OH}{\underset{\bullet}{C}}}\text{-}\phi + O_2 \rightarrow \phi\text{-}\overset{O}{\overset{\|}{C}}\text{-}\phi + HO_2\cdot \qquad (7.168)$$

$$\phi\text{-}\underset{\bullet}{\overset{OH}{\underset{|}{C}}}\text{-}\phi + HO_2\cdot \rightarrow \phi\text{-}\overset{O}{\overset{\|}{C}}\text{-}\phi + H_2O_2 \qquad (7.169)$$

7.9.4.2.3.2. *Benzophenone ketyl radical anion*

BENZOPHENONE KETYL RADICAL ANION (*7.136*) is produced during irradiation of benzophenone in isopropanol–water (1:1) solution in the pH region 10–13:

$$\phi\text{-}\overset{O}{\overset{\|}{C}}\text{-}\phi + H_2O + (CH_3)_2CHOH \overset{+h\nu}{\rightarrow} \phi\text{-}\underset{\bullet}{\overset{O^-}{\underset{|}{C}}}\text{-}\phi + H_3O^+ + (CH_3)_2\overset{\bullet}{C}OH$$

$$(7.170)$$

$$(7.136)$$

The extinction coefficient for the ketyl radical anion has been determined as $\varepsilon = 5.0 \times 10^3$ (at 630 nm).[367]

The ketyl radical anion may dimerize by the following processes:

$$\phi\text{-}\underset{\bullet}{\overset{O^-}{\underset{|}{C}}}\text{-}\phi + \phi\text{-}\overset{OH}{\underset{\bullet}{\underset{|}{C}}}\text{-}\phi \overset{k_{t(1)}}{\longrightarrow} DIMER \qquad k_{t(1)} = 1.1 \times 10^9 \text{ mole}^{-1}\text{ s}^{-1} \qquad (7.171)$$

$$\phi\text{-}\underset{\bullet}{\overset{O^-}{\underset{|}{C}}}\text{-}\phi + \phi\text{-}\underset{\bullet}{\overset{O^-}{\underset{|}{C}}}\text{-}\phi \overset{k_{t(2)}}{\longrightarrow} \begin{array}{c} \overset{O^-}{\underset{|}{C}} \\ \phi\text{-}C\text{-}\phi \\ | \\ \phi\text{-}C\text{-}\phi \\ | \\ O_- \end{array} \qquad (7.172)$$

$$k_t(2) = 1.1 \times 10^5 \text{ mole}^{-1}\text{ s}^{-1}$$

The decay of the ketyl radical anion is second order in radical concentration and dependent critically on the PH of the medium (Fig. 7.8), the lifetime increasing from a few milliseconds at pH 10 to several seconds at pH 13.[367]

The benzophenone ketyl radical anion was characterized by ESR in photo-reduction of the ketone in aqueous and ethanolic isobutylamine and *tert*-butylamine.[867]

7.9.4.2.3.3. *Benzophenone photoreaction with hydrogen peroxide*

The excited triplet state (T_1) of benzophenone reacts with hydrogen peroxide (H_2O_2) by intermolecular hydrogen abstraction to yield mainly hydroxyl $(HO\cdot)$ and hydroperoxy $(HO_2\cdot)$ radicals according to the reactions:[1323]

278

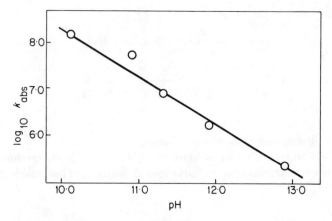

Fig. 7.8 Decay of benzophenone ketyl radical anion in media of varying pH[367]

$$^3(\phi\text{--}\overset{\overset{\displaystyle \dot{O}}{|}}{\underset{\displaystyle \cdot}{C}}\text{--}\phi) + H_2O_2 \rightarrow \left[\phi\text{--}\overset{\overset{\displaystyle OH}{|}}{\underset{\displaystyle \cdot}{C}}\text{--}\phi + HO_2\cdot\right] \rightarrow \left[\begin{matrix}\phi & OOH \\ & C \\ \phi & OH\end{matrix}\right] \quad (7.173)$$

$$\left[\begin{matrix}\phi & OOH \\ & C \\ \phi & OH\end{matrix}\right] \overset{\nearrow \phi\text{--}\overset{\overset{\displaystyle O}{\|}}{C}\text{--}\phi + 2\,HO\cdot \quad (7.174)}{\underset{\searrow \phi\cdot + \phi COOH + HO\cdot \quad (7.175)}{}}$$

If the reaction is carried out in the presence of methyl methacrylate, hydroxyl radicals (HO·) initiate free radical polymerization.

7.9.4.2.3.4. Photoreduction of benzophenone in various solvents

Quantum yields of benzophenone disappearance in various solvents are given in Table 7.13. Three classes of behaviour of benzophenone in the various solvents can be distinguished:[367]

(i) Solvents in which the quantum yield of disappearance is nearly zero, such as water and benzene.

(ii) Solvents in which the quantum yield is somewhat less than or equal to unity and is concentration-independent, as in hexane, toluene, and ethanol.

Table 7.13 Quantum yields (Φ) of disappearance of benzophenone in various solvents at room temperature[367]

Solvent	λ irrad (nm)	Φ
Water	254	0.02
Benzene	366	0
Hexane	254	0.55
Toluene	366	0.45
Isopropanol	254	0.8–1.9
Isopropanol	366	1.48–1.9
Ethanol	254	1.0

Probably all benzophenone molecules in the triplet state (T_1) abstract hydrogen atoms from these solvents. The C–H bond energies in hexane and toluene are about 85 kcal mol^{-1} and 80 kcal mol^{-1}, respectively. The radical (R·) formed from the solvent (RH) may participate in a reversion of the primary hydrogen abstraction reaction:

$$\underset{\underset{\phi-\overset{\displaystyle|}{\underset{\displaystyle\cdot}{C}}-\phi}{}}{\overset{\overset{\displaystyle OH}{\displaystyle|}}{}} + R\cdot \rightarrow \underset{\underset{\phi-C-\phi}{}}{\overset{\overset{\displaystyle O}{\displaystyle\|}}{}} + RH \qquad (7.176)$$

which would result in an overall quantum yield of less than unity.

(iii) Solvents in which the quantum yield is greater than or equal to unity and concentration-dependent, such as isopropanol (cf. reaction 7.178).

It has also been proposed that formation of exciplex between benzophenone molecule and solvent may be responsible for this behaviour.[2154]

The rate constants for the hydrogen abstraction (k_H) from different solvents (cf. equation 7.161) are shown in Table 7.14. In studies of hydrogen abstraction it is common practice to interpret the experimentally observed rate constants k_H per hydrogen atom (primary, secondary, or tertiary). In general, when the number of abstractable hydrogens of a given type (primary, secondary, or tertiary) increases, the observed k_H remains essentially constant. The rate constant k_H differs for

Table 7.14 Absolute rate constants for the hydrogen abstraction (k_H) from different solvents[1335]

Solvent	$k_H \times 10^{-5}$ ($M^{-1} s^{-1}$)
Benzene	0.25
Toluene	1.9
2,2-Dimethylpropane	0.43
2,2,3,3-Tetramethylbutane	0.31
2-Methylpropane	5.7
2,3-Dimethylbutane	9.3
Cyclohexane	7.2
2,2,4,4-Tetramethylpentane	1.7
Isooctane (2,2,4-trimethylpentane)	1.7

abstraction of primary, secondary, or tertiary hydrogen as shown in Table 7.15.[1335]

Table 7.15 Relative rate constants for the hydrogen abstraction (k_H) from different hydrocarbons[1335]

Type of R–H bonl broken	$k_H \times 10^{-5}$ $(\text{M}^{-1}\text{s}^{-1})$
Primary	1
Secondary	40
Tertiary	300
Secondary cyclohexyl	35
Primary benzylic	74
Benzene C–H	0.016

In reaction of ketyl radicals with n-alkanes it has been found that rate constant k_H is a function of the alkane chain length (Fig. 7.9)[3916,3922] Each of the methylene (CH_2) groups in the alkanes are equally reactive towards intermolecular hydrogen abstraction by the excited triplet state of benzophenone. Methyl groups are relatively unreactive. When the excited triplet molecule of benzophenone interacts with an alkane, the lifetime of the interaction is determined by the rate of diffusion away from the encounter pair (cf. Section 5.1.8). Within the encounter complex the excited triplet molecules of benzophenone can come in contact with approximately 14 methylene (CH_2) groups. Rate constants for the intermolecular hydrogen abstraction reaction are four orders of magnitude slower than the diffusion limit.

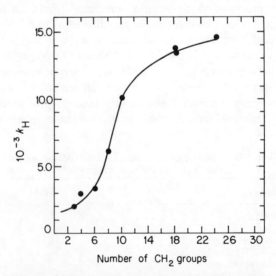

Fig. 7.9 Plot of the second-order rate constant versus number of methylene groups $(CH_2)_n$ for the reaction of benzophenone in its excited triplet state with $CH_3(CH_2)_nCH_3$ in CCl_4[3922]

Benzophenone undergoes photoreduction in benzene but with very low quantum yield ($\Phi = 0.005$).[367,373,2898] The low quantum yield in benzene cannot be due to transfer of electronic energy from triplet benzophenone since the energy of the lowest triplet state of benzene lies above that of benzophenone. The benzophenone triplet lifetime is shorter in benzene (10^{-5}–10^{-6} s) than in other inert solvents (10^{-3} s).

The bond energies with respect to hydrogen-atom abstraction in benzene are relatively high (> 100 kcal mol^{-1}) and these energies are presumably high enough to prevent hydrogen-atom abstraction by the triplet with anything other than very low collisional efficiency.

The very low quantum yield of benzophenone photoreduction (Table 7.13) and the low value of the k_H constant (Table 7.14) indicate a quenching reaction of triplet benzophenone (T_1) by benzene molecules. It has been suggested that the excited triplet state of benzophenone (T_1) forms a charge-transfer complex with benzene.[338, 373, 2898, 3234, 3238, 3781] In such a complex hydrogen abstraction cannot compete with the quenching reaction.

Another explanation is that the excited triplet state of benzophenone (T_1) may be, for an unstable diradical adduct with benzene (*7.137*):[338,339,554, 3171,3172,3234]

$$(7.177)$$

$$(7.137)$$

Formation of this diradical species which absorbs in the same region as the benzophenone triplet itself, and the ketyl radical, can explain why the lifetime of the benzophenone triplet in benzene is shorter than in other solvents.

Toluene has much more readily abstractable hydrogen than does benzene.

Photoreduction of benzophenone in isopropanol (*7.138*) occurs with formation of acetone (*7.138*):[254, 367, 1176, 2555, 2556, 2955, 2956, 2975, 3836]

$$(7.178)$$

$$k_H = 1.28 \times 10^6 \text{ mole}^{-1} \text{ s}^{-1}$$

$$2 \, (CH_3)_2\dot{C}OH \rightarrow (CH_3)_2CO + (CH_3)_2CHOH \qquad (7.179)$$

The isopropanol radical (*7.139*) can be easily trapped with a t-nitrosobutane spin trap and well characterized by ESR spectroscopy.[2227]

The quantum yield of benzophenone photoreduction in isopropanol is concentration-dependent (Table 7.16). The high quantum yield is due to the hydrogen exchange between isopropanol radical (7.139) and benzophenone according to the reaction:

$$(CH_3)_2\dot{C}OH + \phi\text{-}\overset{\overset{\displaystyle O}{\|}}{C}\text{-}\phi \rightarrow (CH_3)_2CO + \phi\text{-}\overset{\overset{\displaystyle OH}{|}}{\underset{\displaystyle \cdot}{C}}\text{-}\phi \qquad (7.180)$$

(7.139)

Table 7.16 Variation of quantum yield of disappearance of benzophenone (Φ) in degassed isopropanol with different solution concentrations during u.v. irradiation (253.7 nm)[367]

Concentration (mole)	Quantum yield (Φ) (mole einstein^{-1})
8×10^{-6}	0.84 ± 0.06
7×10^{-5}	1.02 ± 0.03
5×10^{-4}	1.35 ± 0.05
1×10^{-2}	1.70 ± 0.08
1×10^{-1}	1.90 ± 0.08

During prolonged irradiation of benzophenone in isopropanol an adduct (7.140) is formed, giving rise to increased u.v. absorption between 300 and 370 nm.[1108]

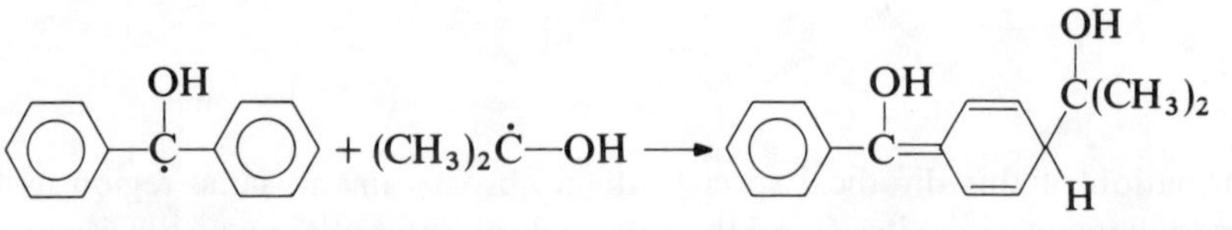

(7.140)   (7.181)

The excited triplet state (T_1) of benzophenone reacts with benzyl methyl ether (7.141) by intermolecular hydrogen abstraction during which ketyl and α-methoxybenzyl radical (7.142) are formed:[2883]

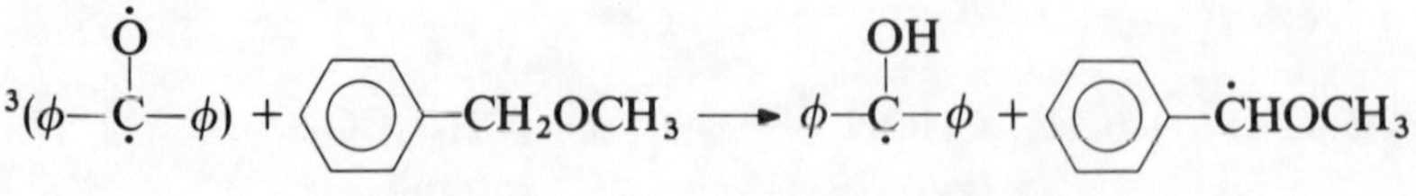

(7.141)   (7.142)   (7.182)

7.9.4.2.3.5. *Benzophenone photoreaction with tetrahydrofurane*
The excited triplet state of benzophenone (T_1) abstracts hydrogen from tetrahydrofurane (7.143) according to the reaction:[2152,2857]

$$^3(\phi-\overset{\overset{\textstyle \cdot}{O}}{\underset{\textstyle \cdot}{C}}-\phi) + \underset{O}{\boxed{}} \xrightarrow{k_H} \phi-\overset{\overset{\textstyle OH}{|}}{\underset{\textstyle \cdot}{C}}-\phi + \underset{O}{\boxed{}}\cdot \qquad (7.183)$$

$$(7.143) \qquad\qquad\qquad (7.144)$$

$$k_H = 3 \times 10^6 \text{ mol}^{-1}\text{ s}^{-1}$$

The tetrahydrofurane radical (7.144) may initiate free radical polymerization of vinyl monomers such as: methyl methacrylate, styrene, and acrylonitrile but not vinyl acetate:[452,1727,2152]

$$\underset{O}{\boxed{}}\cdot + RCH\!=\!CHR' \longrightarrow \underset{O}{\boxed{}}\!\!\overset{}{\underset{CHR\dot{C}HR'}{}} \qquad (7.184)$$

The inability of tetrahydrofurane radicals to initiate polymerization of vinyl acetate is due to the fact that these radicals are not capable of adding to monomer.[452,1727]

Benzophenone derivatives such as 3,3',4,4'-benzophenone tetracarboxylic dianhydride (7.145) and 3,3',4,4'-tetramethoxycarbonyl benzophenone (7.146) are more efficient photoinitiators of methyl methacrylate in tetrahydrofurane than benzophenone itself.[452,1729]

$$(7.145) \qquad\qquad\qquad\qquad (7.146)$$

Efficiency increases in the order benzophenone < (7.146) < (7.145), although the u.v. absorption spectra of these compounds show only minor variations in the region 330–380 nm.

7.9.4.2.3.6. Benzophenone photoreaction with amines

Benzophenone is rapidly photoreduced by primary and secondary aliphatic amines which contain the >CHNH- group to benzophenone ketyl radical (7.148).[738,739,748,859] Primary amines (7.147) led to imines (7.149), whereas secondary amines (7.150) led to alkylimines (7.151):

$$^3(\phi-\overset{\overset{\textstyle O}{\|}}{\underset{\textstyle \cdot}{C}}-\phi) + RR'CHNH_2 \longrightarrow {}^3(EXCIPLEX) \longrightarrow \phi-\overset{\overset{\textstyle OH}{|}}{\underset{\textstyle \cdot}{C}}-\phi + RR'\dot{C}NH_2$$

$$(7.147) \qquad\qquad\qquad\qquad\qquad (7.148) \qquad\qquad (7.185)$$

$$\phi-\overset{\overset{\text{O}}{\|}}{\text{C}}-\phi + \text{RR}'\overset{.}{\text{C}}\text{NH}_2 \longrightarrow \phi-\overset{\overset{\text{OH}}{|}}{\underset{.}{\text{C}}}-\phi + \text{RR}'\text{C}=\text{NH} \qquad (7.186)$$

$$(7.149)$$

$$^3(\phi-\overset{\overset{\text{O}}{\|}}{\text{C}}-\phi) + \text{RR}'\text{CHNHR}'' \longrightarrow \,^3(\text{EXCIPLEX}) \longrightarrow \phi-\overset{\overset{\text{OH}}{|}}{\underset{.}{\text{C}}}-\phi + \text{RR}'\overset{.}{\text{C}}\text{NHR}''$$

$$(7.150) \qquad\qquad\qquad (7.148) \qquad\qquad (7.187)$$

$$\phi-\overset{\overset{\text{O}}{\|}}{\text{C}}-\phi + \text{RR}'\overset{.}{\text{C}}\text{NHR}'' \longrightarrow \phi-\overset{\overset{\text{OH}}{|}}{\underset{.}{\text{C}}}-\phi + \text{RR}'\text{C}=\text{NR}'' \qquad (7.188)$$

$$(7.151)$$

These reactions proceed rapidly in dilute solutions of hydrocarbons by formation of exciplexes (cf. Section 3.5) between the excited triplet state of benzophenone (T_1) and amine.[213, 748, 871, 1297, 1306–1308, 1311, 2238, 2455, 3179]

Formation of benzophenone ketyl radical (7.148) during photoreduction of benzophenone by amines has been supported by ESR studies.[215,865,873] Using t-nitrosobutane (7.131) as a spin trap, several amine radicals have been trapped and characterized by ESR spectroscopy.[2227] A typical ESR spectrum of amino-nitroxide radicals formed during the photoreduction of benzophenone by n-butylamine and isopropyl amine in the presence of t-nitrosobutane is shown in Fig. 7.10.

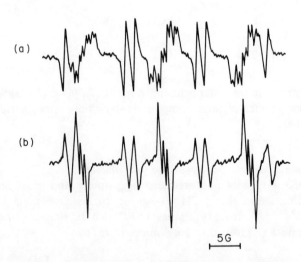

Fig. 7.10 ESR spectrum of nitroxide radicals formed during photolysis of: (a) benzophenone-n-butylamine, and (b) benzophenone-iso-propylamine in solutions containing t-nitrosobutane[2227]

During the photoreduction of benzophenone with aqueous and ethanolic solutions of isobutylamine and *tert*-butylamine benzophenone ketyl radical anion (*7.136*) is formed. This has been well characterized by ESR spectroscopy.[867]

Photoreduction of benzophenone with amines shows less selectivity towards the nature of the hydrogen atom than that observed with alkanes.[739, 2458] The photoreduction is greater in polar solvents than in non-polar solvents.[863]

Studies of the photoreduction of benzophenone with aromatic amines give the following interesting results:

(a) Excited triplet state of benzophenone (T_1) reacts with diphenylamine (*7.152*) in *tert*-butyl alcohol and acetonitrile but not in benzene, leading in the polar solvents to 4-(*N*-phenylamino)phenyldiphenyl methanol (*7.153*) but not to benzopinacol.[2867]

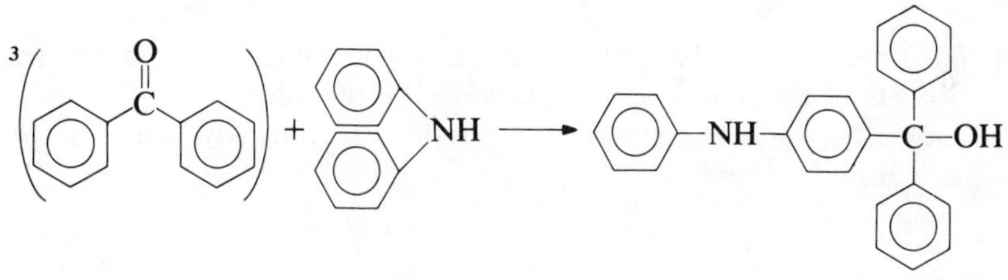

$$(7.152) \qquad\qquad (7.153) \qquad (7.189)$$

(b) The photoreduction of benzophenone by diphenylamine and methyl-diphenylamine can be quenched by triplet quenchers such as naphthalene and ferric dibenzoylmethanate.[863]

(c) The photoreduction of benzophenone with benzhydrol is inhibited by the presence of tertiary amines such as triphenylamine or tri-*p*-tolylamine.[863]

(d) Benzophenone ketyl radical formed during photoreduction of benzophenone with aniline, reacts very fast with aniline radical in reversible reaction:[859]

$$^3(\phi-\overset{O}{\overset{\|}{C}}-\phi) + \phi NH_2 \longrightarrow {}^3(\text{EXCIPLEX}) \longrightarrow \phi-\overset{OH}{\underset{\cdot}{C}}-\phi + \phi\overset{\cdot}{N}H$$

$$(7.190)$$

$$\phi-\overset{OH}{\underset{\cdot}{C}}-\phi + \phi\overset{\cdot}{N}H \longrightarrow \phi-\overset{O}{\overset{\|}{C}}-\phi + \phi NH_2 \qquad\qquad (7.191)$$

(e) Laser photolysis of benzophenone and *N*,*N*-diethylaniline shows that ionic dissociation (7.190) competes with hydrogen abstraction reaction (7.191):[213,2231]

$$\phi-\overset{\overset{\displaystyle O^-}{|}}{\underset{\displaystyle \cdot}{C}}-\phi + \phi-\overset{+}{\underset{\displaystyle \cdot}{N}}(CH_2CH_3)_2$$

$$(7.192)$$

$$^3(\phi-\overset{\overset{\displaystyle O}{\|}}{C}-\phi) + \phi-N(CH_2CH_3)_2 \longrightarrow 3(EXCIPLEX)$$

$$\phi-\overset{\overset{\displaystyle OH}{|}}{\underset{\displaystyle \cdot}{C}}-\phi + \phi-N\overset{\overset{\displaystyle \dot{C}HCH_3}{\diagup}}{\underset{\displaystyle CH_2CH_3}{\diagdown}}$$

$$(7.193)$$

This system may cause both free radicals and ionic polymerization (cf. Section 5.4.6).

(f) It has been proved that potential radicals which initiate free radical polymerizations are amino radicals but not ketyl radicals.[1727, 2030, 2031, 3179]

(g) Oxygen, if present, reacts with amino radicals by the following mechanism:[315,316,2962]

$$R_2\dot{C}-NR_2 + O_2 \longrightarrow R_2\overset{\overset{\displaystyle \dot{O}}{|}}{\underset{\overset{\displaystyle O}{\|}}{C}}-NR_2 \qquad (7.194)$$

$$R_2\overset{\overset{\displaystyle \dot{O}}{|}}{\underset{\overset{\displaystyle O}{\|}}{C}}-NR_2 + R_2CH-NR_2 \longrightarrow R_2\overset{\overset{\displaystyle OOH}{|}}{C}-NR_2 + R_2\dot{C}-NR_2 \qquad (7.195)$$

(h) The photoreduction of benzophenone can also be caused by tetramethylurea and N-methyl-2-pyrrolidone.[1847]

7.9.4.2.3.7. Benzophenone photoreaction with thiols
The excited triplet state of benzophenone (T_1) abstracts hydrogen from thiols (mercaptans) (RSH) and produces THIYL RADICALS (*7.154*):[2566,2570]

$$^3(\phi-\overset{\overset{\displaystyle \dot{O}}{|}}{\underset{\displaystyle \cdot}{C}}-\phi) + RSH \longrightarrow \phi-\overset{\overset{\displaystyle OH}{|}}{\underset{\displaystyle \cdot}{C}}-\phi + RS^{\cdot} \qquad (7.196)$$

$$(7.154)$$

This reaction is enhanced by the relatively low S–H bond dissociation energy. Thiols are relatively good quenchers of the triplet state of benzophenone (T_1)[1469,2650] and acetophenone.[4001]

Thiyl radicals can be trapped with a spin trap *t*-nitrosobutane (*7.131*) and are well characterized by ESR spectroscopy.[2227]

Thiyl radicals may initiate free radical polymerization of vinyl and allyl monomers:[1682,1683,2568,2570]

$$RS\cdot + RCH = CHR' \rightarrow RS\text{–}RCH\text{–}\dot{C}HR' \qquad (7.197)$$

Using polyfunctional thiols (7.155) and bifunctional monomers (*7.156*) it is possible to obtain well-designed crosslinked polymers:[1990,2566]

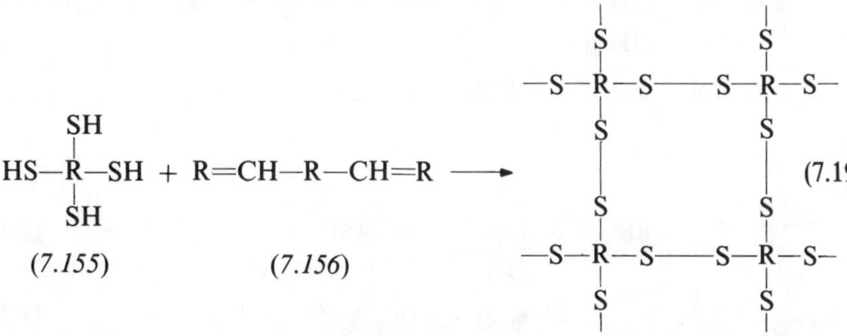

(*7.155*) (*7.156*) (7.198)

When benzophenone was irradiated in the presence of partially deuterized isopropanol (*7.157*) and mercaptan (RSH), isopropanol containing two deuterium atoms (*7.158*) was isolated:[745]

$$\phi\text{—}\underset{\displaystyle}{\overset{\displaystyle O}{\overset{\|}{C}}}\text{—}\phi + CH_3\text{—}\underset{H}{\overset{OD}{C}}\text{—}CH_3 + RSH \xrightarrow{h\nu} \phi\text{—}\overset{O}{\overset{\|}{C}}\text{—}\phi + RSH + CH_3\text{—}\underset{D}{\overset{OD}{C}}\text{—}CH_3 \qquad (7.199)$$

(*7.157*) (*7.158*)

In the absence of mercaptan or in the presence of naphthalene as quencher no product of this type was observed.[744] Since the presence of the mercaptan does not prevent the formation of ketyl radicals, the incorporation of the second deuterium must result from transfer of the following type:

$$^3\left(\phi\text{—}\overset{O}{\overset{\|}{C}}\text{—}\phi\right) + CH_3\text{—}\underset{H}{\overset{OD}{C}}\text{—}CH_3 \longrightarrow \phi\text{—}\overset{OH}{\dot{C}}\text{—}\phi + CH_3\text{—}\overset{OD}{\dot{C}}\text{—}CH_3 \qquad (7.200)$$

$$CH_3\text{—}\overset{OD}{\dot{C}}\text{—}CH_3 + RSSR \longrightarrow CH_3\text{—}\overset{O}{\overset{\|}{C}}\text{—}CH_3 + RS\cdot + RSD \qquad (7.201)$$

$$RSD + CH_3\text{—}\overset{OD}{\dot{C}}\text{—}CH_3 \longrightarrow RS\cdot + CH_3\text{—}\underset{D}{\overset{OD}{C}}\text{—}CH_3 \qquad (7.202)$$

$$2RS \longrightarrow RSSR \qquad (7.203)$$

In general, the mechanism for the inhibition of the photoreduction of benzophenone by mercaptans can be written as:

$$\phi-\overset{O}{\overset{\|}{C}}-\phi + CH_3-\overset{OH}{\overset{|}{\underset{H}{C}}}-CH_3 \xrightarrow{h\nu} \phi-\overset{OH}{\overset{|}{\underset{\cdot}{C}}}-\phi + CH_3-\overset{OH}{\overset{|}{\underset{\cdot}{C}}}-CH_3 \quad (7.204)$$

$$CH_3-\overset{OH}{\overset{|}{\underset{\cdot}{C}}}-CH_3 + \phi-\overset{O}{\overset{\|}{C}}-\phi \longrightarrow \phi-\overset{OH}{\overset{|}{\underset{\cdot}{C}}}-\phi + CH_3-\overset{O}{\overset{\|}{C}}-CH_3 \quad (7.205)$$

$$CH_3-\overset{OH}{\overset{|}{\underset{\cdot}{C}}}-CH_3 + RSH \longrightarrow CH_3-\overset{OH}{\overset{|}{\underset{H}{C}}}-CH_3 + RS\cdot \quad (7.206)$$

$$RS\cdot + \phi-\overset{OH}{\overset{|}{\underset{\cdot}{C}}}-\phi \longrightarrow RSH + \phi-\overset{O}{\overset{\|}{C}}-\phi \quad (7.207)$$

$$\phi-\overset{OH}{\overset{|}{\underset{\cdot}{C}}}-\phi \longrightarrow \phi-\overset{OH}{\overset{|}{\underset{\phi}{C}}}-\overset{OH}{\overset{|}{\underset{\phi}{C}}}-\phi \quad (7.208)$$

$$2RS\cdot \longrightarrow RSSR \quad (7.209)$$

7.9.4.2.3.8. *Benzophenone photoreactions with phenols*

Hindered phenols react with the excited triplet state of benzophenones to yield fuchsone derivatives:[363,364]

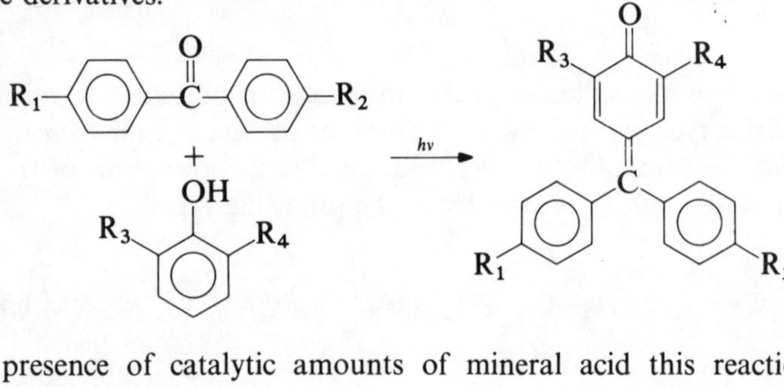

$$(7.210)$$

In the presence of catalytic amounts of mineral acid this reaction yields tetraphenylmethanes (*7.159*):

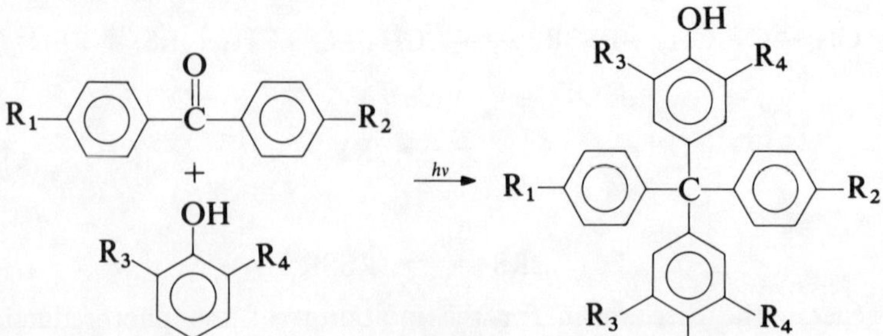

$$(7.159)$$

$$(7.211)$$

These reactions occur by the following mechanism:[363,364]

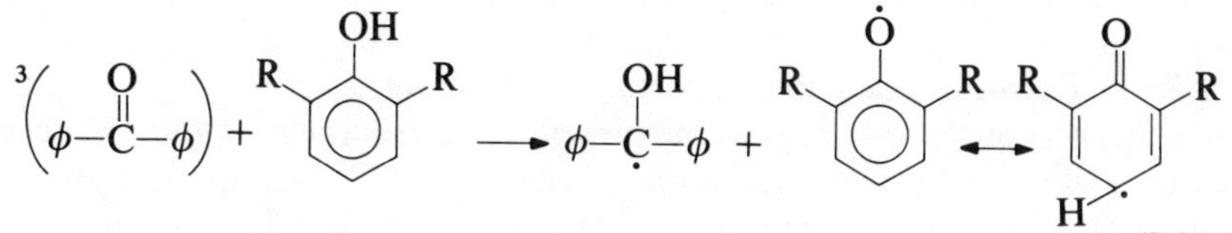

(7.212)

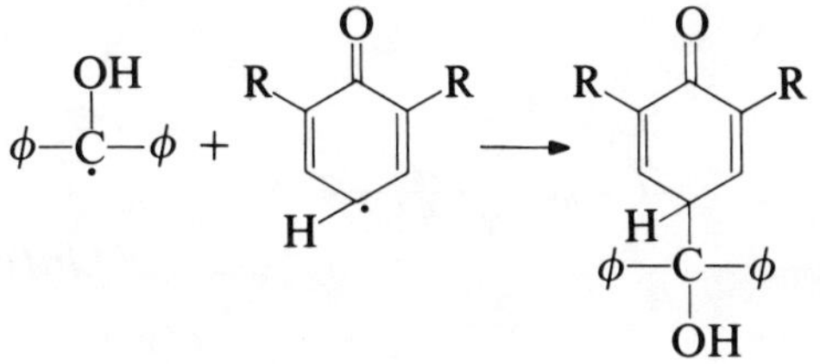

(7.213)

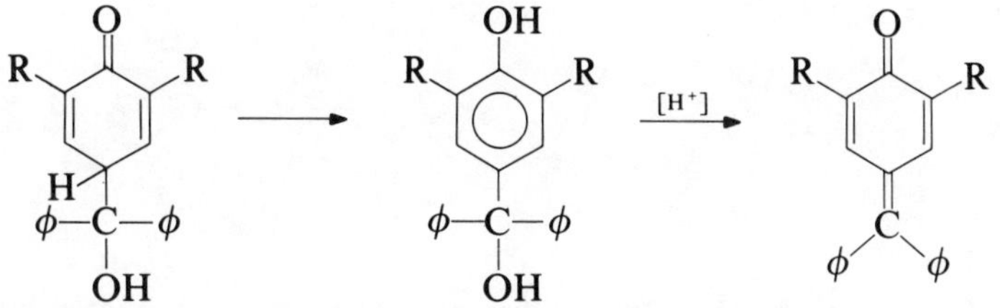

(7.214)

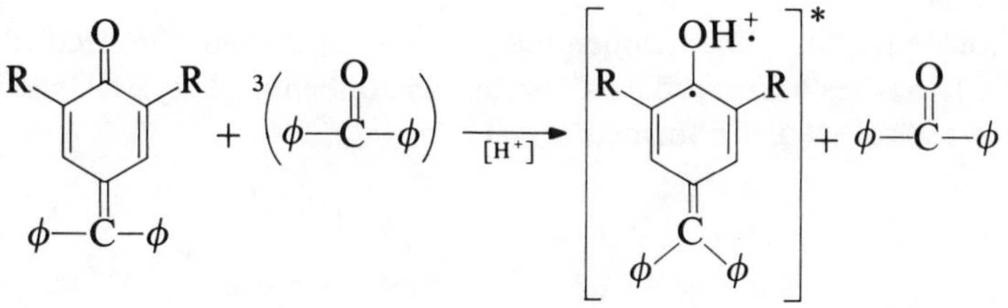

(7.215)

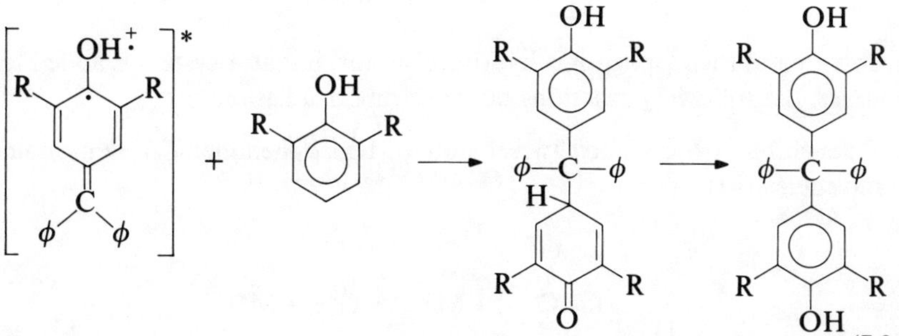

(7.216)

The second step leading to the tetraphenyl methane occurs by energy transfer from the excited triplet state of benzophenone to the fuchsone, resulting in the triplet excitation of fuchsone.

Using *t*-nitrosobutane (7.131) as a spin trap, the phenoxy radicals have been trapped and well characterized by ESR spectroscopy.[2227]

7.9.4.2.3.9. Benzophenone photoreactions with norbornene

Benzophenone photoirradiated in the presence of norbornene (*7.160*) produces oxetane (*7.161*) with high quantum efficiency.[224, 3301] Addition of 0.02 M triethylamine completely suppresses oxetane formation, and benzopinacol (*7.162*) is the only product which is formed:[3179]

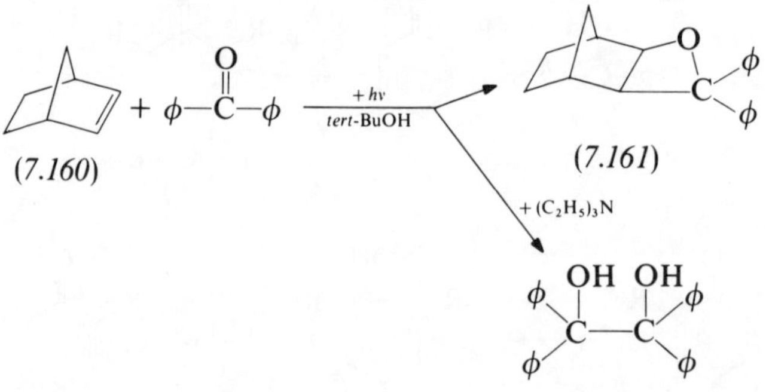

(*7.160*) (*7.161*) (7.217)

(*7.162*) (7.218)

7.9.4.2.3.10. Benzophenone–hydrogen-donors hybrid systems as photoinitiators of polymerization

During photoreduction of benzophenone with hydrogen-donor molecules (RH) e.g. solvent, two types of free radicals, such as benzophenone ketyl radical (*7.163*) and solvent radical (R·), are formed:

$$^3(\phi\text{—}\overset{\overset{\displaystyle O}{\|}}{C}\text{—}\phi) + RH \xrightarrow{k_H} \phi\text{—}\underset{\bullet}{\overset{\overset{\displaystyle OH}{|}}{C}}\text{—}\phi + R\cdot \qquad (7.219)$$

(*7.163*)

If such a hybrid (benzophenone–hydrogen-donor) initiator system is added to a monomer, the following reactions occur during irradiation:

(i) Quenching of the excited triplet state of benzophenone (T_1) by monomer molecules (M):[185, 589, 1554, 2152, 2153, 2459, 2497, 3179]

$$^3(\phi\text{–}\overset{\overset{\displaystyle O}{\|}}{C}\text{–}\phi) + M \overset{k_Q}{\rightarrow} \phi\text{–}\overset{\overset{\displaystyle O}{\|}}{C}\text{–}\phi + M^* \qquad (7.220)$$

The quenching effectiveness (Table 7.17) decreases in the sequence: styrene > α-methylstyrene > vinyl pyrrolidone > methyl methacrylate > acrylonitrile > vinyl acetate.[2152, 2153, 3179] The quenching rate constant (k_Q) for styrene is $k_Q = 3.3 \times 10^9$ mole^{-1} s^{-1}, which indicates that the excited

Table 7.17 Quenching of benzophenone excited triplet
state (T_1) by vinyl monomers in benzene

Monomer	Quenching rate constant (k_Q) $(M^{-1} s^{-1})$	References
Styrene	$(3.3 \pm 0.3) \times 10^9$	
α-Methylstyrene	$(2.7 \pm 0.3) \times 10^9$	
Vinyl pyrrolidone	$(3.6 \pm 0.4) \times 10^8$	
Methyl methacrylate	$(6.9 \pm 0.7) \times 10^9$	2152
Acrylonitrile	$(3.4 \pm 0.3) \times 10^7$	
Vinyl acetate	$(5.4 \pm 0.5) \times 10^6$	
Isobutylene	3.0×10^7	589

triplet state of benzophenone is efficiently quenched by styrene, and polymerization of this monomer is inefficient.[1554, 1555]

(ii) Initiation of polymerization by solvent free radicals (R·):[452, 1727, 2152]

$$M + R\cdot \rightarrow R\text{--}M\cdot \xrightarrow{+M} \text{polymer} \qquad (7.221)$$

Reactivity of R· radicals depends on the resonance, inductive, steric, and electron-repulsion effects in radical initiation of monomers.[2997]
The PROBABILITY (α_R) of formation of radicals (R·) capable of initiating the radical polymerization of monomer (M) is given by equation (7.222) and shown in Fig. 7.11.[2152, 3213]

$$\alpha_R = \frac{k_H[RH]}{k_H[RH] + k_Q[M]} \qquad (7.222)$$

(iii) Initiation of polymerization by ketyl radicals (7.163)

$$M + \phi\text{--}\underset{\overset{|}{\underset{\bullet}{C}}}{\overset{\overset{OH}{|}}{}}\text{--}\phi \xrightarrow{k_i} MH\cdot + \phi\text{--}\overset{\overset{O}{\|}}{C}\text{--}\phi \left[\text{or} \quad \phi\text{--}\underset{\underset{\bullet}{\overset{|}{M}}}{\overset{\overset{OH}{|}}{C}}\text{--}\phi \right] \qquad (7.223)$$

(7.163)

Ketyl radicals may initiate free radical polymerization of several monomers such as vinyl acetate ($k_i = 5.5 \pm 1.5 \times 10^3 \text{ mole}^{-1} \text{s}^{-1}$), acrylonitrile ($k_i = 3.8 \pm 1.0 \times 10^3$ mole^{-1}s^{-1}), methyl methacrylate ($k_i = 9.0 \pm 2.0 \times 10^3$ mole^{-1}s^{-1}) when monomer concentration is sufficiently high (0.4 M).[2152] At lower concentration of methyl methacrylate than 0.4 M ketyl radicals cannot initiate polymerization but act as terminating species.[1727, 2240] With other monomers, such as styrene, α-methylstyrene and vinyl pyrrolidone the situation is that ketyl radicals are not formed, because all of these monomers efficiently quench the triplet state of benzophenone.

Oxygen plays an evident role in the following steps of photoinitiated free radical polymerization:

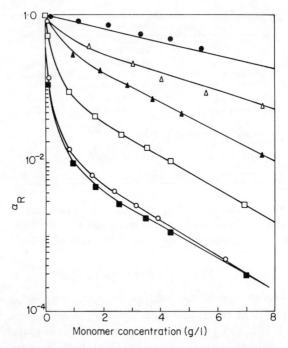

Fig. 7.11 Probability (α_R) for the formation of radicals capable of inducing polymerization versus the monomer concentration according to equation (7.222): (●) vinyl acetate, (△) acrylonitrile, (▲) methyl methacrylate, (□) N-vinyl-3-pyrrolidone, (○) α-methylstyrene, and (■) styrene[2152]

(i) Deactivates the excited triplet state (T_1) of benzophenone.

(ii) Reacts with initiator radicals.[2569]

(iii) Reacts with growing polymer (P·) radical to give polymer peroxy radical (POO·), which is less reactive in propagation:

$$P \cdot + O_2 \rightarrow POO \cdot \qquad (7.224)$$

Polymer peroxy radical abstracts mainly hydrogen from hydrogen-donor sites to produce peroxy groups:

$$POO \cdot + RH \rightarrow POOH + R \cdot \qquad (7.225)$$

The mechanism and kinetics of benzophenone photoinitiated polymerization of unsaturated monomers has been a subject of various publications.[588, 589, 1144, 1148, 1151, 1158, 2087, 2088, 2152, 2154, 2170, 2491, 2498, 3213, 3719]

It has been proposed that the excited triplet state of benzophenone can form, with unsaturated monomer, an exciplex which is further converted to a diradical (*7.164*), CT COMPLEX (*7.165*) or deactivated to the starting compounds (reaction 7.228):[589, 2087, 2088]

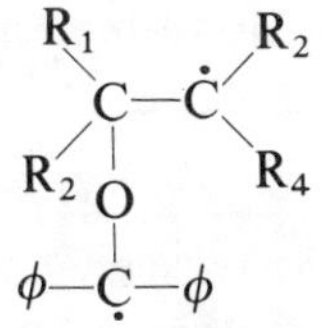

$$\begin{matrix} R_1 & & R_2 \\ & C\!-\!\overset{\textstyle .}{C} & \\ R_2 & O & R_4 \\ & | & \\ & \phi\!-\!\overset{\textstyle .}{C}\!-\!\phi & \end{matrix}$$

(7.226)

(7.164)

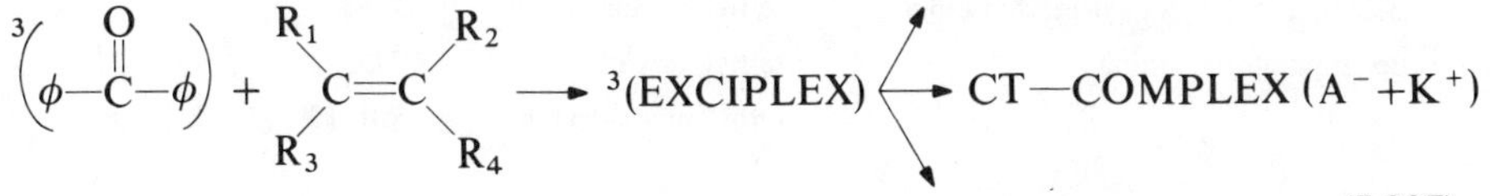

$$3\!\left(\phi\!-\!\overset{\textstyle O}{\overset{\|}{C}}\!-\!\phi\right) + \begin{matrix} R_1 & & R_2 \\ & C\!=\!C & \\ R_3 & & R_4 \end{matrix} \longrightarrow {}^3(\text{EXCIPLEX}) \begin{matrix} \nearrow \\ \longleftrightarrow \text{CT}\!-\!\text{COMPLEX (A}^-\!+\text{K}^+) \\ \searrow \end{matrix}$$

(7.165)   (7.227)

$$\phi\!-\!\overset{\textstyle O}{\overset{\|}{C}}\!-\!\phi + \begin{matrix} R_1 & & R_2 \\ & C\!=\!C & \\ R_3 & & R_4 \end{matrix}$$

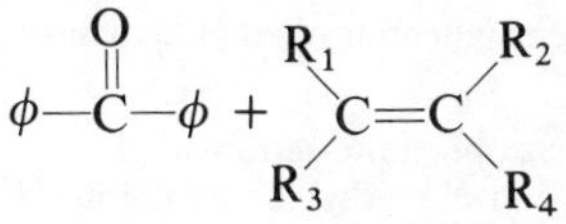

(7.228)

It has been assumed that in the charge-transfer complex (CT complex) the olefine is acting as an electron donor.[589, 2087, 2088] Measurements of photocurrents during the flashing benzophenone–olefine system gave evidence that exciplexes with CT character are generated which precede the formation of free ions.[2154]

An interesting mechanism of benzophenone-photoinitiated polymerization of acrylic monomers in sodium deodecyl sulphate micells has been reported.[1148]

Several benzophenone–hydrogen-donor molecule (RH) systems (Table 7.18) are capable photoinitiate free radical polymerization.

7.9.4.2.3.11. Substituted benzophenones as photoinitiators

Several substituted benzophenones has been used as effective photoinitiators for polymerization of different monomers.

4-Chloromethyl-benzophenone (7.166) photocleavage under u.v. irradiation, yielding 4-benzoylbenzyl radical (7.167) and chlor radical (Cl·):[1846, 3146]

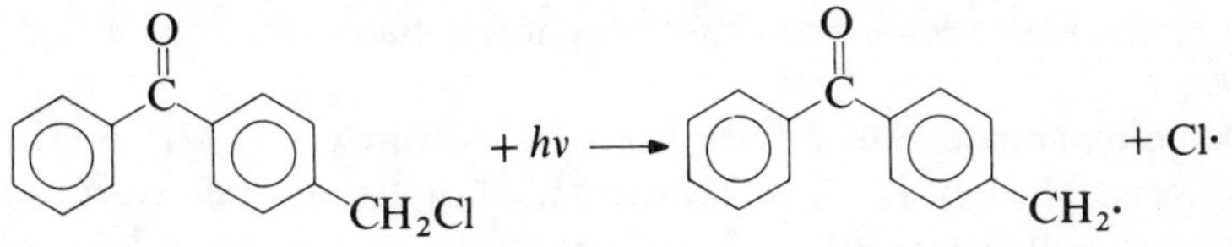

(7.166)   (7.167)   (7.229)

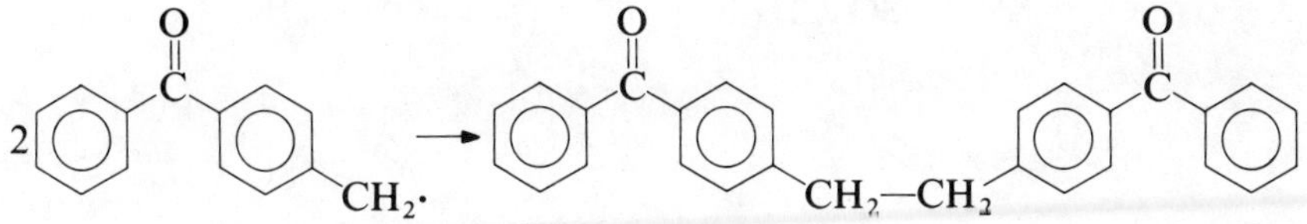

(7.230)

Table 7.18   Examples of benzophenone- hydrogen-donor systems which are capable photoinitiate free radical polymerization

System	Monomer	References
Benzophenone–hydrogen peroxide	Methyl methacrylate	1323
Benzophenone–tetrahydrofurane	Vinyl monomers	452, 1727
Benzophenone–isopropanol	Acrylonitrile	2602
Benzophenone–benzyl methyl ether	Methyl methacrylate	2883
Benzophenone–amines	Methyl acrylate	727, 3179
	Methyl methacrylate	861, 1297, 1307, 2240, 2455, 2458, 2461
	Butyl acrylate	729
	Acrylonitrile	2148
Benzophenone–methyldiethanolamine	Acrylate and methacrylate monomers	417
Benzophenone–tetramethyl- and N-methyl-2-pyrrolidone	Acrylated polyurethanes	1847
Benzophenone–thiols	Vinyl and allyl monomers	1682, 1683, 2568, 2570

Chlor radicals are responsible for initiation of radical polymerization. 4-benzoylbenzyl radicals are less reactive due to resonance stabilization and participate in the initiation of polymerization to a minor extent.

Ketosulphides (*7.168*) photocleavage under u.v. irradiation, yielding 4-ben-zoylbenzyl (*7.169*) and sulphide (*7.170*) radicals:[3146]

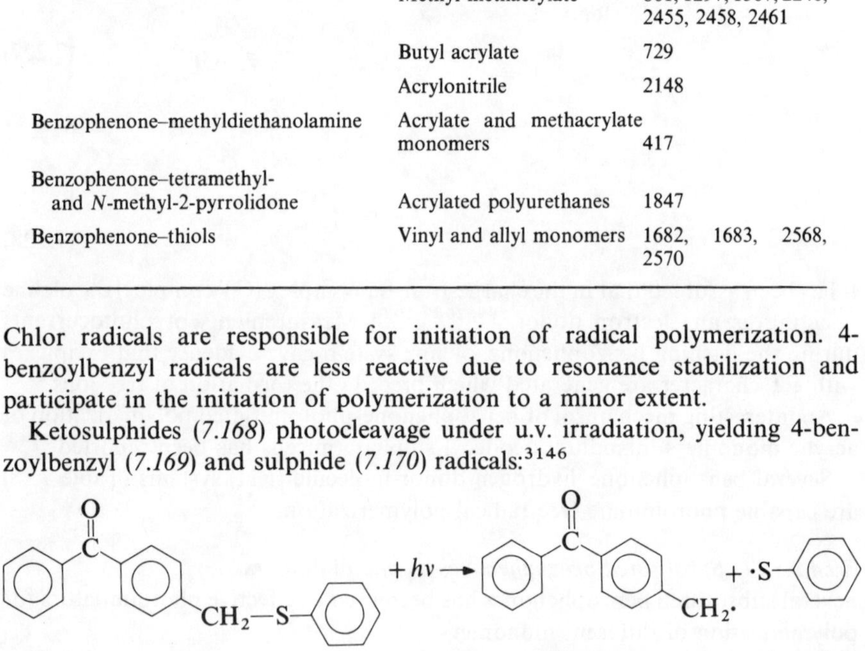

(*7.168*)   (*7.169*)   (*7.170*)

(7.231)

Ketosulphides have been used as effective photoinitiators for polymerization of acrylates.

Methanesulphonate esters (mesylates) of 2-hydroxy- (*7.171*) and 2,4-di-hydroxy-benzophenone (*7.172*) are efficient photoinitiators of polymerization of divinyl ether and diallyl ester, which may occur by free radical or cationic mechanisms.[2891]

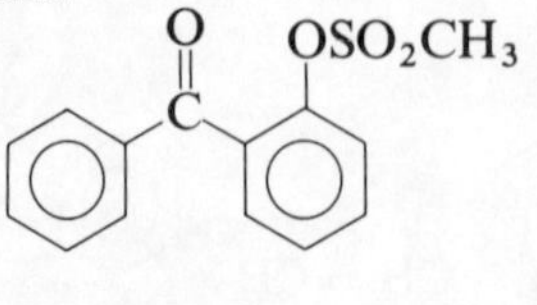

(*7.171*) 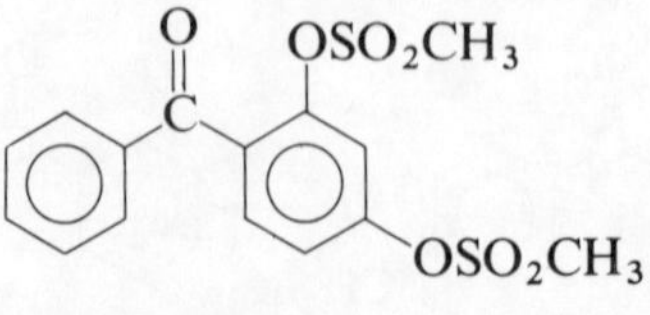   (*7.172*)

Mesylates in the presence of thiols may photoinitiate polymerization of vinyl esters by free radical mechanism. The excited triplet state of mesylate (T_1) abstracts hydrogen from thiols and produces thiyl radicals $(RS\cdot)$ which may initiate polymerization:

(7.232)

Mesylates in the presence of traces of water are photolysed to 2-hydroxybenzophenone (7.173) and methanesulphonic acid (7.174):

(7.173)   (7.174)   (7.233)

The photogenerated methanesulphonic acid initiates cationic polymerization of vinyl ethers by the following mechanism:

$$CH_3SO_3H + CH_2 = CH-OR \rightarrow CH_3-\overset{+}{C}H-OR + CH_3SO_3^- \qquad (7.234)$$

or

$$CH_3-\overset{+}{C}H-OR + CH_2 = CH-OR \rightarrow CH_3-CH-CH_2-\overset{+}{C}H-OR \qquad (7.235)$$

7.9.4.2.3.12. Photochemistry of substituted benzophenones

Substitution at the *ortho*-position (and in some cases at the *para*-position) to the carbonyl group in benzophenone has a remarkable effect on the photochemistry of the molecule. Steric and electronic interaction of the substituted group with the carbonyl group confers stability to u.v. radiation on the compound. For this reason hydroxybenzophenones are used as stabilizers of plastics (cf. Section 15.5).

In all cases where the substituent has a fairly labile hydrogen atom attached to a group in the *ortho*-position, the ketone is photostabile.[591, 2664, 3772]

Thus, pinacolization does not occur with: *ortho* hydroxy-, *ortho* amino-, 2,4-dihydroxy-, 2-monomethylamino-, and 4-amino-benzophenone.

Pinacolization does occur with: 4-dimethylamino-, 4,4'-dichloro-, 4,4'-dimethoxy-, 4-nitro-, and 2-methoxybenzophenone.

The photochemical behaviour of substituted benzophenones can be divided into two groups:[368]

(i) Compounds with no *ortho* substituent or with *ortho* substituents having no hydrogen directly attached to the α-carbon atom. This group has spectra, quantum yield of photolysis, luminescence characteristic, and transient product spectra very similar to those of benzophenone itself. Intersystem crossing (ISC) (cf. Section 1.7) to the triplet state occurs with very high probability, and this state readily abstracts hydrogen from the solvent.

(ii) Compounds having a hydrogen directly attached to carbon substituted at *ortho* position. This group has low quantum yield of photolysis, luminescence is weak or absent, ketyl radicals are not observed on flash photolysis, and they form metastable intermediates of several hours lifetime which are not free radicals (no ESR spectra) and are probably tautomers formed by transfer of a hydrogen atom from the *ortho* substituent to the carbonyl group. A six-membered ring transition state is essential for this tautomerization to compete effectively with hydrogen abstraction from the solvent. The tautomers decay to regenerate the original ketone and react rapidly with oxygen (cf. Section 15.5).

7.9.4.2.3.13. Benzopinacol

Benzopinacol (1,1,2,3-tetraphenyl-1,2-ethanediol) (*7.175*) is easily photolysed by u.v. radiation (250–260 nm) to ketyl radicals (*7.176*):[2697]

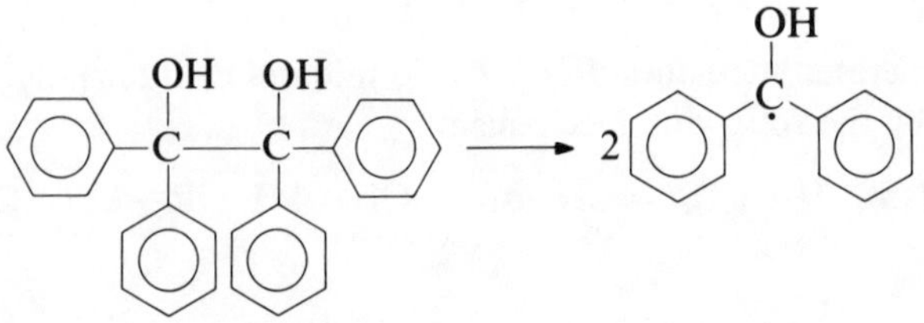

$$(7.236)$$

(*7.175*)   (*7.176*)

In the presence of oxygen, benzophenone and hydroperoxy radicals ($HO_2\cdot$) are further formed (cf. reaction 7.168).

Photolysis of benzopinacol in polymer matrix such as poly(vinyl alcohol) (*7.177*) or polyvinylbutyral (*7.178*), which contain labile hydrogen atoms, provide hydrogen abstraction by ketyl radicals:[2697]

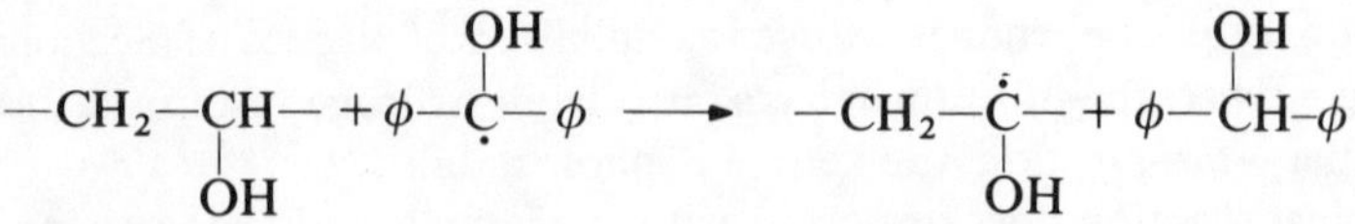

$$(7.237)$$

(*7.177*)

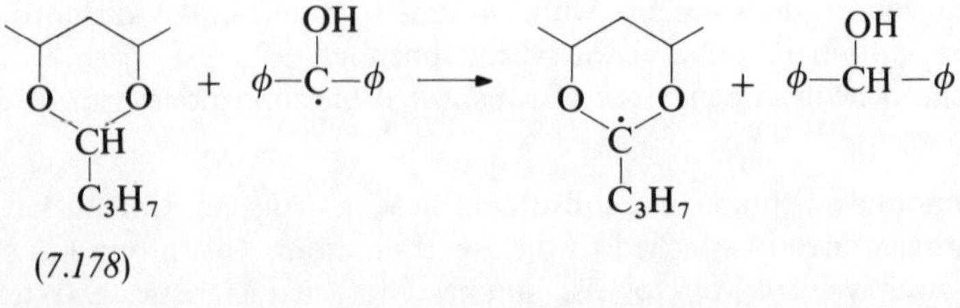

$$(7.238)$$

(*7.178*)

The resulting radicals disappear bimolecularly by disproportionation or recombination, or by reaction with a second ketyl radical with production of benzophenone.

7.9.4.2.3.14. Aminoaromatic ketones

An aminoaromatic ketone such as 4,4′-bis(*N*,*N*-dialkylamino)benzophenone (Michler's ketone) (*7.179*) has both ketone and amine functionality. Aminoaromatic ketones have large molar absorptivities and relatively long triplet lifetime in solutions.[2460, 3235, 3494]

The excited triplet state of Michler's ketone (T_1) abstracts hydrogen from another molecule of Michler's ketone:[740, 2086, 3019]

(*7.179*) (*7.180*) (*7.181*)

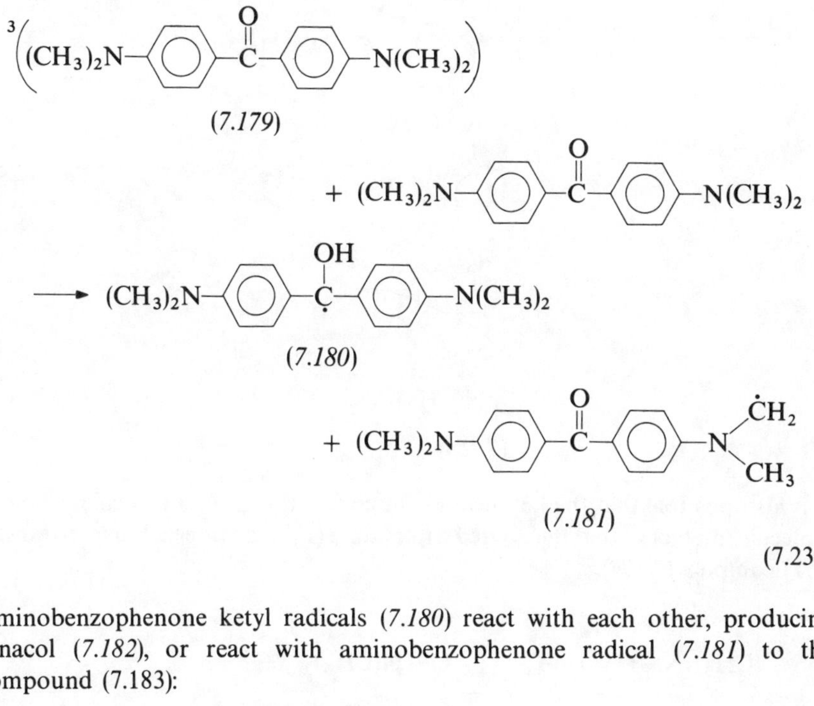

(7.239)

Aminobenzophenone ketyl radicals (*7.180*) react with each other, producing pinacol (*7.182*), or react with aminobenzophenone radical (*7.181*) to the compound (7.183):

(*7.181*) (*7.182*)

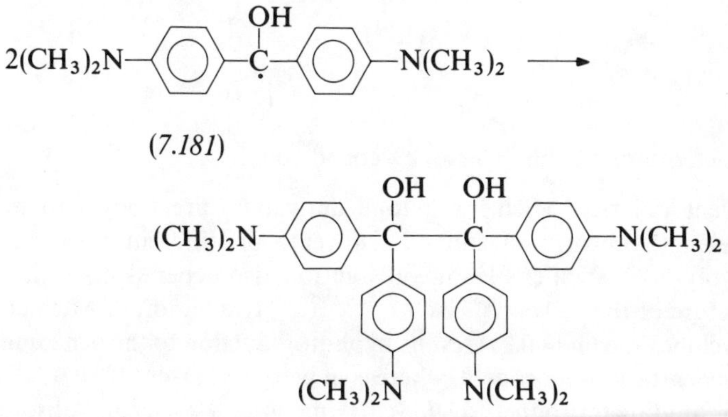

(7.240)

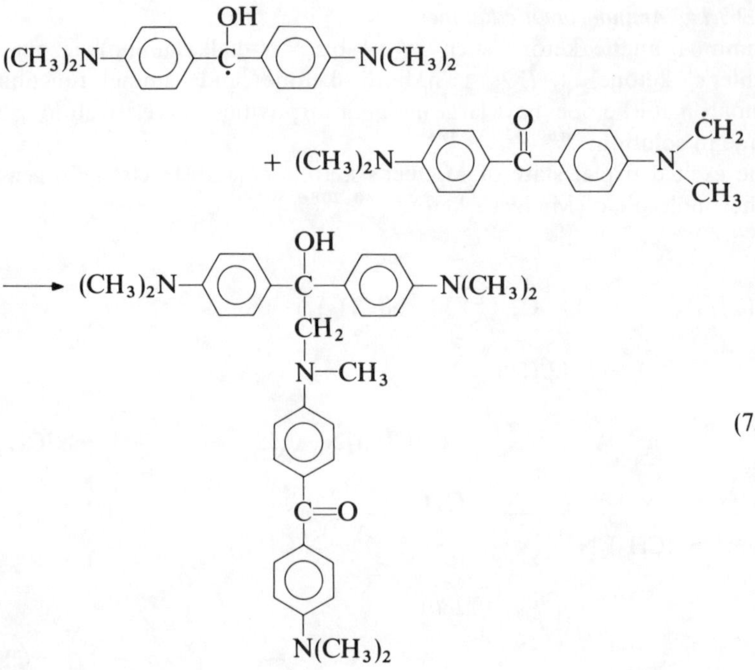

$$(CH_3)_2N—\bigcirc—\overset{OH}{\underset{\cdot}{C}}—\bigcirc—N(CH_3)_2$$

$$+ \; (CH_3)_2N—\bigcirc—\overset{O}{\overset{\|}{C}}—\bigcirc—N\overset{\dot{C}H_2}{\underset{CH_3}{\diagdown}}$$

$$\longrightarrow \; (CH_3)_2N—\bigcirc—\overset{OH}{\underset{CH_2}{C}}—\bigcirc—N(CH_3)_2$$
$$\underset{N—CH_3}{|}$$
$$\bigcirc$$
$$\underset{C=O}{|}$$
$$\bigcirc$$
$$N(CH_3)_2$$

(7.241)

(7.183)

It is also possible that the ketone and amine functionalities in Michler's ketone molecule interact so that the excited triplet state (T_1) is an internal charge-transfer (CT) complex (7.184).[2972]

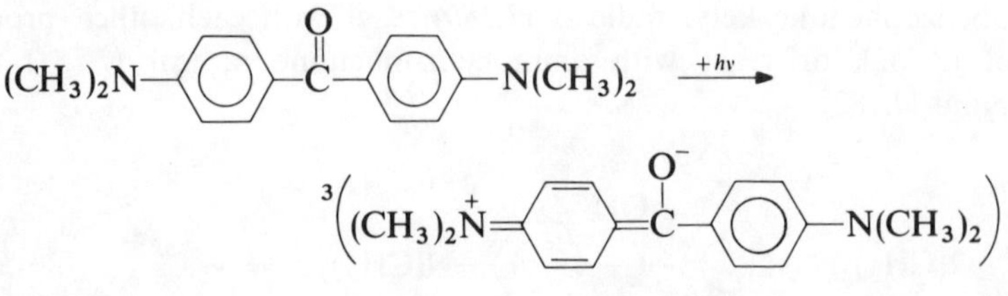

$$(CH_3)_2N—\bigcirc—\overset{O}{\overset{\|}{C}}—\bigcirc—N(CH_3)_2 \xrightarrow{+h\nu}$$

$$^3\left((CH_3)_2\overset{+}{N}=\bigcirc=\overset{O^-}{\underset{}{C}}—\bigcirc—N(CH_3)_2\right)$$

(7.184)   (7.242)

Photoreactivity of Michler's ketone depends on:

(i) Solvent polarity. Michler's ketone derivatives are known to have large wavelength absorption shifts in solvents of different polarity, and the photolysis of Michler's ketone in solution also depends upon the chemical structure of the solvent.[2973, 3235, 3494, 3495] Irradiation of Michler's ketone in cyclohexane ($\Phi = 0.91$) results in photoreduction to the benzopinacol and is concentration-dependent, whereas in benzene solvent ($\Phi = 1.00$) it results in photodimer products (7.183) having long wavelength absorption. In alcohol solvents ($\Phi = 0.08$) it shows little photochemical activity.[3235, 3494]

(ii) Self-quenching (concentration quenching) reactions.[2462]
(iii) Oxygen quenching.[3494, 3495]

A solvent also has an effect on the rate and mechanism of Michler's ketone photoinitiated polymerization of methyl methacrylate (Table 7.19).[2462, 2463] These results indicate that free radical polymerization is mainly initiated by the aminobenzophenone radical (7.181).

Table 7.19 Solvent effects on the polymerization of methyl methacrylate photoinitiated by Michler's ketone[2462]

Solvent	Monomer conversion (%)	Remarks
Benzene	3×10^{-3}	Large amount of polymer formed
Cyclohexane	1×10^{-4}	Very little polymer formed
Methanol	–	No polymer formed

Michler's ketone in mixture with benzophenone (7.185) exhibits PHOTOSYNERGISTIC PROPERTIES. This synergism can only be explained by formation of the triplet exciplex (7.186):[1705, 2086, 2458, 2460, 2463, 2546, 2879, 2881, 3794]

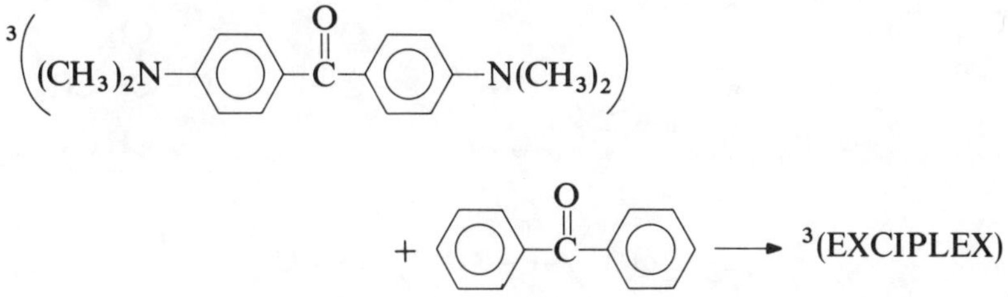

$$^3\left((CH_3)_2N-\text{(ring)}-\overset{O}{\overset{\|}{C}}-\text{(ring)}-N(CH_3)_2 \right)$$

$$+ \quad \text{(ring)}-\overset{O}{\overset{\|}{C}}-\text{(ring)} \quad \longrightarrow \quad ^3(\text{EXCIPLEX})$$

(7.185)   (7.186)

(7.243)

The triplet exciplex may dissociate to aminobenzophenone radical (7.181) and benzophenone ketyl radical (7.187):

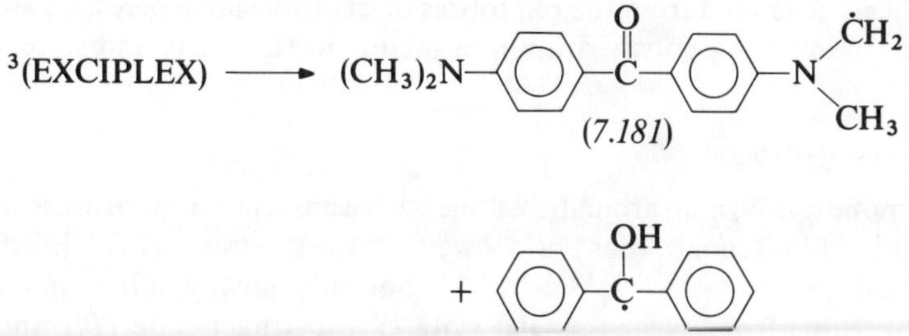

$$^3(\text{EXCIPLEX}) \quad \longrightarrow \quad (CH_3)_2N-\text{(ring)}-\overset{O}{\overset{\|}{C}}-\text{(ring)}-N\overset{\dot{C}H_2}{\underset{CH_3}{}}$$

(7.181)

$$+ \quad \text{(ring)}-\overset{OH}{\underset{\dot{C}}{}}-\text{(ring)}$$

(7.187)   (7.244)

Energy transfer from excited triplet Michler's ketone to benzophenone cannot occur because the triplet energy of benzophenone, $E_T = 69$ kcal mol^{-1}, is higher than the triplet energy of Michler's ketone, $E_T = 61$ kcal mol^{-1}.

The mixture of Michler's ketone with benzophenone has been found to be very effective in the photocuring of pigmented inks, and has been found to be most effective when the two components are intimately mixed before incorporation in the ink vehicle.[1564,1565]

Michler's ketone itself sensitizes photopolymerization of N-vinylsuccinimide in the crystal solid state, and this process can be applied to crystalline matrix photoimaging.[1758]

7.9.4.2.3.15. Benzophenone peresters

Benzophenone peresters such as p-benzoylperoxybenzoic acid *tert*-butyl ester (7.188) undergoes an efficient photodecomposition when irradiated of 366 nm to three types of radicals:[1464,1466,2666,3597]

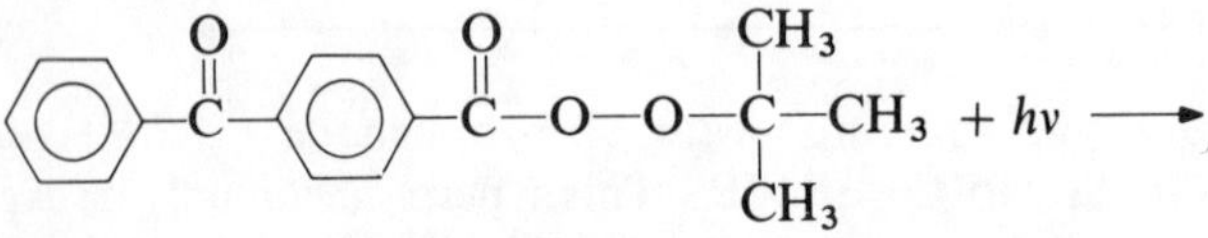

(7.188)

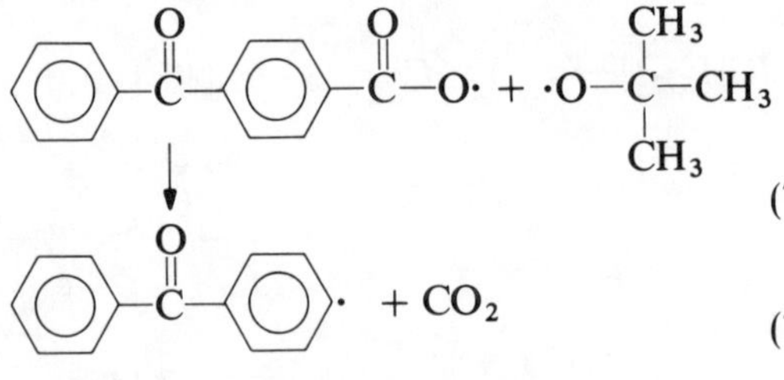

(7.245)

(7.246)

The excited triplet state (T_1) of the benzophenone carbonyl group of the perester is quenched by vinyl monomers (the degree of which depends on the type of monomer used), thus reducing the rate of decomposition of perester when it is used as a photoinitiator.

Radicals generated from the photolysis of benzophenone peresters are highly efficient in initiating polymerization of methyl methacrylate and styrene.[3]

7.9.4.2.4. Fluorenone

Fluorenone (7.189) is an aromatic ketone whose absorption spectrum is shown in Fig. 7.12. Fluorenone has a lowest triplet state (T_1) (π, π^*) $(E_T = 53$ kcal mol$^{-1})$.[587,1570,1719,3989,4008] Solvents greatly affect inter-system crossing (ISC) from excited singlet state (S_1) to triplet state (T_1). Increasing solvent polarity decreases the facility for ISC and simultaneously increases the

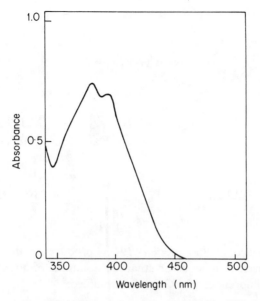

Fig. 7.12 Absorption spectrum of fluorenone[2238]

quantum yield for fluorescence. Thus triplet state activity is maximized in solvents such as benzene and cyclohexane, for which values of triplet yields are 0.93 and 1.03 respectively.[865,873] In contrast, triplet yields in solvents such as alcohols, acetone, and acrylonitrile are substantially less than unity with a concomitant increase in quantum yields for fluorescence.[585,864,1468,2553,3347]

The excited singlet state (S_1) of fluorenone is quenched by amines, and the efficiency increases as the ionization of the amine decreases.[742] The apparent increase in efficiency of quenching when the solvent polarity is increased is due to an increase in lifetime of the singlet state of the ketone caused by the solvent change.[2553]

The triplet state (T_1) of fluorenone is reduced by tertiary amines, and less efficiently by secondary and primary amines.[742,747,862,867,889] The efficiency of photoreduction decreases as the polarity of the solvent is increased due to the drop in efficiency of intersystem crossing (ISC). Fluorenone is not directly photoreduced in alcohol, ether, or alkanes but it is in the presence of amines (e.g. triethylamine) where photoreduction occurs by formation of an exciplex (7.190):[742,743,747,861,862,889,1729,2231,2238,2240]

$$\text{(fluorenone)} + h\nu \longrightarrow {}^{1}\text{(fluorenone)}^* \xrightarrow{\text{ISC}} {}^{3}\text{(fluorenone)}^*$$

(7.189)   (7.247)

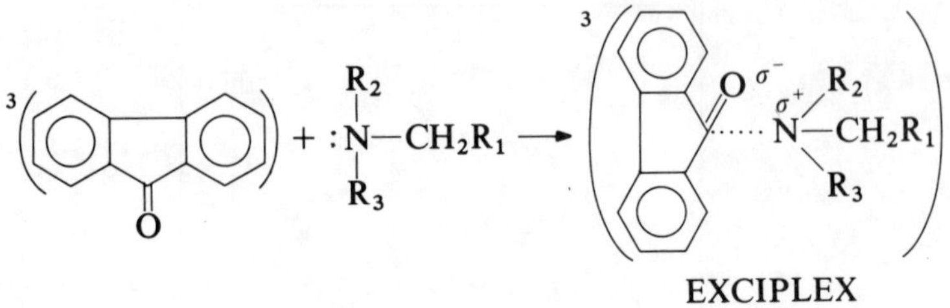

$$3 \quad \text{(fluorenone)} + \overset{R_2}{\underset{R_3}{:N{-}CH_2R_1}} \longrightarrow \left(\,^3 \text{(exciplex structure)} \, \overset{O\ \sigma^-}{\underset{\sigma^+}{}}\cdots\cdots \overset{R_2}{\underset{R_3}{N{-}CH_2R_1}} \right)$$

EXCIPLEX

(7.190)   (7.248)

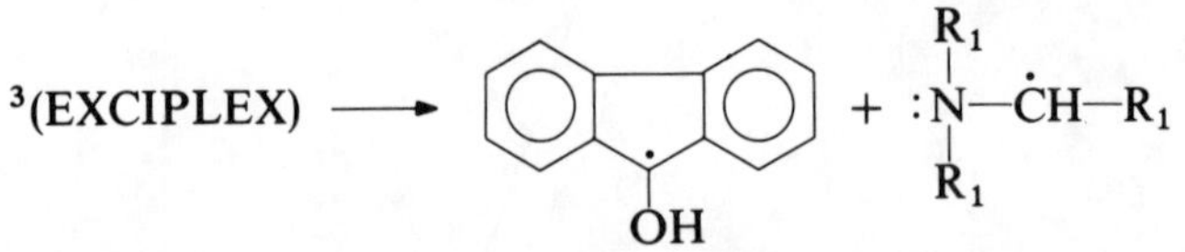

$$\,^3\text{(EXCIPLEX)} \longrightarrow \text{(fluorenyl radical)}\overset{\cdot}{OH} + \overset{R_1}{\underset{R_1}{:N{-}\overset{\cdot}{C}H{-}R_1}}$$

(7.249)

Normally the pinacol of fluorenone (7.191) is produced in the photoreduction reactions. However, if aqueous or ethanolic solutions of aliphatic amines are used, 9-hydroxyfluorenone (7.192) is also obtained:[867]

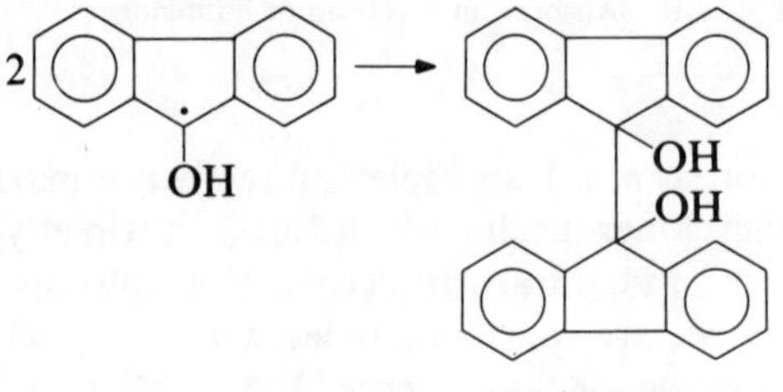

$$2 \ \text{(fluorenyl radical)}\overset{\cdot}{OH} \longrightarrow \text{(pinacol dimer)}\ \overset{OH}{\underset{OH}{}}$$

(7.250)

(7.191)

$$\text{(fluorenyl radical)}\overset{\cdot}{OH} + \overset{R_2}{\underset{R_1}{:N{-}CH_2R_1}} \longrightarrow \text{(fluorene)}\ \underset{H\ \ OH}{} + \overset{R_2}{\underset{R_3}{:N{-}\overset{\cdot}{C}H{-}R_1}}$$

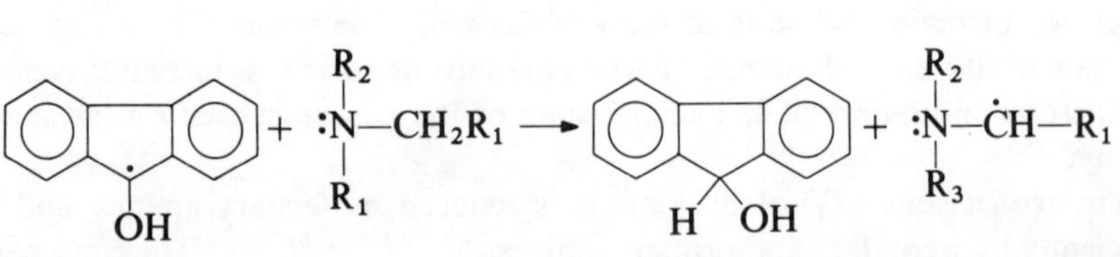

(7.192)   (7.251)

Pure fluorenone does not photoinitiate polymerization of methyl methacrylate and acrylonitrile, but it does in the presence of different amines.[861,2148,2231,2238,2240,2391,3968] Photoinitiation of polymerization by the fluorenone–amine hybrid system is accompanied by consumption of fluorenone.

7.9.4.2.5. Thioxanthone

Thioxanthone (7.193) is an aromatic ketone whose absorption spectrum is shown in Fig. 7.13. Thioxanthone shows two different electronic transitions of the carbonyl group in the excited triplet state depending on solvent polarity: $T_2(n, \pi^*)$

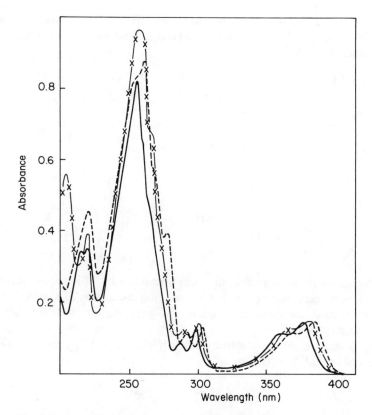

Fig. 7.13 Absorption spectra of: (——) thioxanthone, (– – –) 2-chlorothio-xanthone, and (–×–) 2-isopropylthioxanthone[1369]

in non-polar solvents (e.g. hexane) and $T_1(\pi;\pi^*)$ in polar solvents (e.g. alcohols)[821] (cf. Section 7.9.4.2.1).

The excited triplet state (T_1) is readily deactivated by two processes:[1570, 2194, 3864]

(i) Concentration deactivation (self-quenching)

$$T_1 + S_0 \xrightarrow{\ k_D\ } S_0 + S_0 \qquad (7.252)$$

The specific rate constants for concentration deactivation (k_D) are: $k_D = 6 \times 10^{10}\ \text{mole}^{-1}\,\text{s}^{-1}$ (in isopropanol[680], $k_D = 2.3 \times 10^9\ \text{mole}^{-1}\,\text{s}^{-1}$ (in benzene)[905] and $k_D = 5 \times 10^7\ \text{mole}^{-1}\,\text{s}^{-1}$ (in acetonitrile).[186,3976] The self-quenching mechanism probably involves the sulphur atom, which can act as an electron donor in the formation of a charge-transfer (CT) complex.

(ii) Triplet–triplet annihilation

$$T_1 + T_1 \xrightarrow{\ k_T\ } S_0 + S_0 \quad (k_T = 7 \times 10^9\ \text{mole}^{-1}\,\text{s}^{-1})^{[186]} \qquad (7.253)$$

The triplet state $T_1(\pi, \pi^*)$ ($E_T = 65.5\ kcal\ mol^{-1}$)[821, 1570, 3976] abstracts hydrogen from hydrogen-donor molecules (RH) only and produces a thioxanthone ketyl radical (7.194).

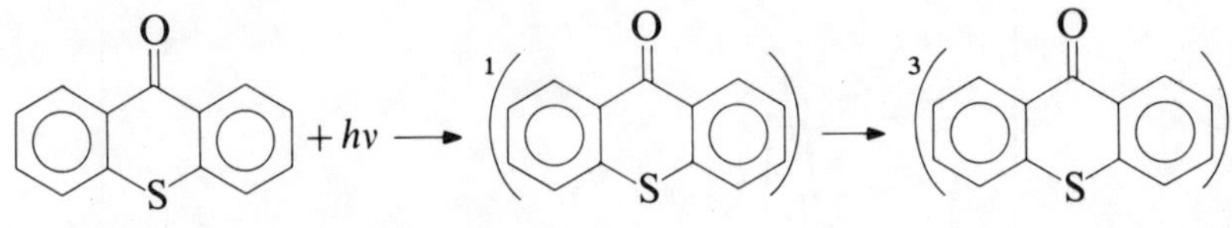

(7.193)   (7.254)

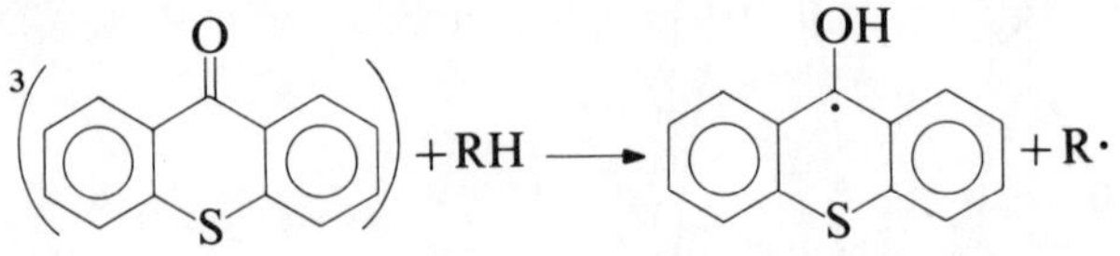

(7.194)

The thioxanthone ketyl radical has been measured by ESR spectroscopy.[3908] Thioxanthone ketyl radicals do not initiate polymerization of vinyl monomers such as methyl methacrylate, styrene, or acrylonitrile, but act rather as terminating species.[186]

Vinyl monomers (M) effectively quench the excited triplet state (T_1) of thioxanthone and its derivatives:

$$T_1 + M \xrightarrow{k_Q} S_0 + M^* \qquad (7.256)$$

The specific rate constant for quenching (k_Q) decreases according to the series:

styrene > N-vinyl-2-pyrrolidone > methyl methacrylate
> acrylonitrile > buthyl-vinyl ether > vinyl acetate

from $k_Q = 6 \times 10^9\ mol^{-1}\ s^{-1}$ to $k_Q = 2 \times 10^4\ mol^{-1}\ s^{-1}$.[186]

Effective photoinitiation of free radical polymerization commonly employs the thioxanthone–amine (hydrogen donor) hybrid system.[186, 2457] The excited triplet state (T_1) of thioxanthone forms, with an amine, the exciplex (7.195) which further dissociates into thioxanthone ketyl radical (7.196) and amine radical (7.197):

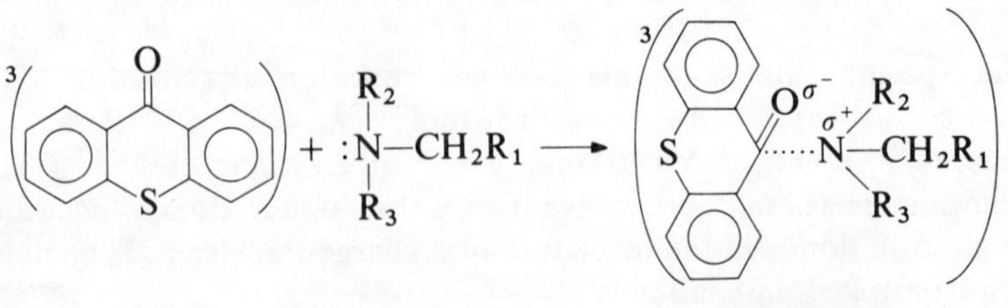

EXCIPLEX
(7.195)   (7.257)

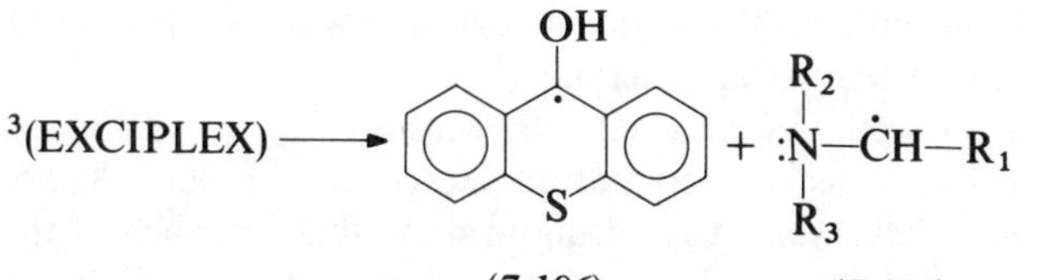

$$^3(\text{EXCIPLEX}) \longrightarrow \underset{(7.196)}{\underset{\text{S}}{\overset{\text{OH}}{\bigcirc\!\!\!\dot{\bigcirc}\!\!\!\bigcirc}}} + \underset{(7.197)}{\overset{\text{R}_2}{\underset{\text{R}_3}{:\!\text{N}\!-\!\dot{\text{C}}\text{H}\!-\!\text{R}_1}}} \qquad (7.258)$$

Amine radicals (*7.197*) have been successfully trapped with 2-methyl-2-nitrosopropane (7.131) and well characterized by ESR spectroscopy.[1369]

Instead of thioxanthone its derivatives, such as 2-chlorothioxanthone (*7.198*), 2-methylthioxanthone (*7.199*), 2-isopropylthioxanthone (*7.200*), have been widely used. These provide the best combination of reactivity with amines, solubility, and absorption spectra.[892]

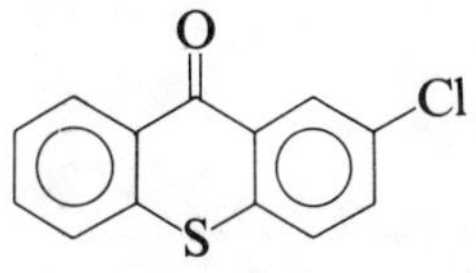

(7.198) 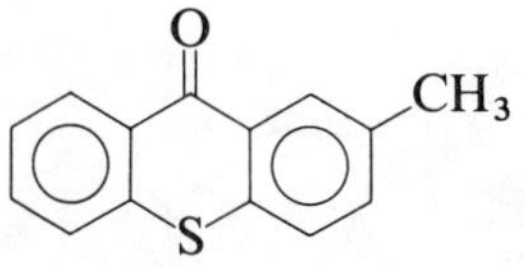   (7.199) 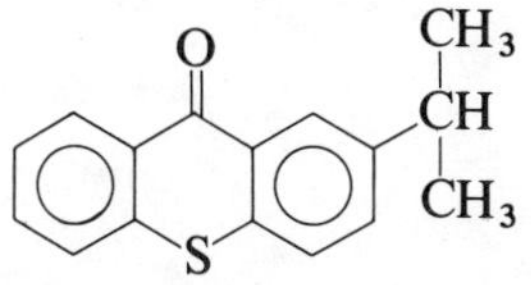   (7.200)

Thioxanthone derivatives proved to be especially useful for titanium dioxide (TiO_2) pigmented coating formulations polymerization.[1705-1707,2457,2696,2731] This is due to the fact that thioxanthone and its derivatives possess a relatively intense absorption around 380 nm (Fig. 7.13), where TiO_2 (anatase) is rather transparent. Optimum film properties were obtained with 2-chlorothioxanthone (*7.198*) and with 2-isopropylthioxanthone (*7.200*) when they were used in mixture with 2-(dimethylamino)ethyl benzoate (*7.201*) or 4-(dimethylamino)ethyl benzoate (*7.202*).[892, 2546, 3131]

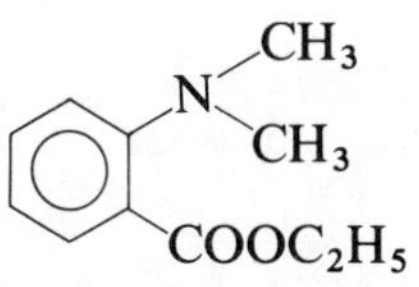

(7.201) 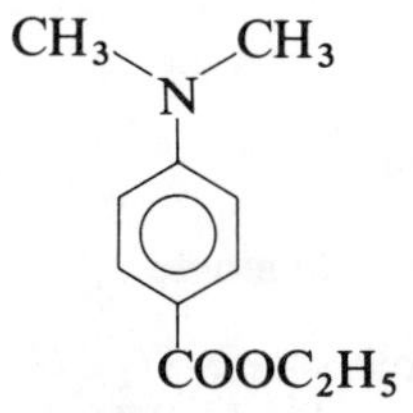   (7.202)

Thioxanthone–amine hybrid systems are only active in photocuring acrylic resins, such as epoxy acrylates, polyester acrylates, and urethane acrylates dissolved in acrylic monomers.[892,1402,1705,2901] Unsaturated polyester resins in styrene cannot be cured with this photoinitiator system. A serious disadvantage of coatings cured by thioxanthone derivatives is that they have a yellow coloration that becomes very pronounced upon exposure to light.

Thioxanthone–amine hybrid system have also been used for photoinitiation of polymerization of methyl methacrylate.[1369]

It has been reported that the excited triplet state (T_1) $(E_T = 65.5 \text{ kcal mol}^{-1})$ of thioxanthone can transfer its excitation energy to quinoline sulphonyl chloride (7.203) $(E_T = 60 \text{ kcal mol}^{-1})$, resulting in homolytic cleavage of the sulphonyl-chloride bond to produce initiating free radicals, capable of polymerization of methyl methacrylate:[2454, 2457]

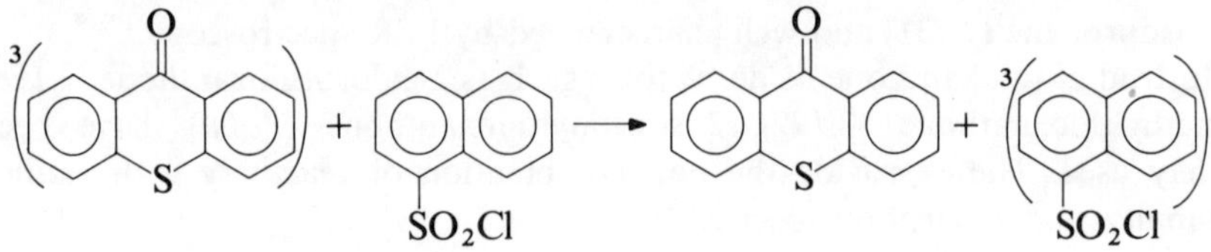

(7.203)   (7.259)

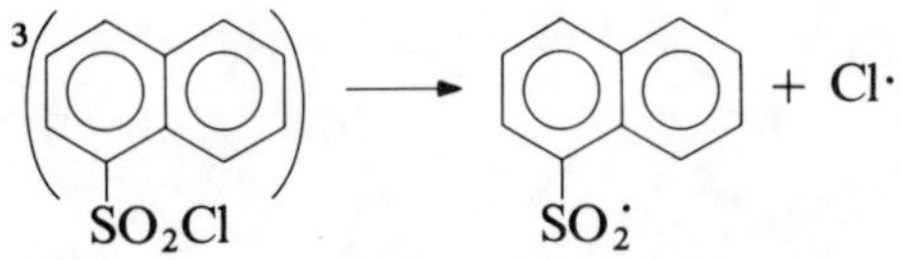

(7.260)

This hybrid photoinitiating system has been used for photocuring of coatings.

7.9.4.2.6. Acridone

Keto-imino compounds such as acridone (7.204) can photoinitiate a free radical polymerization of methyl methacrylate upon irradiation with visible light.[1295]

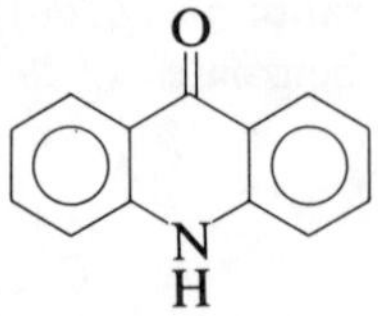

(7.204)

7.9.4.2.7. Aromatic diketones

The excited triplet state (T_1) of benzil (7.205) abstracts hydrogen from a hydrogen-donor molecule (RH) by the following mechanism:[558, 559, 1764, 2550]

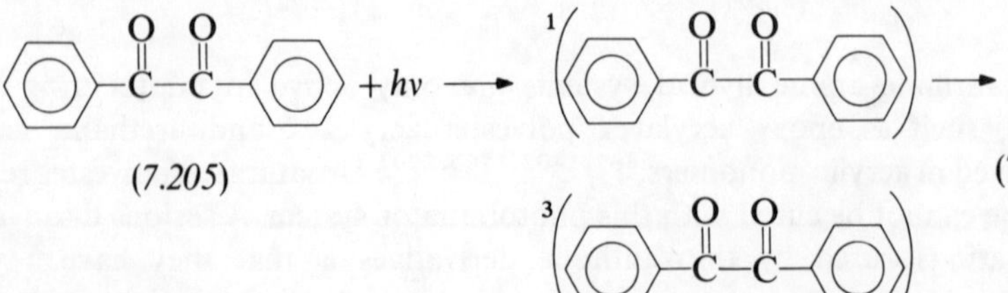

(7.205)   (7.261)

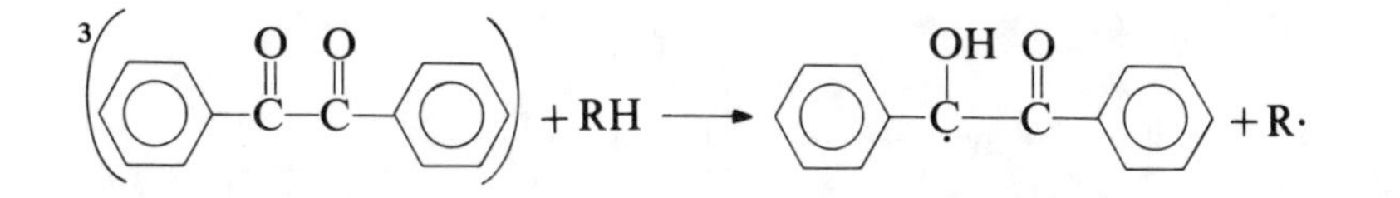

$$(7.206) \qquad\qquad (7.262)$$

The benzil ketyl radical (7.206) can be trapped with spin trap nitroso-compounds and characterized by ESR spectroscopy.[2242]

Vinyl monomers effectively quench the excited triplet state (T_1) of benzil and the rate of polymerisation of methyl methacrylate and styrene is low.[430, 518, 701, 759, 1056, 1235, 1728] The photoinitiating efficiency of benzil can be improved by a factor of three on addition to methyl methacrylate of 10 wt% of tetrahydrofuran.[1728]

The excited triplet state (T_1) of benzil forms, with aniline (7.207), an exciplex (7.208) which is further converted to benzil ketyl radical (7.209) and amine radical (7.210), which may initiate polymerization of methyl methacrylate:[3297]

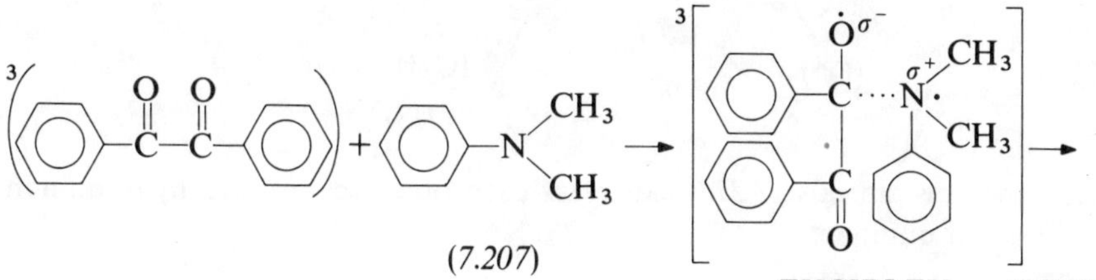

$$\text{EXCIPLEX} \qquad (7.263)$$

$$(7.207) \qquad\qquad (7.208)$$

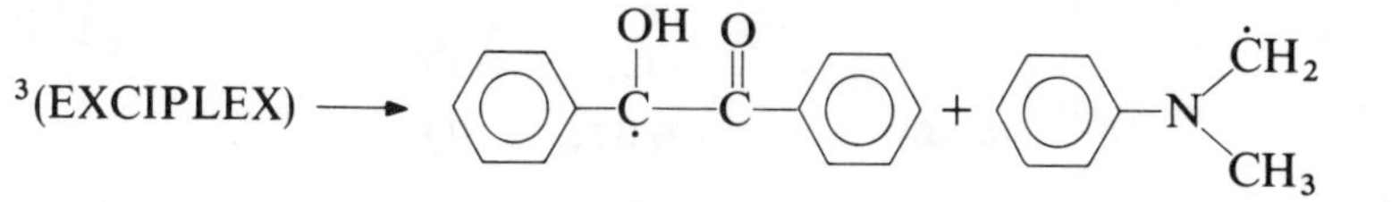

$$(7.209) \qquad\qquad (7.210) \qquad\qquad (7.264)$$

7.10. QUINONES

The photochemistry of p-quinones[537] and o-quinones[3144] has been discussed in detail.

The longest wavelength absorption band for simple quinones (7.211) is in the region 400–500 nm ($\varepsilon = 20$–100) and is due to S_1 (n,π*) transition, or in some cases, e.g. duroquinones, to the corresponding $S_2(\pi,\pi*)$ transition. The intersystem crossing (ISC) from $S_1(n,\pi*)$ to the triplet state $T_1(n,\pi*)$ occurs with efficiency ($\Phi_{ISC} = 0.8$–1.0)[2206, 3887]. The forbidden transition $S_0 \rightarrow T_1$ has been reported for 1,4-benzoquinone in heptane but with very low efficiency.[1576]

The excited triplet state (T_1) of quinone (denoted here as Q*) can be considered as a biradical (7.212):

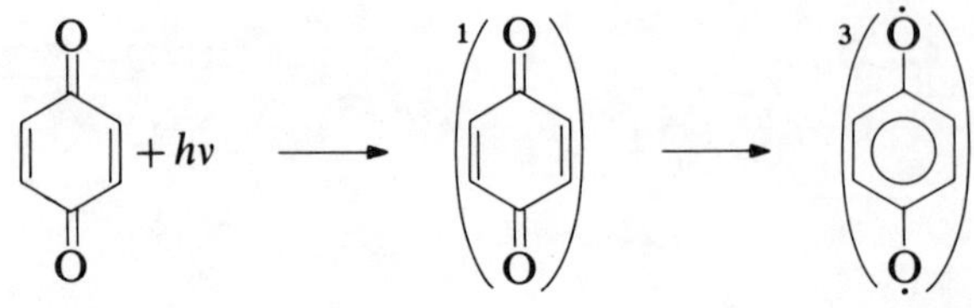

$$(7.265)$$

(Q)

(7.211)

(Q*)

(7.212)

which is very reactive and can abstract hydrogen from a hydrogen-donor molecule (RH) (e.g. alcohols RCH_2OH):[526, 536, 537, 2286, 2287, 3601, 3861]

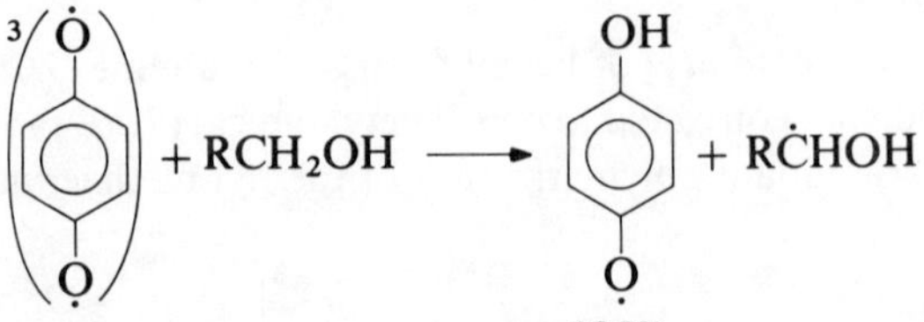

$$(7.266)$$

(Q*)

(QH·)

(7.213)

Semiquinone radicals (7.213) can react each other to produce hydroquinone (7.214) and quinone:

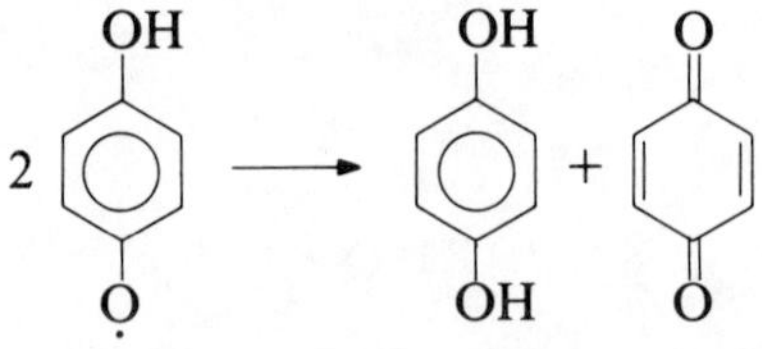

$$(7.267)$$

(QH·)

(QH$_2$)   (Q)

(7.214)

or they can react with oxygen to produce quinone and hydroperoxy radicals ($HO_2 \cdot$):

$$QH \cdot + O_2 \rightarrow Q + HO_2 \cdot \qquad (7.268)$$

Secondary reactions involve free radicals formed:

$$2HO_2 \cdot \rightarrow H_2O_2 + O_2 \qquad (7.269)$$

$$Q + R\dot{C}HOH \rightarrow QH \cdot + RCHO \qquad (7.270)$$

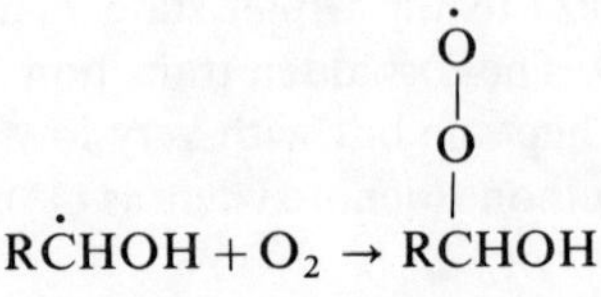

$$(7.271)$$

$$R\dot{C}HOH + O_2 \rightarrow \underset{\underset{|}{O}}{\overset{\dot{O}}{\overset{|}{R}CHOH}}$$

$$\overset{\displaystyle \overset{\cdot}{O}}{\underset{|}{\overset{|}{O}}}$$
$$\underset{|}{\overset{|}{R}CHOH} + HO_2\cdot \rightarrow RCHO + H_2O_2 + O_2 \tag{7.272}$$

$$\overset{\displaystyle \overset{\cdot}{O}}{\underset{|}{\overset{|}{O}}}$$
$$2\underset{|}{RCHOH} \rightarrow 2RCOOH + H_2O_2 \tag{7.273}$$

$$\overset{\displaystyle \overset{\cdot}{O}}{\underset{|}{\overset{|}{O}}}$$
$$\underset{|}{RCHOH} \rightarrow RCHO + HO_2\cdot \tag{7.274}$$

It has been suggested that the excited triplet state (T_1) of quinone can abstract hydrogen from water to produce semiquinone radical $(QH\cdot)$ and hydroxy radical $(HO\cdot)$,[530, 531] but the ionic mechanism is much preferred:

$$Q^* + Q \rightarrow Q^{\overline{\cdot}} + Q^{+}\cdot \tag{7.275}$$

$$Q^{\overline{\cdot}} + Q^{+}_{\cdot} \rightarrow 2Q \tag{7.276}$$

$$H_2O \rightleftarrows HO^- + H^+ \tag{7.277}$$

$$Q^{\overline{\cdot}} + H^+ \rightarrow QH\cdot \tag{7.278}$$

$$2QH\cdot \rightarrow QH_2 + Q \tag{7.279}$$

$$2Q^{\overline{\cdot}} \rightarrow Q^{2-} + Q \tag{7.280}$$

$$Q^{+}_{\cdot} + QH\cdot \rightarrow 2Q + H^+ \tag{7.281}$$

$$Q^{+}_{\cdot} + HO^- \rightarrow Q + HO\cdot \tag{7.282}$$

$$Q + HO\cdot \rightarrow (QOH)\cdot \tag{7.283}$$

$$(QOH)\cdot + Q \rightarrow QOH + QH\cdot \tag{7.284}$$

where:

$Q^{\overline{\cdot}}$ and $Q^{+}_{\cdot}$ are semiquinone anion and semiquinone cation radicals, respectively,

$(QOH)\cdot$ represents the quinone–hydroxy radical adduct, and QOH the hydroquinone derived from it by removal of a nuclear (α- or β-position) hydrogen atom.

An additional series of steps, accounting for the formation of hydrogen peroxide, can be envisaged for systems containing oxygen:

$$Q^{\overline{\cdot}} + O_2 \rightarrow Q + O_2^{\overline{\cdot}} \tag{7.285}$$

$$2O_2^{\overline{\cdot}} \rightarrow O_2^{2-} + O_2 \tag{7.286}$$

$$QH\cdot + O_2 \rightarrow Q + HO_2\cdot \tag{7.287}$$

$$2HO_2\cdot \rightarrow H_2O_2 + O_2 \tag{7.288}$$

$$(QOOH)\cdot + O_2 \rightarrow QOH + HO_2\cdot \tag{7.289}$$

Reactions such as these account for the kinetics observed over the pH range 3–11 (the rate increases with increasing pH), but they do not account for them outside this range. An additional equation

$$Q_.^+ + X \rightarrow Q + X_.^+ \tag{7.290}$$

in which X is not defined, has been introduced[2946] to cover this.

p-Quinone (7.211) can form blue complexes with hydroquinones (7.215), which are known as quinhydrones and have a structure (7.216):[99, 141]

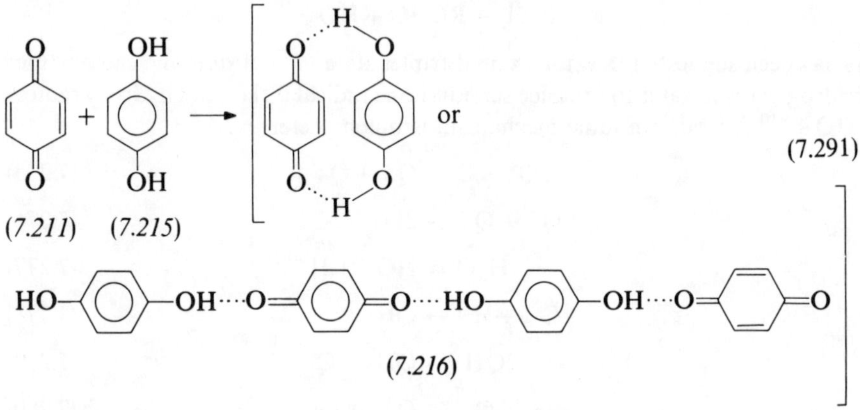

$$\tag{7.291}$$

$$(7.211) \quad (7.215)$$

$$(7.216)$$

Anthraquinone (7.217) under, u.v. irradiation, is excited to the triplet state (T_1) which can be considered as a biradical (7.218):

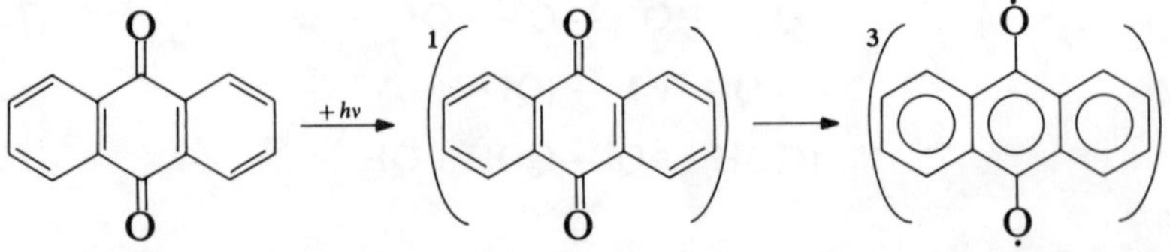

$$\tag{7.292}$$

$$(7.217) \qquad\qquad\qquad\qquad (7.218) \quad (7.292)$$

The triplet excited state of anthraquinone may abstract hydrogen from iso-propanol (7.219) producing semiquinone radical (2.220) and finally hydro-anthraquinone (2.221) by the following reactions:[62, 86−88, 130, 131, 525]

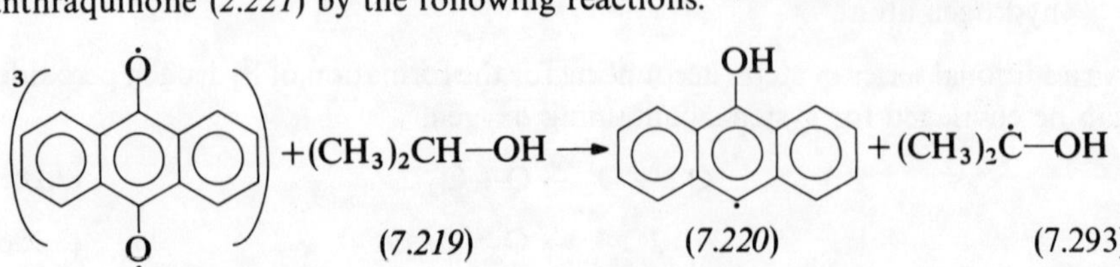

$$(7.219) \qquad\qquad (7.220) \qquad\qquad (7.293)$$

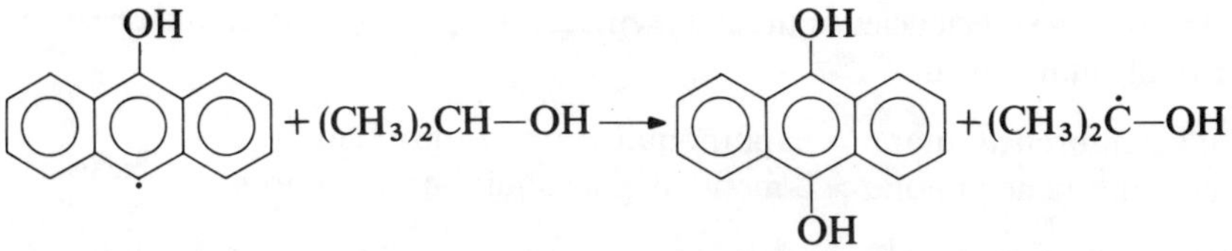

$$(7.221) \qquad\qquad (7.294)$$

Anthraquinone and its derivatives (e.g. 2-*tert*-butylanthraquinone)[452, 757, 1036, 2239, 2286, 2671, 3601, 3860, 3862, 3863, 3887] in the presence of hydrogen donor solvents, such as isopropane–cyclohexane solution[2286–2288] or tetrahydrofurane (7.222),[1729, 2239] are effective photoinitiators of free radical polymerization of methyl methacrylate, acrylonitril, and styrene, but not vinyl acetate. The radical formed from tetrahydrofurane (7.224) can initiate polymerization (reaction 7.296), whereas semiquinone radicals (7.223) act as termination radicals (reactions 7.297 and 7.298):

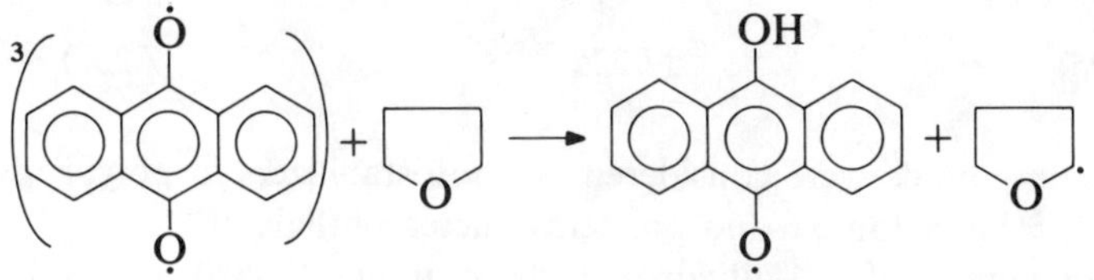

$$(7.295)$$

$$(7.222) \qquad\qquad (7.223) \qquad\qquad (7.224)$$

$$(7.296)$$

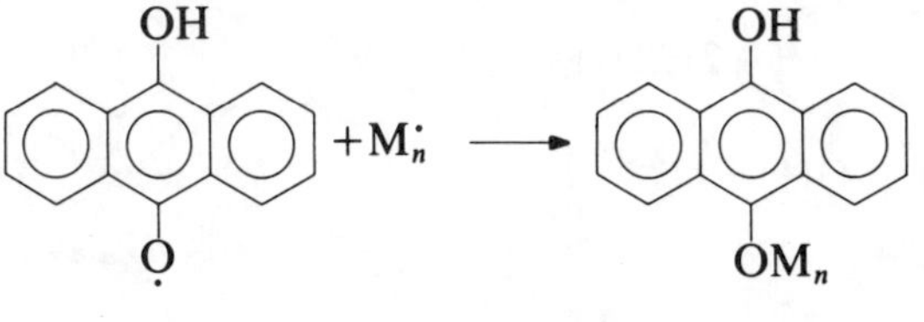

$$(7.297)$$

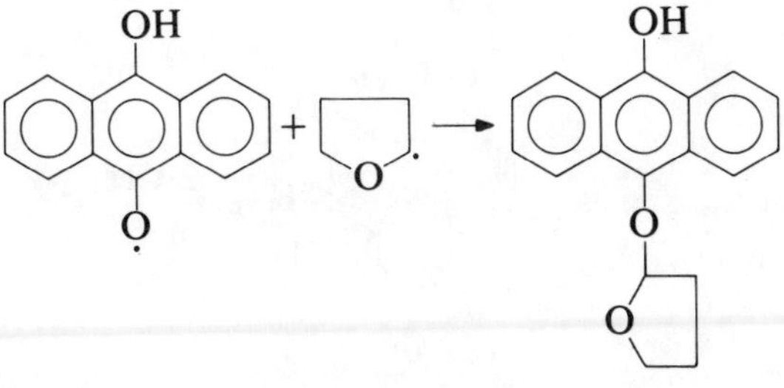

$$(7.298)$$

The relative efficiencies of methyl methacrylate polymerization observed for a series of quinones are:

phenanthrenequinone > naphthoquinone > anthraquinone > acenaphthenequinone > 2-methylnaphthoquinone > chloranil.[452, 2239]

The polymerization of aqueous methyl acrylate by sodium anthraquinone-2-sulphonate in the presence of chloride ions has been explained by the fact that initiating species are chlorine radicals.[207, 208]

2-Aminoanthraquinone (*7.225*) and 2-hydroxyanthraquinone (*7.226*) in 2-propanol produces the radical anion (A$^{\overline{\cdot}}$), whereas 2-amino-3-hydroxy-anthraquinone (*7.227*) gives the semiquinone radical (AH$\cdot$):[105, 129, 130]

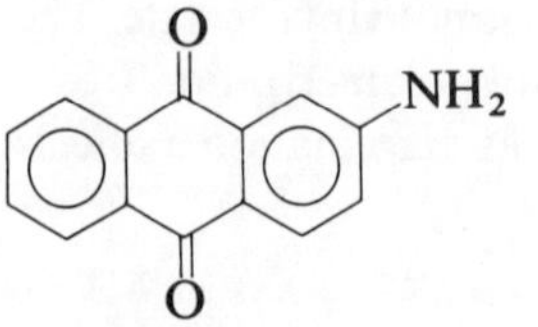

(*7.225*) 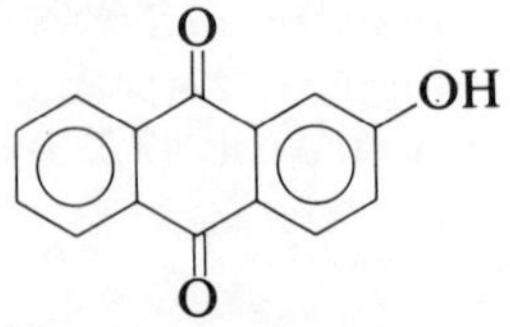　　　　(*7.226*) 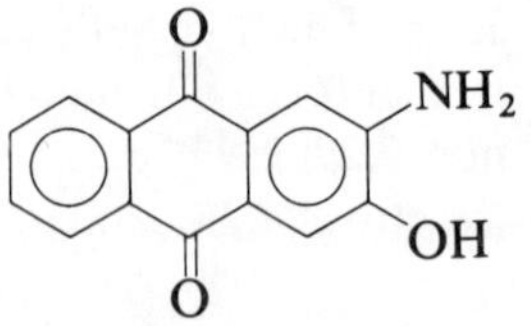　　　　(*7.227*)

These compounds were considered as photostabilizers in polypropylene,[141] polyesters,[85, 130] nylon-6,6 and poly(ethyleneterephthalate).[407]

Photoreduction of 1,4-dihydroxyanthraquinone (*7.228*) and 1,4,5,8-tetra-hydroxyanthraquinone (*7.231*) in isopropanol causes formation of their leuco-forms (*7.229* and *7.230*) and (*7.232* and *7.233*), respectively:[132]

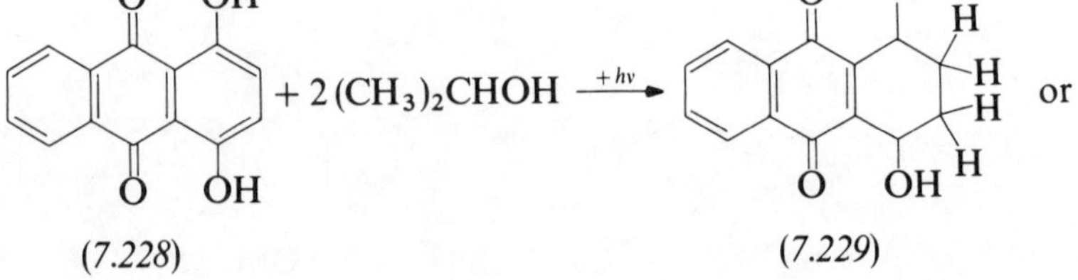

(*7.228*)　　　　　　　　(*7.229*)

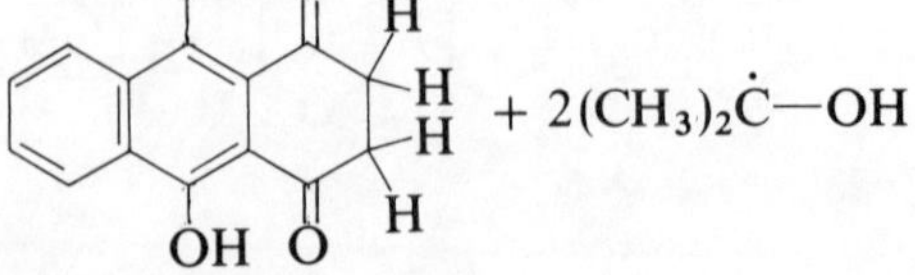

(*7.230*)　　　　　　　　　　　　(*7.299*)

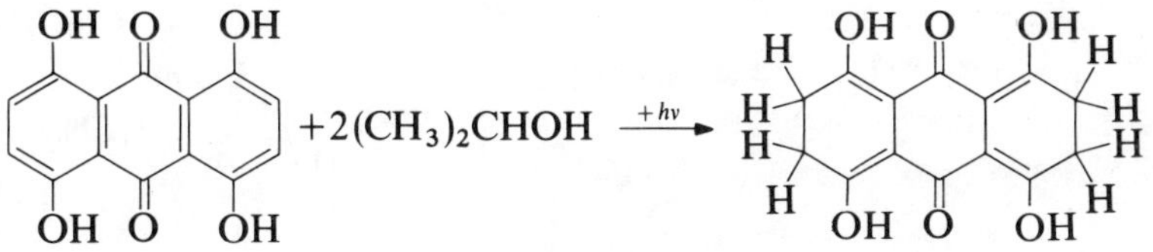

(7.231)   +2(CH₃)₂CHOH $\xrightarrow{+h\nu}$ (7.232)   or

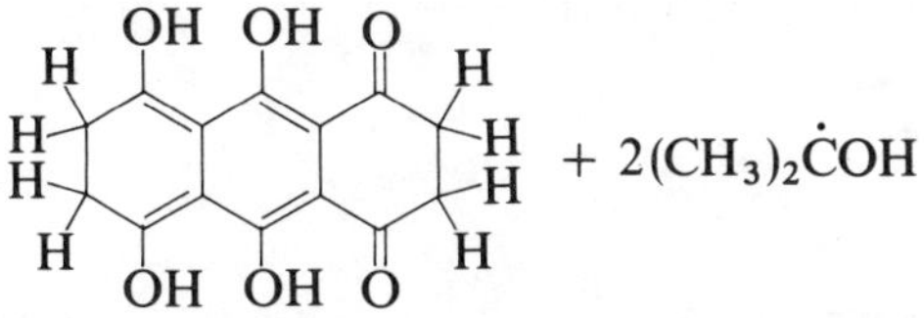

(7.300)

(7.233)   + 2(CH₃)₂ĊOH

7.11. POLYCYCLIC HYDROCARBONS

Polycyclic hydrocarbons such as naphthalene (7.234), anthracene (7.235), phenanthrene (7.236), and pyrene (7.237) can photoinitiate polymerization of methyl methacrylate and acrylonitrile.[319, 597, 600, 2149]

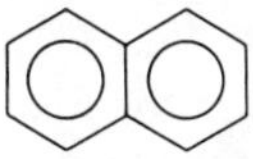

(7.234) 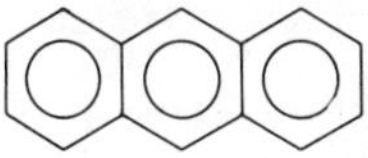   (7.235) 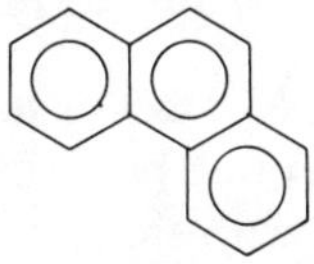   (7.236) 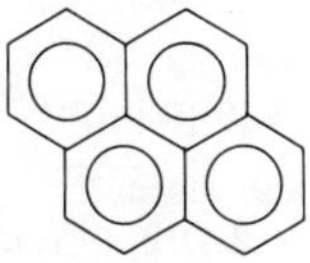   (7.237)

The initiating radicals are formed in decomposition reaction of the mixed exciplex formed between aromatic hydrocarbon in the excited singlet state (S_1) and monomer in the ground state (S_0).[319] Similarly, it was assumed, when using the radiation (356 nm) in the system naphthalene–benzophenone–acrylonitrile, that the initiating species were formed in subsequent reactions of the excited complex (exciplex) of naphthalene in the excited triplet state (T_1) and acrylonitrile in the ground state.[597] Transformation of exciplexes into solvated ion pairs, and subsequent dissociation of the solvated ion pairs into ion radicals, is facilitated in a polar medium.[1407, 2266, 3075]

This reaction is accelerated by the Lewis acids such as $ZnCl_2$, where radicals are formed by the photodecomposition of an excited complex pyrene (Py)-acrylonitrile (AN) and $ZnCl_2$:[598]

$$Py^*(S_1 \text{ or } T_1) + (AN \dots ZnCl_2) \rightarrow (Py \dots AN \dots ZnCl_2)^* \rightarrow$$

$$\rightarrow \text{FREE RADICALS} \quad (7.301)$$

Polycyclic hydrocarbons, e.g. naphthalene, with aliphatic amines, form the exciplex, which is decomposed into free radicals:[2176]

$$(7.302)$$

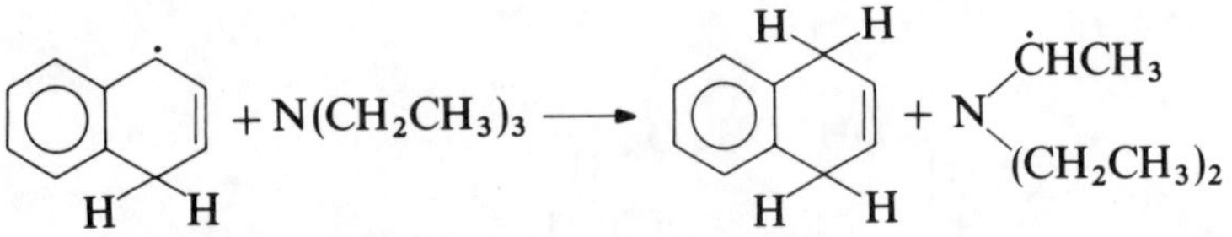

$$(7.238) \qquad (7.239) \qquad (7.303)$$

The main product of this photoreaction is 1,4-dihydronaphthalene (*7.238*) and amino radicals (*7.239*),[748] which may initiate free radical polymerization.

Aromatic hydrocarbons—such as naphthalene, anthracene, phenanthrene— and amines—triethylamine, diethylamine, N,N-dimethylaniline, N-methylaniline, and aniline—were used as a binary photosensitizing system for polymerization of acrylonitrile and methyl methacrylate.[2149]

7.12. AZOCOMPOUNDS

The azo group ($-\ddot{N} = \ddot{N}-$) has low molar absorptivity ($\varepsilon < 1$–10) in the 350 nm region. This weak absorption band arises from (n, π^*) transition of the lone pairs of electrons associated with the azo groups.

During photolysis of 2,2'-azobisisobutyronitrile (2,2'-azobis(2-methyl-propionitrile) (AIBN) (*7.240*) cyanoisopropyl radicals (*7.241*)[1255] are formed. These can be detected by ESR spectroscopy[3394]:

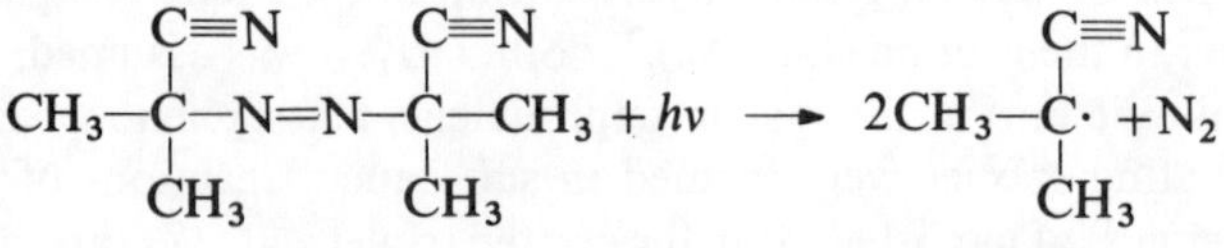

$$(7.304)$$

$$(7.240) \qquad\qquad (7.241)$$

The quantum yield of photodecomposition of azobisisobutyronitrile has been found to be $\Phi = 0.4$ (at 25°C) and $\Phi = 0.6$ (at 45°C).

The cyanoisopropyl radical (*7.241*) can form an isomorphic keteniminyl radical (*7.242*):

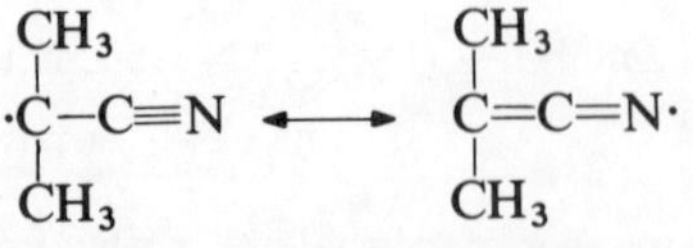

$$(7.305)$$

$$(7.241) \qquad\qquad (7.242)$$

Both radicals can recombine and produce several products:

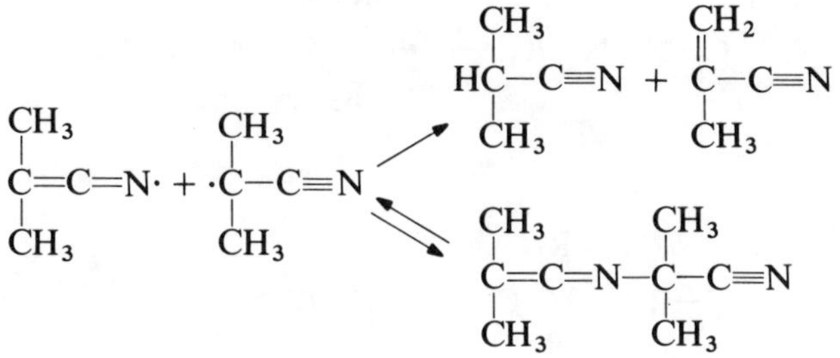

$$
\begin{array}{c}
CH_3 \\
| \\
HC-C\equiv N \\
| \\
CH_3
\end{array}
+
\begin{array}{c}
CH_2 \\
\| \\
C-C\equiv N \\
| \\
CH_3
\end{array}
\qquad (7.306)
$$

$$
\begin{array}{cc}
CH_3 & CH_3 \\
| & | \\
C=C=N\cdot + & \cdot C-C\equiv N \\
| & | \\
CH_3 & CH_3
\end{array}
$$

$$
\begin{array}{cc}
CH_3 & CH_3 \\
| & | \\
C=C=N-C-C\equiv N \\
| & | \\
CH_3 & CH_3
\end{array}
\qquad (7.307)
$$

Azobisisobutyronitrile has been successfully used for photoinitiation of methacrylic acid,[293] methyl methacrylate,[426, 1727, 3392–3394] acrylonitrile,[2148, 2536] and methacrylonitrile.[3392–3394]

Nitrogen which is formed during the photodecomposition of azobisisobutyronitrile has been employed for mechanical rupture of polyamide microcapsules.[2420] The aim of this work was to develop a microcapsulator system which can be induced to release its content, in a controlled manner, by exposure of light.

α-Azo-1-cyclohexacarbonitrile (7.243) irradiated with 350 nm undergoes direct photofragmentation:[3978]

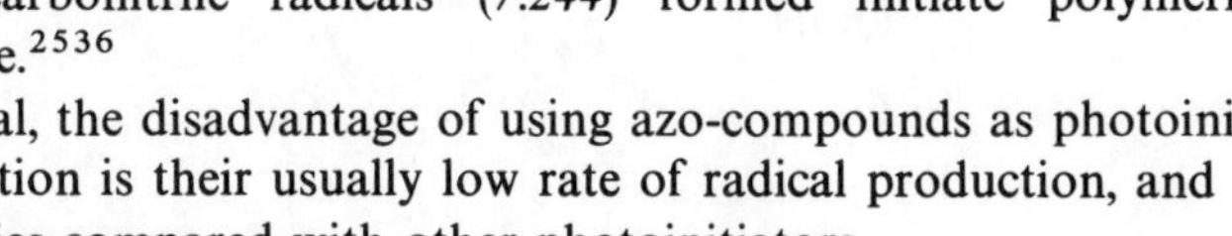

$$
(7.243) \qquad\qquad (7.244) \qquad\qquad (7.308)
$$

Cyclohexacarbonitrile radicals (7.244) formed initiate polymerization of acrylonitrile.[2536]

In general, the disadvantage of using azo-compounds as photoinitiators for polymerization is their usually low rate of radical production, and low molar absorptivities compared with other photoinitiators.

7.13. HYDRAZONES

Hydrazones (7.245), upon u.v. irradiation, are photooxidized to α-azohydroperoxides (7.246):[2211]

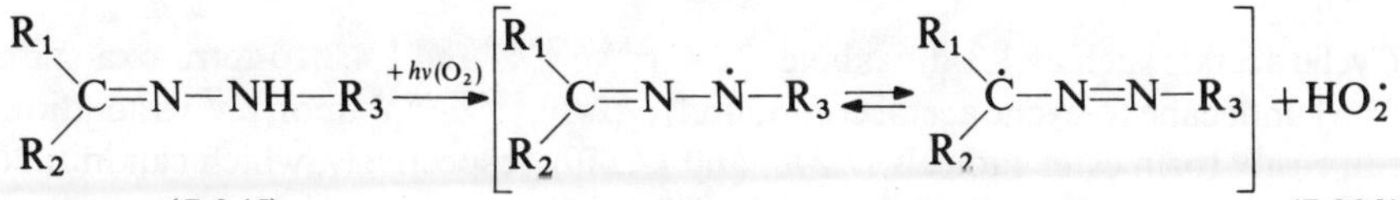

$$
(7.245) \qquad\qquad\qquad\qquad\qquad\qquad\qquad (7.309)
$$

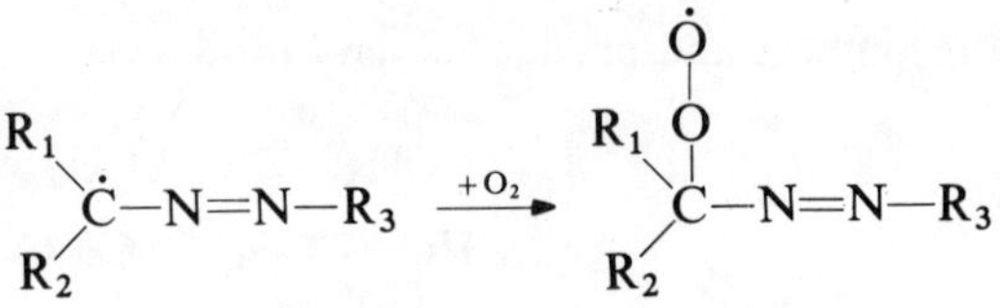

$$R_2R_1\dot{C}-N=N-R_3 \xrightarrow{+O_2} R_2R_1\underset{\underset{\dot{O}}{|}}{\overset{\overset{O}{|}}{C}}-N=N-R_3 \qquad (7.310)$$

$$R_2R_1\underset{\underset{\dot{O}}{|}}{\overset{\overset{O}{|}}{C}}-N=N-R_3 \;+\; R_2R_1C=N-NH-R_3 \longrightarrow$$

$$R_2R_1\overset{OOH}{\underset{|}{C}}-N=N-R_3 \;+\; R_2R_1C=N-\dot{N}-R_3$$

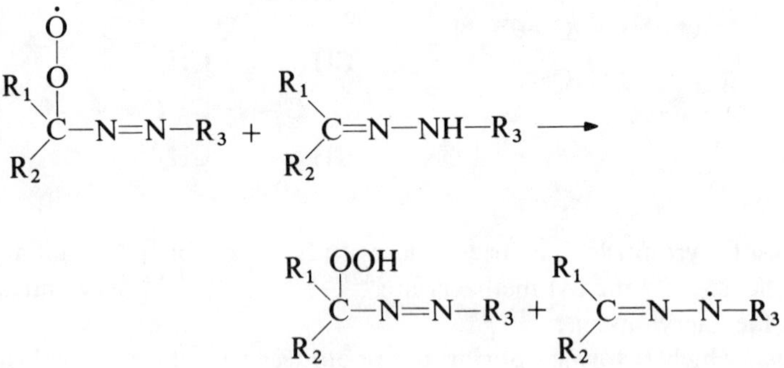

$$(7.246) \qquad\qquad (7.311)$$

α-Azohydroperoxides (*7.246*) are known as a source of hydroxy radicals (HO·) for effective hydroxylation aromatic compounds in anhydrous media.[3596] They are photolysed according to mechanism:

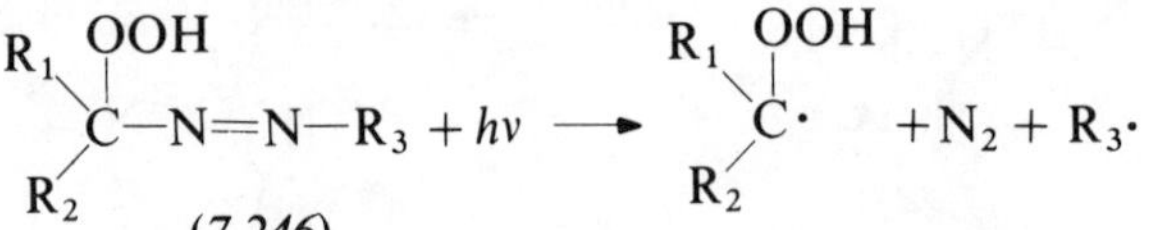

$$R_2R_1\overset{OOH}{\underset{|}{C}}-N=N-R_3 + h\nu \longrightarrow R_2R_1\overset{OOH}{\underset{|}{C}}\cdot \;+ N_2 + R_3\cdot \qquad (7.312)$$

$$(7.246)$$

$$R_2R_1\overset{OOH}{\underset{|}{C}}\cdot \longrightarrow R_2R_1C=O + \cdot OH \qquad (7.313)$$

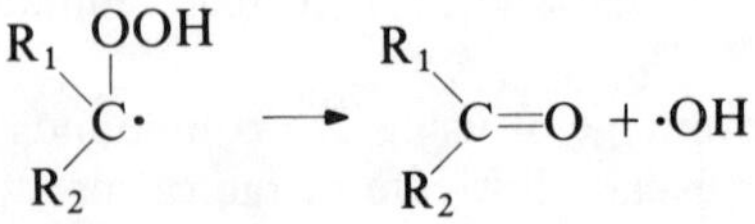

Hydroxy radicals can initiate polymerization of acrylamide and are also effective for image-forming systems.[2211]

7.14. CYCLIC ACETALS

Cyclic acetals such as 1,3-dioxalane (*7.247*)[2843, 2845] and 2,4,8,10-tetra-oxa-spiro [5,5] undecane (dicyclic acetal compound) (*7.249*),[2843, 2848] upon u.v. irradiation, can easily form ester radicals (*7.248*) and (*7.250*), respectively, which can initiate polymerization of styrene and maleic anhydride:

$$\underset{(7.247)}{\begin{array}{c} CH_2-CH_2 \\ O \qquad O \\ C \\ H \quad H \end{array}} \xrightarrow{+h\nu} \begin{array}{c} CH_2-CH_2 \\ O \qquad O \\ \overset{\cdot}{C} \\ H \end{array} + H\cdot \qquad (7.314)$$

$$\underset{}{\begin{array}{c} CH_2-CH_2 \\ O \qquad O \\ \overset{\cdot}{C} \\ H \end{array}} \longrightarrow \underset{(7.248)}{\cdot CH_2CH_2OC\begin{array}{c} H \\ \diagdown \\ O \end{array}} \qquad (7.315)$$

$$\cdot CH_2CH_2OC\overset{H}{\underset{O}{\diagdown}} + \underset{H \quad H}{\begin{array}{c} CH_2-CH_2 \\ O \qquad O \\ C \end{array}} \longrightarrow \underset{H}{\begin{array}{c} CH_2-CH_2 \\ O \qquad O \\ \overset{\cdot}{C} \end{array}} + CH_3CH_2OC\overset{H}{\underset{O}{\diagdown}}$$

or $\qquad\qquad (7.316)$

$$\underset{(7.249)}{\begin{array}{c} H \quad O-CH_2 \quad CH_2-O \quad H \\ C \qquad\qquad C \qquad\qquad C \\ H \quad O-CH_2 \quad CH_2-O \quad H \end{array}} \xrightarrow{+h\nu} \begin{array}{c} H \quad O-CH_2 \quad CH_2-O \quad H \\ C \qquad\qquad C \qquad\qquad \overset{\cdot}{C} \\ H \quad O-CH_2 \quad CH_2-O \end{array} + H\cdot \qquad (7.317)$$

$$\underset{(7.250)}{\begin{array}{c} H \quad O-CH_2 \quad CH_2-O \quad H \\ C \qquad\qquad C \qquad\qquad \overset{\cdot}{C} \\ H \quad O-CH_2 \quad CH_2-O \end{array}} \longrightarrow \begin{array}{c} H \quad O-CH_2 \quad CH_2\cdot \\ C \qquad\qquad C \qquad\qquad H \\ H \quad O-CH_2 \quad CH_2OC\overset{}{\underset{O}{\diagdown}} \end{array} \qquad (7.318)$$

Photopolymerization of maleic anhydride in the presence of cyclic acetals probably occurs by the charge-transfer (CT) complex mechanism.[2843]

Vinyl compounds that contain cyclic acetal group, e.g. 2-vinyl-1,3-dioxalane (7.251),[2847] 2-vinyl-4-hydroxymethyl-1,3-dioxalane,[2846] and 2-methyl-1,3-dioxepane,[1061] are photopolymerized by the following mechanism:

$$\underset{(7.251)}{\begin{array}{c} CH_2=CH \quad O-CH_2 \\ C \\ H \qquad O-CH_2 \end{array}} \xrightarrow{+h\nu} \begin{array}{c} CH_2=CH \quad O-CH_2 \\ \overset{\cdot}{C} \\ O-CH_2 \end{array} + H\cdot \longrightarrow$$

$$\cdot CH_2CH_2OC\underset{O}{\overset{CH=CH_2}{\diagdown}} \qquad (7.319)$$

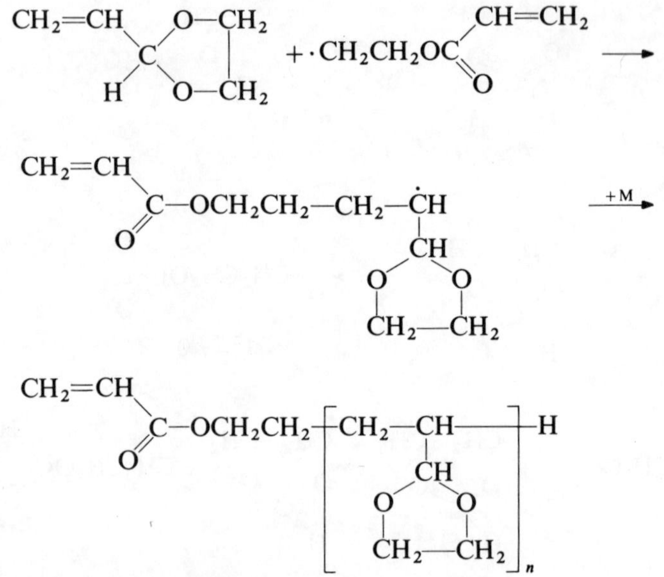

$$CH_2{=}CH\underset{H}{\overset{\displaystyle \,}{\underset{\displaystyle \,}{C}}}\!\!\begin{array}{c}O{-}CH_2\\[2pt]O{-}CH_2\end{array} + \cdot CH_2CH_2OC\!\!\begin{array}{c}CH{=}CH_2\\[2pt]\parallel\\[-2pt]O\end{array} \longrightarrow$$

$$CH_2{=}CH\!\!\begin{array}{c}\\[-4pt]\end{array}\!\!\underset{O}{\overset{\displaystyle \,}{C}}{-}OCH_2CH_2{-}CH_2{-}\overset{\displaystyle \cdot}{C}H \qquad \xrightarrow{\;+M\;}$$

$$CH_2{=}CH\underset{O}{\overset{\displaystyle \,}{C}}{-}OCH_2CH_2{-}\!\!\left[CH_2{-}CH\right]\!{-}H \tag{7.320}$$

Divinyl compounds that contain a dicyclic acetal group, e.g. diallylidene pentaerythritol (*7.252*) can be photopolymerized according to the reaction:[2849]

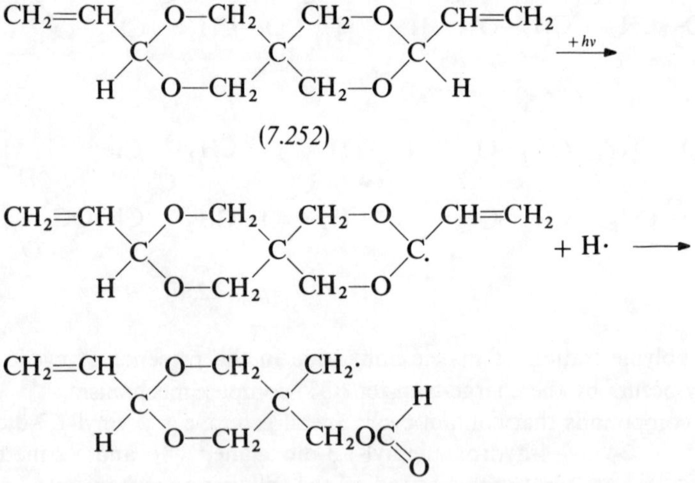

$$(7.252) \qquad \xrightarrow{\;+h\nu\;}$$

$$\cdots \qquad + H\cdot \longrightarrow$$

$$(7.253) \tag{7.321}$$

The radical (*7.253*) is an initiation polymerization radical.

7.15. 1,3-DITHIOLANE

The 1,3-dithiolane (*7.254*), upon u.v. irradiation, undergoes photolysis according to the reaction:[2842]

$$\underset{\underset{CH_2}{\underset{|}{S}}\overset{CH_2-CH_2}{\underset{|}{S}}} \xrightarrow{+h\nu} \cdot CH_2SCH_2CH_2S\cdot \longrightarrow \cdot SCH_2\cdot + \cdot SCH_2CH_2\cdot$$

(7.254)   (7.322)

$$\cdot SCH_2CH_2\cdot \underset{\longleftarrow}{\rightleftharpoons} \underset{S}{CH_2-CH_2}$$ (7.323)

Biradicals formed may give oligomeric products having the structure:

$$H(SH_2)_x - (SCH_2CH_2)_yH \quad \text{where } x:y = 0.76:0.24$$

or copolymerize with methyl methacrylate, acrylonitrile, styrene, and maleic anhydride.[2842]

7.16. PHOTOPOLYMERIZATION INITIATED BY SACCHARIDES

Emulsion polymerization of methyl methacrylate in water (an aqueous micellar solution has been obtained with sodium lauryl sulphate) can be photoinitiated by saccharide compounds such as glucose, cellobiose, fructose, sucrose, or xylose. The mechanism of photoinitiation is not completely understood.[2147, 2494]

7.17. METAL OXIDES

Hydrogen peroxide (H_2O_2), which may photoinitiate polymerization of vinyl monomers (cf. Section 7.3), is formed on the surface of some metallic oxides (ZnO, TiO_2) in the presence of oxygen, water, or organic solvents under u.v. irradiation and is further photolysed into hydroxyl radicals ($HO\cdot$) (cf. Section 15.1.1.2):[2166, 2388]

$$H_2O + \tfrac{1}{2}O_2 \xrightarrow{+h\nu(ZnO)} H_2O_2$$ (7.324)

$$H_2O_2 + h\nu \rightarrow 2HO\cdot$$ (7.325)

The fact that one molecule of oxygen disappears for each molecule of hydrogen peroxide formed indicates, however, that zinc oxide is being partially oxidized also:

$$ZnO + 2H_2O + O_2 \xrightarrow{+h\nu} H_2O_2 + Zn\underset{OOH}{\overset{OH}{<}}$$ (7.326)

This leaves the hydroxyl radicals ($HO\cdot$) bound to the surface of ZnO. The polymerization process initiated on exposure to u.v. irradiation is completed in darkness.[2166]

A TiO_2 semiconductor immersed in solution of styrene, and a supporting electrolyte under u.v. irradiation, initiate photoelectrochemical polymerization.[1197] A platinized suspension of TiO_2 semiconductor powder also initiates photopolymerization of methyl methacrylate.[2127]

Such a photoelectrochemical system is composed of a semiconductor that functions as a working electrode, a metallic counter-electrode, and a solution that contains monomer and supporting electrolyte. The initial event is the absorption of a photon into the semiconductor and the promotion of a photon from the valence to the conduction band. This may be followed by a chemical interaction caused by the injection of electrons into the conduction band, or to the transfer of holes from the valence band to species in solution.

In the photoelectrochemical reaction electron transfer may occur directly, and the reacting molecules need not possess a chromophore-absorbing group. A substance that undergoes irreversible oxidation or reduction can be used to provide a source of free radicals or other active species.

The distinctive feature of a semiconductor is the discrete separation of energy in the valence and conduction bands. When an n-type semiconductor is immersed in solution that contains a redox couple the energy level of the semiconductor (characterized by the Fermi level) and that of the redox couple (characterized by the electrochemical potential) must equilibrate. This process involves the movement of electrons into the bulk of the semiconductor and produces a 'band bending' that erects a potential barrier against the occurrence of the back reaction.

Titanium dioxide (TiO_2) exists in two morphological crystalline forms, anatase and rutile, which exhibit different photocatalytic effects (cf. Section 15.1.1.2).

The TiO_2 (anatase form, but not rutile) is effective in the initiation of the photopolymerization of hydroxyethyl acrylate in the presence of a co-catalyst such as ketone–amine or dyes (methylene blue or eosin).[2890,3877] Anatase pigments coated with alumina or silica are not effective as photosensitizers. Two mechanisms have been proposed:[3877]

(i) During irradiation of TiO_2 with energy higher than 3.0 eV an EXCITON (electron (e^-)/hole (p^+) pair) is formed:

$$TiO_2 + h\nu \rightarrow e^- + p^+ \qquad (7.327)$$

The energy of the exciton is utilized by electron-transfer reactions to produce triplet excited state (T_1) of ketone (or dye) (K*), which subsequently undergoes hydrogen-abstraction from the amine (R_3N) to produce initiator radicals (I·) which initiate free radical polymerization:

$$K + e^- \rightarrow K^{\overline{\cdot}} \qquad (7.328)$$

$$K^{\overline{\cdot}} + p^+ \rightarrow K^* \qquad (7.329)$$

$$K^* + R_3N \rightarrow I· \qquad (7.330)$$

$$I· + monomer \rightarrow polymerization \qquad (7.331)$$

This reaction requires that the energy of the exciton must be sufficient to produce the excited state of ketone (or dye) (T_1). Triplet energies (E_T) for different ketones and dyes are shown in Table 7.20.

Table 7.20   Triplet (E_T) and redox (E_{REDOX}) energies of
ketones and dyes[3877]

Ketone (or dye)	E_T (kcal mol^{-1})*	E_{REDOX} (kcal mol^{-1})*
Methylene blue	34	37
Eosin	43	47
Benzil	53	46
Anthraquinone	62	44
Fluorenone	53	53
Thioxanthone	65.5	62
Benzophenone	69.3	62
Xanthone	74	62

*For conversion energy units, use 1 kcal mol^{-1} = 4.1868 × 10^3 J

(ii) The exciton interacts with ketone (or dye) and amine, respectively, to produce radical ions (K $\dot{-}$) from which initiator radicals (I·) are generated by proton (H·) transfer from the amine to the ketone (or dye) radical ion:

$$K + e^- \rightarrow K\dot{-} \qquad (7.332)$$

$$R_3N + p^+ \rightarrow R_3N\dot{+} \qquad (7.333)$$

$$R_3N\dot{+} + K\dot{-} \rightarrow I· \qquad (7.334)$$

A band model for the mechanism is shown in Fig. 7.14. This mechanism requires that the energy of the exciton be sufficient to generate the radical ions of the ketone (or dye) and amine. Redox energies (E_{REDOX}), based on half-wave reduction potentials of ketone (or dye) and the half-wave oxidation potential of triethanol amine, are presented in Table 7.20. Photosensitization by TiO_2 do not occur when the energy requirement (E_T or E_{REDOX}) is greater than the available energy of the exciton of atanase.

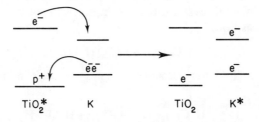

Fig. 7.14   A band model for energy transfer between TiO_2 exciton ($TiO_2*(e^- + p^+)$) and ketone (K)[3877]

7.18. ION PAIR COMPLEXES

The transfer of electrons from an ion to another ion or to the solvent molecule play an important role in the majority of inorganic photoreactions.

Table 7.21 compares the absorption maxima of certain anions and cations, the spectra of which presumably are of the charge transfer type, with the changes of free energy in the reaction:[591]

$$X^n + H^+ \rightarrow X^{n+1} + \tfrac{1}{2}H_2 \qquad (7.335)$$

Table 7.21 Free energies (ΔF^0) of electron transfer and λ_{max} for several ions[591]

Typical ions for electron transfer	ΔF^{0*} (kcal/mol^{-1})†	λ_{max} (nm)	Energy at λ_{max} (kcal mol^{-1})†
F^-	+94	150	191
Cl^-	56	190	150
Br^-	44	195.5	143
I^-	29	232	123
CO^{2+}	42	220	130
Fe^{2+}	18	285	100
V^{2+}	−5	350	81
Cr^{2+}	−9	380	74

* Calculated for reaction $X^n + H^+ \rightarrow X^{n+1} + \tfrac{1}{2}H_2$
† For conversion energy units, use 1 cal mol^{-1} = 4.1868 × 10³ J

The F^- ion is the most electronegative ($\Delta F^0 = +94$ kcal mol^{-1}), the maximum of this absorption band is placed in the short-wave region at 150 nm (corresponding to the energy of 191 kcal einstein^{-1}). Conversely the process:

$$Cr^{2+} + H^+ \rightarrow Cr^{3+} + \tfrac{1}{2}H_2 \qquad (7.336)$$

is exothermic by 10 kcal mol^{-1}, and Cr^{2+} absorption is placed far in the visible region.

A number of observations suggest that hydration water of an ion participates in the electronic transition, e.g.

$$X^n H_2O + h\nu \rightarrow (X^{n+1} H_2O^{-1}) \qquad (7.337)$$

It is assumed that the latter species dissociates or reacts in various ways, for example:

$$(X^{n+1}H_2O^-) + H^+ \rightarrow X^{n+1} + H_2O + H \qquad (7.338)$$

$$(X^{n+1}H_2O^-) \rightarrow X^{n+1} + OH^- + H \qquad (7.339)$$

Cations which are photoreduced through an electron transfer process include Hg^{2+} (180 nm), Cu^{2+} (200 nm), Pb^{2+} (308 nm), Fe^{3+} (230 nm), and Ce^{4+} (320 nm). Aqueous solutions of these cations show a strong absorption. Hydration water apparently donates the electrons:

$$X^n H_2O + h\nu \rightarrow (X^{n-1} H_2O^+) \qquad (7.340)$$

$$(X^{n-1}H_2O^+) + OH^- \rightarrow X^{n-1} + H_2O + OH \qquad (7.341)$$

$$(X^{n-1}H_2O^+) \rightarrow X^{n-1} + OH + H^+ \qquad (7.342)$$

The absorption spectra of the ion pairs of these cations with various anions are shifted towards the visible region and the magnitude of that shift is almost directly proportional to the decrease in electron affinity of the anion. Thus for:[3717]

$$Fe^{3+}Cl^- \quad \lambda_{max} = 320 \text{ nm}$$

$$Fe^{3+}Br^- \quad \lambda_{max} = 380 \text{ nm}$$

$$Fe^{3+}SCN^- \quad \lambda_{max} = 460 \text{ nm}$$

It may be concluded that in such an ion pair the anion donates the electron and the primary act of absorption is, for example:

$$Fe^{3+}X^- + h\nu \rightarrow Fe^{2+}X \qquad (7.343)$$

and it is followed by various reactions of the intermediate compound including its dissociation:

$$Fe^{2+}X \rightarrow Fe^{3+} + X \cdot \qquad (7.344)$$

The difference in the energies of the complex $Fe^{3+}OH^-$ and the completely dissociated and hydrated ions was calculated from thermodynamic measurements;[1072] it is about 40 kcal mol^{-1}. The value obtained from the light absorption maximum corresponds to about 90 kcal mol^{-1}. This difference is by the Franck–Condon principle since, during the optical transition, the interatomic distances remain unchanged. In other words the inter-nuclear distance and the configuration of the solvated shell in $Fe^{2+}OH$ will be the same as that in the ion pair $Fe^{3+}OH^-$. Thus the system is not an equilibrium configuration and the difference of about 50 kcal mol^{-1} should be attributed to the repulsion energy between the nuclei in these states. Consequently the dissociation of the $Fe^{2+} + OH$ complex into $Fe^{3+} + \cdot OH$ may be expected.

Fe_3OH^- effectively photoinitiates decomposition of tert-butyl hydroperoxides (7.255):[424]

$$Fe^{3+}OH^- + h\nu \rightarrow Fe^{2+} + HO \cdot \qquad (7.345)$$

$$(CH_3)_3COOH + \cdot OH \rightarrow (CH_3)_3COO \cdot + H_2O \qquad (7.346)$$

$$(7.255)$$

$FeCl_3$, u.v. irradiated in the presence of alkanes (e.g. 2-methylpentane, 2,4-dimethylpentane), accelerates the formation of alkyl radicals by the following redox catalytic mechanism:[2672]

$$CH_3CH(CH_3)CH_3 + Fe^{3+}Cl^- + h\nu \rightarrow CH_3\dot{C}(CH_3)CH_3 + Fe^{2+} + H^+ + Cl^- \qquad (7.347)$$

Different oxidizing metal salts were used as photoinitiators for polymerization of 2-alkyl-1-vinylimidazoles.[3578]

7.19. METAL CHLORIDES

The polymerization mechanism of vinyl monomers photoinitiated by the $Fe^{3+}Cl^-$ ion pair complex (i.e. $FeCl_3$) can be written as follows:[1071]

$$Fe^{3+}Cl^- + hv \rightarrow Fe^{2+}Cl \tag{7.348}$$

$$Fe^{2+}Cl + monomer\ (M) \rightarrow Fe^{3+} + Cl—M\cdot \tag{7.349}$$

$$Fe^{2+}Cl \rightarrow Fe^{3+} + Cl\cdot \tag{7.350}$$

$$Cl\cdot + M \rightarrow Cl—M\cdot \tag{7.351}$$

$$Cl—M\cdot + nM \rightarrow polymer \tag{7.352}$$

This polymerization takes place in the radiation range of 253.7–310 nm. Wavelengths exceeding 436 nm do not initiate polymerization. The quantum yields with regard to dFe^{2+}/dt depend upon the initiator species. The values of quantum yield are $Fe^{3+}OH^- = 5 \times 10^{-2}$ and $Fe^{3+}Cl^- = 0.13$. The quantum yield with regard to monomer disappearance depends upon the concentration of monomer, intensity of radiation, and type of photoinitiator used.

The photoinitiator $FeCl_3$ has been used for initiation of several vinyl monomers such as methyl methacrylate, methacrylic acid, vinyl acetate, and acrylonitrile in aqueous solutions.[346, 814–819, 1071–1074, 2484, 3471, 3472] $FeCl_3$ also photoinitiates polymerization of acrylonitrile in dimethyl formamide[398] and γ-butyrolactone[399] solutions.

Zinc chloride ($ZnCl_2$) forms CT complexes with vinyl monomers having carbonyl or cyano groups:

$$\tag{7.353}$$

Such charge-transfer (CT) complexes, under u.v. irradiation, photoinitiate polymerization of a monomer (e.g. phenyl acrylate[2295]).

It has been reported that zinc chloride photoinitiates terpolymerization of acrylonitrile, maleic anhydride, and styrene.[599]

Titanium tetrachloride ($TiCl_4$) and zirconocene dichloride were employed as photoinitiators of polymerization of epichlorhydrin and phenyl glycil ether,[1907, 1908] whereas sodium tetrachloroaurate ($NaAuCl_4\cdot 2H_2O$) was employed as photoinitiators of polymerization of acrylamide[1755] and β-propiolactone.[3562] Polymerization of cyclic monomers occurs via ring-opening cationic polymerization.

Metal chlorides such as $CuCl_2$, $CoCl_2$, $MnCl_2$, $FeCl_3$, and other metal salts such as $FeSO_4$ or $Fe(NO_3)_3$ give, with amines, complexes which may photoinitiate polymerization of vinyl monomers.[2783]

7.20. URANIUM SALTS

Uranyl ions (UO_2^{2+}) absorb in the visible region 360–500 nm, and are excited to singlet and triplet states. The energy from the excited triplet state is transferred to a monomer molecule to form an excited triplet state of a monomer. The monomer triplet reacts with another monomer molecule to produce a monomer radical, which initiates polymerization. Uranyl ions are typical photosensitizers. No reduction from U^{6+} to U^{4+} takes place, and this is a direct evidence against an electron transfer mechanism between uranyl ions and monomer.

Uranyl ions were used as photosensitizers of polymerization of methyl methacrylate, acrylonitrile,[2370, 3741, 3742] vinyl chloride (at $-75°C$),[2378] and β-propiolactone.[3164]

7.21. METAL CARBONYLS

The dominant photochemical process for metal carbonyls, e.g. dimanganese decacarbonyl ($Mn_2(CO)_{10}$) and dirhenium decacarbonyl ($Re_2(CO)_{10}$) is a homolytic cleavage of metal–metal bond to produce two metal carbonyl paramagnetic fragments:[1704, 3448, 3793, 3940, 3941]

$$Mn_2(CO)_{10} + h\nu \rightarrow 2Mn(CO)_5 \qquad (7.354)$$

and evolution of carbon oxide:[708, 3448]

$$Mn_2(CO)_{10} + h\nu \rightarrow Mn_2(CO)_9 + CO \qquad (7.355)$$

In the dark a recombination reaction occurs:

$$2Mn(CO)_5 \rightarrow Mn_2(CO)_{10} \qquad (7.356)$$

Photolysis of dimanganese decacarbonyl ($Mn_2(CO)_{10}$) in tetrahydrofurane (THF) occurs in the same way that $Mn(CO)_5$ undergoes a disproportionation reaction, which yields Mn^{2+} and $Mn(CO)_5^-$ according to the scheme:[1699]

$$3Mn_2(CO)_{10} + 12THF \xrightarrow{+h\nu(320\,nm)} 2Mn(THF)_6^{2+} + 4Mn(CO)_5^- + 10CO$$

$$(7.357)$$

With different solvents, different species can be formed.[1701]

A replacement by the ligand on the $Mn(CO)_5^-$ anion occurs[574,575] with manganese carbonyl ($Mn_2(CO)_{10}$), in the presence of nitrogen-donor monomeric or polymeric ligands (L) (e.g. γ-picoline or poly(vinyl pyrrolidone)):

$$3Mn_2(CO)_{10} + 12THF + 4L \xrightarrow{+hv(320\ nm)}$$

$$2Mn(THF)_6^{2+} + 4Mn(CO)_4L^- + 14CO \quad (7.358)$$

It has been found that $Mn_2(CO)_{10}$, $Re_2(CO)_{10}$, and $Os_3(CO)_{12}$, in the presence of suitable chlorine or bromine derivatives such as CCl_4, $CCl_3COOC_2H_5$ or $BrCH_2COOH$, are efficient initiators of free radical polymerization of vinyl monomers such as methyl methacrylate, styrene, vinyl acetate, acrylonitrile,[277,278,285,286,289,295,303,3458,3463] vinyl chloride,[3459,3460,3462] and propylenoxide.[3461] The monomer here plays an important role as coordinating molecule (S). The following mechanism of photoinitiation of polymerization has been proposed:[280-282]

$$Mn_2(CO)_{10} + hv \rightarrow (Mn_2(CO)_{10})^* \quad (7.359)$$

$$(Mn_2(CO)_{10})^* + S \rightarrow [Mn(CO)_5-S-Mn(CO)_5] \rightarrow$$
$$Mn(CO)_5-S\cdot + Mn(CO)_5\cdot \quad (7.360)$$

$$Mn(CO)_5-S\cdot + CCl_4 \rightarrow Mn(CO)_5Cl + \cdot CCl_3 \text{ (slow reaction)} \quad (7.361)$$

$$Mn(CO)_5\cdot + CCl_4 \rightarrow Mn(CO)_5Cl + \cdot CCl_3 \quad (7.362)$$

$$\cdot CCl_3 + \text{monomer} \rightarrow \text{polymerization} \quad (7.363)$$

Manganese pentacarbonyl halide ($Mn(CO)_5Cl$) may subsequently enter into other reactions such as ligand exchange and dimerization.[283,284,286,292,303] In the absence of halides manganese carbonyl ($Mn_2(CO)_{10}$) is ineffective with common vinyl monomers, but $Re_2(CO)_{10}$ gives rise to relatively slow polymerization which has been attributed to hydrogen abstraction.[286, 295] The kinetics of this reaction have been studied in detail.[304, 305]

Halide activity is governed by factors which determine the ease of accepting electrons; thus it increases with multiple substitution, i.e. in the series CH_3Cl < CH_2Cl_2 < $CHCl_3$ < CCl_4, and with introduction of electron-attracting groups into the molecule. Bromine derivatives are much more active than the corresponding compounds of chlorine, and saturated fluorine and iodine compounds are ineffective.

Manganese carbonyl and rhenium carbonyl are both active photoinitiators of free radical polymerization of liquid tetrafluoroethylene ($CF_2=CF_2$) at $-93°C$ and other vinyl monomers (e.g. methyl methacrylate, styrene, and acrylonitrile) containing low concentration of tetrafluoroethylene.[296-298] The initiation involves the following reactions:

$$Mn_2(CO)_{10} + hv \rightarrow 2\ Mn(CO)_5\cdot \quad (7.364)$$

$$Mn(CO)_5\cdot + CF_2 = CF_2 \rightarrow Mn(CO_5)CF_2\dot{C}F_2 \quad (7.365)$$

$$Mn(CO)_5CF_2\dot{C}F_2 + \text{monomer} \rightarrow \text{polymerization} \quad (7.366)$$

The radical $Mn(CO)_5CF_2\dot{C}F_2$ is favoured by the relatively high strength of the

Mn–CF$_2$ bond. Polymeric products, e.g. poly(methyl methacrylate) have metal atoms in their terminal groups, e.g.

$$\underset{\overset{|}{\text{COOCH}_3}}{\overset{\overset{\text{CH}_3}{|}}{\text{Mn(CO)}_5\text{CF}_2\text{CF}_2\text{CH}_2\text{C}\text{----}}}$$

Using metal carbonyls (e.g. Mn$_2$(CO)$_{10}$ or Re$_2$(CO)$_{10}$) and a halogen coinitiator system, several terminating groups can be built in the polymer chain. The nature of these groups may be varied widely. For example Mn$_2$(CO)$_{10}$ and N-bromoacetyldibenz[b,f]azepine (7.256) coinitiator can be used for the preparation of poly(methyl methacrylate) or polystyrene with terminal photoactive groups of N-acyldibenz[b,f]azepine (7.257):[39]

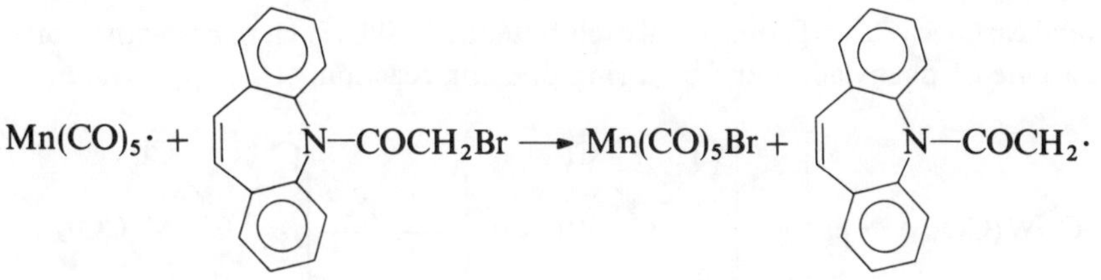

(7.256)   (7.367)

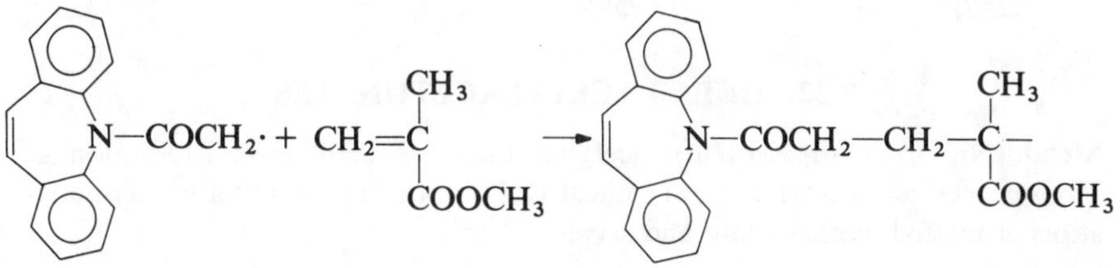

(7.257)   (7.368)

Using manganese carbonyl (Mn$_2$(CO)$_{10}$) and coinitiator (e.g. CBr$_4$) it is possible to synthesize alternating block copolymers,[287,290,299,301,302,306] grafting[243,288] and negative photoresists.[3770]

Radical generation involving metal carbonyls can be achieved not only with organic halogen compounds but also with suitable olefins carrying electron-attracting substituents or acetylene and acetylene dicarboxylic acid and its esters.[278,279,296-298,300] For a coinitiator to possess high activity it must be able to form adequately strong metal–carbon bonds. This can be achieved in the olefinic compounds by virtue of electronegative substituents such as carbonyl, nitryl and fluorine. Steric factors are also important here.[300] In general, Re$_2$(CO)$_{10}$ is more active than Mn$_2$(CO)$_{10}$.

Dimanganese decacarbonyl has also been used as photoinitiator for ring-opening cationic polymerization of epichlorhydrin.[3461]

Tungsten carbonyl ($W(CO)_6$) and molybdenum carbonyl ($Mo(CO)_6$), in the presence of coinitiator carbon tetrachloride, may photoinitiate polymerization of phenylacetylene.[2403,2404] The initiation occurs via formation of metal carbene.[1220] The following mechanism has been proposed:

$$W(CO)_6 + h\nu \rightarrow W(CO)_5 + CO \qquad (7.369)$$

$$W(CO)_5 + CCl_4 \rightarrow \overset{Cl}{\underset{Cl}{\diagdown}}C{=}W(CO)_4 + CO \qquad (7.370)$$

$$\overset{Cl}{\underset{Cl}{\diagdown}}C{=}W(CO)_4 + {-}C{\equiv}C{-} \rightarrow \left[\overset{Cl}{\underset{Cl}{\diagdown}}C{=}W(CO)_6 \cdots \overset{|}{\underset{|}{\overset{C}{\underset{C}{|||}}}} \right] \qquad (7.371)$$

Metal carbene (*7.258*) forms metal cyclobutene (*7.259*), which reacts with another molecule of phenylacetylene by a ring-opening reaction:

$$\left[\overset{Cl}{\underset{Cl}{\diagdown}}C{=}W(CO)_4 \cdots \overset{C}{\underset{C}{|||}}\right] \rightarrow \left[\overset{Cl}{\diagup}\overset{Cl}{\diagdown}\underset{-C=C-}{C{-}W(CO)_4}\right] \xrightarrow{+-C\equiv C-} \left[\overset{Cl}{\diagup}\overset{Cl}{\diagdown}\underset{-C-C-}{\overset{||}{C}}\overset{||}{C}M(CO)_4 \cdots \overset{C}{\underset{C}{|||}}\right]$$

(*7.258*)   (*7.259*)   (*7.372*)

7.22. METAL ACETYLACETONATES

Metal (Ni(II), Co(III), Mn(III)) acetylacetonates (*7.260*) upon irradiation at 365 nm produces acetylacetonyl radical (*7.261*), which may initiate polymerization of methyl methacrylate and styrene:[291, 1909]

$$Me^{III}\left(CH\underset{\underset{O}{\overset{||}{C-CH_3}}}{\overset{\overset{O}{\overset{||}{C-CH_3}}}{\diagup\diagdown}}\right)_3 + h\nu \rightarrow Me^{II}(acac)_2 + CH_3{-}\overset{O}{\overset{||}{C}}{-}\overset{\cdot}{C}H{-}\overset{O}{\overset{||}{C}}{-}CH_3$$

(*7.261*)

(*7.373*)

(*7.260*)

This reaction is accompanied by the photoreduction of metal ion.[1202,1233,1909] Metal acetylacetonates develop a colour characteristic to metal ion; for example $Co(acac)_3$ green and $Fe(acac)_3$ red. Absorption at 365 nm corresponds to metal-to-ligand excitations, whereas the dissociation process is likely to involve ligand-to-metal charge-transfer excited states. Evidently a non-radiative relaxation from the initially excited states to the reacting states is involved, and this may account for the low overall quantum yield of the reaction and low percentage of conversion of monomer into polymer (Table 7.22).

Table 7.22 Conversion (%) of a monomer into polymer in the presence of different metal acetylacetonates $(Me(acac)_n)$[1909]

Monomer	$Me(acac)_n$	Time of irradiation (h)	Conversion (%)
Methyl methacrylate	$\{Ti(acac)_3\}_2 TiCl_6$	5.0	Trace
	$Fe(acac)_3$	15.5	5.94
	$Mn(acac)_3$	3.0	26.6
	$MoO_2(acac)_2$	3.0	5.55
	$Al(acac)_3$	12	16.9
	$Cu(ba)_2$*	3.5	2.09
	$Ni(acac)_2$	5.0	2.87
	$Co(acac)_2$	4.0	9.03
	$Mg(acac)_2$	10	16.8
	$Co(acac)_3$	6.0	28.7
Styrene			
	$Mn(acac)_3$		15.7
	$MoO_2(acac)_2$		0
	$Al(acac)_3$		0.43
	$Cu(ba)_2$		0.39
	$Mg(acac)_2$		0.41
	$Co(acac)_3$		9.48
	$Cr(acac)_3$		1.03
	$Zn(acac)_2$		0

* $Cu(ba)_2$-copper-bis(benzoylacetonate)

In the case of initiation by manganese(III) tris-(1,1,1-trifluoroacetyl)acetonate, the rate of photodissociation is markedly dependent upon the solvent used, in contrast to the manganese(III) acetylacetonate.[218] This phenomenon can be explained on the basis of exciplex formation between metal acetonate and solvent.

The polymerization of methyl methacrylate photoinitiated by tricyanatotripyridineiron(III) complexes has also been described.[3167]

7.23. FERROCENE

Ferrocene (7.262) is a light-stable compound which absorbs light in the ultraviolet and visible range with λ_{max} at 325 and 440 nm. Solutions of ferrocene in cyclohexanone, acetone, and 2-propanol, as well as in dekaline, methanol, and 1-propanol, remain unchanged on irradiation. A photoreaction is observed with CCl_4, CBr_4, $CHCl_3$, CH_2Cl_2, and CCl_3COOH.[3553]

Substituted ferrocenes display a markedly increased photosensitivity, e.g. 1,1-dibenzoylferrocene shows colour changes and decomposes on irradiation in alcohols, ethers, esters, acetone, acetonitrile, and pyridine. Products of these reaction have not been investigated.[3553]

Solution of ferrocene in CCl_4 shows a new band in the absorption spectra due to charge-transfer (CT) complex (7.263) formation:[504]

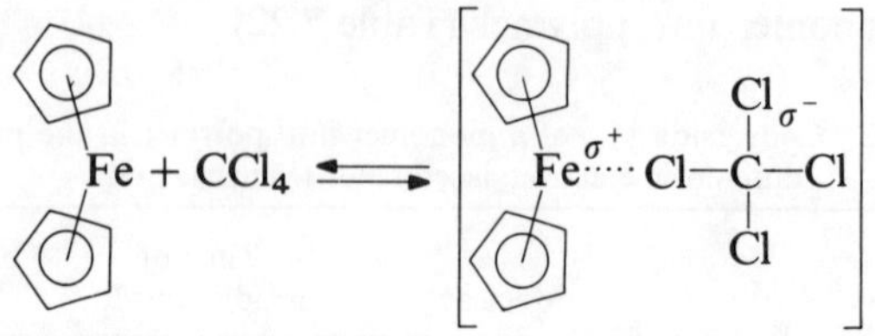

$$(7.374)$$

(7.262)   (7.263)

Irradiation of the complex (7.263) causes formation of trichloromethyl radical ($CCl_3 \cdot$) and ferricenium tetrachloroferrate (7.264):[2094]

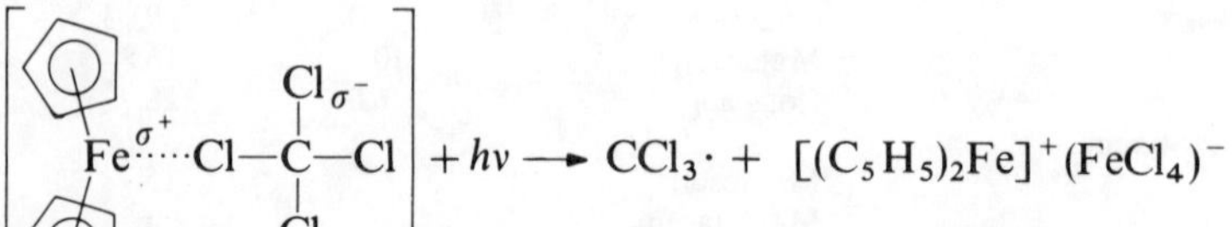

$$(7.375)$$

Several vinyl monomers such as methyl methacrylate, styrene, acrylonitrile, vinyl acetate, etc., easily polymerize under irradiation in the presence of a ferrocene/CCl_4 initiating system.[3577,7636] Polymerization is initiated by trichloromethyl radicals ($CCl_3 \cdot$). Addition of free radical scavenger such as α,α-diphenyl-β-picrylhydrazyl (DPPH) inhibits polymerization.

It has been reported that ferrocene efficiently quenches excited triplet states of triphenylene ($E_T = 66.6$ kcal mol^{-1}) and anthracene ($E_T = 42.6$ kcal mol^{-1}).[1186]

Ferrocene has been also used as light-stabilizer against photodegradation.[3211]

7.24. METAL COMPLEXES

A number of metal complexes under u.v. irradiation may initiate polymerization of vinyl monomers.

Azidopentaaminecobalt(III) chloride (7.265) under u.v. irradiation is photolysed to pentaamine(III) chloride (7.266) with evolution of an azide radical ($N_3 \cdot$):

$$[CO(NH_3)_5N_3]Cl_2 \cdot H_2O + h\nu \rightarrow [Co(NH_3)_5H_2O]Cl_2 + N_3 \cdot \quad (7.376)$$

(7.265)   (7.266)

Azidopentaaminecobalt(III) chloride [2116,2657,2658] and diazidotetraaminecobalt(III)[2115] have been used as photoinitiators of vinyl monomers such as methyl methacrylate, arylamide, acrylonitrile, acrylic acid, and N-vinylpyrrolidone polymerization.

Irradiation of copper(II) amino acid (glutamic acid, serine or vaniline) chelates (7.267) dissolved in water or in organic solvents leads to photoredox reactions. The principal feature of the photochemical reactions of copper(II) complexes is the reduction of the metal centre by electron transfer from the ligand leading to the formation of copper(I) species and of the oxidized ligand radical. The ligand radical (7.268) thus produced undergoes secondary reactions and copper(I) is slowly reoxidized to copper(II) in air-equilibrated solutions.[2654] The following reactions are involved in the photoreduction of bis(valinato) Cu(II) complex (7.267):

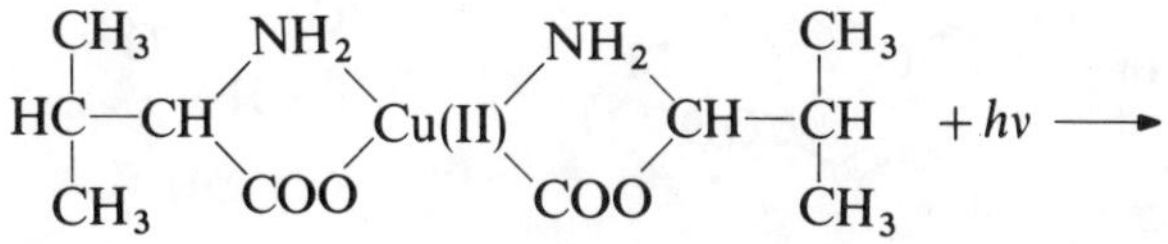

(7.267)

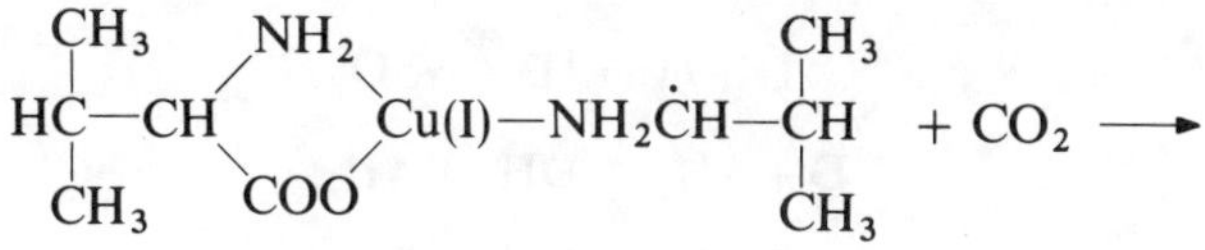

(7.268)　　　　　　　　　　　(7.377)

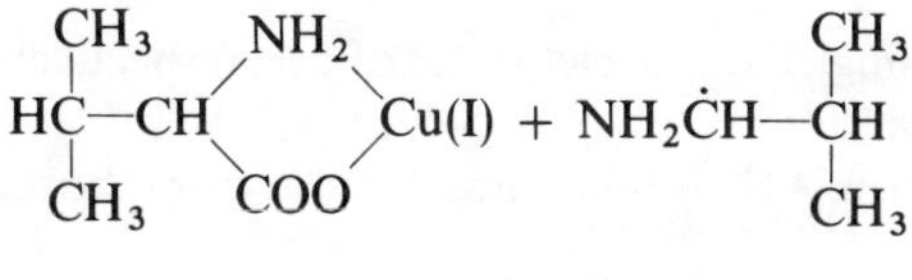

(7.269)

Amino radicals (7.269) formed during the photodecomposition of copper(II) amino chelates may initiate polymerization of vinyl monomers.[1756, 1759, 2655, 2659, 2782]

It has also been reported that diacidobis(ethylenediamine)cobalt(III) complexes photoinitiate polymerization of vinyl monomers.[237]

The iron(III) amino complexes in the presence of CCl_4[2782] and iron(III) tetraphenylporphyrine complex in the presence of CCl_4 and free amines (e.g. ethylenediamine),[1758] photoinitiate polymerization of methyl methacrylate.

Cobalt(III) tetraphenylporphyrine complex in the presence of cholesteryl carbonate (liquid crystal) photoinitiates polymerization of methyl methacrylate and copolymerization of methyl methacrylate and styrene.[978]

Trisoxalatoferrate-arene onium salts have been used as a photoinitiator for polymerization of acrylamide.[340,3161]

7.25. DYES

The photochemistry of dyes is complicated because it is dependent on their chemical structure.

Xanthene dyes (D), e.g. eosin (7.270), in the presence of reducing agent (AH_2), e.g. ascorbic acid (7.271), upon visible light irradiation produce several radicals according to the scheme:[1754,2096,2823]

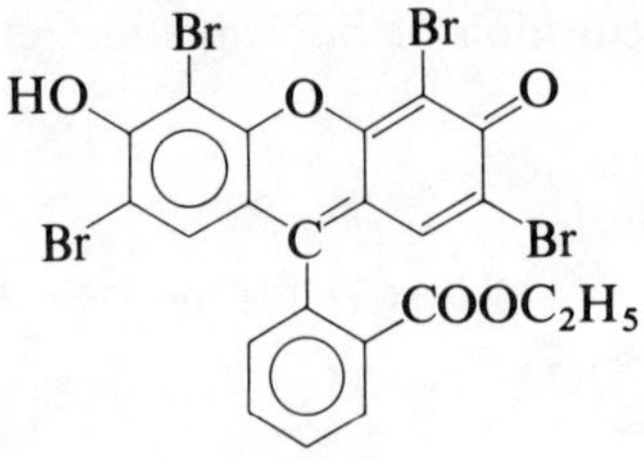

(D)

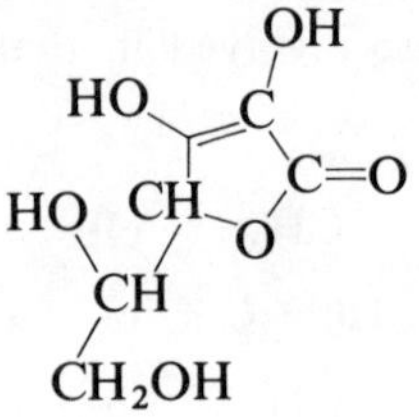

(AH_2)

(7.270)

(7.271)

$$D + hv \rightarrow {}^1D \xrightarrow{ISC} {}^3D \tag{7.378}$$

$$^3D + AH_2 \rightarrow DH \cdot + AH \cdot \tag{7.379}$$

$$2\,DH \cdot \rightarrow DH_2 + D \tag{7.380}$$

$$DH \cdot + AH_2 \rightarrow DH_2 + AH \cdot \tag{7.381}$$

where:

^1D and ^3D are excited singlet and triplet states of dye, respectively; $DH \cdot$ is the semiquinone dye radical;

DH_2 is the leuco dye form; $AH \cdot$ is the radical formed from the reducing agent (AH_2).

There is disagreement in published papers on the mechanism of the initiation of polymerization by dyes in the presence of reducing agents. The following suggestions have been made:

(i) Initiation of polymerization is caused by hydroperoxy radicals ($HO_2 \cdot$) produced by the reaction of the semiquinone dye radical ($DH \cdot$) with dissolved oxygen:[3961]

$$DH \cdot + O_2 \rightarrow D + HO_2 \cdot \tag{7.382}$$

(ii) Initiation of polymerization by a redox system consisting of hydrogen peroxide and reducing agent; the peroxide being generated by re-oxidation of leuco (DH_2) and semiquinone dye radical ($DH \cdot$) by dissolved oxygen.[935]

(iii) Initiation of polymerization by a photoredox process between photoreduced dye and peroxide compound produced during an induction period which precedes polymerization.[3524]

(iv) Initiation of polymerization by hydroxyl radicals (HO ·) formed in hydrogen peroxide/ascorbic acid redox reactions:[2922]

$$AH_2 + H_2O \rightleftarrows AH^- + H_3O^+ \tag{7.383}$$

$$AH^- + H_2O_2 \rightarrow A^{\overline{\cdot}} + H_2O + HO \cdot \tag{7.384}$$

$$AH^- + HO \cdot \rightarrow A^{\overline{\cdot}} + H_2O \tag{7.385}$$

$$2A^{\overline{\cdot}} \rightarrow A + A^{2-} \tag{7.386}$$

where:

AH^- is the mono-ion of the ascorbic acid (AH_2),
$A^{\overline{\cdot}}$ is the radical anion (semireduced form) of ascorbic acid, A is the dehydroascorbic acid.

The excited triplet state of dye (^3D) may react with the mono-ion of the ascorbic acid (AH^-) producing the semiquinone dye radical $(DH \cdot)$ and the radical anion of ascorbic acid $(A^{\overline{\cdot}})$ according to the reaction:

$$^3D + AH^- \rightarrow DH \cdot + A^{\overline{\cdot}} \tag{7.387}$$

The role of oxygen in all of these reaction is still not clear.[651,2376,3310,3807] In the presence of oxygen the semiquinone dye radical $(DH \cdot)$ can produce hydrogen peroxide according to reaction:[2923]

$$DH \cdot + O_2 \rightarrow D + HO_2 \cdot \tag{7.388}$$

$$DH + HO_2 \cdot \rightarrow D + H_2O_2 \tag{7.389}$$

$$HO_2 \cdot + AH^- \rightarrow A^{\overline{\cdot}} + H_2O_2 \tag{7.390}$$

$$2 HO_2 \cdot \rightarrow H_2O_2 + O_2 \tag{7.391}$$

When all the oxygen has been consumed by the above reaction, the dye is reduced to the leuco form (DH_2) either by reaction with the mono-ion of the ascorbic acid (AH^-) or disproportionation of the semiquinone radicals $(DH \cdot)$:

$$DH \cdot + AH^- \rightarrow A^{\overline{\cdot}} + DH_2 \tag{7.392}$$

$$2 DH \cdot \rightarrow D + DH_2 \tag{7.393}$$

Both the $DH \cdot$ and $A^{\overline{\cdot}}$ radicals are stabilized by delocalization of the unpaired electron, and are not sufficiently reactive to act as initiating species for polymerization, e.g. vinyl acetate.[2922,2923]

7.26. POLYMERIC PHOTOINITIATORS (PHOTOSENSITIZERS)

Low molecular photoinitiators (photosensitizers) can be attached to macromolecules (Table 7.23) without destroying their photochemical activity. Polymeric photoinitiators and photosensitizers are also called HETEROGENEOUS PHOTOCATALIZATORS.

Table 7.23   Examples of polymeric photoinitiators (photosensitizers)

Photoinitiator (photosensitizer)	Structure	Application	Reference
Polybenzaldehyde oligomer	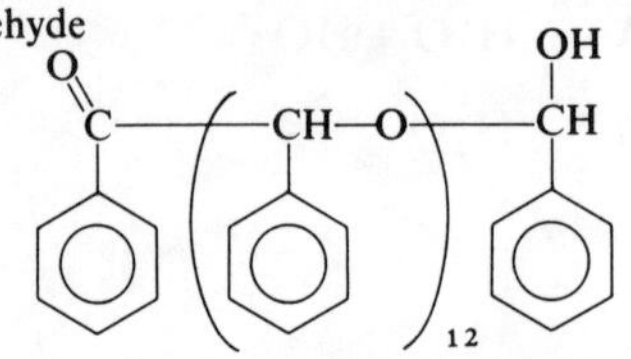	Photosensitizer of isomerization of 1,3-pentadiene and 1,3-cyclohexadiene	941–943
Poly(phenyl vinyl ketone)	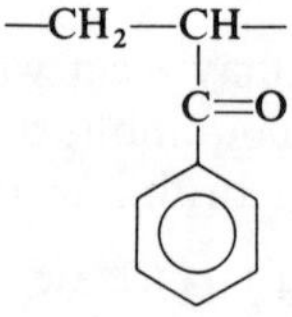	Photosensitizer of isomerization of *trans/cis*-stilbene	1774
Poly(vinyl benzophenone)	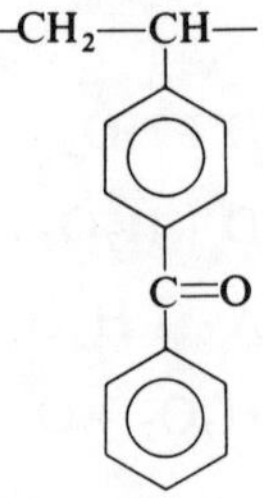	Photoinitiator of grafting	320, 604, 2030, 3489
		Photoinitiator of crosslinking	604, 840, 1333, 1691, 3177
		Photosensitizer of isomerization of *trans/cis*. stilbene	3278, 3709
		Photosensitizer of isomerization of soluble conjugated dienes	485
		Photosensitizer of (2 + 2) cycloaddition of cyclohexadiene to maleic anhydride	484
Poly(4-*N,N*-dimethyl-amino)benzophenone	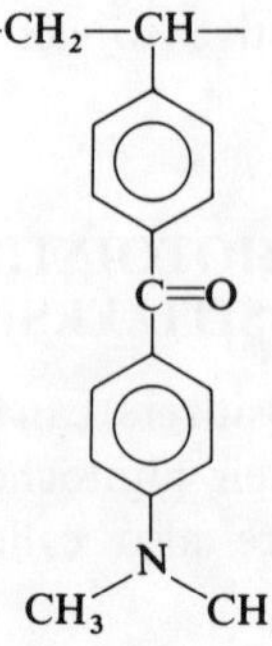	Photosensitizer of isomerization of norbornadiene to quadricyclane	1531, 2026

Table 7.23 (*Contd.*)

Photoinitiator (photosensitizer)	Structure	Application	Reference
Poly(naphthoylstyrene)	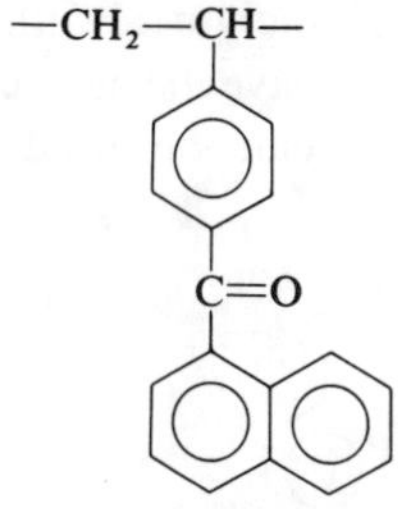 —CH₂—CH— (structure with phenyl ring, C=O, and naphthyl group)	Photosensitizer of isomerization of *trans/cis*-stilbene	1491
Poly(4-acryloxybenzophenone) and copolymers of acrylobenzophenone with methyl methacrylate, methyl acrylate, 1-acrylo-*xy*-2-ethoxyethane, and 2-(*N,N*-dimethylamino)styrene	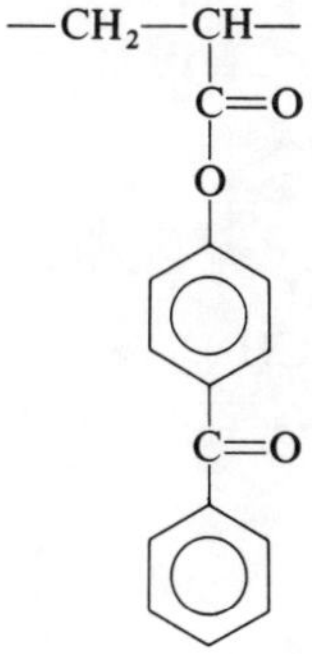 —CH₂—CH— (structure with C=O, O, phenyl ring, C=O, phenyl ring)	Photoinitiators for polymerization of acrylic monomers	604, 1117
Co[2(9,10-anthraquinonyl)/methyl methacrylate]	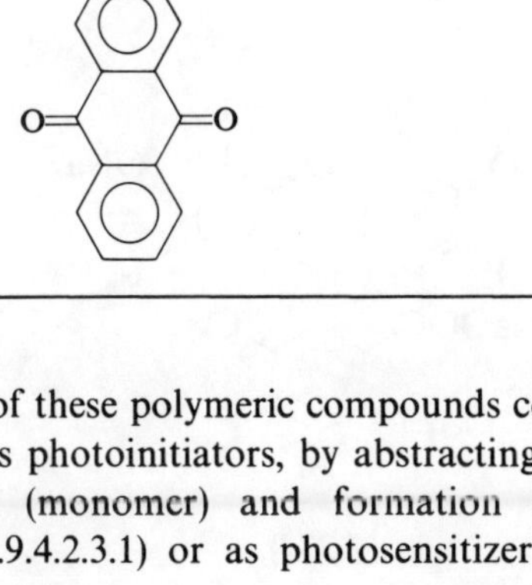 (polymer structure with anthraquinone group)	Photosensitizer of oxidation/reduction of L-ascorbic acid	2620

Some of these polymeric compounds containing a benzophenone side group can act as photoinitiators, by abstracting a hydrogen from a donor hydrogen molecule (monomer) and formation of the polymer ketyl radical (cf. Section 7.9.4.2.3.1) or as photosensitizer (by energy transfer mechanism, cf.

Chapter 2). They exhibit higher efficiency with respect to low molecular benzophenone according to energy migration along the chain, arising from interactions between neighbouring excited groups. Their efficiency depends markedly on the nature of the comonomer.[604, 1117]

Poly(divinylbenzophenone) (7.272) and copolymers with methyl methacrylate, acrylonitrile, and styrene–divinylbenzene, have been used as energy transfer donors.[1465] These polymeric sensitizers can be used for sensitization of

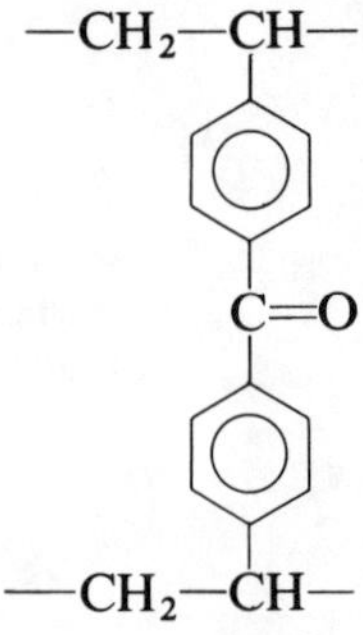

(7.272)

photocyclization or photoisomerization (cf. Table 7.23) of some organic compounds.

Rose Bengal (disodium salt of 3',4',5',6'-tetrachloro-2,4,5,7-tetraiodofluorescein) (7.273) used as photosensitizer in photooxygenation reactions in polar solvents, such as water and methanol, bleaches fast upon irradiation.[2204] Co(styrene/divinylbenzene) supports provide a hydrophobic immobilization centre for Rose Bengal, and these immobilized dyes in polymer matrix (7.274) are more stable to oxidation bleaching:[2400,3201,3532]

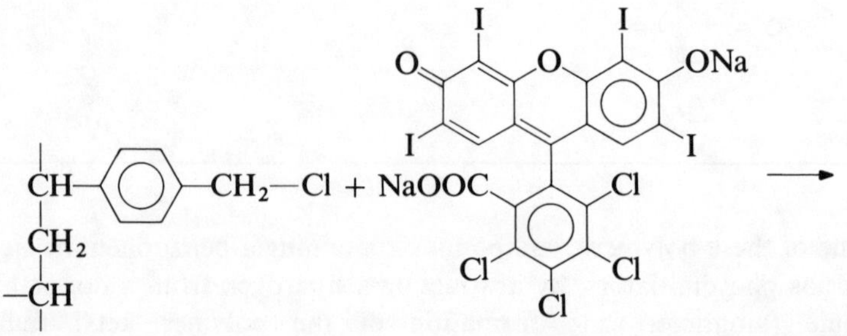

(7.273)

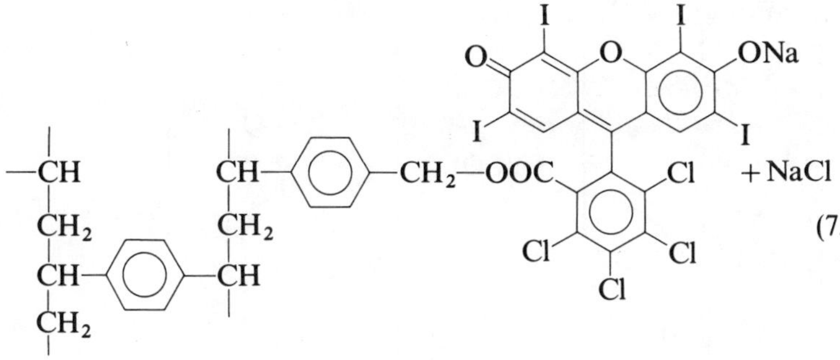

(7.394)

(7.274)

Polymer matrices having free base porphyrins, e.g. tetraphenylporphyrin, tetra-p-methylporphyrin and tetra-p-hydroxyporphyrin, have been synthesized and examined for their photocatalytic activities.[454,2098]

Poly(γ-1-naphthyl-methyl-L- and DL-glutamate) and copolypeptides of γ-1-naphthyl-L-glutamate and γ-1-benzyl-L-glutamate, have been used as photosensitizers for isomerization of 1,2-diphenylcyclopropane.[3710]

Polymer-bound photosensitizers have several advantages over free sensitizers in solution:

(i) They can be used in solvents in which the unbound sensitizer is insoluble and therefore unable to sensitize the reaction efficiently.

(ii) They are significantly more stable towards photodecomposition (photolysis or photobleaching) than are the free sensitizers.

(iii) The polymer-bound sensitizers can be easily removed at the end of reaction by filtration, and can be reused with little or no loss in efficiency.

The serious disadvantage of polymer-based photosensitizers is absorption and scattering of incident radiation by the polymer beads, which complicates the use of covalently bound photosensitizers in studies such as determination of photooxidation quantum yields, in which the intensity of the absorbed radiation must be well known.

7.27. PHOTOCATALYTIC BEHAVIOUR OF POLYMER-ANCHORED Fe(CO)$_n$

Phosphinated styrene-divinylbenzene resin with anchored Fe(CO)$_n$ (7.275) and (7.276) has been shown to serve as a photochemical source of catalytically active species that are capable of initiating alkene (e.g. 1-pentene (7.277) isomerization:[3182]

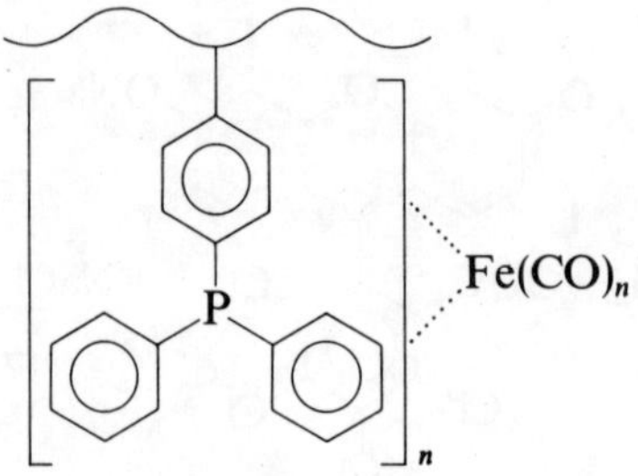

(7.275) $n = 3, 4$

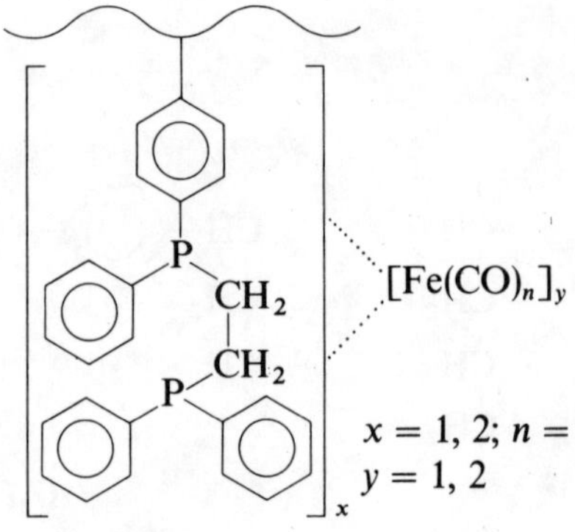

(7.276) $x = 1, 2; n = 3, 4; y = 1, 2$

(7.277)

(7.395)

and reaction of alkene (e.g. 1-pentene) with a silicon hydride (7.278):[3182]

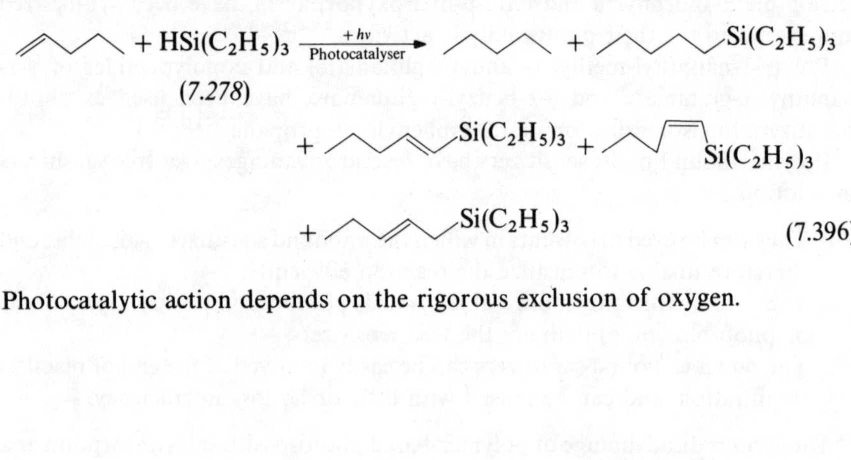

(7.278)

(7.396)

Photocatalytic action depends on the rigorous exclusion of oxygen.

8. Photomodification of polymer surfaces

A number of properties of polymeric materials, such as adhesivity, printability, dyeability, paintability, gloss, anti-static behaviour, anti-fogging, heat stability, scratch resistance, biocompatibility, and many other properties such as chemical reactivities, photoconductivity, anti-thrombic properties, selective permeability, etc. are not determined by bulk but by surface properties.[3560,3573] The importance of surface properties is limited not only to films, but also to fibres and moulded materials. Interface phenomena such as dispersion and adhesivity of fillers in matrix polymers are also relevant to the surface of each component, and of primary importance in composite materials. The correlation of various surface functions to basic surface properties is shown in Table 8.1.

Table 8.1 Correlation between basic surface properties and surface functionalities[3560]

	Practical performance											
	Printability	Adhesivity	Laminateability	Acceptance of writing ink	Paintability	Antistatic surface	Antifogging surface	Prevention of bleed-out	Antistaining surface	Surface hardness	Biocompatibility	Permeability
Hydrophilicity or polar property	0	0	0	0	0	0	0	–	0	–	0	–
Hydrophobicity or non-polar property	–	0	0	–	0	–	–	–	–	–	–	–
Other functionalities*	–	–	–	–	–	–	–	0	–	0	0	0

* 'Other functionalities' include chemical reactivities, photoconductivity, antithrombotic property, specific interactions, selective permeability, and others which are not expressed by polar or non-polar properties

Polymer surface can be modified by a number of photochemical processes such as:

(i) Photografting (cf. Section 8.1).
(ii) Photocycloaddition (cf. Chapter 9).

(iii) Photorearrangement reaction. For example polymers bearing thiocyanato groups (*8.1*), upon u.v. irradiation, undergo photorearrangement of thiocyanate groups (*8.2*):[649]

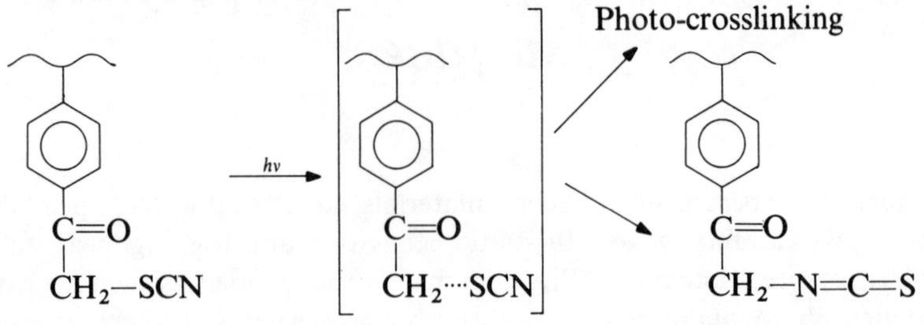

(8.1)

(*8.1*)   (*8.2*)   (8.2)

Further, such u.v.-irradiated film (*8.2*) can be dyed selectively with several kinds of cationic dyes according to the reaction:

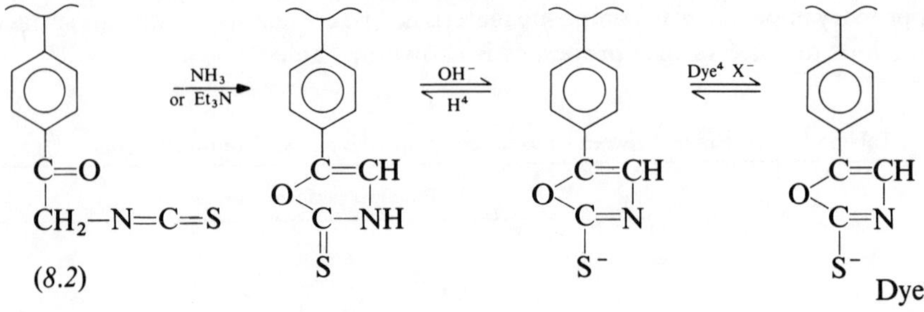

(8.3)

The amount of dye absorption increases with increasing irradiation time.

8.1. PHOTOGRAFTING

PHOTOGRAFTING is a technique utilizing photopolymerization so that the monomer grafts become attached to macromolecules or the surface of other materials. The procedure has been reviewed.[226,227]

The macromolecule can be made to initiate polymerization at points along its chain by generation of radical sites. The sites on the chain may be formed by photoreactions, or by transfer with radicals or radical formers generated in the polymeric matrix. Such techniques offer an opportunity to prepare grafted polymers from monomers and polymers of radically different physical properties, e.g. photografting of a hydrophyllic monomer such as acrylamide onto a hydrophobic polymer such as polyolefins or nylon-6.

Graft copolymerization leads to an appreciable improvement in the existing properties. Grafting can improve weather resistance, wet crease recovery, and dye uptake. In some cases grafting may decrease thermal stability and tear strength.

Photografting of vinylacetate in polytetrafluoroethylene matrix occurs not only at the surface but also in the bulk. The process changes a polymer's surface properties and can be employed in modification of polymeric membranes.[950]

Several examples of photografting of different monomers onto polymers are listed in Table 8.2.

Table 8.2   Examples of grafting

Polymer	Monomer	Initiator	References
Polyethylene	Methyl methacrylate Acrylic acid Methacrylic acid Acrylonitrile Styrene	Benzophenone Anthraquinone Benzoyl peroxide	2761, 2767, 2768, 2827
	Maleimide	Benzophenone	1541
	Acrylamide	Benzophenone	2821
Polypropylene	Acrylic acid	Benzophenone Antraquinone Benzoyl peroxide	2761, 2767, 2768
	Methyl methacrylate Styrene 4-Vinypyridine	Benzoin	201, 202
	Acrylamide	Benzophenone Anthraquinone Benzoyl peroxide	2724, 2725, 3560, 3571, 3573
	p-Styryl diphenyl phosphine	–	1218
Co(ethylene/vinyl acetate)	Binary solid monomer systems, e.g. acenaphthylene–maleimide, acenaphthylene–maleic anhydride	–	1542
Poly(ethylene terephthalate)	Methyl methacrylate	Peroxy diphosphate	2254
	Acrylic acid	Benzophenone Anthraquinone Benzoyl peroxide	2768
Polystyrene	p-Styryl diphenyl phosphine	–	1218
Poly(vinyl chloride)	Acrylamide	Benzophenone	3571
	Epoxy–diacrylate oligomer	2,2-dimethyl-2- hydroxy-acetophenone 1-Benzoyl-cyclo- hexanol	910
	p-Styryl diphenyl phosphine	–	1218
Nylon 6	Methacrylic acid Methyl methacrylate	– –	3714 2589

Table 8.2 (*Contd.*)

Polymer	Monomer	Initiator	References
Nylon 6	Methyl methacrylate	Peroxydiphosphate	2260
	Methyacrylic acid	Benzophenone	1675
	Acrylamide	Fructose	2590
	Styrene	–	1788–1790
Cellulose	Methyl methacrylate	–	1643, 1648
		Hydrogen peroxide	2144
		Metal oxides	2670
		Oxalic acid permanganate	1309
		Potassium persulphate	1310
		Peroxydiphosphate	2253
		Ceric ion	2143, 2763, 2766
		Ferric ion	2151
		Biacetyl	3286
		Benzoin	1143
		Benzophenone	1572
		N-Bromosuccinimide	2252
		Anthraquinone dyes	1236, 1237
		Pyridine–bromine CT complex	2255
		γ-Picoline–bromine CT complex	2262
	Styrene	–	1543
		Benzoin Benzophenone Benzoin ethyl ether	201, 1143
		Uranyl nitrate	201
	Acrylonitrile	Potassium persulfate	1310
		Benzoin Benzophenone	1143
		Anthraquinone dyes	1236, 1237
	Different vinyl monomers	Acetophenone Benzophenone	1573–1575
Aldehyde cellulose	Acrylic acid N-Vinyl-2-pyrrolidone Acrylamide Methyl methacrylate Acrylonitrile Vinyl acetate	–	2759
Cellulose triacetate	Acrylamide	Benzophenone	3571
Cotton	Vinyl monomers	Saccharides	1572
	Methacryl amide	FeCl$_3$	228
	N-Methyl acrylamide	–	1512, 3085

Table 8.2   (*Contd.*)

Polymer	Monomer	Initiator	References
	Glycidyl methacrylate	–	1511
	Vinyl phosphonate oligomer	–	1512
Cellophane	Acrylonitrile Acrylamide	Vinyl ketone	3661
Starch	Water-soluble acrylate monomers	Hydrogen peroxide	3992
	Methyl methacrylate	Potassium permanganate Benzophenone $\alpha,\alpha,$-Dimethoxydeoxybenzoin	1325 2495
Jute fiber	Methyl methacrylate	Ferric sulphate	1324
Wool	Methyl methacrylate	Oxalic acid–Ce^{4+}	1306
	Acrylic acid N-vinyl-2-pyrrolidone	4-(sulphomethyl)benzyl sodium salt	311, 312
	Acrylic monomers	Riboflavine	1668, 2668 2669
Natural rubber	Methyl methacrylate	Benzoin, benzyl, xanthone, 1-chloroanthraquinone Quinolino-bromine	758, 759
		CT complex	1303
	Aqueous acrylamide	Benzophenone	2828
Diethyldithiocarbamated poly(dimethyl siloxane),	Acrylamide, 2-hydroxyethyl methacrylate		
	Methacrylic acid Sodium styrene sulphonate		1763
Dithiocarbamated poly(vinyl chloride)	Methoxyl polyethyleneglycol methacrylate N,N-Dimethyl aminoethyl methacrylate		2534

Photografting has several advantages in comparison with other surface modification process such as liquid–solid chemical treatment, gas–solid treatment, corona discharge treatment, and radiation-induced graft polymerization:[3560]

   (i) Since photoinitiation proceeds via singlet and/or triplet excited states, selective excitation of a desired photoinitiator is feasible.
  (ii) The incident light is absorbed within a thickness of 1–3 μm.
 (iii) Localization of photoinitiator on the polymer–solution boundary is attainable when the reacting solution in contact with a polymer material to be surface-treated is irradiated through the polymer material, if it is transparent.
  (iv) Low-temperature reaction.
   (v) Energy sources for photoreaction, such as mercury lamps, neon lamps, and metal halide lamps,[3023] are comparatively cheap and safe.

(vi) Interactions between the base polymer and the reacting solution are controlled, so as to produce a very thin graft polymer.

Photografting has also several disadvantages, such as:

 (i) The distribution of photochemically produced active species is often inhomogeneous.
(ii) Photodegradation and photooxidation processes[2468,3060] may occur in parallel with photografting reactions.
(iii) Slow reaction rates (irradiation time is still too long for surface treatment of commercial polymers).
(iv) Unfavourable effects of the remaining photoinitiator.
 (v) Photohomopolymerization. When a homopolymer is deposited on the grafted surface, diffusion of monomer and photoinitiator is hindered so that graft polymerization becomes sluggish.
(vi) Oxygen inhibiting of grafting.

8.2. THIN-LAYER SURFACE PHOTOGRAFTING

Thin-layer surface photografting can be made by two techniques:

 (i) Photografting of liquid monomers on solid polymer (solid–liquid system). Photoinitiator can be dissolved in a monomer–solvent mixture or coated on polymer film beforehand.[200,759,893,1237,2761,2827,2828,3204,3571]
(ii) Photografting of monomers in vapour phase on solid polymer on which photoinitiator was coated beforehand (solid–gas system).[1532, 1541, 2670, 2761, 2768, 3286, 3560, 3939]

Surface photografting is sometimes so small that the weight increase of treated polymer cannot be measured gravimetrically, but the results of such a procedure can be detected by other methods, e.g. u.v. and/or i.r. spectroscopy or contact angle measurements (Table 8.3).[3560]

Table 8.3   Examples of hydrophilic treatment of various polymers by surface photografting of acrylamide[3560]

	Contact angle to water, degrees	
	Before treatment	After treatment
Biaxially oriented poly(propylene)	101	35–60
Soft poly(vinyl chloride)	92	30–40
Poly(vinylidene chloride)	75	50–60
1,2- and 1,4-Poly(butadiene)	94	50–60
ABS	98	30–50
Epoxy resin (bisphenol type)	80–90	5–10

Thin-layer surface photografting depends greatly on the kind of solvent used.[2768,3560] The requirements for appropriate solvents are as follows:

(i) The solvent is to be a non-solvent of the base polymer. In order to obtain a thin graft layer the solvent must not swell the base polymer extensively. Slight but definitive interactions are, however, necessary to provide reaction sites for grafting.

(ii) The growing graft chain is not necessarily soluble in the solvent. Good solvent-growing chain interactions are advantageous to assist the propagation of graft chain outside the base polymer surface.

(iii) The solvent has to be inert to the triplet excited state of the photoinitiator. This is particularly important for base polymers which are not very susceptible to radical attack. Low-density polyethylene and polypropylene cannot be surface-grafted from methanol, whereas poly(vinyl chloride) and poly(vinylidene chloride) can.

The choice of photoinitiator is related to the base polymer properties. The photoinitiators other than these which can be excited to triplet state (T_1) are, however, not capable of initiating surface photografting onto polyolefins.

The results of vapor phase photografting of different monomers on polyolefin films in the presence of different photoinitiators (such as benzophenone, anthraquinone, and benzoyl peroxide) are shown in Table 8.4. Graft efficiency was in the range of 60–90 % of each monomer.

Table 8.4 Vapour phase photografting on photoinitiator: (I-1) anthraquinone; (I-2) benzoyl peroxide; (I-3) benzophenone-coated polyolefin films[2761]

| | Percentage grafting | | | | | |
| | Polypropylene | | | Low-density polyethylene | | |
Monomer	(I-1)	(I-2)	(I-3)	(I-1)	(I-2)	(I-3)
Methyl methacrylate	695	654	170	1164	620	146
Acrylic acid	200	241	289	297	251	377
Methacrylic acid	33	33	46	169	16	46
Styrene	3	4	0	2	0	0

The results of liquid phase photografting are shown in Table 8.5.

Figure 8.1 shows the effect of amount of photoinitiator (benzophenone) on photografting of acrylic acid. The percentage graftings for both low-density polyethylene and polypropylene were found to increase with increasing the amount of photoinitiator.

Under conditions where the base polymer is not directly excited by absorbed light, two possibilities of initiating graft polymerization can be considered.

(i) Interaction of excited photoinitiator with the base polymer surface.

(ii) Direct attack of initiating or growing radicals.

Photoinitiators such as benzophenone, substituted benzophenones, fluorenone,

Table 8.5 Liquid phase photographting on photoinitiator (I-1) anthraquinone and (I-2) benzophenone coated low-density polyethylene film in water system[2761]

Monomer	Percentage grafting	
	(I-1)	(I-2)
Methacrylic acid	1120	1350
Acrylic acid	36	491
Acrylic acid*	11	37
Acrylamide	3	21
Acrylonitrile	54	156
Methyl methacrylate*	196	78

*Ethanol:water = 30:70 vol. %

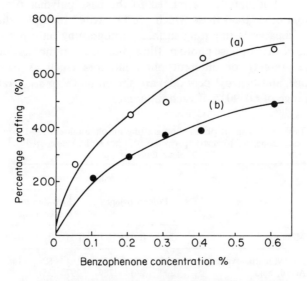

Fig. 8.1 Effect of benzophenone amount (photoinitiator) on liquid phase photografting of acrylic acid on: (a) low-density polyethylene, and (b) polypropylene films[2761]

and benzyl aldehyde and its derivatives are capable of initiating graft polymerization, whereas benzoin isopropyl ether is totally inefficient.

Solvents having labile hydrogen atoms, such as alcohols and amines, exhibit an inhibitory effect on surface grafting. When methanol is used instead of acetone, or a small amount of 2-propanol or amines is added to the reaction system, graft polymerization is inhibited, whereas homopolymerization proceeds rapidly.

The choice of solvent greatly influences the rate of grafting (Fig. 8.2). The degree of solvent–base polymer interaction decides the depth of grafting, and consequently the surface properties of the grafted polymer.[3560,3573] Being nonpolar, polypropylene film has a higher affinity to *n*-hexane or benzene than to acrylonitrile. The affinity of acetone is between the two. Consequently, the

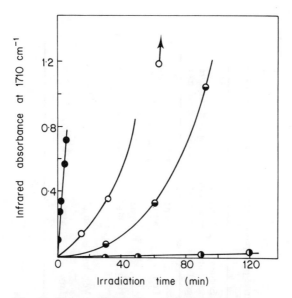

Fig. 8.2 Solvent effect on photografting of acrylic acid onto polypropylene films in different solvents: (●) n-hexane, (○) benzene; (○) acetone, and (○) acetonitrile[3560]

polymer surface will be slightly swollen in contact with these non-polar solvents, and deep grafting can be expected. This indicates that the volume of base polymer available for graft polymerization is larger and the rate tends to be faster. On the other hand, acetonitrile cannot provide grafting sites since the solvent is too incompatible with the polypropylene film. Acetone shows intermediate behaviour. By controlling the solvent–base polymer compatibility it is possible to control the thickness of graft layer.

When a polar monomer is grafted onto a non-polar polymer surface by using a polar solvent, two effects are observed.

(i) An effect which reduces the thickness of the graft layer with decreasing solvent–base polymer when solvent polarity increases.

(ii) An effect which increases solvent-grafted surface interaction with increasing solvent polarity so that the rate of grafting increases with the progression of grafting (Fig. 8.2).

A summary of solvent effects on surface photografting of polar monomers onto nonpolar polymers is shown in Table 8.6.

The facile conversion of a hydrophobic surface by the thin-layer surface photografting has a variety of practical applications in the field of printing, coating, packaging, compositing, biomedical, and agricultural materials.

Photograft copolymerization is a new method for obtaining antithrombogenic heparinized polymers for application in medicine.[2534,2535]

Table 8.6 Summary of solvent effects on surface photografting of polar monomers on to non-polar polymers in the presence of photoinitiator (I) (benzophenone)[3560]

Solvent (S)	S-base polymer interactions	I* + S → radicals	Photografting	Surface concentration of graft chain	Surface modification effect	Proposed structure of grafted polymer
Polar	Weak	Yes	No	—	None	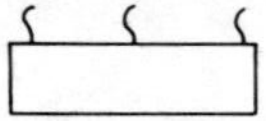
	Weak	No	Yes, but slow	Low	Small	
Moderately polar	Moderate	Yes	No	—	None	
	Moderate	No	Yes, fast to medium	High	Large	
Non-polar	Strong	Yes	Apparently yes (internal polymerization?)	Very low	Small	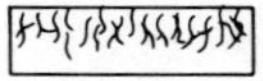
	Strong	No	Yes, very fast	Low	Moderate	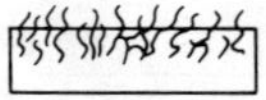

8.3. PHOTOGRAFTING ONTO POLY(DIMETHYLSILOXANES)

Photografting of hydrophyllic vinyl monomers such as acrylamide, N-vinyl-2-pyrrolidonemethacrylic acid, or sodium styrene sulphonate onto the surface of diethyldithiocarbamated poly(dimethylsiloxane) (8.3) occurs by the following mechanism:[1763]

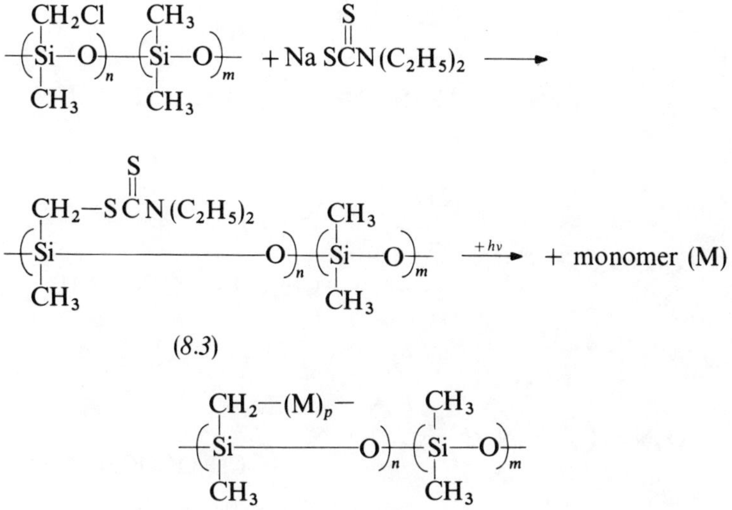

$$(8.3)$$

(grafted polymer)   (8.4)

Surface modification of poly(dimethylsiloxane) typical silicone rubber, is of great importance, particularly in medical applications to influence biocompatibility and thromboresistance.

8.4. PHOTOGRAFTING ONTO CELLULOSE AND ITS DERIVATIVES

Chemical modification of cellulose and cellulosic materials by photografting with suitable monomers improves a variety of properties including tensile strength, resistance to microbiological degradation, abrasion, acid or dye receptivity, wet strength of paper, and adhesion.

Free radicals formed from photolysed photoinitiators ($R\cdot$) can easily abstract hydrogen from hydroxyl groups present in the cellulose backbone, producing cellulose macroradicals (8.4):[226]

$$cell\text{–}OH + R\cdot \rightarrow cell\text{–}O\cdot + RH \qquad (8.5)$$
$$(8.4)$$

A cellulose macroradical can easily initiate graft copolymerization with different monomers (Table 8.2):

$$cell\text{–}O\cdot + monomer\ (M) \rightarrow cell\text{–}O\text{–}(M)_n\text{–} \qquad (8.6)$$

Grafting onto cellulose and its derivatives is affected by wavelength of light,[2145] solvent,[2142] and the photoinitiator used.[2146,2764]

8.5. PHOTOGRAFTING OF ACRYLATED AZO DYES

Several acrylated azo dyes such as:

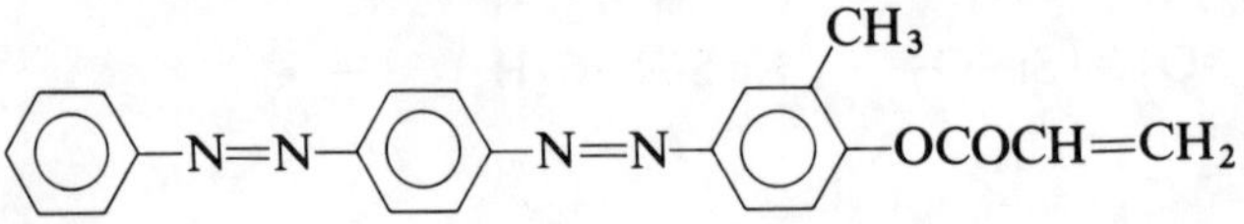

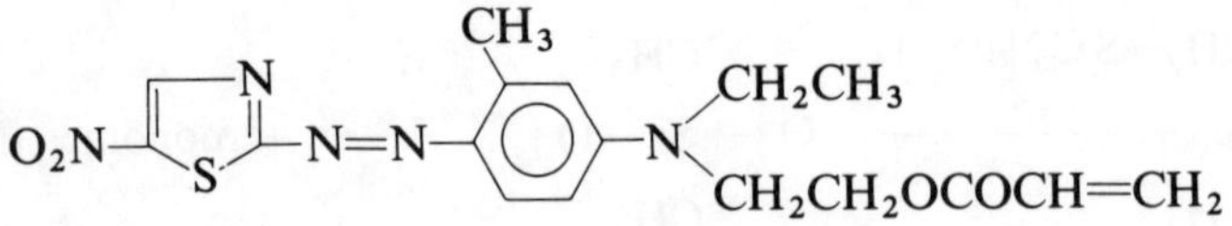

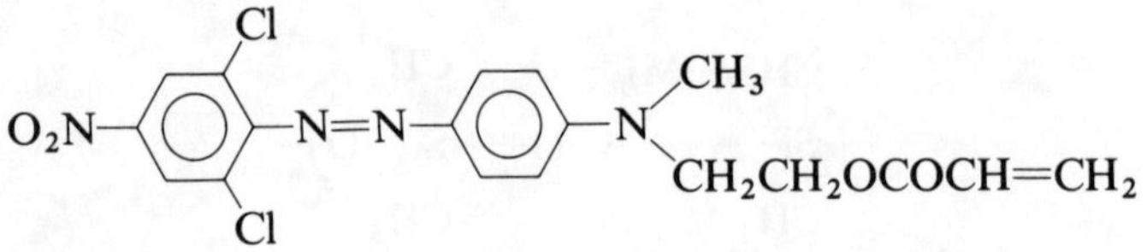

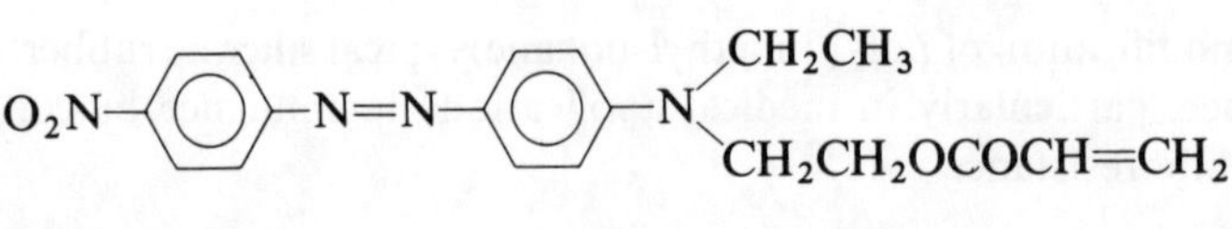

have been successfully grafted on polyamides[386,590], polypropylene,[386, 590, 3288, 3289] poly(ethylene terephthalate),[590] and cellulose.[384]

9. Photocycloaddition reactions

PHOTOCYCLOADDITION (photo-ring forming reaction) of unsaturated monomers (R–X and Y–R) involves a stepwise $(2\pi + 2\pi)$ addition.[989]

(i) Intermolecular photocycloaddition:

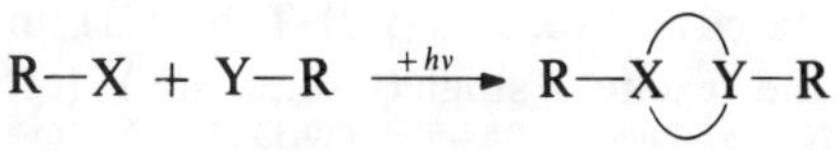

$$R-X + Y-R \xrightarrow{+h\nu} R-XY-R \qquad (9.1)$$

or

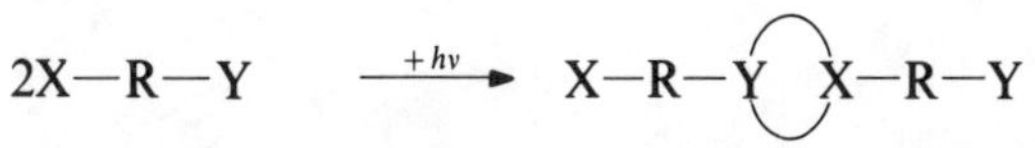

$$2X-R-Y \xrightarrow{+h\nu} X-R-YX-R-Y \qquad (9.2)$$

In contrast to photoinitiated polymerizations, which are chain reactions, photocycloaddition polymerization requires absorption of a photon in each propagation step:

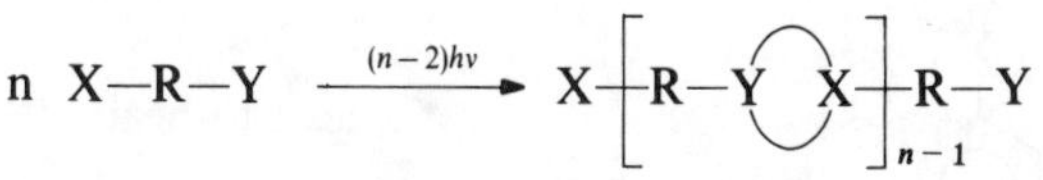

$$n\ X-R-Y \xrightarrow{(n-2)h\nu} X\left[R-YX\right]_{n-1}R-Y \qquad (9.3)$$

(ii) Intramolecular photocycloaddition (where both end groups X and Y are destroyed):

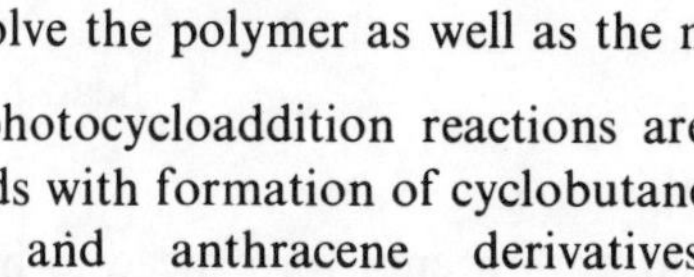

$$X-R-Y \xrightarrow{+h\nu} RXY \qquad (9.4)$$

This process is highly dependent on the length and flexibility of the linkage R. Intramolecular photocycloaddition can occur at any stage of the polymerization, and can involve the polymer as well as the monomer.

Typical examples of photocycloaddition reactions are photodimerization of maleic and fumaric acids with formation of cyclobutane rings,[1408-1410,2932] or 1,3-cyclohexadiene[944] and anthracene derivatives (cf. Section 9.2). Photocycloaddition may also provide formation oligomers and polymers.

9.1. PHOTOCYCLOADDITION OF BENZENE

Irradiation of benzene in the presence of small amounts of cyclooctene (9.1) gives oligomer and polymer (9.2) by photocycloaddition reaction involving the $(2\pi + 3\pi)$ process:[542-544]

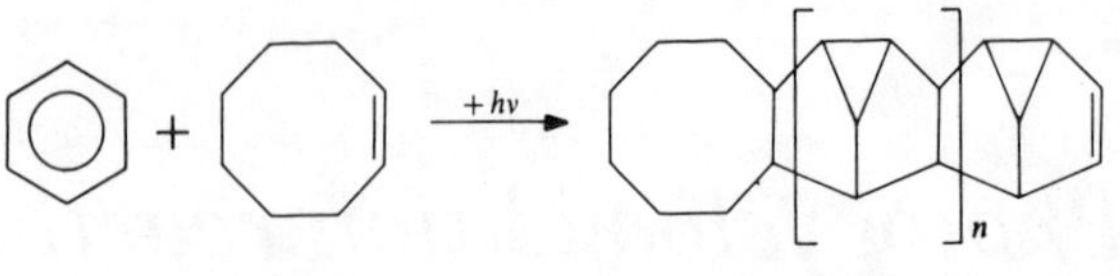

$$(9.5)$$

$$(9.1) \qquad (9.2)$$

A number of intramolecular $(2\pi + 3\pi)$ additions have been reported for substituted benzenes containing olefinic linkages in the side-chain.[1336]

9.2. PHOTOCYCLOADDITION OF ANTHRACENES

Anthracene and substituted anthracenes (9.3) (Table 9.1), under irradiation, form photodimers from the excited singlet state (S_1) (probably via $4\pi + 4\pi$ addition):[499,584,679,736,765,766,956,957,964,970,1104,1403,2664,3375,3565]

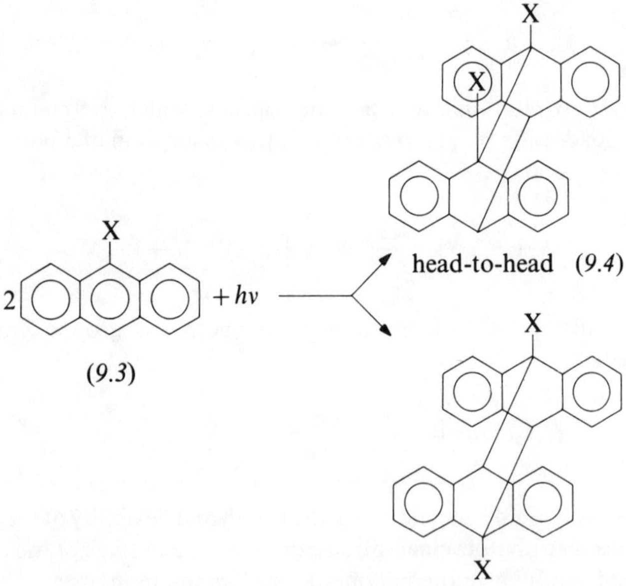

$$(9.6)$$

head-to-head (9.4)

$$2 \text{(9.3)} + h\nu$$

$$(9.7)$$

head-to-tail (9.5)

Most of the 9-substituted anthracenes form head-to-head photodimers (9.4), whereas 9-halogeneanthracenes form head-to-tail photodimers (9.5) (Table 9.1).[2664]

Intermolecular π-electron interaction leads to stabilization of aromatic hydrocarbon dimers of certain configurations with respect to two infinitely isolated monomers.[1551,1824] The condition of stability of these dimers is the occurrence of lowering π-electronic excited states in monomers.[668]

The bis(9-substituted anthracenes) (9.6) polymerizes by photocycloaddition $(4\pi + 4\pi)$:[955,957,959,977,3502,3565,3583,3586,3587]

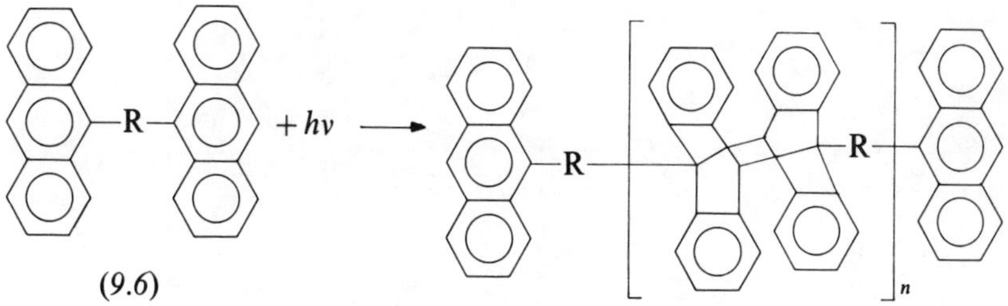

(9.6) (9.8)

Table 9.1 Substituted anthracenes that undergo photodimerization reactions[2664]

Anthracene	Comment
1-Methylanthracene	
2-Methylanthracene	
1-Chloro-4-methylanthracene	
1-Chloroanthracene	
9-Bromoanthracene	Head to tail dimer
9-Bromo-1-chloroanthracene	
Anthracene-9-carboxylic acid	
9-Methylanthracene	
9-Ethylanthracene	
Anthracene-9-carboxaldehyde	Head to head dimer
9-Anthranoic acid, methyl ester	Head to head dimer
9-Hydroxymethylanthracene	Head to head dimer
9-Chloroanthracene	Head to tail dimer
1-Azanthracene	
9-Chloro-1-azanthracene	
2-Phenyl-1-azanthracene	
2-Phenyl-9-chloro-1-azanthracene	
2-Azanthracene	
1,3-Diazanthracene	

Examples of this type of photocycloaddition polymerization is polymerization of bis(anthrylbenzylimidazoles) (9.7) and bis(anthryl–Schiff's base) (9.9) which gives poly(benzylimidazoles) that contain anthracene photodimer (9.8) and (9.10):[2727]

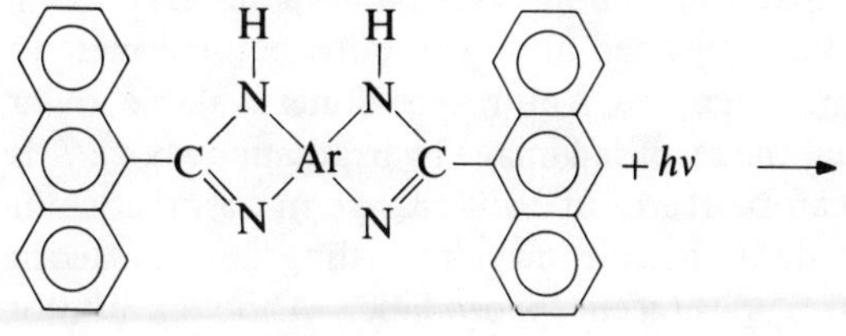

(9.7)

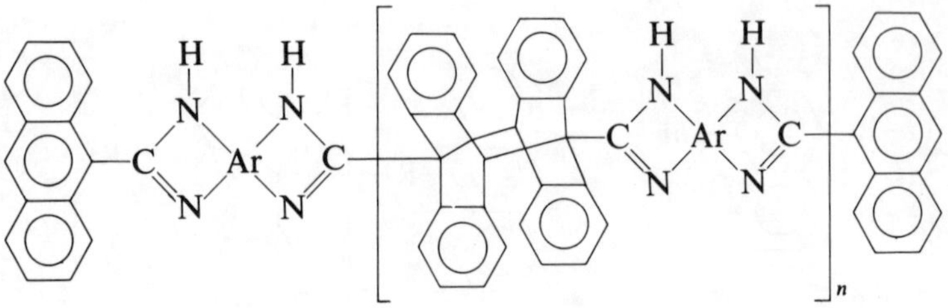

(9.8) (9.9)

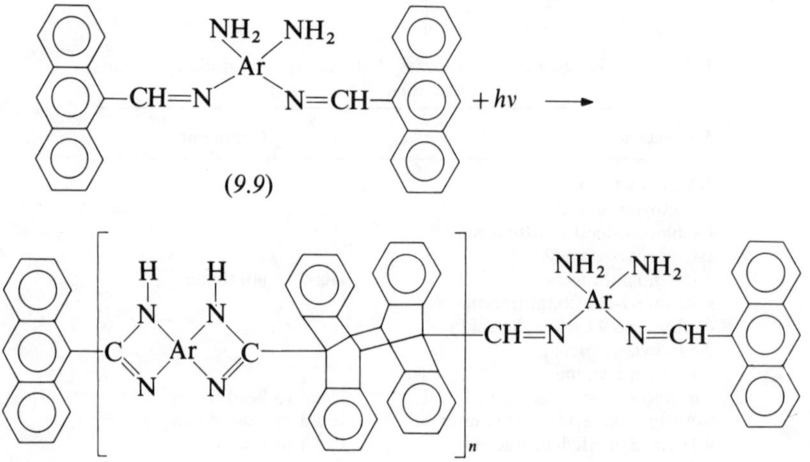

(9.9)

(9.10) (9.10)

where:

Ar =

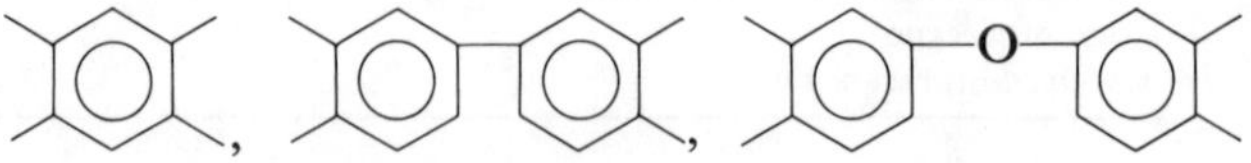

9.2.1. Reversible Photomemory in Polymers

Temperature dependence of the photodimerization of anthryl groups and thermal dissociation at elevated temperatures can be employed in reversible photomemory in polymers operated by dry processes.[676,3566] A process of image recording is conducted between the gloss transition temperature (T_g) and the threshold temperature of the thermal dissociation of the photodimer. By cooling to room temperature below T_g the unreacted anthryl groups become inert to irradiation at the ^1L_a band of anthryl groups. Anthracene dimer is stable under these conditions, which means that the recorded image by irradiation above T_g is fixed by cooling below T_g and can be read out by a change in absorbance or refractive index. To erase the recorded image the image-recording film is heated at 80–100°C to allow dissociation of anthracene photodimer. Although photo-dissociation of the photodimer is also possible, its selective irradiation is not, and recovery of the anthryl groups is much poorer than in thermal processes. The overall processes are presented in Fig. 9.1, and a typical recording is shown in Fig. 9.2. The disadvantages of the polymer with anthryl groups is that the recovery efficiency is 95 % because of the fatigue of repeated recording and erasing, and manipulation in a nitrogen atmosphere is required. A polymer which can have reversible memory must have the functional groups bound to polymer, and these groups have to photodimerize.

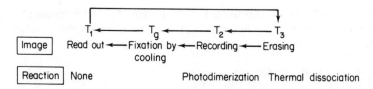

Fig. 9.1 Diagram of processes involved in reversible photomemory in polymers[3566]

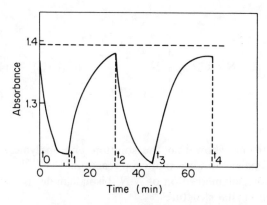

Fig. 9.2 Repetition of photodimerization of polymer-bound anthryl groups and thermal dissociation of the photodimers: t_0-t_1, t_2-t_3 is the light period (dimerization); t_1-t_2 and t_3-t_4 are the dark period (dissociation)[3566]

9.3. PHOTOCYCLOADDITION OF BIS(MALEIMIDES)

Photocycloaddition reactions of N,N'-bis(maleimides) in solution or in crystalline state occurs by a repeated $(2\pi + 2\pi)$ photodimerization process.[457,958,961,968,970,2999,3373]

Upon irradiation of unsubstituted N,N'-polymethylenebismaleimides (9.11) in dilute solution an intramolecular photocycloaddition to diazatetracyclotetracones (9.12) occurs for compounds with chain lengths ranging from three to seven methylene units:

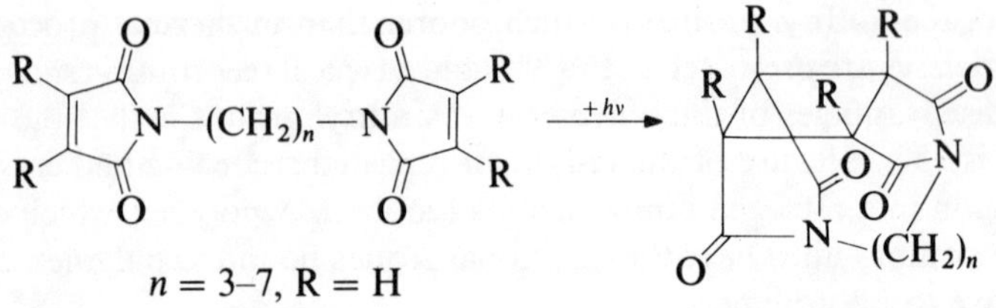

$$n = 3\text{--}7, R = H$$
$$(9.11)$$

$$(9.11)$$

$$(9.12)$$

Upon irradiation of substituted N,N'-polymethylenebismaleimides (9.13) with longer polymethylene chains $(n > 7)$ in dilute solutions photocycloaddition polymerization with formation of cyclobutane rings occurs:

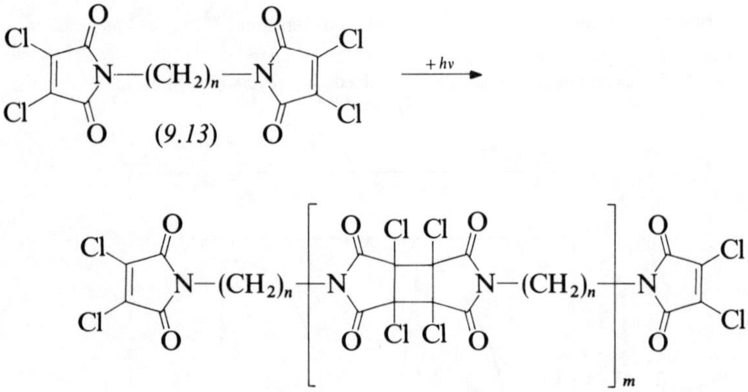

$$(9.13)$$

$$(9.12)$$

Kinetic measurements show that the solution photopolymerization of substituted N,N'-bismaleimides is a multistep reaction.[955,961-963]

Solid-state photopolymerization of N,N'-bismaleimides is controlled by the geometry of their crystal structures.[457]

Polymers with pendant dimethylmaleimide groups (9.14) are highly sensitive photocrosslinkable systems.[457,3373,4009] Crosslinking is achieved by photodimerization of dimethylmaleimide groups according to the reaction:

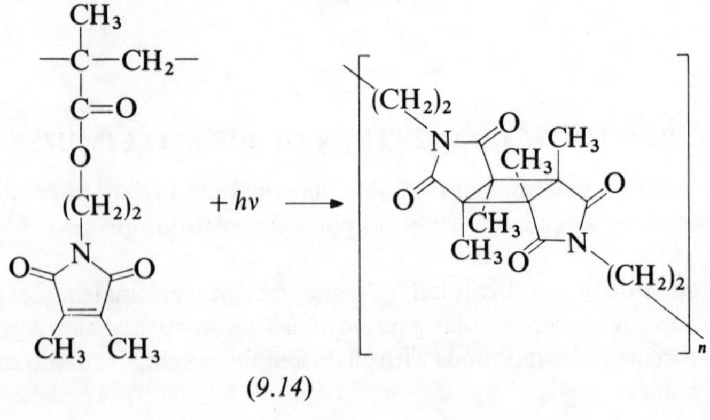

$$(9.14)$$

$$(9.13)$$

This photoreaction can be sensitized effectively by the benzophenone and thioxanthone.

Irradiation of N,N'-bismaleimides in aromatic hydrocarbon solvents in the presence of acetophenone gives copolymers.[1958,2603]

9.4. PHOTOCYCLOADDITION OF DIENONES

Photocycloaddition of dienones, e.g. 4-methyl-4-trichloromethyl-2,5-cyclohexadienone (9.15), upon u.v. irradiation, yields ladder polymers (9.16):[529]

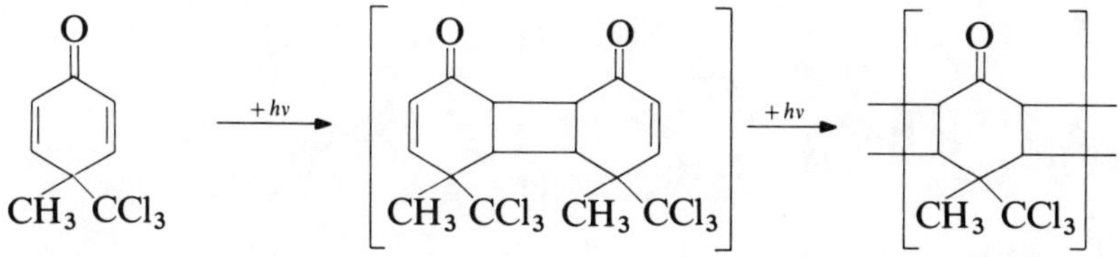

(9.14)

9.5. PHOTOCYCLOADDITION OF BIS(COUMARINES)

Photocycloaddition reaction of bis(coumarines) (9.17) upon u.v. irradiation in the presence of sensitizer (benzophenone) yields polymers (9.18) with molecular weight of approximately 20,000.[957]

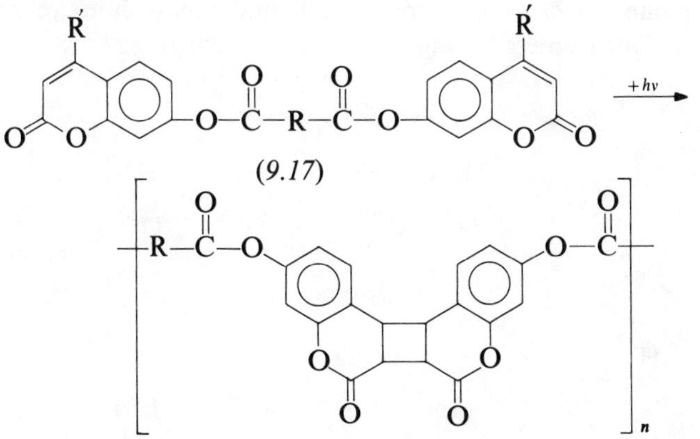

(9.15)

Photodimerization of the coumarine groups takes place after energy transfer from the excited triplet (T_1) of benzophenone.

9.6. PHOTOCYCLOADDITION OF THYMINES

Thymine bases are elements of DNA (deoxyribonucleic acid) (9.19) structure, which absorbs light at 260–265 nm and is photochemically damaged.

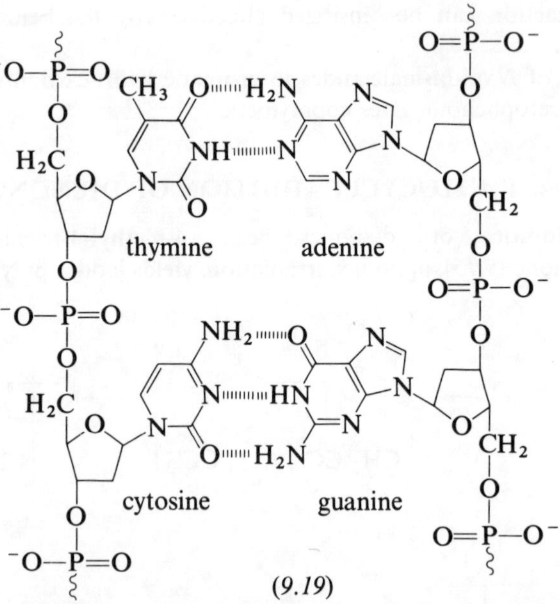

$$(9.19)$$

In DNA the thymine bases are known to interact with neighbouring bases along the polymer chain (stacking interaction), and also with adenine bases in another polymer chain (hydrogen bond interaction) to form double-stranded helical structures.

When thymine (9.20) is u.v. irradiated it undergoes photocyclization with formation of four isomers having a cyclobutane structure:[1109,3739]

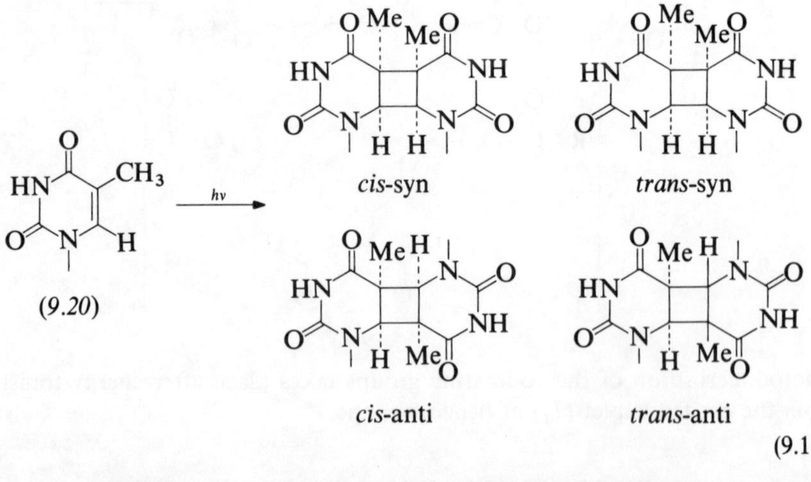

$$(9.16)$$

The mechanism of photodimerization of DNA is even more complicated.[2469,3739,3798] The biological consequences of irradiation are usually ascribed to this type of photoproduct although it is very difficult to relate the killing of cells to the concentration of dimers. However, a wide range of other products are formed, such as hydrates, photoadducts, and crosslinked DNA structures. Photocyloaddition causes mutagenesis in DNA and polynucleotides. The resulting crosslinking of DNA represents a defect in the coding sequence, and hence can be responsible for mutations and cell damage. Which of these reactions is the most important in inducing lethal lesions in the DNA will depend on the particular conditions in which the cell exists while being irradiated.[3799]

Trimethylene-bis(thymine) (9.21) irradiated in a crystal state gives insoluble polymer (9.22):[1174,2265]

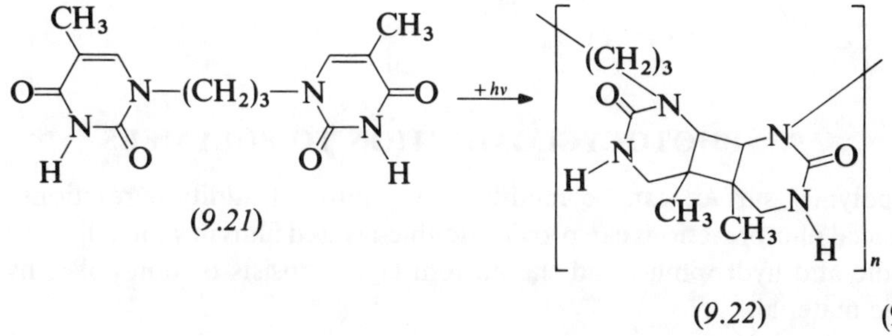

(9.21)　(9.22)　(9.17)

A series of extensive studies on the photocycloaddition of thymine bases covalently bounded as a side-group into polymers (9.23) has been reported:[1757, 2037−2043,2953,3473−3475,3527]

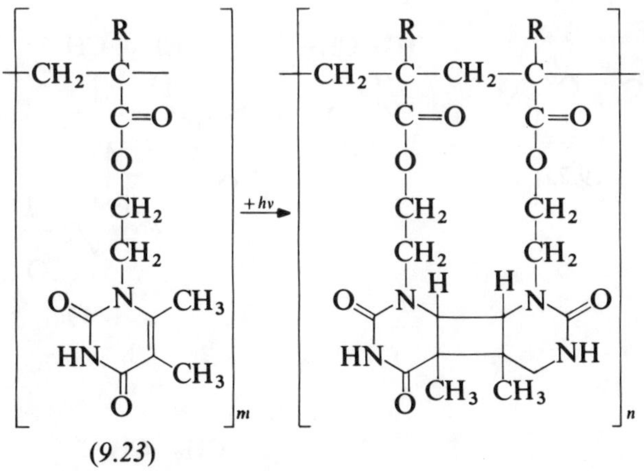

(9.23)　　(9.18)

The quantum efficiency of the photocycloaddition is dependent on the intra-molecular interaction of the thymine bases along the polymer chain, and is greater than those for oligomer model compounds.[2038,2040,2041] Mechanism of photo-dimerization of DNA is complicated.[2469,3739,3798]

Photoinduced dimerization reactions between thymine bases in thymine grafted poly-D- and poly-lysine (9.24) cause the change in helix content.[3475]

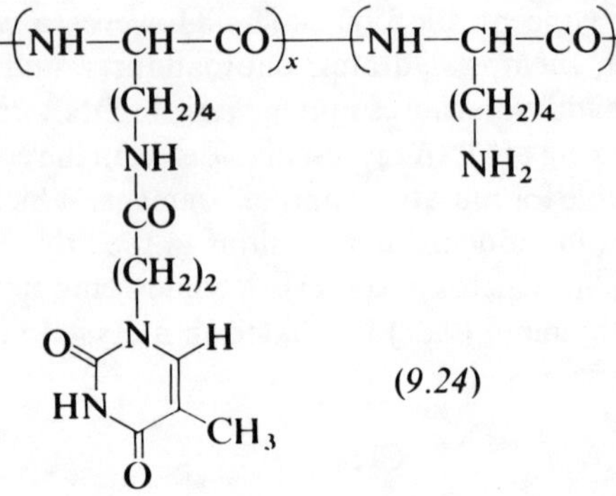

$$\begin{array}{c}
\text{(NH--CH--CO)}_x \quad \text{(NH--CH--CO)}_y \\
\quad\quad | \quad\quad\quad\quad\quad\quad | \\
\quad\quad (CH_2)_4 \quad\quad\quad\quad (CH_2)_4 \\
\quad\quad | \quad\quad\quad\quad\quad\quad | \\
\quad\quad NH \quad\quad\quad\quad\quad NH_2 \\
\quad\quad | \\
\quad\quad CO \\
\quad\quad | \\
\quad\quad (CH_2)_2
\end{array}$$

(9.24)

9.7. PHOTOCYCLOADDITION TO POLYMERS

The polymer surface can be modified by photocycloaddition reactions. The interfacial photoreactions can provide double-layered films in which the top layer is acidic and hydrophilic, and the bottom layer consists of non-polar, hydrophobic material.

9.7.1. Photocycloaddition to Polystyrene

Maleic anhydride (9.25) is photoadded to polystyrene only in chloroform and carbon tetrachloride in the presence of a large excess of acetophenone (sensitizer):[3229]

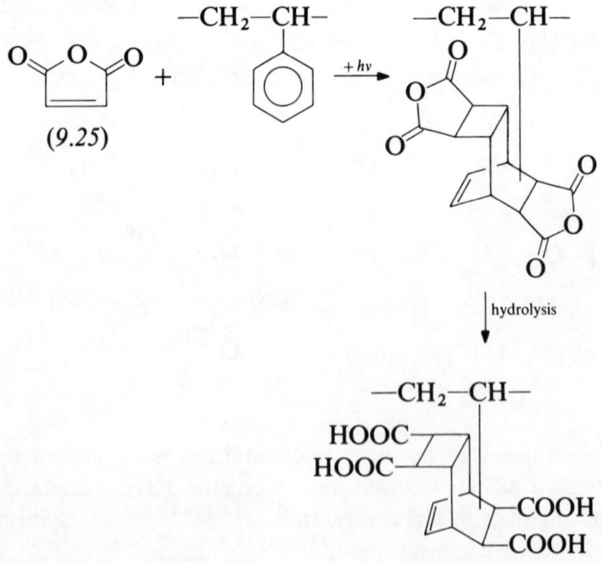

(9.19)

No photoaddition has been observed in cyclohexane.

Cis-cyclooctene (*9.26*) reacts with polystyrene under u.v. irradiation by a 1,3-cycloaddition process, differing in the position of attack on the phenyl ring:[369]

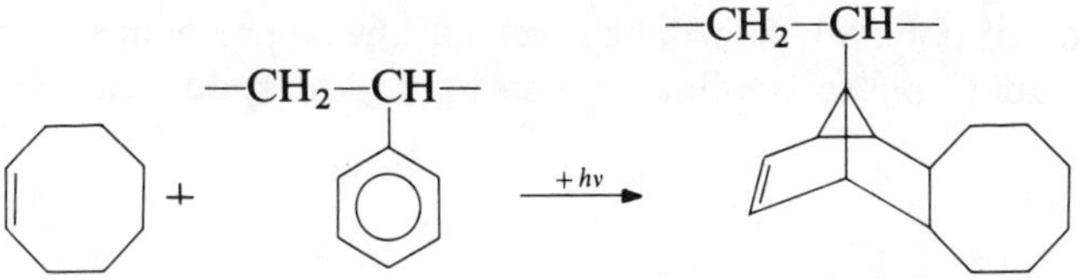

(9.26)   (9.20)

There are different ortho, meta and para positions of 1,3-cycloaddition. The final product formation is influenced by the ionization potential and the electron affinity of the aromatic and the olefinic component relative each other.

9.7.2. Photocycloaddition to Poly(vinylbenzophenone)

A variety of olefins such as isobutene, 2-methyl-2-butene, 2-methyl-2-hexane, and 2,3-dimethyl-2-butene undergo photocycloaddition to poly(4-vinylbenzophenone) and co(styrene/4-vinylbenzophenone) (*9.27*) in benzene solution or in a solid phase to produce polymers whose side-chains contains oxetane rings (*9.28*):[1615]

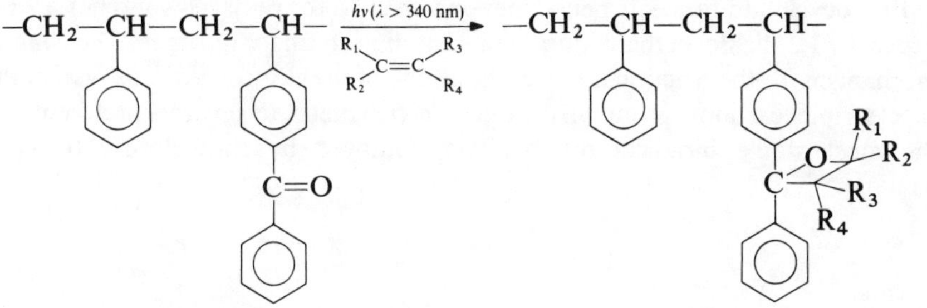

(9.27)   (9.28)   (9.21)

Co(styrene/4-vinylbenzophenone) (*9.27*) reacts with furan (*9.29*) or benzofuran (*9.30*) under u.v. irradiation producing oxetane rings:[3231]

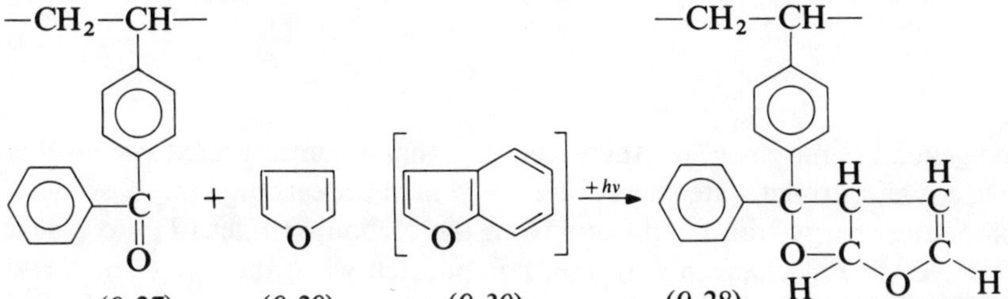

(9.27)   (9.29)   (9.30)   (9.28)

(9.22)

The cycloaddition proceeds by attack of the electron-deficient oxygen of the (n,π^*) triplet state of the aromatic ketone on an electron-rich olefin, followed by ring closure to give an acid-sensitive oxetane ring.

Irradiation of poly(vinylbenzophenone) and the isoprene 'hexamer' squalene leads to rapid photocrosslinking via photocycloaddition and to gel formation.[1615]

9.7.3. Photocycloaddition to Polydienes

Maleic anhydride (9.31) and its derivatives such as 2,3-dichloromaleic anhydride, 2,3-dimethylmaleic anhydride, and dimethyl maleate, photoadd to several polydienes such as poly(1,4-butadiene) (9.32), poly(1,2-butadiene) and poly-isoprene in homogeneous-solution and solid polymer–vapour solution:[1668,1694,3570] .

$$
\begin{array}{ccc}
\underset{(9.31)}{\text{O}\diagdown\text{O}\diagdown\text{O}} & + \quad -\text{CH}_2-\text{CH}=\text{CH}-\text{CH}_2- \quad \xrightarrow{+h\nu} & \underset{(9.23)}{\overset{\overset{\textstyle \text{O}\diagdown\text{O}\diagdown\text{O}}{\overset{\textstyle \text{CH}-\text{CH}}{|\quad\quad|}}}{-\text{CH}_2-\text{CH}-\text{CH}-\text{CH}_2-}}
\end{array}
$$

$$(9.31) \qquad\qquad (9.32) \qquad\qquad\qquad\qquad\qquad (9.23)$$

Photocycloaddition of benzophenone (9.33) to poly(cis-isoprene) (9.34) occurs in 1,2-dichloroethane during u.v. irradiation under nitrogen. The general mechanism for the reaction involves the attack of the electrophyllic oxygen of the (n,π^*) triplet carbonyl group on the electron-rich olefin, to generate preferentially the most stable biradical intermediate, followed by ring closure to yield oxetane:[2687]

$$
\begin{array}{ccc}
\underset{(9.33)}{\text{O}=\text{C}} & + \quad \underset{(9.34)}{-\text{CH}_2-\overset{\overset{\textstyle \text{CH}_3}{|}}{\text{C}}=\text{CH}-\text{CH}_2-} \quad \xrightarrow{+h\nu} &
\end{array}
$$

$$(9.24)$$

The general requirement for this reaction is that the unsaturated system should have a higher triplet state energy level (E_T) than the carbonyl (n,π^*) so that the competing energy transfer process from the carbonyl triplet (T_1) to alkene is inhibited. The mechanism is further complicated when the rate of intersystem crossing from the excited singlet to the triplet is slow in comparison to the rate of addition. In this case the (n,π^*) singlet (S_1) species initiates attack on the system.

9.7.4. Photocycloaddition to Unsaturated Polyesters

Photocycloaddition of phenanthrene (*9.35*) to unsaturated polyesters of fumaric acid (*9.36*) in the presence of sensitizer (thioxanthone) occurs according to the reaction:[1613]

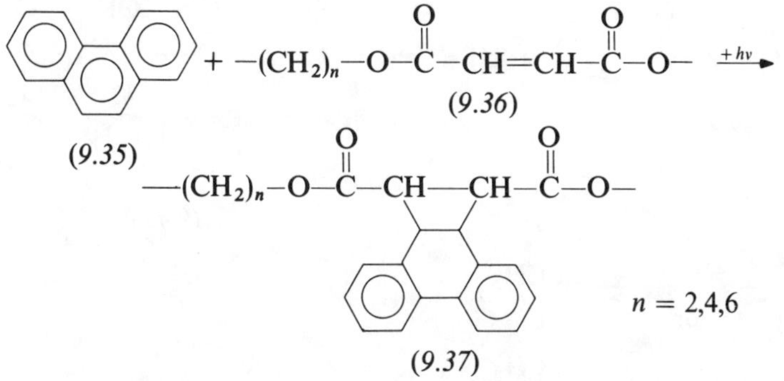

$$(9.25)$$

$$n = 2,4,6$$

(*9.37*)

Cycloaddition proceeds in a variety of solvents; however a high degree of conversion can be obtained only in solvents not possessing readily abstractable hydrogen. In dioxan, tetrahydrofuran, and chloroform, hydrogen abstraction from the solvent by the triplet excited sensitizer (T_1) competes favourably with energy transfer to phenanthrene. As a result, radical chain addition of solvent molecules to the double bonds of the polyester was competitive with the cycloaddition to phenanthrene.

The properties of the polymer cycloadducts (*9.37*) vary considerably, depending on whether a sensitizer is used. Adducts prepared in the unsensitized reaction are much less stable, turning bright yellow–orange on prolonged storage or on brief treatment with methanolic HCl. The instability is attributed to the formation of minor amounts of the oxetane adduct, whose mode of degradation, by analogy with the small-molecule hydrolysis, causes chain scission and regeneration of the phenanthrene.

Diazofluorenone (*9.38*) reacts with unsaturated polyesters (*9.39*) under u.v. radiation at 4–6°C, forming a spirocyclopropane group (*9.40*):[3229]

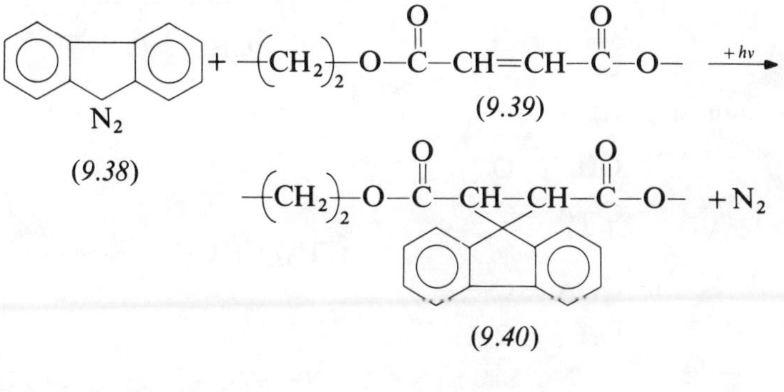

(*9.40*)

$$(9.26)$$

9.8. STEP-GROWTH PHOTOPOLYMERIZATION

Diketones (9.41) in hydrogen donor solvents (RH) (e.g. isopropanol) undergo step-growth photopolymerization to poly(benzopinacols) (9.42):[955,972,974-976,1588,2448,2912]

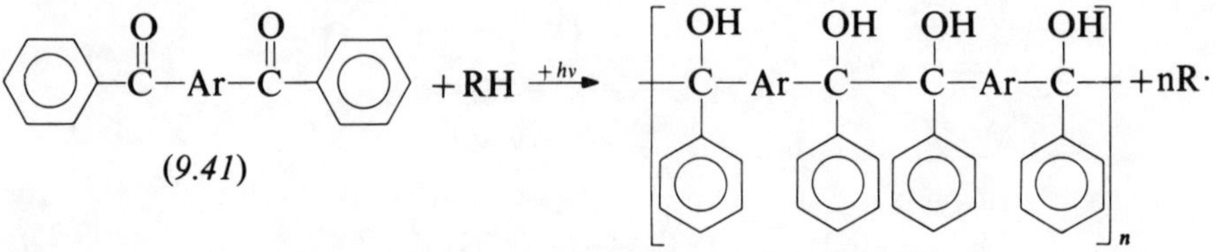

$$
\text{(9.41)} + RH \xrightarrow{+h\nu} \text{(9.42)} + nR\cdot \qquad (9.27)
$$

(9.41)

(9.42)   (9.27)

$$
Ar = \text{(phenyl)} \quad or \quad \text{(phenyl)}-\underset{CH_3}{\overset{CH_3}{C}}-\text{(phenyl)}
$$

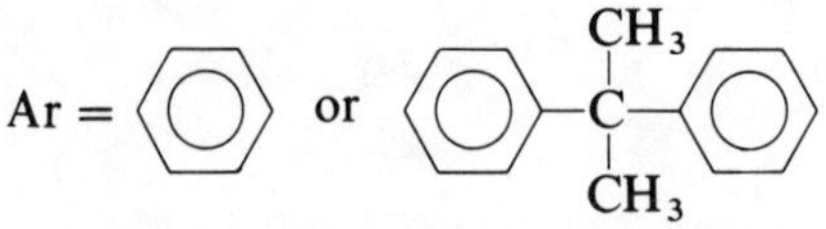

The following step-growth mechanism has been proposed:

(i) Initiation step (I):

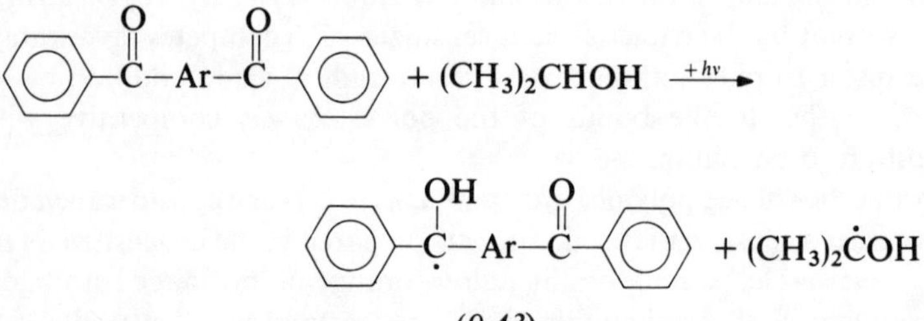

$$
\text{(diketone)} + (CH_3)_2CHOH \xrightarrow{+h\nu}
$$

$$
\text{(9.43)} + (CH_3)_2\dot{C}OH \qquad (9.28)
$$

(9.43)

Two ketyl radicals (9.43) react each other:

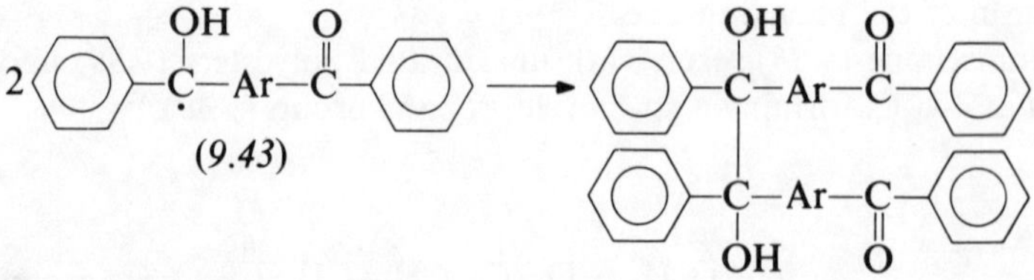

$$
2 \; \text{(9.43)} \longrightarrow \text{(dimer product)}
$$

(9.43)

(ii) Initiation step (II):

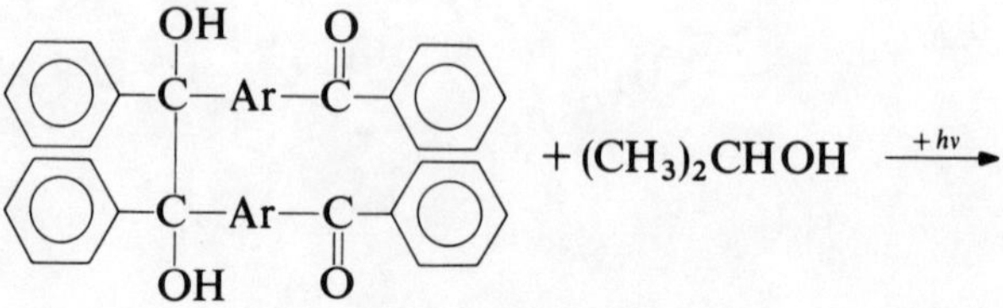

$$
\text{(dimer)} + (CH_3)_2CHOH \xrightarrow{+h\nu}
$$

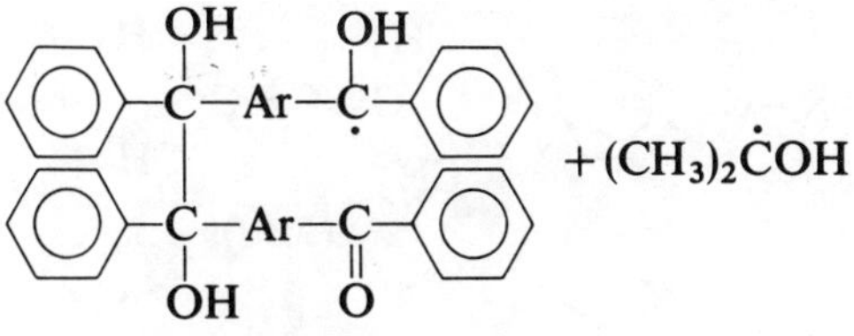

$$\text{(9.44)} \qquad + (CH_3)_2\dot{C}OH$$

(9.29)

Two ketyl radicals (*9.44*) react each other:
(iii) Initiation step (III) occurs analogously to step I or II.

Each following initiation step requires absorption of a photon and recombination of two ketyl radicals formed.

A similar mechanism can be written for step-growth polymerization of hexamethylene bis(ketotriazoles) (*9.45*):[973]

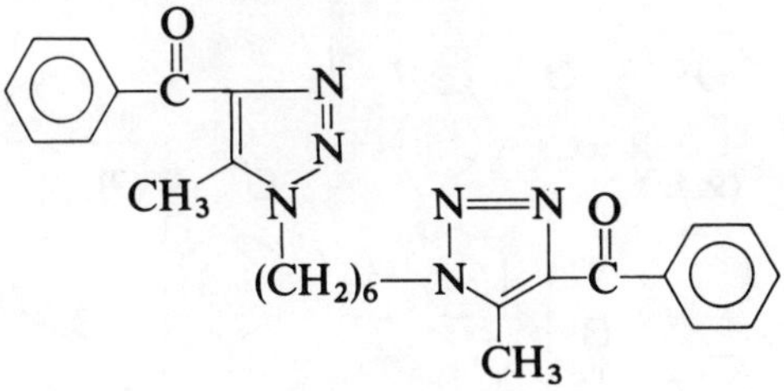

(*9.45*)

Dibenzaldiimines (*9.46*) in alcohol solution in the presence of benzophenone photopolymerize by a step-growth mechanism:[350]

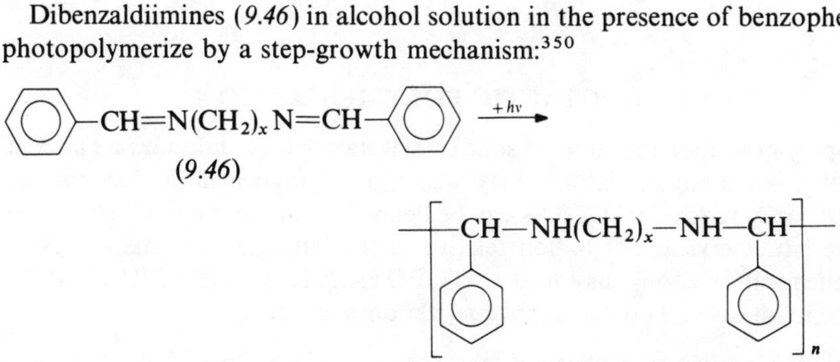

$$\text{(9.46)} \qquad \text{—CH=N(CH}_2)_x\text{N=CH—} \xrightarrow{+h\nu}$$

$$\left[\text{—CH—NH(CH}_2)_x\text{—NH—CH—}\right]_n$$

(9.30)

Photopolymerization of dibenzylbenzenes with metha-dibenzoyl benzene occurs by a similar step-growth mechanism.[1090]

9.8.1. Photocycloaddition of Diketones to Double Bonds

Several series of polymers were prepared in which the chain-growth step was formation of an oxetane ring by photocycloaddition of carbon–carbon double bond to carbonyl group, e.g. bis(benzophenones) (*9.47*) and tetramethyl allene (*9.48*):[198]

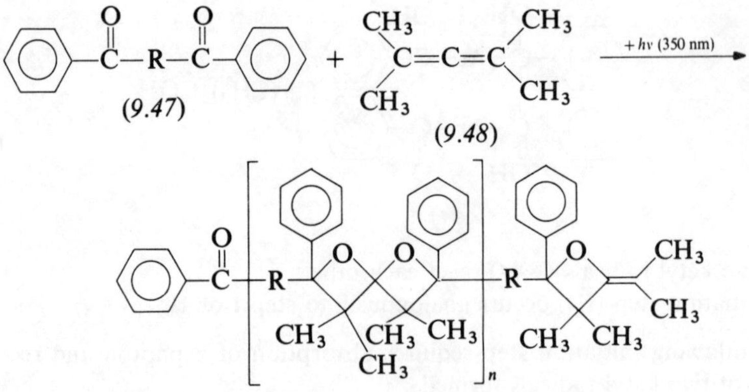

$$(9.31)$$

Another example of this type of reaction is photocycloaddition of bis(benzophenones) (9.47) and furane (9.49) and its derivatives:[199]

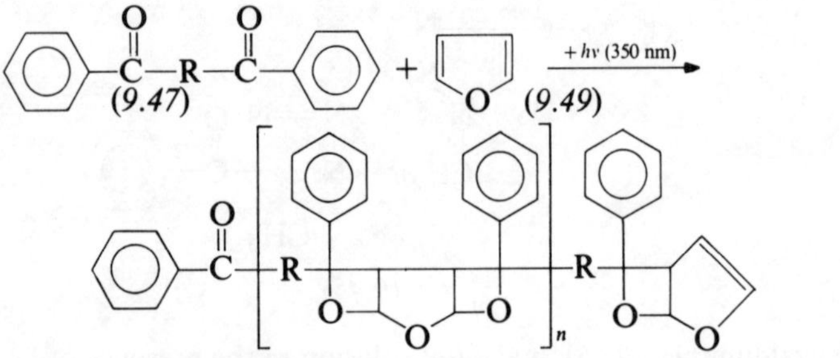

$$(9.32)$$

9.9. TOPOTACTIC POLYMERIZATION

Photopolymerization reaction in a solid crystal state, where molecules are packed in a three-dimensional lattice, may occur as a topochemical reaction. A TOPOCHEMICAL REACTION can be defined as one where products have a definite lattice crystal orientation relative to the lattice of the parent crystal. Topochemical reactions may lead to TOPOTACTIC POLYMERIZATION, but not in all cases. Topotactic polymerization may occur:

(i) Inside a perfect monomer crystal.
(ii) At well-defined sites in crystals, such as packing sequence faults or at dislocations. Lattice defects may play an active role in topochemical reactions by:[515,3831]

   (a) Changing the topology of the molecules localized along the dislocation line and in the stress field of the dislocation as compared to the molecules in the perfect parts of the crystals.
   (b) Preferred trapping of the photoexcitation at or near the defect sites, thus providing a higher probability of occurrence of a chemical event at these sites as compared to the perfect crystal.

(iii) Inside a composite structure of channel inclusion compounds (clathrates), where mobility of host and guest molecules is strictly restricted.

(iv) Along an array of molecules attached or absorbed in a well-defined manner to a rigid molecule in solution (MATRIX POLYMERIZATION).

Because of the requirements for precise geometry and close approach of the reactive or functional groups, only certain crystal forms of the various monomers may undergo topotactic photopolymerization.[1517,1518]

9.9.1. Four-centre Type Photopolymerization

The solid state photopolymerization of monomers containing two conjugate olefinic double bonds (9.50) (Table 9.2) occurs as a topotactic lattice-controlled (FOUR-CENTRE TYPE) polymerization:

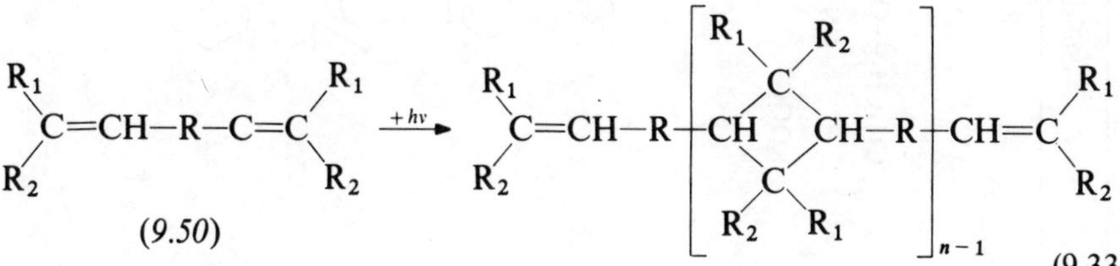

$$(9.33)$$

$$(9.50)$$

Typical examples of such a reaction are photopolymerization of 2,5-distyryl-pyrazine (9.51)[1520,1523,1526,2508,2629,3497] or trans,trans-1,4-bis (β-pyridil-2-vinyl)benzene (9.52)[1745]:

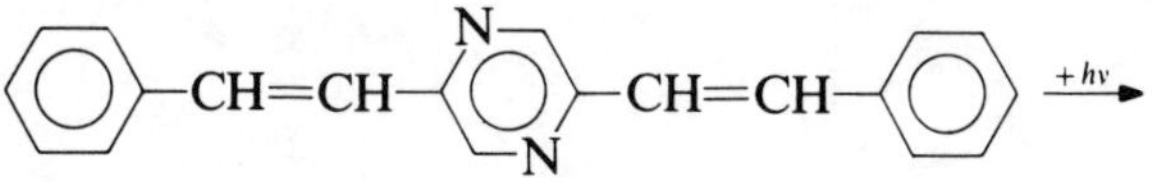

$$(9.51)$$

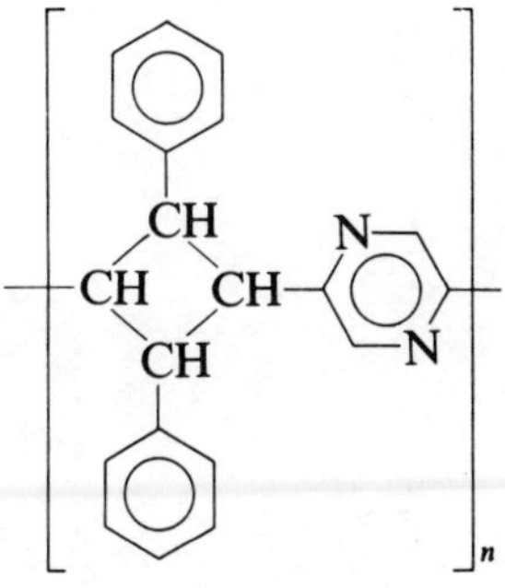

$$(9.34)$$

Table 9.2 Examples of diolefin monomers which undergo four-centre type photopolymerization

Name	Structure	Substituent (R)	References
Bis(β-substituted-vinyl)benzenes	 R—CH=CH——CH=CH—R	R'OOC—	1516, 1517, 1522, 1525, 2700, 3165, 3497
		CH₃CH₂—OCO—C=CH— (CN)	1517, 2647
		ROOC—C=CH— (CN)	2628, 2642
		—NO₂	1516
		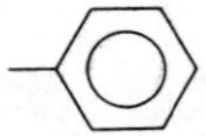	1524, 3504
		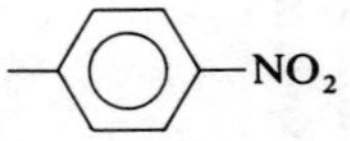	1516, 1517
		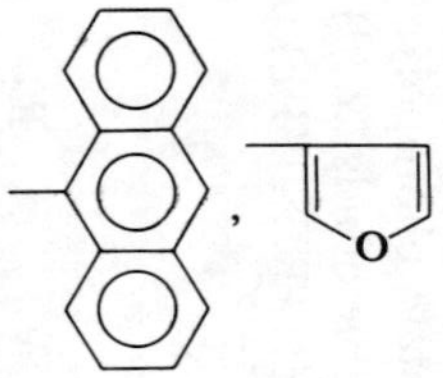	
		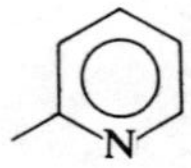	337, 1516, 1524, 1526, 1748, 2644, 3533

Compound	Structure	Substituent	References
	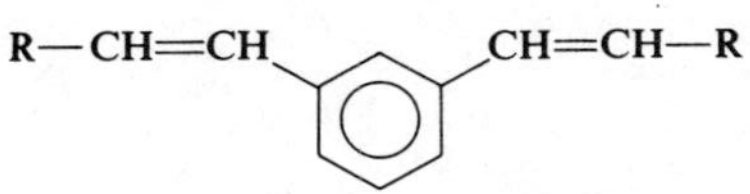 $R-CH=CH-$ (m-phenylene) $-CH=CH-R$	$R'OOC-$	1516, 2539, 2630, 2631, 2638, 3497
β-Cyano-, and bis(β-cyano-β-substituted-vinyl)benzene	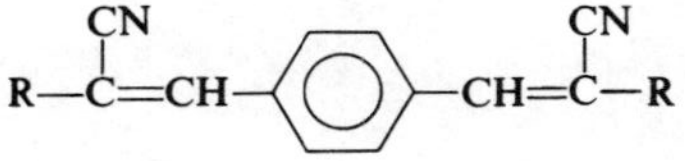 $R-\underset{CN}{C}=CH-$ (p-phenylene) $-CH=\underset{CN}{C}-R$	$-COOCH_3$	2637, 2641
		$-COOCH\underset{CH_3}{\overset{CH_3}{\big<}}$	26, 27, 2629, 2633, 3503, 3736
Bis(α-cyano- β-arylvinyl)benzene	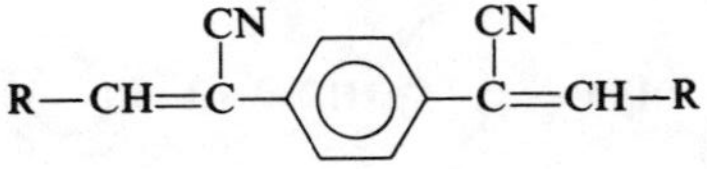 $R-\underset{CN}{C}=CH-$ (p-phenylene) $-CH=\underset{CN}{C}-R$	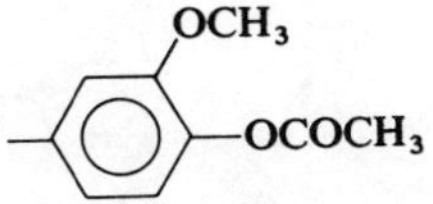 $-OCO-$ (dimethoxy/acetoxy phenyl: OCH_3, $OCOCH_3$)	1625
Bis(β-substituted-vinyl)naphthalene	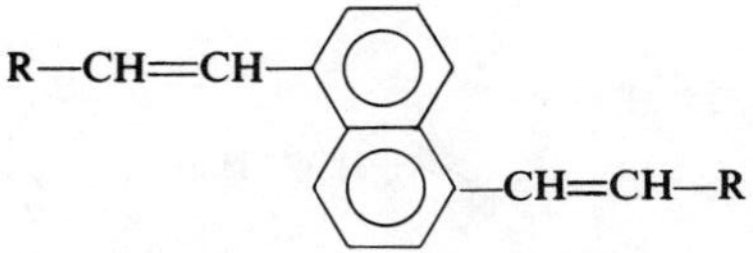 $R-CH=CH-$ (naphthalene) $-CH=CH-R$	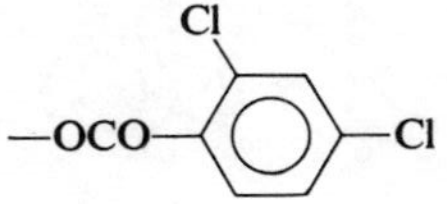 $-OCO-$ (2,4-dichlorophenyl, Cl, Cl)	1064, 2508
Bis(cinnamate)esters	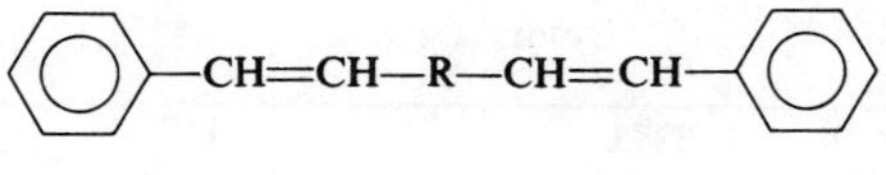 $\phi-CH=CH-R-CH=CH-\phi$	$-OCO-(CH_2)_3-OCO-$	2533, 2632
Bis(cinnamate)amide		$-CONH(CH_2)_6NHCO-$	2632, 3520
Tetra(cinnamate)ester of the pentaerythritol	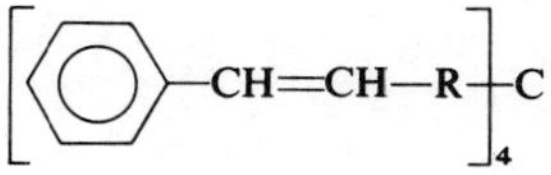 $[\phi-CH=CH-R-]_4 C$	$-OCO-$	827, 1517, 2637

Table 9.2 (Contd.)

Name	Structure	Substituent (R)	References
Bis(styryl)pyrazine	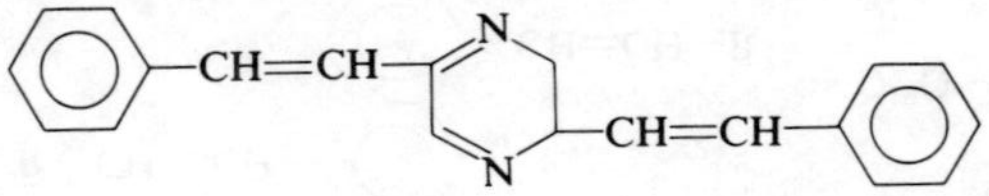		515, 1189, 1516–1521, 1523, 1524, 1526, 1527, 1748, 1951, 2507, 2508, 2632–2637, 2639, 2640, 2643, 2644, 2646, 3185, 3503, 3504, 3533, 3824, 3831
Bis(styryl)-s-triazine	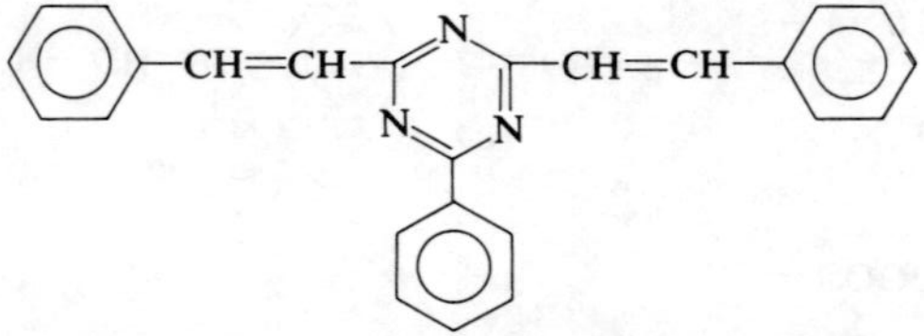		555

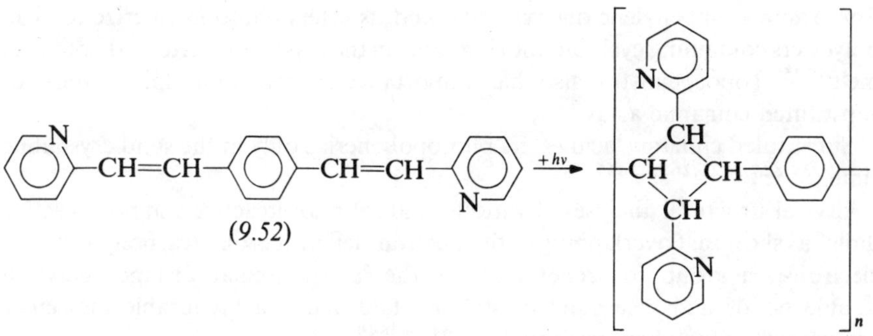

$$\text{(9.52)} \quad \xrightarrow{+h\nu} \quad \text{(9.35)}$$

Four-centre photopolymerization in the crystalline state is a typical TOPOTACTIC POLYMERIZATION.[735,1517,2508,2632,2633,2637,2638,3824]

The fact that isomeric compounds of (9.51) i.e. 1,4-bis(β-pyridyl-3-vinyl)benzene and of (9.52), i.e. 1,4-bis(β-pyridil-4-vinyl)benzene do not photopolymerize indicates the importance of the topochemical arrangement of the double bonds.[1748]

TOPOTAXY is a definite orientation of a polymer crystal with respect to the monomer crystal. The directions of the three axes of the polymer crystal coincide with those of the monomer crystal and the space group of the polymer also agrees with that of monomer. Reactivity of the system is controlled by packing of monomers in a crystalline state[515,735,1517,2508,3831] which may occur by the following mechanism:

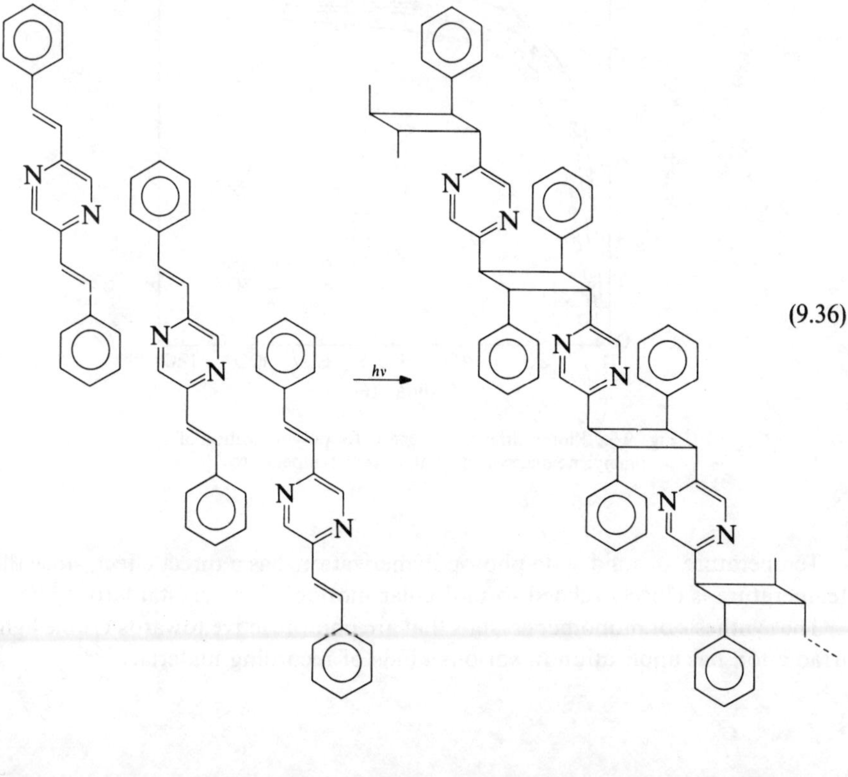

$$\xrightarrow{h\nu} \quad \text{(9.36)}$$

For example p-phenylene diacrylic acid and its esters photopolymerize to yield polyesters containing cyclobutane rings only in the crystalline state, and not in the melt.[1525] Topochemistry also has importance in the photodimerization of substituted cinnamic acids.[737]

Substituted cinnamic acid esters photopolymerize only in the solid crystalline state.[737,827,2533,2632,2637]

Crystal structural analyses of various kinds of photoreactive and non-reactive diolefins show that overlapping of the electron-deficient group (carbonyl) and the electron-rich group (benzene) shortens the intermolecular distance between double bonds and in a solid crystalline state makes a favourable molecular arrangement for polymerization.[2645,2647-2649]

Photopolymerization is usually carried out by dispersing pure monomer crystals in dispersant by stirring with a magnetic stirrer at constant speed. Dispersant is required to provide homogeneous irradiation of the crystal surface.

In the case of p-phenylene diacrylic acids it has been shown that conversion of monomer into polymer occurs rapidly at the initial stage of polymerization and then gradually with irradiation time (Fig. 9.3).[1517,2638] Additives, e.g. photoinitiators, are not involved in the crystalline state of photopolymerization. The surrounding atmosphere generally does not have any significant effect on polymerization kinetics.

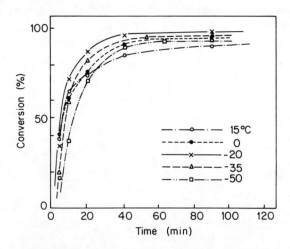

Fig. 9.3 Plot of time vs conversion for polymerization of p-phenylene diacrylic acid at various temperatures[1517]

Temperature, in solid state photopolymerization, has a direct effect, since the temperature is closely related to molecular motion of the crystal lattice.[2638]

The synthesis of monomer crystals that are photoreactive towards visible light irradiation has application in various kinds of recording materials.[2539]

9.9.2. A Photochemical Method to Immobilize Enzymes

Poly(vinyl alcohol) substituted by stilbazolium groups (9.53) shows very high photosensitivity and crosslinks rapidly upon irradiation.[1738,1742,1743] Irradiation of this water-soluble photosensitive polymer yields insolubilized material (9.54) in which the enzymes such as invertase, glucoamylase, or catalase, are entrapped:

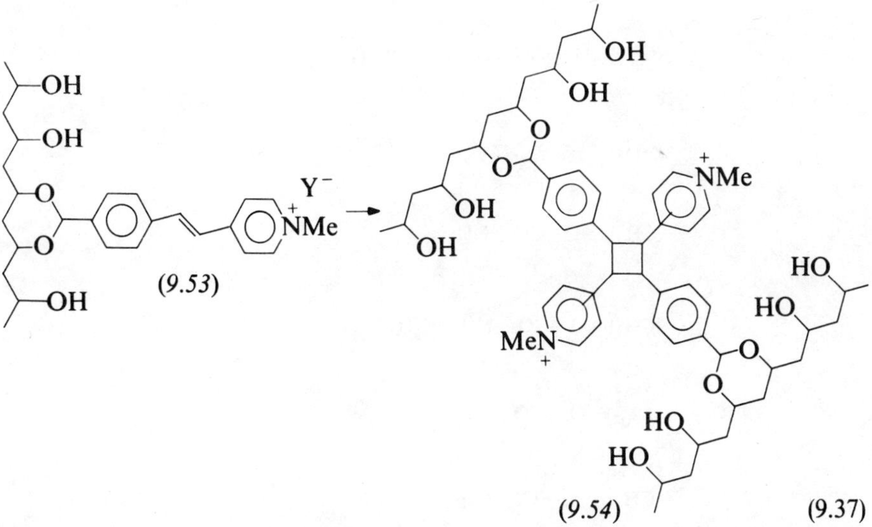

(9.53)   →   (9.54)   (9.37)

Cationic sides of stilbazolium groups immobilize enzymes by coulombic interactions, whereas the crosslinked polymer matrix serves for inclusion of enzyme.[1739] This method requires no covalent bond formation between the matrix and enzyme molecules, and thus maintains the native properties of enzymes.

9.9.3. Photopolymerization of Diacetylenes

The solid state-photopolymerization of diacetylene (9.55) crystals occurs as a topotactic (lattice-controlled) polymerization.[3828, 3829]

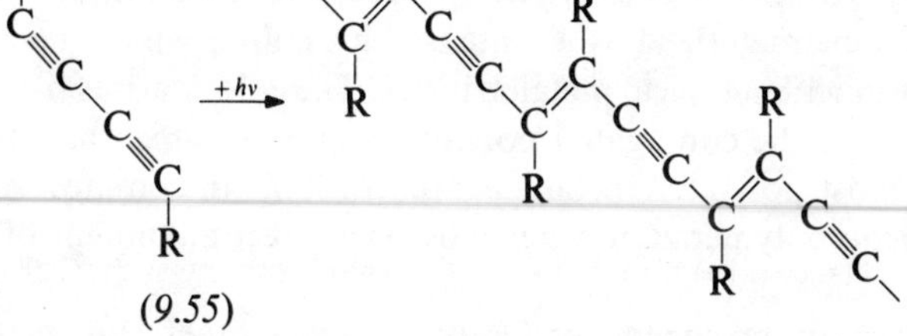

(9.55) $+ h\nu$ →   (9.38)

A list of some substituents (R) is given in Table 9.3.

Table 9.3   List of structure formulae of substituents for
the diacetylenes treated in the text

Substituent (R)

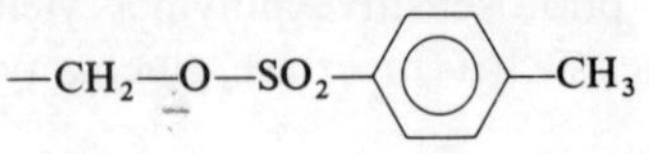

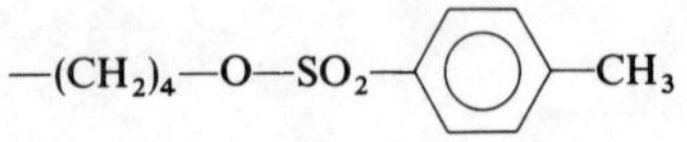

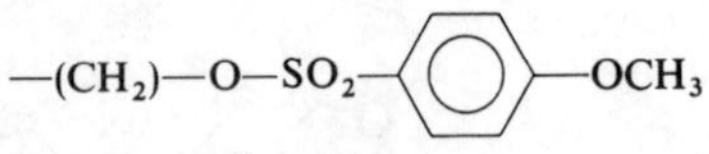

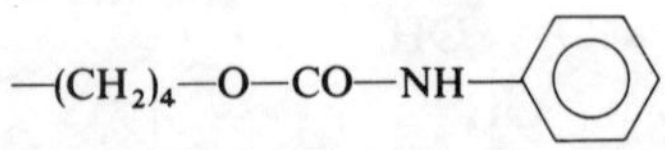

$-CH_2-O-SO_2-\langle\bigcirc\rangle-CH_3$

$-(CH_2)_4-O-SO_2-\langle\bigcirc\rangle-CH_3$

$-(CH_2)-O-SO_2-\langle\bigcirc\rangle-OCH_3$

$-(CH_2)_4-O-CO-NH-\langle\bigcirc\rangle$

$-(CH_2)-O-CO-NH-C_2H_5$

$-(CH_2)-O-CO-NH-\langle\bigcirc\rangle$

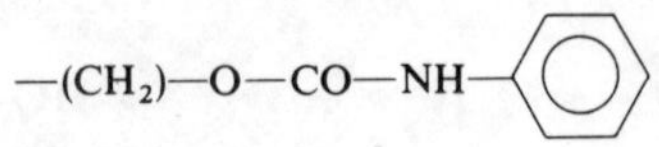

$-(CH_2)_4-O-CO-NH-CH_3$

$-(CH_2)_n-O-CO-NH-CH_2-CO-O-(C_4H_9)$

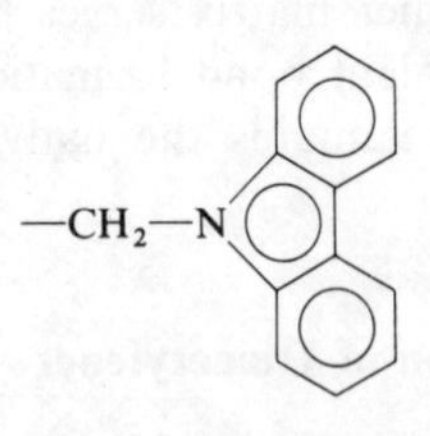

$-CH_2-N\langle\text{carbazole}\rangle$

$R_1: -(CH_2)_9-CH_3$
$R_2: -(CH_2)_8-COOH$

Photopolymerization of diacetylenes proceeds as 1,4-addition to produce a product initially characterized by an intense blue colour which then relaxes to a dark red or green with a typical metallic lustre. The colour arises from the lowest (π,π^*) transition of the conjugated polymer backbone which has its maximum near 600 nm. Most polymeric products are insoluble in common solvents.

This topotactic polymerization depends on correct alignment of the mono-meric units.[327, 453, 664, 1017, 1063, 1065, 1475, 1991, 2547, 2995, 3358, 3825, 3830, 3832] The individual chains grow independently from each other, starting from points randomly distributed throughout the lattice (Fig. 9.4).

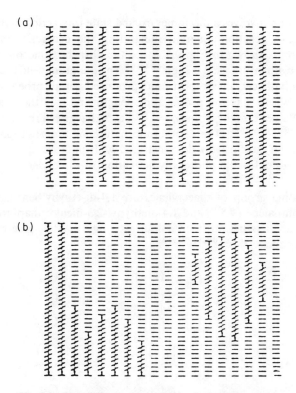

Fig. 9.4 Two different mechanisms of phase transformation in solid-state polymerization: (a) homogeneous reaction in the form of a solid solution, (b) heterogeneous growth by nucleation[3830]

Photopolymerization of bis(*p*-toluene sulphonate) of 2,4-hexadiene-1,6-di-ol (*9.56*), upon u.v. irradiation, is fast and does not show an induction period.[3827] Monomer (*9.56*) can be obtained in high purity in the form of large single crystals by slow evaporation of a concentrated solution in acetone.

$$CH_3\!-\!\langle O \rangle\!-\!\overset{\displaystyle O}{\underset{\displaystyle O}{\overset{\|}{\underset{\|}{S}}}}\!-\!OCH_2\!-\!C\!\equiv\!C\!-\!C\!\equiv\!C\!-\!CH_2O\!-\!\overset{\displaystyle O}{\underset{\displaystyle O}{\overset{\|}{\underset{\|}{S}}}}\!-\!\langle O \rangle\!-\!CH_3$$

(*9.56*)

The photopolymerization process is linear in light intensity and involves formation of the biradical dimer intermediate which has a butatriene structure (*9.57*):

$$\overset{R}{\underset{|}{}}\quad\overset{R}{\underset{|}{}}\,\overset{R}{\underset{|}{}}\quad\overset{R}{\underset{|}{}}$$
$$\cdot C\!=\!C\!=\!C\!=\!C\!-\!C\!=\!C\!=\!C\!=\!C\cdot$$

(*9.57*)

The dimerization involves a monomer excited state interacting with a monomer ground state. A reaction diagram illustrating the energetics of diacetylene polymerization is shown in Fig. 9.5. Since the absorbed photon can initiate a chain reaction, quantum yield is high.[664,666] The quantum yield is defined as the number of poly(diacetylene) repeat units produced per absorbed photon and is equal to nq, where n is the propagation length and q is the chain initiation probability. Values ranging from $nq = 10^{-4}$ to 50 have been reported for various diacetylenes.[664-666,3607] The exothermic nature of the photopolymerization process has been conclusively demonstrated with photoacoustic spectroscopy. The kinetic model for the photopolymerization of diacetylenes is discussed in several papers.[664,666,2995]

An interesting group of diacetylenes are 1,4-di-ethynylbenzene (9.58), 1,4-diethynylnaphthalene (9.59), and 1,4-diethynyl-2,3-dichloronaphthalene (9.60), which upon u.v. irradiation from colourless single crystals give red or brown polymeric products.[3107,3108,3132,3133]

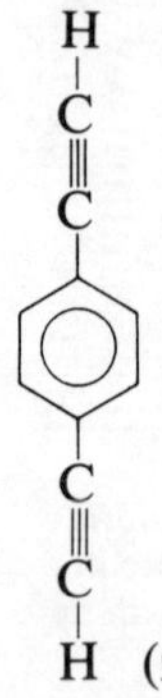

(9.58) 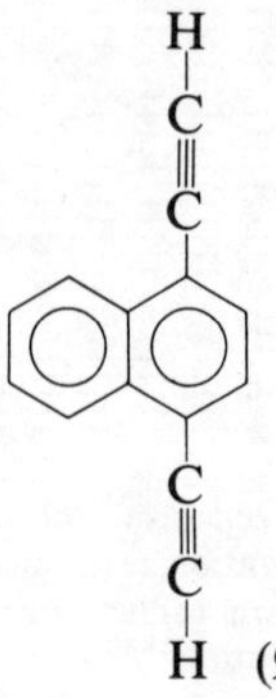   (9.59) 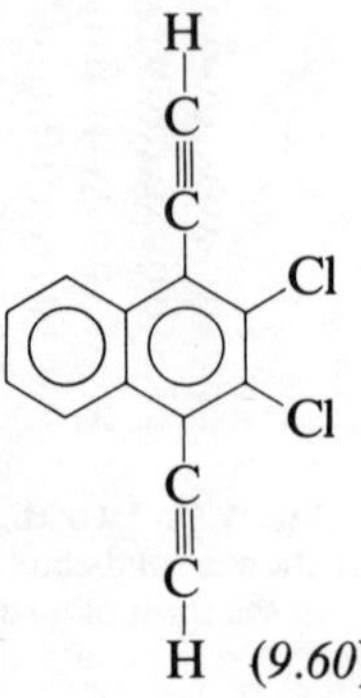   (9.60)

A polymeric product (9.61) obtained from 1,4-diethynylnaphthalene (9.59) has a polyene structure and an average length of about 50 monomeric units.[3132, 3133] Only one ethynyl group participates in the photopolymerization process, whereas the second ethynyl group of the monomer remains unreacted:

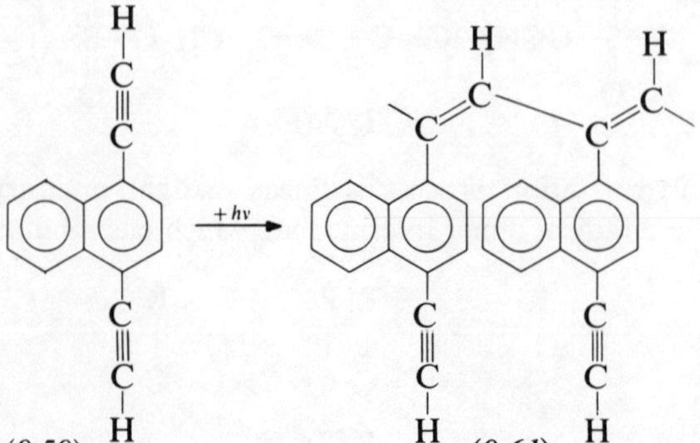

(9.59)   (9.61)   (9.39)

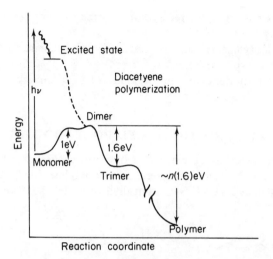

Fig. 9.5   Reaction diagram illustrating the energetics of
diacetylene polymerization[2995]

Several sensitizers, such as phenazine[3602,3608] or certain dyes,[547,548,1152] accelerate photopolymerization of diacetylenes.

Unsaturated diacetylene (diyonic) fatty acids of the general structure (9.62) are known to polymerize either in the solid state, or monolayers and multilayers if irradiated with u.v. radiation.[895,2298,2749,3603,3606,3609]

$$CH_3(CH_2)_{m-1} - C \equiv C - C \equiv C - (CH_2)_n - COOH$$

where: $m = 8$–16 and $n = 0$, 2 and 8

(9.62)

The multilayers consist of a domain structure of bilayers in which the molecules are uniquely oriented.[2298] The polymer chains grow one-dimensionally within the layer plane, restricted in their length by the size of domains.

Various factors influence the photoreactivity, such as:[3604,3605]

(i) Chain length of the acid.
(ii) Nature of the subphase.
(iii) Spreading solvent.
(iv) Additives forming mixed multilayers with the diyonic acids.

Micelles and vesicles built from diacetylenes can be permanently stabilized by photopolymerization.[1420] Diacetylenes can even be incorporated into cell membranes of microorganisms with subsequent polymerization.[1891]

Highly oriented extremely thin polymeric films with potential applications for integrated optics, such as protective coatings or a photoresist, were obtained.[2954,3748]

Because of the dramatic change in colour, as well as mechanical stability, upon polymerization diacetylenes appear to be attractive materials for image tech-

378

nology (such as photographic receptors), information storage, and holography image recording.[3104,3105,3401]

Polymeric diacetylenes can also be used as efficient elements for third harmonic generation in laser spectroscopy.[667,3195,3299]

Poly(diacetylenes) dissolved in organic solvents undergo random chain scission upon u.v. irradiation.[2593]

9.10. PHOTOINDUCED POLYADDITION

Photoinduced polyaddition is a new type of photoreaction in polymeric systems. It occurs between 1,1'-(2,6-naphthalenedicarbonyl)diazirine (9.63) and 1,5-dihydroxynaphthalene (9.64) in tetrahydrofurane solution under u.v. irradiation:[2718]

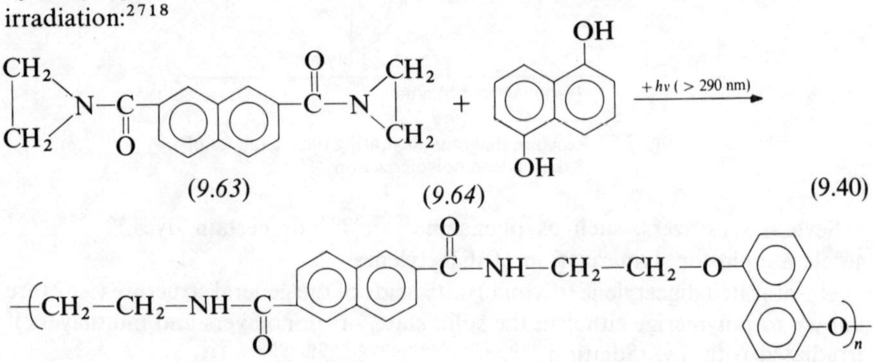

(9.63) (9.64) (9.40)

1,1'-(2,6-naphthalenedicarbonyl)diazirine is also an useful agent for photo-crosslinking of poly(4-hydroxystyrene) (9.65) in film:[2719]

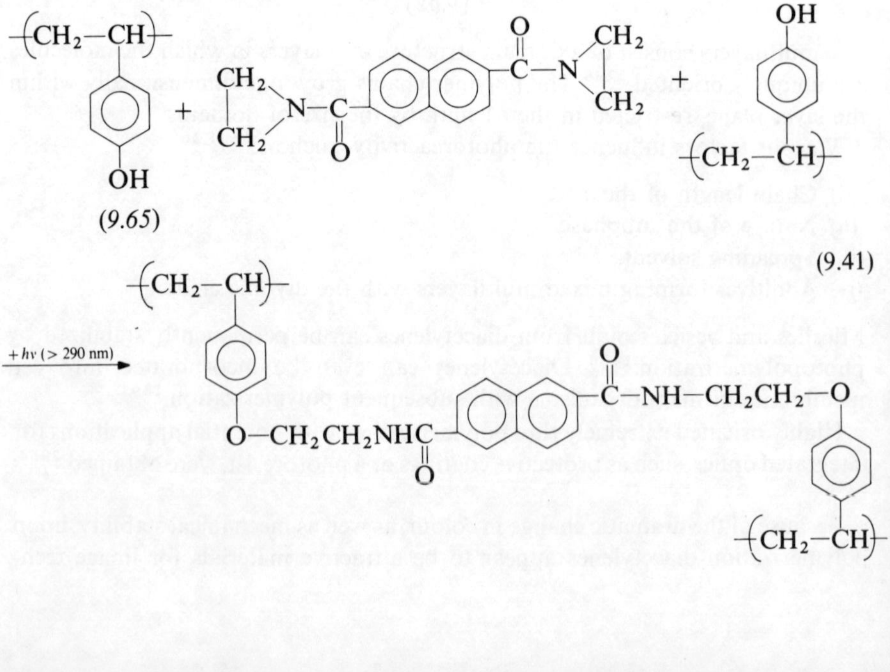

(9.65)

(9.41)

10. Photochromic polymers

PHOTOCHROMISM is defined as a reversible change of a single chemical species between two states having different absorption spectra, such a change being induced in at least one direction by the action of light.[532] Thermal or light-induced reversibility basically differentiates photochromic phenomena from the usual photochemical processes.

In the case of organic compounds, photochromism is related to the following structural modifications:

(i) valence isomerization,
(ii) tautomerism,
(iii) *cis–trans* isomerization,
(iv) ring-opening/ring closure.

PHOTOCHROMIC POLYMERS (PHOTORESPONSIVE POLYMERS) are macromolecules which have covalently connected PHOTOCHROMIC GROUPS,[1768] such as:

(i) Azobenzene groups.[32, 203, 446, 449, 1037, 1038, 1726, 1770, 1773, 1780, 1781, 1794, 1797, 2203, 2322, 2323, 2415–2416, 3319, 3492, 3493, 3732, 3733]

(ii) Stilbene groups.[175, 2512, 2797, 2798]

(iii) Spiropyran groups.[385, 1345, 1770–1772, 1777, 1778, 1974, 1975, 2133, 2135, 2140, 2188–2190, 2483, 2532, 3377, 3378, 3734]

All these groups, under photoirradiation, exhibit *trans–cis* isomerization, which can be studied simply by absorption spectroscopy.

10.1. PHOTOISOMERIZATION OF AZOBENZENE PHOTOCHROMIC POLYMERS

Substituted azobenzene low molecular compounds isomerize from the *trans* (extended stable) from (*10.1*) to the *cis* (compact instable) from (*10.2*) upon irradiation (350–410 nm). After a short time the *cis* form in the dark or under irradiation with $\lambda > 470$ nm is transformed to the *trans* form:[2323, 4004]

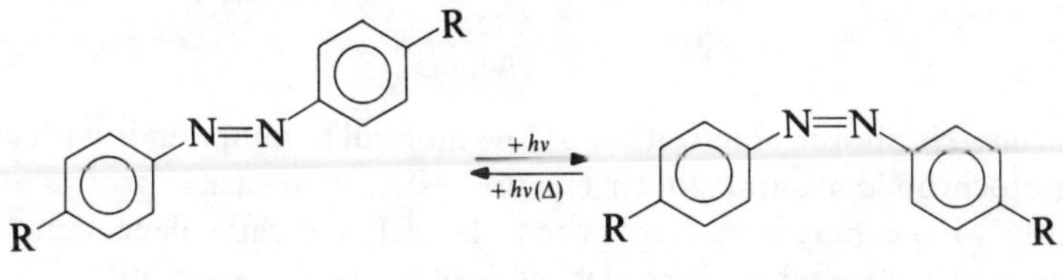

$$(10.1)$$

(*10.1*)　　　　　　　　(*10.2*)

Intense absorption at 350 nm (*trans* form) (due to n, π^* transition), upon irradiation (350–410 nm), decreases and a new peak at 310 nm (*cis* form) (due to π,π^* transition) appears (Fig. 10.1).

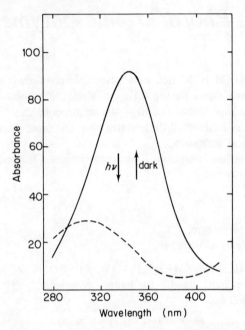

Fig. 10.1 Photochromism of substituted azobenzene low molecular compounds:[2323]

The *trans–cis* photoisomerization of low molecular compounds with azobenzene photochromic group (azobenzene probes) or azobenzene groups attached covalently to the polymers (azobenzene labels) is greatly dependent on the polymer matrix structural properties (cf. Section 10.5).

Photoisomerization of low molecular compounds such as 4-ethoxyazobenzene (*10.3*) or 2,2'-azonaphthalene (*10.4*) in polymer matrix proceeds via first-order kinetics above the glass transition temperature (T_g).[1215] Below T_g, complex kinetics appear to be a result of the multiplicity of different first-order processes.

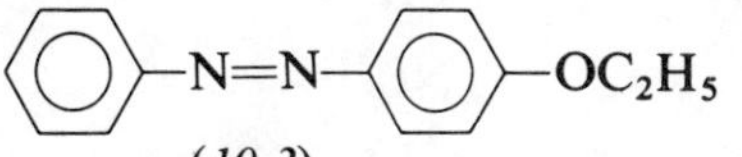

(*10.3*) 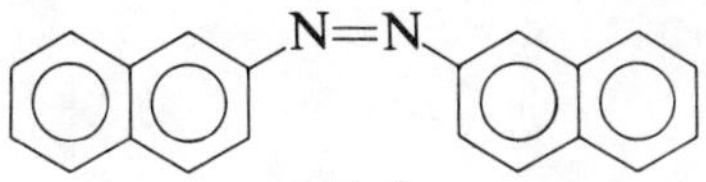   (*10.4*)

Similar behaviour has been observed for the *cis–trans* photoisomerization of azobenzene residues attached as side-groups to copolymers (*10.5*) and (*10.6*) when examined below their glass transition temperature.[1037,1038,1040] In solutions, and in the rubbery state (above T_g), the thermal *cis–trans* recovery follows first-order kinetics, whereas in the glass state (below T_g) some azo groups react much faster and others isomerize much slower.

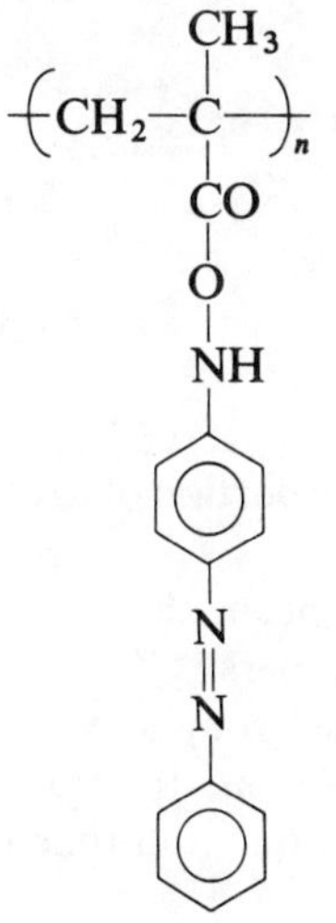

(*10.5*) 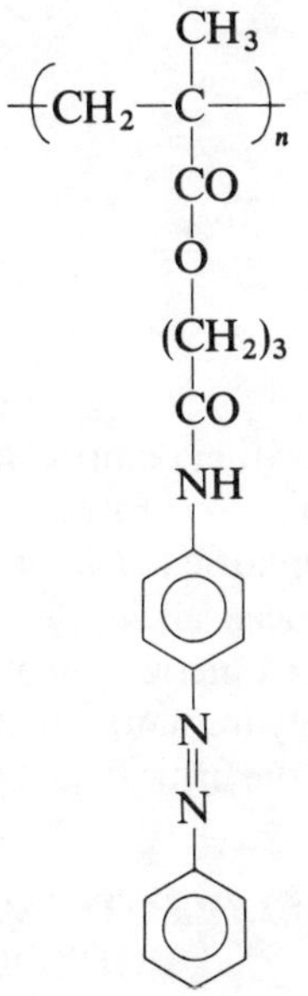   (*10.6*)

Photoisomerization of azobenzene groups in co(4-vinyl-4'-dimethylamino-benzene/styrene) (*10.7*) and co(4-cryloylaminomethylaminoazobenzene/styrene) (*10.8*) occurs 100 times slower in film than in solution.[1931]

Linear polyamides with azobenzene groups in the backbone (*10.9*) undergo conformational changes upon irradiation with light as a consequence of *cis–trans* isomerization of the azobenzene groups.[1770,1773,1780] If each repeating unit of the polymer contains an azobenzene group, and if a major portion of these groups is the *trans* configuration, the polymer chains are extended. The chains,

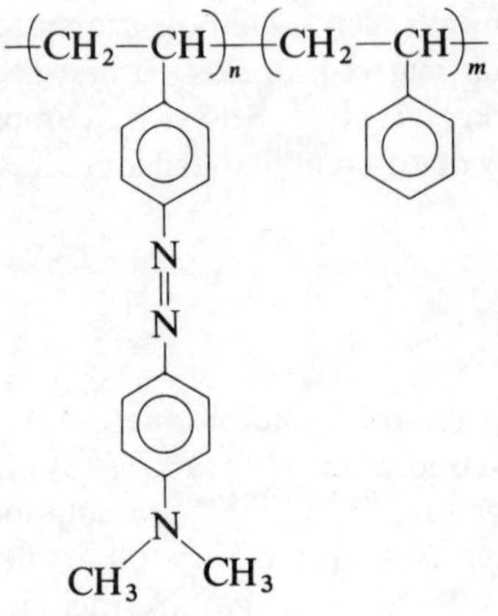

(10.7)

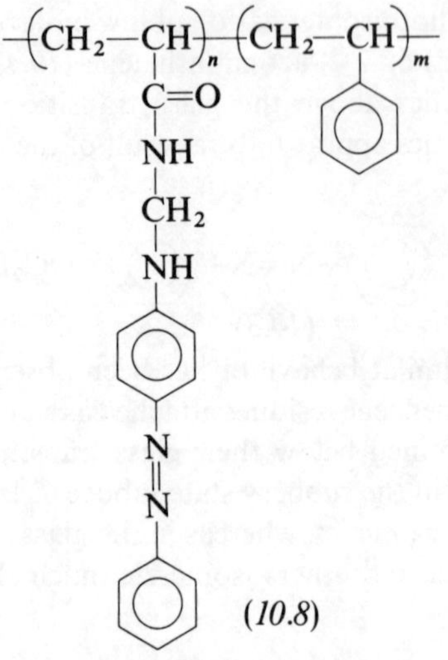

(10.8)

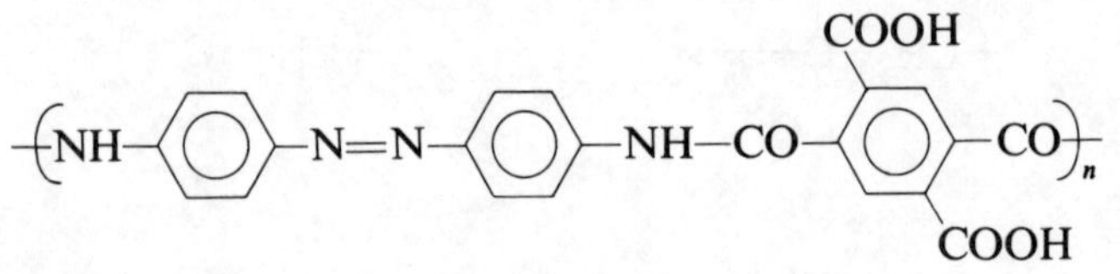

(10.9)

however, form rather compact coils when a major portion of azobenzene groups is converted to the *cis* form.

Azo chromophoric labels such as a molecular probe can be employed for the study of physical ageing in amorphous polymers.[3492,3493] By measuring *trans–cis* photoisomerization of azo photochromic groups covalently bounded to amorphous polyurethanes, it is possible to determine the fraction of the free volume above the critical size at a given temperature and time of ageing.

10.2. PHOTOISOMERIZATION OF STILBENE PHOTOCHROMIC POLYMERS

Stilbene isomerizes from the *trans* (extended stable) form (*10.10*) to the *cis* (compact instable) form (*10.11*) upon irradiation, and this reaction can take place via the lowest triplet state of stilbene:[767]

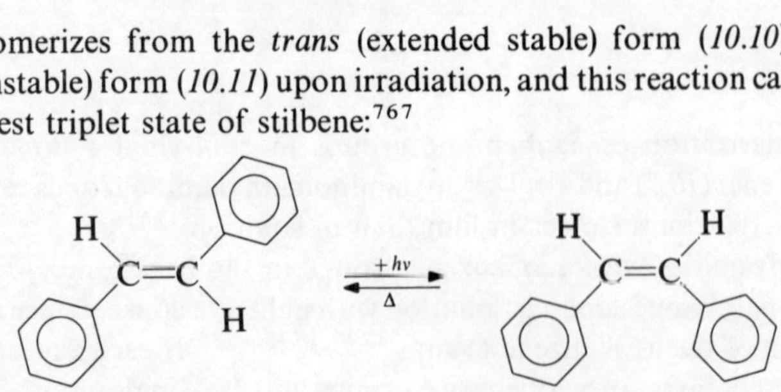

(10.2)

(10.10) (10.11)

A variety of low molecular weight triplet sensitizers (e.g. acetophenone or benzophenone[1487,3173]) and polymeric photosensitizers (cf. Section 7.26) can transfer their excitation energy to the stilbene molecules and triplet–triplet energy transfer may occur with high efficiency, under conditions where only the sensitizer absorbs energy. Stilbene isomerization via direct excitation and photosensitized stilbene isomerization were described in detail elsewhere.[767]

Photoisomerization of *trans–cis* stilbene copolymers, e.g. co(*trans*-N-4-(4'-nitrostyryl)phenyl methacrylamide/2-hydroxyethyl methacrylate) (*10.12*) reaches its steady state faster than did the model compounds (*10.13*):[2512] This was attributed to the lower microenvironmental polarity in the area of the polymer compared with that of the model with solvent.

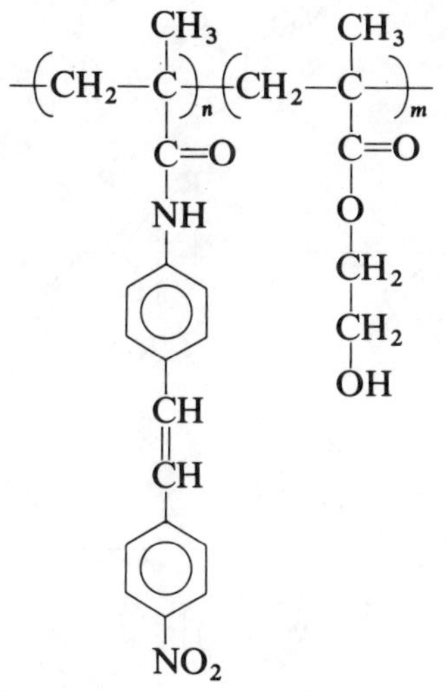

(*10.12*)

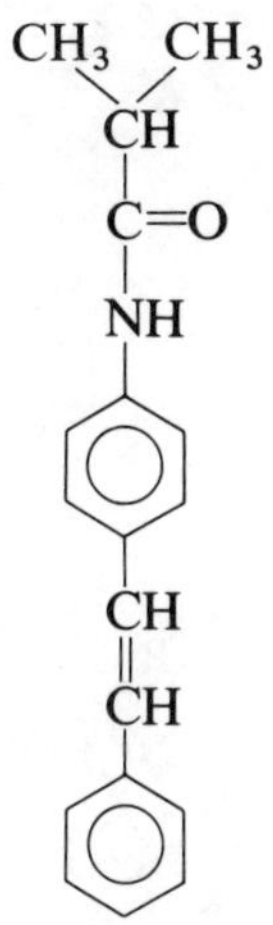

(*10.13*)

The *trans–cis* photoisomerization of co(methyl acrylate/*trans*-4-acryloxystilbene) or co(methyl acrylate/*trans*-vinyl stilbene) produces the disappearance of the dichroic bands associated with the first (π,π^*) electronic transition of the stilbene groups, suggesting the change of the shape of the side-chains induced by light absorption.[175–177]

10.3. PHOTOISOMERIZATION OF SPIROPYRAN PHOTOCHROMIC POLYMERS

Substituted indolinobenzospiropyran low molecular compounds photoisomerize from the 'closed form' (colourless) (*10.14*) to the merocyanine which is an 'open form' (intensively coloured) (*10.15*) upon u.v. irradiation (250–300 nm). The merocyanine form can be reconverted to the spiropyran form upon visible light irradiation, or thermally:[409,419,3162,3907]

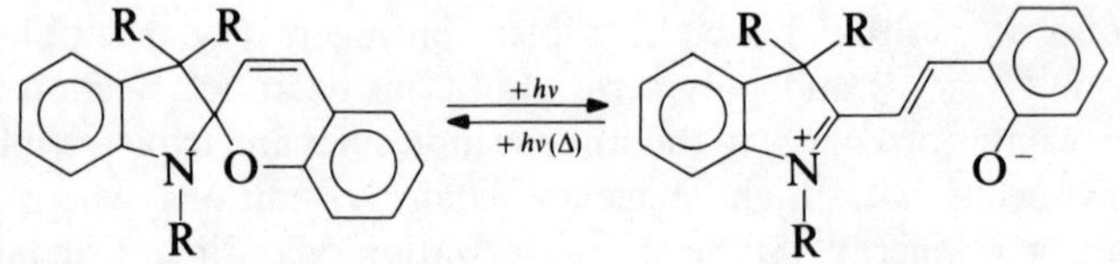

$$(10.3)$$

(10.14)   *(10.15)*

Spiropyran photochromism is complicated by the possible existence of the following eight geometrical isomers:[219, 999, 3162]

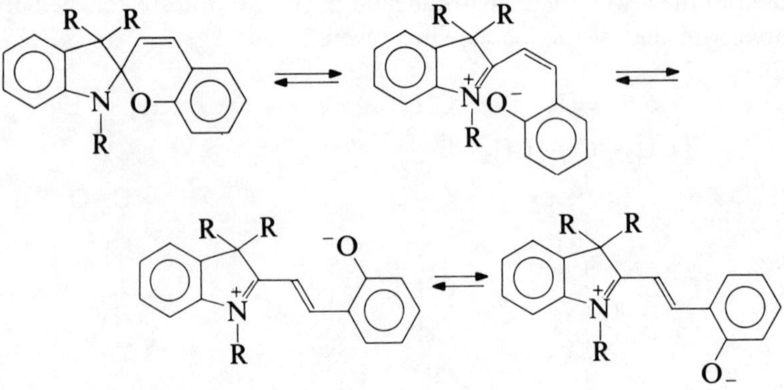

$$(10.4)$$

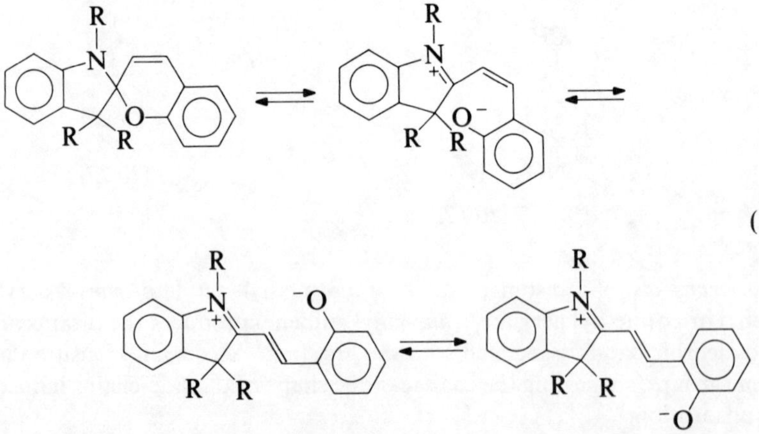

$$(10.5)$$

The ring-opening/ring-closure photoisomerization of low molecular compounds with spiropyran photochromic group (spiropyran probes) or spiropyran groups attached covalently to the polymers (spiropyran labels) is greatly dependent on the polymer matrix structural properties.

The rate of fading of a merocyanine (10.15), obtained by ring-opening of a spiropyran (10.14), is 100–400 times smaller in a poly(methyl methacrylate) film than in a solvent.[1214, 1215]

The rate of fading of a merocyanine also depends on the chemical nature of a polymer matrix. As a result of a strong negative solvatochromism the rate of fading is 8–10 times higher in polystyrene than in poly(methyl methacrylate) and decreases with increasing photochrome concentration in the film.[3743]

The kinetics of fading of merocyanine shows strong deviation from first-order kinetics, when photochrome was dissolved in bulk polymers or linked to the latter as a side-group.[1214, 3378] The decolouration rate can be written as:[3378]

$$[A]_t = [X_1] e^{-k_1 t} + [X_2] e^{-k_2 t} + [X_3] e^{-k_3 t} \qquad (10.6)$$

where:

[X] = contributions of each isomer of merocyanine to the optical density measured at a given time t.

Above the glass transition temperatures (T_g) $(T_g = 55°C$ for (10.16) and $T_g = 56°C$ for (10.17)) the thermal fading (in log(optical density) versus time) (Fig. 10.2) for a polymer (10.16) is very rapid and follows a first-order relationship as in a solution, whereas below T_g the rate is much smaller.

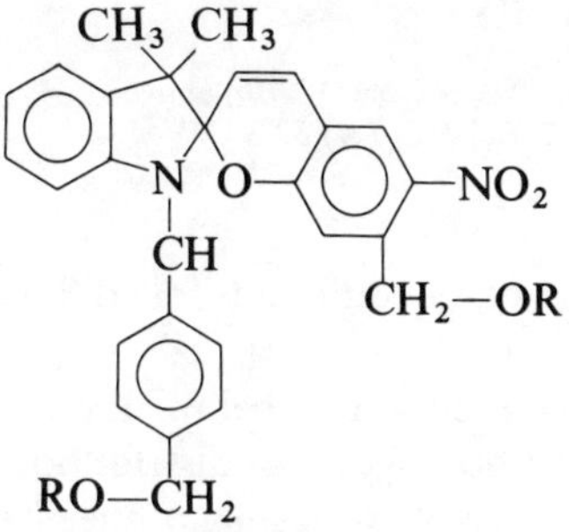

(10.16)

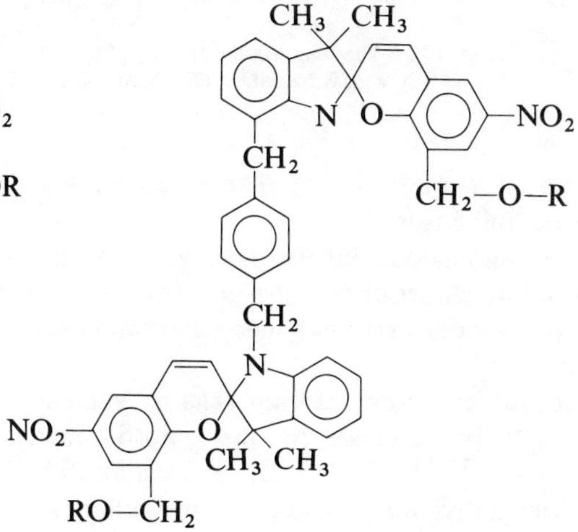

(10.17)

where:

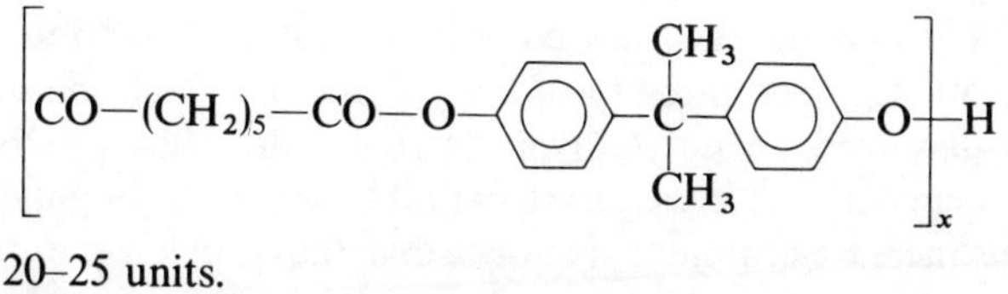

$x = 20$–25 units.

Quenching at low temperature of film colored at 65°C causes thermal fading to become negligible owing to the lack of chain segment mobility below T_g; consequently a permanent image is obtained.

In a rigid polymer matrix the rate of fading is slower because of restrictions of rotation which prevent the merocyanine isomers reaching the necessary confor-

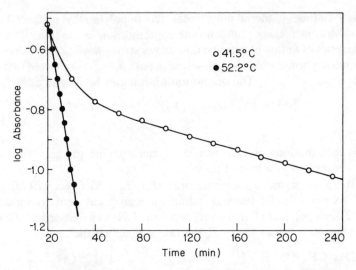

Fig. 10.2 Thermal decoloration of photochromic copolymer (*10.16*) above and below glass transition (T_g) temperature: (○) 41.5°C, (●) 52.2°C[3377]

mation for ring closure. These restrictions may be different for different merocyanine isomers.

The anomalous kinetic behaviour of the merocyanine below the glass transition temperature (T_g) is mainly due to a non-homogeneous distribution of the free volume in the polymer matrix and thus to the physical state of the glassy polymer.[3388]

Substituent bulkiness and chain segment mobility effects are especially strongly pronounced in the photoisomerization of spiropyran photochromes.[3374,3378,3380,3381,3743] Decoloration kinetics in polyester or polystyrene matrix at room temperatures differ considerably for compounds of very similar structure; all contain the same spiropyran photochrome group, but are of different molar values.[3387] By carefully choosing the polymer matrix and the molar size of the photochrome, decoloration can be avoided, i.e. the reversibility of the *cis* ⇄ *trans* equilibrium can be suppressed.

The PHOTORESPONSE, i.e. the coloration under irradiation, depends greatly on the chain segment mobility of the polymer in which the oligomeric photochrome is dissolved. In poly(ethylene terephthalate) the photochemical response is slow below T_g but increases remarkably above T_g. In poly(*n*-propyl methacrylate) this increase around T_g is noticeable, thus much less pronounced; probably on account of the rotation possibilities of the propyloxycarbonyl side-group.

There is considerable difference of opinion in the literature regarding the assignment of the nature of the various transient species and photoproducts observed in the absorption spectra in the visible region of various polymers and copolymers containing the spiropyran groups.[2132,2133,2902]

Nanosecond and picosecond time-resolved spectroscopy of polymers containing spiropyran groups, for the photoinduced aggregation and photoinduced conformation, show changes in spiropyran moieties.[1914-1916]

A number of polymers such as poly(methyl methacrylate), polystyrene,[1345,1778,2133] and copolymers of methyl methacrylate, acrylonitrile, and styrene[1772,1777,2189,2190,3734] with spiropyran photochromes have been synthesized.

Incorporation of the spiropyran groups into side-chains of macromolecules shows evident effects of various length and structure of a spacer on the decoloration of merocyanine.[2188]

Spiropyran groups labelled into the polymer backbone (10.16) and (10.17) show a strong effect of macromolecules on the thermal decoloration process of merocyanine as a function of temperature.[3377,3388]

Stretching of the polymer matrix causes molecular orientation of the photochrome. Stretching of a film of poly(bisphenol-A-pimelate) containing 5 wt% spiropyrane photochrome considerably affects the kinetics of fading (Fig. 10.3).

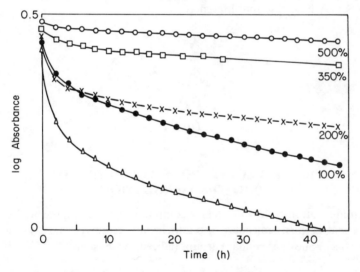

Fig. 10.3 Decoloration of stretched films of poly(bisphenol-A-pimelate) containing 5 wt% spiropyrane photochrome[3378]

The disadvantage of photochromic probes or labels is fatigue phenomenon, i.e. progressive loss of ability to change colour on prolonged exposure to ultraviolet light on repeated irradiation–darkness cycles due to photochemically induced desensitization and oxidation.

Poly(spiropyran methacrylate) shows an interesting link between the crystallization of the atactic homopolymer and spontaneous spiropyran–merocyanine conversion. A polymer up to 40% crystallinity was obtained by slow precipitation

388

or swelling of the amorphous polymer. During this process additional spiropyran side-groups were converted to merocyanine groups. Merocyanine moieties interact and tend to undergo intermolecular stacking, and this is associated with the formation of three-dimensional crystalline domains. As a result, adjacent segments of the polymer chains are brought closer together. This, and the increase of the polarity in the surroundings of the domains, promotes further spiropyran–merocyanine conversion and hence further ordering. This photo-induced crystallization is called ZIPPER CRYSTALLIZATION.[1345]

The photochromic properties of copolymers and copolyesters with spiropyran photochromes have also been interpreted on the basis of the formation of defects within the matrix resulting from the excess of energy on irradiation.[2140]

Much more complex photochromism was observed in the case of naphthospiropyran (e.g. 1,3,3-trimethylindolinonaphthospiropyran) (10.18) in polystyrene matrix.[2532] The kinetics of the decoloration was found to be markedly affected by irradiation time. In a rigid matrix the rate of ring-closure is slower because of restriction of rotation within the coloured forms. Due to these restrictions the isomeric coloured isomers may experience varying degrees of difficulty in achieving the necessary conformation for ring enclosure. These processes may have different lifetimes.

(10.18)

10.4. EFFECT OF POLYMER MATRIX ON PHOTOISOMERIZATION

Photoisomerization of low molecular compounds (probes, guest molecules) in polymer matrix (host molecules) or photochromic groups covalently attached to polymer backbone or as pendant groups (labels) depends on:[1214, 1215, 2137, 2957, 2994, 3738]

(i) Free volume, i.e. the space volume between macromolecules which is available for unhindered isomerization processes.
(ii) Microscopic and macroscopic viscosity.
(iii) Steric interference at a particular environment.

It is well known that quantum yield of photoisomerization of trans (10.10) to cis stilbenes (10.11) decreases with increasing viscosity.[1240] Polymer matrices depending on temperature (below the glass transition temperature (T_g)) reduces the space (free volume) available for rotation of the phenyl groups from trans to cis stilbene configurations.

The free volume is a critical factor in the ability of a guest molecule to isomerize to another form. The photoisomerization kinetics discontinuities during temperature descend near the T_g. The chain segment mobility of the polymer and chain orientations, exert a major influence on photochemical behaviour. For that reason photochromes can be used as a probe for the detection of local chain movements.

When photoisomerizable chromophores are incorporated into the backbone of polymer chains, or into the pendant group, photoisomerization of the chromophores may affect the physical properties of the polymers and the polymer solutions, especially if isomerization involves appreciable changes of a polarity or a geometrical structure.

Photoisomerization of azobenzene photochromes (labels) in a different position in polystyrene such as centre-labelled (10.19), end-labelled (10.20), and side-labelled (10.21), and in the main chain of amorphous polyurethanes, is very sensitive to the volume changes taking place within solid films.[2203,3491,3492]

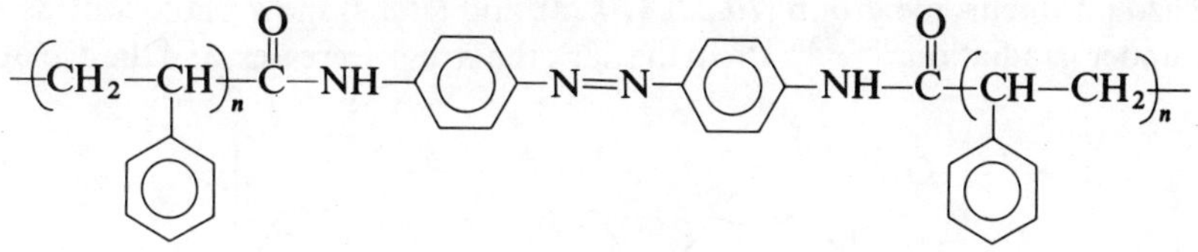

(10.19)

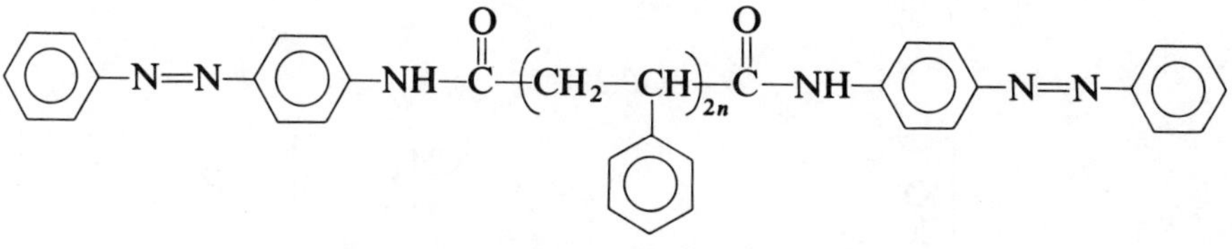

(10.20)

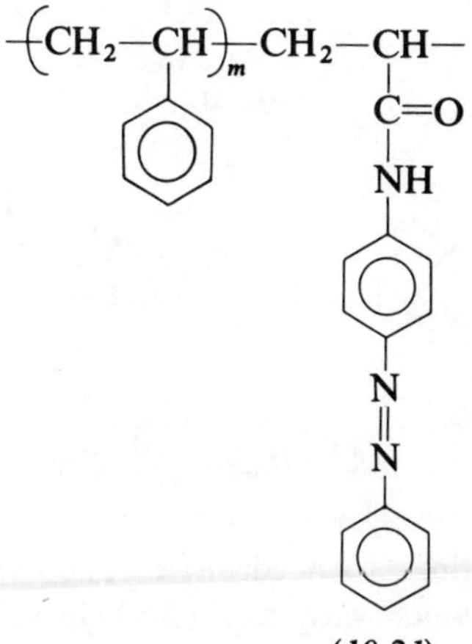

where: $n = 80$, $m = 26$

(10.21)

Photoisomerization in dilute solutions occurs by a single rate process, whereas its initial portion in the solid films may incorporate two separate rate processes. The first is fast, as in dilute solution, and this is followed by a slower one. The fractional amount of the fast process decreases with physical ageing but increases with temperature, plasticization, or glassy deformation.[2203] This fraction is proportional to the number of regions where local free volumes are greater than a critical size necessary for the photoisomerization of the azobenzene groups.

10.5. PHOTOCONTRACTABLE POLYMERS

Incorporation of a photoisomerizable chromophore onto or into macromolecules, during photoisomerization, can cause dimensional changes (PHOTO-INDUCED MECHANICAL CHANGES) of polymer chains, such as shrinkage or dilation.[1037,1039,1118,1214,3384] For example:

(a) Linear photochromic polyamides in which every monomer unit contains an azo photochrome group (10.22), (10.23), and (10.24) show photocontraction under irradiation.[203,336,449] In the dark the stress decreases, and the cycle can

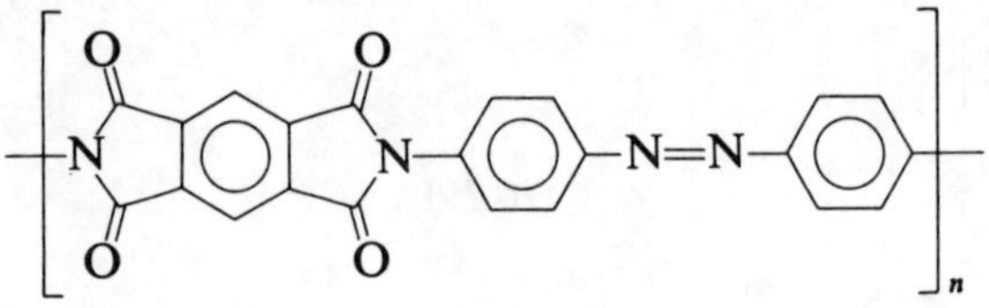

(10.22)

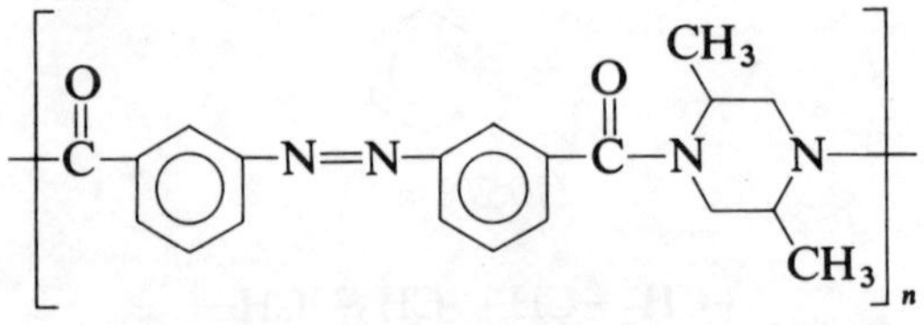

(10.23)

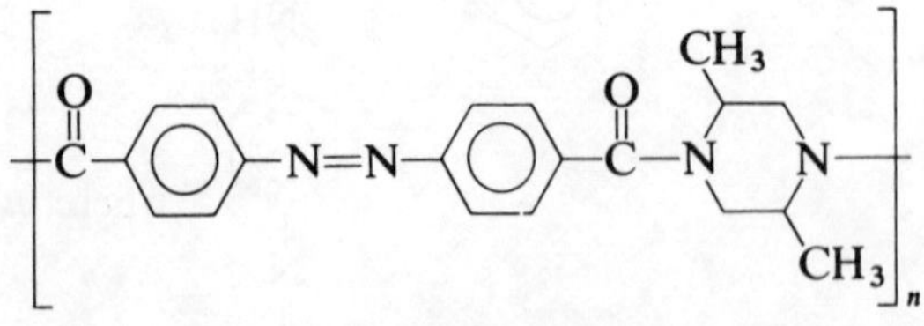

(10.24)

be repeated many times. An idealized photochemical response for a photoresponsive polymer with azo photochromic groups is shown (in Fig. 10.4). When irradiation commences the stress increases, indicating a contraction of the sample length. The stress continues to increase until a

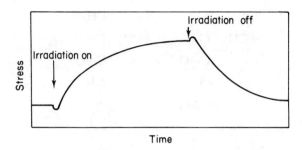

Fig. 10.4 Idealized photomechanical response curve[449]

photostationary state is reached and the stress remains constant until the irradiation ceases. After this the stress decreases, indicating a relaxation of the sample. The process is reversible and may be cycled many times.

(b) Linear polyamide containing a stilbene photochrome in the backbone (*10.25*) shows analogous photomechanical energy conversion.[2797,2798]

$$\left[-NH-\overset{\overset{\displaystyle O}{\|}}{C}-\hspace{-2pt}\bigcirc\hspace{-2pt}-\overset{\overset{\displaystyle O}{\|}}{C}-NH-\bigcirc\atop CH=CH-\bigcirc\right]_n$$

(*10.25*)

(c) Co(β-hydroxyethyl methacrylate/azonaphthol methacrylate) (*10.26*) swollen in water and co(butyl acrylate/methacrylamido-azobenzene) (*10.27*) in dry state under irradiation show photoinduced contraction.[2415,2416]

$$-\left(CH_2-\underset{\underset{\displaystyle OH}{\underset{\displaystyle |}{CH_2}}}{\overset{\overset{\displaystyle CH_3}{|}}{\underset{\underset{\displaystyle CH_2}{|}}{\underset{\displaystyle C=O}{\underset{\displaystyle |}{\underset{\displaystyle O}{|}}}}}C\right)_{0.99}\left(CH_2-\underset{}{\overset{\overset{\displaystyle CH_3}{|}}{C}}\right)_{0.01}$$

(*10.26*)

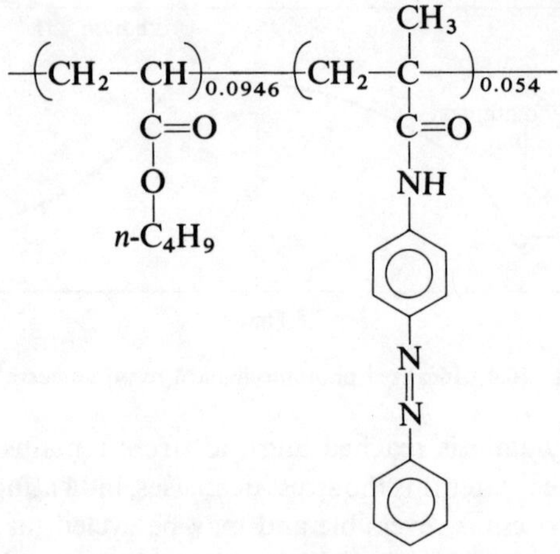

$$\left(CH_2-CH\right)_{0.0946}\left(CH_2-\underset{\underset{\underset{}{CH_3}}{\overset{CH_3}{|}}}{C}\right)_{0.054}$$

(structure with C=O, O, n-C$_4$H$_9$ on left; C=O, NH, phenyl, N=N, phenyl on right)

(10.27)

(d) Reversible photoinduced changes in monolayer surface-pressure have been observed in polymers such as poly(methyl methacrylate), and poly(2-vinylpyridine) mixed with indolinobenzospiropyran low molecular compounds[445,447,3750] or covalently bonded.[1422]

(e) Polyamide gels having triphenylmethane leucoderivatives show, under u.v. irradiation, large deformation of gels.[1775] Triphenylmethane leucoderivatives (10.28) dissociate into ion pairs under u.v. irradiation, with production of intensely coloured triphenyl cations (10.29). The cations thermally recombine with counterions:

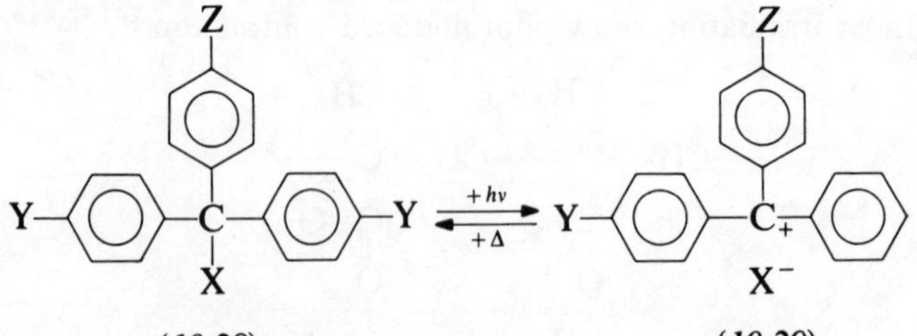

$$Y-\langle\rangle-\underset{\underset{X}{|}}{C}-\langle\rangle-Y \underset{+\Delta}{\overset{+h\nu}{\rightleftarrows}} Y-\langle\rangle-\underset{\underset{X^-}{}}{C_+}-\langle\rangle-Y \qquad (10.7)$$

(10.28)   (10.29)

(f) Co(ethyl acrylate/bis(spiropyran–acrylate)) crosslinks (10.30) show reversible photochemical contraction.[3377,3378,3380,3381,3384] On irradiation of stretched samples at constant temperature, wrinkling (2–3%) takes place, while length recovery takes place in the dark. The cycle can easily be repeated and is reproducible (Fig. 10.5). Poly(methyl methacrylate) and polystyrene-bearing spiropyran photochromes show similar behaviour.[448]

(g) Co(ethyl acrylate/4,4′-dimethylacryloamino-azobenzene) crosslinks (10.31) show similar reversible shrinkage/dilation effects.[1039] Upon irradiation the trans–cis isomerization causes conformational changes of adjacent networks which are considered to be responsible for the photomechanical effect.

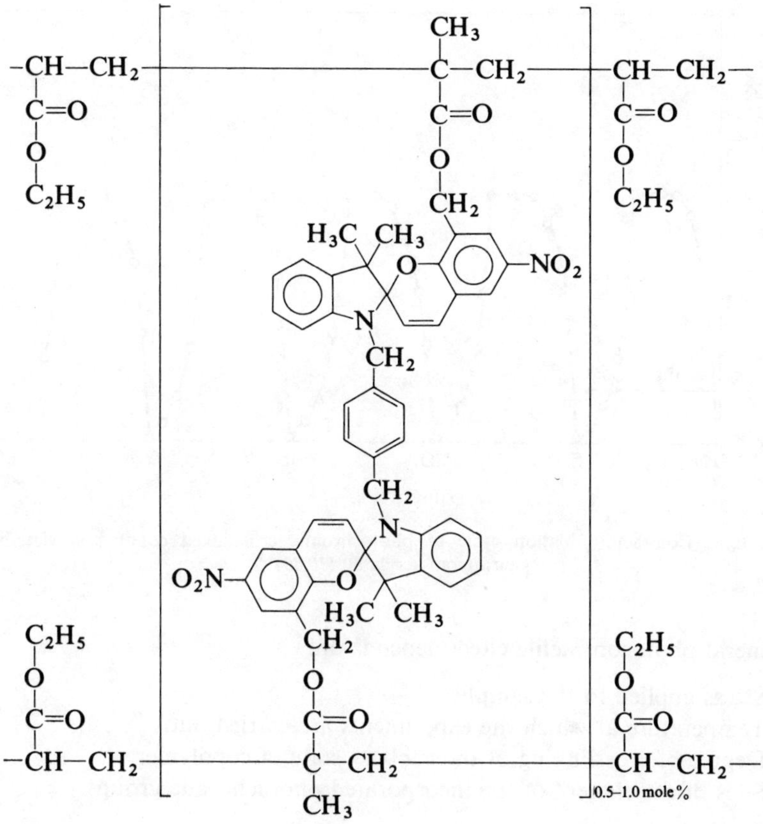

(*10.30*)

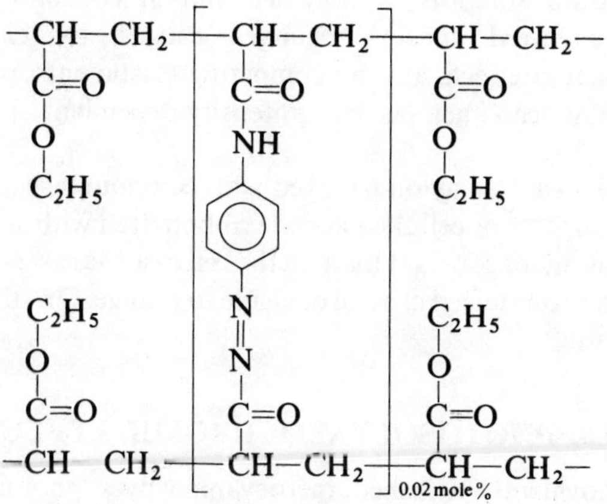

(*10.31*)

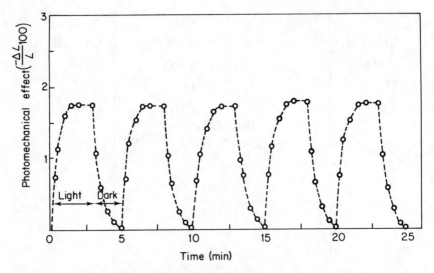

Fig. 10.5 Contraction/dilation cycle of photochromic crosslinked co(ethyl acrylate/bis-(spiropyran-acrylate)) (10.30)[3377]

In general photocontractile effect depends on:

(i) Stress applied to the sample.
(ii) Temperature at which the experiments are carried out.
(iii) Degree of crosslinking of the rubbery state of copolymer.
(iv) Sites of attachment of the incorporated photochromic groups.

Evidence has accumulated showing that other mechanisms may contribute to photochemically induced changes, in addition to the clearly demonstrated photochemical isomerization.[32,3380] Conversion of energy stored in the chromophores, which could lead to local heat effects, was suggested as a possiblity.[2415,2416] Photoirradiation of polymer film at constant length induces previously unrecognized tension changes caused by classical thermal expansion/contraction effects and by competing elastic entropic effects.[2367] A thin polymer film can act as an intensity-dependent photomechanical transducer.[3184]

Several polymers such as nylon 6.6 dyed with compounds containing stilbene photochromic groups[444] or cellulose acetate ribbon dyed with azo dyes[1726] show photocontractions under constant load. In the last case the sample on exposure to sunlight showed a pronounced change of shade to orange. On storage in the dark it recovers its shade.

10.6. PHOTOSOLVATOCHROMIC EFFECT

Polymers with covalently attached merocyanine dyes photochromic groups (10.32) change their absorption spectra with changing solvent polarity.[3734]

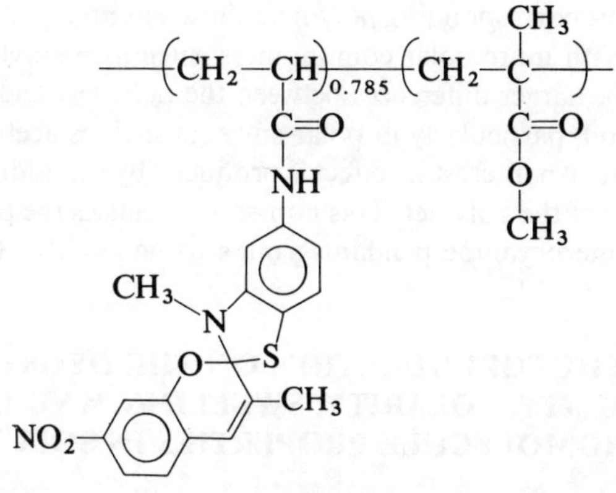

$$(10.32)$$

Thiazolidinebenzospiropyran 'closed form' (10.33), under irradiation, photo-isomerizes to the merocyanine 'open forms' (10.34 and 10.35):

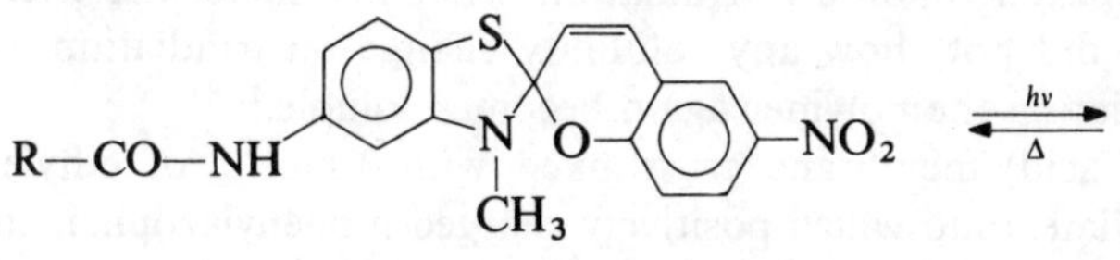

$$\lambda_{max} = 460 \text{ nm}$$

$$(10.33)$$

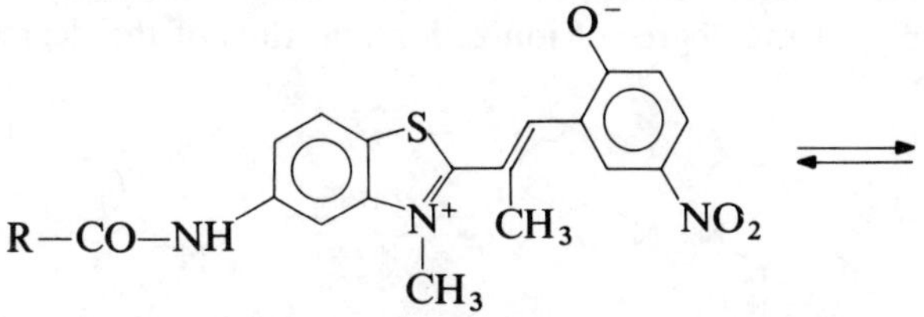

$$\lambda_{max} = 527 \text{ nm}$$

$$(10.34)$$

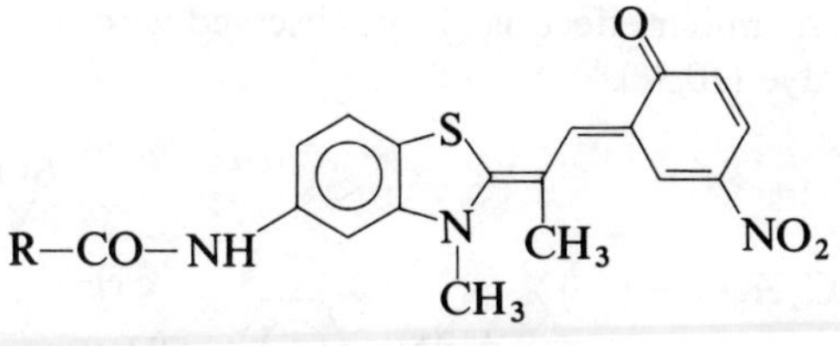

$$\lambda_{max} = 488 \text{ nm}$$

$$(10.35)$$

(10.8)

Model dyes and the corresponding polymers show essentially the same spectra in benzyl alcohol. With more polar comonomers, such as methyl methacrylate or methyl acrylonitrile, larger differences between the polymers and the model compounds were found, particularly in polar solvents such as acetone, chloroform, or tetrahydrofuran. An interesting effect is produced by the addition of hexane to pyridine solutions of the polymer. This non-solvent causes the polymer chains to coil, forcing the merocyanine pendant groups to lie close to the hydrocarbon backbone.

10.7. PHOTOREGULATION ON THE DEGREE OF SOLUBILITY, POLARITY, SWELLING, AND OTHER MACROMOLECULE PROPERTIES IN SOLUTION

Photoisomerization of photochromic groups such as azo- and spiropyran in low molecular compounds, in polymer matrix or attached covalently to macromolecules, can have a PHOTOREGULATION EFFECT on degree of solubility, polarity, viscosity (cf. Section 10.5), and swelling. For example:

(a) Polystyrene with a small number of azobenzene pendant groups becomes insoluble in cyclohexane upon u.v. irradiation, while low molecular weight azobenzene itself did not show any solubility change on irradiation. On visible light irradiation the polymer again becomes soluble.[1781]

(b) Poly(methacrylic acid) membrane crosslinked with 1 mol% of ethylene glycol dimethacrylate, onto which positively charged p-phenylazophenyltrimethylammonium ions were absorbed, shows the photoregulation of the degree of ionization and swelling.[707,3731] Photoisomerization of the *trans* (extended stable) form (10.36) to the *cis* (compact instable) form (10.37) causes a change of the degree of ionization and thus of the degree of swelling:

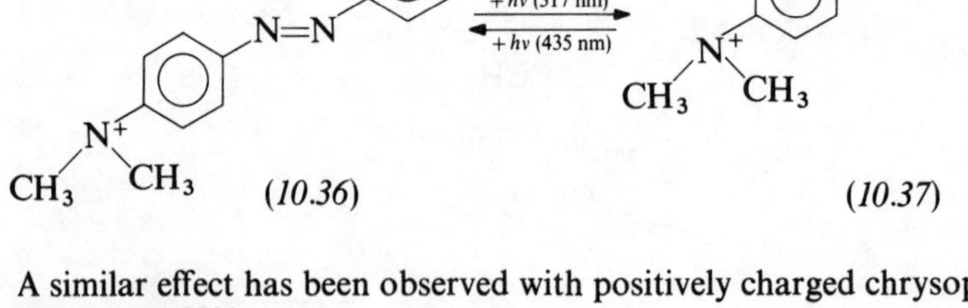

$$+ h\nu \text{ (317 nm)} \rightleftharpoons + h\nu \text{ (435 nm)} \qquad (10.9)$$

A similar effect has been observed with positively charged chrysophenine G dye (10.38).[3732]

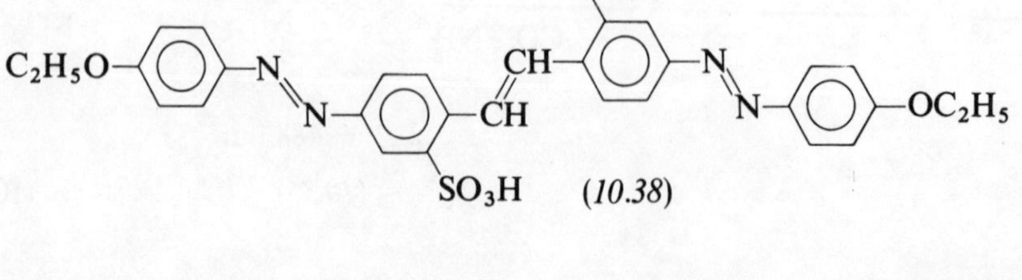

(c) Absorption of photochromic dye such as acid yellow 38(AY) (*10.39*) onto styrene–divinylbenzene copolymer changes reversibly upon irradiation in an aqueous solution system.[2674]

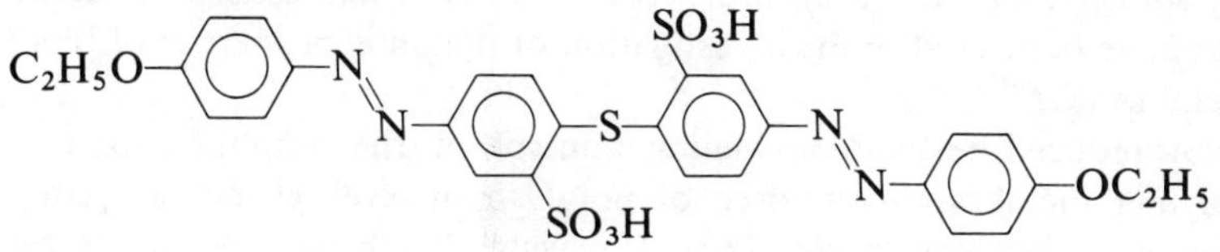

(*10.39*)

(d) In general polymers with covalently bounded azo-photochromic groups exhibit reversible change in wettability which is efficiently increased under irradiation.[1795,2675] The *trans–cis* photoisomerization makes the polymer surface more polar.

(e) Azoaromatic poly(carboxylic acid), co(acrylic acid/p-phenylazoacrylo-nitrile),[1799] poly(methacrylic acid),[1771] and cellulose-2,4-diacetic mem-branes[385,1974,1975] with spiropyran photochromes in aqueous solutions show photoinduced reversible pH change. The pH value of the solution can be reversible, regulated by irradiation and interruption of light in the range of 0.15 pH.[1799]

(f) Photoresponsive polymer membrane prepared by copolymerization of 2-(4-phenylazobenzoyloxy)ethyl methacrylate with ethyl methacrylate (*10.40*) has been employed for photocontrolled diffusibility of a water-soluble organic compound, e.g. p-aminobenzoates.[1794,1797]

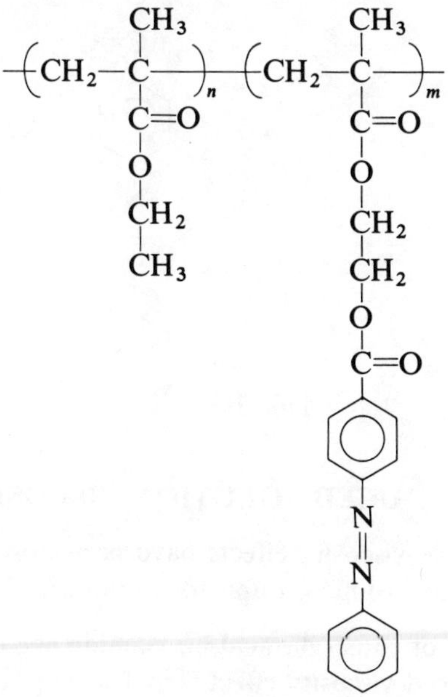

(*10.40*)

(g) The release of NaCl from nylon capsule coated with a lipid bilayer containing an azobenzene photochromic group can be controlled by irradiation.[2775]

(h) Solvent effects on the photochromism of a molecular photochromic probe, e.g. spiropyran, can supply information about the solute/solvent interactions and have been used in the investigation of dynamic properties of biological membranes.[2605]

(i) Photoinduced reversible swelling control of the amphillic azoaromatic polymer membrane composed of poly(2-hydroxyethyl methacrylate) containing azobenzene groups (10.41) in water has been used in practice for separation of water-soluble proteins such as insulin (MW = 600), lyozime (MW = 15,500), chymotrypsin (MW = 23,000), and albumin (MW = 60,000).[1796–1798,1800,1801] The apparatus used for the separation of proteins is shown in Fig. 10.6. Permeation profiles of proteins of various molecular weight through the azoaromatic polymer membrane are presented in Fig. 10.7.

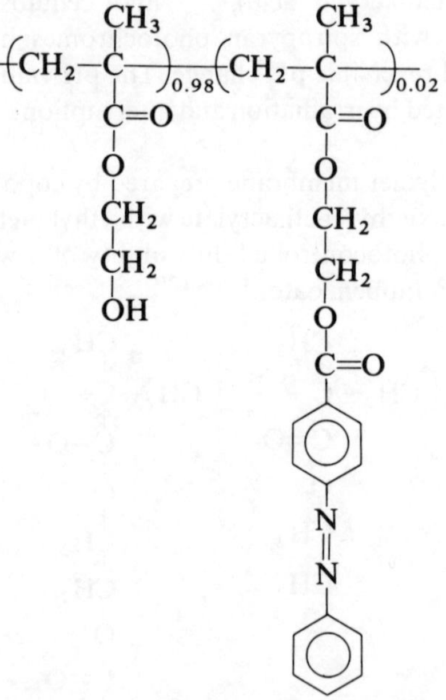

(10.41)

10.8. PHOTOINDUCED SOLUTION VISCOSITY EFFECTS

Photoinduced reversible viscosity effects have been observed in a number of polymers bearing photochromic groups, for example:

(a) Aqueous solutions of co(acrylic acid/acrylamido azo dye monomer) (10.42) show a photoinduced viscosity effect.[2322] The azo group-bearing polymer tends to exist in the dark as hydrophobic agglomerates bearing the less polar

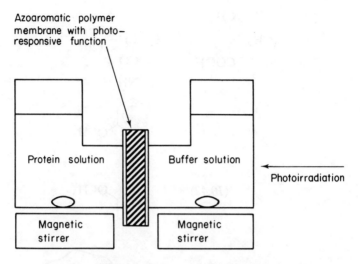

Azoaromatic polymer membrane with photo-responsive function

Protein solution

Buffer solution

Photoirradiation

Magnetic stirrer

Magnetic stirrer

Fig. 10.6   Apparatus used for the separation of proteins[1801]

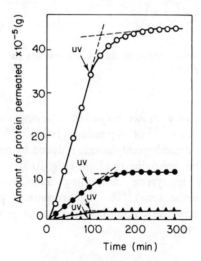

Fig. 10.7   Permeation profiles of proteins: (○) insulin, (●) lysozyme, (△) chymotrypsin, and (▲) albumin; the arrow represents the state of u.v. irradiation of the membrane[1801]

trans form. Irradiation converts the azo groups to the *cis* form, which exerts steric crowding along the polymer chain. The crowding effect causes extension of the polymer backbone to a more rod-like configuration (Fig. 10.8) and thus increases the solution viscosity.

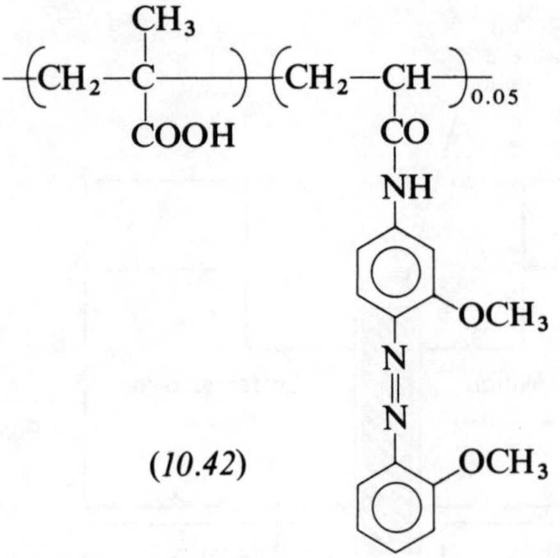

(10.42)

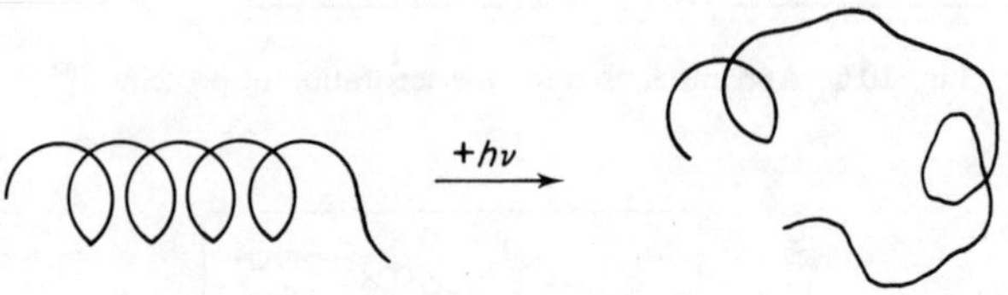

Fig. 10.8 Photoinduced solution viscosity change

(b) The reversible viscosity shows polyamides having the azo photochromic group in the main chain[1773] or as pendant groups.[1781] The viscosity of such polyamides in N,N-dimethylacetylamide reduced by irradiation returned to the initial value after removing the light (Fig. 10.9).

(c) Co(maleic anhydride/styrene containing azo groups), co(methacrylic acid/methacrylamidobenzene),[2416] azoaromatic polyureas,[2161] and poly(methyl methacrylate) containing spiropyran groups[1777,1778] exhibit photoviscosity effects.

10.9. PHOTORESPONSIVE CROWN ETHERS

Photoresponsive crown ethers, that combine within a molecule both a crown ether and a photochromic group, change its conformation in response to photoirradiation, resulting in a change in complexation ability.[982,3312,3320-3324]

The crosslinked polystyrene beads which immobilized one or two terminal azophenoxide groups of an azobenzene-crown (24-crown-8)-azobenzene (10.43) absorbs K^+ and Cs^+ ions in the dark, while they rapidly release them into the solution under u.v. irradiation.[3319] This photosensitive complexation occurs reversibly.

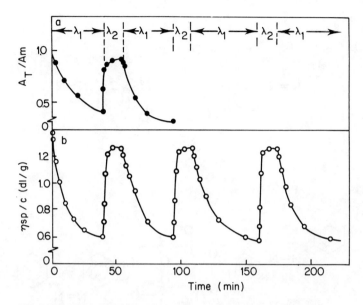

Fig. 10.9 Changes of: (a) content of the *trans* form of azobenzene residues in the polyamide backbone, and (b) viscosity of polyamide containing azobenzene residues in *N*,*N*-dimethylacetamide on alternate irradiation with u.v. (410 > λ_1 > 350 nm) and visible (λ_2 > 470 nm) light[1773]

(*10.43*)

The photoinduced *trans–cis* isomerization of azobenzene photochromes is capable of changing the conformation of the crown ring into a more stretched one which has poor ion-binding ability relative to the 'normal' crown ether, as shown below:

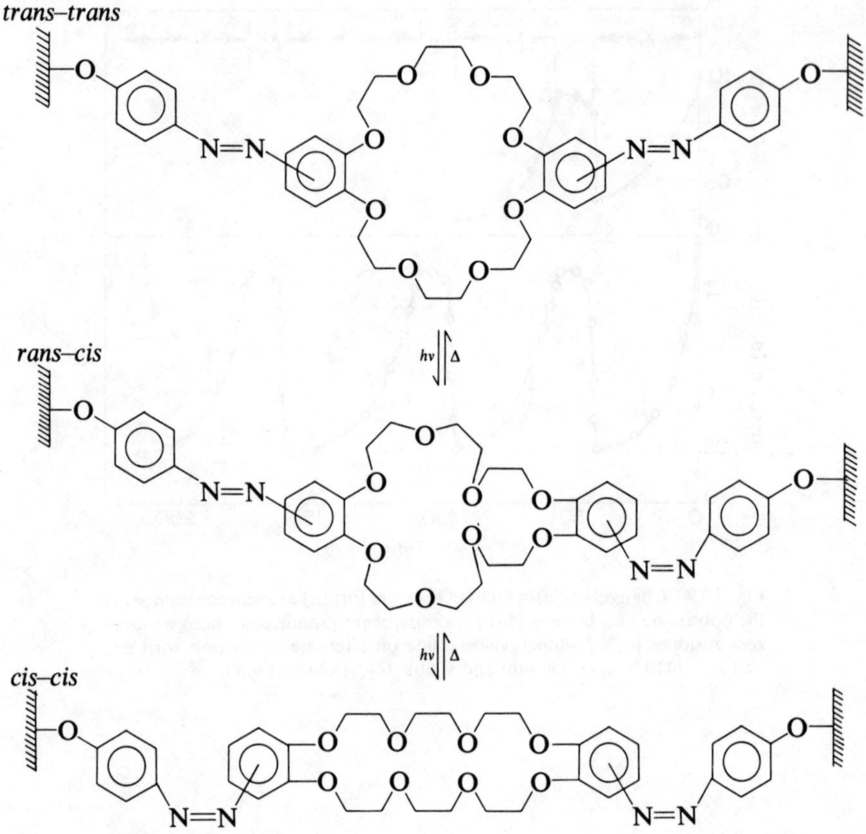

trans–trans

rans–cis

$hv \Big\| \Delta$

$hv \Big\| \Delta$

cis–cis

(10.10)

Photoinduced permeation control of metal cations through a poly(vinyl chloride) membrane containing a photoresponsive crown ether has also been reported.[209, 2160]

These photoresponsive crown ethers can be applied to the photocontrol of solvent extraction and ion transport across membranes.

10.10. MISCELLANEOUS PHOTOCHROMIC POLYMERS

A number of other different photochromic polymers have been investigated intensively, for example:

(a) Reversible viologen cation (V^{2+}) (*10.44*)/viologen radical cation (V^{+}) (*10.45*) (cf. also Chapter 16) development in poly(N-vinyl-2-pyrrolidone) (a typical polar aprotic solid matrix) has been found to be affected by the kinds of viologen cation as well as the paired anion:[1934, 1935]

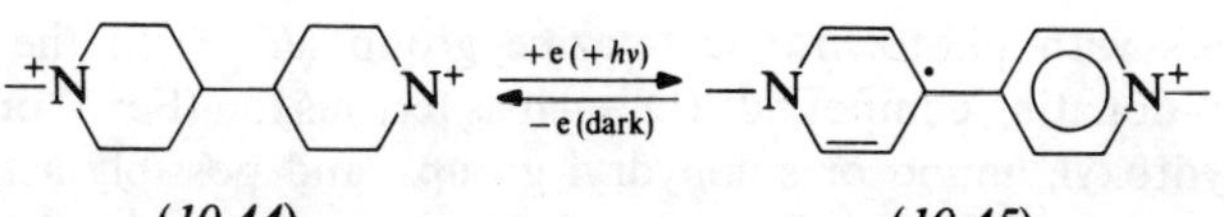

$$\overset{+}{N}\!\!-\!\!\bigcirc\!\!-\!\!\bigcirc\!\!-\!\!\overset{+}{N}{}^{+} \underset{-e\,(dark)}{\overset{+e\,(+h\nu)}{\rightleftarrows}} -N\!\!-\!\!\bigcirc\!\!-\!\!\bigcirc\!\!\overset{\cdot}{\cdot}\!\!-\!\!\overset{+}{N}{}^{+-} \qquad (10.11)$$

$$(10.44) \qquad\qquad\qquad (10.45)$$

The colour thus developed in the film matrix disappears more or less quickly in the absence of u.v. radiation, depending upon the kind of viologen and polymer matrix, as well as the moisture content in the polymer film.

(b) Poly[p-(N,N-dimethylamino)-N-γ-D-glutamanilide] (10.46) cross-linked with 1.5–2.6% of 2,6-bis(bromomethyl)naphthalene in carbon tetrabromide (CBr_4) (acceptor), under u.v. irradiation, exhibits changes in the dimensions of a polyelectrolyte gel, i.e. mechano-optical effect:[247]

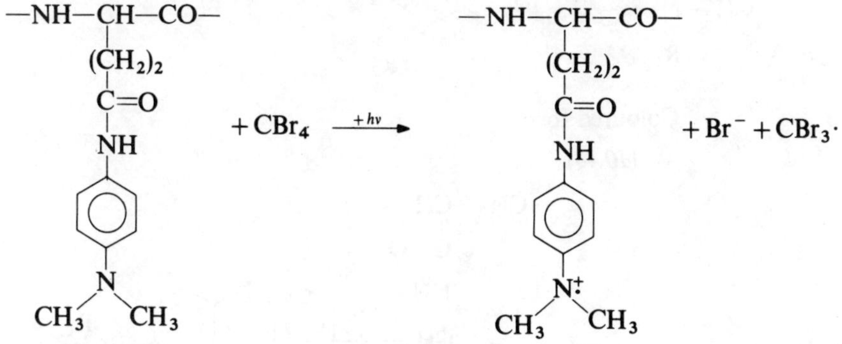

$$(10.46) \qquad\qquad\qquad (10.12)$$

(c) Polymers with a photochromic group, such as mercury thiocarbazonate (10.47), exhibit reversible changes in their colours under irradiation, which is a result of the photodimerization reaction:[1928, 1929]

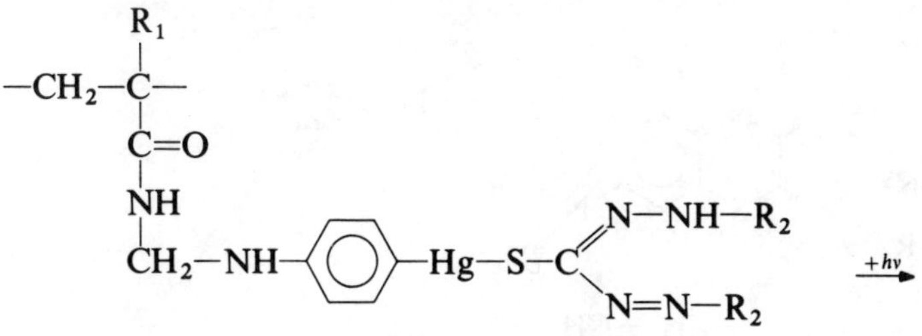

$$(10.47) \qquad\qquad\qquad (10.13)$$

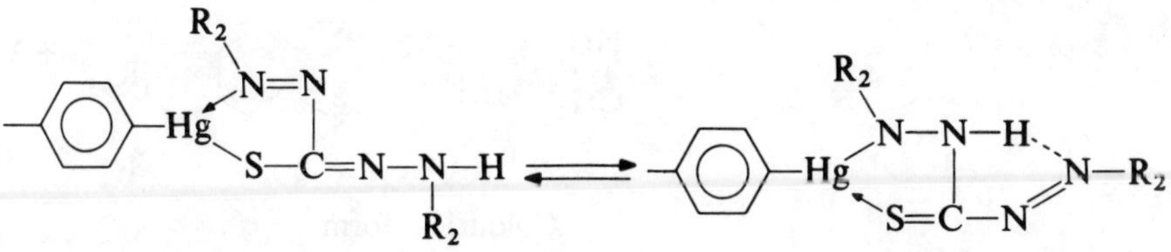

Orange form   Red form

(d) Polymers with photochromic thiazine group (*10.48*) in the presence of electron-donating compounds (D) such as ferrous ion (Fe^{2+}) or compounds with hydroxyl, amino or sulphydryl groups, and possibly activated water molecules, on light irradiation undergo reversible redox reactions into semiquinone polymer (*10.49*) and to a reduced form (*10.50*), which is accompanied by a remarkable change in absorbance around 600 nm:[1928, 1930–1932]

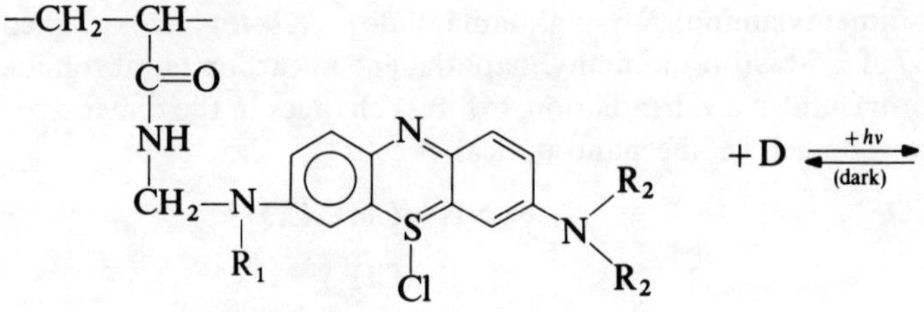

$$-CH_2-CH- \quad \begin{array}{c} | \\ C=O \\ | \\ NH \\ | \\ CH_2-N \\ | \\ R_1 \end{array} \quad \text{[thiazine ring, Cl, } N(R_2)_2\text{]} \quad +D \underset{(dark)}{\overset{+h\nu}{\rightleftharpoons}}$$

Coloured form

(*10.48*)

$$-CH_2-CH- \quad \begin{array}{c} | \\ C=O \\ | \\ NH \\ | \\ CH_2-N \\ | \\ R_1 \end{array} \quad \text{[thiazine ring, } \dot{N}, S, N(R_2)_2\text{]} \quad +A$$

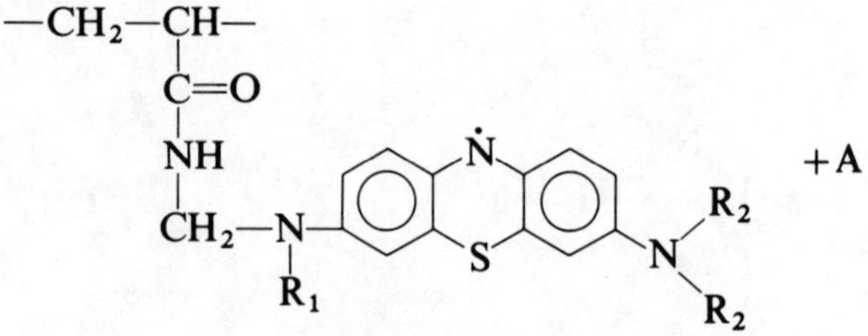

Colourless form

(*10.49*)   (10.14)

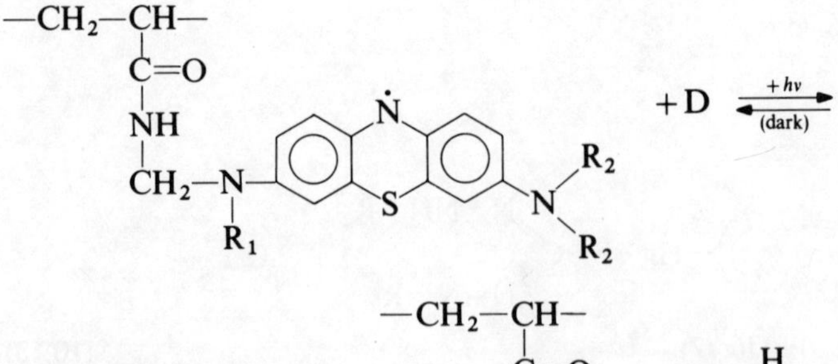

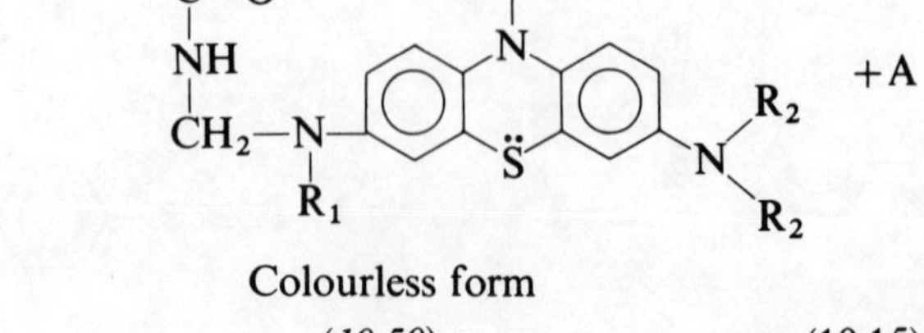

$$-CH_2-CH- \quad \begin{array}{c} | \\ C=O \\ | \\ NH \\ | \\ CH_2-N \\ | \\ R_1 \end{array} \quad \text{[thiazine ring, } \dot{N}, S, N(R_2)_2\text{]} \quad +D \underset{(dark)}{\overset{+h\nu}{\rightleftharpoons}}$$

$$-CH_2-CH- \quad \begin{array}{c} | \\ C=O \\ | \\ NH \\ | \\ CH_2-N \\ | \\ R_1 \end{array} \quad \text{[thiazine ring, } N-H, \ddot{S}, N(R_2)_2\text{]} \quad +A$$

Colourless form

(*10.50*)   (10.15)

where A represents electron-accepting compounds such as ferric ion (Fe^{3+}), compounds with disulphide groups, or oxygen. The reversible colour fading at 600 nm of thiazine polymers under irradiation in the solid state is dependent upon the moisture content in the polymer. Water molecules adsorbed on the active groups in the polymer participate in the reduction of the thiazine groups.

10.11. PHOTOCHROMIC BIOPOLYMERS

10.11.1. Photoinduced Stereochemical Changes

Stereochemical changes in the photochromic molecules incorporated into biopolymer chains can cause conformation variation of the polymer. The photochromic groups act as photoregulated 'switches' changing reversibility from one geometric isomer to another under influence of light.

Polypeptides such as L-p-(phenylazo)phenylalanine,[395,396,1360-1362] poly(L-asparates),[3698-3701,3705-3708] and poly(L-glutamic acid),[709,1674,2948] having azo photochromic groups and poly(L-glutamates) with stilbene photochrome,[1110] can take various conformations whose chiroptical properties are distinctly different from each other under irradiation.

10.11.2. Photoregulation of Enzyme Catalytic Activity

The catalytic activity of immobilized enzymes can be photocontrolled by two methods:[1961-1963,2557,2623]

(i) Direct control of modified enzymes bound on, or entrapped in, an insoluble carrier (e.g. spiropyran–urease in collagen membrane).
(ii) Change of the environmental characteristics around the enzymes. The reversible change of properties of the carrier with light irradiation may cause a change in the activity of the immobilized enzymes.

10.11.3. Photoregulated Biological Processes

Light receptors for photoregulated biological processes seem to consist of photoisomerizable molecules embedded in protein matrix, such as rhodopsin[3784] or phytochromes.[1999] Rhodopsin is composed of protein opsin and a chromophore retinal, which undergoes a light-induced geometrical isomerization from cis to trans form. The isomerization causes a change of conformation of the protein, giving rise to nerve excitation.

11. Photocrosslinking of polymers

PHOTOCROSSLINKING of macromolecules causes formation of three-dimensional network structures, which are insoluble. The crosslinked product which is swollen by the solvent is called GEL. Gel formation occurs after a particular conversion, called the GEL POINT.

Photocrosslinking of a polymer (PH) may occur by the following reactions:[2236]

(i) Photocrosslinking by reaction of two macroradicals (P·) formed during u.v. irradiation of a polymer:

$$P· + P· \rightarrow P-P \tag{11.1}$$

e.g. polypropylene in vacuum[2156] and unsaturated polyesters[2653] upon u.v. irradiation crosslinks.

(ii) Photocrosslinking initiated by the free radicals formed from the photo-decomposition of photoinitiators (I):[2827]

$$I + h\nu \rightarrow R_1· + R_2· \tag{11.2}$$

$$PH + R_1·(or \ R_2·) \rightarrow P· + RH \tag{11.3}$$

$$P· + P· \rightarrow P-P \ (crosslinking) \tag{11.4}$$

A number of different initiators used for photocrosslinking of polymers are collected in Table 11.1.

(iii) Photocrosslinking through photosensitive functional groups attached to macromolecules (photoreactive polymers) (see Section 12).

(iv) Photocrosslinking reactions resulting from the interaction of electron donor and electron acceptor side groups attached to macromolecules. On irradiation photoexcited donor–acceptor complex (exciplex) is formed, in which hydrogen transfer proceeds and results in crosslinking by free radical combination (see Section 11.1).

(v) Photoinitiated cationic crosslinking (cf. Section 11.2).

Photocrosslinked polymers are characterized by:[3022]

(i) NETWORK CHAIN LENGTH, which is the number of chain links between two branch points in the network.

(ii) BRANCH POINT, i.e. the point from which more than two chains radiate.

(iii) DEGREE of BRANCHING, defined as moles of crosslinked monomeric units per total moles of monomeric unit present.

Table 11.1   Examples of photoinitiators for photocrosslinking polymers

Photoinitiator	Polymer	References
Benzophenone	Polyethylene	181, 681, 2827
	Polypropylene	241, 1505
	Polystyrene	3951
	Poly(α-methylstyrene)	1751, 3951
	Co(ethylene/propylene/hexadiene-1,4) (EPDM)	486, 490–493
	Polyenes with polythiols	2570
	Poly(ethylene glycol adipate)	838
Michler's ketone	Poly(1,2-butadiene)	3529
Bis(p-hydroxybenzylidene)acetone	Polyacrylates	520
Sodium benzoate	Poly(vinyl chloride)	3525
Benzoin and its derivatives	Polyethylene	2827
	Polyacrylates	2232
	Unsaturated poly(methacrylic esters)	3383
	Co(ethylene/propylene/hexadiene-1,4) EPDM	486–488, 490, 492
Deoxybenzoin	Polyethylene Co(ethylene/propylene/hexadiene-1,4) EPDM	2827
Fluorenone	Co(methyl methacrylate/aminoethylmethacrylate)	2238
Quinones	Poly(methyl methacrylate)	648
Anthracene	Poly(2,3-epithiopropyl methacrylate)	3673
Metal chelates	Poly(2,3-epithiopropyl methacrylate)	1021
Hexachlorobenzene	Polyethylene	681
	Poly(vinyl chloride)	3367
4-Chlorobenzophenone	Poly(vinyl chloride)	3367
2-Chloroanthraquinone	Poly(vinyl chloride)	3367
	Poly(methyl methacrylate)	2822
Chlorinated polypropylene	Polyethylene	2794
Benzene chromium tricarbonyl 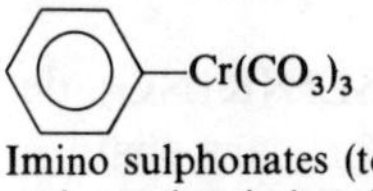	Poly(methyl methacrylate)-co(N-2-hydroxylpropylmethacrylamide)	1032

$$\langle\!\bigcirc\!\rangle\!-\!Cr(CO_3)_3$$

Imino sulphonates (tosyloxyimino
and mesyloxyimino derivatives)

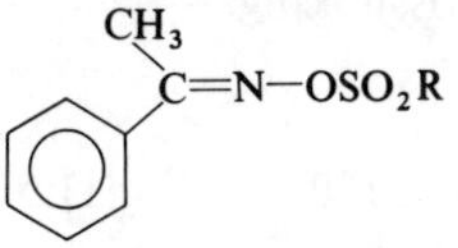

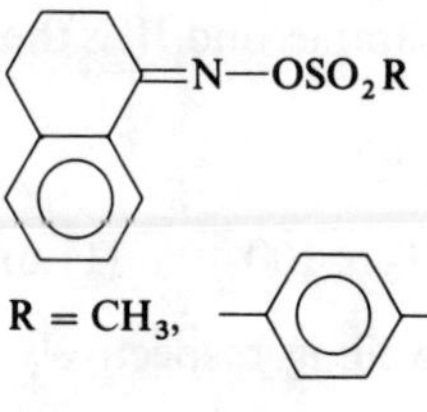

Poly(2,3-epoxy propyl methacrylate)   3326

$$\overset{CH_3}{\underset{}{\bigcirc}}C{=}N{-}OSO_2R$$

$$\bigcirc{=}N{-}OSO_2R$$

$$R = CH_3, \quad -\!\langle\!\bigcirc\!\rangle\!-CH_3$$

Table 11.1   (Contd)

Photoinitiator	Polymer	References
Organic sulphur compounds: Phenacyl phenyl sulphide	Poly(2,3-epoxypropyl methacrylate)	3675

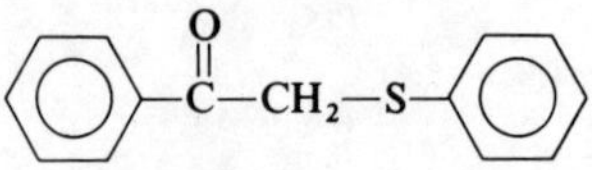

$$\bigcirc\!\!-\!\!\overset{\overset{O}{\|}}{C}\!\!-\!\!CH_2\!\!-\!\!S\!\!-\!\!\bigcirc$$

Phynacyl phenyl sulphoxide

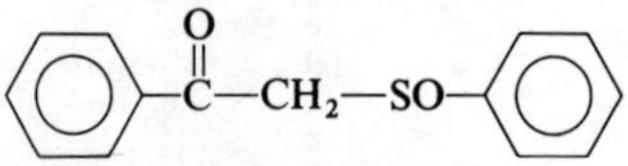

$$\bigcirc\!\!-\!\!\overset{\overset{O}{\|}}{C}\!\!-\!\!CH_2\!\!-\!\!SO\!\!-\!\!\bigcirc$$

Phenacyl phenyl sulphone

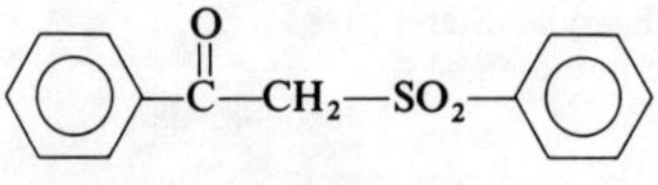

$$\bigcirc\!\!-\!\!\overset{\overset{O}{\|}}{C}\!\!-\!\!CH_2\!\!-\!\!SO_2\!\!-\!\!\bigcirc$$

Diphenyl disulphide

$$\bigcirc\!\!-\!\!S\!\!-\!\!S\!\!-\!\!\bigcirc$$

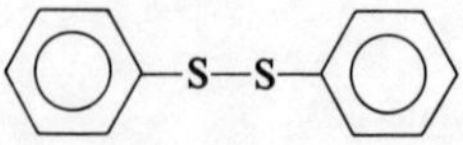

S-phenyl benzenethiosulphonate

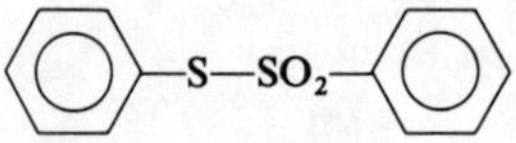

$$\bigcirc\!\!-\!\!S\!\!-\!\!SO_2\!\!-\!\!\bigcirc$$

Diphenyl disulphone

$$\bigcirc\!\!-\!\!SO_2\!\!-\!\!SO_2\!\!-\!\!\bigcirc$$

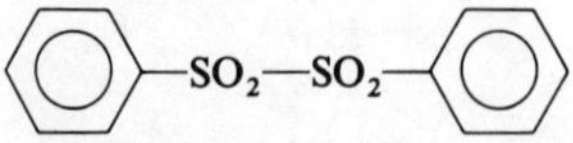		
TiO_2	Polypropylene	2156

(iv) CROSSLINK DENSITY (or the DEGREE OF CROSSLINKING), defined as the number of crosslinked monomeric units per primary chain.

The following methods are used for the study of photocrosslinking:

(i) Measuring the gel fraction (%) in a given solvent:

$$\text{GEL FRACTION } (\%) = (W_0 - W)/W \times 100 \qquad (11.5)$$

where W_0 is the total weight of the partially crosslinked sample, and W is the weight of uncrosslinked soluble fraction.

(ii) Measuring the degree of swelling (%) in a given solvent:

$$\text{DEGREE OF SWELLING } (\%) = (l - l_0)/l_0 \times 100 \qquad (11.6)$$

where l_0 and l are the length of the film before and after swelling, respectively.

(iii) Comparison of i.r. spectra before and after irradiation.[3022]

Mathematical relations formulated to permit calculation of QUANTUM YIELD OF CROSSLINKING (Φ_c)

$$\Phi_c = \frac{\text{Crosslink density}}{\text{Number of quanta absorbed}} \tag{11.7}$$

require data from readily observable quantities such as: polymer molecular weight ($\overline{M}_w$), solution viscosity, gel content, film thickness, radiation absorption coefficient, and initial radiation intensity.[3335]

In practice by measuring the soluble fraction as a function of energy absorbed it is possible to evaluate the quantum yield of crosslinking (Φ_c):[682]

$$S + S^{\frac{1}{2}} = \frac{\rho_0}{q_0} + \frac{1}{\Phi_c \overline{M}_w r} \tag{11.8}$$

where:

S = soluble fraction after absorption of a given number of quanta,

ρ_0 = proportion of main chain units fractured per unit radiation dose,

q_0 = proportion of monomeric units crosslinked per unit radiation dose,

$\overline{M}_w$ = initial weight–average molecular weight,

r = dose in einstein per gram.

A plot $S + S^{\frac{1}{2}}$ against the reciprocal dose yields a straight line which allows calculation of the quantum yield of crosslinking (Fig. 11.1).

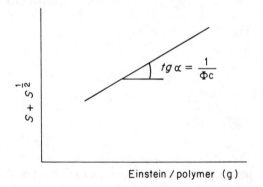

Fig. 11.1 A plot of $S + S^{\frac{1}{2}}$ versus dose of Einsteins per gram of polymer

Photocrosslinking has a wide variety of applications:

(i) Photoresists (cf. Chapter 12).
(ii) Curing (cf. Chapter 13).
(iii) In printing, data storage and retrieval.
(iv) Surface treatment.

Photocrosslinking of linear polymers causes embrittlement and decrease of mechanical properties. In many cases photocrosslinking is followed by chain scission processes.

11.1. PHOTOCROSSLINKING THROUGH DONOR–ACCEPTOR COMPLEXES

This type of photocrosslinking is based on the reaction between electron donor (D) and electron acceptor (A) groups attached to the polymer backbone. On irradiation a photoexcited donor–acceptor complex (exciplex) is formed, in which hydrogen transfer proceeds, giving rise to the production of two radicals, of which the recombination results in crosslinking. This type of photocrosslinking occurs between poly(dimethylaminostyrene) (D) (*11.1*) and poly(vinylbenzophenone) (A) (*11.2*):[3383]

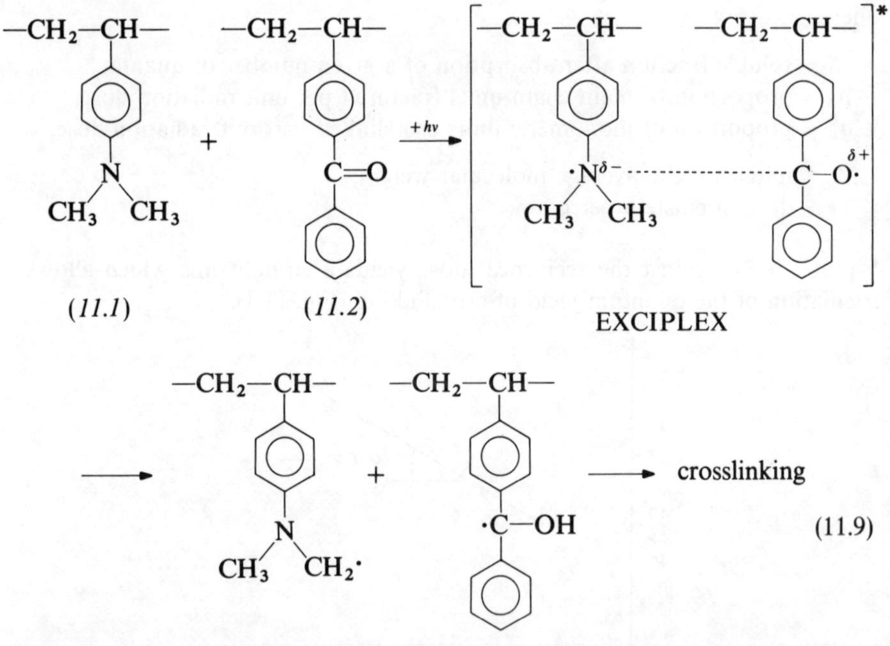

$$D + A \rightleftharpoons [D \ldots A] \tag{11.10}$$

Formation of exciplexes may originate from the excitation of a donor–acceptor complex preexisting in the ground state and/or from the excited triplet state (T_1) of benzophenone with ground state of dimethylaminostyrene units, following the reaction:

$$D + A \rightleftharpoons [D \ldots A] \tag{11.10}$$

$$D + A^* \longleftrightarrow [D^{\overline{\cdot}} + A^{\overset{+}{\cdot}}]^* \tag{11.11}$$

EXCIPLEX

Similar reactions occur during irradiation of co(dimethylaminostyrene/vinyl-benzophenone), and terpolymer(dimethylaminostyrene/vinylbenzophenone/butyl acrylate).[3383]

11.2. PHOTOINITIATED CATIONIC CROSSLINKING

Poly(2-phenylbutadiene) (donor) (*11.3*) in the presence of tetracyanobenzene (acceptor) (*11.4*) in polar solvents under u.v. irradiation is photocrosslinked by cationic mechanism:[1965]

$$-CH_2-\underset{\underset{\bigcirc}{|}}{C}-CH-CH_2- \quad + \quad \underset{CN}{\overset{CN}{\bigcirc}}\overset{CN}{\underset{CN}{}} \quad \xrightarrow{+h\nu}$$

$$(11.3) \qquad\qquad (11.4)$$

$$-CH_2-\underset{\underset{\bigcirc}{|}}{\overset{\cdot}{C}}-\overset{+}{C}H-CH_2- \quad + \quad \underset{CN}{\overset{CN}{\bigcirc}}\overset{CN\overline{\cdot}}{\underset{CN}{}} \qquad (11.12)$$

Poly(2-phenylbutadiene) has an phenol-electron donating group and of double bond in the main chain conjugated with the phenyl group. In the presence of acceptor under u.v. irradiation radical cation is produced in the $C=C$ position. This contributes to the primary process of the crosslinking:

$$-CH_2-\overset{\cdot}{\underset{\underset{\bigcirc}{|}}{C}}-\overset{+}{C}H-CH_2- \quad -CH_2-\underset{\underset{\bigcirc}{|}}{C}-CH-CH_2-$$

$$+ \qquad \longrightarrow \text{ chain propagation}$$

$$(11.13)$$

The termination step of the propagating chain could involve transfer to monomer, combination with the acceptor radical anion, or electron transfer from the acceptor radical anion to the carbonium ion to yield a terminal radical.

The reaction is not quenched appreciably by H_2O, oxygen, or γ-collidine, but is quenched by triethylamine, and cannot be initiated by azoiso-butyronitrile. Small concentrations of H_2O, added before and after irradiation, induce a postcrosslinking process that proceeds to completion within a short time and appears to be proton-initiated.

Similar reactions occur during u.v. irradiation of such polymers as poly[m-(vinyloxyethoxy)styrene] (*11.5*) and poly[m-[vinyl(α-nitrostyrylo-oxyethyl)ether]] (*11.6*) in the presence of an electron acceptor:[578]

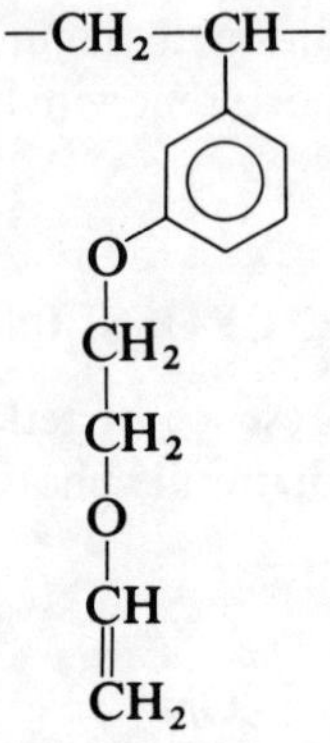

(*11.5*)

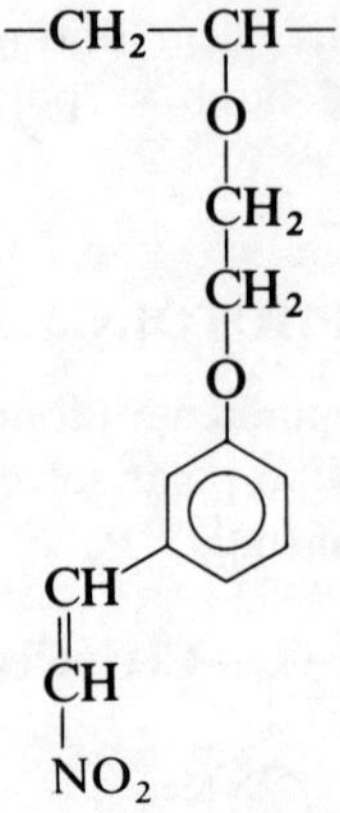

(*11.6*)

Photoinitiated cationic crosslinking has also been reported for the donor–acceptor pairs: poly(2,3-epithiopropyl methacrylate) and 4-*N*,*N*-diethylaminobenzene diazonium tetrafluoroborate[3674] and co(2,3-epithiopropyl methacrylate/methyl methacrylate) and diazonium salts.[1005, 1022]

12. Photoreactive polymers in image formation

The application of photoreactive polymeric systems to image formation has been a subject of many reviews and books.[528, 925, 929–934, 1084, 1093, 1401, 1848, 2113, 2236, 2730, 3050, 3089, 3131, 3298, 3561, 3763, 3892, 3894–3896, 3898, 3899, 3900, 3905, 3998]

IMAGE FORMATION can be obtained using PHOTORESIST COATINGS which change their physical properties (usually solubility) under exposure to light:

(i) Photocrosslinking and photoinitiated polymerization decreases solubility (NEGATIVE-WORKING MODE OF PHOTORESISTS) (Fig. 12.1).
(ii) Photomodification and photodegradation increase solubility (POSITIVE-WORKING MODE OF PHOTORESISTS) (Fig. 12.2).

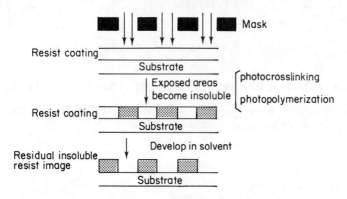

Fig. 12.1   Negative-working mode of photoresists[3900]

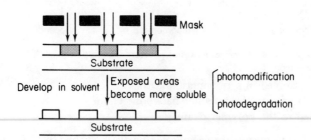

Fig. 12.2   Positive-working mode of photoresists[3900]

413

414

The similarities and differences of the various process steps for positive- and negative-working resists are shown in Fig. 12.3.

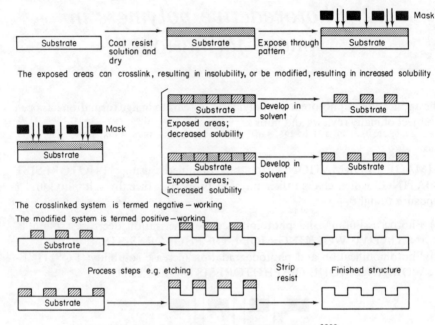

The exposed areas can crosslink, resulting in insolubility, or be modified, resulting in increased solubility

The crosslinked system is termed negative – working
The modified system is termed positive – working

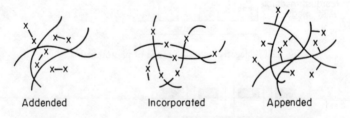

Fig. 12.3 Schematic outline of photoresist use[3900]

12.1. PHOTOCROSSLINKABLE POLYMERIC SYSTEMS

The photoreactive species (group) in a given PHOTOCROSSLINKABLE POLYMERIC SYSTEM can be (Fig. 12.4):

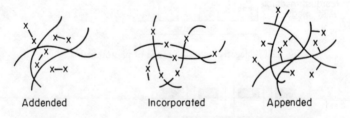

Fig. 12.4 Principal means of photocross-linking polymers[3900]

(i) Simply dissolved in a polymeric matrix.
(ii) Incorporated into the polymer backbone chain.
(iii) Appended to the polymer chain.

Inter- and intramolecular crosslinks can be distinguished by their effect on the insolubilization of the polymer:

(i) INTERMOLECULAR LINKS increase the molecular weight of the polymer and eventually give rise to a three-dimensional network of chains, which constitutes an insoluble gel.

(ii) INTRAMOLECULAR LINKS have no effect on molecular weight and do not contribute to gel formation.

The progress of intermolecular crosslinking may therefore be monitored by the increase in the amount of gel formed during irradiation. A plot of the weight or volume of gel against the radiation dose which produced it is termed the GEL CURVE OF THE PHOTOPOLYMER.[3091]

A point of interest on the gel curve is the minimum radiation dose at which the gel makes its first appearance in the system. This GEL POINT EXPOSURE (I_G) is related in a simple way to the QUANTUM YIELD OF INTER-MOLECULAR CROSSLINK FORMATION (Φ_G).

$$\Phi_G = \frac{\rho}{2.303\varepsilon \, c \, M_w \, I_G} \tag{12.1}$$

where:

ρ = specific gravity of the polymeric material,

ε = molar absorptivity (molecular extinction coefficient) of the reactant at the wavelength of irradiation,

c = concentration of the reactant in the solid film,

M_w = weight—average molecular weight of the polymer.

If the molecular weight (M_w) of the polymer and the optical characterization of the film (ε and c) are known, the quantum yield of intermolecular crosslink formation may be obtained from the gel point exposure of the system.

The sum of the inter- and intramolecular crosslinks produced by a given exposure is measured by the disappearance of reactant from the system. The overall quantum yield (Φ) of the link-producing photoreaction may thus be derived from the rate of change in the absorption spectrum of the film under irradiation. The rate of change of the optical density (D) at the wavelength of chromophore absorption may be expressed in the form:[3091]

$$\frac{dD}{dt} = \Delta\varepsilon \, I_0 \, A \, 10^3 \left(\frac{D - D_\infty}{D} \right) \Phi \tag{12.2}$$

where:

$\Delta\varepsilon$ = difference in molar absorptivity of reactant and product,

I_0 = incident energy flux (einstein cm^{-2} time^{-1}),

A = fraction of the incident energy flux absorbed.

The fraction of intermolecular crosslinks produced on irradiation of the polymer film, which is a direct measure of the degree of interpenetration of the macromolecules in the matrix, is given by the quantum yield ratio (Φ/Φ_G).

The gel point exposure which is required for the determination of the quantum yield of intermolecular crosslinking may be found by extrapolation of the gel curve to zero gel fraction.[3091]

12.2. PHOTOCROSSLINKABLE POLYMERS

PHOTOCROSSLINKABLE POLYMERS are polymers which contain photo-sensitive functional groups which have been attached to or incorporated within a polymer chain, which can become crosslinked under the influence of light.[3895] Examples of such photosensitive functional groups are shown in Table 12.1. Such groups can be attached to many polymers (Table 12.2) in many ways and, depending on its configuration in the polymer chain or where appended to the polymer chain, the light sensitivity and the physical properties of the polymer can be modified.

Rheological and physical properties of photopolymers are relatively un-explored.[415, 451, 675, 1700, 3952]

12.2.1. Photocrosslinkable Polymers with Photosensitive Unsaturated Functional Groups

Polymers having unsaturated double bond (cf. Table 12.1) in a main-chain (*12.1*) or in a pendant group (*12.2*) may, under irradiation, undergo photocrosslinking (reactions 12.3 and 12.5) or photocyclodimerization (reactions 12.4 and 12.6):

$$
\begin{array}{ccc}
& \begin{array}{c} -\overset{|}{C}H-CH- \\ -CH-CH- \\ {\scriptstyle|} \end{array} & (12.3) \\[2em]
\begin{array}{c} -CH=CH- \\ + \\ -CH=CH- \\ (\textit{12.1}) \end{array} \xrightarrow{\ +h\nu\ } & & \\[2em]
& \begin{array}{c} -CH-CH- \\ {\scriptstyle|}\ \ \ {\scriptstyle|} \\ -CH-CH- \end{array} & (12.4)
\end{array}
$$

$$
\begin{array}{ccc}
& \begin{array}{c} -\overset{\top}{C}H\ \ R \\ CH-CH \\ R\ \ \ \ CH- \\ {\scriptstyle\perp} \end{array} & (12.5) \\[2em]
\begin{array}{c} \overset{\top}{\underset{\perp}{\overset{\displaystyle CH}{\underset{\displaystyle CH}{\|}}}}\ {R}\ +\ \overset{\displaystyle CH}{\underset{\displaystyle CH}{\|}}\ {\underset{\perp}{\ }} \\ (\textit{12.2}) \end{array} \xrightarrow{\ +h\nu\ } & & \\[2em]
& \begin{array}{c} \overset{\top}{\ }\ \ R \\ CH-CH \\ CH-CH \\ {\scriptstyle|}\ \ \ {\scriptstyle\perp} \\ R \end{array} & (12.6)
\end{array}
$$

Two groups of this type of polymers, having cinnamate (*12.3*) and chalcone (*12.4*) groups, have been applied in photoresist technology.

Only four of the eleven stereoisomers, which in principle may be formed from two cinnamoyl groups, are observed:

Table 12.1 Photocrosslinkable polymers with different photoreactive (photosensitive functional) groups

Name of group	Structure	λ_{max}	References
Cinnamate	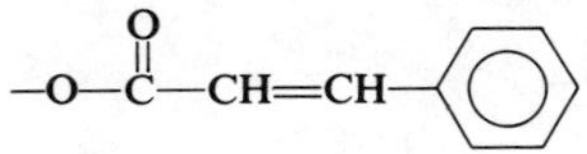 —O—C(=O)—CH=CH—⟨phenyl⟩	250	251–253, 330, 421–423, 495, 813, 1840, 1969, 2027, 2528, 2529, 2712, 2713, 2715, 2716, 2754, 3091, 3102, 3127, 3345, 3642, 3808, 3899
Chalcones (benzylidene-acetophenone)	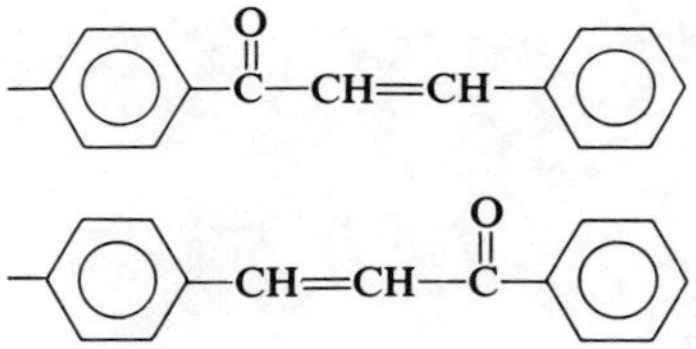 ⟨phenyl⟩—C(=O)—CH=CH—⟨phenyl⟩; ⟨phenyl⟩—CH=CH—C(=O)—⟨phenyl⟩	275	545, 1967, 3127, 3156, 3157, 3715, 3716
Cynnamelidene acetate (styrylacrylate)	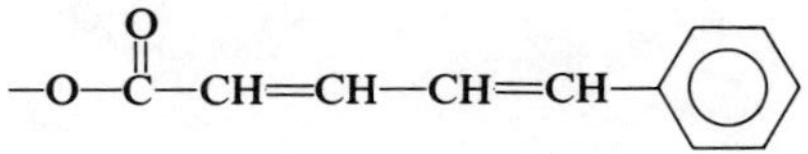 —O—C(=O)—CH=CH—CH=CH—⟨phenyl⟩	304	980, 1968, 1969, 3537, 3539, 3541, 3542, 3953
Chalcone acrylate (4-(2-benzoyl)vinyl cinnamic ester)	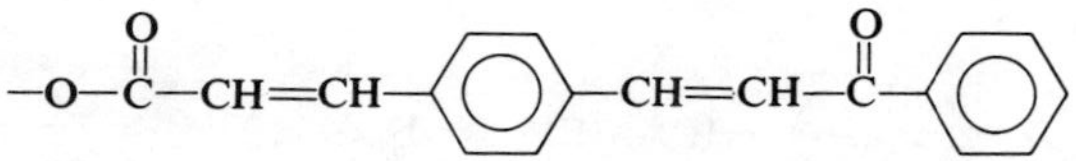 —O—C(=O)—CH=CH—⟨phenyl⟩—CH=CH—C(=O)—⟨phenyl⟩	338	1578, 2711, 2714
Ethyl-p-phenylenebis(acrylate)	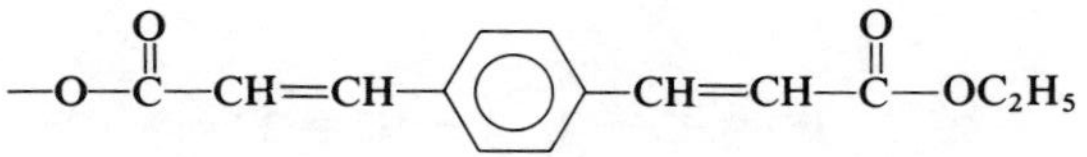 —O—C(=O)—CH=CH—⟨phenyl⟩—CH=CH—C(=O)—OC_2H_5	320	1578
Cinnamylidene malonate	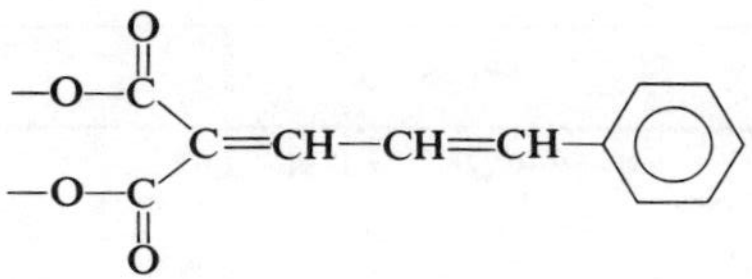 ⟨phenyl⟩—CH=CH—CH=CH—C(—C(=O)—O—)(—C(=O)—O—)	326	3715

417

Table 12.1 (*Contd.*)

Name of group	Structure	λ_{max}	References
p-Phenylene diacrylic ester	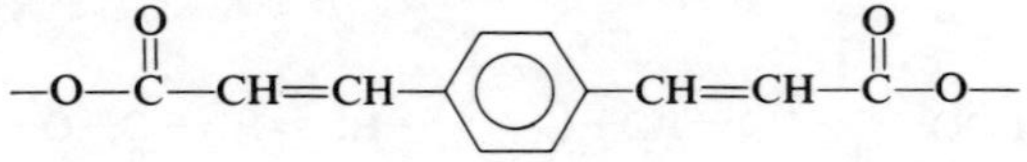	320	471, 1033, 3181, 3894, 3899
Phenol-type polycarbonates and polyesters	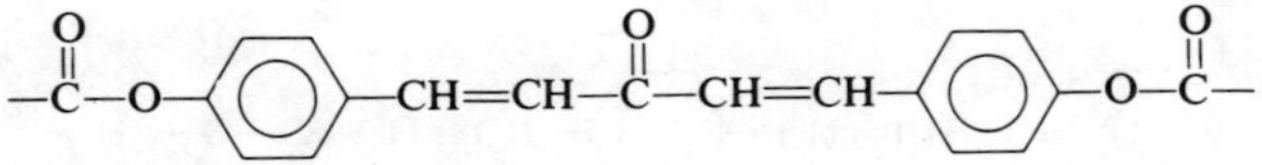	330	730
Analogues			469, 470
β-Furfurylacrylic ester	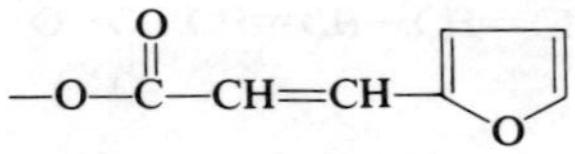		3641
α-Cyanocinnamic ester	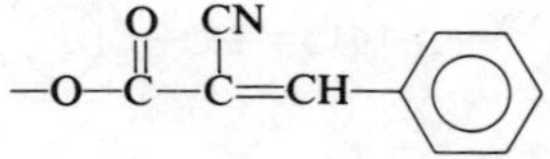		1969, 2710, 3534, 3535
α-Cyano-*β*-styrylacrylic ester (*α*-cyanocynnamelidene acetate)	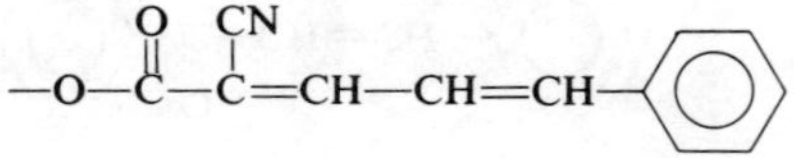		2441, 3142, 3539, 3542
Benzylideno acetones	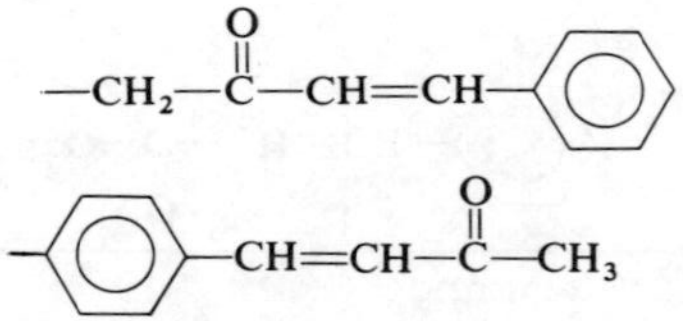		3658

Polyester of *p*-phenylene-bis-(α-cyanobutadiene carboxylic acid)   1835, 2642

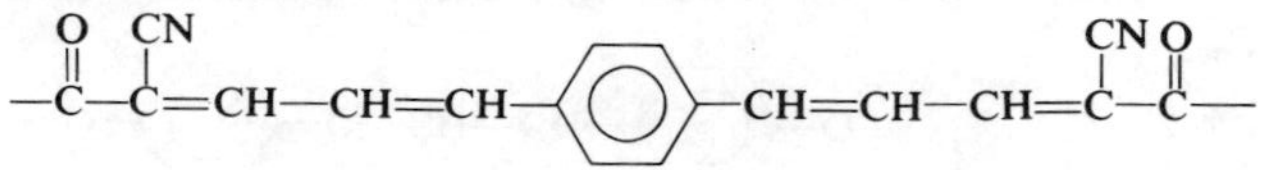

2,5-Diethoxystilbene   305   3467

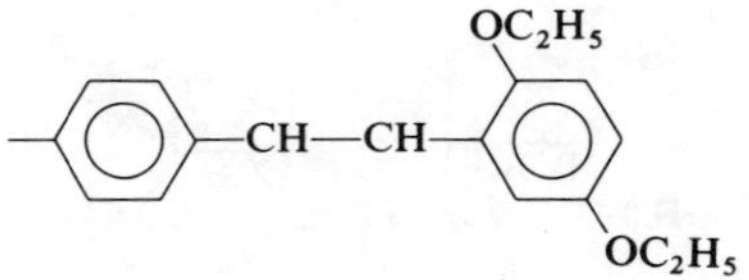

β-Naphthyl-acrylic ester   3540

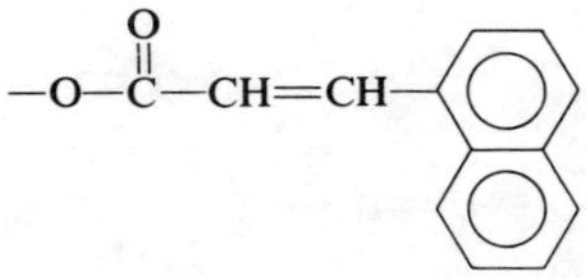

Cinnamelidence-pyruvic ester   3953

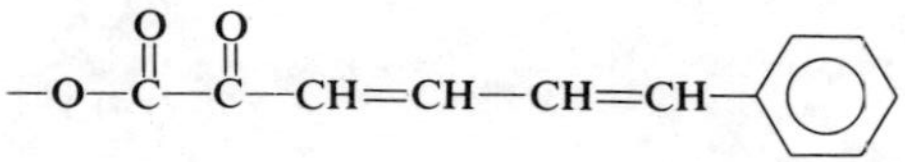

α-Chloro-deoxybenzoin   487

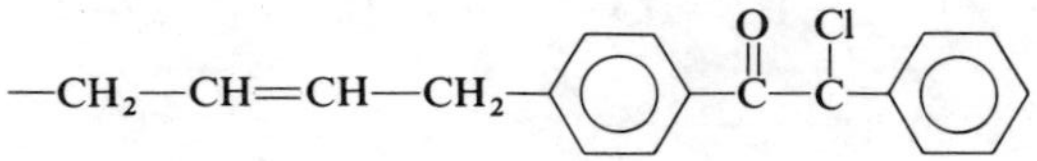

Phenylvinyl   1085

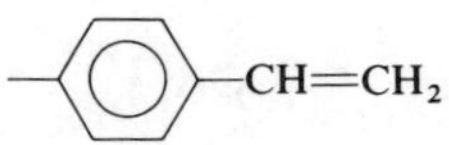

Maleic and fumaric polyesters   3376

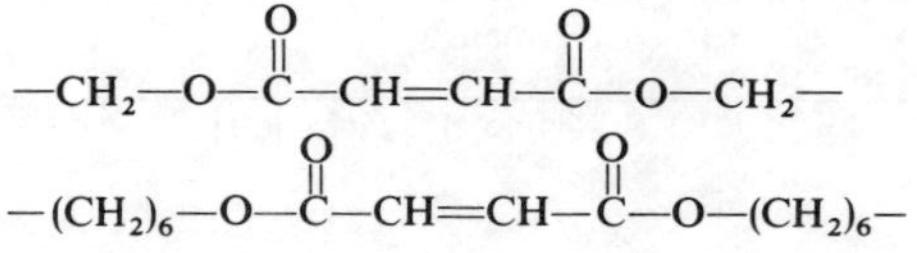

Table 12.1   (Contd.)

Name of group	Structure	λ_{max}	References
Acrylolyl	—C(=O)—CH=CH$_2$		2755
(Vinyloxy)ethoxy	—O—CH$_2$—CH$_2$—O—CH=CH$_2$		2717
Different distyrylketone analogues	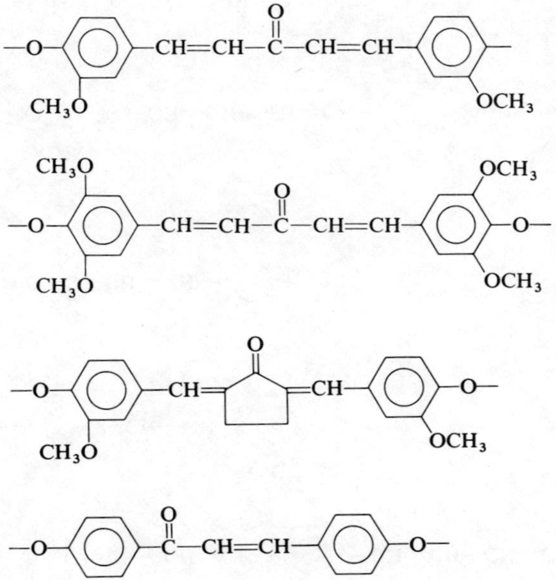		469, 470

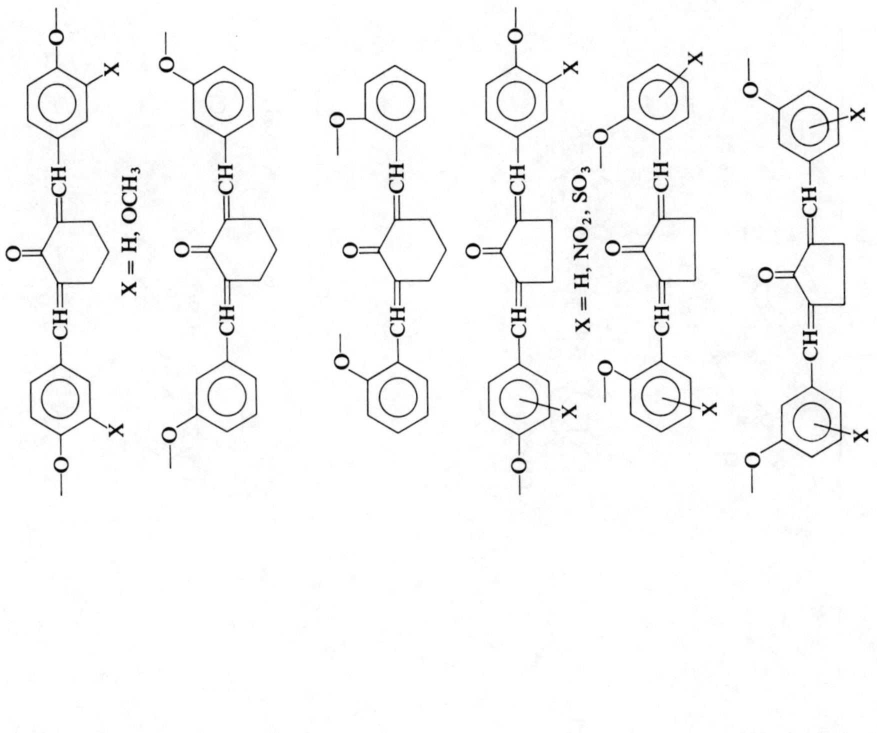

X = H, OCH₃

X = H, NO₂, SO₃

Table 12.1 (Contd.)

Name of group	Structure	λ_{max}	References
1,2-Diphenyl-cyclopropene-4-carboxylate (diphenylcyclopropenoic ester)	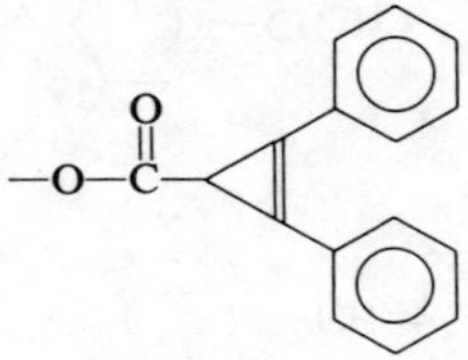	309	903, 906, 2271, 2509
Styrylpyridinium	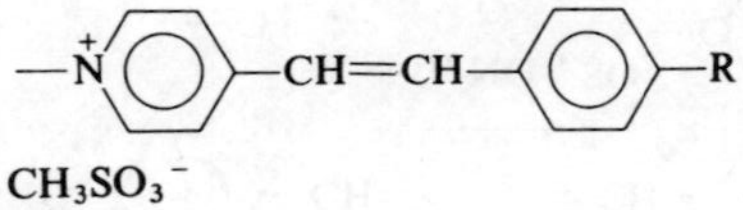 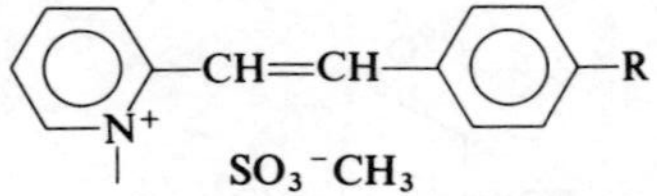	250–680	472, 1333, 1740–1743, 2271, 3897, 3899
4-Styrylquinolinium	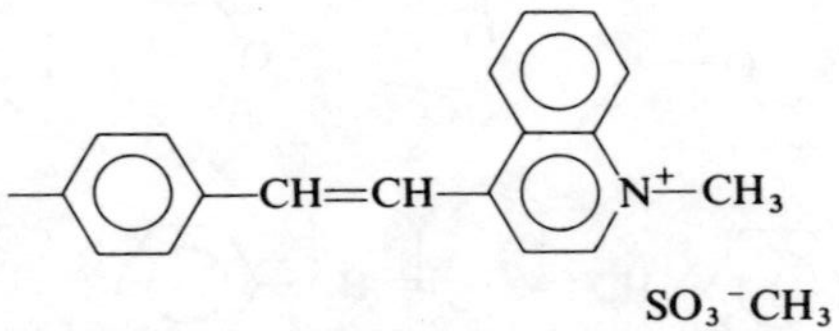		1737
Pyridinium dicyano-methylide	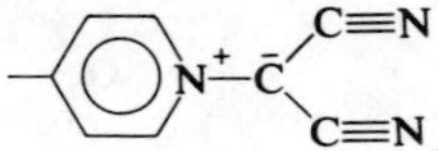		764
Vinylbenzyl-4(1'-pyrenvinyl)pyridinium chloride	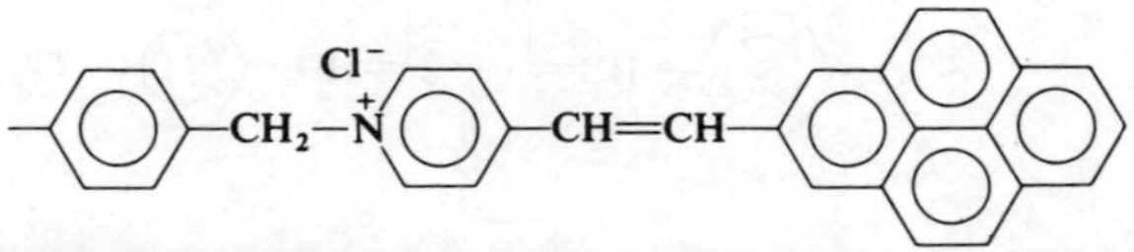		1333

Propargyl ether	$-OCH_2-C\equiv CH$	1972, 1973
Polydiacetylenes	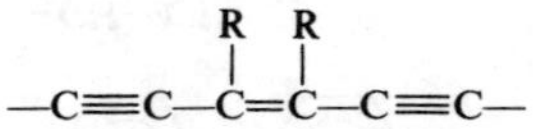	1539, 1540, 2906, 2907, 3826
1,4-Diethynyl-naphthalene and analogues	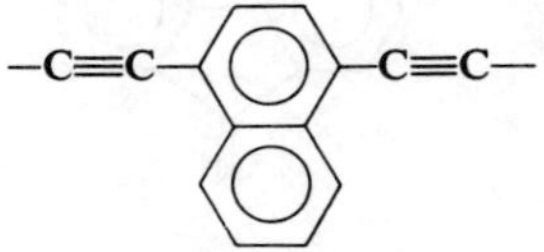	3109, 3132
Bis-propargyl ether of bisphenol (abbreviated to A) and its copolymer with p-diethynylbenzene	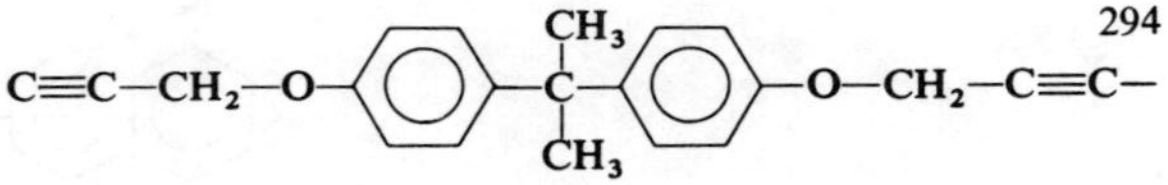	294, 1540
	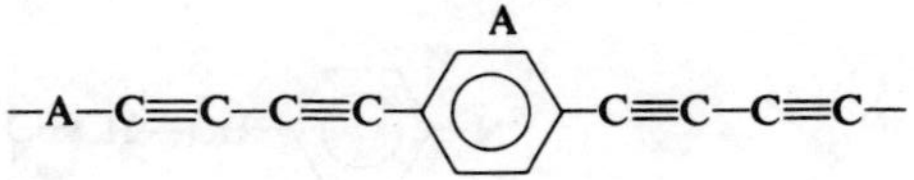	334
Azido-group	$-CH_2-CH-$ N_3	3526
p-Azidophenol	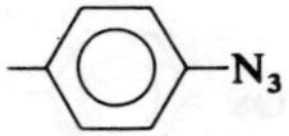	260, 938, 3715
p-Azidobenzoate	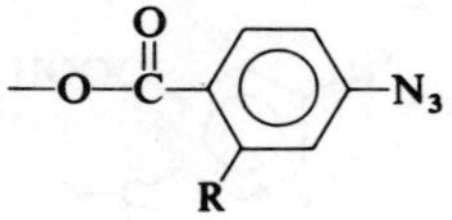 R = H, SO_3H, COOH	274, 2500, 3662, 3666, 3715
Azidophthalate	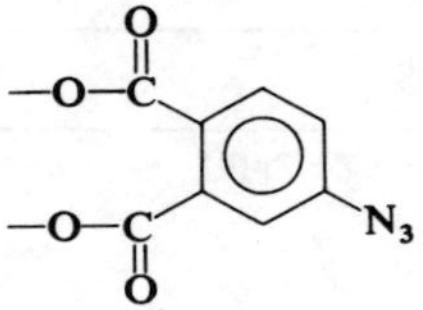	3715

Table 12.1 (Contd.)

Name of group	Structure	λ_{max}	References
p-Azidobenzoic ester	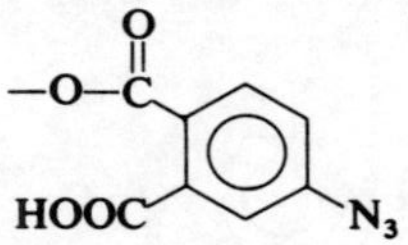		932, 937, 938, 2500
Acid azide	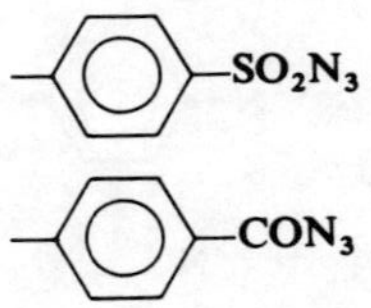		932
p-Azidocynnamic ester	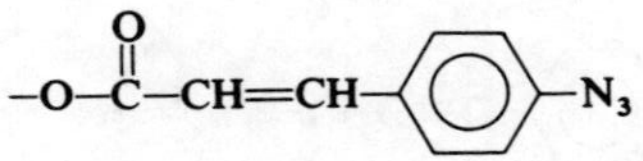		3947, 3948
Naphthylazide	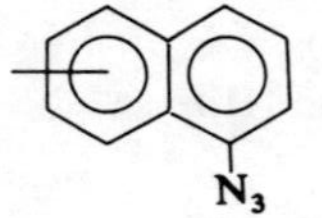		3477
9-Diazofluorenone	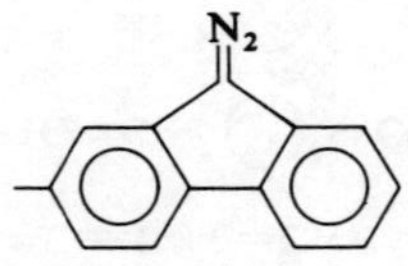		3969
Episulphide	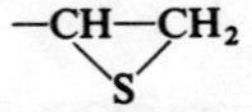	258	1019–1022

Disulphide	$-S-S-$	1697, 1698, 2244
Polysulphide	$-S_n R$	2571
Dithiocarbamate	$-S-\overset{\overset{S}{\|}}{C}-N(C_2H_5)_2$	2622
Xanthate	$-S-\overset{\overset{S}{\|}}{C}-OC_2H_5$	2779
Thiocyanato-acetylstyrene	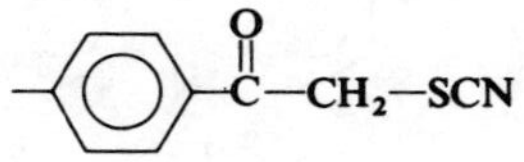 $\overset{\overset{O}{\|}}{C}-CH_2-SCN$ (on benzene ring)	290, 650, 3668, 3669
Polysulphate copolymers	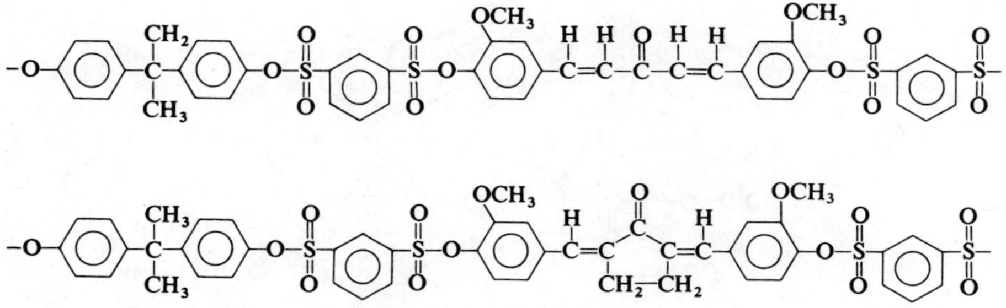	211
Chloroacetic ester	$-O-\overset{\overset{O}{\|}}{C}-CH_2Cl$	2756, 3536
O-acyl oximes	$\overset{\overset{O}{\|}}{C}-O-\overset{\overset{N}{\|}}{C}\overset{R}{\underset{R'}{<}}$	940, 3670

Table 12.1 (Contd.)

Name of group	Structure	λ_{max}	References
Methoxymethylstyrene	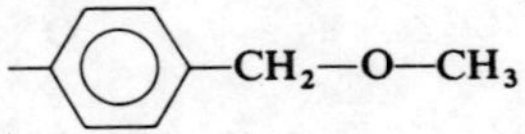		1196
Bromoacetylstyrene	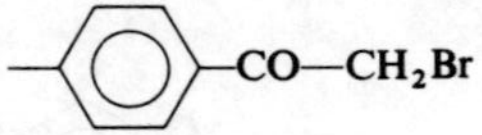		3672
Perester	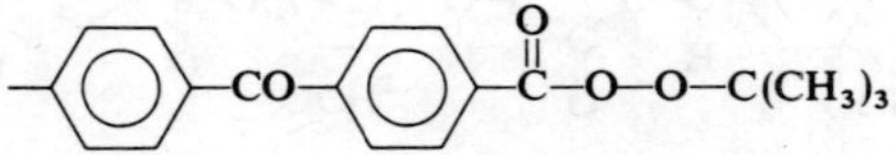		2666
Anthracene	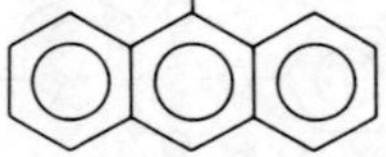		3451, 3565, 3566
N-Carbazolmethyl	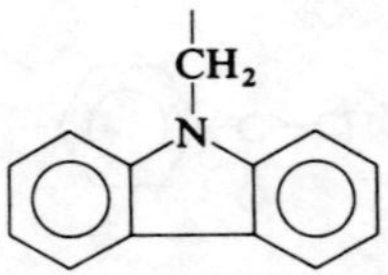		1333
9-Fluorenone	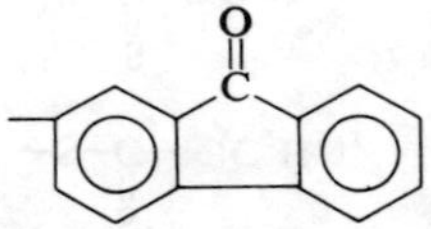		1333
Coumarine	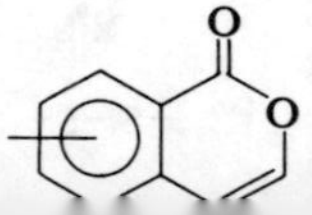		931, 936

Compound	Structure		
1,3-Dioxalane	$-CH-CH_2$, $O-C-R_1$, R_2		2651
Ferrocene		325–440	2097
Oxo-bis(8-quinolyloxy)vanadium(IV)		40	
Pyridine N-oxide		764, 922, 923	
Quinoline N-oxide		922	
p-N,N-dimethyl-aminostyrene N-oxide		923	

Table 12.1 (Contd.)

Name of group	Structure	λ_{max}	References
Pyrazine mono and di N-oxides	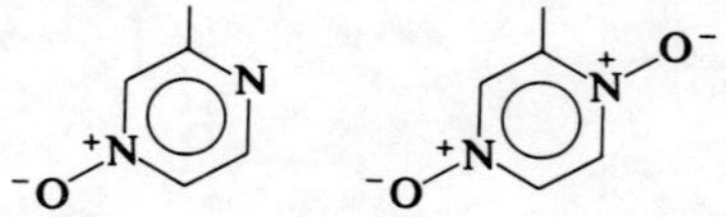		926,
1,2,3-Thiadiazole	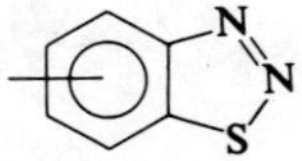	300–360	937
Furoin ether	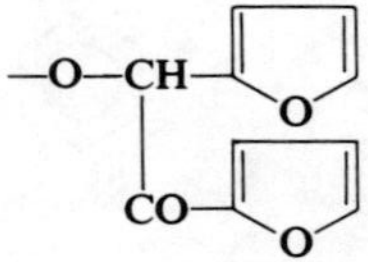		2172
Maleimide and its derivatives	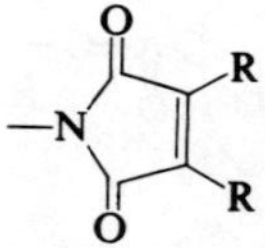 R = H, Cl, CH₃, C₆H₅, CN		411, 457, 1744, 2877, 3373, 3376, 3445, 4009
Dibenzoazepine	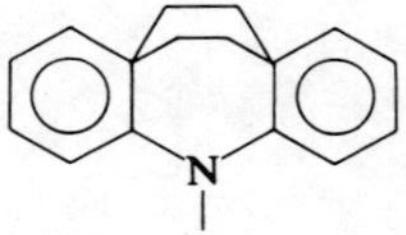		39, 1734

Table 12.2 Linking groups for photocross-linkable (PX) units[3900]

Name	Structure
Poly(vinyl alcohol)esters	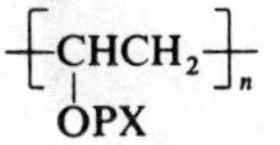 $+\text{CHCH}_2+_n$ OPX
Urethanes	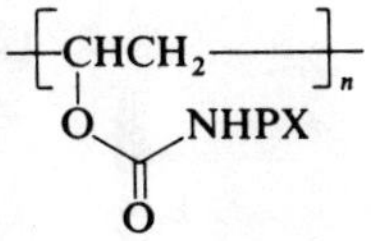 $\left[\text{CHCH}_2\underline{\quad\quad}\right]_n$ O C(=O) NHPX
	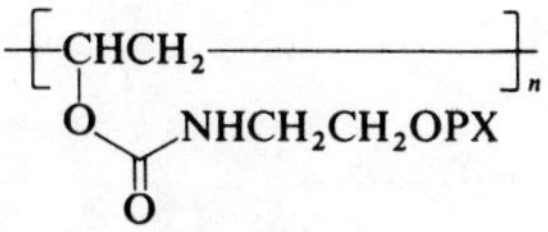 $\left[\text{CHCH}_2\underline{\quad\quad\quad}\right]_n$ O C(=O) NHCH$_2$CH$_2$OPX
Amide	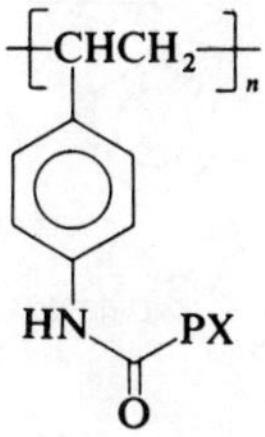 $+\text{CHCH}_2+_n$ with $-C_6H_4-$ HN C(=O) PX
Vinyl ethers	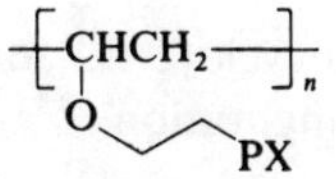 $\left[\text{CHCH}_2\underline{\quad}\right]_n$ O CH$_2$CH$_2$ PX
Bisphenol-epoxy polymers	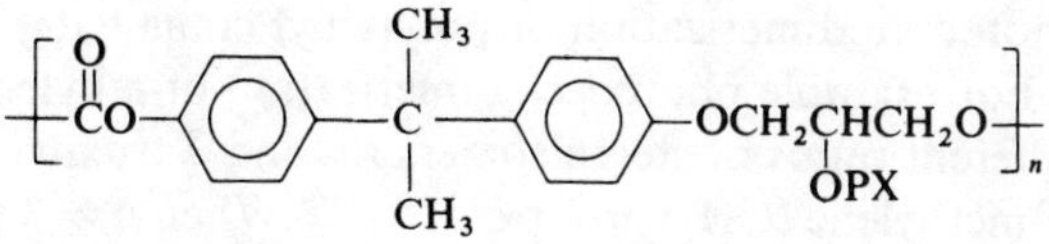 $\left[\overset{O}{\underset{}{C}}O - C_6H_4 - \overset{CH_3}{\underset{CH_3}{C}} - C_6H_4 - OCH_2\overset{OPX}{CH}CH_2O\right]_n$
Polystyryl	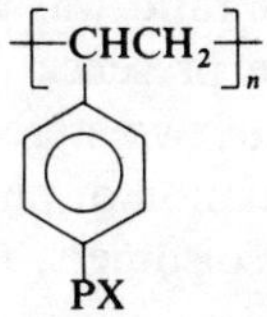 $+\text{CHCH}_2+_n$ with $-C_6H_4-PX$

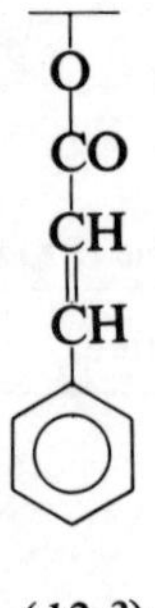

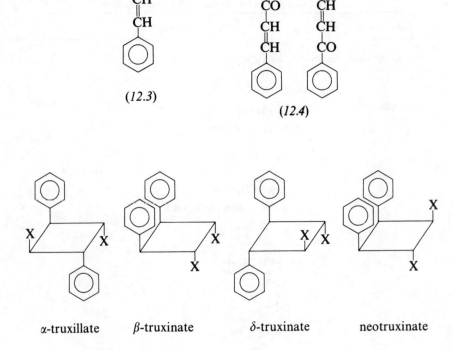

(12.3)

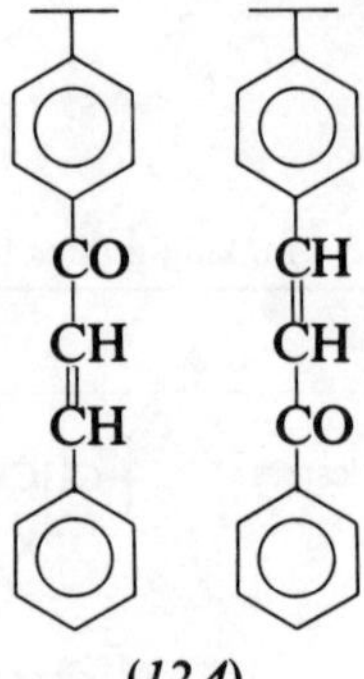

(12.4)

α-truxillate   β-truxinate   δ-truxinate   neotruxinate

The photocyclodimerization (photocyclo(2 + 2)addition) process in amorphous matrice requires stringent steric configuration.[1031] Various low molecular weight cinnamates have been investigated as model compounds for the study of photocyclodimerization of poly(vinyl cinnamate).[1184,1185,3899]

For example photocyclodimerization of ethylene-bis-cinnamate (12.5) shows different ratios of stereoisomers (β- and δ-truxinates) depending on the number of methylene (CH_2) groups.[1184,3899] When $n = 3$ the ratio of β- and δ-truxinate was 4:1, whereas with $n = 4$ or 5, the ratio was 1:4 (reactions 12.7 and 12.8).

Irradiation of phenanthrene in the presence of trans-methyl cinnamate in cyclohexane solution gives a mixture of two stereoisomers, (12.6) and (12.7), in 10:1 ratio respectively. Whereas irradiation of phenanthrene in the presence of cis-methyl cinnamate yields two stereoisomers, (12.8) and (12.9), in the ratio 10:1.[1082]

Attachment of phenanthrene and cinnamate groups to the polymer backbone (12.10) provides faster photocrosslinking, but these copolymers cannot be photosensitized to a speed greater than that obtained for poly(vinyl cinnamate) (12.11).[3899]

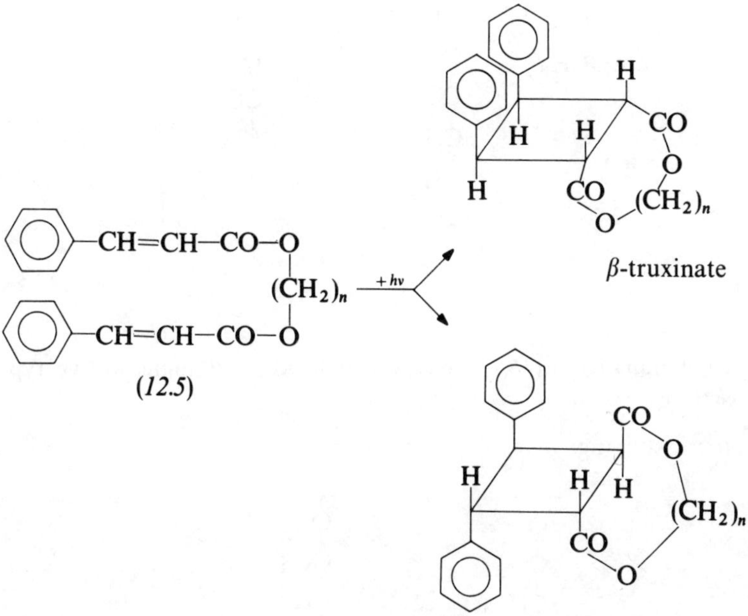

(12.7)

β-truxinate

(12.8)

δ-truxinate

(12.5)

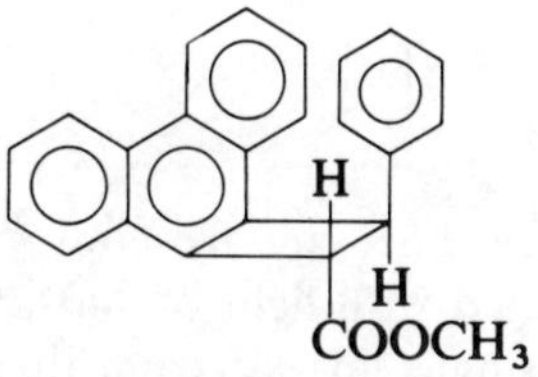

(12.6)

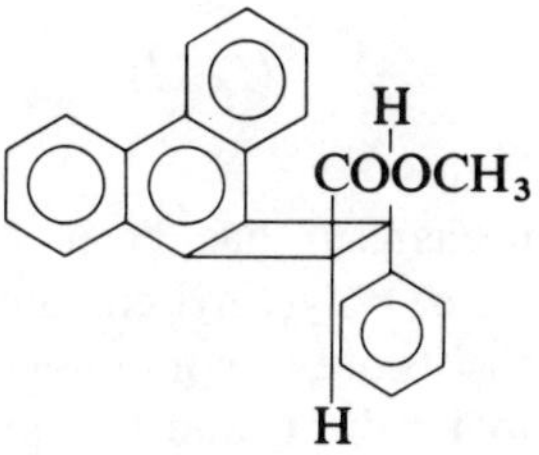

(12.7)

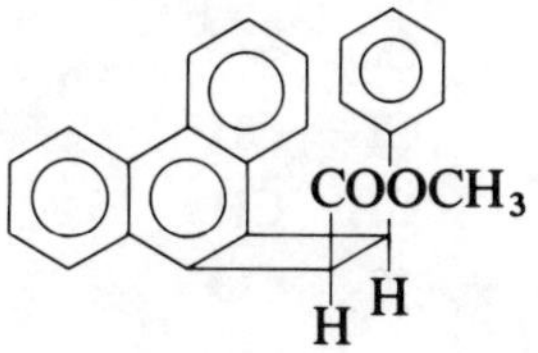

(12.8)

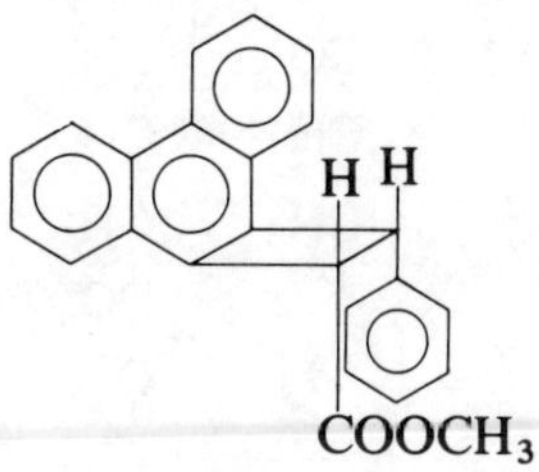

(12.9)

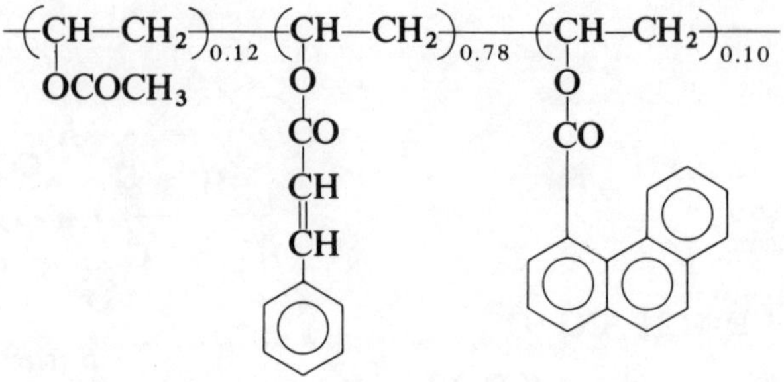

(*12.10*)

Poly(vinyl cinnamate) (*12.11*) may, under irradiation, undergo two types of photoreaction:

(i) Photocrosslinking:

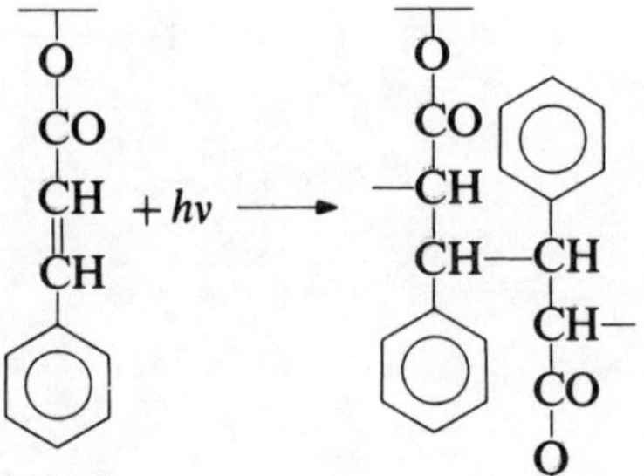

(12.9)

(*12.11*)

This mechanism has been supported by ESR spectroscopy.[2625] ESR spectrum of poly(vinyl cinnamate) irradiated with light of 240–250 nm at 77 K consists of two components: a broad singlet line spectrum attributed to cinnamoyl radicals and the quartet line spectrum to radicals on the main chain, which are produced by hydrogen abstraction reaction.[2020]

(ii) Photocyclodimerization (photocyclo(2 + 2)addition):[1031, 1184, 2529, 2624, 3127, 3414, 3659]

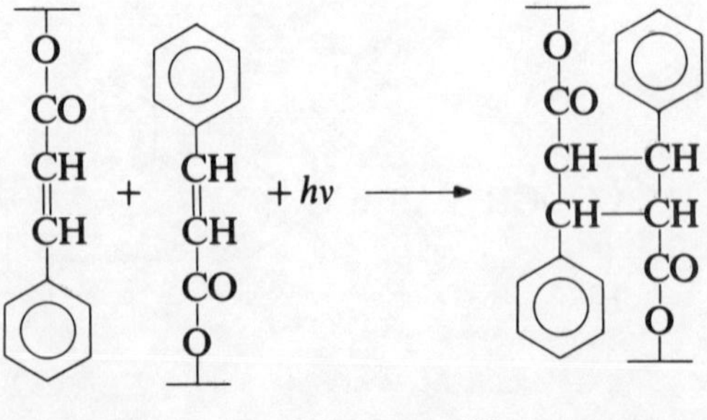

(12.10)

(*12.11*)   (mainly α-truxillic isomer)

Quantum yield of the disappearance of the double bond in poly(vinyl cinnamate) decreases rapidly with irradiation time. Extrapolation to zero time of exposure indicated a quantum yield $\Phi = 0.34$.[2624] The critical hardening value required is approximately 1.3 crosslinks per macromolecule, possessing a degree of polymerization of 1400.[2626] The kinetics of photodimerization of cinnamate groups in polymers have been studied in detail.[253,813,3642] Poly(vinyl cinnamate) does not show any photoreversibility because of very low efficiency of absorption of light by the photodimers of cinnamic acid.[3099]

The electronic structure of the excited states of poly(vinyl cinnamate) has been calculated in detail.[3644] Photoreactions of poly(vinyl cinnamate) can be photosensitized by a number of compounds, which transfer their excitation energy to the cinnamate groups and activate them:[813,3660]

$$S_0 \text{ (sensitizer)} + hv \rightarrow {}^1(\text{sensitizer}) \rightarrow {}^3(\text{sensitizer}) \qquad (12.11)$$

$${}^3(\text{sensitizer}) + S_0(\text{cinnamate}) \rightarrow S_0(\text{sensitizer}) + {}^3(\text{cinnamate}) \qquad (12.12)$$

$${}^3(\text{cinnamate}) \rightarrow \text{photoreactions} \qquad (12.13)$$

Polymers having chalcone-photosensitive groups (Table 12.1) undergo typical singlet excited state (S_1) photoreactions and cannot be photosensitized by the same sensitizers used for poly(vinyl cinnamate).

An especially interesting group of photopolymers are water- and alcohol-soluble styryl–pyridinium polymers, e.g. poly[1-[6-(4-methoxystyryl)-1-methyl-pyridinium-3-yl] ethylene methyl]sulphate (12.12)[472,2271,3899]

(12.12)

Spectral response and absorption maxima for other styryl–pyridinium type polymers are shown in Table 12.3. The rate of photocrosslinking of these polymers depends on structure, solvent used for processing, and degree of reaction.[3899] The water- and alcohol-soluble polymers prepared using anisaldehyde have been found to have relative photographic speeds of from 20,000 to 30,000 times that of unsensitized poly(vinyl cinnamate) (12.11). These polymers become insoluble on exposure to light and are therefore negative working.

Table 12.3 Spectral response and photosensitivity of styryl pyridinium-type polymers[3899]

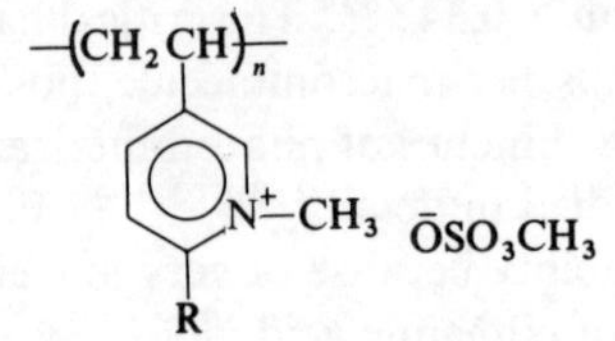

R	λ_{max}	Spectral response
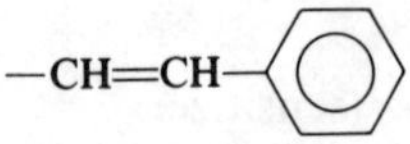 —CH=CH—⬡	344 nm	270–430 nm
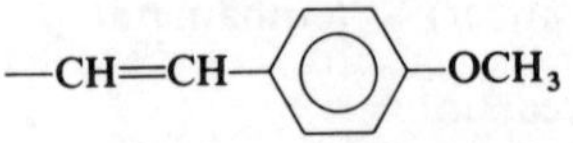 —CH=CH—⬡—OCH$_3$	380 nm	270–480 nm
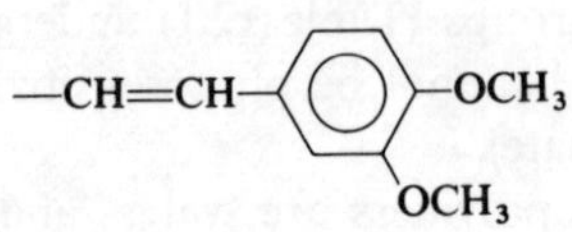 —CH=CH—⬡(—OCH$_3$)(—OCH$_3$)	388 nm	270–560 nm
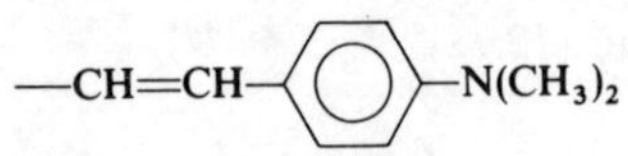 —CH=CH—⬡—N(CH$_3$)$_2$	467 nm	270–630 nm
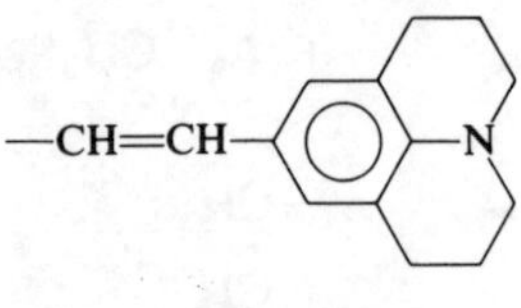 —CH=CH—(julolidine ring system)	486 nm	270–630 nm
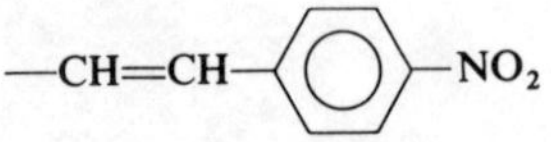 —CH=CH—⬡—NO$_2$	453 nm	270–630 nm
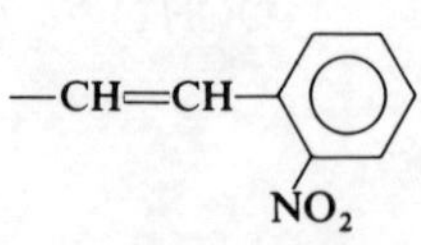 —CH=CH—⬡(—NO$_2$)	345 nm	270–430 nm
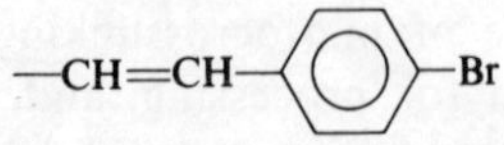 —CH=CH—⬡—Br	351 nm	270–460 nm
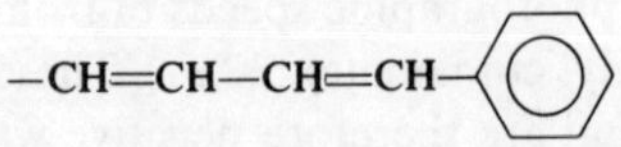 —CH=CH—CH=CH—⬡	375 nm	270–500 nm

Table 12.3 (Contd.)

R	λ_{max}	Spectral response
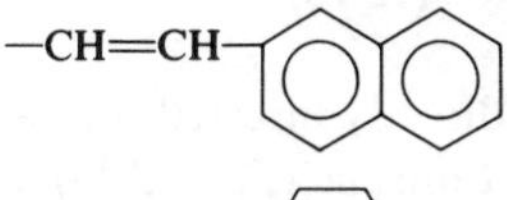 —CH=CH— (naphthalene)	365 nm	270–450 nm
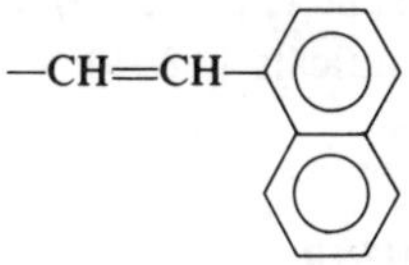 —CH=CH— (biphenyl)	383 nm	270–460 nm
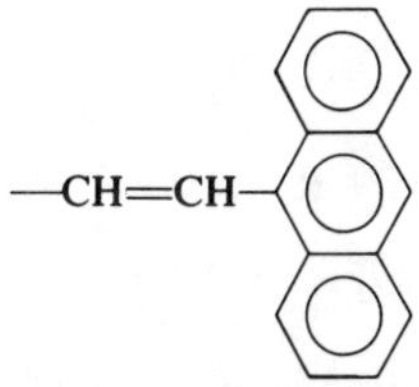 —CH=CH— (terphenyl)	309–408 nm	270–580 nm

By exchange of methoxysulphate counterions with the tetraphenylborate anion, the polymers become organic solvent-soluble. On exposure to light they again become hydrophilic in the exposed areas, and as such are positive working with respect to ink–water discrimination. The conversion to hydrophilicity by light is due to the formation of water-soluble photochemical products from rearrangement of the tetraphenylborate anion.[3899]

The styrylpyridinium-type photosensitive groups undergo typical singlet-type photoreactions and cannot be sensitized by triplet photosensitizers.

12.2.2. Photocrosslinkable Polymers with Photosensitive Azide Groups

Aromatic azides (12.13) are very easily decomposed thermally or photochemically to an excited singlet nitrene (S_1) (12.14), which may then pass by ISC mechanism into a triplet nitrene (T_1) (12.15):[756,3090,3092,3095–3097,3397,3398,3643,3647,3665]

$$\langle \bigcirc \rangle\!-\!N\!-\!\overset{+}{N}\!\equiv\!N^- + h\nu(\Delta) \longrightarrow S_1 \left(\langle \bigcirc \rangle\!-\!\overset{..}{\ddot{N}}:\right)^* + N_2 \longrightarrow T_1 \left(\langle \bigcirc \rangle\!-\!\overset{.}{\ddot{N}}\cdot\right)^*$$

(12.13)   (12.14)   (12.15)

(12.15)

The triplet nitrenes can be formed directly by an energy transfer reaction from photosensitizers.

Nitrenes are unstable, but they are very reactive and may participate in the following reactions:[3093,3386]

(i) Recombination of nitrenes which is spin-allowed for both singlet and triplet state yield azo compounds (12.16):

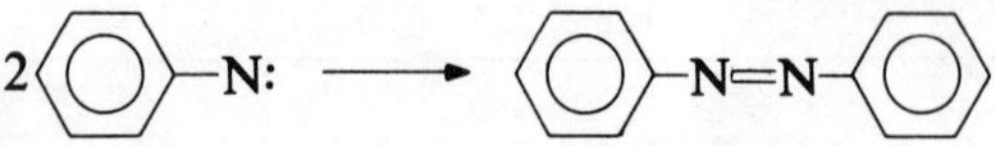

$$2 \left\langle \bigcirc \right\rangle - N: \longrightarrow \left\langle \bigcirc \right\rangle - N=N - \left\langle \bigcirc \right\rangle$$

$$(12.16)$$

(12.16)

(ii) An excited singlet nitrene (S_1) (*12.17*) is directly inserted into a C–H bond (reaction 12.17), whereas an excited triplet nitrene (T_1) (*12.18*) may abstract hydrogen (reaction 12.18) and form amine radicals (*12.19*) which further react by reactions 12.19 and 12.20:

$$S_1(R - \ddot{N}:)^* + RH \rightarrow R-NH-R \qquad (12.17)$$

(12.17)

$$T_1(R-\dot{\ddot{N}}\cdot)^* + RH \rightarrow RNH + R\cdot \qquad (12.18)$$

(12.18)   *(12.19)*

$$R\dot{N}H + RH \rightarrow RNH_2 + R\cdot \qquad (12.19)$$

$$R\dot{N}H + R\cdot \rightarrow R-NH-R \qquad (12.20)$$

(iii) Both singlet and triplet nitrenes add easily to double bonds:

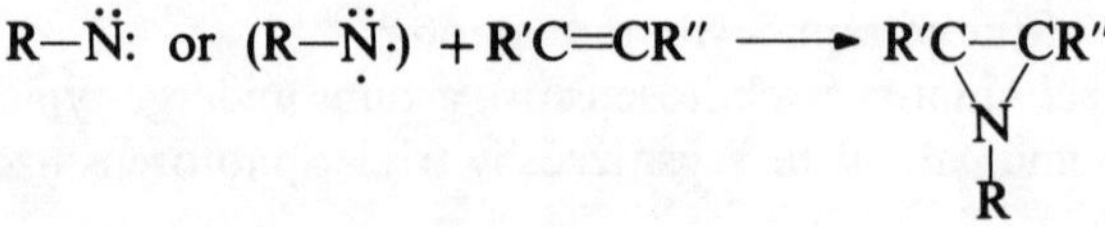

$$R-\ddot{N}: \text{ or } (R-\ddot{N}\cdot) + R'C{=}CR'' \longrightarrow R'C{-}CR''$$
$$\underset{\underset{R}{|}}{\overset{\diagdown \diagup}{N}} \qquad (12.21)$$

Ultraviolet irradiation of bisazides (*12.20*) leads to the formation of a dark-coloured polymer:[482,3230]

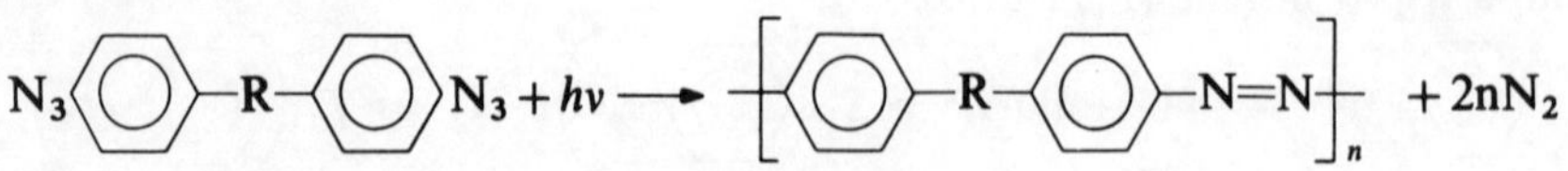

$$N_3 \left\langle \bigcirc \right\rangle - R - \left\langle \bigcirc \right\rangle N_3 + h\nu \longrightarrow \left[\left\langle \bigcirc \right\rangle - R - \left\langle \bigcirc \right\rangle - N=N \right]_n + 2nN_2$$

(12.20)

$$(12.22)$$

Poly(vinyl-4-azidobenzoate) (*12.21*) under irradiation undergoes photocross-linking according to the mechanism (12.23):[3666]

The main products of photolysis of low molecular azides in polymer matrix, such as poly(methyl methacrylate), polystyrene, and polyisoprene, are primary and secondary amines. The yield of secondary amines increases as the rigidity of the polymer matrix increases.[3093]

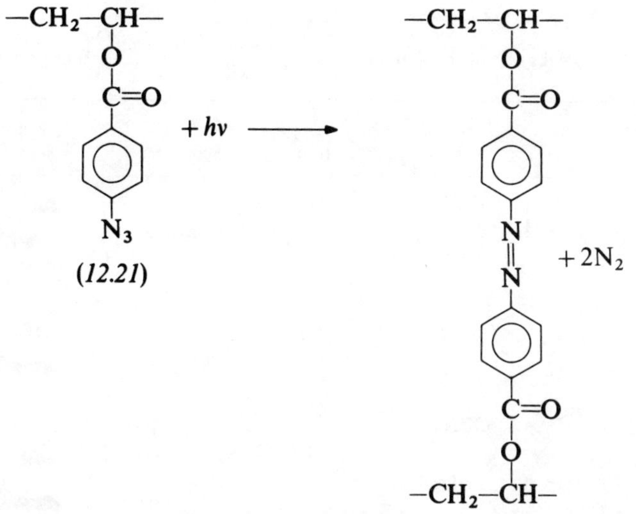

(12.23)

$+2N_2$

12.2.3. Photocrosslinkable Polymeric Systems with Bisarylazides

Cyclized rubber (12.23) obtained from polyisoprene (12.22):

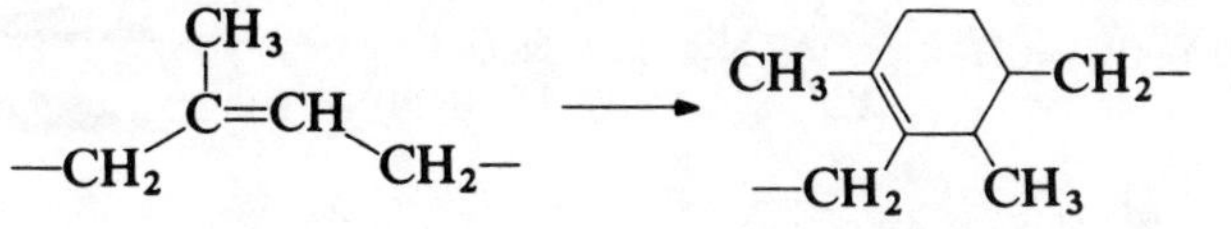

(12.24)

is easily photocrosslinked by bisarylazides (Table 12.4), which absorb light in the range 300–450 nm (Figs 12.5 and 12.6).[1838, 3650] Under irradiation such two-components negative photoresists undergo the following crosslinking reaction:[3905]

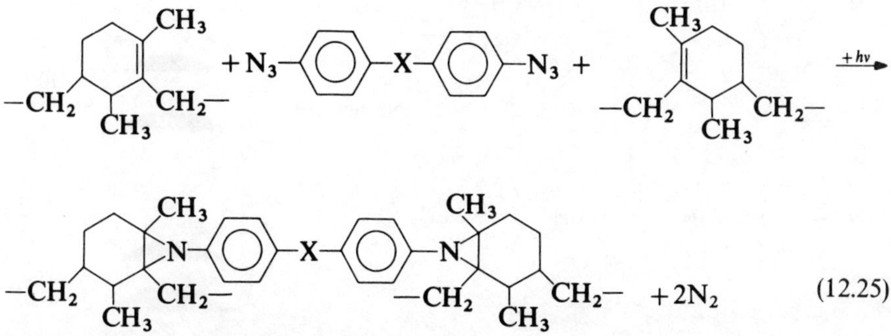

(12.25)

This reaction is efficiently photosensitized by 1-nitropyrene (cf. Table 12.4). The most effective bis-azide compound for the crosslinking of cyclized rubber is 2,6-di(4'-azidobenzal)-4-methylcyclohexanone (12.24).[3317,3650] The activity of this

Table 12.4 Peaks of absorption spectra, spectral, and relative sensitivities for different bisazides[3650]

Bisazide	λ_{max} (nm)	Sensitizer	Spectral sensitivity	RS
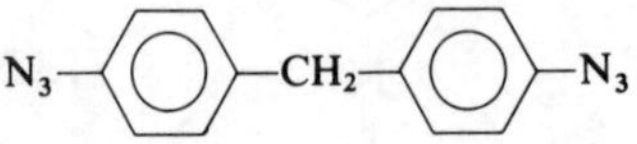 N_3—⬡—CH_2—⬡—N_3	257	None 1-Nitropyrene		1.0 18.4
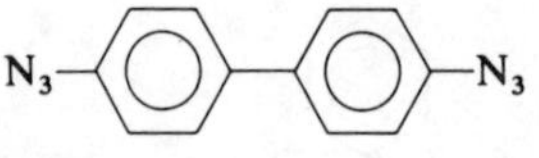 N_3—⬡—⬡—N_3	296	None 1-Nitropyrene		1.3 18.4
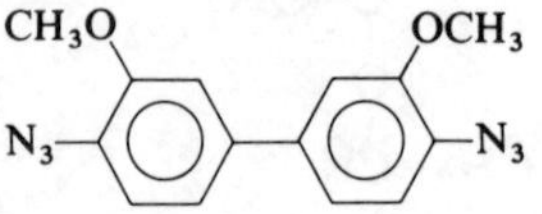 CH_3O, OCH_3 N_3—⬡—⬡—N_3	314	None 1-Nitropyrene		2.7 20.3
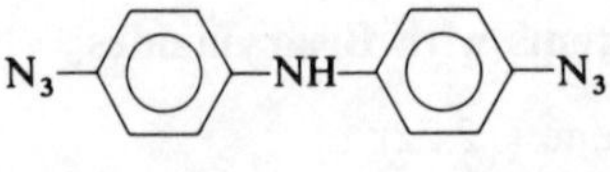 N_3—⬡—NH—⬡—N_3	311	None 1-Nitropyrene		7.5 18.4
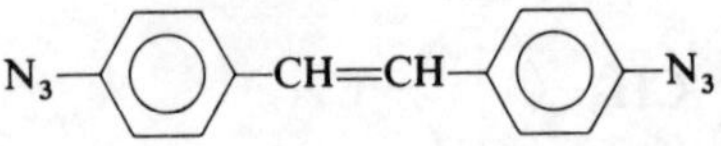 N_3—⬡—CH=CH—⬡—N_3	338	None 1-Nitropyrene		28.0 58.0
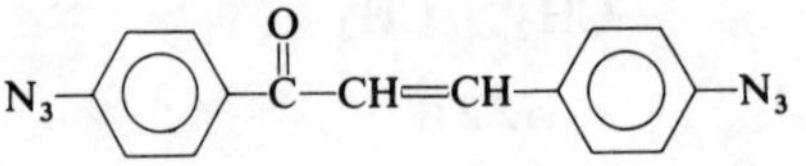 N_3—⬡—C(O)—CH=CH—⬡—N_3	340	None 1-Nitropyrene		28.0 39.0
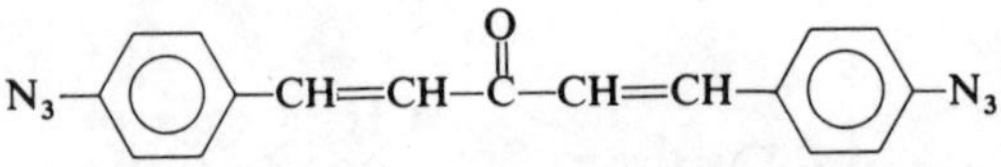 N_3—⬡—CH=CH—C(O)—CH=CH—⬡—N_3	358	None 1-Nitropyrene		10.3 28.0
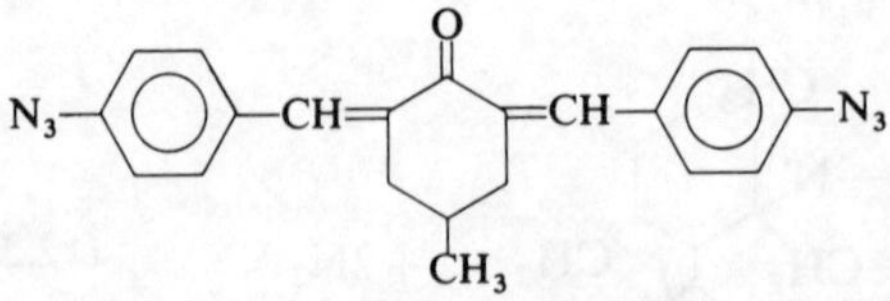 N_3—⬡—CH=C(O)=CH—⬡—N_3 (cyclohexanone, CH_3)	356	None 1-Nitropyrene	 200 300 400 500 600 nm	39.2 84.0

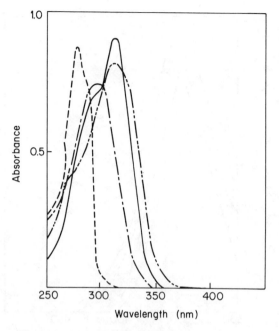

Fig. 12.5 Absorption spectra of different azides:[3650]

(– – –) $N_3-\underset{}{\bigcirc}-CH_2-\underset{}{\bigcirc}-N_3$

(— · —) $N_3-\underset{}{\bigcirc}-\underset{}{\bigcirc}-N_3$

(———) $N_3-\underset{CH_3O}{\bigcirc}-\underset{OCH_3}{\bigcirc}-N_3$

(— ·· —) $N_3-\underset{}{\bigcirc}-NH-\underset{}{\bigcirc}-N_3$

$$N_3-\underset{}{\bigcirc}-CH=\overset{\displaystyle O}{\underset{}{\bigcirc}}=CH-\underset{}{\bigcirc}-N_3$$

$$\underset{CH_3}{}$$

(*12.24*)

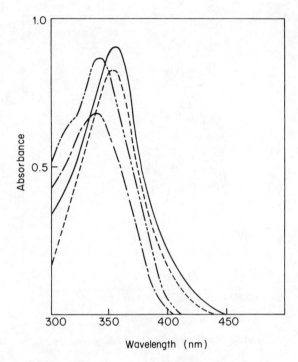

Fig. 12.6 Absorption spectra of different azides:[3650]

$(- \cdot -)$ N$_3$—◯—CH=CH—◯—N$_3$

$(— \cdot\cdot —)$ N$_3$—◯—$\overset{\overset{\text{O}}{\|}}{\text{C}}$—CH=CH—◯—N$_3$

$(——)$ N$_3$—◯—CH=CH—$\overset{\overset{\text{O}}{\|}}{\text{C}}$—CH=CH—◯—N$_3$

$(- - -)$ N$_3$—◯—CH=⬡=CH—◯—N$_3$ (with O and CH$_3$)

azido compound can be effectively photosensitized by other compounds (Table 12.5). Nitro and quinone compounds have a large rate constant for intersystem crossing (ISC) reaction. This means that the energy of excited sensitizer is transferred to bisazide via its lowest triplet state (T_1). The 2′-bromo-1,2-benzanthraquinone photosensitizes bis-azides better than 2′-chloro-1,2-benzanthraquinone. It is well known that a heavy atom increases the efficiency of the ISC reaction.[3650]

Table 12.5 Spectral and relative sensitivities of unsensitized and sensitized 2,6-di(4'-azidobenzal)-4-methylcyclohexanone-cyclized rubber layers[3650]

Sensitizer	Spectral sensitivity	Relative sensitivity
None		39.2
1-Nitropyrene		84.0
1,B-Dinitropyrene		117.6
Cyanoacridine		58.8
	300 600 nm	

Bisarylazide/rubber photoresists are commercially produced under the names KTFR and MX11 (by Kodak), NMR (by Hunt) and Selectilux-N (by Merck). The resolution of these photoresists is limited by the swelling-induced deformation of the expanded (crosslinked) resist patterns during development, and is not sufficient to define patterns smaller than 2 μm size.[921]

Poly(methyl isopropenyl ketone) (12.25), under irradiation in the presence of diazide compounds such as 4,4'-diazidodiphenyl ether and 4,4'-diazidodiphenyl thioether,[3646] or 2,6-di(4'-azidobenzal)-4-methylcyclohexanone (12.24)[2627, 3645, 3979] undergoes the crosslinking reaction:

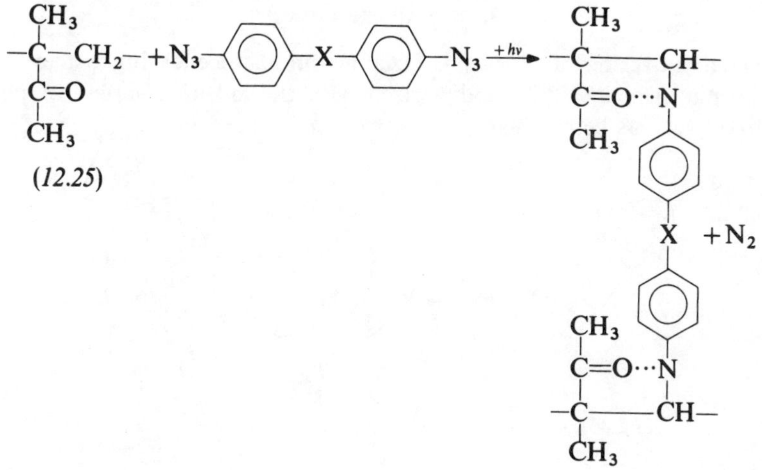

(12.25)

(12.26)

Post-annealing forms hydrogen-bonded products which show a powerful electronic excitation energy quenching effect.[3645] The quencher is more effective than the aromatic compound formed from the azide by post-annealing only.

Poly(methyl isopropenyl ketone)/diazide resists are used as dry film photoresists (cf. Section 12.8.2).

12.2.4. Photocrosslinkable Polymeric Systems with Diazonium Salts

Diazonium salts (12.25) are photochemically reactive (cf. Section 5.4.1). When they are photodecomposed in polymers, e.g. in poly(vinyl alcohol), efficient crosslinking reaction occurs according to the mechanism:[2245,3663,3664]

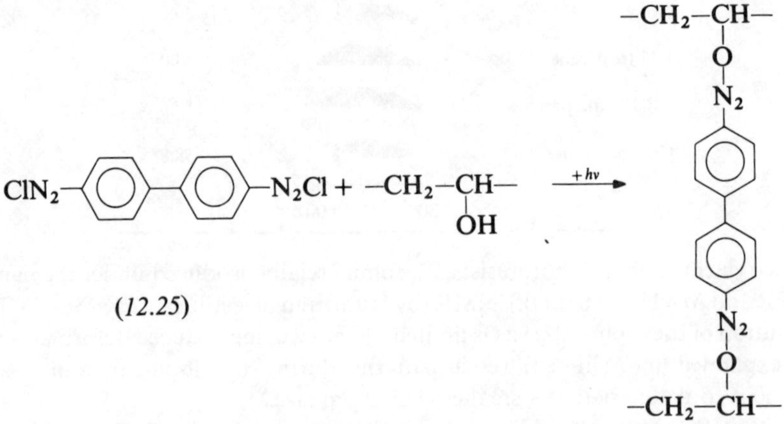

(12.25)

(12.27)

12.2.5. Photocrosslinkable Polymers with Photosensitive Iminostilbene Group

N-acyl derivatives of the dibenz[b,f]azepine (iminostilbene) ring system (12.26) may be photochemically cyclodimerized with the aid of a variety of photosensitizers such as benzophenone or benzyl:[2130]

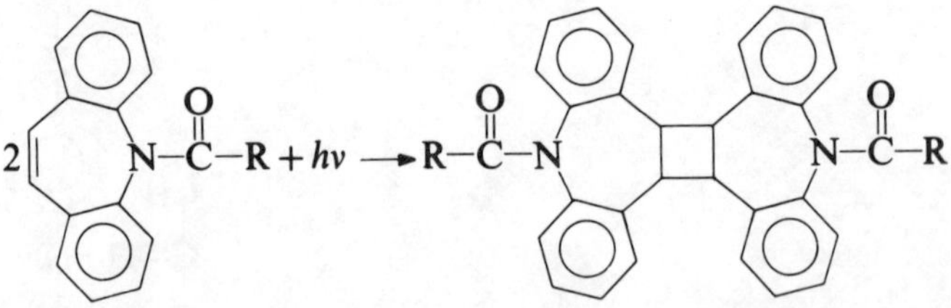

(12.28)

(12.26)

Polymers with the dibenz[b,f]azepine units as pendant groups can be photochemically crosslinked by the above cyclodimerization reaction.[1734]

Photoreactive polymers with terminal iminostilbene groups can be obtained by free radical polymerization of methyl methacrylate, and styrene initiated by free radicals (12.28) formed during the photolysis of N[-4,4'-azo-bis(4-cyanopentanoyl)]-bis-dibenz[b,f]azepine (12.27):[294]

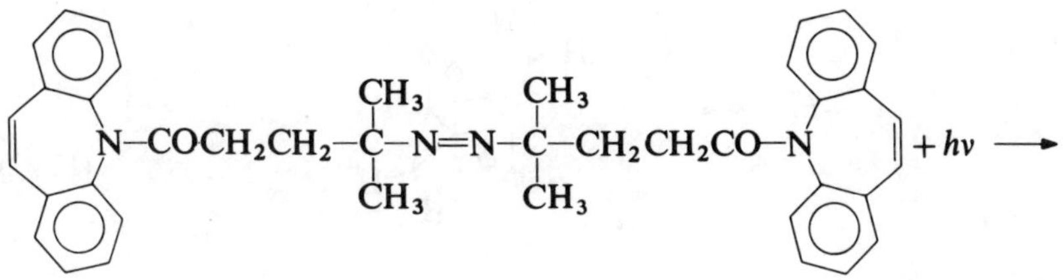

(12.27)

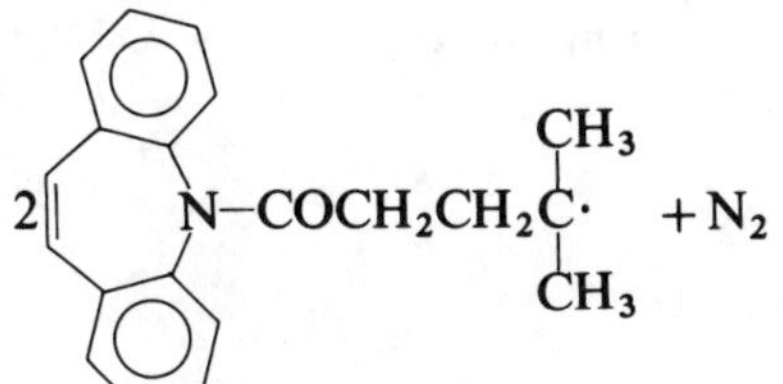

(12.29)

(12.28)

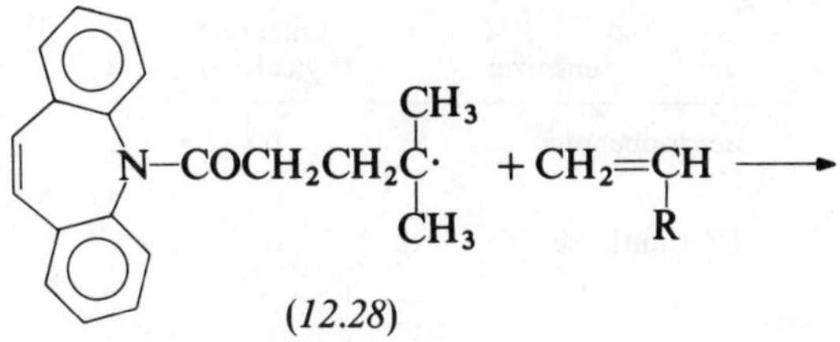

(12.28)

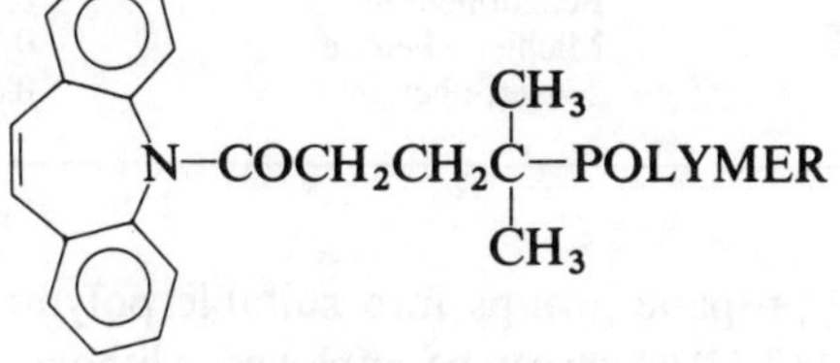

(12.30)

12.2.6. Photocrosslinkable Polymers with Diphenylcyclopropane Group

Direct u.v. irradiation of 1,2-diphenylcyclopropanes (12.29) and (12.30) and 1,2,3-triphenylcyclopropane (12.31) does not give cyclophotodimers.[903,904,906,2748] Sensitized photocyclo $(2+2)$ addition occurs with high quantum yield (Table 12.6) and gives tricyclohexane dimers (major) and cyclopropylcyclopropanes (minor), according to the reactions:

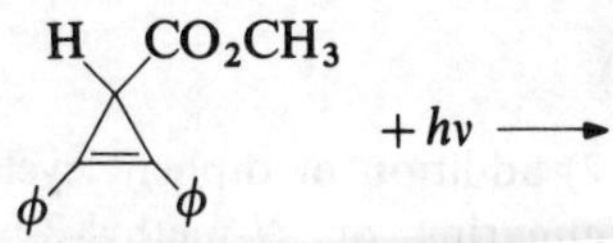

(12.31)

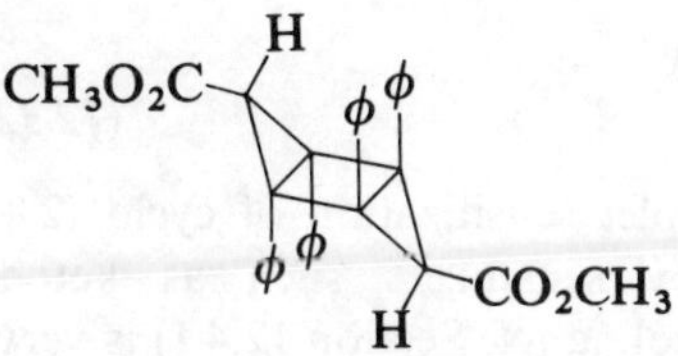

(12.29)

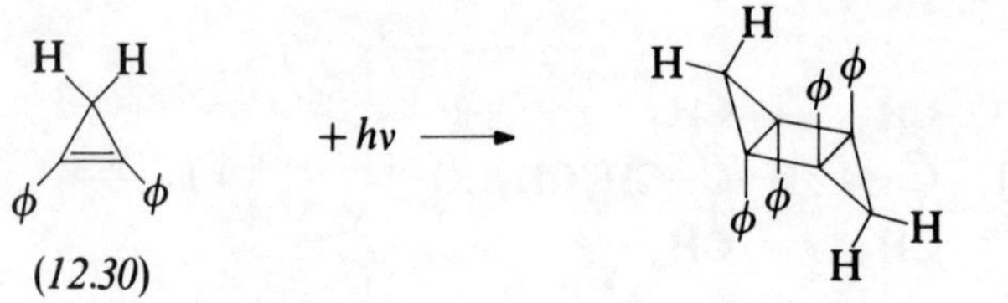

444

(12.32)

(12.30)

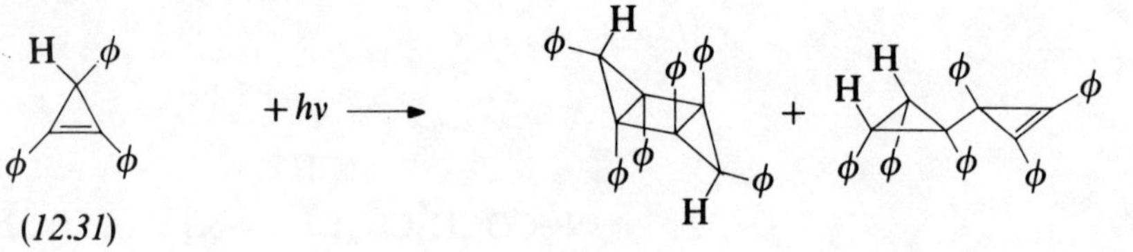

(12.31)

(12.33)

Table 12.6 Quantum yield of cyclopropene $(2+2)$ photodimerization (Φ) in benzene in the presence of different sensitizers[906]

Cyclopropane	Sensitizer	Quantum yield (Φ)
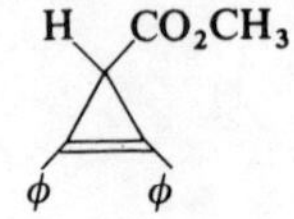	Benzophenone	0.8
	Thioxanthone	0.7
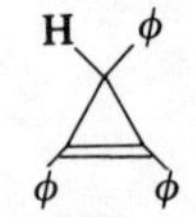	Benzophenone	0.5
	Michler's ketone	0.7
	2-Acetophenone	0.4

Incorporation of cyclopropane groups into suitable polymers gives highly photosensitive copolymer (12.32) which can be employed in lithography.[903,906,2509]

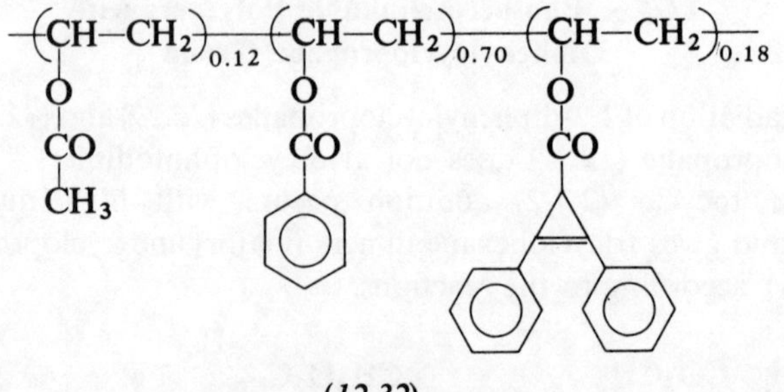

(12.32)

The triplet sensitization of cyclo $(2+2)$ addition of diphenylcyclopropane groups by sensitizers such as ketocoumarins or N-methyl-2-benzoyl-β-naphthiazoline (cf. Section 12.4.1) is very efficient.[903,906,1084,2509]

12.3. PHOTOMODIFICATION OF POLYMER SOLUBILITY

Photomodification of polymer solubility can be obtained by the modification of the solubility of a polymer-bound chromophore group upon light irradiation. The change in functionality modifies the solubility of appended groups, and thus of the polymer in selected solvents.[3900]

Diazoketones (e.g. diazonaphthoquinones (*12.33*)), which are soluble in common solvents but insoluble in aqueous base, upon irradiation produce carbenes (*12.34*), which via Wolf's rearrangement give ketones (*12.35*). The ketone adds water to form an indenecarboxylic acid easily soluble in aqueous base (*12.36*):[503,2868,3905]

(*12.33*) (*12.34*) (*12.35*) (*12.36*)

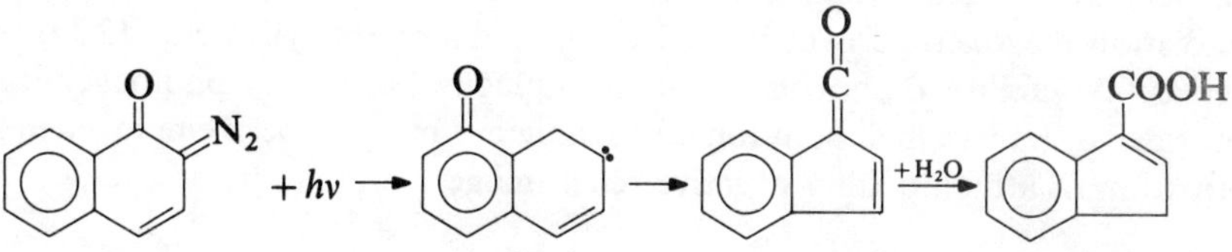

(12.34)

Many compounds having diazoketone chromophores with more complex structures (*12.37–12.41*) have been also investigated:[3906] Unfortunately these materials have practical drawbacks that limit their utility, including synthetic inaccessibility, insolubility, poor bleaching, and thermal instability.

(*12.37*) (*12.38*) (*12.39*)

(*12.40*) (*12.41*)

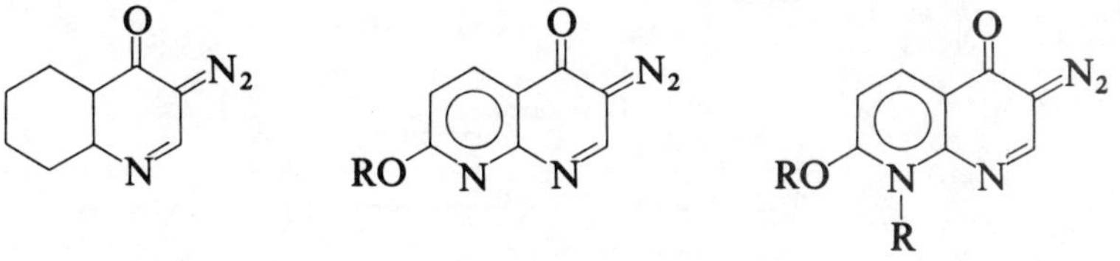

In commercial practice two-component positive resists used are based on novolac resin (*12.42*) and diazoketone (*12.33*).

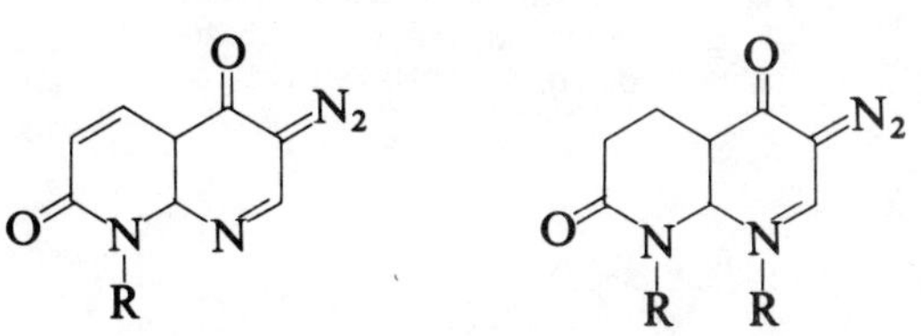

Novolac resins are soluble in common solvents and can be coated from solution to form isotropic, glassy films of high quality. Diazoketone (*12.33*) is also soluble in common organic solvents but is insoluble in aqueous base. In the

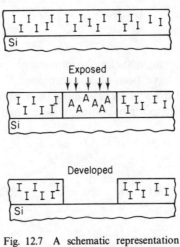

(*12.42*)

formulation of positive photoresists, both polymer and diazoketone are dissolved in ethyl cellulose acetate, diglyme, etc.[503] After coating a required surface, the solvent is removed by heating. The film thus formed contains randomly distributed diazoketone through the novolac matrix (Fig. 12.7). The presence of diazoketone decreases the solubility of novolac resin in aqueous base. Upon irradiation diazoketone is converted to the indenecarboxylic acid (*12.36*) that increases solubility of novolac resin in aqueous base. This photochemically generated difference in dissolution rate in aqueous base has been employed in the formation of a high-quality negative relief image.

Fig. 12.7 A schematic representation of positive resist action in diazo-naphthoquinone-novolac resists:[3905]

(*12.33*)

Base insoluble
sensitizer
(inhibitor) (I)

(*12.36*)

Base soluble
photoproduct
(acid) (A)

Diazidonaphthoquinone/novolac resists are commercially produced under the names AZ1350J (by Shipley), Photoresist 820 (by Kodak), and HPR 204 (by Hunt).

When a polymer-bearing diazoketone group (hydrophobic–non-water-soluble) (*12.43*) is u.v. irradiated it undergoes photorearrangement, resulting in conversion of hydrophobic groups into hydrophilic units (water-soluble) and the image can be developed by treatment with aqueous alkali.[939,2681,3763]

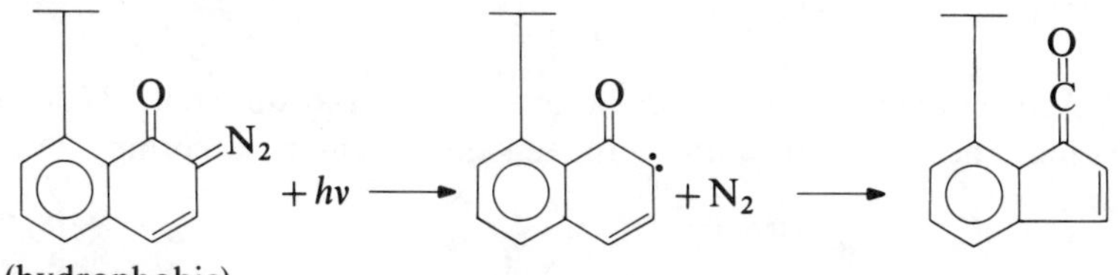

(hydrophobic)

(*12.43*)

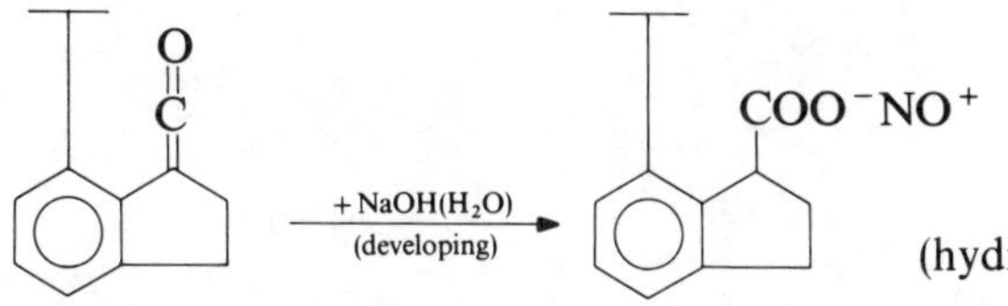

(12.35)

(hydrophilic)

The diazoketone structure is also thermally sensitive. This property has been used to advantage in practical resist applications where the developed image is post-heated to harden the resist image and improve performance.

Aqueous alkali-soluble co(methyl methacrylate/methacrylic acid) is rendered insoluble by the addition of an *o*-nitrobenzyl cholate ester (*12.44*). On irradiation the ester undergoes a photorearrangement to generate *o*-nitrosobenzaldehyde (*12.45*) and carboxylic acid, which are soluble in aqueous base.[3078,3886]

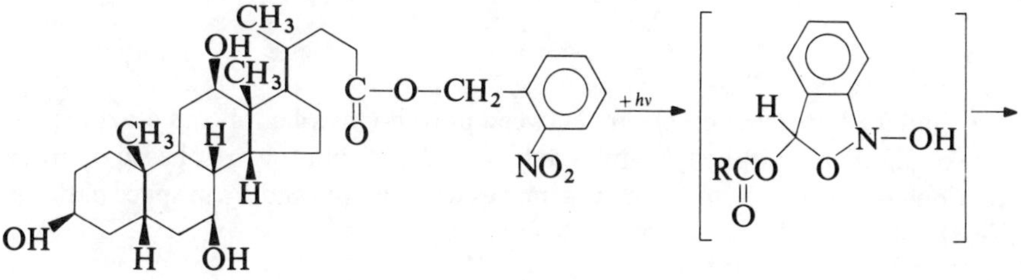

(*12.44*)

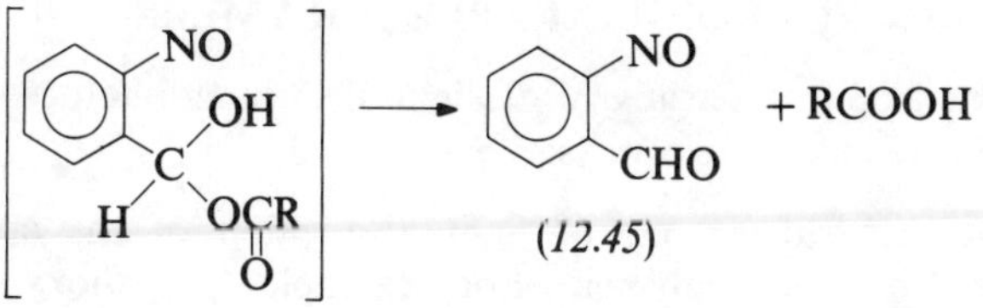

(*12.45*)

$+ RCOOH$ (12.36)

The irradiated regions of the film are then soluble in aqueous alkali.[3077]

Aromatic azides (12.46) react with phenols (12.47) to quinone–imine compounds (12.48):[3667]

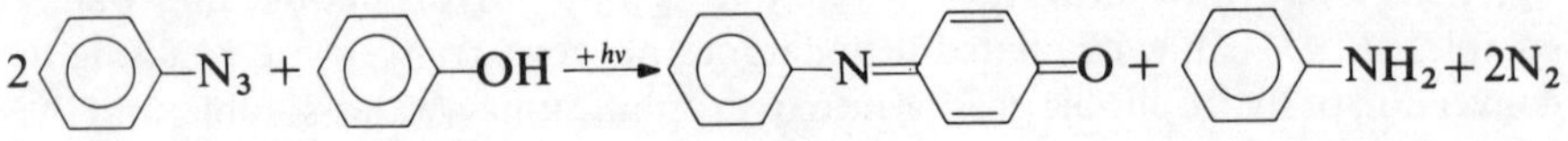

$$(12.46) \qquad (12.47) \qquad\qquad (12.48)$$

(12.37)

A thin layer of phenol resin (12.49) which contains 1-azidopyrene (12.50) under irradiation becomes insoluble in alkaline water according to the reaction:[3649,3667]

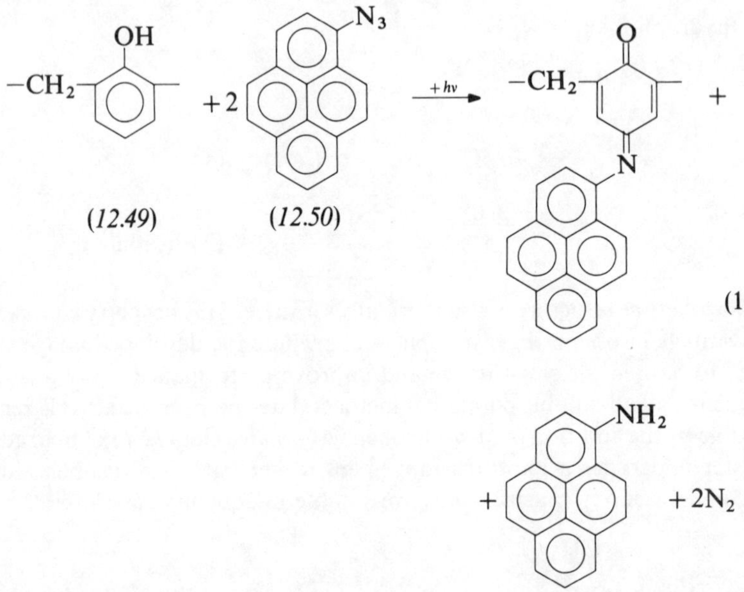

$$(12.49) \qquad (12.50)$$

(12.38)

A similar photoreaction occurs between poly(p-vinylphenol) and 4-azidochalcone or 3,3-diazidodiphenyl sulphone.[1837-1839] The photoresists based on these reactions are useful for the receiving images of 1 μm line and 1 μm space patterns (Fig. 12.8).[2433,2730]

12.4. EXTENSION OF SPECTRAL SENSITIVITY OF PHOTOCROSSLINKABLE POLYMERS

The range of the spectral sensitivity of light-absorbing photocrosslinkable polymers can be extended in two ways:

(i) By modification of the range of light absorption of the crosslinking chromophore in the polymer. For example p-methoxy- and p-

Fig. 12.8 Resist pattern of poly(4-vinyl phenol) containing 3,3-diazido diphenyl sulphone[2730]

dimethylamino-substituted poly(2-vinyl-1-methyl-5-styryl-pyridinium)-methyl sulphate are more susceptible towards insolubilization than the starting polymer.[2271] Their spectral responses (wedge spectrograms) are shown in Fig. 12.9.

Chromophore modification can be accomplished by extension of the unsaturated chain length, e.g. spectral responses of poly(vinyl acetate cinnamate) and poly(vinyl acetate cinnamylidene acetate) are shown in Fig. 12.10).[2270]

(ii) By using spectral sensitizers. Photosensitization reactions occur as a result of singlet or triplet energy transfer from an excited sensitizer (Sens) to chromophore group (X) in a photoreactive polymer:

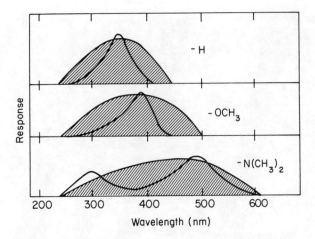

Fig. 12.9 Wedge spectrograms of poly(2-vinyl-1-methyl-5-(*p*-substituted styryl)pyridinium)methosulphate: where R is H, OCH$_3$ and N(CH$_3$)$_2$[3895]

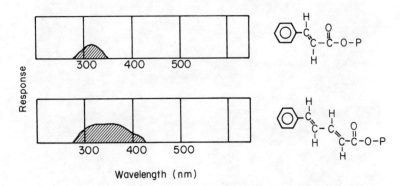

Fig. 12.10 Wedge spectrograms of poly(vinyl cinnamate) and poly(vinyl cinnamylidene acetate)[3895]

$$S_0(\text{Sens}) + h\nu \rightarrow {}^1S(\text{Sens}) \xrightarrow{\text{ISC}} {}^1T(\text{Sens}) \qquad (12.39)$$

$${}^1S(\text{Sens}) + S_0(X) \rightarrow S_0(\text{Sens}) + {}^1S(X) \qquad (12.40)$$

$${}^3S(\text{Sens}) + S_0(X) \rightarrow S_0(\text{Sens}) + {}^1T(X) \qquad (12.41)$$

12.4.1. Photosensitizers for Photocrosslinkable Polymers

Photosensitization of the most photocrosslinkable polymers occurs by the triplet energy transfer from an excited sensitizer molecule to the photocrosslinkable polymers.[813,1084,3899]

The spectral sensitivity and sensitivity values (photographic speed values) for different-type sensitizers in photosensitization of *p*[2-(2-ethylhexyloxycarbonyl)vinyl] cinnamate polymer are compared in Table 12.7.

Table 12.7 Sensitization of p-[2-(2-ethylhexyloxycarbonyl)vinyl] cinnamate type polymer[3899]

Sensitizer	Spectral range (nm)	Sensitivity value
None	270–380	2000
2,6-Bis(4'-ethoxyphenyl)-4-(4'-n-amyloxyphenyl)-thiapyrylium perchlorate	270–520	8000
N-Methyl-2-benzoyl-β-naphthothiazoline	270–400	20,000
4-H-Quinolizine-4-thione	270–480	2800

Efficient triplet sensitizers for photocrosslinkable polymers should have:

(i) High molar absorptivity ($\varepsilon \approx 10^4$) to ensure a considerable absorption of the incident light at a relatively low concentration of the sensitizer in the polymer matrix.

(ii) High efficiency of intersystem crossing (ISC) to the triplet state.

(iii) Triplet energy high enough to ensure an efficient energy transfer to the photoreactive chromophore of the polymer.

(iv) A very small singlet–triplet energy gap, in order to extend the spectral response of the polymer to the longest possible wavelength.

(v) High photochemical stability.

(vi) High solubility in the coating solvent and in the polymer.

The most common sensitizers are aromatic ketones, aromatic nitro compounds, aromatic amines, aromatic hydrocarbons, and a wide variety of heterocyclic compounds.

Special attention has been given to two types of sensitizers, i.e.:

(i) N-methyl-2-benzoyl-β-naphthiazoline (12.51), which may increase sensitivity value of poly(vinyl cinnamate) from 2000 up to 20,000.[1084,3128,3899] This triplet sensitizer[3051] can be used only in the range not exceeding 460 nm. The quantum yield of intersystem crossing (ISC) (Φ_{ISC}) depends on solvent polarity (Table 12.8). The excited molecule of this sensitizer can also transfer its triplet excitation energy to molecular oxygen, during which a singlet oxygen is formed (cf. Section 14.2).[3045,3051,3899]

(12.51)

(ii) Ketocoumarins (12.52) and (12.53) with electron-donating substituents in the 7-position. The structure, absorption maxima, absorptivities, triplet energy, and quantum yield of ISC (Φ_{ISC}) for these triplet sensitizers are shown in Table 12.9.[1084,3900]

Table 12.8 Intersystem crossing quantum yields (Φ_{ISC}) dependence on solvent polarity[1084]

Group	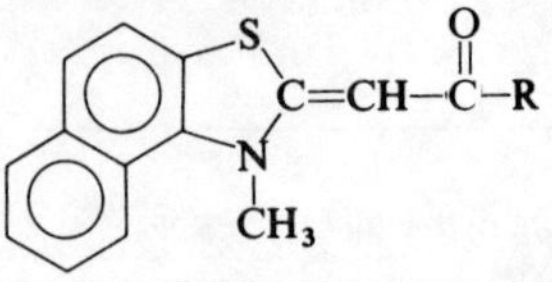		

	Solvent		
R	C_6H_6	$CH_3CO_2C_2H_5$	CH_3CN
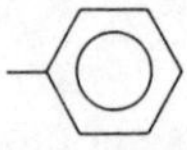	0.24	0.15	0.033
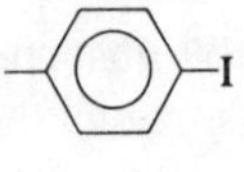	0.34	0.144	0.037
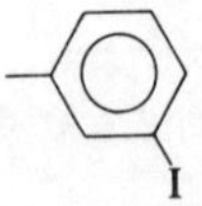	0.15	0.069	0.014
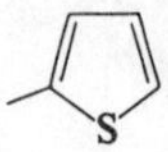	0.6	0.33	0.046
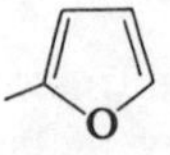		0.375	0.082
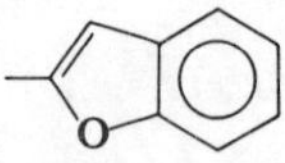	~0.57	0.19	0.018

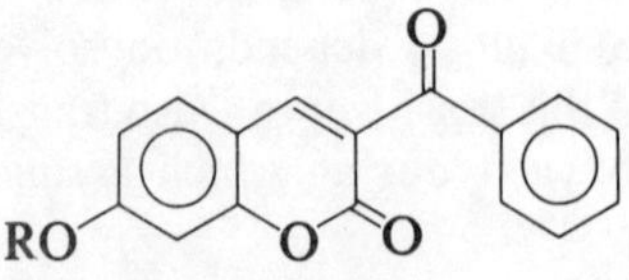

(12.52)

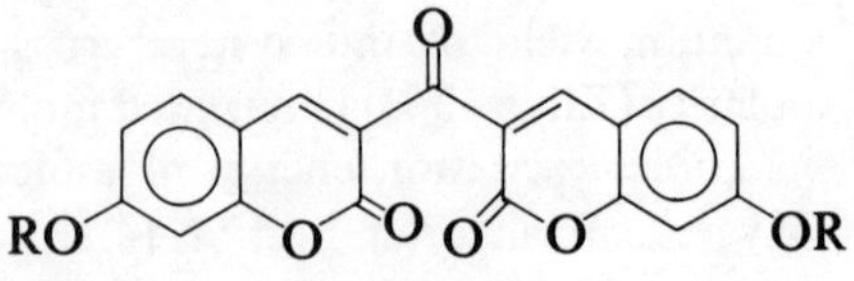

(12.53)

Ketocoumarins are specially effective sensitizers for photocrosslinking polymers with the diphenylcyclopropane group (cf. Section 12.2.6).[906,1084] Ketocoumarins, with a proper combination of some activators such as amines, acetic acid derivatives, and alkoxypyridinium salts, gives quantum yields for radical polymerization much higher than that obtained from the Michler's ketone/benzophenone combinations.[3901] For each class of activators the dependence of the efficiency of polymerization on the redox

Table 12.9 Intersystem crossing quantum yields (Φ_{ISC}) and triplet energies (E_T) of ketocoumarins[1084]

	λ_{max} (nm)	ε	Φ_{ISC}	E_T
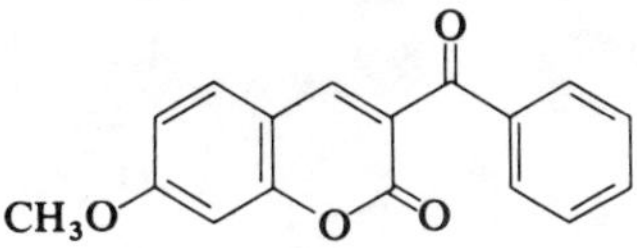	345	19,000	> 0.9	co.59
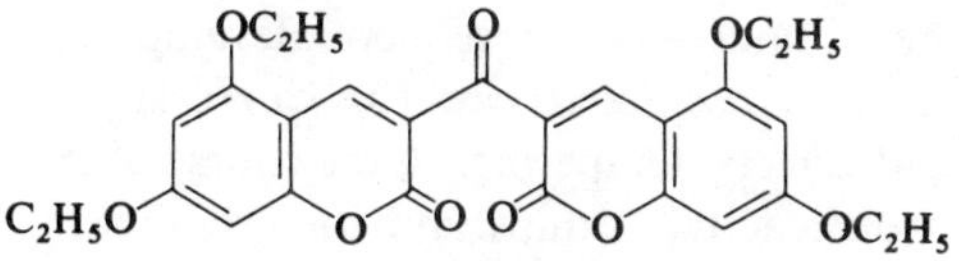	398	46,000	~ 1.0	co.56
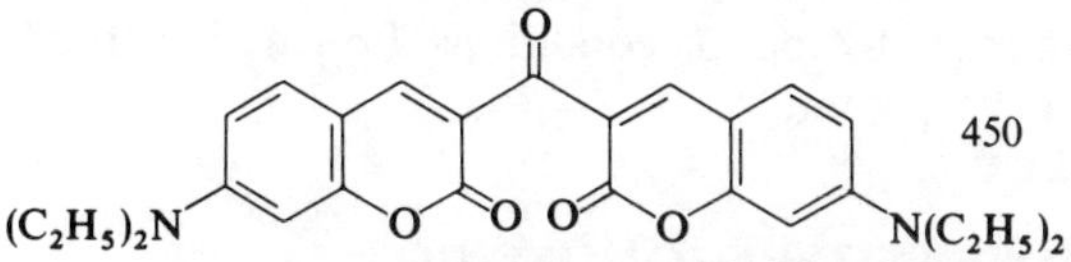	450	92,000	0.94	co.52

properties of the coumarins has been explained in the terms of charge transfer or electron transfer from the activator to the excited ketocoumarin (acetic acid and amine activators, respectively) or electron transfer in the opposite direction (pyridinium salt activators).

12.5. HIGHLY HEAT-RESISTANT PHOTOREACTIVE POLYMERS

Highly heat-resistant polyimides (*12.54*) can be obtained by photochemical ring-closure reactions of soluble polyamides possessing ester-like bound photoreactive groups (*12.55*):[3145]

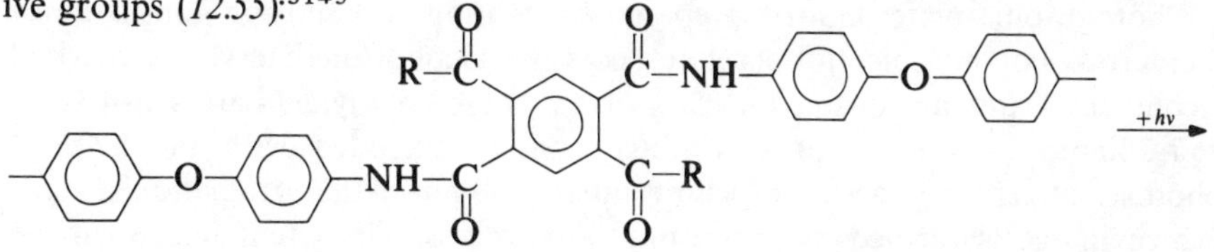

(*12.54*)

$$R = -OCH_2-CH_2-O-\overset{O}{\underset{}{C}}-\overset{CH_3}{\underset{}{C}}=CH_2$$

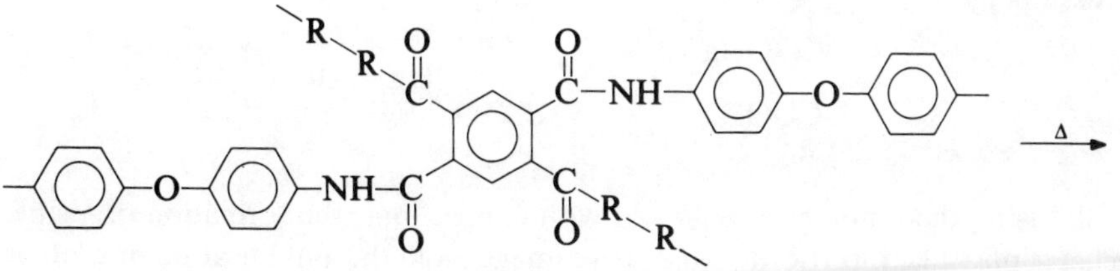

(*12.55*)

$$(12.56) \qquad\qquad\qquad (12.42)$$

The soluble polyamides (12.54), under u.v. irradiation, are photocrosslinked (12.55) and an image pattern can be obtained by subsequent development. Because of their special structure these crosslinked patterns can easily be chemically converted into polyimide patterns by tempering. In the course of this cyclization reaction the remaining photoreactive groups and the crosslinking bridges are set free in the forms of alcohols and polyalcohols. They may be volatilized by selecting the appropriate tempering conditions. Polyimide pattern adhesion to metals and oxides is very good.

12.6. PHOTOSENSITIVE IONOMERS

IONOMERS are ionically crosslinked polymers. The ionic crosslinkage are pairs of ions such as $-COO^- M^+-$ or $-COO^- M^{2+} \ ^-OOC-$ formed by adding mono- or divalent metallic (M) ions, respectively, to a linear polymer with the side chains of carboxyl groups. The strength of the ionic crosslinking depends on the valence of the metallic ion. The ionomer crosslinked by a metallic ion of variable valence on exposure to light may become a new type of photosensitive polymer.[2425,3478,3481,3482,3499-3501]

The mercuric ion, which is produced by photooxidation of mercurous ion, will form crosslinkages if some linear polymer has side-chains of mercurous carboxylic groups.

Photographic plates can be prepared by coating an emulsion consisting of mercurous co(vinyl alcohol/acrylate), co(vinyl alcohol/methacrylate), co(vinyl alcohol/maleate), and water on glass plate. These copolymers are sensitive to 250–340 nm radiation, but this range can be extended with the help of photosensitizers, e.g. acridine yellow up to 600 nm. After irradiation a dark brown image is formed by a printing-out process. The latent image can be amplified with the reducing developers such as hydroquinone or methol. The mercuric ion which is produced by photooxidation of mercurous ion forms crosslinkages:

$$(12.43)$$

By using the ionomer crosslinked with copper ions, the formation of visible relief is possible. Electrically conductive images, and the amplification of a latent image by reducing and oxidizing developments, are also possible.[3479]

12.7. LIQUID-DEVELOPING PHOTORESISTS

Commercially available liquid photoresists are supplied as solutions of the photocrosslinkable polymers in organic or chlorinated solvents with additives such as sensitizers, antioxidants, stabilizers, dyes or pigments, and coating aids.[3900]

Liquid photoresists are employed in the microimage field. Liquid-resist coverage depends on the concentration of solids dissolved in the resist solvent. Thickness can be adjusted by increasing or decreasing the concentration of the solid and by changing the coating parameters. Most coatings are made by spinning the substrate, usually a silicon wafer, while the resist solution flows onto the wafer, or after this has occurred.

Any photoresist formulation, regardless of the type of chemistry involved in its reaction, must fulfil a number of requirements related to physical properties. It must:

(i) Form homogeneous solutions in ecologically acceptable solvents.
(ii) Form smooth, flaw-free coatings with good adhesion in all process steps.
(iii) Be characterized by good solution stability for storage.
(iv) Form tack-free coatings.
(v) Provide good image discrimination. For image discrimination to be achieved the development solvent must remove the soluble portions of the exposed imagery without distorting or swelling the insoluble area. At the same time the contour of the image edges must be controlled to produce virtually vertical walls. In practice, edge definition can be maximized by selection of resist compositions and solvent compositions, times and temperatures. The image quality is controlled by coating and development conditions.
(vi) Resist flow or creep during process at high temperatures, i.e. during etching (plasma etching, reactive-ion etching, ion milling, and ion implantation).
(vii) Form thermally stable images that do not flow. Etching processes can raise the temperature, and resist images can lose dimensional integrity and, as a consequence, flow. This problem can be reduced by chemical modifications of the resist polymer.
(viii) Be removable or strippable after the required fabrication steps.

The preceding requirements depend on critical balancing of polymer or copolymer compositions, which is usually accomplished by adjusting the ratio of non-light-sensitive solubility-controlling units while maintaining the light-sensitivity requirements.

The physical properties required for a photoresist are determined largely by the etching and chemical treatment which the image must withstand. There are two methods of etching:

(i) Wet etching (mainly inorganic acids, organic solvents).
(ii) Dry etching (ion milling, plasma etching, and reactive-ion (ion beams) etching).

Advantages and disadvantages of both methods of etching are listed in Table 12.10. Plasma etching is widely employed commercially.[2910] In this process the selective etching of materials occurs by reaction of substrates with chemically active radicals formed in a glow discharge. The plasma produced contains several reactive species, such as ions, free electrons, and free radicals, which are produced by the plasma decomposition of used gases (O_2, CCl_4, CF_4, etc.). The attack of these reactive species on the substrates (Si, SiO_2, Si_3NO_4, etc.) and metals (Al, Cr, etc.) used in microelectronics forms volatile metal halides (SiF_4) which are removed during etching.

Table 12.10 Advantages and disadvantages of wet and dry etching methods

Method	Advantages	Disadvantages	Specially important resist requirements
Wet etch	Bath process Process well characterized Currently used in production	Chemical handling Space requirements	Ahesion Acid resistance
Dry etch	No chemical Simple process and line pattern improvements Electronic end-point control	Etch discrimination Requires higher temperatures Equipment complicated Process not well understood	Thermal image stability Resistance to active etch agent

12.8. IMAGE FORMATION PHOTOINITIATED POLYMERIZATION SYSTEMS

Image formation photoinitiated polymerizable layers usually consist of the following components:

(i) Polymeric binder such as poly(methyl methacrylate) or poly(vinyl butyral) for holding together the photochemical components.
(ii) Polymerizable monomer such as methylene glycol biacrylate.
(iii) Photoinitiator, e.g. alkyl anthraquinone.
(iv) Thermal stabilizer, e.g. p-methoxyphenol.
(v) Dye, e.g. cyanine dye.

Formulations of such photoinitiated polymerizable layers have resulted in a multitude of patents.

On exposure to light in an image-wise manner the exposed areas are photocrosslinked and then insoluble in solvents. The soluble non-crosslinked areas can be removed with solvent or chemical action to provide relief patterns or images. The crosslinked areas are less easily thermally softened than the unexposed areas. Thus the unexposed and thermally softened areas can be transferred by pressure to a receiving sheet to give a positive image.

Proper choice of the polymer binder and the monomeric acrylate or methacrylate permits control over the ultimate physical properties of exposed and unexposed photoinitiated polymerizable areas. By modifying both the physical properties of the acrylic monomer or prepolymer and the polymeric binder, transfer temperatures can be controlled. Inclusion of hydrophilic groups into the binder can lead to systems which are developable in aqueous washout solvents to give relief images useful for lithographic purposes.

In practice the polymer plays a large role in determining solubility, thermal properties, chemical resistance, developability, image discrimination, physical resistance, and electrical properties of the resist coatings.

The polymer binder or backbone of a photosensitive polymer can influence the photochemical sequences as well as the chemical steps which lead to changes in physical properties.

Each photoresist is characterized by some degree of incident contrast necessary to produce patterns usable for subsequent processing. Depending on substrate properties, the required pattern thickness, and resist edge profiles, conventionally used, the photoresist has a contrast threshold between 85 and 90 % contrast. If the contrast threshold of the resist is reduced, the resolution obtainable with a given optical system is improved, due to the fact that image contrast is a decreasing function of the spatial frequencies present in the image.[1411] Lithography in the production of integrated circuits is predominantly carried out by optical means, which should have high resolution in order to reduce circuit dimensions, improve performance, and increase yield. A new method, designed to produce good-quality patterns with very low contrast illumination through the use of photobleachable materials, has been developed.[1411] In this method a very thin bleachable layer is first applied to the photoresist surface. Following conventional exposure, the layer is removed and the resist developed in the ordinary way. As a result of the presence of the bleachable material (different dyes), the contrast of the illumination that reaches the photoresists is increased. This technique is called the CONTRAST ENHANCEMENT METHOD.

Higher resolution can be obtained by using shorter exposure wavelengths for pattern delineation (deep u.v. lithography). Commercially Xe-Hg or deuterium light sources are employed for exposures. However, applications of excimer lasers have also been reported recently.[1849]

12.8.1. Direct Image System Based on Photopolymerization

Vapour phase photopolymerization of hexachlorobutadiene using u.v. radiation has been employed to obtain negative or positive photoresists.[464, 2164, 2165] Extremely thin (20–100 nm) adherent films, which are electrically and mechanical continuous, can be deposited on the substrate. Films produced at temperatures below 100°C are soluble in acetone, and they can be insolubilized in a pattern by further irradiation at higher temperatures. The soluble part is then removed to produce a negative resist.

To produce positive resists a cross-linked, acetone-insoluble film is deposited by carrying out photopolymerization at about 100°C. The film is further irradiated through a mask in the presence of oxygen for patterned removal of the film by oxidative degradation.

Acrylic monomer (barium acrylate) solution containing phenothiazine dyes (e.g. methylene blue) and a salt of an arylsulphonic acid can photopolymerize giving a direct photopolymer image.[2387, 3152]

Polymerization is initiated by transient free radicals resulting from the redox reaction between the dye–triplet state ($^3D^+$) and the sulphate ion:[2386]

$$D^+ + hv \rightarrow {}^1D^+ \xrightarrow{ISC} {}^3D^+ \tag{12.44}$$

$$^3D^+ + RSO_2^- \rightarrow D\cdot + RSO_2\cdot \tag{12.45}$$

$$RSO_2\cdot + monomer \rightarrow POLYMER \tag{12.46}$$

Desensitization of this reaction is strongly dependent on pH.[2519]

The photopolymer image is due either to the light-scattering properties of polymer or to the refractive index difference within the partially photopolymerized film.

Applications of this method of photopolymerization for rapid-access large-screen display devices[505] and holographic recording[505, 731, 1869, 1870] have been demonstrated.

Crystalline matrix photoimaging is based on photopolymerization of crystalline monomers, e.g. N-vinylsuccinimide, in the presence of such sensitizers as Michler's ketone, benzoin methyl ethane or a photoinitiator specially synthesized for this purpose, such as 4-dimethamino-3'-propionoxy-benzophenone.[1578]

The crystalline coating can be made by spreading of a melted monomer–photoinitiator mixture on a substrate, cooling and crystallizing. After irradiation the unchanged monomer is washed away with a solvent, leaving a polymer image. Photopolymerization does not take place in the crystal lattice but in liquid-like inclusions in a crystalline matrix. The size and compositions of the liquid inclusions are normally determined by the phase relationship of the monomer and photoinitiator. This system is useful for lithography, resists, and other applications.[1578]

12.8.2. Dry-developing Photoresists

A number of systems of dry-developing photoresists are described.

In a dry-film photoresist system a photopolymerizable composition is coated onto a support, e.g. poly(ethylene terephthalate), and can be covered by a protective thin polymer film, such as polypropylene.[3900] Stripping of the protective overcoat exposes the surface of the resist layers, which can be transferred to a work surface such as the copper side of a printed-circuit stock. The resist is imaged through the support by use of a mask, the support is removed, and the image is developed in a suitable solvent. The step requiring solvent handling is development. The use of dry-film resists for fabrication of macrosized circuit boards saves considerable labour in coating, drying, and inspection.

A protected photopolymerization system involving dry-film resists facilitates the use of resist layers much thicker than those possible with photocrosslinkable systems.

Dry-film resists do not lend themselves well to the production of small-geometry imagery because of lower resolution at increased resist thickness and the inability of the resist to conform to topological variations on the device surface. Liquid resists, when coated, conform more easily to surface countours.

A photopolymerizable crosslinkable dry-film system is generally a compatible mixture of three main ingredients:

(i) A preformed polymeric binder, which is non-active and remains chemically unchanged after the process.

(ii) A multifunctional polymerizable monomer, e.g. tri(6-acrylohexyl)-1,3,5-benzenetricarboxylate (12.57).[2546]

$$CH_2 = CHOCO(CH_2)_6OCO \qquad OCO(CH_2)_6OCOCH = CH_2$$

$$(12.57) \qquad OCO(CH_2)_6OCOCH = CH_2$$

(iii) A photoinitiating system, which under irradiation may produce free radicals, radial ions which may initiate polymerization of monomer.

A new concept is the dry-developable negative resist based on poly(methyl isopropenyl ketone) and volatile bis-azides.[3645, 3646] Films of these resists are exposed to u.v. irradiation, baked, and developed with oxygen plasma. Bis-azides in the exposed films are fixed to the polymer and render the films more resistant to oxygen plasma.

Another system of dry-developable resin is based on photoinitiated cationic depolymerization of poly(phthalaldehyde) (12.58) in the presence of onium salts such as diaryliodonium ($Ar_2I^+X^-$) or triarylsulphonium ($Ar_3S^+X^-$) salts:[1819]

$$\left(HC \overset{O}{\diagdown} \overset{O}{\diagup} CH\right)_n \quad \xrightarrow[Ar_2I^+X^-]{+ h\nu} \quad n \overset{CHO\ CHO}{\bigcirc} \qquad (12.47)$$

$$(12.58) \qquad\qquad (12.59)$$

Monomer phthalaldehyde (12.59) formed in this reaction vapourizes, leaving a positive image of the mask.

Photocurable cationic polymerization can also be employed in photoimaging.[784, 785] Epoxy-based photoresists with high resolution have been developed, and used in photography and plastic flexographic printing plates. The photochemistry of photoinitiated cationic polymerization has been described in detail in Section 5.4.

The dry-development technique facilitates an automatic microfabrication process for the semiconductor industry, which promises high process yield, high throughput, and high resolution of the product.

12.8.3. Latent Imaging System Based on
Charge-transfer Complex Formation

Photoreactive systems proceeding through a free radical mechanism are based on the formation of a latent image in two steps:[1016, 1142, 1175, 3388]

(i) under primary irradiation minute amounts of dye are formed at sites exposed to light,
(ii) these trace quantities of dye absorb light energy of much shorter frequency than the initial components, act as photoinitiators of further free radical reactions, and assure the development of the latent image.

Usually such photoreactive systems contain the following components:

(i)   Polymer matrix.
(ii) An activator (e.g. polyhalogen compounds such as carbon tetrabromide, bromoform or benzotribromide).
(iii) Dye-precursor, which may vary considerably e.g. styrylic bases, cyanine bases, diphenyl and triphenyl methane leucoderivatives, or the corresponding bases.

One of such photoreactive systems is based on the following components: poly(N-vinylcarbazol) (12.60) (polymer binder in which the reaction proceeds; it also participates actively in the reaction mechanism), carbon tetrabromide (CBr_4) and di-β-naphthospiropyrane (12.63) in an average molar ratio $5:1:0.1$, respectively.[3388] The following reactions are involved in image formation:

(i) Poly(N-vinylcarbazol) forms, with CBr_4, a charge-transfer (CT) complex (I) (12.61):

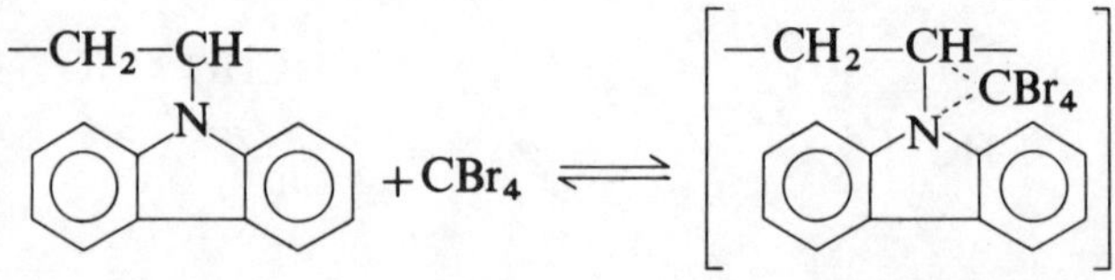

$$(12.48)$$

(12.60)

CT COMPLEX (I)
(12.61)

(ii) Under u.v. irradiation a CT complex (I) (12.61) dissociates to hydrogen bromide (HBr), and $CBr_3 \cdot$ radicals react with poly(N-vinylcarbazole) to poly(tribromomethyl-N-vinylcarbazole) (12.62):

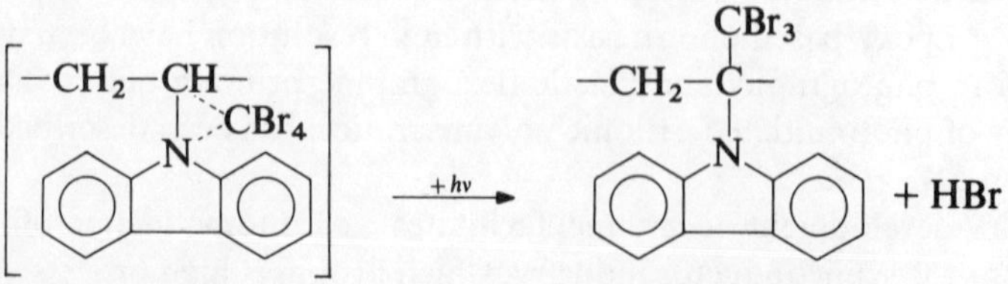

$$(12.49)$$

(12.49)

(12.62)

(iii) Hydrogen bromide reacts with the dye-precursor, i.e. di-β-naphthospiro-pyrane (*12.63*) with formation of a strongly blue-coloured di-β-naphtho-pyrylium bromide (*12.64*):

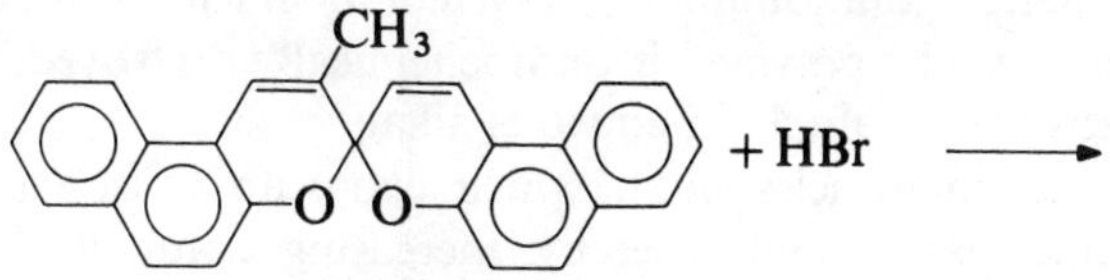

(*12.63*)

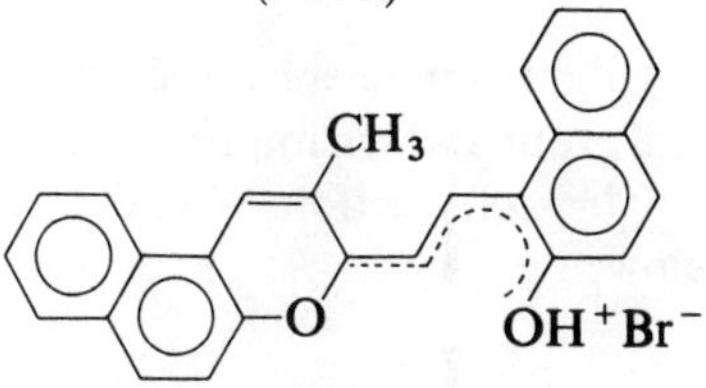

$+ HBr \longrightarrow$

(12.50)

(*12.64*)

(iv) This pyrylium salt (*12.64*) interacts further with CBr_4 with formation of a new CT complex (II):

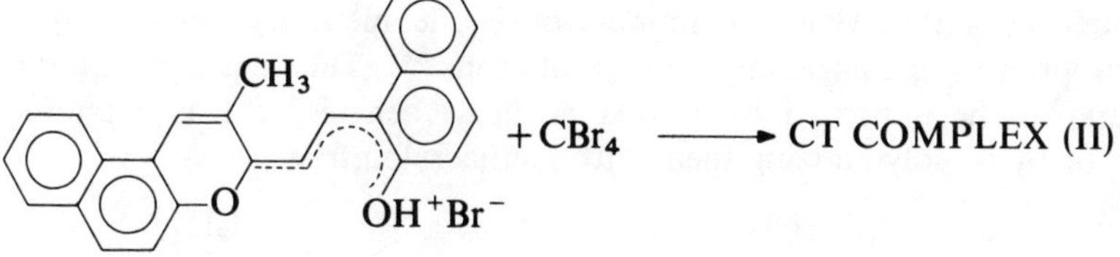

(*12.64*)

$+ CBr_4 \longrightarrow$ CT COMPLEX (II)　(12.51)

(v) During the 'optical development' this CT complex (II) will be photoexcited by exposure at 500–700 nm with renewed production of HBr:

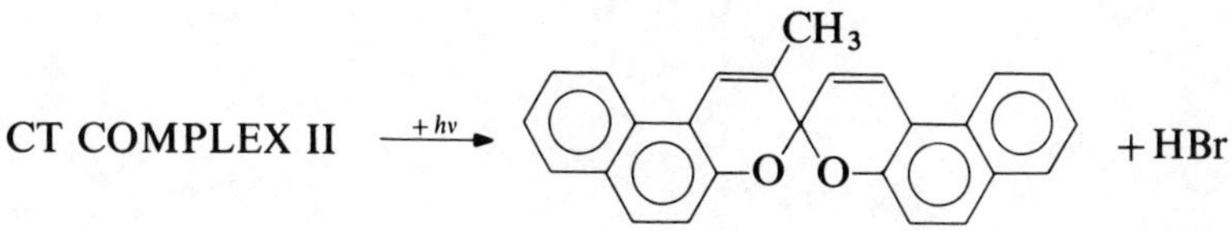

CT COMPLEX II $\xrightarrow{+h\nu}$

$+ HBr$

(12.52)

(vi) This second step results finally in the accumulation of dye at the sites of primary irradiation. Reaction is analogous to reaction (12.50).

In this process heat fixing of the image is assured by removal of the activator (CBr_4) by heating the films at elevated temperature for a short period of time after development.

12.9. POSITIVE-WORKING PHOTORESISTS

Positive-working resists involve either of two different mechanisms:

(i) A polymer matrix–solution inhibitor system in which a solubility inhibitor, dispersed in an acidic polymer, is photochemically destroyed, rendering the exposed areas more soluble in aqueous alkali.

(ii) Photochemical chain scission in which exposure causes a reduction in polymer molecular weight, thereby increasing solubility in a suitable developer solvent.

The decomposition of polymers by light can be enhanced by the incorporation into the polymer chain of functional groups which improve their photosensitivity towards light irradiation[1838,2272,2432] (cf. Section 14.10). The sensitivities of some polymers are compared below:

poly(methyl methacrylate)	1
poly(methyl isopropenyl ketone)	5
poly(butene-1-sulphone)	100

These photodegradable polymers can only be employed in deep-u.v. (200–300 nm) lithographic techniques.

The sensitivity of poly(methyl methacrylate) can be increased by incorporation of a small amount of oxime groups. The N–O bond is photochemically labile. This co(methyl methacrylate/3-oximino-2-butanone methacrylate) (*12.65*) is capable of producing images of 1 μm resolution.[3886] The sensitivity of the copolymers can be improved by adding *p-t*-butyl benzoic acid as a photosensitizer, or by terpolymerizing them with methacrylonitrile.

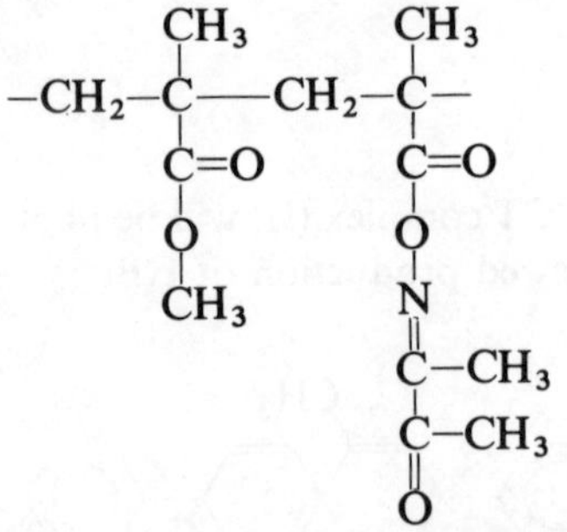

(*12.65*) 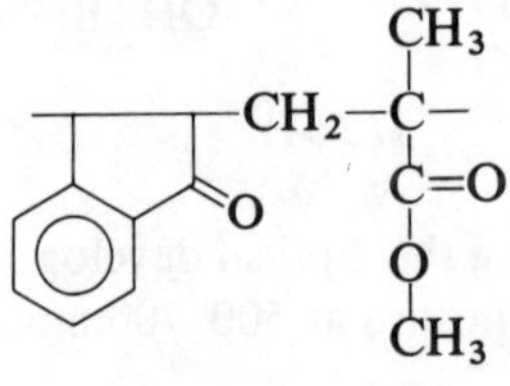   (*12.66*)

The copolymer of methyl methacrylate and indene (*12.66*) shows increased sensitivity. Resolutions of 0.75 μm lines and spaces have been achieved.[3886]

12.10. PHOTOREACTIVE POLYMERS—INDUSTRIAL APPLICATIONS

The industrial application of photoreactive polymers is very broad and increases continuously from year to year. Photoresists are used in the reprographic and the graphic-arts fields (PHOTOFABRICATION PROCESSES) (Fig. 12.11). Photofabrication is well known as a manufacturing technique for making parts or decorating metal surfaces through the use of a photographically produced master pattern, a photosensitive resist, and an etching or plating bath. The process consists of five basic steps:[934]

(i) Preparation of the artwork.
(ii) Photography of the artwork reduced to its final size with a process camera.
(iii) Preparation of a metal support including cleaning, coating with the light-sensitive photoresist material, and drying.
(iv) Exposure of the coating with ultraviolet radiation in contact with the photographic master.
(v) Etching of the material wherein the etchant acts as a cutting or contouring tool that chemically dissolves those portions of the support that are not protected by the exposed and developed photoresist (cf. Section 12.7).

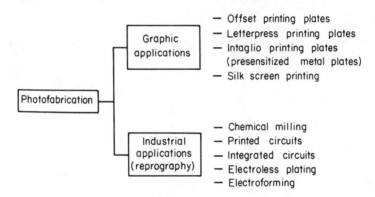

Fig. 12.11   Application fields of photofabrication techniques[934]

Printing plate production is a major market for photopolymers. Cost-effectiveness and high productivity of polymers have resulted in wide acceptance of letterpress plates for newspapers, lithographic plates in commercial printing, and flexographic plates for flexible films. Development of compressible thick flexoplates has opened up a potentially large market in printing of corrugated board. Flexography is also entering the newspaper field.

The printed circuits industry continues to favour dry-film photoresists for their excellent definition and high automation capability.

Ultraviolet screen inks used as solder masks, etch resists, and plating resists, continue to replace solvent-based screen inks.

Photopolymer systems polymerize and/or crosslink by either free radical or ionic mechanisms under irradiation. Bases for these systems include acrylics, alone and in combination with poly(vinyl alcohol), cinnamates, and allylics. The products may be in the form of liquids, solutions, or solid polymer compositions sandwiched between plastic and/or metallic substrates. Photopolymerization of these systems is enhanced by incorporation of sensitizers.

The use of photoresists[503,528,925,1848] and subsequent processing steps is repeated many times in the manufacture of a typical integrated circuit, which may contain as many as 10,000 electronic components on a silicon wafer which is only a few millimetres square. Presently, the image resolution of such photoresists is $2~\mu m$.

The fabrication of an integrated circuit is a complex process which depends on the construction of a precise pattern of electronic components, for example, metal-oxide semiconductor devices. Critical distances in commercially available integrated circuits are often measured in micrometres.

13. Photocuring

PHOTOCURING can be defined as a light-initiated polymerization of unsaturated ink and coating compositions. Photocuring has been the subject of several publications.[603,1144,2232,2880,2882,3131] Photocurable coatings are employed for:

(i) Coating of metal containers, wood and paper, fibers, and floor tiles. Because of their high crosslink density, pigmented and clear coatings are usually quite resistant to natural weathering and to chemical attack from acidic pollutants or solvents. They also exhibit remarkable properties both from optical (brilliance, transparency) and mechanical points of view (hardness, impact and abrasion-resistance, tensile strength, etc.).

(ii) Coatings for laser vision, 'video discs', which are based on acrylate monomers.[2055] The photopolymerization process imposes some very exact requirements on the layer in which the information is stored, such as high curing rate[3730] and low viscosity. The coating on the disc should have high dimensional stability of the pit pattern, good adhesion of the substrate, durability, lack of odour, etc.

(iii) Formation of acrylic pressure adhesives directly on to the backing material.[3300]

(iv) Special applications in fine art conservation (restoration and preservation).[1100,1219]

Depending on the substrate to be coated, coating thickness, and the specific coating formulation, photochemical curing requires from one-half to one-twentieth of the energy used in conventional curing methods.[3226] In addition since liquid monomers are used in photochemical curing, there is no loss of these petrochemically based materials. At the same time, solvent elimination results in a reduction of air pollution and makes this process ecologically attractive.[3227]

A potential way to overcoming the air pollution problem associated with solvent-based coatings exists in the use of solventless, 100 % solid coatings. One common variety of solventless resins consists of solutions of organic prepolymers containing aliphatic unsaturation in copolymerizable organic monomers. Examples of such resins are, for example, solutions of unsaturated polyesters in styrene or vinyl toluene, or solutions of polymers containing esters of acrylic or methacrylic acid in acrylate or methacrylate monomers. Since no solvent has to be evaporated from such systems prior to cure, little if any pollution might be expected from such resins, particularly if such compositions could be cured in a very short time at low temperatures.

In general, four sources of air pollution can be considered:

(i) pollution occuring during application (rolling, spraying, etc.);
(ii) pollution prior to cure, due to evaporation;
(iii) pollution during cure;
(iv) pollution after cure, for example from heating or post-baking of the u.v.-cured resins.

13.1. ULTRAVIOLET-CURABLE COATING SYSTEMS

Ultraviolet-curable coating systems consist of several components, such as:[351,1421,1700,2028,2297,2558,3131,3737]

(i) Polymerizable monomers for film-forming.
(ii) Photoinitiators, which upon radiation produce free radicals or cation radicals that can initiate monomer polymerization.
(iii) Diluents for viscosity reduction.
(iv) Plasticizing diluents, which are incorporated in some systems.
(v) Pigments.
(vi) Dyestuffs for colouring.
(vii) Thermal free radical inhibitor (pot-life stabilizer antioxidant).
(viii) Different additives (for flow, slip, mist, wetting, dispersion control, etc.).

In commercially available u.v.-curable systems most of the ingredients, such as specific resins, are closely guarded proprietary secrets. Resin types currently employed by different companies are:[451,787,805,1232,2394,3131,3813]

(i) unsaturated polyesters,
(ii) acrylated polyesters,
(iii) acrylated epoxy esters,
(iv) acrylated isocyanates,
(v) acrylated triazines (melamines),
(vi) acrylated polyesters,
(vii) thiol/ene systems,
(viii) cationic-cured epoxy systems,
(ix) aminoplasts cured by photoliberated acids.

The use of bis-, tris-, or even tetra-functional monomers, together with suitable binders, leads to crosslinked photopolymer coatings with good physical properties.[918] They become very hard and solvent-resistant on polymerization, owing to formation of crosslinked cage-like structures. An interpenetrating network (*13.1*) is formed within but independent of the binder (reaction 13.1).

Characteristics of u.v. curined coatings and their advantages are shown in Table 13.1. This table compares film thickness, pigments, sensitivity to yellowing, comments on curing conditions, and typical resins used in various applications.

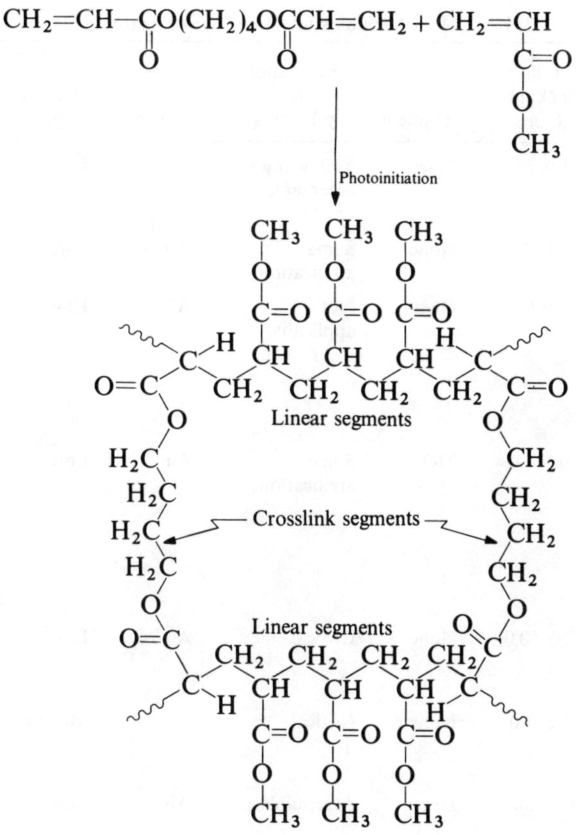

$$CH_2{=}CH{-}\underset{\underset{O}{\|}}{C}O(CH_2)_4O\underset{\underset{O}{\|}}{C}CH{=}CH_2 + CH_2{=}CH$$

$$\begin{array}{c} C{=}O \\ | \\ O \\ | \\ CH_3 \end{array}$$

| Photoinitiation

(13.1)

(*13.1*)

While photoinitiated free radical polymerization is limited to vinyl monomers susceptible to free radical addition, other monomers such as lactones, epoxides, cyclic ethers, sulphides, and acetals, are potential coating materials using cationic photoinitiators (cf. Section 5.4).

Another advantage is that the photoinitiated cationic polymerization is insensitive towards oxygen. In contrast, photoinitiated free radical polymerizations requires blanketing by an inert atmosphere to eliminate oxygen inhibition of polymerization. This is costly and cumbersome, and is therefore undesirable in an industrial process.

The primary component in the u.v.-curable coating system is the photoinitiator, which upon radiation produces free radicals that can initiate monomer polymerization. The most commonly employed photosensitizers (cf. Chapter 7) are:[576, 1232, 1241, 2296, 2458, 2881, 2891, 2892]

  (i) benzoin ethers,
 (ii) benzophenones (+ amine),
(iii) thioxanthones (+ amine),

Table 13.1 Characteristics of u.v. curied coatings[1232]

	Film thickness (mm)	Pigment	Resistance to yellowing	Atmosphere	Curing speed	Typical composition
Wood fillers	1.0–5.0	Filler	Yellowimg acceptable	Air/N$_2$	Low	Polyester styrene
Paper coatings (Clear)	0.1–0.5	None	Some applications	Air/N$_2$	High	Epoxy acrylate
Inks	0.1	Organic	Not applicable	Air	High	Epoxy acrylate, aromatic urethane acrylate
White pigmented inks and coatings	0.3–1.0	TiO$_2$	Some applications	Air	Low	Epoxy acrylate, aromatic urethane acrylate
Floor coatings	2.0–10.0	None	Critical	Air/N$_2$	Low	Aliphatic urethane acrylate
Clear exterior protective coatings	0.5–3.0	None	Critical	Air	Medium	Aliphatic urethane acrylate
Electronics	2.0–200.0	Dyes	Acceptable	Air	Low	Epoxy acrylate

(iv) ketals,

(v) acetophenones,

Each of these classes has certain performance characteristics which account for their commercial acceptance (Table 13.2).

Examples of some ink and coating formulations are given in Table 13.3.

A typical photoreactive coating used to produce flexible circuitry consists of:[3747]

Monomer   300 g

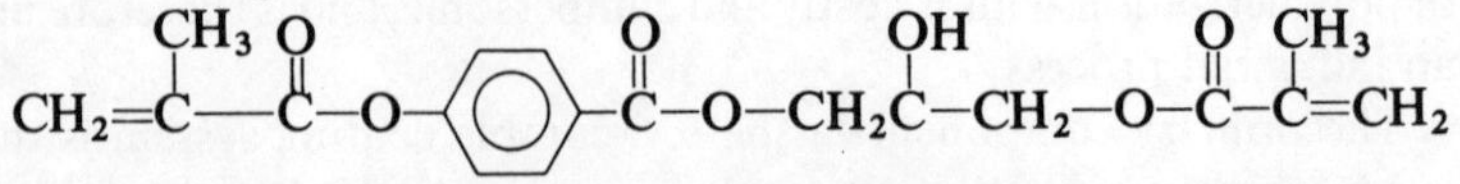

$$CH_2=C{-}\overset{CH_3}{|}{-}\overset{O}{\overset{\|}{C}}{-}O{-}\bigcirc{-}\overset{O}{\overset{\|}{C}}{-}O{-}CH_2\overset{OH}{\overset{|}{C}}{-}CH_2{-}O{-}\overset{O}{\overset{\|}{C}}{-}\overset{CH_3}{\overset{|}{C}}=CH_2$$

Binder   250 g   Poly(methyl methacrylate–co-ethyl acrylate–co-methacrylic acid) (50:34:16)

Photoinitiator 35 g   Benzophenone
                 5 g   Michler's ketone

Stabilizer   0.2 g   p-Methoxyphenol

Coating solvent 490 g   2-Ethoxyethane

469

Table 13.2 Photoinitiators properties and applications[1232]

Photoinitiators	Physical properties			Applications						
	Physical state	Colour	Solubility	Polyester-styrene wood filler, composites	Paper varnishes	Inks	White pigmented	Floor coatings	Exterior	Electronics
Benzoin ethers										
Isopropyl benzoin ethers	Solid	White	Good	●						
Isobutyl benzoin ether mixture)	Liquid	Colourless	Good	●						
Benzophenones										
Benzophenone	Solid	White-amber	Good		●	●				●
Michler's ketone	Solid	Yellow-blue	Fair			●	●			●
Thioxanthones										
Chlorothioxanthone	Solid	Yellow	Poor			●	●			
Isopropylthioxanthone	Solid	Yellow	Fair			●	●			
Dodecylthioxanthone	Liquid	Brown	Good			●	●			
Ketals										
Benzyl dimethyl ketal	Solid	White	Good	●	●	●	●			●
Acetophenone diethyl ketal	Liquid	Colourless	Good					●	●	
Acetophenones										
α-Hydroxy, cyclohexyl phenyl ketone	Solid	White	Good		●			●	●	
2-Hydroxy, 2-methyl phenyl propanone	Liquid	Colourless	Good		●			●	●	

Table 13.3   Some photocuring formulations[603]

Lithographic ink

35.0	oligomer
37.0	multifunctional monomer
8.0	photoinitiator
5.0	wax compound
2.5	monoacrylate monomer
0.5	inhibitor
12.0	colorant
100.0	

Clear varnish

20.0	oligomer
53.0	multifunctional acrylate
5.0	photoinitiator
0.2	slip agent
19.5	monoacrylate monomer
0.3	inhibitor
2.0	accelerator
100.0	

White can coating

15.0	oligomer
12.5	multifunctional acrylate
25.0	monoacrylate monomer
5.0	photoinitiator
0.5	inhibitor
2.0	accelerator
40.0	titanium dioxide
100.0	

Particle-board filler

12.0	oligomer
10.0	multifunctional acrylate
9.7	monoacrylate monomer
1.0	photoinitiator
0.3	inhibitor
23.0	calcium carbonate
44.0	talc
100.0	

The above formulation, when coated and dried, gives tack-free coatings which can be rolled up and stored without danger of adhesion between the coated side and the reverse side of a flexible support. This is accomplished by careful balancing of the nature of the binder and the monomer and their ratio to each other.

In opaque or coloured coatings there is competition between photoinitiator and pigment in that the photoinitiator must:[1232,3131]

(i) Absorb light energy in the same region of the absorption spectrum as the pigment. In this case the molar absorptivity (molar absorptivity (ε) of the photoinitiator (Table 13.4) must be large, or a large concentration of photoinitiator is necessary to effect photopolymerization.

Table 13.4   Photoinitiator molecular absorptivity (ε) coefficients at selected wavelengths[1232]

Photoinitiators	Molecular absorptivity		
	260 nm	360 nm	405 nm
Benzoin ethers			
Isopropyl benzoin ethers	11.379	50	~0
Isobutyl benzoin ether (mixture)	*	*	~0
Benzophenones			
Benzophenone	14.922	51	~0
Michler's ketone	8.040	37,500	1,340
Thioxanthones			
Chlorothioxanthone	*	3,944	197
Isopropylthioxanthone	*	5,182	102
Dodecylthioxanthone	*	3,620	127
Ketals			
Benzil dimethyl ketal	9,740	97	~0
Acetophenone diethyl ketal	5,775	19	~0
Acetophenones			
α-Hydroxy, cyclohexyl phenyl ketone	3,170	18	~0
2-Hydroxy, 2-methyl phenyl propanone	2,710	9	~0

* Not measured

(ii) Not absorb light energy in the same region of the absorption spectrum as the pigment, or have a different absorption spectrum. In this case, the molar absorptivity (ε) of the photoinitiator may be small, and a smaller concentration of photoinitiators can be used.

In each of these cases the thickness of the coating films is very important (inks 2 to 10 micrometres versus a fluid coating of 0.5–2 millimetres) on the amount of light energy absorbed by the photoinitiator.[3877] The amounts of pigment, dispersion, and reflectance of the paint film are also important for light absorption by the photoinitiator (cf. Section 15.1).

The most commonly employed pigments are

(i) white pigments (TiO_2),
(ii) colour organic pigments,
(iii) carbon black,
(iv) extender pigments which absorb little or no visible light (SiO_2, $CaCO_3$).

The presence of pigments can have profound effects on u.v. curable coatings and inks, and cannot be thought of as inert additives. The following effects of pigmentation should be taken into consideration:[1232,2457,3737,3877]

(i) Light scattering (internal or external reflectance) and penetration of energy.
(ii) Refractive index and wavelength of light absorption of the pigment.
(iii) Free radical catalytic activity.
(iv) Particle size and degree of dispersion (viscosity effects).

(v) Amount of pigment and film thickness in relation to hiding power and effects on cure speed.

(vi) Photophysical properties of pigments (cf. Section 15.1.1).

(vii) Photochemical involvement of the pigment in the process.

In the absence of pigment, absorption of a u.v.-curable system is a function of:

(i) Radiation intensity and spectrum.

(ii) Molar absorptivity and concentration of photoinitiator.

(iii) Molar absorptivity of other components of a given u.v.-curable resin.

(iv) Film thickness.

In the case of u.v.-curable free radical polymerization systems the radiation should be absorbed at any layers within the film. Since free radicals have a short lifetime, and since viscosity of the coating will increase as polymerization proceeds, the diffusion path of free radicals is limited to a maximum of 0.1 μm.

The degree of scattering is dependent on:

(i) Difference in refractive index between pigment and vehicle.

(ii) Particle size of pigment.

(iii) Pigment concentration.

(iv) Wavelengths of radiation.

There are three effects of scattering on radiation entering a film:

(i) Increased reflectance.

(ii) Effective path length in the film is increased.

(iii) Radiation within the film is diffused by the scattering so that reflection back into the film from the underside of the top surface increases due to the effect of angle of incidence on the percentage reflectance.

Due to scattering, and the fact that pigments are in polymer, calculation of the absorption by photoinitiator in the presence of a pigment is very complex.[3876, 3877]

Absorption of radiation is affected by the substrate over which a coating or ink is applied. As the concentration of photoinitiator is increased, total absorption of radiation increases and the fraction absorbed by photoinitiator increases. The optimum concentration of photoinitiator depends on a wide range of variables, such as:[728, 857, 1564]

(i) Wavelength energy distribution of the source.

(ii) Absorption spectrum of the photoinitiator.

(iii) Absorption and scattering coefficient of the pigment as a function of wavelength.

(iv) Film thickness.

(v) Substrate reflectance.

Figure 13.1 shows the effect of benzoin concentration and thickness on percentage polymerization. Since obtaining complete surface cure is often the

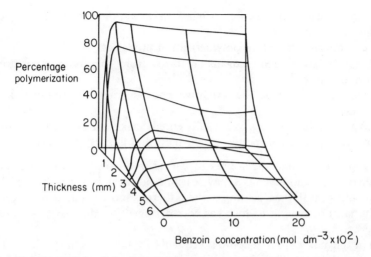

Fig. 13.1 Effect of benzoin concentration and thickness on percentage polymerization[728]

overriding problem, high concentrations of photoinitiator are used to the detriment of radiation transmission, and hence cure of the lower layers. The degree of below-surface curing will affect scratch resistance, hardness, and adhesion to the substrate. A high concentration of photoinitiator would also be expected to produce lower molar mass polymer and to leave high levels of initiator by-products in the cured film, giving further depletion of properties.

Many pigments interfere with u.v. curing due to reduction in absorption of radiation by the photoinitiator, particularly at the bottom of films.[1626] The following list of variables, has been proposed to minimize the effects:[3877]

(i) Select pigments having the least u.v. absorption, that fulfil other requirements.

(ii) Except for very thin films select pigments with the lowest possible scattering coefficients.

(iii) Avoid pigments having high absorption coefficients in wavelength regions where absorption of the pigments is at a minimum (in the case of white pigments this is usually in the very near u.v.).

(iv) Use the u.v. source with high intensity (e.g. lasers[909, 911]) in wavelength regions of comparatively low absorption by pigment and high absorption by photoinitiator.

(v) Optimize photoinitiator concentration for the particular system. This concentration will vary with pigment, monomer, film thickness, substrate, and radiation source.

(vi) If possible, utilize a highly reflective substrate.

The pigment effects and the effect of sample thickness on u.v. curing is usually overshadowed by oxygen inhibition at the surface. Oxygen has the following effects on u.v. curing processes:[1232]

(i) It can quench the excited singlet (S_1) and/or triplet (T_1) states of photoinitiator.

(ii) It can retard free radical polymerization processes.[919, 3869]

(iii) It reacts with free radicals formed from photodecomposition of photoinitiators.[2569]

(iv) It may participate in the oxidative photodegradation and/or yellowing processes.[1711]

(v) Oxygen inhibition leads to partially cured, 'tacky' surfaces, which results in reduction of such properties as hardness and impact resistance.

Some applications lend themselves to curing in an inert atmosphere such as nitrogen or argon; however, this may be economically impractical for most large-scale facilities. In thinner coatings, where there is more contact with oxygen, a photoinitiator or synergist which has been designed to reduce oxygen inhibition may be necessary.

One of the major approaches to successful formulation is the judicious use of the proper inhibitors. The additives (e.g. hindered phenols, cf. Section 15.13) must be used in trace quantities, and give a commercially acceptable package stability. At the same time the addition must not impede the cure.

Coatings and inks must possess appropriate rheological properties to permit application, and the desired flow after application. Many conventional coatings are applied with 50 % or more solvent. The flow properties of a heterogeneous dispersion system such as pigmented ink or coating are a function of the volume of the dispersed phase.[3676, 3875] As the volume of dispersed phase increases, viscosity increases slowly until the packing of the particles causes interference with flow. In general, viscosity begins to increase rapidly as the volume of dispersed phase approaches 50%. Most polymerization reactions result in a reduction in volume. If the surface of a film polymerizes before the lower layers, wrinkling can result.

13.2. SURFACE PROTECTION BY PHOTOCURING

Most of the industrial polymers used for outdoor applications undergo degradation (cf. Section 14.7) when they are exposed to the combined action of sunlight, oxygen from air, humidity, and acid pollutants. This process develops primarily at the surface of the polymer and leads to substantial modification in the appearance (colour, gloss) of the irradiated material.

One of the possible methods of protecting polymers against natural weathering consists of applying organic coatings that offer a good resistance both to u.v. light and chemical attack. These coatings are usually highly crosslinked polymers which can be conveniently obtained by photopolymerization of multifunctional monomers and oligomers. Besides improving the surface properties of the coated materials, this treatment may provide, by its light-screening effect, a good protection for polymers which are very susceptible to photodegradation; for instance, poly(vinyl chloride), one of the leading thermoplastics.

Ultraviolet curable systems used as varnishes, printing inks, or plastics and paper coatings often contain urethane–acrylate oligomers,[2921, 2963, 3154] which, after cure, remain flexible enough to ensure a good adhesion to the support.

The other main type of u.v. coatings used are based on epoxy-acrylates, which exhibit excellent surface hardness and impact resistance characteristics.[910] Unfortunately the adhesion of these epoxy resins onto several polymeric materials, and on poly(vinyl chloride) is often poor, thus reducing their effectiveness as protective coatings.

Since epoxy–acrylate coatings absorb u.v. radiation of wavelengths below 300 nm, which were shown to be very harmful to poly(vinyl chloride), they are expected to act as an effective u.v. filter, thus preventing the photodegradation of poly(vinyl chloride). This protective action of the coatings is clearly apparent on Fig. 13.2, which shows u.v. absorption spectra of poly(vinyl chloride) film and an epoxy–acrylate coated poly(vinyl chloride) film, initially and after 60 and 90 min exposure to light (cf. Section 15.10).

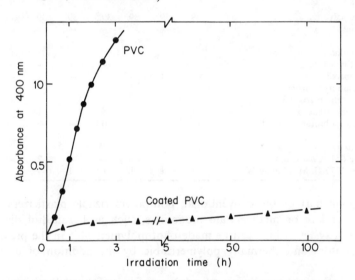

Fig. 13.2 Kinetics of the discoloration of a 100 μm poly(vinyl chloride) film (●) and of an epoxy-acrylate-coated poly(vinyl chloride) film (▲) upon exposure to u.v. light[910]

13.3. DUAL U.V./THERMALLY CURABLE PLASTISOLS

PLASTISOLS are dispersions of solid polymeric resins in non-volatile organic plasticizers. They range in viscosity from pourable liquids to heavy pastes, and can be fused by heating.

Photoreactive, thermally curable plastisol compositions are made by mixing a thermoplastic (preferably poly(vinyl chloride)), a methacrylate, a thermal initiator, a photoinitiator, and a conventional plasticizer (e.g. phthalates, adipates, and phosphates).[2567] Examples of dual u.v./thermally curable plastisol compositions are shown in Table 13.5. Short exposure of these compositions to u.v. radiation results in a tack-free skin cure. Heating after u.v. irradiation gives simultaneous crosslinking and fusion.

Table 13.5   Dual u.v./thermally curable plastisol formulations[2567]

Formulations	A	B	C	D	E	F	G	H
				(in weight percent)				
Reactive monomers								
Trimethylolpropane trimethacrylate	20	20	20	19.9	20	20	—	34
Hexanediol diacrylate	—	—	—	—	—	—	20	—
Polymeric resins								
Poly(vinyl chloride)	39	39	39	39.5	—	39	39	—
Vinylchloride/vinyl acetate copolymer	—	—	—	—	39	—	—	—
Methyl methacrylate/butyl methacrylate copolymer	—	—	—	—	—	—	—	12
Plasticizers								
Adipate polyester	39	39	39	39.5	39	39	39	—
Phthalate	—	—	—	—	—	—	—	51
Thermal initiators								
Benzoyl peroxide	—	—	—	—	—	—	—	1.5
Benzopinacol	—	1	—	—	—	—	—	—
Azobisisobutyronitrile	—	—	1	—	—	—	—	—
1,1-Bis(t-butylperoxy)-3,3,5-trimethylcyclohexane	1	—	—	—	1	1	1	—
Tetramethylthiuram disulfide	—	—	—	0.1	—	—	—	—
Photoinitiators								
Irgacure 651 (Ciba-Geigy)	—	—	—	—	—	2	—	—
Darocur 1173 (E.M. Chemicals)	1	1	1	1	1	—	1	1.5

Replacing part of the conventional, non-polymerizable plasticizers with a reactive acrylate or methacrylate monomers, such as hexanediol diacrylate, results in a system which can be made photopolymerizable in the presence of photoinitiators and thermally polymerizable by the addition of a thermal initiator.

These dual u.v./thermally curable plastisols are useful as adhesives, sealants, encapsulants, and in many other applications.

13.4. PHOTODEGRADATION OF U.V./CURED COATINGS

Relatively few fundamental studies report the problem of photodegradation and weathering of u.v. cured coatings.[161, 397, 465, 910, 1256, 1339, 1711, 2033, 2035, 2791, 2998, 3057, 3209, 3276]

The detrimental effects of sunlight on u.v. cured coatings consist of:

(i) Discoloration (mainly yellowing), more or less pronounced, depending on the chemical structure of the oligomer and photoinitiator used.[1232, 1339, 1709, 1710, 3209]

(ii) A loss of the mechanical and surface properties, that results from chain scission, crosslinking, and oxidation processes.

In order to improve the durability of the coating adequate light-stabilizers must be added.[2035] The photostabilization of u.v. cured organic coatings have been the subject of few reports.[520, 1708-1711, 3057, 3411] It is also important that, during outdoor exposure, some coatings may lose their adhesion to the support so that even a well-stabilized polymer may become ineffective in protecting substrates against weathering for extended periods. In the case of organic polymeric coatings, photografting is one possible way to overcome this difficulty.

The most reliable data on the weatherability of coatings are obtained by subjecting specimens to natural outdoor exposure.[2033, 3276] (cf. Section 14.7). This testing practice has several drawbacks:

(i) Years of natural exposure may be necessary to reveal weatherability differences between good and very good coatings.

(ii) The choice of a site and details of the testing procedure can significantly affect test results. For example, coatings perform differently at test sites in Florida (under humid conditions) than they do in Arizona (under dry conditions).

(iii) Weather fluctuations at test locations from season to season and year to year.

These difficulties have necessitated many attempts at producing an artificial weather environment in the laboratory, in which the deterioration of coating or polymer samples is accelerated.[3060] Commercial equipment is available.[3023]

14. Photodegradation of polymers

PHOTODEGRADATION, which includes such processes as chain scission, crosslinking, and secondary oxidative reactions, has been the subject of several books[43,162,1046,1260,1535,1857,1858,2187,2468,3060,3216,3331] and reviews.[104, 631-633, 1270, 1271, 1583, 2800, 3002, 3003, 3006, 3017, 3032, 3061, 3065, 3068, 3070, 3218, 3247, 3452, 3844, 3883, 3884]

Several theoretical treatments of polymer chain scission and crosslinking by u.v. radiation have been published.[683,1120,1856,2439,3335,3338,3340] Photodegradation data on various polymer films have been examined within the framework of the derived mathematical expressions.[1120, 3336, 3337, 3342] These papers demonstrate the importance of taking into account radiation intensity decreases with depth in the polymer system in the analysis of the changes in molecular weight averages and molecular weight distributions in irradiated optically thick polymer films (i.e. those in which a substantial portion of incident light is absorbed). The effect of uniformity of the intensity of activating radiation, and the changes in absorption produced by the reaction on the kinetics of photolysis of a solute in unstirred[2006] and well-stirred[3339] solutions, have been examined. In the unstirred system the case involving light absorption by the photolabile solute only (i.e. photolysis products and solvents were assumed to be transparent) was considered. Photolysis of solute in the well-stirred system was analysed for the general case in which solute, products, and solvent all absorbed light. However only the light absorbed by the solute led to its photolysis.[3341]

A major problem in studying polymer photooxidation is the low concentration of oxidation products in the early stages of reaction. In extensive photooxidation, volatile products make interpretation of the result difficult, due to complex secondary reactions.

Experimental investigation of photodegradation involved the use of various highly sophisticated instruments: u.v./vis. and i.r. spectrometers, and u.v./vis. derivative spectrometers,[51, 80, 154, 2558, 3189] electron spin resonance (ESR),[2480, 3062, 3653, 3654] electron spectroscopy for chemical application (ESCA),[715, 717,719-722,724,725,986,987,1647,2596,2914-2916] fluorescence and phosphorescence spectrometers,[3023] nano- and pico-second spectrometers,[3023] and many other instruments such as mass spectrometers, chromatographs, electron microscopes, instruments for determining molecular weight and molecular weight distribution, etc.[3022]

Photoacoustic spectroscopy is a new technique which has recently been applied for early detection of photooxidation of polymers[2291] and used in studying the natural weathering of polymers.[8-11,24,189,2445]

Better knowledge of the degradation of polymers would not only be of academic interest but also would be of practical importance. It could improve stabilization by the development of a more suitable and less unstable polymer structures, or by the introduction of more effective stabilizers.

14.1. PHOTOOXIDATIVE DEGRADATION OF POLYMERS BY FREE RADICAL MECHANISM

PHOTOOXIDATIVE DEGRADATION OF POLYMERS, which includes such processes as chain scission, crosslinking, and secondary oxidative reactions, occurs by the free radical mechanism (similar to thermal oxidation) in the following steps:[2468,3060]

(i) Photoinitiation step:
  (a) Internal and/or external chromophoric groups (cf. Section 14.2) absorb light and produce low molecular radicals ($R\cdot$) and /or polymeric radicals ($P\cdot$).
  (b) Energy absorbed by a given chromophoric groups can be transferred to another group (energy transfer process), which further dissociate into free radicals.
  (c) Energy absorbed by a given chromophoric group can be accumulated at a given bond (by an energy migration process), and further dissociate into free radicals.
(ii) Propagation step:
  (a) Subsequent reactions of low molecular free radicals ($R\cdot$) and polymer alkyl radicals ($P\cdot$) in a chain process like abstraction of hydrogen from polymer molecule (PH):

$$PH + R\cdot(P\cdot) \rightarrow P\cdot + RH(PH) \tag{14.1}$$

  (b) Reactions of polymer alkyl radicals with oxygen, during which polymer alkylperoxy radicals ($POO\cdot$) are formed:

$$P\cdot + O_2 \rightarrow POO\cdot \tag{14.2}$$

  (c) Abstraction of hydrogen from the same or another polymer molecule by polymer alkylperoxy radical, with formation of a hydroperoxy group:

$$POO\cdot + PH \rightarrow POOH \tag{14.3}$$

  (d) Photodecomposition of hydroperoxide groups with formation of polymer alkyloxy ($PO\cdot$), polymer alkylperoxy ($POO\cdot$), and hydroxyl ($HO\cdot$) radicals:

$$POOH + h\nu \rightarrow PO\cdot + \cdot OH \tag{14.4}$$

$$2POOH + h\nu \rightarrow PO\cdot + POO\cdot + H_2O \tag{14.5}$$

(e) Abstraction of hydrogen from the same or another polymer molecule by polymer alkyloxy radical with formation of hydroxyl groups:

$$PO\cdot + PH \rightarrow POH + P\cdot \tag{14.6}$$

(f) Disproportionation reaction (β-scission process) of polymer alkoxy radicals with formation of end-aldehyde groups and end-polymer alkyl radicals:

$$PO\cdot \rightarrow -C\underset{\diagdown H}{\overset{\diagup O}{}} + \cdot CH_2- \tag{14.7}$$

(g) Rearrangement reactions, addition reactions, etc.

(iii) Termination step:

    (a) Subsequent reactions of free radicals ($P\cdot$, $POO\cdot$, and $PO\cdot$) among each other, resulting in the crosslinking.

    (b) Reactions of polymer free radicals with low molecular radicals, resulting in the termination reactions.

Formation of propagating radicals ($POO\cdot$) by reaction (14.2) in semi-rigid polymer system is very low (ca. 1%).[2436] because of efficient polymer alkyl radicals ($P\cdot$) recombination before O_2 reacts with them.

Polymer alkylperoxy radicals ($POO\cdot$) are strongly resonance-stabilized, and they can exist at room temperature in polymer matrix.[3062] Most of them show asymmetric singlet-line ESR spectra. They are relatively selective electrophylic species abstracting tertiary hydrogen in preference to secondary or primary bonds.

Polymer alkylperoxy radicals ($POO\cdot$) are reported to be photolysed in vacuum and in air[894,1226,2516,2518,2689] to polymer alkyl radicals ($P\cdot$). In solid polymers, particularly at low temperatures, low radical mobility greatly affects the termination reaction.[948] The cross-termination between polymer alkyl and polymer alkylperoxy radicals should be more rapid than the self-termination of two polymer alkylperoxy radicals from liquid-phase data.[1676] Photoinduced polymer alkyl radical production will lead to a cross-termination reaction only if a polymer alkylperoxy radical is sufficiently close to interact with it before the polymer alkyl radical is reconverted to a polymer alkylperoxy radical by oxygen. When radicals are generated at low temperatures the separation of a radical pair will, on average, be less than if radical generation took place at higher temperatures or if an annealing step led to radical dispersion.

The probability of a polymer alkyl radical being photochemically generated close to a polymer alkylperoxy radical is slight, and the rate of photoinduced radical combination will be less than that of polymer alkylperoxy radical photolysis because the majority of the photochemically generated polymer alkyl radicals will simply be scavenged by oxygen.

Polymer alkylperoxy radicals ($POO\cdot$) recombine in the cage and only a very small number (ca. 0.5%) escape the recombination process to produce hydroperoxide groups ($POOH$) (Reaction 14.3).

Polymeric hydroperoxides (POOH) are the dominant source of free radicals.[79, 624, 625, 1438, 1762, 2198, 2515, 2518, 2689, 2691] They are formed extensively during the thermal processing (Fig. 14.1)[654] (cf. Section 14.3.1).

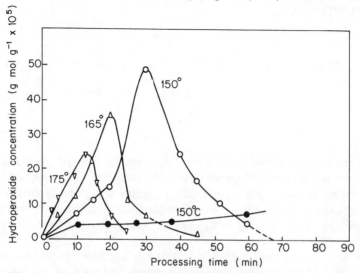

Fig. 14.1 The effect of processing temperature on hydroperoxide formation in low-density polyethylene[654]

The light quanta produced by sun irradiation are energetically sufficient to cleave PO–OH and also P–OOH, but hardly POO–H bonds, which have the dissociation energies of 42 kcal mol^{-1} (PO–OH), 70 kcal mol^{-1} P–OOH and 90 kcal mol^{-1} POO–H.[405, 2578] The large difference in bond dissociation energy between PO–OH and P–OOH mean that reaction with formation PO· and ·OH radicals will predominate during light irradiation. Hydroperoxide groups are transparent at wavelength > 340 nm and they have very low molar absorptivity (molar extinction coefficient) ($\varepsilon = 10$–150 mol^{-1} cm^{-1}) at wavelengths λ = 340 nm. The O–O bond has no low-lying stable excited states, and the potential energy surfaces of the first excited states are dissociative.[1077] The quantum yield of hydroperoxide groups photocleavage in the near ultraviolet is close to 1.0.[632] The most probable mechanism of photodecomposition of OOH groups occurs via an energy transfer processes (cf. Chapter 2) from the excited carbonyl[1270, 1440, 2610, 3697, 3792] or aromatic hydrocarbon[491, 1079, 1263, 1270, 1271, 1276, 2690, 3543] groups (donors) to hydroperoxy groups (acceptors):

$$(\text{donor})^* + -\text{OOH}(\text{acceptor}) \rightarrow \text{donor} + (-\text{OOH})^* \qquad (14.8)$$

$$(-\text{OOH})^* \rightarrow -\text{O·} + \text{·OH} \qquad (14.9)$$

A likely mechanism for this energy transfer is the formation of an encounter complex of finite lifetime between the excited carbonyl group and the ground state hydroperoxide group (or peroxide) as an intermediate in the quenching process. Coupling of the carbonyl n,π* electronic energy and the peroxy vibrational energy then occurs in the charge-transfer (CT) intermediate (exciplex) and leads to subsequent deactivation of the n,π^* species. The resultant vibrationally excited peroxy system might undergo bond dissociation to yield the radicals ($-O\cdot$ and $\cdot OH$):

$$^3(>CO) + P-OOH \rightarrow \left(\begin{array}{c} O \\ \| \\ >C \\ \vdots \\ O-P \\ | \\ OH \end{array}\right)^* \rightarrow (POOH)^* \rightarrow PO\cdot + \cdot OH \qquad (14.10)$$

(vibrational excitation)

Experimental data obtained for the rate of disappearance of hydroperoxy groups during the ketone-sensitized photolysis of *tert*-butylhydroperoxide and of *cis*-polyisoprene hydroperoxide suggest that a significant portion of energy transferred from the excited carbonyl groups to the hydroperoxide groups results in direct scission of the peroxy linkage by an energy transfer mechanism.[1436, 2688-2691]

Using nanosecond laser flash spectroscopy it has been shown that excited polyketones and their copolymers can abstract hydrogen of hydroperoxide groups:[3446]

$$(>C = O)^* + POOH \rightarrow >\dot{C}-OH + PO_2^\cdot \qquad (14.11)$$

It has also been suggested that the hydroperoxide group may form a complex with the carbonyl group (14.1), which under u.v. radiation decomposes by a 'non-free radical' mechanism (without production of any free hydroxyl radical):[1274, 1275]

$$\begin{array}{c} \sigma^+H\cdots O^{\sigma-} \\ | \qquad \| \\ \sigma^-O\cdots C^{\sigma+} \\ / \quad \backslash \\ P-O \quad P \quad P \end{array} \xrightarrow{+h\nu} \begin{array}{c} H \quad O \\ | \quad \| \\ + \quad O-C \\ \qquad \backslash P \\ P-O-P \end{array} \qquad (14.12)$$

(14.1)

During the photothermal oxidation of polymers the thermal decomposition (thermolysis) of hydroperoxy groups increases with increasing temperature, and at temperatures exceeding 140–160°C no hydroperoxide group can exist.

It should also be considered that a free radical can initiate hydroperoxy group decomposition:

$$POOH + R\cdot \begin{array}{c} \nearrow PO\cdot + ROH \\ \searrow POO\cdot + RH \end{array} \qquad (14.13)$$

$$POOH + R\cdot \rightarrow P\cdot + ROOH \rightarrow P\cdot + RO\cdot + \cdot OH \qquad (14.14)$$

The final results of oxidative degradation of polymers are decreased molecular weight, increased molecular weight distribution, formation of crosslinked structures, and formation of a variety of oxidation products such as aldehydes, ketones, carboxylic acids, esters, peroxides, endoperoxides, peracids, peroxyesters, and γ-lactones.[355,612,614,615,633,692,3020,3060,3509,3844]

The chain scission reaction observed during photooxidative degradation of polymers may occur in:

(i) The photoinitiation step, as a direct photodissociation of a given bond, or ketone group in the backbone photolysis (by the Norrish Type I and type II reactions, cf. Section 14.3.1).

(ii) The propagation step as a disproportionation reaction of polymer alkoxy radicals (PO.) (β-scission process. cf. Reaction 14.7).

Crosslinking reactions are the result of a termination step. The statistical theories of chain scission and crosslinking of polymers have been reviewed elsewhere.[835]

The extent of MAIN-CHAIN SCISSION(S) can be calculated with the equation:[180, 2678]

$$S = \frac{w}{M_{n(0)}} \left(\frac{M_{n(0)}}{M_{n(t)}} - 1 \right)$$
(14.15)

where:

w = weight of the irradiated sample (grams),

$M_{n(0)}, M_{n(t)}$ = initial and final number-average molecular weights after irradiation time (t).

The QUANTUM YIELD FOR THE MAIN-CHAIN SCISSION (Φ_s), which is the number of scissions occurring per quantum absorbed, is given by the following equation:

$$\Phi_s = \frac{c}{M_{n(0)}} \left(\frac{d(M_{n(0)}/M_{n(t)} - 1)}{d(I_a t)} \right)$$
(14.16)

where:

c = concentration of a polymer (in solution),

I_a = intensity of light absorbed by a polymer.

The thermal and processing history of fabricated articles such as films and fibres has tremendous effect on the photostability. The conditions to which polymers are subjected during processing are much more severe than those experienced by the polymer during storage. The high temperature normally involved, coupled with the inevitable presence of small amounts of oxygen even in a commercial processing operation and the powerful shearing process to which the polymer is subjected in a screw extruder, are responsible for producing free radicals, and for their oxidation to hydroperoxyradicals.

The effects of processing conditions on photooxidative stability of polymers have been studied in detail on the following polymers:

polyolefins,[624,653,654,656,658,1223,1730,2479,3250] (cf. Section 14.3.1).

polystyrene,[1265] polyethylene/polystyrene blends[3258], co(styrene/acrylonitrile),[1264] and poly(vinyl chloride).[3252, 3266-3268]

The useful lifetimes of polymers and plastics depend on the progress of photooxidative degradation processes.[3924] As a result of these processes physical and mechanical properties change, and are manifest in:

(i) Discoloration.
(ii) Formation of cracks and bubbles on the surface.
(iii) Embrittlement.
(iv) Loss of tensile properties.
(v) Increased electrical conductivity.

The useful lifetime of polymers and plastics also depend on other factors such as:

(i) Chemical structure of a polymer and its photoreactivity.
(ii) Presence and concentration of external and/or internal chromophoric groups (impurities and abnormal groups).
(iii) Physical structure of a polymer (polymer morphology, chain stiffness, crystallinity, etc.).
(iv) Chemical structure and photoreactivity of additives such as pigments, dyes, photostabilizers, antioxidants, thermostabilizers, plasticizers, etc. (cf. Chapter 15).
(v) Environmental condition and weathering conditions (cf. Section 14.7)

Photodegradation and photooxidative degradation processes can be reduced by:

(i) Prevention of the photoinitiation step by application of photostabilizers.
(ii) Interruption of the propagation step by using free radical scavengers and antioxidants (destruction of hydroperoxy groups (OOH)).
(iii) Employing rapid and efficient termination processes.

Despite the many papers published on the photodegradation and photooxidation of polymers, it is still difficult to thoroughly understand these processes because samples and experimental conditions are often different and no quantitative correlation is possible between results obtained by different authors.

14.2. LIGHT-ABSORBING IMPURITIES IN POLYMERS

Industrially produced polymers contain a number of light-absorbing impurities produced in side-reactions during polymerization, processing, and storage. These impurities can be divided into two groups:[3060, 3250]

(i) Internal impurities which contain chromophoric groups, introduced into macromolecules during polymerization, processing and storage; this includes:
    (a) Hydroperoxides (polyolefins, polystyrene, poly(vinyl chloride), polydienes).
    (b) Carbonyls (polyolefins, polystyrene, poly(vinyl chloride), polydienes).
    (c) Unsaturated bonds (C=C) (polyolefins, polystyrene, poly(vinyl chloride).

(d) Catalyst residue (polypropylene, high-density polyethylene, poly(vinyl chloride)).

(e) Charge-transfer (CT) complexes with oxygen (polyolefins, polystyrene).

(ii) External impurities, which may contain chromophoric groups, are:

(a) Traces of solvents, catalyst, etc.

(b) Compounds from polluted urban atmosphere and smog,[1371, 3066, 3677], e.g. polynuclear hydrocarbons such as naphthalene and anthracene in polypropylene,[612, 3250] and polybutadiene.[3040]

(c) Additives (pigments, dyes, thermal stabilizers, antioxidants, photo-stabilizers, etc.).

(d) Traces of metals and metal oxides from processing equipment and containers. The extrusion, milling, chopping, and compounding steps involved in polymer processing can all introduce traces or even particles of metals such as Fe, Ni, or Cr into the polymer.

Anomalous structural units within the polymer molecule can greatly affect photochemical properties of a polymer and the degradation mechanism. Most abnormal groups originate in the initiation step of free radical polymerization.[3404] The free radicals from added initiator, or combination of initiators, add to monomer and are thereby incorporated as chain ends of polymer molecules.

The thermal dissociation of benzoyl peroxide (*14.2*) gives both benzoyloxy (*14.3*) and phenyl radicals (*14.4*):

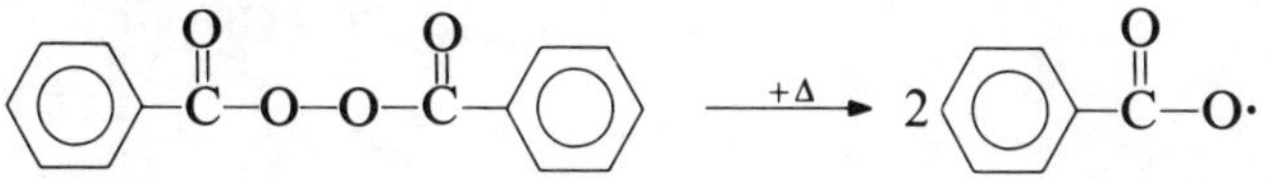

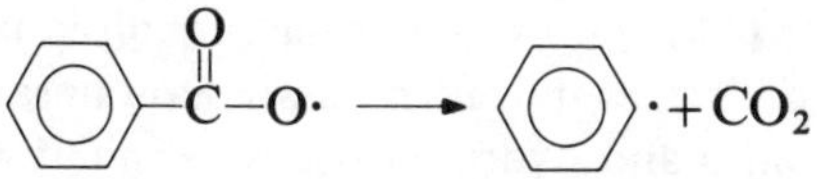

$$\text{(14.17)}$$

$$\text{(14.18)}$$

The resultant polymer molecules therefore contain either ester or phenyl residues as end-group.[2543] The benzylic methylene is susceptible to a variety of hydrogen abstraction and oxidation reactions, which may result in reduced stability of the polymer molecules.

Peresters, e.g. perbenzoate (*14.5*) dissociate to acyloxy (*14.6*) and alkoxy (*14.7*) radicals or, in the case of peroxydicarbonates and peroxalates (*14.8*), to alkoxy radicals (*14.9*) and carbon dioxide:

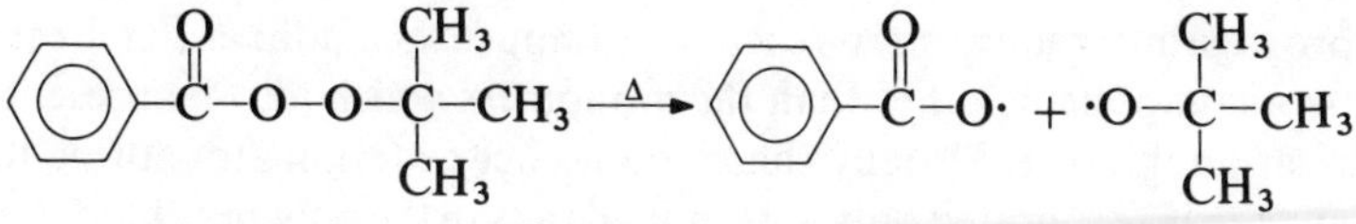

$$\text{(14.19)}$$

$$CH_3-\underset{\underset{CH_3}{|}}{\overset{\overset{CH_3}{|}}{C}}-O-O-\overset{\overset{O}{\|}}{C}-\overset{\overset{O}{\|}}{C}-O-O-\underset{\underset{CH_3}{|}}{\overset{\overset{CH_3}{|}}{C}}-CH_3 \xrightarrow{+\Delta} 2CH_3-\underset{\underset{CH_3}{|}}{\overset{\overset{CH_3}{|}}{C}}-O\cdot + 2CO_2$$

(*14.8*) (*14.9*) (14.20)

The alkoxy and acyloxy radicals may further decompose to alkyl or aryl radicals. The *tert*-butoxy radical (*14.9*) derived from decomposition of *tert*-butyl peroxalate (*14.8*) can abstract hydrogen from monomers as well as undergoing double-bond addition:[1414, 3116, 3117, 3192]

$$CH_2=\underset{\underset{COOCH_3}{|}}{\overset{\overset{\dot{C}H_2}{|}}{C}} + (CH_3)_3C-OH$$

(*14.10*) (14.21)

$$CH_3-\underset{\underset{CH_3}{|}}{\overset{\overset{CH_3}{|}}{C}}-O\cdot + CH_2=\underset{\underset{COOCH_3}{|}}{\overset{\overset{CH_3}{|}}{C}} \xrightarrow{+\Delta}$$

(*14.9*)

$$CH_2=\underset{\underset{\dot{C}OOCH_2}{|}}{\overset{\overset{CH_3}{|}}{C}} + (CH_3)_3C-OH$$

(*14.11*) (14.22)

$$(CH_3)_3CO-CH_2-\underset{\underset{COOCH_3}{|}}{\overset{\overset{CH_3}{|}}{\dot{C}}}$$

(14.23)

If the radicals (*14.10*) or (*14.11*) propagate or are involved in chain termination reactions, the polymer molecules thus formed will have unsaturated end-groups. The use of other peroxides and hydroperoxides as initiators will result in the incorporation of alkoxy, peroxy, or hydroxy end-groups, depending on the initiator and reaction conditions.

During the thermal and/or photodecomposition of azobisisobutyronitrile (AIBN) two radicals are formed: cyano-isopropyl radical and keteniminyl radical (cf. Section 7.12), which initiate polymerization and form end-groups.[429] The keteniminyl end-groups are thermally labile and could undergo dissociation during subsequent treatments of the polymer. Generally it should be noted that AIBN-initiated polymers are usually more photostable than those prepared with peroxide initiators.[250]

The propagating radical may react with impurities, additives, and other by-products or components other than the monomers which may be present in the polymerization mixture. Usually these species act as terminators or as transfer agents, and are incorporated only as terminal units in the polymer chain. A variety of species may, however, be incorporated by copolymerization.

Oxygen is normally an effective inhibitor or chain terminator of radical polymerization but can form peroxy-oligomers (*14.12*) with methyl methacrylate, styrene, and other vinyl monomers:[2438]

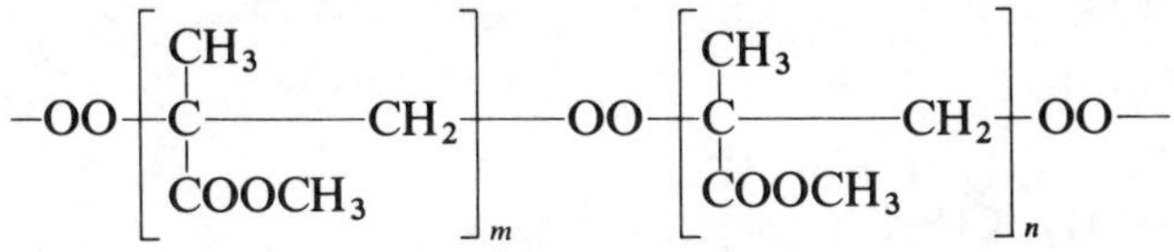

$$-OO\left[\begin{array}{c} CH_3 \\ | \\ C\!-\!\!-\!\!-\!\!-CH_2 \\ | \\ COOCH_3 \end{array}\right]_m\!\!-OO\left[\begin{array}{c} CH_3 \\ | \\ C\!-\!\!-\!\!-\!\!-CH_2 \\ | \\ COOCH_3 \end{array}\right]_n\!\!-OO-$$

$$(14.12) \quad m,n = 1\text{--}3$$

The chain peroxide linkage (*14.13*) can be formed during free-radical polymerization of styrene where there is no rigorous exclusion of oxygen from the reaction:[560,1206,1252,2219,3845]

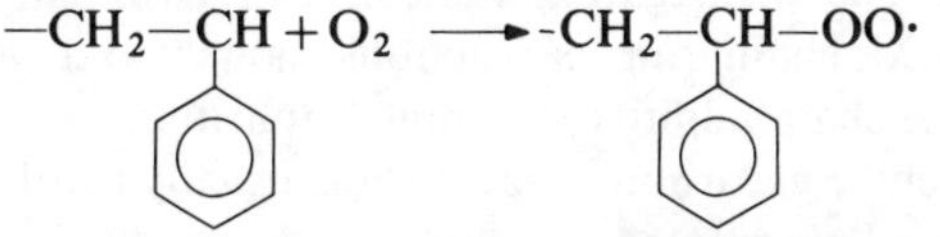

$$-CH_2-\overset{\cdot}{C}H + O_2 \longrightarrow -CH_2-CH-OO\cdot$$

(14.24)

$$-CH_2-CH-OO\cdot + CH_2{=}CH \longrightarrow -CH_2-CH-O-O-CH_2-\overset{\cdot}{C}H$$

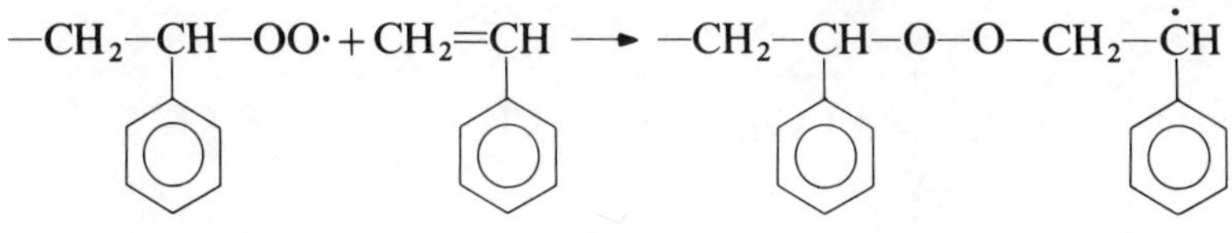

$$(14.13) \qquad (14.25)$$

During u.v. irradiation and/or under heating, peroxide linkages can disproportionate to acetophenone (*14.14*) and hydroxyl (*14.15*) end-groups:

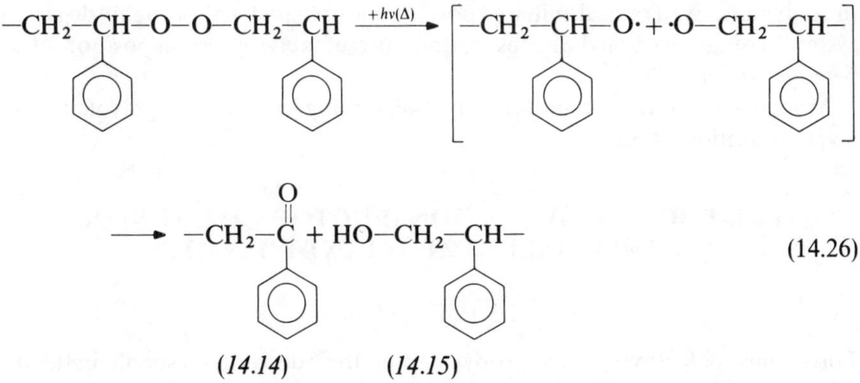

$$-CH_2-CH-O-O-CH_2-CH \xrightarrow{+h\nu(\Delta)} \left[-CH_2-CH-O\cdot + \cdot O-CH_2-CH-\right]$$

$$\longrightarrow -CH_2-\overset{O}{\overset{\|}{C}} + HO-CH_2-CH-$$

(14.26)

$$(14.14) \qquad (14.15)$$

If the polymerization is made in benzene, there is the possibility that solvent molecules may be incorporated in the polymer. The propagation radical of poly(vinyl acetate) (*14.16*) can react with benzene and form dihydroxybenzenoid adducts (*14.17*):[2913]

$$-CH_2-\underset{\underset{OCOCH_3}{|}}{C}\cdot \quad + \quad \bigcirc \quad \longrightarrow \quad -CH_2-\underset{\underset{OCOCH_3}{|}}{\overset{\overset{H\ H}{|}}{C}}\langle \odot \rangle \quad \xrightarrow{+CH_2=CH-OCOCH_3}$$

(*14.16*)

$$-CH_2-CH\overset{\overset{H}{\diagup}}{\underset{\underset{OCOCH_3}{|}}{\diagdown}}\overset{H}{\diagdown}\underset{\underset{OCOCH_3}{|}}{CH_2-CH-} \tag{14.27}$$

(*14.17*)

The termination process of polymerization may also result in the formation of abnormal groups. The propagating radicals can terminate by combination, disproportionation, or chain transfer reactions; namely to solvent, to initiator, or monomer, polymer, chain additive, or some impurities.

Transfer to solvent, e.g. carbon tetrachloride, usually involves the abstraction of a halogen from a halogenated solvent:

$$-CH_2-\underset{\underset{Y}{|}}{\overset{\overset{X}{|}}{C}}\cdot + CCl_4 \quad \longrightarrow \quad -CH_2-\underset{\underset{Y}{|}}{\overset{\overset{X}{|}}{C}}-Cl + \cdot CCl_3 \tag{14.28}$$

The solvent-derived radicals ($CCl_3\cdot$) can initiate further chain polymerization.

Transfer to additives or impurities involves the abstraction of hydrogen by polymer radicals. The radicals derived from the transfer agent then initiate further polymerization. For example, the carbonyl fragments derived from aldehyde impurities or ketonic groups can promote u.v.-initiated degradation of the polymers. Even radical-initiated polystyrene prepared in thoroughly degassed systems contain carbonyl groups, formed presumably by decomposition of α-peroxy species.[1250,1252]

More details on the formation of light-absorbing impurities in polyolefins are given in Section 14.3.1.

14.3. PHOTODEGRADATION (PHOTOOXIDATION) OF COMMERCIALLY IMPORTANT POLYMERS

14.3.1. Polyolefins

Polyolefins (polyethylene, polypropylene, and their derivatives) should in theory be photooxidatively stable on the basis of their structure; commercial polymers are in fact not so. Several reviews[54,1227,1438,3250] and books[54,2468,3060] discuss the problem of polyolefin instability towards u.v. irradiation. The photooxidative degradation of different polyolefins has been a subject of many experimental publications (Table 14.1).

Table 14.1 Photodegradation and photooxidative degradation of polyolefins—references

Name	Structure	References
Polyethylene	$-CH_2-CH_2-$	46, 54, 182, 183, 220, 400, 556, 697, 1227, 1437, 1732, 2054, 3071, 3250, 3251
Polypropylene	$-CH_2-CH-$ $\qquad\;\; CH_3$	46, 79, 81, 82, 84, 222, 238, 266, 612–615, 624, 625, 1225, 1382, 1959, 3071, 3251, 3626, 3754, 3986
Polyisobutylene	CH_3 $-CH_2-C-$ $\qquad CH_3$	360, 639, 643, 3056
Poly(1-butene)	$-CH_2-CH-$ $\qquad\;\; CH_2$ $\qquad\;\; CH_3$	3353
Poly(1-butenesulfone)	$-CH_2-CH-SO_2-$ $\qquad\;\; CH_2$ $\qquad\;\; CH_3$	1666
Poly(4-methylpent-1-ene)	$-CH_2-CH-$ $\qquad\;\; CH_2$ $\qquad\;\; CH$ $CH_3 \quad CH_3$	113
Co(ethylene/propylene)		1272, 1274, 2284
Co(ethylene/propylene/1,4-hexadiene) (EPDM)		489, 490
Co(ethylene/propylene/diene)		126
Co(ethylene/vinyl acetate)		3364, 3366
Co(ethylene/methyl methacrylate)		3625
Polyethylene/poly(methyl methacrylate) blends		1377, 1383
Polyethylene/polystyrene blends		3258
Polyisobutylene/poly(methyl methacrylate) blends		2099
Model compounds: squalene and 2,6,10,15,19,23-hexamethylhydrotetracosane		2421

Commercial polyolefins may contain a number of internal and/or external impurities which can be formed during the following stages:[3250]

(i) Polymer manufacture (unsaturation, catalyst residues, hydroperoxide, and carbonyl groups by adventitious oxidation).

(ii) Processing and fabrication (hydroperoxides and carbonyl groups formed under high-temperature oxidation, transition metal ions from machinery or compounding ingredients).

(iii) Environmental exposure (polycyclic hydrocarbons from atmospheric pollution, carbonyl groups by photolysis of hydroperoxides, unsaturation by photolysis of ketones, transition metal ions (particularly iron and copper)).

The photoinitiation step of the photooxidative degradation of polyolefins has been a long-standing subject of some controversy. Several internal and external impurities are believed to be responsible for this step: hydroperoxides,[79,169,183, 606,612,1228,1338] carbonyls,[612,624,1338,1440] unsaturated bonds,[81,84,183,3147] metal catalyst residue,[82,181,612,632,2479] and traces of polynuclear aromatic hydrocarbons.[612,632]

It is known that commercial polyolefins always contain small amounts of olefinic unsaturations of which vinyl (*14.18*) is the more important in high-density polyethylene (HDPE) and vinylidene (*14.19*) in low-density polyethylene (LDPE):[3147]

$$
\begin{array}{cc}
-\mathrm{CH}-\mathrm{CH_2}- & -\mathrm{C}-\mathrm{CH_2}- \\
\hspace{0.3em}| & \hspace{0.3em}\| \\
\hspace{0.3em}\mathrm{CH} & \hspace{0.3em}\mathrm{CH_2} \\
\hspace{0.3em}\| & \\
\hspace{0.3em}\mathrm{CH_2} & (14.19) \\
\\
(14.18) &
\end{array}
$$

These groups, like the unsaturation in polybutadiene polymers, become involved in oxidative reactions during a typical processing operation, and hydroperoxides (OOH) can be measured in LDPE quite early during processing.[183,654] The rate of formation of hydroperoxides increases with increasing temperature (Fig. 14.1), indicating that they are formed thermally rather than mechanochemically. The maximum hydroperoxide concentration achieved is also highest at the lowest processing temperature (Fig. 14.1), consistent with the known decreasing stability of hydroperoxides with increasing temperature.

The processing time has an evident influence on the formation and disappearance of vinylidene, hydroperoxide, and carbonyl groups (Fig. 14.2).[182,183,654, 3250] Ketonic carbonyls increase in an auto-accelerating mode[182,183] and vinylidene, which is initially the most important unsaturated group present, decays rapidly after an induction period. Vinylidene decrease is associated with the rapid formation and rapid decay of hydroperoxide groups. The formation of conjugated carbonyl (*14.20*) in the later stages of oxidation in place of vinylidene (*14.19*) indicates that vinylidene decay is associated with the oxidation of the allylic group:

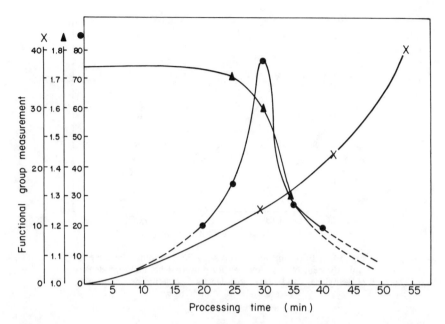

Fig. 14.2 Change in functional group concentrations during processing of low-density polyethylene: (▲) vinylidene intensity, (×) carbonyl intensity, and (○) hydroperoxide $(g\ mol^{-1} \times 10^6)$[3250]

$$-CH_2-\underset{\underset{CH_2}{\|}}{C}-CH_2-CH_2- \xrightarrow{+O_2} -CH_2-\underset{\underset{CH_2}{\|}}{\overset{OOH}{C}}-CH-CH_2-$$

$$(14.19)$$

$$(14.29)$$

$$-CH_2-\underset{\underset{CH_2}{\|}}{\overset{OOH}{C}}-CH-CH_2- \xrightarrow{\Delta} -CH_2-\underset{\underset{CH_2}{\|}}{C}-\overset{O}{\overset{\|}{C}}-CH_2- \ +H_2O$$

$$(14.20)$$

$$(14.30)$$

Processing time has an effect on the time to embrittlement (Fig. 14.3). The time to embrittlement of polyolefins during photooxidation is not directly related to the concentration of carbonyl groups formed (Fig. 14.4).[1730,1731,2479,3247] This finding is based on the assumption that no carbonyl groups are responsible for the initiation of photodegradation in polyolefins. Long-chain dialkyl ketones[1504, 3250,3938] are not photoinitiators for oxidative degradation of polypropylene, whereas aromatic ketones are.[181,1504]

492

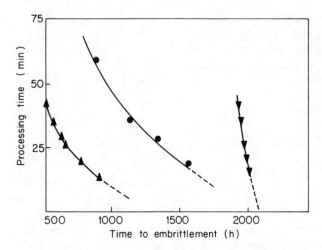

Fig. 14.3 Effect of melt processing in air upon the time to embrittlement of: (▲) low-density polyethylene, (▲) high-density polyethylene, and (●) low-density polyethylene containing $Fe(acac)_3(1 \times 10^{-4}$ mol/100 g)[3250]

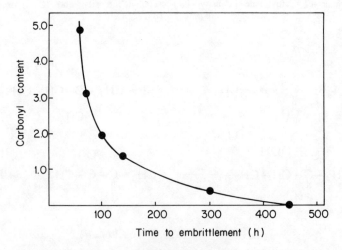

Fig. 14.4 Relationship between carbonyl content and embrittlement time of high-density polyethylene[3250]

Photooxidation of LDPE, which was subjected to a processing treatment, leads to a very rapid decay of vinylidene groups and to the auto-accelerating formation of ketonic carbonyl groups (Fig. 14.5).

Allylic hydroperoxides (*14.21*) formed from thermal oxidation of vinylidene groups (*14.19*) are the most important photoinitiating groups initially present in

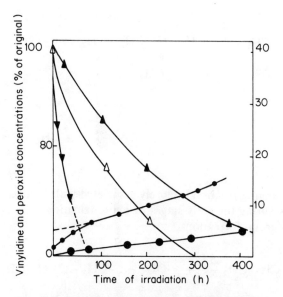

Fig. 14.5 Effect of irradiation time on functional group concentration in low-density polyethylene: vinylidene intensity (▲) 5 hours and (△) 30 hours of processing, carbonyl (●) 5 hours and (○) 30 hours of processing, and hydroperoxide (▲) 30 hours of processing[3250]

thermally treated polyolefins.[183] The thermal and/or photodecomposition of hydroperoxy groups leads to:

(i) A chain scission reaction, during which end-aldehyde groups (*14.22*) are formed:

$$-CH_2-\underset{\underset{CH_2}{\|}}{\overset{\overset{OOH}{|}}{C}}-CH-CH_2- \longrightarrow -CH_2-\underset{\underset{CH_2}{\|}}{\overset{\overset{\overset{\cdot}{O}}{|}}{C}}-CH-CH_2- + \cdot OH$$

(*14.21*)

(14.31)

$$-CH_2-\underset{\underset{CH_2}{\|}}{\overset{\overset{\overset{\cdot}{O}}{|}}{C}}-CH-CH_2- \longrightarrow -CH_2-\underset{\underset{CH_2}{\|}}{\overset{}{C}}-CHO + \cdot CH_2-$$

(*14.22*)

(14.32)

(ii) A crosslinking reaction, which occurs only during u.v. irradiation, but not in thermal oxidation:

$$-CH_2-\overset{\overset{\displaystyle \cdot O}{|}}{\underset{\underset{\displaystyle CH_2}{\|}}{C}}-CH-CH_2- \ + \ -CH_2-\underset{\underset{\displaystyle CH_2}{\|}}{C}-CH_2-CH_2- \ \rightarrow \ -CH_2-\overset{\overset{\displaystyle \cdot}{|}}{\underset{\underset{\underset{\underset{\underset{\displaystyle CH_2}{|}}{C}}{-CH_2-\overset{\|}{\underset{CH_2}{}}}}{\underset{\displaystyle O}{|}}}{C}}-CH-CH_2-$$

$$-CH_2-\overset{\overset{\displaystyle \cdot}{|}}{C}-CH-CH_2-$$
$$| $$
$$CH_2$$
$$|$$
$$O$$
$$|$$
$$-CH_2-\underset{\underset{\displaystyle CH_2}{\|}}{C}-CH-CH_2- \qquad (14.33)$$

This crosslinking reaction is closely associated with changes in mechanical properties during the early stages of photooxidation.

(iii) Carbonyl group formation:

$$-CH_2-\overset{\overset{\displaystyle \cdot O}{|}}{\underset{\underset{\displaystyle CH_2}{\|}}{C}}-CH-CH_2- \ + \ \cdot OH \ \rightarrow \ -CH_2-\underset{\underset{\displaystyle CH_2}{\|}}{C}-\overset{\overset{\displaystyle O}{\|}}{C}-CH_2- \ + \ H_2O \qquad (14.34)$$

The hydroperoxide photolysis and chain scission are the most important initiating processes occurring in the photooxidation of polyolefins.[79, 169, 183, 606, 1228]

The role of charge-transfer (CT) complexes between polyolefins and oxygen,[46, 612, 3655] and singlet oxygen $(1/O_2)$ in the formation of hydroperoxide groups in polyolefins has been considered, but no evident proof was obtained.[621, 632, 1119, 1954, 3634]

Commercial polyolefins may contain some catalyst residues. For example, commercial isotactic polypropylene is polymerized from a heterogeneous organo–aluminium–titanium complex (Ziegler–Natta process), or less frequently from metallic oxides of Cr, V, or Mo bonded to inert support (e.g. Phillips process).

The process employing an unsupported Ti catalyst can leave considerable amounts of Ti and Al mixed with or bonded to the polymer. For example polypropylene film commercially produced can contain as many as 14 different metals at the concentrations: Ti—15 ppm, Al—25 ppm, Fe—8.5 ppm, Mg—2 ppm, Mn—0.03 ppm, Pb—0.08 ppm, Cr—0.09 ppm, Mo—0.5 ppm, Ni—0.2 ppm, Zn—1 ppm, Na > 25 ppm, Cu—0.4 ppm, Ca—3.5 ppm, and Sn—0.08 ppm.[3106] The precise chemical composition of these catalyst residues has yet to be established, but they probably include TiO_2, $OTiCl_2$, $OTiCl(OR)$, $TiCl_2(OR)_2$, $TiCl_4$, or $Ti(OR)_4$ (where R may be alkyl or polymer molecule).[711] Most Ti compounds absorb in the u.v. region up to ca. 350 nm,[710, 711, 1765], whereas Al_2O_2 only absorbs at 300 nm and is probably unimportant as a photoinitiator in comparison to Ti.

The u.v. stability of polypropylene decreases with increasing ash content (mainly Ti and Al). Ti residues may also lead to rapid destruction of antioxidants in the melt and act by catalysing the decomposition of hydroperoxide groups under u.v. irradiation.[265] Transition metals from complexes with hydroperoxide groups, which can be initiating species.[443, 3245, 3651] In a thin metal-free surface layer, no oxidation products appear in oxidized polyethylene.[723] It is also well known that transition metals accelerate the autooxidation of hydrocarbons, which may be present in very low concentrations,[3718] and also accelerate the hydrogen peroxide photolysis.[2347] The effects of typical pro-oxidant transition metal ion complexes on the u.v. stability of LDPE are shown in Table 14.2. It is clear that cobalt and iron in soluble form have a catastrophic effect on the u.v. stability of LDPE.[2806, 3250] Moreover, stability is affected in proportion to the effect of the metal ion on the carbonyl index of the polymer as a result of oxidation during processing.

Table 14.2 Effect of transition metal acetylacetonates on the oxidation of low-density polyethylene during processing and on u.v. irradiation[3250]

Additive	Concentration (mol/100 g)	Carbonyl index (before exposure)	Time to embrittlement (h)
Fe(acac)$_3$	2×10^{-3}	10.22	120
	1×10^{-3}	4.64	216
	0.7×10^{-5}	1.10	260
Co(acac)$_3$	2×10^{-3}	43.57	0
	1.4×10^{-4}	22.82	50
Ni(acac)$_2$	2×10^{-3}	0.51	1968
Mn(acac)$_2$.2H$_2$O	2×10^{-3}	0.64	600
	1.7×10^{-4}	2.70	510
Ce(acac)$_3$	2×10^{-3}	0.48	700
Zn(acac)$_2$	1×10^{-3}	0.38	1370
None	–	0.15	2100

Both thermal and photoinitiated oxidations of polypropylene have been reported to occur preferentially in the surface layers of thick samples, although interpretations differ as to the cause.[481, 625, 626, 1224, 1536, 3151] Oxygen diffusion control has been proposed as the factor limiting the oxidation of the sample interior,[481, 1419] but in the case of photooxidation of polypropylene the concentration of light-absorbing impurities (cf. Section 14.2) in the surface layers seems a more likely explanation.[626, 632]

The oxidation of solid polypropylene occurs only in the amorphous phase.[3925] During crystallization of an impure molten polymer material the impurities are concentrated in the interlamellar amorphous regions at the spherulic boundaries.[1992, 2588]

The crystalline phase has been reported[1333] to be impermeable to oxygen, and the radicals formed within the crystallites are thought to migrate to defects or phase boundaries where oxygen is available.[3285] The crystalline lattice of polypropylene is known to be highly imperfect, particularly when the melt is cooled rapidly, as occurs in commercial film and fibre production.[3306] The crystal structure produced under these rapid cooling conditions is partially oxygen-permeable.

Permeability of semicrystalline polymers is reduced by increase in orientation of the amorphous phase, which greatly hinders the diffusion of the penetrant.[3749] Permeability is rapidly reduced during the drawing of polyethylene,[2929] whereas it is less orientation-dependent in the case of drawing polypropylene films.[753]

Increases in crystallinity decrease the permeability of most semicrystalline polyolefins by reducing the volume available to the penetrant, and by limiting the mobility of the amorphous segments nearby.

For a semicrystalline polymers such as polypropylene, photostabilizers and antioxidants are concentrated in the amorphous regions of the polymer where photooxidation reactions occur.

Secondary cage radical recombination in the photooxidation of solid polypropylene, can considerably limit the photooxidative propagation reaction.[1228] Most peroxy radical pairs formed as a result of hydroperoxide photolysis in the solid polymer terminate after only a few propagation steps (secondary cage recombination). The small proportion of peroxy radicals escaping the secondary cage propagate with very long kinetic chain length to produce the bulk of photoinitiating hydroperoxides so that even relatively inefficient radical scavenging of these intermediates can result in a significant reduction in the rate of hydroperoxide formation.

The progressive oxidative backbone scission of intercrystalline tie molecules leads to crack formation in the surface zone and causes a rapid decrease in residual elongation with increasing irradiation time. Massive embrittlement occurs at hydroperoxy group (OOH) concentration $1–5 \times 10^{-2}$ M, i.e. 0.04–0.3 reactions on the tertiary C–H sites.[624] The strain greatly enhances the photodegradation of polyethylene.[394] Photooxidative processes have a tremendous effect on the decreasing of mechanical properties (tensile strength, elongation at break) of polyolefins.[2864, 3000]

Polyolefins which contain a carbonyl group in the backbone (14.23) are photolysed primarily by the NORRISH TYPE I scission reaction, during which carbon monoxide and two macroradicals are formed, whereas if they contain carbonyl end-group (14.24) they are photolysed by a NORRISH TYPE II scission to give acetone and an unsaturated polymer chain end (reactions 14.35–14.40).[182, 356, 467, 527, 612, 624, 1349, 1442, 3357]

A new mechanism of chain scission in co(ethylene/propylene), in which free radicals formed from the photolysis of the hydroperoxy group ($PO\cdot$ and $\cdot OH$) react with neighbouring ketone groups, has been proposed (reactions 14.41 and 14.42).[1272, 1274]

Formation and disappearance of different functional groups during u.v. irradiation of polyolefins can be followed easily by measuring i.r. spectra. The most significant changes observed during u.v. irradiation of LDPE is a rapid decrease of vinylidene groups (887 cm^{-1}) and formation of carbonyl groups

$$-CH_2\overset{\underset{\displaystyle H}{|}}{\overset{\displaystyle CH_3}{C}}CH_2\overset{\displaystyle O}{\overset{\|}{C}}CH_2\overset{\underset{\displaystyle H}{|}}{\overset{\displaystyle CH_3}{C}}$$

(*14.23*)

$\xrightarrow[\text{Norrish I}]{\geqslant 90\%}$ $-\overset{\underset{\displaystyle H}{|}}{\overset{\displaystyle CH_3}{C}}CH_2\cdot + O=\overset{\displaystyle \cdot}{C}CH_2\overset{\displaystyle CH_3}{C}-\longrightarrow CO + \cdot CH_2\overset{\underset{\displaystyle H}{|}}{\overset{\displaystyle CH_3}{C}}-$ (14.35)

$\xrightarrow[\text{Norrish II}]{\leqslant 10\%}$ $\left.\begin{array}{l} -CH=C\overset{\displaystyle CH_3}{\underset{\displaystyle H}{}} \\[2ex] -CH_2C\overset{\displaystyle CH_2}{\underset{\displaystyle H}{}} \end{array}\right\} + CH_2=\overset{\displaystyle OH}{\overset{|}{C}}CH_2\overset{\underset{\displaystyle H}{|}}{\overset{\displaystyle CH_3}{C}}$ (14.36)

$\longrightarrow CH_3\overset{\displaystyle O}{\overset{\|}{C}}CH_2\overset{\underset{\displaystyle H}{|}}{\overset{\displaystyle CH_3}{C}}-$ (14.37)

$$-CH_2\overset{\underset{\displaystyle H}{|}}{\overset{\displaystyle CH_3}{C}}CH_2\overset{\displaystyle O}{\overset{\|}{C}}\overset{}{\underset{\displaystyle CH_3}{}}$$

(*14.24*)

$\xrightarrow[\text{Norrish I}]{\sim 15\%}$ $-CH_2\overset{\underset{\displaystyle H}{|}}{\overset{\displaystyle CH_3}{C}}CH_2\cdot + CH_3\overset{\displaystyle \cdot}{C}O \longrightarrow CH_3CHO$ (14.38)

$\longrightarrow CH_3\cdot + CO$ (14.39)

$\xrightarrow[\text{Norrish II}]{\sim 85\%}$ $\left.\begin{array}{l} -CH=C\overset{\displaystyle CH_3}{\underset{\displaystyle H}{}} \\[2ex] -CH_2C\overset{\displaystyle CH_2}{\underset{\displaystyle H}{}} \end{array}\right\} + CH_2=\overset{\displaystyle OH}{\overset{|}{C}}CH_3$ (14.40)

$\downarrow$

CH_3COCH_3

$$POOH + h\nu \longrightarrow PO\cdot + \cdot OH \qquad (14.41)$$

$$\begin{array}{l} PO\cdot \quad O=C \\ \\ HO\cdot \quad O=C \end{array} \longrightarrow \begin{array}{l} O=C \\ OP \\ OH \\ O=C \end{array} \qquad (14.42)$$

(1721 cm^{-1}), carboxylic groups (1185 cm^{-1}), and end-unsaturated groups (909 cm^{-1}) (Fig. 14.6).

The presence of carbonyl and peroxidic groups formed in the thermal-, photo- and outdoor degradation of polyolefins can be measured by luminescence (fluorescence and phosphorescence) spectroscopy.[57,89-92,95,108,122,133,135,139,2805,2807,2813]

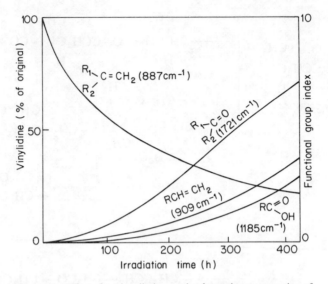

Fig. 14.6 Effect of u.v. irradiation on the change in concentration of functional groups in low-density polyethylene[183]

Computer simulation of photooxidation kinetics of polyolefins[1272,1449] can predict the final concentration of different functional groups and products in photodegraded samples.

14.3.2. Polydienes

Photodegradation (photooxidative degradation) of polydienes and their copolymers and blends has been the subject of many experimental publications (Table 14.3). Polydienes contain unsaturated double bonds in the main chain such as poly(1,4-butadiene) or polyisoprene, or in pendant groups such as poly(1,2-butadiene). These are the polymers most susceptible to photooxidative degradation.

The photoinitiation step during the photodegradation of polybutadiene (14.25) includes formation of polymer chain-end radicals (14.26) by the dissociation of a weak $-CH_2-CH_2-$ bond:

$$-CH_2-CH=CH-CH_2-CH_2-CH=CH-CH_2- \xrightarrow{+h\nu}$$

$$(14.25)$$

$$-CH_2-CH=\overset{\cdot}{C}H-CH_2\cdot + \cdot CH_2-CH=CH-CH_2-$$

$$(14.26) \hspace{3cm} (14.43)$$

or abstraction of hydrogen from polymer molecule by a free radical ($R\cdot$) and formation of polymer allyl radical (14.27):

Table 14.3 Photodegradation and photooxidative degradation of different elastomers—references

Name	Structure	References
1,4-Polybutadiene	$-CH_2-CH=CH-CH_2-$	353, 355, 640, 642, 949, 1283, 1498, 1830, 2030, 2198, 2334, 3024, 3030, 3711
Polyisoprene	$-CH_2-C=CH-CH_2-$ $\quad\quad\;\; \mid$ $\quad\quad\;\; CH_3$	
1,4-Polypiperylene	CH_3 $\mid$ $-CH-CH=CH-CH_2-$	641, 642
1,2-Polybutadiene	$-CH_2-CH-$ $\quad\quad\;\; \mid$ $\quad\quad\;\; CH$ $\quad\quad\;\; \parallel$ $\quad\quad\;\; CH_2$	355, 1353, 1911, 2334
Polynorbornene	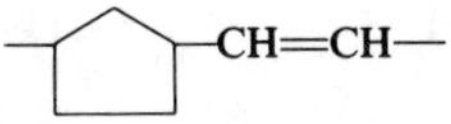 $-CH=CH-$	3945
Co(isobutylene/butadiene) (butyl rubber)	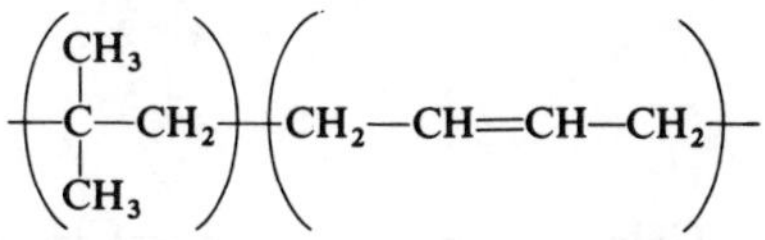	3355
Co(ethylene/propylene/1,4-hexadiene) (EPDM)	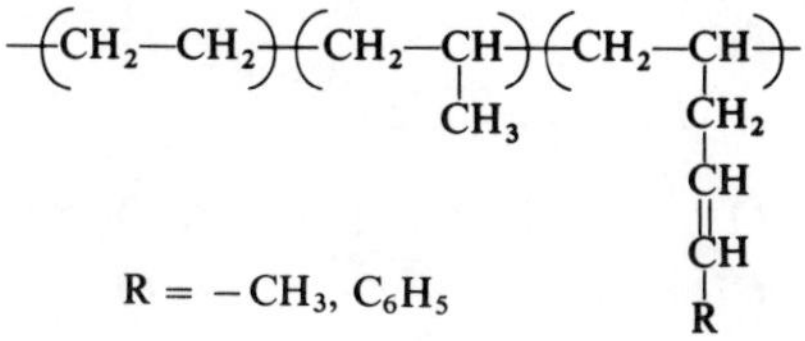 $R = -CH_3, C_6H_5$	489–491
Co(butadiene/acrylonitrile)		3363
Co(acrylonitrile/butadiene/styrene) (ABS)		29, 30, 263, 501, 883, 885, 1257, 1278, 1279, 1994, 2101, 3315, 3316, 3263
Co(butadiene/styrene)/ polyethylene blends		1281
Co(butadiene/styrene)/poly(methyl methacrylate) blends		3263
Co(acrylonitrile/butadiene/styrene) (ABS)/poly(methyl acrylate) blends		3263
Polybutadiene modified polystyrene		1282–1285

$$-CH_2-CH=CH-CH_2- + R\cdot \rightarrow -CH_2-CH=CH-\dot{C}H- + RH \quad (14.44)$$
$$(14.27)$$

Formation of this type of radicals has been confirmed by ESR spectroscopy.[640–642,3062]

Both radicals (14.26) and (14.27) are further oxidized to polymer peroxy (POO·) and oxy (PO·) radicals, and in consequence give hydroperoxy, cycloperoxy, α-, and β-unsaturated carbonyl, aldehyde groups:[353,355,356,949,1283,1498,2468,3030,3060]

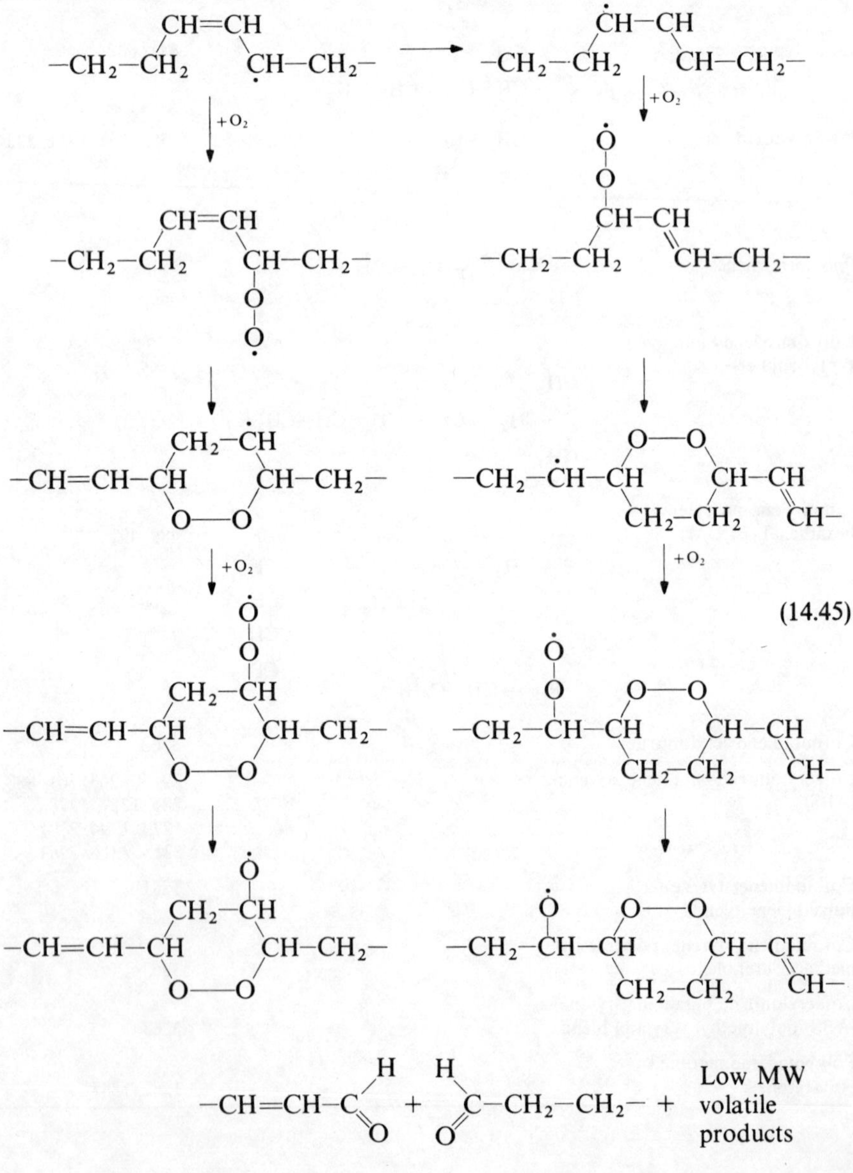

$$(14.45)$$

$$-CH=CH-C\overset{H}{\underset{O}{\diagdown}} \quad + \quad \overset{H}{\underset{O}{\diagup}}C-CH_2-CH_2- \quad + \quad \text{Low MW volatile products}$$

Direct excitation of a double bond, or by an energy transfer mechanism, can provide formation of biradicals (*14.28*) and further secondary reactions such as:[3030,3060]

(i) *Cis–trans* isomerization:

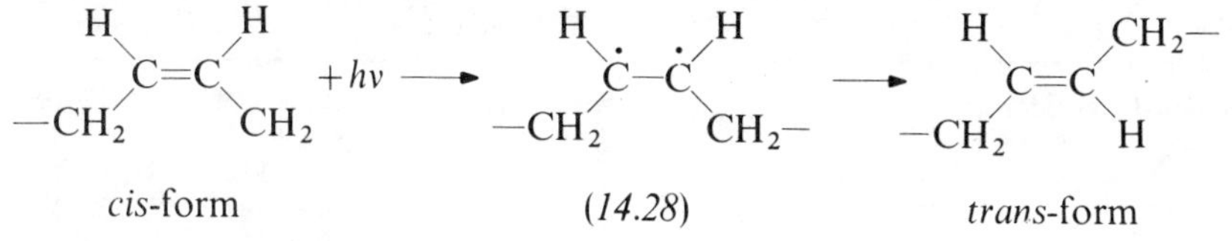

cis-form (*14.28*) trans-form

(14.46)

(ii) Formation of cycloperoxides:

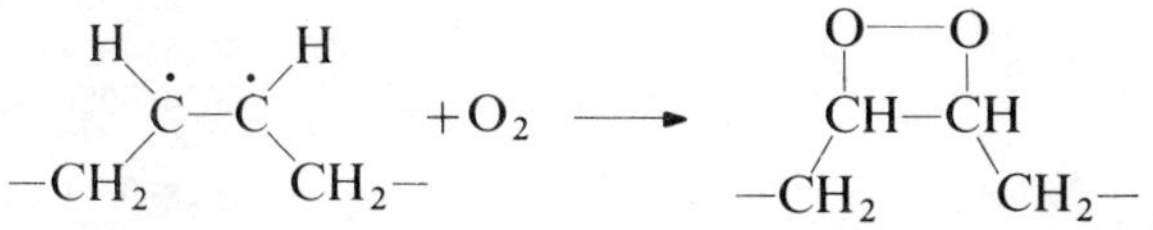

(14.47)

(iii) Formation of cyclopropyl groups:

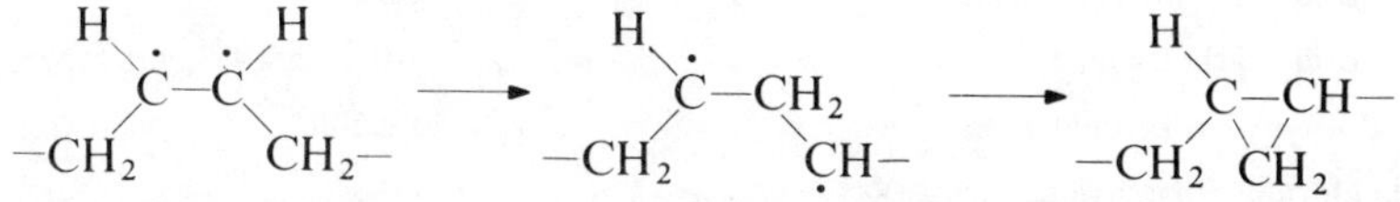

(14.48)

Polydienes may also react with singlet oxygen $^1O_2$ by the ENE mechanism which has been separately discussed in Section 14.4.

Photooxidative degradation of poly(1,2-butadiene) occurs by analogous reactions in which pendant unsaturated groups are involved.[355,1911,2334]

A specific photoreaction for polymers with pendant unsaturated groups is the photocyloaddition of double bonds in the 1,6-diene arrangement:[1353]

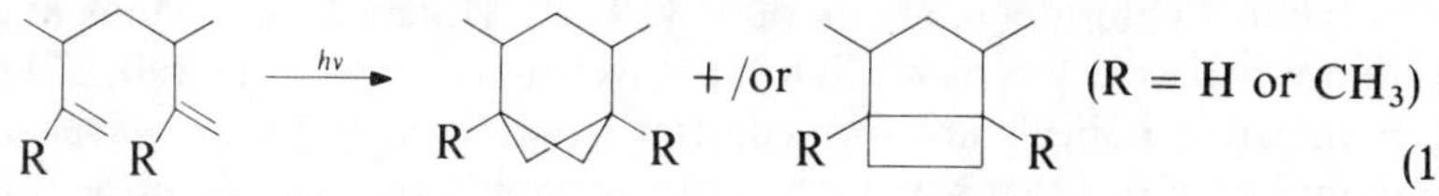

(14.49)

It is generally agreed that all copolymers or blends which contain as one component polydiene structures (Table 14.3) are very susceptible to both thermal oxidations and photooxidation. For example the styrene–butadiene block copolymer initiates photooxidative degradation of polyethylene in blends of these polymers due to the presence of unsaturation in butadiene.[1281]

Rubber vulcanizates possess many types of different crosslinking bonds depending on curing agents. Sulphur crosslinked rubber is photodegraded at 400–500 nm irradiation, whereas cumyl peroxide crosslinked rubber is photodegraded with irradiation below 400 nm.[2561,2577,3965]

14.3.3. Poly(Vinyl Chloride)

Photodegradation (photooxidative degradation) and photodehydrochlorination of poly(vinyl chloride) and its copolymers and blends have been discussed in a number of reviews,[506,912,2851,2854,3067,3427] and books,[2468,2855,3060] and were the subject of many experimental publications (Table 14.4).

Table 14.4 Photodegradation, photooxidative degradation, and photodehydrochlorination of poly(vinyl chloride) and its analogue—references

Name	Structure	References

Poly(vinyl chloride)	$-CH_2-CH-$
	       $\mid$
	       Cl
		10, 267, 508–510, 770, 908, 913, 914, 916, 1322, 1327–1329, 1331, 1332, 1467, 1875, 1925, 2004, 2095, 2179, 2180, 2381, 2396–2398, 2423, 2572–2574, 2581, 2801, 2852, 2858, 2859, 2861, 2875, 3021, 3026, 3028, 3047, 3067, 3070, 3087, 3189, 3262, 3268, 3332, 3361, 3456, 3628, 3711–3713, 3744

Chlorinated poly(vinyl chloride)		916
Co(vinyl chloride/vinylidene chloride)		2826
Co(vinyl chloride/carbon monoxide)		510, 512, 1983, 1984, 3745
Co(vinyl chloride/vinyl ketone)		1585
Poly(vinyl chloride)/polyethylene blends		1280
Poly(vinyl chloride)/polybutadiene blends		2101, 2200
Poly(vinyl chloride)/co(acrylonitrile/butadiene) (NBR) blends		3366
Poly(vinyl chloride)/co(acrylonitrile/butadiene/ styrene) (ABS) blends		3262, 3264, 3265

The poly(vinyl chloride) contains only, C–C, C–H, and C–Cl, bonds and is therefore not expected to absorb light of wavelength longer than 190–220 nm. The fact that free radicals are formed after irradiation of longer wavelength (220–370 nm) indicates that some kinds of chromophores must be present in a polymer.[1331,1467,2004,2396,2398,2854,3427] The light and heat instability of poly(vinyl chloride) must be caused by structural abnormalities that are present to varying extents in different types of commercially available polymer samples, such as unsaturated end-groups, branch points (tertiary-bonded chlorine atoms), random unsaturation (allylic chlorine)[762] (Table 14.5), and oxidized structures such as hydroperoxide groups[762,3252,3266–3268] and carbonyl groups.[760,762,996,1853]

It is generally accepted that photodegradation[267,510,913,914,1327–1329,1331,1332,2180,2381,2398,2572,2851,2852,2853,3026,3047,3262,3268,3332] and photothermal degradation[1467,2574,3744] occur with discolouration of the polymer due to the polyene formation (Fig. 14.7):

$$\left(CH_2-\underset{\underset{Cl}{\mid}}{CH}\right)_n + h\nu \longrightarrow \left(CH=CH\right)_x\left(CH_2-\underset{\underset{Cl}{\mid}}{CH}\right)_y + xHCl \tag{14.50}$$

Table 14.5 Structural defects in poly(vinyl chloride)

Unsaturated groups	Possible transformation structures	
	Long branches	Abnormal internal structures
$-CH_2-CHCl-CCl=CH_2$ $-CH_2-CHCl-CH=CH_2$	$-CH_2-CH-CH_2-$ $\mid$ CHCl	$-CH_2-CCl=CH-CH_2-$ $-CH_2-CH=CH-CH_2-$
$-CH_2-CHCl-CH_2-CCl=CH_2$	Cl $\mid$ $-CH_2-C-CH_2-$ $\mid$	
$-CH_2-CHCl-CH_2-CH=CH_2$	CH_2-	$-CH_2-CHCl-CH_2-CCl=CH-CHCl-CH_2$
	$-CH_2-CH-CH_2-$ $\mid$ CH_2-	

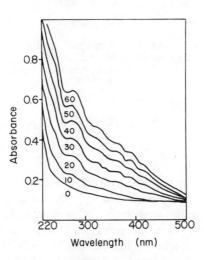

Fig. 14.7 Absorption spectra of irradiated poly(vinyl chloride) film at different irradiation times (0–60 min)[3189]

These yellow–brown-coloured polyene structures consist of 2–14 conjugated bonds, and their formation can be followed by measuring u.v./vis. absorption spectra[913,3189] or u.v./vis. derivative spectra (Fig. 14.8).[3189]

Hydrogen chloride (HCl) formed during photodehydrochlorination of poly(vinyl chloride) can further react with polyene structures and cause their partial bleaching according to the reactions:[914,2851,2861]

$$-(CH=CH)_n + HCl \underset{+h\nu}{\overset{+h\nu}{\rightleftharpoons}} -(CH=CH)_{n-1}-CH_2-\underset{\mid}{CH} \quad (14.51)$$
$$Cl$$

504

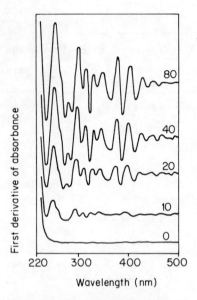

Fig. 14.8 Second derivative spectra of the same poly(vinyl chloride) film as in Fig. 14.7 at different irradiation times (0–60 min)[3189]

$$-(CH=CH)_n + HCl \; \underset{+hv}{\overset{+hv}{\rightleftarrows}} \; -(CH=CH)_x - CH_2 - \underset{Cl}{CH} - (CH=CH)_{n-x} -$$

(14.52)

There is an increase in the rate of dehydrochlorination of poly(vinyl chloride) in nitrogen with increasing temperature.[1328,3087] Dehydrochlorination occurs predominantly in a thin surface layer.[1327] Quantum yield of this reaction (Φ_{HCl} = 0.015) has been found to be independent of the extent of the dehydrochlorination reaction.[913] The fraction of incident light absorbed by the irradiated films increases continuously from ca. 5% initially to about 60% after 1 hour of irradiation (Fig. 14.9).

Large amounts of HCl are evolved during the photooxidation of poly(vinyl chloride), at a rate which increases with irradiation time (Fig. 14.10). This autoaccelerating process has been generally assumed to result from the photolysis of the oxidation products, mainly hydroperoxides and/or carbonyl groups.[509,3262] Hydroperoxy groups can be photodecomposed by an energy transfer mechanism from excited polyene groups:[913,1331]

$$-(CH=CH)_n + hv \longrightarrow -(CH=CH)_n^*$$

(14.53)

$$-(CH=CH)_n^* + POOH \longrightarrow -(CH=CH)_n + PO\cdot + \cdot OH$$

(14.54)

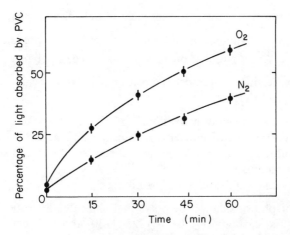

Fig. 14.9 Fraction of incident light absorbed by a 30 μm
poly(vinyl chloride) film as a function of irradiation time[913]

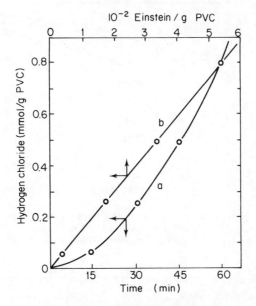

Fig. 14.10 Amounts of hydrogen chloride evolved
in the photooxidation of poly(vinyl chloride) film
as a function of time (curve *a*) and as function of
the number of photon absorbed by the polymer
(curve *b*)[913]

On the other hand it has been proved that the accelerated kinetics of HCl
evolution results exclusively from the increased absorption of light by the
irradiated polymer, since a plot of the amount of HCl evolved, as a function of the
number of photons absorbed by the polymer, yields a straight line (Fig. 14.10).[913]

This slope of this line is the quantum yield of dehydrochlorination (Φ_{HCl}).

$$\Phi_{HCl} = \frac{\text{Number of HCl molecules evolved/g polymer}}{\text{Number of photons absorbed/g polymer}} \qquad (14.55)$$

The role of oxygen in the photodecomposition of poly(vinyl chloride) is complex. It has been found that oxygen:

(i) Decreases the energy of activation.[2875]
(ii) Accelerates the rate of the dehydrochlorination reaction.[267,913,1332]
(iii) Causes bleaching by attacking the polyene structures and shortening the length of conjugation:[508,2095,2859,3044]
(iv) Promotes chain scission and crosslinking.[267,913,3911]

As a result of photooxidation of poly(vinyl chloride), hydroperoxide (OOH) (3450 cm^{-1}) and carbonyl ($>$CO) (1725, 1730, and 1770 cm^{-1}) groups are formed (Fig. 14.11).[267, 913, 2572, 3026, 3047, 3744]

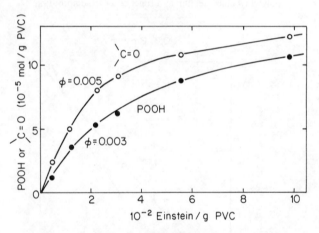

Fig. 14.11 Concentration of carbonyl and hydroperoxide groups as a function of the number of photons absorbed by the polymer in the photooxidation of poly(vinyl chloride) films[913]

The major oxidation products involve primarily the peroxy radical (*14.29*) formed in the initiation step:

$$-(CH=CH)_n-\underset{\underset{Cl}{|}}{CH}-CH_2- + h\nu \longrightarrow -(CH=CH)_n-\overset{\cdot}{C}H-CH_2-+\cdot Cl$$

$$(14.56)$$

$$-(CH=CH)_n-\overset{\cdot}{C}H-CH_2- +O_2 \longrightarrow -(CH=CH)_n-\overset{\overset{\displaystyle\cdot O}{|}}{\underset{}{\overset{\displaystyle O}{|}}}{C}H-CH_2-$$

$$(14.29) \qquad (14.57)$$

and the α-chloroperoxy radical (*14.30*) resulting from the abstraction of a tertiary hydrogen by peroxy radicals:[913]

$$-CH_2-\underset{Cl}{CH}-CH_2\underset{Cl}{CH}- \;+POO\cdot \longrightarrow -CH_2-\underset{Cl}{\overset{\cdot}{C}}-CH_2-\underset{Cl}{CH}- \;+POOH \tag{14.58}$$

$$-CH_2-\underset{Cl}{\overset{\cdot}{C}}-CH_2-\underset{Cl}{CH}- \;+O_2 \longrightarrow -CH_2-\underset{Cl}{\overset{\overset{\cdot}{O}}{\overset{|}{\underset{|}{O}}}{C}}-CH_2-CH_2-\underset{Cl}{CH}-$$

$$(14.30) \tag{14.59}$$

Formation of the polymer peroxy radicals in poly(vinyl chloride) can be measured by ESR spectroscopy, as a single line asymmetric spectrum.[3026,3067]

The chain scission reaction occurs by the disproportionation reaction of the polymer alkyloxy radical (*14.31*) or polymer chloroalkyloxy radical (*14.32*), which are formed from the photodecomposition of related hydroperoxy groups:

$$-CH_2-\overset{OOH}{\underset{}{CH}}-CH_2-\underset{Cl}{CH}- \;+h\nu \longrightarrow -CH_2-\overset{\overset{\cdot}{O}}{\underset{}{CH}}-CH_2-\underset{Cl}{CH}- \;+\cdot OH$$

$$(14.31) \tag{14.60}$$

$$-CH_2-\overset{\overset{\cdot}{O}}{\underset{}{CH}}-CH_2-\underset{Cl}{CH}- \longrightarrow -CH_2-\overset{O}{\underset{H}{C}} \;+\cdot CH_2-\underset{Cl}{CH}- \tag{14.61}$$

$$-CH_2-\underset{Cl}{\overset{OOH}{\underset{}{C}}}-CH_2-\underset{Cl}{CH}- \longrightarrow -CH_2-\underset{Cl}{\overset{\overset{\cdot}{O}}{\underset{}{C}}}-CH_2-\underset{Cl}{CH}- \;+\cdot OH$$

$$(14.32) \tag{14.62}$$

$$-CH_2-\underset{Cl}{\overset{\overset{\cdot}{O}}{\underset{}{C}}}-CH_2-\underset{Cl}{CH}- \longrightarrow -CH_2-\underset{Cl}{\overset{O}{C}} \;+\cdot CH_2-\underset{Cl}{CH}- \tag{14.63}$$

The average number of chain scissions increases linearly with the number of photons absorbed (Fig. 14.12). The quantum yield of chain scission calculated from the slope of this straight line is $\Phi = 1.7 \times 10^{-4}$.[913]

Gel permeation chromatography (PVC) analysis shows a broadening of the molecular weight distribution of the photooxidized poly(vinyl chloride) towards both the lower and the higher molecular weights, thus indicating that crosslinking competes with the chain scission process. The crosslinking process occurs not

508

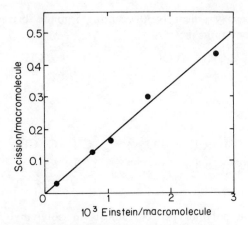

Fig.14.12 Chain scission formation as a function of
the number of photons absorbed by the polymer in
the photooxidation of poly(vinyl chloride) films[913]

only in solid phase[913] but also in solution.[267] The quantum yield for crosslinking
of the solid film $\Phi = 5 \times 10^{-4}$.

The mechanism of photooxidative degradation of poly(vinyl chloride) is not
yet completely understood. For that reason attempts were made to study the
photooxidation of model compounds such as 2,4-dichloropentane and 2,3-
dichlorobutane.[2179, 2573]

When poly(vinyl chloride) films were cast from tetrahydrofurane solution it
was observed that solvent residues retained in a polymer sample have a
tremendous effect on the relative concentration of conjugated polyenes produced,[1329, 2423, 2801] and on photooxidative processes.[1925, 2381, 3047, 3048]
Photodegradation of poly(vinyl chloride) in tetrahydrofurane proceeds twice as
fast as in 1,2-dichloroethane,[267] and photobleaching of coloured polyene
structures is more noticeable.[2859] The effect of tetrahydrofurane on the
photooxidative degradation has been studied in detail.[1329, 1722, 1723,
1925, 2381, 3047, 3048] Tetrahydrofurane (*14.33*) is very susceptible to oxidation,
during which hydroperoxytetrahydrofurane (*14.34*) is formed. Under u.v.
irradiation this is photodecomposed to tetrahydrofurane oxy radicals (*14.35*) (cf.
Section 6.4) which may abstract hydrogen from poly(vinyl chloride) by the
following reactions:[3047, 3048]

$$\underset{(14.33)}{\boxed{}\,_O} + O_2 \longrightarrow \underset{(14.34)}{\boxed{}\,_O{-}OOH} \tag{14.64}$$

$$\boxed{}\,_O{-}OOH \xrightarrow{\,+h\nu\,} \underset{(14.35)}{\boxed{}\,_O{-}O\cdot + \cdot OH} \tag{14.65}$$

$$\underset{O}{\overset{O}{\Box}}{-}O\cdot\ +\ -CH_2-\underset{Cl}{CH}- \longrightarrow \underset{O}{\overset{O}{\Box}}{-}OH$$

$$-\overset{\cdot}{C}H-\underset{Cl}{CH}-\ \left(-CH_2-\underset{Cl}{\overset{\cdot}{C}}-\right)+H_2O$$

$$\text{or}$$

(14.66)

Several commercial additives employed in the poly(vinyl chloride) industry such as pigments (TiO_2, $CaCO_3$),[868,3365] thermal stabilizers,[379, 380,383,761,2575,3027] antioxidants,[1141] lubricants,[3362] and plasticizers[1329,3893] under u.v. irradiation in a polymer matrix may be partially photolysed into free radicals which may further initiate radical dehydrochlorination and oxidation reactions. For that reason it is highly recommended that poly(vinyl chloride) plastics used outdoors, specially as coatings for metals and in building industry, be tested for accelerated ageing, which may be caused by different additives.

Poly(vinyl chloride) can be photochlorinated to the higher content of chlorine in polymer molecules.[420,917,2103,2346,3505] This reaction proceeds by a free radical mechanism:

$$Cl_2 + h\nu \longrightarrow 2Cl\cdot$$

(14.67)

$$-CH_2-\underset{Cl}{CH}-\ +Cl\cdot\ \longrightarrow\ -\overset{\cdot}{C}H-\underset{Cl}{CH}-\ \left(\underset{\text{or}}{-CH_2-\underset{Cl}{\overset{\cdot}{C}}-}\right)+HCl$$

(14.68)

$$-\overset{\cdot}{C}H-\underset{Cl}{CH}-\ \left(\underset{\text{or}}{-CH_2-\underset{Cl}{\overset{\cdot}{C}}-}\right)+Cl\cdot\ \longrightarrow\ -\underset{Cl}{CH}-\underset{Cl}{CH}-\ \left(\underset{\text{or}}{-CH_2-\underset{Cl}{\overset{Cl}{C}}-}\right)$$

(14.69)

Photodegradation of chlorinated poly(vinyl chloride) proceeds with much more efficient evolution of HCl, leading to long chlorinated polyene sequences, which are responsible for the rapid coloration of the irradiated polymer:[915]

$$-\underset{Cl}{CH}-\underset{Cl}{CH}-\ +h\nu\ \longrightarrow\ -\underset{Cl}{C}=CH-\ +HCl$$

(14.70)

Chain scission and crosslinking occur simultaneously, both in the presence and in the absence of oxygen. By reaction with polyenyl radicals, oxygen slightly reduces the efficiency of the dehydrochlorination, while at the same time promoting the formation of carbonyl and hydroperoxide groups in the polymer chain.

14.3.4. Polystyrene

Photodegradation (photooxidative degradation) of polystyrene and its anal-ogues has been reviewed,[1268,2468,3060,3844] and was a subject of many exper-imental publications (Table 14.6). In spite of this great effort the mechanism of photodegradation of polystyrene is still very controversial and not yet clear.

Polystyrene has a strong absorption spectrum (Fig. 14.13) due to the presence of chromophoric phenyl groups. It can also form a charge-transfer (CT) contact complex with oxygen, which has a long-tail extended absorption up to 340 nm.[2746,2747,3955]

The fluorescence spectrum of polystyrene film (Fig. 3.8) consists a broad and structureless band which is due to excimer formation (cf. Section 3.2.1). The 'mirror-image' fluorescence of the phenyl groups is not observed in any of the solid-state spectra of polystyrene. Depending on the excitation wavelength and sample history, most of polystyrene spectra in solution show several fluorescence bands: 298/310 nm (monomeric styrene); 325 nm (excimer); and 338, 352, 372 and 385–410 nm (traces of fluorescent impurities).[851,2067]

The excimer emissions quickly disappear during the photooxidation of polystyrene sample (Fig. 14.14). A new broad band from 380 to 550 nm is formed and attributed to the formation of polyene structures in a backbone. Excimer fluorescence does not disappear when the polystyrene sample is irradiated in vacuum. The most likely explanation is that energy is transferred from the excited phenyl groups to oxidation products instead of migration to excimer sites.[1262,1263]

The phosphorescence spectrum of polystyrene (Fig. 14.15) has been attributed to the presence of acetophenone-type end-groups (14.38).[560,1250-1252,1262,1263,1268,2067] The intensity of this spectrum increases during irradiation time.

A number of impurities and irregularities in polystyrene, such as hydro-peroxide groups, aromatic carbonyl groups, unsaturated bonds, and chain peroxide linkages (cf. Section 14.2), can be responsible for the photoinitiation of free radical oxidation of polystyrene.

Ultraviolet irradiation of polystyrene films evidently changes their u.v./vis. (Fig. 14.16) and i.r. (Fig. 14.17) absorption spectra.[1262,2122,3033] The u.v./vis. absorption spectra show the formation of different possible structures, such as acetophenone, dienes, trienes, and tetraenes.[1262] Formation of polyenes from unsaturated chain end-groups occurs through excitation by an energy transfer mechanism from the excited phenyl groups in polystyrene.[323] The excited unsaturated end-group then decomposes with evolution of hydrogen and formation of a conjugate diene (14.36):

$$-CH-CH_2-C=CH_2^* \longrightarrow -CH=CH-CH=CH_2 + H_2$$

(14.71)

(14.36)

Table 14.6 Photodegradation and photooxidative degradation of polystyrenes—references

Name	Structure	References
Polystyrene	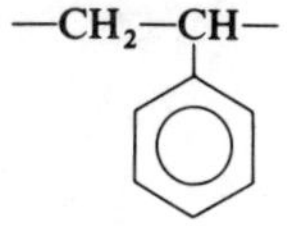 $-CH_2-CH-$ (phenyl)	13, 14, 774, 833, 1157, 1252, 1261–1263, 1391–1394, 1692 1820, 2117, 2122, 2134, 2141, 2219, 2220, 2332, 2831, 2914, 3033, 3059, 3083, 3325, 3627, 3838, 3840, 3841, 3936
Poly(α-methylstyrene)	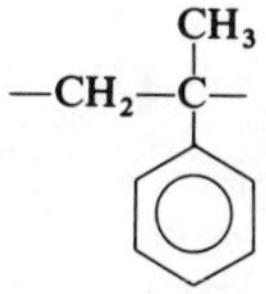 $-CH_2-\overset{CH_3}{\underset{\;}{C}}-$ (phenyl)	352, 774, 1160, 1404, 1752, 3450, 3838
Poly(o,m,p-methylstyrenes)	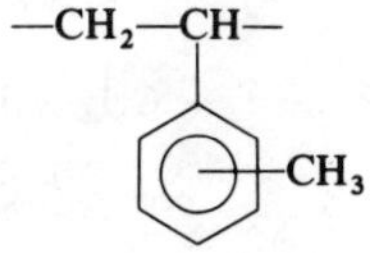 $-CH_2-CH-$ (phenyl)$-CH_3$	3838, 3849, 3854
Poly(p-ethylstyrene)	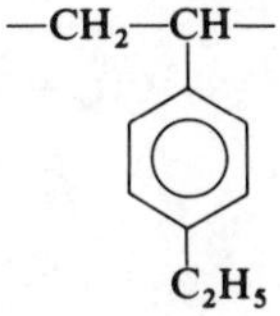 $-CH_2-CH-$ (phenyl)$-C_2H_5$	3857
Poly(p-isopropylstyrene)	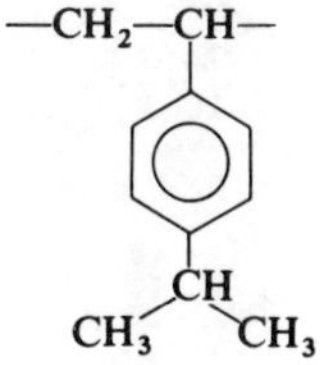 $-CH_2-CH-$ (phenyl) $CH-(CH_3)(CH_3)$	3852
Poly(p-tert-butylstyrene)	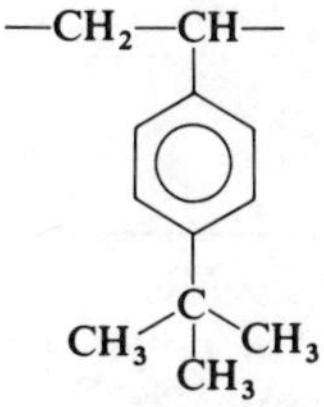 $-CH_2-CH-$ (phenyl) $C(CH_3)(CH_3)(CH_3)$	3853
Poly(p-methoxystyrene)	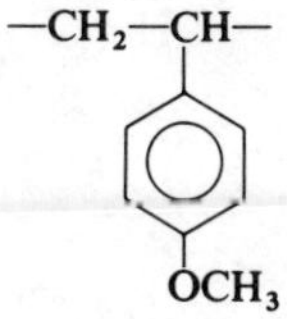 $-CH_2-CH-$ (phenyl)$-OCH_3$	3848, 3855

Table 14.6 (*Contd.*)

Name	Structure	References
Poly(*p*-chloro, bromo, fluoro-amino, nitro-styrenes)	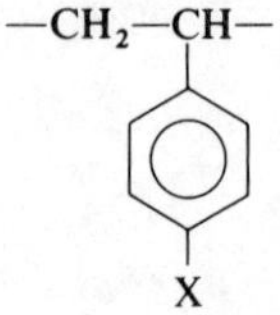 $-CH_2-CH-$ phenyl ring with X $X = Cl, Br, F,$ NH_2, NO_3	3847, 3851
Poly(*p*-bromoacetylstyrene)	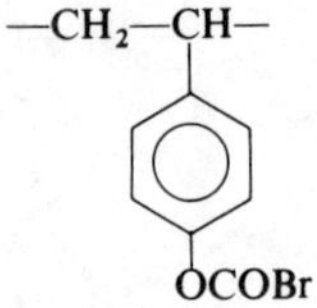 $-CH_2-CH-$ phenyl ring with OCOBr	3672
Polystyrene sulfone	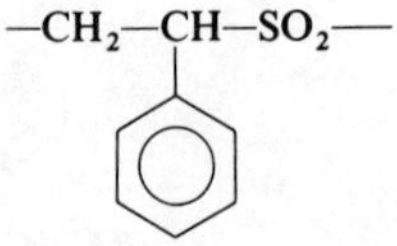 $-CH_2-CH-SO_2-$ phenyl ring	1666
Co(styrene/oxygen)	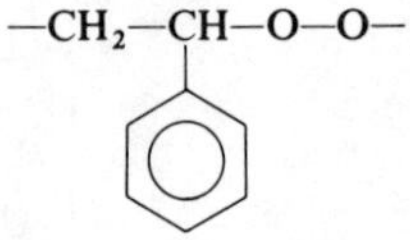 $-CH_2-CH-O-O-$ phenyl ring	3845
Co(styrene/methyl methacrylate)		3623
Co(styrene/vinyl ketone)		1584
Co(styrene/acrylonitrile)		1264
Polystyrene/poly(vinylacetophenone) blends		3855, 3856
Polystyrene/poly(2,6-dimethyl-1,4-phenylene oxide)		1873
Polystyrene/co(acrylonitrile/butadiene/styrene) blends		2101, 3263
Polybutadiene-modified polystyrene (high-impact polystyrene)		1282–1285

Conjugated double-bonds in the polymer backbone (*14.37*) can also be formed under u.v. irradiation in vacuum:[1392]

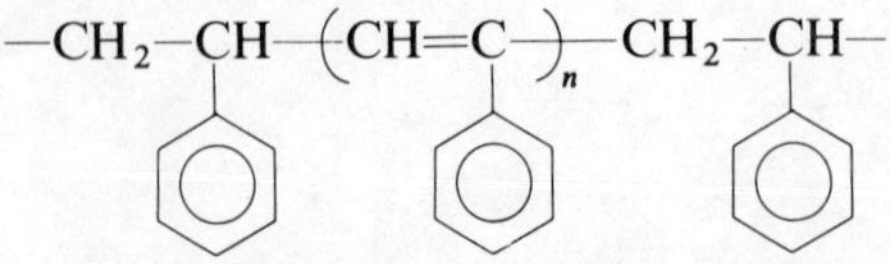

$$-CH_2-CH-(CH=C)_n-CH_2-CH-$$

(each carbon bearing a phenyl ring)

(*14.37*)

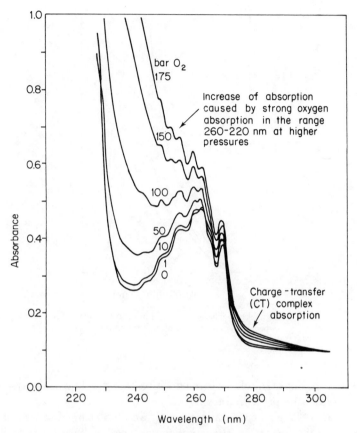

Fig. 14.13 Absorption spectrum of polystyrene films: (——) in vacuum, and (———) in the presence of oxygen at different pressures

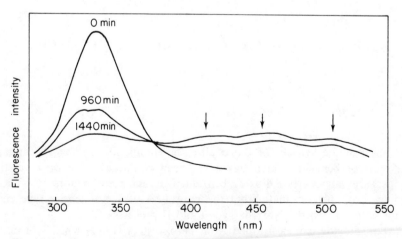

Fig. 14.14 Change of fluorescence spectra of polystyrene upon irradiation in oxygen (600 Torr)[1262]

514

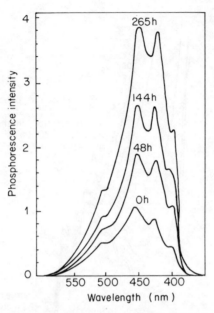

Fig. 14.15 Change of phosphorescence
spectra of polystyrene upon irradiation[560]

It is clear that polyenes should be formed with a higher yield when polystyrene is
u.v. irradiated in the presence of oxygen.

The increase of i.r. absorption at 1685 cm^{-1} and 1730 cm^{-1} has been assigned
respectively to the formation of acetophenone-type (*14.38*) and chain-aliphatic
ketones (*14.39*) and (*14.40*):[349, 1262, 2122]

(*14.38*) (*14.39*) (*14.40*)

Formation of the ketone (*14.39*) requires the elimination of phenyl ring and
formation of benzene, which has never been confirmed. The only volatile
products are water, carbon dioxide, benzaldehyde, and acetophenone.[1262] These
products probably originate from the photolysis of acetophenone-type end-
groups according to the Norrish Type I and II processes.

Carbon dioxide can be formed from peroxyesters (*14.41*) which are intermediates in the oxidation process:[1262]

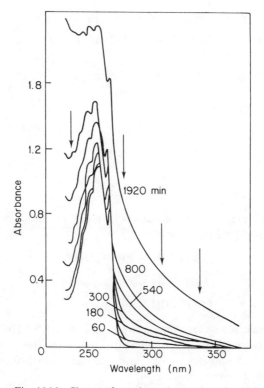

Fig. 14.16 Change of u.v. absorption spectra of polystyrene film irradiated at 254 nm at 600 Torr. Arrows
correspond to absorption maxima expected for acetophenone, dienes, trienes, and tetraenes,
respectively[1262]

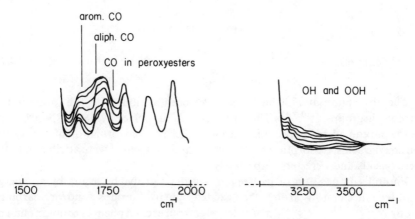

Fig. 14.17 Change of i.r. absorption spectra of polystyrene film irradiated at 254 nm at
600 Torr[1262]

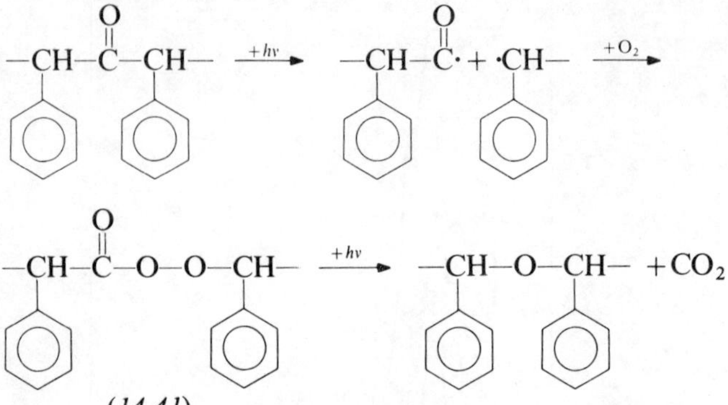

(14.72)

(14.41)

Water and polymer–hydroxyl groups (POH) are the results of decomposition of polymer hydroperoxides (POOH):[1262]

$$POOH \rightarrow PO\cdot + \cdot OH \qquad (14.73)$$

$$PH + \cdot OH \rightarrow P\cdot + H_2O \qquad (14.74)$$

$$PO\cdot + PH \rightarrow POH + P\cdot \qquad (14.75)$$

The quantum yield for water elimination is low, $\Phi = 9.5 \times 10^{-3}$.

Acetophenone-type end-groups can be formed by the thermal decomposition and/or photodecomposition of the tertiary hydroperoxide groups (14.42):[1262,1271]

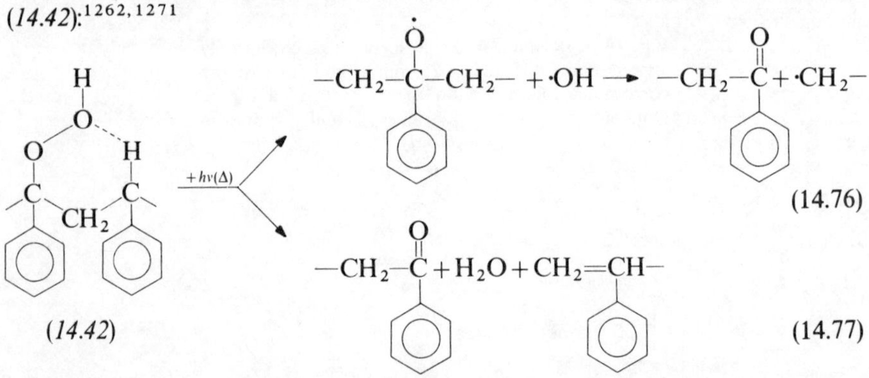

(14.42)

(14.76)

(14.77)

The absorption at 1730 cm^{-1} has also been attributed to the formation of carboxylic groups.[1820,2831] These results were supported by ESCA spectroscopy of the photooxidized polystyrene surface which showed formation of peroxy and carboxylic acids. Absorption at 1774 cm^{-1} and 1760 cm^{-1} were attributed to the peroxyacids and peresters respectively.

Photodecomposition of hydroperoxy groups probably occurs by an energy transfer mechanism from the excited phenyl groups and/or carbonyl groups.[1263,1270,1271,1276] Energy transfer from excited phenyl groups in cumene (donors) to cumene hydroperoxides (acceptors) provide evidence for an energy transfer process which may occur as it does in polystyrene.[1276]

ESCA spectroscopy has been successfully applied to study of photooxidation processes on the surface of polystyrene films.[717,722,2596,2914]

In order to explain the photooxidation mechanism of polystyrene a number of low-molecular model compounds have been investigated; e.g.: 2-phenyl-butane,[2331,2332] 3-phenylpentane,[2327] linear aromatic ketones such as 4-methyl-1,7-diphenyl-3-heptanone, 2,8-diphenyl-4-nonanone, 5-methyl-2,8-diphenyl-4-nonanone, and cyclic ketones such as 1-phenyl-2(3-phenyl-propionyl)cyclopentane and 1-3(phenylbutyryl)-2(2-phenylpropyl)cyclo-pentane[468] photooxidation reactions.

One unresolved problem is a strong yellowing of polystyrene samples observed during u.v. irradiation. There are many alternative theories of the mechanism of yellowing, but all concern the formation of different chromophoric groups (Table 14.7) which are responsible for this phenomenon.

Photooxidation processes of polystyrene are accompanied by changes in molecular weight distribution resulting from chain scission and cross-linking reactions:[833]

The quantum yields of chain scission and crosslinking in vacuum are $\Phi = 2.4 \times 10^{-3}$ and $\Phi = 3.9 \times 10^{-4}$ respectively, whereas corresponding values in oxygen are $\Phi = 5.6 \times 10^{-3}$ and $\Phi = 9.4 \times 10^{-4}$.

Photooxidation of polystyrene samples causes changes in dielectric properties and wettability.[1161,1404,1405,3839–3841,3850]

Photodegradation of polystyrene in solution depends greatly on the type of solvent used.[357,477,1015,2120,2220,2993,3188,3215,3515,3932,3933–3936] The quantum yields for random chain scission of polystyrene in different solvents differ remarkably (Table 14.8).[2993] There is particularly a high quantum yield change in the presence of carbon tetrachloride (Fig. 14.18). With oxygen-free solution no degradation was observed.[357,2993,3188] A possible mechanism relates to formation of charge-transfer (CT) complexes between polystyrene and solvent, e.g. chloroform.[3215,3515]

Chain scission occurs probably by free radical mechanism, which is initiated by free radicals (ClO · and Cl ·) formed from the photolysis of phosgene (14.43). This is a result of photooxidation of carbon tetrachloride and chloroform:[3933]

$$2\ CCl_4 + O_2 \xrightarrow{h\nu} 2\ O{=}C\underset{Cl}{\overset{Cl}{\big<}} + 2\ Cl_2 \qquad (14.78)$$

$$2\ CHCl_3 + O_2 \xrightarrow{h\nu} 2\ O{=}C\underset{Cl}{\overset{Cl}{\big<}} + 2\ HCl \qquad (14.79)$$

$$O{=}C\underset{Cl}{\overset{Cl}{\big<}} + h\nu \rightarrow 2\ ClO\cdot + 2\ Cl\cdot \qquad (14.80)$$

(14.43)

518

Table 14.7 Different chromophoric groups which can be formed during photooxidative degradation of polystyrene

Name of group	Structure	References
Acetophenone	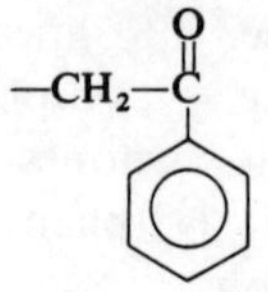	323, 349, 1262, 1271, 1392, 1393, 2122, 2218
Benzal acetophenone (chalcone)	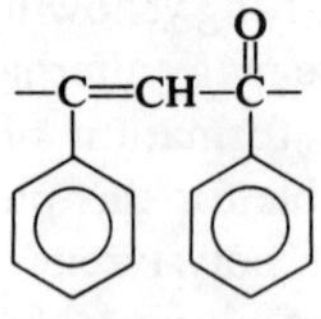	2141, 3785–3788
Diacarbonyl	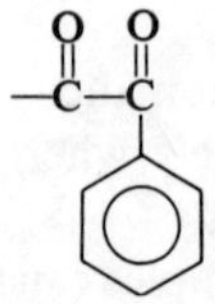	4000
Conjugated dialdehydes	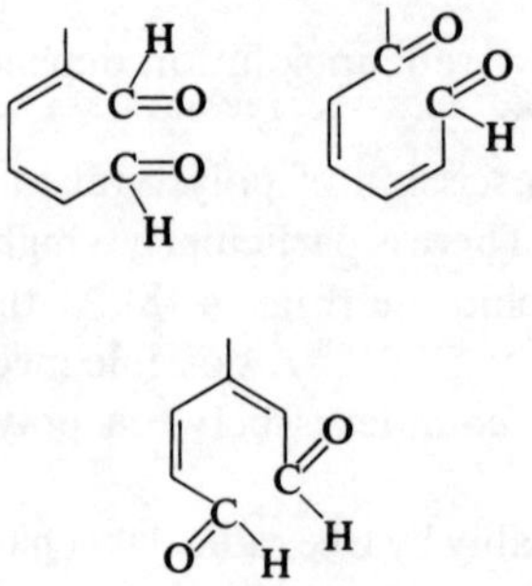	3033
Conjugated double-bonds	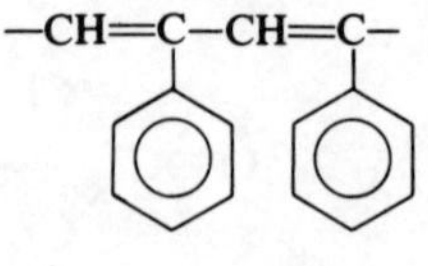	323, 1392–1394
Stilbene	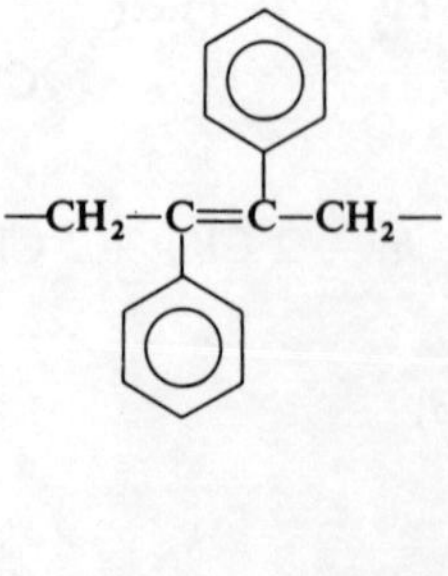	2067

Table 14.7 (*Contd.*)

Name of group	Structure	References
Carbonyl	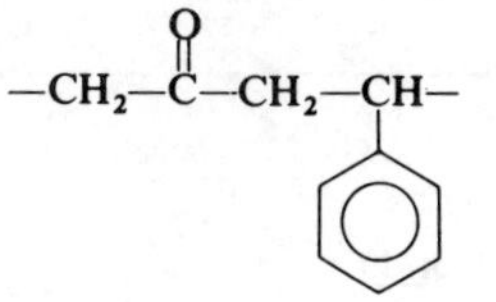	1262, 2331
	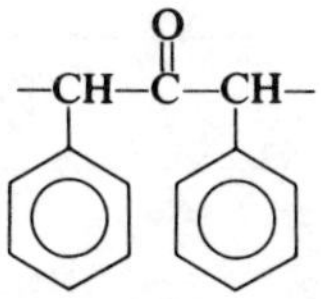	1262, 2331
Peroxy esters	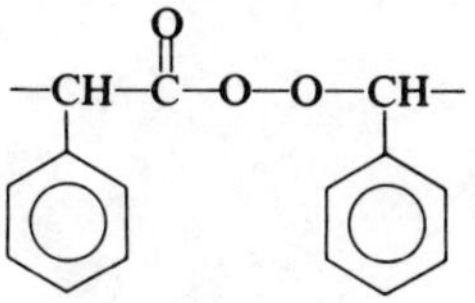	1262
Quinomethane	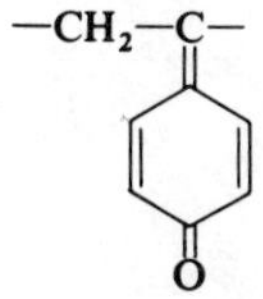	15
Aromatic acid	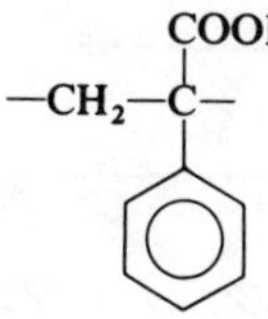	1820
	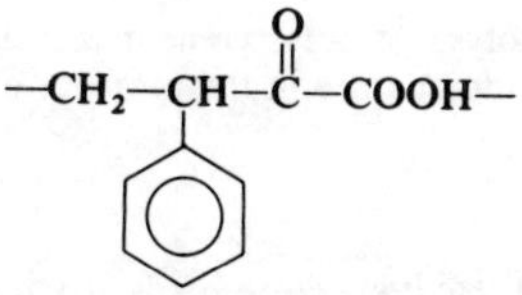	1268
Hydroxyl	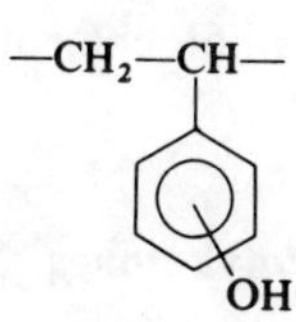	2333, 3842, 3844

520

Table 14.8   Quantum yields for random scission of polystyrene (Φ_s) photolysed in solution in air[2993]

Solvent	$\Phi_s \times 10^4$
Cyclohexane	2
Dioxane	7
Benzene	34
Methylene chloride	44
Chloroform	440
Carbon tetrachloride	1130

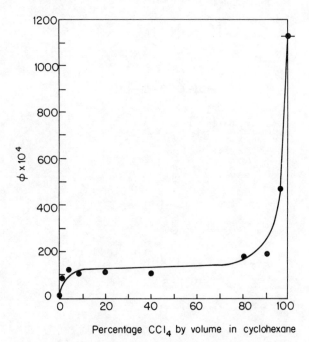

Percentage CCl_4 by volume in cyclohexane

Fig. 14.18   Photolysis of polystyrene in a cyclohexane–carbon tetrachloride mixture in the presence of air[2993]

The photochlorination of polystyrene in solution involves replacement of hydrogen atoms at α- and β-positions along the chain, and in further stage addition of chlorine to phenyl ring. The photochlorination is accompanied by the chain scission of polymer.[427,1479,1872]

14.3.5. Polyketones

The photochemistry and photodegradation of aliphatic and aromatic poly-ketones and their copolymers have been the subject of many publications

(Table 14.9). All of these polyketones contain a carbonyl group as a pendant group.

Photodegradation of polyketones of a general structure (14.44) and its copolymers proceeds by the Norrish Type I and the Norrish Type II reactions by the following scheme:[310, 826, 1047, 1350, 1445 – 1447, 2301, 2610, 3538, 3929]

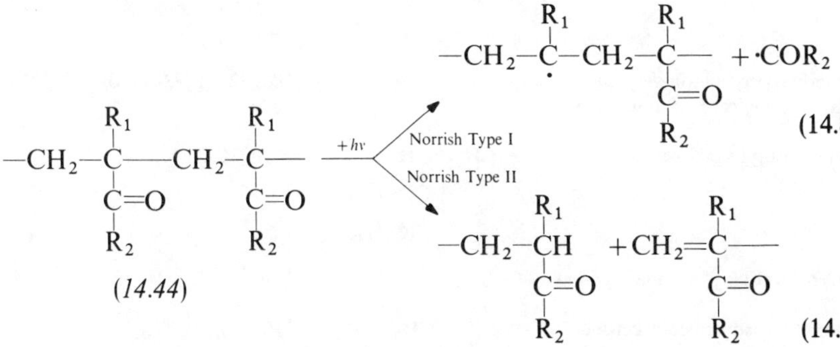

$$-CH_2-\overset{\underset{\displaystyle C=O}{R_1}}{\underset{\displaystyle R_2}{C}}-CH_2-\overset{\underset{\displaystyle C=O}{R_1}}{\underset{\displaystyle R_2}{C}}- \quad \xrightarrow{+h\nu} \quad \begin{array}{l} \text{Norrish Type I} \\ \\ \text{Norrish Type II} \end{array}$$

(14.44)

$$-CH_2-\overset{R_1}{\underset{\cdot}{C}}-CH_2-\overset{\underset{\displaystyle C=O}{R_1}}{\underset{\displaystyle R_2}{C}}- \quad +\cdot COR_2 \qquad (14.81)$$

$$-CH_2-\overset{\underset{\displaystyle C=O}{R_1}}{\underset{\displaystyle R_2}{CH}} \quad +CH_2=\overset{\underset{\displaystyle C=O}{R_1}}{\underset{\displaystyle R_2}{C}}- \qquad (14.82)$$

$R_1 = H, CH_3, R_2 = CH_3-C(CH_3)$

Only the Norrish Type II reaction leads to direct scission of the main chain. The quantum yields for the Norrish Type I and the Norrish Type II reactions for poly(methyl vinyl ketone) in dioxane solution under 313 nm irradiation were determined as $\Phi = 0.025$ and $\Phi = 0.04$, respectively.[1445, 1446] The quantum yield for chain scission decreased with the extent of degradation, and was attributed to a repolymerization initiated by radicals produced by the Type I process, by addition to the terminal double-bonds created by the Type II scission.

Polymer alkyl radicals (14.45) in the presence of oxygen can be oxidized to polymer alkylperoxy radical (POO·) (14.46), which can abstract hydrogen from the same or neighbouring polymer molecule (PH) and form hydroperoxy groups (POOH) (14.47), which, under heat and/or light, are decomposed to polymer alkyloxy radical (PO·) (14.48). This may disproportionate by the β-scission process, and leads to the main chain scission (reaction 14.86):

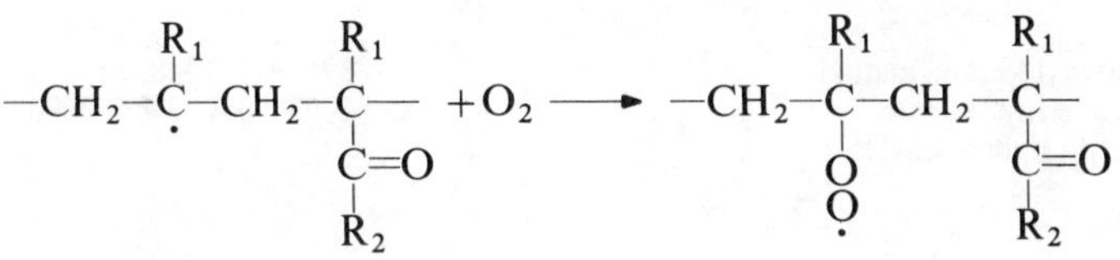

$$-CH_2-\overset{R_1}{\underset{\cdot}{C}}-CH_2-\overset{\underset{\displaystyle C=O}{R_1}}{\underset{\displaystyle R_2}{C}}- \quad +O_2 \longrightarrow -CH_2-\overset{\underset{\displaystyle \underset{\cdot}{O}}{\underset{\displaystyle O}{R_1}}}{C}-CH_2-\overset{\underset{\displaystyle C=O}{R_1}}{\underset{\displaystyle R_2}{C}}-$$

(14.45)   (14.46)   (14.83)

$$-CH_2-\overset{\underset{\displaystyle \underset{\cdot}{O}}{\underset{\displaystyle O}{R_1}}}{C}-CH_2-\overset{\underset{\displaystyle C=O}{R_1}}{\underset{\displaystyle R_2}{C}}- \quad +PH \longrightarrow -CH_2-\overset{\underset{\displaystyle \underset{\displaystyle H}{\overset{\displaystyle O}{O}}}{R_1}}{C}-CH_2-\overset{\underset{\displaystyle C=O}{R_1}}{\underset{\displaystyle R_2}{C}}- \quad +P\cdot$$

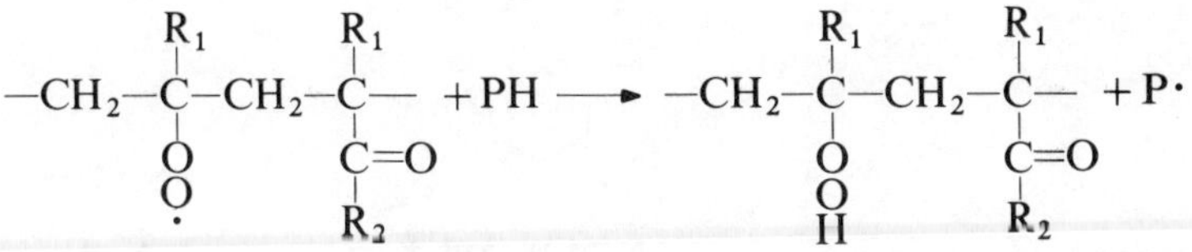

(14.47)   (14.84)

Table 14.9 Photochemistry (including photodegradation and photooxidative degradation) of polyketones—references

Name	Structure	References
Poly(methyl vinyl ketone)	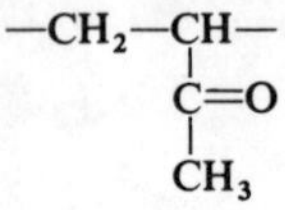 $-CH_2-CH-$ $C=O$ CH_3	180, 826, 1445–1447, 2301, 3929
Co(methyl vinyl ketone/styrene (α-methylstyrene, carbon monoxide))		180, 825, 1378, 1970, 1971, 2283, 2679
Poly(*tert*-butyl vinyl ketone)	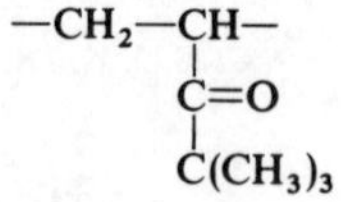 $-CH_2-CH-$ $C=O$ $C(CH_3)_3$	2610, 3538
Co(*tert*-butyl vinyl ketone/styrene)		2610
Poly(methyl isopropenyl ketone)	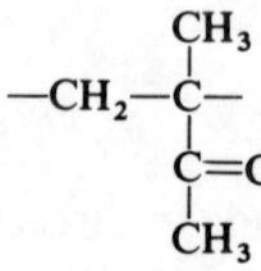 CH_3 $-CH_2-C-$ $C=O$ CH_3	310, 826, 3929
Co(methyl isopropenyl ketone/acrylonitrile (methyl methacrylate, styrene, carbon monoxide))		36, 1378, 1443, 2283, 2690, 3480
Poly(chloromethyl vinyl ketone)	 $-CH_2-CH-$ $C=O$ CH_2Cl	2836
Poly(phenyl vinyl keton)	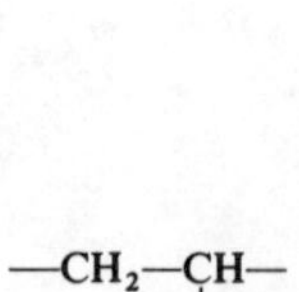 $-CH_2-CH-$ $C=O$ ⬡	343, 358, 359, 826, 839, 841, 843, 1047, 1087–1089, 1350, 1687, 2023, 2049, 2341, 3174, 3371, 3514
Co(phenyl vinyl ketone/methyl methacrylate (vinyl acetate, styrene, vinyl naphthalene))		825, 995, 1348, 1350, 1545, 1971, 2049, 2050, 2345, 2690, 3327
Poly(*p*-substituted phenyl vinyl ketones)	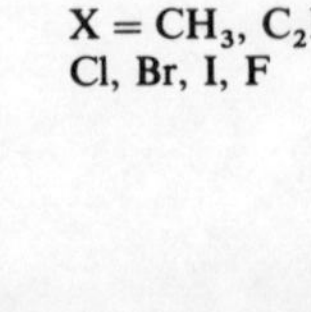 $-CH_2-CH-$ $C=O$ ⬡ X $X = CH_3, C_2H_5,$ Cl, Br, I, F	1689, 1693, 2335, 2339, 2343

Table 14.9   (*Contd.*)

Name	Structure	References
Poly(o-tolylyl vinyl ketone)	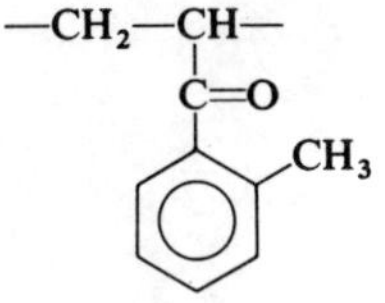	342–344, 3197
Poly(p-methoxyphenyl vinyl ketone) and its copolymers	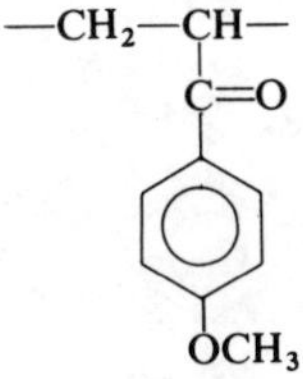	1687, 1693, 2336–2338, 2342
Poly(3,4-dimethoxyphenyl vinyl ketone)	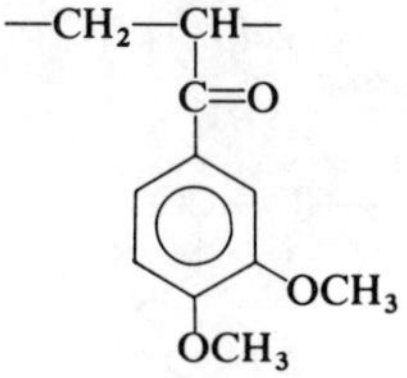	2337
Poly(4′-acetylphenyl vinyl ketone) and its copolymers	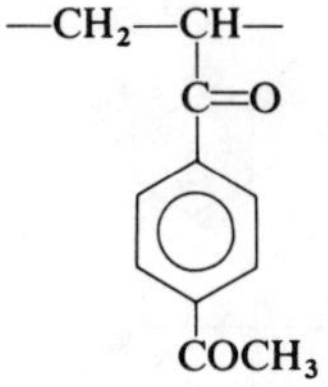	2335
Poly(phenyl isopropenyl ketone) and its copolymers	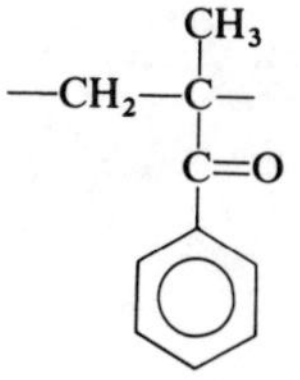	2489, 2611, 2612, 2849, 3480
Poly(vinyl acetophenone) and its copolymers	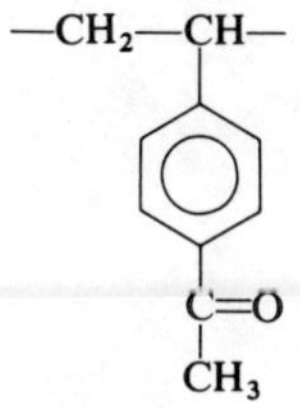	3148, 3843, 3846, 3856

Table 14.9   (*Contd.*)

Name	Structure	References
Poly(*p*-trifluoro vinyl acetophenone)	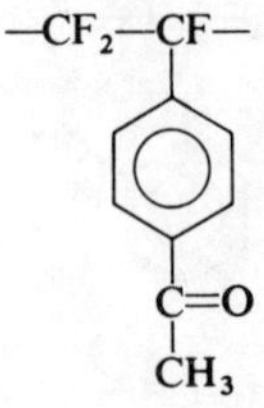 $-CF_2-CF-$ ... $C=O$, CH_3	233
Poly(vinyl benzophenone)	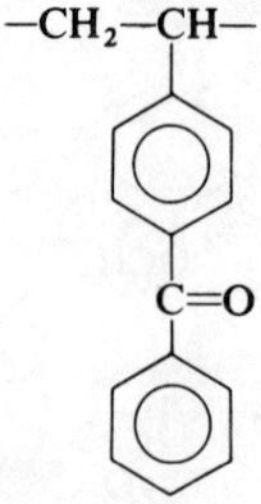 $-CH_2-CH-$ ... $C=O$	840, 2082, 3177, 3214
Poly(*p*-trifluoro vinyl benzophenone)	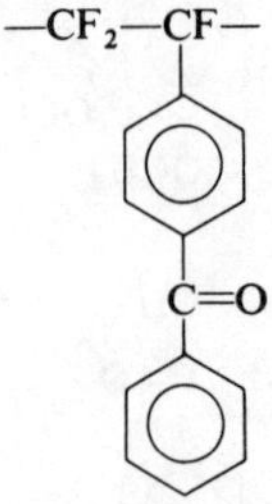 $-CF_2-CF-$ ... $C=O$	233
Other polyketones and their copolymers		647, 1693, 2337, 2344, 2665

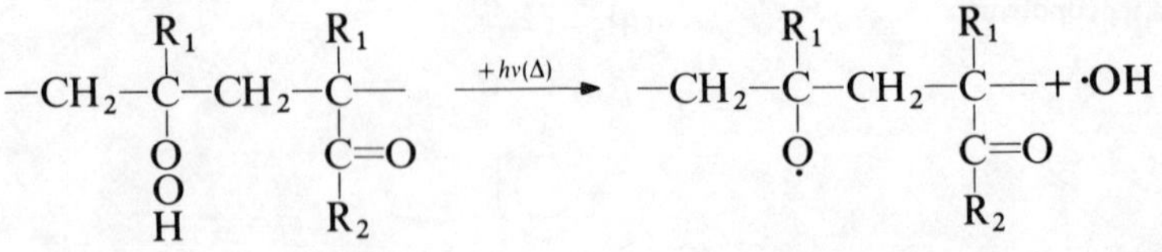

$$-CH_2-\overset{R_1}{\underset{\overset{O}{\underset{H}{O}}}{C}}-CH_2-\overset{R_1}{\underset{\overset{C=O}{R_2}}{C}}- \xrightarrow{+h\nu(\Delta)} -CH_2-\overset{R_1}{\underset{\overset{O}{\cdot}}{C}}-CH_2-\overset{R_1}{\underset{\overset{C=O}{R_2}}{C}}- + \cdot OH$$

$$(14.48) \qquad\qquad (14.85)$$

$$-CH_2-\overset{R_1}{\underset{\overset{O}{\cdot}}{C}}-CH_2-\overset{R_1}{\underset{\overset{C=O}{R_2}}{C}}- \longrightarrow -CH_2-\overset{R_1}{\underset{O}{C}} + \cdot CH_2-\overset{R_1}{\underset{\overset{C=O}{R_2}}{C}}-$$

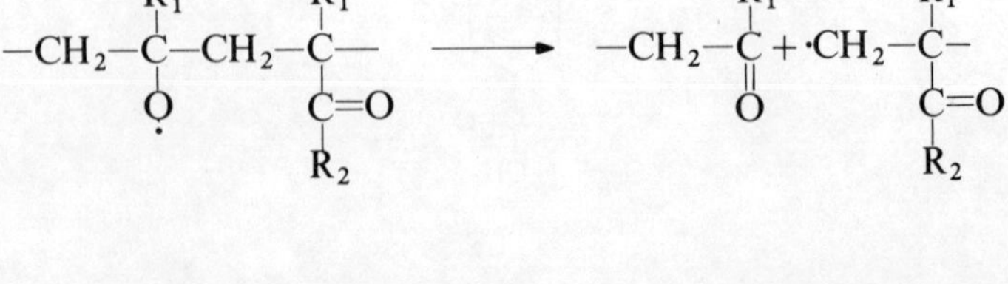

$$(14.86)$$

Poly(phenyl vinyl ketone) (referred to also as poly(acrylophenone)) (Table 14.9) is very reactive photochemically because very rapid intersystem crossing (ISC) from the excited singlet state to the triplet state causes nearly all of the photophysical and photochemical processes to occur from the triplet state.[358, 359, 826, 841, 1047, 1087–1089, 1350, 2049, 2341, 3371]

In the case of poly(phenyl vinyl ketone) a very efficient energy transfer occurs along the polymer chain. Triplet lifetime is of the order of 55–100 ns.[343,358] In the case of co(acrylophenone/vinyl monomers (especially with the excess of styrene)), the phenyl vinyl ketone group can be isolated from one another, and instead of all energy transfer being along the polymer chain, an efficient chain scission reaction can be observed.[1348, 1350, 1997] As a result of the Norrish Type II reaction a number of double-bonds formed at the ends of chains, which can act as quenchers for energy migration. As the degradation proceeds, more and more of these quenching groups are built in the polymer chain itself and excitation can be quenched both internally and from intra- and intermolecular contacts in solution.[1047]

Biradical intermediates formed during photolysis of poly(phenyl vinylketone) are highly reactive and capable of undergoing intermolecular reactions.[3771] The absorbance due to these biradicals was measured at 450 nm and the spectrum was similar to that observed for small aromatic ketones. The addition of radical scavengers such as atomic oxygen, NO and SO$_2$, or *tert*-butyl nitrite to the solution increased the rate of biradical decay. These biradicals are also efficient initiators of free radical polymerization of methyl methacrylate.

Poly(*o*-tolylvinyl ketone) (*14.49*) and copolymers containing *o*-tolyl vinyl-ketone groups are quite photostable polymers, and this stability has been attributed to the reversible photoenolization process:[342,343,3197]

$$(14.87)$$

These processes are well known in the case of small molecules. Typically, two triplet states are observed corresponding to the syn (*14.50*) and anti (*14.51*) conformations of the tolyl group with the respect to the carbonyl chromophore:[1471,3372,3776]

$$-CH_2-CH- \qquad -CH_2-CH-$$

(structures)

syn- anti-
(14.50) (14.51)

The syn conformer (14.50) usually has a very short triplet lifetime (sub-nanosecond range) due to the rapid enolization process. On the other hand the lifetime of the triplet for the anti conformation (14.51) is controlled by the rate of bond rotation to the syn conformer (typically 10–50 ns).[414,3776] When γ-hydrogen atoms are available, the Norrish Type II process competes with the rotationally controlled decay of the anti conformation. In the case of poly(o-tolylvinyl ketone) there is no evidence for Norrish Type II fragmentation. It is conceivable that the energy is transferred along the polymer, through anti conformations of the substituents, until the energy sink, a syn conformation, is reached. From here, photoenolization takes place. In other words, energy migration appears to be substantially faster than either bond rotation or the Norrish Type II reaction.

Photodegradation of poly(vinyl acetophenone) (14.52) and its copolymers proceeds by the Norrish Type I reaction:[3148,3843,3846,3855]

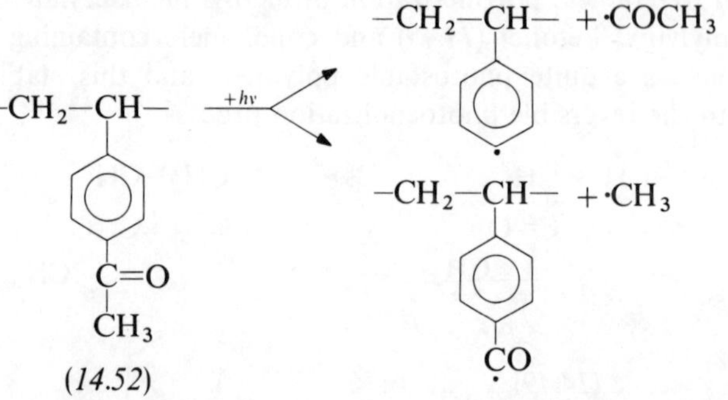

$$-CH_2-CH- \ +\cdot COCH_3 \tag{14.88}$$

$$-CH_2-CH- \ +\cdot CH_3 \tag{14.89}$$

(14.52)

The secondary reactions include decomposition, abstraction, addition, and oxidation of initially formed radicals. Both acetyl (14.53) and polymeric carbonyl radical (14.54) can undergo further decomposition as follows:

$$CH_3CO\cdot \ \longrightarrow \ CH_3\cdot + CO \tag{14.90}$$

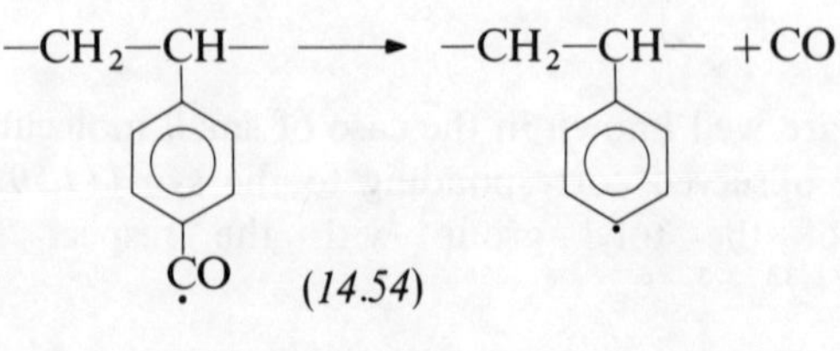

(14.53)

$$-CH_2-CH- \ \longrightarrow \ -CH_2-CH- \ + CO \tag{14.91}$$

(14.54)

or abstract hydrogen from the polymer (PH) molecules:

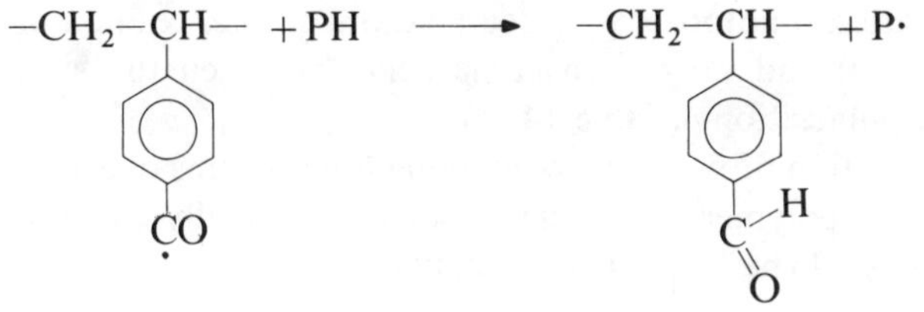

$$-CH_2-CH- \ +PH \ \longrightarrow \ -CH_2-CH- \ +P\cdot$$

(14.92)

$$-CH_2-CH- \ +PH \ \longrightarrow \ -CH_2-CH- \ +P\cdot$$

(14.93)

The importance of ketone groups present in polymers as internal impurities or formed as a result of oxidation processes has been presented in Section 14.1.

14.3.6. Polymers Containing Keto Groups in the Backbone

Polymers of this type can be synthesized by copolymerization of a monomer, e.g. ethylene,[1428, 1513, 1514, 1580, 1581] styrene,[1981, 1984] or aliphatic acryloketones with carbonyl oxide (CO).[2283] They are photodecomposed by the Norrish Type I and II processes:

$$-CH_2\overset{O}{\overset{\|}{C}}\cdot + \cdot CH_2CH_2CH_2CH_2-$$

(14.94)

$$-CH_2\overset{O}{\overset{\|}{C}}CH_2CH_2CH_2CH_2- \ +h\nu$$

Norrish Type I

Norrish Type II

$$-CH_2-\overset{O}{\overset{\|}{C}}CH_3 + CH_2{=}CH-CH_2-$$

(14.95)

The Norrish Type II reaction, an intramolecular elimination, appears to be independent of temperature and phase and is not quenched by atmospheric oxygen. In the glass transition temperatures (T_g), however, the Norrish Type II reaction is inhibited, probably due to restrictions of the freedom of internal motion of the polymer. The Norrish Type I reaction produces free radicals and is temperature-dependent. At 120°C the two processes make approximately equal contributions to a total quantum yield for reaction of about $\Phi = 0.05$. However, at ambient temperature the Norrish Type II process accounts for the major part of the chemical reaction.[1349,1513]

Polymers of this type have found a practical application as photodegradable polymers with controlled lifetime (cf. Section 14.10). Polyesters containing keto groups in the backbone also belong to this type of polymer.[984,2964]

14.3.7. Polyacrylates and Polymethacrylates

Photodegradation (photooxidative degradation) of ester containing polymers of poly(acrylic acid) and poly(methacrylic acid) has been the subject of many experimental publications (Table 14.10).

Photodegradation of polymers containing ester groups of a general structure (14.55) and its copolymers under u.v. irradiation (usually 254 nm) proceeds by both the Norrish Type I and Type II reactions:[985]

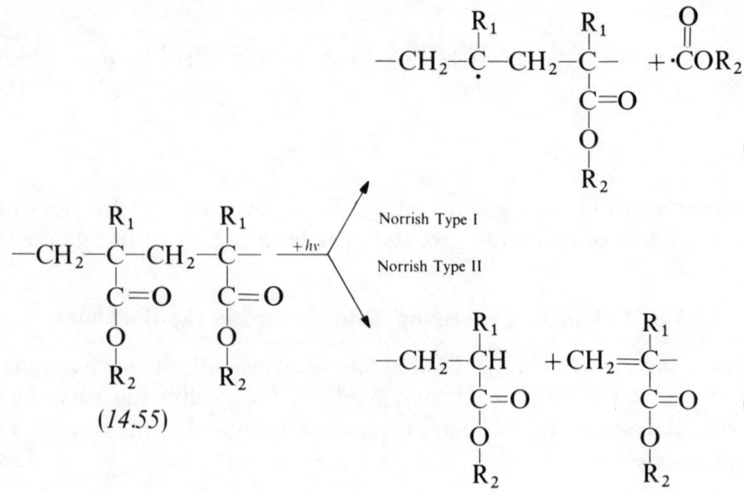

$$-CH_2-\overset{\underset{\displaystyle |}{R_1}}{\underset{\displaystyle \bullet}{C}}-CH_2-\overset{\underset{\displaystyle |}{R_1}}{\underset{\displaystyle \underset{\displaystyle O}{\overset{\displaystyle \|}{C=O}}}{C}}-\quad +\cdot\overset{\displaystyle O}{\overset{\displaystyle \|}{C}}OR_2 \qquad (14.96)$$

$$-CH_2-\overset{R_1}{\underset{\underset{\underset{R_2}{O}}{C=O}}{C}}-CH_2-\overset{R_1}{\underset{\underset{\underset{R_2}{O}}{C=O}}{C}}- \xrightarrow{+h\nu}$$

(14.55)

Norrish Type I

Norrish Type II

$$-CH_2-\overset{R_1}{\underset{\underset{\underset{R_2}{O}}{C=O}}{CH}}\quad +CH_2=\overset{R_1}{\underset{\underset{\underset{R_2}{O}}{C=O}}{C}}- \qquad (14.97)$$

$R_1 = H, CH_3$

$R_2 = CH_3, C_2H_5, C_4H_9$

In addition to these two Norrish reactions, the photolysis of the ester side-chain is also possible:

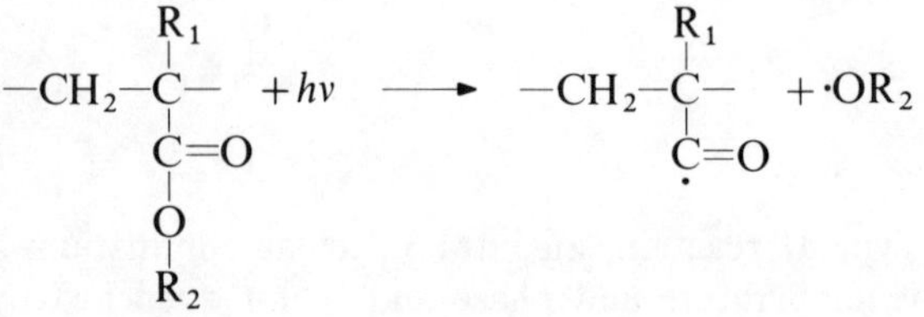

$$-CH_2-\overset{R_1}{\underset{\underset{\underset{R_2}{O}}{C=O}}{C}}-\quad +h\nu \longrightarrow -CH_2-\overset{R_1}{\underset{\underset{\bullet}{C=O}}{C}}-\quad +\cdot OR_2 \qquad (14.98)$$

$$-CH_2-\overset{R_1}{\underset{C=O}{C}}-\quad \longrightarrow -CH_2-\overset{R_1}{\underset{\bullet}{C}}-\ + CO \qquad (14.99)$$

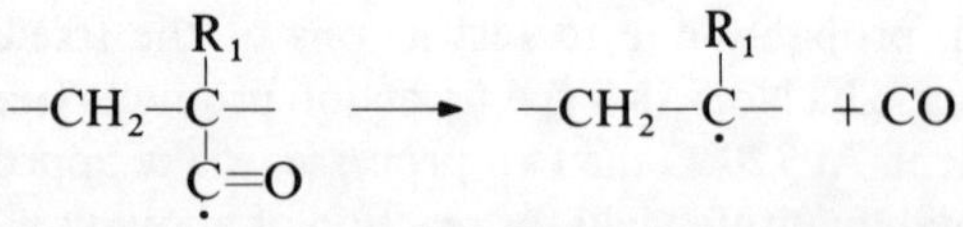

The quantum yield for the chain scission $\Phi = 0.05$, and for the ester side chain cleavage $\Phi = 0.01$.[1456] In the presence of oxygen photooxidative degradation occurs by an analogous process, as shown in Section 14.3.5. At higher temperatures the polymers depolymerize, giving off monomer by the unzipping reaction:[832,1384]

Table 14.10 Photodegradation and photooxidative degradation of polyacrylates—references

Name	Structure	References
Poly(methyl acrylate)	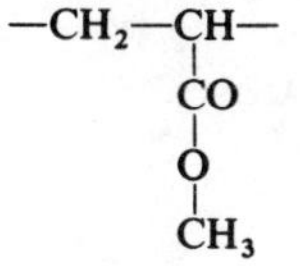 $-CH_2-CH-$ $\quad\quad\;\; CO$ $\quad\quad\;\; O$ $\quad\quad\;\; CH_3$	16, 594, 1153, 1154, 1159, 1379, 1389, 2450, 2453, 2473, 2579, 3455
Poly(ethyl acrylate)	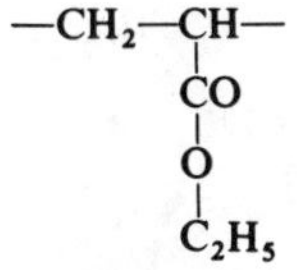 $-CH_2-CH-$ $\quad\quad\;\; CO$ $\quad\quad\;\; O$ $\quad\quad\;\; C_2H_5$	2451–2453
Poly(n-butyl acrylate)	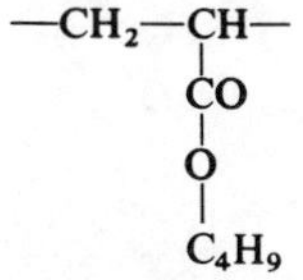 $-CH_2-CH-$ $\quad\quad\;\; CO$ $\quad\quad\;\; O$ $\quad\quad\;\; C_4H_9$	2292, 2451–2453
Poly(phenyl acrylate)	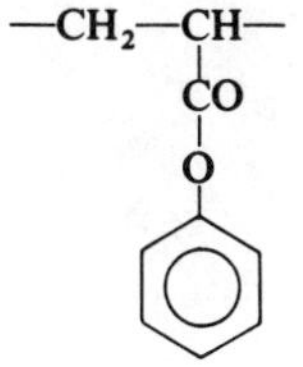 $-CH_2-CH-$ $\quad\quad\;\; CO$ $\quad\quad\;\; O$ (phenyl)	2282
Poly(naphthyl acrylate)	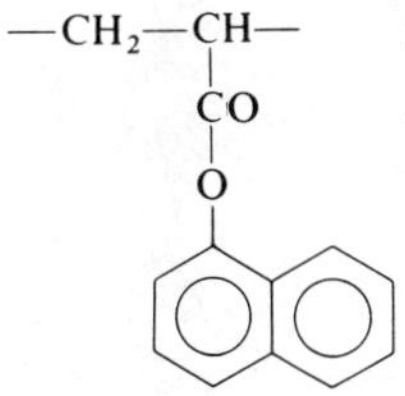 $-CH_2-CH-$ $\quad\quad\;\; CO$ $\quad\quad\;\; O$ (naphthyl)	2490
Poly(methacrylic acid)	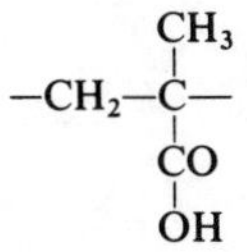 $\quad\quad\;\; CH_3$ $-CH_2-C-$ $\quad\quad\;\; CO$ $\quad\quad\;\; OH$	347, 702, 1862, 1863, 1866
Poly(methyl methacrylate)	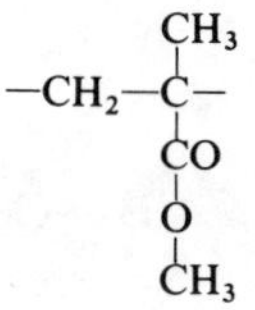 $\quad\quad\;\; CH_3$ $-CH_2-C-$ $\quad\quad\;\; CO$ $\quad\quad\;\; O$ $\quad\quad\;\; CH_3$	359, 361, 844, 985, 1158, 1226, 1376, 1388, 1390, 1456, 1926, 2134, 2337, 2340, 2363, 2395, 2878, 3450, 3622
Co(methyl methacrylate/methyl acrylate (maleic anhydride, α-chloroacrylonitrile))		1376, 1377, 1380, 1381, 1386, 1387

Table 14.10 (*Contd.*)

Name	Structure	References
Poly(*N*-dimethyl-β-amino-ethyl methacrylate)	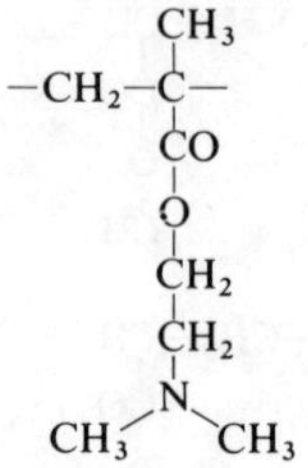	1861
Poly(naphthyl methacrylate)	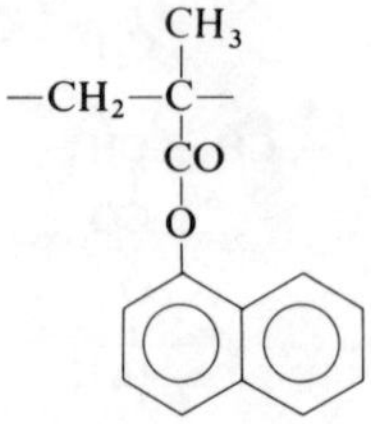	2722
Co(naphthyl methacrylate/butyl methacrylate)		2720, 2722
Poly(methacryl esters of *p*-acrylated 2-phenoxy-ethanols)	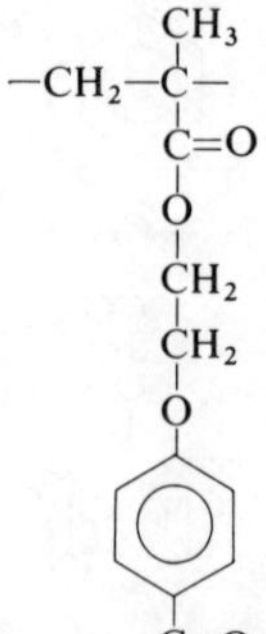	1691
Poly(methyl methacrylate)/polypropylene blends		1382, 1383

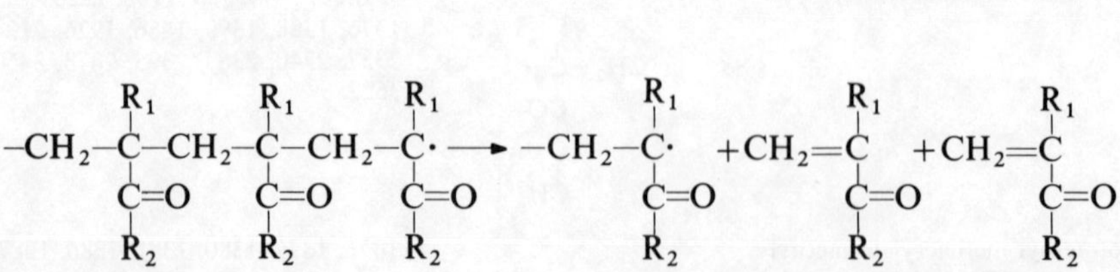

$$(14.100)$$

A stochastic model for predicting the life of photolytically degraded poly(methyl methacrylate) film has been proposed.[2395]

Poly(phenyl acrylate) (14.56)[2282] and poly(naphthyl acrylate)[2490] undergo PHOTO-FRIES REARRANGEMENT:

(14.101)

(14.56)　　　　　(14.57)　　　　　(14.58)

The two main products, the ortho and para hydroxy substituted phenyl (naphthyl) ketone groups (14.57) and (14.58), are formed. The total quantum yield of the photo-Fries rearrangement is high ($\Phi = 0.4$) and nearly independent of whether the reaction was carried out in solution or in the solid state. When cooled below the glass transition temperature ($T_g = 50°C$) no change in quantum yield of either product was observed down to the γ transition, below which the quantum yields decreases. These results suggest that the photo-Fries rearrangement may occur even in very limited free volume in a solid polymer matrix. The relatively small rotation of the phenyl or acyl group within the cage should require very little excess free volume. For the formation of the para hydroxy substituted phenyl group, which requires a larger rotation, a slightly higher activation energy is needed.[2282]

14.3.8. Polycarbonates

In comparison with other polymers, polycarbonates are more stable towards photodegradation and their mechanical properties change slowly during long exposure outdoors. This extended photostability of polycarbonates is a result of the photo-Fries rearrangement reactions.

The mechanism of this reaction observed during the u.v. irradiation of poly(aryl esters) (cf. Section 14.3.9) and polycarbonates and their model compounds, has been a subject of numerous investigations.[1, 17, 391, 393, 714, 716–720, 749, 882, 890, 891, 1078, 1259, 1340, 1455, 1458, 1482, 1670, 1712, 1713, 1917, 2351, 2474, 2506, 2558, 2591, 2592, 3115, 3180, 3624]

For example photorearrangement of bis(4-tert-butylphenyl)carbonate (14.59) in alcoholic solution occurs with formation of salicylate derivative (14.60), dihydroxybenzophenone (14.61), and 4-tert-butylphenol (14.62):[387,1482,1670]

In the case of polycarbonates (14.63) the photo-Fries rearrangement mechanism is a very similar to that described for model compounds:

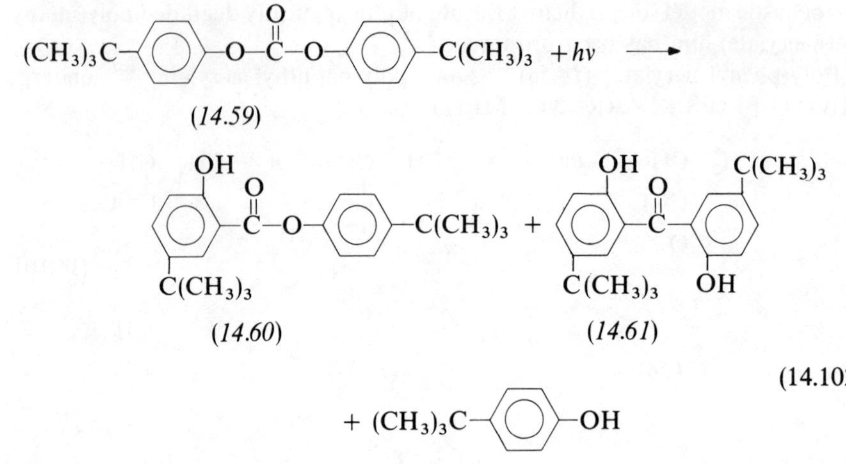

(14.59) (14.60) (14.61) (14.62)

(14.102)

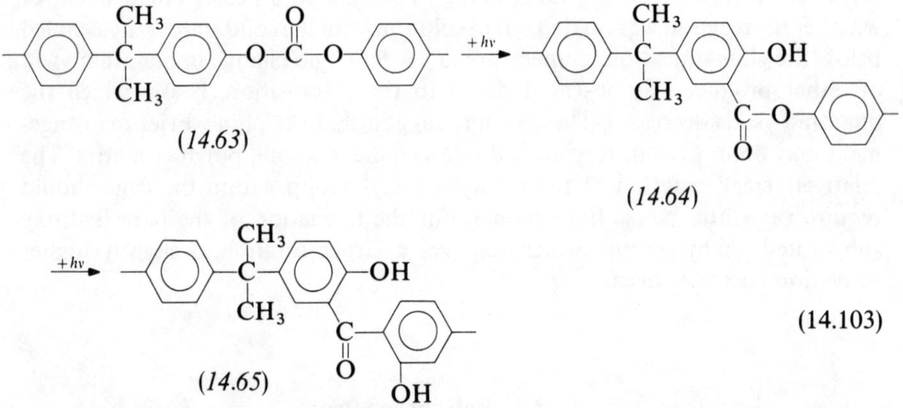

(14.63) (14.64) (14.65)

(14.103)

The salicylate (*14.64*) and *o*-hydroxybenzophenone (*14.65*) structures formed in the backbone may have a photostabilizing effect on the polycarbonate photodegradation. The mechanism of these photostabilizing effects has been given in Sections 15.3 and 15.5.

Independently of the photo-Fries rearrangement, main chain scission occurs by the following reaction:[1713,3115]

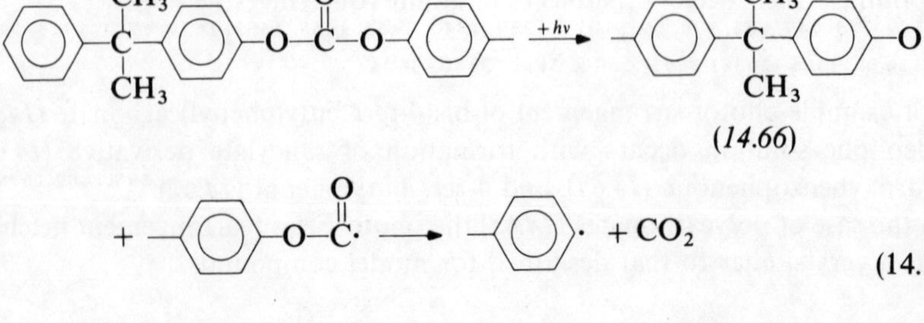

(14.66)

(14.104)

The polymer end-phenoxy radical (14.66) may abstract hydrogen from the methyl groups which may cause initiation of a new chain scission reaction by another mechanism:

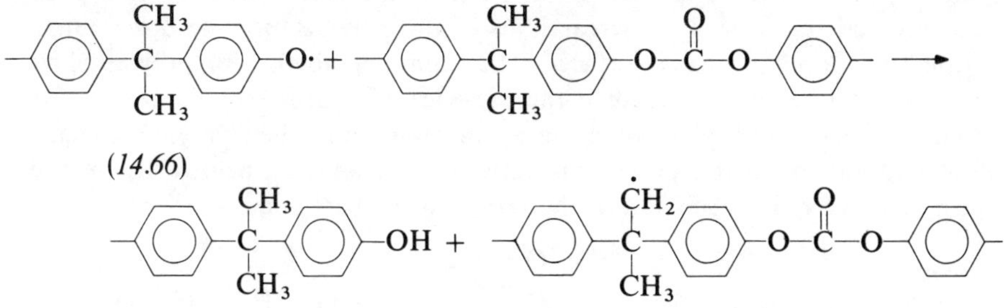

(14.66)

(14.105)

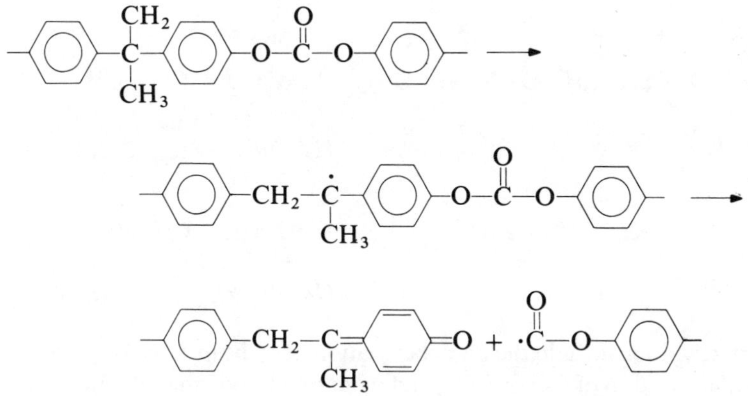

(14.106)

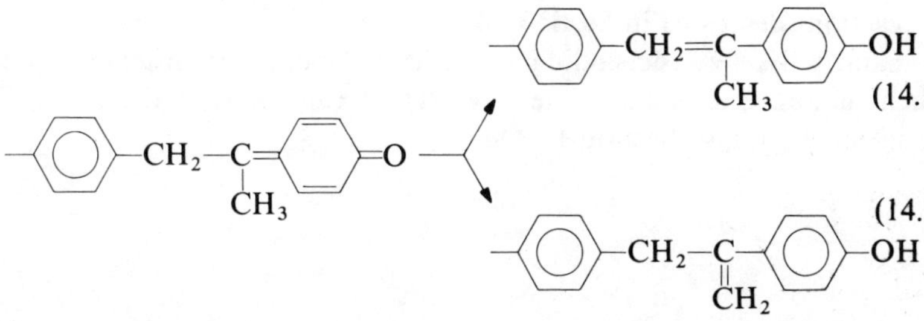

(14.107)

(14.108)

The final mechanism is more complicated because of coupling different radicals to each other.

Electron spectroscopy for chemical application (ESCA) has been successfully applied to the study of the photochemistry of polycarbonates.[717,719,720,1078]

14.3.9. Polyesters

The photodegradation of aliphatic unsaturated polyesters causes extensive chain scission, crosslinking, and yellowing.[2330, 3060, 3069, 3761]

534

Much more interest has been concentrated on the photodegradation of an aromatic polyester such as poly(ethylene terephthalate) (*14.67*),[112,450,595, 695,696,896-900,2382,2499,2818,2869,2917,3434,3436,3437,3728,3789] and its copolymers.[37,927,3468] On exposure to u.v. irradiation fibres tend to lose their elasticity and break easily, whereas films of the polymer become discoloured, brittle, and develop crazed surfaces. The photodegradation of poly(ethylene terephthalate) has been related to the presence of light-absorbing groups, i.e. aromatic ester groups, which are inherent in the polymer chain. Photodegradation of this polymer is initiated by irradiation below 315 nm and occurs by direct chain scission of the bonds in the backbone:

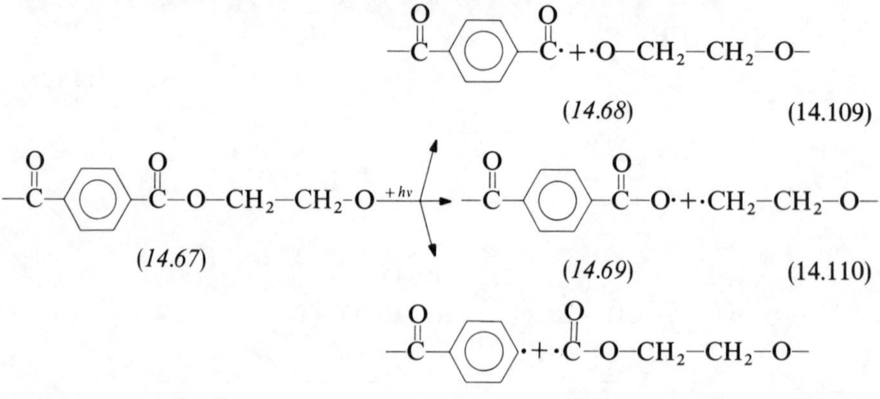

$$-\overset{O}{\underset{\parallel}{C}}-\!\!\!\!\bigcirc\!\!\!\!-\overset{O}{\underset{\parallel}{C}}\cdot + \cdot O - CH_2 - CH_2 - O -$$

(*14.68*)　　　　　　　　(14.109)

$$-\overset{O}{\underset{\parallel}{C}}-\!\!\!\!\bigcirc\!\!\!\!-\overset{O}{\underset{\parallel}{C}}-O-CH_2-CH_2-O-\xrightarrow{+h\nu}$$

(*14.67*)

$$-\overset{O}{\underset{\parallel}{C}}-\!\!\!\!\bigcirc\!\!\!\!-\overset{O}{\underset{\parallel}{C}}-O\cdot + \cdot CH_2 - CH_2 - O -$$

(*14.69*)　　　　　　　　(14.110)

$$-\overset{O}{\underset{\parallel}{C}}-\!\!\!\!\bigcirc\!\!\!\!\cdot + \cdot \overset{O}{\underset{\parallel}{C}}-O-CH_2-CH_2-O-$$

(*14.70*)　　　　　　　　(14.111)

The secondary reactions include the decarboxylation of radicals (*14.68*), (*14.69*), and (*14.70*) with evolution of CO or CO_2, and oxidation to polymer alkylperoxy (POO·), polymer alkyloxy (PO·), polymer hydroperoxide groups (POOH) and other reactions described in Section 14.3.5.

Aromatic polyesters, depending on their structure, for instance poly(9,9-bis(phenyl)fluorenone isophthalate) (*14.71*),[2311] can undergo the photo-Fries rearrangement (cf. also Section 14.3.8):

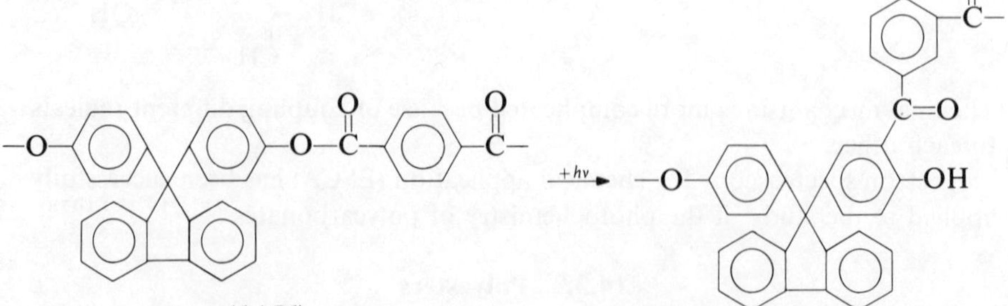

(*14.71*)

(14.112)

It has been suggested that thin coatings of a number of phenyl polyesters can protect polymers ordinarily sensitive to u.v. radiation.[2366] After photo-Fries rearrangement polymers contain a number of *o*-hydroxy-benzophenone groups, which can act as an inner photostabilizer and protect coated substrate from photodegradation.

Photodegradation of poly(ethylene-2,6-naphthate) (*14.72*) occurs by a mechanism analogous to that presented for poly(ethylene terephthalate).[144, 694, 2844] This polymer, like poly(ethylene terephthalate), has excellent physical and chemical properties, but under prolonged u.v. irradiation shows significant deterioration in its mechanical properties.

(*14.72*)

Polyesters with incorporated *O*-acylated aminoketone groups (*14.73*) undergo a very rapid photodegradation when irradiated in methylene chloride with light of 365 nm.[940] In the presence of vinyl or acrylic monomers, block copolymers can be obtained.

(*14.73*)

14.3.10. Polyamides

Photodegradation (photooxidative degradation) of polyamides (mainly Nylon 6,6) which results in a chain scission, crosslinking, and yellowing, has been investigated in detail.[12,58,95,133,136,143,154,1028,1066,1586,2374,2375,2467,2559, 2895, 2981, 3086, 3434–3436, 3452, 3520, 3966, 3967]

The thermal history of nylon has a marked influence on the subsequent photostability.[95,133,136,143] Polyamides contain some impurities which are responsible for their fluorescence and phosphorescence emission.[95,127, 133,134,137,142,143,145,1195] The phosphorescent impurities in nylon polymers have been identified as carbonyl groups conjugated with ethylenic unsaturations.[137,143,145] During irradiation these α,β-unsaturated carbonyl groups are photolysed initially by an isomerization process similar to that described in Section 14.3.1. The β,γ-unsaturated carbonyl groups formed are then photolysed by the Norrish Type I and II reactions.

The dominant chain scission process caused by u.v. irradiation occurs at the amide linkage:

$$-CH_2-CH_2-\overset{\overset{\textstyle O}{\|}}{C}-NH-CH_2-CH_2- + h\nu \longrightarrow$$

$$-CH_2-CH_2-\overset{\overset{\textstyle O}{\|}}{C}\cdot + \cdot NH-CH_2-CH_2-$$

$$(14.113)$$

$$-CH_2-CH_2-\overset{\overset{\textstyle O}{\|}}{C}\cdot \longrightarrow -CH_2-CH_2\cdot + CO$$

$$(14.114)$$

The polymer alkyl end radical (*14.74*) and polymer amino radical (*14.76*) may abstract hydrogens from the same or adjacent polymer chain and form a new polymer alkyl radical (*14.75*):

$$-CH_2-CH_2\cdot + -CH_2-CH_2-\overset{\overset{\textstyle O}{\|}}{C}-NH-CH_2-CH_2- \longrightarrow$$

(*14.74*)

$$-CH_2-CH_3 + -CH_2-\overset{\textstyle \cdot}{C}H-\overset{\overset{\textstyle O}{\|}}{C}-NH-CH_2-CH_2-$$

(*14.75*)

$$(14.115)$$

$$-CH_2-CH_2-NH\cdot + -CH_2-CH_2-\overset{\overset{\textstyle O}{\|}}{C}-NH-CH_2-CH_2- \longrightarrow$$

(*14.76*)

$$-CH_2-CH_2-NH_2 + -\overset{\textstyle \cdot}{C}H-CH_2-\overset{\overset{\textstyle O}{\|}}{C}-NH-CH_2-CH_2-$$

(*14.75*)

$$(14.116)$$

All of these radicals can be oxidized and form polymer alkylperoxy radicals (POO·), and consequently polymer hydroperoxide groups (POOH), which can be thermally and/or photochemically decomposed to polymer alkyloxy radicals (PO·) which can be disproportionated by the β-scission process (cf. Reaction 14.7).

Several other polyamides such as poly(undecanamides),[222,3547] poly(N-chloroamides),[307,1454] and poly(1,3-phenylene isophthalamide) (polyaramides)[610,611,619] have also been investigated.

Photodegradation of polyaramides (*14.77*) in the absence of oxygen occurs mainly by the photo-Fries rearrangement, during which 2-amino (*14.78*) and 4-aminobenzophenone (*14.79*) groups are formed along the backbone:[611]

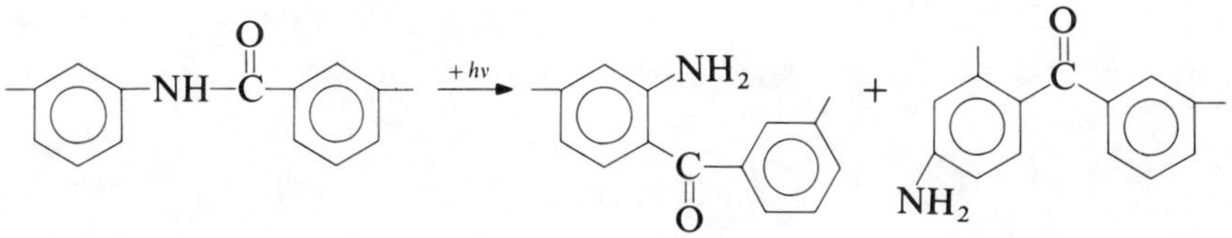

(*14.77*) (*14.78*) (*14.79*)

(14.117)

The photo-Fries reaction is diminished by the presence of oxygen and, instead of aminobenzophenone products, polymer phenyl nitroso groups (*14.81*) and polymer phenyl hydroperoxy groups (*14.82*) are formed:

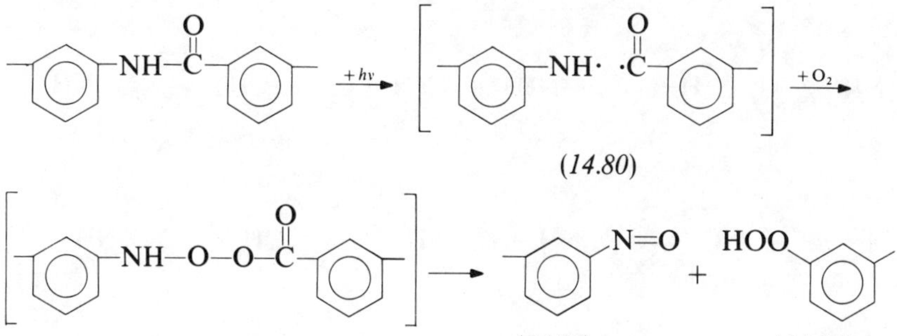

(*14.80*)

(*14.81*) (*14.82*)

(14.118)

The pair of radicals formed in a cage (*14.80*) can also recombine or crosslink. A polymer phenylacyl radical which escapes off the cage decomposes to the polymer phenyl radical and carbon monoxide.

14.3.11. Polyurethanes

The mechanism of photodegradation (photooxidative degradation) of polyurethanes depends on their structure,[8, 11, 106, 693, 1207–1210, 1213, 1590–1592, 1658, 1697, 1698, 2799, 2803, 2804, 2809, 2811, 2815, 3232] but in general is similar to that of polyamides (cf. Section 14.3.10).

Polyurethanes based on 4,4′-diphenylmethane diisocyanate (*14.83*) exhibit intense yellowing during u.v. irradiation.[106]

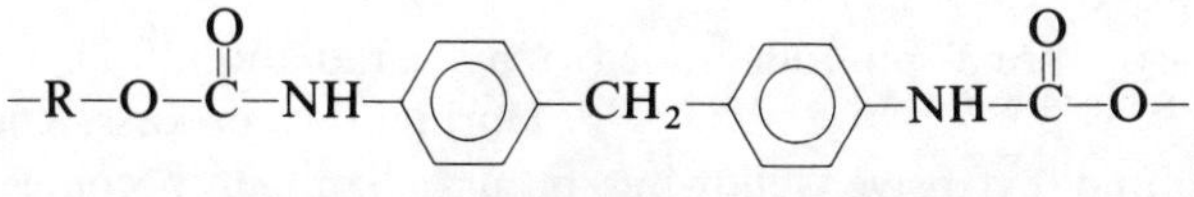

(*14.83*)

Small amounts of benzophenone type (*14.84*) chromophore are responsible for this. They are present in the polymer and can be photooxidized to diquinone imide structure (*14.85*) by the following mechanism:[106]

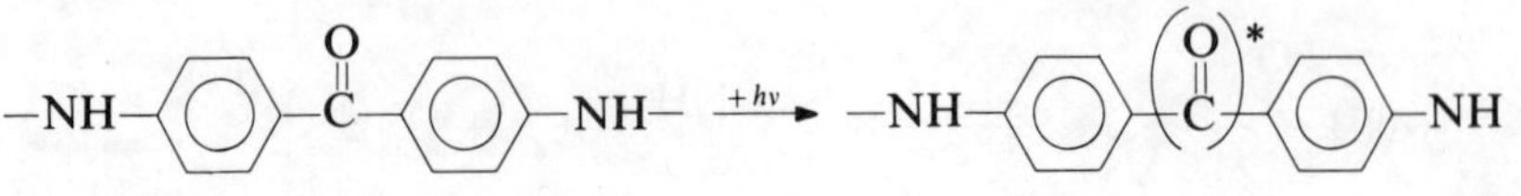

$$-NH-\bigcirc-\overset{\overset{O}{\parallel}}{C}-\bigcirc-NH-\xrightarrow{+h\nu}-NH-\bigcirc-\overset{\overset{(O)^*}{\parallel}}{C}-\bigcirc-NH$$

(14.84)   (14.119)

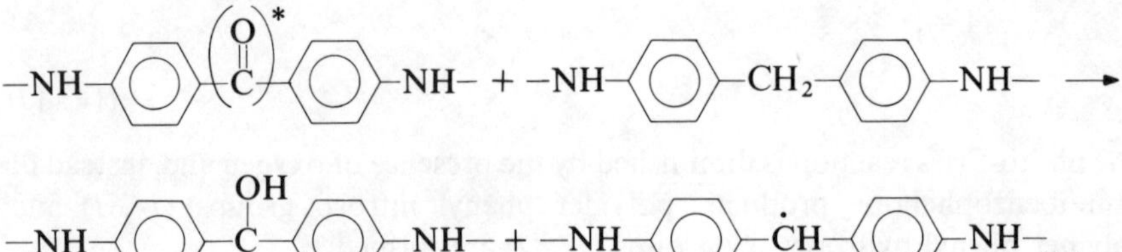

$$-NH-\bigcirc-\overset{(O)^*}{\underset{}{C}}-\bigcirc-NH- + -NH-\bigcirc-CH_2-\bigcirc-NH- \longrightarrow$$

$$-NH-\bigcirc-\overset{\overset{OH}{|}}{\underset{\cdot}{C}}-\bigcirc-NH- + -NH-\bigcirc-\dot{C}H-\bigcirc-NH-$$

(14.120)

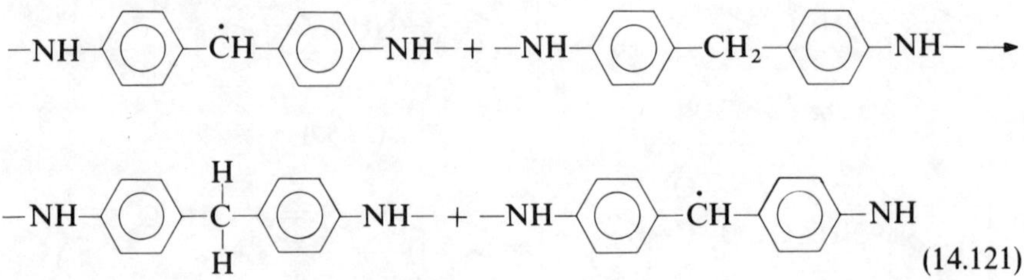

$$-NH-\bigcirc-\dot{C}H-\bigcirc-NH- + -NH-\bigcirc-CH_2-\bigcirc-NH- \longrightarrow$$

$$-NH-\bigcirc-\overset{\overset{H}{|}}{\underset{H}{C}}-\bigcirc-NH- + -NH-\bigcirc-\dot{C}H-\bigcirc-NH$$

(14.121)

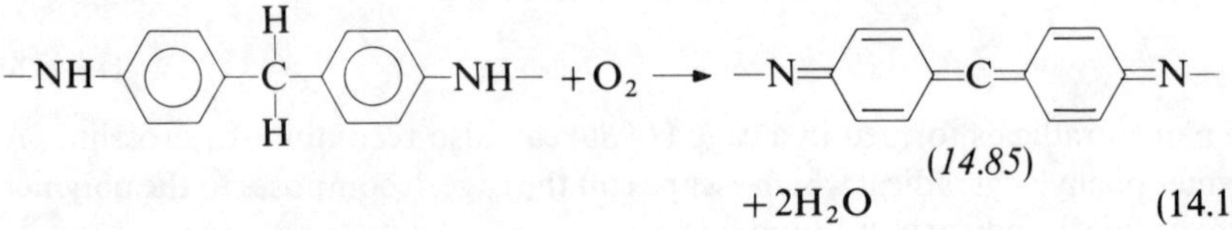

$$-NH-\bigcirc-\overset{\overset{H}{|}}{\underset{H}{C}}-\bigcirc-NH- + O_2 \longrightarrow -N=\bigcirc=C=\bigcirc=N-$$

(14.85)

$$+2H_2O$$ (14.122)

Replacement of the methylene hydrogens with methyl groups inhibits photo-yellowing by preventing the diurethane bridges from oxidation to quinone–imide structures.

14.3.12. Polysulphones

Polysulphones are photodegraded by radiation in the range 300–350 nm.[109,879,881,886,1258,1259,2916] During this process, chain scission, crosslinking, and extensive yellowing occurs. Samples become brittle with decreased strength and elongation at break.

Photolysis of polysulphone prepared from the polycondensation of biphenols with 4,4'-dichlorodiphenylsulphone (14.86) in the absence of oxygen results in the following bond-scission processes:

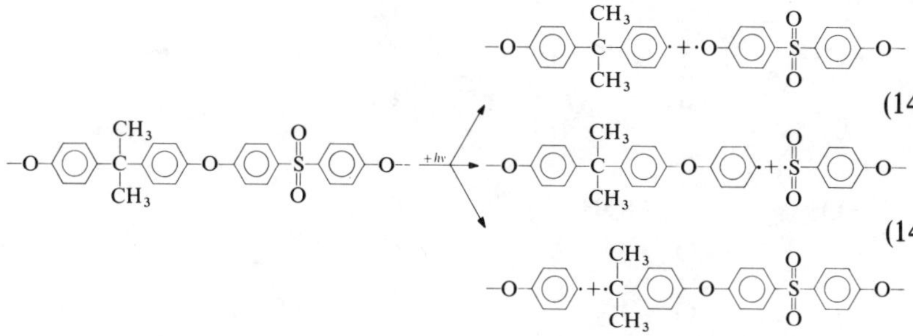

(14.123)

(14.124)

(14.125)

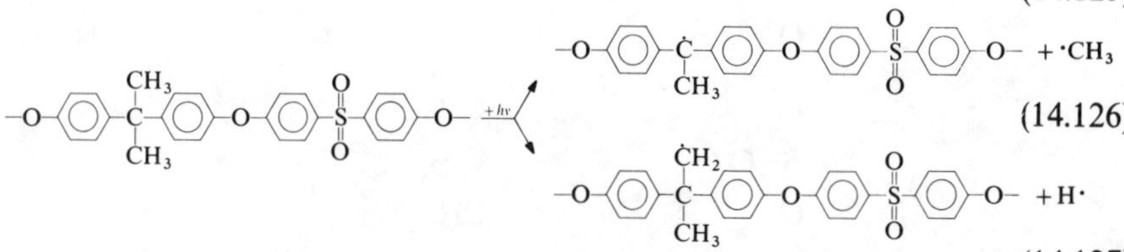

(14.126)

(14.127)

In the presence of oxygen most of the formed polymer radicals can be oxidized, by analogous reactions as described in the previous sections.

Sulphur dioxide, which is formed by scission of carbon-sulphur linkage:

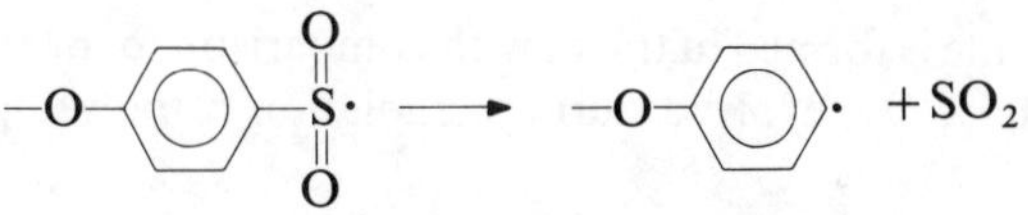

(14.128)

may decompose polymeric hydroperoxide groups (POOH) into polymeric peroxides (POOP) and polymeric esters of sulphuric acid ($POSO_2OH$):[1258,1259]

$$POOH + SO_2 \rightarrow POSO_2OH \tag{14.129}$$

$$2\ POSO_2OH \rightarrow POSO_2OP + H_2SO_4 \tag{14.130}$$

$$POOH + POSO_2OH \rightarrow POOP + POSO_2OH \tag{14.131}$$

$$POOH + POSO_2OH \rightarrow POOP + H_2SO_4 \tag{14.132}$$

The formation of oligomeric sulphonic acids (*14.87*) probably occurs by the following mechanism:

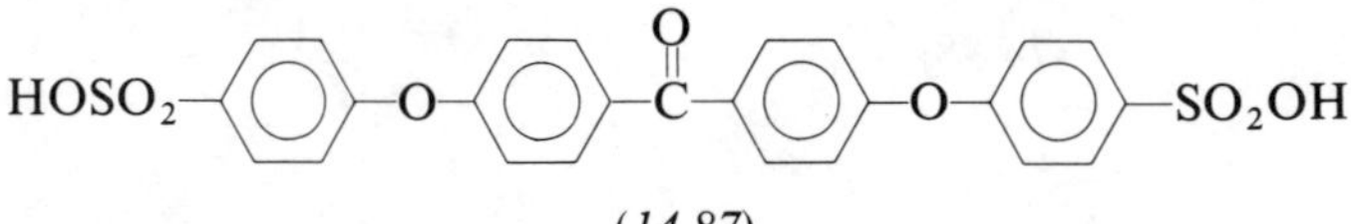

$$HOSO_2 - \bigcirc - O - \bigcirc - \overset{\overset{O}{\parallel}}{C} - \bigcirc - O - \bigcirc - SO_2OH$$

(*14.87*)

(14.133)

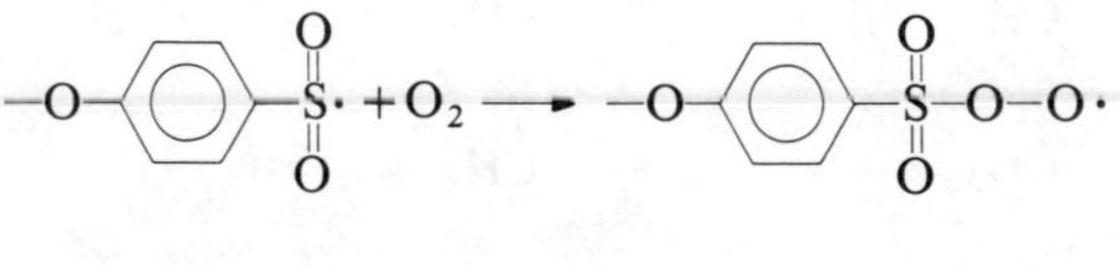

$$-O-\langle\bigcirc\rangle-\overset{\overset{O}{\|}}{\underset{\|}{S}}-O-O\cdot + -\langle\bigcirc\rangle-\overset{CH_3}{\underset{CH_3}{\overset{|}{C}}}-\langle\bigcirc\rangle- \longrightarrow$$

$$-O-\langle\bigcirc\rangle-\overset{\overset{O}{\|}}{\underset{\|}{S}}-OOH + -\langle\bigcirc\rangle-\overset{\overset{\cdot}{C}H_2}{\underset{CH_3}{\overset{|}{C}}}-\langle\bigcirc\rangle- \qquad (14.134)$$

$$-O-\langle\bigcirc\rangle-\overset{\overset{O}{\|}}{\underset{\|}{S}}-OOH \xrightarrow{+h\nu} -O-\langle\bigcirc\rangle-\overset{\overset{O}{\|}}{\underset{\|}{S}}-O\cdot + \cdot OH \qquad (14.135)$$

$$-O-\langle\bigcirc\rangle-\overset{\overset{O}{\|}}{\underset{\|}{S}}-O\cdot + -\langle\bigcirc\rangle-\overset{CH_3}{\underset{CH_3}{\overset{|}{C}}}-\langle\bigcirc\rangle- \longrightarrow$$

$$-O-\langle\bigcirc\rangle-\overset{\overset{O}{\|}}{\underset{\|}{S}}-OH + -\langle\bigcirc\rangle-\overset{\overset{\cdot}{C}H_2}{\underset{CH_3}{\overset{|}{C}}}-\langle\bigcirc\rangle- \qquad (14.136)$$

The sulphur dioxide is formed in traces with comparison to the two main gaseous products, CO and CO_2, evolved during irradiation of polysulphones.

14.3.13. Polysiloxanes

Poly(dimethylsiloxane) (*14.88*) should not theoretically absorb light above 300 nm. It nevertheless deteriorates on exposure to solar radiation. The direct photocleavage of the Si–O– bond is rather improbable because the dissociation energy of the Si–O bond is far too high. The main reaction observed is a photodissociation of side-chains followed further by cross-linking:[928]

$$
\underset{(14.88)}{\overset{CH_3}{\underset{CH_3}{\overset{|}{-\underset{|}{Si}-O-}}}}
\underset{\searrow}{\overset{\nearrow}{}}
\begin{array}{l}
\overset{\cdot}{-Si}-O- + \cdot CH_3 \qquad (14.137)\\
\;\;|\\
\;\;CH_3\\[2mm]
\overset{\cdot}{C}H_2\\
\;\;|\\
-Si-O- + H\cdot \qquad (14.138)\\
\;\;|\\
\;\;CH_3
\end{array}
$$

$$
\begin{array}{l}
CH_3\\
|\\
-Si-O-\\
\cdot\\
CH_2\\
|\\
-Si-O-\\
|\\
CH_3
\end{array}
\longrightarrow
\begin{array}{l}
CH_3\\
|\\
-Si-O-\\
\\
CH_2\\
|\\
-Si-O-\\
|\\
CH_3
\end{array}
\qquad (14.139)
$$

Several gaseous products, such as hydrogen, methane, and ethane, are evolved.

Polysiloxanes which contain as a pendant group a phenyl group are more susceptible towards crosslinking reaction.[1802, 1803, 3240, 3343] Polysiloxanes with phenyldisilanyl groups (*14.89*) under u.v. irradiation are photocleaved, with formation of very reactive silyl radicals (*14.90*) which can easily abstract hydrogen from methyl groups:[1802, 1803]

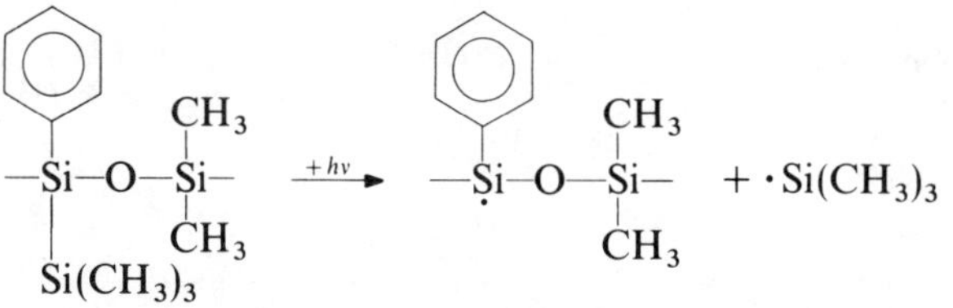

$$ \xrightarrow{+h\nu} \qquad (14.140) $$

(*14.89*)   (*14.90*)

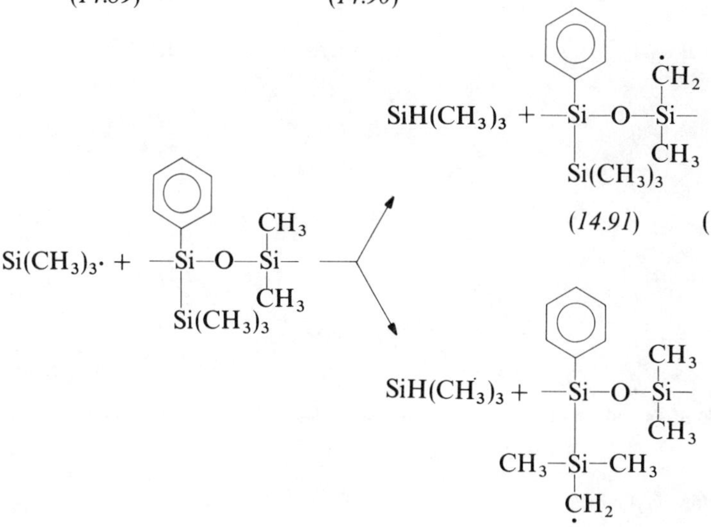

$$ (14.141) $$

(*14.91*)   (14.141)

$$ (14.142) $$

(*14.92*)

All polymeric radicals (*14.90*), (*14.91*), and (*14.92*) effectively formed crosslinks. This process has been proposed as a new class of positive deep u.v. resist[1802,1803] (cf. Chapter 12).

14.3.14. Other Polymers

Photodegradation (photooxidative degradation) of many other polymers[2468, 3060] (Table 14.11) is not described here, because of space limitations. In most cases reactions involved in photodegradation of these polymers are very similar to that described in this Chapter.

Table 14.11 Photodegradation and photooxidative degradation of other polymers—references

Name	Structure	References
Polyoxymethylene	$-CH_2-O-$	120, 1385, 1702, 1996, 2723
Polyacetaldehyde	CH_3 $\|$ $-CH-O-$	2389
Poly(2,6-dimethyl-1,4-phenylene oxide)	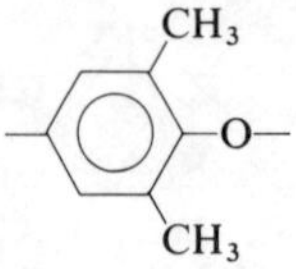	119, 673, 880, 1874, 1997, 2916, 3656, 3795
Epoxides	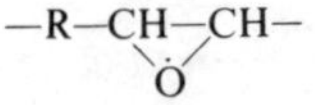 $-R-CH-CH-$ $\diagdown O \diagup$	60, 375–378, 1253, 1259, 1995
Poly(vinyl alcohol)	$-CH_2-CH-$ $\|$ OH	210, 3971
Poly(vinyl acetate)	$-CH_2-CH-$ $\|$ O $\|$ CO $\|$ CH_3	550–552, 837, 1266, 1267, 2513, 3652
Poly(vinl butyral)	$-CH_2-CH-$ $\|$ O $\|$ CO $\|$ C_3H_7	1913, 2513, 3088
Polyacrylonitrile and copolymers	$-CH_2-CH-$ $\|$ CN	36, 1595, 1859, 1865, 3436
Poly(α-chloroacrylonitrile)	Cl $\|$ $-CH_2-C-$ $\|$ CN	1595
Polymers and copolymers bearing o-acyloxime groups	$-CH_2-CH-$ $\|$ CO $\|$ $O \quad R$ $\diagdown$ $N=C$ $\diagdown R$	940, 3670
Poly(vinylpyrrolidone)	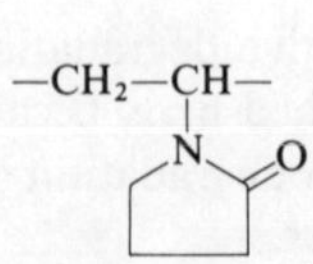 $-CH_2-CH-$ (pyrrolidone ring with N and =O)	1867, 1868, 2440

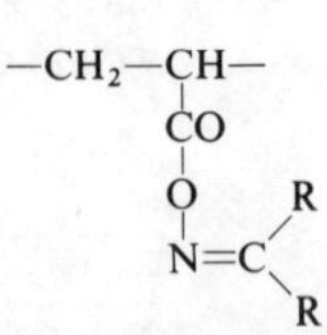

Table 14.11   (Contd.)

Name	Structure	References
Poly(N-vinyl carbazole)	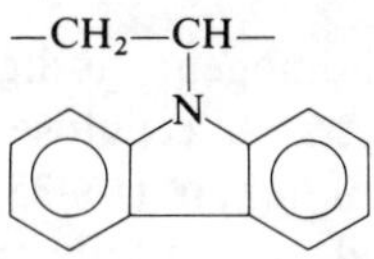	772
Polyacenaphthylene	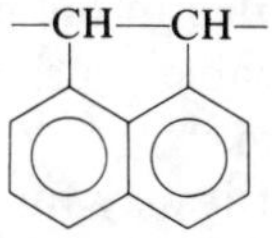	3420, 3431
Poly(dimethylacrylimide)	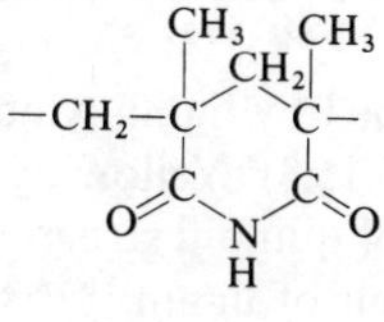	1594
Poly(acrylic anhydride)	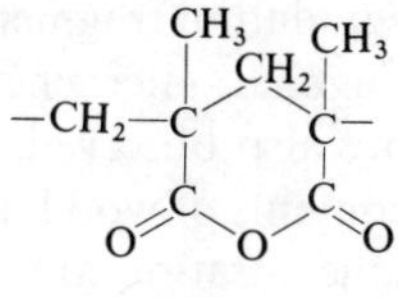	1593
Poly(alkoxanes)	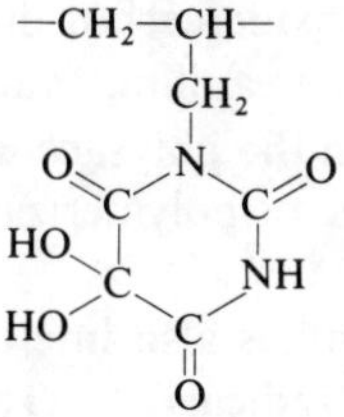	1060
Poly(benzoxazoles)	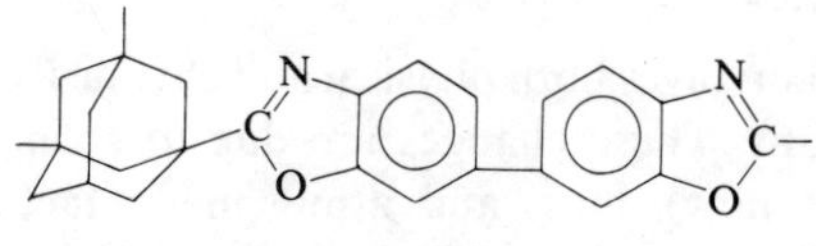	979
Poly(diacetylenes)	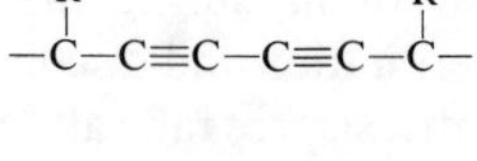	2521, 2593, 2749, 3073
Poly(organophosphazanes)	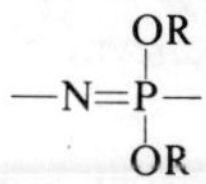	476, 1341, 1342, 1500, 1596, 2750

Poly(N-vinyl carbazole)

$-CH_2-CH-$

Polyacenaphthylene

$-CH-CH-$

Poly(dimethylacrylimide)

$CH_3 \quad CH_3$
CH_2
$-CH_2-C \quad C-$
$O=C \quad C=O$
N
H

Poly(acrylic anhydride)

$CH_3 \quad CH_3$
CH_2
$-CH_2-C \quad C-$
$O=C \quad C=O$
O

Poly(alkoxanes)

$-CH_2-CH-$
CH_2
$O=C \quad C=O$
N
HO
$\quad C \quad NH$
HO
C
O

Poly(diacetylenes)

$\quad R \quad\quad\quad R$
$-C-C≡C-C≡C-C-$

Poly(organophosphazanes)

OR
$-N=P-$
OR

14.3.15. Photodegradation (Photooxidation) of Natural Polymers

14.3.15.1. Photodegradation of wood

Wood is a complex and inhomogeneous lignocellulosic polymer composite, which contains a fibrous structure of cellulose (43 %), hemicellulose (28–38 %), a three-dimensional network of lignin (16–33 %), a small amount (2–8 %) of extractable organic compounds, and a mineral water.[1644,2364]

Wood absorbs light very well, due to the presence of several chromophoric groups,[1653] but u.v. radiation is unable to penetrate into wood more than 75 μm and visible light to only 200 μm.[1651,1654,1655] Photooxidative degradation occurs initially at the surface, but it may further attack the underlying layers. Photooxidation of the wood surface results in yellowing, loss of gloss, roughening, and checking.[677,2221,2222]

The light is mainly absorbed by lignin which contains several chromophoric groups[3729] (cf. Section 14.3.15.3). Yellowing of lignin components has been attributed to the formation of quinoid structures, and begins with the oxidation of the phenolic structural unit of lignin.[1653,2131]

During photooxidative degradation fragments of wood structure, such as cell wall, are destroyed and diagonal microcheck, expansion of aperture, and degeneration of pit domes are also observed.[2526]

The methoxyl and lignin contents of wood during u.v. irradiation are reduced, and acidity and carbonyl concentration are increased. Several low molecular products—such as carbon monoxide, carbon dioxide, hydrogen, water, methanol, formaldehyde, organic acids, vanillin, and syringaldehyde—are formed. Chemical changes also occur in the polysaccharide molecules of wood. Loss of strength and decrease of degree of polymerization and α-cellulose content have been observed.[1642]

Hemicellulose in paper products also interacts with lignin. Xylan, upon u.v. irradiation, photolyses to free radicals.[1630] The photolytic scission of hemicellulose chain may occur at the end of the molecule as well as in the middle of the chain. Yellowing of the paper upon light irradiation is due to the photooxidation of hemicellulose.[1645,3134]

Weathering of wood involves both physical and chemical changes in its surface and near-surface regions. These changes are due to solar radiation, moisture (dew, rain, humidity, snow), heat, and atmospheric factors (oxygen, singlet oxygen, ozone, air pollutants, NO_x, SO_2, etc.).[677, 1092, 1646, 1649, 1650, 2527]

Free radicals generated in wood during the weathering process play an important role in surface deterioration and discoloration. It has been found that wood does not contain any intrinsic free radicals.[1656] Free radicals are produced in wood only when it is subject to various environmental agents such as u.v./vis. radiation, certain chemicals, fungi or enzymes, and certain physical and chemical stresses.[1652,1654]

Water is a polar liquid which may penetrate into, and swell, the cell walls, leading to the reduction of hydrogen bonding. Due to the high polarity of water molecules, they may attract or interact with free radicals. It has been reported that moisture has a significant bearing on formation and decay of free radicals in cellulose exposed to u.v. irradiation. Moisture content in the range of 5–7 % led to a significant inhibition of radical formation, but when the moisture content was lower or higher than this critical range, water appeared to promote free radical formation.[1629,1652]

Electron spectroscopy for chemical application (ESCA) has been successfully applied to the study of oxidized wood surfaces.[1647]

Wood surfaces can be protected against photodegradation by impregnation with metal ions of inorganic salts, such as ferric ions and chromium ions.[677,1091] Complexes between guiacol and catechol and inorganic salts are formed. Probably inorganic salts blocks free phenolic hydroxyl groups in lignin, which are reactive centres in the photooxidation of lignin. Application of ferric chloride, chromium trioxide, and copper chromate treatments increases the weathering resistance of wood.[677,1091,1918] The substitution of hydroxyl groups in wood by acetylation or methylation and carbonyl groups with phenylhydrazine does not prevent photooxidation reactions.[1918]

14.3.15.2. Cellulose

Photodegradation (photooxidative degradation) of cellulose and its esters have been reviewed.[333,334,952,2945,3060]

Three types of chromophoric groups present in cellulose (*14.93*) are responsible for the light absorption (Fig. 14.19):[478,1630]

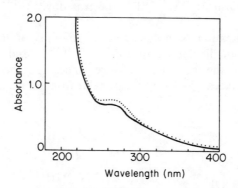

Fig. 14.19 Ultraviolet absorption spectrum of cellulose film: (——) before and (– – –) after 2 h irradiation (254 nm)[1630]

(i) Internal chromophore groups (mainly of acetal or glucosidic linkages) present in original cellulose materials.
(ii) Internal chromophoric groups (such as carbonyl, carboxyl, and hydroperoxide groups) formed in mechanically and/or chemically processed

cellulosic materials, e.g. the ginning process for lint cotton, the pulping process for wood cellulose, and the dissolution process for cellophane.

(iii) External impurities containing some chromophoric groups or other impurities, e.g. metals and metal salts and oxides.[245]

The photochemical reaction of cellulose does not occur in vacuum at wavelengths longer than 340 nm.[1630] In the presence of oxygen a photooxidative degradation occurs, via free radical mechanisms:[479,480,828,829,953,954, 1145,1571,1629,1630,1632-1640,1643,1644,1648,1652,2052,2251,2496,2760,2945,3082,3143]

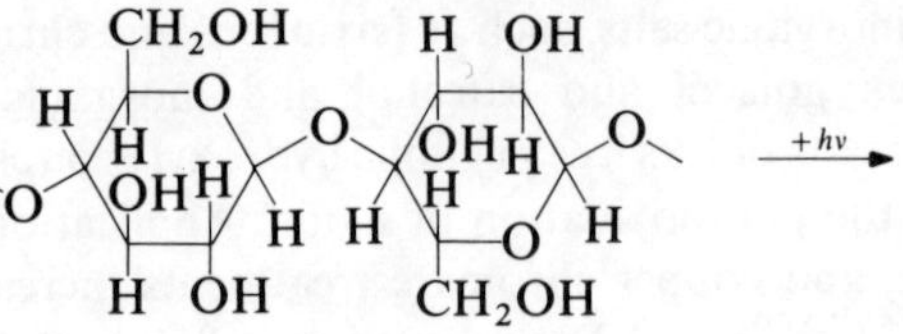

(14.143)

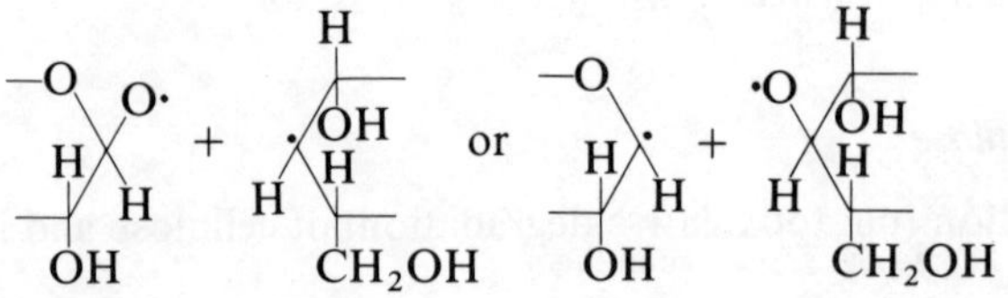

Formation of cellulose radicals during the photocleavage of the glucosic bond has been reconfirmed by ESR spectroscopy.[1630,1632-1634,1637-1640, 1643,1644,1648,2945]

During u.v. irradiation of cellulose small amounts of hydrogen, carbon monoxide, and carbon dioxide are formed.[479,480,1638] Considering the chemical structure of cellulose, the abstraction of hydrogen occurs at these carbons, where a small amount of delocalizing energy is available from the neighbouring oxygen atoms:[1638]

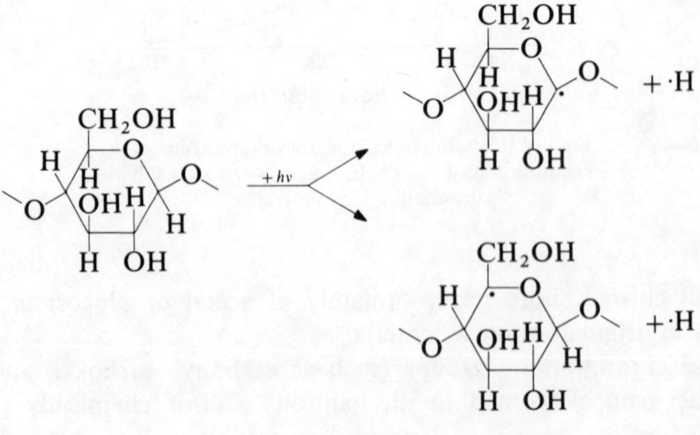

(14.144)

(14.145)

These are also other possibilities that the dehydrogenation can take place:

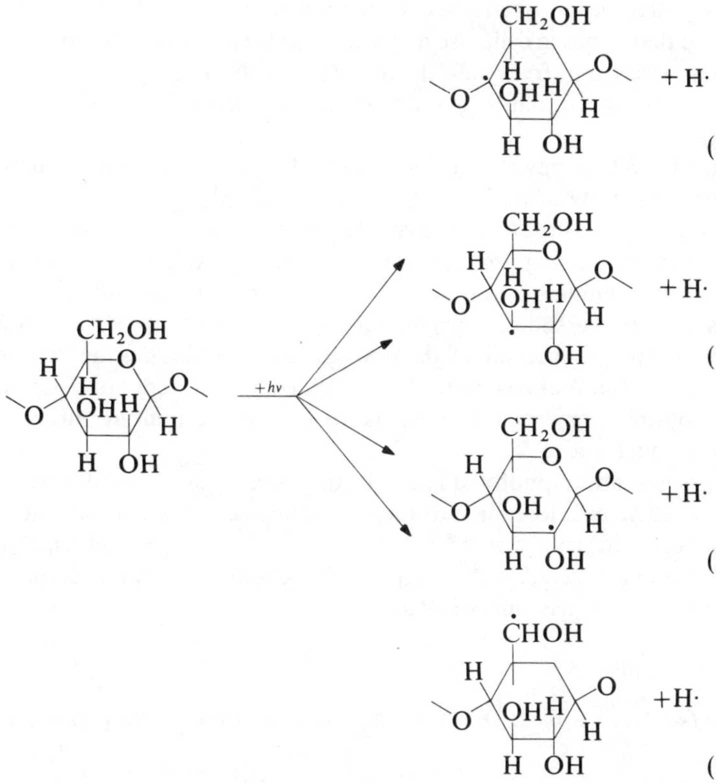

$$\text{CH}_2\text{OH} \quad \cdots \quad + \text{H} \cdot \tag{14.146}$$

$$\quad + \text{H} \cdot \tag{14.147}$$

$$\quad + \text{H} \cdot \tag{14.148}$$

$$\quad + \text{H} \cdot \tag{14.149}$$

Upon u.v. irradiation the photocleavage of C–OH and C–CH$_2$OH bonds may also occur, yielding a number of a new radicals such as HO·, HO–CH$_2$· and polymer radicals (P·).[1638,2760] Polymer radicals can be further oxidized to polymer peroxy radicals (POO·), polymer hydroperoxy groups (POOH), which can be thermally and/or photochemically cleavaged into polymer oxy radicals (PO·) and hydroxy radicals (HO·).

Photooxidation of cellulose is affected by the presence of moisture/water. The water content in cellulose is about 5–7 %. Water may play an important role in the formation of free radicals, which subsequently lead to the degradation of cellulose.[1629,1652]

The photodegradability of cellulose is heavily dependent upon its morphology.[1639] Free radicals are formed exclusively in the amorphous regions of cellulose and are probably limited to the surface layers of the polymer. Hydrogen bonding in cellulose is different in different lattice types, and this may influence free radical formation in unit cells in different lattice types.

During photooxidation of cellulose polymer peroxy radicals (POO·) are formed. This was confirmed by ESR spectroscopy.[1642,2656]

With increasing lignin content in cellulose the concentration of free radicals formed from cellulose decreases. The presence of lignin in cellulose has an inhibiting effect upon free-radical formation in cellulose.[1633,1644] It has been concluded that lignin in cellulose materials affects energy transfer. In spite of the inhibiting effect on free radical formation, yellowing of lignin-containing cellulose due to the formation of quinoid structures in lignin is a most undesirable property.

Electron spectroscopy for chemical application (ESCA) has been successfully applied to the study of the oxidation of cellulose.[724]

It is almost impossible to isolate cellulose in its pure form, because mechanical and chemical processing are responsible for partial oxidation of a sample.

Several photoinitiators (or photosensitizers) present or added to cellulose accelerate its photooxidative degradation (cf. Table 14.15). Metallic ions may form the complexes absorbing light at longer wavelengths. During irradiation of such complexes several new radicals are formed, which were not detected when other photoinitiators such as benzoperoxide, azobisisobutyronitrile, or benzophenone were added.[1635]

Photodegradation (photooxidative degradation) of other cellulose derivatives such as methyl cellulose in dioxane,[3439] cellulose nitrate in ethylacetate,[712] cellulose di- and tri-acetate,[687,1641,1899,2125,2514] hydroxymethylpropyl cellulose,[874] benzyl cellulose,[828] naphthoyl cellulose,[2736] and cyanated cellulose,[828,829] have been investigated.

14.3.15.3. Lignin

Lignin (14.94) has a basic structure that contains twelve coniferyl units:

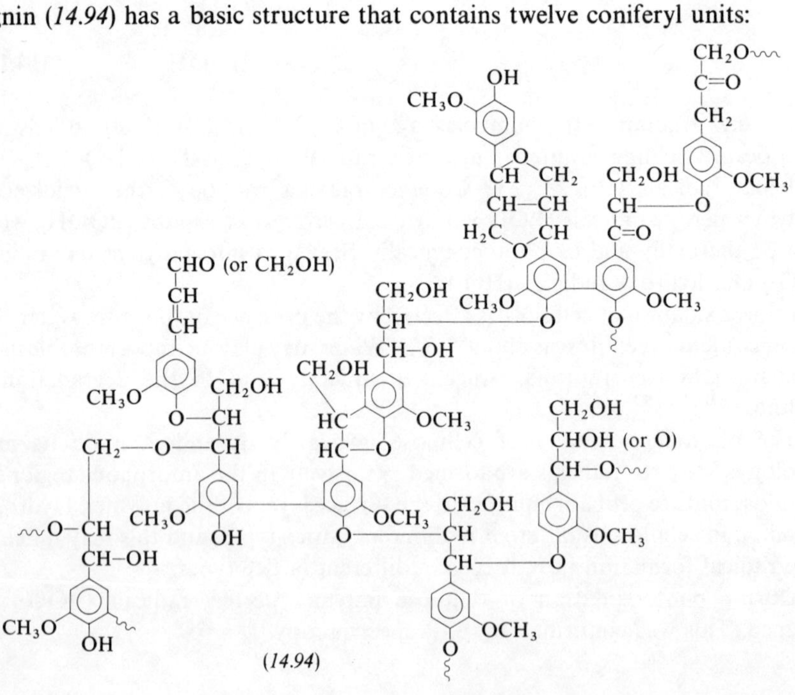

(14.94)

Lignin is an excellent light absorber; its absorption spectrum (Fig. 14.20) extends with tail to over 400 nm.

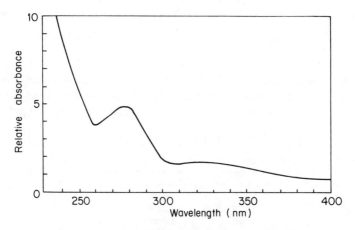

Fig. 14.20   Ultraviolet absorption spectrum of lignin[1644]

Lignin always contains a small amount of stable phenoxy radicals which may be generated during sample preparation, handling, or storage in the presence of light. During u.v. irradiation of lignin several other free radicals are formed, which were detected by ESR spectroscopy.[1644,2656] Because of the very complicated lignin structure (*14.94*) it is extremely difficult to identify free-radical sites formed.

The photochemistry of lignin is not well known.[540,1243,1244,1334,1644,2223,2299,3058] It is generally agreed that the phenoxy radicals (*14.95*) are the major intermediate formed in photoirradiated lignin. Phenoxy radicals may react further by analogous reactions as described for hindered phenols (cf. Section 15.13):

$$\text{(14.95)} \longrightarrow \text{(14.96)} + \cdot OCH_3 \qquad (14.150)$$

The final result is formation of ortho-quinoid structures (*14.96*) which are responsible for yellowing of lignin.

14.3.15.4. Photodegradation of wool

The exposure of wool keratin to sunlight results in a number of physical and chemical changes.[2692]

Wool and hair fibres contain mainly keratin, which consists of polypeptide chains bound by salt linkages between the functional groups of the amino acids (*14.97*) and by cystine (S–S) linkages (*14.98*):

550

$$
\begin{array}{c}
\overset{|}{CO} \\
CHCH_2COO^- \quad \overset{+}{H_3}N(CH_2)_4\overset{|}{CH} \\
\overset{|}{NH} \qquad\qquad\qquad \overset{|}{NH}
\end{array}
\qquad
\begin{array}{c}
\overset{|}{CO} \\
CHCH_2-S-S-CH_2\overset{|}{CH} \\
\overset{|}{NH} \qquad\qquad\qquad \overset{|}{CO}
\end{array}
$$

$$(14.97) \qquad\qquad\qquad (14.98)$$

The amino acid composition of wool fibres is given in Table 14.12. For different animal fibres, the cystine content of the keratin varies, but it is higher than in any other protein. There are differences in the structural position of the amino acids in the keratin between hair from different animal species, and also along the fibres.

Table 14.12 The amino acid composition of wool[3060]

Type of side chain	Acid	Mol. %
Hydrocarbon (or hydrogen)	Glycine	10.5
	Alanine	5.4
	Valine	4.8
	Leucine	9.8
	β-Phenylalanine	2.6
Hydroxy	Serine	11.0
	Threonine	6.2
	Tyrosine	3.3
Acidic	Aspartic acid	6.1
	Glutamic acid	12.4
Basic	Lysine	2.2
	Arginine	8.4
	Histidine	1.0
Sulphur-containing	Cystine	10.3
	Methionine	0.4
Heterocyclic	Proline	4.6
	Tryptophane	1.0

The absorption spectrum of wood (Fig. 14.21) depends very much on its origin, i.e. composition of different amino acids. The absorption between 250 and 300 nm is due essentially to the presence of the amino acids tyrosine (14.99) and

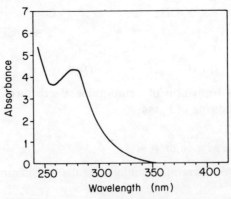

Fig. 14.21 Ultraviolet absorption spectrum of merino-wool keratin[2693]

tryptophan (*14.100*), with minor contributions from cystine (*14.101*) and phenylalanine (*14.102*).[2692,2693]

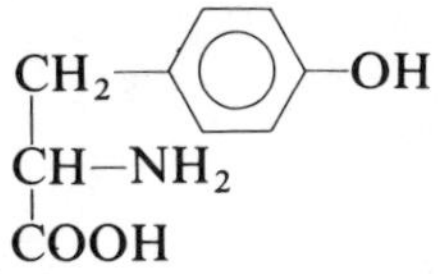

CH$_2$—⟨○⟩—OH
CH—NH$_2$
COOH

(*14.99*)

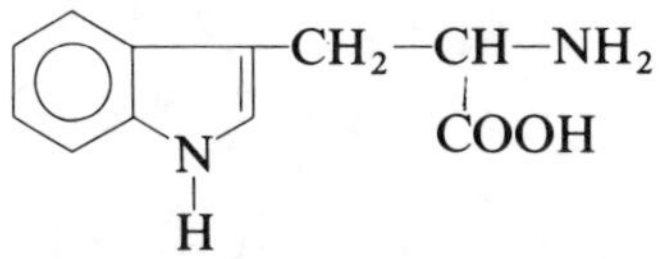

⟨○⟩—CH$_2$—CH—NH$_2$
N   COOH
H

(*14.100*)

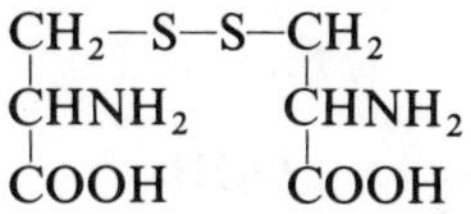

CH$_2$—S—S—CH$_2$
CHNH$_2$ CHNH$_2$
COOH   COOH

(*14.101*)

CH$_2$—⟨○⟩
CH—NH$_2$
COOH

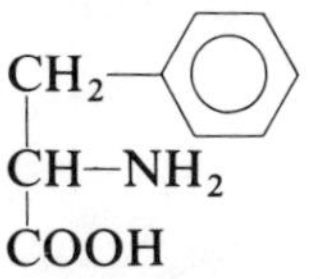

(*14.102*)

Under u.v. and even sun radiation a wool becomes yellow. The degree of yellowing depends very much on the wavelength of radiation used.[2213,2264] Much more extensive yellowing occurs in summer than in winter months.

Oxygen plays an essential role in the yellowing of wool.[2693] Irradiation of dry wool in the absence of oxygen produces a green colour, probably due to the formation of thiyl free radicals. On exposure to atmospheric oxygen these radicals decay, and the green colour of the wool turns to yellow.[2264] The presence of water vapour enhances yellowing, thus wet wool yellows much more rapidly than dry wool.[2523]

Wool exhibits a week fluorescence which can origin from tryptophan (*14.100*).[1289,2109] The disulphide bonds present in wool are responsible for at least 50 % quenching of wool's fluorescence.[2224] The phosphorescence observed in wool also originates from tryptophan.[1289,2109,2225,2693]

During photooxidative degradation of wool several amino acids are formed and main photocleavage occurs between the side amino acid groups and the main chain.[242,1628,2505,2692] Determination of free radicals formed by ESR spectroscopy is difficult, due to a lack of distinct hyperfine structure of the resulting spectra.[2870–2872,3183]

The most widely accepted theory of wool yellowing proposes the photooxidation of the tryptophan groups (*14.100*) to yellow colored. N-formylkynurenine (*14.103*) and kynurenine (*14.104*) (reaction 14.151):[1627,2692,3060]
Up until now the mechanism of wool photooxidative degradation has not been fully elucidated.

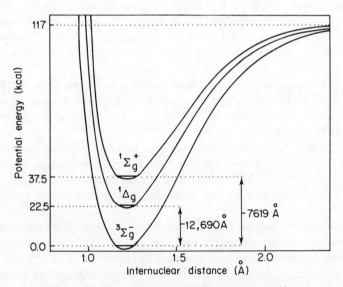

(14.100)

(14.103)

(14.151)

(14.104)

14.4. PHOTOOXIDATIVE DEGRADATION OF POLYMERS
BY SINGLET OXYGEN MECHANISM

Molecular oxygen exists in the three lowest-lying states as triplet state ($^3\Sigma_g^-$), singlet oxygen $^1O_2(^1\Delta_g)$, and $^1O_2(^1\Sigma_g^+)$ (Fig. 14.22)[3060, 3063, 3200, 3806]

Fig. 14.22 Energy diagram of different oxygen species

The isolated molecules of $^1O_2(^1\Delta_g)$ at low pressure undergo a spontaneous transition to the ground state ($^3\Sigma_g^-$) for which the half-lifetime is 45 min:

$$^1O_2(^1\Delta_g) \rightarrow O_2(^3\Sigma_g^-) + h\nu \ (12,700 \ \text{Å}) \tag{14.152}$$

Collisions with other molecules can shorten this lifetime in two ways:

(i) They can induce an electric-dipole transition at the same wavelength.
(ii) They can induce a radiationless transition to the ground state.

For example, in the presence of 1 atm of O_2 the radiative half-life becomes 10 min and the non-radiative half-life becomes 14 ms. This makes the dark process almost 10^4 times faster than the radiative process. The ability of molecules to quench singlet oxygen in the gas phase and in solution, and the practical implications of this process, will be presented in Section 15.15. From many quenching studies it is now known that in the gas phase at 1 atm the lifetime of singlet oxygen can vary between 1 and 10^{-5} s, depending on the nature of the gas. In solution the lifetime varies between 1 ms (in Freon) and 2 μs (in water) (Table 14.13).[2486, 2487]

Table 14.13   Relative lifetimes of $^1O_2$ (τ) in different solvents

Solvent or solvent mixture	τ (s)
H_2O	2.0×10^{-6}
$H_2O + CH_3OH$ (1:1)	3.5×10^{-6}
D_2O	2.0×10^{-5}
CH_3OH	5.0×10^{-6}
CH_3CH_2OH	5.6×10^{-6}
Benzene	1.25×10^{-5}
Cyclohexane	1.7×10^{-5}
Toluene	2.0×10^{-5}
Iso-octane	2.0×10^{-5}
Dioxane	3.2×10^{-5}
$CHCl_3$	6.0×10^{-5}
$CHCl_3 + CH_3OH$ (9:1)	2.6×10^{-5}
Chlorobenzene	9.1×10^{-5}
CH_2Cl_2	1.05×10^{-4}
CS_2	2.0×10^{-4}
CCl_4	7.0×10^{-4}
CF_3Cl (Freon 11)	1.0×10^{-3}
CCl_2FCClF_2 (Freon 113)	2.2×10^{-3}

The most common methods for the generation of $^1O_2$ are:[3025, 3060]

(i) Microwave discharge method.[1452, 1952, 2360, 3036] In the case of gas-phase oxidation of polymers, the amount of hydroperoxy groups formed on the polymer surfaces even after long exposure times to $^1O_2$ (50–200 h) is sometimes too low to be measurable by standard experimental methods. The failure of $^1O_2$ molecules generated in the gas phase to react with the polymer surfaces results from non-elastic collisional deactivation.

(ii) Chemical reactions, e.g. thermal decomposition of ozone–triphenyl-phosphite adduct (14.105):[1953]

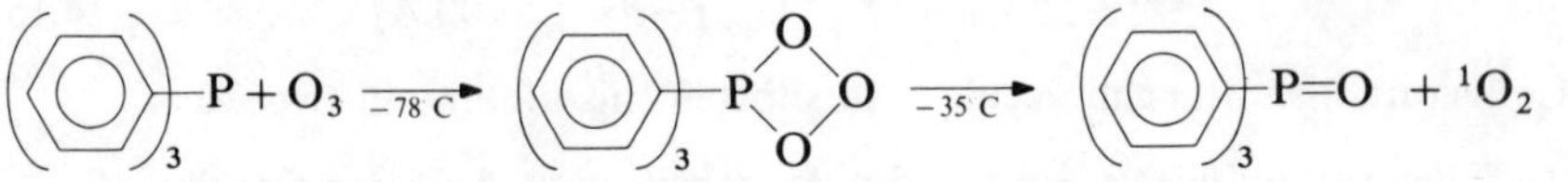

$$(14.105) \qquad\qquad (14.153)$$

(iii) Photosensitization by dyes.[1024, 1029, 1412, 1448, 3037, 3038, 3098, 4010] The dye molecule (Methylene Blue, Rose Bengal, Rhodamine B, or $\alpha,\beta,\gamma,\sigma$-tetraphenyl-porphirine (D) absorb light and after excitation to the singlet (^{1}D) or triplet (^{3}D) states transfer their excitation energy by an energy transfer mechanism to oxygen:

$$D + h\nu \rightarrow {}^1D \xrightarrow{\text{ISC}} {}^3D \qquad\qquad (14.154)$$

$$^1D(^3D) + O_2 \rightarrow D + {}^1O_2 \qquad\qquad (14.155)$$

These processes in many cases are accompanied by reactions in which free radicals are formed and oxidation–reduction processes and/or light-induced fading occurs (cf. Section 7.25).[3405] Unfortunately a number of dyes photosensitize (or photoinitiate) degradation and oxidative degradation of polymers by reactions in which singlet oxygen does not take place.[124, 2226, 3038, 3060]

Singlet oxygen can also be generated in many other reactions, for example polynuclear aromatic hydrocarbons such as 9,10-substituted anthracenes (14.106) can transfer their excitation energy to oxygen and produce $^1O_2$, which reacts with a donor molecule giving 9,10-endoperoxide (14.107):[645, 768]

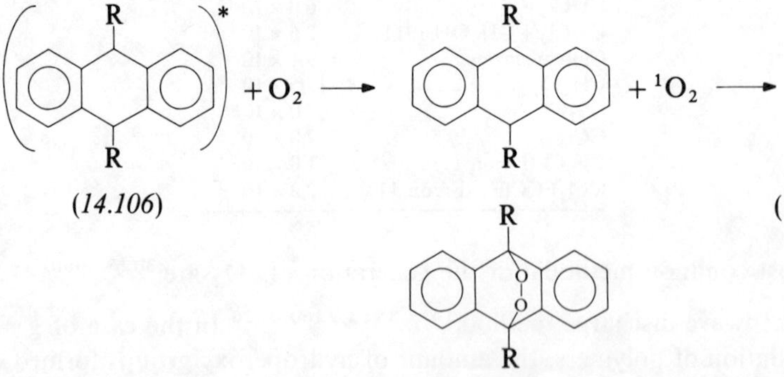

$$(14.106) \qquad\qquad\qquad (14.156)$$

$$(14.107)$$

Under u.v. irradiation (or even thermally) 9,10-endoperoxide is decomposed with evolution of $^1O_2$, or transformed to anthraquinone (14.108):

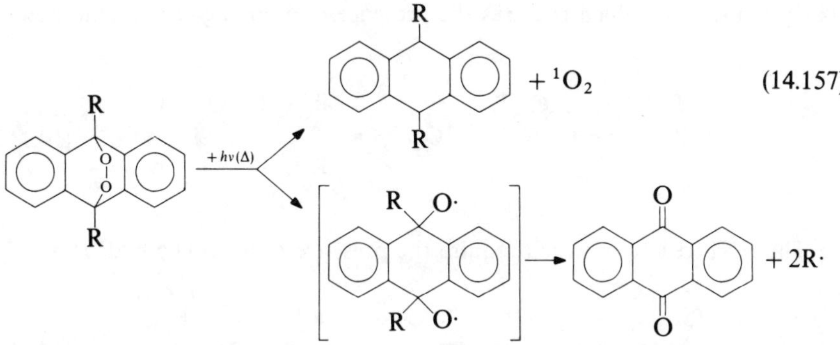

$$+ {}^1O_2 \qquad (14.157)$$

$$\to \quad + 2R\cdot$$

$$(14.108) \qquad (14.158)$$

These reactions may play some important role in photosensitized (photoinitiated) oxidation of a number polymers (cf. Table 14.15) by polycyclic hydrocarbons. Co(styrene/9,10-di-p-styrylanthracene), employing the above reactions, can be used as a carrier of singlet oxygen.[3141]

In photosensitized reactions and in biological processes various types of oxygen species such as oxygen radical anion (O_2^{-}, also called the SUPEROXIDE ANION), hydroxy (HO·), and hydroperoxy (HO$_2$·) radicals, hydrogen peroxide (H_2O_2), oxygen atoms (O), and ozone (O_3) may be formed, by interconversion reactions, into one other:[3130, 3350]

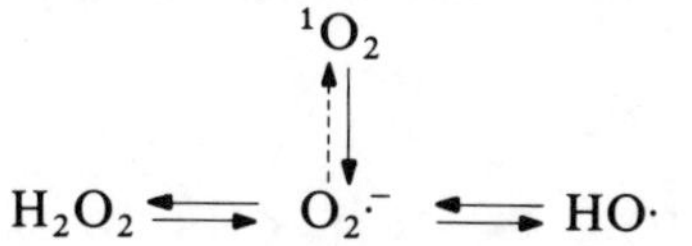

$$H_2O_2 \rightleftharpoons O_2^{-} \rightleftharpoons HO\cdot \qquad \overset{{}^1O_2}{} \qquad (14.159)$$

$$H_2O_2 \rightleftharpoons HO\cdot \rightleftharpoons O_2^{-} \qquad \overset{{}^1O_2}{} \qquad HO_2\cdot \qquad (14.160)$$

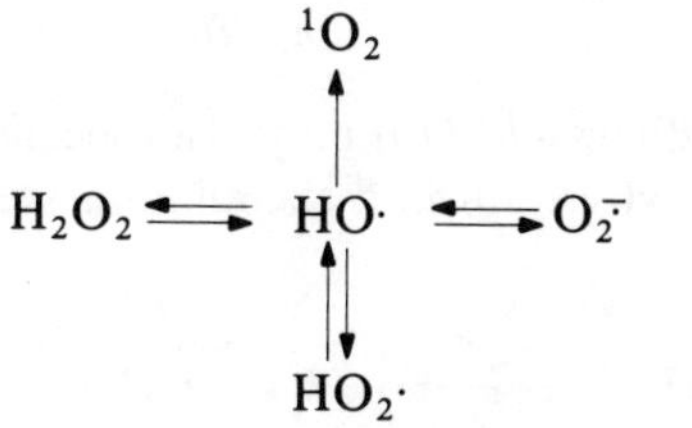

$$H_2O_2 \leftarrow O_3 \rightarrow O_2^{-} \qquad \overset{{}^1O_2}{} \qquad O \qquad HO\cdot \qquad (14.161)$$

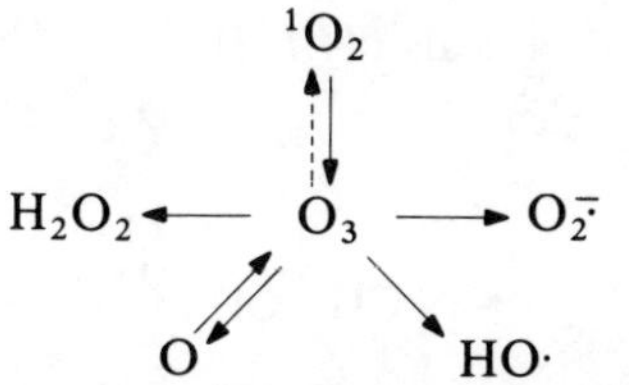

All these oxygen species may participate in effective oxidation processes of polymers.

The two principal reactions characteristic of singlet oxygen are:[1102, 1121, 1125, 1986–1988, 3063]

(i) Cycloaddition, which requires the presence of a conjugated double bond:

$$\begin{array}{c} R\quad\quad R \\ | \quad\quad | \\ R-C\quad\quad C-R \\ \diagdown\;C-C\;\diagup \\ | \quad | \\ R \quad R \end{array} + {}^1O_2 \longrightarrow \begin{array}{c} R\quad O-O\quad R \\ | \quad\diagdown\;\diagup\quad | \\ C\quad\quad C \\ \diagup\quad C=C\quad\diagdown \\ R \quad | \quad | \quad R \\ R\quad R \end{array} \qquad (14.162)$$

(ii) The ENE reaction, which require the presence of an allylic hydrogen:

$$\begin{array}{c} R \\ \diagdown \\ CH-CH=C \\ \diagup\quad\quad\quad\diagdown \\ R\quad\quad\quad R \end{array} + {}^1O_2 \longrightarrow \begin{array}{c} \quad\quad\quad OOH \\ R\quad\quad\quad | \quad R \\ \diagdown\quad\quad\diagup \\ C=CH-C \\ \diagup\quad\quad\diagdown \\ R\quad\quad\quad R \end{array} \qquad (14.163)$$

The ENE reaction may occur by one of the following reactions:

(i) Through an initially formed peroxide intermediate, (14.109):[1102, 1986, 1988]

$$-\underset{|}{C}H-CH=\underset{|}{C}- + {}^1O_2 \longrightarrow -\underset{|}{C}H-\overset{O^+O^-}{\underset{|}{C}H-\underset{|}{C}-} \longrightarrow -\underset{|}{C}=CH-\overset{OOH}{\underset{|}{C}-}$$

$$(14.109) \qquad\qquad\qquad\qquad (14.164)$$

(ii) Via the formation of a four-membered ring dioxetane intermediate (14.110):[1102, 1986, 1988]

$$-\underset{|}{C}H-CH=\underset{|}{C}- + {}^1O_2 \longrightarrow -\underset{|}{C}H-\overset{O-O}{\underset{|}{C}H-\underset{|}{C}-} \longrightarrow -\underset{|}{C}=CH-\overset{OOH}{\underset{|}{C}-}$$

$$(14.110) \qquad\qquad\qquad\qquad (14.165)$$

Dioxetane intermediates (14.110) may also decompose with the exclusive formation of carbonyl groups in a chemiluminescence reaction:[2443, 2444]

$$-\underset{|}{C}H-\overset{O-O}{\underset{|}{C}H-\underset{|}{C}-} \longrightarrow -\underset{|}{C}H-\overset{O}{\overset{\|}{C}H} + -\overset{O}{\overset{\|}{C}-} + h\nu \qquad (14.166)$$

(iii) Via the formation of diradicals (14.111):[1492, 1493, 3150]

$$-\underset{|}{C}H-CH=\underset{|}{C}- + {}^1O_2 \longrightarrow -\underset{|}{C}H-\overset{\dot{O}}{\underset{|}{\overset{O}{\underset{|}{\dot{C}H-C-}}}} \longrightarrow -\underset{|}{C}=CH-\overset{OOH}{\underset{|}{C}-}$$

$$(14.111) \qquad\qquad\qquad\qquad (14.167)$$

Reaction of singlet oxygen with olefins is a fast reaction with activation enthalpies below $5\,kcal\,mol^{-1}$ The relative reactivities of simple olefins towards ${}^1O_2$ are in the order: tetra- > tri- > disubstituted, with vinyl or monosubstituted double-bonds being essentially unreactive.[1351, 1352, 1589] Various theoretical models have been applied in order to explain the ENE reaction.[2300, 3438]

Singlet oxygen reactions with polymers have been carefully reviewed.[596,627, 628,1354,1957,2352,2360,3006,3025,3035,3037,3060-3064,3253] It is generally agreed that only polymers which contain unsaturated bonds as elements of their structures (e.g. polydienes) or in the form of abnormal structures (internal impurities, e.g. polyolefins, poly(vinyl chloride) Table 14.14)) may react with singlet oxygen by an ENE reaction.

For example polyisoprene (*14.112*) reaction with $^1O_2$ leads to the formation of three types of hydroperoxides: [516,596,1354,1357,1828,1830,2582,2688,2689, 2691,3135,4007]

$$-CH_2-\underset{\underset{OOH}{|}}{\overset{\overset{CH_3}{|}}{C}}-CH=CH_2- \qquad (14.168)$$

$$-CH_2-\overset{\overset{CH_3}{|}}{C}=CH-CH_2- \ + ^1O_2 \quad\longrightarrow\quad -CH=\overset{\overset{CH_3}{|}}{C}-\underset{\underset{OOH}{|}}{C}H-CH_2- \qquad (14.169)$$

(*14.112*)

$$-CH_2-\overset{\overset{CH_2}{||}}{C}-\underset{\underset{OOH}{|}}{C}H-CH_2- \qquad (14.170)$$

Reaction of $^1O_2$ with unsaturated bonds in polymeric systems has been confirmed by the study of oxidation of model compounds.[623,634, 652,1589,2689,2691,3549-3552]

Depending upon the reaction conditions, $^1O_2$ oxidation of polydienes may be accompanied by extensive degradation and/or crosslinking. [1356,2517,2689, 3025,3030,3036-3039,3046,3946]

It has also been reported that wood,[1647] cellulose,[1024,1029,1222, 1412,1413,1642,1650] and lignin[541,1242,1243] are susceptible towards $^1O_2$ oxidation. Subsequent work[623] using α-amylose which was exposed to $^1O_2$ did not confirm reaction of singlet oxygen with cellulose.

Wool is also partially oxidized with the $^1O_2$.[2418,2692,2693]

Fulvic and humic acids, which are present in human skin, hair, and brain (substantia nigra); in microorganisms and insect dyes such as melanins; in the soil; and in environmental pollution as parts of smoke, may react with singlet oxygen. However, the mechanism of these reactions is complicated and not fully understood.[3368,3369]

The main importance of singlet oxygen oxidation is the formation of hydroperoxide groups. Even formation of a few OOH groups and their further decomposition by light and/or heat can initiate extensive free-radical oxidation (cf. Section 14.1).

Table 14.14 Polymers, copolymers, and polyblends which contain double-bonds in their structure which can react with singlet oxygen ($^1O_2$)

Group	Structure	References
A: Polymers which have unsaturated double bonds in the repeat units that form the backbone structure of a polymer or in pendant groups.		
Polybutadiene	$-CH_2-CH=CH-CH_2-$	354, 516, 628, 1355, 1828, 1830, 1952, 1953, 1955, 2582, 3025, 3030, 3036, 3046, 4007
Polyisoprene	$\begin{array}{c} CH_3 \\ | \\ -CH_2-C=CH-CH_2- \end{array}$	516, 1354, 1357, 1828, 1830, 2582, 2688, 2689, 2691, 3135, 4007
1,2-Poly(1,4-hexadiene)	$\begin{array}{c} -CH_2-CH- \\ | \\ CH_2 \\ | \\ CH \\ \| \\ CH \\ | \\ CH_2 \\ | \\ CH_3 \end{array}$	1354, 1358
1,4-Poly(2,3-dimethyl-1,3-butadiene)	$\begin{array}{c} CH_3 \ CH_3 \\ | \quad | \\ -CH_2-C=C-CH_2- \end{array}$	1354
Polynorbornene	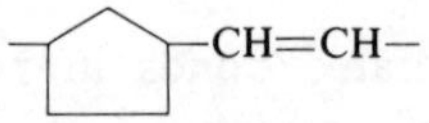 $-CH=CH-$	3945, 3946
Co(ethylene/propylene/ ethylidenenorbornene) (EPDM)	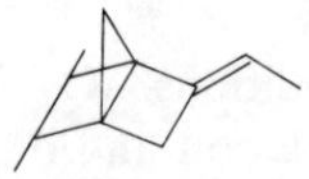	1014, 2741, 2742, 3937
Polyconjugated polymers, e.g. poly(9,10-dianthracene-9,10-diylidine)	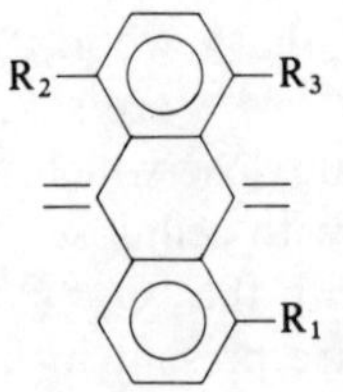 $\ R_1=R_2=R_3=H$ $R_1=R_2=Cl, R_3=H$ $R_1, R_3=Cl, R_2=H$	3225
Co(acrylonitrile/butadiene)		4007, 4010
Co(styrene/butadiene) (SBR)		3039, 4007, 4010
Co(acrylonitrile/butadiene/ styrene) (ABS)		1953, 4010

Table 14.14   (*Contd.*)

Group	Structure	References
Co(styrene/acrylonitrile) (SAN) grafted on polybutadiene		1953
Co(methyl methacrylate/ acrylonitrile/styrene) grafted on polybutadiene		4010
Polybutadiene/poly(vinyl chloride) blends		2200

B: Polymers which have unsaturated bonds in the form of internal impurities

Polyolefins	$-CH_2-CH_2-CH=CH_2$	627, 632, 986, 1954, 2518, 3634
Polystyrenes	$-CH=CH-CH_2-CH-$	580, 986, 3030
Poly(vinyl chloride) (cf. Table 14.5)	$-(CH=CH)_n-CH_2-CH-$ $\hspace{1.5cm}\vert$ $\hspace{1.4cm}Cl$	1332, 2180, 3044, 3047, 3067

14.5. PHOTOINITIATED DEGRADATION

Photodegradation (photooxidative degradation) of polymers (PH) can be photochemically initiated by free radicals formed from the photolysis of a photoinitiator added to a polymer sample:

$$\text{Photoinitiator} + hv \rightarrow \text{free radicals (R·)} \quad \text{(Initiation step)} \quad (14.171)$$

$$PH + R· \rightarrow P· + RH \quad \text{(Propagation step)} \quad (14.172)$$

The propagation and termination steps involved in photooxidative degradation were discussed in Section 14.1.

Most photoinitiated reactions are described in literature as 'photosensitized degradation', but according to the definition of photoinitiators and photosensitizers given in Chapter 7, this nomenclature is wrong. The word 'photosensitized' should only be used for reactions in which the energy transfer process from donor (sensitizer) to acceptor (polymer) occurs without photolysis of a donor into free radicals. In some cases a chemical compound (e.g. benzophenone or dyes) may, depending upon the conditions of the experiment and type of polymer investigated, behave as photoinitiator and/or photosensitizer.

The dye-initiated ('dye-sensitized') photodegradation of fibres is called PHOTOTENDERING in the textile industry.

A number of photoinitiators were employed for the initiation of degradation of many different polymers (Table 14.15).

Table 14.15 Examples of photoinitiators employed for the degradation of different polymers

Photoinitiator	Polymer	References
Metal oxides (ZnO, SnO, CuO, CuO$_2$, Fe$_3$O$_4$, SnO$_2$, etc.)	Natural rubber	179
	Cellulose	1899
Metal oxides (TiO$_2$ and ZnO)	Polyolefins	220, 868, 1337, 1765, 2214
	Polyamides	138, 2853, 3546
	Poly(vinyl chloride)	869, 3365
Metal sulphides (CdS)	Polyethylene	2924
Metal chlorides (LiCl)	Polyamides	2202
Metal chlorides (FeCl$_3$)	Polyolefins	2673, 2762, 2967
	Poly(methyl methacrylate)	2966, 2967
	Poly(vinyl chloride)	1910, 2853
	Poly(ethylene glycols)	3751
	Cellulose	1631, 1632, 1634, 1635, 1639
Tetraethylammoniumtetrachloro-ferrate (C$_2$H$_5$)$_4$NFeCl$_4$	Poly(methyl methacrylate)	2966, 2967
Metal acetylacetonates	Polyethylene	181, 2765, 3250
	Poly(vinyl chloride)	1880, 2802
	Polyurethanes	693, 2804
Metal stearates	Polyolefins	2479, 2805, 2806, 3557, 3651
Metal isopropylxanthates	Poly(phenylene oxide)	672, 673
Metal complexes (ferrocene, copper salicylate, copper phthalocyanine)	Poly(vinyl chloride)	2178, 2180, 2181
Metal complexes (tris-α-thio-picolin-anilide)-cobalt(III)	Polyisobutylene	670
Carbon black	Polyethylene	223
Aliphatic ketones	Polyethylene	1879
	Poly(vinyl chloride)	2359
	Poly(ethylene glycols)	246
	Polyisoprene	3012
Alkyl–aryl ketones	Poly(vinyl chloride)	1509
	Polystyrene	1273
	EPDM	490
Benzophenone	Polyolefins	241, 490, 1505
	EPDM	490
	Poly(vinyl chloride)	869, 1599, 2856, 2857, 2860
	Polystrene	832, 1268, 1273, 2530, 3627, 3951
	Poly(α-methylstyrene)	3951

Table 14.15   (*Contd.*)

Photoinitiator	Polymer	References
	Polydienes	949, 951, 3011, 3018, 3043
	Poly(ethylene glycol adipate)	838
	Polyurethanes	2804
	Cellulose	1631, 1640, 2493
Thiobenzophenone	Polydienes	3008, 3009
Benzoin and its derivatives	EPDM	490
	Poly(vinyl chloride)	2857
Peroxides	Poly(vinyl chloride)	1330
	Polystyrene	832, 2119, 2744
	Polydienes	2198
	Cellulose	1631, 1640, 2493
3,4-Dimethoxybenzoic acid	Poly(methyl isopropenyl ketone)	3648
Azobisisobutyronitrile	Polystyrene	2220, 3841
	Poly(α-methylstyrene)	1752
	Poly(p-methylstyrene)	3854
	Polydienes	3015
	Polydiacetylenes	2593
	Cellulose	1631, 1640
Polycyclic hydrocarbons (naphthalene, anthracene, acenaphthene, pyrene, and phenanthrene)	Polyolefins	238, 612, 627, 632, 2959 3528, 3626, 3657
	Poly(vinyl chloride)	1330, 2860
	Polystyrene	2136, 2738, 2739, 3052
	Poly(vinyl acetate)	4006
	Polydienes	3040
Quinones, anthraquinones	Polyolefins	141, 1504, 3557
	Polystyrene	3034
	Polydienes	3013, 3043
	Polyamides	148
	Polyurethanes	2804
Triphenyl methyl cation	Poly(α-methylstyrene)	1753
1,6-diphenyl-1,3,5-hexatriene and 1,3-diphenyl-1,3-butadiene	Polyethylene	646, 2379
Amines	Polyolefins	3508
	Polydienes	2437
	Cellulose nitrate	713
N-methyl-2-benzoyl-β-naphthazoline	Polyethylene	1912
	Polydienes	3045

Table 14.15   (Contd.)

Photoinitiator	Polymer	References
Hydrazines	Polydienes	2437
Nitrosyl chloride	Polyolefins	2032
Nitrosocompounds	Polydienes	3001
Chloronitrosocompounds	Polydienes	3004, 3005, 3007, 3010, 3014, 3031, 3049
1,2-dibromotetrachloroethane, carbon tetrabromide, 1,1,1,3-tetrabromononane, 1,1,1,3-tetrabromophenylpropane	Poly(styrene)	3188
Luminol/H_2O_2, peroxide/H_2O_2	Lignine	1010
Different dyes	Poly(vinyl alcohol)	1919
	Polydienes	2517, 3038
	Polyamides	147, 148, 1066, 2093
	Cellulose	335, 336, 874, 888, 1023, 1025–1027, 2442
	Biopolymers	381
Oxidized and chlorinated polypropylene	Polyethylene	2794

The photochemistry of most of these photoinitiators has been presented in Chapter 7 and other sources.[3016, 3018, 3020, 3025, 3060]

Careful investigation of photoinitiated (or photosensitized) degradation of polymers is important, because most commercial polymers and plastics materials contain different external and/or internal impurities which may initiate polymer degradation.

It has lately been reported that a number of commercial additives such as pigments, thermal stabilizers, antioxidants, lubricants, and plasticizers used in poly(vinyl chloride) final products may be photolysed under u.v. irradiation to free radicals, which may initiate degradation of a polymer (cf. Section 14.3.3).

14.6. MODIFICATION OF POLYMER SURFACES BY FAR-ULTRAVIOLET RADIATION

The section of the electromagnetic spectrum which extends from 200 to 150 nm is formally defined as far-ultraviolet radiation (cf. Section 1.1). The narrow band which lies between 200 and 180 nm is a practical and highly effective region for the photochemical modification of polymers for the following reasons:[3421–3426]

(i) In this region nearly all organic molecules (with the exception of saturated aliphatic hydrocarbons and fluorocarbons) absorb far u.v.-region photons (typical molar absorptivity $\varepsilon = 0.5$–1.0×10^4 litre mol^{-1} cm^{-1}). As a result this radiation penetrates organic polymers to only ≈ 300 nm before 96% of its intensity is absorbed.

(ii) The radiative lifetimes of the excited states are exceedingly short, typically 0.1 ns. Since fluorescence is rarely observed, lifetimes for bond-breaking must be of the order of 1–10 ps.

(iii) Quantum yields for bond-breaking are of the order of 0.1–1.0, which is 10 to 50-fold larger than in the mid- or near-u.v. regions.

(iv) Since nearly every photon is effective in breaking a bond, the polymer is split up to this depth into smaller fragments (Fig. 14.23). The broken fragments readily recombine unless the presence of air or oxygen traps the free fragments.

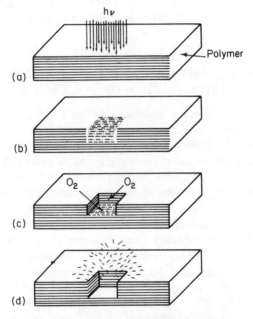

Fig. 14.23   Schematic representation of action of far-ultraviolet radiation on polymer surface[3422]

(v) Overall reaction is faster, and the bulk of the material is unaffected.

(vi) From a practical point of view the 180–200 nm region of the far-ultraviolet is a convenient one, since fused silica freely transmits these wavelengths, water vapour has hardly any absorption, and the oxygen in air absorbs weakly.

Under laser (pulsed) light intensities in the far-u.v. range (185 and 193 nm) photons are absorbed in a thin surface layer and numerous bond breaks occur. The excess energy of the photon above the bond dissociation energy will reside in the fragments as translational, rotational, and vibrational energy. Further, the fragments occupy a larger specific volume than the polymer chain that they replace. The result is that the photoproducts are ejected from the surface of the film or 'ablate'. Hence this phenomenon is called ABLATIVE PHOTODEGRADATION. Material transport out of the irradiated area takes place in ca. 10 μs. As a result very little or no heating of the substrate occurs. The photochemistry for a given pulse of laser light is limited to the depth to which the incident radiation is absorbed. The bulk of the material, i.e. material at a depth of 1 μm or more, is unaffected by the light. Successive pulses continue the process until the polymer in the optical path is photochemically removed. This process is fundamentally different from the decomposition of polymers by visible or infrared radiation from lasers, where the heating effect of the photon is the essential cause of degradation of the polymer[3422, 3424-3426]

The etching of polymer surfaces with laser far-u.v. radiation has been proved with poly(methyl methacrylate), poly(ethylene terephthalate), polycarbonate, and polyimide,[3421, 3422, 3424-3426] and many functional photoresists.[197]

14.7. WEATHERING AND ENVIRONMENTAL DEGRADATION OF POLYMERS

WEATHERING of polymers involves physical, mechanical, and chemical changes in their surfaces and near the surface regions. It may also include the whole polymer sample. These changes are due to the solar energy; moisture (dew, rain, humidity, snow); heat; and atmospheric oxidants (ozone, atomic oxygen, and singlet oxygen); and air pollutants (sulphur dioxide, nitrogen oxides, polycyclic hydrocarbons, etc.). Other factors which may contribute are mechanical factors such as mechanical stress, impact of rain drops or of hail, freezing in winter or gusts of wind loaded with dust, etc.[878,887,1217,3060,3066,3139]

Most importantly, polymers and plastics were tested for their WEATHERABILITY.[887] This term refers to the ability of the plastic to withstand treatment under complex and variable conditions which occur under weathering. Weathering of commercially important polymers such as poly-olefins,[650,812,2201,2864,3755] poly(vinyl chloride),[326,2424,2485,2574-2576,2864] polystyrene,[2864] co(styrene/butadiene)[1008,2789] co(acrylonitrile/butadiene/ styrene) (ABS),[883-885] different polymer blends,[1099,2099] coatings (cf. Section 13.4), and natural polymer wood (cf. Section 14.3.15.1) can be given as examples of such studies.

Air pollutants such as sulphur dioxide, nitrogen dioxide, polycyclic hydrocarbons, and photochemical oxidants such as ozone, atomic oxygen, and singlet oxygen may cause serious deterioration of polymers and plastics exposed to sun radiation. All of these pollutants participate in very complicated photochemical reactions resulting in the formation of smog.[1371,3677] These air pollutants are mainly the result of burning various types of fuels: coal oil, and gas for heating of buildings, generation of electricity and gasoline for transport, etc. The use of gasoline and diesel fuel produces many hydrocarbons, much nitrogen oxide, and almost all the carbon monoxide. Coal and oil combustion in industry and domestic heating produces most of the sulphur dioxide. The role of this pollutant on the photodegradation of polymers has been reviewed in detail.[1217,3060,3066,3243]

Sulphur dioxide, one of the most important air pollutants, may react with polymer oxy radicals (PO·) formed during the thermal decomposition and/or photodecomposition of polymer hydroperoxy groups (POOH) by reactions which lead to the formation of sulphate groups (14.113):[1860,1864,2531,3106]

$$POOH + h\nu \longrightarrow PO· + ·OH \tag{14.173}$$

$$PO· + SO_2 \longrightarrow PO-\overset{\overset{\displaystyle O}{\|}}{\underset{\underset{\displaystyle O}{\|}}{S}}· \tag{14.174}$$

$$PO-\overset{\overset{\displaystyle O}{\|}}{\underset{\underset{\displaystyle O}{\|}}{S}}· + ·OH \longrightarrow PO-\overset{\overset{\displaystyle O}{\|}}{\underset{\underset{\displaystyle O}{\|}}{S}}-OH \tag{14.175}$$

$$(14.113)$$

These reactions are accompanied by chain scission and crosslinking of a polymer.

Polycyclic hydrocarbons may photoinitiate free radical oxidative degradation of polymers (cf. Section 14.5) or participate in the generation of singlet oxygen ($^1O_2$) by an energy transfer mechanism (cf. Section 14.4).

Singlet oxygen ($^1O_2$), which can be produced in complicated photochemical reactions in smog, may react with unsaturated polymers and produce hydroperoxide groups.

Discussion of photochemical activity of other air pollutants and photochemical oxidants such as ozone, atomic oxygen is omitted because of space limitations.

14.8. PHOTOINITIATED CATIONIC DEPOLYMERIZATION

Onium salt photoinitiators such as diaryliodonium ($Ar_2I^+X^-$) or triarylsulphonium ($Ar_3S^+X^-$) salts can be used as photoinitiators of cationic depolymerization of some polymers:[784,785]

(i) Poly(phthalaldehyde) (14.114):[1819]

$$(14.114) \qquad\qquad (14.115) \tag{14.176}$$

The monomer phthalaldehyde (14.115) evaporates. This system has been used in the dry-developable resin technique (cf. Section 12.8.2).

(ii) Polycarbonates (*14.116*):

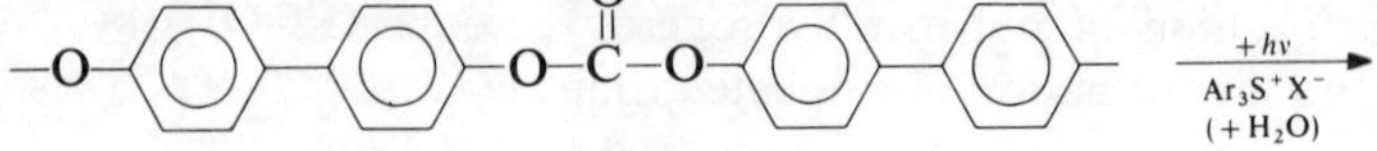

(*14.116*)

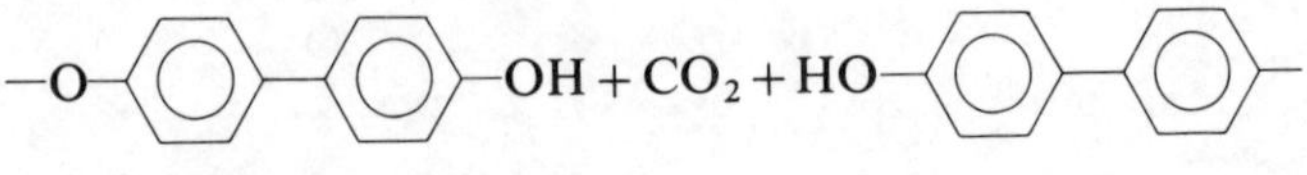

$$-O-\bigcirc-\bigcirc-OH + CO_2 + HO-\bigcirc-\bigcirc-$$

(14.177)

The acid produced from the photodecomposition of onium salts in the presence of water catalyses chain scission of hydrolytically sensitive polycarbonates.

The mechanism of photolysis of onium salts is presented in Section 5.4.3.

14.9. PHOTODEGRADATION THROUGH CHARGE-TRANSFER INTERACTION

The *p*-methylphenol (*p*-cresol) (*4.117*) photosensitizes dehydrochlorination of poly(vinyl chloride) by the formation of an excited charge-transfer complex (exciplex) (*14.118*) (cf. Section 3.5):[1141,1599]

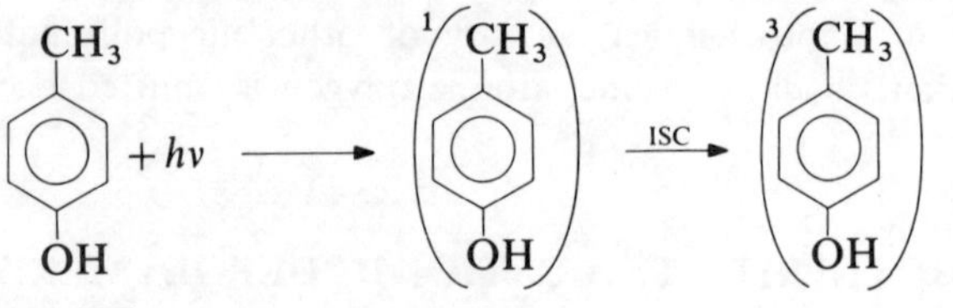

(14.178)

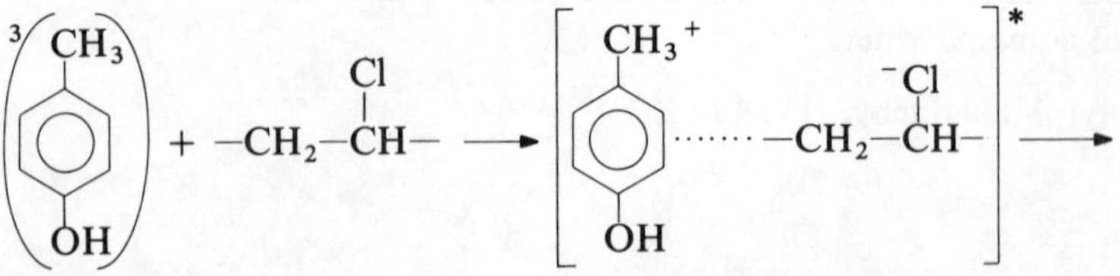

EXCIPLEX
(*14.118*)

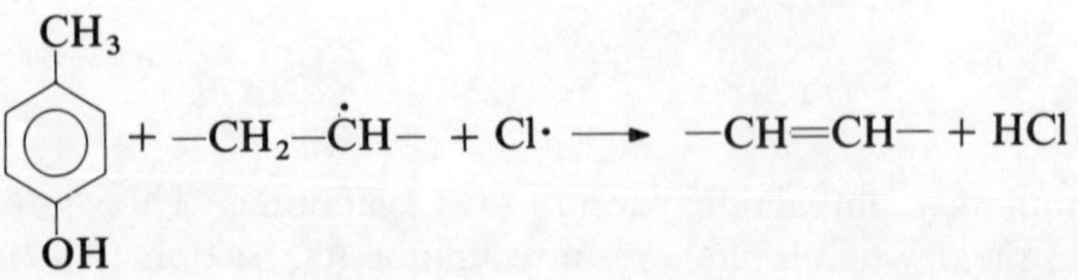

(14.179)

Similarly poly(α-methylstyrene) (*14.119*) can form, with tetracyanobenzene (*14.120*) or pyrromellitic anhydride, a charge-transfer (CT) complex (*14.121*) in polar solvents, which after light irradiation is excited to the exciplex:[1966]

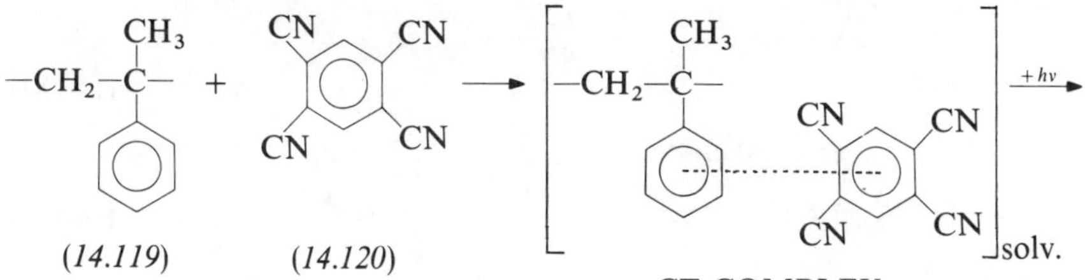

(*14.119*)   (*14.120*)

CT-COMPLEX
(*14.121*)

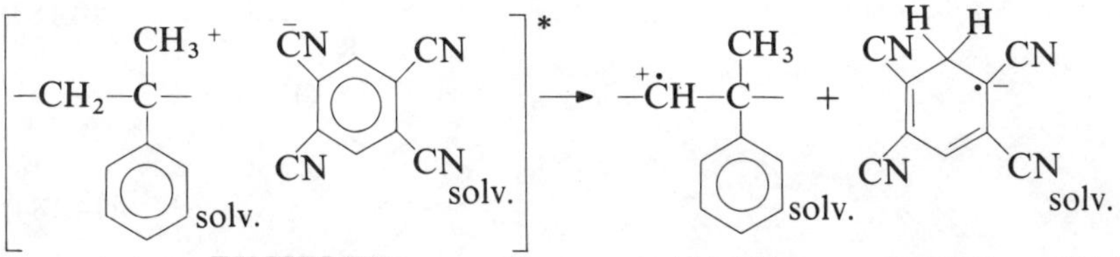

EXCIPLEX   (*14.122*)   (*14.123*)   (14.180)

The photoexcited CT complex (exciplex) dissociates into the solvated radical cation of poly(α-methylstyrene) (*14.122*) and solvated radical anion of tetracyanobenzene (*14.123*).

14.10. PHOTODEGRADABLE POLYMERS

Plastics contribute considerably to the solid waste disposal problem and disposal of the solid plastics waste from homes, offices, and factories is already a major problem in big cities. Solid waste disposal and litter are among the many problems that arise from the relationship between man and his environment. Plastics materials are particularly bad pollutants, due to their ever-growing production and to their high degree of nondegradability. There are several possible processes for the disposal of solid waste, e.g. incineration, pyrolysis, composting, destructive distillation, and recovery of the base elements. The use of plastics materials capable of transforming themselves into products which can reenter the biological life cycle appears to be one of the best solutions to the disposal problem.[3556] The problem of degradable plastics can be partially solved by application of specially designed photodegradable polymers.[755,1432, 1433, 1583, 2584, 3060, 3248, 3249, 3419]

Present attempts to develop plastics with reduced outdoor stability are based on:

(i) Synthesis of polymers with light-sensitive groups, e.g.:
    (a) Aliphatic ketone groups located predominantly in the main chain or in pendant (side) groups.[36,180,1513,1580,1692,2085,2964,3480] The primary

reactions of these carbonyl groups include the Norrish Type I and/or Norrish Type II processes:

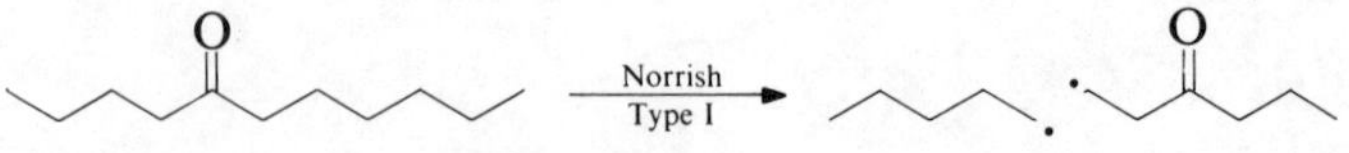

(14.181)

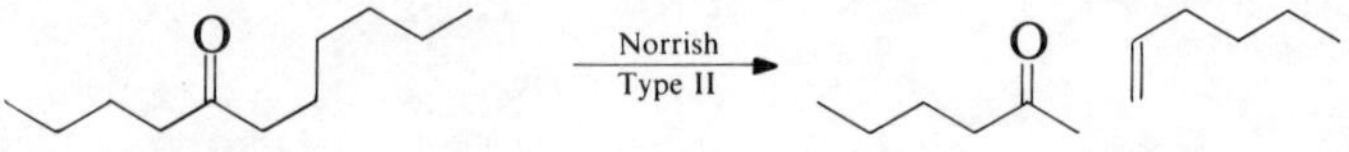

(14.182)

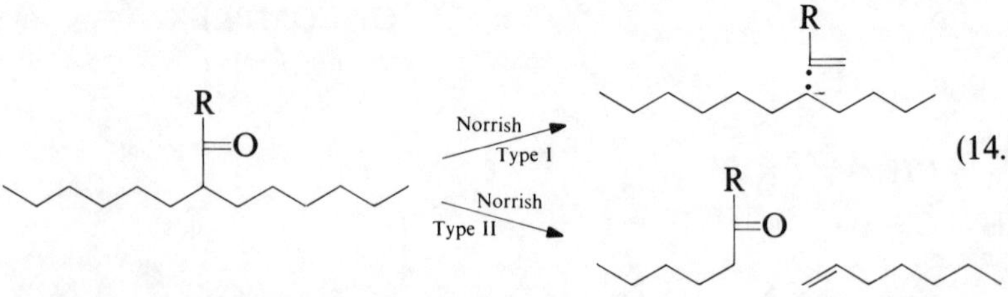

(14.183)

(14.184)

The Norrish Type I reaction can result in crosslinking, but the Norrish Type II reactions always result in polymer chain cleavage. For a Norrish Type II reaction to occur, a hydrogen located at the α-position to the keto group is necessary.

Polymers with light-sensitive carbonyl groups can be synthesized, for example by polymerization of carbon monoxide with vinyl ketone[180] or vinyl aldehyde,[2085] or copolymerization of styrene with benzalacetophenone[1692] or methyl methacrylate (or vinyl acetate) with methyl isopropyl ketone.[36] A list of available ketone polymers and copolymers is given in Table 14.9.

Ketone polymers have been developed on an industrial scale by Eco Plastics Ltd,[1433, 1583, 3419] and carbon monoxide copolymers by Du Pont, Eastman Kodak, and Ethylene Plastique.[1583]

(c) o-Nitrobenzaldehyde acetal groups.[2931] This photopolymer (14.124) under u.v. light is photorearranged to glycols (14.125) and o-nitrosobenzoate ester derivatives (14.126) by a multistep degradation process which occurs by the following mechanism (reactions 14.185 and 14.186):

(d) Oxime (14.127)[2986] or phenacylamine (14.128)[1696,1733] groups in the main chain:

(e) Diaryliodonium group in the pendant (14.129) or in the main (14.130) chain of a macromolecule:[785,796,3949,3950]

Polymers containing diaryliodonium salt groups have absorption below 300 nm and are rapidly photolysed when u.v.-irradiated according to a mechanism similar to that described for the photoinitiation of cationic/radical polymerization (cf. Section 5.4.6).

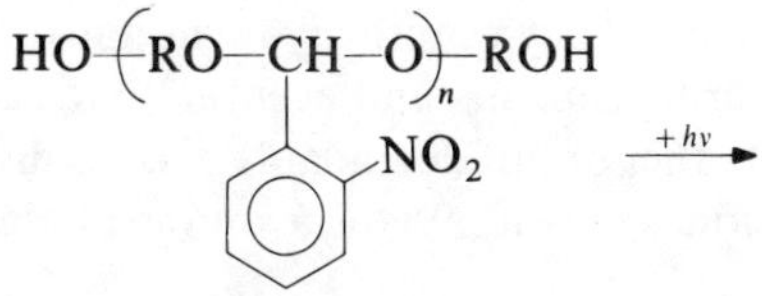

$$HO-(RO-CH-O)_n-ROH$$

at the CH position: NO_2 substituent on benzene ring

(14.124)

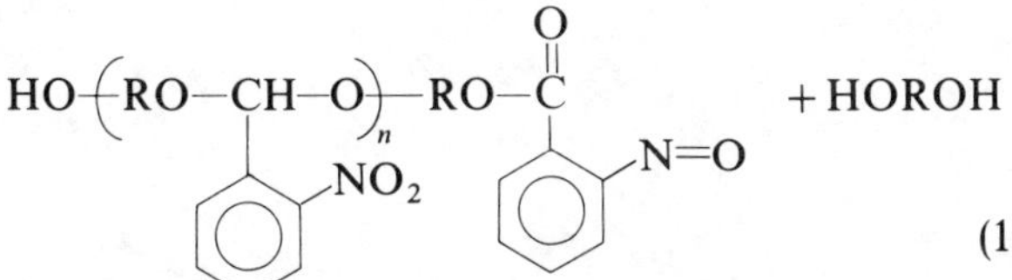

$$\xrightarrow{+h\nu}$$

$$HO-(RO-CH-O)_n-RO-\overset{O}{\underset{\|}{C}} \quad + HOROH$$

with NO_2 on one ring and $N=O$ on the carbonyl-bearing ring

(14.185)

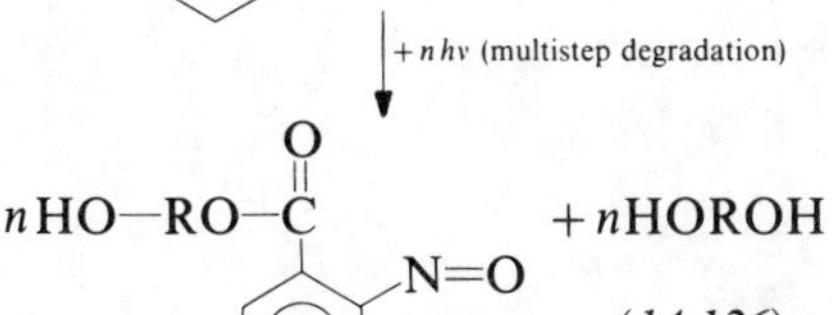

$$\downarrow +nh\nu \text{ (multistep degradation)}$$

$$n\,HO-RO-\overset{O}{\underset{\|}{C}} \quad + n\,HOROH$$

with $N=O$ substituent on ring

(14.126)　(14.186)

(14.125)

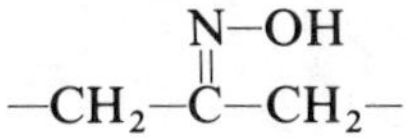

$$-CH_2-\overset{N-OH}{\underset{\|}{C}}-CH_2-$$

(14.127)

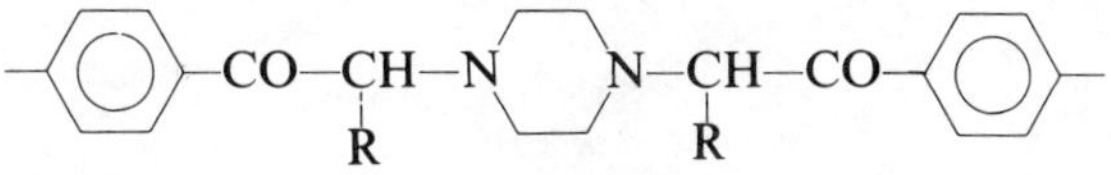

$$-\bigcirc-CO-\underset{R}{CH}-N\bigcirc N-\underset{R}{CH}-CO-\bigcirc-$$

(14.128)

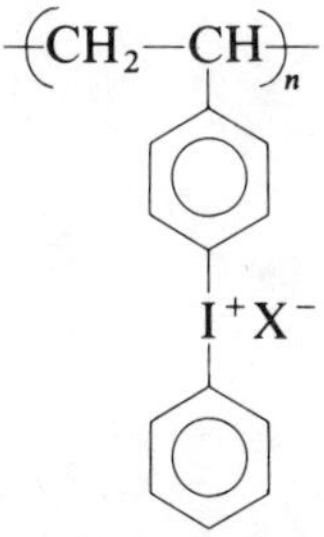

$$+(CH_2-CH)_n$$

with benzene ring bearing I^+X^- and phenyl

(14.129)

$$+\left(\bigcirc-\overset{X^-}{\underset{}{I^+}}-\bigcirc-(CH_2)_3-\bigcirc\right)_n$$

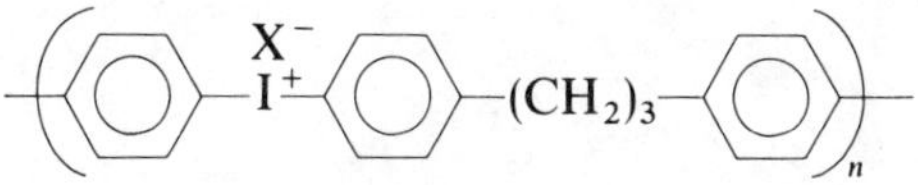

(14.130)

(ii) Modification of polymers by addition of another polymer. For example polyethylene film containing radiation-modified atactic polypropylene[2793] a more light-sensitive polymer, was used as a photodegradable plastic for agriculture.[2792,3159]

(iii) Application of photoinitiators (or photosensitizers) which, added to polymers or plastics, photoinitiate (or photosensitize) their degradation.[1111, 3060, 3244] A number of photoinitiators are listed in Table 14.15.

Such photodegradable plastics, under sun irradiation, lose their physical and mechanical properties, become brittle, and are finally broken down into small particles by natural erosion, rain, wind, etc. They eventually form powder which passes into the soil, is attacked by microorganisms and eventually reenters the biocycle. A controversial question is whether or not degradable plastics disappear without leaving intermediates and residues which may be a source of air, water, and soil pollution.

15. Photostabilization of polymers

The PHOTOSTABILIZATION of polymers involves the retardation or elimination of photochemical process in polymers and plastics.

ULTRAVIOLET STABILIZERS are additives to plastics and other polymer materials which prevent the photodegradation caused by ultraviolet light, present in sunlight and various kinds of artificial light sources. The use outdoors of plastics, fibres, and elastomers is possible owing to the wide application of different ultraviolet stabilizers.

A number of books[43, 54, 1260, 1534, 1535, 1857, 2128, 2187, 2468, 3060, 3218, 3761] and reviews[172, 458, 614, 684, 1347, 1348, 1426, 1430, 1431, 1470, 1494, 1562, 1563, 1583, 1602, 1993, 2064, 2800, 3196, 3218, 3247, 3250, 3332, 3881, 3884] discuss photostabilization mechanisms and photostabilizer properties.

The practical problem in the study of photostabilization of polymers is a voluminous patent literature on these products, whereas a highly fragmented and often contradictory scientific literature has attempted to answer the mechanistic questions. Although various researchers have studied narrow aspects of photodegradation or photostabilization, a clear understanding of these reactions requires knowledge of the photochemistry and photophysics of photodegradation and photostabilization, and knowledge of the nature and concentration of all important species involved in stabilization of the solid polymer through its photooxidative lifetime.

Most ultraviolet stabilizers are easily incorporated into plastics. They can be added on a hot mill or in a Banbury. The lower-melting absorbers are especially easy to disperse in this manner.

In compositions that contain plasticizers, dissolving the ultraviolet stabilizer is often a more convenient method of addition, and ensures complete dispersion of the absorber.

Other methods of addition include distributing or dry-blending the absorber on the resin powder or granules before processing. In a few cases, ultraviolet stabilizers may be applied to finished articles from a non-aqueous solution or from an emulsion.

Ultraviolet stabilizers are more effective in clear, transparent plastics than in opaque plastics. Since light penetrates only a short distance into opaque plastics, any degradation by ultraviolet radiation is necessarily close to the surface. For an absorber to provide the best protection, therefore, its highest concentration should be in this surface volume of light penetration.

The photostabilizing action of light stabilizers depends greatly on the prior processing history.[54,97,657]

Stabilizer mobility and compatibility also play a role in determining stabilizer effectiveness.

Stabilizers should also be resistant to:

(i) Thermal and chemical decomposition.
(ii) Photolysis and photooxidation processes (attack by different free radicals, such as R·, RO·, ROO·, HO·, etc.).
(iii) Reaction with atomic oxygen (O), singlet oxygen ($^1O_2$) and ozone (O_3).

The amount of an absorber required to provide economical protection in a plastic is governed by several factor such as:

(i) Thickness of the plastic.
(ii) Tolerance of colour.
(iii) Added cost of ultraviolet absorber.
(iv) Effect of high concentrations of absorber in plastics.
(v) Compatibility of the absorber in the plastic.
(vi) Ultraviolet–visible light sensitivity of the polymer towards degradation.

Optimum protection with minimum side-effects must always be determined for each individual application. The hydroxybenzophenones, for example, have good heat stability and fair to very good compatibility in most plastics, whereas in amine-type plastics (e.g. nylon) benzophenones become yellower in colour.

In general the concentration of stabilizer added is a function of the thickness of the plastic. The thinner the plastic, the higher will be the concentration of ultraviolet absorber necessary for adequate protection. The effect of light is greatest on the surface of a plastic. In a thin film the surface layer is a much larger fraction of the total film than it is for a thick film. Therefore it plays a more important role in the total properties of the thin film.

Any type of additive (photostabilizer, antioxidant, thermal stabilizer, etc.) must be evenly distributed, which requires that it be compatible with the polymer.[684] Stabilizers are usually more compatible with the amorphous part of a polymer, and are substantially excluded from the spherulites or dendrides of semi-crystalline polymers.[592] Furthermore, oxygen diffusion is higher in the amorphous region. It is clear that much of the photooxidative reaction and stabilization occurs in the amorphous polymer region, and high stabilizer concentrations are usually present in this phase. Although an overall amount of about 1 wt% of photostabilizer is used, this is often concentrated to 10–40% in the amorphous portion of the polymer.[3615]

A significant amount of stabilizers can be lost from plastics due to exudation, volatilization, solvent extraction during fabrication, and their long-term use. The problem is most severe for articles with a high surface area to volume ratio, such as fibres and films. For instance, during polypropylene fibre formation, stabilizers are evaporated around the extrusion die, leading to a decrease in the real content in the polymer.

Formation of a thin surface of the stabilizer on the surface of the polymer would improve the effectiveness of an additive,[1537,2349] but abrasion of the polymer, or washing, lead to removal of the protective layer and thus to a reduction in the effectiveness of incompatible stabilizers.

Washing with water,[2348] or dry cleaning in organic solvents,[2349] decreases the content of low molecular weight stabilizers in the polymer.

Extraction of polypropylene films stabilized with 2-hydroxy-4-octyloxybenzophenone removed 20 and 70% with perchloroethylene and methanol, respectively.[1107] Perchloroethylene cleaning (dry cleaning) of polyester and polyamide fibres stabilized with 2,2'-dihydroxy-4,4'-dimethoxybenzophenone extracted as much as 20% of the stabilizer from the fibres.[3457]

As the molecular weight increases, the volatility of stabilizers decreases during storage,[1013] thermal treatment at 60–120°C,[1013] or at processing temperatures (280°C).[274] For example 2-hydroxy-4-alkoxybenzophenones with long alkoxy side-chains are more compatible with polyethylene and more effective than stabilizers with short side-chains.[1843]

The bonding of stabilizers into a chain of the polymer, or the use of higher molecular weight (oligomeric or polymeric) stabilizers (cf. Section 15.9) can solve the problem of the loss of stabilizers in the polymer, but will not protect against their reduction in concentration by radical reactions. Therefore the use of higher molecular weight stabilizers, where the molecular weight prevents volatilization from the polymer and reduces migration to the polymer surface, is specially suitable for semicrystalline polymers. In addition, the decrease in stabilizer content may result from migration of stabilizers to the polymer surface due to incompatibility.[1013] The volatility of stabilizers has been studied elsewhere.[1612, 2961] Additives migration in polymers has been successively determined by fluorescence techniques.[2036,3796]

All effective u.v. stabilizers are progressively destroyed during long-term use (5–10 years).[615] Several u.v. stabilizers are almost completely destroyed well before the end of the photooxidative lifetime of the polymer. For example polyethylene containing 1 wt% of 4-dodecyloxy-2-hydroxybenzophenone showed no exudation or surface cracking after 5 years, while plates containing 1 wt% of 2-hydroxy-4-methoxybenzophenone showed exudation and cracked after 2 years of weathering.[684] Most photostabilizers are photostable in polymer film in the absence of oxygen, but are destroyed by oxidative processes.

Ultraviolet stabilizers are destroyed in polymer matrix by the following processes:

(i) Photolysis and/or thermolysis.

(ii) Reactions with low molecular radicals and macroradicals formed during photooxidative degradation of polymers, e.g. peroxy radicals (ROO·).[615, 1603]

(iii) Photografting to polymer, e.g. 2-hydroxy-4-octyloxybenzophenone has been found to be photografted to polypropylene.[3756]

(iv) Reactions with singlet oxygen, e.g. phenolic stabilizers.[633,2431,3175]

574

Effectiveness of photostabilizers (and antioxidants) depends on physical factors such as:[262,614]

  (i) Volatility (evaporation).[455,1538,2682,2961]
 (ii) Diffusivity (mobility).[1843]
(iii) Compatibility.[1843]
 (iv) Solubility.[3457]
  (v) Dispersion in the amorphous phase.
 (vi) Morphology of polymer (amorphous and/or crystalline phase).
(vii) Orientation state.
(viii) Surface-to-volume ratio.
 (ix) Synergistic/antagonistic effects with other additives.[56,3060]

When the observed effect of a mixture of two or more additives (e.g. photostabilizer and antioxidant) is not additive, but greater than would be expected from the additivity law, the phenomenon of SYNERGISM is observed. In the opposite case, when the observed effect is weaker, ANTAGONISM between the components of the mixture is said to occur (Fig. 15.1). The mechanism of synergistic/antagonistic effects is not well known. Several mechanisms have been proposed, which are different for different components mixed together.

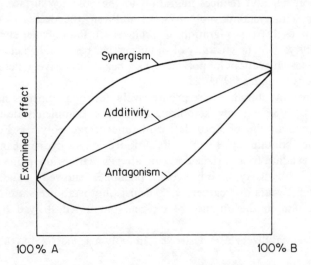

Fig. 15.1 Synergism and antagonism effects

Ultraviolet stabilizers can be classified according to their ability to act as:

(i) Absorbers to reduce the number of photons absorbed by internal and external chromophores present in a polymer (u.v. absorbers and light screeners).

(ii) Compounds which may deactivate excited states (singlet and/or triplet) of chromophoric groups present in a polymer (quenchers).

(iii) Compounds which can decompose hydroperoxide groups before they are photolysed by absorbed photons (decomposers).

(iv) Compounds which can react with free radicals and thus interrupt degradative chain processes (scavengers).

(v) Compounds which can react or deactivate singlet oxygen ($^1O_2$) (singlet oxygen quenchers).

Ultraviolet stabilizers that are available commercially can be classified by chemical structure:

(i) Pigments (cf. Section 15.1).
(ii) Chelates (cf. Section 15.2).
(iii) Salicylates (cf. Section 15.3).
(iv) Salicylanilides (cf. Section 15.4).
(v) Hydroxybenzophenones (cf. Section 15.5).
(vi) Hydroxyphenylbenzotriazoles (cf. Section 15.6).
(vii) Methal xanthates (cf. Section 15.7).

15.1. LIGHT SCREENERS

LIGHT SCREENERS reduce the amount of u.v. radiation reaching the chromophore groups in polymers by the following mechanisms:

(i) Radiation reflection and/or scattering.
(ii) Radiation absorption.

Reflective or opaque pigment particles belong to the light screeners acting as light shields. Not all polymers are protected by the same degree by a given pigment, because of differences in the absorption spectra of pigments and polymers. Other important factors are the dispersion of pigments and their effect on physical properties such as electrical properties, which are affected by the addition of pigments. The chemical nature of pigment is also very important; for instance iron oxide protects polyolefins but rapidly catalyses the decomposition of poly(vinyl chloride) on exposure to u.v. radiation.

15.1.1. Photochemistry of Pigments

PIGMENTS are insoluble inorganic or mineral and organic compounds of complex structure (Table 15.1), which, used as additives, are incorporated into polymers, coating, inks, etc. for:

(i) Cost reduction.
(ii) Reinforcement.
(iii) Hardening.
(iv) Improving slip and skid resistance.

Table 15.1 Examples of inorganic and organic pigments[2053]

Inorganic pigments

Titanium dioxide	Ultramarine blue
Iron oxides	Chromium oxide
Cadmium yellow	Molybdate yellow
Cadmium red	

Organic pigments

Name	Structure	Commercial name
Carbon black		
Phthalocyanine blue	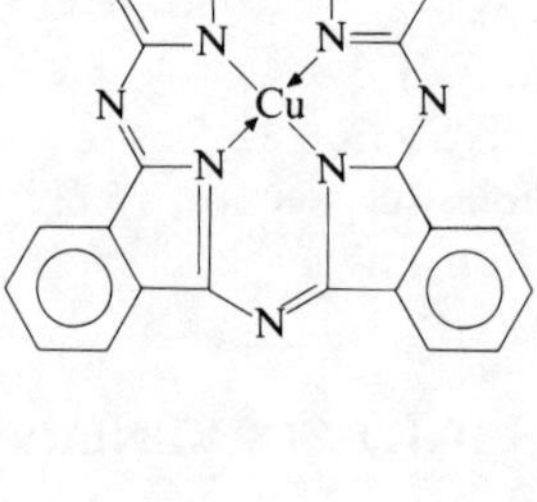	Pigment Blue 15:3
Dioxazine	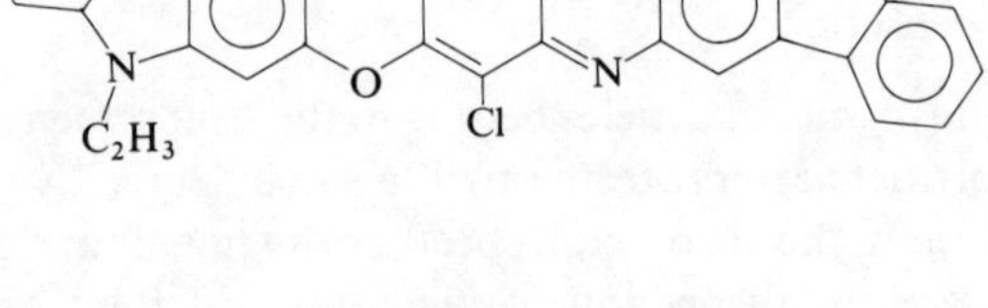	
Tetrachloroisoindolines	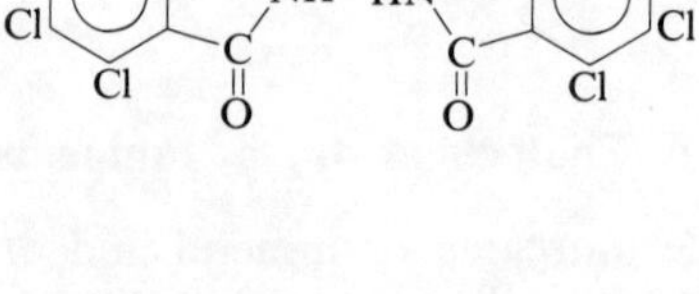	Pigment Yellow 109, 110
Perylene	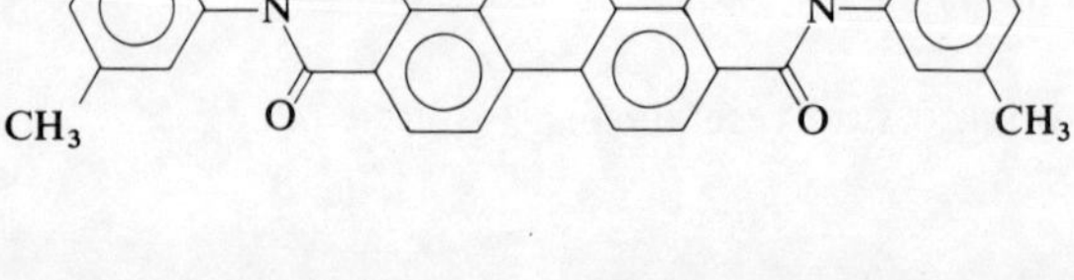	Pigment Red 149

Table 15.1 (*Contd.*)

Flavanthron

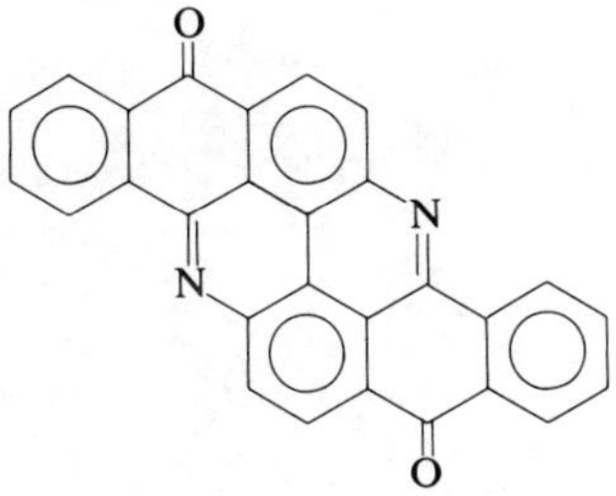

Thioindigo

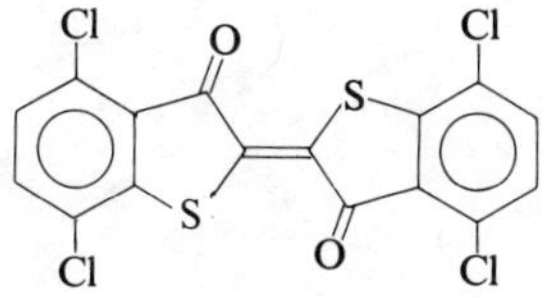

Pigment Red 88

Azo condensation

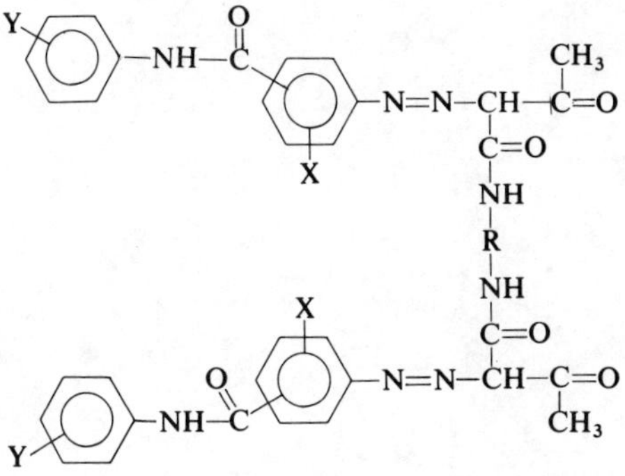

Pigment Yellow 93, 94, 95

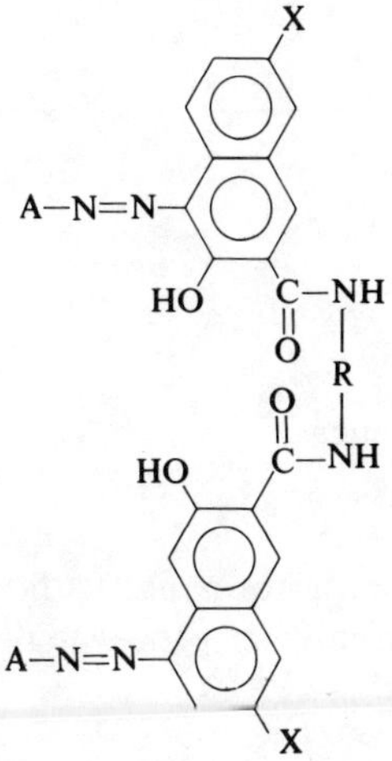

Pigment Orange 31
Pigment Red 144, 166

Table 15.1 (*Contd.*)

Diarylide   Pigment Yellow 83

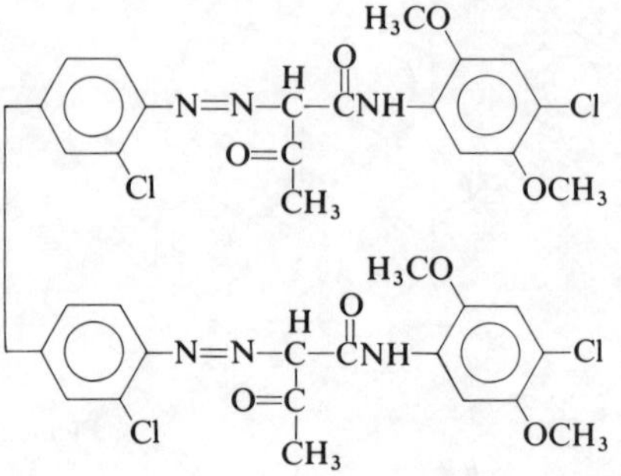

Quinacridone

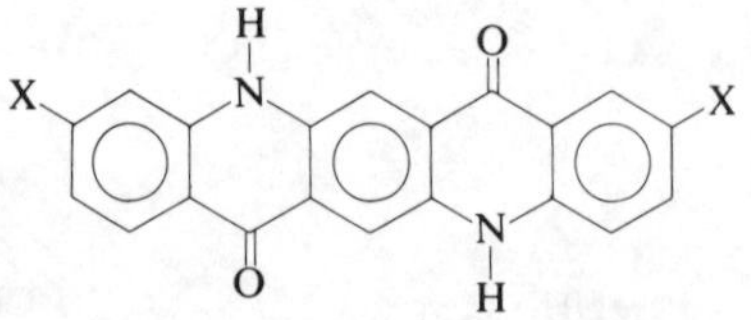

Quinophthalone

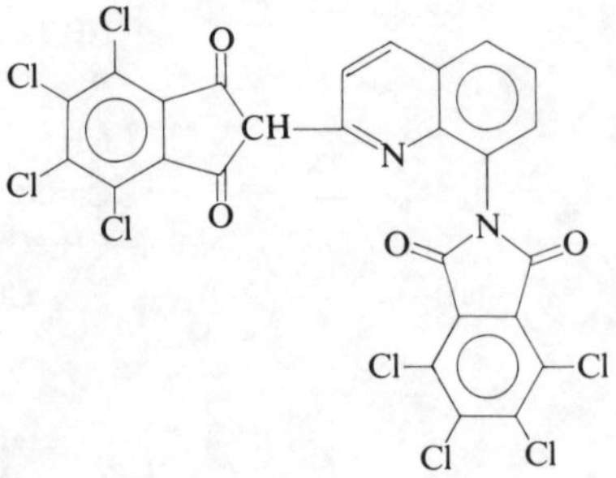

  (v) Flow properties.
 (vi) Colour effects.
 (vii) Storage stability.
(viii) Gloss and flatting efficiency.

Pigments should be:

  (i) Light-stable for long-term performance without fading.
 (ii) Heat-stable to withstand polymer processing conditions.
(iii) Migration-resistant.
(iv) Cheap and non-toxic.

Light-absorbing properties, photochemical behaviour, and the nature of pigment fading, determine the practical application of pigments in polymers.[55,123,124,2053] Pigments may also influence polymer morphology because of its incompatibility with polymer matrix.

The influence of pigments in polymer photostability is not completely understood. If an absorbing pigment is introduced into a polymer, it acts as an inner screen for the photoproducts. If these products are not photooxidized, they accumulate in the polymer matrix. Since pigments act as highly absorbing additives, photooxidative phenomena are limited mainly to the surface of samples.

Inorganic and organic pigments have some favourable or adverse effects on polymer photostability.[55,159,2053,3430,3725] Phthalocyanine blues and greens have been found to exhibit good photoprotection,[55,74,159,2053,3725] whereas yellow, orange, and red organic pigments and dyes have been shown to have deleterious effects.[55,2053,3430]

The effectiveness of pigment on light stability of low-density polyethylene was:[2541,2559]

Chrome green > phthalocyanine green > ultramarine blue > cadmium yellow
In polypropylene the order of effectiveness was different:[3430]
Rutile (TiO_2) > cadmium red > cadmium yellow > phthalocyanine blue > iron oxide red.

Studies on the photochemical activity of pigments in commercial polymers have been mainly concerned with titanium dioxide (TiO_2) (cf. Section 15.1.1.2).

Some pigments have negative influence on polymer light stability, such as:

 (i) Photocatalytic effect on polymer degradation (e.g. TiO_2).
 (ii) Sensitization of singlet oxygen formation, which can react with unsaturated polymers and form hydroperoxide groups.
(iii) Selective absorption of other stabilizing additives (e.g. carbon black).
(iv) Chemical interaction with stabilizers (e.g. carbon black).
 (v) Formation of polymer-harmful products from light fading.
(vi) Semiconductor phenomena.

15.1.1.1. Carbon black

The classification of carbon black of various types has been based traditionally on the manufacturing process. The following classes of carbon black are present, distinguished according to the range of average particle size expressed in Å:

 (i) Furnace black (170–700 Å).
 (ii) Impingement black (100–270 Å).
(iii) Thermal black (1500–5000 Å).
(iv) Lamp black (500–900 Å).
 (v) Acetylene black (350–500 Å).

The chemistry of carbon black is highly irregular. Particles are electrostatically aggregated. Aggregate size is dependent upon the wetting property of the carbon in the solvent system.[997]

The surface of carbon black contains various functional groups primarily originated by oxidation. This largely arises from incomplete pyrolysis of petroleum feed, as well as natural oxidation of carbon black. Figure 15.2 shows a model of carbon black with different functional groups such as carbonyl, hydroxyl, quinone, ether, etc. Carbonyl functionality in the form of quinone groups on the surface of carbon black was established by chemical and spectroscopic techniques.[1481,3469,3470]

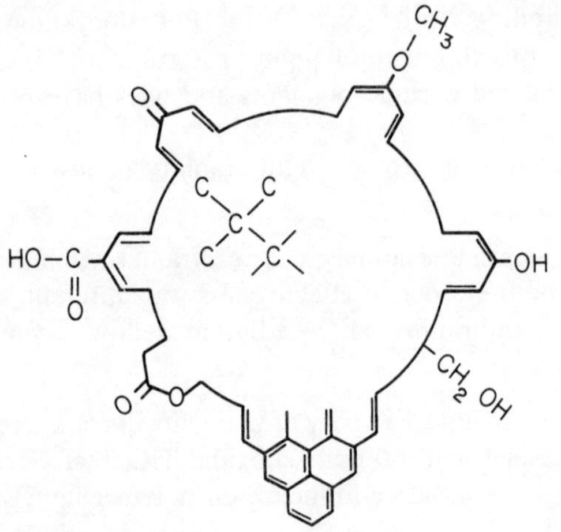

Fig. 15.2 Carbon black model illustrating possible functional groups[678]

Surface quinone groups of carbon black (15.1) can participate in the deactivation of free radicals, e.g. ketyl radicals (15.2) (or monomer radicals) in the following mechanism:[678]

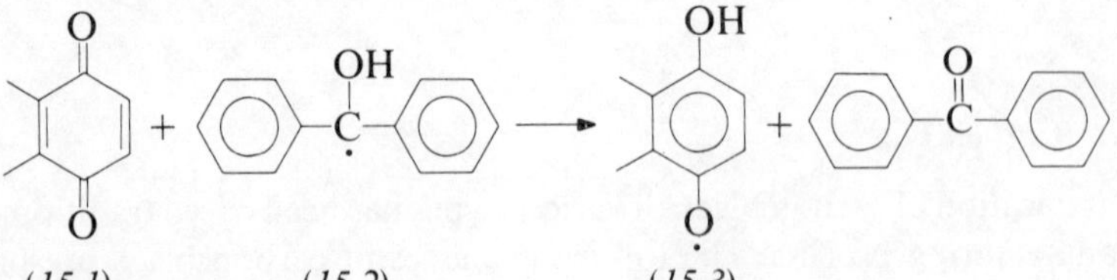

$$(15.1)$$

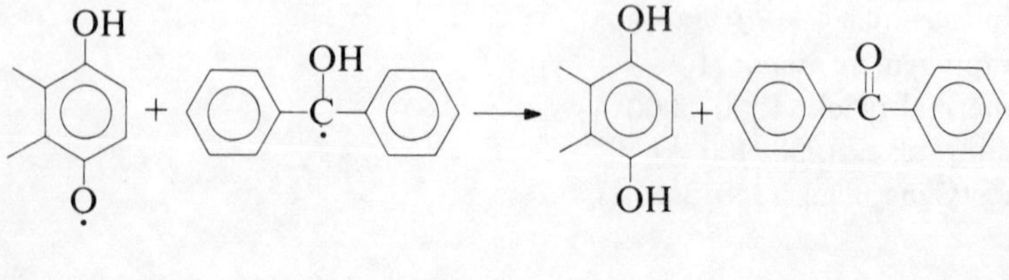

$$(15.2)$$

$$\text{(structure with OH groups)} + O_2 \longrightarrow \text{(structure with O groups)} + H_2O_2 \qquad (15.3)$$

The semiquinone radical (15.3) is relatively stabilized by the adjacent aromaticity present in carbon black.

The behaviour of carbon black in the presence of polymerizable monomers is extremely complex. In solution polymerization, as well as in bulk polymerization, carbon black has been shown to accelerate,[1002, 2477, 2772] as well as retard,[1002, 2477, 2586, 2770, 2771] free radical polymerization. This trend is dependent upon the initiator of polymerization, type of polymerization, and the solvent system used in the reaction. This problem also exists in the production of u.v. cure inks.[325, 678, 1221]

Introduction of small quantities (< 1 wt %) of carbon black in branched polyethylene induces a photothermal degradation at temperatures $> 70°C$.[223]

Carbon black absorbs u.v. radiation more efficiently than conventional colour pigments, and is one of the most efficient light stabilizers for polymers, e.g. polyethylene.[3924, 3926]

15.1.1.2. Titanium dioxide

Titanium dioxide (TiO_2) exists in two morphological crystalline forms—anatase and rutile—which exhibit different photocatalytic effects on polymers.[100, 104, 107, 111, 117, 124, 221, 222, 1115, 1116, 1129, 2053, 2249, 3485] Anatase and rutile have different absorption spectra (Fig. 15.3). TiO_2 is an extrinsic semiconductor of the n-type and the distances between the valence and the conduction

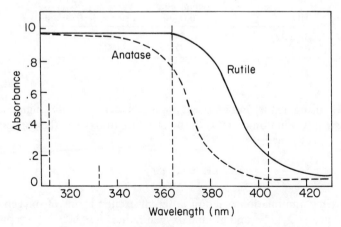

Fig. 15.3 Ultraviolet absorption spectra of: (---) anatase and (——) rutile[2901]

band are: anatase 3.29 eV (385 nm), rutile 3.05 eV (385 nm). With the absorption of a quantum of higher energy than 3.3 eV or 3.05 eV, an electron is lifted out of the valence band and into the conduction band, thereby forming an EXCITON (Fig. 15.4).[3766] The fact that rutile is more stable is explained by the more rigid bond of its surface hydroxyl groups.[456] Although rutile more readily forms excitons, the positive hole reacts slower with the hydroxyl anions (OH$^-$).

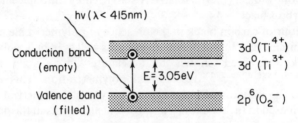

Fig. 15.4   Schematic of energy band model of the rutile form
of TiO$_2$[3766]

A measure of the energy of the electron/hole pair may be derived from emission studies, since the wavelength of emission corresponds to the energy released on electron/hole recombination. Emission studies have been reported on: TiO$_2$ molecules in a neon matrix at 4K, single crystal rutile at 77K, and polycrystalline anatase and rutile powders at 77K (Table 15.2). These results indicate that the available energy of the electron/hole pairs of photoexcited anatase is substantially greater than that of rutile.[3877]

Table 15.2   Emission data from TiO$_2$[3877]

TiO$_2$	Modification temperature (kelvin)	Emission max (nm)	$E_{emission}$ (kcal/mol)
Anatase	77	530	54
Rutile	77	850	34

Manufacturing history has an effect on luminescence of TiO$_2$ pigments.[67,116] During u.v. irradiation of TiO$_2$ in the presence of oxygen, the oxygen radical anion (O$_2$$^-$) is formed:[104,107,2890,3485,3764]

$$TiO_2 + O_2 \xrightarrow{+h\nu} TiO_2{}^+ \cdot + O_2{}^- \cdot \qquad (15.4)$$

In a reversible annihilation reaction different excited forms of oxygen such as singlet oxygen ($^1O_2$),[2887] and atomic oxygen radical (O·),[994,1577] are produced:

$$TiO_2^+ \cdot + O_2^- \cdot \rightarrow TiO_2 + {}^1O_2 \qquad (15.5)$$

On the TiO_2 surface, in the presence of water, hydroxyl groups $Ti^{4+}OH^-$ are formed (Fig. 15.5).[3766] Under light irradiation with energy higher than 3.0 eV the following reactions may occur:

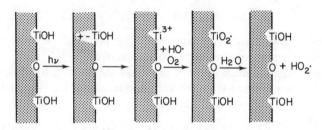

Fig. 15.5 Surface reactions on TiO_2 generate deleterious free radicals[3766]

(i) Formation of an exciton (electron (e^-)/hole (p^+) pair):

$$Ti^+OH^- + h\nu \rightarrow e^- + p^+ \qquad (15.6)$$

(ii) The exciton reacts further instantaneously with a surface hydroxide ion, and a Ti^{4+} ion of the lattice forms a hydroxyl radical (HO·) and Ti^{3+} ion:[63,431,456,807,2890,3766,3767]

$$p^+ + Ti^{4+}OH^- \rightarrow Ti^{4+} + HO\cdot \qquad (15.7)$$

$$e^- + Ti^{4+} \rightarrow Ti^{3+} \qquad (15.8)$$

ESR spectroscopy studies have demonstrated formation of hydroxyl radicals on the surface of TiO_2 (Fig. 15.6).[2890,3766]

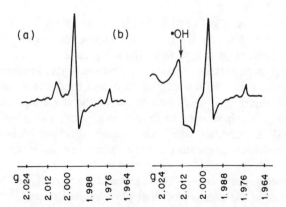

Fig. 15.6 ESR signal of hydroxyl radical (HO·) observed during u.v. irradiation of anatase/water: (a) before irradiation, (b) after irradiation[3766]

Irradiation of a suspension of atanase in normal primary alcohols (methanol, ethanol, or n-butanol) in nitrogen atmosphere, led to the formation of Ti^{3+} ions and corresponding aldehyde in solution:[577,2377] The hydroxyl radicals formed are responsible for the reaction:

$$2HO\cdot + RCH_2OH \rightarrow RCHO + 2H_2O \tag{15.9}$$

During irradiation of isopropanol in the presence of TiO_2 acetone is formed by a reaction analogous to that above:[1765,1767]

(iii) The next step is the addition of an atmospheric oxygen molecule, which takes over the electron from the Ti^{3+} and turns into the superoxide radical ion ($O_2^-\cdot$):

$$Ti^{3+} + O_2 \rightarrow Ti^{4+} + O_2^-\cdot \tag{15.10}$$

(iv) This is followed by the important reaction in which water is consumed. The water reacts in the form of its dissociation products:

$$O_2^-\cdot + H^+ \rightarrow HO_2\cdot \tag{15.11}$$

$$Ti^{4+} + OH^- \rightarrow Ti^{4+}OH^- \tag{15.12}$$

As a result of this reaction hydroperoxy radicals ($HO_2\cdot$) are formed. Hydroxyl ions (HO^-) remain on the TiO_2 surface to participate again in the cycle.

The reactive hydroxyl ($HO\cdot$) and hydroperoxy ($HO_2\cdot$) radicals may further react with polymer and initiate its degradation. For example it has been reported that the effect of moisture on photodegradation of Nylon 66 is much greater in the presence than in the absence of TiO_2.[2312]

This effect is particularly important in destruction of coatings pigmented with TiO_2 and known as CHALKING, during which the polymer surface gradually erodes away.[3764,3766,3767,4002] Similar photocatalytic oxidation can be caused by zinc oxide (ZnO).

According to the ASTM D 658 definition, CHALKING is the phenomenon manifested in paint films by the presence of loose removable powder, evolved from the film itself, at or just beneath the surface.

With anatase pigments destruction by photocatalytic oxidation cycle takes place more rapidly than by u.v. degradation, and therefore governs the overall picture of chalking (Fig. 15.7). After weathering the anatase particles remain on the paint surface in what might be described as 'pits'. Rutile pigment particles therefore stand on 'pedestals' which have remained behind because the pigment particles cast a shadow preventing the u.v. radiation from reaching the paint in these regions.[3766]

The difference in photoactivity between anatase and rutile has been attributed to the difference in the energies of their photoexcited states.[138,140,144,150] The anatase form of TiO_2 has a photocatalytic effect on the degradation of such polymers as polyolefins,[63,64,66,100,138,140,150,868,990,1767,2193,2250] polyamides,[111,138,144,1030,2250,2311,2559,3103,3547,3548,3555,4002] and poly(vinyl

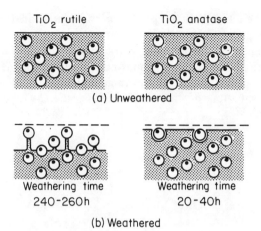

TiO$_2$ rutile   TiO$_2$ anatase

(a) Unweathered

Weathering time   Weathering time
240-260h   20-40h

(b) Weathered

Fig. 15.7 Schematic of the degradation processes during the weathering of binders pigmented with TiO$_2$[3766]

chloride);[3365,3766] whereas the rutile form of TiO$_2$ has a stabilizing effect on photodegradation.

Anatase (but not rutile) has been found to quench the long-lived phosphorescence emission from polyolefins.[140,150] The following electron transfer mechanism between triplet excited (T_1) carbonyl group and TiO$_2$ has been proposed:[3877]

$$>C=O \xrightarrow{+h\nu} (>C=O)^* \text{ (triplet)} \tag{15.13}$$

$$(>C=O)^* + TiO_2 \text{ (anatase)} \rightarrow >C-O^{\overline{\cdot}} + TiO_2^+ \cdot \tag{15.14}$$

The carbonyl radical anion may further abstract hydrogen from a polymer.

It has been also reported that TiO$_2$ catalyses the photodecomposition of *tert*-butyl hydroperoxide and hydroperoxides formed during the oxidation of isooctane.[2193]

To retard the photocatalytic effect many commercial TiO$_2$ pigments are usually coated with one or several layers of inorganic barrier coatings such as alumina, silica, zinc oxide, or manganese acetate.[1767,3765] Alumina and silica, which are strong electron-acceptors, probably attach to the surface of hydroxyl groups:[1703]

$$\overset{|||}{Ti} \overset{H}{\underset{|||}{:O:}} \longrightarrow \overset{H}{\underset{H}{\overset{O}{\underset{|}{\overset{|}{Al}}}}} :OH \tag{15.15}$$

$$\overset{|||}{Ti} \overset{H}{\underset{|||}{:O:}} \longrightarrow \overset{H}{\underset{H}{\overset{O}{\underset{|}{\overset{|}{Al}}}}} -O-\overset{H}{\underset{H}{\overset{O}{\underset{|}{\overset{|}{Si}}}}} -OH \tag{15.16}$$

This can reduce the activity to a few per cent of that of uncoated anatase pigment, but does not completely eliminate it.

The role of manganese, cobalt, zinc, and cerium acetates (Table 15.3) can be attributed to their effectiveness to bond metal acetates on the surface of the pigments, which serve as sites for the rapid annihilation of excitons by the preferential participation of the metal acetate, thereby avoiding reactions involving titanium:[1767]

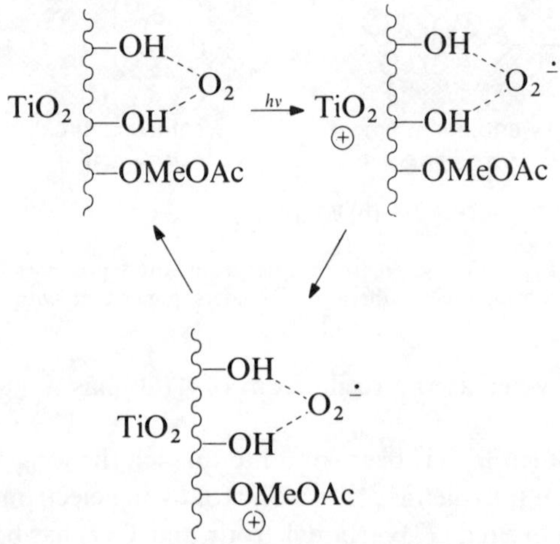

$$(15.17)$$

$$(15.18)$$

The rutile TiO_2 pigment does not show photoactivity as high as the anatase TiO_2, and it is normally not coated with an inorganic barrier.

It has been reported that TiO_2 pigments exhibit antagonistic or synergistic effects with several additives such as photostabilizers and/or antioxidants.[100,990,3725] Nickel chelates are more effective photostabilizers than hydroxybenzophenones in the presence of TiO_2 pigments. This difference in stabilizing efficiency appears to be due to more efficient quenching of the excited state of anatase by nickel chelate than by the hydroxybenzophenone.[150]

The hindered piperidines were found to be ineffective in protecting the polypropylene against the photocatalytic effect of anatase.[100,150,1767] The photocatalytic effect of rutile in polyolefins has been found to be significantly reduced by the addition of phenolic antioxidants.[65,66,1767,2214]

15.2. QUENCHERS

QUENCHERS deactivate excited states (singlet and/or triplet) of chromophoric groups in polymers before bond scission can occur by the following mechanisms;[3881,3884]

(i) Chemical and/or physical deactivation.
(ii) Energy transfer process (cf. Chapter 2).

Table 15.3 Photoactivity of anatase TiO_2 coated with metal acetates[1767]

Metal acetate coating (3 wt %)	Quantum yield of photoactivity in isopropyl alcohol (Φ)
None	1.0
Silver	0.9
Thallium	0.9
Gallium	0.9
Ferric	0.6
Lead	0.6
Rubidium	0.5
Strontium	0.5
Aluminum	0.5
Lanthanum	0.5
Zirconium	0.4
Uranyl	0.4
Potassium	0.4
Samarium	0.4
Praseodymium	0.3
Niobium	0.3
Neodymium	0.3
Cupric	0.3
Magnesium	0.3
Barium	0.3
Yttrium	0.3
Sodium	0.3
Lithium	0.2
Chromic	0.2
Stannous	0.2
Didymium	0.2
Nickelous	0.2
Calcium	0.2
Cerous	0.02
Zinc	0.01
Cobaltous	0.01
Manganous	<0.001

The quenching process is successful only if the quencher molecule is within quenching distance of the excited (singlet and/or triplet) of the chromophoric group within the latter's lifetime. High diffusion constants in a polymer matrix, i.e. good mobility of quencher and/or excited chromophoric group and long lifetime of excited chromophoric group, may therefore enlarge the apparent action sphere of the quencher. In other words it may lower the concentration of quencher necessary to observe a certain effect. The action sphere can be calculated for each specific case, e.g. a freely diffusing quencher of molecular weight 500 and a fixed excited chromophoric group with an active diameter of 5 Å.[1563] In Fig. 15.8 the quencher concentration is plotted as a function of the mean square displacement ($\bar{x}^2$), i.e. the concentration needed to allow quenching within the lifetime (τ_s) in a polymer matrix characterized by a diffusion constant D. Each curve represents a specific distance (R_Q) at which the energy transfer becomes operative.

588

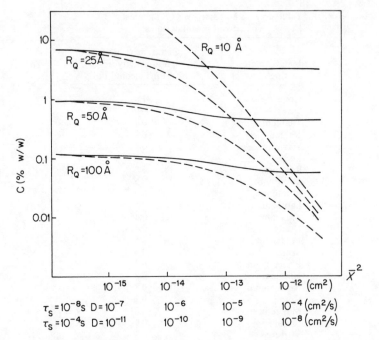

Fig. 15.8 Concentration of quencher versus mean displacement[1563]

The dashed lines give the best possible case (unit probability), i.e. the biggest enlargement of quencher action sphere, which results if each quencher is surrounded by excited chromophoric groups. Under these circumstances any movement of the quencher, independent of direction, would leave to successful encounter.

The unbroken lines reflect a situation in which excited chromophoric group concentration is equal to or even lower than quencher concentration. In this case the direction of quencher diffusional movement–toward or away from the excited chromophoric group–starts to play a role.

The concentration presented in Fig. 15.8 can be considered from the point of view of light stabilizer concentrations actually used in practice, which range from 0.1 to 0.5 % weight by weight. This figure demonstrates that at practical additive levels only quenchers with an operational mode effective at or above 50 Å can be expected to deactivate excited states of chromophoric group efficiently. In other words, only long-range energy transfer can be expected to contribute to excited-state deactivation with the usual half-life of singlets and triplets.

Photostabilization of polymers by the energy transfer mechanism depends on the relative energy levels of the triplet of the polymer and the photostabilizer (Fig. 15.9). The triplet energies of some commercial photostabilizers are listed in Table 15.4.[353] This table shows that the commercial photostabilizers exhibit a wide range of triplet energies, which are highest for salicylate and nickel chelate, but are lowest for hydroxybenzophenones and hydroxyphenylbenzotriazoles. The last

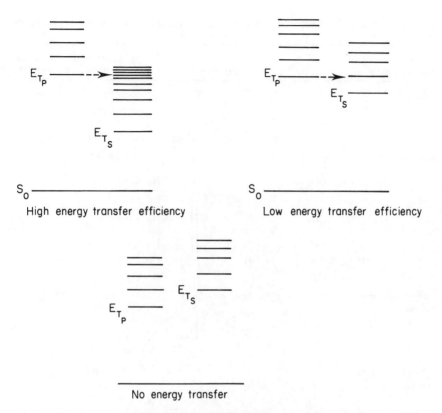

High energy transfer efficiency   Low energy transfer efficiency

No energy transfer

Fig. 15.9   Diagram of energy transfer between polymer and stabilizer for different triplet
energy levels: (E_{T_P}) polymer and (E_{T_S}) stabilizer

compounds with the lowest triplet energies should act as quenchers by
triplet–triplet energy transfer, but they are rather involved in other mechanisms
of photostabilization (cf. Sections 15.5 and 15.6). Nickel chelates, which are very
effective quenchers of the triplet state of carbonyl groups in polyolefins, are less
effective in the quenching of the triplet state of the carbonyl group in poly(vinyl
ketone).[19,20,620,700,1450,1453,2740]

A number of very controversial papers have been published discussing the role
of photostabilizers in the quenching of the triplet state of polymers. The results
suggest that it is necessary to extend the investigation into triplet quenching in
polymers on the basis of more fundamental research.

15.2.1. Nickel Chelates

Nickel chelates are not thermally stable up to 300°C; they form black nickel
sulphide. A number of nickel chelates of different structures (Table 15.5) have
found   industrial   application   for   stabilization   of   polyolefins.[68–70,75,76,93,128,149,268,527,614,615,630,1450,1452,3054,3060,3250]   Nickel

Table 15.4 Examples of triplet energies of some of commercially available photostabilizers (adopted from Ref. 353)

Chemical name	Structure	Commercial name	E_T (kcal mol^{-1})
Salicylate	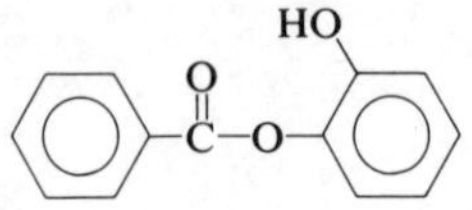	RMB	80.8
Nickel chelate	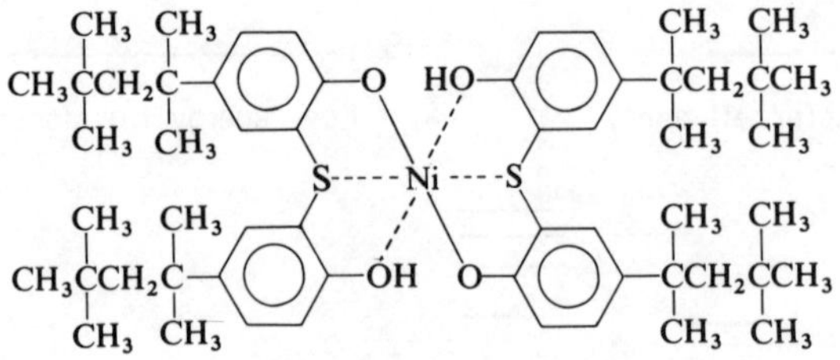	Ferro AM 101	77.8
o-Hydroxy-benzo-phenone	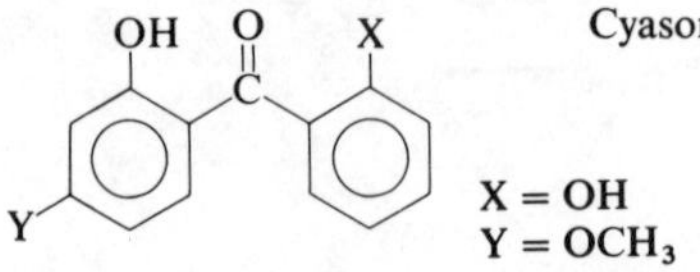 X = OH Y = OCH$_3$	Cyasorb UV24	67.3
o-Hydroxy-benzo-triazoles	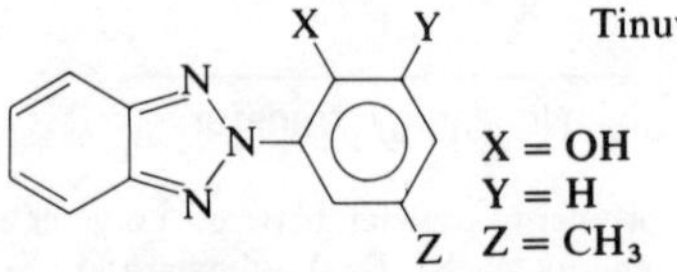 X = OH Y = H Z = CH$_3$	Tinuvin P	63.0

chelates were also tested for photostabilization of polyisobutylene,[671] polybutadiene,[2197] polystyrene,[1249,1506] poly(vinyl chloride),[1286,1287] poly(2,6-dimethyl-1,4-phenyl oxide),[669] and polyurethanes.[2812]

Nickel chelates can photostabilize a polymer by one or more of the following mechanisms:

(i) Quenching of the excited states of carbonyl groups (ketones) (energy transfer mechanism).[19,20,150,527,620,700,1450,1452,1453,1507,2740] (cf. Chapter 2).
(ii) Quenching of singlet oxygen ($^1O_2$) (cf. Section 15.15).[72,614,633,1450,1452,1506]
(iii) Decomposing of hydroperoxides (OOH).[633,2215,3053-3055,3247]
(iv) Scavenging of free radicals (alkylperoxy (ROO·) and alkyloxy (RO·).[68-70,615,629,1677,2310,3055,3718]

Nickel chelates show a synergistic effect in the presence of nickel stearates, and an antagonistic effect with calcium stearate. The mechanism of this interaction has been proposed as some degree of metal exchange with certain stearates.[71,77]

Table 15.5 Examples of commercially available nickel chelate photostabilizers for polyolefins

Name	Structure	Commercial name
syn-Methyl-2-hydroxy-4-methylphenyl-N-alkylketimine	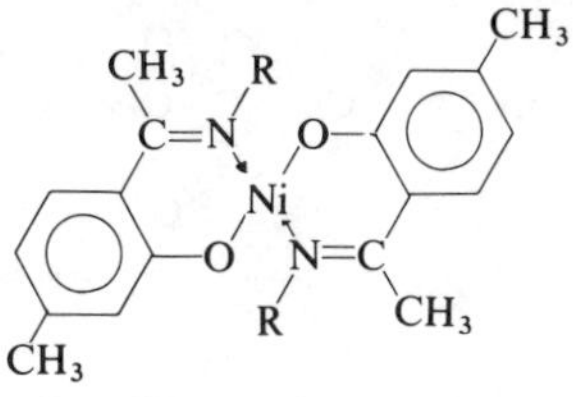 R = OH, n − C$_4$H$_9$, C$_6$H$_5$	
Bis[2,2-thiobis-4-(1,1,3,3-tetramethylbutyl)phenolato]nickel	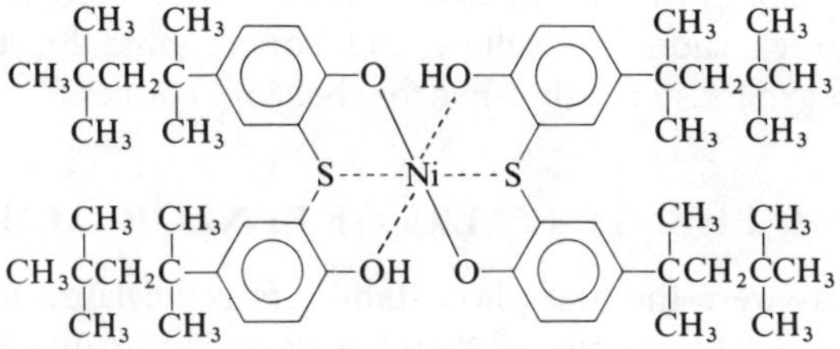	Ferro AM101
[2,2-Thiobis-4-(1,1,3,3-tetramethylbutyl)phenolato-n-butylamine]nickel	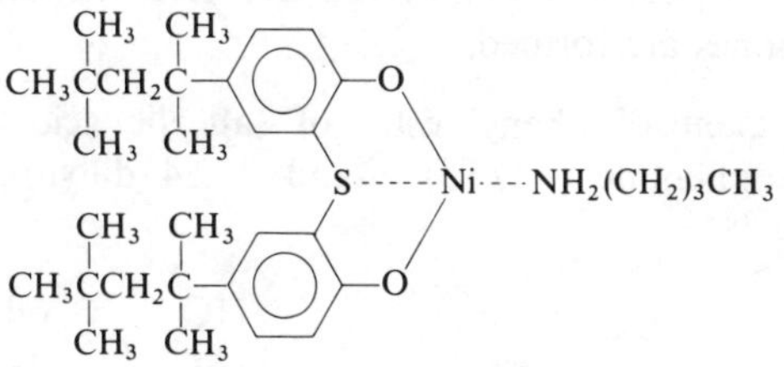	Cyasorb UV 1084
Dibutyldithiocarbamate nickel	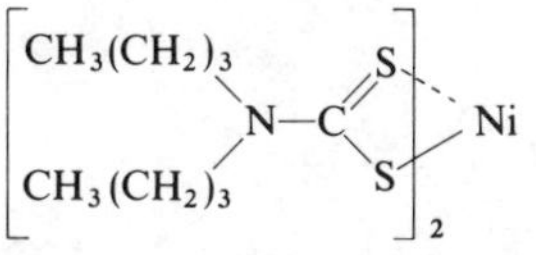	NBC
Bis[o-alkyl-3,5-di-t-butyl-4-hydroxybenzyl]phosphonate nickel	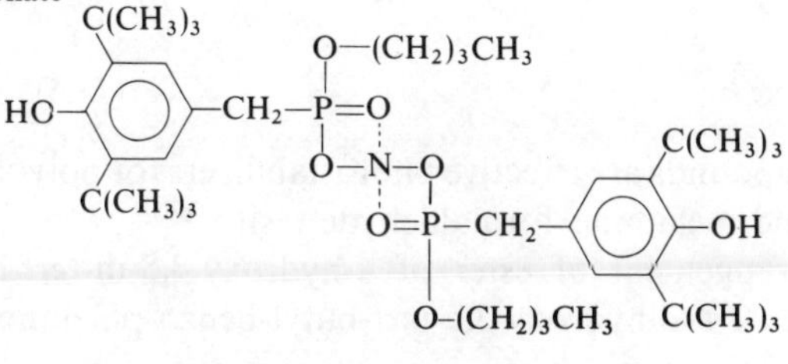	Irgastab 2001

Table 15.5 (Contd.)

Name	Structure	Commercial name
Bis[(1-phenyl-3-methyl)-4-docanolyl-5-pyrazolate] nickel	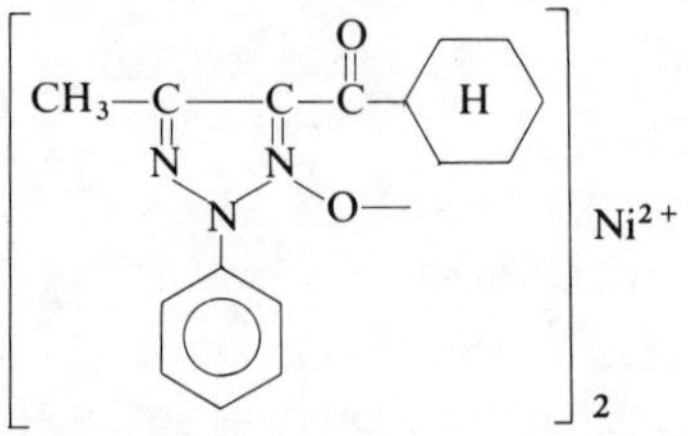	Sanduvor

Nickel chelates also show an interaction with different pigments such TiO_2 (anatase and rutile), cadmium yellow and copper phthalocyanine.[74] Nickel chelates are also synergistic with o-hydroxybenzophenones.[2478]

15.3. PHENYL ESTERS OF BENZOIC ACID

These compounds were the first photostabilizers (salicylates) used technically, and they are still available because of their low price. Unfortunately most of them turn yellow on exposure to ultraviolet light, and this limits their use as photostabilizers for colourless and transparent plastics. The yellowing of these compounds is due to the PHOTOFRIES REARRANGEMENT in which hydroxybenzophenones are formed.

(i) Photorearrangement of phenyl ester of salicylic acid (15.4) gives 2,2'-dihydroxybenzophenone (15.5) and 2,4'-dihydroxybenzophenone (15.6):[684, 1899, 2682]

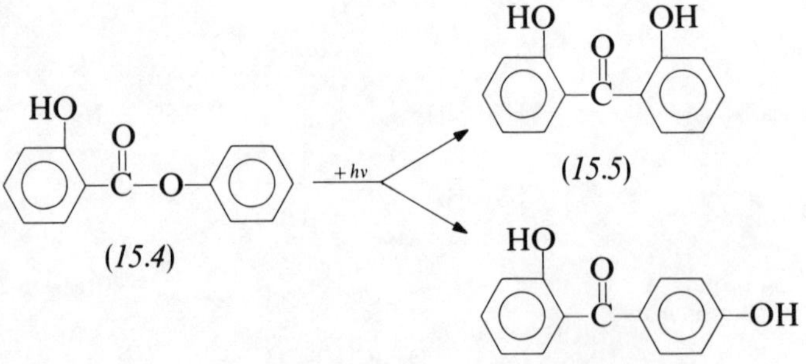

(15.19)

(15.5)

(15.4)

(15.6)   (15.20)

These compounds are effective photostabilizers for polyolefins,[2862] cellulose acetate, and melamine–formaldehyde resins.[3210]

(ii) Photorearrangement of ester of 4-hydroxy-3,5-tert-butyl-benzoic acid (15.7) gives 2,4'-dihydroxy-3,5-tert-butyl-benzo-phenone (15.8):[763,1562]

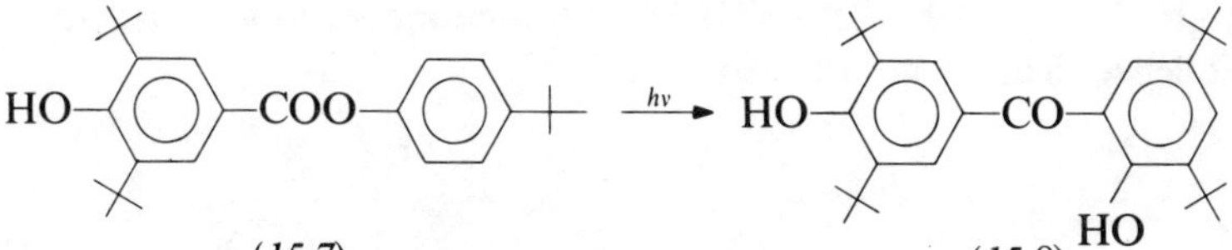

(15.7) (15.8)

(15.21)

These compounds are effective photostabilizers for polyolefins.[615,1562] It has been reported lately that nickel and zinc salts of 4-hydroxy-3,5-di-*tert*-butylbenzoic acid can be used as photostabilizers in polypropylene.[78]

More information available on practical application of phenyl esters of benzoic acid are published elsewhere.[3060]

15.4. SALICYLANILIDES

o-Hydroxysalicylanilides can form three different intramolecular hydrogen bonding depending on solvent polarity:[1041]

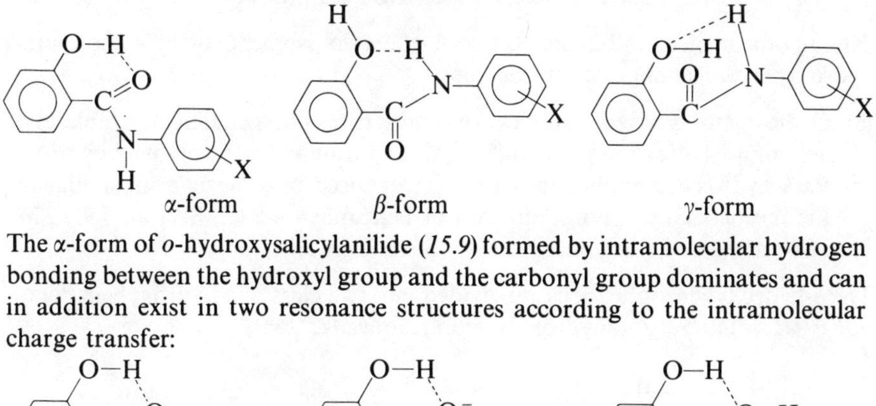

α-form β-form γ-form

The α-form of *o*-hydroxysalicylanilide (*15.9*) formed by intramolecular hydrogen bonding between the hydroxyl group and the carbonyl group dominates and can in addition exist in two resonance structures according to the intramolecular charge transfer:

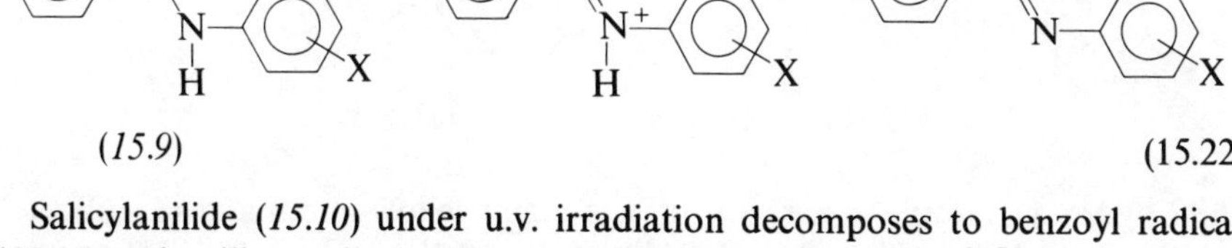

(*15.9*) (15.22)

Salicylanilide (*15.10*) under u.v. irradiation decomposes to benzoyl radical (*15.11*) and aniline radicals (*15.12*) according to the reaction:[2174]

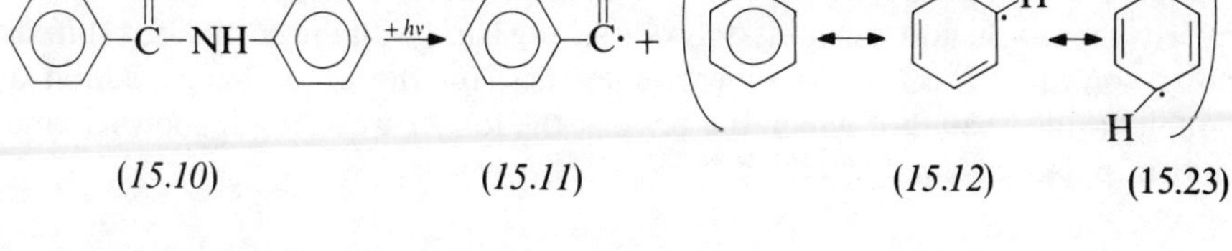

(*15.10*) (*15.11*) (*15.12*) (15.23)

594

As a result of FRIES REARRANGEMENT, o-aminobenzophenone (15.13) and p-aminobenzophenone are formed (15.14):

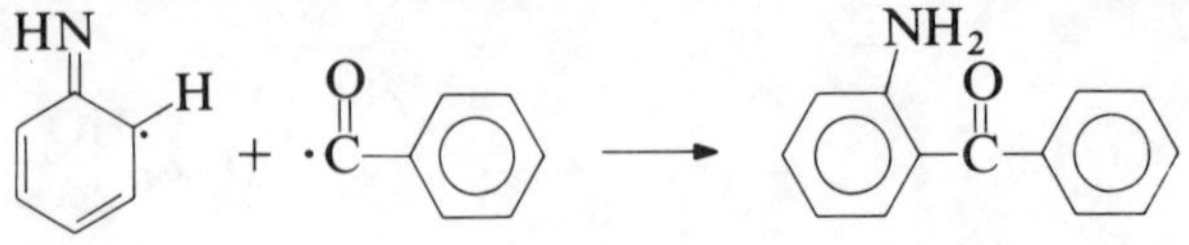

(15.24)

(15.13)

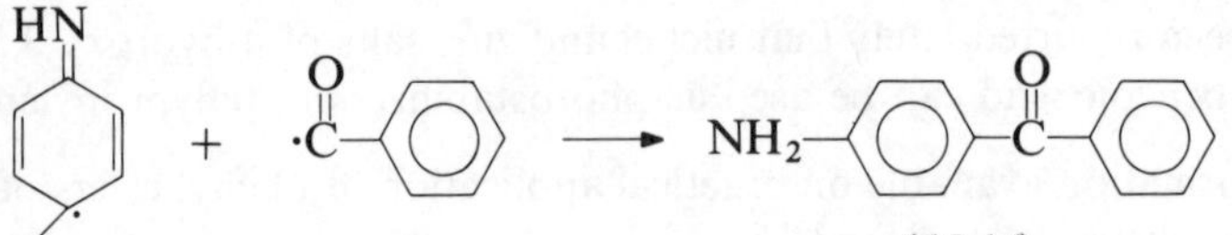

(15.25)

(15.14)

Salicylanilides show a remarkable effect on the photostabilization of natural rubber.[2013]

15.5. HYDROXYBENZOPHENONES

The photochemical difference between the two isomers, ortho- and para-hydroxybenzophenones, is very distinct:

(a) Ortho-hydroxybenzophenones are in most cases photochemically stable, and are employed for the photostabilization of a number of polymers.[1205,1426]
(b) Para-hydroxybenzophenones are photoreduced by a mechanism similar to the mechanism of photoreduction of benzophenone (cf. Section 7.9.4.2.3), and are used as photoinitiators (cf. Section 14.5).

The o-hydroxybenzophenones are divided into two classes: 2-hydroxybenzophenones (15.15) and 2,2'-dihydroxybenzophenones (15.16).

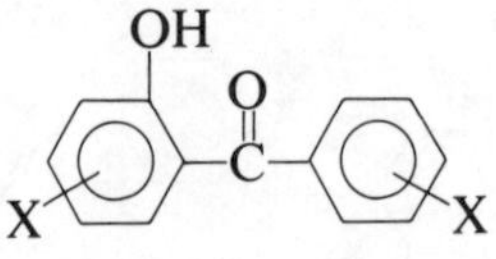

(15.15) 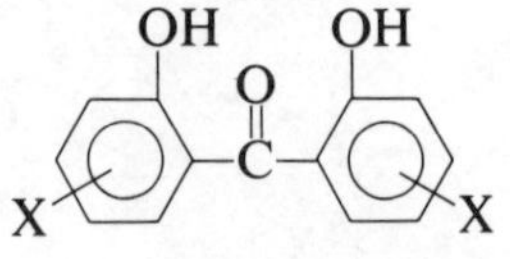   (15.16)

The 2-hydroxybenzophenones usually have a faint yellow colour and an ability to stabilize polymers to light below 370 nm. The 2,2'-dihydroxybenzophenones absorb light effectively up to 400 nm and also have a faint yellow colour. They photostabilize polymers to light below 400 nm.

Figure 15.10 shows the spectra of some hydroxybenzophenones. The p-hydroxybenzophenones exhibit only one strong band with the normal red shift in polar solvents. The o-hydroxybenzophenones, on the other hand, exhibit a double-headed band, of which the peak at the longer wavelength shows a blue shift in polar solvents.[988,1561,1686,2183,2184]

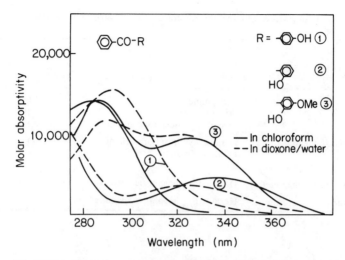

Fig. 15.10 Absorption spectra of *p*-hydroxybenzophenones in chloroform and water/dioxane[1561]: (1) *p*-hydroxybenzophenone, (2) *o*-hydroxybenzophenone, (3) 2-hydroxy-4-methoxybenzophenone

The visual hypsochromic effect of the methoxy group in para-position to the carbonyl group is due to a reduction of the band-width of the very flat long wavelength peak of the simple *o*-hydroxybenzophenone. The short wavelength peak exhibits the normal batochromic effect expected from the introduction of an auxochrome.[1561] These results indicate the formation of a strong hydrogen bond. Substitution in the para-position by a methoxy group increases the strength of the hydrogen bond. The stable position of the proton is at the phenolic oxygen atom in the ground state, and in the first excited triplet state at the carbonyl group. This behaviour is very pronounced in the case of *p*-methoxy-*o*-hydroxybenzophenone.

Rigid, polar solvents are able to disrupt the intramolecular hydrogen bond in *o*-hydroxybenzophenones and allow population of the low-lying triplet level.[2208] The quantum yield for the photoreaction of various benzophenones in amine solvents is shown in Table 15.6. These data reflect the fact that solvation by hexamethylphosphoric triamide efficiently disrupts intramolecular hydrogen bonding involving the 2-hydroxy substituent, allowing population of the reactive (n,π^*) triplet state.[1766]

The big differences in quantum yield in hexamethylphosphoric triamide relative to those obtained in butylamine and triethylamine probably reflect the differences in the ability of the various solvents to solvate the hydroxybenzophenone, rather than differences in their electron-donating ability. Photoreduction in hexamethylphosphoric triamide (*15.17*) then proceeds by electron transfer from a lone pair on a hexamethylphosphoric triamide nitrogen, followed by proton transfer and, in part, radical combination according to the following scheme:[1766]

Table 15.6 Quantum yields (Φ) for the photoreaction of various benzophenones in amine solvents[1766]

Compound	Solvent	Quantum yield (Φ)
Benzophenone	Butylamine	1.0
	Triethylamine	0.19
	Hexamethylphosphoric triamide	0.17
2-Hydroxybenzophenone	Hexamethylphosphoric triamide	0.11
	Triethylamine	1.6×10^{-3}
	Butylamine	7.5×10^{-5}
2,4-Dihydroxybenzophenone	Hexamethylphosphoric triamide	3.7×10^{-4}
2,4,4'-Trihydroxybenzophenone	Hexamethylphosphoric triamide	7.5×10^{-4}

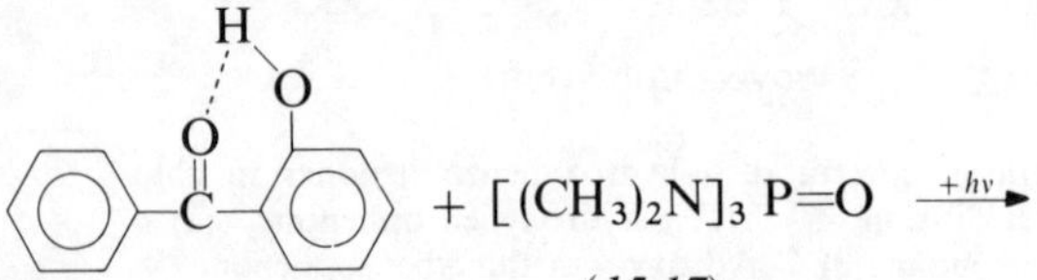

$$+ [(CH_3)_2N]_3\,P{=}O \xrightarrow{+h\nu}$$

(15.17)

$$+ [(CH_3)_2N]_3\,P{\rightarrow}O \longrightarrow$$

$$+ [(CH_3)_2N]_2\,\overset{O}{\overset{\uparrow}{P}}{-}\overset{+}{\overset{\cdot}{N}}(CH_3)_2 \longrightarrow$$

$$+ [(CH_3)_2N]_2\,\overset{O}{\overset{\uparrow}{P}}{-}N\overset{CH_3}{\underset{CH_2\cdot}{}} \longrightarrow$$

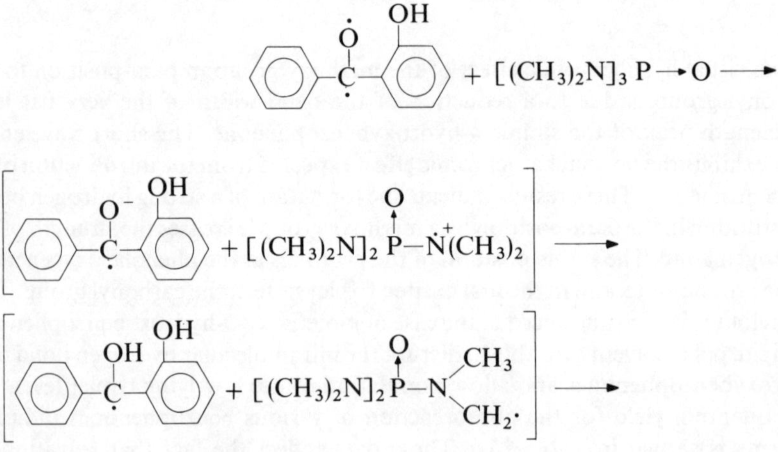

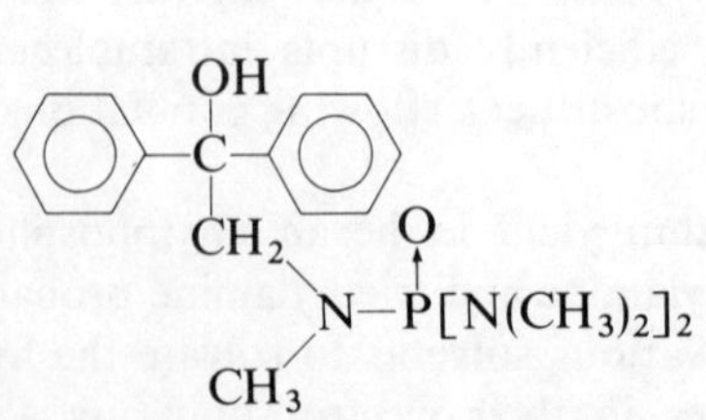

(15.26)

The quantum yield of 7×10^{-3} for the photoreduction of o-hydroxybenzophenone by direct irradiation in isopropyl alcohol[1344] is much lower than the quantum yield of 1.0 for the photoreduction of benzophenone, (cf. Section 7.9.4.2.3), indicating the relative inefficiency of the alcohol in solvating o-

hydroxybenzophenone. The quantum yield for the photoreduction of *o*-hydroxybenzophenone in hexamethylphosphoric triamide ($\Phi = 0.11$) is similar to that reported for the reduction of benzophenone in triethylamine.[738] Either hexamethylphosphoric triamide solvated *o*-hydroxy-benzophenone with very high efficiency, or the electron-transfer process, involves the excited singlet state (S_1) of *o*-hydroxybenzophenone and occurs in hexamethylphosphoric triamide at a rate which competes very favourably with singlet excited state (S_1) deactivation. The low quantum yield for *o*-hydroxy-benzophenones having 4-hydroxy substituents (Table 15.6) indicates that deactivation of the benzophenone excited singlet state (S_1), by loss of a proton to hexamethylphosphoric triamide and formation of an unreactive charge-transfer complex, competes favourably with the intersystem crossing (ISC) process.

The photostabilizing mechanism of *o*-hydroxybenzophenones has been the subject of many investigations.[56, 196, 368, 685, 1459, 1489, 1563, 1657, 1725, 2068, 2069, 2183–2185, 2208, 2501, 2751]

The most probable mechanism involved in a photostabilization is the change of energy of absorbed photon to the intramolecular proton transfer during which a 'quinone structure' is formed. This reaction may occur by one of four proposed cycles:

(i) *o*-Hydroxybenzophenone molecule in the ground state (S_0) absorbs a photon and is excited to the singlet (S_1) charge-transfer state:[2068,2069]

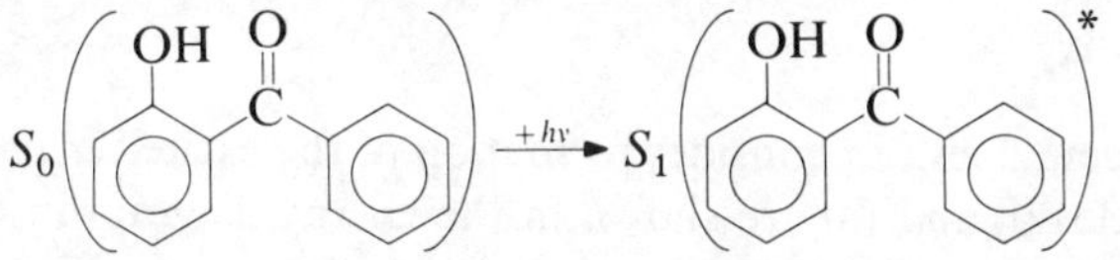

$$(15.27)$$

In this excited singlet state (S_1) the proton of the hydroxyl group is transferred to the carbonyl group, during which 'quinone structure' (possibly of n,π* character) is formed:

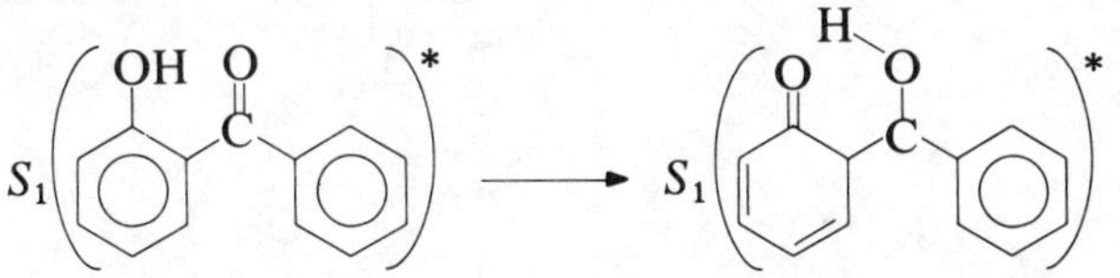

$$(15.28)$$

The excited singlet state (S_1) of 'quinone structure' pass by intersystem crossing (ISC) process to the excited triplet state (T_1):

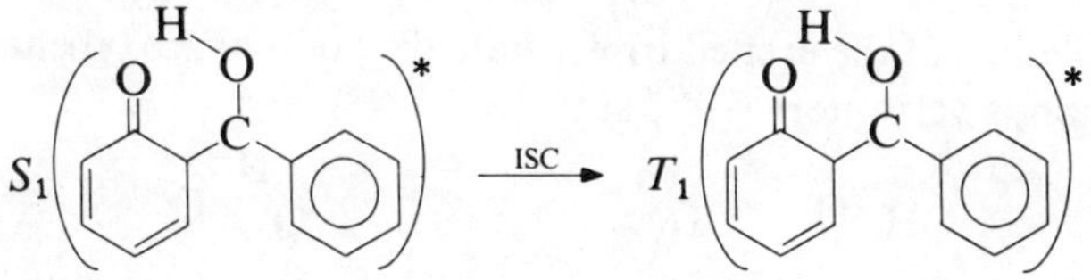

$$(15.29)$$

Since in the excited triplet state (T_1) the stable structure is the 'ketone structure', the proton tunnels back:

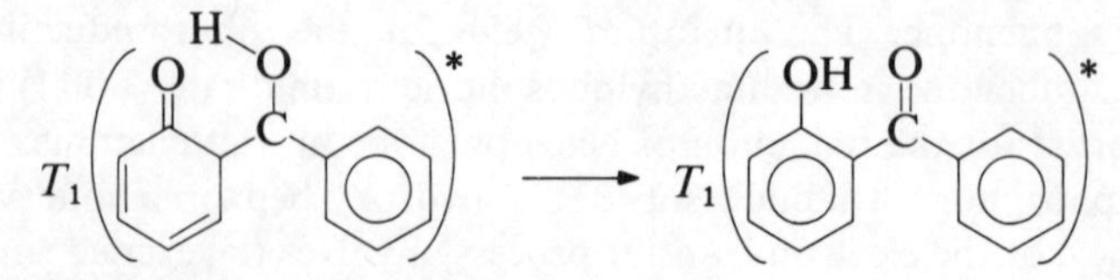

$$T_1 \left(\text{structure} \right)^* \longrightarrow T_1 \left(\text{structure} \right)^* \qquad (15.30)$$

The intersystem crossing (ISC) process leads the excited triplet state (T_1) of 'ketone structure' directly to the ground state (S_0):

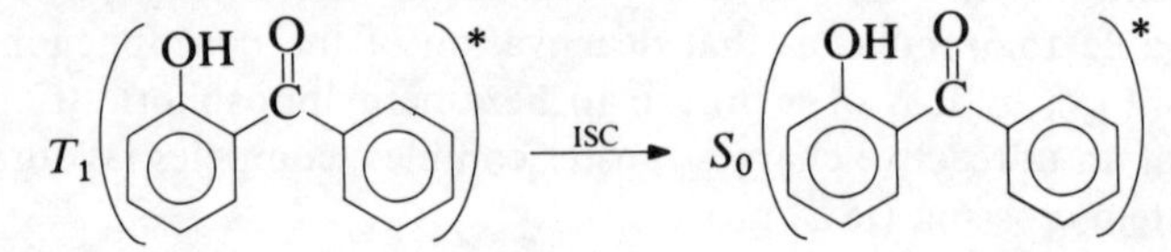

$$T_1 \left(\text{structure} \right)^* \xrightarrow{\text{ISC}} S_0 \left(\text{structure} \right)^* \qquad (15.31)$$

(ii) Another explanation postulates that the rotation of the hydroxyphenyl group in the excited singlet state (S_1) of the 'quinone structure' enhances the Franck–Condor factor between the excited singlet state (S_1) and the ground state (S_0), thus leading to a high rate for the internal conversion (IC) process to the ground state:[1561-1563]

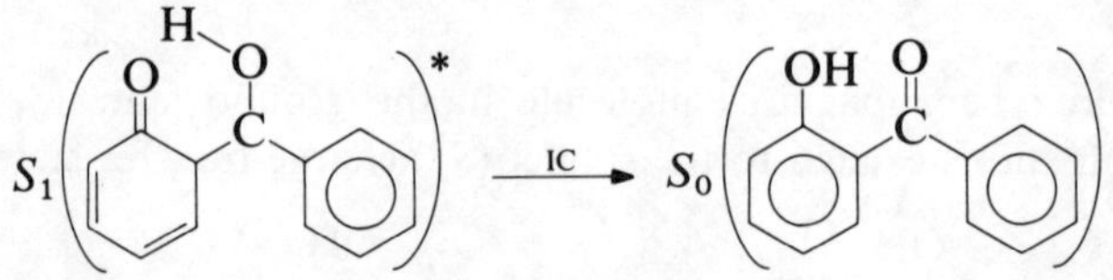

$$S_1 \left(\text{structure} \right)^* \xrightarrow{\text{IC}} S_0 \left(\text{structure} \right) \qquad (15.32)$$

This scheme does not consider formation of the excited triplet states (T_1). Both cycles (i) and (ii) are shown in the energy diagram in Fig. 15.11.

(iii) This cycle has been proposed on the basis of nanosecond flash spectroscopy:[196,685,1459,1725,2501]

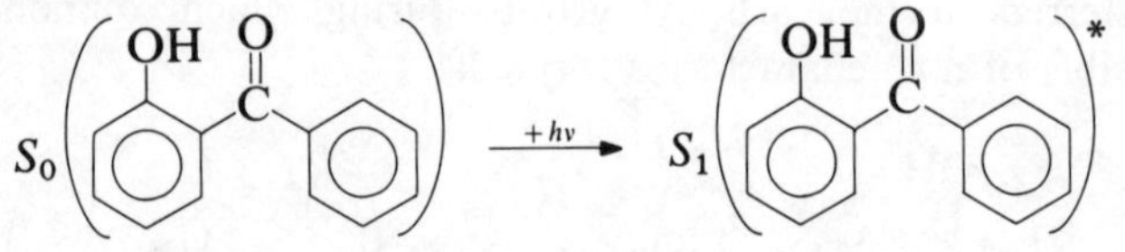

$$S_0 \left(\text{structure} \right) \xrightarrow{+h\nu} S_1 \left(\text{structure} \right)^* \qquad (15.33)$$

$$S_1 \left(\text{structure} \right)^* \xrightarrow{\text{ISC}} T_1 \left(\text{structure} \right)^* \qquad (15.34)$$

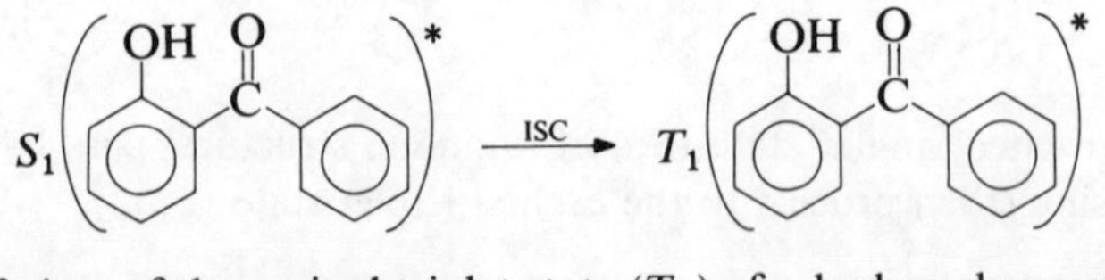

The lifetime of the excited triplet state (T_1) of o-hydroxybenzophenone in solution is very short (10^{-8} s).[2208]

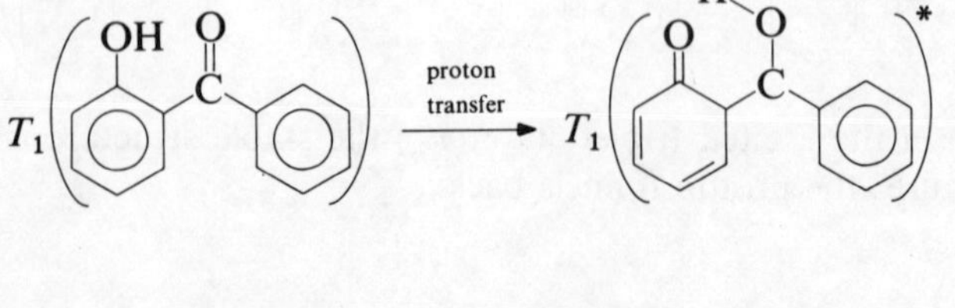

$$T_1 \left(\text{structure} \right) \xrightarrow[\text{transfer}]{\text{proton}} T_1 \left(\text{structure} \right)^* \qquad (15.35)$$

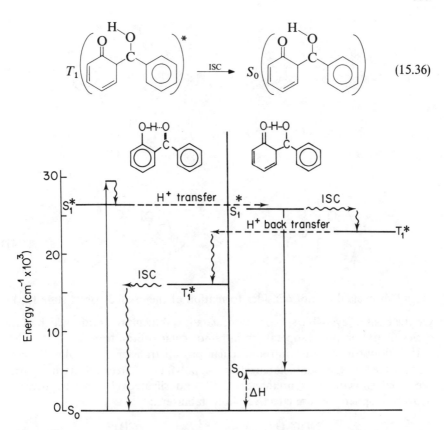

$$T_1 \left(\text{structure} \right)^* \xrightarrow{\text{ISC}} S_0 \left(\text{structure} \right) \tag{15.36}$$

Fig. 15.11 Energy diagram of photophysical processes involved in the photostabilizing mechanism of ortho-hydroxybenzophenones[2068]

The lifetime of the "quinone structure" in the ground state (S_0) is shorter than 10^{-9} sec.[1459]

$$S_0 \left(\text{structure} \right) \longrightarrow S_0 \left(\text{structure} \right) \tag{15.37}$$

(iv) This cycle is an alternative proposition based on picosecond flash spectroscopy:[2501]

$$S_0 \left(\text{structure} \right) \xrightarrow{+ h\nu} S_1 \left(\text{structure} \right)^* \tag{15.38}$$

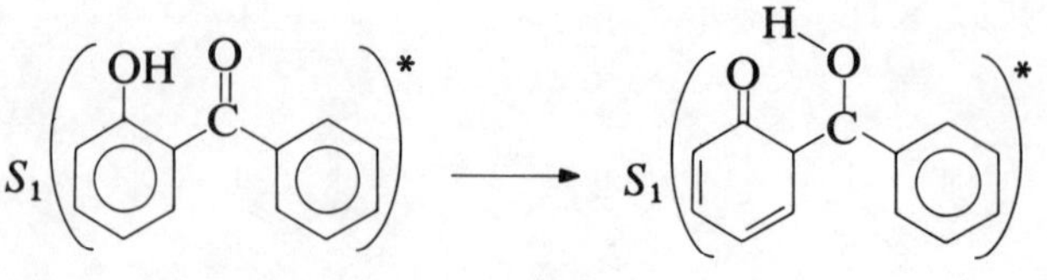

$$S_1 \left(\overset{OH}{\underset{}{\bigcirc}} \overset{O}{\underset{C}{}} \bigcirc \right)^* \longrightarrow S_1 \left(\overset{H}{\underset{O}{}} \overset{O}{\underset{C}{}} \bigcirc \right)^* \qquad (15.39)$$

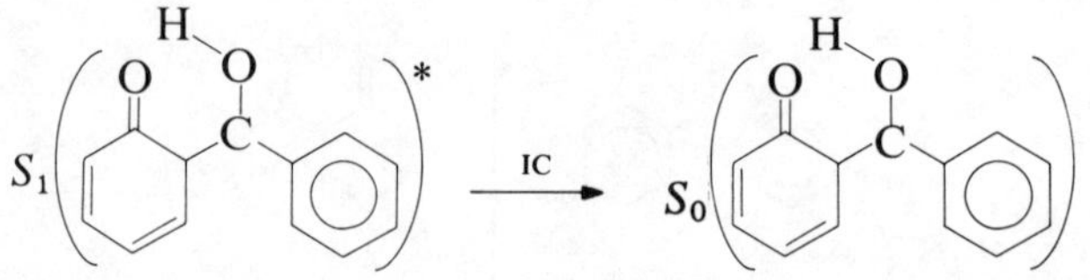

$$S_1 \left(\overset{H}{\underset{O}{}} \overset{O}{\underset{C}{}} \bigcirc \right)^* \xrightarrow{\text{IC}} S_0 \left(\overset{H}{\underset{O}{}} \overset{O}{\underset{C}{}} \bigcirc \right) \qquad (15.40)$$

$$S_0 \left(\overset{H}{\underset{O}{}} \overset{O}{\underset{C}{}} \bigcirc \right) \longrightarrow S_0 \left(\overset{OH}{\underset{}{\bigcirc}} \overset{O}{\underset{C}{}} \bigcirc \right) \qquad (15.41)$$

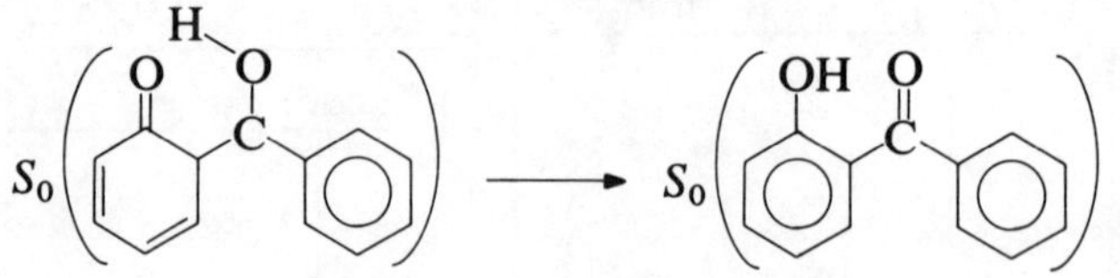

This cycle does not consider formation of the excited triplet state (T_1).

On the basis of available spectroscopic data it is difficult to decide, which of the cycles (i)–(iv) proposed, operate in intramolecular proton transfer.

It is doubtful whether intramolecular proton transfer cycles alone can be responsible for the photostabilization of polymers. Lately results have been presented showing that urethane (*15.18*) and silyated (*15.19*) derivatives of hydrobenzophenones are effective photostabilizers:[2791]

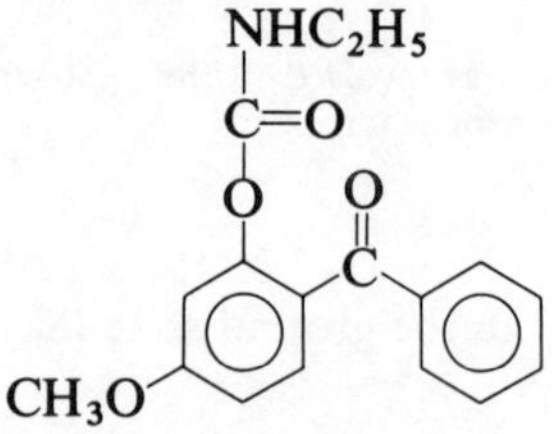

(*15.18*) 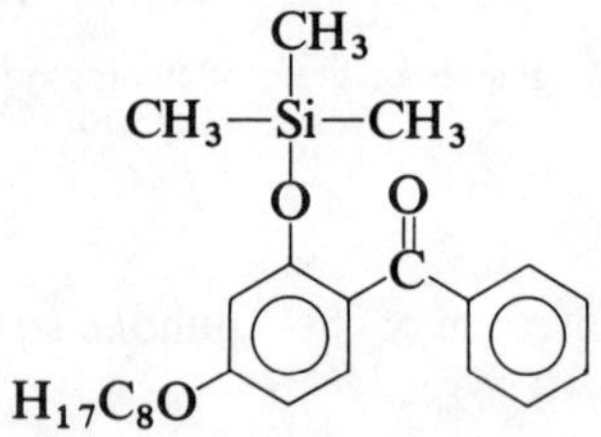   (*15.19*)

These compounds cannot photostabilize by the proposed mechanism of intramolecular proton transfer.

Several other propositions have been given in order to explain photostabilization properties of o-hydroxybenzophenones. These compounds can also:

(i) Act as light screeners.[656,1451,3999]

(ii) Quench excited states of chromophoric groups (e.g. carbonyl group) present in polymers by an energy transfer mechanism.[94,1249,1453,1690,3752]

(iii) Act as free radical scavengers.[656,3254,3752,3756]

(iv) Act as hydroperoxy group (OOH) decomposers. It has been found that o-hydroxybenzophenones are photodecomposed more rapidly in processed polypropylene compared with unprocessed polymer.[629] The rate of photo-decomposition correlates exactly with initial hydroperoxide concentrations in the polymer.[102] These result can be explained by the mechanism in which polymer oxy (PO·) and hydroxy (HO·) radicals produced in the photolysis of hydroperoxides abstract a hydrogen atom of the hydroxyl group in o-hydroxybenzophenone:[659]

$$POOH + hv \longrightarrow PO· + ·OH \qquad (15.42)$$

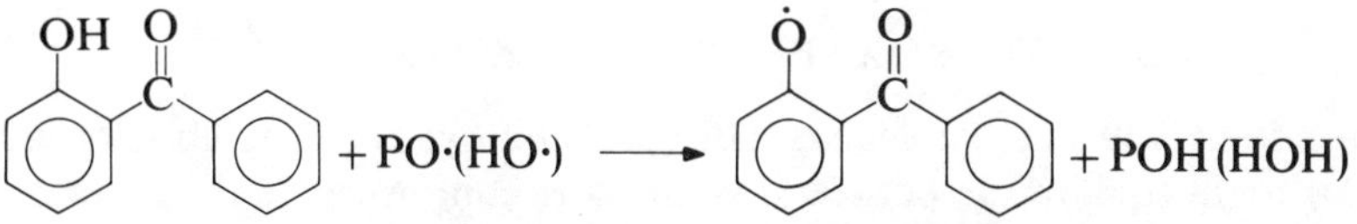

$$(15.20)$$
$$(15.43)$$

The radical (15.20) is no longer capable of intramolecular hydrogen bonding.

It has been reported that 4-methoxy (15.21) and 4-n-octoxy (15.22) derivatives of o-hydroxybenzophenone exhibit higher stabilization effects that pure o-hydroxybenzophenones.[102,151] This effect has been explained by the fact that during u.v. radiation these groups may photolyse to give corresponding alkyl radicals:

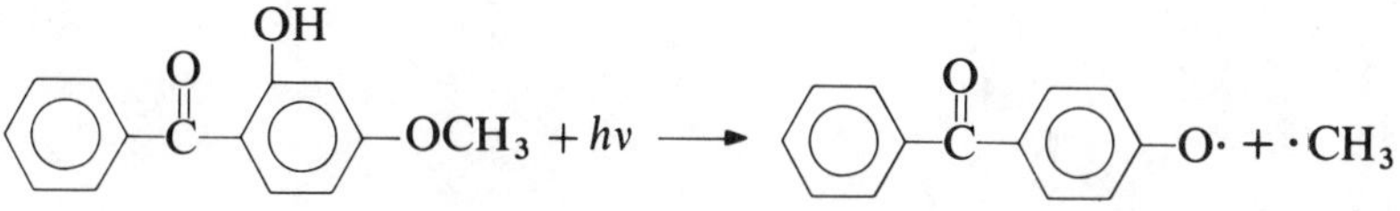

(15.21)
$$(15.44)$$

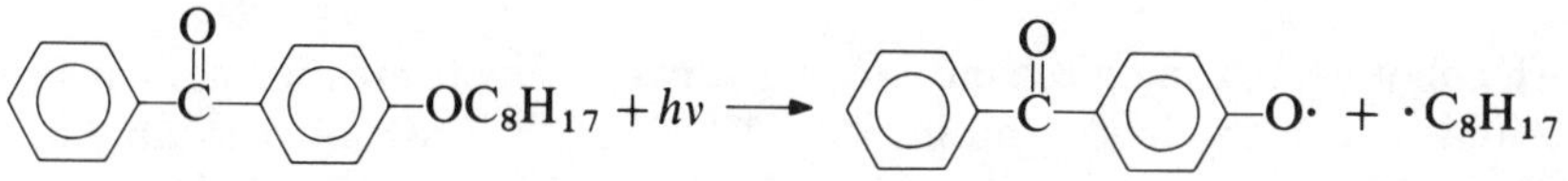

(15.22)
$$(15.45)$$

The smaller methyl radicals (·CH$_3$) may move easily out of the polymer cage, whereas the larger octyl radicals (·C$_8$H$_{17}$) may not. In the latter case radical recombination could therefore increase the light stabilities of the 4-methoxy and 5-n-octoxy derivatives of o-hydroxybenzophenones.

The list of different o-hydroxybenzophenone derivatives proposed as photo-stabilizers mentioned in patents or offered for trial is extensive, and includes hundreds of different compounds.

From 0.5 to 1.0 wt % of monohydroxybenzophenones or 0.1–0.2 wt % of the dihydroxybenzophenones have been found to be effective in the light stabiliz-ation of polyolefins,[102,152,268,633,684,1843,3756] poly(methyl methacrylate),[2340] polystyrene,[1249,2960] poly(vinyl chloride),[1286,1287,2396] co(acrylonitrile/

butadiene/styrene)(ABS),[2102] polyamides,[3866] polyurethanes,[2812] cellulose acetate and melamine–formaldehyde resin,[3210] and wool.[644,2522,3140]

o-Hydroxybenzophenones show synergistic effects with nickel chelates.[2478] 2-Hydroxy-4-octyloxybenzophenone (15.22) synergizes with both peroxide-decomposing antioxidants (metal dithiocarbamates) and chain-breaking antioxidants (hindered phenols (e.g. Irganox 1076)).[656]

15.6. HYDROXYPHENYLBENZOTRIAZOLES

Absorption spectra measurements indicate that o-hydroxyphenylbenzotriazoles (15.23) are in equilibrium between two or more conformers in the ground state (S_0). Two conformers absorbing at 302 and 340 nm may be stabilized by a combination of intra- and intermolecular hydrogen bonding:[56,1460,1724,1725]

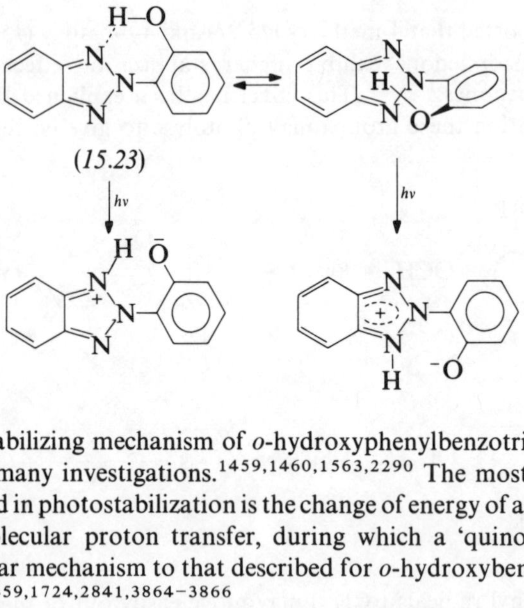

$$(15.46)$$

$$(15.47)$$

The photostabilizing mechanism of o-hydroxyphenylbenzotriazoles has been the subject of many investigations.[1459,1460,1563,2290] The most probable mechanism involved in photostabilization is the change of energy of absorbed photon to the intramolecular proton transfer, during which a 'quinone structure' is formed (a similar mechanism to that described for o-hydroxybenzophenones, cf. Section 15.5):[1459,1724,2841,3864-3866]

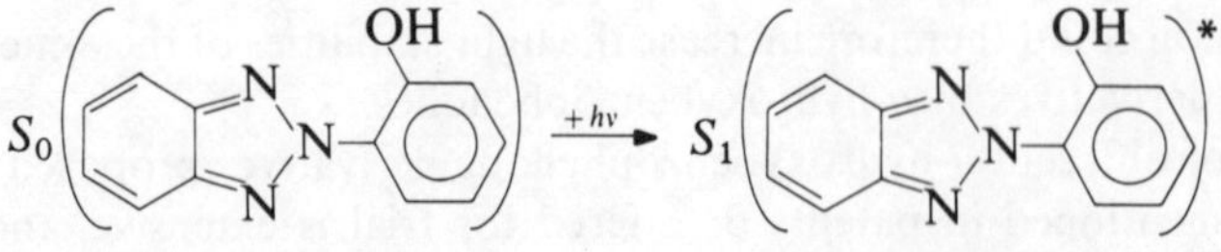

$$(15.48)$$

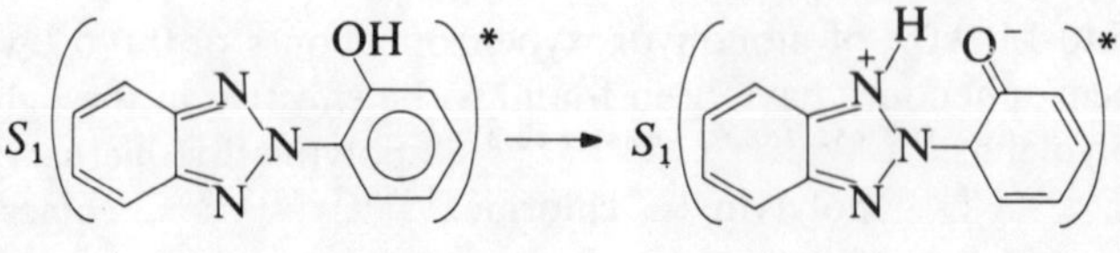

$$(15.49)$$

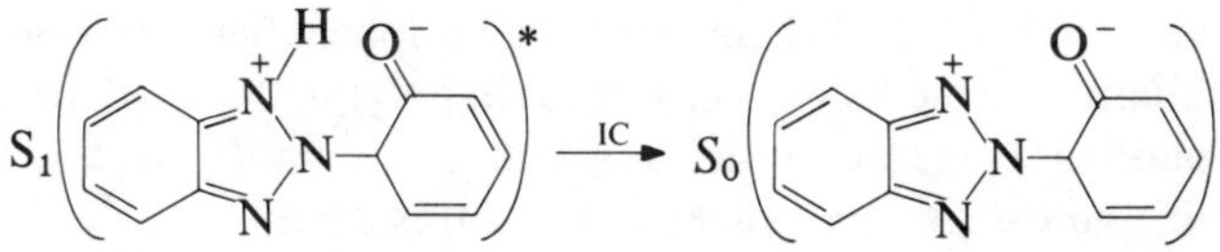

$$S_1 \left(\text{[structure with } \overset{H}{\underset{}{N^+}}, O^-] \right)^* \xrightarrow{\text{IC}} S_0 \left(\text{[structure with } N^+, O^-] \right) \quad (15.50)$$

$$S_0 \left(\text{[structure with } \overset{H}{\underset{}{N^+}}, O^-] \right) \longrightarrow S_0 \left(\text{[structure with } OH] \right) \quad (15.51)$$

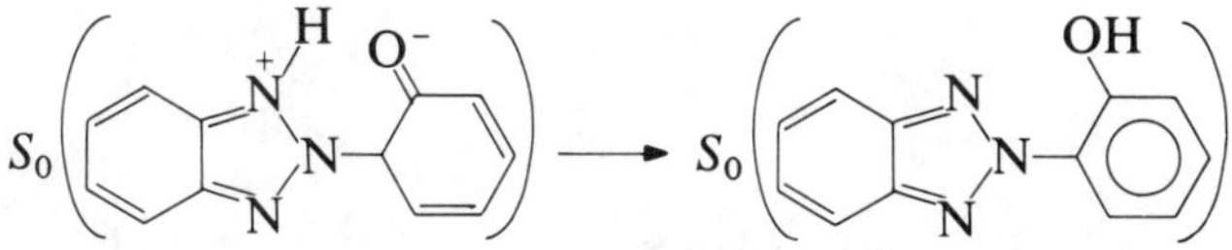

The lifetime of the 'quinone structure' in the ground state (S_0) is less than 1.0 ns.[1459]

The 2-(2'-hydroxy-5'-methylphenyl)benzotriazole (Tinuvin P) also shows phosphorescence in ethanol at 77K.[1453,2036]

It is doubtful whether the intramolecular proton transfer cycle alone can be responsible for the photostabilization of polymers by o-hydroxyphenyl-benzotriazoles. Recent results have been presented indicating that urethane (15.24) and sylilated (15.25) derivatives of o-hydroxyphenylbenzotriazoles are effective photostabilizers of coatings:[2790,2791]

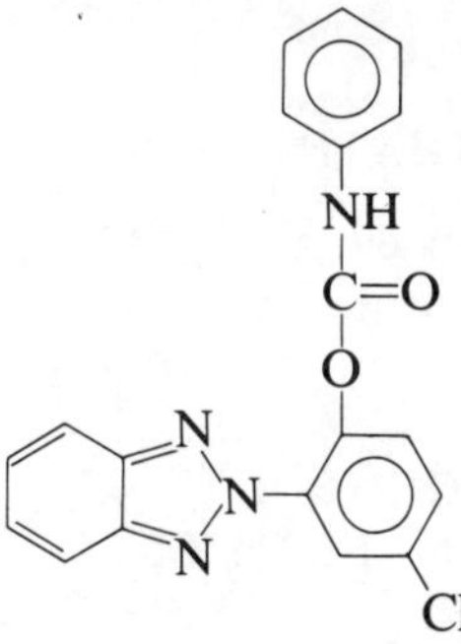

$$(15.24) \qquad\qquad (15.25)$$

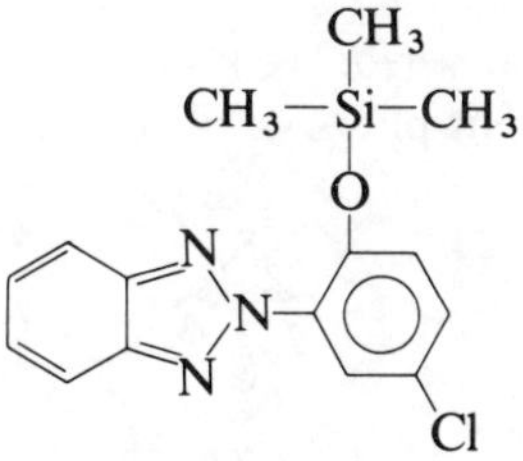

The marked effect of processing on the photodecomposition of o-hydroxyphenylbenzotriazole appears to be due to the formation of polymer hydroperoxide groups (P–OOH) which thermally decompose to polymer oxy (PO·) and hydroxyl (HO·) radicals. Both radicals may abstract hydrogen atoms from the hydroxyl group in o-hydroxyphenylbenzotriazole:[98]

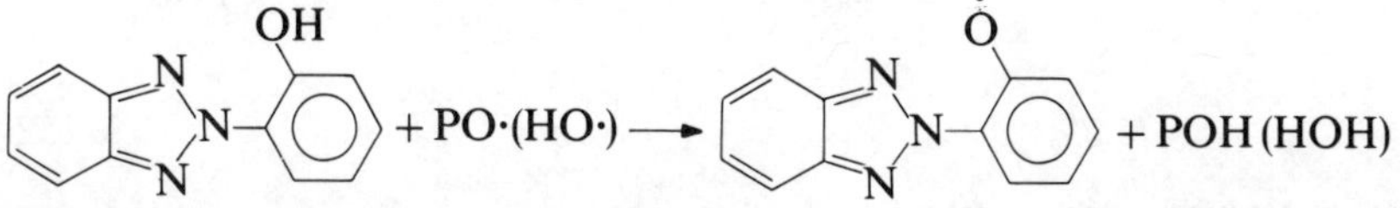

$$(15.52)$$

This reaction has also been confirmed by reaction of o-hydroxyphenylbenzo-triazole with low molecular peroxy radicals in the dark.[1603,1605,1609]

The acetate ester of 2-(2′-hydroxy-5′-methylphenyl)benzotriazole (*15.26*), upon u.v. irradiation in methylene chloride or in poly(methyl methacrylate) films, undergoes photocleavage reaction in which (Tinuvin P) (*15.27*) is regenerated.[2290] No photo-Fries rearrangement was observed:

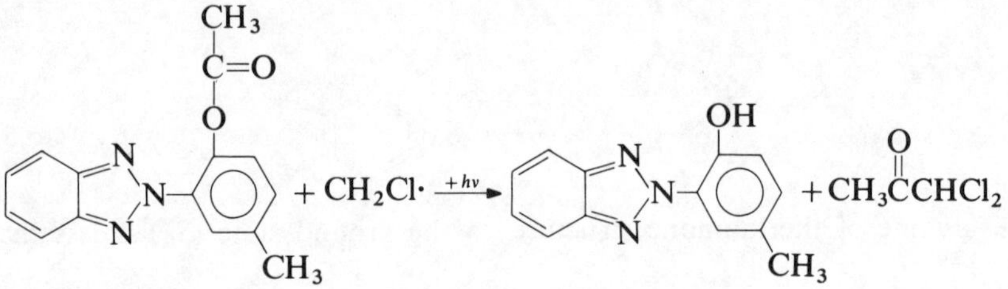

$$(15.26) \qquad\qquad (15.27) \qquad\qquad (15.53)$$

o-Hydroxyphenylbenzotriazoles have been found to be effective in the light stabilization of polypropylene,[268,1451] polystyrene,[1249,2960] co(acrylonitrile/butadiene/styrene) (ABS),[2100] poly(vinyl chloride),[1286,1287] polybutadiene,[212] cellulose acetate and melamine–formaldehyde resin,[3210] and wool.[2228,2522,3809]

15.7. METAL XANTHATES

Metal xanthates (*15.28*), known as good antioxidants,[165,168–171] can also be used as efficient photostabilizers for polypropylene.[166,662]

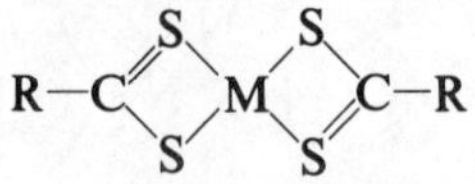

$R = OC_8H_{17}$
$M = Ni, Fe, Cu, Co$

$$(15.28)$$

The iron xanthate (*15.29*) is very unstable to u.v. irradiation, and its photolysis may occur by the following mechanism:[166,3259]

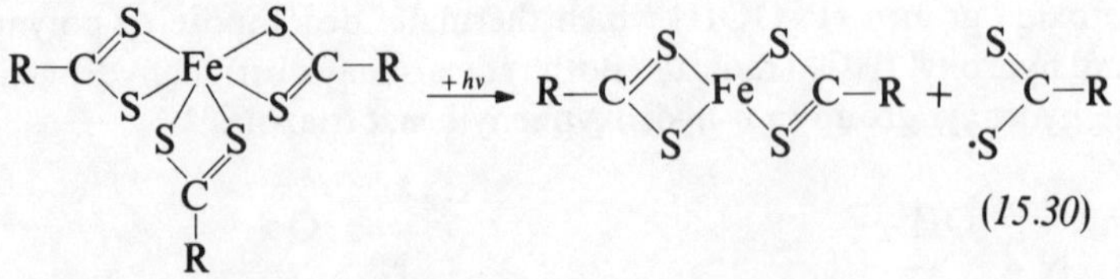

$$(15.30)$$

$$(15.29) \qquad\qquad\qquad (15.54)$$

The disulphide radical (thiyl radical, xantogen radical) (*15.30*) can abstract hydrogen from a polymer molecule (PH) and in this way produce polymer alkyl radical (P·), which is susceptible to oxidation to polymer alkylperoxy radical (PO·) and further to the formation of polymer hydroperoxide (POOH):

$$R-C\overset{\displaystyle S}{\underset{\displaystyle S\cdot}{\Big\langle}} + PH \longrightarrow R-C\overset{\displaystyle S}{\underset{\displaystyle SH}{\Big\langle}} + P\cdot \tag{15.55}$$

(15.30)

$$P\cdot \to O_2 \to POO\cdot \longrightarrow POOH + P\cdot \tag{15.56}$$

Disulphide radicals can recombine and form dixantogen (15.31), which is an effective antioxidant responsible for hydroperoxide decomposition according to the reaction:[165]

$$R-C\overset{\displaystyle S}{\underset{\displaystyle S\cdot}{\Big\langle}} + \overset{\displaystyle S}{\underset{\displaystyle \cdot S}{\Big\rangle}}C-R \longrightarrow R-C\overset{\displaystyle S}{\underset{\displaystyle S-S}{\Big\langle}}\overset{\displaystyle S}{\Big\rangle}C-R \tag{15.57}$$

(15.31)

$$2\,R-C\overset{\displaystyle S}{\underset{\displaystyle S-S}{\Big\langle}}\overset{\displaystyle S}{\Big\rangle}C-R + 2\,POOH \longrightarrow 2\,R-C\overset{\displaystyle S}{\underset{\displaystyle SO\cdot}{\Big\langle}} + 2\,PO\cdot + H_2O \tag{15.58}$$

With an excess of hydroperoxides, formation of sulphuric acid has been observed.[166]

Light stability of xanthate complexes of nickel, cobalt, and copper is high, whereas the light stability of iron complex and corresponding dixantogen (15.31) is low.[166] Xanthates of nickel and cobalt photostabilize through a combination of complementary mechanisms such as radical scavenging, peroxide decomposition, and u.v. screening, whereas iron xanthate and dixantogen function almost exclusively as peroxide decomposers.[166]

15.8. NEW PHOTOSTABILIZERS

A number of papers and many patents recommending various chemical compounds as efficient photostabilizers have been published in the past decade. The majority of these compounds have not, however, found wide application on an industrial scale. Some new propositions are presented below:

(a) Cetyl-3,5-di-*tert*-butyl-4-hydroxybenzoate (commercially available as Cyasorb UV 2908) (15.32) has been shown to be an effective photostabilizer in high-density polyethylene,[159] and polypropylene.[156,158]

$$HO-\overset{\displaystyle \times}{\underset{\displaystyle \times}{\bigcirc}}-COOC_{16}H_{33}$$

(15.32)

(b) α-Cyano-β,β-diphenylacrylate esters (*15.33*) (commercially available under the trade name N-35, produced by GAF Co, USA) has been recommended for applications in polystyrene, acrylics, and surface coatings, but is useless in polyolefins.[3196]

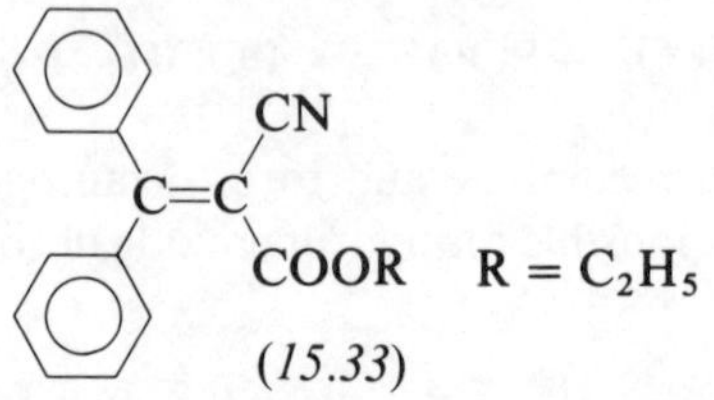

$$R = C_2H_5$$

(*15.33*)

(c) Copolymer (*15.36*) obtained from reaction of *p*-xylylene (*15.34*) with *N*-oxyl biradical (2,2,6,6-tetramethylpiperidinoxy-4-spiro-2'-(1',3'-dioxane)-5'-spiro-5''-(1'',3''-dioxane)-2''-spiro-4'''-(2''',2''',6''',6'''-tetramethyl piperidinoxy) (*15.35*) has been proposed as an effective photostabilizer for polypropylene:[1190-1194]

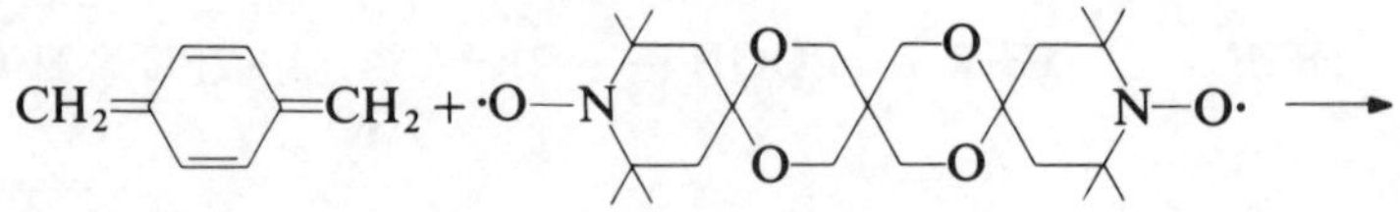

(*15.34*)　　　　　　　　　　(*15.35*)

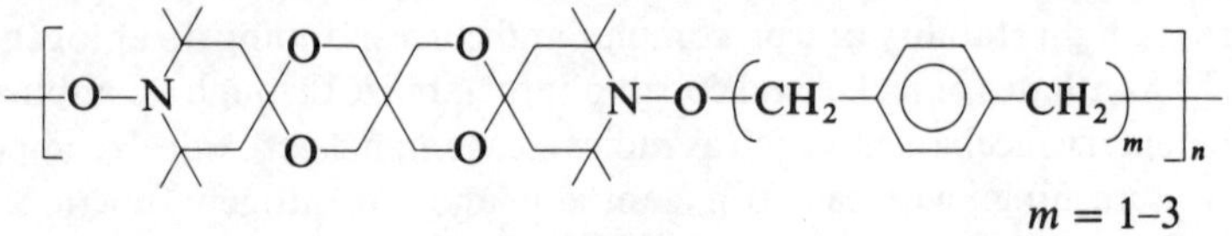

(*15.36*)

$$m = 1-3$$

(15.59)

(d) 4-(1-imidazolyl)phenol (*15.37*) is an effective photostabilizer in polybutadiene.[2326]

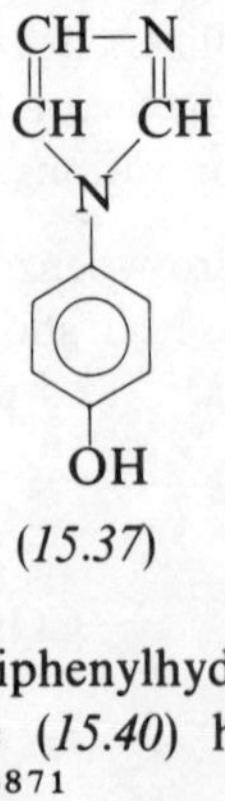

(*15.37*)

(e) Nitrosobenzene (*15.38*), diphenylhydroxylamine (*15.39*), and 2-methylnaphtho(1,8-di)triazine (*15.40*) have been proposed as photostabilizers for polybutadiene.[3871]

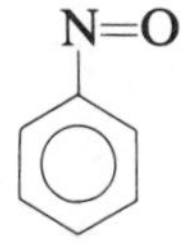

(15.38) 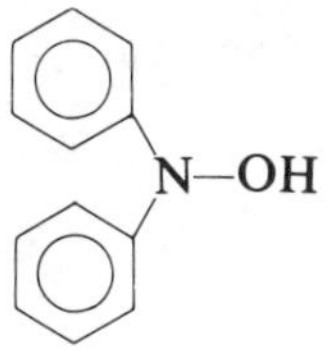   (15.39) 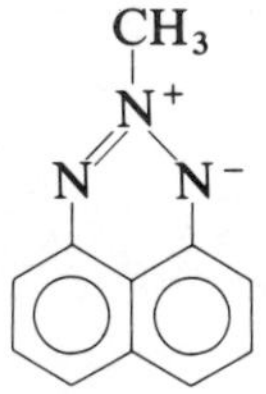   (15.40)

(f) A number of azo compounds such as azobenzene and azonaphthalene derivatives have been found to be effective light stabilizers for butyl rubber, polyurethanes, and isoprene/styrene and butadiene/styrene block copolymers.[1018,2812,3761] The stabilization mechanism probably occurs according to the quinone–hydrazone tautomerism, which proceeds via the following scheme:[1610]

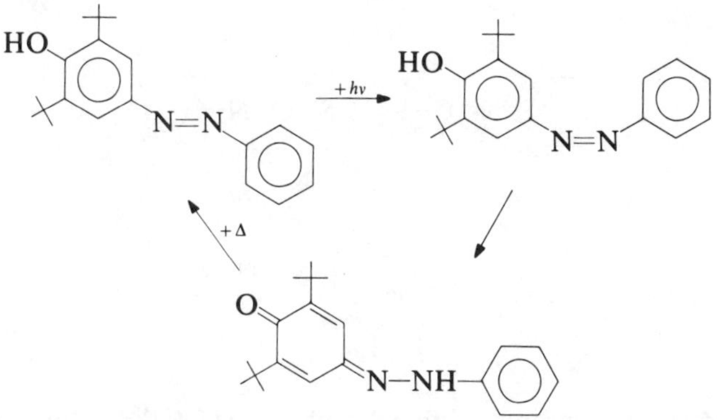

(15.60)

(15.61)

(g) The cerous chloride (CeCl$_3$)–poly(vinyl chloride) complex has been proposed as an effective u.v. screener for polycarbonates.[2051]

(h) Metal (Co^{2+}, Mn^{2+}, Ni^{2+}) salts of 2-[(1-hydroxy-naphthalenyl)carbonyl]-benzoic acid (15.41) can be considered as u.v. stabilizers for polybutadiene.[3042]

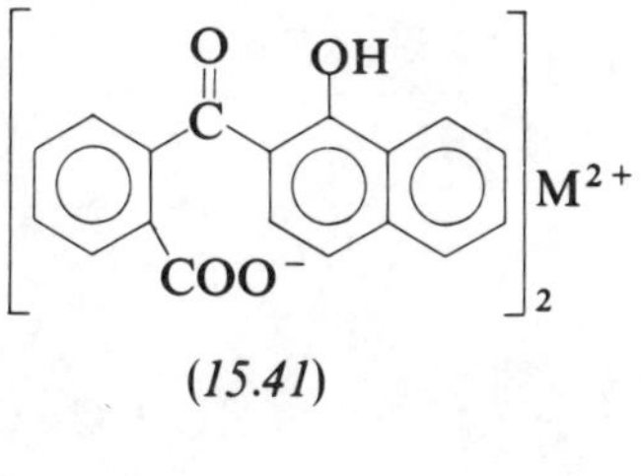

(15.41)

(i) p-Benzoquinone–tin polycondensates (15.42) act as photostabilizers for polybutadiene.[3970]

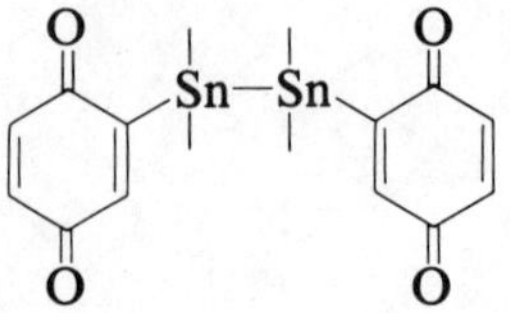

608

(15.42)

(j) New metal chelates such as nickel(II) (α-thiopicoline anilide) *(15.43)* and manganese(III) (α-thiopicoline anilide) *(15.44)* have been proposed as photostabilizers in poly(2,6-dimethyl-1,4-phenylene oxide)[669] and polyisobutylene,[671] respectively:

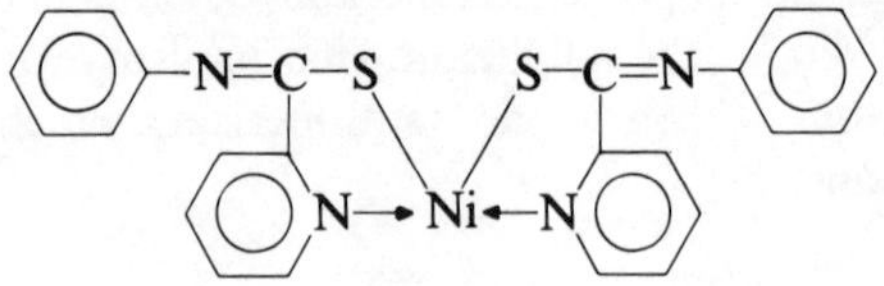

(15.43)

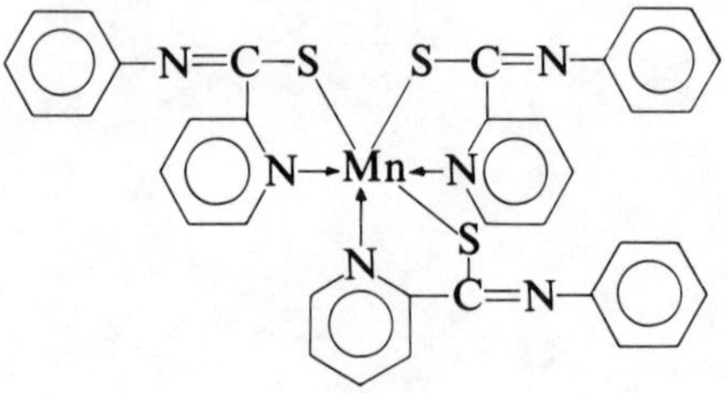

(15.44)

Cuprous di-isopropyl dithiophosphate *(15.45)*[3335] and copper(II) bis(pyrrole-isopropanolamine *(15.46)*[3354] have been shown to be effective light stabilizers for co(butadiene/styrene) (SBR):

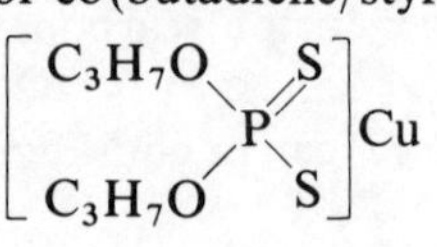

(15.45)

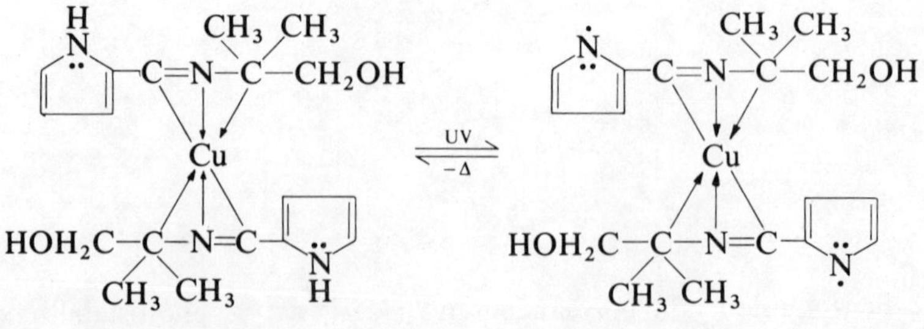

(15.46)   (15.62)

This brief survey cannot present all the trends in research and development of new photostabilizers. A complete report in this field is considered by the author as a separate publication.

15.9. POLYMERIC PHOTOSTABILIZERS

Ultraviolet absorbing groups can be attached to polymers without destroying the stabilizer's ability to absorb light and dissipate the energy in a non-harmful form, by the following methods:[262, 3611, 3613, 3760]

(i) Radical polymerization and copolymerization of vinyl monomers containing photostabilizing groups such as: salicylate ester groups,[261, 1106, 1834, 3614] o-hydroxybenzophenone groups,[259, 260, 1107, 1459, 1461, 1604, 1657, 1933, 2340, 2348, 2349, 2537, 2810, 2814, 2951, 2952, 3302, 3612, 3615, 3864] o-hydroxyphenylbenzotriazole groups,[1460, 2280, 2281, 3411, 3987, 3988] and α-cyano-β,β-diphenylacrylate ester group.[3487, 3488]

(ii) Radical polymerization with a radical initiator molecule attached to an ultraviolet absorber.

(iii) Ring-opening polymerizations and polymerizations with coordination catalysts.[2348, 2350]

(iv) Introduction of difunctional ultraviolet stabilizers into condensation polymers.

(v) Polymer modification reactions by photo-Fries pendant group rearrangement,[2780] or main chain rearrangement.[390, 393, 749, 2366, 2780]

(vi) Grafting of monomers or compounds with functional groups, bearing photostabilizing groups onto polymers.[562, 1105, 1279, 2810, 2814, 3302, 3902]

On the basis of the information available on polymeric photostabilizers, the following conclusions can be drawn:[262]

(i) Polymeric u.v. stabilizers are less susceptible to solvent extraction, exudation, and evaporation from plastics than low molecular weight stabilizers and most oligomeric stabilizers.

(ii) Oligomeric u.v. stabilizers prepared by the condensation of two 2-hydroxybenzophenone molecules with formaldehyde are more resistant to exudation and extraction than stabilizers with long aliphatic side-chains.

(iii) Oligomeric u.v. stabilizers prepared with long side-chains and copolymers containing a low percentage of polymerizable u.v. stabilizers are more compatible with polymers than many low molecular weight stabilizers. Compatible copolymers which contain low percentages of u.v. stabilizer repeat units are more effective than incompatible polymeric u.v. stabilizers.[1107]

(iv) Introduction of a u.v. stabilizing group into a polymer chain can decrease the mechanical properties of the plastic by decreasing polymer crystallinity.[3615] It may be possible to covalently incorporate stabilizers into polymers without affecting the crystallinity of the polymer by attaching the stabilizers as polymer end-groups.

(v) Copolymers which contain low percentages of u.v. stabilizers repeat units can affect the physical properties of the copolymers by decreasing the crystallinity of copolymers. It is preferable to stabilize highly crystalline homopolymers such as polyethylene with a large amount of copolymer containing a high concentration of u.v. stabilizer repeat units.

(vi) The advantages of using polymeric u.v. stabilizers for the protection of plastics against oxidative photodegradation are most evident in samples with high surface area to volume ratios, which are processed at high temperatures, exposed to solvents, and used or tested over a long period of time. In short-term tests, u.v. stabilizer compatibility is most important for polymer protection.

(vii) For thick plastic articles it may be advantageous to coat the sample with a film of a copolymer containing stabilizer units rather than mixing the copolymeric stabilizer into the plastic. The comonomer should be chosen to give the copolymeric u.v. stabilizer good adhesion to the plastic substrate.

(viii) For thick plastic articles it may also be advantageous to use a u.v. stabilizer with a compatible side-chain, for example a long aliphatic side-chain for polyolefins, which can continuously exude from the plastic and replenish the stabilizer lost at the surface.

(ix) A broad range of polymer properties can be obtained by copolymerization of monomeric u.v. stabilizers with the appropriate comonomers. Consequently the copolymers may be used in a wide variety of applications.

15.10. SURFACE PROTECTION BY PHOTOGRAFTING

One of the possible methods of protecting polymers against natural weathering consists of applying organic coatings that offer a good resistance to both u.v. light and chemical attack. These coatings are usually highly crosslinked polymers which can be conveniently obtained by photopolymerization of multifunctional monomers or oligomers. For example coatings based on epoxy-acrylates are rather resistant to weathering, and exhibit excellent surface hardness and impact resistance characteristics can be used for protection of light-instable poly(vinyl chloride).[910] (cf. Section 13.2)

Soluble polyesters based on bisphenols and dicarboxylic acids were employed to coat plastics. A thin skin on the surface of the polyester coating photo-Fries rearranges on exposure to u.v. radiation, and the skin protects the substrate and lower layers of the coating from degradation. As the skin of a rearranged polymer is slowly destroyed, the radiation is able to penetrate deeper into the polymer, causing the lower layers of polyester to rearrange. This type of reaction is described as a 'SELF-HEALING PROCESS'.[749]

Polymer photostabilization can also be achieved by surface grafting.[561]

15.11. HINDERED AMINES

Hindered amine light stabilizers (abbreviated to HALS) belong to a new group of products developed during the last decade on the industrial scale. (Table 15.7).

Table 15.7 Examples of hindered piperidines, some of them commercially available

Name	Structure	Commercial name
4-Hydroxy-2,2,6,6-tetramethyl-piperidine	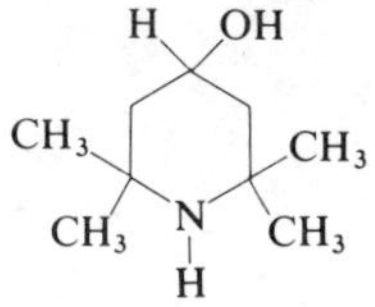	
4-Hydroxy-1,2,2,6,6-pentamethyl-piperidine	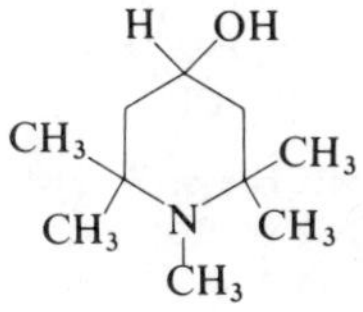	
4-Benzoyloxy-2,2,6,6-tetramethyl-piperidine	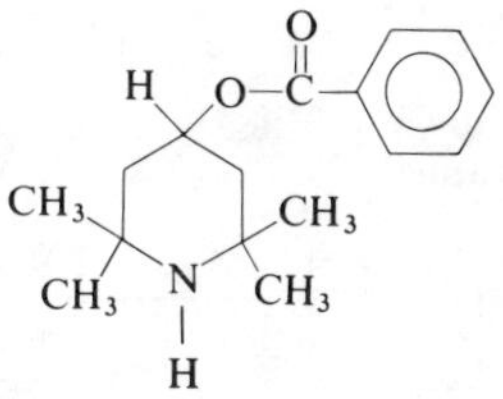	Sanol 774
Bis(2,2,6,6-tetramethyl-4-piperidinyl)sebacate	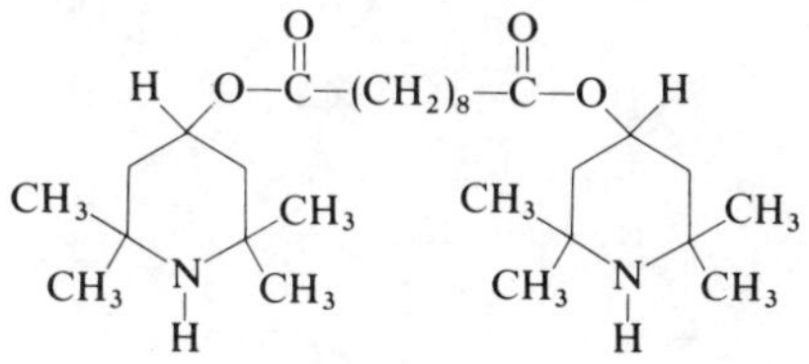	Tinuvin 770
Bis(1,2,2,6,6-pentamethyl-4-piperidinyl) sebacate	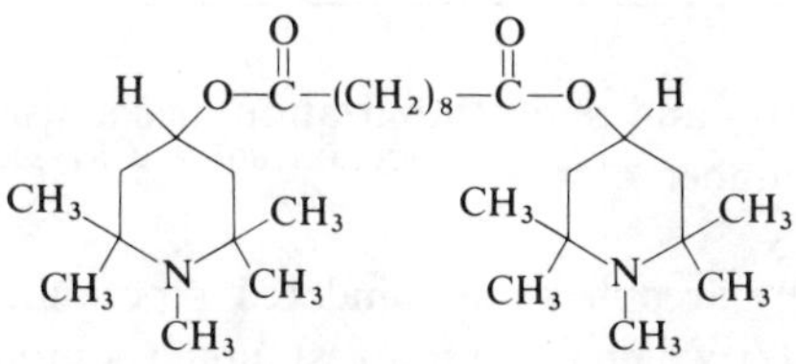	Tinvuin 292
Bis(1,2,2,6,6-pentamethyl-4-piperidinyl)-2-n-butyl-2-(3,5-di-$tert$-butyl-4-hydroxybenzyl)malonate	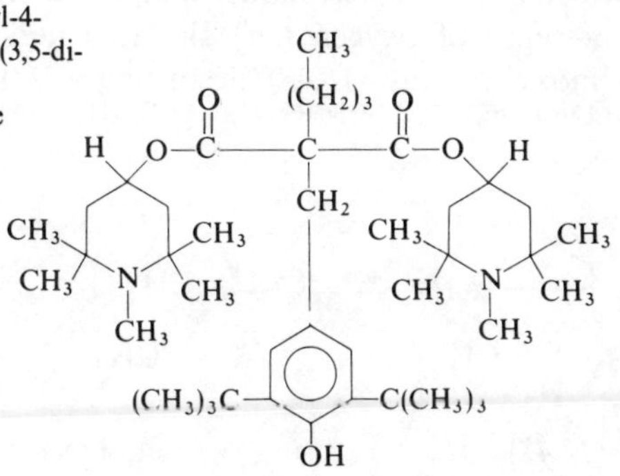	Tinuvin 144

Table 15.7 *(Contd.)*

Name	Structure	Commercial name
Polyster of succinnic acid with *N-β*-hydroxy-ethyl-2,2-6,6-tetramethyl-4-hydroxy-piperidine	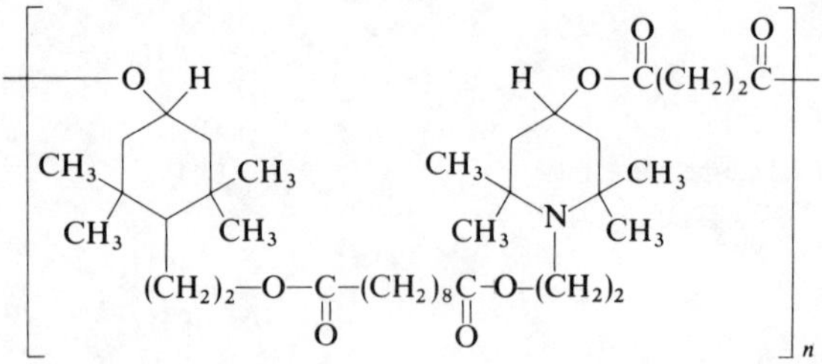	Tinuvin 622

| Poly[2-*N,N'*-di(2,2,6,6-tetramethyl-4-piperi-dinyl)hexanediamino-4-1-amino-1,1,3,3-tetramethyl]symtriazine | 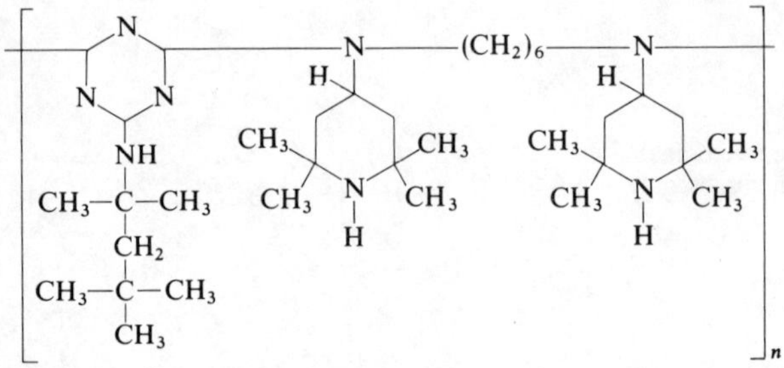 | Chimassorb 944 |

Photochemistry and photostabilization mechanisms of HALS have been the subject of a number of reviews.[50,389,606,614,615,768,1098,1563,1608,3280,3333,3410,3884,3885]

It is generally accepted that hindered piperidine (e.g. 2,2,6,6-tetra-methyl-piperidine (*15.47*), which is the simplest model compound of HALS), during u.v. irradiation in the presence of oxygen (air) and free radicals (R·), produces hindered a piperidinoxy radical (*15.48*) according to the reaction:[257,614,615,633,660,746,1095,1256,1396,1397,1562,1964,2598,3175,3280,3282,3314,3334,3410,3476,3519,3753]

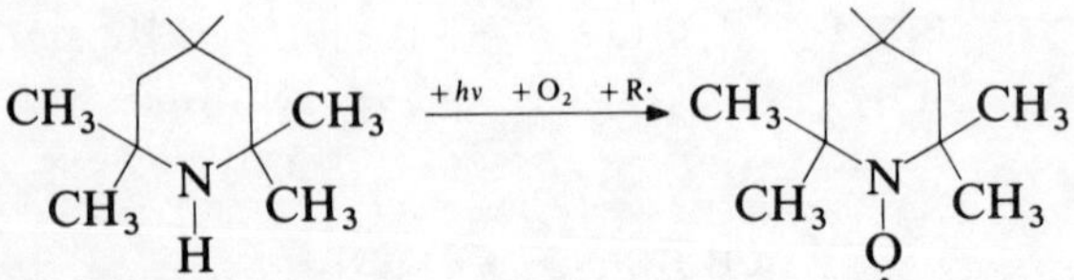

(15.63)

Note: for simplicity the word 'hindered' will be omitted in the remainder of this Section.

The generation of piperidinoxy radicals (*15.48*) can be detected by electron spin resonance spectroscopy (ESR).[47,54,96,97,99,101,115,125,149,270,609,615,635,1130, 1606,1608,3112,3334] Piperidinoxy radicals belong to the class of stable organic free radicals and show triplet ESR spectra with equal components of equal intensities[3062] (Fig. 15.12). The mechanism of the formation of piperidinoxy radicals (*15.48*) is not completely clear. It has been proposed that a charge-transfer (CT) complex formed between piperidine and oxygen can be an intermediate step in the formation of piperidinoxy radicals:[2329]

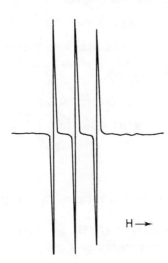

$H \longrightarrow$

Fig. 15.12   ESR spectrum of piperidinoxy radical

$$R\cdot + \cdot OH \longrightarrow ROH \qquad (15.65)$$

Formation of CT complexes between oxygen and several amines (*15.49*) is well established.[3637] On the other hand the role of the free radical (R·) in this mechanism is not well understood.

Piperidinoxy radicals (*15.48*) react with alkyl radicals (R·) to give a substituted hydroxylpiperidine (*15.50*):

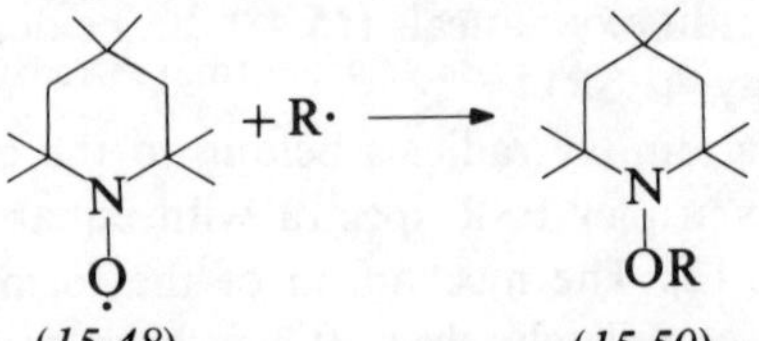

$$(15.66)$$

The rate constant for this reaction is a factor of ~ 20 slower than for a competitive reaction between alkyl radicals and oxygen:[1396,3884]

$$R \cdot + O_2 \rightarrow ROO \cdot \qquad (15.67)$$

Alkylperoxy radicals (ROO·) do not react with piperidinoxy radicals, but they can react with substituted hydroxylpiperidine (*15.50*) to regenerate piperidinoxy radicals (*15.48*):[606,1396,1608,3314,3476]

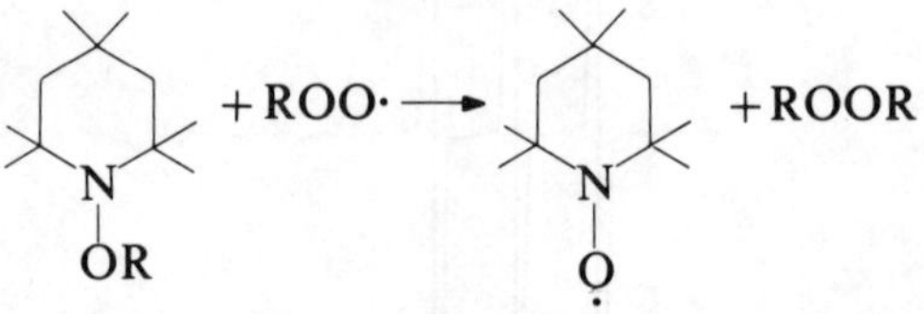

$$(15.68)$$

Similar reactions occur for the polymer alkyl radical (P·) and polymer peroxy radicals (POO·).

Substituted hydroxylpiperidines (*15.51*) are thermally unstable, slowly decomposing even at room temperature by a complex mechanism:[1395,1678,3884]

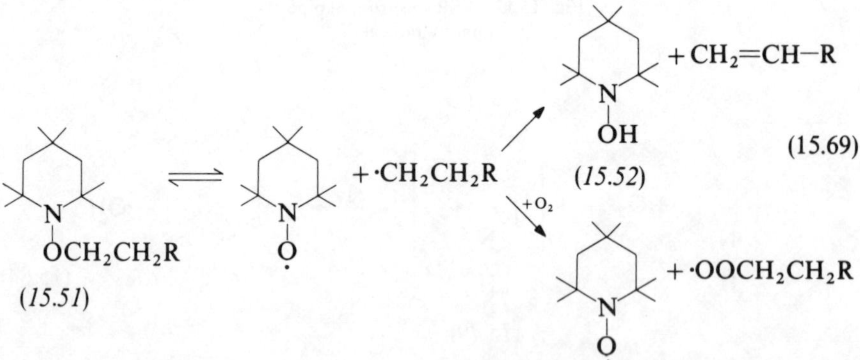

$$(15.69)$$

$$(15.70)$$

Reaction (15.69) occurs at only 1/50 of the rate of reaction (15.70) in the presence of air in the liquid phase. In the solid polymer reaction (15.69) may be favoured over the peroxy regeneration step. Hydroxylpiperidine (*15.52*), which is formed from reaction (15.69), reacts further with peroxy radicals (POO·):[1395]

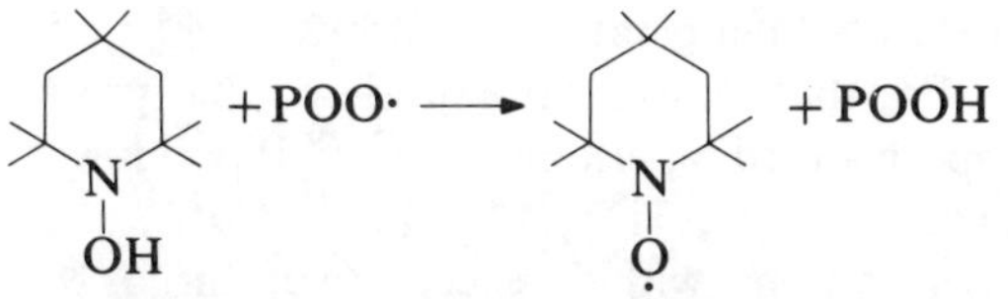

$$+ POO \cdot \longrightarrow \qquad + POOH \qquad (15.71)$$

(15.52)

Piperidines (15.47) and piperidinoxy radicals (15.48) participate in the decomposition (or photodecomposition) of hydroperoxy groups (OOH) by the following mechanisms, in which formation of charge-transfer (CT) complexes (15.53) and (15.54) has been considered:[48, 54, 160, 257, 271, 535, 605, 1095, 1097, 1098, 1397, 1426, 3280–3283, 3352, 3996]

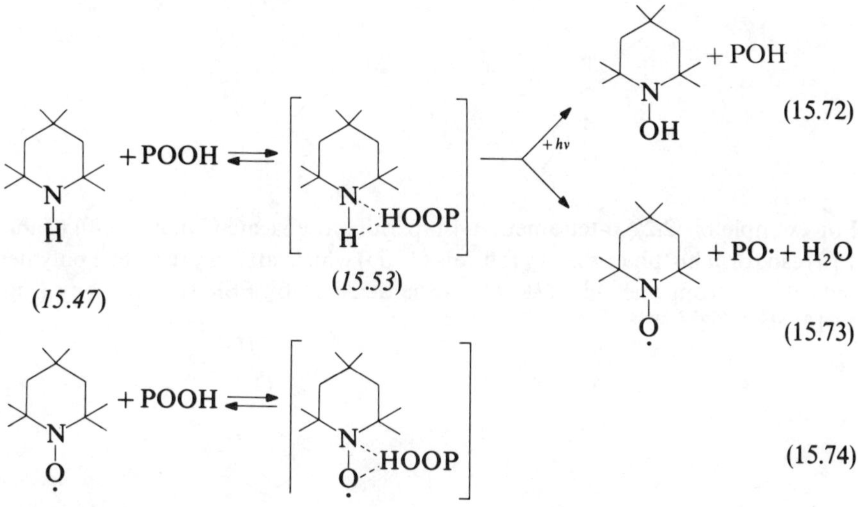

$$+ POH \qquad (15.72)$$

$$+ PO \cdot + H_2O \qquad (15.73)$$

$$(15.74)$$

(15.47)   (15.53)

(15.48)   (15.54)

Formation of CT complexes between piperidines and hydroperoxy groups is supported by the fact that several amines can form hydrogen-bonded associates with hydroperoxides.[2830, 3637]

Reaction (15.73) raises the local concentration of piperidinoxy radicals in regions where polymer hydroperoxy groups photocleavage. Under these conditions the piperidinoxy radicals would then effectively compete with oxygen for the polymer alkyl radicals (P·) and stabilize polymer. Evidence of this mechanism is based on infrared studies[50,54,3280] and observation of an increase in piperidine adsorption from solution by oxidized polypropylene films containing higher concentrations of hydroperoxide groups.[605] This reaction has been also verified by study of the reactions between piperidine (Tinuvin 770) and t-butylhydroperoxide[3283] and di-tert-butylperoxide.[2329] In general hydroperoxy group decomposition in oxidized polypropylene at room temperature occurs very slowly.[632] Similar observations have been reported on studies of primary, secondary, and tertiary substituted hydroperoxide decomposition by piperidines and piperidinoxy radicals.[1097,3410] Piperidines decompose hydroperoxide groups rapidly at elevated temperatures ($> 100°C$).[52,54,271,605,635,657,660,1098,2329,3280,3283,3885] For that reason it is considered that processing of polymers can in part decompose added piperidines.[657,2329] Piperidines may also complex with metal ion species (Zn^{2+}, Ti^{3+}, V^{4+}, Fe^{2+}, Co^{2+}, and Fe^{3+}) producing insoluble products in benzene (with the exception of Zn^{2+}).[1080,1081] Metal traces in the form of their salts from processing equipment may also be responsible for decreasing amounts of piperidines added to a polymer. It has been considered that the metal–piperidine complex may participate in the stabilization of decomposition of hydroperoxide groups.[1080]

Piperidinoxy radicals can react with polymer alkyl radicals ($P\cdot$):[606, 614, 635, 1012, 1606–1608, 2680, 3314, 3334]

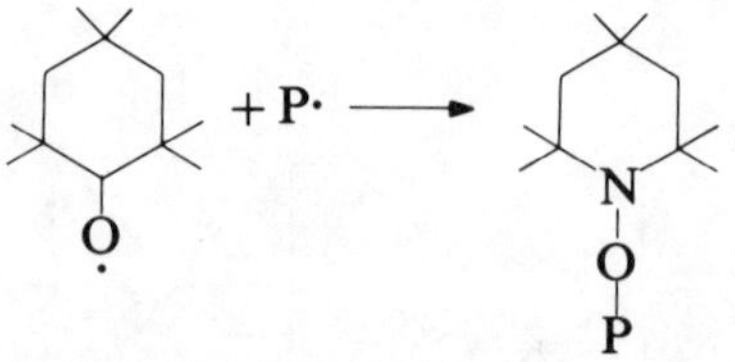

$$\tag{15.75}$$

For example bis(2,2,6,6-tetramethyl-4-piperidinyl)sebacate (Tinuvin 770) (Table 15.7) can form bis(piperidinoxy radicals (*15.55*) which, after grafting to a polymer radical ($P\cdot$) from one side, (*15.56*) can be detected by ESR spectroscopy (Fig. 15.13):[115,149,609,1606,1608]

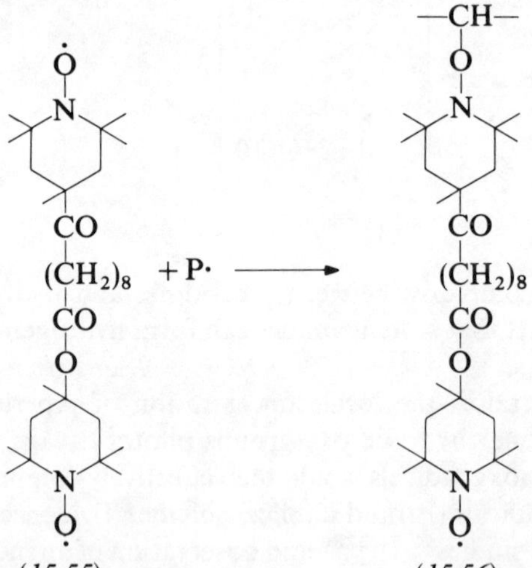

$$\tag{15.76}$$

(*15.55*)　　　　　　　　　(*15.56*)

It has also been proposed that piperidinoxy radicals in polypropylene can produce ethylenic unsaturation:[160,257,660]

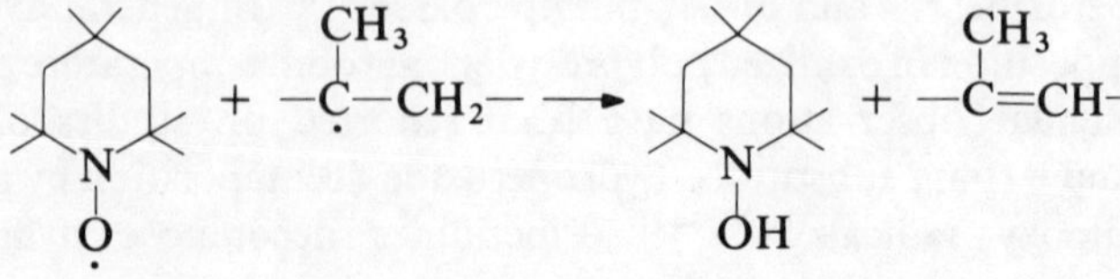

$$\tag{15.77}$$

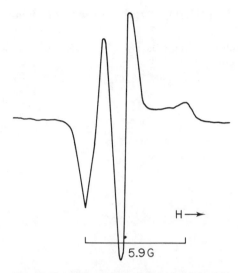

Fig. 15.13   ESR spectrum of piperidinoxy radical
grafted onto polypropylene film[1606]

Reactions occurring during processing of polypropylene can be responsible for a decrease in the piperidinoxy radical concentration.[97,657]

One of many other suggestions is that the piperidine and piperidinoxy radicals may cause a conversion of α,β to β,γ unsaturated ketone groups in photooxidized polyolefins, and in this way contribute to polymer stabilization:[52,110,115,121,146,149]

$$-CH_2-CH{=}CH-\overset{\overset{\displaystyle O}{\|}}{C}- \xrightarrow{+h\nu} -CH{=}CH-CH_2-\overset{\overset{\displaystyle O}{\|}}{C}- \qquad (15.78)$$

The photostabilization mechanism and its kinetics[606-608,614,616,617,1396,3996] are complicated because piperidine, piperidinoxy radicals, hydroxylpiperidine, and substituted hydroxylpiperidines all participate in stabilization reactions. Computer simulation has been employed to test the action of piperidine stabilizers.[1272]

Hindered piperidines are effective light stabilizers in thick layers, films, and fibres.[410,1425,2268,2269,3633] A wide range of structural types of different hindered piperidines have appeared in the patent literature and technical literature. The most commercially used hindered piperidines are listed in Table 15.7. They were mainly used for the photostabilization of poly-olefins[45,52,54,70,83,96, 103,149,153,269, 270,615,635,1012,1272, 1396,1427,1607,1608, 3724,3757] and co(ethylene/propylene/ethylidenenorbornene) (EPDM),[2743] poly-dienes,[669,1497,2328,3963,3964] poly(vinyl chloride),[2854] and acrylic/melamine coatings.[1256] Mono(hindered piperidines) are less efficient as light stabilizers than bis(hindered piperidines) (such as Tinuvin 144, Tinuvin 292, or Tinuvin

770, cf. Table 15.7), probably largely because of migration and volatilization.[615, 635, 1094, 3724, 3964]

Polymeric piperidinoxy radicals (15.57) would also be considered for the stabilization of polymers.[1960,2169]

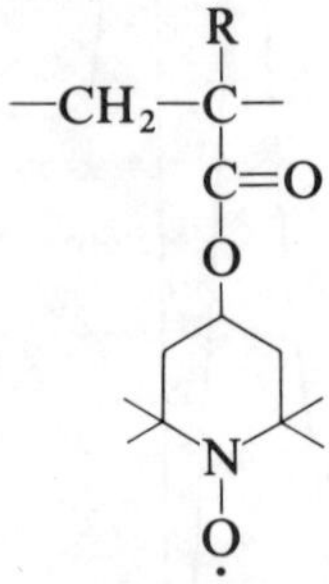

$$-CH_2-\underset{\underset{C=O}{\underset{|}{|}}}{\overset{\overset{R}{|}}{C}}-$$

(15.57)

15.11.1. Other Aspects of Photochemistry of Hindered Piperidines

Piperidines and piperidinoxy radicals may react with different organic compounds under u.v. irradiation.

Piperidinoxy radicals are capable of oxidizing aliphatic alcohols to ketones by the following reaction:[1204]

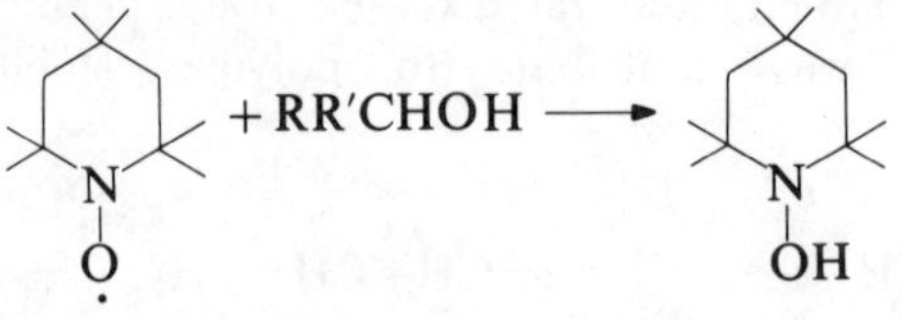

$$\text{(structure)} + RR'CHOH \longrightarrow \text{(structure)} + RCOR' + H\cdot \tag{15.79}$$

Piperidines and N-methyl substituted piperidines (15.58) are ineffective quenchers of the excited singlet (S_1) and triplet state (T_1) of carbonyl groups, whereas piperidinoxy radicals have this ability.[94,392,1094,1563,3280]

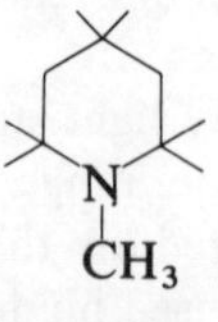

(15.58)

The quenching rate constants (k_q) determined for several hindered piperidines show that introduction of a methyl group instead of a hydrogen atom at the nitrogen atom enhances quenching.[2551]

Irradiation of diisopropyl ketone under oxygen in the presence of 4-benzyloxy-2,2,6,6-tetramethylpiperidine (Sanol 774) (15.59) results in formation of acetone, isobutyric acid, and small amounts of isopropanol. At the same time piperidine (15.59) is quantitatively oxidized to corresponding piperidinoxy radical (15.61):[1095-1098]

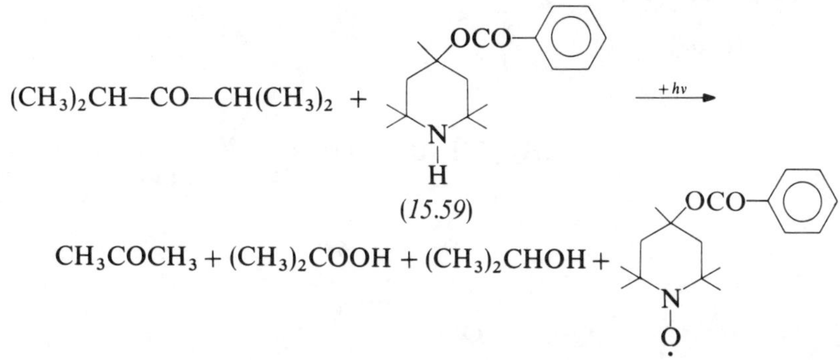

$$(CH_3)_2CH-CO-CH(CH_3)_2 \ + \ \text{(15.59)} \ \xrightarrow{+h\nu}$$

$$CH_3COCH_3 + (CH_3)_2COOH + (CH_3)_2CHOH + \ \text{(15.60)} \qquad (15.80)$$

It has been suggested that the isopropylacylperoxy radical (15.61) is responsible for the oxidation of piperidine to piperidinoxy radical (15.60):

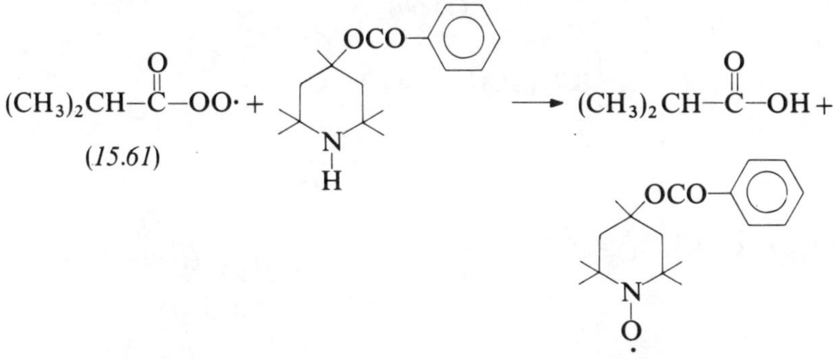

$$(CH_3)_2CH-\overset{O}{\overset{\|}{C}}-OO\cdot \ + \ \text{(15.59)} \ \longrightarrow \ (CH_3)_2CH-\overset{O}{\overset{\|}{C}}-OH \ + \ \text{(15.60)}$$

$$(15.61) \qquad\qquad\qquad (15.81)$$

Irradiation of diisopropyl ketone in the presence of piperidinoxy radical (15.60) and oxygen gives a high yield of acetone and isobutyric acid (15.66).[1098] During this reaction concentration of piperidinoxy radical is practically unchanged. Two possible forms of this reaction have been proposed.

(i) Formation of charge-transfer complexes (15.63) and (15.64) between iso-propylperoxy (15.62) and isopropylacylperoxy (15.61) radicals and piperi-dinoxy radicals (15.60): (reactions 15.82 and 15.83)
   In the next step the CT complexes reacts with the remaining isopropylacyl-peroxy (15.61) or isopropyl (15.65) radicals to form acetone and isobutyric acid (15.66): (reactions 15.84 and 15.85)
(ii) Formation of hydroxypiperidine isopropyl ether (15.67), which further undergoes cleavage by the isopropylacylperoxy radical (15.61) with formation of acetone and isobutyric acid, and liberation of the piperidinoxy radical (15.60): (reactions 15.86 and 15.87).

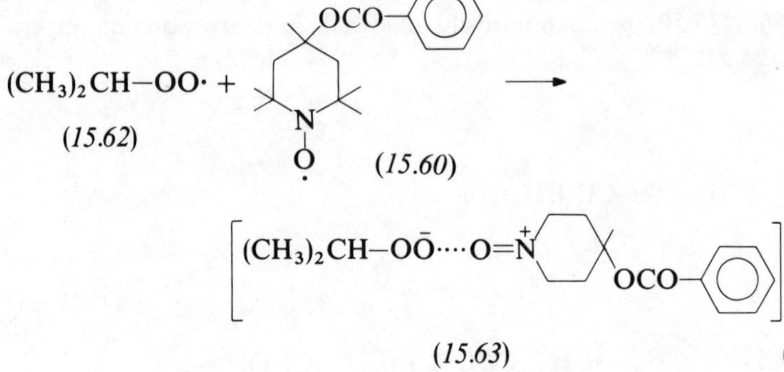

$$(CH_3)_2CH-OO\cdot +\ \text{(15.60)} \longrightarrow$$

(15.62)

$$\left[(CH_3)_2CH-O\bar{O}\cdots O=\overset{+}{N} \diagdown\!\!\diagup OCO\!\!-\!\!\bigcirc \right]$$

(15.63)   (15.82)

$$(CH_3)_2CH-\overset{O}{\overset{\|}{C}}-OO\cdot +\ \text{(15.60)} \longrightarrow$$

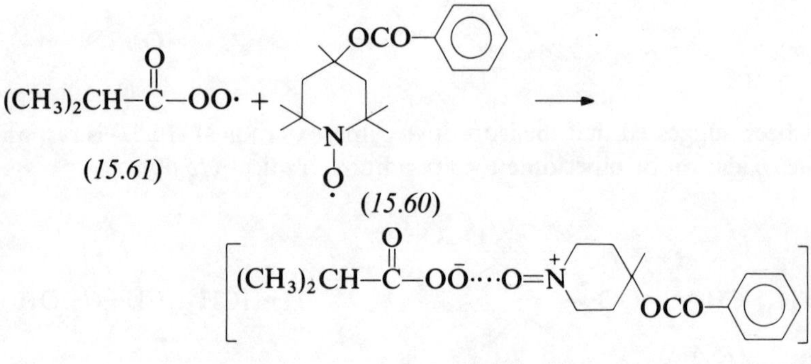

(15.61)

$$\left[(CH_3)_2CH-\overset{O}{\overset{\|}{C}}-O\bar{O}\cdots O=\overset{+}{N} \diagdown\!\!\diagup OCO\!\!-\!\!\bigcirc \right]$$

(15.64)   (15.83)

$$\left[(CH_3)_2CH-O\bar{O}\cdots O=\overset{+}{N} \diagdown\!\!\diagup OCO\!\!-\!\!\bigcirc \right] + (CH_3)_2CH-\overset{O}{\overset{\|}{C}}-OO\cdot \longrightarrow$$

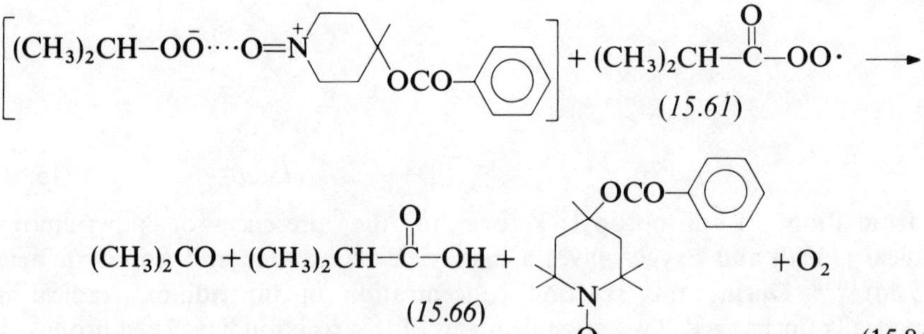

(15.61)

$$(CH_3)_2CO + (CH_3)_2CH-\overset{O}{\overset{\|}{C}}-OH +\ \text{OCO}\!\!-\!\!\bigcirc + O_2$$

(15.66)   (15.84)

$$\left[(CH_3)_2CH-\overset{O}{\overset{\|}{C}}-O\bar{O}\cdots O=\overset{+}{N} \diagdown\!\!\diagup OCO\!\!-\!\!\bigcirc \right] + (CH_3)_2CH\cdot \longrightarrow$$

(15.65)

$$(CH_3)_2CO + (CH_3)_2CH-\overset{O}{\overset{\|}{C}}-OH +\ \text{OCO}\!\!-\!\!\bigcirc + O_2$$

(15.66)   (15.85)

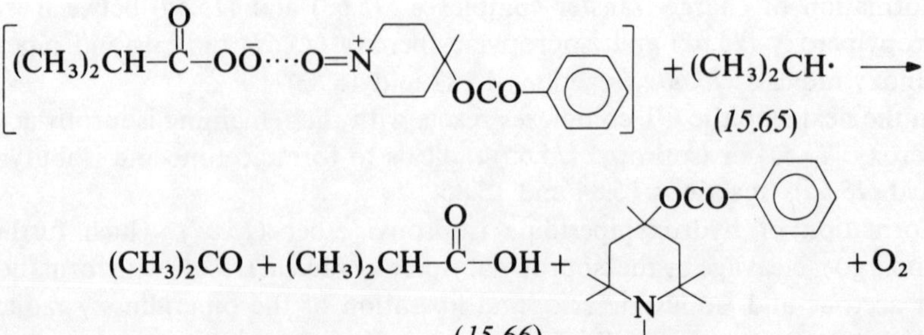

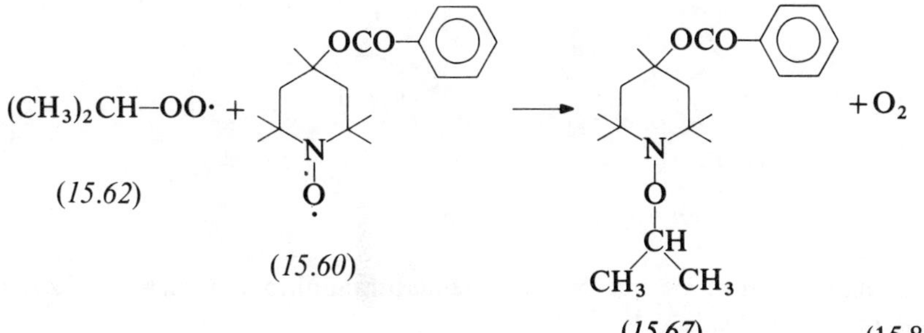

(*15.62*) (*15.60*) (*15.67*) (15.86)

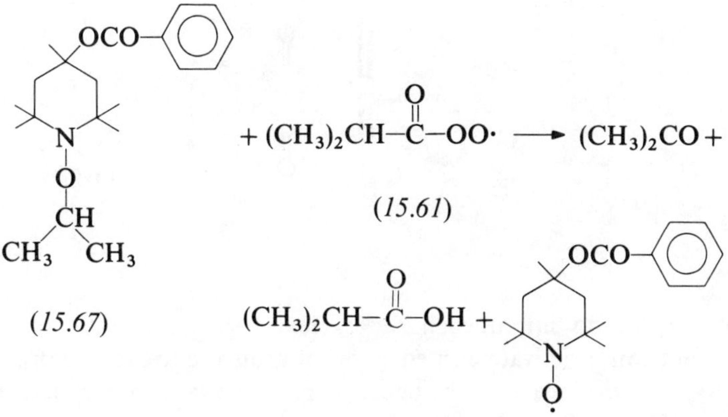

(*15.67*) (*15.61*) (*15.60*) (15.87)

These mechanisms are a little controversial, because it has been reported that the peroxy radical (ROO·) does not react with piperidinoxy radicals.[535,2677]

It has been reported that benzophenone accelerates formation of piperidinoxy radicals from piperidines under u.v. irradiation.[2329] The mechanism of this reaction is not very clear, because benzophenone ketyl radicals (*15.68*) are not capable of abstracting hydrogen from piperidine[83,101] but they can react with piperidinoxy radicals according to the reaction:[47,83,98]

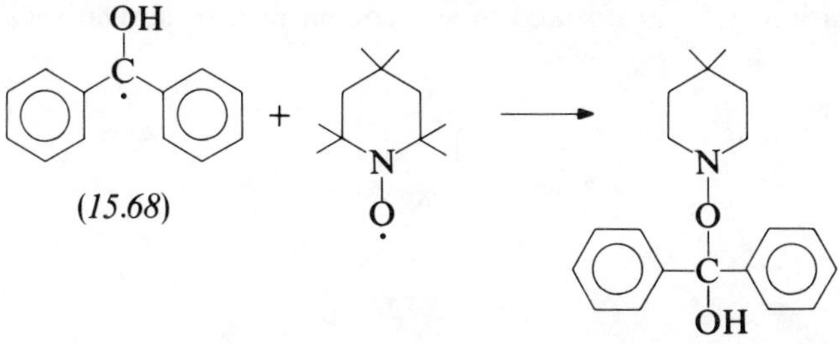

(*15.68*) (15.88)

Piperidinoxy radicals added to *p*-quinone inhibit formation of hydroquinone, probably by reaction with semiquinone (*15.69*) radicals:[99]

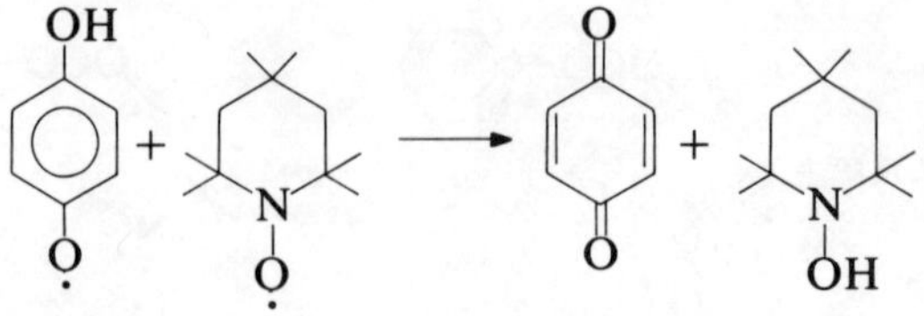

(15.89)

(15.69)

A similar reaction occurs between semianthraquinone radicals (15.70) and piperidinoxy radicals:[48,49,52,83]

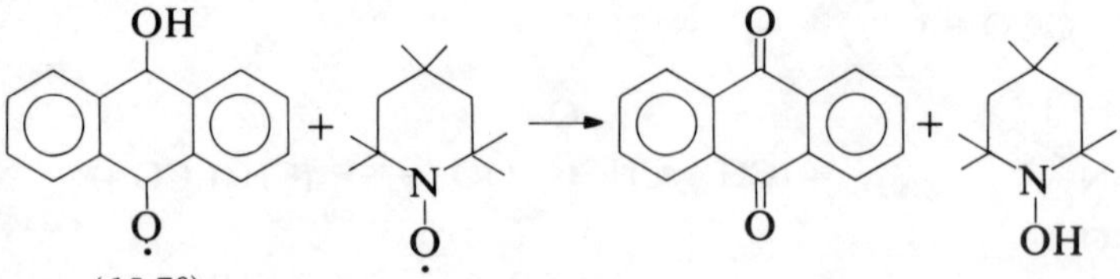

(15.70)

(15.90)

Piperidinoxy radicals do not quench the excited triplet state (T_1) of anthraquinone,[79] but can deactivate excited states of aromatic hydrocarbons.[3811]

Piperidinoxy radicals react with benzoyl peroxide and form nitrones (15.71):[2542]

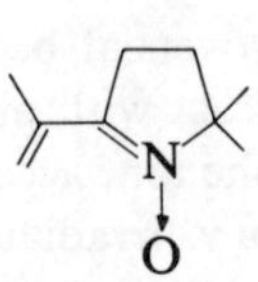

(15.71)

It has been reported that piperidinoxy radicals, during thermal oxidation and photooxidation, are transformed to subsequent nitroso derivatives (15.72):[3269]

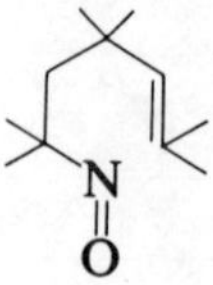

(15.72)

Bis(2,2,6,6-tetramethyl-4-piperidinyl)sebacate (Tinuvin 770) (15.73) reacts with 2-hydroxyanthraquinone (15.74) yielding a red-orange bis[N-(2-anthraquinonyl)2,2,6,6-tetramethyl-4-piperidinyl]sebacate (15.75):[101]

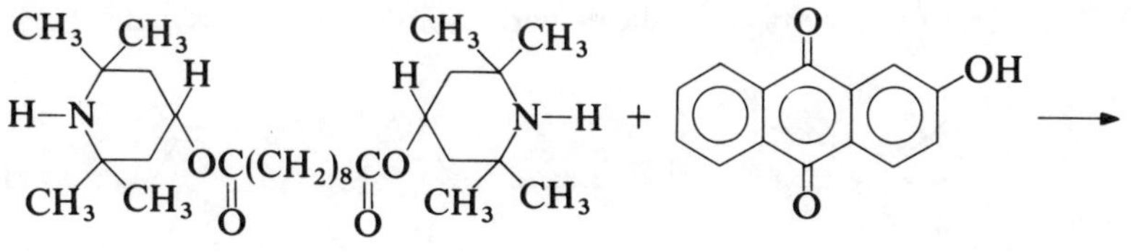

(15.73)   (15.74)

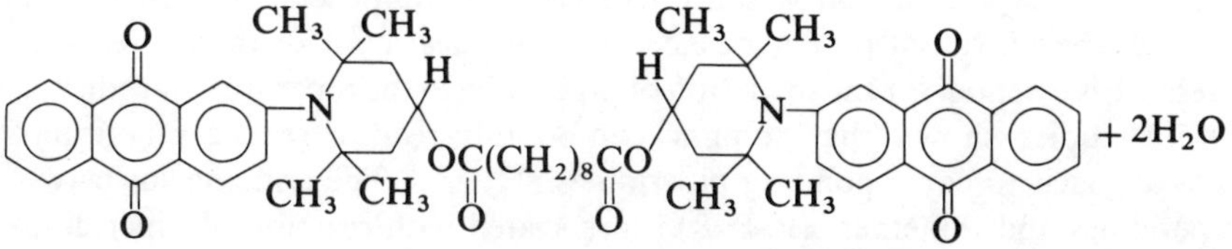

(15.75)   (15.91)

Piperidines do not react with 2,2-diphenyl-1-picrylhydrazyl (DPPH), which is an effective free radical scavenger,[3410] whereas they do react with 2,4,6-tri-*t*-butyl-phenoxy radicals.[3283]

Piperidines may also react with singlet oxygen ($^1O_2$) but this reaction is not very effective (cf. Section 15.15).

15.12. INTERACTION OF HINDERED PIPERIDINES WITH DIFFERENT ADDITIVES

Despite the high efficiency of these stabilizers they often interact unfavourably with many other additives used in commercial polymers and plastics. The synergistic/antagonistic effects depend on the structure of additive and on the structure of piperidine derivatives. In this field many controversial results are published on the synergistic/antagonistic activity for a given piperidine–additive system. This can be the result of different experimental conditions in which such effects were evaluated.

The antagonism between piperidines and phenolic antioxidants[44,45,49,52,53,97,98,118,146,657,2328,3633] may be associated with the following processes:

(i) Inhibition of hydroperoxide formation by the antioxidant thus preventing mechanism given by reactions (15.73).[45,49,52,146,148]

(ii) Oxidation of the hindered phenol (*15.76*) to an active cyclohexadienonyl radicals (*15.77*):[44, 49, 52, 53, 2328]

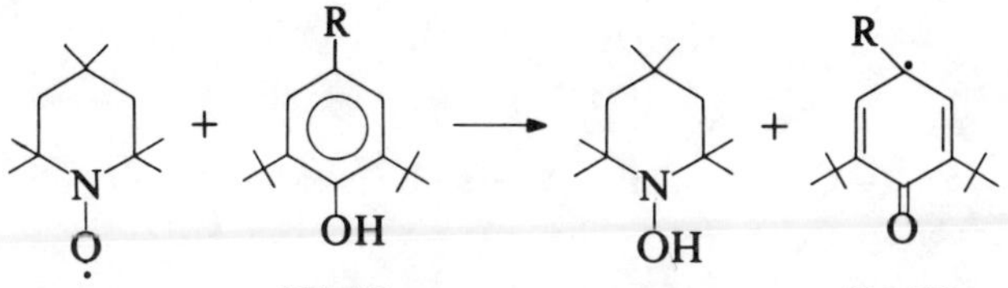

(15.92)

(15.76)   (15.77)

(iii) Reaction of piperidinoxy radicals with cyclohexadienonyl radicals:[2328]

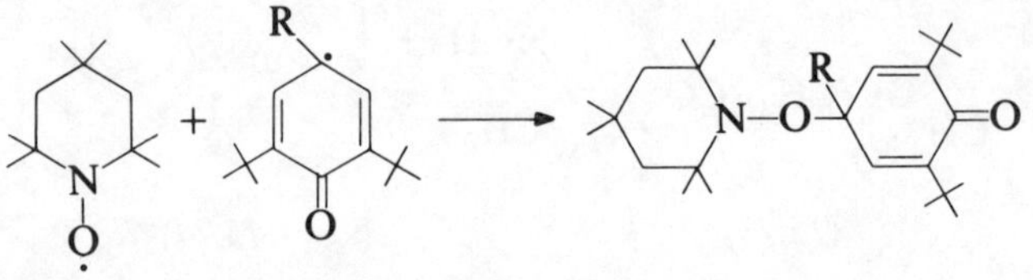

$$(15.93)$$

Metal dialkyl dithiocarbamates (peroxide decomposers) are antagonistic towards the activity of piperidines, because they destroy hydroperoxides, which presence is essential for the oxidation of piperidine to the piperidinoxy radical.[657] Other studies shown that sulphur compounds readily reduce piperidinoxy radicals back to corresponding piperidines.[2480, 2599] The antagonism between piperidines and thioethers (R–S–R) is associated with reaction of piperidinoxy radicals with sulphenyl radicals (15.78) which gives inactive sulphonpiperidines (15.79):[3280]

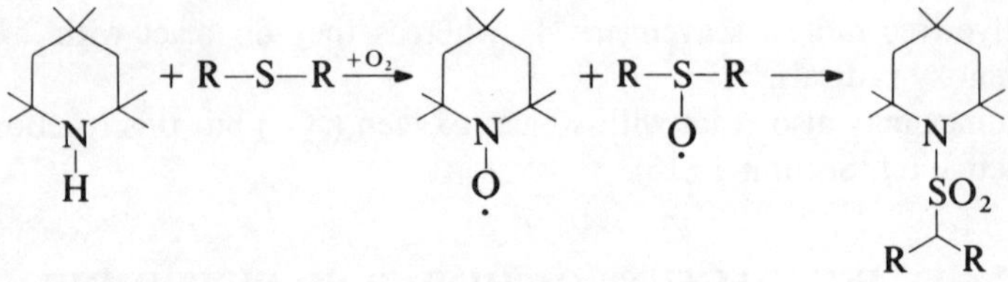

$$(15.94)$$

$$(15.78) \qquad (15.79)$$

Piperidinoxy-substituted metal xanthates (15.80) and piperidinoxy-substituted disulphides (15.81) show antagonism under u.v. irradiation in comparison to the simple alkyl xanthates (15.82) and corresponding disulphides (15.83):[167]

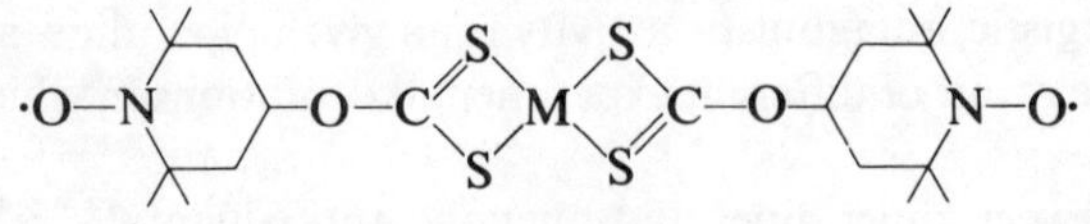

$$(15.80)$$

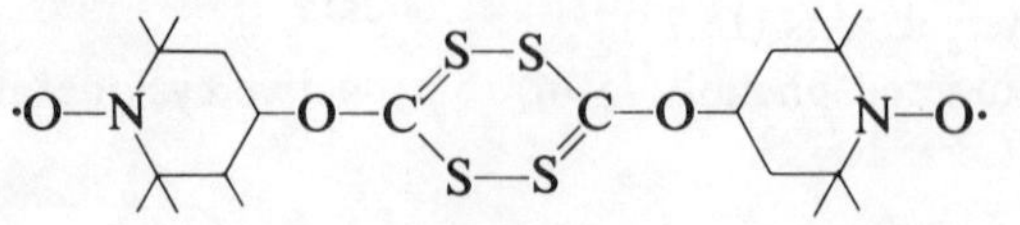

$$(15.81)$$

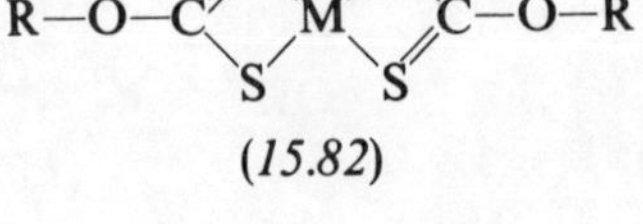

$$(15.82)$$

$$R-O-C\underset{S-S}{\overset{S-S}{\diagdown}}C-O-R \qquad R = C_8H_{17}$$

(15.83)

Interaction of piperidines with *o*-hydroxybenzophenones,[45,101,269,270,1426, 1608] *o*-hydroxyphenylbenzotriazoles,[45,98,269] nickel chelates,[45,53,73,97,269,270] and 1,4-diazabiscyclo(2,2,2) octane (DABCO)[45] is not synergistic but nevertheless greater than that of the original piperidine alone.

Piperidines can form complexes with coloured pigments such as phthalocyanine, chromic oxide, and cadmium yellow.[155] Interaction of piperidines with pigments, often enhanced, is associated with the ability of the pigment to absorb the additives into its surface, thereby increasing the effect of any potential interactions.[100,155]

15.13. ANTIOXIDANTS

Antioxidants are classified into two groups according to their protection mechanisms:

(i) Kinetic chain-breaking antioxidants.
(ii) Peroxide decomposers.

Their stabilization mechanism on polymer oxidative degradation has been the subject of several reviews.[2978,2979,3060,3243,3246,3255,3257]

Kinetic chain-breaking antioxants include a number of compounds such as hindered amines (HALS) (cf. Section 15.11), hindered phenols, and many others able to produce stable radicals which may act as free radical scavengers for polymer alkyl (P·), polymer alkylperoxy (POO·) and polymer alkyloxy (PO·) radicals, as well for a low molecular radicals such as R·, ROO·, RO·, HO·, etc.

The radical chain-breaking reaction is, in general, a complicated process in which phenoxy radicals *(15.84)* formed during the oxidation of phenols, and many other radicals formed after different transformation reactions, act as free radical scavengers for radicals formed during thermal and/or photooxidative degradation of polymers.[157,389,2328,2429,2978–2980,3175,3519,3997]

The stability and reactivity of phenoxy radicals is determined by steric effects of the substituents R_1, R_2, and R_3,[178,2371,3868] and by the extent of delocalization of the unpaired electron.[3110,3111] In the series of phenoxy radicals the lifetime increases as steric hindrance in the ortho- and para-position increases:

(15.84)

(15.95)

Radical lifetime →

← Radical reactivity

Phenoxy radicals with small substituents (R_1 and R_2) at the ortho-positions are less stable than those with bulky groups because of the ease of dimerization. A bulky substituent such as a tertiary butyl group in the ortho- or para-position to the phenolic carbon has a significant steric hindrance effect, thus minimizing dimerization. The following order of steric hindrance is valid:[549]

$$\textit{tert}\text{-butyl} > \text{benzyl} > \text{alkyl} > \text{methyl} > \text{hydrogen}$$

The formation and concentration of phenoxy radicals can be determined by ESR spectroscopy.[1009, 3062] The spin concentration measurements can provide a basis for comparing the radical scavenging activity of different antioxidant systems. The concentration of radicals formed from antioxidant cannot be measured directly within the bulk polymer, because of their low concentration in relation to the formed polymer radicals. With a model system composed of the antioxidant in solution with a known concentration of initiating radicals it is possible to measure ESR spectra of antioxidant radicals and their concentration.[1009]

The hindered phenoxy radicals (15.84) depending on substituents can be more or less stable (even over a period of 400 hours), because they can isomerize into more resonance-stabilized cyclohexadienonyl radicals (15.85) and (15.86):

$$R_1 \overset{\cdot O}{\underset{R_3}{\bigcirc}} R_2 \longrightarrow R_1 \overset{O}{\underset{\overset{\cdot}{R_3}}{\bigcirc}} R_2 \rightleftharpoons R_1 \overset{O}{\underset{R_3}{\bigcirc}} \overset{\cdot}{R_2} \tag{15.96}$$

$$(15.84) \qquad (15.85) \qquad (15.86)$$

The rate of isomerization is of the order of magnitude 10^8 litres $mol^{-1} s^{-1}$.[699, 1004]

In secondary reactions cyclohexadienonyl radicals, depending on the steric effects of R_1, R_2, and R_3 groups, undergo a C–C coupling reaction to cyclohexadienone dimers (15.87):[549, 2422, 2769, 3279]

$$R_1 \overset{O}{\underset{\overset{\cdot}{R_3}}{\bigcirc}} R_2 + R_1 \overset{O}{\underset{\overset{\cdot}{R_3}}{\bigcirc}} R_2 \longrightarrow O = \overset{R_2 \quad\quad R_1}{\underset{R_1 \quad\quad R_2}{\bigcirc \overset{R_3}{\underset{R_3}{}} \bigcirc}} = O \tag{15.97}$$

$$(15.87)$$

The C–C coupling reaction occurs for cresol ($R_1 = H$ and $R_3 = CH_3$),[1902] for 2,4- or 2,6-dialkyl phenols,[1671, 1672, 2019, 2372, 3810, 3977] and for 2,4,6-trisubstituted phenols which have a non-bulky and readily cleavable substituent in the ortho- or para-position.[178]

The cyclohexadienones formed from disubstituted phenols may, under u.v. irradiation, intermolecularly isomerize to 4,4'-biphenyldiols (*15.88*), which can be further oxidized to diphenylquinones (*15.89*):[1672]

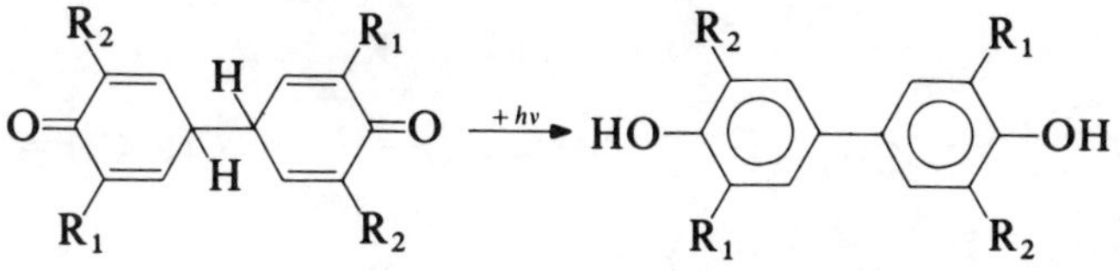

$$(15.98)$$

(*15.88*)

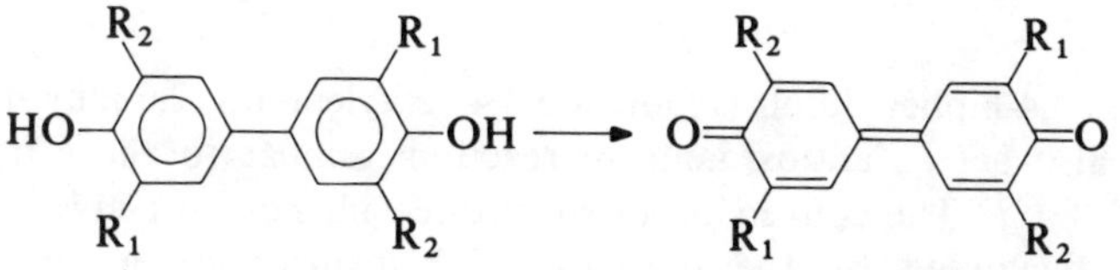

$$(15.99)$$

(*15.89*)

Phenols which have a methyl group in the para-position (*15.90*) may, in the presence of alkyloxy (RO·) and/or alkylperoxy (ROO·) radicals, isomerize to 4,4'-ethylene-bis(2,6-alkylphenol) (*15.91*) according to the following mechanism:[332,1747]

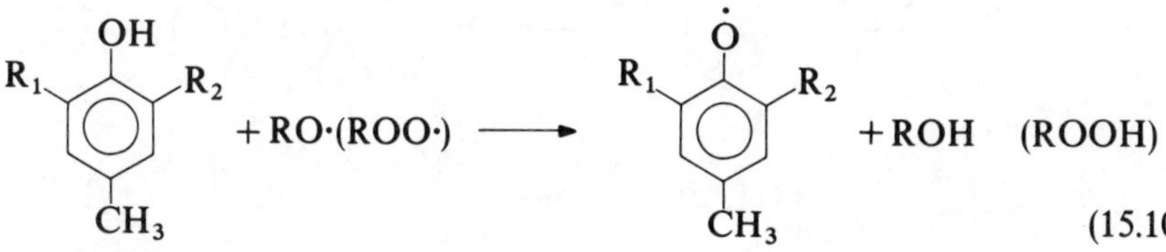

$$(15.100)$$

(*15.90*)

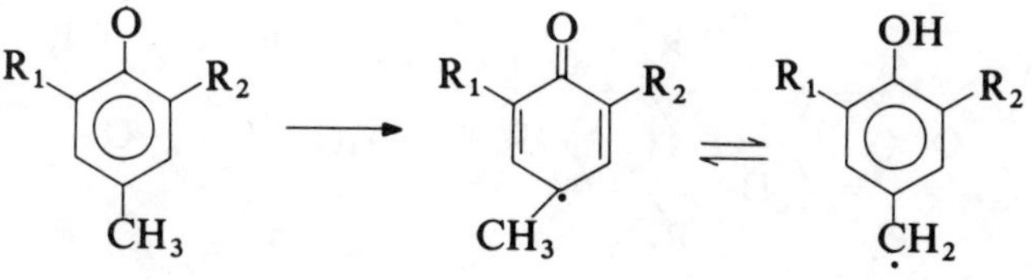

$$(15.101)$$

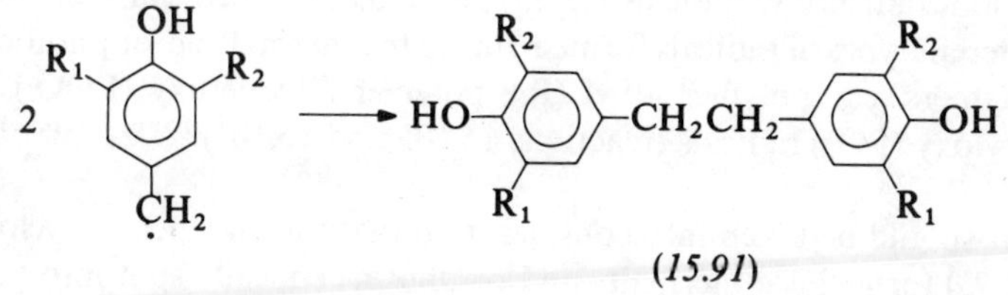

(*15.91*)

$$(15.102)$$

During the processing of polymers formation of stilbenequinone compounds (*15.92*) has been observed by the oxidation process of 4,4'-ethylene-bis(2,6-alkylphenol) (*15.91*):[332,1747,2157,3273]

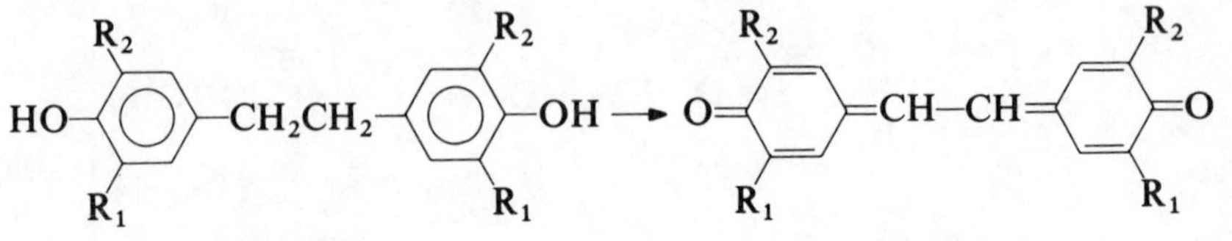

$$(15.91) \qquad (15.92)$$

(15.103)

Substituted 4,4-biphenyldiols (*15.88*) and 4,4'-ethylene-bis (2,6-alkyl-phenols) (*15.91*) can also act as antioxidants in reactions similar to those described previously.[2978,3255] The conversion of substituted phenols to active oxidation products is facilitated by the presence of hydroperoxide groups in the polymer.[2157,3273]

Phenoxy radicals (*15.84*) may also react with cyclohexadienonyl radicals (*15.85*) (C–O coupling) according to the reaction:

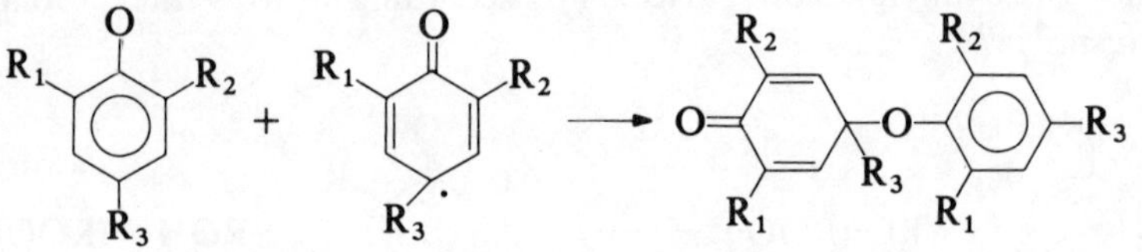

$$(15.84) \qquad (15.85) \qquad (15.104)$$

In the case of disubstituted phenols (where R_3 = H, e.g. 2,6-di-*tert*-butylphenol), under u.v. irradiation, intermolecular isomerization may occur:[1672,1902]

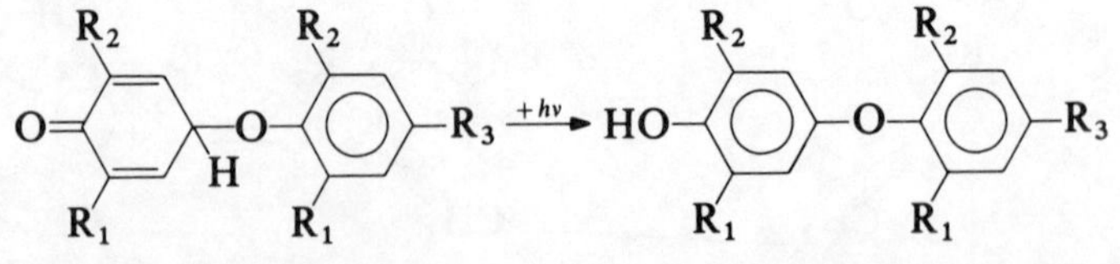

(15.105)

It is generally accepted that the chain-breaking reaction involves mainly cyclohexadienonyl radicals (*15.85*) and (*15.86*), which can react (scavenge) with different types of radicals formed during the thermal and/or photodegradation of polymers, e.g. polymer alkyl (P·), polymer alkylperoxy (POO·), and polymer alkyloxy (PO·) radicals (reactions 15.105 and 15.106):[1566, 2328, 2978, 2979, 3041, 3255]

It should be taken into consideration that the alkylperoxycyclohexadienones (*15.93*) formed are thermally and/or photochemically split into a new group of oxy radicals (*15.94*) (reaction 15.108):

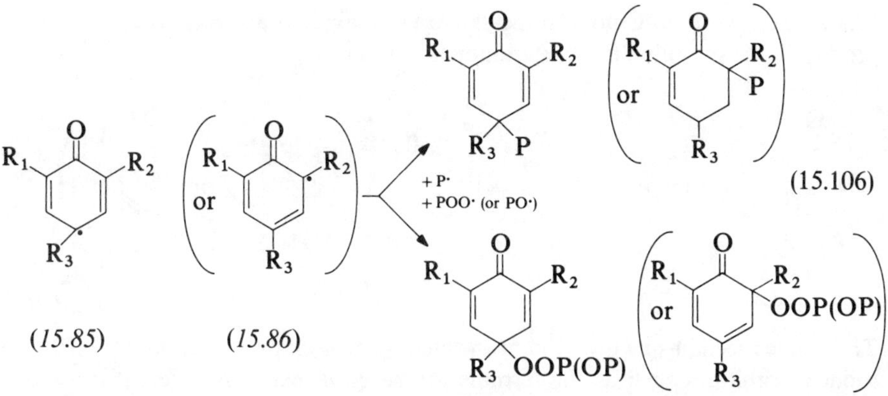

$$ (15.106) $$

$$ (15.107) $$

(15.85)   (15.86)

(15.93)

(15.93)   (15.94)

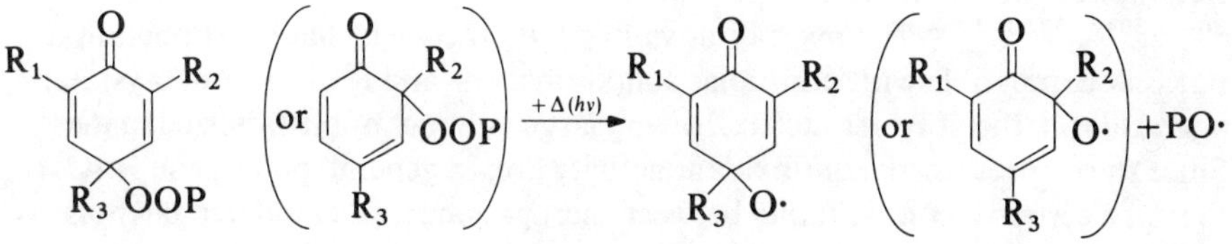

$$ (15.108) $$

$+\Delta (h\nu)$

$+PO\cdot$

Cyclohexadienonyl radicals (15.85) and (15.86) may also react with oxygen and form cyclohexadienonyl-peroxy radicals (15.95):

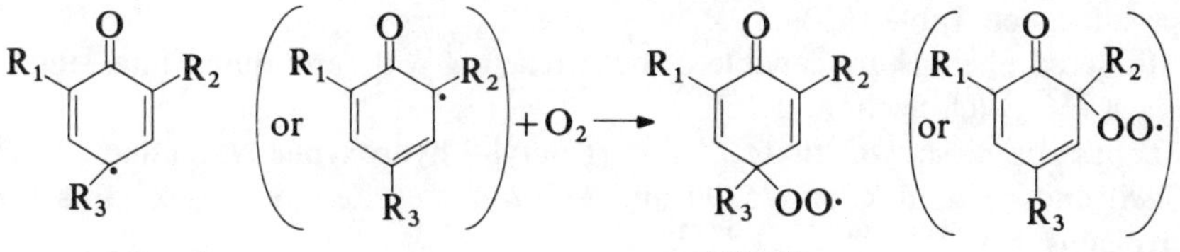

(15.85)   (15.86)   (15.95)

$+O_2 \longrightarrow$

$$ (15.109) $$

Both radicals (15.94) and (15.95) may further initiate free radical oxidation (they are pro-oxidants) and/or degradation and/or crosslinking of polymers.

Alkyloxycyclohexadienone radicals (15.94) can undergo α-cleavage reaction with formation of corresponding quinones (15.96) and alkyl radicals:

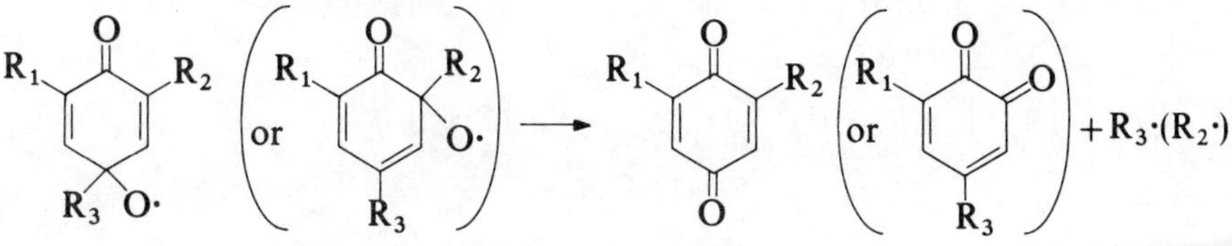

(15.94)   (15.96)

$+R_3\cdot (R_2\cdot)$

$$ (15.110) $$

Alkyloxycyclohexadienone radicals (*15.94*) may also abstract hydrogen and produce corresponding hydroxyquinones (*15.97*):

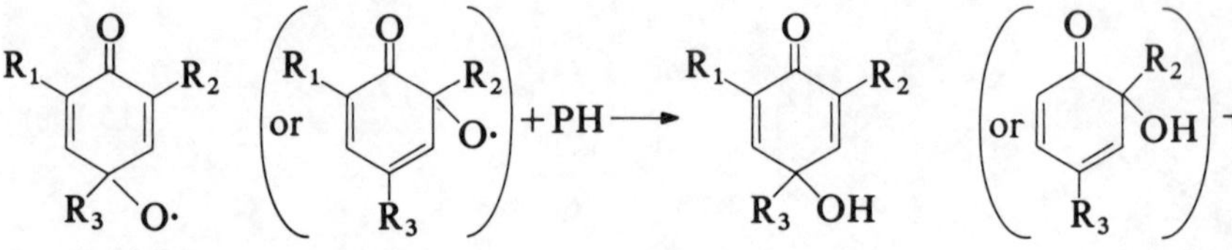

(*15.94*)   (*15.97*)

(15.111)

The above mechanisms involved in the chain-breaking processes do not include other possibilities such as the termination reaction between different types of radicals formed, transformation of different products, etc.

Hindered phenols have been carefully investigated, as antioxidants against free-radical photooxidation of polymers.[614, 1566, 2118, 2121, 2123, 2124, 2745, 2978, 3041, 3043, 3758, 3964] The main disadvantage in practice is that most phenols are not very compatible with non-polar solid polymers, and tend to form crystalline aggregates in the polymer matrix, leaving large volumes of unstabilized material. Since the centres for the antioxidant activity are, in general, polar groups (OH), there is a certain incompatibility between inert polymers and hindered phenols. In practice this effect can be overcome by incorporating hydrophobic groups into the phenol molecules. Examples of some commercially available antioxidants on the base of hindered phenols are shown in Table 15.8.

The stabilizing efficiency of hindered phenols can be enhanced by coupling them with other compounds, e.g. with imidazolyl groups[2326] or with hindered piperidines (cf. Table 15.7).

Hindered phenols are capable of both reacting with, and quenching, singlet oxygen ($^1O_2$) (cf. Section 15.15).

It has been shown that 3,5-di-*tert*-butyl-4-hydroxyphenyl-methane-thiol (*15.98*) and its analogues (*15.99*) and (*15.100*) are effective antioxidants for polyolefins:[1086, 2100, 3261, 3270–3273]

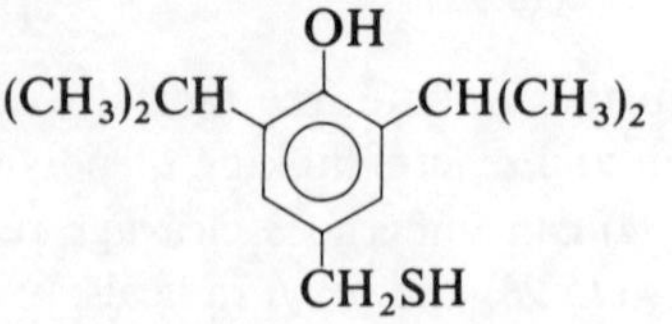

(*15.98*) 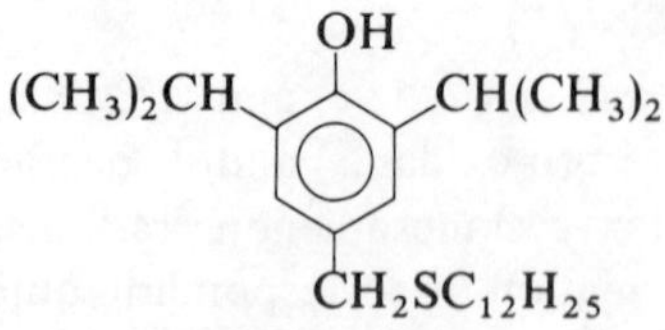   (*15.99*)

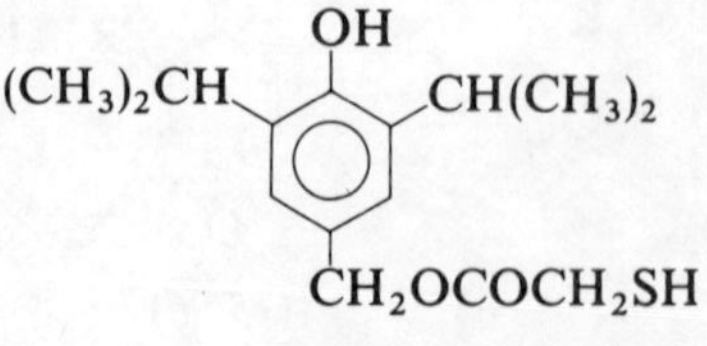

(*15.100*)

Table 15.8 Examples of commercially available phenolic antioxidants[157]

Name		Commercial name
Cetyl-3,5-di-*tert*-butyl-4-hydroxybenzoate	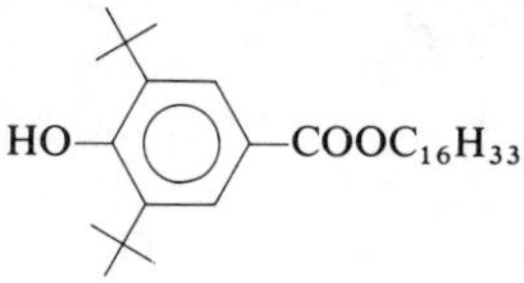	Cyasorb UV 2908
4-Hydroxy-3,5-di-*tert*-butylbenzoic acid	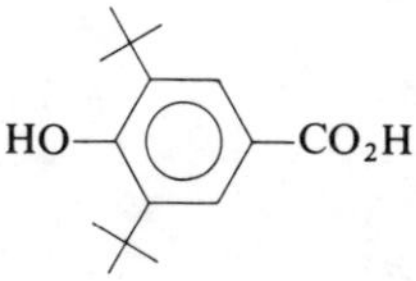	Shell Acid
1,3,5-tris(4-*tert*-butyl-3-hydroxy-2,6-dimethylbenzyl)	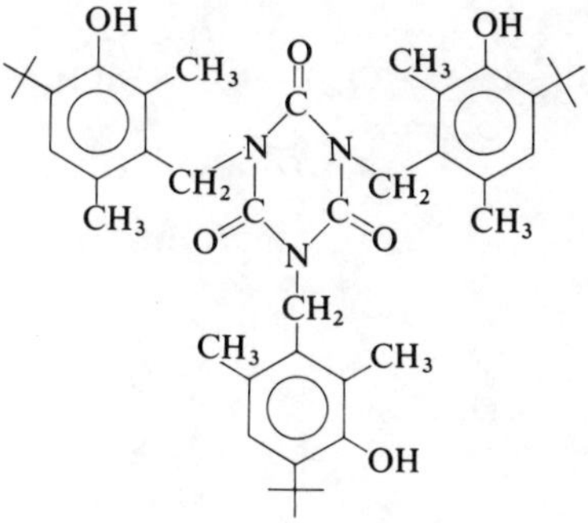	Cyanox 1790
Stearyl-3-(3,5-di-*tert*-butyl-4-hydroxyphenyl)proprionate	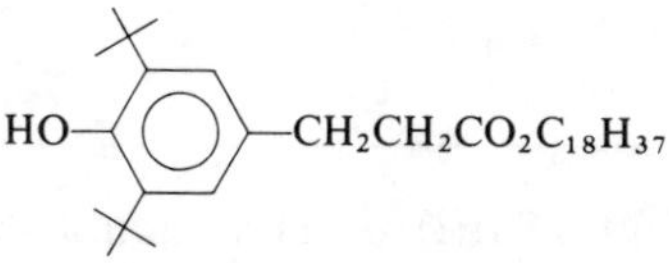	Irganox 1076
Pentaerithrityl tetrabis (3,5-di-*tert*-butyl-4-hydroxyphenyl)	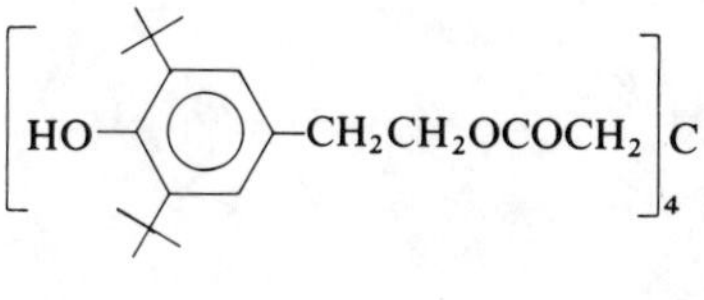	Iraganox 1010
2,6-di-*tert*-butyl-4-methylphenol	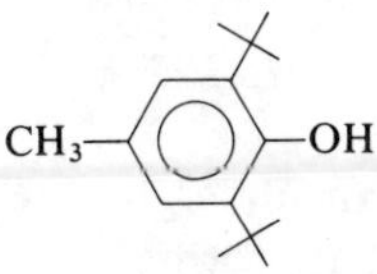	Topanol OC

Table 15.8   (Contd.)

Name		Commercial name
1,3,5-tris(3,5-di-*tert*-butyl-4-hydroxybenzyl)isocyanurate	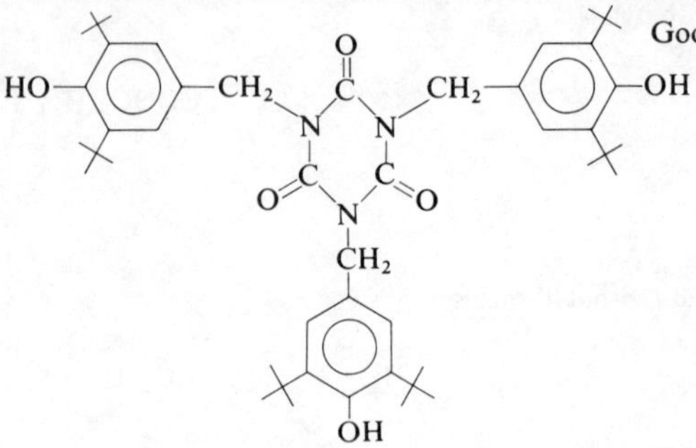	Goodrite 3114
Tris[3-(3′,5′-di-tert-butyl-4-hydroxybenzyl)-2″-acetohexyl]isocyanurate	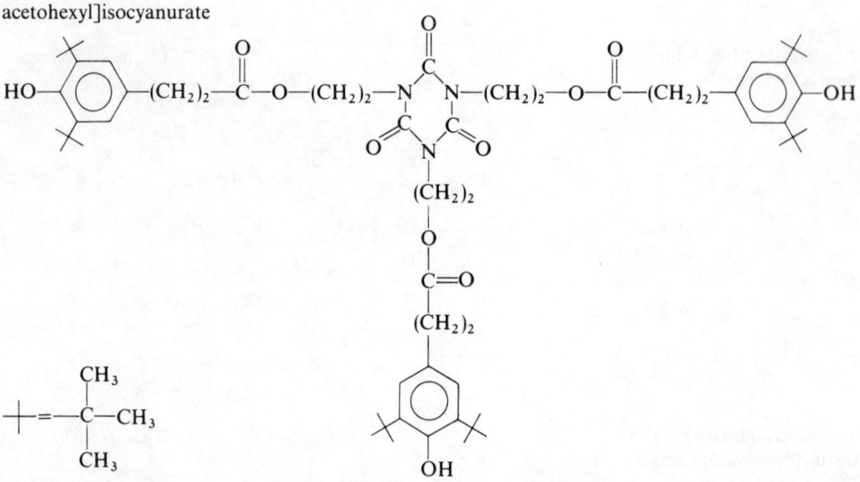	Goodrite 3125

Addition of antioxidant (*15.101*) to olefinic double-bonds suggests that the chain-breaking reactions occur predominantly through the thiol radical (*15.103*) rather than the phenoxy radical (*15.102*):[3261]

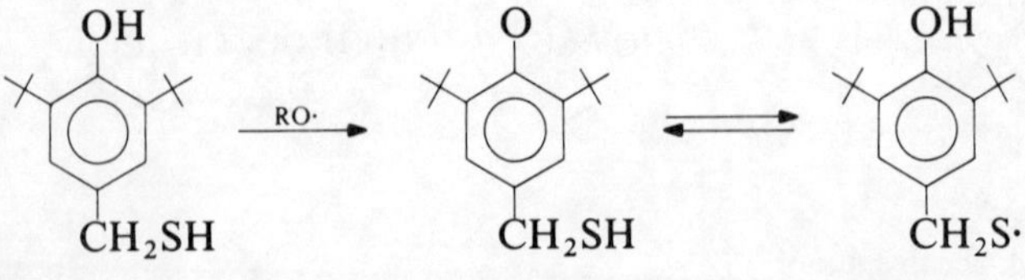

The phenoxy radical (*15.102*) can undergo a reaction during which quinone methide (*15.104*) and ṠH radical are formed:

(*15.102*) (*15.104*)

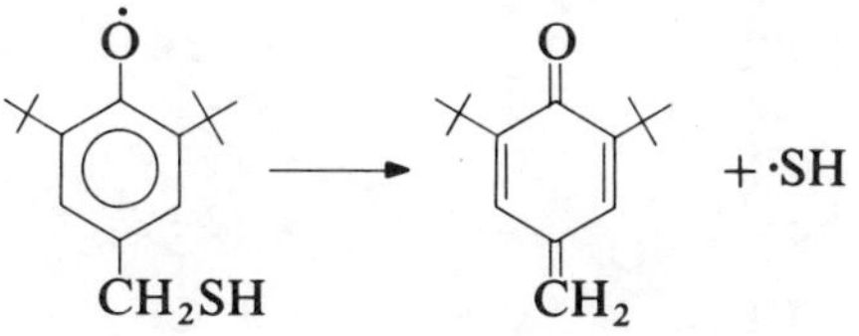

(15.113)

The thiol radical (*15.103*) can react with hydroperoxy groups (OOH) by the following mechanism:[3261]

(*15.103*) (*15.105*)

(15.114)

(15.115)

(*15.106*)

(15.116)

(*15.107*)

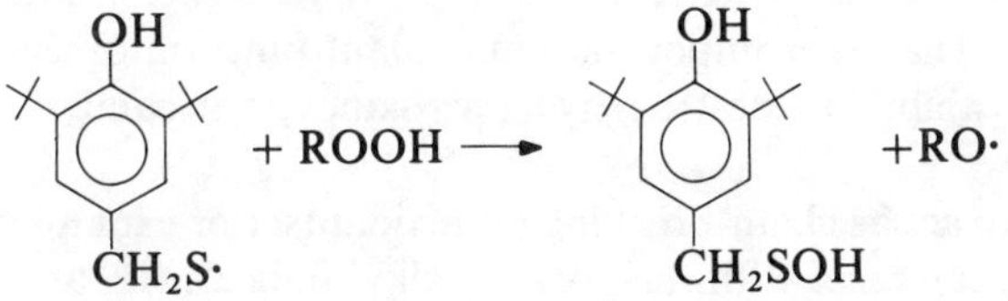

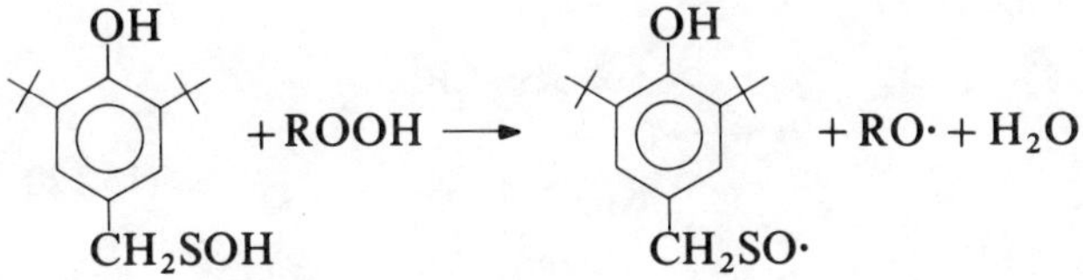

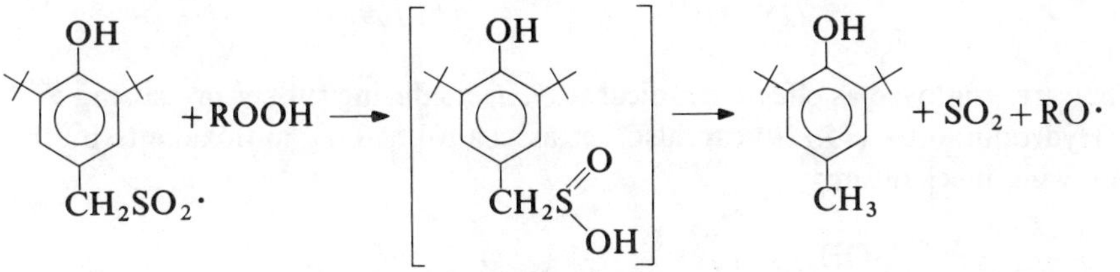

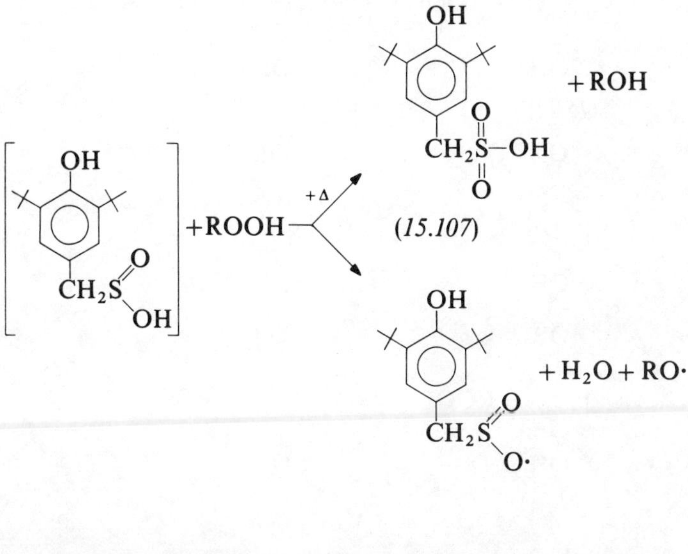

(15.117)

(15.118)

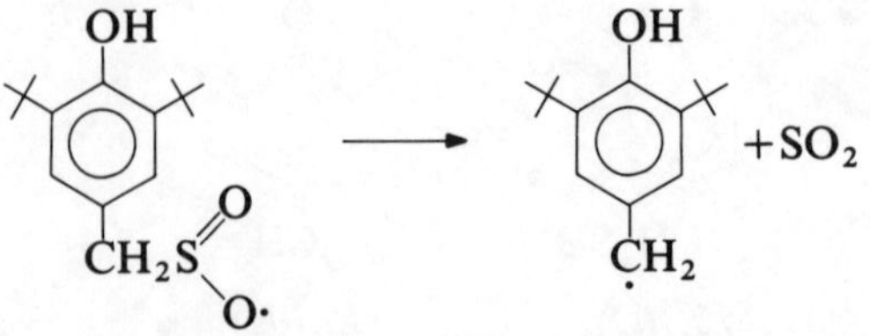

$$+ SO_2 \qquad (15.119)$$

In this reaction a number of products are formed, such as sulphenic acids (15.105),[217,218] sulphinic acids (15.106),[216,1721] sulphonic acids (15.107),[216,1721] and sulphur dioxide.[1720] These are effective radical scavengers and participate in the antioxidation mechanism. The other important antioxidant function of the acidic sulphur species is the ability to destroy hydroperoxides in a catalytic process.[1624,3256]

Many other compounds can act as chain-breaking antioxidants. For example substituted quinones (15.108) can react with the polymer alkyl radicals (P·) and produce phenoxy radicals (15.109):

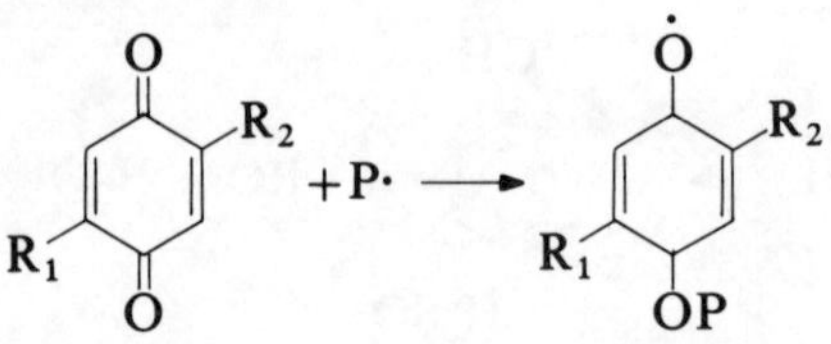

$$+ P· \longrightarrow \qquad (15.120)$$

(15.108)   (15.109)

They are employed as effective radical scavengers during rubber processing.[3812]

Hydroquinones (15.110) can also act as chain-breaking antioxidants by the following mechanism:[3255]

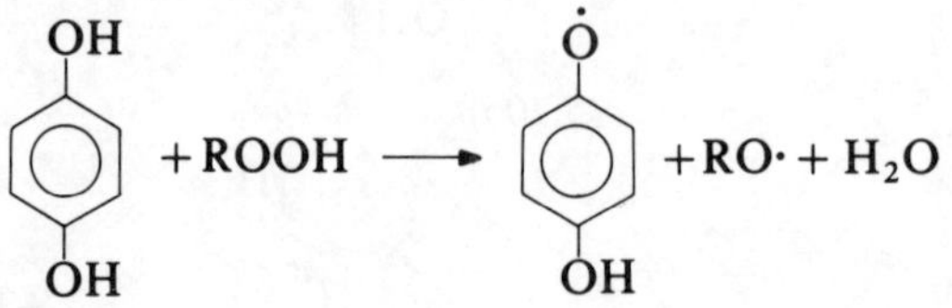

$$+ ROOH \longrightarrow \qquad + RO· + H_2O \qquad (15.121)$$

(15.110)

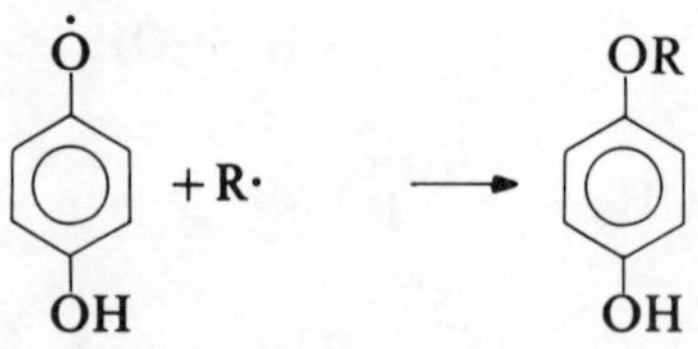

$$+ R· \longrightarrow \qquad (15.122)$$

$$+ ROO· \longrightarrow \qquad + ROOH \qquad (15.123)$$

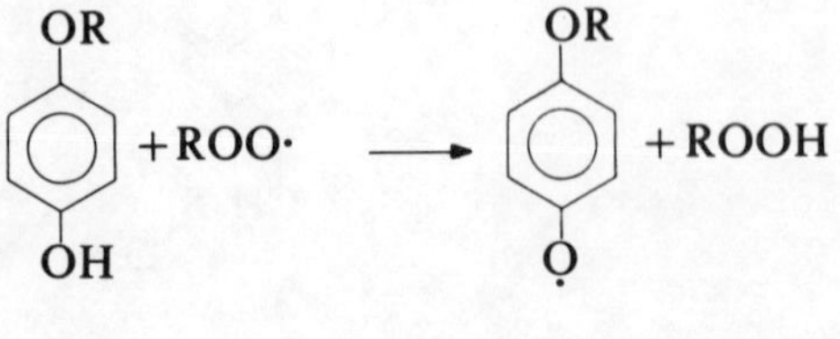

Hydroxygalvinol (15.111) reacts with alkylperoxy (ROO·) radicals, producing a stable galvinoxy radical (15.112):[661,3257]

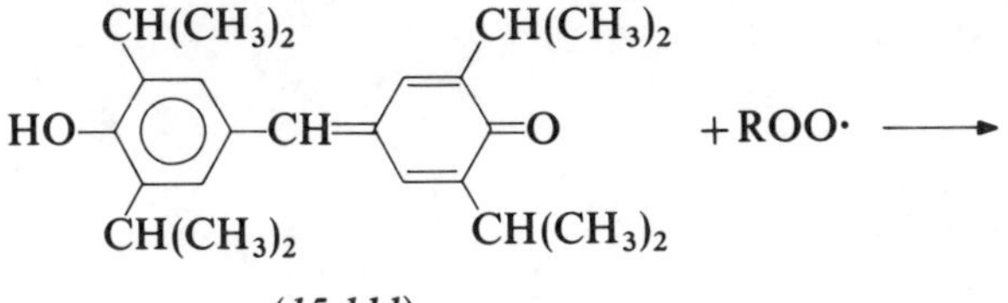

(15.111)

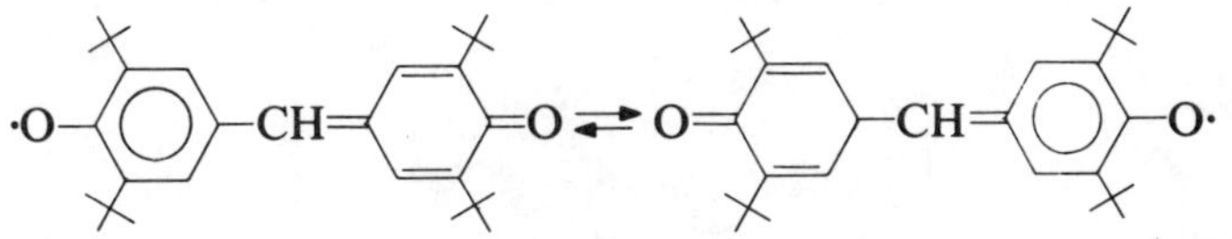

(15.112)   (15.124)

The hydroxygalvinol/galvinoxy radical redox couple is an effective thermal antioxidant and processing stabilizer for polypropylene.[256-258,1566,3255]

Diarylamines (15.113) are also effective alkyloxy (RO·) and alkylperoxy (ROO·) radical scavengers in a process in which nitroxy radicals (15.114) and finally hydroxylamine (15.115) are formed:[3255]

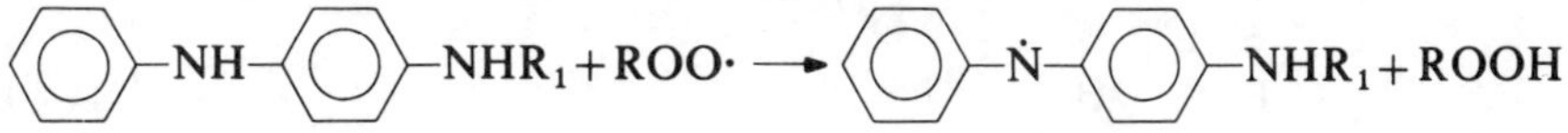

(15.113)   (15.125)

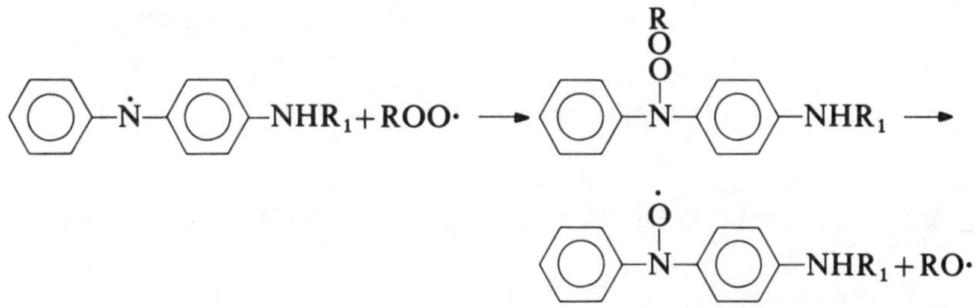

(15.114)   (15.126)

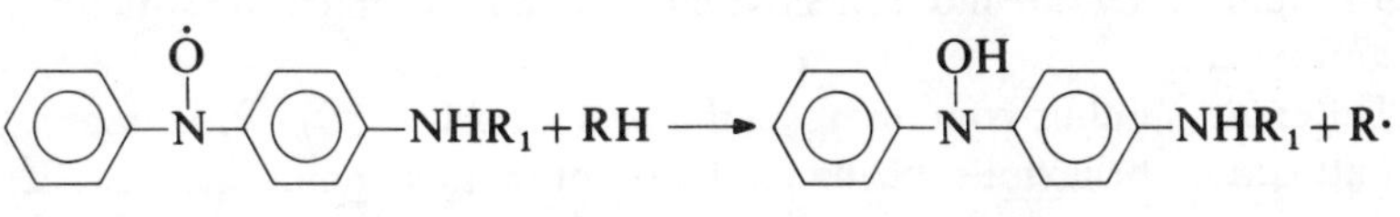

(15.115)   (15.127)

Hindered piperidines (cf. Section 15.11) can also be considered as effective chain-breaking antioxidants.

Hindered nitrones (15.116) react with alkylperoxy radicals (ROO·) producing a corresponding nitroxy radicals (15.117)[661]

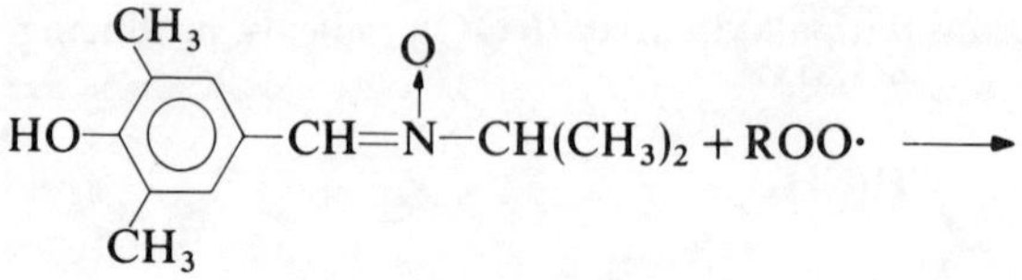

(*15.116*)

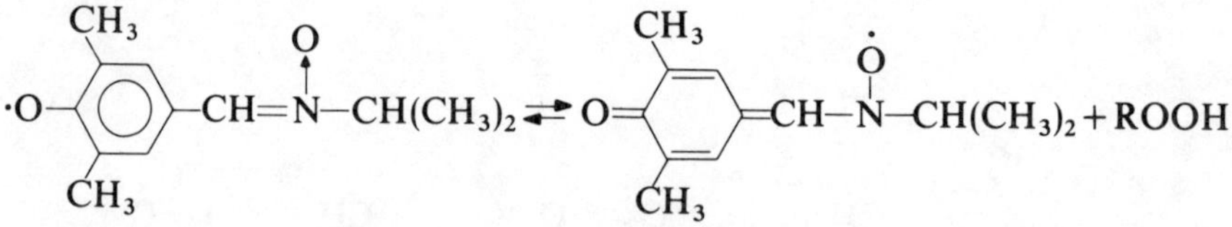

(*15.117*)   (15.128)

Another group of antioxidants react as decomposers of hydroperoxides to non-radical products, for example:

(i) Metal complexes, e.g. 2-hydroxyacetophenone oxime (*15.118*):[3055]

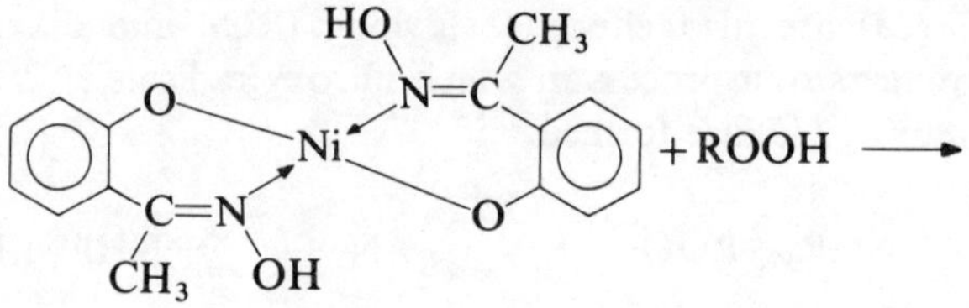

(15.129)

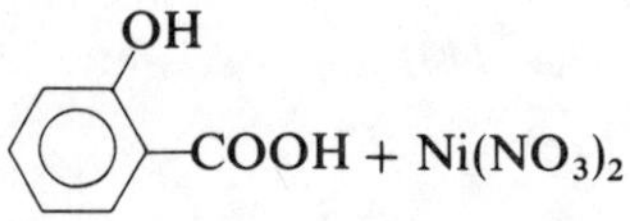

(*15.118*)

(ii) Phosphite esters (*15.119*):[1716-1718,3260]

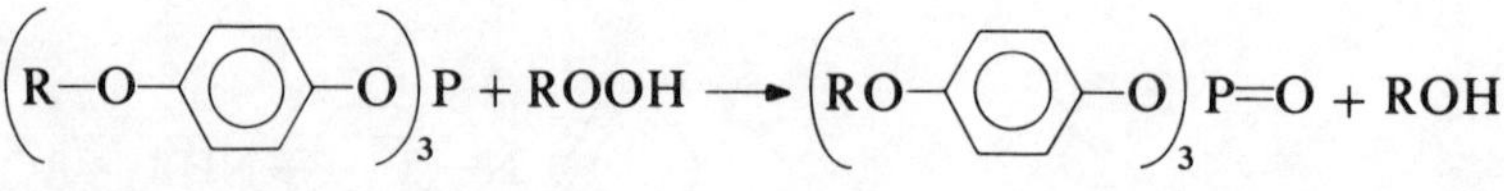

(*15.119*)   (15.130)

Phosphite esters are more effective when used in combination with phenolic antioxidants.

(iii) Different sulphur-containing stabilizers (Table 15.9). Prooxidant and antioxidant behaviour during oxidation of monosulphides (*15.120*) can be presented by the following reactions:[3255]

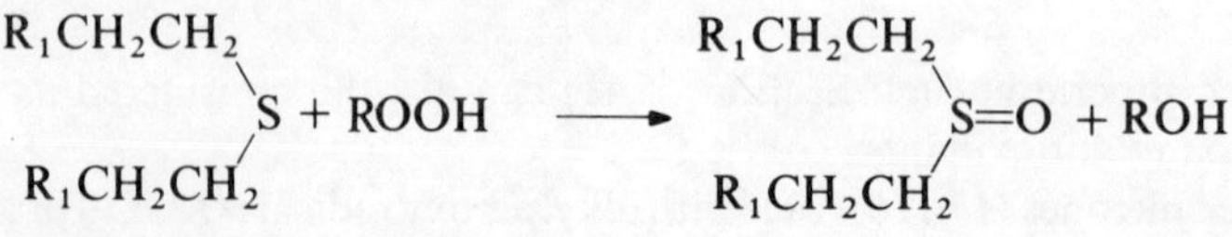

(*15.120*)   (15.131)

$$R_1CH_2CH_2 \atop R_1CH_2CH_2 {\Large\diagdown \atop \diagup} S{=}O \longrightarrow R_1CH_2CH_2SOH + R_1CH{=}CH_2 \qquad (15.132)$$

$$R_1CH_2CH_2SOH + ROOH \underset{R_1CH_2CH_2S\diagup_{OH}^{\diagup\!\diagup O}}{\overset{R_1CH_2CH_2SO\cdot + RO\cdot + H_2O}{\bigg<}}$$

(15.133)

(15.134)

$$2\ R_1CH_2CH_2SOH \longrightarrow R_1CH_2CH_2\overset{O}{\overset{\|}{S}}{-}SCH_2CH_2R_1 \qquad (15.135)$$

$$R_1CH_2CH_2\overset{O}{\overset{\|}{S}}{-}SCH_2CH_2R_1 \longrightarrow R_1CH_2CH_2\overset{O}{\overset{\|}{S}}\cdot + \cdot SCH_2CH_2R_1 \qquad (15.136)$$

$$2\ R_1CH_2CH_2\overset{O}{\overset{\|}{S}}\cdot \longrightarrow R_1CH_2CH_2\overset{O}{\overset{\|}{S}}{-}\overset{O}{\overset{\|}{S}}{-}CH_2CH_2R_1 \qquad (15.137)$$

$$2\ R_1CH_2CH_2S\cdot \longrightarrow R_1CH_2CH_2S{-}SCH_2CH_2R_1 \qquad (15.138)$$

For other sulphur-containing stabilizers (Table 15.9) reactions involved in decomposition of hydroperoxides can be even more complicated. The application of antioxidants[1105] is optimized according to their activity, synergistic/antagonistic effects with other additives (e.g. with photostabilizers[56, 1829]), compatibility with the polymer, and rate of loss from the polymer by volatilization. For the last reason attempts were made to synthesize polymeric antioxidants (cf. Section 15.14).

The oxidation process in polymers, which greatly affects their photostability, may be very slow at room temperature but is accelerated by heat, light, metal catalysts, organic catalyst, ozone, and mechanical deformations. The processing of polymers is usually carried out at elevated temperatures which accelerate the oxidation process, e.g. the extrusion process for polyolefins uses temperatures in the range 200–300°C (cf. Section 14.3.1). Many antioxidants (and other additives) decompose at these temperatures and lose their stabilizing properties. Most commercial antioxidants are effective mainly at room temperature and up to 150°C. From this point of view it is important that the antioxidants are stable and non-volatile at higher temperatures, and do not lose their effectiveness through side-reactions.

The mechanism of action of antioxidants during photooxidation of polymers is similar to that described in this chapter. Antioxidants should be tested for possible photolysis reactions and photoinitiating effects. For these reasons antioxidants should be stable to light and not photoreactive. It seems that the introduction of almost any antioxidant or other additives may cause complicated photochemical reactions to occur in a polymer.

Table 15.9 Examples of sulphur-containing antioxidants

Name	Structure	References
Monosulphides	R_1SR_2	216–218, 3255, 3307, 3308
Thiophenols	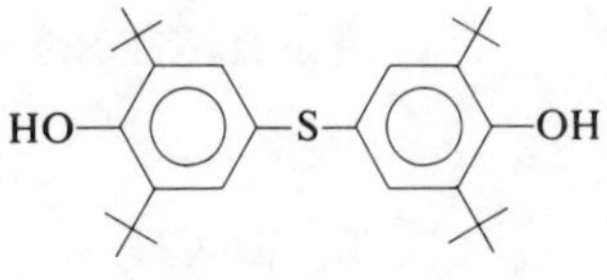	1876, 1877
Mercaptobenzothiazoles	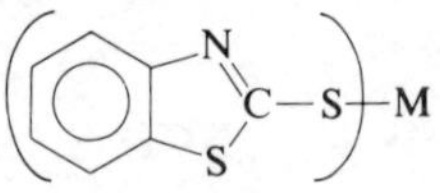	1721, 1746
Mercaptobenzoimidazoles	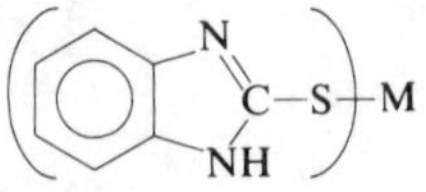	3255
Xanthates	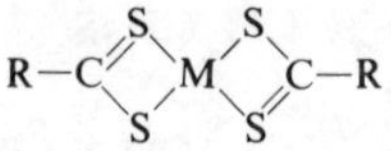	165, 168–171, 173
Dithiocarbamates	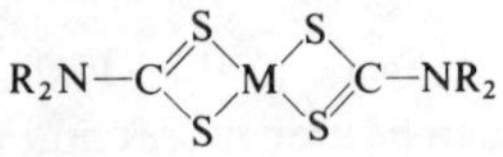	655, 1450, 1452, 1624, 3055
Dithiophosphates	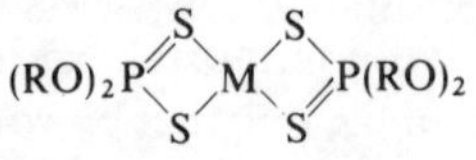	2950

15.14. POLYMERIC ANTIOXIDANTS

A number of polymeric antioxidants or polymers containing a chemically bonded antioxidant, such as 2,6-di-*tert*-butyl phenol group, have been reported.[21, 511, 1067, 1105, 1587, 1956, 2158, 2538]

After u.v. irradiation of co(styrene/2,6-di-*tert*-butyl-4-vinyl-phenol) (*15.121*) a stable polymeric radical is formed. This exhibits an asymmetric singlet line ESR spectrum (Fig. 15.14).[21]

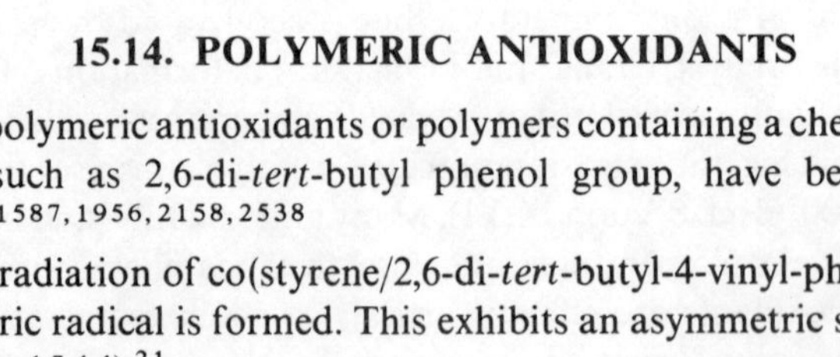

(*15.121*)

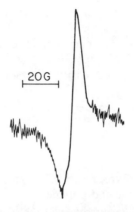

Fig. 15.14 ESR spectrum of phenoxy radicals produced by u.v. irradiation of a co(styrene/2,6-di-*tert*-butyl-4-vinyl-phenol)[21]

The radical concentration is proportional to the concentration of phenolic group in the copolymer. The half-life of the polymeric phenoxy radical is: in darkness under nitrogen 20 days, in darkness and air 14 days, and in light and air 10 days.

Incorporation of a bonded 2,6-di-*tert*-butyl phenolic moiety in polystyrene does not serve as an effective antioxidant against photooxidation.[21] The reason for this is probably the limited mobility of polymeric phenoxy radical in a polymer matrix.

15.15. SINGLET OXYGEN QUENCHERS

Singlet oxygen ($^1O_2$) can react with polymers which contain unsaturated bonds by ENE reaction and form hydroperoxide groups (OOH) (cf. Section 14.4).

Deactivation of $^1O_2$ by a quencher (Q) occurs by two mechanisms:[388, 3025, 3440–3442]

(i) Chemical quenching:

$$Q + {}^1O_2 \xrightarrow{k_1} QO_2 \tag{15.139}$$

(ii) Physical quenching:

$$Q + {}^1O_2 \xrightarrow{k_2} Q + O_2 \tag{15.140}$$

The rate constant of quenching (k_q), depending on the reaction, can be: $k_q = k_1$, or $k_q = k_2$, or $k_q = k_1 + k_2$. The quenching constants for most quenchers were determined in solutions. Experimental measurement of quenching rate constant in solid polymers, which is diffusion-controlled, is difficult.[3025] These results show that $^1O_2$ quenchers in the polymer matrix are less efficient, and quencher activity is spread over a much narrower range than found in the liquid phase.[582,614]

The β-carotene (15.122) and other carotenoids deactivate $^1O_2$ by physical and chemical quenching.[1075, 1122, 1124, 1125, 1216, 3155, 3284] The rate constant of quenching (k_q) for β-carotene is 3×10^{10} liter mol^{-1} s^{-1}.[1125]

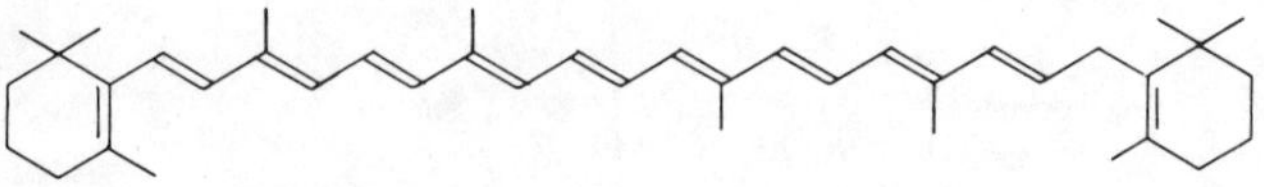

(15.122)

Carotenes may also react with free radicals. They are effective antioxidants in the photooxidation of polystyrene and polydienes.[2737,2739,3029,3033]

Polycyclic hydrocarbons, despite the fact that they can produce singlet oxygen by an energy transfer mechanism (cf. Section 14.4), are effective quenchers of $^1O_2$ which reacts with them forming endoperoxides. The most effective $^1O_2$ quencher is rubrene (15.123) (orange-coloured), which is oxidized to 9,10-endoperoxide (15.124) (colourless) with a rate constant for quenching $k_q = 7.3 \times 10^7$ liter mol^{-1} s^{-1}.[389, 618, 2551, 2552, 3443, 3909]

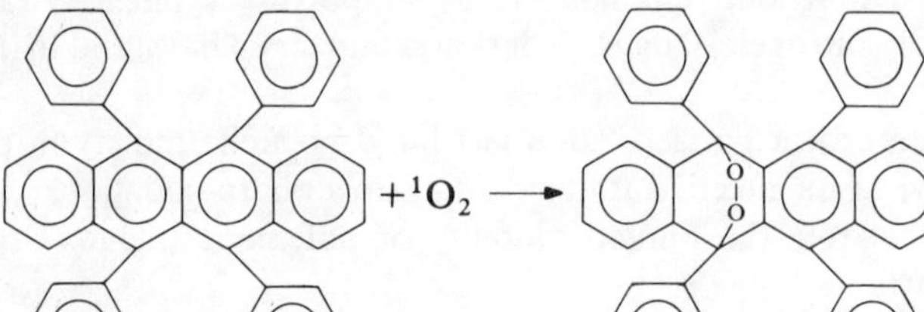

(15.141)

(15.123)   (15.124)

Some polycyclic hydrocarbons, such as 1,2,5,6-dibenzanthracene, chrysene, and phenanthrene, exhibit enhanced fluorescence caused by singlet oxygen, which occurs via a singlet oxygen feedback mechanism in the following steps:[2000-2003]

(i) Singlet oxygen is generated by an energy transfer mechanism from a donor triplet state (^3D) to oxygen:

$$^3D + O_2 \rightarrow D + {^1O_2} \qquad (15.142)$$

(ii) Diffusion of the $^1O_2$ to the site of an acceptor triplet state (^3A).

(iii) Energy transfer from $^1O_2$ to the acceptor (^3A) via formation of an OXCIPLEX ($^3A \ldots {^1O_2}$), and generation of the first excited singlet state of the acceptor molecule (^1A):

$$^1O_2 + {^3A} \rightarrow ({^3A} \ldots {^1O_2}) \rightarrow {^1A} + O_2 \qquad (15.143)$$

(iv) Emission of enhanced fluorescence from (^1A):

$$^1A \rightarrow A + \text{fluorescence} \qquad (15.144)$$

The luminescence generated by this mechanism is SINGLET OXYGEN ANNIHILATION FLUORESCENCE. Efficient deactivation must occur at a rate that greatly exceeds that expected for collisional quenching of singlet oxygen if there are to be sufficient ground-state oxygen molecules to explain the continued quenching of the acceptor triplets. The feedback mechanism, however, does provide an efficient $^1O_2$ deactivation process. The singlet oxygen feedback mechanism can be considered as a new way for the development of a new class of effective $^1O_2$ quenchers for use as stabilizers of polymer systems.

One of the most effective $^1O_2$ quenchers is the 1,3-diphenylisobenzofuran (referred also as 2,5-diphenyl-3,4-isobenzofuran) (*15.125*).[22,374,475,2199, 2417,2419,2486–2488,3186,3351,3444,3909,3993] This compound can quench $^1O_2$ physically, or react with it to ozonide of 1,2-diphenyl-3,4-benzocyclodiene (*15.126*):

$$(15.145)$$

(*15.125*)　　　　　　　(*15.126*)

The rate constant of quenching (k_q) for 1,3-diphenylisobenzofuran is 8×10^8 liter mol^{-1} s^{-1}.[475,2199]

2,6-di-*tert*-butyl-4-methylphenol has been found to be photooxidized in carbon tetrachloride, hexane, or in polystyrene matrix by the addition of anthracene.[2745] The reaction probably occurs by singlet oxygen ($^1O_2$) oxidation, which is generated in the energy transfer mechanism between the excited triplet state (T_1) of anthracene and molecular oxygen. The rate constant of quenching (k_q) for 2,6-di-*tert*-butylphenol is 10^6 liter mol^{-1} s^{-1}.

Hindered phenols (*15.127*) are capable of both reacting with and quenching singlet oxygen ($^1O_2$), depending on their substituents (R_1, R_2, and R_3) and the reaction conditions.[2427,2430,3518,3600] On the basis of the rate effects and the detection of the phenoxy radicals the following mechanism for $^1O_2$ oxidation of hindered phenols has been proposed (reactions 15.146 and 15.147):[1123, 1126, 2430, 3163, 3518, 3519, 3600]

Further reactions of phenoxy radicals (*15.128*) and cyclohexadienonyl hydroperoxide (*15.129*) are described in detail in Section 15.13.

Hindered piperidines (*15.130*) may react with singlet oxygen (chemical quenching of $^1O_2$) but this reaction is not very effective, whereas hindered piperidinoxy radicals (*15.132*) are more efficient physical quenchers of $^1O_2$.[45,272,392,1094,1563,3410,3964] The main reaction of piperidines with $^1O_2$ probably occurs via a charge-transfer (CT) complex (*15.131*) and results in the formation of piperidinoxy radicals (*15.132*):[272,392,1831,2304,2305,2544,2551,3112]

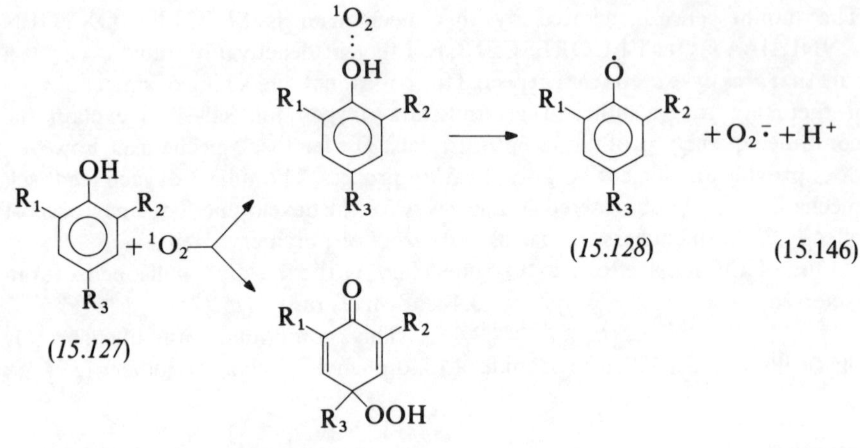

$$^1O_2$$

(15.128)   (15.146)

(15.127)

(15.129)   (15.147)

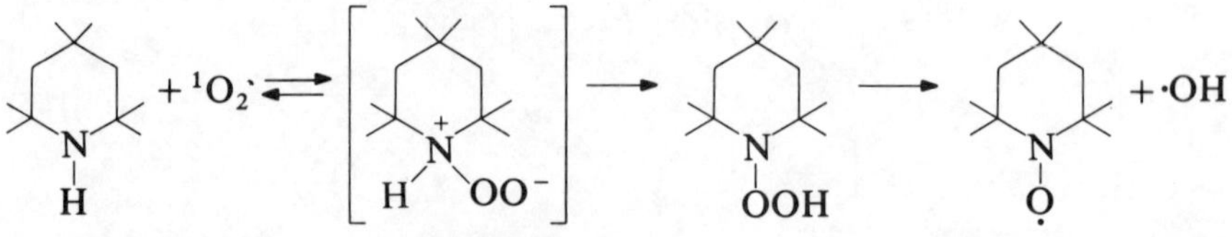

(15.130)   CT COMPLEX
(15.131)   (15.132)   (15.148)

The rate constants for quenching (k_q) of $^1O_2$ by different hindered piperidines are given in Table 15.10.[2551] These k_q values show that introduction of a methyl group instead of hydrogen at the nitrogen atom inhibits quenching. This suggests that the nitrogen atom and $^1O_2$ molecule must come into close proximity in order to form a charge-transfer (CT) complex.

Nickel chelates are efficient quenchers of singlet oxygen.[388,389,618,622, 627,1119,1198,1199,1450,1452,2197,2215,2554,3042,3882,4010,4011] The rate constants for quenching (k_q) by some of the nickel chelates are listed in Table 15.11.[2554] Diamagnetic nickel chelates are excellent quenchers with rate constants which vary from the diffusion-controlled rate limit to about 1/10 of this value. Paramagnetic complexes quench $^1O_2$ at 1/100 to 1/1000 of this value, or less. Diamagnetic nickel complexes have coordinating atoms arranged in square-planar geometry. Singlet oxygen has a relatively unhindered approach to the nickel atom from above or below the plane formed by the corresponding atoms, and can easily interact with an orbital perpendicular to this plane. In para-magnetic complexes the approach to the nickel atoms is blocked by the tetrahedral or octahedral arrangements of the ligands.

The factors which determine the relative position of a complex within each group are less obvious. The best quenchers have nickel coordinated to sulphur atoms, and throughout the diamagnetic groups the order $S_4 > N_4 > N_2O_2$ is generally followed. This corresponds to a nephelauxetic series, indicating that

Table 15.10   Rate constants of quenching (k_q) of singlet oxygen
($^1O_2$) by piperidines[2551]

Piperidine	k_q (liter mol^{-1} s^{-1})
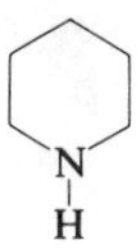 piperidine (N–H)	5.8×10^6
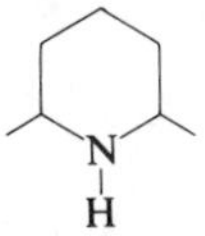 2,6-dimethylpiperidine (N–H)	3.0×10^6
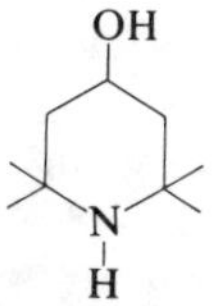 2,2,6,6-tetramethyl-4-hydroxypiperidine (N–H, OH)	$<2 \times 10^5$
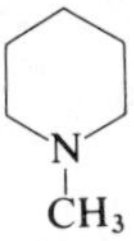 N-methylpiperidine (N–CH$_3$)	5.3×10^7
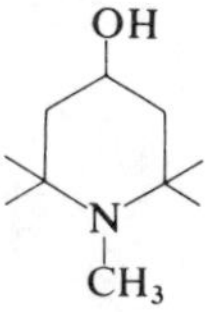 2,2,6,6-tetramethyl-4-hydroxy-N-methylpiperidine (N–CH$_3$, OH)	9.2×10^7
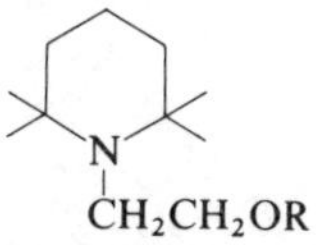 2,2,6,6-tetramethyl-N-(CH$_2$CH$_2$OR)piperidine R = OH, OCH$_3$	$<2 \times 10^5$

Table 15.11   Rate constants of quenching (k_q) of singlet oxygen ($^1O_2$) by
different metal chelates[2554]

Chelate	Magnetic properties	K_q (liter mol^{-1} s^{-1})
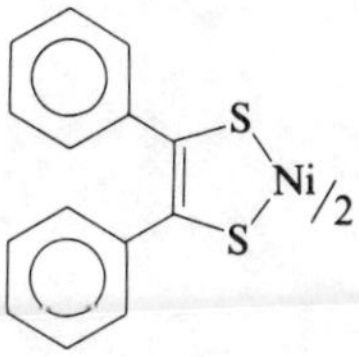 bis(1,2-diphenylethylenedithiolato)nickel (Ni, S–S) $_{1/2}$	Diamagnetic	1.1×10^{10}

Table 15.11 (*Contd.*)

Chelate	Magnetic properties	K_q (liter mol^{-1} s^{-1})
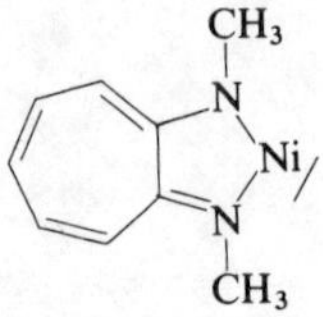 CH$_3$ / N / N / Ni / N / CH$_3$ (triazole-cycloheptatriene)	Diamagnetic	6.1×10^9
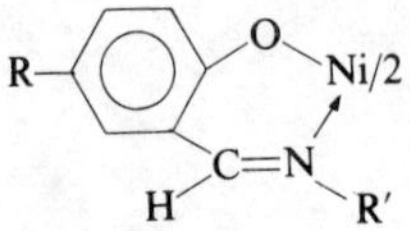 R— O—Ni/2, C=N—R', H; R = Br, n-C$_4$H$_9$	Diamagnetic	3.7×10^9
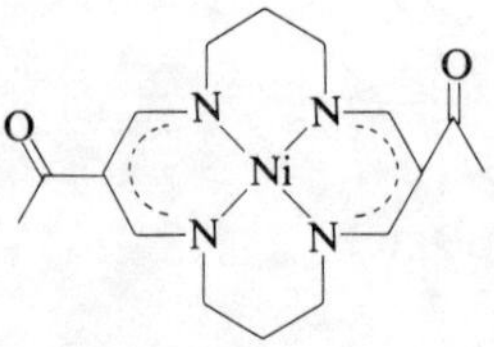 O=, Ni, N N N N tetraazamacrocycle with acetyl groups	Diamagnetic	1.6×10^9
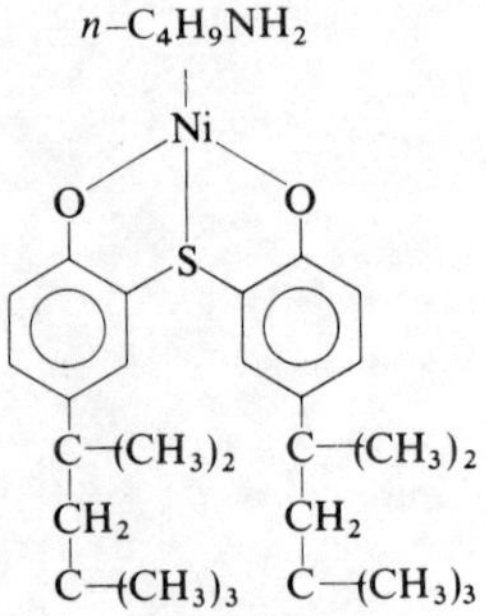 n-C$_4$H$_9$NH$_2$, Ni, O S O, C-(CH$_3$)$_2$ C-(CH$_3$)$_2$, CH$_2$ CH$_2$, C-(CH$_3$)$_3$ C-(CH$_3$)$_3$	Paramagnetic	1.7×10^8
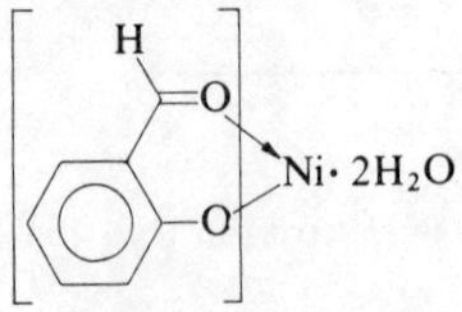 [H, O, Ni · 2H$_2$O, O] (salicylaldehyde)	Paramagnetic	4.6×10^7
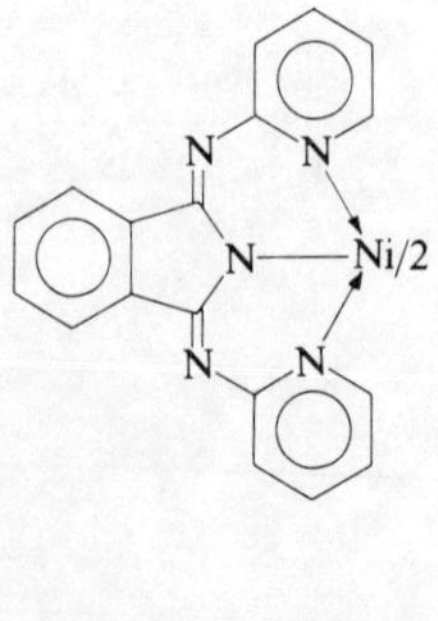 N N, N—Ni/2, N (bis-pyridyl phthalazine)	Paramagnetic	1.6×10^8

Table 15.11 (*Contd.*)

Chelate	Magnetic properties	K_q (liter mol^{-1} s^{-1})
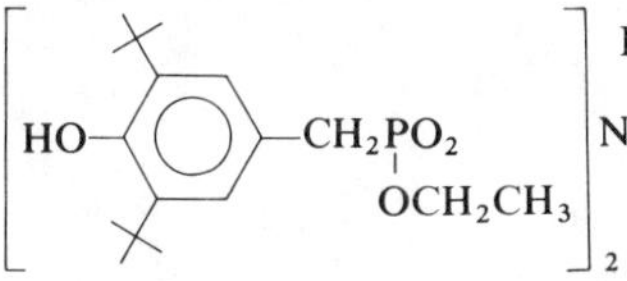	Paramagnetic	2.2×10^7
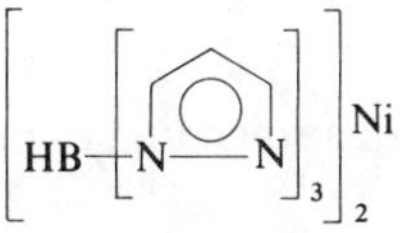	Paramagnetic	2.1×10^6

singlet oxygen quenching rates (k_q) increase as the polarizability of the ligands increases. This observation is consistent with quenching occurring by interaction of an orbital on nickel with one of $^1O_2$, followed by electronic energy transfer.[1199]

Metal chelates may also react simultaneously as decomposers of hydroperoxides, free radical scavengers, and quenchers of excited singlet (S_1) and/or triplet (T_1) of different chromophoric groups.[2215]

In conclusion, it should be stressed that singlet oxygen oxidation of polymers is always simultaneously accompanied by a free radical oxidation mechanism. The best approach to protecting polymers from both of these processes is by adding stabilizers that can both quench singlet oxygen (chemically or physically) and react with free radicals present in a free radical oxidation.

16. Polymeric systems in solar energy conversion

16.1. PRINCIPLES OF CHEMICAL CONVERSION OF SOLAR ENERGY

Principles of solar energy conversion and possibilities of application of polymeric systems in these processes have been reviewed.[752,1529,1946,2968]

Water decomposition into H_2 and O_2 is one of the promising solutions of the photochemical conversion systems of solar energy, and can be carried out in three ways:

(i) Photoredox reactions.
(ii) Heterogenous photocatalytic reactions.
(iii) Photoelectrolysis in photogalvanic cells.

These three methods are closely related and merge into one other.

Water photolysis may theoretically occur by one of these three processes:

(i) One-electron process into free radicals ($H\cdot$ and $\cdot OH$):

$$H_2O + h\nu \rightarrow H\cdot + \cdot OH \tag{16.1}$$

which requires more than 2.52 eV (Fig. 16.1).
(ii) Two-electron process into H_2O_2 and hydrogen:

$$H_2O + h\nu \rightarrow H_2O_2 + H_2 \tag{16.2}$$

which requires 1.74 eV.
(iii) Four-electron process into hydrogen and oxygen:

$$2\,H_2O + h\nu \rightarrow 2\,H_2 + O_2 \tag{16.3}$$

which requires 1.23 eV for each of the four electron transfers involved in this mechanism. From the thermodynamical point of view this is the most favourable reaction.

A schematic presentation of solar spectral distribution and the theoretical threshold energies and wavelengths for dissociation of water by one-, two-, and four-electron processes is given in Fig. 16.2. The light of any wavelength shorter than the threshold should be capable of inducing the reaction. However, all excess photon energy will be transformed into heat.

Conversion of solar energy via photochemical redox processes relies on electron-transfer reactions between the excited state of light-absorbing species (redox couple S^+_{red}/S_{red} which can be reduced, or redox couple S^+_{ox}/S_{ox} which

646

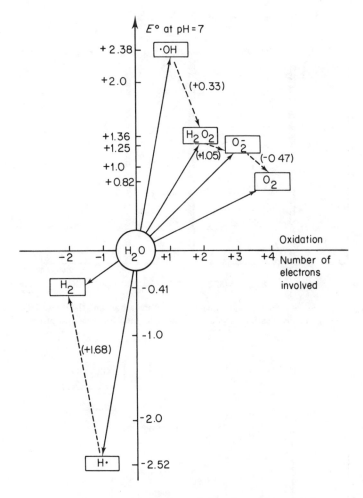

Fig. 16.1 Redox potentials for the various redox reactions of the water
molecule[2247]

can be oxidized) and another substance which is either oxidized or reduced, may
occur by:[2968]

(i) Absorption of photon by the redox couple S^+_{red}/S_{red} raises one of its
electrons to a higher energy level, and the molecule is a better electron-
donor in its excited state $(S^+_{red}/S_{red})^*$ than in its ground state. Its reducing
potential is raised. Electronic excitation of the reducing redox couple
S^+_{red}/S_{red} makes it a better reducer or lowers the redox potential of the
couple by an amount corresponding to the energy absorbed (Fig. 16.3).

(ii) Absorption of photon by the redox couple S^+_{ox}/S_{ox} creates an electron
'hole' in the original molecular energy level and the molecule is a better
electron- acceptor in its excited state $(S^+_{ox}/S_{ox})^*$ than in its ground state. Its

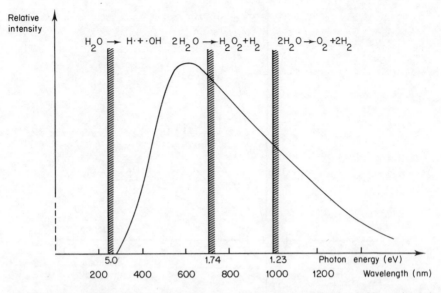

Fig. 16.2   Schematic representation of solar spectra, distribution, and the threshold energies and wavelengths for dissociation of water by one-, two-, and four-electron processes[2247]

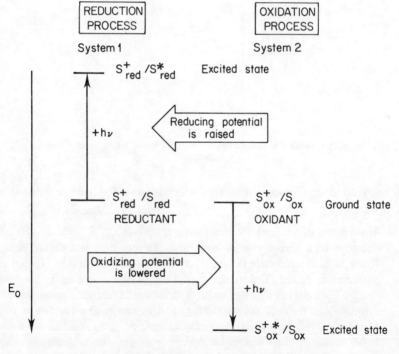

Fig. 16.3   Relative potentials of a redox couple S^+/S

oxidizing potential is lowered. Electronic excitation of the oxidizing redox couple S^+_{ox}/S_{ox} makes it a better oxidizer, or raises the redox potential of the couple by an amount corresponding to the energy absorbed (Fig. 16.3).

The four-electron process requires the following redox potentials (Fig. 16.1):

$$2H_2O \rightarrow O_2 + 4e^- + 4H^+ \qquad E^0 = +0.82 \text{ eV} \qquad (16.4)$$

$$2H^+ + 2e^- \rightarrow H_2 \qquad E^0 = -0.41 \text{ eV} \qquad (16.5)$$

For simultaneous photoreduction and photooxidation of water, the redox schemes for the two systems are shown in Fig. 16.4:[2968]

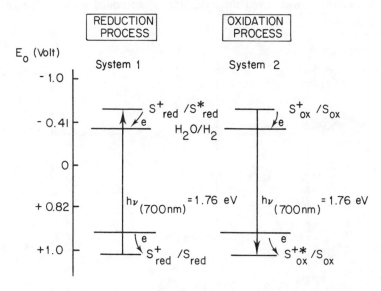

Fig. 16.4 Redox schemes for the simultaneous photooxidation and photoreduction of water: System 1: light absorbed by the reductant; System 2: Light absorbed by the oxidant[769]

(i) Light is absorbed by reductant (S^+_{red}/S_{red})

$$S^+_{red}/S_{red} + h\nu \rightarrow (S^+_{red}/S_{red})^* \qquad (16.6)$$

Considering a redox couple S^+_{red}/S_{red} with a redox potential in the ground state of 1.05 eV, the light absorption of 700 nm (1.76 eV) produces redox excited state $(S^+_{red}/S_{red})^*$ with potential -0.71 eV (1.05 − 1.76 eV). Such a situation makes it energetically possible for the couple in its ground state (S^+_{red}/S_{red}) to oxidize water, and in its excited state $(S^+_{red}/S_{red})^*$ to reduce water, using a wavelength of 700 nm, provided the redox potential of the couple in its ground state lies between 0.81 and 1.34 eV ($-0.41 + 1.73$ eV).

(ii) Light is absorbed by oxidant (S^+_{ox}/S_{ox})

$$S^+_{ox}/S_{ox} + h\nu \rightarrow (S^+_{ox}/S_{ox})^* \qquad (16.7)$$

650

Considering a redox couple S^+_{ox}/S_{ox} with a redox potential in the ground state of -0.66 eV, the light absorption of 700 nm (1.76 eV) produces redox excited state $(S^+_{ox}/S_{ox})^*$ with potential 1.1 eV $(-0.66 + 1.76$ eV$)$. Such a situation makes it energetically possible for the couple in its ground state (S^+_{ox}/S_{ox}) to reduce water, and in its excited state $(S^+_{ox}/S_{ox})^*$ to oxidize water, using a wavelength of 700 nm, provided the redox potential of the couple in its ground state lies between -0.42 eV and -0.95 eV $(0.82 - 1.76$ eV$)$.

From the energetics point of view, much more effective systems are:

(i) Reduction of water and the oxidation of an added donor (D) molecule (Fig. 16.5).

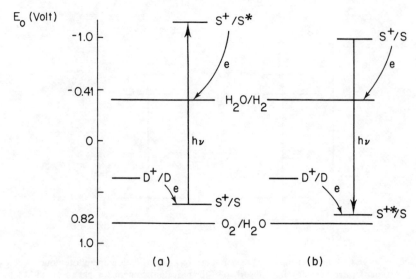

Fig. 16.5 Redox schemes for the photoreduction of water using a sacrificial donor D: (a) System 1, (b) System 2[769]

(ii) Oxidation of water and the reduction of an added acceptor (A^+) molecule (Fig. 16.6).

In these both systems two photons are required for each electron transferred.

In practice the following components have been used for the photoreduction of water:[1946,2968]

(i) Sensitizer (S) (which can form a redox couple S^+_{red}/S_{red} or S^+_{ox}/S_{ox}),
(ii) acceptor (A),
(iii) donor (D),
(iv) catalysts (cat),

and a number of reactions result, two of which are important to considered:

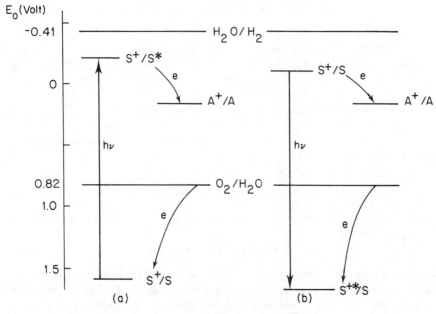

Fig. 16.6 Redox schemes for the photooxidation of water using a sacrificial acceptor A^+: (a) System 1, (b) System 2[769]

(i) Reduction of water by a system of components such as sensitizer (S) (S^+_{red}/S_{red}) and intermediate electron acceptor (A):

$$S + hv \rightarrow S^* \qquad (16.8)$$

$$S^* + A \rightarrow S^+ + A^- \qquad (16.9)$$

$$S^+ + D \rightarrow S + D^+ \qquad (16.10)$$

$$A^- + H_2O \xrightarrow{cat} A + OH^- + \tfrac{1}{2}H_2 \qquad (16.11)$$

(ii) Reduction of water by a system of components such as sensitizer (S) (S^+_{ox}/S_{ox}) and intermediate electron donor (D):

$$S + hv \rightarrow S^* \qquad (16.12)$$

$$S^* + D \rightarrow S^- + D^+ \qquad (16.13)$$

$$S^- + A \rightarrow S + A^- \qquad (16.14)$$

$$A^- + H_2O \xrightarrow{cat} A + OH^- + \tfrac{1}{2}H_2 \qquad (16.15)$$

Quantum yield for photoreactions involving energy or electron transfer is dependent on temperature.[1892]

16.2. PHOTOREDUCTION OF WATER

Several organic compounds and metallic organic complexes (Table 16.1) were successfully employed as photosensitizers.

Tris(2,2'-bipyridyl)ruthenium(II) complex (Table 16.1) (abbreviated to $Ru(bpy)_3^{2+}$) is one of the most promising photosensitizers in solar energy conversion process.[275,474,663,780,1340,1400,1937,1945,1946,2034,2247,3207,3496,3531,3690,3874] Its absorption maximum of 452 nm in water (Fig. 16.7) is near the peak of the solar spectrum (ca. 500 nm, Fig. 16.2) and the molar absorptivity (ε) is fairly high ($\varepsilon = 1.38 \times 10^4$). Irradiation of $Ru(bpy)_3^{2+}$ with light below 560 nm results in the formation of relatively long-lived ($\tau = 0.6\ \mu s$ in water at 25°C) charge-transfer excited triplet state $(Ru(bpy)_3^{2+})^*$:[2661]

$$Ru(bpy)_3^{2+} + h\nu \rightarrow (Ru(bpy)_3^{2+})^* \tag{16.16}$$

The excited state $(Ru(bpy)_3^{2+})^*$ is potentially capable of reducing water to hydrogen:[946,1510]

$$(Ru(bpy)_3^{2+})^* + H_2O \rightarrow Ru(bpy)_3^{3+} + \tfrac{1}{2}H_2 + OH^- \tag{16.17}$$

For the $Ru(bpy)_3^{2+}$ complex to function in a catalytic role, the $Ru(bpy)_3^{3+}$ produced when being generated must be rapidly reconverted to the initial ground state $Ru(bpy)_3^{2+}$. This can be accomplished through its reduction by hydroxide ion according to the equation:[780,2247]

$$Ru(bpy)^{3+} + OH^- \rightarrow Ru(bpy)_3^{2+} + \tfrac{1}{2}O_2 + \tfrac{1}{2}H_2O \tag{16.18}$$

The redox potentials of the $Ru(bpy)_3^{2+}$ complexes are shown in Fig. 16.8. In practice $Ru(bpy)_3^{2+}$ as a photosensitizer for water photolysis itself is useless, because multi-electron processes, instead of four-electron processes occur. The more complicated systems must be used for photoreduction of water, e.g. $Ru(bpy)_3^{2+}-MV^{2+}-EDTA$-colloidal platinum:[1400, 1946, 2247]

(i) The first step is the excitation of ruthenium complex $(Ru(bpy)_3^{2+})$ by light:

$$Ru(bpy)_3^{2+} + h\nu \rightarrow (Ru(bpy)_3^{2+})^* \tag{16.19}$$

(ii) The photoexcited molecule of the ruthenium complex $(Ru(bpy)_3^{2+})^*$ further reduces the methyl viologen (1,1'-dimethyl-4,4'-dipyridinium dichloride (16.1), abbreviated here as MV^{2+}) by an electron transfer mechanism, which is very fast ($k = 10^8 - 10^9\ \text{M}^{-1}\text{s}^{-1}$) to the viologen cation radical ($MV^+\cdot$) (16.2):[2702]

$$(Ru(bpy)_3^{2+})^* + MV^{2+} \underset{\substack{\text{back electron} \\ \text{transfer}}}{\longrightarrow} Ru(bpy)_3^{3+} + MV^+\cdot \tag{16.20}$$

Methyl viologen (16.1) considered to function as an electron mediator or electron relay.[1045,1669,2112]

$$CH_3-\overset{+}{N}\bigcirc-\bigcirc\overset{+}{N}-CH_3 \qquad CH_3-\overset{+}{N}\bigcirc-\bigcirc(\cdot)N-CH_3$$

(16.1)   (16.2)

Table 16.1   Examples of organic and metallic organic complexes employed as photosensitizers

Sensitizer	Type	References
Tris(2,2'-bipyridil)ruthenium(II) complex (abbreviated to Ru(bpy)$_3^{2+}$) 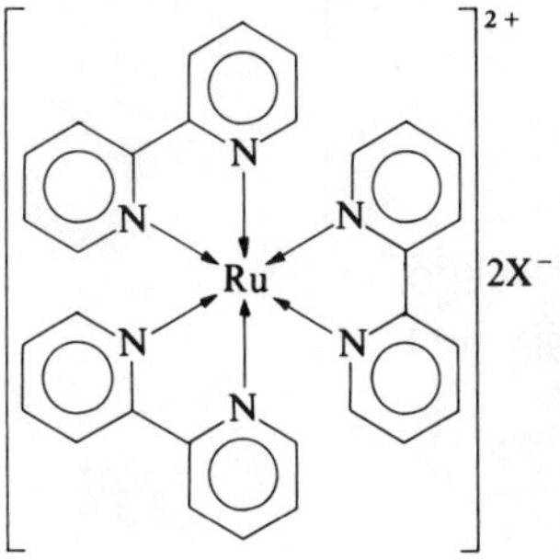	S^+_{red}/S_{red}	275, 276, 474, 663, 780, 945, 946, 1400, 1937, 1945, 1946, 2034, 2247, 2661, 3207, 3496, 3531, 3690, 3874, 3943

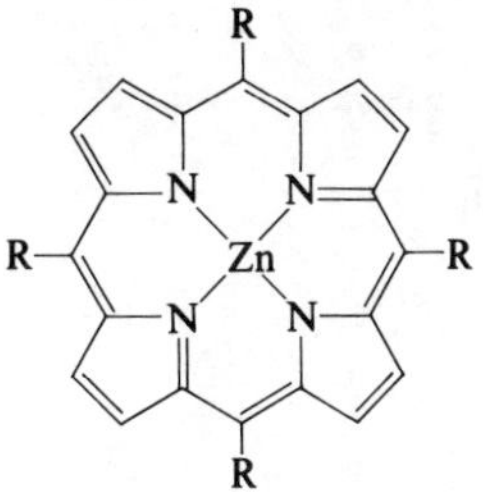 Porphyrin derivatives	S^+_{red}/S_{red}	992, 1495, 1496, 1660, 1924, 2435, 2470, 2785, 2786, 3873

where R = *N*-methylpyridinium chloride

Chlorophyll

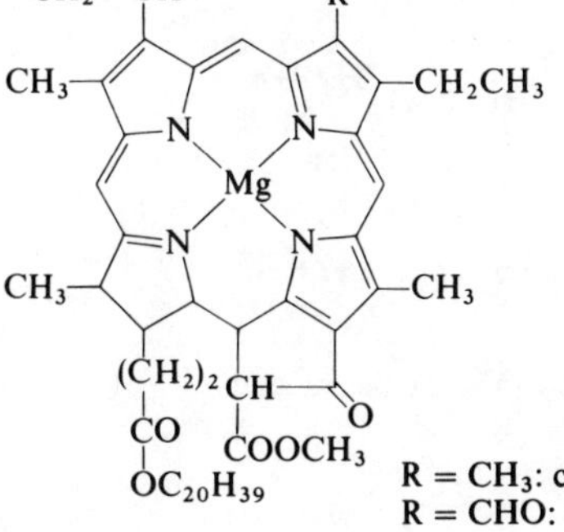

R = CH$_3$: chlorophyll *a*
R = CHO: chlorophyll *b*

Proflavin 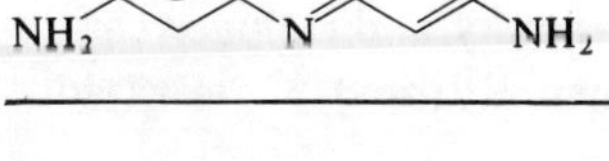	S^+_{ox}/S_{ox}	382, 1415, 1923, 2112, 2126

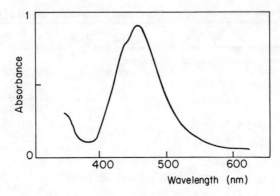

Fig. 16.7 Refractive spectrum of the tris(bipyridyl)-ruthenium(II) chloride[3638]

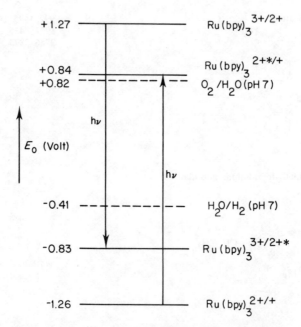

Fig. 16.8 Redox potentials of $Ru(bpy)_3^{2+}$ complexes and water[2247]

The viologen cation radical ($MV^{+}\cdot$) has an absorption with maximum at 620 nm (Fig. 16.9) and gives a single-line ESR spectrum with $g = 1.967$ and $\Delta H_{msl} = 13.5$ gauss (at 25°C) (Fig. 16.10).

(iii) The back electron transfer reaction occurs very rapidly, and viologen cation radical ($MV^{+}\cdot$) can be accumulated only when $Ru(bpy)_3^{3+}$ is rapidly

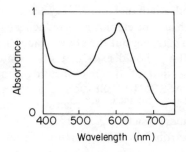

Fig. 16.9 Visible absorption spectrum of cation radical of methyl viologen $(MV^{+\cdot})$[1939]

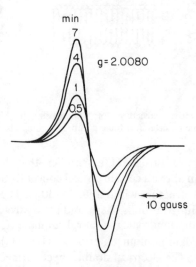

Fig. 16.10 ESR spectrum of cation radical of methyl viologen $(MV^{+\cdot})$ formed on cellulose under irradiation with visible light *in vacuo*. The numbers represent reaction time in minutes[1945]

reduced by a reducing agent such as ethylene diaminetetra-acetic acid (*16.3*) (EDTA):[752, 1400, 2034, 3530]

$$Ru(bpy)_3^{3+} + EDTA \rightarrow Ru(bpy)_3^{2+} + EDTA^+ \qquad (16.21)$$

$$
\begin{array}{cc}
H^+\,{}^-OOC-CH_2 & CH_2-COO^-\,H^+ \\
\diagdown & \diagup \\
N-CH_2-CH_2-N & \\
\diagup & \diagdown \\
H^+\,{}^-OOC-CH_2 & CH_2-COO^-\,H^+
\end{array}
$$

(*16.3*)

Triethylamine and triethanolamine,[31] cysteine or polymer-bound imi-nodiacetic acid can also be employed as reducing agents.[1946, 2167] In order to have succees in carrying out the above reaction, $[Ru(bpy)_3{}^{3+}-EDTA]$ and $[MV^{+}\cdot-EDTA]$ should exist in the separate phases. This can be effected by application of the microheterogeneous conversion system (MOLECULAR ASSEMBLIES) such as micelles,[41, 42, 533, 1398, 1714, 1715, 2368, 2580, 2581, 3074, 3783, 3931] microemulsions,[2044, 3904] or bi-layer membrane liposomes (vesicles)[593, 1127, 1128, 1761, 1946, 3610] which provide unidirectional electron flow. The structure of such molecular assembly is shown in Fig. 16.11.

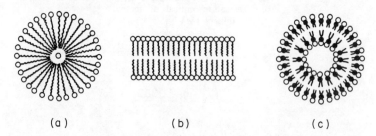

Fig. 16.11 Illustrative structure of: (a) micell, (b) bilayer membrane, (c) liposome (vesicle). (O) Ionic group, (⌁) long alkyl group[3690]

MICELLES are surfactant assemblies that form spontaneously in aqueous solution above a certain critical concentration.[1101, 2302, 3544] They have a spherical structure with radius 15–30 Å. The polar head groups are exposed to the aqueous bulk phase, and the hydrocarbon tails protrude in the interior. Compounds such as lauryl sulphate $(CH_3(CH_2)_{11}SO_4{}^-Na^+)$ or cetyltrimethylammonium chloride $(CH_3(CH_2)_{15}N^+(CH_3)_3Cl^-)$ are used as micelles. The electrical double-layer formed around ionic micelles (Fig. 16.12) is very important for unidirectional electron flow. This electrical double-layer gives surface potentials of 150 mV and provides a microscopic barrier for prevention of thermal back-transfer of the electrons. When both $Ru(bpy)_3{}^{2+}$ and MV^{2+} are dissolved in water, the micelle serves as a carrier for the reduced $MV^+\cdot$.

In a case where the amphilic N-tetradecyl-N'-methyl viologen (*16.4*) has been employed, the oxidized form (*16.4*) has hydrophilic properties, and the reduced form (*16.5*) has hydrophobic properties:[539, 1400]

$$CH_3-\overset{+}{N}\langle\bigcirc\rangle-\langle\bigcirc\rangle\overset{+}{N}-(CH_2)_{13}-CH_3 \xrightarrow{+e^-}$$

(*16.4*)

$$CH_3-N\langle\odot\rangle-\langle\bigcirc\rangle\overset{+}{N}-(CH_2)_{13}-CH_3$$

(*16.5*)   (16.22)

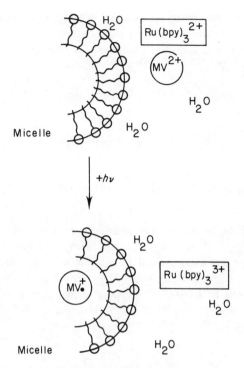

Fig. 16.12   Principles of charge and compounds
separation by cationic micelles

Due to its hydrophilic character the oxidized form (*16.4*) is present mainly in the aqueous phase, and does not associate with the micelle aggregates (e.g. cetyltrimethylammonium chloride). When electron transfer occurs, the reduced form (*16.5*) acquires hydrophobic properties and is rapidly soluble in the micelle aggregate (Fig. 16.12). The oxidized sensitizer is prevented from approaching the micelles by the positive surface charge. Such micelle assemblies decreases the rate constant of the back electron transfer at least by 500.[538]

(iv) The viologen cation radical ($MV^{+}\cdot$) is thermodynamically capable of reducing water, but the reaction does not occur spontaneously. It must be catalysed by colloidal platinum (Pt):

$$2\,MV^{+}\cdot + 2\,H^{+} \xrightarrow{\text{Pt}} 2\,MV^{2+} + H_2 \tag{16.23}$$

The reduction potential of $MV^{2+}/MV^{+}\cdot$ is low (-0.44 eV at pH 7). The role of the platinum catalyst in this reaction is probably that of microelectrode.[1400] Colloidal platinum must be stabilized by water-soluble polymers such as poly(vinyl alcohol),[2046,2047] poly(vinyl pirrolidone),[1399, 1949,2046,2702,3187,3631] or polyacrylamide.[31] The activity of such catalysts increases remarkably in colloidal dispersion.

(v) By employing two catalysts, colloidal platinum for reduction and ruthenium dioxide (RuO_2) powder for oxidation, it is possible to carry out combined reduction and oxidation of water simultaneously, according to the reactions:[1400, 1920, 1949, 2045, 2048, 2248]

$$(Ru(bpy)^{2+})^* + MV^{2+} \longrightarrow Ru(bpy)_3^{3+} + MV^{+\cdot} \qquad (16.24)$$

$$2MV^{+\cdot} + 2H^+ \xrightarrow{Pt} 2MV^{2+} + H_2 \qquad (16.25)$$

$$4Ru(bpy)_3^{3+} + 4OH^- \xrightarrow{RuO_2} 4Ru(bpy)_3^{2+} + O_2 + 2H_2O \quad (16.26)$$

Ruthenium dioxide (RuO_2) can also be supported on co(styrene/maleic anhydride).[1920, 1921] All of these reactions involved in photochemical water splitting are summarized in a cyclic scheme shown in Fig. 16.13. Instead of RuO_2 other catalysts have been investigated such as TiO_2 doped with Pt as well as RuO_2,[1980] or polynuclear metal complexes such as Prussian blue (ferrous potassium hexacyanoferrate(III)) ($KFe[Fe(CN)_6]$).[1942]

16.3. POLYMERIC PHOTOSENSITIZERS

In order to explore polymeric synthesizers for solar energy conversion co(styrene/vinyl-2,2′-bipyridine) (16.6)[1200, 1201] and co(styrene/styryl-2,2′-dipyridine) (16.7)[601] were synthesized. These substances can form complexes with transition metal salts, e.g. ruthenium:[1936, 1938–1940, 1946, 1947, 1950, 3486]

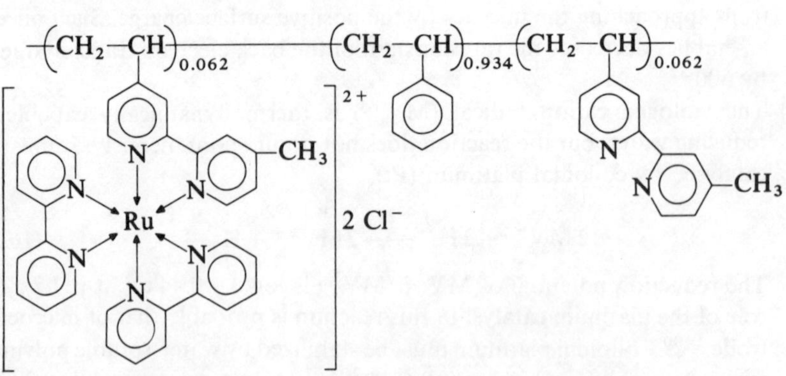

(16.6)

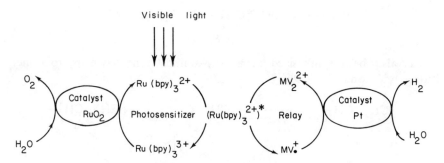

Fig. 16.13 Cyclic scheme for reactions involved in photochemical water splitting[2247]

$$\left(CH_2-CH\right)_{0.020} \left(CH_2-CH\right)_{0.378} \left(CH_2-CH\right)_{0.165} \left(CH_2-CH\right)_{0.847}$$

(16.7)

Absorption and emission spectra of a polymeric complex (16.6) in dimethylformamide are shown in Fig. 16.14.

The $Ru(bpy)_3^{2+}$ group can be separated from the phenyl ring in polystyrene beads by an oligo(oxyethylene) spacer (16.8):[3638]

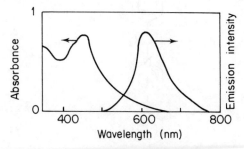

Fig. 16.14 Absorption and emission spectra of polymeric complex (16.6) in dimethylformamide[1939]

$$\text{Poly(styryl-}(OCH_2CH_2)_n - Ru(bpy)_3{}^{2+})2Cl^-$$

$$(16.8)$$

It has also been synthesized as a co(p-aminostyrene/N-vinylpyrrolidone)-$Ru(bpy)_3{}^{2+}$ complex (16.9):[2168]

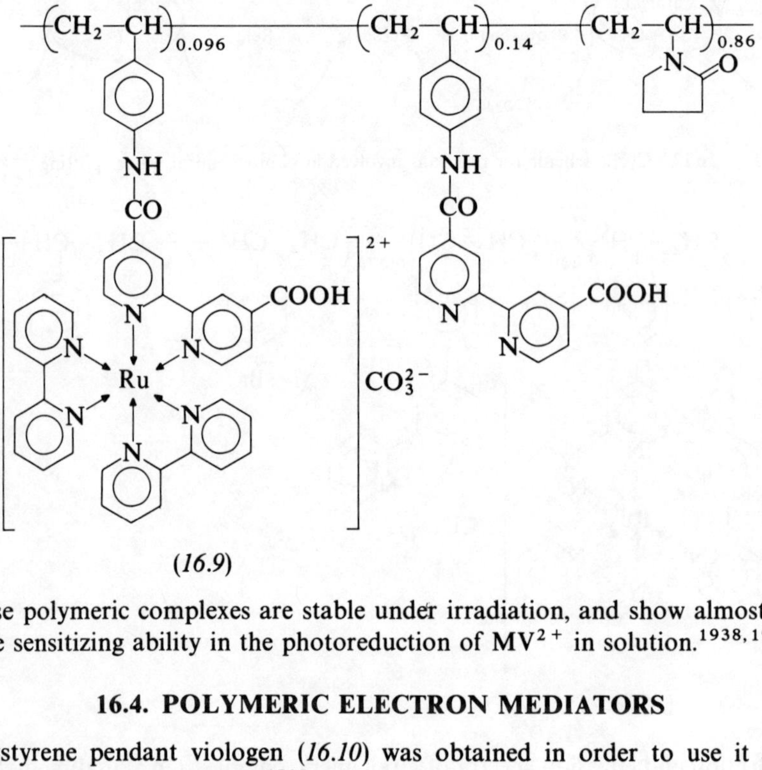

$$(16.9)$$

These polymeric complexes are stable under irradiation, and show almost the same sensitizing ability in the photoreduction of MV^{2+} in solution.[1938,1947]

16.4. POLYMERIC ELECTRON MEDIATORS

Polystyrene pendant viologen (16.10) was obtained in order to use it as a polymeric electron mediator.[2702]

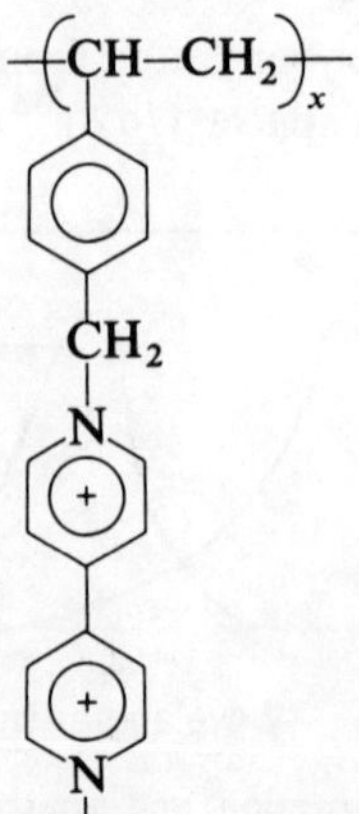

$$(16.10)$$

Another type of polymeric electron mediator is a co(styrene/acrylamide) containing pendant viologen groups.[31] These polymeric electron mediators can stabilize Pt colloids themselves.

A polystyrene–Ru(bpy)$_3$$^{2+}$ complex has been covalently linked to viologen (16.11) units in order to obtain polymeric system which consists of photochemical centre and electron mediator.[2426]

(16.11)

Such polymeric structures show high charge separation efficiency in a viologen cation radical in the presence of EDTA as reducing agent.

16.5. POLYMERIC CARRIERS OF PHOTOCHEMICAL REACTION COMPONENTS

A polymer solid matrix can work as a carrier of photochemical reaction components. The irradiation of cellulose paper [1937, 1944, 1945] or cellophane[2524] after adsorbing Ru(bpy)$_3$$^{2+}$, MV^{2+}, and EDTA induced rapid formation of MV$^+$· in solid phase.

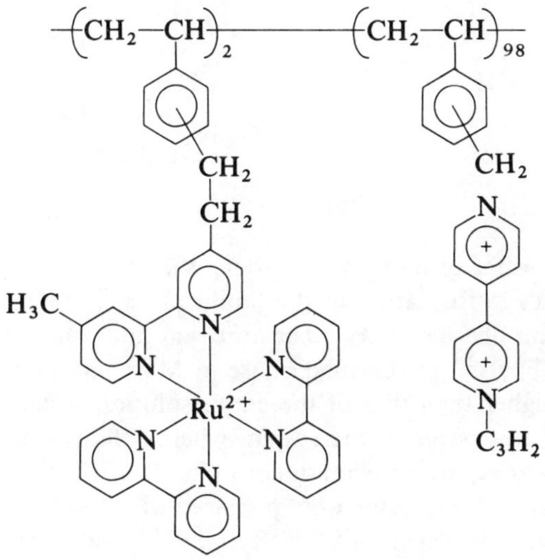

The reducing power of the MV$^+$· formed in a solid phase can be transferred to the liquid phase. Ru(bpy)$_3$$^{2+}$ and MV^{2+} were adsorbed in water-swollen polymer beads containing iminodiacetic acid (16.12) groups.[2167] The irradiation of the beads induced rapid formation of MV$^+$· in the solid state through the electron relay of. During the irradiation of beads in water in the presence of oxygen, hydrogen peroxide was formed by interfacial electron transfer from MV$^+$· formed in the solid phase to O$_2$ in the liquid phase.[2167]

A photoinduced electron relay system at solid–liquid interface was also constructed by utilizing polymer pendant Ru(bpy)$_3$$^{2+}$. The irradiation of mixture

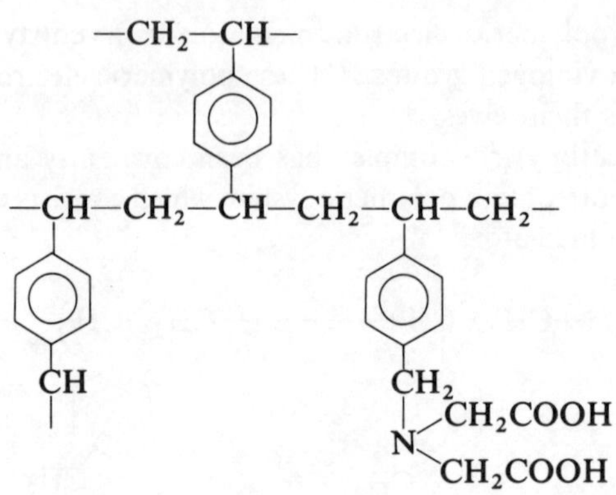

$$-CH_2-CH-$$

(polymer structure)

$$-CH-CH_2-CH-CH_2-CH-CH_2-$$

$$CH$$

$$CH_2$$
$$N \big\langle {}^{CH_2COOH}_{CH_2COOH}$$

(16.12)

of EDTA and the water-insoluble polystyrene–Ru(bpy)$_3{}^{2+}$ complex containing MV^{2+} induced MV$^{+}\cdot$ formation in the liquid phase.[1939] The proton and Pt catalyst in the liquid phase gives H$_2$. The apparent rate constant of the electron transfer from Ru(bpy)$_3{}^{2+}$ in the solid phase to MV^{2+} in the liquid phase was estimated to be higher than that of the entire solution system. Photoinduced electron relay did not occur in the system where a film of polymer pendant Ru(bpy)$_3{}^{2+}$ complex separates two aqueous phases of EDTA and MV^{2+}.

Crosslinked poly(styrene) beads were prepared with Ru(bpy)$_3{}^{2+}$ anchored at an oligo(oxyethylene) spacer group.[3638] Ry(bpy)$_3{}^{2+}$ and its dicinnamate were immobilized in a membrane prepared from cinnamate of poly(vinyl alcohol) by photocrosslinking, and the immobilized complex sensitized the photoreduction of MV^{2+}.[1982]

16.6. CONVERSION OF SOLAR ENERGY BY POLYMERIC PHOTOGALVANIC CELLS

When a photochemical process in solution gives a photoresponse at the electrode, the system can form a PHOTOGALVANIC CELL. The photoinduced redox reaction is typical for photogalvanic cells.

The best-known photoredox system is thionine (16.13) (abbreviated to TH$^+$) and ferrous ion. The photoexcited TH$^+$ is reduced by Fe^{2+} to give TH$_2{}^{+}\cdot$ cation radical and Fe^{3+}:

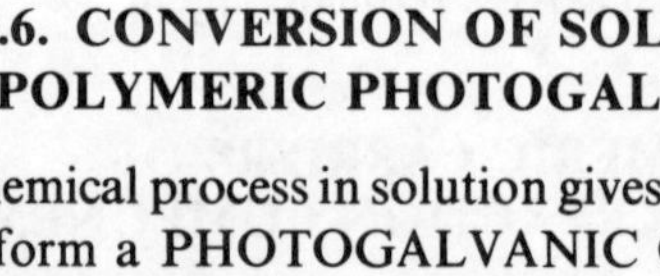

$$NH_2 \quad S \quad \overset{+}{N}H_2$$

(16.13)

$$TH^+ + Fe^{2+} + H^+ \underset{dark}{\overset{+h\nu}{\rightleftharpoons}} TH_2{}^{+}\cdot + Fe^{3+} \qquad (16.27)$$

(Violet)   (colourless)

During irradiation the aqueous violet solution is decolorized, whereas when the light is cut off, the colour again returns to violet. This reversible photoredox system gives a photopotential when irradiated in a photochemical cell composed of light and dark chambers (Fig. 16.15).[1943]

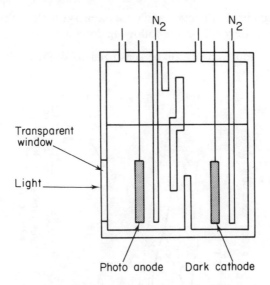

Fig. 16.15   Photogalvanic cell composed of light and dark chambers[1946]

A photogalvanic cell operating on a photoredox system, i.e. polycation polymer pendant thionine (TH^+) (*16.14*) and Fe^{2+} (reducing agent), has been constructed:[34, 3313]

$$-\left(OCHCH_2\right)_n \underline{\hspace{3cm}} \left(OCHCH_2\right)_{1-n} -$$

(*16.14*)

A photoredox system can also be constructed from thionine and polyamine,[1941] and polymers containing $Ru(bpy)_3^{2+}$ and viologen as reducing agent.[1936, 1940, 1948, 2862]

16.7. APPLICATION OF POLYMERS IN SOLAR ENERGY STORAGE

Solar energy can be stored by organic or organometallic substances which can rapidly isomerize to an energy-rich unstable isomer under irradiation.[1892–1895, 2195, 3202] The energy storage scheme involves a closed cycle of reactants:

$$A \underset{+\Delta}{\overset{+h\nu}{\rightleftharpoons}} B \tag{16.28}$$

which absorb light in one direction and releases energy in the reverse. Selection of reactants A and B depends on the following criteria:[1893, 1895]

(i) Reactant A must absorb or be sensitized to sunlight (300–700 nm, Fig. 16.16).

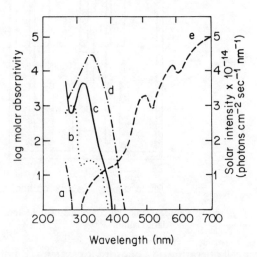

Fig. 16.16 Absorption spectra of: (a) norbornadiene, (b) acetophenone, (c) o-benzoyloxybenzaldehyde, (d) 4-(N,N,-dimethyl-amino)benzophenone, (e) solar radiance spectrum[1531]

(ii) Ensure that photochemical reactions occur in one direction A → B (reaction should be photochromic).

(iii) Reactant B should not absorb solar radiation or be involved in energy transfer with a sensitizer.

(iv) The quantum efficiency of photoreaction A → B should be near unity.

(v) Thermal reaction (B → A) should have a large negative enthalpy.

(vi) Photoproduct B should be kinetically stable (i.e. revert to A negligibly at ambient temperatures). Ideally a catalyst should induce back-reaction at a controllable rate and temperatures.

(vii) Chemical components A and B must survive a large number of energy storage and reversion cycles.

(viii) Chemicals A and B must be inexpensive, available in large quantity, readily handled, and relatively non-toxic.

The best-known system to store solar energy is photoisomerization of norbornadiene (16.15) to quadricyclane (16.16):[1531, 2175, 3241]

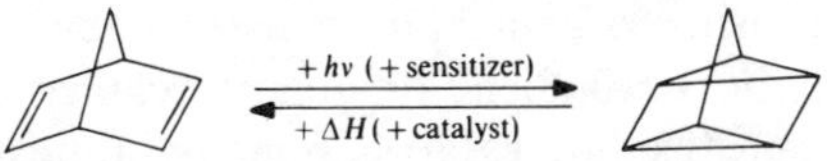

$$\text{(16.15)} \quad \xrightleftharpoons[+ \Delta H \, (+ \text{catalyst})]{+ h\nu \, (+ \text{sensitizer})} \quad \text{(16.16)} \tag{16.29}$$

Advantages of such a solar energy storage system are:

(i) Norbornadiene is readily available and comparatively inexpensive.
(ii) Quadricyclane has a high storage capacity ($\Delta H = 21$ kcal mol^{-1}).[1906]
(iii) Both norbornadiene and quadricyclane are liquids.
(iv) Conditions exist for which both steps in the cyclical process can be virtually quantitative.

The system also has several disadvantages, such as:

(i) Norbornadiene ($E_T = 70$ kcal mol^{-1}) has no absorption in the visible region (Fig. 16.16) and sensitizers such as aromatic ketones (e.g. benzophenone, fluorenone, or biacetyl) are required for norbornadiene isomerization by visible light.[321,1365,1366,1488,1490,2175,2601]
(ii) The reverse reaction of quadricyclane to norbornadiene, which releases heat, must be catalysed, e.g. by Co(II)tetraphenylporhyrin.[1611]
(iii) Both sensitizer and catalyst must be immobilized on porous polymers; ketones in the form of polyketones[1530, 1531, 2175] (e.g. poly(4-N,N-dimethyl-amino)benzophenone) (16.17) and Co(II)tetraphenylporphyrin adsorbed on silica gel or anchored to porous polystyrene (16.18).[1531, 2026] An additional problem is that the polyketones (e.g. poly(vinylacetophenone)) are photodegraded by sunlight[3843] (cf. Section 14.3.5).

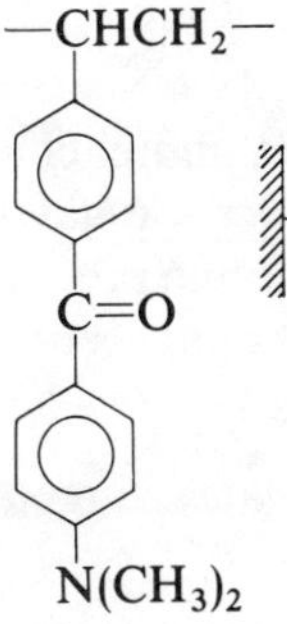

$Z = $ —CO—, $R = CO_2CH_3$
$Z = SO_2$—, $R = SO_3CH_3$

(16.17)

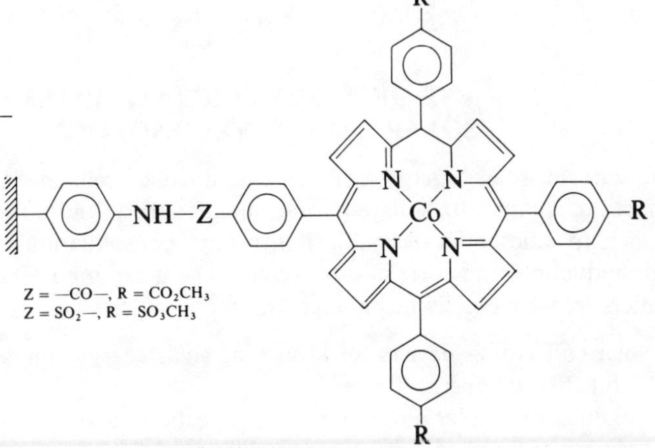

(16.18)

Low molecular sensitizers and catalysts may produce undesirable photoadducts with either norbornadiene or quadricyclane. Usually the sensitizing activity of the sensitizer or catalyst remains almost unchanged through immobilization, but is sometimes decreased, depending on their structure. Activity decrease can be observed after several numbers of recyclic norbornadiene/quadricyclane system.

A flow type model of solar energy storage device, utilizing polymer pendant sensitizer and catalysts (Fig. 16.17), was constructed and operated.[1531,1893] Comparison of insoluble polymeric vs soluble monomeric sensitizers for the photoisomerization of norbornadiene to quadricyclane is given in Table 16.2. All polymers show relatively lower quantum efficiency than monomeric photosensitizers.[1531]

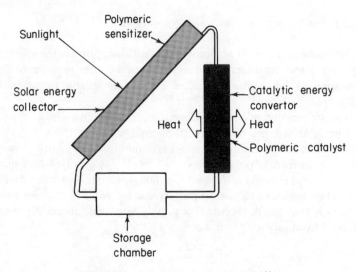

Fig. 16.17   Solar energy storage device[1531]

16.8. USE OF POLYMERIC MATERIALS IN SOLAR ENERGY TECHNOLOGY

Most current solar energy collectors and solar cells need more efficient production, simpler installation and maintenance, and lower capital cost. Polymers (plastics) offer design flexibility, parts consolidation, rapid processing, weight reduction, and corrosion resistance. The most important applications of polymers in solar energy technologies are:[636, 1238]

(i) Solar collectors—devices for absorbing solar energy and converting it into useful thermal energy.[636, 1239]

(ii) Luminescence solar concentrators (cf. Section 16.9).

(iii) Encapsulating photovoltaic cells.[28,806,2279,2502]

(iv) Optical components such as Fresnel lenses, light conductors, etc.

Table 16.2 Comparison of insoluble polymeric vs soluble monomeric sensitizers for the photo-isomerization of norbornadiene to quadricyclane[1531]

Sensitizer	Structure	Quantum efficiency for production of quadricyclane
Acetophenone	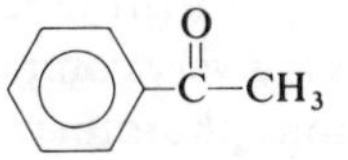	0.9
Poly(acetophenone)	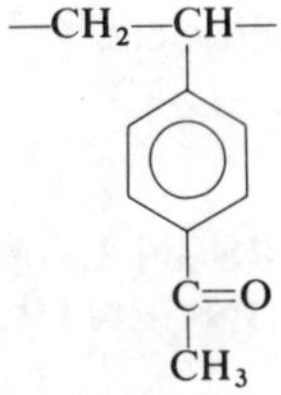	0.26
o-Benzyloxybenzaldehyde	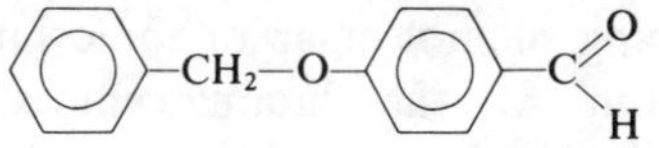	0.3
Poly(o-benzyloxybenzaldehyde)	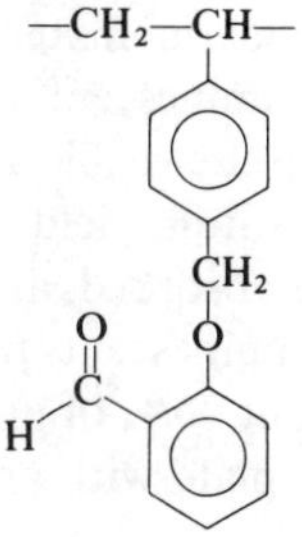	0.24
4-(N,N-dimethylamino)benzophenone	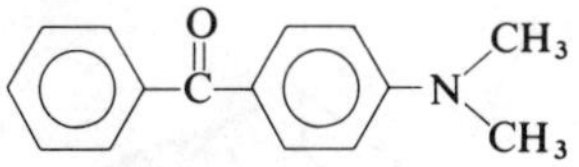	0.5
Poly[4-(N,N,-dimethylamino)benzophenone]	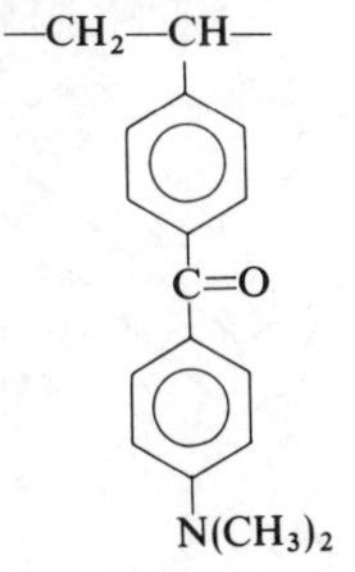	0.55

(v) Conventional applications[636,1238] such as adhesives, coatings,[2909,3928] sealants,[3928] moisture barriers, electrical and thermal insulation, and structural frames.

Polymers (plastics) employed in solar technologies must not only withstand u.v. irradiation (photodegradative oxidation processes) but must have suitable mechanical properties and durability to maintain operation under the combined effects of sunlight, wind, hail, rain, and atmospheric corrosion pollutants, for several year (5–20 years). Environmental corrosion of polymers in solar technologies applications is a very complex problem, and includes: photodegradation, thermal degradation, photothermal degradation, weathering, biological (soil) degradation, change of physical (optical) and mechanical properties, dimensional stability, permeability, etc.[2482] Effective methods for photostabilization are especially required.

16.9. POLYMERS IN LUMINESCENT SOLAR CONCENTRATORS

The operation of luminescent solar concentrators is based on the idea of light pipe-trapping of molecular or ionic luminescence. This trapped light can be coupled out of the luminescent solar concentrator into photovoltaic cells.[328, 329, 4003] The operation of a planar solar concentrator is shown in Fig. 16.18. A transparent polymeric material (e.g. poly(methyl methacrylate)) is impregnated with guest luminescent absorbers (e.g. organic dye molecules) having strong absorption in the visible and u.v. regions of the sun spectrum, and also having an efficient quantum yield of emission. Solar photons entering the upper face of the plate are absorbed, and luminescent photons then emitted. A large number of these luminescent photons are trapped by total internal reflection, for example about 75 % of an isotropic emission can be trapped in a poly(methyl methacrylate) plate with an index of refraction of 1.49.

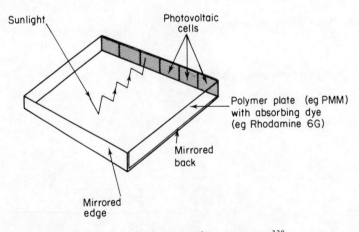

Fig. 16.18   A luminescent solar concentrator[328]

The dyes used (e.g. Rhodamine 6G or Coumarine-6) should be compatible with the polymer matrix (most dyes are polar, whereas poly(methyl methacrylate) is non-polar).[3507] The essential concentration required is 10^{-3}–10^{-4} M of dye. If the dye concentration increases the self-absorption processes lower quantum efficiency of the emitted light (quenching of luminescence). Dyes should not be photolysed (photobleached).

References

1. Abbås, K., *Appl. Polym. Symp.*, No. 35, 345 (1979).
2. Abdul-Rasoul, F. A. M., Ledwith, A., and Yagci, Y., *Polymer*, **19**, 1219 (1978).
3. Abu-Abdoun, I. I., Thijs, L., and Neckers, D. C., *J. Polym. Sci.*, A1, **21**, 3,129 (1983).
4. Abu-Abdoun, I. I., Thijs, L., and Neckers, D. C., *Macromolecules*, **17**, 282 (1984).
5. Abuin, E. A., Lissi, E. A., Gargallo, L., and Radic, D., *Europ. Polym. J.*, **15**, 373 (1979).
6. Abuin, E. A., Lissi, E. A., Gargallo, L., and Radic, D., *Europ. Polym. J.*, **16**, 793, 1023 (1980).
7. Abuin, E. A., Lissi, E. A., Gargallo, L., and Radic, D., *Europ. Polym. J.*, **20**, 105 (1984).
8. Abu-Zeid, M. E., Marafi, M. A., Nofal, E. E., and Anani, A. A., *J. Photochem.*, **18**, 347 (1982).
9. Abu-Zeid, M. E., Nofal, E. E., Abdul-Rasoul, F. A., Marafi, M. A., Mahmoud, D. G. S., and Ledwith, A., *J. Appl. Polym. Sci.*, **28**, 2317 (1983).
10. Abu-Zeid, M. E., Nofal, E. E., Marafi, M. A., Tahseen, L. A., Rasoul, F. A. A., and Ledwith, A., *J. Appl. Polym. Sci.*, **29**, 2431 (1984).
11. Abu-Zeid, M. E., Nofal, E. E., Tahseen, L. A., Abdul-Rasoul, F. A., and Ledwith, A., *J. Appl. Polym. Sci.*, **29**, 2443 (1984).
12. Achhammer, B. G., Reinhart, F. W., and Kline, G. M., *J. Res. Nat. Bur. Stand.*, **46**, 391 (1951). (1951).
13. Achhammer, B. G., Reiney, M. J., and Reinhart, F. W., *J. Res. Nat. Bur. Stand.*, **47**, 116 (1951).
14. Achhammer, B. G., Reiney, M. J., Wall, L. A., and Reinhart, F. W., *J. Polym. Sci.*, **8**, 555 (1952).
15. Achhammer, B. G., Reiney, M. J., Wall, L. A., and Reinhart, F. W., *Nat. Bur. Stand. Circ.*, **525**, 205 (1953).
16. Ackerman, L., and McGill, W. J., *South Afr. Chem. Inst.*, **26**, 82 (1973).
17. Adam, W., *J. Chem. Soc. Chem. Commun.*, **1974**, 289.
18. Adam, W., and Cilento, G. (eds), *Chemical and Biological Generation of Excited States*, Academic Press, New York, 1982.
19. Adamczyk, A., and Wilkinson, F., *J. Chem. Soc. Faraday II*, **68**, 2031 (1972).
20. Adamczyk, A., and Wilkinson, F., *J. Appl. Polym. Sci.*, **18**, 1225 (1974).
21. Adams, D., and Braun, D., *J. Polym. Sci.*, B, **18**, 629 (1980).
22. Adams, D. R., and Wilkinson, F., *J. Chem. Soc. Faraday Trans.*, **68**, 586 (1972).
23. Adams, G. E., and Wilson, R. L., *J. Chem. Soc. Faraday Trans.*, **69**, 719 (1973).
24. Adams, M. G., *Progr. Anal. Atom. Spectr.*, **5**, 153 (1982).
25. Adamson, A. W., Waltz, W. L., Zinato, E., Watts, D. W., Fleischauer, F. D., and Lindholm, R. D., *Chem. Rev.*, **68**, 541 (1968).
26. Addadi, L., Van Mil, J., and Lahav, M., *Origins Life*, **11**, 107 (1981).
27. Addadi, L., Van Mil, J., and Lahav, M., *J. Amer. Chem. Soc.*, **104**, 3422 (1982).
28. Addeo, A., Bonadies, V., Carfagna, C., Guerra, G., and Moscheiti, A., *Solar Energy*, **30**, 421 (1983).
29. Adeniyi, J. B., *Europ. Polym. J.*, **20**, 291 (1984).
30. Adeniyi, J. B., and Kolawole, E. G., *Europ. Polym. J.*, **20**, 43 (1984).
31. Ageishi, K., Endo, T., and Okawara, M., *J. Polym. Sci.*, A1, **19**, 1085 (1981).
32. Agolini, F., and Gay, F. P., *Macromolecules*, **3**, 349 (1970).
33. Al-Abidin, K. M. Z., and Jones, R. G., *J. Chem. Soc. Faraday I*, **75**, 774 (1979).
34. Albery, W. J., Fould, A. W., Hall, K. J., Hillman, A. R., *J. Electr. Soc.*, **127**, 654 (1980).
35. Alexander, I. J., and Scott, R. J., Europ. Pat. Appl. 14.293 (Scott Bader Ltd) (1980).
36. Alexandru, L., and Guillet, J. E., *J. Polym. Sci.*, A1, **13**, 483 (1975).
37. Alexandru, L., and Guillet, J. E., *J. Polym. Sci.*, A1, **14**, 2791 (1976).
38. Alexandru, L., and Somersall, A. C., *J. Polym. Sci.*, A1, **15**, 2013 (1977).
39. Alimoglu, A. K., Bamford, C. H., Ledwith, A., and Mulik, S. U., *Macromolecules*, **10**, 1081 (1977).
40. Aliwi, S. M., and Bamford, C. H., *Polymer*, **18**, 375 (1977).

671

41. Alkaitis, S. A., Beck, G., and Grätzel, M., *J. Amer. Chem. Soc.*, **97**, 5723 (1975).
42. Alkaitis, S. A. and Grätzel, M., *J. Amer. Chem. Soc.*, **98**, 3549 (1976).
43. Allara, D. L., and Hawkins, W. L. (eds), *Stabilization and Degradation of Polymers*, Advances in Chemistry Series, No. 169, American Chemical Society, Washington, D.C., 1978.
44. Allen, N. S., *Makromol. Chem. Rapid Commun.*, **1**, 235 (1980).
45. Allen, N. S., *Polym. Degrad. Stabil.*, **2**, 129 (1980).
46. Allen, N. S., *Polym. Degrad. Stabil.*, **2**, 155 (1980).
47. Allen, N. S., *Polym. Degrad. Stabil.*, **2**, 179 (1980).
48. Allen, N. S., *Polym. Degrad. Stabil.*, **2**, 269 (1980).
49. Allen, N. S., *Makromol. Chem.*, **181**, 2413 (1980).
50. Allen, N. S., in *Developments in Polymer Photochemistry* (Allen, N. S., ed.), Vol. 2, Applied Science Publishers, Ltd, London, 1981, p. 239.
51. Allen, N. S., *Polym. Photochem.*, **1**, 43 (1981).
52. Allen, N. S., *Polym. Photochem.*, **1**, 243 (1981).
53. Allen, N. S., *Polym. Degrad. Stabil.*, **3**, 73 (1981).
54. Allen, N. S. (ed.), *Degradation and Stabilization of Polyolefins*, Applied Science Publishers, Ltd, London, 1983.
55. Allen, N. S., in *Degradation and Stabilization of Polyolefins* (Allen, N. S., ed.), Applied Science Publishers, Ltd, London, 1983, p. 337.
56. Allen, N. S., *Polym. Photochem.*, **3**, 167 (1983).
57. Allen, N. S., *Polym. Degrad. Stabil.*, **6**, 193 (1984).
58. Allen, N. S., *Polym. Degrad. Stabil.*, **8**, 55 (1984).
59. Allen, N. S., Binkley, J. P., Parsons, B. J., Phillips, G. O., and Tennent, N. H., in *Photophysics of Synthetic Polymers* (Phillips, D., and Roberts, A. J., eds), Science Reviews Ltd, Northwood, England, 1982, p. 128.
60. Allen, N. S., Binkley, J. P., Parsons, B. J., Phillips, G. O., and Tennent, N. H., *Polym. Photochem.*, **2**, 97 (1982).
61. Allen, N. S., Binkley, J. P., Parsons, B. J., Phillips, G. O., and Tennent, N. H., *Polym. Photochem.*, **2**, 389 (1982).
62. Allen, N. S., Binkley, J. P., Parsons, B. J., Phillips, G. O., and Tennent, N. H., *Dyes Pigments*, **4**, 11 (1983).
63. Allen, N. S., Bullen, D. J., and McKellar, J. F., *Chem. Ind. (London)*, **1977**, 797.
64. Allen, N. S., Bullen, D. J., and McKellar, J. F., *J. Materials Sci.*, **12**, 1320 (1977).
65. Allen, N. S., Bullen, D. J., and McKellar, J. F., *J. Materials Sci.*, **13**, 2692 (1978).
66. Allen, N. S., Bullen, D. J., and McKellar, J. F., *J. Materials Sci.*, **14**, 759 (1979).
67. Allen, N. S., Bullen, D. J., and McKellar, J. F., *J. Materials Sci.*, **14**, 1941 (1979).
68. Allen, N. S., Chirinos-Padron, A., and Appleyard, J. H., *Polym. Degrad. Stabil.*, **4**, 223 (1982).
69. Allen, N. S., Chirinos-Padron, A., and Appleyard, J. H., *Polym. Degrad. Stabil.*, **5**, 29 (1983).
70. Allen, N. S., Chirinos-Padron, A., and Appleyard, J. H., *Polym. Degrad. Stabil.*, **5**, 55 (1983).
71. Allen, N. S., Chirinos-Padron, A., and Appleyard, J. H., *Polym. Degrad. Stabil.*, **5**, 323 (1983).
72. Allen, N. S., Chirinos-Padron, A., and Appleyard, J. H., *Polym. Degrad. Stabil.*, **6**, 31 (1984).
73. Allen, N. S., Chirinos-Padron, A., and Appleyard, J. H., *Polym. Degrad. Stabil.*, **6**, 149 (1984).
74. Allen, N. S., Chirinos-Padron, A., and Appleyard, J. H., *Polym. Degrad. Stabil.*, **9**, 15 (1984).
75. Allen, N. S., Chirinos-Padron, A., and Appleyard, J. H., *Polym. Photochem.*, **5**, 333 (1984).
76. Allen, N. S., Chirinos-Padron, A., and Appleyard, J. H., *Europ. Polym. J.*, **20**, 433 (1984).
77. Allen, N. S., Chirinos-Padron, A., Appleyard J. H., and Henman, T. J., *Polym. Degrad. Stabil.*, **5**, 105 (1983).
78. Allen, N. S., Chirinos-Padron, A., and Taylor, C., *Europ. Polym. J.*, **20**, 1031 (1984).
79. Allen, N. S., Fatinikun, K. O., *Polym. Degrad. Stabil.*, **3**, 327 (1980/81).
80. Allen, N. S., Fatinikun, K. O., and Henman, T. J., *Chem. Ind. (London)*, **1981**, 119.
81. Allen, N. S., Fatinikun, K. O., and Henman, T. J., *Polym. Degrad. Stabil.*, **4**, 59 (1982).
82. Allen, N. S., Fatinikun, K. O., and Henman, T. J., *Europ. Polym. J.*, **19**, 551 (1983).
83. Allen, N. S., Fatinikun, K. O., Luc-Gardette, J., and Lemaire, J., *Polym. Degrad. Stabil.*, **3**, 243 (1981).
84. Allen, N. S., Fatinikun, K. O., Luc-Gardette, J., and Lemaire, J., *Polym. Degrad. Stabil.*, **4**, 95 (1982).
85. Allen, N. S., and Harwood B., *Polym. Degrad. Stabil.*, **4**, 319 (1982).
86. Allen, N. S., Harwood, B., and McKellar, J. F., *J. Photochem.*, **9**, 559 (1978).
87. Allen, N. S., Harwood, B., and McKellar, J. F., *J. Photochem.*, **10**, 187 (1979).
88. Allen, N. S., Harwood, B., and McKellar, J. F., *J. Photochem.*, **10**, 193 (1979).

672

89. Allen, N. S., Homer, J., and McKellar, J. F., *Chem. Ind. (London)*, **1976**, 692.
90. Allen, N. S., Homer, J., and McKellar, J. F., *J. Appl. Polym. Sci.*, **20**, 2553 (1976).
91. Allen, N. S., Homer, J., and McKellar, J. F., *J. Appl. Polym. Sci.*, **21**, 2261 (1977).
92. Allen, N. S., Homer, J., and McKellar, J. F., *J. Appl. Polym. Sci.*, **21**, 3147 (1977).
93. Allen, N. S., Homer, J., and McKellar, J. F., *J. Appl. Polym. Sci.*, **22**, 611 (1978).
94. Allen, N. S., Homer, J., and McKellar, J. F., *Makromol. Chem.*, **179**, 1575 (1978).
95. Allen, N. S., Homer, J., McKellar, J. F., and Phillips, G. O., *Brit. Polym. J.*, **7**, 11 (1975).
96. Allen, N. S., Kotecha, J., Luc-Gardette, J., and Lemaire, J., *Polym. Commun.*, **25**, 235 (1984).
97. Allen, N. S., Luc-Gardette, J., and Lemaire, J., *Polym. Photochem.*, **1**, 111 (1981).
98. Allen, N. S., Luc-Gardette, J., and Lemaire, J., *Polym. Degrad. Stabil.*, **3**, 199 (1981).
99. Allen, N. S., Luc-Gardette, J., and Lemaire, J., *Polym. Degrad. Stabil.*, **4**, 261 (1982).
100. Allen, N. S., Luc-Gardette, J., and Lemaire, J., *Dyes Pigments*, **3**, 295 (1982).
101. Allen, N. S., Luc-Gardette, J., and Lemaire, J., *J. Appl. Polym. Sci.*, **27**, 2761 (1982).
102. Allen, N. S., Luc-Gardette, J., and Lemaire, J., *Polym. Photochem.*, **3**, 251 (1983).
103. Allen, N. S., Luc-Gardette, J., and Lemaire, J., *Polym. Degrad. Stabil.*, **8**, 133 (1984).
104. Allen, N. S., and McKellar, J. F., *Chem. Soc. Rev.*, **4**, 544 (1975).
105. Allen, N. S., and McKellar, J. F., *J. Photochem.*, **5**, 317 (1976).
106. Allen, N. S., and McKellar, J. F., *J. Appl. Polym. Sci.*, **20**, 1441 (1976).
107. Allen, N. S., and McKellar, J. F., *Brit. Polym. J.*, **9**, 302 (1977).
108. Allen, N. S., and McKellar, J. F., *Polymer*, **18**, 986 (1977).
109. Allen, N. S., and McKellar, J. F., *J. Appl. Polym. Sci.*, **21**, 1129 (1977).
110. Allen, N. S., and McKellar, J. F., *Chem. Ind. (London)*, **1977**, 537.
111. Allen, N. S., and McKellar, J. F., *Macromol. Rev.*, **13**, 241 (1978).
112. Allen, N. S., and McKellar, J. F., *Makromol. Chem.*, **179**, 523 (1978).
113. Allen, N. S., and McKellar, J. F., *J. Appl. Polym. Sci.*, **22**, 625 (1978).
114. Allen, N. S., and McKellar, J. F., *J. Appl. Polym. Sci.*, **22**, 2085 (1978).
115. Allen, N. S., and McKellar, J. F., *J. Appl. Polym. Sci.*, **22**, 3277 (1978).
116. Allen, N. S., and McKellar, J. F., *Chem. Ind. (London)*, **1978**, 907.
117. Allen, N. S., and McKellar, J. F., *Photochemistry of Man-Made Polymers*, Applied Science Publishers, London, 1979.
118. Allen, N. S., and McKellar, J. F., *Plast. Rubber Molec. Appl.*, **5**, 170 (1979).
119. Allen, N. S., and McKellar, J. F., *Macromol. Chem.*, **180**, 2875 (1979).
120. Allen, N. S., and McKellar, J. F., *Polym. Degrad. Stabil.*, **1**, 47 (1979).
121. Allen, N. S., and McKellar, J. F., *Polym. Degrad. Stabil.*, **1**, 205 (1979).
122. Allen, N. S., and McKellar, J. F., in *Developments in Polymer Degradation* (Grassie, N., ed.), Vol. 2, Applied Science Publishers, Ltd, London, 1979, p. 129.
123. Allen, N. S., and McKellar, J. F., *Photochemistry of Dyed and Pigmented Polymers*, Applied Science Publishers, Ltd, London, 1980.
124. Allen, N. S., and McKellar, J. F., in *Photochemistry of Dyed and Pigmented Polymers* (Allen, N. S., and McKellar, J. F., eds), Applied Science Publishers, Ltd, London, 1980, p. 247.
125. Allen, N. S., and McKellar, J. F., *Europ. Polym. J.*, **16**, 553 (1980).
126. Allen, N. S., McKellar, J. F., Arnaud, R., and Lemaire, J., *Makromol. Chem.*, **179**, 2103 (1978).
127. Allen, N. S., McKellar, J. F., and Chapman, C. B., *J. Appl. Polym. Sci.*, **20**, 1717 (1976).
128. Allen, N. S., McKellar, J. F., and Homer, J. F., *J. Appl. Polym. Sci.*, **22**, 611 (1978).
129. Allen, N. S., McKellar J. F., and Moghaddam, B. M., *Chem. Ind. (London)*, **1979**, 214.
130. Allen, N. S., McKellar, J. F., and Moghaddam, B. M., *J. Photochem.*, **11**, 101 (1979).
131. Allen, N. S., McKellar, J. F., and Moghaddam, B. M., *J. Appl. Chem., Biotechnol.*, **29**, 119 (1979).
132. Allen, N. S., McKellar, J. F., Moghaddam, B. M., and Phillips, G. O., *Chem. Ind. (London)*, **1979**, 593.
133. Allen, N. S., McKellar, J. F., and Phillips, G. O., *J. Polym. Sci.*, A1, **12**, 1233 (1974).
134. Allen, N. S., McKellar, J. F., and Phillips, G. O., *J. Polym. Sci.*, A1, **12**, 2623 (1974).
135. Allen, N. S., McKellar, J. F., and Phillips, G. O., *J. Polym. Sci.*, B, **12**, 253 (1974).
136. Allen, N. S., McKellar, J. F., and Phillips, G. O., *J. Polym. Sci.*, B, **12**, 477 (1974).
137. Allen, N. S., McKellar, J. F., and Phillips, G. O., *J. Polym. Sci.*, A1, **13**, 2857 (1975).
138. Allen, N. S., McKellar, J. F., Phillips, G. O., and Chapman, C. B., *J. Polym. Sci.*, B, **12**, 723 (1974).
139. Allen, N. S., McKellar, J. F., Phillips, G. O., and Wood, D. G. M., *J. Polym. Sci.*, A1, **12**, 2647 (1974).

140. Allen, N. S., McKellar, J. F., Phillips, G. O., and Wood, D. G. M., *J. Polym. Sci.*, B, **12**, 241 (1974).
141. Allen, N. S., McKellar, J. F., and Protopapas, S. A., *J. Appl. Polym. Sci.*, **22**, 1451 (1978).
142. Allen, N. S., McKellar, J. F., and Wilson, D., *J. Photochem.*, **6**, 73 (1976).
143. Allen, N. S., McKellar, J. F., and Wilson, D., *J. Photochem.*, **6**, 337 (1976).
144. Allen, N. S., McKellar, J. F., and Wilson, D., *J. Photochem.*, **7**, 319 (1977).
145. Allen, N. S., McKellar, J. F., and Wilson, D., *J. Polym. Sci.*, A1, **15**, 2973 (1977).
146. Allen, N. S., McKellar, J. F., and Wilson, D., *Chem. Ind. (London)*, **1978**, 887.
147. Allen, N. S., McKellar, J. F., and Wilson, D., *J. Photochem.*, **7**, 405 (1977).
148. Allen, N. S., McKellar, J. F., and Wilson, D., *Makromol. Chem.*, **179**, 269 (1978).
149. Allen, N. S., McKellar, J. F., and Wilson, D., *Polym. Degrad. Stabil.*, **1**, 205 (1979).
150. Allen, N. S., McKellar, J. F., and Wood, D. G. M., *J. Polym. Sci.*, A1, **13**, 2319 (1975).
151. Allen, N. S., Muhder, M., and Green, P., *Polym. Degrad. Stabil.*, **7**, 83 (1984).
152. Allen, N. S., Muhder, M., and Green, P., *Polym. Degrad. Stabil.*, **9**, 145 (1984).
153. Allen, N. S., and Parkinson, A., *Polym. Degrad. Stabil.*, **4**, 161 (1982).
154. Allen, N. S., and Parkinson, A., *Polym. Degrad. Stabil.*, **4**, 239 (1982).
155. Allen, N. S., and Parkinson, A., *Polym. Degrad. Stabil.*, **5**, 189 (1983).
156. Allen, N. S., Parkinson, A., Loffelman, F. F., and Sussi, P. V., *Polym. Degrad. Stabil.*, **5**, 241 (1983).
157. Allen, N. S., Parkinson, A., Loffelman, F. F., and Sussi, P. V., *Polym. Degrad. Stabil.*, **5**, 403 (1983).
158. Allen, N. S., Parkinson, A., Loffelman, F. F., and Sussi, P. V., *Angew. Makromol. Chem.*, **116**, 203 (1983).
159. Allen, N. S., Parkinson, A., Loffelman, F. F., Rauhut, M. M., and Sussi, P. V., *Polym. Degrad. Stabil.*, **7**, 153 (1984).
160. Allen, N. S., Parkinson, A., Luc-Gardette, A., and Lemaire, J., *Polym. Degrad. Stabil.*, **5**, 135 (1983).
161. Allen, N. S., Robinson, P. J., White, N. J., and Skelhorne, G. G., *Europ. Polym. J.*, **20**, 13 (1984).
162. Allen, N. S., and Rabek, J. F., *"New Trends in the Photochemistry of Polymers"*, Elsevier Applied Sci. Publ., London, 1985.
163. Allen, P. E. M., and Patrick, C. R., *Makromol. Chem.*, **48**, 89 (1961).
164. Allerhand, A., and Heilstone, R. K., *J. Chem. Phys.*, **56**, 3718 (1972).
165. Al-Malaika, S., Chakraborty, K. B., and Scott, G., in *Developments in Polymer Stabilization* (Scott, G., ed.), Vol. 6, Applied Science Publishers, Ltd, London, 1983, p. 73.
166. Al-Malaika, S., Huczkowski, P., and Scott, G., *Polymer*, **25**, 1006 (1984).
167. Al-Malaika, S., Kok, L., and Scott, G., *Polym. Commun.*, **25**, 233 (1984).
168. Al-Malaika, S., and Scott, G., *Europ. Polym. J.*, **16**, 503 (1980).
169. Al-Malaika, S., and Scott, G., *Europ. Polym. J.*, **16**, 709 (1980).
170. Al-Malaika, S., and Scott, G., *Europ. Polym. J.*, **19**, 241 (1983).
171. Al-Malaika, S., and Scott, G., *Polym. Degrad. Stabil.*, **5**, 415 (1983).
172. Al-Malaika, S., and Scott, G., in *Degradation and Stabilization of Polyolefins* (Allen, N. S., ed.), Applied Science Publishers, Ltd, London, 1983, p. 283.
173. Al-Malaika, S., Scott, G., and Huczkowski, P., *Polym. Degrad. Stabil.*, **7**, 95 (1984).
174. Almgren, M., Griesner, F., and Thomas, J. K., *J. Amer. Chem. Soc.*, **101**, 279 (1979).
175. Altomare, A., Carlini, C., Ciardelli, F., Solaro, R., Houben, J. L., and Rosato, N., *Polymer*, **24**, 95 (1983).
176. Altomare, A., Carlini, C., and Solaro, R., *Polymer*, **23**, 1355 (1982).
177. Altomare, A., Ciardelli, C. C., Solaro, R., and Rosato, N., *J. Polym. Sci.*, A1, **22**, 1267 (1984).
178. Altwicker, E. R., *Chem. Ind. (London)*, **1967**, 475.
179. Ambelang, J. C., Kline, R. H., Lorentz, O. M., Parks, C. R., Wandelin, C., and Shelton, J. R., *Rubb. Chem. Technol.*, **36**, 1497 (1963).
180. Americ, Y., and Guillet, J. E., *Macromolecules*, **4**, 375 (1971).
181. Amin, M. U., and Scott, G., *Europ. Polym. J.*, **10**, 1019 (1974).
182. Amin, M. U., Scott, G., Tillekeratne, L. M. K., *Europ. Polym. J.*, **11**, 85 (1975).
183. Amin, M. U., Tillekeratne, L. M. K., and Scott, G., *Europ. Polym. J.*, **11**, 85 (1976).
184. Amirzadeh, G., Kuhlmann, R., and Schnabel, W., *J. Photochem.*, **10**, 133 (1979).
185. Amirzadeh, G., Kuhlmann, R., and Schnabel, W., *Proc. IUPAC Makro Mainz.*, **1**, 335 (1979).
186. Amrizadeh, G., and Schnabel, W., *Makromol. Chem.*, **182**, 2821 (1981).
187. Amrani, F., Hung, J. M., and Morawetz, H., *Macromolecules*, **13**, 649 (1980).

674

188. Azumi, J., and Azumi, H., *Bull. Chem. Soc. Japan*, **39**, 2317 (1966).
189. Anani, A., Mobasher, A., and Rasoul, F. A., *J. Appl. Polym. Sci.*, **29**, 1491 (1984).
190. Ander, P., and Mohmoudhagh, M. K., *Macromolecules*, **15**, 213 (1982).
191. Anderson, J. C., and Reeje, C. B., *Tetrahed. Lett.*, **1962**, 1.
192. Anderson, R., and Prausnitz, J. M., *J. Chem. Phys.*, **39**, 1225 (1963).
193. Anderson, R. A., Birch, D. J. S., Davidson, K., Imhof, R. E., and Soutar, I., in *Photophysics of Synthetic Polymers* (Phillips, D., and Roberts, A. J., eds), Science Reviews, Ltd, Northwood, England, 1982, p. 140.
194. Anderson, R. A., Reid, R. F., and Soutar, I., *Europ. Polym. J.*, **15**, 925 (1979).
195. Anderson, R. M., Reid, R. F., and Soutar, I., *Europ. Polym. J.*, **16**, 945 (1980).
196. Anderson, R. W. Jr, Damaschen, D. E., Scott, G. W., and Talley, L. D., *J. Chem. Phys.*, **71**, 1134 (1978).
197. Andrew, J. E., Dyer, P. E., Forster, D., and Key, P. H., *Appl. Phys. Lett.*, **48**, 717 (1983).
198. Andrews, D. J., and Feast, W. J., *J. Polym. Sci.*, A1, **14**, 319 (1976).
199. Andrews, D. J., and Feast, W. J., *J. Polym. Sci.*, A1, **14**, 331 (1976).
200. Ang, C. H., Davis, N. P., Garnett, J. L., and Yen, N. T., *Radiat. Phys. Chem.*, **9**, 831 (1977).
201. Ang, C. H., Garnett, J. L., Levot, R., and Long, M. A., *J. Macromol. Sci. Chem.*, A, **17**, 87 (1982).
202. Ang, C. H., Garnett, J. L., Levot, R., Long, M. A., and Yen, N. T., *J. Polym. Sci.*, B, **18**, 471 (1980).
203. Angolini, F., and Gay, F. P., *Macromolecules*, **3**, 349 (1970).
204. Angus, H. J. F., McDonald, J., and Bryce-Smith, D., *J. Chem. Soc.*, **1960**, 2003.
205. Anufrieva, E. V., and Gotlib, Yu. Yu., *Adv. Polym. Sci.*, **40**, 1 (1981).
206. Anufrieva, E. V., Gotlib, Yu. Yu., Krakovyak, M. G., Skorokhodov, S. S., Sheveleva, T. V., *Dokl. Akad, Nauk SSSR*, **194**, 1108 (1970).
207. Anwaruddin, Q., and Santappa, M., *J. Polym. Sci.*, **35**, 361 (1967).
208. Anwaruddin, Q., and Santappa, M., *J. Polym. Sci.*, A1, **7**, 1315 (1969).
209. Anzai, J., Ueno, A., Sasaki, H., Shimokawa, K., and Osa, T., *Makromol. Chem. Rapid Commun.*, **4**, 731 (1983).
210. Aoki, H., Uehara, M., Suzuki, T., and Yoshida, A., *Europ. Polym. J.*, **16**, 571 (1980).
211. Arcesi, J. A., Rauner, F. J., and Williams, J. L. R., *J. Appl. Polym. Sci.*, **15**, 513 (1971).
212. Arct, J., Dul, M., Rabek, J. F., and Rånby, B., *Europ. Polym. J.*, **17**, 1041 (1981).
213. Arimitsu, S., and Masuhara, H., *Chem. Phys. Lett.*, **22**, 543 (1973).
214. Arimitsu, S., Masuhara, H., Magata, N., and Tsubomura, H., *J. Phys. Chem.*, **19**, 1255 (1975).
215. Arimitsu, S., and Tsubomura, H., *Bull. Chem. Soc. Japan*, **44**, 2288 (1971).
216. Armstrong, C., Husbands, M. J., and Scott, G., *Europ. Polym. J.*, **15**, 241 (1979).
217. Armstrong, C., Plant, M. A., and Scott, G., *Europ. Polym. J.*, **11**, 161 (1975).
218. Armstrong, C., and Scott, G., *J. Chem. Soc.*, **1971**, 1747.
219. Arnaud, J., Wippler, C., and Beau d'Angeres, F., *J. Chim. Phys.*, **64**, 1165 (1967).
220. Arnaud, R., and Lemaire, J., in *Developments in Polymer Degradation* (Grassie, N., ed.), Vol. 2, Applied Science Publishers, Ltd, London, 1979, p. 159.
221. Arnaud, R., and Lemaire, J., in *Developments in Polymer Degradation* (Grassie, N., ed.), Vol. 2, Applied Science Publishers, London, 1979, p. 247.
222. Arnaud, R., and Lemaire, J., in *Developments in Polymer Photochemistry* (Allen, N. S., ed.), Vol. 2, Applied Science Publishers, Ltd, London, 1981, p. 135.
223. Arnaud, R., Lemaire, J., Quemner, J., and Roche, G., *Europ. Polym. J.*, **12**, 499 (1976).
224. Arnold, D. R., Hinman, R. L., and Glick, A. H., *Tetrahed. Lett.*, **1964**, 1425.
225. Arnold, D. R., and Maroulis, A. J., *J. Amer. Chem. Soc.*, **98**, 5931 (1976).
226. Arthur, J. C. Jr, in *Developments in Polymer Photochemistry* (Allen, N. S., ed.), Vol. 1, Applied Science Publishers, Ltd, London, 1980, p. 69.
227. Arthur, J. C. Jr, in *Developments in Polymer Photochemistry* (Allen, N. S., ed.), Vol. 2, Applied Science Publishers, Ltd, London, 1981, p. 39.
228. Arthur, J. C., and Hinojosa, O., *Appl. Polym. Symp.*, No. 26, 147 (1975).
229. Asai, D., Okada, A., Kondon, S., and Tsuda, K., *J. Macromol. Sci. Chem.*, A, **18**, 1011 (1982).
230. Asai, M., and Tazuke, S., *Macromolecules*, **6**, 818 (1973).
231. Asai, M., Tazuke, S., and Okamura, S., *J. Polym. Sci.*, A1, **12**, 45 (1974).
232. Asai, M., Tazuke, S., Okamura, S., Ohno, T., and Kato, S., *Chem. Lett.*, **1973**, 993.
233. Asai, N., and Neckers, D. C., *J. Org. Chem.*, **45**, 2903 (1980).
234. Asanuma, T., Gotoh, T., Tsuchida, A., Yamamoto, M., and Nishijima, Y., *J. Chem. Soc. Chem. Commun.*, **1977**, 485.

675

235. Asanuma, T., Yamamoto, M., and Nishijima, Y., *J. Chem. Soc. Chem. Comm.*, **1975**, 56.
236. Asanuma, T., Yamamoto, M., and Nishijima, Y., *J. Chem. Soc. Chem. Comm.*, **1975**, 608.
237. Aslam, M., Anwaruddin, Q., and Natarajan, L. V., *Polym. Photochem.*, **5**, 41 (1984).
238. Aspler, J. S., Carlsson, D. J., and Wiles, D. M., *Macromolecules*, **9**, 691 (1976).
239. Aspler, J. S., and Guillet, J. E., *Macromolecules*, **12**, 1082 (1979).
240. Aspler, J. S., Hoyle, C. E., and Guillet, J. E., *Macromolecules*, **11**, 925 (1978).
241. Asquith, R. S., Gardner, K. L., Geehan, T. G., and McNally, G., *J. Polym. Sci.*, B, **15**, 435 (1973).
242. Asquith, R. S., Hirst, L., and Rivett, D. E., *Appl. Polym. Symp.*, No. 18, 333 (1971).
243. Ashworth, J., Bamford, C. H., and Smith, E. G., *Pure Appl. Chem.*, **30**, 25 (1972).
244. Atkinson, B., *Experentia*, **14**, 272 (1958).
245. Attala, R. H., and Nagel, S. C., *J. Chem. Soc. Chem. Comm.*, **1972**, 1049.
246. Augustyniak, W., and Wojtczak, J., *J. Polym. Sci.*, A1, **18**, 1339 (1980).
247. Aviram, A., *Macromolecules*, **11**, 1275 (1978).
248. Avivi, P., and Weinreb, A., *J. Chem. Phys.*, **27**, 716 (1957).
249. Avouris, P., Kordas, J., and El-Bayoumi, A., *Chem. Phys. Lett.*, **26**, 373 (1974).
250. Ayrey, G., *Chem. Rev.*, **63**, 645 (1963).
251. Azuma, C., Hashizume, E., Sanui, K., Ogata, N., *J. Appl. Polym. Sci.*, **28**, 543 (1983).
252. Azuma, C., Sanui, K., and Ogata, N., *J. Appl. Polym. Sci.*, **25**, 1273 (1980).
253. Azuma, C., Sanui, K., and Ogata, N., *J. Appl. Polym. Sci.*, **27**, 2065 (1982).
254. Bäckström, H. L. J., Appelgren, K. L., and Niklasson, R. J. V., *Acta Chem. Scand.*, **19**, 1555 (1965).
255. Bagdasaryan, K. S., and Sinitsyna, A. A., *J. Polym. Sci.*, **52**, 31 (1961).
256. Bagheri, R., Chakraborty, K. B., and Scott, G., *Chem. Ind. (London)*, **1980**, 865.
257. Bagheri, R., Chakraborty, K. B., and Scott, G., *Polym. Degrad. Stab.*, **4**, 1 (1982).
258. Bagheri, R., Chakraborty, K. B., and Scott, G., *Polym. Degrad. Stab.*, **5**, 197 (1983).
259. Bailey, D., Tirrell, D., Pinazzi, C., and Vogel, O., *Macromolecules*, **11**, 312 (1978).
260. Bailey, D., Tirrell, D., and Vogl, O., *J. Macromol. Sci. Chem.*, **A12**, 661 (1968).
261. Bailey, D., Tirrell, D., and Vogl, O., *J. Polym. Sci.*, A1, **14**, 2725 (1976).
262. Bailey, D., and Vogl, O., *J. Macromol. Sci. Rev. Macromol. Chem.*, C, **14**, 267 (1976).
263. Bair, H. E., Boyle, D. J., and Kelleher, P. G., *Polym. Eng. Sci.*, **20**, 995 (1980).
264. Bal, T. S., Cox, A., Kemp, T. J., and Pinot de Moira, P., *Polymer*, **21**, 423 (1980).
265. Balaban, L., Majer, J., Ryšavy, D., and Šlama, Z., *Kunstoffe*, **60**, 341 (1970).
266. Balaban, L., Majer, J., Vesely, K., *J. Polym. Sci.*, C, **22**, 1059 (1969).
267. Balandier, M., and Decker, C., *Europ. Polym. J.*, **14**, 995 (1978).
268. Balint, G., Kelen, T., Rehak, A., and Tüdös, F., *React. Kinet. Catal. Lett.*, **4**, 467 (1976).
269. Balint, G., Kelen, T., Tüdös, F., and Rehak, A., *Polym. Bull.*, **1**, 647 (1979).
270. Balint, G., Rockenbauer, A., Kelen, T., Tüdös, F., and Jokay, L., *Polym. Photochem.*, **1**, 139 (1981).
271. Ball, S., and Bruice, T. C., *J. Amer. Chem. Soc.*, **102**, 6498 (1980).
272. Ballardini, R., Beggatio, G., Bortolus, P., Faucitano, A., Buttafana, A., and Gratani, F., *Polym. Degrad. Stabil.*, **7**, 41 (1984).
273. Ballester, M., Riera, J., Castaner, J., and Casulleras, M., *Tetrahed. Lett.*, **1978**, 643.
274. Balogh, A., Durmis, J., Holčik, J. and Karvaš, M., *Plasty Kaučok*, **14**, 204 (1977).
275. Balzani, V., Bolletta, F., Gandofi, M. T., and Maestri, M., *Top Current Chem.*, **75**, 1 (1978).
276. Balzani, V., Moggi, L., Manfrin, M. F., Bolletta, F., Laurence, G. S., *Chem. Rev.*, **15**, 321 (1975).
277. Bamford, C. H., *Europ. Polym. J.*, Suppl., **1969**, 1.
278. Bamford, C. H., *Pure Appl. Chem.*, **34**, 173 (1973).
279. Bamford, C. H., *Polymer*, **17**, 321 (1976).
280. Bamford, C. H., and Al-Lamee, K. G., *J. Chem. Soc. Faraday Trans. I*, **80**, 2175 (1984).
281. Bamford, C. H., and Al-Lamee, K. G., *J. Chem. Soc. Faraday Trans. I*, **80**, 2187 (1984).
282. Bamford, C. H., Al-Lamee, K. G., and Konstantinov, C. J., *J. Chem. Soc. Faraday Trans. I*, **73**, 1406 (1977).
283. Bamford, C. H., Bingham, J., and Block, H., *Trans. Faraday Soc.*, **66**, 2612 (1970).
284. Bamford, C. H., Burley, J. W., and Goldbeck, M., *J. Chem. Soc. Dalton Trans.*, **1972**, 1846.
285. Bamford, C. H., Crowe, P. A., Hobbs, J., and Wayne, R. P., *Proc. Rny. Soc., (London)* A, **292**, 153 (1966).
286. Bamford, C. H., Crowe, P. A., and Wayne, R. P., *Proc. Roy. Soc. (London)*, A, **284**, 445 (1965).
287. Bamford, C. H., Duncan, F. J., Reynolds, R. J., and Seddon, J. D., *J. Polym. Sci.*, C, **23**, 419 (1968).

676

288. Bamford, C. H., Dyson, R. W., and Eastmond, G. C., *J. Polym. Sci.*, C, **16**, 2425 (1967).
289. Bamford, C. H., Dyson, R. W., and Eastmond, G. C., *Polymer*, **10**, 885 (1969).
290. Bamford, C. H., Eastmond, G. C., Woo, J., and Richards, F. H., *Polymer*, **2**, 643 (1982).
291. Bamford, C. H., and Ferrar, A. N., *J. Chem. Soc. Faraday Trans. I*, **68**, 1243 (1972).
292. Bamford, C. H., Hobbs, J., and Wayne, R. P., *Chem. Commun.*, **1965**, 469.
293. Bamford, C. H., Jenkins, A. D., and Ward, K. C., *J. Polym. Sci.*, **48**, 37 (1960).
294. Bamford, C. H., Ledwith, A., and Yagci, Y., *Polymer*, **19**, 354 (1978).
295. Bamford, C. H., and Mahud, M. U., *J. Chem. Soc. Faraday Trans. I*, **68**, 762 (1972).
296. Bamford, C. H., and Mullik, S. U., *Polymer*, **14**, 38 (1973).
297. Bamford, C. H., and Mullik, S. U., *J. Chem. Soc. Faraday Trans. I*, **69**, 1127 (1973).
298. Bamford, C. H., and Mullik, S. U., *J. Chem. Soc. Faraday Trans. I*, **71**, 625 (1975).
299. Bamford, C. H., and Mullik, S. U., *Polymer*, **17**, 94 (1976).
300. Bamford, C. H., and Mullik, S. U., *J. Chem. Soc. Faraday Trans. I*, **73**, 1260 (1977).
301. Bamford, C. H., and Mullik, S. U., *Polymer*, **19**, 948 (1978).
302. Bamford, C. H., Mullik, S. U., and Puddephatt, R. J., *J. Chem. Soc. Faraday Trans. I*, **71**, 2213 (1975).
303. Bamford, C. H., and Paprotny, J., *Polymer*, **13**, 208 (1972).
304. Bamford, C. H., and Sakamoto, I., *J. Chem. Soc. Faraday Trans. I*, **70**, 330 (1974).
305. Bamford, C. H., and Sakamoto, I., *J. Chem. Soc. Faraday Trans. I*, **70**, 344 (1974).
306. Bamford, C. H., and Xiao-Zu Han, *Polymer*, **22**, 1299 (1981).
307. Banihashemi, A., and Schulz, R. C., *Makromol. Chem.*, **179**, 855 (1978).
308. Banks, E., Okamoto, Y., and Ueba, Y., *J. Appl. Polym. Sci.*, **25**, 359 (1980).
309. Barber, J., (ed.), *Photosynthesis in Relation to Model Systems*, Elsevier, 1979.
310. Bargon, J., *J. Polym. Sci.*, A1, **16**, 2747 (1978).
311. Barker, P. S., Bottom, R. A., Guthrie, J. T., and Beddows, C. G., *Polym. Photochem.*, **2**, 87 (1982).
312. Barker, P. S., Guthrie, J. T., Davis, M. J., Godfrey, A. A., and Green, P. N., *J. Appl. Polym. Sci*, **26**, 521 (1981).
313. Barkowski, D., Mersch, W., and Dörfmüller, T., *J. Polym. Sci.*, A1, **20**, 953 (1982).
314. Barltrop, J. A., and Coyle, J. D., *Excited States in Organic Photochemistry*, Wiley, New York, 1975.
315. Bartholomew, R. F., and Davidson, R. S., *J. Chem. Soc.*, C, **1971**, 2342.
316. Bartholomew, R. F., Davidson, R. S., and Howell, M. J., *J. Chem. Soc.*, C, **1971**, 2804.
317. Bartholomew, R. F., Davidson, R. S., Lambeth, P. F., McKellar, J. F., and Turner, P. H., *J. Chem. Soc. Perkin II*, **1972**, 577.
318. Barton, J., and Čapek, I., *Makromol. Chem.*, **182**, 3513 (1981).
319. Barton, J., Čapek, I., and Hrdlovic, P., *J. Polym. Sci.*, A1, **13**, 2671 (1975).
320. Barton, J., Čapek, I., Susoliak, O., and Juranicova, V., *Makromol. Chem.*, **179**, 2997 (1978).
321. Barwise, A. J. G., German, A. A., Leyland, R. L., Smith, P. G., and Rodgers, M. A. J., *J. Amer. Chem. Soc.*, **100**, 1814 (1978).
322. Basile, L. J., *J. Chem. Phys.*, **36**, 2204 (1962).
323. Basile, L. J., *Trans. Faraday Soc.*, **60**, 1702 (1964).
324. Basile, L. J., *Trans. Faraday Soc.*, **62**, 3163 (1965).
325. Bassemir, R. W., *Amer. Ink Maker*, **1974**, Dec., 33.
326. Bassewitz, K., and Menzel. G., *Angew. Makromol. Chem.*, **47**, 201 (1975).
327. Bässler, H., *Adv. Polym. Sci.*, **63**, 2 (1984).
328. Batchelder, J. S., Zewail, A. H., and Cole, T., *Appl. Opt.*, **18**, 3090 (1979).
329. Batchelder, J. S., Zewail, A. H., and Cole, T., *Appl. Opt.*, **20**, 3733 (1981).
330. Battais, A., Boutevin, B., Maliszkiewicz, M., and Vial-Reveillon, F., *Europ. Polym. J.*, **19**, 499 (1983).
331. Bauer, D. R., Brayman, J. T., and Pecora, R., *Macromolecules*, **8**, 443 (1975).
332. Bauer, R. H., and Coppinger, G. M., *Tetrahedron*, **19**, 1201 (1963).
333. Baugh, P. J., in *Developments in Polymer Photochemistry* (Allen, N. S., ed.), Vol. 2, Applied Science Publishers Ltd, London, 1981, p. 165.
334. Baugh, P. J., and Phillips, G. O., In *Cellulose and Cellulosic Derivatives* (Bikales, N., and Segal, L., eds), Vol. 5, Wiley, New York, 1971, p. 1047.
335. Baugh, P. J., Phillips, G. O., and Worthington, N. W., *J. Soc. Dyers Colour.*, **85**, 241 (1969).
336. Baugh, P. J., Phillips, G. O., and Worthington, N. W., *J. Soc. Dyers Colour.*, **86**, 19 (1970).
337. Baughan, R. H., *J. Appl. Phys.*, **42**, 4579 (1971).
338. Baum, E., and Norman, R. O. C., *J. Chem. Soc.*, B, **1968**, 227.
339. Baum, E., and Norman, R. O. C., *J. Chem. Soc.*, B, **1968**, 749.

677

340. Bauman, H., Strehmel, B., and Timpe, H. J., *Polym. Photochem.*, **4**, 233 (1984).
341. Baumgärtel, H., and Zimmermann, Z., *Z. Naturforsch.*, B, **18**, 406 (1963).
342. Bays, J. P., Encinas, M. V., and Scaiano, J. C., *Macromolecules*, **12**, 348 (1979).
343. Bays, J. P., Encinas, M. V., and Scaiano, J. C., *Macromolecules*, **13**, 815 (1980).
344. Bays, J. P., Encinas, M. V., and Scaiano, J. C., *Polymer*, **21**, 283 (1980).
345. Baysal, B., and Tobolsky, A. V., *J. Polym. Sci.*, **8**, 529 (1952).
346. Baxendale, J. H., Evans, M. G., and Park, G. S., *Trans. Faraday Soc.*, **42**, 155 (1946).
347. Baxendale, J. H., and Thomas, J. K., *Trans. Faraday Soc.*, **54**, 1515 (1958).
348. Baxendale, J. H., and Wilson, J. A., *Trans. Faraday Soc.*, **53**, 344 (1957).
349. Beachell, H. C., and Smiley, L. H., *J. Polym. Sci.*, A1, **5**, 1635 (1967).
350. Beak, P., and Payet, C. R., *J. Org. Chem.*, **35**, 3281 (1970).
351. Bean, A. J., and Bassemir, R. W., in *UV Curing: Science and Technology*, (Pappas, S. P., ed.), Vol. 1, Stanford Technology Marketing Crop., Norwalk, CT, 1978, p. 187.
352. Beavan, S. W., Beck, G., and Schnabel, W., *Europ. Polym. J.*, **14**, 385 (1978).
353. Beavan, S. W., Hackett, P. A., and Phillips, D., *Europ. Polym. J.*, **10**, 925 (1974).
354. Beavan, S. W., Hargreaves, J. S., and Phillips, D., *Adv. Photochem.*, **11**, 207 (1979).
355. Beavan, S. W., and Phillips, D., *Europ. Polym. J.*, **10**, 593 (1974).
356. Beavan, S. W., and Phillips, D., *J. Photochem.*, **3**, 349 (1974/75).
357. Beavan, S. W., and Schnabel, W., *Macromolecules*, **11**, 782 (1978).
358. Beck, G., Dobrowolski, G., Kiwi, J., and Schnabel, W., *Macromolecules*, **8**, 9 (1975).
359. Beck, G., Kiwi, J., Lindenau, D., and Schnabel, W., *Europ. Polym. J.*, **10**, 1069 (1974).
360. Beck, G., Lindenau, D., and Schnabel, W., *Europ. Polym. J.*, **11**, 761 (1975).
361. Beck, G., Lindenau, D., and Schnabel, W., *Macromolecules*, **10**, 135 (1977).
362. Beck, S., Hallam, A., and North, A. M., *Polymer*, **20**, 1177 (1979).
363. Becker, H. D., *J. Org. Chem.*, **32**, 2115 (1967).
364. Becker, H. D., *J. Org. Chem.*, **32**, 2124 (1967).
365. Becker, H. D., *Pure Appl. Chem.*, **54**, 1589 (1982).
366. Beckett, A., Osborne, A., and Porter, G., *Trans. Faraday Soc.*, **60**, 873 (1964).
367. Beckett, A., and Porter, G., *Trans. Faraday Soc.*, **59**, 2038 (1963).
368. Beckett, A., and Porter, G., *Trans. Faraday Soc.*, **59**, 2051 (1963).
369. Beckewitz, F., and Heusinger, H., *J. Polym. Sci.*, A1, **21**, 2763 (1983).
370. Bednar, B., Morawetz, H., and Shafer, J. A., *Macromolecules*, **17**, 1634 (1984).
371. Belicar, J., *Radiat. Cur.*, **4**, 15 (1977).
372. Bell, E. R., Rust, F. F., and Vaughan, W. E., *J. Amer. Chem. Soc.*, **72**, 337 (1950).
373. Bell, J. A., Linschitz, H., *J. Amer. Chem. Soc.*, **85**, 528 (1963).
374. Bell, J. A., and MacGillvray, J. D., *J. Chem. Educ.*, **51**, 677 (1974).
375. Bellenger, V., Bouchard, C., Claveirolle, P., and Verdu, J., *Polym. Photochem.*, **1**, 69 (1981).
376. Bellenger, V., and Verdu, J., *J. Appl. Polym. Sci.*, **28**, 2599 (1983).
377. Bellenger, V., and Verdu, J., *J. Appl. Polym. Sci.*, **28**, 2677 (1983).
378. Bellenger, V., and Verdu, J., *Polym. Photochem.*, **5**, 295 (1984).
379. Bellenger, V., Verdu, J., and Carette, L. B., *J. Macromol. Sci. Chem.*, A, **17**, 1149 (1982).
380. Bellenger, V., Verdu, J., Carette, L. B., Vymazalova, Z., and Vymazal, Z., *Polym. Degrad. Stabil.*, **4**, 303 (1982).
381. Bellin, J. S., *Photochem. Photobiol.*, **8**, 383 (1968).
382. Bellin, J. S., Alexander, R., and Mahoney, R. D., *Photochem. Photobiol.*, **17**, 17 (1973).
383. Bellenger, V., Verdu, J., Carette, L. B., Unpublished results (cf. 379).
384. Bellobono, I. R., Calgari, S., Leonardi, M. C., Selli, E., and Paglia, E. D., *Angew. Makromol. Chem.*, **100**, 135 (1981).
385. Bellobono, I. R., Giovanardi, S., Marcandalli, B., Calgari, S., and Nosari, D., *Polym. Photochem.*, **4**, 59 (1984).
386. Bellobono, I. R., Tolusso, F., Selli, E., Calgari, S., and Berlin, A., *J. Appl. Polym. Sci.*, **26**, 619 (1981).
387. Belluš, D., *Adv. Photochem.*, **8**, 109 (1971).
388. Belluš, D., in *Singlet Oxygen: Reactions with Organic Compounds and Polymers* (Rånby, B., and Rabek, J. F., eds), Wiley, Chichester, 1978, p. 61.
389. Belluš, D., *Adv. Photochem.*, **11**, 105 (1979).
390. Belluš, D., and Hrdlovič, P., *Chem. Rev.*, **67**, 599 (1967).
391. Belluš, D., Hrdlovič, P., and Manašek, Z., *J. Polym. Sci.*, B, **4**, 1 (1966).
392. Belluš, D., Lind, H., and Wyatt, J. F., *J. Chem. Soc. Chem. Comm.*, **1972**, 1199.

678

393. Belluš, D., Manašek, Z., Hrdlovič, P., and Slama, P., *J. Polym. Sci.*, C, **16**, 267 (1967).
394. Benachour, D., and Rogers, C. E., in *Photodegradation and Photostabilization of Coatings* (Pappas, S. P., and Winslow, F. H., eds), ACS Symp. Ser., No. 151., Amer. Chem. Soc., Washington, D.C., 1981, p. 263.
395. Benedetti, E., and Goodman, M., *Biochem.*, **7**, 4242 (1968).
396. Benedetti, E., Kossoy, A., Falxa, M. L., and Goodman, M., *Biochem.*, **7**, 4234 (1968).
397. Bendaikha, T., and Decker, T., *J. Radiat. Curing*, **1984**, Apr., 6.
398. Bengough, W. I., and Carson, T. G., *Europ. Polym. J.*, **6**, 1459 (1970).
399. Bengough, W. I., and Ross, I. C., *Trans. Faraday Soc.*, **62**, 2251 (1966).
400. Benham, J. V., and Pullukat, T. S., *J. Appl. Polym. Sci.*, **20**, 3295 (1976).
401. Bennett, R. G., and Kellog, R., in *Progress in Reaction Kinetics*, (Porter, G., ed.), Vol. 4, Pergamon Press, London, 1966.
402. Bennett, R. G., Schwenker, R. P., and Kellog, R., *J. Chem. Phys.*, **41**, 3040 (1967).
403. Bensasson, R. V., Ronfard-Haret, J. C., Land, E. J., and Webber, S. E., *Chem. Phys. Lett.*, **68**, 438 (1979).
404. Benson, S. W., *J. Chem. Phys.*, **40**, 1007 (1964).
405. Benson, S. W., *J. Chem. Educ.*, **42**, 501 (1965).
406. Benson, S. W., and Boss, J., *J. Chem. Phys.*, **29**, 546 (1958).
407. Bentley, P., McKellar, J. P., and Phillips, G. O., *Rev. Progr. Colour. Rel. Topics*, **5**, 33 (1974).
408. Bentz, J. P., Beyl, J. P., Beinert, G., and Weill, G., *Europ. Polym. J.*, **11**, 711 (1975).
409. Bercovici, T., Heiligman-Rim, R., and Fischer, E., *Mol. Photochem.*, **1**, 23 (1969).
410. Berger, J., *Kunstoffe Fortschr.*, **2**, 79 (1976).
411. Berger, J., *Polymer*, **25**, 1629 (1984).
412. Berger, M., Goldblatt, I. L., and Steel, C., *J. Amer. Chem. Soc.*, **95**, 1717 (1973).
413. Bergmann, F., and Hirsberg, Y., *J. Amer. Chem. Soc.*, **65**, 1429 (1943).
414. Bergmark, W. R., Beckmann, B., and Lindenberger, W., *Tetrahed. Lett.*, **1971**, 2259.
415. Berkower, I. J., Beutel, J., and Waiker, P., *Photogr. Sci. Eng.*, **12**, 283 (1968).
416. Berlman, I. B., Goldschmidt, C. R., Stein, G., Tomkiewicz, Y., and Weinreb, A., *Chem. Phys. Lett.*, **4**, 338 (1969).
417. Berner, G., Kirchmayer, R., and Rist, G., *J. Oil Colour Chem. Assoc.*, **61**, 105 (1978).
418. Berner, G., Puglisi, J., Kirchmayer, R., and Rist, G., *J. Radiat. Curing*, **1979**, Apr., 2.
419. Bertelson, T., *Techn. Chem.*, **3**, 45 (1971).
420. Berticat, P., *J. Chim. Phys.*, **79**, 887 (1967).
421. Bertrains, H., Boutevin, B., Maliszkiewicz, M., and Vernet, J. L., *Europ. Polym. J.*, **18**, 779 (1982).
422. Bertrains, H., Boutevin, B., Maliszkiewicz, M., and Vernet, J. L., *Europ. Polym. J.*, **18**, 785 (1982).
423. Bertrains, H., Boutevin, B., Maliszkiewicz, M., Pietrasanta, Y., and Vernet, J. L., *Europ. Polym. J.*, **18**, 791 (1982).
424. Betts, A. T., and Uri, N., *Chem. Ind.*, **25**, 512 (1967).
425. Bevington, J. C., *Proc. Roy. Soc.*, A, **239**, 420 (1957).
426. Bevington, J. C., Melville, H. W., and Taylor, R. P., *J. Polym. Sci.*, **12**, 449 (1954).
427. Bevington, J. C., and Ratti, L., *Europ. Polym. J.*, **8**, 1105 (1972).
428. Bevington, J. C., and Toole, J., *J. Polym. Sci.*, **28**, 413 (1958).
429. Bevington, J. C., and Troth, H. G., *Trans. Faraday Soc.*, **58**, 186 (1966).
430. Bhaduri, R., and Aditya, S., *Colloid. Polym. Chem.*, **256**, 659 (1972).
431. Bickley, R. I., and Jayanti, R. K., *Faraday Discuss. Soc.*, **58**, 194 (1974).
432. Biddle, D., and Chapoy, L. L., *Polym. Photochem.*, **5**, 129 (1984).
433. Biddle, D., and Chapoy, J. L., *Macromolecules*, **17**, 1751 (1984).
434. Biddle, D., and Nordström, T., *Arkiv Kemi*, **22**, 359 (1970).
435. Bilen, C. S., and Morantz, D. J., *Polymer*, **17**, 109 (1976).
436. Birks, J. B., *Progr. Reaction Kinetics*, **5**, 181 (1970).
437. Birks, J. B., *Rep. Progr. Phys.*, **38**, 903 (1975).
438. Birks, J. B., (ed.), *Organic Molecular Photophysics*, Vol. 1 and Vol. 2, Wiley, London, 1973 and 1975.
439. Birks, J. B., and Christophoru, L. G., *Proc. Roy. Soc.*, A, **277**, 571 (1964).
440. Birks, J. B., and Conte, J. C., *Proc. Roy. Soc.*, A, **303**, 85 (1974).
441. Birks, J. B., Dyson, D. J., and King, T. A., *Proc. Roy. Soc.*, A, **277**, 270 (1964).
442. Birks, J. B., Dyson, D. J., and Munro, I. H., *Proc. Roy. Soc.*, A, **275**, 575 (1963).
443. Black, J. F., *J. Amer. Chem. Soc.*, **100**, 527 (1978).

679

444. Blair, H. S., and Kaulaw, T., *Polymer*, **21**, 1475 (1980).
445. Blair, H. S., and McArdle, C. B., *Polymer*, **25**, 999 (1984).
446. Blair, H. S., and McArdle, C. B., *Polymer*, **25**, 1347 (1984).
447. Blair, H. S., and Pogue, H. T., *Polymer*, **20**, 99 (1979).
448. Blair, H. S., and Pogue, H. T., *Polymer*, **23**, 779 (1982).
449. Blair, H. S., Pogue, H., and Ridran, H. I., *Polymer*, **21**, 1195 (1980).
450. Blais, P., Day, M., Wiles, D. M., *J. Appl. Polym. Sci.*, **17**, 1895 (1973).
451. Blanding, J. M., Osborn, C. L., and Watson, S. L., *J. Radiat. Curing*, **5**, 13 (1978).
452. Block, H., Ledwith, A., and Taylor, A. R., *Polymer*, **12**, 271 (1971).
453. Bloor, D., Koski, L., Stevens, G. C., Preston, F. H., and Ando, D. J., *J. Mater. Sci.*, **10**, 1678 (1975).
454. Blossey, E. C., Neckers, D. C., Thayer, A. L., and Schaap, A. P., *J. Amer. Chem. Soc.*, **95**, 5820 (1973).
455. Blumberg, M., Boss, C. R., and Chien, J. C. W., *J. Appl. Polym. Sci.*, **9**, 3837 (1965).
456. Boehm, H. P., *Z. Anorg. Allg. Chem.*, **352**, 156 (1967).
457. Boens, N., DeSchrywer, F. C., and Smets, G., *J. Polym. Sci.*, A1, **13**, 201 (1975).
458. Boettner, E. A., *Kunstof. Rundschau*, **18**, 57 (1971).
459. Bokobza, L., Jasse, B., and Monnerie, L., *Europ. Polym. J.*, **13**, 921 (1977).
460. Bokobza, L., Jasse, B., and Monnerie, L., *Europ. Polym. J.*, **16**, 715 (1980).
461. Bokobza, L., and Monnerie, L., *Polymer*, **22**, 235 (1981).
462. Bokobza, L., Pajot-Augy, E., Monnerie, L., Bouas-Laurent, H., and Castellan, A., *Polymer*, **22**, 1309 (1981).
463. Bokobza, L., Pajot-Augy, E., Monnerie, L., Castellan, A., and Bouas-Laurent, H., *Polym. Photochem.*, **5**, 191 (1984).
464. Bolon, D. A., and Kunz, C. O., *Polym. Eng. Sci.*, **12**, 109 (1972).
465. Bolon, D. A., Lucas, G. M., Olson, D. R., and Webb, K. K., *J. Appl. Polym. Sci.*, **25**, 543 (1980).
466. Bolt, J. D., and Turro, N. J., *J. Phys. Chem.*, **85**, 4029 (1981).
467. Bonkowski, J. E., *Text. Res. J.*, **39**, 243 (1969).
468. Bonnans-Plaisance, C., Casals, P. F., and Levesque, G., *Makromol. Chem.*, **182**, 2523 (1981).
469. Borden, D. G., *J. Appl. Polym. Sci.*, **22**, 239 (1978).
470. Borden, D. G., Unruh, C. C., and Merrill, S. H., U.S. Pat. 966,296 (Aug. 12, 1964).
471. Borden, D. G., Unruh, C. C., and Merrill, S. H., U.S. Pat. 3,453,237 (July 1, 1969), and U.S. Pat. 3,647,444 (Mar. 7, 1972).
472. Borden, D. G., and Williams, J. L. R., *Makromol. Chem.*, **178**, 3035 (1977).
473. Borer, A., Kirchmayr, R., and Rist., *Helv. Cim. Acta*, **61**, 305 (1978).
474. Borgarello, E., Kiwi, J., Pelizzetti, E., Visca, M., and Grätzel, M., *Nature*, **298**, 158 (1981).
475. Bortolus, P., Delltone, S., and Beggiato, G., *Europ. Polym.*, **13**, 185 (1977).
476. Bortolus, P., Minto, F., Beggiato, G., and Lora, S., *J. Appl. Polym. Sci.*, **24**, 285 (1979).
477. Bortolus, P., Minto, F., Loras, S., Gleria, M., and Beggiato, G., Polym. Photochem., **4**, 45 (1984).
478. Bos, A., *J. Appl. Polym. Sci.*, **16**, 2567 (1972).
479. Bos, A., *J. Polym. Sci.*, A1, **12**, 2283 (1974).
480. Bos, A., and Buchman, A. S., *J. Polym. Sci.*, A1, **11**, 833 (1973).
481. Boss, C. R., and Chien, J. C. W., *J. Polym. Sci.*, A1, **4**, 1543 (1966).
482. Bössler, H. H., and Schulz, R. C., *Makromol. Chem.*, **158**, 113 (1972).
483. Boudevska, H., Broutchkov, C., and Astrough, H., *Europ. Polym. J.*, **8**, 737 (1983).
484. Bourdelande, J. L., Font, T., and Sanchez-Ferrando, F., *Tetrahed. Lett.*, **21**, 3805 (1980).
485. Bourdelande, J. L., Font, T., and Sanchez-Ferrando, F., *Polym. Photochem.*, **2**, 383 (1982).
486. Bousquet, J. A., Donnet, J. B., Faure, J., Fouassier, J. P., Haidar, B., and Vidal, A., *J. Polym. Sci.*, **A1**, 17 (1979).
487. Bousquet, J. A., Donnet, J. B., Faure, J., Fouassier, J. P., Haidar, B., and Vidal, A., *J. Polym. Sci.*, A1, **18**, 765 (1980).
488. Bousquet, J. A., Faure, J., Fouassier, J. P., Donnet, J. P., Haidar, B., and Vidal, A., *J. Polym. Sci.*, A1, **17**, 1685 (1979).
489. Bousquet, J. A., and Fouassier, J. P., *J. Photochem.*, **20**, 197 (1982).
490. Bousquet, J. A., and Fouassier, J. P., *Polym. Degrad. Stabil.*, **5**, 113 (1983).
491. Bousquet, J. A., and Fouassier, J. P., *J. Polym. Sci.*, A1, **22**, 3865 (1984)
492. Bousquet, J. A., Fouassier, J. P., Haidar, B., Vidal, A., Donnet, J. B., and Faure, J., *C.R. Acad. Sci. Paris*, **283C**, 613 (1976).

680

493. Bousquet, J. A., Hajdar, B., Fouassier, J. P., and Vidal, A., *Europ. Polym. J.*, **19**, 135 (1983).
494. Boustead, I., *Europ. Polym. J.*, **6**, 731 (1970).
495. Boutevin, B., and Maliszkiewicz, M., *Makromol. Chem.*, **184**, 977 (1983).
496. Bowen, E. J., *Trans. Faraday Soc.*, **50**, 97 (1954).
497. Bowen, E. J. (ed.), *Chemiluminescence in Chemistry*, Van Nostrand, London, 1968.
498. Bowen, E. J., and Metcalf, W. S., *Proc. Roy. Soc.*, A, **206**, 437 (1951).
499. Bowen, E. J., and Tanner, D. W., *Trans. Faraday Soc.*, **51**, 475 (1955).
500. Bowie, W. T., and Feldman, M. R., *J. Amer. Chem. Soc.*, **99**, 4721 (1977).
501. Boyle, D. J., and Gesner, B. D., *J. Appl. Polym. Sci.*, **12**, 1193 (1963).
502. Box, H. C., Budzinski, E. E., and Freund, H. G., *J. Amer. Chem. Soc.*, **92**, 5305 (1970).
503. Bramwell, F. B., Zadjura, R. E., Paley, C., and Farhrenholtz, S. R., *J. Chem. Educ.*, **55**, 403 (1978).
504. Brand, J. C. D., and Snedden, W., *Trans. Faraday Soc.*, **53**, 894 (1957).
505. Brault, R. C., Jenney, J. A., Margerum, J. D., Miller, L. J., and Rust, J. B., *Image Techn.*, **13**, 13 (1971).
506. Braun, D., *Pure Appl. Chem.*, **53**, 549 (1981).
507. Braun, D., and Becker, K. H., *Angew. Makromol. Chem.*, **6**, 186 (1969).
508. Braun, D., and Bender, R. F., *Europ. Polym. J.*, Suppl. **1969**, p. 269.
509. Braun, D., and Kull, S., *Angew. Makromol. Chem.*, **85**, 79 (1980).
510. Braun, D., and Kull, S., *Angew. Makromol. Chem.*, **86**, 171 (1980).
511. Braun, D., and Meier, B., *Makromol. Chem.*, **175**, 791 (1974).
512. Braun, D., and Wolf, M., *Angew. Makromol. Chem.*, **70**, 71 (1978).
513. Braun, H., and Förster, Th., *Ber. Bunsenges. Phys. Chem.*, **70**, 1091 (1966).
514. Braun, H., and Förster, Th., *Z. Phys. Chem.*, **78**, 40 (1972).
515. Braun, H. G., and Wegner, G., *Makromol. Chem.*, **184**, 1103 (1983).
516. Breck, A. K., Taylor, C. L., Russell, K. E., and Wan, J. K. S., *J. Polym. Sci.*, A1, **12**, 1505 (1974).
517. Breitenbach, J. W., and Derkosch, J., *Makromol. Chem.*, **81**, 530 (1950).
518. Breitenbach, J. W., and Frittum, H., *J. Polym. Sci.*, **29**, 565 (1958).
519. Breitenbach, J. W., Sommer, F., and Unger, G., *Monatsh. Comm.*, **101**, 32 (1970).
520. Breskman, E. L., and Pappas, S. P., *J. Coatings Technol.*, **48**, 35 (1976).
521. Breslow, R., *Chem. Soc. Rev.*, **1**, 533 (1972).
522. Breslow, R., Baldwin, S., Flechtner, T., Kalicky, P., Liu, S., and Washburn, W., *J. Amer. Chem. Soc.*, **95**, 3251 (1973).
523. Breslow, R., and Winnik, M., *J. Amer. Chem. Soc.*, **91**, 3083 (1969).
524. Brewer, T., *J. Amer. Chem. Soc.*, **93**, 775 (1971).
525. Bridge, N. K., and Porter, G., *Proc. Roy. Soc.*, A, **244**, 259 (1958).
526. Bridge, N. K., and Reed, M., *Trans. Faraday Soc.*, **56**, 1796 (1960).
527. Briggs, P. J., and McKellar, J. F., *J. Appl. Polym. Sci.*, **12**, 1825 (1968).
528. Brinckman, E., Delzenne, G., Poot, A., and Willems, J., *Unconventional Imaging Processes*, Focal Press, London, 1978.
529. Brisimitzakis, A. C., and Karydas, A. C., *J. Polym. Sci.*, B, **21**, 565 (1983).
530. Broadbent, A. D., *Chem. Commun.*, **1967**, 382.
531. Broadbent, A. D., and Newton, P. R., *Can. J. Chem.*, **50**, 381 (1972).
532. Brown, G. L., *Photochromism*, Techniques in Chemistry, Vol. 3, Wiley, New York, 1971.
533. Brown, G. M., Brunschwig, B. S., Greutz, C., Endicott, S., and Suttin, W., *J. Amer. Chem. Soc.*, **101**, 1298 (1979).
534. Brown, K., and Soutar, I., *Europ. Polym. J.*, **10**, 433 (1974).
535. Brownlie, J. T., and Ingold, K. U., *Can. J. Chem.*, **45**, 2427 (1967).
536. Bruce, J. M., *Quart. Rev.*, **21**, 405 (1967).
537. Bruce, J. M., in *The Chemistry of the Quinoid Compounds* (Patai, S., ed.), Wiley, 1974, p. 465.
538. Brugger, P. A., Infelta, P. P., Braun, A. M., and Grätzel, M., *J. Amer. Chem. Soc.*, **103**, 320 (1981).
539. Brugger, P. A., and Grätzel, M., *J. Amer. Chem. Soc.*, **102**, 2461 (1980).
540. Brunow, G., and Eriksson, B., *Acta Chem. Scand.*, **25**, 2779 (1971).
541. Brunow, G., Forsskåhl, I., Grönlund, A. C., Lindström, G., and Nyberg, K., in *Singlet Oxygen: Reactions with Organic Compounds and Polymers*, (Rånby, B., and Rabek, J. F., eds), Wiley, Chichester, 1978, p. 311.
542. Bryce-Smith, D., *Pure Appl. Chem.*, **34**, 193 (1973).
543. Bryce-Smith, D., and Gilbert, A., *Chem. Commun.*, **1966**, 643.

681

544. Bryce-Smith, D., Gilbert, A., and Orger, B. H., *Chem. Commun.*, **1966,** 512.
545. Brzozowski, Z. K., Jozwiak, B., and Kielkiewicz, J., *Appl. Polym. Symp.*, No. 35,337 (1979).
546. Buback, M., Choe, C. R., and Franck, E. U., *Makromol. Chem.*, **185,** 1665 and 1685 (1984).
547. Bubeck, C., Tieke, B., and Wegner, G., *Ber. Bunsenges. Phys. Chem.*, **86,** 499 (1982).
548. Bubeck, C., Tieke, B., and Wegner, G., *Mol. Cryst. Liq. Cryst.*, **96,** 109 (1983).
549. Buchachenko, A., *Stable Radicals*, Consultant Bureau, New York, 1965.
550. Buchanan, K. J., and McGill, W. J., *Europ. Polym. J.*, **16,** 309 (1980).
551. Buchanan, K. J., and McGill, W. J., *Europ. Polym. J.*, **16,** 313 (1980).
552. Buchanan, K. J., and McGill, W. J., *Europ. Polym. J.*, **16,** 319 (1980).
553. Buettner, A. V., *J. Phys. Chem.*, **68,** 3253 (1964).
554. Buettner, A. V., and Dedinas, J., *J. Phys. Chem.*, **75,** 187 (1971).
555. Bührer, H. G., Aeschbach, R., Phillipou, T., Parnaud, J. J., and Elias, H. G., *Makromol. Chem.*, **13,** 157 (1972).
556. Buil, J., and Verdu, J., *Europ. Polym. J.*, **15,** 389 (1979).
557. Bullock, A. T., Cameron, G. G., and Smith, P. M., *J. Phys. Chem.*, **77,** 1635 (1973).
558. Bunbury, D. L., and Chuang, T. T., *Can. J. Chem.*, **47,** 2045 (1969).
559. Bunbury, D. L., and Wang, C. T., *Can. J. Chem.*, **46,** 1473 (1968).
560. Burchill, P. J., and George, G. A., *J. Polym. Sci.*, B, **12,** 497 (1974).
561. Burchill, P. J., and Pinkerton, D. M., *J. Polym. Sci.* Symposia, No. 55, 185 (1976).
562. Burchill, P. J., and Pinkerton, D. M., *Polym. Degrad. Stabil.*, **2,** 239 (1980).
563. Burkhart, R. D., *Macromolecules*, **9,** 234 (1976).
564. Burkhart, R. D., *Chem. Phys.*, **46,** 11 (1980).
565. Burkhart, R. D., *Macromolecules*, **16,** 820 (1983).
566. Burkhart, R. D., *J. Phys. Chem.*, **87,** 1566 (1983).
567. Burkhart, R. D., and Abla, A. A., *J. Phys. Chem.*, **86,** 468 (1982).
568. Burkhart, R. D., and Aviles, R. G., *J. Phys. Chem.*, **83,** 1897 (1979).
569. Burkhart, R. D., and Aviles, R. G., *Macromolecules*, **12,** 1073 (1979).
570. Burkhart, R. D., and Aviles, R. G., *Macromolecules*, **12,** 1078 (1979).
571. Burkhart, R. D., Aviles, R. G., and Magrini, K., *Macromolecules*, **14,** 91 (1981).
572. Burkhart, R. D., and Lonson, E. R., *Chem. Phys. Lett.*, **54,** 85 (1978).
573. Burnett, M., and North, A. M., *Transfer and Storage of Energy by Molecules*, Wiley, London, 1979.
574. Busetto, C., Mattucci, A. M., Cernia, E., *J. Appl. Polym. Sci.*, **25,** 395 (1980).
575. Busetto, C., Mattucci, A. M., Cernia, E., Guizzi, G., and Belluco, U., *Inorg. Chem. Acta*, **34,** 155 (1979).
576. Bush, R. W., Ketley, A. D., Morgan, C. R., and Whitt, D. G., *J. Radiat. Curing*, **7,** 20 (April) (1980).
577. Buss, A. D., Malati, M. A., and Atkinson, R., *J. Oil Colour. Chem. Assoc.*, **59,** 369 (1976).
578. Butler, G. B., and Ferree, W. I., *J. Polym. Sci.*, Symposia, No. 56, 397 (1977).
579. Butty, E., and Suppan, P., *Polym. Photochem.*, **5,** 171 (1984).
580. Butyagin, G. P., Ivanov, V. B., and Shlyapintokh, V. Ya, *Vysokomol. Soedin.*, B, **18,** 827 (1976).
581. Buxton, G., and Wilmarth, W. K., *J. Phys. Chem.*, **67,** 2835 (1963).
582. Bystritskaya, E. V., Karpukhin, O. N., and Karpovitch, T. S., *Dokl. Akad. Nauk SSSR*, **235,** 607 (1977).
583. Cabanes, W. R., and Lin, L. C., *J. Polym. Sci.*, A1, **22,** 857 (1984).
584. Calas, R., Lalsnde, R., Fauger, J. G., and Moulines, M. F., *Bull. Soc. Chem. France*, **1965,** 119.
585. Caldwell, R. A., *Tetrahed Lett.*, **1969,** 2121.
586. Caldwell, R. A., Creed, D., and Ohita, H., *J. Amer. Chem. Soc.*, **96,** 2994 (1974).
587. Caldwell, R. A., and Gajewski, R. P., *J. Amer. Chem. Soc.*, **93,** 532 (1971).
588. Caldwell, R. A., and James, S. P., *J. Amer. Chem. Soc.*, **91,** 5184 (1969).
589. Caldwell, R. A., Sorocool, G. W., and Gajewski, R. P., *J. Amer. Chem. Soc.*, **95,** 2549 (1973).
590. Calgari, S., Selli, E., and Bellobono, R., *J. Appl. Polym. Sci.*, **27,** 527 (1982).
591. Calvert, J. G., and Pitts, J. N., *Photochemistry*, Wiley, New York, 1966.
592. Calvert, P. D., and Ryan, T. G., *Polymer*, **19,** 611 (1978).
593. Calvin, M., *Int. J. Energy Res.*, **3,** 73 (1979).
594. Cameron, G. G., and Kane, D. R., *Makromol. Chem.*, **109,** 194 (1967).
595. Campbell, D., Monteith, L. K., and Turner, D. T., *J. Polym. Sci.*, A1, **8,** 2703 (1970).
596. Canva, G. P., and Canva, J. J., *Rubber J.*, **1971,** Sept., 36.
597. Čapek, I., and Barton, J., *J. Polym. Sci.*, A1, **12,** 327 (1974).

682

598. Čapek, I., and Barton, J., *J. Polym. Sci.*, A1, **17**, 937 (1979).
599. Čapek, I., and Barton, J., *Makromol. Chem.*, **182**, 3505 (1981).
600. Čapek, I., Barton, J., and Danciger, J., *J. Polym. Sci.*, A1, **17**, 943 (1979).
601. Card, R. J., and Neckers, D. C., *J. Amer. Chem. Soc.*, **99**, 7733 (1977).
602. Carlblom, L. H., and Pappas, S. P., *J. Polym. Sci.*, A1, **15**, 1381 (1977).
603. Carlick, D. J., in *Encyclopedia of Polymer Science and Technology*, (Mark, H. F., and Bikales, N. M., eds), Supplement, Vol. 1, Wiley, New York, 1976, p. 367.
604. Carlini, C., Ciardelli, F., Donati, D., and Gurzoni, F., *Polymer*, **24**, 599 (1983).
605. Carlsson, D. J., Chan, K. H., Durmis, J., and Wiles, D. M., *J. Polym. Sci.*, A1, **20**, 575 (1982).
606. Carlsson, D. J., Chan, K. H., Garton, A., and Wiles, D. M., *Pure Appl. Chem.*, **52**, 289 (1980).
607. Carlsson, D. J., Chan, K. H., Jensen, J. P. T., Wiles, D. M., and Durimis, J., in *Polymer Additives* (Kresta, J. E., ed.), Plenum Press, New York, 1984, p. 35.
608. Carlsson, D. J., Chan, K. H., and Wiles, D. M., *Amer. Chem. Soc., Org. Coat. Plast. Chem., Prepr.* **42**, 555 (1980).
609. Carlsson, D. J., Chan, K. H., and Wiles, D. M., in *Photodegradation and Photostabilization of Coatings* (Pappas, S. P., and Winslow, F. H., eds), ACS Symp. Ser. No. 151, ACS Washington, D.C., 1981, p. 51.
610. Carlsson, D. J., Gan, L. H., Parnell, R. D., and Wiles, D. M., *J. Polym. Sci.*, **11**, 683 (1973).
611. Carlsson, D. J., Gan, L. H., and Wiles, D. M., *J. Polym. Sci.*, A1, **16**, 2353 (1978).
612. Carlsson, D. J., Garton, A., and Wiles, D. M., *Macromolecules*, **9**, 695 (1976).
613. Carlsson, D. J., Garton, A., and Wiles, D. M., *J. Appl. Polym. Sci.*, **21**, 2963 (1977).
614. Carlsson, D. J., Garton, A., and Wiles, D. M., in *Developments in Polymer Stabilization* (Scott, G., ed.), Vol. 1, Applied Science Publishers, Ltd, London, 1979, p. 219.
615. Carlsson, D. J., Grattan, D. W., Suprunchuk, T., and Wiles, D. M., *J. Appl. Polym. Sci.*, **22**, 2217 (1978).
616. Carlsson, D. J., Grattan, D. W., and Wiles, D. M., *Amer. Chem. Soc., Org. Coat. Plast. Chem., Prepr.*, **39**, 628 (1978).
617. Carlsson, D. J., Grattan, D. W., and Wiles, D. M., *Polym. Degrad. Stabil.*, **1**, 69 (1979).
618. Carlsson, D. J., Mendenhall, G. D., Suprunchuk, T., and Wiles, D. M., *J. Amer. Chem. Soc.*, **94**, 8960 (1972).
619. Carlsson, D. J., Parnell, R. D., and Wiles, D. M., *J. Polym. Sci.*, B, **11**, 149 (1973).
620. Carlsson, D. J., Sproule, D. E., and Wiles, D. M., *Macromolecules*, **5**, 659 (1972).
621. Carlsson, D. J., Suprunchuk, T., and Wiles, D. M., *J. Appl. Polym. Sci.*, **16**, 675 (1972).
622. Carlsson, D. J., Suprunchuk, T., and Wiles, D. M., *J. Polym. Sci.*, B, **11**, 61 (1973).
623. Carlsson, D. J., Suprunchuk, T., and Wiles, D. M., *J. Polym. Sci.*, B, **14**, 193 (1976).
624. Carlsson, D. J., and Wiles, D. M., *Macromolecules*, **2**, 587 (1969).
625. Carlsson, D. J., and Wiles, D. M., *J. Polym. Sci.*, B, **8**, 419 (1970).
626. Carlsson, D. J., and Wiles, D. M., *Macromolecules*, **4**, 174 and 179 (1971).
627. Carlsson, D. J., and Wiles, D. M., *J. Polym. Sci.*, B, **11**, 759 (1973).
628. Carlsson, D. J., and Wiles, D. M., *Rubber Chem. Technol.*, **49**, 991 (1974).
629. Carlsson, D. J., and Wiles, D. M., *J. Polym. Sci.*, A1, **12**, 2217 (1974).
630. Carlsson, D. J., and Wiles, D. M., *Macromolecules*, **7**, 259 (1974).
631. Carlsson, D. J., and Wiles, D. M., *J. Radiat. Curing*, **2**, April, 2 (1975).
632. Carlsson, D. J., and Wiles, D. M., *J. Macromol. Sci. Rev. Macromol. Chem.*, C, **14**, 65 (1976).
633. Carlsson, D. J., and Wiles, D. M., *J. Macromol. Sci. Rev. Macromol. Chem.*, C, **14**, 155 (1976).
634. Carlsson, D. J., and Wiles, D. M., *J. Polym. Sci.*, A1, **14**, 493 (1976).
635. Carlsson, D. J., and Wiles, D. M., *Polym. Degrad. Stabil.*, **6**, 1 (1981).
636. Carroll, W. F., and Schissel, P., in *Polymers in Solar Energy Utilization* (Gebelein, C. G., Williams, D. J., and Deanin, R. D., eds), ACS Symp. Ser. No. 220, ACS Washington, D.C., 1983, p. 3.
637. Carruthers, R. A., Crellin, R. A., and Ledwith, A., *Chem. Commun.*, **1969**, 252.
638. Carsey, T. P., Findley, G. L., and McGlynn, S. P., *J. Amer. Chem. Soc.*, **101**, 4502 (1979).
639. Carstensen, P., *Dansk. Kemi*, **49**, 97 (1968).
640. Carstensen, P., *Makromol. Chem.*, **135**, 219 (1970).
641. Carstensen, P., *Makromol. Chem.*, **142**, 145 (1971).
642. Carstensen, P., in *ESR Application to Polymer Research* (Kinell, P. O., Rånby, B., and Runnström-Reio, V., eds), Almqvist & Wiksell, Stockholm, 1973, p. 159.
643. Carstensen, P., and Rånby, B., in *Radiation Research*, North-Holland, Amsterdam, 1967, p. 297.
644. Cegarra, J., Ribe, J., and Miro, P., *J. Soc. Dyers Colour.*, **88**, 293 (1972).

645. Čeppan, M., Lapčik, L., Liška, M., and Pelikan, P., *Europ. Polym. J.*, **16**, 607 (1980).
646. Cernia, E., Mantovani, E., Marconi, W., Mazzei, M., Palladino, N., and Zanobi, A., *J. Appl. Polym. Sci.*, **19**, 15 (1975).
647. Cernia, E., Marconi, W., Palladino, N., and Bacchin, P., *J. Appl. Polym. Sci.*, **18**, 2085 (1974).
648. Cha, Y. S., Tsunooka, M., and Tanaka, M., *J. Polym. Sci.*, A1, **22**, 2973 (1984).
649. Cha, Y. S., Tsunooka, M., Tanaka, M., and Konishi, H., *J. Appl. Polym. Sci.*, **29**, 2941 (1984).
650. Chan, M. G., and Hawkins, W. L., *Polym. Prepr. Amer. Chem. Soc., Div. Polym. Chem.*, **9**, 1638 (1968).
651. Chaberek, S., and Allen, R. J., *J. Phys. Chem.*, **69**, 647 (1965).
652. Chaineaux, J., and Tanielian, C., in *Singlet Oxygen: Reactions with Organic Compounds and Polymers* (Rånby, B. and Rabek, J. F., eds), Wiley, Chichester, 1978, p. 164.
653. Chakraborty, K. B., and Scott, G., *Polymer*, **18**, 98 (1977).
654. Chakraborty, K. B., and Scott, G., *Europ. Polym. J.*, **13**, 731 (1977).
655. Chakraborty, K. B., and Scott, G., *Europ. Polym. J.*, **13**, 865 (1977).
656. Chakraborty, K. B., and Scott, G., *Europ. Polym. J.*, **13**, 1007 (1977).
657. Chakraborty, K. B., and Scott, G., *Chem. Ind. (London)*, **1978**, 237.
658. Chakraborty, K. B., and Scott, G., *Polym. Degrad. Stabil.*, **1**, 37 (1979).
659. Chakraborty, K. B., and Scott, G., *Europ. Polym. J.*, **15**, 35 (1979).
660. Chakraborty, K. B., and Scott, G., *Polymer*, **21**, 252 (1980).
661. Chakraborty, K. B., and Scott, G., *J. Polym. Sci.*, B, **22**, 553 (1984).
662. Chakraborty, K. B., Scott, G., and Poyner, W. R., *Polym. Degrad. Stabil.*, **8**, 1 (1984).
663. Chan, S. F., Chou, M., Creutz, C., Matsubara, T., and Sutin, N., *J. Amer. Chem. Soc.*, **103**, 369 (1981).
664. Chance, R. R., and Patel, G. N. *J. Polym. Sci.*, A2, **16**, 859 (1978).
665. Chance, R. R., Patel, G. N., and Witt, J. D., *J. Chem. Phys.*, **71**, 206 (1979).
666. Chance, R. R., and Shand, M. L., *J. Chem. Phys.*, 948 (1980).
667. Chance, R. R., Shand, M. L., Hogg, C., and Silbey, R., *Phys. Rev.*, B, **22**, 3540 (1980).
668. Chandra, A. K., *Chem. Phys. Lett.*, **5**, 229 (1970).
669. Chandra, R., *Polym. Photochem.*, **3**, 367 (1983).
670. Chandra, R., and Handa, S. P., *J. Appl. Polym. Sci.*, **27**, 1945 (1982).
671. Chandra, R., and Handa, S. P., *Polym. Photochem.*, **3**, 391 (1983).
672. Chandra, R., and Singh, B. P., *Europ. Polym. J.*, **18**, 199 (1982).
673. Chandra, R., Singh, B. P., Singh, S., and Handa, S. P., *Polymer*, **22**, 523 (1981).
674. Chandross, E. A., and Dempster, C. J., *J. Amer. Chem. Soc.*, **92**, 3586 (1970).
675. Chang, C. T., *Photogr. Sci. Eng.*, **23**, 311 (1979).
676. Chang, J. M., and Aklonis, J. J., *J. Polym. Sci.*, B, **21**, 999 (1983).
677. Chang, S. T., Hon, D. N. S., and Feist, W. C., *Wood Fiber*, **14**, 104 (1982).
678. Chang, Y. C., *Photogr. Sci. Eng.*, **23**, 301 (1979).
679. Chapman, O. L., and Lee, K., *J. Org. Chem.*, **34**, 4166 (1969).
680. Chapman, O. L., and Wampfler, G., *J. Amer. Chem. Soc.*, **91**, 5390 (1969).
681. Charlesby, A., Grace, C. S., and Pilkington, F. B., *Proc. Roy. Soc.*, A, **268**, 205 (1962).
682. Charlesby, A., and Pinner, S. H., *Proc. Roy. Soc.*, A, **249**, 367 (1959).
683. Charlesby, A., and Thomas, D. K., *Proc. Roy. Soc.*, A, **269**, 104 (1962).
684. Chaudet, J. H., Newland, G. C., Patton, H. W., and Tamblyn, J. W., *SPE Trans.*, **1**, 26 (1961).
685. Chaudet, J. H., and Tamblyn, J. W., *SPE Trans.*, **1**, 57 (1961).
686. Chen, C., *J. Polym. Sci.*, A1, **3**, 1107 and 1127 (1964).
687. Chen, C. S. H., Jankowski, S., and Brother, A., *Adv. Chem. Ser.*, No. 66, 240 (1967).
688. Chen, D. T. L., and Morawetz, H., *Macromolecules*, **9**, 463 (1976).
689. Chen, D. T. L., and Morawetz, H., *Macromolecules*, **15**, 1445 (1982).
690. Chen, H. L., and Morawetz, H., *Europ. Polym. J.*, **19**, 923 (1983).
691. Chen, T. S., and Thomas, J. K., *J. Polym. Sci.*, A1, **17**, 1103 (1979).
692. Cheng, H. N., Schilling, F. C., and Bovey, F. A., *Macromolecules*, **9**, 363 (1976).
693. Cheu, E. L., and Osawa, Z., *J. Appl. Polym. Sci.*, **19**, 2947 (1975).
694. Cheung, P. S. R., and Roberts, C. W., *J. Polym. Sci.*, B, **17**, 227 (1979).
695. Cheung, P. S. R., Dellinger, J. A., Stuckey, W. C., and Roberts, C. W., in *Photodegradation and Photostabilization of Coatings* (Pappas, S. P., and Winslow, F. H., eds), ACS Symp. Ser., No. 151, ACS Washington, D.C., 1981, p. 239.
696. Cheung, P. S. R., Roberts, C. W., and Wagener, K. B., *J. Appl. Polym. Sci.*, **24**, 1809 (1979).
697. Chew, C. H., Gan, L. M., and Scott, G., *Europ. Polym. J.*, **13**, 361 (1977).
698. Chien, J. C. W., *J. Phys. Chem.*, **69**, 4317 (1965).

684

699. Chien, J. C. W., and Boss, C. R., *J. Polym. Sci.*, A1, **5**, 1683 (1967).
700. Chien, J. C. W., and Conner, W. P., *J. Amer. Chem. Soc.*, **90**, 1001 (1968).
701. Chinmayanaandam, B. R., and Melville, H. W., *Trans. Faraday Soc.*, **53**, 73 (1965).
702. Chou, C. H., and Jellinek, H. H. G., *Can. J. Chem.*, **42**, 522 (1964).
703. Christensen, J. E., Jacobine, A. F., and Scanio, C. J. V., *Radiat. Curing*, **8**, March, 1 (1981).
704. Chryssomallis, G., and Drickamer, H. G., *J. Chem. Phys.*, **71**, 4817 (1979).
705. Chu, B., and Kubota, K., *Macromolecules*, **14**, 1838 (1981).
706. Chu, D. Y., and Thomas, J. K., *Macromolecules*, **17**, 2142 (1984).
707. Chuang, J. C., DeSorgo, M., and Prins, W., J. Mechanochem. *Cell Mobility*, **2**, 105 (1973).
708. Churg, P., Hermann, H., Grevels, F. W., Schaffner, K., *J. Chem. Soc., Chem. Commun.*, **1984**, 785.
709. Ciardelli, F., Chiellini, E., Carlini, C., Pieroni, O., Salvadori, P., and Menicagli, R., *J. Polym. Sci. Polym. Symposia*, No. 62, 143 (1978).
710. Cicchetti, O., *Adv. Polym. Sci.*, **7**, 70 (1970).
711. Cicchetti, O., and Gratini, F., *Europ. Polym. J.*, **8**, 561 (1972).
712. Claesson, S., and Wettermark, G., *Arkiv Kemi*, **17**, 355 (1961).
713. Claesson, S., Palm, G., and Wettermark, G., *Arkiv Kemi*, **17**, 579 (1961).
714. Clark, D. T., *Pure Appl. Chem.*, **54**, 415 (1982).
715. Clark, D. T., Dilks, A., and Thomas, H. R., in *Developments in Polymer Degradation* (Grassien, N., ed.), Vol. 1, Applied Science Publishers, Ltd, London, 1977, p. 87.
716. Clark, D. T., and Munro, H. S., *Polym. Degrad. Stabil.*, **4**, 83 (1982).
717. Clark, D. T., and Munro, H. S., *Polym. Degrad. Stabil.*, **4**, 441 (1982).
718. Clark, D. T., and Munro, H. S., *Polym. Degrad. Stabil.*, **5**, 23 (1983).
719. Clark, D. T., and Munro, H. S., *Polym. Degrad. Stabil.*, **5**, 227 (1983).
720. Clark, D. T., and Munro, H. S., *Polym. Degrad. Stabil.*, **8**, 195 (1984).
721. Clark, D. T., and Munro, H. S., *Polym. Degrad. Stabil.*, **8**, 213 (1984).
722. Clark, D. T., and Munro, H. S., *Polym. Degrad. Stabil.*, **9**, 63 (1984).
723. Clark, D. T., and Peeling, J., *Polym. Degrad. Stabil.*, **3**, 177 (1981).
724. Clark, D. T., and Stephenson, P. J., *Polym. Degrad. Stabil.*, **4**, 185 (1982).
725. Clark, D. T., and Thomas, H. R., *J. Polym. Sci.*, A1, **16**, 791 (1978).
726. Clark, D. T., Litt, A. D., and Steel, C., *J. Amer. Chem. Soc.*, **91**, 5143 (1969).
727. Clarke, S. R., and Shanks, R. A., *J. Macromol. Sci. Chem.*, A, **14**, 69 (1980).
728. Clarke, S. R., and Shanks, R. A., *Polym. Photochem.*, **1**, 103 (1981).
729. Clarke, S. R., and Shanks, R. A., *J. Macromol. Sci. Chem.*, A, **17**, 77 (1982).
730. Clecak, N. J., Fox, R. J., and Moreau, W. M., in Soc. Plast. Eng. Regional Technol. Conf., Ellenville, N. Y., Oct. 24, 1973, Abstr., p. 43.
731. Close, D. M., Jacobson, A. D., Margerum, J. D., Brault, R. G., and McClung, F. J., *Appl. Phys. Lett.*, **14**, 159 (1969).
732. Close, G. L., and Paulson, D. J., *J. Amer. Chem. Soc.*, **92**, 7229 (1970).
733. Coates, J. R., Sullivan, R. J., and Ott, J. B., *J. Phys. Chem.*, **63**, 589 (1959).
734. Cocivera, M., and Trozzolo, A. M., *J. Amer. Chem. Soc.*, **92**, 1772 (1970).
735. Cohen, M. D., *Angew. Chem.*, **87**, 439 (1975).
736. Cohen, M. D., Ludmer, A., and Yakhot, V., *Chem. Phys. Lett.*, **38**, 398 (1976).
737. Cohen, M. D., and Schmidt, G. M. J., *J. Chem. Soc.*, **1964**, 1996.
738. Cohen, S. G., and Baumgarten, R. J., *J. Amer. Chem. Soc.*, **89**, 3471 (1967).
739. Cohen, S. G., and Chao, H. M., *J. Amer. Chem. Soc.*, **90**, 165 (1968).
740. Cohen, S. G., and Cohen, J. I., *J. Amer. Chem. Soc.*, **89**, 164 (1967).
741. Cohen, S. G., and Cohen, J. I., *Israel J. Chem.*, **6**, 757 (1968).
742. Cohen, S. G., and Gutenplan, J. B., *Tetrahed. Lett.*, **1968**, 5353.
743. Cohen, S. G., and Gutenplan, J. B., *Tetrahed. Lett.*, **1969**, 2125.
744. Cohen, S. G., Laufer, D. A., and Sherman, W. V., *J. Amer. Chem. Soc.*, **86**, 3060 (1964).
745. Cohen, S. G., Orman, S., and Laufer, D. A., *J. Amer. Chem. Soc.*, **84**, 3905 (1962).
746. Cohen, S. G., Ostberg, B. E., Sparow, D. B., and Blout, E. L., *J. Polym. Sci.*, **3**, 264 (1948).
747. Cohen, S. G., and Pearsons, G., *J. Amer. Chem. Soc.*, **92**, 7603 (1970).
748. Cohen, S. G., Parola, A., and Parsons, G. H., *Chem. Rev.*, **73**, 141 (1973).
749. Cohen, S. M., Young, R. H., and Markart, A. H., *J. Polym. Sci.*, **9**, 3263 (1971).
750. Cohen, W. D., *Rec. Trav. Chim.*, **39**, 243 (1920).
751. Collart, P., Demeyer, K., Toppet, S., and DeSchryver, F. C., *Macromolecules*, **16**, 1390 (1983).
752. Connolly, J. S., *Photochemical Conversion and Storage of Solar Energy*, Academic Press, New York, 1981.

753. Connor, W. P., and Schertz, G. L., *SPE Trans.*, **1963,** 186.
754. Cook, W. D., J. Macromol., *Sci. Chem.*, A, **17,** 99 (1982).
755. Cooney, J. D., and Wiles, D. M., *Soc. Plast. Eng. Techn. Pap.*, **20,** 420 (1974).
756. Coope, J. A. R., Farmer, S. B., Gardner, G. L., and McDowell, C. A., *J. Chem. Phys.*, **42,** 54 (1965).
757. Cooper, H. R., *Trans. Faraday Soc.*, **62,** 2865 (1966).
758. Cooper, W., and Fielden, M., *J. Polym. Sci.*, **28,** 442 (1958).
759. Cooper, W., Vaugham, G., Miller, S., and Fielden, M., *J. Polym. Sci.*, **34,** 651 (1953).
760. Cooray, B. B., and Scott, G., *Europ. Polym. J.*, **16,** 169 (1980).
761. Cooray, B. B., and Scott, G., *Polym. Degrad. Stabil.*, **2,** 35 (1980).
762. Cooray, B. B., and Scott, G., *Polym. Degrad. Stabil.*, **3,** 127 (1981).
763. Coppinger, G. M., and Bell, E. R., *J. Phys. Chem.*, **70,** 3479 (1966).
764. Cottard, J. J., Loucheux, C., Lablanche-Combier, A., *J. Appl. Polym. Sci.*, **26,** 1233 (1981).
765. Cowan, D. O., *J. Amer. Chem. Soc.*, **94,** 6779 (1972).
766. Cowan, D. O., and Drisko, R. L., *J. Amer. Chem. Soc.*, **89,** 3068 (1967).
767. Cowan, D. O., and Drisko, R. L., *Elements of Organic Photochemistry*, Plenum Press, New York, 1976.
768. Cowell, G. W., and Pitts, J. N. Jr., *J. Amer. Chem. Soc.*, **90,** 1106 (1968).
769. Coyle, J. D., Hill, R. R., and Roberts, D. R., *Light, Chemical Change and Life*, The Open University Press, Milton Keynes, 1982.
770. Cox, W. C., Crawford, D. J., and Peill, P. L. D., *J. Appl. Polym. Sci.*, **14,** 611 (1970).
771. Cox, A., Kemp, T. J., Payne, D. R., and Pinot de Moira, P., *J. Photogr. Sci.*, **25,** 208 (177).
772. Cozzens, R. F., in *Photodegradation and Photostabilization of Coatings*, (Pappas, S. P., and Winslow, F. H., eds), ACS Symp. Ser., No. 151, ACS Washington, D.C., 1981, p. 137.
773. Cozzens, R. F., and Cox, R. B., *J. Chem. Phys.*, **50,** 1532 (1969).
774. Cozzens, R. F., Monitz, W. B., and Fox, R. B., *J. Chem. Phys.*, **48,** 581 (1968).
775. Craubner, H., *J. Polym. Sci.*, Al, **18,** 2011 (1980).
776. Craubner, H., *J. Polym. Sci.*, A1, **20,** 1935 (1982).
777. Creed, D., and Caldwell, R. A., *J. Amer. Chem. Soc.*, **96,** 7369 (1974).
778. Creed, D., Caldwell, R. A., Ohta, H., and DeMarco, D.C., *J. Amer. Chem. Soc.*, **99,** 277 (1977).
779. Creed, D., Caldwell, R. A., and Melton, L. A., *J. Amer. Chem. Soc.*, **98,** 622 (1976).
780. Creutz, C., and Sutin, N., *Proc. Natl. Acad. Sci.*, U.S., **72,** 2858 (1975).
781. Crivello, J. V., *Chemtech,* **10,** 624 (1980).
782. Crivello, J. V., in *UV Curing: Science and Technology* (Pappas, S. P., ed.), Vol. 1, Stanford Technology Marketing Corp., Norwalk, CT, 1978, p. 24.
783. Crivello, J. V., in *Developments in Polymer Photochemistry* (Allen, N. S., ed.), Vol. 2, Applied Science Publishers, Ltd, London, 1981, p. 1.
784. Crivello, J. V., *Ann. Rev. Mater. Sci.*, **13,** 173 (1983).
785. Crivello, J. V., *Polym. Eng. Sci.*, **23,** 953 (1983).
786. Crivello, J. V., *Adv. Polym. Sci.*, **62,** 1 (1984).
787. Crivello, J. V., and Conlon, D. A., *J. Polym. Sci.*, A1, **21,** 1785 (1983).
788. Crivello, J. V., and Lam, J. H. W., *J. Polym. Sci.*, Polym. Symp., No. 56, 383 (1976).
789. Crivello, J. V., and Lam, J. H. W., *Macromolecules*, **10,** 1307 (1977).
790. Crivello, J. V., and Lam, J. H. W., *J. Polym. Sci.*, A1, **16,** 2441 (1978).
791. Crivello, J. V., and Lam, J. H. W., *J. Polym. Sci.*, A1, **17,** 759 (1979).
792. Crivello, J. V., and Lam, J. H. W., *J. Polym. Sci.*, A1, **17,** 977 (1979).
793. Crivello, J. V., and Lam, J. H. W., *J. Polym. Sci.*, A1, **17,** 1047 (1979).
794. Crivello, J. V., and Lam, J. H. W., *J. Polym. Sci.*, A1, **17,** 1059 (1979).
795. Crivello, J. V., and Lam, J. H. W., *J. Polym. Sci.*, A1, **17,** 2877 (1977).
796. Crivello, J. V., and Lam, J. H. W., *J. Polym. Sci.*, A1, **17,** 3845 (1979).
797. Crivello, J. V., and Lam, J. H. W., *J. Polym. Sci.*, A1, **18,** 1021 (1980).
798. Crivello, J. V., and Lam, J. H. W., *J. Polym. Sci.*, A1, **18,** 2677 (1980).
799. Crivello, J. V., and Lam, J. H. W., *J. Polym. Sci.*, A1, **18,** 2697 (1980).
800. Crivello, J. V., Lam, J. H. W., Moore, J. E., and Schroeter, S. H., *J. Radiat. Curing,* **5,** Jan., 2 (1978).
801. Crivello, J. V., Lam, J. H. W., and Volante, C. N., *J. Radiat. Curing,* **4,** Mar., 2 (1977).
802. Crivello, J. V., and Lee, J. L., *Macromolecules*, **14,** 1142 (1981).
803. Crivello, J. V., and Lee, J. L., *Polym. Photochem.*, **2,** 219 (1982).
804. Crivello, J. V., and Lee, J. L., *Macromolecules*, **16,** 864 (1983).

686

805. Crivello, J. V., Lee, J. L., and Conlon, D. A., *J. Radiat. Curing*, **10,** Jan., 6 (1983).
806. Cuddihy, E. F., Coulbert, C. D., Willis, P., Baum, B., Garcia, A., and Minning, C., in *Polymers in Solar Energy Utilization* (Gebelein, C. G., Williams, D. J., and Deanin, R. J., eds), ACS Symp. Ser., No. 220, ACS Washington, D.C., 1983, p. 353.
807. Cundall, E. B., Hulme, B., Rudham, R., and Salim, M. S., *J. Oil Colour Chem. Assoc.*, **61,** 351 (1978).
808. Cundall, R. B., and Pereira, L. C., *Chem. Phys. Lett.*, **18,** 371 (1971).
809. Cundall, R. B., Robinson, D. A., and Pereira, L. C., *Adv. Photochem.*, **10,** 148 (1977).
810. Cunibert, C., and Perico, A., *Europ. Polym. J.*, **13,** 369 (1977).
811. Cunibert, C., and Perico, A., *Europ. Polym. J.*, **16,** 887 (1980).
812. Cunliffe, A. V., and Davis, A., *Polym. Degrad. Stabil.*, **4,** 17 (1982).
813. Curme, H. G., Natale, C. C., and Kelly, D. J., *J. Phys. Chem.*, **71,** 767 (1967).
814. Dainton, F. S., and Eaton, R. S., *J. Polym. Sci.*, **39,** 313 (1959).
815. Dainton, F. S., and James, D. G. L., *Dis. Faraday Soc.*, **14,** 244 (1953).
816. Dainton, F. S., and James, D. G. L., *J. Polym. Sci.*, **39,** 299 (1959).
817. Dainton, F. S., and Jones, R. G., *Trans. Faraday Soc.*, **63,** 1512 (1967).
818. Dainton, F. S., Seaman, P. H., James, D. G. L., and Eaton, R. S., *J. Polym. Sci.*, **34,** 209 (1959).
819. Dainton, F. S., and Tordoff, M., *Trans. Faraday Soc.*, **53,** 666 (1957).
820. Dale, R. E., and Eisinger, J., *Proc. Natl. Acad. Sci. U.S.*, **73,** 271 (1976).
821. Dalton, J. C., and Montgomery, F. C., *J. Amer. Chem. Soc.*, **96,** 6230 (1971).
822. Dalton, J. C., Dawes, D. S., Turro, N. J., Weiss, D. S., Barltrop, J. A., and Coyle, J. D., *J. Amer. Chem. Soc.*, **93,** 7213 (1971).
823. Dalton, J. C., and Turro, N. J., *Ann. Rev. Phys. Chem.*, **21,** 499 (1970).
824. Dammers-de-Klerk, A., *Mol. Phys.*, **1,** 141 (1958).
825. Dan, E., and Guillet, J. E., *Macromolecules*, **6,** 230 (1973).
826. Dan, E., Somersall, A. C., and Guillet, J. C., *Macromolecules*, **6,** 228 (1973).
827. Danusso, F., Ferruti, P., Moro, A., Tieghi, G., and Zocchi, M., *Polymer*, **18,** 161 (1977).
828. Daruwalla, E. H., Moonim, S. M., and Arthur, J. C. Jr, *Text. Res. J.*, **42,** 592 (1972).
829. Daruwalla, E. H., Moonim, S. M., and Tamboskar, P. K., Cell. Chem. Technol., **6,** 695 (1972).
830. Das, P. K., and Scaiano, J. C., *Macromolecules*, **14,** 693 (1981).
831. Dauben, W. G., Salem, L., and Turro, N. J., *Acc. Chem. Res.*, **8,** 41 (1975).
832. David, C., in *Comprehensive Chemical Kinetics* (Bamford, C. H., and Tipper, C. F. eds), Vol. 14, Elsevier, Amsterdam, 1975, p. 1.
833. David, C., Baeyens-Volant, D., Delaunois, G., Lu Vinh, Q., Piret, W., and Geuskens, G., *Europ. Polym. J.*, **14,** 501 (1978).
834. David, C., Baeyens-Volant, D., and Geuskens, G., *Europ. Polym. J.*, **12,** 71 (1976).
835. David, C., and Baeyens-Volant, D., *Europ. Polym. J.*, **14,** 29 (1977).
836. David, C., Baeyens-Volant, D., and Piens, D., *Europ. Polym. J.*, **16,** 413 (1980).
837. David, C., Borsu, M., and Geuskens, G., *Europ. Polym. J.*, **6,** 959 (1970).
838. David, C., Camera, O., and Geuskens, G., *Polymer*, **18,** 198 (1977).
839. David, C., Demarteau, W., and Geuskens, G., *Polymer*, **8,** 497 (1967).
840. David, C., Demerteau, W., and Geuskens, G., *Polymer*, **10,** 21 (1969).
841. David, C., Demarteau, W., and Geuskens, G., *Europ. Polym. J.*, **6,** 537 (1970).
842. David, C., Demarteau, W., and Geuskens, G., *Europ. Polym. J.*, **6,** 1397 (1970).
843. David, C., Demarteau, W., and Geuskens, G., *Europ. Polym. J.*, **6,** 1405 (1970).
844. David, C., Fuld, D., Geuskens, G., and Charlesby, A., *Europ. Polym. J.*, **5,** 641 (1969).
845. David, C., Lempereur, M., and Geuskens, G., *Europ. Polym. J.*, **8,** 417 (1972).
846. David, C., Lempereur, M., and Geuskens, G., *Europ. Polym. J.*, **9,** 1315 (1973).
847. David, C., Lempereur, M., and Geuskens, G., *Europ. Polym. J.*, **10,** 1181 (1974).
848. David, C., Naegelen, V., Piret, W., and Geuskens, G., *Europ. Polym. J.*, **11,** 569 (1975).
849. David, C., Piens, M., and Geuskens, G., *Europ. Polym. J.*, **8,** 1019 (1972).
850. David, C., Piens, M., and Geuskens, G., *Europ. Polym. J.*, **8,** 1291 (1972).
851. David, C., Piens, M., and Geuskens, G., *Europ. Polym. J.*, **9,** 533 (1973).
852. David, C., Piens, M., and Geuskens, G., *Europ. Polym. J.*, **12,** 621 (1976).
853. David, C., Putman, N., Lempereur, M., and Geuskens, G., *Europ. Polym. J.*, **8,** 409 (1972).
854. David, C., Putman-de Lavareille, N., and Geuskens, G., *Europ. Polym. J.*, **10,** 617 (1974).
855. David, C., Putman-de Lavareille, N., and Geuskens, G., *Europ. Polym. J.*, **12,** 71 (1976).
856. David, C., Putman-de Lavareille, N., and Geuskens, G., *Europ. Polym. J.*, **13,** 15 (1977).
857. Davidson, H. R., and Hemmendinger, H., *J. Opt. Soc. Amer.*, **56,** 1102 (1966).
858. Davidson, J. A., and Abrahamson, E. W., *Photochem. Photobiol.*, **15,** 403 (1972).

859. Davidson, R. S., *Chem. Commun.*, **1966,** 575.
860. Davidson, R. S., and Goodin, J. W., *Europ. Polym. J.*, **18,** 589 (1982).
861. Davidson, R. S., and Goodin, J. W., *Europ. Polym. J.*, **18,** 597 (1982).
862. Davidson, R. S., and Lambeth, P. F., *Chem. Commun.*, **1967,** 1265.
863. Davidson, R. S., and Lambeth, P. F., *Chem. Commun.*, **1968,** 511.
864. Davidson, R. S., and Lambeth, P. F., *Chem. Commun.*, **1969,** 1098.
865. Davidson, R. S., Lambeth, P. F., McKellar, J. F., Turner, P. H., and Wilson, R., *Chem. Commun.*, **1969,** 732.
866. Davidson, R. S., Lambeth, P. F., and Santhanam, M., *J. Chem. Soc. Perkin II*, **1972,** 2351.
867. Davidson, R. S., Lambeth, P. F., Yonis, F. A., and Wilson, J., *J. Chem. Soc.*, C, **1969,** 2203.
868. Davidson, R. S., and Meek, R. R., *Europ. Polym. J.*, **17,** 163 (1981).
869. Davidson, R. S., and Meek, R. R., *Polym. Photochem.*, **2,** 1 (1982).
870. Davidson, R. S., and Orton, P. S., *J. Chem. Soc. Chem. Commun.*, **1974,** 209.
871. Davidson, R. S., and Santhanam, M., *J. Chem. Soc. Perkin II*, **1972,** 2355.
872. Davidson, R. S., and Steiner, P. R., *J. Chem. Soc. Commun.*, **1971,** 1115.
873. Davidson, R. S., and Wilson, R., *J. Chem. Soc.,* B, **1970,** 71.
874. Davies, D. H., and Dixon, D., *J. Appl. Polym. Sci.*, **16,** 2449 (1972).
875. Davies, D. H., Phillips, D. C., and Smith, J. D. B., *J. Polym. Sci.*, A1, **10,** 3253 (1972).
876. Davies, D. H., Smith, J. D. B., and Phillips, D. C., *Macromolecules*, **6,** 163 (1973).
877. Davies, P. B., and North, A. M., *Proc. Chem. Soc. (London)*, **1964,** 141.
878. Davis, A., in *Developments in Polymer Degradation* (Grassie, N., ed.) Applied Science Publishers, Ltd, London, 1977, p. 279.
879. Davis, A., Deane, G. W. H., and Diffey, B. C., *Nature*, **261,** 169 (1976).
880. Davis, A., Deane, G. W. H., Gordon, D., Howell, G. U., and Ledbury, K. J., *J. Appl. Polym. Sci.*, **20,** 1165 (1976).
881. Davis, A., and Gardiner, D., *Polym. Degrad. Stabil.*, **4,** 145 (1982).
882. Davis, A., Golden, J. H., *J. Macromol. Sci., Rev. Macromol. Chem.*, C, **3,** 49 (1969).
883. Davis, A., and Gordon, D., *J. Appl. Polym. Sci.*, **18,** 1159 (1974).
884. Davis, A., and Gordon, D., *J. Appl. Polym. Sci.*, **18,** 1173 (1974).
885. Davis, A., and Gordon, D., *J. Appl. Polym. Sci.*, **18,** 1181 (1974).
886. Davis, A., Howes, B. V., Ledbury, K. J., and Pearce, P. J., *Polym. Degrad. Stabil.*, **1,** 121 (1979).
887. Davis, A., and Sims, D., *Weathering of Polymers*, Applied Science Publishers, Ltd, London, 1983.
888. Davis, A. K., Gee, G. A., McKellar, J. F., and Phillips, G. O., *Chem. Ind.*, **1973,** 431.
889. Davis, G. A., Carapellucci, D. A., Szoc, K., and Gressler, J. D., *J. Amer. Chem. Soc.*, **91,** 2264 (1969).
890. Davis, G. A., and Golden, J. H., *J. Chem. Soc.*, B, **1968,** 426.
891. Davis, G. A., Golden, J. H., *J. Chem. Soc., B*, **1968,** 40.
892. Davis, M. J., Doherty, J., Godfrey, A. A., Green, P. N., Young, J. R. A., and Parrish, M. A., *J. Oil. Colour Chem. Assoc.*, **61,** 256 (1978).
893. Davis, N. P., Garnett, J. L., and Urquhart, R. C., *J. Polym. Sci.* Symp., No. 55, 287 (1976).
894. Davydov, E. Ya., Margolin, A. L., Pariskii, G. P., Postnikov, L. M., and Toptygin, D. Ya., *Dokl. Akad. Nauk SSSR*, **243,** 1475 (1978).
895. Day, D. R., and Ringsdorf, H., *J. Polym. Sci.*, B, **16,** 205 (1978).
896. Day, M., and Wiles, D. M., *J. Polym. Sci.*, B, **9,** 665 (1971).
897. Day, M., and Wiles, D. M., *Can. J. Chem.*, **49,** 2916 (1971).
898. Day, M., and Wiles, D. M., *J. Appl. Polym. Sci.*, **16,** 175 (1972).
899. Day, M., and Wiles, D. M., *J. Appl. Polym. Sci.*, **16,** 191 (1972).
900. Day, M., and Wiles, D. M., *J. Appl. Polym. Sci.*, **16,** 203 (1972).
901. Deb, P. C., *Europ. Polym. J.*, **11,** 31 (1975).
902. Deb, P. C., and Meyerhoff, G., *Europ. Polym. J.*, **10,** 709 (1974).
903. DeBoyer, C. D., *J. Polym. Sci.*, B, **11,** 25 (1973).
904. DeBoyer, C. D., and Breslow, R., *Tetrahed. Lett.*, **1967,** 1033.
905. DeBoyer, C. D., and Schlessinger, R. H., *J. Amer. Chem. Soc.*, **94,** 655 (1972).
906. DeBoyer, C. D., Wadsworth, D. H., and Perkins, W. C., *J. Amer. Chem. Soc.*, **95,** 861 (1973).
907. Debye, P. J., *Trans. Electrochem. Soc.*, **82,** 265 (1942).
908. Decker, C., *J. Appl. Polym. Sci.*, **20,** 3321 (1976).
909. Decker, C., *Polym. Photochem.*, **3,** 131 (1983).
910. Decker, C., *J. Appl. Polym. Sci.*, **28,** 97 (1983).

688

911. Decker, C., *J. Polym. Sci.*, A1, **21**, 2451 (1983).
912. Decker, C., in *Degradation and Stability of Poly(vinyl chloride)* (Owen, E. D., ed.), Elsevier Applied Science, London, 1984, p. 81.
913. Decker, C., and Balandier, M., *Polym. Photochem.*, **1**, 221 (1981).
914. Decker, C., and Balandier, M., *Europ. Polym. J.*, **18**, 1085 (1982).
915. Decker, C., and Balandier, M., *Makromol. Chem.*, **183**, 1263 (1982).
916. Decker, C., and Balandier, M., *Polym. Photochem.*, **5**, 267 (1984).
917. Decker, C., Balandier, M., and Faure, J., *J. Macromol. Sci. Chem.*, A, **16**, 1463 (1981).
918. Decker, C., and Bendaikha, T., *Europ. Polym. J.*, **20**, 753 (1984).
919. Decker, C., Faure, J., Fizet, M., and Rychla, L., *Photogr. Sci. Eng.*, **23**(3), 137 (1979).
920. Decker, C., and Fizet, M., *Makromol. Chem. Rapid Commun.*, **1**, 637 (1980).
921. Deckert, C. A., and Ross, D. L., *J. Electrochem.*, **127**, 45C (1980).
922. Decout, J. L., Lablache-Combier, A., and Loucheux, C., *J. Polym. Sci.*, A1, **18**, 2371 (1980).
923. Decout, J. L., Lablache-Combier, A., and Loucheux, C., *J. Polym. Sci.*, A1, **18**, 2391 (1980).
924. Dedinas, J., *J. Phys. Chem.*, **75**, 181 (1971).
925. Deforest, W. S., *Photoresists: Materials and Process*, McGraw-Hill, New York, 1975.
926. Deledalle, P., Lablache-Combier, A., and Loucheux, C., *J. Appl. Polym. Sci.*, **29**, 125 (1984).
927. Dellinger, J. A., and Roberts, C. W., *J. Appl. Polym. Sci.*, **26**, 321 (1981).
928. Delman, A. D., Landy, M., and Simms, B. B., *J. Polym. Sci.*, A1, **7**, 3375 (1969).
929. Delzenne, G. A., *Ind. Chim. Belg.*, **30**, 679 (1965).
930. Delzenne, G. A., *Ind. Chim. Belg.*, **39**, 249 (1969).
931. Delzenne, G. A., *Europ. Polym. J.*, Suppl., **1969**, 55.
932. Delzenne, G. A., *J. Macromol Sci. Rev. Polym. Technol.*, D, **1**, 185 (1971).
933. Delzenne, G. A., *Makromol. Chem.*, Suppl., **2**, 169 (1979).
934. Delzenne, G. A., in *Encyclopedia of Polymer Science and Technology* (Mark, H. F., and Bikales, N. M., eds), Suppl. 1, Wiley, N. Y., 1976, p. 401.
935. Delzenne, G. A., Dewinter, W., Toppet, S., and Smets, G., *J. Polym. Sci.*, A, **2**, 1069 (1964).
936. Delzenne, G. A., and Laridon, U., *36 Congress Intern. de Chimie Industrielle*, Bruxelles, 1966, p. 373.
937. Delzenne, G. A., and Laridon, U., *Ind. Chem. Belg.*, **34**, 395 (1969).
938. Delzenne, G. A., and Laridon, U., *J. Polym. Sci.*, Polym. Symp., No. 22, 1149 (1969).
939. Delzenne, G. A., and Laridon, U., U.S. Pat., 3,502,470 (May 24, 1970).
940. Delzenne, G. A., Laridon, U., and Peeters, H., *Europ. Polym. J.*, **6**, 933 (1970).
941. De Maré, G. R., Fontaine, M. C., and Goldfinger, P., *J. Org. Chem.*, **33**, 2528 (1968).
942. De Maré, G. R., Fox, J. R., Termonia, M., and Tshibangila, B., *Europ. Polym. J.*, **12**, 119 (1976).
943. De Maré, G. R., and Fox, J. R., *Europ. Polym. J.*, **17**, 315 (1981).
944. De Maré, G. R., Goldfinger, P., Huybrechts, G., Jonas, E., and Toth, M., *Ber. Bunsenges, Physik Chem.*, **73**, 867 (1969).
945. Demas, J. N., and Adamson, A. W., *J. Amer. Chem. Soc.*, **93**, 3042 (1971).
946. Demas, J. N., and Crosby, G. A., *J. Amer. Chem. Soc.*, **93**, 2841 (1971).
947. Demyer, K., Vanderauweraer, M., Aerts, J., and DeSchryver, F. C., *J. Chim. Phys.*, **77**, 493 (1980).
948. Denisov, E. T., *Usp. Khim.*, **47**, 1090 (1978).
949. De Paoli, M. A., *Europ. Polym. J.*, **19**, 761 (1983).
950. De Paoli, M. A., Tamashito, I., and Galembeck, F., *J. Polym. Sci.*, B, **17**, 391 (1979).
951. De Paoli, M. A., and Schulz, G. W., *Polym. Bull.*, **8**, 437 (1982).
952. Desai, R. L., *Pulp Pap. Mag. Canad.*, **69**, T322 (1968).
953. Desai, R. L., and Shields, J. A., *Makromol. Chem.*, **122**, 134 (1969).
954. Desai, R. L., and Shields, J. A., *J. Colloid Interface Sci.*, **31**, 585 (1969).
955. DeSchryver, F. C., *Pure Appl. Chem.*, **34**, 213 (1973).
956. DeSchryver, F. C., *Makromol. Chem.*, Suppl., **3**, 85 (1979).
957. DeSchryver, F. C., Anand, L., Smets, G., and Switten, J., *J. Polym. Sci.*, B, **9**, 777 (1971).
958. DeSchryver, F. C., Bhardwaj, I., and Put, J., *Angew. Chem.*, **81**, 224 (1969).
959. DeSchryver, F. C., Boens, N., Huybrechts, J., Daemen, J., De Brackeleiré, M., *Pure Appl. Chem.* **49**, 237 (1977).
960. DeSchryver, F. C., Boens, N., and Put, J., *Adv. Photochem.*, **10**, 359 (1977).
961. DeSchryver, F. C., Boens, N., and Smets, G., *J. Polym. Sci.*, A1, **10**, 1687 (1972).
962. DeSchryver, F. C., Boens, N., and Smets, G., *J. Amer. Chem. Soc.*, **69**, 6463 (1974).
963. DeSchryver, F. C., Boens, N., and Smets, G., *Macromolecules*, **7**, 339 (1974).

964. DeSchryver, F. C., De Brackeleiré, M., Toppet, S., and van Schoor, M., *Tetrahed. Lett.*, **1973,** 1253.
965. DeSchryver, F. C., Demeyer, K., and Toppet, S., *Macromolecules,* **16,** 89 (1983).
966. DeSchryver, F. C., Demeyer, K., and Toppet, S., *Macromolecules,* **16,** 1589 (1983).
967. DeSchryver, F. C., Demeyer, K., van der Auweraer, M., and Quanten, E., *Ann. N. Y. Acad. Sci.*, **93,** 366 (1981).
968. DeSchryver, F. C., Feast, W. J., and Smets, G., *J. Polym. Sci.*, A1, **8,** 1989 (1970).
969. DeSchryver, F. C., Moens, L., van der Auweraer, M., Boens, N., Monnerie, L., and Bokobza, L., *Macromolecules,* **15,** 64 (1982).
970. DeSchryver, F. C., and Put, J., *Ind. Chim. Belg.*, **37,** 1107 (1972).
971. DeSchryver, F. C., Vandendriessche, J., Toppet S., Demeyer, K., and Boens, N., *Macromolecules,* **15,** 406 (1982).
972. DeSchryver, F. C., Thien, T. V., and Smets, G., *J. Polym. Sci.*, B, **9,** 425 (1971).
973. DeSchryver, F. C., Thien, T. V., and Smets, G., *J. Polym. Sci.*, A1, **13,** 215 (1975).
974. DeSchryver, F. C., Thien, T. V., Toppet, S., and Smets, G., *J. Polym. Sci.*, A1, **13,** 227 (1975).
975. DeSchryver, F. C., Tran Van, T., and Smets, G., *J. Polym. Sci.*, B, **9,** 425 (1971).
976. DeSchryver, F. C., Tran Van, T., and Smets, G., *J. Polym. Sci.*, A1, **13,** 215 (1975).
977. DeSchryver, F. C., Tran Van, T., Toppet, S., and Smets, G., *J. Polym. Sci.*, A1, **13,** 227 (1975).
978. Desande, D. D., and Aravindakshan, P., *Polym. Photochem.*, **4,** 295 (1984).
979. Despax, B., Paillous, N., Lattes, A., and Paillous, A., *J. Polym. Sci.*, A1, **18,** 593 (1980).
980. Dessauer, R., and Looney, C., *Photogr. Sci. Eng.*, **23,** 287 (1979).
981. Destor, C., Langevin, D., and Rondelez, F., *J. Polym. Sci.*, B, **16,** 229 (1978).
982. Desvergne, J. P., and Bouas-Lanrent, H., *J. Chem. Soc., Chem. Commun.*, **1978,** 403.
983. Dexter, D. L., *J. Chem. Phys.*, **21,** 836 (1953).
984. Dhanraj, J., and Guillet, J. E., *J. Polym. Sci.*, Polym. Symp., No. 23, 433 (1968).
985. Dickens, B., Martin, J. W., and Waksman, D., *Polymer,* **25,** 706 (1984).
986. Dilks, A., *J. Polym. Sci.*, A1, **19,** 1319 (1981).
987. Dilks, A., and Clark, D. T., *J. Polym. Sci.*, A1, **19,** 2847 (1981).
988. Dilling, W. L., *J. Org. Chem.*, **31,** 1045 (1966).
989. Dilling, W. L., *Chem. Revs,* **83,** 1 (1983).
990. Dills, W. L., and Reeve, T. B., *Plast. Technol.*, **16,** 50 (1970).
991. Dimitrov, D. I., Kumanova, B. K., and Bisolnakov, N., *Zh. Prikl. Khim.*, **46,** 1803 (1973).
992. Dilung, I. I., and Kapinus, E. R., *Russ. Chem. Rev.*, **47,** 43 (1978).
993. Dixon, G. D., *J. Polym. Sci.*, A1, **12,** 1717 (1974).
994. Djeghri, N., Formenti, M., Juillet, F., and Teichner, S., *J. Faraday Discuss. Chem. Soc.*, **58,** 185 (1974).
995. Dobrowolski, G., Kiwi, J., and Schnabel, W., *Europ. Polym. J.*, **12,** 657 (1976).
996. Dolezal, B., and Stepek, J., *Chem. Promysl.*, **10,** 381 (1960).
997. Donnet, J. B., and Voet, A., *Carbon Black*, Dekker, New York, 1976.
998. Dorfman, L. W., and Salsburg, Z. W., *J. Amer. Chem. Soc.*, **73,** 255 (1951).
999. Dougherty, R. C., *J. Amer. Chem. Soc.*, **93,** 7187 (1971).
1000. Drake, R. C., Christensen, R. L., and Phillips, D., *Polym. Photochem.*, **5,** 141 (1984).
1001. Drake, R. C., Phillips, D., Roberts, A. J., and Soutar, I., in *Photophysics of Synthetic Polymers* (Phillips, D., and Roberts, A. J., eds), Science Reviews Ltd, Northwood, England, 1982, p. 150.
1002. Draus, G., Gruver, J. T., and Rollman, K. W., *J. Polym. Sci.*, **36,** 564 (1959).
1003. Drickamer, H. G., *Ann. N.Y. Acad. Sci.*, **371,** 261 (1981).
1004. Dubinskij, V. Z., Rosinskij, V. A., Miller, V. B., and Shlapnikova, I. A., *Izv. Akad. Nauk SSSR*, **1975,** 1180.
1005. Dubois, J. C., Eranian, A., and Datamanti, I., *Proc. Symp. Electron Ion Beam Sci. Technol.* 8th Intern. Conf., **78,** (5), 303 (1978).
1006. Dubois, J. T., and Van Hemert, I., *J. Chem. Phys.*, **40,** 923 (1964).
1007. Duncan, F. J., Trotman-Dickenson, A. F., *J. Chem. Soc.*, **1962,** 4672.
1008. Dunn, P., and Hart, S. J., *J. Inst. Rubber Ind.*, **3,** 1 (1969).
1009. Dunn, T. S., Williams, E. E., and Williams, J. L., *J. Polym. Sci.*, **20,** 1599 (1982).
1010. Duran, N., and Mansilla, H., *J. Macromol. Sci. Chem.*, A, **21,** 1467 (1984).
1011. Durden, A., and Crosby, D. G., *J. Org. Chem.*, **30,** 1684 (1965).
1012. Durmis, J., Carlsson, D. J., Chan, K. H., and Wiles, D. M., *J. Polym. Sci.*, B, **19,** 549 (1981).
1013. Durmis, J., Karvaš, M., Čaučik, P., and Holčik, J., *Europ. Polym. J.*, **11,** 219 (1975).
1014. Duynstee, E. F. J., and Mevis, M. E. A. H., *Europ. Polym. J.*, **8,** 1375 (1972).

690

1015. Easton, M. J., and MacCallum, J. R., *Polym. Degrad. Stabil.*, **3**, 229 (1981).
1016. Eaton, D. F., *Photogr. Sci. Eng.*, **23**, 150 (1979).
1017. Eckhard, H., Prusik, T., and Chance, R. R., *Macromolecules*, **16**, 732 (1983).
1018. Efremkin, A. F., and Ivanov, V. B., *Polym. Photochem.*, **4**, 179 (1984).
1019. Egawa, H., Ishikawa, M., Tsunooka, M., Ueda, T., and Tanaka, M., *J. Polym. Sci.*, A1, **21**, 479 (1983).
1020. Egawa, H., Nonaka, T., Hurusawa, T., Tsunooka, M., and Tanaka, M., *J. Polym. Sci.*, A1, **21**, 2597 (1983).
1021. Egawa, H., Nonaka, T., Tokushige, K., Tsunooka, M., and Tanaka, M., *J. Polym. Sci.*, A1, **21**, 1233 (1983).
1022. Egawa, H., Tsunooka, M., Ueda, T., and Tanaka, M., *J. Polym. Sci.*, B, **19**, 201 (1981).
1023. Egerton, G. S., *J. Soc. Dyers Colour.*, **63**, 161 (1947).
1024. Egerton, G. S., *Brit. Polym. J.*, **3**, 63 (1971).
1025. Egerton, G. S., and Assaad, N. E. N., *Chem. Ind.*, **1967**, 2112.
1026. Egerton, G. S., and Assaad, N. E. N., *J. Soc. Dyers Colour.*, **83**, 85 (1967).
1027. Egerton, G. S., Assaad, N. E. N., and Uffindell, N. D., *J. Soc. Dyers Colour.*, **83**, 409 (1967).
1028. Egerton, G. S., Assaad, N. E. N., and Uffindell, N. D., *Chem. Ind.*, **1967**, 1172.
1029. Egerton, G. S., and Morgan, A. G., *J. Soc. Dyes Colour.*, **87**, 268 (1971).
1030. Egerton, G. S., and Shah, K. M., *Text. Res. J.*, **38**, 130 (1968).
1031. Egerton, P. L., Pitts, E., and Reiser, A., *Macromolecules*, **14**, 95 (1981).
1032. Egerton, P. L., Reiser, A., Shaw, W., and Wagner, H. M., *J. Polym. Sci.*, A1, **17**, 3315 (1973).
1033. Egerton, P. L., Trigg, J., Hyde, E. M., and Reiser, A., *Macromolecules*, **14**, 100 (1981).
1034. Eichler, C. P., Hertz, C. P., Naito, I., and Schnabel, W., *J. Photochem*, **12**, 225 (1980).
1035. Eichler, C. P., Hertz, C. P., and Schnabel, W., *Angew. Makromol. Chem.*, **91**, 39 (1980).
1036. Eigenmann, G., *Helv. Chim. Acta*, **46**, 804 (1963).
1037. Eisenbach, C. D., *Makromol. Chem.*, **179**, 2489 (1978).
1038. Eisenbach, C. D., *Polym. Bull.*, **1**, 517 (1979).
1039. Eisenbach, C. D., *Polymer*, **21**, 1175 (1980).
1040. Eisenbach, C. D., *Ber. Bunsenges. Phys. Chem.*, **84**, 680 (1980).
1041. El-Azmirly, M. A., Morsi, S. E., Issa, R. M., and Barsoum, S., *Europ. Polym. J.*, **11**, 95 (1975).
1042. Ellinger, L. P., *Chem. Ind.*, **1963**, 1982.
1043. Ellinger, L. P., *Adv. Macromol. Chem.*, **1**, 169 (1968).
1044. Elmgren, H., *J. Polym. Sci.*, B, **18**, 815 (1980).
1045. Elotson, R. M., and Edsberg, R. L., *Can. J. Chem.*, **33**, 646 (1957).
1046. Emanuel, H. M., and Buchachenko, A. L., *Chemical Physics of Polymer Ageing*, Nauka, Moskva, 1982 (in Russian).
1047. Encina, M. V., Funabashi, K., and Scaiano, J. C., *Macromolecules*, **12**, 1167 (1979).
1048. Encina, M. V., and Lissi, E. A., *J. Photochem.*, **3**, 237 (1974).
1049. Encina, M. V., and Lissi, E. A., *J. Photochem.*, **4**, 321 (1975).
1050. Encina, M. V., and Lissi, E. A., *J. Photochem.*, **6**, 173 (1976/1977).
1051. Encina, M. V., and Lissi, E. A., *J. Polym. Sci.*, A1, **17**, 1645 (1979).
1052. Encina, M. V., and Lissi, E. A., *J. Photochem*, **20**, 153 (1982).
1053. Encina, M. V., and Lissi, E. A., *J. Polym. Sci.*, A1, **21**, 2157 (1983).
1054. Encina, M. V., Lissi, E. A., and Abarca, M. T., *J. Polym. Sci.*, A1, **17**, 19 (1979).
1055. Encina, M. V., Lissi, E. A., Gargallo, L., Radic, D., and Olea, A. F., *Macromolecules*, **17**, 2261 (1984).
1056. Encina, M. V., Lissi, E. A., Kochi, V., and Elorza, E., *J. Polym. Sci.*, A1, **20**, 73 (1982).
1057. Encina, M. V., Nogales, A., and Lissi, E. A, *J. Photochem.*, **4**, 75 (1975).
1058. Encina, M. V., Riviera, M., and Lissi, E. A., *J. Polym. Sci.*, A1, **16**, 1709 (1978).
1059. Encina, M. V., Soto, H., and Lissi, E. A., *J. Photochem.*, **3**, 467 (1974).
1060. Endo, T., Fujiwara, E., and Okawara, M., *J. Polym. Sci.*, B, **16**, 211 (1978).
1061. Endo, T., Okawara, M., Bailey, W. J., Azuma, K., Nate, K., and Yokono, H., *J. Polym. Sci.*, B, **21**, 373 (1983).
1062. Engel, P. S., *J. Amer. Chem. Soc.*, **92**, 6074 (1970).
1063. Enkelman, V., *Adv. Polym. Sci.*, **63**, 92 (1984).
1064. Enkelman, V., Kapp, H., and Meyer, W., *Acta Crystal.*, B, **34**, 2350 (1978).
1065. Enkelman, V., and Wegner, G., *Angew. Chem.*, **89**, 432 (1977).
1066. Ershov, Yu. A., and Dovbii, Eu. V., *J. Appl. Polym. Sci.*, **21**, 1511 (1977).
1067. Evans, B. V., and Scott, G., *Europ. Polym. J.*, **10**, 453 (1974).
1068. Evans, D. F., *J. Chem. Soc.*, **1953**, 345.

1069. Evans, D. F., *J. Chem. Soc.*, **1957,** 1351.
1070. Evans, D. F., *J. Chem. Soc.*, **1960,** 1736.
1071. Evans, M. G., Santappa, M., and Uri, N., *J. Polym. Sci.*, **7,** 243 (1951).
1072. Evans, M. G., and Uri, N., *Trans. Faraday Soc.*, **45,** 224 (1949).
1073. Evans, M. G., and Uri, N., *Nature*, **164,** 404 (1949).
1074. Evans, M. G., and Uri, N., *J. Soc. Dyers Colour.*, **65,** 709 (1949).
1075. Evans, N. A., and Leaver, I. H., *Austr. J. Chem.*, **27,** 1797 (1974).
1076. Everett, A. J., and Minkoff, G. J., *Trans. Faraday Soc.*, **49,** 410 (1953).
1077. Evleth, E. M., *J. Amer. Chem. Soc.*, **98,** 1637 (1976).
1078. Factor, A., and Chu, M. L., *Polym. Degrad. Stabil.*, **2,** 203 (1980).
1079. Fahrenholtz, A. M., *J. Amer. Chem. Soc.*, **93,** 151 (1971).
1080. Fairgrieve, S. P., and MacCallum, J. R., *Polym. Commun.*, **25,** 44 (1984).
1081. Fairgrieve, S. P., and MacCallum, J. R., *Polym. Degrad. Stabil.*, **8,** 107 (1984).
1082. Farid, S., in VIII Int. Conf. Photochem., Edmonton, Canada, Aug. 7, 1975, Abstract Y-1.
1083. Farid, S., and Brown, K. A., *Chem. Commun.*, **1976,** 564.
1084. Farid, S., Martic, P. A., Daly, R. C., Thomson, D. R., Specht, J. P., Hartman, S. E., and Williams, J. L. R., *Pure Appl. Chem.*, **51,** 249 (1979).
1085. Farrall, M. J., Alexis, M., and Trecarten, M., *Polymer*, **24,** 114 (1983).
1086. Farzaliev, V. M., Fernando, W. S. E., and Scott, G., *Europ. Polym. J.*, **14,** 785 (1978).
1087. Faure, J., *Pure Appl. Chem.*, **49,** 487 (1977).
1088. Faure, J., Fouassier, J. P., and Lougnot, D. J., *J. Photochem.*, **5,** 13 (1976).
1089. Faure, J., Fouassier, J. P., Lougnot, D. J., and Salvin, R., *Nouv. J. Chim.*, **1,** 15 (1977).
1090. Feast, W. J., and Spanomanolis, C., *Polym. Photochem.*, **1,** 285 (1981).
1091. Feist, W. C., *Chemtech.*, **8,** 160 (1978).
1092. Feist, W. C., and Hon, D. N. S., *Adv. Chem. Ser.*, **207,** 401 (1984).
1093. Feit, E. D., in *UV Curing: Science and Technology* (Pappas, S. P., ed.), Vol. 1, Stanford Technology Marketing Corp., Norwalk, CT, 1978, p. 230.
1094. Felder, B., and Schumacher, R., *Angew. Makromol. Chem.*, **31,** 35 (1973).
1095. Felder, B., Schumacher, R., and Sitek, F., *Helv. Chim. Acta*, **63,** 132 (1980).
1096. Felder, B., Schumacher, R., and Sitek, F., *Chem. Ind. (London)*, **4,** 155 (1980).
1097. Felder, B., Schumacher, R., and Sitek, F., *Am. Chem. Soc. Org. Coat. Plast. Chem. Prepr.*, **42,** 561 (1980).
1098. Felder, B., Schumacher, R., and Sitek, F., in *Photodegradation and Photostabilization of Coatings* (Pappas, S. P., and Winslow, F. H., eds), ACS Symp. Ser., No. 151, ACS Washington, D.C., 1981, p. 65.
1099. Feldman, D., *J. Appl. Polym. Sci.*, **26,** 3493 (1981).
1100. Feller, R. L., Curran, M., and Bailie, C., in *Photodegradation and Photostabilization of Coatings* (Pappas, S. P., and Winslow, F. H., eds), ACS Symp. Ser., No. 151, ACS Washington, D.C., 1981, p. 183.
1101. Fendler, J. H., and Fendler, E. J., *Catalysis in Micellar and Macromolecular Systems*, Academic Press, New York, 1975.
1102. Fenical, W., Kearns, D. R., and Radlick, P., *J. Amer. Chem. Soc.*, **91,** 3396 (1969).
1103. Ferguson, J., Castellan, A., Desvergne, P., and Bouas-Laurent, A., *Chem. Phys. Lett.*, **78,** 446 (1981).
1104. Ferguson, J., and Mau, A. W., *Mol. Phys.*, **27,** 377 (1976).
1105. Fernando, W. S. E., and Scott, G., *Europ. Polym. Sci.*, **16,** 971 (1980).
1106. Fertig, J., Goldberg, A. I., and Skoultchi, M., *J. Appl. Polym. Sci.*, **9,** 903 (1965).
1107. Fertig, J., Goldberg, A. I., and Skoultchi, M., *J. Appl. Polym. Sci.*, **10,** 663 (1966).
1108. Filipescu, N., and Minn, F. L., *J. Amer. Chem. Soc.*, **90,** 1544 (1968).
1109. Fischer, G. J., and Johns, H. E., in *Photochemistry and Photobiology of Nucleic Acids* (Wang, S. Y., ed.), Vol. 1, Academic Press, New York, 1976, p. 225.
1110. Fissi, A., Houben, J. L., Rosato, N., Pieroni, O., and Ciardelli, F., *Macromol. Chem. Rapid Comm.*, **3,** 29 (1982).
1111. Fitton, S. L., Howard, R. N., and Williamson, G. R., *Brit. Polym. J.*, **2,** 217 (1970).
1112. Fitzgibbon, P. D., and Frank, C. W., *Macromolecules*, **14,** 1650 (1981).
1113. Fitzgibbon, P. D., and Frank, C. W., *Bull. Amer. Phys. Soc.*, **26,** 367 (1981).
1114. Fitzgibbon, P. D., and Frank, C. W., *Macromolecules*, **15,** 733 (1982).
1115. Fitzky, H. G., Kampf, G., Klaren, A., and Voltz, H. G., *Amer. Chem. Soc. Div. Org. Coat. Plast. Chem. Prepr.*, **42,** 660 (1980).
1116. Fitzky, H. G., Kampf, G., and Voltz, H. G., *Progr. Org. Coat.*, **3,** 233 (1974).

692

1117. Flamigni, L., Barigelletti., F., and Bortolus, P., *Europ. Polym. J.*, **20**, 171 (1984).
1118. Flannery, J. B., *J. Amer. Chem. Soc.*, **90**, 5660 (1968).
1119. Flood, J., Russell, K. E., and Wan, J. K., Macromolecules, **6**, 669 (1973).
1120. Flynn, J. H., *J. Polym. Sci.*, **27**, 83 (1958).
1121. Foote, C. S., *Acc. Chem. Res.*, **1**, 104 (1968).
1122. Foote, C. S., Chang, Y. C., and Denny, R. W., *J. Amer. Chem. Soc.*, **92**, 5216 (1970).
1123. Foote, C. S., Ching, T. Y., and Geller, G. C., *Photochem. Photobiol.*, **20**, 511 (1974).
1124. Foote, C. S., and Denny, R. W., *J. Amer. Chem. Soc.*, **90**, 6233 (1968).
1125. Foote, C. S., Denny, R. W., Weaver, L., Chang, Y. C., and Peters, J., *Ann. N.Y. Acad. Sci.*, **171**, 139 (1970).
1126. Foote, C. S., Thomas, M., and Ching, T. Y., *J. Photochem.*, **5**, 172 (1976).
1127. Ford, W. E., Otvos, J. W., and Calvin, M., *Nature*, **274**, 507 (1978).
1128. Ford, W. E., Otvos, J. W., and Calvin, M., *Proc. N.Y. Acad. Sci.*, **76**, 3590 (1979).
1129. Fornes, R. E., Gilbert, R. D., and Stowe, B. S., *Polym. Plast. Technol. Eng.*, **3**, 159 (1974).
1130. Forrester, A. R., Hay, J. M., and Thomson, R. H., *Organic Chemistry of Stable Free Radicals*, Academic Press, New York, 1968.
1131. Förster, T., *Amer. Phys.*, **2**, 55 (1948).
1132. Förster, T., *Z. Naturforsch.*, **4**, A, 321 (1949).
1133. Förster, T., in *Modern Quantum Chemistry* (Sinanoğlu, O., ed.), Vol. 3, Academic Press, New York, 1965.
1134. Förster, T., *Discuss. Faraday Soc.*, **27**, 7 (1959).
1135. Förster, T., *Z. Elektrochem.*, **64**, 157 (1960).
1136. Förster, T., in *Comparative Effects of Radiation* (Burton, M., Kirby-Smith, J., and Magee, J. L., eds), Wiley, New York, 1960.
1137. Förster, T., *Fluoreszenz Organische Verbindungen*, Vandenhoech and Ruprech, Göttingen, 1961.
1138. Förster, T., *Faraday Discuss. Chem. Soc.*, **27**, 7 (1969).
1139. Förster, T., Lieber, C. O., Seidel, H. P., and Weller, A., *Z. Phys. Chem.*, **39**, 265 (1963).
1140. Foster, R. *Organic Charge-Transfer Complexes*, Academic Press, London, 1969.
1141. Foster, R. J., Whitling, P. H., Mellor, J. M., and Phillips, D., *J. Appl. Polym. Sci.*, **22**, 1129 (1978).
1142. Fotland, R. A., *J. Photogr. Sci.*, **18**, 33 (1970).
1143. Fouassier, J. P., in *Graft Copolymerization of Lignocellulosic Fibers* (Hon, D. N. S., ed.), ACS Symp. Ser., No. 187, ACS Washington, D.C., 1982, p. 83.
1144. Fouassier, J. P., *J. Chim. Phys.*, **80**, 339 (1983).
1145. Fouassier, J. P., Freytag, L. J., and Viallier, P., *Bull. Sci. Int., Text. France*, **11**, 81 (1982).
1146. Fouassier, J. P., Jacques, P., and Lougnot, D. J., *J. Radiat, Curing*, **10**, 9 (1983).
1147. Fouassier, J. P., Jacques, P., Lougnot, D. J., and Pilot, T., *Polym. Photochem.*, **5**, 57 (1984).
1148. Fouassier, J. P., and Lougnot, D. J., *Polym. Photochem.*, **3**, 79 (1983).
1149. Fouassier, J. P., Lougnot, D. J., and Faure, J., *C.R. Acad. Sci.*, **284**, C 643 (1977).
1150. Fouassier, J. P., Lougnot, D. J., and Faure, J., *Makromol. Chem.*, **179**, 437 (1978).
1151. Fouassier, J. P., and Riviere, D., *Polym. Photochem.*, **3**, 29 (1983).
1152. Fouassier, J. P., Tieke, B., and Wegner, G., *Israel J. Chem.*, **18**, 227 (1979).
1153. Fourie, J., and McGill, W. J., *South Afr. J. Chem.*, **32**, 63 (1979).
1154. Fourie, J., and McGill, W. J., *South Afr. J. Chem.*, **32**, 156 (1979).
1155. Fox, R. B., *Pure Appl. Chem.*, **30**, 87 (1972).
1156. Fox, R. B., in *Photochemical Processes in Polymer Chemistry*, (DeSchryver, F. C., and Smets, G., eds), Butterworths, London, 1973, p. 235.
1157. Fox, R. B., Isaacs, L. G., Saalfeld, F. E., and McDowell, M. V., NRL Report 6284 (1965).
1158. Fox, R. B., Isaacs, L. G., and Stokes, S. J., *J. Polym. Sci.*, A, **1**, 1079 (1963).
1159. Fox, R. B., Isaacs, L. G., Stokes, S., and Kagarise, R. E., *J. Polym. Sci.*, A, **2**, 2085 (1964).
1160. Fox, R. B., and Price, T. R., *J. Polym. Sci.*, A, **3**, 2303 (1965).
1161. Fox, R. B., Price, T. R., and Cain, D. S., *Adv. Chem. Ser.*, No. 87, 72 (1968).
1162. Fox, R. B., Price, T. R., Cozzens, R. F., and Echols, W. H., *Macromolecules*, **7**, 937 (1974).
1163. Fox, R. B., Price, T. R., and Cozzens, R. F., *J. Chem. Phys.*, **54**, 79 (1971).
1164. Fox, R. B., Price, T. R., Cozzens, R. F., and McDonald, J. R., *J. Chem. Phys.*, **57**, 534 (1972).
1165. Fox, R. B., Price, T. R., Cozzens, R. F., and McDonald, J. R., *J. Chem. Phys.*, **57**, 2284 (1972).
1166. Frank, C. W., *J. Chem. Phys.*, **61**, 2015 (1974).
1167. Frank, C. W., *Macromolecules*, **8**, 305 (1975).
1168. Frank, C. W., *Plast. Compound.*, **1981**, Jan/Feb., p. 67.

693

1169. Frank, C. W., and Gashgari, M. A., *Macromolecules*, **12**, 163 (1979).
1170. Frank, C. W., and Gashgari, M. A., *Polym. Prepr. Amer. Chem. Soc. Div. Polym. Chem.*, **21**, 42 (1980).
1171. Frank, C. W., and Gashgari, M. A., *Ann. N.Y. Acad. Sci.*, **366**, 387 (1981).
1172. Frank, C. W., Gashgari, M. A., Chutikamontham, P., and Haverly, V. J., in *Structure and Properties of Amorphous Polymers* (Walton, A. G., ed.), Elsevier, New York, 1980, p. 187.
1173. Frank, C. W., and Harrah, L. A., *J. Chem. Phys.*, **61**, 1526 (1974).
1174. Frank, J. K., and Paul, I. C., *J. Amer. Chem. Soc.*, **95**, 2324 (1973).
1175. Fränz, K., Maslowski, S., and Vollmer, H. P., *Z. Angew. Phys.*, **31**, 116 (1971).
1176. Franzen, V., *Ann. Chem.*, **633**, 1 (1960).
1177. Frazer, I. M., MacCallum, J. R., and Moran, K. T., *Europ. Polym. J.*, **20**, 425 (1984).
1178. Fredrickson, G. H., Andersen, H. C., and Frank, C. W., *J. Chem Phys.*, **79**, 3572 (1983).
1179. Fredrickson, G. H., Andersen, H. C., and Frank, C. W., *Macromolecules*, **16**, 1456 (1983).
1180. Fredrickson, G. H., Andersen, H. C., and Frank, C. W., *Macromolecules*, **17**, 54 (1984).
1181. Fredrickson, G. H., Andersen, H. C., and Frank, C. W., *Macromolecules*, **17**, 1496 (1984).
1182. Fredrickson, G. H., and Frank, C. W., *Macromolecules*, **16**, 572 (1983).
1183. Fredrickson, G. H., and Frank, C. W., *Macromolecules*, **16**, 1198 (1983).
1184. Freedman, M., Mahadgen, Y., Rennert, J., Soloway, S., and Waltchen, I. *Org. Prep. Proc.*, **1**, 267 (1969).
1185. Frings, R. B., and Schnabel, W., *Polym. Photochem.*, **3**, 325 (1983).
1186. Fry, A. J., Liu, R. S. H., and Hammond, G. S., *J. Amer. Chem. Soc.*, **88**, 4781 (1966).
1187. Fuhr, K., *Polym. Paint Colour J.*, **167**, 672 (1977).
1188. Fuhrmann, J., and Leicht, R., *Colloid Polym. Sci.*, **258**, 631 (1980).
1189. Fujishige, S., and Hasegawa, M., *J. Polym. Sci.*, A1, **7**, 2037 (1969).
1190. Fujita, T., Kurumada, T., Toda, T., and Yoshioka, T., *J. Polym. Sci.*, B, **19**, 609 (1981).
1191. Fujita, T., and Yoshioka, T., *J. Polym. Sci.*, B, **18**, 549 (1980).
1192. Fujita, T., and Yoshioka, T., *J. Polym. Sci.*, A1, **20**, 1639 (1982).
1193. Fujita, T., Yoshioka, T., and Soma, N., *J. Polym. Sci.*, B, **16**, 515 (1978).
1194. Fujita, T., Yoshioka, T., and Soma, N., *J. Polym. Sci.*, A1, **18**, 3253 (1980).
1195. Fujiwara, Y., and Zeronian, S. H., *J. Appl. Polym. Sci.*, **26**, 3729 (1981).
1196. Fukuda, H., and Nakashima, Y., *J. Polym. Sci.*, A1, **21**, 1423 (1983).
1197. Funt, B. L., and Tan, S. R., *J. Polym. Sci.*, A1, **22**, 605 (1981).
1198. Furue, H., and Russell, K. E., in *Singlet Oxygen: Reactions with Organic Compounds and Polymers* (Rånby, B., and Rabek, J. F., eds), Wiley, Chichester, 1978, p. 316.
1199. Furue, H., and Russell, K. E., *Can. J. Chem.*, **56**, 1595 (1978).
1200. Furue, M., Sumi, K., and Nozakura, S., *Polym. Prepr. Japan*, **29**, 280 (1980).
1201. Furue, M., Sumi, K., and Nozakura, S., *Polym. Prepr. Japan*, **30**, 405 and 415 (1981).
1202. Gafney, H. D., and Lintvedt, R. L., *J. Amer. Chem. Soc.*, **92**, 728 (1970).
1203. Gali, G., Solaro, R., Chiellini, E., Fernyhough, A., and Ledwith, A., *Macromolecules*, **16**, 502 (1983).
1204. Ganem, B., *J. Org. Chem.*, **40**, 1998 (1975).
1205. Gantz, G. M., and Summer, W. G., *Text. Res. J.*, **27**, 244 (1957).
1206. Garcia Rubio, L. H., Rao, N., and Patel, R. D., *Macromolecules*, **17**, 1998 (1984).
1207. Gardette, J. L., and Lemaire, J., *Makromol. Chem.*, **182**, 2723 (1981).
1208. Gardette, J. L., and Lemaire, J., *Makromol. Chem.*, **183**, 2415 (1982).
1209. Gardette, J. L., and Lemaire, J., *Polym. Degrad. Stabil.*, **6**, 135 (1984).
1210. Gardette, J. L., and Lemaire, J., *Makromol. Chem.*, **185**, 467 (1985).
1211. Gardette, J. L., and Phillips, D., *Polym. Comm.*, **25**, 366 (1981).
1212. Gardette, J. L., Abuin, E. B., and Lissi, E. A., *Scientia*, **1978**, 11.
1213. Gargallo, L., and Morawetz, H., *Macromolecules*, **14**, 1838 (1981).
1214. Garlund, Z. G., *J. Polym. Sci.*, B, **6**, 57 (1968).
1215. Garlund, Z. G., and Laverty, J. J., *J. Polym. Sci.*, B, **7**, 719 (1969).
1216. Garner, A., and Wilkinson, F., in *Singlet Oxygen: Reactions with Organic Compounds and Polymers*, (Rånby, B., and Rabek, J. F., eds), Wiley, Chichester, 1978, p. 48.
1217. Garner, D. P., and Stahl, G. A., (eds) *The Effect of Hostile Environments on Coatings and Plastics*, ACS Symp. Ser., No. 229, Washington, DC., 1983.
1218. Garnett, J. L., Levot, R., and Long, M. A., *J. Polym. Sci.*, A1, **9**, 23 (1981).
1219. Garnett, J. L., and Major, G., *J. Radiat. Curing*, **9**, Jan, 4 (1982).
1220. Garner, F., Krausz, P., and Rudler, H., *J. Organomet. Chem.*, **186**, 77 (1980).
1221. Garret, M. D., *Amer. Ink Maker*, **1974**, Nov. 22.

694

1222. Garston, B., *J. Soc. Dyers Colour.*, **96**, 535 (1980).
1223. Garton, A., Carlsson, D. J., and Wiles, D. M., *J. Appl. Polym. Sci.*, **21**, 2963 (1977).
1224. Garton, A., Carlsson, D. J., and Wiles, D. M., *J. Polym. Sci.*, A1, **16**, 33 (1978).
1225. Garton, A., Carlsson, D. J., and Wiles, D. M., *Macromolecules,* **12**, 1071 (1979).
1226. Garton, A., Carlsson, D. J., and Wiles, D. M., *J. Polym. Sci.*, A1, **18**, 3245 (1980).
1227. Garton, A., Carlsson, D. J., and Wiles, D. M., in *Developments in Polymer Photochemistry*, (Allen, N. S., ed.), Vol. 1, Ch. 4, Applied Science Publishers, Ltd, London, 1980.
1228. Garton, A., Carlsson, D. J., and Wiles, D. M., *Makromol. Chem.*, **181**, 1841 (1980).
1229. Gashgari, M. A., and Frank, C. W., *Macromolecules*, **14**, 1558 (1981).
1230. Gatechair, L. R., and Pappas, S. P., *Org. Coat. Appl. Polym. Sci. Proc.*, **46,** 183 ACS Meeting, Las Vegas, March 25–April 2, 1982.
1231. Gatechair, L. R., and Pappas, S. P., in *Initiation of Polymerization*, (Bailey, F. E., Jr, ed.), ACS Symp. Ser., No. 212, ACS Washington, D.C., 1983, p. 173.
1232. Gatechair, L. R., and Wostratzky, D., *J. Radiat. Curing*, **10**, March 4 (1983).
1233. Gafney, H. D., Lintvedt, R. L., and Jaworiwsky, I. S., *Inorg. Chem.*, **9**, 728 (1970).
1234. Gaur, H. A., Groenenboom, C. J., Hageman, H. J., Hokvoort, G. T. M., Oosterhoff, P., Overeem, T., Polman, R. J., and van der Werf, S., *Makromol. Chem.*, **185**, 1795 (1984).
1235. Gaylord, N. G., and Dixit, S. S., *J. Polym. Sci.*, B, **9**, 823 (1971).
1236. Geacitov, N. V., Stannett, V., and Abrahamson, E. W., *Makromol. Chem.*, **36**, 62 (1959).
1237. Geacitov, N. V., Stannett, V., Abrahamson, E. W., and Hermans, J. J., *J. Appl. Polym. Sci.*, **3**, 54 (1960).
1238. Gebelein, C. G., Williams, D. J., and Deanin, R. D., (eds,) *Polymers in Solar Energy Utilization*, ACS Symp. Ser., No. 220, ACS Washington, D.C., 1983.
1239. Geddes, K. A., and Deanin, R. D., *Chemtech*, **1982**, Dec., 736.
1240. Gegiou, D., Muszket, K. A., and Fischer, E., *J. Amer. Chem. Soc.*, **90**, 12 (1968).
1241. Gehlhaus, J., and Ohngemach, J., *J. Radiat. Curing*, **10**, July, 20 (1983).
1242. Gellerstedt, G., Kringstadt, K., and Lindfors, E. L., in *Singlet Oxygen: Reactions with Organic Compounds and Polymers* (Rånby, B., and Rabek, J. F., eds), Wiley, Chichester, 1978, p. 302.
1243. Gellerstedt, G., and Pettersson, E. L., *Acta Chem. Scand.*, B, **29**, 1005 (1975).
1244. Gellerstedt, G., and Pettersson, E. L., *Sven. Papperstids.*, **80**, 15 (1977).
1245. Gelles, R., and Frank, C. W., *Macromolecules*, **15**, 741 (1982).
1246. Gelles, R., and Frank, C. W., *Macromolecules*, **15**, 747 (1982).
1247. Gelles, R., and Frank, C. W., *Macromolecules*, **15**, 1486 (1982).
1248. Gelles, R., and Frank, C. W., *Macromolecules*, **16**, 1448 (1983).
1249. George, G. A., *J. Appl. Polym. Sci.*, **18**, 117 (1974).
1250. George, G. A., *J. Appl. Polym. Sci.*, **18**, 419 (1974).
1251. George, G. A., and Hodgeman, D. K. C., *Europ. Polym. J.*, **13**, 63 (1976).
1252. George, G. A., and Hodgeman, D. K. C., *J. Polym. Sci.*, C, **55**, 195 (1976).
1253. George, G. A., Sacher, R. E., and Sprouse, J. F., *J. Appl. Polym. Sci.*, **21**, 2241 (1977).
1254. Georgescauld, D., Desmasez J., Lapouyade, R., Babeau, M., and Winnik, M., *Photochem. Photobiol.*, **31**, 539 (1980).
1255. Gerlock, J. L., and Bauer, D. R., *J. Polym. Sci.*, A1, **22**, 447 (1984).
1256. Gerlock, J. L., Van Oene, H., and Bauer, D. R., *Europ. Polym. J.*, **19**, 11 (1983).
1257. Gesner, B. D., *J. Appl. Polym. Sci.*, **9**, 3701 (1965).
1258. Gesner, B. D., and Kelleher, P. G., *J. Appl. Polym. Sci.*, **12**, 1199 (1968).
1259. Gesner, B. D., and Kelleher, P. G., *J. Appl. Polym. Sci.*, **13**, 2183 (1969).
1260. Geuskens, G., ed. *Degradation and Stabilization of Polymers*, Wiley, New York, 1975.
1261. Geuskens, G., in *Developments in Polymer Degradation* (Grassie, N., ed.) Vol. 3, Applied Science Publishers, Ltd, London, 1981, p. 207.
1262. Geuskens, G., Baeyens-Volant, D., Delaunois, G., Lu-Vinh, Q., Piret, W., and David, C., *Europ. Polym. J.*, **14**, 291 (1978).
1263. Geuskens, G., Baeyens-Volant, D., Delaunois, G., Lu-Vinh, Q., Piret, W., and David, C., *Europ. Polym. J.*, **14**, 299 (1978).
1264. Geuskens, G., and Bastin, P., *Polym. Degrad. Stabil.*, **4**, 111 (1982).
1265. Geuskens, G., Bastin, P., Lu-Vinh, Q, and Rens, M., *Polym. Degrad. Stab.*, **3**, 295 (1980/81).
1266. Geuskens, G., Borsu, M., and David, C., *Europ. Polym. J.*, **8**, 883 (1972).
1267. Geuskens, G., Borsu, M., and David, C., *Europ. Polym. J.*, **8**, 1347 (1972).
1268. Geuskens, G., and David, C., in *Degradation and Stabilization of Polymers* (Geuskens, G., ed.), Halsted Press, New York, 1975, p. 113.
1269. Geuskens, G., and David, C., *Pure Appl. Chem.*, **49**, 479 (1977).

1270. Geuskens, G., and David, C., *Pure Appl. Chem.*, **51**, 233 (1979).
1271. Geuskens, G., and David, C., *Pure Appl. Chem.*, **51**, 2385 (1979).
1272. Geuskens, G., Debie, F., Kabamba, M. S., and Nedelkos, G., *Polym. Photochem.*, **5**, 313 (1984).
1273. Geuskens, G., Delaunois, G., Lu-Vinh, Q., Piret, W., and David, C., *Europ. Polym. J.*, **18**, 387 (1982).
1274. Geuskens, G., and Kamba, M. S., *Polym. Degrad. Stabil.*, **4**, 69 (1982).
1275. Geuskens, G., and Kamba, M. S., *Polym. Degrad. Stabil.*, **5**, 399 (1983).
1276. Geuskens, G., and Lu-Vinh, Q., *Europ. Polym. J.*, **18**, 307 (1982).
1277. Geweda, E., Prochorov, J., *Chem. Phys. Lett.*, **30**, 155 (1975).
1278. Ghaemy, M., and Scott, G., *Polym. Degrad. Stabil.*, **3**, 233 (1981).
1279. Ghaemy, M., and Scott, G., *Polym. Degrad. Stabil.*, **3**, 253 (1981).
1280. Ghaffar, A., Sadrmohaghegh, C., and Scott, G., *Polym. Degrad. Stabil.*, **3**, 341 (1981).
1281. Ghaffar, A., Sadrmohaghegh, C., and Scott, G., *Europ. Polym. J.*, **17**, 941 (1981).
1282. Ghaffar, A., Scott, A., and Scott, G., *Europ. Polym. J.*, **11**, 271 (1975).
1283. Ghaffar, A., Scott, A., and Scott, G., *Europ. Polym. J.*, **12**, 615 (1976).
1284. Ghaffar, A., Scott, A., and Scott, G., *Europ. Polym. J.*, **13**, 83 (1977).
1285. Ghaffar, A., Scott, A., and Scott, G., *Europ. Polym. J.*, **13**, 89 (1977).
1286. Ghatge, N. D., and Vernekar, S. P., *Angew. Makromol. Chem.*, **20**, 165 (1971).
1287. Ghatge, N. D., and Vernekar, S. P., *Angew. Makromol. Chem.*, **20**, 175 (1971).
1288. Ghiggino, K. P., Archibald, D. A., and Thistlewaite, P. J., *J. Polym. Sci.*, B, **18**, 673 (1980).
1289. Ghiggino, K. P., Nicholls, C. H., and Pailthorpe, M. T., *J. Photochem.*, **4**, 155 (1975).
1290. Ghiggino, K. P., Roberts, A. J., and Phillips, D., *J. Photochem.*, **9**, 301 (1978).
1291. Ghiggino, K. P., Roberts, A. J., and Phillips, D., *Adv. Polym. Sci.*, **40**, 69 (1981).
1292. Ghiggino, K. P., Wright, R. D., and Phillips, D., *Chem. Phys. Lett.*, **53**, 552 (1978).
1293. Ghiggino, K. P., Wright, R. D., and Phillips, D., *Europ. Polym. J.*, **14**, 567 (1978).
1294. Ghiggino, K. P., Wright, R. D., and Phillips, D., *J. Polym. Sci.*, A2, **16**, 1499 (1978).
1295. Ghosh, A. K., Biwas, S., and Banerjee, A. N., *J. Makromol. Sci. Chem.*, A, **20**, 927 (1983).
1296. Ghosh, A. K., and Banerjee, A. N., *J. Makromol. Sci. Chem.*, A, **21**, 1253 (1984).
1297. Ghosh, P., and Bandyopadhyay, A. R., *Europ. Polym. J.*, **20**, 1117 (1984).
1298. Ghosh, P., and Banerjee, A. N., *J. Polym. Sci.*, A1, **12**, 375 (1974).
1299. Ghosh, P., and Banerjee, A. N., *J. Polym. Sci.*, A1, **15**, 203 (1977).
1300. Ghosh, P., and Banerjee, A. N., *J. Polym. Sci.*, A1, **16**, 633 (1978).
1301. Ghosh, P., and Banerjee, A. N., *J. Macromol. Sci. Chem.*, A, **16**, 1385 (1981).
1302. Ghosh, P., and Banerjee, A. N., *J. Macromol. Sci. Chem.*, A, **18**, 615 (1982).
1303. Ghosh, P., and Banerjee, A. N., *J. Macromol. Sci. Chem.*, A, **17**, 1273 (1982).
1304. Ghosh, P., and Biswas, S., *Makromol. Chem.*, **182**, 1985 (1981).
1305. Ghosh, P., and Biswas, S., *J. Macromol. Sci. Chem.*, A, **18**, 503 (1982).
1306. Ghosh, P., Chowdhury, D. K., and Bandyopadhyay, A. R., *J. Macromol. Sci. Chem.* A, **20**, 549 (1983).
1307. Ghosh, P., and Ghosh, R., *Europ. Polym. J.*, **17**, 545 (1981).
1308. Ghosh, P., and Ghosh, R., *Europ. Polym. J.*, **17**, 817 (1981).
1309. Ghosh, P., and Ghosh, T. K., *J. Macromol. Sci. Chem.*, A, **17**, 847 (1982).
1310. Ghosh, P., and Ghosh, T. K., *J. Macromol. Sci. Chem.*, A, **18**, 361 (1982).
1311. Ghosh, P., and Ghosh, T. K., *J. Macromol. Sci. Chem.*, A, **19**, 525 (1983).
1312. Ghosh, P., and Ghosh, T. K., *J. Polym. Sci.*, A1, **22**, 2295 (1984).
1313. Ghosh, P., and Jana, S., *J. Macromol. Sci. Chem.*, A, **17**, 743 (1982).
1314. Ghosh, P., Jana, S., and Biswas, S., *Europ. Polym. J.*, **16**, 93 (1980).
1315. Ghosh, P., Jana, S., and Biswas, S., *J. Polym. Sci.*, A1, **21**, 3347 (1983).
1316. Ghosh, P., and Mitra, P. S., *J. Polym. Sci.*, A1, **13**, 921 (1975).
1317. Ghosh, P., and Mitra, P. S., *J. Polym. Sci.*, A1, **14**, 981 (1976).
1318. Ghosh, P., and Mitra, P. S., *J. Polym. Sci.*, A1, **15**, 1743 (1977).
1319. Ghosh, P., and Mitra, P. S., and Banerjee, A. N., *J. Polym. Sci.*, A1, **11**, 2021 (1973).
1320. Ghosh, P., and Mukherji, N., *Europ. Polym. J.*, **15**, 797 (1979).
1321. Ghosh, P., and Mukherji, N., *Europ. Polym. J.*, **17**, 541 (1981).
1322. Ghosh, P., and Mukherji, N., *Europ. Polym. J.*, **19**, 493 (1983).
1323. Ghosh, P., Mukhopadhyay, G., and Ghosh, R., *Europ. Polym. J.*, **16**, 457 (1980).
1324. Ghosh, P., and Paul, S. K., *J. Macromol. Sci. Chem.*, A, **20**, 169 (1983).
1325. Ghosh, P., and Paul, S. K., *J. Macromol. Sci. Chem.*, A, **20**, 261 (1983).
1326. Ghosh, P., Sengupta, P. K., and Mukherjee, N., *J. Polym. Sci.*, A1, **17**, 2119 (1979).

696

1327. Gibb, W. H., and MacCallum, J. R., *Europ. Polym. J.*, **7**, 1231 (1971).
1328. Gibb, W. H., and MacCallum, J. R., *Europ. Polym. J.*, **8**, 1233 (1972).
1329. Gibb, W. H., and MacCallum, J. R., *Europ. Polym. J.*, **9**, 77 (1973).
1330. Gibb, W. H., and MacCallum, J. R., *J. Polym. Sci.*, C, No. 40, 9 (1973).
1331. Gibb, W. H., and MacCallum, J. R., *Europ. Polym. J.*, **10**, 529 (1974).
1332. Gibb, W. H., and MacCallum, J. R., *Europ. Polym. J.*, **10**, 533 (1974).
1333. Gibson, H. W., Bailey, F. C., and Chu, J. Y. C., *J. Polym. Sci.*, A1, **17**, 777 (1979).
1334. Gierer, J., and Lin, S. T., *Svensk Papperstidn.*, **75**, 233 (1972).
1335. Giering, L., Berger, M., and Steel, C., *J. Amer. Chem. Soc.*, **96**, 953 (1974).
1336. Gilbert, A., and Taylor, G. N., *Chem. Comm.*, **1979**, 229.
1337. Ginhac, J. M., Arnaud, R., and Lemaire, J., *Makromol. Chem.*, **182**, 1229 (1981).
1338. Ginhac, J. M., Gardette, J. L., Arnaud, R., and Lemaire, J., *Makromol. Chem.*, **182**, 1017 (1981).
1339. Gismondi, T. E., *J. Radiat. Curing*, **11**, April, 14 (1984).
1340. Gleria, M., Minto, F., Beggiato, G., and Bortolus, P., *J. Chem. Soc. Chem. Commun.*, **1978**, 285.
1341. Gleria, M., Minto, F., Lora, S., and Bortolus, P., *Europ. Polym. J.*, **15**, 671 (1971).
1342. Gleria, M., Minto, F., Lora, S., Bortolus, P., and Ballardini, R., *Macromolecules*, **14**, 687 (1981).
1343. Gochanour, C. R., and Fayer, M. D., *J. Phys. Chem.*, **85**, 1989 (1981).
1344. Godfrey, T. S., Porter, G., and Suppan, P., *Discuss, Faraday Soc.*, **1965**, 194.
1345. Goldburt, E., Shvartman, F., Fishman, S., and Krongauz, V., *Macromolecules*, **17**, 1225 (1984).
1346. Goldenberg, M., Emert, J., and Morawetz, H., *J. Amer. Chem. Soc.*, **100**, 7171 (1978).
1347. Golemba, F. J., and Guillet, J. E., *J. Paint Technol.*, **41**, 315 (1969).
1348. Golemba, F. J., and Guillet, J. E., *Soc. Plast. Eng.*, **26**, 88 (1970).
1349. Golemba, F. J., and Guillet, J. E., *Macromolecules*, **5**, 63 (1972).
1350. Golemba, F. J., and Guillet, J. E., *Macromolecules*, **5**, 212 (1972).
1351. Gollnick, K., *Adv. Photochem.*, **6**, 2 (1968).
1352. Gollnick, K., *Adv. Chem. Ser.*, **77**, 78 (1968).
1353. Golub, M. A., *Pure Appl. Chem.*, **30**, 105 (1972).
1354. Golub, M. A., *Pure Appl. Chem.*, **52**, 305 (1980).
1355. Golub, M. A., Gemmer, E. V., and Rosenberg, M. L., *Adv. Chem. Ser.*, **169**, 11 (1957).
1356. Golub, M. A., and Rosenberg, M. L., *J. Polym. Sci.*, A1, **18**, 2543 (1980).
1357. Golub, M. A., Rosenberg, M. L., and Gemmer, R. V., *Rubber Chem. Technol.*, **50**, 704 (1977).
1358. Golub, M. A., Rosenberg, M. L., and Gemmer, R. V., *J. Polym. Sci.*, A1, **17**, 3751 (1979).
1359. Gomer, R., and Noyes, W. A., *J. Amer. Chem. Soc.*, **72**, 101 (1950).
1360. Goodman, M., and Benedetti, E., *Biochem.*, **7**, 4226 (1968).
1361. Goodman, M., Falxa, M. L., *J. Amer. Chem. Soc.*, **89**, 3863 (1966).
1362. Goodman, M., and Kossoy, A., *J. Amer. Chem. Soc.*, **88**, 5010 (1966).
1363. Gordon, M., and Ware, W. R., (eds), *The Exciplex*, Academic Press, New York, 1975.
1364. Gorin, S., and Monnerie, L., *J. Chem. Phys.*, **67**, 869 (1970).
1365. Gorman, A. A., and Leyland, R. L., *Tetrahed. Lett.*, **1972**, 5345.
1366. Gorman, A. A., Leyland, R. L., Rodgers, M. A., and Smith, P. G., *Tetrahed. Lett.*, **1973**, 5086.
1367. Gorse, R. A., and Volman, D. H., *J. Photochem.*, **1**, 1 (1972).
1368. Gösele, V., Hauser, M., Klein, U. K. A., and Frey, R., *Chem. Phys. Lett.*, **34**, 519 (1975).
1369. Göthe, S., Thesis, Royal Institute of Technology, Stockholm.
1370. Gotoh, T., Yamamoto, M., and Nishijima, Y., *J. Polym. Sci.*, B, **17**, 143 (1979).
1371. Gould, R. F. (ed.), *Photochemical Smog and Ozone Reactions*, ACS Symp. Ser., No. 113, ACS Washington, D.C., 1972.
1372. Govindjee, R., and Govindjee, G., *Sci. Amer.*, **231**, No. 6 (1974).
1373. Grabowski, Z. R., Rotkiewicz, K., Siemiarczuk, A., Cowley, D. J., and Baumann, W., *Nouv, J. Chim.*, **3**, 443 (1979).
1374. Graley, M., Reiser, A., Roberts, A. J., and Phillips, D., in *Photophysics of Synthetic Polymers* (Phillips, D., and Roberts, A. J., eds), Sciences Reviews Ltd, Northwood, England, 1982, p. 26.
1375. Graley, M., Reisser, A., Roberts, A. J., and Phillips, D., *Macromolecules*, **14**, 1752 (1981).
1376. Grassie, N., in *Photochemical Processes in Polymer Chemistry*, (DeSchryver, F. C., and Smets, G., eds), Butterworth, London, 1973, p. 247.
1377. Grassie, N., and Davidson, A. J., *Polym. Degrad. Stabil.*, **3**, 25 (1981).
1378. Grassie, N., and Davidson, A. J., *Polym. Degrad. Stabil.*, **3**, 45 (1981).

1379. Grassie, N., and Davis, T. I., *Makromol. Chem.*, **175**, 2657 (1974).
1380. Grassie, N., and Holmes, A. S., *Polym. Degrad. Stabil.*, **3**, 145 (1981).
1381. Grassie, N., and Leeming, W. B. H., *Europ. Polym. J.*, **11**, 809 (1975).
1382. Grassie, N., and Leeming, W. B. H., *Europ. Polym. J.*, **11**, 819 (1975).
1383. Grassie, N., and Leeming, W. B. H., in *UV Light Induced Reactions in Polymers* (Labana, S. S., ed.), ACS Symp. Ser., No. 25, ACS Washington, D.C., 1976, p. 367.
1384. Grassie, N., and Melville, H. W., *Proc. Roy. Soc.*, A, **199**, 14 (1949).
1385. Grassie, N., and Roche, R. S., *Makromol. Chem.*, **112**, 34 (1964).
1386. Grassie, N., Scotney, A., Jenkins, R., and Davis, T. I., *Chem. Zvesti*, **26**, 208 (1972).
1387. Grassie, N., Scotney, A., and Davis, T. I., *Makromol. Chem.*, **176**, 963 (1975).
1388. Grassie, N., Scotney, A., and MacKinnon, L., *J. Polym. Sci.*, A, **15**, 251 (1977).
1389. Grassie, N., and Torrance, B. J. D., *J. Polym. Sci.*, A1, **6**, 3303, 3315 (1968).
1390. Grassie, N., Torrance, B. D. J., and Colford, J. G., *J. Polym. Sci.*, A1, **7**, 1425 (1969).
1391. Grassie, N., and Weir, N. A., *J. Appl. Polym. Sci.*, **9**, 963 (1965).
1392. Grassie, N., and Weir, N. A., *J. Appl. Polym. Sci.*, **9**, 975 (1965).
1393. Grassie, N., and Weir, N. A., *J. Appl. Polym. Sci.*, **9**, 987 (1965).
1394. Grassie, N., and Weir, N. A., *J. Appl. Polym. Sci.*, **9**, 999 (1965).
1395. Grattan, D. W., Carlsson, D. J., Howard, J. A., and Wiles, D. M., *Can. J. Chem.*, **57**, 2834 (1979).
1396. Grattan, D. W., Carlsson, D. J., and Wiles, D. M., *Polym. Degrad. Stabil.*, **1**, 69 (1979).
1397. Grattan, D. W., Reddoch, A. H., Carlsson, D. J., and Wiles, D. M., *J. Polym. Sci.*, B, **16**, 143 (1978).
1398. Grätzel, M., in *Micellization and Microemulsions* (Mittal, K. C., ed.), Vol. 2, Plenum Press, New York, 1977, p. 531.
1399. Grätzel, M., *Ber. Bunsenges. Phys. Chem.*, **84**, 981 (1980).
1400. Grätzel, M., in *Photochemical Conversion and Storage of Solar Energy*, (Connolly, J. S., ed.), Academic Press, New York, 1981, p. 131.
1401. Green, G. E., Stark, B. P., and Zahir, S. A., *J. Macromol. Sci. Rev. Macromol. Chem.*, C, **21**, 187 (1981/82).
1402. Green, P. N., Young, J. R. A., and Davis, M. J., *Paint Manufac.*, **48**, 32 (1978).
1403. Greene, F. D., Misrock, S. L., and Wolfe, J. R. Jr., *J. Amer. Chem. Soc.*, **77**, 3852 (1955).
1404. Greenwood, R., and Weir, N., *Makromol. Chem.*, **176**, 2041 (1975).
1405. Greenwood, R., and Weir, N. A., in *Ultraviolet Light Induced Reactions in Polymers* (Labana, S. S., ed.), ACS Symp. Ser., No. 25, ACS, Washington, D.C., 1976, p. 220.
1406. Grellmann, K. H., and Scholz, H. G., *Chem. Phys. Lett.*, **62**, 64 (1979).
1407. Grellmann, K. H., Watkins, A. R., and Weller, A., *J. Phys. Chem.*, **76**, 469 (1972).
1408. Griffin, G. W., Bassinski, J. E., and Peterson, L. I., *J. Amer. Chem. Soc.*, **84**, 1012 (1961).
1409. Griffin, G. W., Bassinski, J. E., and Vellturo, A. F., *Tetrahed. Lett*, **3**, 13 (1960).
1410. Griffin, G. W., Vellturo, A. F., and Furukawa, F., *J. Amer. Chem. Soc.*, **83**, 2725 (1961).
1411. Griffing, B. F., and West, P. R., *Polym. Eng. Sci.*, **23**, 947 (1983).
1412. Griffiths, J., and Hawkins, C., *J. Soc. Dyers Colour.*, **89**, 173 (1973).
1413. Griffiths, J., and Hawkins, C., *Polymer*, **17**, 1114 (1976).
1414. Griffiths, P. G., Rizzardo, E., and Solomon, D. H., *J. Macromol. Sci. Chem.*, A, **17**, 45 (1982).
1415. Grimaldi, J. J., Boileau, S., and Lehn, J. M., *Nature*, **265**, 229 (1977).
1416. Gritter, R. J., and Sabatino, E. C., *J. Org. Chem.*, **29**, 1965 (1964).
1417. Grodel, M., and Polacki, Z., *Z. Naturforsch.*, **27a**, 713 (1972).
1418. Groenenboom, C. J., Hageman, H. J., Overeem, T., and Weber, A. J. M., *Makromol. Chem.*, **183**, 281 (1982).
1419. Gromov, B. A., Zubov, Y. A., Kiryushkin, S. G., Marin, A. P., Rapport, N. Yu, Selikhova, V. T., and Shlyapintokh, Y. A., *Vyskomol. Soedin.*, B, **15**, 580 (1973).
1420. Gros, L., Ringsdorf, H., and Shupp, H., Angew. Chem., **93**, 311 (1981).
1421. Gruber, G. W., in *UV Curing: Science and Technology*, (Pappas, S. P., ed.), Vol. 1, Stanford Technology Marketing Corp., Norwalk, CT, 1978, p. 172.
1422. Gruel, H., Vilanove, R., and Rondelez, F., *Phys. Rev. Lett.*, **44**, 590 (1980).
1423. Guarino, A., *J. Photochem.*, **11**, 273 (1979).
1424. Gueron, M., Eisinger, J., and Shulman, R. G., *J. Chem. Phys.*, **47**, 4077 (1967).
1425. Gugumus, F., *Kunst. Plast.*, **22**, 11 (1975).
1426. Gugumus, F., in *Developments in Polymer Stabilization* (Scott, G., ed.), Vol. 1, Applied Science Publishers, Ltd, London, 1979, p. 261.
1427. Gugumus, F., in *Polymer Additives* (Kresta, J. E., ed.), Plenum Press, New York, 1984, p. 17.

698

1428. Guillet, J. E., *Pure Appl. Chem.*, **30,** 135 (1972).
1429. Guillet, J. E., *Naturwissensch.*, **59,** 503 (1972).
1430. Guillet, J. E., *Pure Appl. Chem.*, **36,** 127 (1973).
1431. Guillet, J. E., *Chim. Industrie Gen. Chim.*, **106,** 977 (1973).
1432. Guillet, J. E., ed. *Polymers and Ecological Problems*, Plenum, New York, 1973.
1433. Guillet, J. E., in *Polymers and Ecological Problems* (Guillet, J. E., ed.), Plenum, New York, 1973, p. 1.
1434. Guillet, J. E., *Polym. Eng. Sci.*, **14,** 482 (1974).
1435. Guillet, J. E., in *Degradation and Stabilization of Polymers* (Geuskens, G., ed.) Applied Science Publishers, Ltd, London, 1975.
1436. Guillet, J. E., *Pure Appl. Chem.*, **49,** 249 (1977).
1437. Guillet, J. E., *Adv. Chem. Ser.*, **169,** 1 (1978).
1438. Guillet, J. E., in *Stabilization and Degradation of Polymers* (Allara, D. L., and Hawkins, W. L., eds), ACS Symp. Series, No. 169, ACS Washington, D.C., 1978, p. 1.
1439. Guillet, J. E., *Polym. Prep. Amer. Chem. Soc. Div. Polym. Chem.*, **20,** 395 (1979).
1440. Guillet, J. E., *Pure Appl. Chem.*, **52,** 285 (1980).
1441. Guillet, J. E., *Polymer Photophysics and Photochemistry*, Cambridge University Press, Cambridge, 1985.
1442. Guillet, J. E., Dhanraj, J., Golemba, F. J., and Hartley, G. H., *Adv. Chem. Ser.*, **85,** 272 (1968).
1443. Guillet, J. E., Houvenaghel-Defoort, B., Kilp, T., Turro, N. J., Steinmetler, H. C., and Schuster, G., *Macromolecules*, **7,** 942 (1974).
1444. Guillet, J. E., Hoyle, C. E., and MacCallum, J. R., *Chem. Phys. Lett.*, **54,** 337 (1978).
1445. Guillet, J. E., and Norrish, R. G. W., *Nature*, **173,** 626 (1954).
1446. Guillet, J. E., and Norrish, R. G. W., *Proc. Roy. Soc.*, A, **233,** 153 (1955).
1447. Guillet, J. E., and Norrish, R. G. W., *Proc. Roy. Soc.*, A, **233,** 172 (1955).
1448. Guillet, J. E., Sherren, J., Gharapetian, H. M., and MacInnis, W. K., *J. Photochem.*, **25,** 501 (1984).
1449. Guillet, J. E., Somersall, A. C., and Gordon, J. W., in *Polymers in Solar Energy Utilization* (Gebelein, R. D., Williams, D., and Deanin, R. D., eds), ACS Symp. Ser., No. 220, ACS Washington, D.C., 1983, p. 217.
1450. Guillory, J. P., and Becker, R. S., *J. Polym. Sci.*, A1, **12,** 993 (1974).
1451. Guillory, J. P., and Cook, C. F., *J. Polym. Sci.*, A1, **9,** 1529 (1971).
1452. Guillory, J. P., and Cook, C. F., *J. Polym. Sci.*, A1, **11,** 1927 (1973).
1453. Guillory, J. P., and Cook, C. F., *J. Amer. Chem. Soc.*, **95,** 4885 (1973).
1454. Gunster, E. J., and Schulz, R. C., *Makrmol. Chem.*, **181,** 289 (1980).
1455. Gupta, A., Liang, R., Moacanin, J., Goldbeck, R., and Kliger, D., *Macromolecules*, **13,** 262 (1980).
1456. Gupta, A., Liang, R., Tsay, F. D., and Moacanin, J., *Macromolecules*, **13,** 1696 (1980).
1457. Gupta, A., Liang, R., Moacanin, J., Kliger, D., Goldbeck, R., Horowitz, J., and Miskowski, V. M., *Europ. Polym. J.*, **17,** 485 (1981).
1458. Gupta, A., Rembaum, A., and Moacanin, J., *Macromolecules*, **11,** 1285 (1978).
1459. Gupta, A., Scott, G. W., Kliger, D., in *Photodegradation and Photostabilization of Coatings* (Pappas, S. P., and Winslow, F. H., eds), ACS Symp. Ser. No. 151, ACS Washington, D.C., 1981, p. 27.
1460. Gupta, A., Scott, G. W., Kliger, D., and Vogl, O., in *Polymers in Solar Energy Utilization* (Gebelein, C. G., Williams, D. J., and Deanin, R. D., eds), ACS Symp. Ser., No. 220, ACS Washington, D.C., 1983, p. 293.
1461. Gupta, A., Yavrouian, A., Stefano, S. D., Merritt, C. D., and Scott, G. W., *Macromolecules*, **13,** 821 (1980).
1462. Gupta, I., Gupta, S. N., and Neckers, D. C., *J. Polym. Sci.*, A1, **20,** 147 (1982).
1463. Gupta, M. C., Gupta, A., Horowitz, J., and Kliger, D., *Macromolecules*, **15,** 1372 (1982).
1464. Gupta, S. N., Gupta, I., and Neckers, D. C., *J. Polym. Sci.*, A1, **103** (1981).
1465. Gupta, S. N., Thijs, L., and Neckers, D. C., *Macromolecules*, **13,** 1037 (1980).
1466. Gupta, S. N., Thijs, L., and Neckers, D. C., *J. Polym. Sci.*, **19,** 855 (1981).
1467. Gupta, V. P., and Pierre, L. E. S., *J. Polym. Sci.*, A1, **17,** 931 (1979).
1468. Guttenplan, J. B., and Cohen, S. G., *Tetrahed. Lett.*, **1969,** 2125.
1469. Guttenplan, J. B., and Cohen, S. G., *J. Org. Chem.*, **38,** 2001 (1973).
1470. Gysling, H., and Heller, H. J., *Kunststoffe*, **51,** 18 (1961).
1471. Haag, R., Wirtz, J., and Wagner, P. J., *Helv. Chim. Acta.*, **60,** 2595 (1977).
1472. Haan, S. W., and Zwanzig, R., *J. Chem. Phys.*, **68,** 1879 (1978).

1473. Haas, E., Katchalski-Katzir, E., and Steinberg, I. Z., *Biophysics*, **17**, 11 (1978).
1474. Haas, E., Wilchek, M., Katchalski-Katzir, E., and Steinberg, I. Z., *Proc. Natl. Acad. Sci. USA*, **72**, 1807 (1975).
1475. Hadicke, E., Mez, E. C., Kauch, H., Wegner, G., and Kaiser, J., *Angew. Chem.*, **83**, 253 (1971).
1476. Hageman, H. J., Makromol. *Chem. Rapid Commun.*, **2**, 517 (1981).
1477. Hageman, H. J., van der Maeden, F. P. B., and Janssen, P. C. G. M., *Makromol. Chem.*, **180**, 2531 (1979).
1478. Halary, J. L., Ubrich, J. M., Nunzi, J. M., Monnerie, L., and Stein, R. S. *Polymer*, **25**, 956 (1984).
1479. Haldon, R. A., and Hay, J. N., *J. Polym. Sci.*, A1, **5**, 2297 (1967).
1480. Hallensleben, M. L., *Europ. Polym. J.*, **9**, 227 (1973).
1481. Hallum, J. V., and Drushel, H. V., *J. Phys. Chem.*, **62**, 110 (1958).
1482. Hama, Y., and Shinohara, K., *J. Polym. Sci.*, **8**, 651 (1970).
1483. Hamanoue, K., Teranishi, H., Okamoto, M., Furukawa, Y., Tagawa, S., and Tabata, Y., *J. Polym. Sci.*, A1, **18**, 91 (1980).
1484. Hamity, M., and Scaiano, J., *J. Photochem.*, **4**, 229 (1975).
1485. Hammond, G. S., Baker, W. P., and Moore, W. M., *J. Amer. Chem. Soc.*, **83**, 2795 (1961).
1486. Hammond, G. S., and Leermakers, P. A., *J. Amer. Chem. Soc.*, **84**, 207 (1962).
1487. Hammond, G. S., Saltiel, J., Lamola, A. A., Turro, N. J., Bradshaw, J. B., Cowen, D. O., Cousell, R. C., Vogt, V., and Dakton, C., *J. Phys. Chem.*, **86**, 3197 (1964).
1488. Hammond, G. S., Turro, N. J., and Fisher, A., *J. Amer. Chem. Soc.*, **83**, 4674 (1961).
1489. Hammond, G. S., Turro, N. J., and Leermakers, P. A., *J. Phys. Chem.*, **66**, 1144 (1962).
1490. Hammond, G. S., Wyatt, P., DeBoer, C. D., and Turro, N. J., *J. Amer. Chem. Soc.*, **86**, 2532 (1964).
1491. Hammond, H. A., Doty, J. C., Laakso, T. M., and Williams, J. L. R., *Macromolecules*, **3**, 711 (1970).
1492. Harding, L. B., and Goddard, W. A., III, *J. Amer. Chem. Soc.*, **99**, 4520 (1977).
1493. Harding, L. B., and Goddard, W. A., III, *J. Amer. Chem. Soc.*, **102**, 439 (1980).
1494. Hardy, W. B., in *Developments in Polymer Degradation* (Allen, N. S., ed.), Vol. 2, Applied Science Publishers, Ltd, London, 1982, p. 287.
1495. Harel, Y., and Manassen, J., *J. Amer. Chem. Soc.*, **99**, 5817 (1977).
1496. Harel, Y., Manasse, J., and Levanon, H., *Photochem. Photobiol.*, **23**, 337 (1976).
1497. Hargreaves, J. S., and Phillips, D., *Europ. Polym. J.*, **15**, 119 (1979).
1498. Hargreaves, J. S., and Phillips, D., *J. Polym. Sci.*, A1, **17**, 1711 (1979).
1499. Hargreaves, J. S., and Webber, S. E., *Macromolecules*, **15**, 424 (1982).
1500. Hargreaves, J. S., and Webber, S. E., *Polym. Photochem.*, **2**, 359 (1982).
1501. Hargreaves, J. S., and Webber, S. E., *Macromolecules*, **16**, 1017 (1983).
1502. Hargreaves, J. S., and Webber, S. E., *Macromolecules*, **17**, 235 (1984).
1503. Hargreaves, J. S., and Webber, S. E., *Macromolecules*, **17**, 1741 (1984).
1504. Harper, D. J., and McKellar, J. F., *Chem. Ind. (London)*, **1972**, 848.
1505. Harper, D. J., and McKellar, J. F., *J. Appl. Polym. Sci.*, **17**, 3503 (1973).
1506. Harper, D. J., and McKellar, J. F., *J. Appl. Polym. Sci.*, **18**, 1233 (1974).
1507. Harper, D. J., McKellar, J. F., and Turner, P. H., *J. Amer. Chem. Soc.*, **18**, 2805 (1974).
1508. Harrah, L. A., *J. Chem. Phys.*, **56**, 385 (1972).
1509. Harriman, A., Rockett, B. W., and Poyner, W. R., *J. Chem. Soc. Perkin II*, **1974**, 485.
1510. Harrington, R. W., Hager, G. D., and Crosby, G. A., *Chem. Phys. Lett.*, **21**, 487 (1973).
1511. Harris, J. A., and Arthur, J. C. Jr., *J. Appl. Polym. Sci.*, **24**, 1767 (1979).
1512. Harris, J. A., Keating, E. J., and Goynes, W. R., *J. Appl. Polym. Sci.*, **25**, 2295 (1980).
1513. Hartley, G. H., and Guillet, J. E., *Macromolecules*, **1**, 165 (1968).
1514. Hartley, G. H., and Guillet, J. E., *Macromolecules*, **1**, 413 (1968).
1515. Harwood, H. J., and Ritchey, W. M., *J. Polym. Sci.*, B, **2**, 601 (1964).
1516. Hasegawa, M., *Chem. High Polym.*, **27**, 337 (1970).
1517. Hasegawa, M., *Adv. Polym. Sci.*, **42**, 1 (1982).
1518. Hasegawa, M., *Chem. Revs*, **83**, 507 (1983).
1519. Hasegawa, M., Nakanishi, H., and Yurugi, T., *Chem. Lett.*, **1975**, 497.
1520. Hasegawa, M., Nakanishi, H., and Yurugi, T., *J. Polym. Sci.*, A1,**16**, 2113 (1978).
1521. Hasegawa, M., Nakanishi, H., Yurugi, T., and Ishida, K., *J. Polym. Sci.*, B, **12**, 57 (1974).
1522. Hasegawa, M., and Shiba, S., *J. Phys. Chem.*, **86**, 1490 (1982).
1523. Hasegawa, M., and Suzuki, Y., *J. Polym. Sci.*, B, **5**, 813 (1967).
1524. Hasegawa, M., Suzuki, Y., Nakanishi, H., and Nakanishi, F., *Progr. Polym. Sci. Japan*, **5**, 143 (1973).

700

1525. Hasegawa, M., Suzuki, F., Nakanishi, H., and Suzuki, Y., *J. Polym. Sci.*, B, **6**, 293 (1968).
1526. Hasegawa, M., Suzuki, Y., Suzuki, F., and Nakanishi, H., *J. Polym. Sci.*, A1, **7**, 743 (1969).
1527. Hasegawa, M., Suzuki, Y., and Tamaki, T., *Bull. Chem. Soc. Japan*, **43**, 3020 (1970).
1528. Haszeldine, R. N., *J. Chem. Soc.*, **1953**, 3761.
1529. Hautala, R. R., King, R. B., and Kutal, C. (eds), *Solar Energy: Chemical Conversion and Storage*, The Humana Press, 1979.
1530. Hautala, R. R., and Little, J. L., *Adv. Chem. Ser.*, No. 184, 1 (1980).
1531. Hautala, R. R., Little, J., and Swett, E., *Solar Energy*, **19**, 503 (1977).
1532. Haward, G. J., Kim, S. R., and Peters, R. H., *J. Soc. Dyers Colour.*, **85**, 468 (1969).
1533. Hawkins, W. L., *SPE Trans.*, **1964**, 187.
1534. Hawkins, W. L. (ed.), *Polymer Stabilization*, Wiley, New York, 1972.
1535. Hawkins, W. L., *Polymer Degradation and Stabilization*, Springer Verlag, Berlin, 1984.
1536. Hawkins, W. L., Matreyek, W., and Winslow, F. H., *J. Polym. Sci.*, **41**, 1 (1959).
1537. Hawkins, W. L., Worthington, M. A., and Matreyek, W., *J. Appl. Polym. Sci.*, **3**, 227 (1960).
1538. Hawkins, W. L., Worthington, M. A., and Matreyek, W., *Ind. Eng. Chem., Prod. Res. Develop.*, **1**, 241 (1962).
1539. Hay, A. S., Bolon, D. A., Leimer, K. R., and Clark, R. F., *J. Polym. Sci.*, B, **8**, 97 (1970).
1540. Hay, A. S., Bolon, D. A., and Leimer, K. R., *J. Polym. Sci.*, A1, **8**, 1022 (1970).
1541. Hayakawa, K., Kawase, K., and Yamakata, H., *J. Polym. Sci.*, A1, **12**, 2603 (1974).
1542. Hayakawa, K., Kawase, K., and Yamakata, H., *J. Polym. Sci.*, A1, **17**, 3337 (1979).
1543. Hayashi, K., *J. Radiat, Curing*, **7**, 11 (1980).
1544. Hayashi, K., and Irie, M., *Pure Appl. Chem.*, **34**, 259 (1973).
1545. Hayashi, K., Irie, M., Kiwi, J., and Schnabel, W., *Polymer, J.*, **9**, 41 (1977).
1546. Hayashi, K., Irie, M., and Hayashi, K., *Polym. J.*, **3**, 762 (1972).
1547. Hayashi, K., and Irie, M., in *Photochemical Processes in Polymer Chemistry* (DeSchryver, F. C., and Smets, G., eds), Butterworths, London, 1973, p. 259.
1548. Hayashi, K., Irie, M., and Yamamoto, Y., *J. Polym. Sci.*, C, No. 56, 173 (1976).
1549. Hayashi, T., and Maeda, K., *Bull. Chem. Soc. Japan*, **42**, 3509 (1969).
1550. Hayashi, T., Maeda, K., and Shida, T., *Bull. Chem. Soc. Japan*, **43**, 652 (1970).
1551. Hayashi, T., Suzuki, T., Mataga, N., Sakata, Y., and Misumi, S., *J. Phys. Chem.*, **81**, 420 (1977).
1552. Heine, H. G., *Tetrahed. Lett.*, **1972**, 4755.
1553. Heine, H. G., Hartmann, W., Kory, D. R., Magyar, J. G., Hoyle, C. E., McVey, J. K., and Lewis, F. D., *J. Org. Chem.*, **39**, 691 (1974).
1554. Heine, H. G., Rosenkranz, H. J., and Rudolph, H., *Angew. Chem.*, **84**, 1032 (1972).
1555. Heine, H. G., Rosenkranz, H. J., and Rudolph, H., *Angew. Chem. Ind., Ed. Engl.*, **11**, 974 (1972).
1556. Heine, H. G., Rudolph, H., and Rosenkranz, H. J., *Appl. Polym. Symp.*, C, No. 26, 157 (1975).
1557. Heine, H. G., and Traenchner, H. J., *Progr. Org. Coatings*, **3**, 115 (1975).
1558. Heisel, F., and Laustriat, G., *J. Chim. Phys.*, **66**, 1881 (1969).
1559. Heisel, F., and Laustriat, G., *J. Chim. Phys.*, **66**, 1895 (1969).
1560. Heller, A., *Mol. Photochem.*, **1**, 257 (1969).
1561. Heller, H. J. *Europ. Polym. J.*, Suppl., **1969**, 105.
1562. Heller, H. J., and Blattmann, H. R., *Pure Appl. Chem.*, **30**, 145 (1972).
1563. Heller, H. J., and Blattmann, H. R., *Pure Appl. Chem.*, **36**, 141 (1979).
1564. Hencken, G., *Fabre Lack*, **81**, 916 (1975).
1565. Hencken, G., *Amer. Ink Maker*, **56**, 57 (1978).
1566. Henman, T. J., in *Developments of Polymer Stabilization* (Scott, G., ed.), Vol. 1., Applied Science Publishers, Ltd, London, 1979, p. 39.
1567. Heppel, G. E., *Photochem. Photobiol.*, **4**, 7 (1965).
1568. Herkstroeter, W. G., and Hammond, G. S., *J. Amer. Chem. Soc.*, **88**, 4769 (1966).
1569. Herkstroeter, W. G., Jones, L. B., and Hammond, G. S., *J. Amer. Chem. Soc.*, **88**, 4777 (1966).
1570. Herkstroeter, W. G., Lamola, A. A., and Hammond, G. S., *J. Amer. Chem. Soc.*, **86**, 4537 (1964).
1571. Hernadi, S., *Cellul. Chem. Technol.*, **9**, 31 (1975).
1572. Herold, R., and Fouassier, J. P., *Angew. Makromol. Chem.*, **86**, 109 (1980).
1573. Herold, R., and Fouassier, J. P., *Angew. Makromol. Chem.*, **86**, 123 (1980).
1574. Herold, R., and Fouassier, J. P., *Makromol. Chem. Rapid Commun.*, **2**, 699 (1981).
1575. Herold, R., and Fouassier, J. P., *Angew. Makromol. Chem.*, **97**, 137 (1981).
1576. Herre, W., and Weis, P., *Spectrochem. Acta*, A, **29**, 205 (1973).
1577. Hermann, J. M., Disdier, J., and Pichat, P., *J. Catal.*, **60**, 369 (1979).

1578. Hertler, W. R., McCartin, P. J., Merrill, J. R., Nacci, G. R., and Nebe, W. J., *Photogr. Sci. Eng.*, **23**, 297 (1979).
1579. Herz, C. P., and Eichler, J., *Farbe Lack*, **85**, 933 (1979).
1580. Heskins, M., and Guillet, J. E., *Macromolecules*, **1**, 97 (1968).
1581. Heskins, M., and Guillet, J. E., *Macromolecules*, **3**, 224 (1970).
1582. Heskins, M., and Guillet, J. E., *Photochemistry of Macromolecules*, Plenum Press, New York, 1970, p. 39.
1583. Heskins, M., and Guillet, J. E., in *Encyclopedia of Polymer Science and Technology*, Suppl. (Mark, H. F., and Bikales, N. M., eds), Wiley, New York, 1976, p. 378.
1584. Heskins, M., McAneney, T. B., and Guillet, J. E., in *Ultraviolet Light Induced Reactions in Polymers* (Labana, S. S., ed.), ACS Symp. Ser., No. 125, ACS Washington, D.C., 1976, p. 281.
1585. Heskins, M., Reid, W. J., Pinchin, D. J., and Guillet, J. E., in *Ultraviolet Light Induced Reactions in Polymers* (Labana, S. S., ed.), ACS Symp. Ser., No. 125, ACS Washington, D.C., 1976, p. 272.
1586. Heuvel, H. M., and Lind, K. C. J. B., *J. Polym. Sci.*, A2, **8**, 401 (1970).
1587. Hewgill, F. R., and Smith, W. T., *J. Polym. Sci.*, B, **14**, 463 (1976).
1588. Higgins, J., Johannes, A. H., Jones, J. F., Schultz, R., McCombs, D. A., and Menon, C. S., *J. Polym. Sci.*, A1, **8**, 1987 (1970).
1589. Higgins, R., Foote, C. S., and Cheng, H., *Adv. Chem. Ser.*, **77**, 102 (1972).
1590. Hippe, Z., and Jablonski, H., *Polimery*, **12**, 203 (1967).
1591. Hippe, Z., and Jablonski, H., *Polimery*, **12**, 261 (1967).
1592. Hippe, Z., and Jablonski, H., *Polimery*, **13**, 208 (1968).
1593. Hiraoka, H., *Macromolecules*, **9**, 359 (1976).
1594. Hiraoka, H., *Macromolecules*, **10**, 719 (1977).
1595. Hiraoka, H., and Lee, W. Y., *Macromolecules*, **11**, 622 (1978).
1596. Hiraoka, H., Lee, W. Y., Welsh, L. W. Jr., and Allen, R. W., *Macromolecules*, **12**, 753 (1979).
1597. Hirayama, F., *J. Chem. Phys.*, **42**, 3163 (1965).
1598. Hirayama, F., Basile, L. J., and Kikuchi, C., *Mol. Cryst.*, **4**, 83 (1968).
1599. Hirayama, F., Foster, R. J., Mellor, J. M., Whitling, P. H., Grant, K. R., and Phillips, D., *Europ. Polym. J.*, **14**, 679 (1978).
1600. Hirota, K., Yamamoto, M., and Nishijima, Y., *Rep. Progr. Polym. Phys. Japan*, **16**, 509 (1972).
1601. Hirt, R. C., Schmitt, R. G., Searle, N. D., Sullivan, A. P., *J. Opt. Soc. Amer.*, **50**, 706 (1960).
1602. Hirt, R. C., Searle, N. Z., and Schmitt, R. G., *SPE Trans.*, **1961**, Jan., 21.
1603. Hodgeman, D. K. C., *J. Polym. Sci.*, B, **16**, 161 (1976).
1604. Hodgeman, D. K. C., *Polym. Degrad. Stabil.*, **1**, 155 (1979).
1605. Hodgeman, D. K. C., *J. Macromol. Chem. Ed.*, A, **14**, 173 (1980).
1606. Hodgeman, D. K. C., *J. Polym. Sci.*, A1, **18**, 533 (1980).
1607. Hodgeman, D. K. C., *J. Polym. Sci.*, A1, **19**, 807 (1981).
1608. Hodgeman, D. K. C., in *Development of Polymer Degradation* (Grassie, N., ed.), Vol. 4, Applied Science Publishers, Ltd, London, 1982, p. 189.
1609. Hodgeman, D. K. C., and Gelbert, E. P., *J. Polym. Sci.*, A1, **18**, 1105 (1980).
1610. Hofer, E., *Z. Naturforsch.*, **35b**, 233 (1980).
1611. Hogeveen, H., and Voger, H. C., *J. Amer. Chem. Soc.*, **89**, 2486 (1967).
1612. Holčik, J., Karvaš, M., Kassovicova, D., and Durmis, J., *Europ. Polym. J.*, **12**, 173 (1976).
1613. Holden, D. A., and Guillet, J. E., *J. Polym. Sci.*, **17**, 15 (1979).
1614. Holden, D. A., and Guillet, J. E., in *Developments in Polymer Photochemistry* (Allen, N. S., ed.), Vol. 1, Applied Science Publishers, Ltd, London, 1980, p. 27.
1615. Holden, D. A., and Guillet, J. E., *J. Polym. Sci.*, A1, **18**, 565 (1980).
1616. Holden, D. A., and Guillet, J. E., *Macromolecules*, **13**, 289 (1980).
1617. Holden, D. A., and Guillet, J. E., *Macromolecules*, **15**, 1475 (1982).
1618. Holden, D. A., Kovarova, J., Guillet, J. E., Engel, D., Rhein, Th., and Schulz, R. C., *Europ. Polym. J.*, **19**, 1071 (1983).
1619. Holden, D. A., Ng, D., and Guillet, J. E., *Br. Polym. J.*, **14**, 159 (1982).
1620. Holden, D. A., Ren, X. X., and Guillet, J. E., *Macromolecules*, **17**, 1500 (1984).
1621. Holden, D. A., Rendall, W. A., and Guillet, J. E., *Ann. N. Y. Acad. Sci.*, **366**, 11 (1981).
1622. Holden, D. A., Sheppard, S. E., and Guillet, J. E., *Macromolecules*, **15**, 1481 (1982).
1623. Holden, D. A., Wang, P. Y. K., and Guillet, J. E., *Macromolecules*, **13**, 295 (1980).
1624. Holdsworth, J. D., Scott, G., and Williams, D., *J. Chem. Soc.*, **1964**, 4692.
1625. Holm, M. J., and Zlenty, F. B., *J. Polym. Sci.*, A1, **10**, 1311 (1972).
1626. Holman, R., and Rubin, H., *J. Oil Colour. Chem. Assoc.*, **67**, 189 (1978).

702

1627. Holt, L. A., and Milligan, B., *J. Text. Ind.*, **67**, 269 (1976).
1628. Holt, L. A., and Milligan, B., *Text. Res. J.*, **47**, 620 (1977).
1629. Hon, D. N. S., *J. Polym. Sci.*, A1, **13**, 955 (1975).
1630. Hon, D. N. S., *J. Polym. Sci.*, A1, **13**, 1347 (1975).
1631. Hon, D. N. S., *J. Polym. Sci.*, A1, **13**, 1933 (1975).
1632. Hon, D. N. S., *J. Polym. Sci.*, A1, **13**, 2363 (1975).
1633. Hon, D. N. S., *J. Polym. Sci.*, A1, **13**, 2641 (1975).
1634. Hon, D. N. S., *J. Polym. Sci.*, A1, **13**, 2653 (1975).
1635. Hon, D. N. S., *J. Appl. Polym. Sci.*, **19**, 2789 (1975).
1636. Hon, D. N. S., *J. Polym. Sci.*, A1, **14**, 14 (1976).
1637. Hon, D. N. S., *J. Polym. Sci.*, A1, **14**, 225 (1976).
1638. Hon, D. N. S., *J. Polym. Sci.*, A1, **14**, 2497 (1976).
1639. Hon, D. N. S., *J. Polym. Sci.*, A1, **14**, 2513 (1976).
1640. Hon, D. N. S., *J. Macromol. Sci. Chem.*, A1, **10**, 1175 (1976).
1641. Hon, D. N. S., *J. Polym. Sci.*, A1, **15**, 725 (1977).
1642. Hon, D. N. S., *J. Polym. Sci.*, A1, **17**, 441 (1979).
1643. Hon, D. N. S., *J. Appl. Polym. Sci.*, **23**, 3591 (1979).
1644. Hon, D. N. S., in *Developments in Polymer Degradation* (Grassie, N., ed.), Vol. 3, Applied Science Publishers, Ltd, London, 1981, p. 229.
1645. Hon, D. N. S., in *Preservation of Paper and Textiles of Historic and Artistic* (Williams, J. C., ed.), Vol. 2, ACS Symp. Ser., No. 193, ACS, Washington, D.C., 1981, p. 119.
1646. Hon, D. N. S., *J. Appl. Polym. Sci.*, **37**, 845 (1983).
1647. Hon, D. N. S., *J. Appl. Polym. Sci.*, **29**, 2777 (1984).
1648. Hon, D. N. S., and Chan, H. C., in *Graft Copolymerization of Lignocellulosic Fibers* (Hon, D. N. S., ed.), ACS Symp. Ser., No. 187, ACS Washington, D.C., 1982, p. 101.
1649. Hon, D. N. S., and Chang, S. T., *J. Polym. Sci.*, A1, **22**, 2227 (1984).
1650. Hon, D. N. S., Chang, S. T., and Feist, W. C., *Wood Sci. Technol.*, **16**, 193 (1982).
1651. Hon, D. N. S., and Feist, W. C., *Wood Fiber*, **12**, 121 (1980).
1652. Hon, D. N. S., and Feist, W. C., *Wood Sci.*, **14**, 41 (1981).
1653. Hon, D. N. S., and Glasser, W., *Polym. Plast. Techn. Eng.*, **12**, 159 (1979).
1654. Hon, D. N. S., and Ifju, G., *Wood Sci.*, **11**, 118 (1978).
1655. Hon, D. N. S., and Ifju, G., *Wood Sci.*, **11**, 121 (1978).
1656. Hon, D. N. S., Ifju, G., and Feist, W. C., *Wood Fiber*, **12**, 121 (1980).
1657. Hon, S. Y., Hetherington, W. J. III., Koreschowski, G. M., and Eisenthal, K. S., *Chem. Phys. Lett.*, **68**, 282 (1979).
1658. Hong, S. I., Kurosaki, T., and Okawara, M., *J. Polym. Sci.*, A1, **10**, 3405 (1972).
1659. Hong, S. I., Kurosaki, T., and Okawara, M., *J. Polym. Sci.*, A1, **12**, 2553 (1974).
1660. Hopf, F. R., and Whitten, D. G., in *Porphyrines and Metalloporhyrines* (Smith, K. M., ed.), Elsevier, Amsterdam, 1975, p. 667.
1661. Horie, K., and Mita, I., *Polym. J.*, **9**, 201 (1977).
1662. Horie, K., and Mita, I., *Macromolecules*, **11**, 1175 (1978).
1663. Horie, K., and Mita, I., *Chem. Phys. Lett.*, **93**, 61 (1982).
1664. Horie, K., and Mita, I., *Europ. Polym. J.*, **11**, 1037 (1984).
1665. Horie, K., Moroshita, K., and Mita, I., *Macromolecules*, **17**, 1746 (1984).
1666. Horie, K., and Schnabel, W., *Polym. Photochem.*, **2**, 419 (1982).
1667. Horie, K., Schnabel, W., Mita, I., and Ushiki, H., *Macromolecules*, **14**, 1422 (1981).
1668. Horie, K., Nishijima, Y., Fujimoto, T., Ozawa, K., Kujirai, C., and Nakamura, Y., *Appl. Polym. Symp.*, **18**, 513 (1971).
1669. Hornbaugh, J. E., Sanquist, J. E., Burris, R. H., and Orme-Johnson, W. H., *Biochem.*, **15**, 2633 (1976).
1670. Horspool, W. M., and Pauson, P. L., *J. Chem. Soc.*, **1965**, 5162.
1671. Horswill, E. C., and Ingold, K. U., *Can. J. Chem.*, **44**, 263 (1966).
1672. Horswill, E. C., and Ingold, K. W., *Can. J. Chem.*, **44**, 269 (1966).
1673. Houben, J. L., Natucci, B., Solaro, R., Collela, O., Chiellini, E., and Ledwith, A., *Polymer*, **19**, 811 (1978).
1674. Houben, J. L., Pieroni, O., Fissi, A., and Ciardelli, F., *Biopolym.*, **17**, 799 (1978).
1675. Howard, G. J., Kim, S. R., and Peters, R. H., *J. Soc. Dyers Colour.*, **85**, 468 (1969).
1676. Howard, J. A., *Adv. Free Radical Chem.*, **4**, 49 (1972).
1677. Howard, D. A., Ohkatsu, Y., Chenier, J. H. B., and Ingold, K. U., *Can. J. Chem.*, **51**, 1543 (1973).
1678. Howard, J. A., and Tait, J. C., *J. Org. Chem.*, **43**, 4279 (1978).

1679. Hoyle, C. E., and Guillet, J. E., *J. Polym. Sci.*, B, **16**, 185 (1978).
1680. Hoyle, C. E., and Guillet, J. E., *Macromolecules*, **11**, 221 (1978).
1681. Hoyle, C. E., and Guillet, J. E., *Macromolecules*, **12**, 956 (1979).
1682. Hoyle, C. E., Hensel, R. D., and Grubb, M. B., *J. Polym. Sci.*, A1, **23**, 1865 (1984).
1683. Hoyle, C. E., Hensel, R. D., and Grubb, M. B., *Polym. Photochem.*, **4**, 69 (1984).
1684. Hoyle, C. E., Nemzek, T. L., Mar, A., and Guillet, J. E., *Macromolecules*, **11**, 429 (1978).
1685. Hoytink, G. J., *Acc. Chem. Res.*, **2**, 114 (1969).
1686. Hrdlovič, P., Belluš, D., and Lazar, M., *Coll. Czech. Chem. Commun.*, **33**, 59 (1968).
1687. Hrdlovič, P., Daneček, J., Berek, D., and Lukáč, I., *Europ. Polym. J.*, **13**, 123 (1977).
1688. Hrdlovič, P., and Lukáč, I., *J. Polym. Sci.*, C, No. 47, 319 (1974).
1689. Hrdlovič, P., and Lukáč, I., in *Developments in Polymer Degradation* (Grassi, N., ed.) Vol. 4, Applied Science Publishers, Ltd, London, 1982, p. 101.
1690. Hrdlovič, P., Lukáč, I., and Manašek, Z., *Chem. Zvesti*, **26**, 433 (1972).
1691. Hrdlovič, P., Lukáč, I., and Zvara, I., *Europ. Polym. J.*, **17**, 1121 (1981).
1692. Hrdlovič, P., Lukáč, I., Zvara, I., Kuličkova, M., and Berek, D., *Europ. Polym. J.*, **16**, 651 (1980).
1693. Hrdlovič, P., Trekoval, J., and Lukáč, I., *Europ. Polym. J.*, **15**, 229 (1979).
1694. Hrdlovič, P., Záhumenský, I., Lukáč, I., and Sláma, P., *J. Polym. Sci.*, A1, **16**, 877 (1978).
1695. Hrstka, J., and Klimovic, J., *Polym. Bull.*, **7**, 631 (1982).
1696. Huang, S., J., and Byrne, C. A., *J. Appl. Polym. Sci.*, **27**, 2467 (1982).
1697. Huang, W. N., and Aklonis, J. J., *J. Macromol. Sci. Phys.*, B, **13**, 291 (1977).
1698. Huang, W. N., and Aklonis, J. J., *J. Macromol. Sci. Phys.*, B, **15**, 45 (1978).
1699. Hudson, A., Lappert, M. F., Macquitty, J. J., Nicholson, B. K., Zainal, H., Luchurst, C., Zannoni, S., Bratt, S. W., and Symons, M. C. R., *J. Organomet. Chem.*, **110**, C5 (1976).
1700. Huemmer, T. E., *J. Radiat. Curing*, **1**, 9 (1974).
1701. Huffadine, A. S., Peake, B. M., Robinson, B. H., Simpson, J., and Souwson, P. A., *J. Organomet. Chem.*, **125**, 391 (1976).
1702. Hughes, O. R., and Coard, L. C., *J. Polym. Sci.*, A1, **7**, 1861 (1969).
1703. Hughes, W., *Tenth FATIPEC Congress*, Verlag Chemie, Weinheim, 67 (1970).
1704. Hughey, J. L., Anderson, C. P., and Meyer, T. J., *J. Organomet. Chem.*, **125**, C49 (1977).
1705. Hulme, B. E., Hird, M. S., and Marron, J. J., SME Tech. Paper FC76-492 (1976).
1706. Hult, A., and Rånby, B., *Amer. Chem. Soc. Polym. Prepr.*, **25**(1), 329 (1984).
1707. Hult, A., and Rånby, B., in *Materials for Microlithography* (Thompson, L. F., Wilson, C. G., Frecht, J. M. F., eds) ACS Symp. Ser. No. 206, ACS Washington, D.C., 1984, p. 457.
1708. Hult, A., and Rånby, B., *Polym. Degrad. Stabil.*, **9**, 1 (1984).
1709. Hult, A., and Rånby, B., *Polym. Degrad. Stabil.*, **8**, 75 (1984).
1710. Hult, A., and Rånby, B., *Polym. Degrad. Stabil.*, **8**, 89 (1984).
1711. Hult, A., Yuan, Y. Y., and Rånby, B., *Polym. Degrad. Stabil.*, **8**, 241 (1984).
1712. Humphrey, J. S., Jr., and Rolber, R. S., *Mol. Photochem.*, **3**, 35 (1971).
1713. Humphrey, J. S., Schultz, A. R., and Jaquiss, D. B. G., *Macromolecules*, **6**, 305 (1973).
1714. Humphrey-Baker, R., Grätzel, M., Tundo, P., and Pelizetti, E., *Angew. Chem. Ind. Ed.* **18**, 63 (1979).
1715. Humphrey-Baker, R., Moroi, Y., Grätzel, M., Pelizetti, E., and Tundo, P., *J. Amer. Chem. Soc.*, **102**, 3689 (1980).
1716. Humphris, K. J., and Scott, G., *J. Chem. Soc. Perkin Trans. II.*, **1973**, 831.
1717. Humphris, K. J., and Scott, G., *Pure Appl. Chem.*, **36**, 163 (1973).
1718. Humphris, K. J., and Scott, G., *J. Chem. Soc. Perkin Trans. II*, **1974**, 617.
1719. Hunter, T. F., *Trans. Faraday Soc.*, **66**, 300 (1970).
1720. Husbands, M. J., and Scott, G., *Europ. Polym. J.*, **15**, 249 (1979).
1721. Husbands, M. J., and Scott, G., *Europ. Polym. J.*, **15**, 879 (1979).
1722. Huson, M. G., and McGill, W. J., *South Afr. J. Chem.*, **31**, 91 (1978).
1723. Huson, M. G., and McGill, W. J., *South Afr. J. Chem.*, **31**, 95 (1978).
1724. Huston, A. L., Scott, G. W., and Gupta, A., *J. Chem. Phys.*, **76**, 4978 (1982).
1725. Huston, A. L., and Scott, G. W., *Proc. Intern. Soc. Eng.*, **322**, 215 (1982).
1726. Husy, H., Merian, E., and Schetty, G., *Textile Res. J.*, **39**, 94 (1969).
1727. Hutchinson, J., Lambert, M. C., and Ledwith, A., *Polymer*, **14**, 250 (1973).
1728. Hutchinson, J., and Ledwith, A., *Polymer*, **14**, 405 (1973).
1729. Hutchinson, J., and Ledwith, A., *Adv. Polym. Sci.*, **14**, 49 (1974).
1730. Hutson, G. V., and Scott, G., *Chem. Ind. (London)*, **1972**, 725.
1731. Hutson, G. V., and Scott, G., *Europ. Polym. J.*, **10**, 45 (1974).

704

1732. Huvet, A., Phillippe, J., and Verdu, J., *Europ. Polym. J.*, **14**, 709 (1978).
1733. Hyatt, J. A., *J. Org. Chem.*, **37**, 1254 (1072).
1734. Hyde, P., Kricka, L. J., and Ledwith, A., *J. Polym. Sci.*, B, **11**, 415 (1973).
1735. Ibemesi, J. A., Kinsinger, J. B., and Ashraf El-Bayoumi, *J. Polym. Sci.*, A1, **18**, 879 (1980).
1736. Ibrahim, T. M., Al-Lamee, K. G., and Adam, G. A., *Polym. Photochem.*, **4**, 13 (1984).
1737. Ichimura, K., *Polym. Progr. Polym. Sci. Japan*, **29**(2), 217 (1980).
1738. Ichimura, K., *J. Polym. Sci.*, A1, **20**, 1411 (1982).
1739. Ichimura, K., *J. Polym. Sci.*, A1, **22**, 2817 (1984).
1740. Ichimura, K., and Watanabe, S., *Chem. Lett.*, **1978**, 1289.
1741. Ichimura, K., and Watanabe, S., *J. Polym. Sci.*, B, **18**, 613 (1980).
1742. Ichimura, K., and Watanabe, S., *J. Polym. Sci.*, A1, **18**, 891 (1980).
1743. Ichimura, K., and Watanabe, S., *J. Polym. Sci.*, A1, **20**, 1419 (1982).
1744. Ichimura, K., Watanabe, S., and Ochi, H., *J. Polym. Sci.*, B, **14**, 207 (1976).
1745. Iguchi, M., Nakamishi, H., and Hasegawa, M., *J. Polym. Sci.*, A1, **6**, 1055 (1968).
1746. Ingham, F. A. A., Scott, G., and Stuckey, J. E., *Europ. Polym. J.*, **11**, 783 (1975).
1747. Ingold, K. U., *Can. J. Chem.*, **41**, 2816 (1963).
1748. Iguchi, M., Nakanishi, H., and Hasegawa, M., *J. Polym. Sci.*, A1, **6**, 1055 (1968).
1749. Iinuma, F., Mikawa, H., and Shirota, Y., *Macromolecules*, **14**, 1747 (1981).
1750. Ikeda, M., Teramoto, Y., and Yasutake, M., *J. Polym. Sci.*, A1, **16**, 1175 (1978).
1751. Ikeda, T., Kawaguchi, K., Yamaoka, H., and Okamura, S., *Macromolecules*, **11**, 735 (1978).
1752. Ikeda, T., Okamura, S., and Yamaoka, H., *J. Polym. Sci.*, A1, **15**, 2971 (1977).
1753. Ikeda, T., Shimizu, H., Yamaoka, H., and Okamura, S., *Macromolecules*, **10**, 1417 (1977).
1754. Imamura, M., and Koizumi, M., *Bull. Chem. Soc. Japan*, **29**, 899 (1957).
1755. Imamura, M., Asai, M., Tazuke, S., and Okamura, *Makromol. Chem.*, **168**, 91 (1973).
1756. Inaki, Y., Ishiyama, N., and Takemoto, K. *Angew. Chem.*, **27**, 175 (1972).
1757. Inaki, Y., Suda, Y., Kita, Y., and Takemoto, K., *J. Polym. Sci.*, A1, **19**, 2519 (1981).
1758. Inaki, Y., Takahashi, M., Kameo, Y., and Takemoto, K., *J. Polym. Sci.*, A1, **16**, 399 (1978).
1759. Inaki, Y., Takahashi, M., and Takemoto, K., *J. Macromol. Sci. Chem.*, A, **9**, 1133 (1975).
1760. Inbar, S., Linschitz, H., and Cohen, S. G., *J. Amer. Chem. Soc.*, **103**, 1084 (1981).
1761. Infelta, P. P., Fendler, J. H., and Grätzel, M., *J. Amer. Chem. Soc.*, **102**, 1479 (1980).
1762. Ingold, K. U., *Acc. Chem. Res.*, **2**, 1 (1969).
1763. Inoue, H., and Kohama, S., *J. Appl. Polym. Sci.*, **29**, 877 (1984).
1764. Inoue, H., Takido, S., Somemiya, T., and Nomura, Y., *Tetrahed. Lett.*, **1975**, 2755.
1765. Irick, G. Jr., *J. Appl. Polym. Sci.*, **16**, 2387 (1972).
1766. Irick, G. Jr., and Newland, G. C., *Tetrahed. Lett.*, **1970**, 4151.
1767. Irick, G. Jr., Newland, G. C., and Wang, R. H. S., in *Photodegradation and Photostabilization of Coatings* (Pappas, S. P., and Winslow, F. H., eds), ACS Symp. Ser., No. 151, ACS Washington, D.C., 1981, p. 147.
1768. Irie, M., in *Molecular Models of Photoresponsiveness* (Montagnolo, G., and Erlander, B. F., eds), Plenum Press, New York, 1983, p. 291.
1769. Irie, M., and Hayashi, K., *Polym. Sci. Japan*, **8**, 105 (1975).
1770. Irie, M., and Hayashi, K., *J. Macromol. Sci. Chem.*, A, **13**, 511 (1979).
1771. Irie, M., Hayashi, K., and Menju, A., *Polym. Photochem.*, **1**, 233 (1981).
1772. Irie, M., Hayashi, K., and Smets, G., Photogr. Sci. Eng., **23**, 191 (1979).
1773. Irie, M., Hirano, Y., Nashimoto, S., and Hayashi, K., *Macromolecules*, **14**, 262 (1981).
1774. Irie, S., Irie, M., Yamamoto, Y., and Hayashi, K., *Macromolecules*, **8**, 424 (1975).
1775. Irie, M., and Kungwatchakun, D., *Makromol. Chem. Rapid Commun.*, **5**, 829 (1985).
1776. Irie, M., Masuhara, M., Hayashi, K., and Mataga, N., *J. Phys. Chem.*, **78**, 341 (1974).
1777. Irie, M., Menju, A., and Hayashi, K., *Macromolecules*, **12**, 1176 (1979).
1778. Irie, M., Menju, A., Hayashi, K., and Smets, G., *J. Polym. Sci.*, **17**, 29 (1979).
1779. Irie, M., Sasaoka, S., Yamamoto, Y., and Hayashi, K., *J. Polym. Sci.*, A1, **17**, 815 (1979).
1780. Irie, M., and Schnabel, W., *Macromolecules*, **14**, 1426 (1981).
1781. Irie, M., and Tanaka, H., *Macromolecules*, **16**, 210 (1983).
1782. Irie, M., Tomimoto, S., and Hayashi, K., *J. Polym. Sci.*, B, **8**, 585 (1970).
1783. Irie, M., Tomimoto, S., and Hayashi, K., *J. Phys. Chem.*, **76**, 1419 (1972).
1784. Irie, M., Tomimoto, S., and Hayashi, K., *J. Polym. Sci.*, A1, **10**, 3235 (1972).
1785. Irie, M., Tomimoto, S., and Hayashi, K., *J. Polym. Sci.*, A1, **10**, 3243 (1972).
1786. Irie, M., Yamamoto, Y., and Hayashi, K., *J. Macromol. Sci. Chem.*, A, **9**, 817 (1975).
1787. Irie, M., Yamamoto, Y., and Hayashi, K., *Pure Appl. Chem.*, **49**, 455 (1977).
1788. Ishibashi, H., *Kobunshi Kagaku*, **23**, 620 (1966).

1789. Ishibashi, H., *Kobunshi Kagaku*, **25**, 481 (1968).
1790. Ishibashi, H., and Tamaxi, H., *Kobunshi Kagaku*, **24**, 171 (1967).
1791. Ishida, I., Sondo, S., and Tsuda, K., *Macromol. Chem.*, **178**, 3221 (1977).
1792. Ishida, I., Takahashi, H., Sato, H., and Tsubomura, H., *J. Amer. Chem. Soc.*, **92**, 275 (1970).
1793. Ishida, H., Takahashi, H., and Tsubomura, H., *Bull. Chem. Soc. Japan*, **43**, 3130 (1970).
1794. Ishihara, K., Hamada, N., Hiraguri, Y., and Shinohara, I., *Makromol. Chem. Rapid Commun.*, **5**, 459 (1984).
1795. Ishihara, K., Hamada, N., Kato, S., and Shinohara, I., *J. Polym. Sci.*, A1, **21**, 1551 (1983).
1796. Ishihara, K., Hamada, N., Kato, S., and Shinohara, I., *J. Polym. Sci.*, A1, **22**, 121 (1984).
1797. Ishihara, K., Hamada, N., Kato, S., and Shinohara, I., *J. Polym. Sci.*, A1, **22**, 881 (1984).
1798. Ishihara, K., Kato, S., Hamada, N., and Shinohara, I., *J. Polym. Sci.*, Al, **21**, 1551 (1983).
1799. Ishihara, K., Matsuo, T., Tsunemitsu, K., Shinohara, I., and Negishi, N., *J. Polym. Sci.*, A1, **22**, 3687 (1984).
1800. Ishihara, K., Okazaki, A., Negishi, N., Shinohara, I., Okano, T., Kataoka, K., and Sakurai, Y., *J. Appl. Polym. Sci.*, **27**, 239 (1982).
1801. Ishihara, K., and Shinohara, I., *J. Polym. Sci.*, B, **22**, 515 (1984).
1802. Ishikawa, M., Hongzhi, N., Matsusaki, K., Nate, K., Inoue, T., and Yokono, H., *J. Polym. Sci.*, B, **22**, 669 (1984).
1803. Ishikawa, M., Imamura, N., Mioyshi, N., and Kumada, M., *J. Polym. Sci.*, B, **21**, 657 (1983).
1804. Ishii, T., Handa, T., and Matsunaga, S., *Makromol. Chem.*, **177**, 283 (1976).
1805. Ishii, T., Handa, T., and Matsunaga, S., *Makromol. Chem.*, **178**, 2351 (1977).
1806. Ishii, T., Handa, T., and Matsunaga, S., *Macromolecules*, **11**, 40 (1978).
1807. Ishii, T., Handa, T., and Matsunaga, S., *J. Polym. Sci.*, A2, **17**, 811 (1979).
1808. Ishii, T., Matsunaga, S., and Handa, T., *Macromol. Chem.*, **177**, 283 (1976).
1809. Itagaki, H., Horie, K., and Mita, I., *Europ. Polym. J.*, **12**, 1201 (1983).
1810. Itagaki, H., Horie, K., and Mita, I., *Macromolecules*, **16**, 1395 (1983).
1811. Itagaki, H., Obukata, N., Okamoto, A., Horie, K., and Mita, I., *Chem. Phys. Lett.*, **78**, 143 (1981).
1812. Itagaki, H., Obukata, N., Okamoto, A., Horie, K., and Mita, I., *J. Amer. Chem. Soc.*, **104**, 4469 (1982).
1813. Itagaki, H., Okamoto, A., Horie, K., and Mita, I., *Chem. Phys. Lett.*, **78**, 143 (1981).
1814. Itagaki, H., Okamoto, A., Horie, K., and Mita, I., *Europ. Polym. J.*, **18**, 885 (1982).
1815. Itaya, A., Okamoto, K., and Kusabayashi, S., *Bull. Chem. Soc. Japan*, **49**, 2037 (1976).
1816. Itaya, A., Okamoto, K., and Kusabayashi, S., *Bull. Chem. Soc. Japan*, **49**, 2082 (1976).
1817. Itaya, A., Okamoto, K., and Kusabayashi, S., *Bull. Chem. Soc. Japan*, **51**, 79 (1978).
1818. Itaya, A., Okamoto, K., Masuhara, H., Ikeda, N., Mataga, N., and Kusabayashi, S., *Macromolecules*, **15**, 1213 (1982).
1819. Ito, H., and Willson, C. G., *Polym. Eng. Sci.*, **23**, 1012 (1983).
1820. Ito, M., and Porter, R. S., *J. Appl. Polym. Sci.*, **27**, 4471 (1982).
1821. Ito, S., Yamamoto, M., and Nishijima, Y., *Rep. Progr. Polym. Phys. Japan*, **19**, 421 (1976).
1822. Ito, S., Yamamoto, M., and Nishijima, Y., *Rep. Progr. Polym. Phys. Japan*, **20**, 481 (1977).
1823. Itoh, M., *J. Amer. Chem. Soc.*, **96**, 7390 (1974).
1824. Itoh, M., Fuke, K., and Kobyashi, S., *J. Chem. Phys.*, **72**, 1417 (1980).
1825. Itoh, M., Furuya, S., and Okamoto, T., *Bull. Chem. Soc. Japan*, **50**, 2509 (1977).
1826. Itoh, M., Mimura, T., *Chem. Phys. Lett.*, **24**, 551 (1974).
1827. Itoh, M., Mimura, T., Usui, H., and Okamoto, T., *J. Amer. Chem. Soc.*, **95**, 4388 (1973).
1828. Ivanov, V. B., Kuznecova, M. I., Angert, L. G., and Shlyapintokh, V. Ya., *Dokl. Akad. Nauk SSSR*, **228**, 1144 (1976).
1829. Ivanov, V. B., Lozovskaya, E. L., and Shlyapintokh, V. Ya., *Polym. Photochem.*, **2**, 55 (1982).
1830. Ivanov, V. B., and Shlyapintokh, V. Ya., *J. Polym. Sci.*, **16**, 899 (1978).
1831. Ivanov, V. B., Shlyapintokh, V. Ya., Kvostach, O. M., Shapiro, A. B., and Rozantsev, E. G., *J. Photochem.*, **4**, 313 (1975).
1832. Ivanova, T. V., Mokeeva, G. A., and Sveshukov, B. Ya., *Opt. Spectrosc.*, **12**, 325 (1962).
1833. Iwai, K., Itoh, Y., Fure, M., and Nozakura, S. I., *J. Polym. Sci.*, A1, **21**, 2439 (1983).
1834. Iwasaki, M., Tirrell, D., and Vogl, O., *J. Polym. Sci.*, A1, **18**, 2755 (1980).
1835. Iwata, K., Hagiwara, T., and Matsuzawa, H., *J. Polym. Sci.*, B, **22**, 215 (1984).
1836. Iwaya, Y., and Tazuke, S., *Macromolecules*, **15**, 396 (1982).
1837. Iwayanagi, T., Hashimoto, M., Nonogaki, S., Koibuchi, S., and Makino, D., *Polym. Eng. Sci.*, **23**, 935 (1983).
1838. Iwayanagi, T., Kohashi, T., and Nonogaki, S., *J. Electrochem. Sci.*, **127**, 2759 (1980).

706

1839. Iwayanagi, T., Kohashi, T., Nonogaki, S., Matsuzawa, T., Douta, H., and Yanazawa, H., *IEEE Trans. Electron Devices*, ED-28, 1306 (1981).
1840. Izawa, T., Nishikubo, N., and Takahashi, E., *Makromol. Chem.*, **184**, 2297 (1983).
1841. Jablonski, A., Z. *Phys.*, **16**, 236 (1935).
1842. Jachowicz, J., and Morawetz, H., *Macromolecules*, **15**, 828 (1982).
1843. Jackson, R. A., Oldland, S. R. D., and Pajaczkowski, A., *J. Appl. Polym. Sci.*, **12**, 1297 (1968).
1844. Jacob, N., Balakrishnan, S., and Reddy, M. P., *J. Phys. Chem.*, **81**, 17 (1977).
1845. Jacobi, M., and Henne, A., *J. Radiat. Curing*, **10**, Oct., 16 (1983).
1846. Jacobine, A. F., Lipka, B., and Scanio, C. J. V., *J. Radiat. Curing*, **10**, July, 33 (1983).
1847. Jacobine, A. F., and Scanio, J. V., *J. Radiat. Curing*, **10**, July, 26 (1983).
1848. Jacobson, K. I., and Jacobson, R. E., *Imaging Systems*, Focal Press, London, 1976.
1849. Jain, K., Rice, S., and Lin, B. J., *Polym. Eng. Sci.*, **23**, 1019 (1983).
1850. Jakus, C., and Smets, G., *Macromolecules*, **15**, 435 (1982).
1851. Jarry, J. P., and Monnerie, L., *Macromolecules*, **12**, 927 (1979).
1852. Jarry, J. P., and Monnerie, L., *J. Polym. Sci.*, A2, **18**, 1879 (1980).
1853. Jaschig, W., *Kunstoffe*, **52**, 458 (1962).
1854. Jassim, A. N., MacCallum, J. R., and Moran, K. T., *Europ. Polym. J.*, **19**, 909 (1983).
1855. Jassim, A. N., MacCallum, and T. M. Shepherd, T. M., *Europ. Polym. J.*, **17**, 125 (1981).
1856. Jellinek, H. H. G., *J. Polym. Sci.*, **62**, 281 (1962).
1857. Jellinek, H. H. G., *Aspects of Degradation and Stabilization of Polyolefines*, Elsevier, Amsterdam, 1978.
1858. Jellinek, H. H. G., *Degradation and Stabilization of Polymers*, Elsevier, Amsterdam, 1983.
1859. Jellinek, H. H. G., and Bastien, I. J., *Can. J. Chem.*, **39**, 2056 (1961).
1860. Jellinek, H. H. G., and Chaudhuri, A., *J. Polym. Sci.*, A1, **10**, 1773 (1972).
1861. Jellinek, H. H. G., and Chou, C. H., *J. Macromol. Sci. Chem.*, A, **4**, 255 (1970).
1862. Jellinek, H. H. G., and Lipovac, S. N., *J. Macromol. Chem.*, **1**, 773 (1966).
1863. Jellinek, H. H. G., and Lipovac, S. N., *J. Polym. Sci.*, C, No. 22, 621 (1969).
1864. Jellinek, H. H. G., and Pavlinec, J., *Photochemistry of Macromolecules*, Plenum Press, New York, 1971, p. 91.
1865. Jellinek, H. H. G., and Schlueter, W. A., *J. Appl. Polym. Sci.*, **3**, 206 (1960).
1866. Jellinek, H. H. G., and Wang, I., *Kolloid Z.Z. Polym.*, **202**, 1 (1965).
1867. Jellinek, H. H. G., and Wang, L. C., *J. Macromol. Sci. Chem.*, A, **2**, 781 (1968).
1868. Jellinek, H. H. G., and Wang, L. C., *J. Macromol. Sci. Chem.*, A, **2**, 1353 (1968).
1869. Jenney, J. A., *J. Opt. Soc. Amer.*, **60**, 1155 (1970).
1870. Jenney, J. A., *J. Opt. Soc. Amer.*, **61**, 1116 (1971).
1871. Jenkins, A. D., and Ledwith, A. (eds), *Structure Reactivity and Mechanism in Polymer Chemistry*, Wiley, London, 1974.
1872. Jenkins, R. K., Byrd, N. R., and Lister, J. L., *J. Appl. Polym. Sci.*, **12**, 2059 (1968).
1873. Jensen, J. P. T., and Kops, J., *J. Polym. Sci.*, A1, **18**, 2737 (1980).
1874. Jerussi, R. A., *J. Polym. Sci.*, A1, **9**, 2009 (1971).
1875. Jiráčková-Audouin, L., Bellnger, V., and Verdu, J., *Polym. Photochem.*, **5**, 283 (1984).
1876. Jiráčkova-Audouin, L., Bory, J. F., Farrenq, J. F., and Verdu, J., *Polym. Degrad. Stabil.*, **6**, 17 (1984).
1877. Jiráčkova-Audouin, L., Verdu, J., and Pospisil, J., *Polym. Degrad. Stabil.*, **7**, 67 (1984).
1878. Jirkovsky, J., Fojtik, A., and Becker, H. G. O., *Coll. Czech. Chem. Commun.*, **46**, 1560 (1981).
1879. Joffe, Z., and Rånby, B., in *ESR Applications to Polymer Research* (Kinell, P. O., Rånby, B., and Runnström-Reio, V., eds), Almqvist-Wiksell, Stockholm, 1973, p. 171.
1880. Joffe, Z., and Rånby, B., *J. Appl. Polym. Sci., Polym. Symp.*, **35**, 307 (1979).
1881. Johnson, D. H., and Tobolsky, A. V., *J. Amer. Chem. Soc.*, **74**, 938 (1952).
1882. Johnson, G. E., *J. Chem. Phys.*, **61**, 3002 (1974).
1883. Johnson, G. E., *J. Chem. Phys.*, **62**, 4697 (1975).
1884. Johnson, G. E., *Macromolecules*, **13**, 145 (1980).
1885. Johnson, G. E., *Macromolecules*, **13**, 839 (1980).
1886. Johnson, G. E., and Good, I. A., *Macromolecules*, **15**, 409 (1982).
1887. Johnson, P. C., and Offen, H. W., *J. Chem. Phys.*, **55**, 2945 (1971).
1888. Johnson, P. C., and Offen, H. W., *J. Chem. Phys.*, **56**, 1638 (1972).
1889. Johnson, P. C., and Offen, H. W., *Chem. Phys. Lett.*, **18**, 258 (1973).
1890. Johnson, P. C., and Offen, H. W., *J. Chem. Phys.*, **59**, 801 (1973).
1891. Johnston, D. S., Sanghera, S., Pons, M., and Chapman, D., *Biochem. Biophys. Acta*, **602**, 57 (1980).

1892. Jones, G. II, and Butler, R. J., *Solar Energy*, **29**, 579 (1982).
1893. Jones, G. II, Chiang, S. H., and Xuan, P. T., *J. Photochem.*, **10**, 1 (1979).
1894. Jones, G. II, Reinhart, T. G., and Bergmark, W. R., *Solar Energy*, **20**, 241 (1978).
1895. Jones, G. II, Xuan, P. T., and Chiang, S. H., in *Solar Energy, Chemical Conversion and Storage* (Hautala, R. R., King, B. B., and Kutal, C., eds), Humana Press, New York, 1979.
1896. Jones, P. F., and Siegel, S., *J. Phys. Chem.*, **50**, 1134 (1969).
1897. Jones, R. R., and Karimian, R., *Polymer*, **21**, 832 (1980).
1898. Jones, R. G., and Khalid, N., *Europ. Polym. J.*, **18**, 285 (1982).
1899. Jortner, J., *J. Polym. Sci.*, **37**, 199 (1959).
1900. Jortner, J., and Sokolov, U., *Nature*, **190**, 1003 (1961).
1901. Jortner, J., and Sokolov, U., *J. Phys. Chem.*, **65**, 1633 (1961).
1902. Joschek, H. W., and Miller, S. J., *J. Amer. Chem. Soc.*, **88**, 14 (1966).
1903. Joshi, M. G., *J. Appl. Polym. Sci.*, **26**, 3945 (1981).
1904. Joshi, M. G., and Rodriguez, F., *J. Appl. Polym. Sci.*, **27**, 3151 (1982).
1905. Joshi, M. G., and Rodriguez, F., *J. Appl. Polym. Sci.*, **29**, 1345 (1984).
1906. Kabakoff, D. S., Bünzli, J. C. G., Oth, J. F. M., Hammond, W. B., and Berson, J. A., *J. Amer. Chem. Soc.*, **97**, 1510 (1975).
1907. Kaeriyama, K., *Makromol. Chem.*, **153**, 229 (1972).
1908. Kareiyama, K., *J. Polym. Sci.*, A1, **14**, 1547 (1976).
1909. Kaeriyama, K., and Shimura, T., *Makromol. Chem.*, **167**, 429 (1973).
1910. Kagiya, V. T., in Conf. on 'Degradability of Polymers and Plastics', Inst. Electr. Eng., London, 1973, p. 8/1.
1911. Kagiya, V. T., and Takemoto, K., *J. Macromol. Sci. Chem.*, A, **10**, 795 (1976).
1912. Kagiya, V. T., Takemoto, K., Shi, H., and Rabek, J. F., *J. Polym. Sci.*, B, **16**, 619 (1978).
1913. Kaiser, W. D., and Wagner, H., *Korrosion*, **7**, 3 (1976).
1914. Kalisky, Y., Orlowski, T. E., and Williams, D. J., *J. Phys. Chem.*, **87**, 5333 (1983).
1915. Kalisky, Y., and Williams, D. J., *Chem. Phys. Lett.*, **91**, 77 (1983).
1916. Kalisky, Y., and Williams, D. J., *Macromolecules*, **17**, 292 (1984).
1917. Kalmus, C. E., and Hercules, D. M., *J. Amer. Chem. Soc.*, **96**, 449 (1974).
1918. Kalnins, M. A., *J. Appl. Polym. Sci.*, **29**, 105 (1984).
1919. Kolontarov, I. Ya., Kostina, N. V., and Kiseleva, N. N., *J. Polym. Sci.*, A1, **12**, 939 (1974).
1920. Kalyanasundaram, K., and Grätzel, M., *Ang. Chem. Int. Ed.*, **18**, 701 (1979).
1921. Kalyanasundaram, K., and Grätzel, M., *Angew. Chem.*, **91**, 759 (1979).
1922. Kalyanasundaram, K., Mićić, O., Promauro, E., and Grätzel, M., *Helv. Chim. Acta*, **62**, 2432 (1979).
1923. Kalyanasundaram, K., and Grätzel, M., *J. Chem. Soc., Chem. Comm.*, **1979**, 1137.
1924. Kalyanasundaram, K., and Grätzel, M., *Helv. Chim. Acta*, **63**, 4781 (1980).
1925. Kamal, M. R., El-Kaissy, M. M., and Avedesian, M. M., *J. Appl. Polym. Sci.*, **16**, 83 (1972).
1926. Kambe, H., Watanabe, H., Nagatomo, S., and Itoh, Y., *Polymer*, **18**, 1063 (1977).
1927. Kamijo, T., Irie, M., and Hayashi, K., *Bull. Chem. Soc. Japan*, **51**, 3286 (1978).
1928. Kamogawa, H., *J. Appl. Polym. Sci.*, **13**, 1883 (1969).
1929. Kamogawa, H., *J. Polym. Sci.*, **9**, 335 (1971).
1930. Kamogawa, H., and Hasegawa, H., *J. Appl. Polym. Sci.*, **17**, 745 (1973).
1931. Kamogawa, H., Kato, M., and Sugiyama, H., *J. Polym. Sci.*, A1, **6**, 2967 (1968).
1932. Kamogawa, H., Katsuta, S., and Nanasawa, M., *J. Appl. Polym. Sci.*, **27**, 1621 (1982).
1933. Kamogawa, H., Nanasawa, M., and Uehara, Y., *J. Polym. Sci.*, B, **15**, 675 (1977).
1934. Kamogawa, H., Masui, T., and Amemiya, S., *J. Polym. Sci.*, A, **22**, 383 (1984).
1935. Kamogawa, H., Masui, T., and Nanasawa, M., *Chem. Lett.*, **1980**, 1145.
1936. Kaneko, M., Moriya, S., Yamada, A., Yamamoto, H., and Oyama, N., *Electrochem. Acta*, **29**, 115 (1984).
1937. Kaneko, M., Motoyoshi, J., and Yamada, A., *Nature*, **285**, 468 (1980).
1938. Kaneko, M., Nemoto, S., Yamada, A., and Kurimura, Y., *Ing. Chim. Acta*, **44**, L289 (1980).
1939. Kaneko, M., Ochiai, M., Kinosita, K. Jr., and Yamada, A., *J. Polym. Sci.*, A1, **20**, 1011 (1982).
1940. Kaneko, M., Ochiai, M., and Yamada, A., *Makromol. Chem. Rapid. Commun.*, **3**, 299 (182).
1941. Kaneko, M., Sato, S., and Yamada, A., *Makromol. Chem.*, **179**, 1277 (1978).
1942. Kaneko, M., Takabayashi, N., and Yamada, A., *Chem. Lett.*, **1982**, 1647.
1943. Kaneko, M., and Yamada, A., *J. Phys. Lett.*, **81**, 1213 (1977).
1944. Kaneko, M., and Yamada, A., *Makromol. Chem.*, **182**, 1111 (1981).
1945. Kaneko, M., and Yamada, A., *Photochem. Photobiol.*, **33**, 793 (1981).
1946. Kaneko, M., and Yamada, A., *Adv. Polym. Sci.*, **55**, 2 (1984).

708

1947. Kaneko, M., Yamada, A., and Kurimura, Y., *Ing. Chim. Acta*, **45**, L73 (1980).
1948. Kaneko, M., Yamada, A., Oyama, N., and Yamaguchi, S., *Makromol. Chem. Rapid Commun.*, **3**, 769 (1982).
1949. Kaneko, M., Yamada, A., Tsuchida, E., and Kurimura, Y., *Polym. Prep. Japan*, **31**, 1665 (1982).
1950. Kaneko, M., Yamada, A., Tsuchida, E., and Kurimura, Y., *J. Polym. Sci.*, B, **20**, 593 (1982).
1951. Kanetsuna, H., Hasegawa, M., Mitsuhashi, S., Kurita, T., Sasaki, K., Maeda, K., Obata, H., and Hatakeyama, T., *J. Polym. Sci.*, A2, **8**, 1027 (1970).
1952. Kaplan, M. L., and Kelleher, P. G., *Science*, **169**, 1206 (1970).
1953. Kaplan, M. L., and Kelleher, P. G., *J. Polym. Sci.*, **8**, 3163 (1970).
1954. Kaplan, M. L., and Kelleher, P. G., *J. Polym. Sci.*, B, **9**, 565 (1971).
1955. Kaplan, M. L., and Kelleher, P. G., *Rubber Chem. Technol.*, **45**, 423 (1972).
1956. Kaplan, M. L., Kelleher, P. G., Bebbington, G. H., and Hartless, R. L., *J. Polym. Sci.*, B, **11**, 357 (1973).
1957. Kaplan, M. L., and Trozzolo, A. M., in *Singlet Oxygen* (Wasserman, H. H., and Murray, R. W., eds), Academic Press, New York, 1979, p. 575.
1958. Kardush, N., and Stevens, M. P., *J. Polym. Sci.*, A1, **10**, 1093 (1972).
1959. Karpukhin, O. N., and Slobodetskaya, E. M., *J. Polym. Sci.*, A1, **17**, 3687 (1979).
1960. Karrer, F. E., *Makromol. Chem.*, **181**, 595 (1980).
1961. Karube, I., Nakamoto, Y., Namba, K., and Suzuki, S., *Biochem. Biophys. Acta*, **429**, 975 (1976).
1962. Karube, I., Nakamoto, Y., and Suzuki, S., *Biochem. Biophys. Acta*, **445**, 774 (1976).
1963. Karube, I., Suzuki, S., Nakamoto, Y., and Nishida, M., *Biotech. Bioeng.*, **19**, 1549 (1972).
1964. Katbab, A., and Scott, G., *Chem. Ind.*, **14**, 573 (1980).
1965. Kato, K., Okamura, S., and Yamaoka, H., *J. Polym. Sci.*, B, **14**, 211 (1976).
1966. Kato, K., Sasaki, H., Okamura, S., and Yamaoka, H., *J. Polym. Sci.*, B, **18**, 197 (1980).
1967. Kato, M., *J. Polym. Sci.*, B, **7**, 605 (1969).
1968. Kato, M., Hasegawa, M., and Ichijo, T., *J. Polym. Sci.*, B, **8**, 263 (1970).
1969. Kato, M., Ichijo, T., Ishii, K., and Hasegawa, M., *J. Polym. Sci.*, A1, **9**, 2109 (1971).
1970. Kato, M., and Yamazaki, M., *Makromol. Chem.*, **177**, 3455 (1976).
1971. Kato, M., and Yoneshige, Y., *Makromol. Chem.*, **164**, 159 (1973).
1972. Kato, M., and Yoneshige, Y., *J. Polym. Sci.*, A1, **17**, 79 (1979).
1973. Kato, M., and Yoneshige, Y., *J. Rad. Curing*, **7**, Oct., 23 (1980).
1974. Kato, S., Aizawa, M., and Suzuki, S., *J. Membrane Sci.*, **1**, 289 (1976).
1975. Kato, S., Aizawa, M., and Suzuki, S., *J. Membrane Sci.*, **2**, 39 (1977).
1976. Kauffmann, H. F., *Makromol. Chem.*, **180**, 2649 (1979).
1977. Kauffmann, H. F., *Makromol. Chem.*, **180**, 2665 (1979).
1978. Kauffmann, H. F., *Makromol. Chem.*, **180**, 2681 (1979).
1979. Kauffmann, H. F., Breitenbach, T. W., and Olaj, O. F., *J. Polym. Sci.*, A1, **11**, 737 (1973).
1980. Kawai, T., and Sakata, T., *Chem. Phys. Lett.*, **72**, 87 (1980).
1981. Kawai, W., *J. Polym. Sci.*, A1, **15**, 1479 (1977).
1982. Kawai, W., *Kobunshi Ronbunshu*, **37**, 303 (1980).
1983. Kawai, W., and Ichihashi, T., *J. Polym. Sci.*, A1, **12**, 201 (1974).
1984. Kawai, W., and Ichihashi, T., *J. Polym. Sci.*, A1, **12**, 1041 (1974).
1985. Kawasaki, M., Ibuki, T., Iwasaki, M., and Takezaki, Y., *J. Chem. Phys.*, **59**, 2075 (1973).
1986. Kearns, D. R., *J. Amer. Chem. Soc.*, **91**, 655 (1969).
1987. Kearns, D. R., *Chem. Rev.*, **71**, 395 (1971).
1988. Kearns, D. R., Fenical, W., and Radlick, P., *Ann. N.Y. Acad. Sci.*, **171**, 34 (1970).
1989. Kearwall, A., and Wilkinson, F., in *Transfer and Storage of Energy by Molecules* (Burnell, M., and North, M., eds), Vol. 1, Wiley, London, 1969, p. 94.
1990. Kehr, C. L., and Wszlock, W. R., *Prepr. Div. Organ. Coating Plast. Chem.*, ACS, **33**(1), 295 (1973).
1991. Keiser, J., Wegner, G., and Fischer, W. E., *Israel J. Chem.*, **10**, 157 (1972).
1992. Keith, H. D., and Padden, F. J. Jr., *J. Appl. Phys.*, **35**, 1270 (1961).
1993. Kelen, T., *Polymer Degradation*, Van Nostrand-Reinhold, New York, 1983.
1994. Kelleher, P. G., Boyle, D. J., and Gesner, B. D., *J. Appl. Polym. Sci.*, **11**, 1731 (1967).
1995. Kelleher, P. G., and Gesner, B. D., *J. Appl. Polym. Sci.*, **13**, 9 (1969).
1996. Kelleher, P. G., and Jassie, L. B., *J. Appl. Polym. Sci.*, **9**, 2501 (1965).
1997. Kelleher, P. G., Jassie, L. B., and Gesner, R. D., *J. Appl. Polym. Sci.*, **11**, 137 (1967).
1998. Kemula, W., and Grabowska, A., *Bull. Acad. Polon. Sci. Ser. Sci. Chem.*, **8**, 525 (1960).

1999. Kendrick, R. E., and Spruit, C. J. P., *Photochem. Photobiol.*, **26**, 201 (1977).
2000. Kenner, R. D., and Khan, A. U., *Chem. Phys. Lett.*, **36**, 643 (1975).
2001. Kenner, R. D., and Khan, A. U., *J. Chem. Phys.*, **64**, 1877 (1976).
2002. Kenner, R. D., and Khan, A. U., *Chem. Phys. Lett.*, **45**, 76 (1977).
2003. Kenner, R. D., and Khan, A. U., *J. Chem. Phys.*, **67**, 1605 (1977).
2004. Kenyon, A. S., Natl. Bur. Stand., Circ. No. 525, 81 (1953).
2005. Kerr, J. A., *Chem. Rev.*, **66**, 465 (1966).
2006. Kessler, H. C. Jr., *J. Phys. Chem.*, **71**, 2736 (1967).
2007. Ketley, A. D., and Tsao, J. H., *J. Radiat. Curing*, **6**, Feb., 22 (1979).
2008. Kettle, G. J., and Soutar, I., *Europ. Polym. J.*, **14**, 895 (1978).
2009. Kettle, G. J., and Soutar, I., *Polym. Photochem.*, **1**, 123 (1981).
2010. Keyanpour-Rad, M., and Ledwith, A., *Polym. Bull.*, **1**, 337 (1979).
2011. Keyanpour-Rad, M., Ledwith, A., Hallam, A., North, A. M., Breton, M., Hoyle, C., and Guillet, J. E., *Macromolecules*, **11**, 1114 (1978).
2012. Keyanpour-Rad, M., Ledwith, A., and Johnson, G. E., *Macromolecules*, **13**, 222 (1980).
2013. Khalifa, W. M., El-Azmerly, M. A., Morsi, S. E., and Barsoum, S., *J. Appl. Polym. Sci.*, **23**, 2225 (1979).
2014. Kharash, M. S., and Friedlander, H. N., *J. Org. Chem.*, **14**, 239 (1949).
2015. Kharash, M. S., Jensen, E. V., and Urry, W. H., *J. Amer. Chem. Soc.*, **69**, 110 (1947).
2016. Kharash, M. S., Jerome, J. J., and Urry, W. H., *J. Org. Chem.*, **15**, 966 (1950).
2017. Kharash, M. S., Reinmuth, O., and Urry, W. H., *J. Amer. Chem. Soc.*, **69**, 1105 (1957).
2018. Kharash, M. S., Simon, E., and Nudenberg, W., *J. Org. Chem.*, **18**, 328 (1953).
2019. Kharash, M, S., and Yoshi, B. S., *J. Org. Chem.*, **22**, 1439 (1957).
2020. Kikuchi, S., and Nakamura, K., in *Prepr. Soc. Plast. Eng. Reg. Techn. Conf.*, Ellenville, New York, 6 Nov. 1967, p. 175.
2021. Kilp, T., and Guillet, J. E., *Macromolecules*, **14**, 1680 (1981).
2022. Kilp, T., Guillet, J. E., Galin, J. C., and Roussel, R., *Macromolecules*, **15**, 980 (1982).
2023. Kilp, T., Guillet, J. E., Merle-Aubery, L., and Merle, Y., *Macromolecules*, **15**, 60 (1982).
2024. Kim, N., and Webber, S. E., *Macromolecules*, **13**, 1233 (1980).
2025. Kim, N., and Webber, S. E., *Macromolecules*, **15**, 430 (1982).
2026. King, R. B., and Sweet, E. M., *J. Org. Chem.*, **44**, 385 (1979).
2027. Kinoshita, A., and Namariyama, Y., *J. Appl. Polym. Sci.*, **28**, 1255 (1983).
2028. Kinstle, J. F., *Paint Varnish Prod.*, **17**, June, 3 (1973).
2029. Kinstle, J. F., *J. Radiat. Curing*, **1**, Feb., 2 (1974).
2030. Kinstle, J. F., and Watson, S. L. Jr., *J. Radiat. Curing*, **2**, Jan., 7 (1975).
2031. Kinstle, J. F., and Watson, S. L. Jr., *J. Radiat. Curing*, **3**, Feb., 2 (1976).
2032. Kinstle, J. F., and Watson, S. L. Jr., *Macromolecules*, **15**, 894 (1982).
2033. Kinmonth, F. A. Jr., and Norton, J. E., *J. Coating Technol.*, **49**, 37 (1977).
2034. Kirch, M., Lehn, J. M., and Sauvage, J. P., *Helv. Chim. Acta*, **62**, 1345 (1979).
2035. Kirchmayr, R., Berner, G., Husler, R., and Rist, G., *Farbe u. Lack*, **88**, 910 (1982).
2036. Kirkbright, G. F., Narayanaswamy, R., and West, T. S., *Anal. Chem. Acta*, **52**, 237 (1970).
2037. Kita, Y., Futugawa, H., Inaki, Y., and Takemoto, K., *Polym. Bull.*, **2**, 195 (1980).
2038. Kita, Y., Inaki, Y., and Takemoto, K., *J. Polym. Sci.*, A1, **18**, 427 (1980).
2039. Kita, Y., Uno, T., Inaki, Y., and Takemoto, K., *Nucleic Acid Res.*, **6**, 63 (1979).
2040. Kita, Y., Uno, T., Inaki, Y., and Takemoto, K., *J. Polym. Sci.*, **19**, 477 (1981).
2041. Kita, Y., Uno, T., Inaki, Y., and Takemoto, K., *J. Polym. Sci.*, A1, **19**, 1733 (1981).
2042. Kita, Y., Uno, T., Inaki, Y., and Takemoto, K., *J. Polym. Sci.*, A1, **19**, 2347 (1981).
2043. Kita, Y., Uno, T., Inaki, Y., and Takemoto, K., *J. Polym. Sci.*, A1, **19**, 3315 (1981).
2044. Kiwi, J., and Grätzel, M., *J. Amer. Chem. Soc.*, **100**, 6314 (1978).
2045. Kiwi, J., and Grätzel, M., *Chimia*, **33**, 289 (1979).
2046. Kiwi, J., and Grätzel, M., *Nature*, **281**, 657 (1979).
2047. Kiwi, J., and Grätzel, M., *J. Amer. Chem. Soc.*, **101**, 7214 (1979).
2048. Kiwi, J., and Grätzel, M., *Angew. Chem. Int. Ed.*, **17**, 860 (1979).
2049. Kiwi, J., and Schnabel, W., *Macromolecules*, **8**, 430 (1975).
2050. Kiwi, J., and Schnabel, W., *Macromolecules*, **9**, 468 (1976).
2051. Klein, A. J., Yu, H., and Yen, W. M., *J. Appl. Polym. Sci.*, **26**, 2381 (1981).
2052. Kleinert, T. N., *Papier (Darmstadt)*, **24**, 207 (1970).
2053. Klemchuk, P. P., *Polym. Photochem.*, **3**, 1 (1983)
2054. Klinshpont, E. R., Milinchuk, V. K., and Vasilenko, V. V., *Polym. Photochem.*, **4**, 329 (1984).

710

2055. Kloosterboer, J. G., and Lippits, G. J. M., *J. Radiat. Curing*, **11**, Jan., 10 (1984).
2056. Kloosterboer, J. G., van de Hei, G. M. M., Gossink, R. G., and Dortant, G. C. M., *Polym. Comm.*, **25**, 322 (1984).
2057. Kloosterboer, J. G., van de Hei, G. G. M., Boots, H. M. M., *Polym. Comm.*, **25**, 354 (1981).
2058. Klöpffer, W., *Ber. Bunseng. Phys. Chem.*, **73**, 864 (1969).
2059. Klöpffer, W., *J. Chem. Phys.*, **50**, 2337 (1969).
2060. Klöpffer, W., *J. Chem. Phys.*, **50**, 1689 (1969).
2061. Klöpffer, W., *Chem. Phys. Lett.*, **4**, 193 (1969).
2062. Klöpffer, W., *Chem. Phys. Lett.*, **7**, 137 (1970).
2063. Klöpffer, W., *Kunstoffe*, **61**, 533 (1971).
2064. Klöpffer, W., *Kunstoffe*, **60**, 385 (1970).
2065. Klöpffer, W., *Ber. Bunseng. Phys. Chem.*, **74**, 693 (1970).
2066. Klöpffer, W., in *Organic Molecular Photophysics* (Birks, J. B., ed.), Vol. 1, Wiley, London 1973, p. 357.
2067. Klöpffer, W., *Europ. Polym. J.*, **11**, 203 (1975).
2068. Klöpffer, W., *J. Polym. Sci.*, Polym. Symp., No. 57, 205 (1976).
2069. Klöpffer, W., *Adv. Photochem.*, **10**, 311 (1977).
2070. Klöpffer, W., *Spectrosc. Lett.*, **11**, 863 (1978).
2071. Klöpffer, W., *Ann. N.Y. Acad. Sci.*, **366**, 373 (1981).
2072. Klöpffer, W., *Chem. Phys.*, **57**, 75 (1981).
2073. Klöpffer, W., in *Electric Properties of Polymers* (Mort, J., and Pfisteri, G., eds), Wiley, New York, 1982, p. 161.
2074. Klöpffer, W., and Bauser, H., *Z. Phys. Chem.*, **101**, 25 (1976).
2075. Klöpffer, W., and Fischer, D., *J. Polym. Sci.* Polym. Symp., No. 40, 43 (1973).
2076. Klöpffer, W., Fischer, D., and Naundorf, G., *Macromolecules*, **10**, 450 (1977).
2077. Klöpffer, W., and Liptay, W., *Z. Naturforsch.*, A, **25**, 1091 (1970).
2078. Klöpffer, W., Turro, N. J., Chow, M. F., and Noguchi, Y., *Chem. Phys. Lett.*, **54**, 157 (1978).
2079. Knapczyk, J. W., Lubinowski, J. J., and McEwen, W. E., *Tetrahed. Lett.*, **1972**, 3739.
2080. Knapczyk, J. W., and McWeen, W. E., *J. Org. Chem.*, **35**, 2539 (1970).
2081. Knibbe, H., and Weller, A., *Z. Physik Chem.*, **55**, 99 (1967).
2082. Knoesel, R., and Weill, G., *Polym. Photochem.*, **2**, 167 (1982).
2083. Ko, M., Sato, T., and Otsu, T., *J. Polym. Sci.*, A1, **12**, 2943 (1974).
2084. Ko, M., Sato, T., and Otsu, T., *J. Polym. Sci.*, A1, **7**, 3329 (1969).
2085. Kobayashi, J., Kinoshita, Y., Ide, F., and Nakatsuka, K., *Kobunshi Kagaku*, **28**, 437 (1971).
2086. Koch, T. H., and Jones, A. H., *J. Amer. Chem. Soc.*, **92**, 7503 (1970).
2087. Kochevar, I. E., and Wagner, P. J., *J. Amer. Chem. Soc.*, **92**, 5742 (1970).
2088. Kochevar, I. E., and Wagner, P. J., *J. Amer. Chem. Soc.*, **94**, 3859 (1972).
2089. Koch, J. K., and Krusic, P. J., *J. Amer. Chem. Soc.*, **91**, 3940 (1969).
2090. Kochi, J. K., and Rataliff, M. A., *J. Org. Chem.*, **36**, 3112 (1971).
2091. Kodaira, T., and Hayashi, K., *J. Polym. Sci.*, B, **9**, 907 (1971).
2092. Kodaira, T., Hayashi, K., and Ohnishi, T., *Polym. J.*, **4**, 1 (1973).
2093. Koenig, H. S., and Roberts, C. W., *J. Appl. Polym. Sci.*, **19**, 1847 (1975).
2094. Koerner von Gustorf, E., and Grewels, F. W., *Fortsch. Chem. Forsch.*, **13**, 366 (1969).
2095. Koh, N. P., Marchal, C., and Verdu, J., *Anal. Chem*, **51**, 1000 (1079).
2096. Koizumi, M., and Usui, Y., *Mol. Photochem.*, **57**, 4 (1972).
2097. Kojima, K., Iwabuchi, S., Nakahira, T., Uchiyama, T., and Koshiyama, Y., *J. Polym. Sci.*, B, **14**, 143 (1976).
2098. Kojima, K., Nakahira, T., Kosuge, Y., Saito, M., and Iwabuchi, S., *J. Polym. Sci.*, B, **19**, 193 (1981).
2099. Kolawole, E. G., *Europ. Polym. J.*, **20**, 629 (1984).
2100. Kolawole, E. G., and Adeniyi, J. B., *Europ. Polym. J.*, **18**, 469 (1982).
2101. Kolawole, E. G., and Agaboola, M. O., *J. Appl. Polym. Sci.*, **27**, 2317 (1982).
2102. Kolawole, E. G., and Scott, G., *J. Appl. Polym. Sci.*, **26**, 2581 (1981).
2103. Kolinsky, M., Doskocilova, D., Drahoradova, E., Schneider, B., Stokr, J., and Kuska, V., *J. Polym. Sci.*, A1, **9**, 791 (1971).
2104. Köller, H., Rabold, G. P., Weiss, K., Mukherjee, T. K., *Proc. Chem. Soc.*, **1964**, 332.
2105. Kondo, S., Itoh, S., and Tsuda, K., *J. Macromol. Sci. Chem.*, A, **20**, 433 (1983).
2106. Kondo, S., Kondo, Y., and Tsuda, K., *J. Polym. Sci.*, B, **21**, 217 (1983).
2107. Kondo, S., Muramatsu, M., Senga, M., and Tsuda, K., *J. Polym. Sci.*, A1, **22**, 1187 (1984).

711

2108. Kondo, S., Muramatsu, M., and Tsuda, K., *J. Macromol. Sci. Chem.*, A, **19**, 999 (1983).
2109. Konev, S. V., *Fluorescence and Phosphorescence of Proteins and Nucleic Acids*, Plenum Press, New York, 1967.
2110. Konishi, H., Shinagawa, Y., Azuma, A., Okano, T., and Kiji, J., *Makromol. Chem.*, **183**, 2941 (1982).
2111. Kornis, G., and de Mayo, P., *Can. J. Chem.*, **42**, 2822 (1964).
2112. Koryakin, B. V., Dzhabiev, T. S., and Shilov, A. E., *Dokl. Akad. Nauk. SSSR*, **298**, 620 (1977).
2113. Kosar, J., *Light-Sensitive Systems: Chemistry and Application of Nonsilver Halide Photographic Processes*, Wiley, New York, 1965.
2114. Kossanyi, J., and Furth, B., *L'Actualité Chim.*, **1974**, 2.
2115. Kothandaraman, H., and Santappa, M., *Polym. J.*, **2**, 148 (1971).
2116. Kothandaraman, H., Srinivasan, K. S. V., and Santappa, M., *J. Polym. Sci.*, Al, **10**, 3685 (1972).
2117. Kovačevic, V., Bravar, M., and Hace, D., *Polym. Photochem.*, **4**, 393 (1984).
2118. Kowal, J., *Polym. Degrad. Stabil.*, **7**, 175 (1984).
2119. Kowal, J., and Nowakowska,L., *Polymer*, **20**, 1003 (1979).
2120. Kowal, J., and Nowakowska, M., *Polymer*, **23**, 281 (1982).
2121. Kowal, J., and Nowakowska, M., *Makromol. Chem.*, **183**, 1701 (1982).
2122. Kowal, J., Nowakowska, M., and Waligora, B., *Polymer*, **19**, 1313 (1978).
2123. Kowal, J., and Waligora, B., *Makromol. Chem.*, **179**, 707 (1978).
2124. Kowal, J., and Waligora, B., *Makromol. Chem.*, **179**, 713 (1978).
2125. Kozmina, O. P., Dubyaga, V. P., Belyakov, V. K., and Zaichukova, W. A., *Europ. Polym. J.*, **5**, 447 (1969).
2126. Krasna, A. I., *Photochem. Photobiol.*, **29**, 267 (1979).
2127. Kraeutler, B., Reiche, H., and Bard, A. J., *J. Polym. Sci.*, B, **17**, 535 (1979).
2128. Kresta, J. E., *Polymer Additives*, Plenum Press, New York, 1984.
2129. Krentz, F. H., *Trans. Faraday Soc.*, **51**, 172 (1955).
2130. Kricka, L. J., Lambert, M. C., and Ledwith, A., *J. Chem. Soc. Perkin Trans. I*, **1974**, 52.
2131. Kringstad, K. P., and Lin, S. Y., *Tappi*, **53**, 2296 (1970).
2132. Krongauz, V. A., *Israel J. Chem.*, **18**, 304 (1979).
2133. Krongauz, V. A., and Goldburt, E. S., *Macromolecules*, **14**, 1382 (1981).
2134. Kryszewski, M., Galeski, A., and Milczarek, P., *J. Polym. Sci.*, B, **14**, 365 (1976).
2135. Kryszewski, M., and Nadolski, B., *Pure Appl. Chem.*, **49**, 511 (1977).
2136. Kryszewski, M., and Nadolski, B., in *Singlet Oxygen:Reactions with Organic Compounds and Polymers* (Rånby, B., and Rabek, J. F., eds), Wiley, Chichester, 1978, p. 244.
2137. Kryszewski, M., Nadolski, B., Imhof, R. E., North, A. M., and Pethrick, R. A., *Makromol. Chem.*, **183**, 1257 (1982).
2138. Kryszewski, M., Wandelt, B., Birch, D. J. S., Imhof, R. E., North, A. M., and Pethrick, R. A., *Polymer*, **23**, 924 (1982).
2139. Kryszewski, M., Wandelt, B., Birch, D. J. S., Imhof, R. E., North, A. M., and Pethrick, R. A., *Polym. Commun.*, **24**, 73 (1983).
2140. Kryszewski, M., Zapiens, D., and Nadolski, B., *J. Polym. Sci.*, Al, **11**, 2423 (1973).
2141. Kubica, J., and Waligora, B., *Europ. Polym. J.*, **13**, 325 (1977).
2142. Kubota, H., Murata, Y., and Ogiwara, Y., *J. Polym. Sci.*, Al, **11**, 485 (1978).
2143. Kubota, H., and Ogiwara, Y., *J. Appl. Polym. Sci.*, **14**, 2611 (1970).
2144. Kubota, H., and Ogiwara, Y., *J. Appl. Polym. Sci.*, **14**, 2879 (1970).
2145. Kubota, H., and Ogiwara, Y., *J. Appl. Polym. Sci.*, **15**, 2767 (1971).
2146. Kubota, H., and Ogiwara, Y., *J. Appl. Polym. Sci.*, **16**, 337 (1972).
2147. Kubota, H., and Ogiwara, Y., *J. Appl. Polym. Sci.*, **20**, 1405 (1976).
2148. Kubota, H., and Ogiwara, Y., *J. Appl. Polym. Sci.*, **27**, 2683 (1982).
2149. Kubota, H., and Ogiwara, Y., *J. Appl. Polym. Sci.*, **28**, 2425 (1983).
2150. Kubota, H., and Ogiwara, Y., *Polym. Photochem.*, **4**, 317 (1984).
2151. Kubota, H., Ogiwara, Y., and Matsuzaki K., *J. Polym. Sci.*, Al, **12**, 2809 (1974).
2152. Kuhlmann, R., and Schnabel, W., *Polymer*, **17**, 419 (1976).
2153. Kuhlmann, R., and Schnabel, W., *Angew. Makromol. Chem.*, **57**, 195 (1977).
2154. Kuhlmann, R., and Schnabel, W., *J. Photochem.*, **7**, 287 (1977).
2155. Kuhlmann, R., and Schnabel, W., *Polymer*, **18**, 1163 (1977).
2156. Kujirai, C., Hashiya, S., Furuno, H., and Terada, N., *J. Polym. Sci.*, Al, **6**, 589 (1968).
2157. Kularatne, K. W. S., and Scott, G., *Europ. Polym. J.*, **14**, 835 (1978).
2158. Kularatne, K. W. S., and Scott, G., *Europ. Polym. J.*, **15**, 827 (1979).

712

2159. Kulevsky, N., Wang, C. T., and Stenberg, V. I., *J. Org. Chem.*, **34**, 1345 (1969).
2160. Kumano, A., Niwa, O., Kaiyama, T., Takayanagi, M., Kano, K., and Shinkai, S., *Chem. Lett.*, **1983**, 1327.
2161. Kumar, G. S., De Pra, P., Zhang, K., and Neckers, D. C., *Macromolecules*, **17**, 2463 (1984).
2162. Kunetake, T., Aso, C., Ito, K., *Makromol. Chem.*, **97**, 10 (1966).
2163. Kunieda, A., Kondo, S., and Tsuda, K., *J. Polym. Sci.*, B, **12**, 395 (1974).
2164. Kunz, C. O., Long, P. G., and Wright, A. N., *Polym. Eng. Sci.*, **12**, 209 (1972).
2165. Kunz, C. O., and Wright, A. N., *J. Chem. Soc. Trans. Faraday*, **68**, 140 (1971).
2166. Kuriacose, J. C., and Markham, M. C., *J. Phys. Chem.*, **65**, 2232 (1961).
2167. Kurimura, Y., Nagashima, M., Takato, K., Tsuchida, E., Kaneko, M., and Yamada, A., *J. Phys. Chem.*, **86**, 2432 (1982).
2168. Kurimura, Y. Shinozaki, N., Ito, F., Uratani, Y., Shigehara, K., Tsuchida, E., Kanekom, M., and Yamada, A., *Bull. Chem. Soc. Japan*, **55**, 380 (1982).
2169. Kurosaki, T., Lee, K. W., Okawara, M., *J. Polym. Sci.*, Al, **10**, 3295 (1972).
2170. Kurowski, S. R., and Morrison, H., *J. Amer. Chem. Soc.*, **94**, 507 (1972).
2171. Kurpiewski, T., and McDowell, J. R., SME Tech. Pap. Ser. FC78-509 (1978).
2172. Kurusu, Y., Nishiyama, H., and Okawara, M., *Makromol. Chem.*, **133**, 49 (1970).
2173. Kusai, M. Z., Al-Abidin, Z., and Jones, R. G., *J. Chem. Soc. Faraday II*, **75**, 774 (1979).
2174. Küster, B., and Herlinger, H., *Angew. Makromol. Chem.*, **40/41**, 265 (1974).
2175. Kutal, C., Schwendiman, D. P., and Grutsch, P. A., *Solar Energy*, **19**, 651 (1977).
2176. Kuzmin, M. G., and Gusewa, L. N., *Chem. Phys. Lett.*, **3**, 71 (1969).
2177. Kuzmin, M. G., Sadovskii, N. A., and Soboleva, I. V., *J. Photochem.*, **23**, 27 (1983).
2178. Kwei, K. R. S., *J. Appl. Polym. Sci.*, **12**, 1543 (1968).
2179. Kwei, K. P. S., *J. Polym. Sci.*, Al, **7**, 237 (1969).
2180. Kwei, K. P. S., *J. Polym. Sci.*, Al, **7**, 1075 (1969).
2181. Kwei, K. P. S., and Kwei, T. K., *J. Appl. Polym. Sci.*, **12**, 1551 (1968).
2182. Kyle, B. R. M., and Kilp, T., *Polymer*, **25**, 989 (1984).
2183. Kysel, O., and Zahradnik, R., *Coll. Czech. Chem. Commun.*, **35**, 3030 (1970).
2184. Kysel, O., Zahradnik, R., Bellus, D., and Sticzay, T., *Coll. Czech. Chem. Commun.*, **35**, 3191 (1970).
2185. Kysel, O., Zahrdnik, R. and Pakula, B., *Coll. Czech. Chem. Commun.*, **35**, 3020 (1970).
2186. Labana, S. S., *J. Macromol. Sci. Rev. Macromol. Chem.*, C, **11**, 299 (1974).
2187. Labana, S. S., ed, *Ultraviolet Light Induced Reactions in Polymers*, ACS Symp. Ser., No. 25, ACS Washington, D.C., 1976.
2188. Labský, J., Kálal, J., and Mikes, F., *Polym. Photochem.*, **2**, 289 (1982).
2189. Labský, J., Konopecky, I., Nespórek, S., and Kálal, J., *Europ. Polym. J.*, **17**, 309 (1981).
2190. Labský, J., Mikes, F., and Kálal, J., *Polym. Bull.*, **2**, 785 (1980).
2191. Lachish, U., Anderson, R. W., and Williams, D. J., *Macromolecules*, **13**, 1143 (1980).
2192. Lachish, U., and Williams, D. J., *Macromolecules*, **13**, 1322 (1980).
2193. Lacoste, J., Arnaud, R., and Lemaire, J., *J. Polym. Sci.*, Al, **22**, 3885 (1984).
2194. Lai, T., and Lim, E. C., *Phys. Lett.*, **73**, 2244 (1980).
2195. Laird, T., *Chem. Ind.*, **1978**, 186.
2196. Lakowicz, J. R., *Principle of Fluorescence Spectroscopy*, Plenum Press, New York, 1984.
2197. Lala, D., and Rabek, J. F., *Polym. Degrad. Stabil.*, **3**, 383 (1980/81).
2198. Lala, D., and Rabek, J. F., *Europ. Polym. J.*, **17**, 7 (1981).
2199. Lala, D., Rabek, J. F., and Rånby, B., *Europ. Polym. J.*, **16**, 735 (1980).
2200. Lala, D., Rabek, J. F., and Rånby, B., *Polym. Degrad. Stabil.*, **3**, 307 (1980/81).
2201. La Mantia, F. P., *Europ. Polym. J.*, **20**, 993 (1981).
2202. La Mantia, F. P., and Acierno, D., *Polym. Photochem.*, **4**, 271 (1984).
2203. Lamarre, L., and Sung, C. S. P., *Macromolecules*, **16**, 1729 (1983).
2204. Lamberts, J. J. M., and Neckers, D. C., *Z. Naturforsch.*, **39b**, 474 (1984).
2205. Lamola, A. A., in *Techniques of Organic Chemistry* (Leermakers, P. A., and Weissberger, A., eds), Vol. 14, Interscience, New York, 1969, p. 17.
2206. Lamola, A. A., and Hammond, G. S., *J. Chem. Phys.*, **43**, 2129 (1965).
2207. Lamola, A. A., Leermakers, P. A., Byers, G. W., and Hammond, G. S., *J. Amer. Chem. Soc.*, **87**, 2322 (1965).
2208. Lamola, A. A., and Sharp, L. J., *J. Phys. Chem.*, **70**, 2634 (1966).
2209. Landi, V. R., and Heidt, L. J., *J. Phys. Chem.*, **73**, 2361 (1969).
2210. Lanza, E., Berghmans, H., and Smets, G., *J. Polym. Sci.*, A2, **11**, 95 (1973).
2211. Laridon, U., Delzenne, G., and Conix, A., *Photogr. Sci. Eng.*, **9**, 364 (1965).

713

2212. Lashkov, A. I., and Ermolaev, V. L., *Opt. Sepectr.*, **22,** 462 (1967).
2213. Launer, H. F., *Text. Res. J.*, **35,** 395 (1965).
2214. Laurenson, P., Arnand, R., Lemaire, J., Quemner, J., and Roche, G., *Europ. Polym. J.*, **14,** 129 (1978).
2215. Laver, H. S., in *Developments in Polymer Stabilization* (Scott, G., ed.), Vol. 1, Applied Science Publishers, Ltd, London, 1979, p. 167.
2216. Law, K. Y., *Polymer*, **25,** 399 (1984).
2217. Law, K. Y., and Loufty, R. O., *Polymer*, **24,** 439 (1983).
2218. Lawrence, J. B., and Weir, N. A., *Chem. Commun.*, **1966,** 273.
2219. Lawrence, L. B., and Weir, N. A., *J. Polym. Sci.,* **11,** 105 (1973).
2220. Lawrence, L. B., and Weir, N. A., *J. Appl. Polym. Sci.*, **18,** 1821 (1974).
2221. Leary, G. J., *Tappi*, **50,** 17 (1967).
2222. Leary, G. J., *Tappi*, **51,** 257 (1968).
2223. Leary, G. J., *J. Chem. Soc. Perkin Trans.*, II, **1972,** 640.
2224. Leaver, I. H., *Photochem. Photobiol.*, **21,** 197 (1975).
2225. Leaver, I. H., *Photochem. Photobiol.*, **27,** 439 (1978).
2226. Leaver, I. H., in *Photochemistry of Dyed and Pigmented Polymers* (Allen, N. S., and McKellar, J. F., eds), Applied Science Publishers, Ltd, Lodnon, 1980, p. 161.
2227. Leaver, I. H., and Ramsay, G. C., *Tetrahedron*, **25,** 5669 (1969).
2228. Leaver, I. H., Waters, P. J., and Evans, N. A., *J. Polym. Sci.*, Al, **17,** 1531 (1979).
2229. Ledger, M. B., and Porter, G., *J. Chem. Soc. Faraday Trans.*, **68,** 539 (1972).
2230. Ledwith, A., *Acc. Chem. Res.*, **5,** 133 (1972).
2231. Ledwith, A., *J. Oil Colour. Chem. Assoc.*, **59,** 157 (1976).
2232. Ledwith, A., *Pure Appl. Chem.*, **49,** 431 (1977).
2233. Ledwith, A., *Polymer*, **19,** 1217 (1978).
2234. Ledwith, A., *Makromol. Chem. Suppl.*, **3,** 348 (1979).
2235. Ledwith, A., *Pure Appl. Chem.*, **54,** 549 (1982).
2236. Ledwith, A., in *Developments in Polymerization* (Haward, R. N. ed.), Vol. 3, Applied Science Publishers, Ltd, London, 1982, p. 55.
2237. Ledwith, A., Al-Kass, S., Sherrington, D. C., and Bonner, P., *Polymer*, **22,** 143 (1981).
2238. Ledwith, A., Bosley, J. A., and Purbick, M., *J. Oil Colour Chem. Assoc.*, **61,** 95 (1978).
2239. Ledwith, A., Ndaalio, G., and Taylor, A. R., *Macromolecules*, **8,** 1 (1975).
2240. Ledwith, A., and Purbick, M., *Polymer*, **14,** 521 (1973).
2241. Ledwith, A., Rawley, N. J., and Walker, S. M., *Polymer*, **22,** 435 (1981).
2242. Ledwith, A., Russel, P. J., and Sutcliffe, L. H., *J. Chem. Soc. Perkin Trans.*, *II,* **1972,** 1925.
2243. Lee, C. H., Waddell, W. H., and Casassa, E. F., *Macromolecules*, **18,** 1021 (1981).
2244. Lee, C. Y. C., and Aklonis, J. J., ACS Symp. Ser. No. 95, ASC Washington, D.C., 1979, p. 219.
2245. Lee, W. E., Calvert, J. G., and Malmberg, E. W., *J. Amer. Chem. Soc.*, **83,** 1928 (1961).
2246. Leermakers, P. A., Byers, G. W., Lamola, A. A., and Hammond, G. S., *J. Amer. Chem. Soc.*, **85,** 2670 (1963).
2247. Lehn, J. M., in *Photochemical Conversion and Storage of Solar Energy* (Connolly, J. S., ed.), Academic Press, New york, 1981, p. 161.
2248. Lehn, J. M., Sauvage, J. P., and Ziessel, R., *Nuov. J. Chim.*, **3,** 429 (1979).
2249. Lemaire, J., *Pure Appl. Chem.*, **54,** 1667 (1982).
2250. Lemaire, J., and Arnaud, R., *Polym. Photochem.*, **5,** 243 (1984).
2251. Le Nest, J. F., and Silvy, J., *C.R. Acad. Sci., Ser.*, C, **271,** 102 (1970).
2252. Lenka, S., and Dash, M., *J. Macromol. Chem. Sci.*, A, **18,** 1141 (1982).
2253. Lenka, S., and Nayak, P. L., *J. Appl. Polym. Sci.*, **27,** 3787 (1982).
2254. Lenka, S., and Nayak, P. L., *J. Polym. Sci.*, Al, **21,** 1871 (1983).
2255. Lenka, S., Nayak, P. L., and Dash, M., *Polym. Photochem.*, **3,** 109 (1983).
2256. Lenka, S., Nayak, P. L., and Dash, M., *Polym. Photochem.*, **4,** 1 (1984).
2257. Lenka, S., Nayak, P. L., and Dash, M., *Polym. Photochem.*, **4,** 83 (1984).
2258. Lenka, S., Nayak, P. L., and Nayak, S. K., *J. Macromol. Sci. Chem.*, A, **20,** 835 (1983).
2259. Lenka, S., Nayak, P. L., and Nayak, S. K., *J. Polym. Sci.*, Al, **22,** 429 (1984).
2260. Lenka, S., Nayak, P. L., and Pradhan, A. K., *Polym. Photochem.*, **3,** 357 (1983).
2261. Lenka, S., Nayak, P. L., and Ray, S., *Polym. Photochem.*, **4,** 167 (1984).
2262. Lenka, S., Nayak, P. L., and Tripathy, A. K., *J. Appl. Polym. Sci.*, **27,** 1853 (1982).
2263. Lenka, S., Nayak, P. L., and Tripathy, A. K., *J. Polym. Sci.*, Al, **20,** 2097 (1982).
2264. Lennox, F. G., King, M. G., Leaver, I. H., Ramsay, G. G., and Sauvage, W. F., *Appl. Polym. Symp.*, No. 18, 353 (1971).

714

2265. Leonard, N. J., McCredie, R. S., Louge, M. W., and Cundall, R. L., *J. Amer. Chem. Soc.*, **95**, 2320 (1973).
2266. Leonard, N. J., and Weller, A., *Z. Elektrochem.*, **67**, 791 (1963).
2267. Leonov, M. R., Redoshkin, B. A., and Suslov, V. A., *Zh. Obsch. Khim.*, **32**, 3959 (1962).
2268. Leu, K. W., *Plastics News*, **1975**, April, 10.
2269. Leu, K. W., Gugumus, F., Linhart, H., and Wieber, A., *Plaste Kauts.*, **24**, 408 (1977).
2270. Leubner, G. W., Unruh, C. C., US Pat. 3,257,664 (21 June 1966).
2271. Leubner, G. W., Williams, J. L. R., and Unruh, C. C., US Pat. 2,811,510 (29 Oct. 1957).
2272. Levine, A. W., Kaplan, M., and Piliniak, E. S., *Polym. Eng. Sci.*, **14**, 518 (1974).
2273. Levine, R., and Jortner, J., eds, *Molecular Energy Transfer*, Wiley New York, 1976.
2274. Lewis, F. D., Hoyle, C. H., Magyar, J. G., Heine, H. G., and Hartman, W., *J. Org. Chem.*, **40**, 488 (1975).
2275. Lewis, F. D., Johnson, R. W., and Kory, D. R., *J. Amer. Chem. Soc.*, **96**, 6100 (1974).
2276. Lewis, F. D., Lauterbach, R. T., Heine, H. G., Hartmann, W., and Rudolph, H., *J. Amer. Chem. Soc.*, **97**, 1519 (1975).
2277. Lewis, F. D., and Magyar, J. G., *J. Amer. Chem. Soc.*, **95**, 5973 (1973).
2278. Lewis, F. M., and Matheson, M. S., *J. Amer. Chem. Soc.*, **71**, 747 (1949).
2279. Lewis, K. J., in *Polymers in Solar Energy Utilization* (Gebelein, C. G., Williams, D. J., and Deanin, R. J., eds), ACS Symp. Ser., No. 220, ACS Washington, D.C., 1983, p. 367.
2280. Li, S., Albertsson, A. C., Gupta, A., Bassett, W. Jr., and Vogl, O., *Monatshefte Chem.*, **115**, 853 (1984).
2281. Li, S., Gupta, A., Albertsson, A. C., Bassett, W. Jr., and Vogl, O., *Polym. Bull.*, **12**, 237 (1984).
2282. Li, S. K. L., and Guillet, J. E., *Macromolecules*, **10**, 840 (1977).
2283. Li, S. K. L., and Guillet, J. E., *J. Polym. Sci.*, A1, **18**, 2221 (1980).
2284. Li, S. K. L., and Guillet, J. E., *Macromolecules*, **17**, 41 (1984).
2285. Li, S. K. L., and Guillet, J. E., *Polym. Photochem.*, **4**, 21 (1984).
2286. Li, Z. T., Kubota, H., and Ogiwara, Y., *J. Macromol. Sci. Chem.*, A, **18**, 483 (1982).
2287. Li, Z. T., Kubota, H., and Ogiwara, Y., *J. Appl. Polym. Sci.*, **27**, 1465 (1982).
2288. Li, Z. T., Kubota, H., and Ogiwara, Y., *Polym. Photochem.*, **3**, 435 (1983).
2289. Li, X. B., Winnik, M. A., and Guillet, J. E., *Macromolecules*, **16**, 992 (1983).
2290. Li, X. B., Winnik, M. A., and Guillet, J. E., *J. Polym. Sci.*, A1, **21**, 1263 (1983).
2291. Liang, R. H., Coulter, D. R., Dao, C., and Gupta, A., in *Polymers in Solar Energy Utilization* (Gebelein, C. G., Williams, D. J., and Deanin, R. D., eds), ACS Symp. Ser., No. 220, ACS Washington, D.C., 1983, p. 265.
2292. Liang, R. H., Tsay, F. D., and Gupta, A., *Macromolecules*, **15**, 974 (1982).
2293. Liao, T. P., and Morawetz, H., *Macromolecules*, **13**, 1228 (1980).
2294. Liao, T. P., Okamoto, Y., and Morawetz, H., *Macromolecules*, **12**, 535 (1979).
2295. Liaw, D. J., and Lin, R. S., *Polym. Photochem.*, **5**, 23 (1984).
2296. Lieberman, R. A., *J. Radiat. Curing*, **3**, Aug., 13 (1981).
2297. Lieberman, R. A., *J. Radiat. Curing*, **11**, Jan., 22 (1984).
2298. Liesner, G., Tieke, B., and Wegner, G., *Thin Solid Films*, **68**, 77 (1980).
2299. Lin, S. Y., and Kringstad, K., *Tappi*, **53**, 1675 (1970).
2300. Lindberg, J. J., and Pyykö, P., in *Singlet Oxygen: Reactions with Organic Compounds and Polymers* (Rånby, B., and Rabek, J. F., eds), Wiley, Chichester, 1978, p. 224.
2301. Lindenau, D., Beavan, S. W., Beck, G., and Schnabel, W., *Europ. Polym. J.*, **13**, 819 (1977).
2302. Lindman, B., and Wennerström, H., *Topics Current Chem.*, **87**, 3 (1980).
2303. Lindsell, W. E., Robertson, F. C., and Soutar, I., *Europ. Polym. J.*, **17**, 203 (1981).
2304. Lion, Y., Delmelle, M., and Van de Vorst, A., *Nature*, **263**, 442 (1976).
2305. Lion, Y., Gandin, E., and Van de Vorst, A., *Photochem. Photobiol.*, **31**, 305 (1980).
2306. Lissi, E. A., and Encina, M. V., *J. Polym. Sci.*, **17**, 2791 (1979).
2307. Lissi, E. A., Encina, M. V., and Abarca, M. T., *J. Polym. Sci.*, A1, **17**, 19 (1979).
2308. Lissi, E. A., Garrido, J., and Zanocco, A., *J. Polym. Sci.*, B, **12**, 391 (1984).
2309. Lissi, E. A., and Zanocco, *J. Polym. Sci.*, A1, **21**, 2197 (1983).
2310. Liston, T. V., Ingersoll, H. G., and Adams, J. Q., *Amer. Chem. Soc. Div. Petrol., Prepr.*, **14**, (4), A83 (1969).
2311. Lo, J., Lee, S. N., and Pearce, E. M., *J. Appl. Polym. Sci.*, **29**, 35 (1984).
2312. Lock, L. M., and Frank, G. C., *Text. Res. J.*, **43**, 502 (1973).
2313. Long, R. C., and Walsh, T. D., *J. Amer. Chem. Soc.*, **89**, 3943 (1967).
2314. Longworth, J. W., *Biopolym.*, **4**, 1131 (1966).
2315. Longworth, J. W., and Bovey, F. A., *Biopolym.*, **4**, 1115 (1966).

2316. Loutfy, R. O., *Macromolecules*, **14**, 270 (1981).
2317. Loutfy, R. O., *J. Polym. Sci.*, A2, **20**, 825 (1982).
2318. Loutfy, R. O., and Law, K. Y., *J. Phys. Chem.*, **84**, 2803 (1980).
2319. Loutfy, R. O., and Teegarden, D. M., *Macromolecules*, **16**, 452 (1983).
2320. Lougnot, D. J., and Fouassier, J. P., *Macromol. Chem. Rapid Comm.*, **4**, 11 (1983).
2321. Lougnot, D. J., Merlin, A., Salvin, R., Fouassier, J. P., and Faure, J., *Il Nuovo Cimento*, **63B**, 284 (1981).
2322. Lovrien, R., *Proc. U.S. Natl. Acad. Sci.*, **57**, 236 (1967).
2323. Lovrien, R., and Waddington, J. C. B., *J. Amer. Chem. Soc.*, **86**, 2315 (1964).
2324. Lower, S. K., and El-Sayed, M. A., *Chem. Rev.*, **66**, 199 (1966).
2325. Loyons, E. H. Jr., and Dickinson, R. G., *J. Amer. Chem. Soc.*, **57**, 443 (1935).
2326. Lucki, J., Rabek, J. F., and Rånby, B., *Polym. Bull.*, **1**, 563 (1979).
2327. Lucki, J., Rabek, J. F., and Rånby, B., *J. Appl. Polym. Sci.*, Appl. Polym. Symp., No. 35, 275 (1979).
2328. Lucki, J., Rabek, J. F., and Rånby, B., *Polym. Photochem.*, **5**, 351 (1984).
2329. Lucki, J., Rabek, J. F., Rånby, B., and Dai, G. S., *Polym. Photochem.*, **5**, 385 (1984).
2330. Lucki, J., Rabek, J. F., Rånby, B., and Ekström, C., *Europ. Polym. J.*, **17**, 919 (1981).
2331. Lucki, J., and Rånby, B., *Polym. Degrad. Stabil.*, **1**, 1 (1979).
2332. Lucki, J., and Rånby, B., *Polym. Degrad. Stabil.*, **1**, 165 (1979).
2333. Lucki, J., and Rånby, B., *Polym. Degrad. Stabil.*, **1**, 251 (1979).
2334. Lucki, J., Rånby, B., and Rabek, J. F., *Europ. Polym. J.*, **15**, 1101 (1979).
2335. Lukáč, I., Chmela, Š., and Hrdlovič, P., *J. Polym. Sci.*, A1, **17**, 2893 (1979).
2336. Lukáč, I., and Hrdlovič, P., *Europ. Polym. J.*, **11**, 767 (1975).
2337. Lukáč, I., and Hrdlovič, P., *Europ. Polym. J.*, **14**, 339 (1978).
2338. Lukáč, I., and Hrdlovič, P., *Europ. Polym. J.*, **15**, 533 (1979).
2339. Lukáč, I., and Hrdlovič, P., *Polym. Photochem.*, **2**, 277 (1982).
2340. Lukáč, I., Hrdlovič, P., Maňasek, Z., and Belluš, D., *Europ. Polym. J.*, Suppl., 1969, p. 523.
2341. Lukáč, I., Hrdlovič, P., Maňásek, Z., and Belluš, D., *J. Polym. Sci.*, A1, **9**, 69 (1971).
2342. Lukáč, I., Moravčik, M., and Hardlovič, P., *J. Polym. Sci.*, A1, **12**, 1913 (1974).
2343. Lukáč, I., Pilka, J., Kuličková, M., and Hrdlovič, P., *J. Polym. Sci.*, A1, **15**, 1645 (1977).
2344. Lukáč, I., Zvara, I., and Hrdlovič, P., *Europ. Polym. J.*, **18**, 427 (1982).
2345. Lukáč, I., Zvara, T., Hrdlovič, P., and Maňásek, Z., *Chem. Zvesti*, **26**, 404 (1972).
2346. Lukas, R., Kolinsky, M., and Doskocilova, D., *J. Polym. Sci.*, A1, **17**, 2691 (1979).
2347. Lunak, S., Veprek-Siska, J., *Z. Naturforsch.*, **36B**, 654 (1981).
2348. Luston, J., Guniš, J., and Maňásek, Z., *J. Macromol. Sci. Chem.*, A, **7**, 587 (1973).
2349. Luston, J., Maňásek, Z., and Samuhelová, Z., *J. Appl. Polym. Sci.*, **24**, 2005 (1979).
2350. Luston, J., Schubertova, N., and Maňásek, Z., *J. Polym. Sci.*, Polym. Symp., No. 40, 33 (1972).
2351. Lyons, A. R., Symons, M. C., and Yandell, J. K., *J. Chem. Soc. Faraday Trans., II*, **1971**, 495.
2352. MacCallum, J. R., in *Developments in Polymer Degradation* (Grassie, N., ed.), Vol. 1, Applied Science Publishers, Ltd, London, 1977, p. 237.
2353. MacCallum, J. R., in *Ann. Rep. Progr. Chem. Sect. A. Phys, Inorg. Chem.*, **75**, 99 (1978).
2354. MacCallum, J. R., *Europ. Polym. J.*, **17**, 209 (1981).
2355. MacCallum, J. R., *Polymer*, **23**, 175 (1982).
2356. MacCallum, J. R., in *Photophysics of Synthetic Polymers* (Phillips, D., and Roberts, A. J., eds), Sciences Reviews Ltd., Northwood, England, 1982, p. 39.
2357. MacCallum, J. R., Hoyle, C. E., and Guillet, J. E., *Macromolecules*, **13**, 1647 (1980).
2358. MacCallum, J. R., El-Sayed, F. E., Pomery, P. J., and Shepherd, M., *J. Chem. Soc., Faraday Trans., II*, **75**, 79 (1979).
2359. MacCallum, J. R., and Gibb, W. H., *Europ. Polym. J.*, **10**, 529 (1974).
2360. MacCallum, J. R., and Rankin, C. T., *Makromol. Chem.*, **175**, 2477 (1974).
2361. MacCallum, J. R., and Rudkin, A. L., *Nature*, **266**, 338 (1977).
2362. MacCallum, J. R., and Rudkin, A. L., *Europ. Polym. J.*, **17**, 953 (1981).
2363. MacCallum, J. R., and Schoff, C. F., *Trans. Faraday Soc.*, **67**, 2372 and 2383 (1971).
2364. MacGregor, E. A., and Greenwood, C. T., *Polymers in Nature*, Wiley, New York, 1980.
2365. Maeda, K., Hayashi, T., and Moringa, M., *Bull. Chem. Soc. Japan*, **37**, 1563 (1964).
2366. Maerov, S. B., *J. Polym. Sci.*, A, **3**, 487 (1965).
2367. Maerov, S. B., Avakian, P., and Matheson, R. R., *J. Appl. Polym. Sci.*, **19**, 2797 (1984).
2368. Maestri, M., Infelta, P. P., and Grätzel, M., *J. Chem. Phys.*, **69**, 1522 (1978).
2369. Maguire, W. J., and Pink, R. C., *Trans. Faraday Soc.*, **63**, 1097 (1967).
2370. Mahadevian, V., and Santappa, M., *Current Sci.*, **29**, 91 (1960).

716

2371. Mahoney, L. R., *Angew. Chem.*, **81,** 555 (1969).
2372. Mahoney, L. R., and Weiner, S. A., *J. Amer. Chem. Soc.*, **94,** 585 (1972).
2373. Mahraj, U., Winnik, M. A., Dors, B., and Schäfer, H. J., *Macromolecules*, **12,** 905 (1979).
2374. Mai, Y. W., and Cotterell, B., *J. Appl. Polym. Sci.*, **27,** 4885 (1982).
2375. Mai, Y. W., Head, D. R., Cotterell, B., and Roberts, B. W., *J. Mater. Sci.*, **15,** 3057 (1980).
2376. Maiti, S., Saha, M. K., and Palit, S. R., *Makromol. Chem.*, **127,** 224 (1969).
2377. Malati, M. A., and Seager, N. J., *J. Oil Colour Chem. Assoc.*, **64,** 231 (1981).
2378. Manson, J. A., Iobst, S. A., and Acosta, R., *J. Polym. Sci.*, A1, **10,** 179 (1972).
2379. Mantovani, E., Mazzei, M., Robertiello, A., and Zanobi, A., *J. Appl. Polym. Sci.*, **21,** 589 (1977).
2380. Mao, T. J., and Eldred, R. J., *J. Polym. Sci.*, A1, **5,** 1741 (1967).
2381. Marchal, J. C., *J. Macromol. Sci. Chem.*, A, **12,** 609 (1978).
2382. Marcotte, F. B., Campbell, D., and Cleaveland, J. A., *J. Polym. Sci.*, A1, **5,** 481 (1967).
2383. Marek, M., *J. Polym. Sci.*, Polym. Symp., No. 56, 149 (1976).
2384. Marek, M., and Toman, L., *J. Polym. Sci.*, Polym. Symp., No. 42, 339 (1973).
2385. Mark, M., Toman, L., and Pilar, J., *J. Polym. Sci.*, A1, **13,** 1565 (1975).
2386. Margerum, J. D., Lackner, A. M., Little, M. J., and Petrusis, C. T., *J. Phys. Chem.*, **75,** 3066 (1971).
2387. Margerum, J. D., Miller, L. J., and Rust, J. B., *Photogr. Sci. Eng.*, **9,** 40 (1969).
2388. Markham, M. C., and Laidler, J., *J. Phys. Chem.*, **57,** 363 (1953).
2389. Marsh, D. G., *J. Polym. Sci.*, A1, **14,** 3013 (1976).
2390. Marsh, D. G., and Heicklen, J., *J. Amer. Chem. Soc.*, **88,** 269 (1966).
2391. Marsh, D. G., Yanus, J. F., and Pearson, J. M., *Macromolecules*, **8,** 427 (1975).
2392. Martic, P. A., Daly, R. C., Williams, J. L. R., and Farid, S. Y., *J. Polym. Sci.*, B, **15,** 295 (1977).
2393. Martic, P. A., Daly, R. C., Williams, J. L. R., and Farid, S. Y., *J. Polym. Sci.*, B, **17,** 305 (1979).
2394. Martin, B. M., in *UV Curing: Science and Technology* (Pappas, S. P., ed.) Vol. 2, Technology Marketing Co., Norwalk, CT, 1984, p. 107.
2395. Martin, J. W., *J. Appl. Polym. Sci.*, **29,** 777 (1984).
2396. Martin, K. G., *Brit. Polym. J.*, **4,** 53 (1972).
2397. Martin, K. G., and Tilley, R. I., *Brit. Polym. J.*, **1,** 213 (1969).
2398. Martin, K. G., and Tilley, R. I., *Brit. Polym. J.*, **3,** 36 (1971).
2399. Martin, K. G., and Kalantar, A. H., *J. Phys. Chem.*, **72,** 2265 (1968).
2400. Martinez-Utrilla, R., Botija, J. M., and Sastre, R., *Polym. Bull.*, **12,** 119 (1984).
2401. Martinez-Utrilla, R., Catalina, F., and Sastre, R., *Polym. Photochem.*, **4,** 361 (1984).
2402. Masuda, T., Kuwane, Y., and Higashimura, T., *J. Polym. Sci.*, A1, **20,** 1043 (1982).
2403. Masuda, T., Kuwane, Y., Yamamoto, K., and Nigashimura, T., *Polym. Bull.*, **2,** 823 (1980).
2404. Masuda, T., Yamamoto, K., and Higashimura, T., *Polymer*, **23,** 1663 (1982).
2405. Masuhara, H., and Mataga, N., *J. Lumin.*, **24/25,** 511 (1981).
2406. Masuhara, H., Ohwada, S., Mataga, N., Itaya, A., Okamoto, K., and Kusabayashi, S., *J. Phys. Chem.*, **84,** 2363 (1980).
2407. Masuhara, H., Ohwada, S., Seki, Y., Mataga, N., Sato, K., and Tazuke, S., *Photochem. Photobiol.*, **32,** 9 (1980).
2408. Masuhara, H., Shioyama, H., Mataga, N., Inoue, T., Kitamura, N., Tanabe, T., and Tazuke, S., *Macromolecules*, **14,** 1738 (1981).
2409. Masuhara, H., Tamai, N., Mataga, N., DeSchryver, F. C., and Van den Driessche, J., *J. Amer. Chem. Soc.*, **105,** 7256 (1983).
2410. Masuhara, H., Tamai, N., Inoue, K., and Mataga, N., *Chem. Phys. Lett.*, **91,** 109 (1982).
2411. Masuhara, H., Tanaka, J. A., Mataga, N., DeSchryver, F. C., and Collart, P., *Polym. J.*, **15,** 915 (1983).
2412. Masuhara, H., Van den Driessche, J., Demyer, K., Boens, N., and DeSchryver, F. C., *Macromolecules*, **15,** 1471 (1982).
2413. Mataga, N., and Murata, Y., *J. Amer. Chem. Soc.*, **91,** 3144 (1969).
2414. Matějka, L., and Dušek, K., *Makromol. Chem.*, **182,** 3223 (1981).
2415. Matějka, L., Dušek, K., and Ilavsky, M., *Polym. Bull.*, **1,** 659 (1979).
2416. Matějka, L., Ilavsky, M., Dušek, K., and Wichterle, O., *Polymer*, **22,** 1511 (1981).
2417. Matheson, I. B. C., and Lee, J., *Chem. Phys. Lett.*, **7,** 475 (1970).
2418. Matheson, I. B. C., and Lee, J., *Photochem. Photobiol.*, **29,** 879 (1979).
2419. Matheson, I. B. C., Lee, J., Yamanashi, B. S., and Wolbarsht, M. L., *J. Amer. Chem. Soc.*, **96,** 3343 (1974).
2420. Mathiowitz, E., Raziel, A., Cohen, M. D., and Fischer, E., *J. Appl. Polym. Sci.*, **26,** 809 (1981).

2421. Mathur, A. B., Kumar, U., and Mathur, G. N., *Polym. Photochem.*, **2**, 161 (1982).
2422. Matisova-Rychla, L., Ambrovic, P., Kulickova, N., and Rychly, J., *J. Polym. Sci.*, Polym. Symp., No. 57, 181 (1976).
2423. Matsumoto, T., Mune, I., and Watanai, S., *J. Polym. Sci.*, A1, **7**, 1609 (1969).
2424. Matsumoto, S., Ohshima, H., and Hasuda, Y., *J. Polym. Sci.*, A1, **22**, 869 (1984).
2425. Matsumoto, K., Sugita, K., Yokoyama, F., and Suzuki, S., *Graphic Arts of Japan*, **1974**, (4), 7.
2426. Matsuo, T., Sakamoto, T., Takuma, K., Sakurai, K., and Ohsako, T., *J. Phys. Chem.*, **85**, 1277 (1982).
2427. Matsuura, T., Nishinaga, N., Yoshimura, N., Arai, T., Okamura, K., and Saito, I., *Tetrahed. Lett.*, **1969**, 1673.
2428. Matsuura, T., and Omura, K., *Chem. Commun.*, **1966**, 127.
2429. Matsuura, T., Omura, K., and Nakashima, R., *Bull. Chem. Soc. Japan*, **28**, 1358 (1965).
2430. Matsuura, T., Yoshimura, N., Nishinaga, A., and Saito, I., *Tetrahed. Lett.*, **1969**, 1669.
2431. Matsuura, T., Yoshimura, N., Nishinaga, A., and Saito, I., *Tetrahed.*, **28**, 4933 (1972).
2432. Matsuzawa, T., and Tomioka, H., *IEEE Trans. Electron Device Lett.*, **EDL-2**, 90 (1981).
2433. Matsuzawa, T., and Tomioka, H., *IEEE Trans. Electron Devices*, **ED-28**, 1306 (1981).
2434. Matsuyama, Y., and Tazuke, S., *Polym. J.*, **8**, 481 (1976).
2435. Mauzerall, D., in *Porphyrins*, (Dolphin, D., ed.), Vol. 5, Academic Press, New York, 1978, p. 53.
2436. Mayo, F. R., *Macromolecules*, **11**, 942 (1978).
2437. Mayo, F. R., Heller, J., Walrath, R. L., and Irwin, K. C., *Rubber Chem. Technol.*, **41**, 289 (1968).
2438. Mayo, F. R., Miller, A. A., and Russell, G. A., *J. Amer. Chem. Soc.*, **80**, 2500 (1958).
2439. Maxim, L. D., and Kuist, C. H., *Off. Dig.*, **36**, 723 (1964).
2440. Mazzocchi, P. H., Danisi, F., and Thomas, J. J., *J. Polym. Sci.*, B, **13**, 737 (1975).
2441. M'Bon, G., Roucox, C., Lablanche-Combier, A., and Loucheux, C., *J. Appl. Polym. Sci.*, **29**, 651 (1984).
2442. McAlpine, E., and Sinclair, R. S., *Textile Res. J.*, **47**, 283 (1977).
2443. McCapra, F., *Quart. Rev. (London)*, **20**, 485 (1966).
2444. McCapra, F., *Chem. Commun.*, **1968**, 155.
2445. McClelland, D. F., *Anal. Chem.*, **55**, 89 (1983).
2446. McCloskey, C. M., and Bond, J., *Ind. Eng. Chem.*, **47**, 2125 (1955).
2447. McClure, D. S., and Hanst, P. L., *J. Chem. Phys.*, **23**, 1722 (1955).
2448. McCombes, D. A., Menon, C. S., and Higgins, J., *J. Polym. Sci.*, A1, **9**, 1261 (1971).
2449. McDonald, J. R., Echols, W. E., Price, T. R., and Fox, R. B., *J. Chem. Phys.*, **57**, 1746 (1972).
2450. McGill, W. J., *J. Appl. Polym. Sci*, **19**, 2781 (1975).
2451. McGill, W. J., and Ackerman, L., *J. Polym. Sci.*, A1, **12**, 1541 (1974).
2452. McGill, W. J., and Ackerman, L., *J. Polym. Sci.*, A1, **12**, 2697 (1974).
2453. McGill, W. J., and Ackerman, L., *J. Appl. Polym. Sci.*, **19**, 2773 (1975).
2454. McGinniss, V. D., *ACS Coating and Plastics Preprints*, **35**, (1), 118 (1975).
2455. McGinniss, V. D., *J. Radiat-Curing*, **2**, Jan., 3 (1975).
2456. McGinniss, V. D., *SME Tech. Pap (Ser.)*, FC76-486 (1976).
2457. McGinniss, V. D., in *Ultraviolet Light Induced Reactions in Polymers*, (Labana, S. S., ed.), ACS Symp. Ser. No. 25, ACS Washington, D.C., 1976, p. 135.
2458. McGinniss, V. D., *Photogr. Sci. Eng.*, **23**, 124 (1979).
2459. McGinniss, V. D., in *Developments in Polymer Photochemistry*, (Allen, N. S., ed.), Vol. 3, Applied Science Publishers, Ltd, London, 1982, p. 1.
2460. McGinness, V. D., and Dusek, D. M., *Polym. Prepr. Amer. Chem. Soc. Div. Polym. Chem.*, **15** (1), 480 (1974).
2461. McGinniss, V. D., and Kay, A., *Polym. Eng. Sci.*, **17**, 478 (1977).
2462. McGinniss, V. D., Prouder, T., Kuo, C., and Gallopo, A., *Macromolecules*, **11**, 393 (1978).
2463. McGinniss, V. D., Prouder, T., Kuo, C., and Gallopo, A., *Macromolecules*, **11**, 405 (1978).
2464. McGinniss, V. D., and Stevenson, D. R., *Polym. Prepr. Amer. Chem. Soc. Div. Polym. Chem.*, **15** (2), 302 (1974).
2465. McInally, I., Reid, R. F., Rutherford, H., and Soutar, I., *Europ. Polym. J.*, **15**, 723 (1979).
2466. McInally, I., Soutar, I., and Steedman, W., *J. Polym. Sci.*, A1, **15**, 2511 (1977).
2467. McKellar, J. F., and Allen, N. S., *Macromol. Rev.*, **13**, 241 (1978).
2468. McKellar, J. F., and Allen, N. S., *Photochemistry of Man-Made Polymers*, Applied Science Publishers, Ltd, 1979.
2469. McLaren, A. D., and Shugar, D., *Photochemistry of Proteins and Nucleic Acids*, Pergamon Press, New York, 1964.

718

2470. McLendon, G., and Miller, D. S., *J. Chem. Soc. Chem. Commun.*, **1980**, 533.
2471. McMillan, G. R., *J. Amer. Chem. Soc.*, **83**, 3018 (1961).
2472. McMillan, G. R., and Wiljan, M. J. H., *Can. J. Chem.*, **36**, 1227 (1958).
2473. McNiell, I. C., Ackerman, L., Gupta, S. N. Zulfiqar, M., and Zulfiqar, S., *J. Polym. Sci.*, A1, **15**, 2381 (1977).
2474. McRae, J. A., and Symons, M. C. R., *J. Chem. Soc.*, B, **1968**, 428.
2475. McVey, J. K., Shold, D. M., and Yang, N. C., *J. Chem. Phys.*, **65**, 3375 (1976).
2476. Meagher, J. F., and Heicklen, J., *J. Photochem.*, **3**, 455 (1975).
2477. Medalia, A. I., Hagopain, E., and Hall, J. P., *J. Polym. Sci.*, **57**, 693 (1962).
2478. Melchore, J. A., *Ind. Eng. Chem. Prod. Res. Dev.*, **1**, 232 (1962).
2479. Mellor, D. C., Moir, A. B., and Scott, G., *Europ. Polym. J.*, **9**, 219 (1973).
2480. Melnikov, M. Ya., Lipskerova, E. M., Mikhalik, O. M., and Fock, N. V., *Polym. Photochem.*, **2**, 133 (1982).
2481. Melville, H. W., Robb, J. C., and Tutton, R. C., *Disc. Faraday Soc.*, **14**, 150 (1953).
2482. Mendelson, M. A., Navish, F. W. Jr., Luck, R. M., and Yeoman, F. A., in *Polymers in Solar Energy Utilization*, (Gebelein, C. G., Williams, D. J. and Deanin, R. D., eds), ACS Symp. Ser., No. 220, ACS Washington, D.C., 1983, p. 39.
2483. Menju, A., Hayashi, K., and Irie, M., *Macromolecules*, **14**, 755 (1981).
2484. Menon, C. C., and Santappa, M., *Can. J. Chem.*, **35**, 1267 (1957).
2485. Menzel, G., *Angew. Makromol. Chem.*, **47**, 181 (1975).
2486. Merkel, P. B., and Kearns, D. R., *J. Amer. Chem. Soc.*, **94**, 1029 (1972).
2487. Merkel, P. B., and Kearns, D. R., *J. Amer. Chem. Soc.*, **94**, 7244 (1972).
2488. Merkel, P. B., Nilsson, R., and Kearns, D. R., *J. Amer. Chem. Soc.*, **94**, 1030 (1972).
2489. Merle, Y., Merle-Aubry, L., and Guillet, J. E., *Macromolecules*, **17**, 288 (1984).
2490. Merle-Aubry, L., Holden, D. A., Merle, Y., and Guillet, J. E., *Macromolecules*, **13**, 1138 (1980).
2491. Merlin, A., and Fouassier, J. P., *J. Polym. Sci.*, B, **17**, 709 (1979).
2492. Merlin, A., and Fouassier, J. P., *Polymer*, **21**, 1363 (1980).
2493. Merlin, A., and Fouassier, J. P., *Angew. Makromol. Chem.*, **86**, 109 (1980).
2494. Merlin, A., and Fouassier, J. P., *J. Polym. Sci.*, A1, **19**, 2357 (1981).
2495. Merlin, A., and Fouassier, J. P., *Makromol. Chem.*, **182**, 3053 (1981).
2496. Merlin, A., and Fouassier, J. P., *Angew. Makromol. Chem.*, **108**, 185 (1982).
2497. Merlin, A., Lougnot, D. J., and Fouassier, J. P., *Polym. Bull.*, **2**, 847 (1980).
2498. Merlin, A., Lougnot, D. J., and Fouassier, J. P., *Europ. Polym. J.*, **17**, 755 (1981).
2499. Merrill, R. G., and Roberts, C. W., *J. Appl. Polym. Sci.*, **21**, 2745 (1977).
2500. Merrill, S. H., and Unruh, C. C., *J. Appl. Polym. Sci.*, **7**, 273 (1963).
2501. Merrit, C., Scott, G. W., Gupta, A., and Yavrouian, A., *Chem. Phys., Lett.*, **69**, 169 (1980).
2502. Metz, P. D., Teoh, H., Vanderhart, D. L., and Wilhelm, W. G., in *Polymers in Solar Energy Utilization* (Gebelein, C. G., Williams, D. J., and Deanin, R. J., eds) ACS Symp. Ser., No. 220, ACS Washington, D.C., 1983, p. 421.
2503. Mezger, T., and Cantow, H. J., *Makromol. Chem. Rapid Commun.*, **4**, 313 (1983).
2504. Mezger, T., and Cantow, H. J., *Polym. Photochem.*, **5**, 49 (1984).
2505. Meybeck, A., and Meybeck, J., *Photochem. Photobiol.*, **6**, 355 (1967).
2506. Meyer, J. W., and Hammond, G. S., *J. Amer. Chem. Soc.*, **94**, 2219 (1972).
2507. Meyer, W., Lieser, G., and Wegner, G., *Makromol. Chem.*, **178**, 631 (1977).
2508. Meyer, W., Lieser, G., and Wegner, G., *J. Polym. Sci.*, A1, **16**, 1365 (1978).
2509. Mijovic, M. V., Beynon, P. J., Shaw, T. J., Petrak, K., Reiser, A., Roberts, A. J., and Phillips, D., *Macromolecules*, **15**, 1464 (1982).
2510. Mikeš, F., Morawetz, H., and Dennis, K. S., *Macromolecules*, **13**, 969 (1980).
2511. Mikeš, F., Morawetz, H., and Dennis, K. S., *Macromolecules*, **17**, 60 (1984).
2512. Mikeš, F., Strop, P., and Kala, J., *Makromol. Chem.*, **175**, 2375 (1974).
2513. Mikhalik, M. Ya., Seropegina, Ya, N., Melnikov, M. Ya., and Fock, N. V., *Europ. Polym. J.*, **17**, 1011 (1981).
2514. Mikhelev, Yu. A., Guseva, L. N., Rogova, L. S., and Toptygin, D. Ya., *Polym. Photochem.*, **2**, 457 (1982).
2515. Milinchuk, V. K., *Vysokomol. Soedin.*, **7**, 1293 (1965).
2516. Milinchuk, V. K., and Klinshpont, E. R., *J. Polym. Sci., Polym. Symp.*, No. 40, 1 (1973).
2517. Mill, T., Irvin, K. C., and Mayo, F. R., *Rubber Chem. Technol.*, **41**, 296 (1976).
2518. Mill, T., Richardson, H., and Mayo, F. R., *J. Polym. Sci.*, A1, **11**, 2899 (1973).
2519. Miller, L. J., Margerum, J. D., Rust, J. B., Brault, R. G., and Lackner, A. M., *Macromolecules*, **7**, 179 (1974).

2520. Miller, L. J., and Nort, A. M., *J. Chem. Soc. Faraday*, II, **71**, 1233 (1975).
2521. Miller, M. A., and Wegner, G., *Makromol. Chem.*, **185**, 1727 (1984).
2522. Milligan, B., and Holt, L. A., *Polym. Degrad. Stabil.*, **5**, 339 (1983).
2523. Milligan, B., and Tucker, D. J., *Text. Res. J.*, **32**, 634 (1962).
2524. Milosavljevic, B. H., and Thomas, J. K., *Macromolecules*, **17**, 2244 (1984).
2525. Mimura, T., and Itoh, M., *J. Amer. Chem. Soc.*, **98**, 1095 (1976).
2526. Miniutti, V. P., *Forest Prod. J.*, **14** 571 (1964).
2527. Miniutti, V. P., *J. Paint Technol.*, **45**, 27 (1973).
2528. Minsk, L. M., and Van Deusen, W. P. (Eastman Kodak Co.,) US Pat. 2,690,906 (1954).
2529. Minsk, L. M., Smith, J. C., Van Deusen, W. P., and Wright, J. R., *J. Appl. Polym. Sci.*, **2**, 302 (1959).
2530. Mita, I., Takagi, T., Horie, K., and Shindo, Y., *Macromolecules*, **17**, 2256 (1984).
2531. Mitchell, J., and Perkins, L. R., *Adv. Polym. Symp.*, **4**, 167 (1967).
2532. Miura, M., Hayashi, T., Akutsu, F., and Nagakubo, K., *Polymer*, **19**, 348 (1978).
2533. Miura, M., Kitami, T., and Nagakubo, K., *J. Polym. Sci.*, B, **6**, 463 (1968).
2534. Miyama, H., Fuji, N., Shimazaki, Y., and Ikeda, K., *Polym. Photochem.*, **3**, 445 (1983).
2535. Miyama, H., Harumiya, N., Mori, Y., and Tanazawa, H., *J. Biomed. Mater. Res.*, **11**, 251 (1977).
2536. Miyama, H., Harumiya, H., and Takeda, A., *J. Polym. Sci.*, A1, **10**, 943 (1972).
2537. Mizutani, Y., and Kusumoto, K., *J. Appl. Polym. Sci.*, **19**, 713 (1975).
2538. Mizutani, Y., Kusumoto, K., and Hisano, S., *Bull. Chem. Soc. Japan*, **44**, 1134 (1971).
2539. Mizutano, T., Hattori, S., and Tawata, M., *J. Opt. Soc. Amer.*, **67**, 1651 (1977).
2540. Mlezina, J., and Cermak, V., *Farbe Lack*, **79**, 1153 (1973).
2541. Mlinac, M., Rolich, J., and Bravar, M., *J. Polym. Sci.*, Polym. Symp., No. 57, 161 (1976).
2542. Moad, G., Rizzardo, G., and Solomon, D. H., *Tetrahed. Lett.*, **1981**, 1165.
2543. Moad, G., Rizzardo, E., and Solomon, D. H., *J. Macromol. Sci. Chem.*, A, **17**, 51 (1982).
2544. Moan, J., and Wold, E., *Nature*, **279**, 450 (1979).
2545. Mochel, W. E., Crandall, J. L., and Peterson, J. H., *J. Amer. Chem. Soc.*, **77**, 494 (1955).
2546. Molaire, M. F., *J. Polym. Sci.*, A1, **20**, 847 (1982).
2547. Mondong, R., and Bassler, H., *Chem. Phys. Lett.*, **78**, 371 (1981).
2548. Monitz, W. B., Sojka, S. A., Poranski, C. F. Jr., and Birkle, D. L., *J. Amer. Chem. Soc.*, **100**, 7940 (1978).
2549. Monnerie, L., in *Photophysics of Synthetic Polymers* (Phillips, D., and Roberts, A. J., eds), Science Reviews Ltd, Northwood, England, 1982, p. 70.
2550. Monroe, B. M., *Adv. Photochem.*, **8**, 77 (1970).
2551. Monroe, B. M., *J. Phys. Chem.*, **81**, 1861 (1977).
2552. Monroe, B. M., *J. Phys. Chem.*, **82**, 15 (1978).
2553. Monroe, B. M., and Groff, R. P., *Tetrahed. Lett.*, **1973**, 3955.
2554. Monroe, B. M., and Mrowca, J. J., *J. Phys. Chem.*, **83**, 591 (1979).
2555. Monroe, B. M., and Weiner, S. A., *J. Amer. Chem. Soc.*, **91**, 450 (1969).
2556. Monroe, B. M., Weiner, S. A., and Hammond, G. S., *J. Amer. Chem. Soc.*, **90**, 1913 (1968).
2557. Montagnoli, G., Pieroni, O., and Suzuki, S., *Polym. Photochem.*, **3**, 279 (1983).
2558. Moore, J. E., in *Photodegradation and Photostabilization of Coatings,* (Pappas, S. P., and Winslow, F. H., eds), ACS Symp. Ser., No. 151, ACS Washington, D.C., 1981, p. 97.
2559. Moore, R. F., *Polymer*, **4**, 493 (1963).
2560. Moore, W. M., Hammond, G. S., and Fass, R. P., *J. Amer. Chem. Soc.*, **83**, 2789 (1961).
2561. Morand, M. J., *Rubber, Chem. Technol.*, **39**, 537 (1966).
2562. Morawetz, H., *Science*, **203**, 405 (1979).
2563. Morawetz, H., *Ann. N.Y. Acad. Sci.*, **366**, 404 (1981).
2564. Morawetz, H., *Polym. Eng. Sci.*, **23**, 689 (1983).
2565. Morawetz, H., and Amrani, F., *Macromolecules*, **11**, 281 (1978).
2566. Morgan, C. R., *J. Radiat. Curing*, **1**, March, 11 (1974).
2567. Morgan, C. R., *J. Radiat,. Curing*, **10**, Oct., 8 (1983).
2568. Morgan, C. R., and Ketley, A. D., *J. Radiat. Curing*, **7**, Apr., 10 (1980).
2569. Morgan, C. R., and Kyle, D. R., *J. Radiat. Curing*, **10**, Oct., 4 (1983).
2570. Morgan, C. R., Magnotta, F., and Ketley, A. D., *J. Polym. Sci.*, A1, **15**, 627 (1977).
2571. Mori, K., Harada, H., and Nakamura, T., *J. Polym. Sci.*, A1, **12**, 2497 (1974).
2572. Mori, F., Koyama, M., and Oki, Y., *Angew. Makromol. Chem.*, **64**, 89 (1977).
2573. Mori, F., Koyama, M., and Oki, Y., *Angew. Makromol. Chem.*, **68**, 137 (1978).
2574. Mori, F., Koyama, M., and Oki, Y., *Angew. Makromol. Chem.*, **75**, 113 (1979).

2575. Mori, F., Koyama, M., and Oki, Y., *Angew. Makromol. Chem.*, **75**, 123 (1979).
2576. Mori, F., Koyama, M., and Oki, Y., *Angew. Makromol. Chem.*, **75**, 223 (1979).
2577. Morichaud, J. P., *Rev. Gen. Cautchouc*, **43**, 1615 (1966).
2578. Morison, J. D., and Nicholson, A. J. G., *J. Chem. Phys.*, **20**, 1021 (1952).
2579. Morito, K., and Suzuki, S., *J. Appl. Polym. Sci.*, **16**, 2947 (1972).
2580. Moroi, Y., Braun, A. M., and Grätzel, M., *J. Amer. Chem. Soc.*, **101**, 567 (1979).
2581. Moroi, Y., Infelta, P. P., and Grätzel, M., *J. Amer. Chem. Soc.*, **101**, 573 (1979).
2582. Morsi, S. E., Khalifa, W. M., Zaki, A., Al-Sayed, M. A., and Etaiw, S. H., *Polymer*, **22**, 942 (1981).
2583. Mort, J., and Pfister, G., *Electronic Properties of Polymers*, Wiley, New York, 1982.
2584. Mortillaro, L., *Mater. Plast. Elastom.*, **8**, 836 (1972).
2585. Morton, M., *J. Chem. Educ.*, **50**, 740 (1973).
2586. Moustafa, A. B., and Sakr, M., *J. Polym. Sci.*, A1, **14**, 2327 (1976).
2587. Movaghar, J. N., Birks, J. B., and Nagvi, K. R., *Proc. Phys. Soc. London*, **91**, 449 (1967).
2588. Moyer, J. D., and Ochs, R. J., *Science*, **142**, 1316 (1963).
2589. Mukherjee, A. K., and Goel, H. R., *J. Macromol. Sci. Chem.*, A, **17**, 545 (1982).
2590. Mukherjee, A. K., Sachdev, H. S., and Gupta, A., *J. Appl. Polym. Sci.*, **28**, 2125, and 2217 (1983).
2591. Mullen, P. A., and Searle, N. Z., *J. Appl. Polym. Sci.*, **13**, 2183 (1967).
2592. Mullen, P. A., and Searle, N. Z., *J. Appl. Polym. Sci.*, **14**, 765 (1970).
2593. Muller, M. A., and Wegner, G., *Makromol. Chem.*, **185**, 1727 (1984).
2594. Mungall, W. S., Martin, C. L., and Borgeson, G. C., *Macromolecules*, **8**, 934 (1875).
2595. Munk, A. U., and Scott, J. F., *Nature*, **177**, 587 (1956).
2596. Munro, H. S., Clark, D. T., and Peeling, J., *Polym. Degrad. Stabil.*, **9**, 185 (1984).
2597. Murahashi, S., Nozakura, S., Umhera, A., and Obala, K., *Kobunshi Kagaku*, **21**, 625 (1964).
2598. Murayama, K., Morimura, S., and Yoshika, T., *Bull. Chem. Soc. Japan*, **42**, 1640 (1969).
2599. Murayama, K., and Yoshioka, T., *Bull. Chem. Soc. Japan*, **42**, 1942 (1969).
2600. Murov, S. L., *Handbook of Photochemistry*, Dekker, New York, 1973.
2601. Murov, S., and Hammond, G. S., *J. Phys. Chem.*, **72**, 3797 (1968).
2602. Murthy, C. P., Sethuram, B., and Rao, T. N., *J. Macromol. Sci. Chem.*, A, **21**, 739 (1984).
2603. Musa, Y., and Stevens, M. P., *J. Polym. Sci.*, A1, **10**, 319 (1972).
2604. Mustroph, H., Epperlein, J., and Pietsch, H., *Photogr. Sci. Eng.*, **24**, 196 (1980).
2605. Nadolski, B., Uznanski, P., and Kryszewski, M., *Makromol. Chem. Rapid Commun.*, **5**, 327 (1984).
2606. Nagaro, I., Nishihara, K., and Sakota, N., *J. Polym. Sci.*, A1, **12**, 785 (1974).
2607. Nagata, I., Li, R., Banks, E., and Okamoto, Y., *Macromolecules*, **16**, 903 (1983).
2608. Nagata, I., and Morawetz, H., *Macromolecules*, **14**, 87 (1981).
2609. Nagata, I., and Okamoto, Y., *Macromolecules*, **16**, 749 (1983).
2610. Naito, I., Imamura, K., Shintomi, H., Okamura, K., and Kinoshita, A., *Polym. Photochem.*, **4**, 149 (1984).
2611. Naito, I., Koga, K., Kinoshita, A., and Schnabel, W., *Europ. Polym. J.*, **16**, 109 (1980).
2612. Naito, I., Kuhlmann, R., and Schnabel, W., *Polymer*, **20**, 165 (1979).
2613. Nakahira, T., Ishizuka, S., Iwabuchi, S., and Kojima, K., *Makromol. Chem. Rapid Commun.*, **1**, 437 (1980).
2614. Nakahira, T., Ishizuka, S., Iwabuchi, S., and Kojima, K., *Macromolecules*, **15**, 1217 (1982).
2615. Nakahira, T., Ishizuma, S., Iwabuchi, S., and Kojima, K., *Macromolecules*, **16**, 297 (1983).
2616. Nakahira, T., Maruyama, I., Iwabuchi, S., and Kojima, K., *Makromol. Chem.*, **180**, 1853 (1979).
2617. Nakahira, T., Minami, C., Iwabuchi, S., and Kojima, K., *Makromol. Chem.*, **180**, 2245 (1979).
2618. Nakahira, T., Sakuma, T., Iwabuchi, S., and Kojima, K., *J. Polym. Sci.*, A1, **20**, 1863 (1982).
2619. Nakahira, T., Sasaoka, T., Iwabuchi, S., and Kojima, K., *Makromol. Chem.*, **183**, 1239 (1982).
2620. Nakahira, T., Shinomiya, T., Fukimoto, T., Iwabuchi, S., and Kojima, K., *Europ. Polym. J.*, **14**, 317 (1978).
2621. Nakahira, T., Yuasa, S., Tanihira, K., Yokoyama, I., Iwabuchi, S., and Kojima, K., *Makromol. Chem. Rapid Commun.*, **5**, 483 (1984).
2622. Nakai, T., and Okawara, M., *Makromol. Chem.*, **108**, 95 (1967).
2623. Nakamoto, Y., Karube, T., Terewaki, S., and Suzuki, S., *J. Solid-Phase Biochem.*, **1**, 143 (1976).
2624. Nakamura, K., and Kikuchi, S., *J. Chem. Soc. Japan*, **87**, 930 (1966).
2625. Nakamura, K., and Kikuchi, B., *Bull. Chem. Soc. Japan*, **40**, 2684 (1967).

2626. Nakamura, K., Sakata, T., and Kikuchi, S., *Bull. Chem. Soc. Japan*, **41**, 1765 (1968).
2627. Nakane, H., Yokota, A., Yamamoto, S., and Kanai, W., *Polym. Eng. Sci.*, **23**, 1050 (1983).
2628. Nakanishi, F., Arai, M., and Nakanishi, H., *J. Polym. Sci.*, A1, **22**, 2095 (1984).
2629. Nakanishi, F., and Hasegawa, M., *J. Polym. Sci.*, A1, **8**, 2151 (1970).
2630. Nakanishi, F., Nakanishi, H., Hasegawa, M., and Yamada, Y. J., *J. Polym. Sci.*, A1, **13**, 2499 (1975).
2631. Nakanishi, F., Nakanishi, H., Kato, M., Tawata, M., and Hattori, S., *J. Appl. Polym. Sci.*, **26**, 3505 (1981).
2632. Nakanishi, H., Hasegawa, M., and Sasada, Y., *J. Polym. Sci.*, A2, **10**, 1537 (1972).
2633. Nakanishi, H., Hasegawa, M., and Sasada, Y., *J. Polym. Sci.*, A2, **15**, 173 (1977).
2634. Nakanishi, H., Hasegawa, M., and Sasada, Y., *J. Polym. Sci.*, B, **17**, 459 (1979).
2635. Nakanishi, H., Hasegawa, M., and Yurugi, T., *J. Polym. Sci.*, A1, **14**, 2079 (1976).
2636. Nakanishi, H., Jones, W., and Parkinson, G. M., *Acta Crystallogr.*, B, **35**, 3102 (1979).
2637. Nakanishi, H., Jones, W., Thomas, J. M., Hasegawa, M., and Lees, W. L., *Proc. Roy. Soc. London*, A, **396**, 307 (1980).
2638. Nakanishi, H., Nakanishi, F., Suzuki, Y., and Hasegawa, M., *J. Polym. Sci.*, A1, **11**, 2503 (1973).
2639. Nakanishi, H., Nakano, N., and Hasegawa, M., *J. Polym. Sci.*, B, **8**, 755 (1970).
2640. Nakanishi, H., Parkinson, G. M., Jones, W., Thomas, J. M., and Hasegawa, M., *Israel J. Chem.*, **18**, 261 (1979).
2641. Nakanishi, H., and Sasada, Y., *Acta Crystallogr.*, B, **34**, 332 (1978).
2642. Nakanishi, H., Suzuki, M., and Nakanishi, F., *J. Polym. Sci.*, B, **20**, 653 (1982).
2643. Nakanishi, H., Suzuki, Y., Suzuki, F., and Hasegawa, M., *J. Polym. Sci.*, A1, **7**, 753 (1969).
2644. Nakanishi, H., Ueno, K., Hasegawa, M., and Sasada, Y., *Chem. Lett.*, **1972**, 301.
2645. Nakanishi, H., Ueno, K., Hasegawa, M., Sasada, Y., and Yurugi, T., *Amer. Chem. Soc. Div. Org. Coat. Plast. Pap.*, **35**(1), 75 (1975).
2646. Nakanishi, H., Ueno, K., and Sasada, Y., *Acta Crystallogr.*, B, **32**, 3352 (1976).
2647. Nakanishi, H., Ueno, K., and Sasada, Y., *Acta Crystallogr.*, B, **34**, 2036 (1978).
2648. Nakanishi, H., Ueno, K., and Sasada, Y., *Acta Crystallogr.*, B, **34**, 2209 (1978).
2649. Nakanishi, H., Ueno, K., and Y., *J. Polym. Sci.*, A1, **16**, 767 (1978).
2650. Nakasaki, M., *J. Chem. Soc. Japan, Pure Chem. Sect.*, **74**, 403, 405, 518 (1953).
2651. Nakashima, Y., and Fukuda, H., *J. Polym. Sci.*, A1, **17**, 245 (1979).
2652. Nakata, T., Tokumaru, K., and Simamura, O., *Tetrahed. Lett.*, **1967**, 3303.
2653. Nal, E. H. S., and Smets, G., *Polym. Photochem.*, **5**, 93 (1984).
2654. Namasivayam, C., and Natarajan, P., *J. Polym. Sci.*, A1, **21**, 1371 (1983).
2655. Namasivayam, C., and Natarajan, P., *J. Polym. Sci.*, A1, **21**, 1385 (1983).
2656. Nanassy, A. J., and Desai, R. L., *J. Appl. Polym. Sci.*, **15**, 2245 (1971).
2657. Natarajan, L. V., and Santappa, M., *J. Polym. Sci.*, B, **5**, 357 (1967).
2658. Natarajan, L. V., and Santappa, M., *J. Polym. Sci.*, A1, **6**, 3245 (1968).
2659. Natarajan, P., Chandrasekaran, K., and Santappa, M., *J. Polym. Sci.*, A1, **14**, 455 (1976).
2660. Natsume, T., Nishimura, M., Fujimatsu, M., Shimizu, M., Shirota, Y., Hirata, S., Kusabayshi, S., and Mikawa, H., *Polym. J.* (*Japan*), **1**, 181 (1970).
2661. Navon, G., and Sutin, N., *Inorg. Chem.*, **13**, 2159 (1974).
2662. Nayak, P. L., Lenka, S., and Mishra, M. K., *J. Polym. Sci.*, **19**, 2457 (1981).
2663. Neckers, D. C., *Tetrahed. Lett.*, **1965**, 1889.
2664. Neckers, D. C., *Mechanistic Organic Photochemistry*, Reinhold, New York, 1967.
2665. Neckers, D. C., *Chem. Technol.*, **8**, 108 (1978).
2666. Neckers, D. C., *J. Radiat. Curing*, **10**, Apr., 19 (1983).
2667. Neckers, D. C., and Abu-Abdoun, I. I., *Macromolecules*, **17**, 2468 (1984).
2668. Needles, H. L., *J. Polym. Sci.*, B, **5**, 595 (1967).
2669. Needles, H. L., *J. Appl. Polym. Sci.*, **12**, 1557 (1968).
2670. Needles, H. L., and Alger, K. W., *J. Appl. Polym. Sci.*, **19**, 2207 (1975).
2671. Neely, W. C., and Dearman, H. H., *J. Chem. Phys.*, **44**, 1302 (1966).
2672. Negishi, A., and Ogiwara, Y., *J. Appl. Polym. Sci.*, **25**, 1095 (1980).
2673. Negishi, A., Ogiwara, Y., and Ozawa, Z., *J. Appl. Polym. Sci.*, **23**, 2953 (1978).
2674. Negishi, N., Ishihara, K., and Shinohara, I., *J. Polym. Sci.*, B, **19**, 593 (1981).
2675. Negishi, N., Ishihara, K., Shinohara, I., Okano, M., Kataoka, K., and Sakurai, Y., *Makromol. Chem. Rapid Commun.*, **2**, 95 (1981).
2676. Nielson, R. D. M., Soutar, I., and Steedman, N., *J. Polym. Sci.*, A2, **15** 617 (1977).

722

2677. Neiman, M. B., and Rosantsev, E., *Bull. Acad. Sci. SSSR, Chem. Ser.*, **1964**, 1095.
2678. Nemzek, T. L., and Guillet, J. E., *Macromolecules*, **10**, 94 (1977).
2679. Nenkov, G., Stoyanov, A., Bagdansaliev, T., and Kabaivanov, V., *Europ. Polym. J.*, **20**, 19 (1984).
2680. Nethsinghe, L. P., and Scott, G., *Europ. Polym. J.*, **20**, 213 (1984).
2681. Neugebauer, W., Enderman, F., and Reichel, M. K., U.S. Pat. 3,201,239 (17 Aug. 1965).
2682. Newland, G. C., and Tamblyn, J. W., *J. Appl. Polym. Sci.*, **8**, 1949 (1964).
2683. Ng, D., and Guillet, J. E., *Macromolecules*, **14**, 405 (1981).
2684. Ng, D., Guillet, J. E., *Macromolecules*, **15**, 724 (1982).
2685. Ng, D., and Guillet, J. E., *Macromolecules*, **15**, 728 (1982).
2686. Ng, D., Yoshiki, K., and Guillet, J. E., *Macromolecules*, **16**, 568 (1983).
2687. Ng, H. C., and Guillet, J. E., *Macromolecules*, **10**, 866 (1977).
2688. Ng, H. C., and Guillet, J. E., in *Singlet Oxygen: Reactions with Organic Compounds and Polymers* (Rånby, B., and Rabek, J. F., eds), Wiley, Chichester, 1978, p. 278.
2689. Ng, H. C., and Guillet, J. E., *Macromolecules*, **11**, 929 (1978).
2690. Ng, H. C., and Guillet, J. E., *Macromolecules*, **11**, 937 (1978).
2691. Ng, H. C., and Guillet, J. E., *Photochem. Photobiol.*, **28**, 571 (1978).
2692. Nichols, C. H., in *Developments in Polymer Photochemistry* (Allen, N. S., ed.), Vol. 1, Applied Science Publishers, Ltd, London, 1980, p. 125.
2693. Nichols, C. H., and Pailthorpe, M. T., *J. Text. Inst.*, **67**, 397 (1976).
2694. Nicolaus, W., Hesse, A., and Schulz, D., *Plastverarbeiter*, **31**, 723 (1980).
2695. Nielson, R. D. M., Soutar, I., and Steedman, W., *Macromolecules*, **10**, 1193 (1977).
2696. Niessner, C. T., *Radiat. Curing*, **3**, Aug, 25 (1981).
2697. Nijst, G., and Smets, G., *Makromol. Chem. Rapid Commun.*, **2**, 481 (1981).
2698. Nishihara, K., Suzuoka, A., and Sakota, N., *J. Polym. Sci.*, A1, **11**, 2315 (1973).
2699. Nishihara, T., and Kaneko, M., *Makromol. Chem.*, **124**, 84 (1969).
2700. Nishii, M., and Hayashi, Y., *Ann. Rev. Materials Sci.*, **5**, 135 (1975).
2701. Nishijima, A., Teramoto, A., Yamamoto, M., and Hiratsuka, S., *J. Polym. Sci.*, A2, **5**, 23 (1967).
2702. Nishijima, T., Nagamura, T., and Matsuo, T., *J. Polym. Sci.*, B, **19**, 65 (1981).
2703. Nishijima, Y., *J. Polym. Sci.*, Polym. Symp., No. 31, 352 (1970).
2704. Nishijima, Y., *J. Macromol. Sci. Phys.*, B, **8**, 389 (1973).
2705. Nishijima, Y., Mitani, K., Katayama, S., and Yamamoto, M., *Rep. Progr. Polym. Phys. Japan*, **13**, 421 (1970).
2706. Nishijima, Y., Sasaki, Y., Tsujisaki, M., and Yamamoto, M., *Rep. Progr. Polym. Phys. Japan*, **15**, 453 (1972).
2707. Nishijima, Y., Takedono, S., and Oku, S., *Rep. Progr. Polym. Phys. Japan*, **9**, 497 (1966).
2708. Nishijima, Y., and Yamamoto, M., *Polym. Preprints, Amer. Chem. Soc., Div. Polym. Chem.*, **20**(1), 391 (1979).
2709. Nishijima, Y., Yamamoto, M., Katayama, S., Hirota, K., Sasaki, Y., and Tsujisaki, M., *Rep. Progr. Polym. Phys. Japan*, **15**, 445 (1972).
2710. Nishikubo, T., Ichijo, T., and Tanaka, T., *J. Appl. Polym. Sci.*, **18**, 2009 (1974).
2711. Nishikubo, T., Iizawa, T., and Hasegawa, M., *J. Polym. Sci.*, B, **19**, 113 (1981).
2712. Nishikubo, T., Iizawa, T., and Saito, Y., *J. Polym. Sci.*, A1, **21**, 2291 (1983).
2713. Nishikubo, T., Iizawa, T., Takahashi, E., and Udagawa, A., *Makromol. Chem. Rapid Commun.*, **5**, 131 (1984).
2714. Nishikubo, T., Iizawa, T., Taneichi, H., and Hasegawa, M., *Polym. Prepr. Polym. Soc. Japan*, **29**(2), 214 (1980).
2715. Nishikubo, T., Iizawa, T., and Yamada, M., *J. Polym. Sci.*, B, **19**, 177 (1981).
2716. Nishikubo, T., Iizawa, T., Yamada, M., and Tsuchiya, K., *J. Polym. Sci.*, A1, **21**, 2025 (1983).
2717. Nishikubo, T., Iizawa, T., Yoshinaga, A., and Nitta, M., *Makromol. Chem.*, **183**, 789 (1982).
2718. Nishimoto, S., Izukawa, T., Haruta, Y., and Kagiya, T., *J. Polym. Sci.*, B, **22**, 199 (1984).
2719. Nishimoto, S., Izukawa, T., Haruta, Y., and Kagiya, T., *J. Polym. Sci.*, B., **22**, 323 (1984).
2720. Nishimoto, S., Yamamoto, K., and Kagiya, T., *Macromolecules*, **15**, 720 (1982).
2721. Nishimoto, S., Yamamoto, K., and Kagiya, T., *Macromolecules*, **15**, 1180 (1982).
2722. Nishimoto, S., Yamamoto, K., and Kagiya, T., *Polym. Photochem.*, **5**, 231 (1984).
2723. Nishimura, O., and Osawa, Z., *Polym. Photochem.*, **1**, 191 (1981).
2724. Nito, K., Suzuki, S., Miyasaka, K., and Ishikawa, K., *Polym. Prep. Japan*, **21**, 171 (1980).
2725. Nito, K., Suzuki, S., Miyasaka, K., and Ishikawa, K., *J. Appl. Polym. Sci.*, **27**, 637 (1982).
2726. Nitzan, A., Jortner, J., and Rentzepis, P. M., *Chem. Phys. Lett.*, **8**, 445 (1971).

723

2727. Niume, K., Toyofuku, K., Toda, F., Uno, K., Hasegawa, M., and Iwakura, Y., *J. Polym. Sci.*, A1, **20,** 663 (1982).
2728. Nobbs, J. H., Bower, D. I., and Ward, I. M., *J. Polym. Sci.*, A1, **17,** 259 (1979).
2729. Nomura, H., and Miyahara, Y., *Polym. J.*, **8,** 30 (1976).
2730. Nonogaki, S., *Polym. Eng. Sci.*, **23,** 985 (1983).
2731. Noomen, A., *J. Oil Color Chem. Assoc.*, **64,** 347 (1981).
2732. Norrish, R. G. W., and Searby, M. H., *Proc. Roy Soc. (London)*, A, **237,** 464 (1956).
2733. North, A. M., *Brit. Polym. J.*, **7,** 119 (1975).
2734. North, A. M., and Ross, D. A., *J. Polym. Sci.*, Polym. Symp., No. 55, 259 (1976).
2735. North, A. M., and Soutar, I., *J. Chem. Soc. Faraday Trans.*, **68,** 1101 (1972).
2736. North, D. A., Harrop, R., Phillips, G. O., and Baugh, P. J., *Polym. Photochem.*, **2,** 309 (1982).
2737. Nowakowska, M., in *Singlet Oxygen: Reactions with Organic Compounds and Polymers* (Rånby, B., and Rabek, J. F., eds), Wiley, Chichester, 1978, p. 254.
2738. Nowakowska, M., *Makromol. Chem.*, **179,** 2953 and 2959 (1978).
2739. Nowakowska, M., *Makromol. Chem.*, **181,** 1013 and 1021 (1980).
2740. Nowakowska, M., *Poly. Photochem.*, **3,** 243 (1983).
2741. Nowakowska, M., *Polym. Photochem.*, **4,** 307 (1984).
2742. Nowakowska, M., *Polym. Photochem.*, **4,** 347 (1984).
2743. Nowakowska, M., *Polym. Degrad. Stabil.*, **7,** 205 (1984).
2744. Nowakowska, M., and Kowal, J., *Polymer*, **21,** 545 (1980).
2745. Nowakowska, M., and Kowal, J., *Europ. Polym. J.*, **16,** 615 (1980).
2746. Nowakowska, M., Kowal, J., and Waligora, B., *Polymer*, **19,** 1317 (1978).
2747. Nowakowska, M., Najbar, J., and Waligora, B., *Europ. Polym. J.*, **12,** 387 (1975).
2748. Obata, N., and Moritani, I., *Tetrahed. Lett.*, **1966,** 1503
2749. O'Brien, D. F., Whitesides, T. H., and Klingbiel, R. T., *J. Polym. Sci.*, A1, **19,** 95 (1981).
2750. O'Brien, J. P., Ferara, W. T., and Allock, H. R., *Macromolecules*, **12,** 108 (1979).
2751. O'Conell, E. J., *J. Amer. Chem. Soc.*, **90,** 6550 (1968).
2752. Odani, H., *Bull. Inst. Chem. Res. Kyoto Univ., Japan*, **51,** 351 (1973).
2753. Ogata, N., and Azuma, C., *J. Polym. Sci.*, A1, **9,** 1759 (1971).
2754. Ogata, N., Azuma, C., and Kurukami, S., *J. Polym. Sci.*, B, **17,** 1849 (1979).
2755. Ogata, N., Azuma, C., and Saito, K., *J. Polym. Sci.*, A1, **11,** 3029 (1973).
2756. Ogata, N., Sanui, K., and Tanaka, H., *Polym. Prepr. Amer. Chem. Soc., Div., Polym. Chem.*, **29**(2), 218 (1980).
2757. Ogata, Y., and Sawaki, Y., *J. Org. Chem.*, **41,** 373 (1976).
2758. Ogata, Y., Tomizawa, K., and Furuta, K., in *Chemistry of Functional Groups Peroxides*, (Patai, S., ed.), Wiley, Chichester, 1983, p. 712.
2759. Ogiwara, Y., Hayase, K., and Kubota, H., *J. Appl. Polym. Sci.*, **23,** 1 (1979).
2760. Ogiwara, Y., Hon, N. S., and Kubota, H., *J. Appl. Polym. Sci.*, **18,** 2057 (1974).
2761. Ogiwara, Y., Kanda, M., Takumi, M., and Kubota, H., *J. Polym. Sci.*, B, **19,** 457 (1981).
2762. Ogiwara, Y., Kimura, Y., Osawa, Z., and Kubota, H., *J. Polym. Sci.*, **15,** 1667 (1977).
2763. Ogiwara, Y., and Kubota, H., *J. Polym. Sci.*, A1, **6,** 3119 (1968).
2764. Ogiwara, Y., and Kubota, H., *J. Polym. Sci.*, A1, **9,** 2549 (1971).
2765. Ogiwara, Y., Kubota, H., and Kimura, Y., *J. Polym. Sci.*, A1, **16,** 2865 (1978).
2766. Ogiwara, Y., Kubota, H., and Ogiwara, Y., *J. Polym. Sci.*, A1, **6,** 3119 (1968).
2767. Ogiwara, Y., Takumi, M., and Kubota, H., *J. Appl. Polym. Sci.*, **27,** 3743 (1982).
2768. Ogiwara, Y., Torikoshi, K., and Kubota, H., *J. Polym. Sci.*, B, **20,** 17 (1982).
2769. Ohkatsu, Y., Haruna, T., and Osa, T., *J. Macromol. Sci. Chem.*, **11,** 1975 (1977).
2770. Ohkita, K., Shimoura, M., and Tsujita, T., *Carbon*, **16,** 156 (1976).
2771. Ohkita, K., Tsubokawa, N., and Saitoh, E., *Carbon*, **16,** 41 (1978).
2772. Ohkita, K., Tsubokawa, N., Saitoh, E., Noda, E., and Takashina, N., *Carbon*, **13,** 443 (1975).
2773. Ohno, A., Kito, N., and Kawase, N., *J. Polym. Sci.*, **10,** 133 (1972).
2774. Ohita, H., Creed, D., Wine, P. H., Caldwell, R. A., and Mecton, L. A., *J. Amer. Chem. Soc.*, **98,** 2002 (1976).
2775. Okahata, Y., Lim, H., and Hachiya, S., *Makromol. Chem. Rapid Commun.*, **4,** 303 (1983).
2776. Okamoto, K., Kusabayashi, S., and Mikawa, H., *Bull. Chem. Soc. Japan*, **46,** 2613 (1973).
2777. Okamoto, Y., Ueba, Y., Dzhanibekov, N. F., and Banks, E., *Macromolecules*, **14,** 17 (1981).
2778. Okamoto, Y., Ueba, Y., Nagata, I., and Banks, E., *Macromolecules*, **14,** 807 (1981).
2779. Okawara, M., *J. Org. Chem.*, **30,** 2025 (1965).
2780. Okawara, M., Endo, T., Kurusu, Y., in *Progress in Polymer Science in Japan* (Imahori, K., and Iwakura, Y., eds), Vol. 4, Wiley, New York, 1972, p. 128.

724

2781. Okieimen, E. F., *Polymer*, **22,** 1737 (1981).
2782. Okimoto, T., Takahashi, M., Inaki, Y., and Takemoto, K., *J. Polym. Sci.*, B, **12,** 121 (1974).
2783. Okimoto, T., Takahashi, M., Inaki, Y., and Takemoto, K., *Angew. Makromol. Chem.*, **38,** 81 (1974).
2784. Okubo, T., and Turro, N. J., *J. Phys. Chem.*, **85,** 4034 (1981).
2785. Okura, I., and Kim-Thuan, N., *J. Molec. Catalysis*, **6,** 227 (1979).
2786. Okura, I., and Kim-Thuan, N., *J. Chem. Soc. Chem. Commun.*, **1980,** 84.
2787. Olaj, O. F., Breitenbach, J. W., and Kaufman, H. F., *J. Polym. Sci.*, B, **9,** 883 (1971).
2788. Olea, A. F., Encinas, M. V., and Lissi, E. A., *Macromolecules*, **15,** 1111 (1982).
2789. Oldfield, D., *J. Macromol. Sci. Chem.*, A1, **17,** 273 (1982).
2790. Olson, D. R., *J. Appl. Polym. Sci.*, **28,** 1159 (1983).
2791. Olson, D. R., and Schroeter, S. H., *J. Appl. Polym. Sci.*, **22,** 2165 (1978).
2792. Omichi, H., and Hagiwara, M., *Polym. Photochem.*, **1,** 15 (1981).
2793. Omichi, H., Hagiwara, M., and Araki, K., *Makromol. Chem.*, **180,** 1923 (1979).
2794. Omichi, H., Hagiwara, M., Asano, M., and Araki, K., *J. Appl. Polym. Sci.*, **24,** 2311 (1979).
2795. Omura, K., and Matsuura, T., *Tetrahed. Lett.*, **1970,** 255.
2796. Orgel, L. E., and Mulliken, R. S., *J. Amer. Chem. Soc.*, **79,** 4839 (1957).
2797. Osada, Y., and Katsumura, E., *Makromol. Chem. Rapid Comunn.*, **2,** 241 (1981).
2798. Osada, Y., and Katsumura, E., *Makromol. Chem. Rapid Commun.*, **2,** 411 (1981).
2799. Osawa, Z., in *Developments in Polymer Degradation* (Allen, N. S., ed.) Vol. 2, Applied Science Publishers, Ltd, London, 1982, p. 209.
2800. Osawa, Z., in *Degradation and Stabilization of Polymers* (Jellinek, H. H. G., ed.), Elsevier, Amsterdam, 1983, p. 162.
2801. Osawa, Z., and Aiba, M., *Polym. Photochem.*, **2,** 339 (1982).
2802. Osawa, Z., and Aiba, M., *Polym. Photochem.*, **2,** 447 (1982).
2803. Osawa, Z., Cheu, E. L., and Ogiwara, Y., *J. Polym. Sci.*, A1, **13,** 535 (1975).
2804. Osawa, Z., Cheu, E. L., and Magashima, K., *J. Polym. Sci.*, A1, **15,** 445 (1977).
2805. Osawa, Z., Kurisu, N., and Kuroda, H., *Polym. Degrad. Stabil.*, **3,** 265 (1981).
2806. Osawa, Z., Kurisu, N., Nagashima, K., and Nakano, K., *J. Appl. Polym. Sci.*, **23,** 3583 (1979).
2807. Osawa, Z., and Kuroda, H., *J. Polym. Sci.*, B, **20,** 577 (1982).
2808. Osawa, Z., Kuroda, H., and Kobyashi, Y., *J. Appl. Polym. Sci.*, **29,** 2843 (1984).
2809. Osawa, Z., Magashima, K., Ohshima, H., and Chew, E. L., *J. Polym. Sci.*, A1, **17,** 409 (1979).
2810. Osawa, Z., Matsui, K., and Ogiwara, Y., *J. Macromol. Sci. Chem.*, A, **1,** 581 (1967).
2811. Osawa, Z., and Nagashima, K., *Polym. Degrad. Stabil.*, **1,** 311 (1979).
2812. Osawa, Z., Nagashima, K., Ohshima, H., and Cheu, E. L., *J. Polym. Sci.*, A1, **17,** 409 (1979).
2813. Osawa, Z., and Nakano, K., *J. Polym. Sci.*, Polym. Symp., No. 57, 267 (1976).
2814. Osawa, Z., Suzuki, M., and Ogiwara Y., *J. Macromol. Sci. Chem.*, A, **5,** 275 (1971).
2815. Osawa, Z., Tajima, E., Yanagisawa, T., and Suzuki, K., in *Polymer Additives* (Kresta, J. E., ed.), Plenum Press, New York, 1984, p. 49.
2816. Osborn, C. L., *J. Radiat. Curing*, **3,** Feb., 2 (1976).
2817. Osborn, C. L., and Sander, M. R., *ACS Div. Org. Coat. Plast. Chem., Preprints*, **34**(1), 660 (1974).
2818. Osborne, K. R., *J. Polym. Sci.*, **38,** 357 (1959).
2819. Ostberg, G. L., and LeRoy, D. J., *Can. J. Chem.*, **29,** 333 (1951).
2820. Oster, G., *Nature (London)*, **173,** 300 (1954).
2821. Oster, G., Brit. Pat., 856, 884, Dec. 21, 1960.
2822. Oster, G., *J. Polym. Sci.*, B, **2,** 1181 (1964).
2823. Oster, G., and Adelman, A. H., *J. Amer. Chem. Soc.*, **78,** 913 (1956).
2824. Oster, G., Geacitov, N., and Cassen, T., *J. Opt. Soc. Amer.*, **58,** 1217 (1968).
2825. Oster, G., and Nishijima, Y., *Fortsch. Hochpolym. Forsch.*, **3,** 313 (1964).
2826. Oster, G., Oster, G. K., and Kryszewski, M., *J. Polym. Sci.*, **57,** 937 (1962).
2827. Oster, G., Oster, G. K., and Moroson, H., *J. Polym. Sci.*, **34,** 671 (1959).
2828. Oster, G., and Shibata, O., *J. Polym. Sci.*, **26,** 233 (1957).
2829. Oster, G., and Yang, N., *Chem. Rev.*, **68,** 125 (1968).
2830. Oswald, D. A., Noel, P., and Stephenson, A. J., *J. Org. Chem.*, **26,** 3969 (1961).
2831. Otocka, E. P., Curran, S., and Porter, R. S., *J. Appl. Polym. Sci.*, **28,** 3227 (1983).
2832. Otsu, T., *J. Polym. Sci.*, **21,** 559 (1956).
2833. Otsu, T., and Kyriyama, A., *Polym. Bull.*, **11,** 135 (1984).
2834. Otsu, T., and Nayatami, K., *Makromol. Chem.*, **27,** 149 (1958).
2835. Otsu, T., Nayatani, K., Muto, I., and Imai, M., *Makromol. Chem.*, **27,** 142 (1958).

725

2836. Otsu, T., Tanaka, H., and Wasaki, H., *Polymer*, **20**, 55 (1979).
2837. Otsu, T., and Yoshida, M., *Makromol. Chem. Rapid Commun.*, **3**, 127 (1982).
2838. Otsu, T., and Yoshida, M., *Polym. Bull.*, **7**, 197 (1982).
2839. Otsu, T., Yoshida, M., and Kyriyama, A., *Polym. Bull.*, **7**, 45 (1982).
2840. Otsu, T., and Yoshioka, M., Makromol. *Chem. Rapid Commun.*, **3**, 127 (1982).
2841. Otterstedt, J. A., *J. Chem. Phys.*, **58**, 5716 (1973).
2842. Ouchi, T., and Aso, M., *J. Polym. Sci.*, **17**, 2639 (1979).
2843. Ouchi, T., and Azuma, T., *Europ. Polym. J.*, **18**, 809 (1982).
2844. Ouchi, T., Hosoi, M., and Matsumoto, F., *J. Appl. Polym. Sci.*, **20**, 1983 (1976).
2845. Ouchi, T., Nakamura, S., Hamada, M., and Oiwa, M., *J. Polym. Sci.*, A1, **13**, 455 (1975).
2846. Ouchi, T., Nakamura, S., and Oiwa, M., *J. Macromol. Sci. Chem.*, A, **11**, 1651 (1977).
2847. Ouchi, T., Nakamura, T., and Yonekura, K., *J. Polym. Sci.*, A1, **17**, 147 (1979).
2848. Ouchi, T., Sato, C., and Kemeoka, M., *Techn. Rept of Kansai Univ., Japan*, **22**, 93 (1980).
2849. Ouchi, T., Tatsumi, M., and Oiwa, M., *J. Polym. Sci.*, A1, **17**, 3979 (1979).
2850. Owen, E. D., in *Developments in Polymer Photochemistry* (Allen, N. S., ed.) Vol. 1, Applied Science Publishers, Ltd, London, 1980, p. 1.
2851. Owen, E. D., *in Photodegradation and Photostabilization* of Coatings, (Pappas, S. P., and Winslow, F. H., eds), ACS Symp. Ser., No. 151, ACS Washington, D.C., 1981, p. 217.
2852. Owen, E. D., in *Photophysics of Synthetic Polymers* (Phillips, D., and Roberts, A. J., eds), Science Reviews, Ltd, Northwood, England, 1982, p. 97.
2853. Owen, E. D., *Org. Coat. Appl. Polym. Sci. Proc.*, **46**, 641 (1982), Las Vegas, ACS Meeting, 1982.
2854. Owen, E. D., in *Developments in Polymer Degradation* (Allen, N. S., ed.), Applied Science Publishers, Ltd, London, 1982, p. 165.
2855. Owen, E. D., ed., *Degradation and Stabilization of Poly(vinyl chloride)*, Elsevier, London, 1984.
2856. Owen, E. D., and Bailey, R. J., *J. Polym. Sci.*, A1, **10**, 113 (1972).
2857. Owen, E. D., and Pasha, I., *J. Appl. Polym. Sci.*, **25**, 2417 (1980).
2858. Owen, E. D., and Read, R. L., *Europ. Polym. J.*, **15**, 41 (1979).
2859. Owen, E. D., and Read, R. L., *J. Polym. Sci.*, A1, **17**, 2719 (1979).
2860. Owen, E. D., and Williams, J. I., *J. Polym. Sci.*, A1, **11**, 905 (1973).
2861. Owen, E. D., and Williams, J. I., *J. Polym. Sci.*, A1, **12**, 1933 (1974).
2862. Oyama, N., Yamaguchi, S., Kaneko, M., and Yamada, A., *J. Electroanal. Chem.*, **139**, 215 (1982).
2863. Ozawa, T., Sukegawa, S., and Masaki, K., *Chem. High Polym.*, **17**, 367 (1960).
2864. Pabiot, J., and Verdu, J., *Polym. Eng. Sci.*, **21**, 32 (1981).
2865. Pac, C., and Sakurai, H., *Tetrahed. Lett.*, **1968**, 1865.
2866. Pac, C., and Sakurai, H., *Chem. Commun.*, **1969**, 20.
2867. Pac, C., Sakurai, H., and Tosa, T., *Chem. Commun.*, **1970**, 1311.
2868. Pacansky, J., and Lyerla, J. R., *IBM J. Res. Develop.*, **23**, 42 (1979).
2869. Pacifici, J. G., and Straley, J. M., *J. Polym. Sci.*, B, **7**, 7 (1969).
2870. Pailthorpe, M. I., and Nicholls, C. H., *Photochem. Photobiol.*, **14**, 135 (1971).
2871. Pailthorpe, M. T., and Nicholls, C. H., *Radiat. Res.*, **49**, 112 (1972).
2872. Pailthorpe, M. T., and Nicholls, C. H., *Photochem. Photobiol.*, **15**, 465 (1972).
2873. Pajot-Augy, E., Bokobza, L., Monnerie, L., Castellan, A., Bouas-Laurent, H., and Millet, C., *Polymer*, **24**, 117 (1983).
2874. Pajot-Augy, E., Bokobza, L., Monnerie, L., Castellan, A., Laurent, H. B., *Macromolecules*, **17**, 1490 (1984).
2875. Palma, G., and Carenza, M., *J. Appl. Polym. Sci.*, **14**, 1737 (1970).
2876. Pan, S. S., and Morawetz, H., *Macromolecules*, **13**, 1157 (1980).
2877. Panda, S. P., *J. Sci. Ind. Res.*, **35**, 560 (1976).
2878. Panke, D., and Wunderlich, W., *J. Appl. Polym. Sci., Appl. Polym. Symp.*, No. 35, 321 (1979).
2879. Pappas, S. P., *Progr. Org. Coat.*, **2**, 333 (1974).
2880. Pappas, S. P., ed., *UV Curing: Science and Technology*, Vol. 1, Stamford Technology Marketing, Norwalk, CT, 1978.
2881. Pappas, S. P., *Radiat. Curing*, **8**, March, 28 (1981).
2882. Pappas, S. P., ed., *UV Curing: Science and Technology*, Vol. 2, Stamford Technology Marketing, Norwalk, CT, 1984.
2883. Pappas, S. P., and Asmus, R A., *J. Polym. Sci.*, A1, **20**, 2643 (1983).
2884. Pappas, S. P., and Chattopadhyay, A. K., *J. Amer. Chem. Soc.*, **95**, 6484 (1973).
2885. Pappas, S. P., and Chattopadhyay, A. K., *J. Polym. Sci.*, B, **13**, 483 (1975).

726

2886. Pappas, S. P., Chattopadhyay, A. K., and Carlblom, L. H., in *Ultraviolet Light Induced Reactions in Polymers* (Labana, S. S., ed.), ACS Symp. Ser., No. 25, ACS Washington, D.C., 1976, p. 12.

2887. Pappas, S. P., and Fiher, R. M., *J. Paint Technol.*, **46**, 65 (1974).

2888. Pappas, S. P., Gatechair, L. R., and Jilek, J. H., *J. Polym. Sci.*, A1, **22**, 77 (1984).

2889. Pappas, S. P., and Jilek, J. H., *Photogr. Eng. Sci.*, **23**, 140 (1979).

2890. Pappas, S. P., and Kühhirt, W., *J. Paint Technol.*, **47**, 42 (1975).

2891. Pappas, S. P., and Lam, C. W., *J. Radiat. Curing*, **7**, Jan., 2 (1980).

2892. Pappas, S. P., and McGinniss, V. D., in *UV Curing: Science and Technology* (Pappas, S. P., ed.), Vol. 1, Stamford Technology Marketing, Norwalk, CT, 1978, p. 1.

2893. Pappas, S. P., Pappas, B. C., Gatechair, L. R., Jilek, J. H., and Schnabel, W., *Polym. Photochem.*, **5**, 1 (1984).

2894. Pappas, S. P., Pappas, B. C., Gatechair, L. R., Schnabel, W., *J. Polym. Sci.*, A1, **22**, 69 (1984).

2895. Pariiskii, G. B., Postnikov, L. M., Toptygin, D. Ya., and Davydow, E. Ya., *J. Polym. Sci.*, Polym. Symp., No. 42, 1287 (1973).

2896. Parker, C. A., *Photoluminescence of Solutions*, Elsevier, New York, 1968.

2897. Parker, C. A., and Hatchard, C. G., *Trans. Faraday Soc.*, **59**, 284 (1963).

2898. Parker, C. A., and Joyce, T. A., *Chem. Commun.*, **1968**, 749.

2899. Parker, C. A., and Joyce, T. A., *Trans. Faraday Soc.*, **65**, 2823 (1969).

2900. Parker, C. A., and Rees, W. T., *Analyst*, **87**, 83 (1962).

2901. Parrish, M. A., *J. Oil Colour Chem. Assoc.*, **60**, 474 (1977).

2902. Parashutkin, A. A., and Krongauz, V. P., *Mol. Photochem.*, **6**, 437 (1974).

2903. Pasch, N. F., and Webber, S. E., *Chem. Phys.*, **16**, 361 (1976).

2904. Pasch, N. F. and Webber, S. E., *Macromolecules*, **11**, 727 (1978).

2905. Pasch, N. P., McKenzie, R. D., and Webber, S. E., *Macromolecules*, **11**, 733 (1978).

2906. Patil, A. O., Deshpande, D. D., and Talwar, S. S., *Polymer*, **22**, 434 (1981).

2907. Patil, A. O., Deshpande, D. D., Talwar, S. S., and Biswas, A. B., *J. Polym. Sci.*, A1, **19**, 1155 (1981).

2908. Patterson, L. K., Porter, G., and Topp, M. R., *Chem. Phys. Lett.*, **7**, 612 (1970).

2909. Paul, D. F., and Gumbs, R. W., *J. Appl. Polym. Sci.*, **21**, 959 (1977).

2910. Paulsen, R. G., *J. Vac. Sci. Technol.*, **14**, 266 (1977).

2911. Pavlik, J. O., Plooard, P. I., Somersall, A. C., and Guillet, J. E., *Can. J. Chem.*, **51**, 1435 (1973).

2912. Pearson, D. E., and Thiemann, P. D., *J. Polym. Sci.*, A1, **8**, 2103 (1970).

2913. Peebles, L. H., Jr., Clarke, J. T., and Stockmayer, W. H., *J. Amer. Chem. Soc.*, **82**, 4780 (1960).

2914. Peeling, J., and Clark, D. T., *Polym. Degrad. Stabil.*, **3**, 97 (1980/81).

2915. Peeling, J., and Clark, D. T., *Polym. Degrad. Stabil.*, **3**, 177 (1980/81).

2916. Peeling, J., and Clark, D. T., *J. Appl. Polym. Sci.*, **26**, 3761 (1981).

2917. Peeling, J., Courval, G., and Jazzar, M. S., *J. Polym. Sci.*, A1, **22**, 419 (1984).

2918. Pekcan, O., Winnik, M. A., and Croucher, M. D., *J. Coloid Inter, Sci.*, **95**, 420 (1983).

2919. Pekcan, O., Winnik, M. A., Egan, L., and Croucher, M. D., *Macromolecules*, **16**, 699 (1983).

2920. Pekcan, O., Winnik, M. A., and Croucher, M. D., *Polym. Sci.*, B, **21**, 1011 (1983).

2921. Pelgrims, J., *J. Oil. Color Chem. Assoc.*, **61**, 111 (1978).

2922. Pemberton, D. R., and Johnson, A. F., *Polymer*, **25**, 529 (1984).

2923. Pemberton, D. R., and Johnson, A. F., *Polymer*, **25**, 536 (1984).

2924. Penot, G., Arnaud, R., and Lemaire, J., *Polym. Photochem.*, **2**, 39 (1982).

2925. Perkins, W. C., *J. Radiat. Curing*, **8**, 16 (1981).

2926. Perrin, F. C., *C.R. Acad. Sci., Ser. C.*, **178**, 1978 (1924).

2927. Perrin, F. C., *Ann. Phys.*, **12**, 169 (1929).

2928. Perrins, N. C., and Simons, J. P., *Trans. Faraday Soc.*, **65**, 390 (1969).

2929. Peterlin, A., *Pure Appl. Chem.*, **39**, 239 (1974).

2930. Petropoulos, C. C., *J. Polym. Sci.*, A, **2**, 69 (1964).

2931. Petropoulos, C. C., *J. Polym. Sci.*, A1, **15**, 1637 (1977).

2932. Petterson, L. T., Hager, R. G., Vellturo, A. F., and Griffin, G. W., *J. Org. Chem.*, **32**, 1018 (1967).

2933. Petterson, R. C., Dimaggio, A., Herberts, A. L., Haley, T. J., Mykytka, J. P., and Sarkar, I. M., *J. Org. Chem.*, **36**, 631 (1971).

2934. Pfordte, K., *J. Prakt. Chem.*, **5**, 196 (1957).

2935. Phillips, D., and Roberts, A. J., eds, *Photophysics of Synthetic Polymers*, Science Reviews, Northwood, England, 1982.

2936. Phillips, D., Roberts, A. J., Rumbles, G., and Soutar, I., *Macromolecules*, **16**, 1597 (1983).

2937. Phillips, D., Roberts, A. J., and Soutar, I., *J. Polym. Sci.*, A2, **18**, 2401 (1980).
2938. Phillips, D., Roberts, A. J., and Soutar, I., *J. Polym. Sci.*, B, **18**, 123 (1980).
2939. Phillips, D., Roberts, A. J., and Soutar, I., *Europ. Polym. J.*, **17**, 101 (1981).
2940. Phillips, D., Roberts, A. J., and Soutar, I., *Polymer*, **22**, 293 (1981).
2941. Phillips, D., Roberts, A. J., and Soutar, I., *Polymer*, **22**, 427 (1981).
2942. Phillips, D., Roberts, A. J., and Soutar, I., *J. Polym. Sci.*, A2, **20**, 411 (1982).
2943. Phillips, D., Roberts, A. J., and Soutar, I., *Macromolecules*, **16**, 1593 (1983).
2944. Phillips, D., and Rumbles, G., *Polym. Photochem.*, **5**, 153 (1984).
2945. Phillips, G. O., Baugh, P. J., McKellar, J. F., and von Sontag, C., in *Cellulose Chemistry and Technology* (Arthur, J. C., ed.), 1977, p. 313.
2946. Phillips, G. O., Worthington, N. W., McKellar, J. F., and Sharpe, R. R., *J. Chem. Soc.*, A, **1969**, 767.
2947. Philphott, M. R., *J. Chem. Phys.*, **63**, 485 (1975).
2948. Pieroni, O., Houben, J. L., Fissi, A., Constantino, P., and Ciardelli, F., *J. Amer. Chem. Soc.*, **102**, 5913 (1980).
2949. Pilar, J., Toman, L., and Marek, M., *J. Polym. Sci.*, A1, **14**, 2399 (1976).
2950. Pimblott, J. G., Scott, G., and Stuckey, J. E., *J. Appl. Polym. Sci.*, **19**, 865 (1975).
2951. Pinazzi, P., and Fernandez, A., *Makromol. Chem.*, **167**, 147 (1973).
2952. Pinazzi, P., and Fernandez, A., *Makromol. Chem.*, **168**, 19 (1973).
2953. Pitha, J., *Polymer*, **19**, 425 (1977).
2954. Pitt, C. W., and Nalpita, L. M., *Thin Solid Films*, **68**, 101 (1980).
2955. Pitts, J. N. Jr., Johnson, H. W., and Kuwana, T., *J. Phys. Chem.*, **66**, 2456 (1962).
2956. Pitts, J. N. Jr., Letsinger, R. L., Taylor, R. P., Patterson, J. M., Rectenwald, G., and Martin, R. B., *J. Amer. Chem. Soc.*, **81**, 1068 (1959).
2957. Pitts, J. N. Jr., Wan, J. K. S, and Schuck, E. A., *J. Amer. Chem. Soc.*, **86**, 3606 (1964).
2958. Pitts, J. N. Jr., and Wan, J. K. S., in *Chemistry of the Carbonyl Group* (Patai, S. ed.), Interscience Publishers, London, 1966, p. 825.
2959. Pivovarov, A. P., Gak, Y. V., and Lukovnikov, A. F., *Vysokomol. Soedin.*, A, **13**, 2110 (1971).
2960. Pivovarov, A. P., Pivovarova, T. S., and Lukovnikov, A. F., *Vyskomol. Soedin.*, A, **15**, 661 (1973).
2961. Plant, M. A., and Scott, G., *Europ. Polym. J.*, **7**, 1173 (1971).
2962. Plews, G., and Phillips, R., *J. Coat. Technol.*, **51**, 69 (1979).
2963. Plews, G., and Phillips, R., *J. Coat. Technol.*, **51**, 648 (1979).
2964. Plooard, P. T., and Guillet, J. E., *Macromolecules*, **5**, 405 (1972).
2965. Plummer, V. T., and Carson, R. K., *Science*, **166**, 1141 (1969).
2966. Pokholok, T. V., Parysky, G. B., Zaitseva, N. I., and Toptygin, D. Ya., *Europ. Polym. J.*, **20**, 609 (1984).
2967. Pokholok, T. V., Zaitseva, N. I., Parysky, G. B., and Toptygin, D. Ya., *Polym. Photochem.*, **2**, 429 (1982).
2968. Porter, G., in *Light, Chemical Change and Life* (Coyle, J. D., Hill, R. R., and Roberts, D. R., eds), The Open University Press, Milton Keynes, England, 1982, p. 362.
2969. Porter, G., in *Light, Chemical Change and Life* (Coyle, J. D., Hill, R. R., and Roberts, D. R., eds), The Open University Press, Milton Keynes, England, 1982, p. 355.
2970. Porter, G., and Archer, M., *Interdiscip. Sci. Rev.*, **1**, 119 (1976).
2971. Porter, G., and Suppan, P., *Proc. Chem. Soc.*, **1964**, 191.
2972. Porter, G., and Suppan, P., *Pure Appl. Chem.*, **9**, 499 (1964).
2973. Porter, G., and Suppan, P., *Trans. Faraday Soc.*, **61**, 1664 (1965).
2974. Porter, G., and Wilkinson, F., *Proc. Roy. Soc.*, A, **264**, 1 (1961).
2975. Porter, G., and Wilkinson, F., *Trans. Faraday Soc.*, **57**, 1686 (1961).
2976. Porter, G., and Windsor, M. W., *Proc. Roy. Soc.*, A, **245**, 238 (1958).
2977. Pospelova, M. A., Belokon, V. G., and Gurman, V. S., *Dokl. Akad. Nauk. SSR*, **212**, 414 (1973).
2978. Pospišil, J., in *Developments in Polymer Stabilization* (Scott, G., ed.), Vol. 1. Applied Science Publishers, Ltd, London, 1979, p. 1.
2979. Pospišil, J., *Adv. Polym. Sci.*, **36**, 70 (1980).
2980. Pospišil, J., in *Developments in Polymer Photochemistry* (Allen, N. S., ed.), Vol. 2, Applied Science Publishers, Ltd, London, 1981, p. 53.
2981. Postnikov, C. M., Margolin, A. L., Shlyapintokh, V. Ya., Vichutinskaya, E. V., *J. Polym. Sci.*, Polym. Symp., No. 42, 1275 (1973).
2982. Potashnik, R., Goldsmidt, C. R., and Ottolenghi, M., *Chem. Phys. Lett.*, **9**, 424 (1971).

728

2983. Powell, R. C., J. Chem. Phys., 55, 1871 (1971).
2984. Powell, R., and Kim, Q., J. Lumin., 6, 351 (1975).
2985. Pownall, H. J., and Huber, J. R., J. Amer. Chem. Soc., 93, 6429 (1971).
2986. Pozzi, V., Silvers, A. E., Giuffre, L., and Cernia, E., J. Appl. Polym. Sci., 19, 923 (1975).
2987. Pratte, J. F., Noyes, W. A., Jr., and Webber, S. E., Polym. Photochem., 1, 1 (1981).
2988. Pratte, J. F., and Webber, S. E., Macromolecules, 15, 417 (1982).
2989. Pratte, J. F., and Webber, S. E., J. Phys. Chem., 87, 449 (1983).
2990. Pratte, J. F., and Webber, S. E., Macromolecules, 16, 1188 (1983).
2991. Pratte, J. F., and Webber, S. E., Macromolecules, 16, 1193 (1983).
2992. Pratte, J. F., and Webber, S. E., Macromolecules, 17, 2116 (1984).
2993. Price, T. R., and Fox, R. B., J. Polym. Sci., 4, 771 (1966).
2994. Priest, W. J., and Sifain, M. M., J. Polym. Sci., A1, 9, 3161 (1971).
2995. Prock, A., Shand, M. L., and Chance, R. R., Macromolecules, 15, 238 (1982).
2996. Pryce, A., J. Oil Color Chem. Assoc., 59, 166 (1976).
2997. Pryor, W. A., Free Radicals, McGraw-Hill, New York, 1966.
2998. Puglisi, J., and Vigeant, F., Radiat. Curing, 7, 3 (1980).
2999. Put, J., and DeSchryver, F. C., J. Amer. Chem. Soc., 95, 137 (1973).
3000. Raab, M. Kotulak, L., Kolarik, J., and Pospišil, J., J. Appl. Polym. Sci., 27, 2457 (1982).
3001. Rabek, J. F., J. Polym. Sci., B, 2, 557 (1964).
3002. Rabek, J. F., Polimery, 9, 128 (1964).
3003. Rabek, J. F., Polimery, 9, 221 (1964).
3004. Rabek, J. F., J. Appl. Polym. Sci., 9, 2121 (1965).
3005. Rabek, J. F., Chem. Stosow., 9, 433 (1965).
3006. Rabek, J. F., Wiadom. Chem., 20, 291, 355, 435 (1966).
3007. Rabek, J. F., Chem. Stosow., 10, 331 (1966).
3008. Rabek, J. F., J. Polym. Sci., A1, 4, 1311 (1966).
3009. Rabek, J. F., Polimery, 11, 266 (1966).
3010. Rabek, J. F., J. Polym. Sci., Polym. Symp., No. 16, 949 (1967).
3011. Rabek, J. F., Chem. Stosow., 11, 53 (1967).
3012. Rabek, J. F., Chem. Stosow., 11, 73 (1967).
3013. Rabek, J. F., Chem. Stosow., 11, 89 (1967).
3014. Rabek, J. F., Chem. Stosow., 11, 169 (1967).
3015. Rabek, J. F., Chem. Stosow., 11, 183 (1967).
3016. Rabek, J. F., Photochem. Photobiol., 7, 5 (1968).
3017. Rabek, J. F., Polimery, 10, 258 (1971).
3018. Rabek, J. F., Pure Appl. Chem., 8, 29 (1971).
3019. Rabek, J. F. in Materials of 3rd Techn. Conf. on Photopolymers, Soc. Plast. Eng. Mid. Hudson Sect., Ellenvile, 1973, p. 27.
3020. Rabek, J. F., in Comprehensive Chemical Kinetics (Bamford, C. H., and Tipper, C. F., eds), Vol. 14 Elsevier, Amsterdam, 1975, p. 425.
3021. Rabek, J. F., in Ultraviolet Light Induced Reactions in Polymers (Labana, S. S., ed.), ACS Symp. Ser., No. 25, ACS Washington, D.C., 1976, p. 255.
3022. Rabek, J. F., Experimental Methods in Polymer Chemistry, Wiley, Chichester, 1980.
3023. Rabek, J. F., Experimental Methods in Photochemistry and Photophysics, Wiley, Chichester, 1982.
3024. Rabek, J. F., in Polymer Additives (Kresta, J. E., ed.), Plenum Press, New York, 1984.
3025. Rabek, J. F., in Singlet Oxygen (Frimer, A. A., ed.), Vol. 4, CRC Press, Boca Raton, 1985, p. 1.
3026. Rabek, J. F., Canbäck, G., Lucki, J., and Rånby, B., J. Polym. Sci., A1, 14, 1447 (1976).
3027. Rabek, J. F., Canbäck, G., and Rånby, B., J. Appl. Polym. Sci., 21, 2211 (1977).
3028. Rabek, J. F., Canbäck, G., and Rånby, B., J. Appl. Polym. Sci., Appl. Polym. Symp., No. 35, 299 (1979).
3029. Rabek, J. F., and Lala, D., J. Polym. Sci., B, 18, 427 (1980).
3030. Rabek, J. F., Lucki, J., and Rånby, B., Europ. Polym. J., A1, 15, 1089 (1979).
3031. Rabek, J. F., and Rabek, T. I., J. Appl. Polym. Sci., 7, S37 (1963).
3032. Rabek, J. F., and Rånby, B., in ESR Applications to Polymer Research (Kinell, P. O., Rånby, B., and Runnstrnl-Reio, V., eds), Almqvist-Wiksel, Stockholm, 1972, p. 201.
3033. Rabek, J. F., and Rånby, B., J. Polym. Sci., A1, 12, 273 (1974).
3034. Rabek, J. F., and Bånby, B., J. Polym. Sci., A1, 12, 295 (1974).
3035. Rabek, J. F., and Rånby, B., Polym. Eng. Sci., 15, 40 (1975).
3036. Rabek, J. F., and Rånby, B., J. Polym. Sci., A1, 14, 1463 (1976).

729

3037. Rabek, J. F., and Rånby, B., *Photochem. Photobiol.*, **28**, 557 (1978).
3038. Rabek, J. F., and Rånby, B., *Photochem. Photobiol.*, **30**, 133 (1979).
3039. Rabek, J. F., and Rånby, B., *J. Appl. Polym. Sci.*, **23**, 2481 (1979).
3040. Rabek, J. F., and Rånby, B., *Rev. Rouman. Chem.*, **7**, 1045 (1980).
3041. Rabek, J. F., Rånby, B., and Arct, J., *Polym. Degrad. Stabil.*, **5**, 65 (1983).
3042. Rabek, J. F., Rånby, B., Arct, J., and Golubski, Z., *Europ. Polym. J.*, **18**, 81 (1982).
3043. Rabek, J. F., Rånby, B., Arct, J., and Liu, R., *J. Photochem.*, **25**, 519 (1984).
3044. Rabek, J. F., Rånby, B., Östensson and Flodin, P., *J. Appl. Polym. Sci.*, **24**, 2408 (1979).
3045. Rabek, J. F., Rämme, G., Canbäck, G., Rånby, B., and Kagiya, V. T., *Europ. Polym. J.*, **15**, 339 (1979).
3046. Rabek, J. F., Shur, Y. J., and Rånby, B., in *Singlet Oxygen: Reactions with Organic Compounds and Polymers* (Rånby, B., and Rabek, J. F., eds), Wiley, Chichester, 1978, p. 264.
3047. Rabek, J. F., Shur, Y. J., and Rånby, B., *J. Polym. Sci.*, A1, **13**, 1285 (1975).
3048. Rabek, J. F., Skowronski, T. A., and Rånby, B., *Polymer*, **21**, 226 (1980).
3049. Rabek, J. F., Zabrzewski, J., and Zielinski, S., *Nukleonika*, **12**, 265 (1967).
3050. Rabek, J. F., and Zaleskii, A., *Polimery*, **9**, 437, 499, 555 (1970).
3051. Rämme, G., *J. Photochem.*, **9**, 439 (1978).
3052. Rämme, G., and Rabek, J. F., *Europ. Polym. J.*, **13**, 855 (1977).
3053. Ranaweera, R. P. R., and Scott, G., *J. Polym. Sci.*, B, **13**, 71 (1975).
3054. Ranaweera, R. P. R., and Scott, G., *Europ. Polym. J.*, **12**, 591 (1976).
3055. Ranaweera, R. P. R., and Scott, G., *Europ. Polym. J.*, **12**, 825 (1976).
3056. Rånby B., nd Carstensen, P., *Adv. Chem. Ser.*, No. 66, 256 (1967).
3057. Rånby, B., and Hult, A., in *Organic Coatings Science and Technology* (Parfitt, G. D., and Patsis, A. V., eds), Vol. 7, Dekker, New York, 1984, p. 137.
3058. Rånby, B., Kringstad, K., Cowling, E. B., and Lins, S. Y., *Acta Chem. Scand.*, **23**, 3257 (1969).
3059. Rånby, B., and Lucki, J., *Pure Appl. Chem.*, **52**, 295 (1980).
3060. Rånby, B., and Rabek, J. F., *Photodegradation, Photooxidation and Photostabilization of Polymers*, Wiley, London, 2975.
3061. Rånby, B., and Rabek, J. F., in *Ultraviolet Light Induced Reactions in Polymers* (Labana, S. S., ed.), ACS Symp. Ser., No. 25, ACS Washington, D.C., 1976, p. 391.
3062. Rånby, B., and Rabek, J. F., *ESR Spectroscopy in Polymer Research*, Springer Verlag, Berlin, 1977.
3063. Rånby, B., and Rabek, J. F., eds, *Singlet Oxygen, Reactions with Organic Compounds and Polymers*, Wiley, Chichester, 1978.
3064. Rånby, B., and Rabek, J. F., in *Singlet Oxygen: Reactions with Organic Compounds and Polymers* (Rånby, B., and Rabek, J. F., eds), Wiley, Chichester, 1978, p. 211.
3065. Rånby, B., and Rabek, J. F., *J. Appl. Polym. Sci.*, Appl. Polym. Symp., No. 35, 243 (1979).
3066. Rånby, B., and Rabek, J. F., in *The Effects of Hostile Environment on Coatings and Plastics* (Garner, D. P., and Stahl, G. A., eds), ACS Symp. Ser., No. 229, ACS Washington, D.C., 1983, p. 291.
3067. Rånby, B., Rabek, J. F., and Canbäck, G., *J. Macromol. Sci. Chem.*, A, **12**, 587 (1978).
3068. Rånby, B., Rabek, J. F., and Joffe, Z., in Proc. Conf. on *Degradability of Polymers and Plastics*, The Plastic Institute, London, No. 27/28, 1973, p. 3/1.
3069. Rånby, B., Rabek, J. F., and Lucki, J., in *Physicochemical Aspects of Polymer Surfaces* (Mittal, K. L., ed.), Vol. 1, Plenum Press, New York, 1983, p. 283.
3070. Rånby, B., Rabek, J. F., Shur, Y. J., Joffe, Z., and Wikström, K., in Proc. Conf. on *Degradation and Stabilization of Polymers*, Brussels, 13–14 Sept. 1974, p. 35.
3071. Rånby, B., and Yoshida, H., *J. Polym. Sci.*, Polym. Symp., No. 12, 263 (1966).
3072. Rastogi, R. P., and Nigam, R. Y., *Trans. Faraday Soc.*, **55**, 2005 (1959).
3073. Rau, N. O., Schulz, R. C. *Makromol. Chem. Rapid Commun.*, **5**, 725 (1984).
3074. Razem, B., Wong, M., and Thomas, J. K., *J. Amer. Chem. Soc.*, **100**, 1629 (1978).
3075. Rehm, D., and Weller, A., Israel, *J. Chem.*, **8**, 259 (1970).
3076. Redpath, A. E. C., and Winnik, M. A., *Polymer*, **24**, 1286 (1983).
3077. Reichmanis, E., Gooden, R., Wilkins, C. W. Jr., and Schonhorn, H., *J. Polym. Sci.*, A1, **21**, 1075 (1983).
3078. Reichmanis, E., Wilkins, C. W. Jr., and Chandros, E. A., *J. Vac. Sci. Technol.*, **19**, 1338 (1981).
3079. Reid, R. F., and Soutar, I., *J. Polym. Sci.*, B, **15**, 153 (1977).
3080. Reid, R. F., and Soutar, I., *J. Polym. Sci.*, A2, **16**, 231 (1978).
3081. Reid, R. F., and Soutar, I., *J. Polym. Sci.*, A2, **18**, 457 (1980).
3082. Reine, A. H., and Arthur, J. C. Jr., *Text. Res. J.*, **40**, 90 (1970).

730

3083. Reiney, M. J., Tryon, M., and Achhammer, B. G., *Bur. Stand. J. Res.*, **51**, 155 (1953).
3084. Reinhard, G. D., and Lardy, H. A., *Biochem.*, **19**, 1484 (1980).
3085. Reinhardt, R. M., Arthur, J. C. Jr., and Muller, C. I., *J. Appl. Polym. Sci.*, **24**, 1739 (1979).
3086. Reinisch G., Jaeger, W., and Ulbert, K., *Faserforsch. Textiltechn.*, **19**, 363 (1968).
3087. Reinisch R. F., Gloria, H. R., and Androes, G. M., *Photochemistry of Macromolecules*, Plenum Press, New York, 1970, p. 185.
3088. Reinöhl, V., Sedlar, J., and Navratil, M., *Polym. Photochem.*, **1**, 165 (1981).
3089. Reiser, A., *J. Chim. Phys.*, **77**, 469 (1980).
3090. Reiser, A., Bowen, G., and Horne, R. J., *Trans. Faraday Soc.*, **62**, 3162 (1966).
3091. Reiser, A., and Egerton, P. L., *Macromolecules*, **12**, 670 (1979).
3092. Reiser, A., and Froster, V., *Nature*, **208**, 682 (1965).
3093. Reiser, A., Leyshon, L. J., and Johnson, L., *Trans. Faraday Soc.*, **67**, 2389 (1971).
3094. Reiser, A., Leyshon, L., Saunders, D., Mijnvic, M. J., Bright, A., and Bogie, J., *J. Amer. Chem. Soc.*, **94**, 2414 (1972).
3095. Reiser, A., and Marlye, R., *Trans. Faraday Soc.*, **64**, 1806 (1967).
3096. Reiser, A., Wagner, H., and Bowen, G., *Tetrahed Lett.*, **1966**, 2635.
3097. Reiser, A., Wagner, H. M., Marley, R., and Bowen, G., *Trans. Faraday Soc.*, **63**, 2403 (1967).
3098. Rembold, M. W., and Kramer, H. E. A., *J. Soc. Dyers Color.*, **94**, 12 (1978).
3099. Rennert, J., *Photogr. Sci. Eng.*, **15**, 60 (1971).
3100. Renschler, C. L., and Faulkner, L. R., *J. Amer. Chem. Soc.*, **104**, 3315 (1982).
3101. Rentzepis, P. M., and Busch, G. E., *Mol. Photochem.*, **4**, 353 (1972).
3102. Reynold, G. A., Laakso, T. M., Borden, D. G., and Williams, J. L. R., U.S. Pt., 3,748,131 (24 July, 1973).
3103. Richards, D. P., and Bovenizer, G. W., *J. Paint Technol.*, **44**, 90 (1972).
3104. Richter, K. H., Güttler, W., and Schwoerer, M., *Chem. Phys. Lett.*, **92**, 4 (1982).
3105. Richter, K. H., Güttler, W., and Schwoerer, M., *Appl. Phys.*, A, **32**, 1 (1983).
3106. Richters, P., *Macromolecules*, **3**, 262 (1970).
3107. Ried, W., *Angew. Chem.*, **76**, 933 (1964).
3108. Ried, W., *Angew. Chem.*, **76**, 973 (1964).
3109. Ried, W., and Wesselborg, K. M., *J. Prakt. Chem.*, **12**, 306 (1961).
3110. Rieker, A., and Kessler, H., *Z. Naturforsch.*, **21b**, 939 (1966).
3111. Rieker, A., Müller, E., and Becker, W., *Z. Naturforsch.*, **17b**, 718 (1962).
3112. Rigo, A., Argese, ., Stevanto, R., Orsega, E. P., and Viglino, P., *Inorg. Chem. Acta*, **24**, 171 (1977).
3113. Rippen, F. Kaufmann, G., and Klöpffer, W., *Chem. Phys.*, **52**, 165 (1980).
3114. Rippen, G., and Klöpffer, W., *Ber. Bunsenges. Phys. Chem.*, **83**, 437 (1979).
3115. Rivaton, A., Sallet, D., and Lemaire, J., *Polym. Photochem.*, **3**, 463 (1983).
3116. Rizzardo, E., and Solomon, D. H., *Polym. Biul.*, **1**, 529 (1979).
3117. Rizzardo, E., and Solomon, D. H., *J. Macromol. Sci. Chem.*, A, **13**, 1005 (1979).
3118. Robb, W., *Inoran. Chem.*, **6**, 382 (1967).
3119. Robbins, W. K., and Eastman, R. H., *J. Amer. Chem. Soc.*, **92**, 6076 (1970).
3120. Roberts, A. J., in *Photophysics of Synthetic Polymers* (Phillips, D., and Roberts, A. J., eds), Sbience Reviews Ltd, Northwood, England, 1982, p. 111.
3121. Roberts, A. J., and Soutar, I., in *Polymer Photophysics*, (Phillips, D., ed), Chapman and Hall Ed., London, 1985, p. 221.
3122. Roberts, A. J., Cureton, C. G., and Phillips, D., *Chem. Phys. Lett.*, **72**, 554 (1980).
3123. Roberts, A. J., O'Connor, D. V., and Phillips, D., *Ann. N.Y. Acad. Sci.*, **366**, 109 (1981).
3124. Roberts, A. J., and Phillips, D., *Macromolecules*, **15**, 678 (1982).
3125. Roberts, A. J., Phillips, D., Abdul-Rasould, F. A. M., and Ledwith, A., *J. Chem. Soc. Faraday Trans.*, I, **77**, 2725 (1981).
3126. Roberts, A. J., Phillips, D., Chapoy, L. L., and Biddle, D., *Chem. Phys. Lett.*, **103**, 271 (1984).
3127. Robertson, E. M., Van Deusen, W. P., and Minsk, L. M., *J. Appl. Polym. Sci.*, **2**, 308 (1959).
3128. Robertson, E. M., and West, W., U.S. Pat. 2,732,301 (24 Jan., 1956).
3129. Rock, W. P., *Brit. Dent. J.*, **136**, 455 (1974).
3130. Rodgers, M. A. J., and Powers, E. L., *Oxygen and Oxy-Radicals in Chemistry and Biology*, Academic Press, New York, 1981.
3131. Roffey, C. G., *Photopolymerization of Surface Coatings*, Wiley, Chichester, 1982.
3132. Rohde, O., and Wegner, G., *Makromol. Chem.*, **179**, 1999 (1978).
3133. Rohde, O., and Wegner, G., *Makrolol. Chem.*, **179**, 2013 (1978).
3134. Rollingson, S. M., *Tappi*, **38**, 625 (1955).

3135. Rooney, M. L., *Chem. Ind.* (*London*), **1982,** 197.
3136. Roots, J., and Nyström, B., *Europ. Polym. J.,* **15,** 1127 (1979).
3137. Roots, J., and Nyström, B., *Polym. Commun.,* **25,** 166 (1984).
3138. Roquitte, B. C., *J. Phys. Chem.,* **70,** 1334, 2699 (1966).
3139. Rosato, D. V., and Schwartz, R. J., *Environmental Effect on Plastics Materials,* Wiley-Interscience, New York, Vol. 1–2.
3140. Rose, W. G., Walden, M. K., and Moore, J. E., *Text. Res. J.,* **31,** 495 (1961).
3141. Rosenthal, I., and Acher, A. J., *Israel J. Chem.,* **12,** 897 (1974).
3142. Roucoux, C., Loucheux, C., and Lablache-Combier, A., *J. Appl. Polym. Sci.,* **26,** 1221 (1981).
3143. Rowell, R. M., Feist, W. C., and Ellis, W. D., *Wood Sci.,* **13,** 202 (1981).
3144. Rubin, M. B., *Topics Current Chem.,* **13,** 251 (1969).
3145. Rubner, R., Ahne, H., Kühn, E., and Kolodlie, J. G., *Photogr. Sci. Eng.,* **23,** 303 (1979).
3146. Rudolph, H., Rosenkrantz, H. J., and Heine, H. G., *J. Appl. Polym. Sci.,* Appl. Polym. Symp., No. 26, 157 (1975).
3147. Rugg, F. M., Smith, J. J., and Waterman, L. H., *J. Polym. Sci.,* **11,** 1 (1953).
3148. Rujimethabhas, M., and Weir, N. A., *Europ. Polym. J.,* **19,** 779 (1983).
3149. Rumbles, G., in *Photophysics of Synthetic Polymers,* (Phillips, D., and Roberts, A. J., eds), Science Reviews Ltd., Northwood, England, 1982, p. 5.
3150. Russeau, G., LePerchec, P., and Conia, J. M., *Tetrahed. Lett.,* **1977,** 2517.
3151. Russell, C. A., and Pascale, J. V., *J. Appl. Polym. Sci.,* **7,** 959 (1963).
3152. Rust, J. B., Miller, L. J., and Margerum, J. D., *Polym. Eng. Sci.,* **9,** 40 (1969).
3153. Rutherford, H., and Soutar, I., *J. Polym. Sci.,* A2, **15,** 2213 (1977).
3154. Rybny, C., *J. Oil Color Chem. Assoc.,* **61,** 179 (1978).
3155. Sachikiko, I., Suong, B. H., and Takeo, S., *Tetrahed.,* **1969,** 279.
3156. Sadafule, D. S., and Panda, S. P., *J. Appl. Polym. Sci.,* **24,** 511 (1979).
3157. Sadafule, D. S., and Panda, S. P., *Polym. Photochem.,* **2,** 13 (1982).
3158. Sadhir, R. K., Smith, J. D. B., and Castle, P. M., *J. Polym. Sci.,* A1, **21,** 1315 (1983).
3159. Sadrmohagegh, G., Scott, G., and Setoudeh, E. *Polym. Degrad. Stabil.,* **3,** 469 (1980/81).
3160. Saegusa, T., and Matsumoto, S., *J. Polym. Sci.,* **6,** 1559 (1968).
3161. Sahul, K., Natarajan, L. V., and Aaruddin, Q., *J. Polym. Sci.,* B, **15,** 603 (1977).
3162. Sahyun, M. R. V., *Photogr. Sci. Eng.,* **16,** 63 (1972).
3163. Saito, I., Imura, M., and Matssura, T., *Tetrahed.,* **1972,** 5307.
3164. Sakamoto, M., Hayashi, K., and Okamura, S., *J. Polym. Sci.,* B, **3,** 205 (1965).
3165. Sakamoto, M., Huy, S., Nakanishi, H., Nakanishi, F., Yurugi, T., and Hasegawa, M., *Chem. Lett.,* **1981,** 99.
3166. Sakota, N., Kisihiue, K., Shimada, S., and Minoura, Y., *J. Polym. Sci.,* A1, **12,** 1787 (1974).
3167. Sakota, N., Takahashi, K., and Nishihara, K., *Makromol. Chem.,* **161,** 173 (1972).
3168. Salmassi, A., Eichler, J., Herz, C. P., and Schnabel, W., *Polym. Photochem.,* **2,** 209 (1982).
3169. Salmassi, A., and Schnabel, W., *Polym. Photochem.,* **5,** 215 (1984).
3170. Salomon, G. A., and Noyes, R. M., *J. Amer. Chem. Soc.,* **84,** 672 (1962).
3171. Saltiel, J., Curtis, H. C., and Jones, B., *Mol. Photochem.,* **2,** 331 (1970).
3172. Saltiel, J., Curtis, H. C., Metts, L., Miley, J. W., Winterle, J., and Wrighton, M., *J. Amer. Chem. Soc.,* **92,** 410 (1970).
3173. Saltiel, J., and Megarity, E. D., *J. Amer. Chem. Soc.,* **94,** 2742 (1972).
3174. Salvin, R., Meybeck, J., and Faure, J., *Makromol. Chem.,* **178,** 2275 (1977).
3175. Samsonova, L. V., Taims, L., and Pospišil, J., *Angew. Makromol. Chem.,* **65,** 197 (1977).
3176. Sanchez, G., Knoesel, R., and Weill, G., *Europ. Polym. J.,* **14,** 485 (1978).
3177. Sanchez, G., Weill, G., and Knosel, R., *Makromol. Chem.,* **179,** 131 (1978).
3178. Sander, M. R., and Osborn, C. L., *Tetrahed. Lett.,* **1974,** 415.
3179. Sander, M. R., Osborn, C. L., and Trecker, D. J., *J. Polym. Sci.,* A1, **10,** 3173 (1972).
3180. Sander, M. R., Trecker, D. J., and Hedaya, E., *J. Amer. Chem. Soc.,* **90,** 7249 (1968).
3181. Sandhu, M. A., Savage, D. J., and Martin, T. W., *Photogr. Sci. Eng.,* **33,** 159 (1979).
3182. Sander, R. D., Austin, R. G., Wrighton, M. S., Honnick, W. D., and Pittman, L. U., Jr., in *Interfacial Photoprocesses, Energy Conversion and Synthesis* (Wrighton, M. S., ed.), ACS Symp. Ser., No. 184, ACS Washington, D.C., 1980, p. 13.
3183. Santus, R., and Grossweiner, L. I., *Photochem. Photobiol.,* **15,** 101 (1972).
3184. Sarbolouki, M. N., and Fedors, R. F., *J. Polym. Sci.,* B, **17,** 629 (1979).
3185. Sasada, Y., Shimanouchi, H., Nakanishi, H., and Hasegawa, M., *Bull. Chem. Soc. Japan,* **44,** 1262 (1971).

732

3186. Sasaki, T., Kanematsu, K., Hayakawa, K., and Sugira, M., *J. Amer. Chem. Soc.*, **75**, 355 (1975).
3187. Sasse, W. H. F., in *Solar Power and Fuels* (Bolton, J. R., ed.), Academic Press, New York, 1977, p. 277.
3188. Sastre, R., and Gonzales, F., *Polym. Photochem.*, **1**, 153 (1981).
3189. Sastre, R., Martinez, G., Castillo, F., and Millan, J. L., *Makromol. Chem. Rapid Commun.*, **5**, 541 (1984).
3190. Sato, T., Abe, M., and Otsu, T., *Makromol. Chem.*, **180**, 1165 (1979).
3191. Sato, T., and Obase, H., *Tetrahed. Lett.*, **1967**, 1633.
3192. Sato, T., and Otsu, T., *Makromol. Chem.*, **178**, 1941 (1977).
3193. Saunders, D., and Heicklen, J., *J. Phys. Chem.*, **70**, 1950 (1966).
3194. Saunders, D. S., and Winnik, M. A., *Macromolecules*, **11**, 25 (1978).
3195. Sauteret, C., Hermann, J. P., Frey, R., Pradere, F., Ducuing, J., Baughman, R. H., and Chance, R. R., *Phys. Rev. Lett.*, **36**, 956 (1976).
3196. Savides, C., *SPE J.*, **29**, 38 (1973).
3197. Scaiano, J. C., Bays, J. P., and Encinas, M. V., in *Photodegradation and Photostabilization of Coatings* (Pappas, S. P., and Winslow, F. H., eds), ACS Symp. Ser. No. 151, ACS Washington, D.C., 1981, p. 19.
3198. Scaiano, J. C., and Selwyn, J. C., *Macromolecules*, **14**, 1723 (1981).
3199. Scaiano, J. C., and Wubbeis, G. G., *J. Amer. Chem. Soc.*, **103**, 640 (1981).
3200. Schaap, A. P., ed., *Singlet Molecular Oxygen*, Dowden, Hutchinson Ross, Inc., Stroudsburg, USA, 1976.
3201. Schaap. A. P., Thayer, A. L., Blossey, E. C., and Neckers, D. C., *J. Amer. Chem. Soc.*, **97**, 3741 (1975).
3202. Scharf, von H. D., Fleischhauer, J., Schleker, W., and Weitz, R., *Angew. Chem.*, **91**, 696 (1979).
3203. Schenck, G. O., Cziesla, M., Eppinger, K., Matthias, G., and Pope, M., *Tetrahed. Lett.*, **1967**, 193.
3204. Schindler, A., Gratzl, M., and Platt, K. L., *J. Polym. Sci.*, A1, **15**, 1541 (1977).
3205. Schlessinger, S. I., *Polym. Eng. Sci.*, **14**, 513 (1974).
3206. Schlesinger, S. I., *Photogr. Sci. Eng.*, **18**, 387 (1974).
3207. Schmehl, R. H., Whitesell, L. G., and Whitten, D. G., *J. Amer. Chem. Soc.*, **103**, 3761 (1981).
3208. Schmidle, C. J., *Amer. Ink Maker*, **55**, 41,44,103 (1977).
3209. Schmidt, S. R., *J. Radiat, Curing*, **11**, Apr. 19 (1984).
3210. Schmitt, R. G., and Hirt, R. G., *J. Polym. Sci.*, **61**, 361 (1962).
3211. Schmitt, R. G., and Hirt, R. G., *J. Appl. Polym. Sci.*, **7**, 1565 (1963).
3212. Schnabel, W., *Pure Appl. Chem.*, **51**, 2373 (1979).
3213. Schnabel, W., *Photogr. Sci. Eng.*, **23**, 154 (1979).
3214. Schnabel, W., *Makromol. Chem.*, **180**, 1487 (1979).
3215. Schnabel, W., *Polym. Eng. Sci.*, **20**, 688 (1980).
3216. Schnabel, W., *Polymer Degradation: Principles and Practical Applications*, Hanser Int., München, 1981.
3217. Schnabel, W., in *Photophysics of Synthetic Polymers* (Phillips, D., and Roberts, A. J., eds), Science Reviews, Ltd., Northwood, England, 1982, p. 55.
3218. Schnabel, W., in *Developments in Polymer Degradation* (Allen, N. S., ed.), Vol. 2, Applied Science Publishers, Ltd, London, 1982, p. 237.
3219. Schnepp, O., and Levy, M., *J. Amer. Chem. Soc.*, **84**, 172 (1962).
3220. Schneider, F., *Z. Naturforsch.*, **24a**, 863 (1969).
3221. Schneider, F., and Springer, J., *Makromol. Chem.*, **146**, 181 (1971).
3222. Schneider, F., *Naturforsch.*, A, **24a**, 863 (1969).
3223. Schönberg, A., Fateen, A. K., and Omran, S. M. A. R., *J. Amer. Chem. Soc.*, **78**, 1224 (1956).
3224. Schönberg, A., and Mustafa, A., *J. Chem. Soc.*, **1944**, 67.
3225. Schopov, I., and Jossifov, C., *Makromol. Chem.*, **180**, 2037 (1979).
3226. Schroeter, S. H., *Amer. Chem. Soc. Div. Org. Coat. Plast. Chem.*, New York, 146th Meeting, Aug. 1972, Vol. 32 (2), p. 401.
3227. Schroeter, S. H., and Moore, J. C., *Amer. Chem. Soc., Div. Org. Coat. Plast. Chem.*, New York, 164th Meeting, Aug. 1972, Vol. 32(2), p. 404.
3228. Schulman, S. G., *Fluorescence and Phosphorescence Spectroscopy: Physicochemical Principles and Practice*, Pergamon Press, Oxford, 1977.
3229. Schulz, R. C., *Pure Appl. Chem.*, **34**, 305 (1973).
3230. Schulz, R. C., and Büssler, H. H., *Umschau*, **1971**, 673.

3231. Schulz, R. C., Rohe, L., and Adler, H., *Europ. Polym. J.*, Suppl., **1969**, p. 309.
3232. Schultze, H., *Makromol. Chem.*, **172,** 57 (1973).
3233. Schumb, W. C., Satterfield, C. N., and Wentworth, R. L., *Hydrogen Peroxide*, Reinhold, New York, 1955.
3234. Schuster, D. I., and Brizzolara, D. F., *J. Amer. Chem. Soc.*, **92,** 4359 (1970).
3235. Schuster, D. I., Goldstein, M. D., and Bane, P., *J. Amer. Chem. Soc.*, **99,** 187 (1977).
3236. Schuster, D. I., and Weil, T. M., *J. Amer. Chem. Soc.*, **95,** 4091 (1973).
3237. Schuster, D. I., and Weil, T. M., *Mol. Photochem.*, **6,** 69 (1974).
3238. Schuster, D. I., Weil, T. M., and Halpern, A. M., *J. Amer. Chem. Soc.*, **94,** 8248 (1972).
3239. Schuster, D. I., Weil, T. M., and Topp, M. R., *Chem. Commun.*,**1971**, 1212.
3240. Schwartz, A., Weisbrook, J. B., and Turner, D. T., *Macromolecules*, **14,** 216 (1981).
3241. Schwendiman, D. P., and Kutal, C., *J. Amer. Chem. Soc.*, **99,** 5677 (1977).
3242. Schweitzer, D., Colpa, J. P., Behnke, J., Hausser, K. H., Haenel, M., and Staab, H. A., *Chem. Phys.*, **11,** 373 (1975).
3243. Scott, G., *Atmospheric Oxidation and Antioxidants*, Elsevier, London, 1965.
3244. Scott, G., *Plast. Rubb. Text.*, **1,** 361 (1970).
3245. Scott, G., *Brit. Polym. J.*, **3,** 24 (1871).
3246. Scott, G., *Pure Appl. Chem.*, **30,** 267 (1972).
3247. Scott, G., *Macromol. Chem.*, **8,** 319 (1973).
3248. Scott, G., in *Polymers and Ecological Problems* (Guillet, J. E., ed.), Plenum Press, New York, 1973, p. 27.
3249. Scott, G., *J. Oil. Colour Chem. Assoc.*, **56,** 521 (1973).
3250. Scott, G., in *Ultraviolet Light Induced Reactions with Polymers* (Labana, S. S., ed.), ACS Symp. Ser., No. 25, ACS Washington, D.C., 1976, p. 340.
3251. Scott, G., in *Developments in Polymer Degradation* (Grassie, N., ed.), Vol. 1, Applied Science Publishers, Ltd, London, 1977, p. 205.
3252. Scott, G., *Polym. Plast. Technol. Eng.*, **11,** 1 (1978).
3253. Scott, G., in *Singlet Oxygen: Reactions with Organic Compounds and Polymers* (Rånby, B., and Rabek, J. F., eds), Wiley, Chichester, 1978, p. 230.
3254. Scott, G., *Pure Appl. Chem.*, **52,** 365 (1980).
3255. Scott, G., in *Developments in Polymer Stabilization* (Scott, G., ed.), Vol. 4, Applied Science Publishers, Ltd, London, 1981, p. 1.
3256. Scott, G., in *Developments in Polymer Stabilization* (Scott, G., ed.), Vol. 6, Applied Science Publishers, Ltd, London, 1983, p. 29.
3257. Scott, G., *Polym. Eng. Sci.*, **24,** 1007 (1984).
3258. Scott, G., and Setoudeh, E., *Europ. Polym. J.*, **18,** 901 (1982).
3259. Scott, G., Sheena, H. H., and Harriman, A. M., *Europ. Polym. J.*, **14,** 1071 (1978).
3260. Scott, G., and Smith, K. V., *Europ. Polym. J.*, **14,** 39 (1978).
3261. Scott, G., and Suharto, R., *Europ. Polym. J.*, **20,** 139 (1984).
3262. Scott, G., and Tahan, M., *Europ. Polym. J.*, **11,** 535 (1975).
3263. Scott, G., and Tahan, M., *Europ. Polym. J.*, **13,** 981 (1977).
3264. Scott, G., and Tahan, M., *Europ. Polym. J.*, **13,** 989 (1977).
3265. Scott, G., and Tahan, M., *Europ. Polym. J.*, **13,** 997 (1977).
3266. Scott, G., Tahan, M., and Vyvoda, J., *Chem. Ind.* (*London*), **1976,** 903.
3267. Scott, G., Tahan, M., and Vyvoda, J., *Europ. Polym. J.*, **14,** 377 (1978).
3268. Scott, G., Tahan, M., and Vyvoda, J., *Europ. Polym. J.*, **14,** 1021 (1978).
3269. Scott, G., Tedder, J. M., and Walton, J. C., *J. Chem. Soc. Perkin II*, **1980,** 260.
3270. Scott, G., and Yusoff, M. F., *Polym. Degrad. Stabil.*, **2,** 321 (1980).
3271. Scott, G., and Yusoff, M. F., *Polym. Degrad. Stabil.*, **3,** 13 (1980).
3272. Scott, G., and Yusoff, M. F., *Polym. Degrad. Stabil.*, **3,** 53 (1980).
3273. Scott, G., and Yusoff, M. F., *Europ. Polym. J.*, **16,** 497 (1980).
3274. Scott, G. E., and Senogles, E., *J. Macromol. Sci. Rev. Macromol. Chem.*, **9,** 49 (1973).
3275. Scott, G. E., and Senogles, E., *J. Macromol. Sci. Chem.*, **8,** 753 (1974).
3276. Scott, J. L., *J. Cat. Technol.*, **49,** 27 (1977).
3277. Searby, M. H., and Norrish, R. G. W., *Proc. Roy. Soc.*, A, **220,** 464 (1956).
3278. Searle, R., Williams, J. L. R., Doty, J. C., DeMeyer, D. E., Merrill, S. H., and Laalso, T. M., *Makromol. Chem.*, **107,** 246 (1967).
3279. Sedlar, J., Kovarova, J., and Pospišil, J., *Polym. Photochem.*, **1,** 25 (1981).
3280. Sedlar, J., Marchal, J., and Petruj, J., *Polym. Photochem.*, **2,** 175 (1982).
3281. Sedlar, J., Petruj, J., Pac, J., and Navratil, M., *Polymer*, **21,** 5 (1980).

734

3282. Sedlar, J., Petruj, J., Pac, J., and Zahradinckova, A., *Europ. Polym. J.*, **16**, 659 (1980).
3283. Sedlar, J., Petruj, J., Pac, J., and Zahradnickova, A., *Europ. Polym. J.*, **16**, 663 (1980).
3284. Seely, G. K., and Meyer, T. H., *Photochem. Photobiol.*, **13**, 27 (1971).
3285. Seguchi, T., and Tomka, N., *J. Phys. Chem.*, **77**, 40 (1972).
3286. Seiber, R. P., and Needles, H. G., *J. Appl. Polym. Sci.*, **19**, 2187 (1975).
3287. Seidel, H. P., and Sellinger, B. K., *Aust. J. Chem.*, **18**, 977 (1965).
3288. Selli, E., Bellobono, I. R., Tolusso, F., and Calgari, S., *Ann. Chim. (Roma)*, **71**, 147 (1981).
3289. Selli, E., Bellobono, I. R., Calgari, S., and Berlin, A., *J. Soc. Dyers Colour.*, **97**, 438 (1981).
3290. Selwyn, J. C., Scaiano, J. C., *Polym. Commun.*, **21**, 1365 (1980).
3291. Semerak, S. N., and Frank, C. W., *Polym. Prepr., Amer. Chem. Soc., Div. Polym, Chem.*, **22**, 314 (1981).
3292. Semerak, S. N., and Frank, C. W., *Macromolecules*, **14**, 443 (1981).
3293. Semerak, S. N., and Frank, C. W., *Adv. Chem. Ser.*, No. 203, 757 (1983).
3294. Semerak, S. N., and Frank, C. W., *Adv. Polym. Sci.*, **54**, 31 (1983).
3295. Semerak, S. N., and Frank, C. W., *Macromolecules*, **17**, 1148 (1984).
3296. Sengupta, P. K., and Bevington, J. C., *Polymer*, **14**, 527 (1973).
3297. Sengupta, P. K., and Modak, S. K., *J. Macromol. Sci. Chem.*, A, **20**, 789 (1983).
3298. Senih, G. A., and Florin, R. E., *J. Macromol. Sci. Rev. Macromol. Chem.*, C, **24**, 240 (1984).
3299. Shand, M. L., Chance, R. R., and Silbey, R., *Chem. Phys. Lett.*, **64**, 448 (1979).
3300. Shanks, R. A., and Clarke, S. R., *Polym. Photochem.*, **3**, 157 (1983).
3301. Sharf, D., and Korte, F., *Tetrahed. Lett.*, **1963**, 821.
3302. Sharma, Y. N., Naqvi, M. K., Gawande, P. S., and Bhardwaj, I. S., *J. Appl. Polym. Sci.*, **27**, 2605 (1982).
3303. Shea, K. J., Okhata, Y., and Dougherty, T. K., *Macromolecules*, **17**, 296 (1984).
3304. Sheeham, J. C., and Wilson, R. M., *J. Amer. Chem. Soc.*, **86**, 5277 (1964).
3305. Sheeham, J. C., Wilson, R. M., and Oxford, W. A., *J. Amer. Chem. Soc.*, **93**, 7222 (1971).
3306. Sheehan, W. C., and Cole, T. R., *J. Appl. Polym. Sci.*, **8**, 2359 (1964).
3307. Shelton, J. R., *Rubb. Chem. Technol.*, **47**, 949 (1974).
3308. Shelton, J. R., and Harrington, E. R., *Rubb. Chem. Technol.*, **49**, 147 (1976).
3309. Shelton, J. R., and Uzelmeier, C. W., *J. Org. Chem.*, **35**, 1576 (1970).
3310. Sherriff, A. I. M., and Santappa, M., *J. Pooym. Sci.*, A, **3**, 3131 (1965).
3311. Shiah, T. Y. J., and Morawetz, H., *Macromolecules*, **17**, 792 (1984).
3312. Shiga, M., Takagi, M., and Ueno, K., *Chem. Lett.*, **1980**, 1021.
3313. Shigehara, K., Sano, H., and Tsuchida, E., *Makromol. Chem.*, **179**, 1531 (1978).
3314. Silov, J. B., and Denisov, E. I., *Vyskomol. Soedin.*, A, **16**, 2313 (1974).
3315. Shimada, J. I., and Kabuki, K., *J. Appl. Polym. Sci.*, **12**, 655 (1968).
3316. Shimada, J. I., and Kabuki, K., *J. Appl. Polym. Sci.*, **12**, 671 (1968).
3317. Shimizu, S., and Bird, G., *J. Electrochem. Soc.*, **124**, 1394 (1977).
3318. Shinitzky, M., Dianoux, A. C., Gitler, C., and Weber, G., *Biochem.*, **10**, 2106 (1971).
3319. Shinkai, S., Kinda, H., Ishihara, M., and Manabe, E. O., *J. Polym. Sci.*, A1, **21**, 3525 (1983).
3320. Shinkai, S., Minami, T., Kusano, Y., and Manabe, O., *J. Amer. Chem. Soc.*, **104**, 1967 (1982).
3321. Shinkai, S., Nakaji, T., Nishida, Y., Ogawa, T., and Manabe, O., *J. Amer. Chem. Soc.*, **102**, 5860 (1980).
3322. Shinkai, S., Nakaji, T., Ogawa, T., Shigematsu, K., and Manabe, O., *J. Amer. Chem. Soc.*, **103**, 111 (1981).
3323. Shinkai, S., Ogawa, T., Nakaji, T., Kusano, Y., and Manabe, O., *Tetrahedron Lett.*, **35**, 4569 (1979).
3324. Shinkai, S., Shigematsu, K., Kusano, Y., and Manabe, O., *J. Chem. Soc., Perkin, I.*, **1981**, 3279.
3325. Shiono, T., Niki, E., and Kamiya, Y., *Bull. Chem. Soc. Japan*, **51**, 3290 (1978).
3326. Shirai, M., Masuda, T., Tsunooka, M., and Tanaka, M., *Makromol. Chem. Rapid Commun.*, **5**, 689 (1984).
3327. Shirai, S., Seno, M., Ishii, M., and Asahara, T., *J. Appl. Polym. Sci.*, **20**, 2429 (1976).
3328. Shirota, Y., Kawai, K., Yamamoto, N., Tada, K., Shida, T., Mikawa, H., Tsubomura, H., *Bull. Chem. Soc. Japan*, **45**, 2883 (1972).
3329. Shirota, Y., Matsumoto, A., and Mikawa, H., *Polym. J.*, **3**, 643 (1972).
3330. Shirota, Y., Tada, K., Shimazu, M., Kusabayashi, S., and Mikawa, H., *Chem. Commun.*, **1970**, 1110.
3331. Shlyapintokh, V. Ya., *Photochemical Reactions and Stabilization of Polymers* (in Russian), Khimia, Moscow, 1979.

3332. Shlyapintokh, V. Ya., in *Developments in Polymer Photochemistry*, (Allen, N. S., ed.), Vol. 2, Applied Science Publishers, London, 1981, p. 215.
3333. Shlyapintokh, V. Ya., and Ivanov, V. B., in *Developments in Polymer Stabilization* (Scott, G., ed.), Vol. 5, Applied Science Publishers, Ltd, London, 1982, p. 41.
3334. Shlyapintokh, V. Ya., Ivanov, V. B., Khvostach, O. M., Shapiro, A. B., and Rozantsev, E. G., *Dokl. Akad. Nauk SSSR*, **225**, 1132 (1975).
3335. Shultz, A. R., *J. Chem. Phys.*, **29**, 200 (1958).
3336. Shultz, A. R., *J. Polym. Sci.*, **47**, 267 (1960).
3337. Shultz, A. R., *J. Phys. Chem.*, **65**, 967 (1961).
3338. Shultz, A. R., *J. Appl. Polym. Sci.*, **10**, 353 (1966).
3339. Shultz, A. R., *J. Polym. Sci.*, Polym. Symp., No. 25, 115 (1968).
3340. Shultz, A. R., in *Durability of Macromolecular Materials* (Eby, R. K., ed.), ACS Symp. Ser., No. 95, ACS Washington, D.C., 1979, p. 29.
3341. Shultz, A. R., *J. Polym. Sci.*, A2, **22**, 1753 (1984).
3342. Shultz, A. R., and Leahy, S. M., *J. Appl. Polym. Sci.*, **5**, 64 (1961).
3343. Siegel, S., Champetier, R. J., and Calloway, A. R., *J. Polym. Sci.*, A1, **4**, 2107 (1966).
3344. Sienicki, K., and Bojarski, C., *Polym. Photochem.*, **4**, 435 (1984).
3345. Sierocka, M., Lyk, B., Paczkowski, J., Zakrzewski, A., and Wrzeszczynski, A., *Polym. Photochem.*, **4**, 207 (1984).
3346. Silbert, L. S., in *Organic Peroxides* (Swen, D., ed.), Vol. 2, Wiley, New York, 1971, p. 678.
3347. Singer, L. A., *Tetrahed. Lett.*, **1969**, 923.
3348. Singer, L. A., and Bartlett, P. D., *Tetrahed. Lett.*, **1964**, 1887.
3349. Singer, L. A., Brown, R. E., and Davis, G. A., *J. Amer. Chem. Soc.*, **95**, 8638 (1973).
3350. Singh, A., *Photochem. Photobiol.*, **28**, 429 (1978).
3351. Singh, A., McIntyre, N. R., and Koroll, G. W., *Photochem. Photobiol.*, **28**, 595 (1978).
3352. Singh, H., and Tedder, J. M., *J. Chem. Soc. Chem. Comm.*, **1981**, 70.
3353. Singh, R. P., *Polym. Bull.*, **5**, 443 (1981).
3354. Singh, R. P., *Polym. Photochem.*, **2**, 331 (1982).
3355. Singh, R. P., and Chandra, R., *Polym. Photochem.*, **2**, 257 (1982).
3356. Sisido, M., Imanishi, Y., and Higashimura, T., *Macromolecules*, **12**, 975 (1979).
3357. Sitek, F., and Guillet, J. E., *J. Polym. Sci.*, Polym. Symp., No. 57, 343 (1976).
3358. Sixl, H., *Adv. Polym. Sci.*, **63**, 51 (1984).
3359. Skilton, P. F., and Ghiggino, K. P., *Polym. Photochem.*, **5**, 179 (1984).
3360. Skilton, P. F., Sakurovs, R., and Ghiggino, K. P., *Polym. Photochem.*, **2**, 409 (1982).
3361. Skowronski, T., and Rabek, J. F., *Polimery*, **10**, 441 (1971).
3362. Skowronski, T., Rabek, J. F., and Rånby, B., *Polymer*, **24**, 1189 (1983).
3363. Skowronski, T., Rabek, J. F., and Rånby, B., *Polym. Degrad. Stabil.*, **5**, 173 (1983).
3364. Skowronski, T., Rabek, J. F., and Rånby, B., *Polym. Photochem.*, **3**, 341 (1983).
3365. Skowronski, T., Rabek, J. F., and Rånby, B., *Polym. Degrad. Stabil.*, **8**, 37 (1984).
3366. Skowronski, T., Rabek, J. F., and Rånby, B., *Polym. Eng. Sci.*, **24**, 276 (1984).
3367. Skowronski, T., Rabek, J. F., and Rånby, B., *Polym. Photochem.*, **5**, 77 (1984).
3368. Slawinski, J., Puzyna, W., Slawinska, D., *Photochem. Photobiol.*, **28**, 459 (1978).
3369. Slawinska, S., and Michalska, T., in *Singlet Oxygen: Reactions with Organic Compounds and Polymers* (Rånby, B., and Rabek, J. F., eds), Wiley, Chichester, 1978, p. 294.
3370. Sloan, J. P., Tedder, J. M., and Walton, J. C., *J. Chem. Soc. Faraday Trans.*, I, **69**, 1143 (1973).
3371. Small, R. D. Jr., and Scaiano, J. C., *Macromolecules*, **11**, 840 (1978).
3372. Small, R. D. Jr., and Scaiano, J. C., *J. Amer. Chem. Soc.*, **99**, 7713 (1979).
3373. Smets, G. J., *J. Prakt. Chem.*, **313**, 546 (1971).
3374. Smets, G., *Pure Appl. Chem.*, **30**, 1 (1972).
3375. Smets, G., *Pure Appl. Chem.*, **42**, 509 (1975).
3376. Smets, G., in Proc. Intern. Symp. on Macromolecules, Rio de Janeiro, 26 July, 1974 (1975).
3377. Smets, G., *J. Polym. Sci.*, A1, **13**, 2223 (1975).
3378. Smets, G., *Adv. Polym. Sci.*, **50**, 17 (1983).
3379. Smets, G., Aerts, A., van Erum, J., *Polymer, J.*, **12**, 539 (1980).
3380. Smets, G, Bracken, J., and Irie, M., *Pure Appl. Chem.*, **50**, 845 (1978).
3381. Smets, G., De Blauwe, F., *Pure Appl. Chem.*, **39**, 225 (1974).
3382. Smets, G., De Winter, W., and Delzenne, G., *J. Polym. Sci.*, **55**, 767 (1961).
3383. Smets, G., El Hamouly, S. N., and Oh, T. J., *Pure Appl. Chem.*, **56**, 439 (1984).
3384. Smets, G., and Evans, G., *Pure Appl. Chem.*, **8**, 357 (1953).
3385. Smets, G., and Matsumoto, S., *J. Polym. Sci.*, A1, **14**, 2983 (1976).

736

3386. Smets, G., Nijst, C., Schmitz-Smets, M., and Somers, A., *J. Polym. Sci.*, Polym. Symp., No. 67, 83 (1980).
3387. Smets, G., and Schmitz-Smets, M., in *Structural Orders in Polymers* (Ciardelli, F., and Giusti, P., eds), Pergamon Press, Oxford, 1981, p. 221.
3388. Smets, G., Thoen, J., and Aerts, A., *J. Polym. Sci.*, Polym. Symp., No. 51, 119 (1975).
3389. Smirnova, T. Ya., Gladyshev, G. P., and Rafikov, S. R., *Dokl. Akad. Nauk SSSR*, **170**, 118 (1966).
3390. Smith, B. A., *Macromolecules*, **15**, 469 (1982).
3391. Smith, B. A., Samulski, E. T., Yu, L. P., and Winnik, M. A., *Phys. Rev. Lett.*, **52**, 45 (1984).
3392. Smith, P., Gilman, L. B., and De Lorenzo, R. A., *J. Magn. Res.*, **10**, 179 (1973).
3393. Smith, P., and Stevens, R. D., *J. Phys. Chem.*, **76**, 3141 (1972).
3394. Smith, P., Stevens, R. D., and Gilman, L. B., *J. Phys. Chem.*, **79**, 2688 (1975).
3395. Smith, R. C., and Wyard, S. J., *Nature*, **191**, 896 (1961).
3396. Smith, R. N., Smith, R. H., and Young, D. A., *Inorg. Chem.*, **5**, 145 (1966).
3397. Smolinski, G., Snyder, L. C., and Wasserman, E., *Rev. Mod. Phys.*, **35**, 576 (1963).
3398. Smolinsky, G., Wasserman, E., and Yager, W. A., *J. Amer. Chem. Soc.*, **84**, 3220 (1962).
3399. Smoluchowski, M., *Z. Phys. Chem.*, **92**, 129 (1917).
3400. Snelling, D. R., *Chem. Phys. Lett.*, **2**, 346 (1968).
3401. Sohn, J. E., Garito, A. F., Desai, K. N., Narang, R. S., and Kuzyk, M., *Makromol. Chem.*, **180**, 2975 (1979).
3402. Sojka, S. A., Poranski, C. F. Jr., and Monitz, W. B., *J. Amer. Chem. Soc.*, **97**, 5953 (1975).
3403. Solly, R. K., and Benson, S. W., *J. Amer. Chem. Soc.*, **93**, 1592 (1971).
3404. Solomon, D. H., *J. Macromol. Sci. Chem.*, A, **17**, 337 (1982).
3405. Somer, G., and Green, M. E., *Photochem. Photobiol.*, **17**, 179 (1973).
3406. Somersall, A. C., Dan, E., and Guillet, J. E., *Macromolecules*, **7**, 233 (1974).
3407. Somersall, A. C., and Guillet, J. E., *Macromolecules*, **5**, 410 (1972).
3408. Somersall, A. C., and Guillet, J. E., *Macromolecules*, **6**, 218 (1973).
3409. Somersall, A. C., and Guillet, J. E., *J. Macromol. Sci. Rev. Macromol. Chem.*, **13**, 135 (1975).
3410. Son, P. N., *Polym. Degrad. Stabil.*, **2**, 295 (1980).
3411. Song. Z., Rånby, B., Gupta, A., Borsig, E., and Vogl, O., *Polym. Bull.*, **12**, 245 (1984).
3412. Sonntag, C. V., *Fortsch. Chem. Forsch.*, **13**, 333 (1969).
3413. Sonntag, C. V., and Schuchmann, H. P., *Adv. Photochem.*, **10**, 59 (1977).
3414. Sonntag, F. I., and Srinivasan, R., *Tech. Pap., Reg. Techn. Conf. Soc. Plast. Eng., Mid-Hudson Sect.*, 1967, p. 163.
3415. Soutar, I., *Ann. N.Y. Acad. Sci.*, **336**, 24 (1981).
3416. Soutar, I., in *Developments in Polymer Photochemistry* (Allen, N. S., ed.), Vol.3, Applied Science Publisher's Ltd, London, 1982, p. 125.
3417. Soutar, I., Philips, D., Roberts, A. J., and Rumbles, G., *J. Polym. Sci.*, A2, **20**, 1759 (1982).
3418. Speed, R., and Selinger, B., *Aust. J. Chem.*, **22**, 9 (1969).
3419. Spencer, L. R., Heskins, M., and Guillet, J. E., in *Proc. 3rd Intern. Biodegradation Symp.* (Sharpley, J. M., and Kaplan, A. M., eds), Applied Science Publishers, Ltd, London, 1976, p. 753.
3420. Springer, J., Ueberreiter, K., and Wenzel, R., *Makromol. Chem.*, **96**, 134 (1966).
3421. Srinivasan, R., *Polymer*, **23**, 1863 (1982).
3422. Srinivasan, R., *J. Radiat. Curing*, **10**, Oct., 12 (1983).
3423. Srinivasan, R., *J. Vac. Sci., Tech.*, B, **1**, 923 (1983).
3424. Srinivasan, R., and Braken, B., *J. Polym. Sci.*, A1, **22**, 2601 (1984).
3425. Srinivasan, R., and Leigh, W. J., *J. Amer. Chem. Soc.*, **104**, 6784 (1982).
3426. Srinivasan, R., Mayne-Banton, V., *Appl. Phys. Lett.*, **41**, 576 (1982).
3427. Starnes, W. H. Jr., in *Photodegradation and Photostabilization of Polymers*, (Pappas, S. P., and Winslow, F. H., eds), ACS Symp. Ser. 151, ACS Washington, D.C., 1981, p. 197.
3428. Steenken, S., Schuchmann, H. P., and von Sonntag, C., *J. Phys. Chem.*, **79**, 763 (1975).
3429. Steiner, R. F., *Excited States in Biopolymers*, Plenum Press, New York, 1983.
3430. Steinlin, F., and Saar, W., *Melliand Textilberichte* II, **1980**, 941.
3431. Stelter, T., and Springer, J., *Makromol. Chem.*, **185**, 1719 (1984).
3432. Stenberg, V. I., Olson, R. D., Wang, C. T., and Kulevski, N., *J. Org. Chem.*, **32**, 3327 (1967).
3433. Stenberg, V. I., Wang, C. T., and Kulevsky, N., *J. Org. Chem.*, **35**, 1774 (1970).
3434. Stephenson, C. V., Lacey, J. C. Jr., and Wilcox, W. S., *J. Polym. Sci.*, **55**, 477 (1961).
3435. Stephenson, C. V., Moses, B. C., Burks, R. E. Jr., Coburn, W. C. Jr., and Wilcox, W. S., *J. Polym. Sci.*, **55**, 465 (1961).

737

3436. Stephenson, C. V., Moses, B. C., and Wilcox, W. S., *J. Polym. Sci.*, **55**, 451 (1961).
3437. Stephenson, C. V., and Wilcox, W. S., *J. Polym. Sci.*, A1, **1**, 2741 (1963).
3438. Stephenson, L. M., Gridina, M. J., and Orfanopoulos, M., *Acc. Chem. Res.*, **13**, 419 (1980).
3439. Steurer, E., *Z. Phys. Chem*, **47**, 127 (1940).
3440. Stevens, B., *Acc. Chem. Res.*, **6**, 90 (1973).
3441. Stevens, B., in *Singlet Oxygen: Reactions with Organic Compounds and Polymers* (Rånby, B., and Rabek, J. F. eds), Wiley, Chichester, 1978, p. 54.
3442. Stevens, B., and Algar, B. E., *Ann. N. Y. Acad. Sci.*, **171**, 50 (1970).
3443. Stevens, B., and Ors, J. A., *J. Phys. Chem.*, **80**, 2164 (1976).
3444. Stevens, B., Ors, J. A., and Pinsky, L. M., *Chem. Phys. Lett.*, **27**, 355 (1975).
3445. Stevens, M. P. in *Tech. Pap. Rep. Techn. Conf. SPE*, **1977**, 17. CA. **86**, 56111 (1977).
3446. Stewart, L. C., Carlsson, D. J., Wiles, D. M., and Scaiano. J. C., *J. Amer. Chem. Soc.*, **105**, 3605 (1983).
3447. Stief, L. J., and DeCarlo, V. J., *J. Chem. Phys.*, **50**, 1234 (1969).
3448. Stigman, A. E., and Tyler, D. R., *Acc. Chem. Res.*, **17**, 61 (1984).
3449. Stockmayer, W. H., and Matsud, K., *Macromolecules*, **5**, 766 (1972).
3450. Stokes, S., and Fox, R. B., *J. Polym. Sci.*, **56**, 507 (1962).
3451. Štolka, M., *Macromolecules*, **8**, 8 (1975).
3452. Stowe, B. S., Fornes, R. E., and Gilbert, R. D., *Polym. Plast. Tech. Eng.*, **3**, 159 (1974).
3453. Strachan, A. N., and Blacet, F. I., *J. Amer. Chem. Soc.*, **77**, 5254 (1955).
3454. Strambini, G., and Galley, W. C., *J. Chem. Phys.*, **63**, 3467 (1975).
3455. Strauss, S., and Madorsky, S. L., *J. Res. Nat. Bur. Stand.*, **50**, 165 (1953).
3456. Strelkova, D. O., Fedoseva, G. T., and Minser, K. C., *Vysokomol. Soedin.*, A, **18**, 2064 (1976).
3457. Strobel, A. F., and Catino, S. C., *Ind. Eng. Chem., Prod. Res. Dev.*, **1**, 241 (1962).
3458. Strohmeier, W., *Z. Naturforsch*, B, **19**, 655 (1964).
3459. Strohmeier, W., *Z. Naturforsch.*, B, **22**, 98 (1967).
3460. Strohmeier, W., *Z. Naturforsch.*, B, **22**, 113 (1967).
3461. Strohmeier, W., and Barbeau, C., *Makromol. Chem.*, **81**, 86 (1965).
3462. Strohmeier, W., and Grübel, H., *Z. Naturforsch.*, B, **22**, 553 (1967).
3463. Strohmeier, W., and Hartman, P., *Z. Naturforsch.*, B, **19**, 882 (1964).
3464. Stryer, L., *J. Mol. Biol.*, **13**, 482 (1965).
3465. Stryer, L., *Science*, **162**, 526 (1968).
3466. Stryer, L., and Haugland, R. P., *Proc. Nat. Acad. Sci. U.S.*, **58**, 719 (1967).
3467. Stuber, F. A., Ulrich, H., Rao, D. V., and Sayigh, A. A. R., *J. Appl. Polym. Sci.*, **13**, 2247 (1969).
3468. Stuckey, W. C., and Roberts, C. W., *J. Appl. Polym. Sci.*, **26**, 701 (1981).
3469. Studebaker, M. L., Huffman, E. W. D., Wolfe, A. C., and Nabors, L. G., *Ind. Eng. Chem.*, **48**, 162 (1956).
3470. Studebaker, M. L., and Rinehart, R. W., *Rubb. Chem. Technol.*, **45**, 106 (1972).
3471. Subramanyan, R. V., and Santappa, M., *Current Sci.*, **24**, 230 (1955).
3472. Subramanyan, R. V., and Santappa, M., *Current Sci.*, **25**, 218 (1956).
3473. Suda, Y., Inaki, Y., and Takemoto, K., *J. Polym. Sci.*, A1, **21**, 2813 (1983).
3474. Suda, Y., Inaki, Y., and Takemoto, K., *Nucleic Acid Res.*, **10**, 169 (1983).
3475. Suda, Y., Inaki, Y., and Takemoto, K., *J. Polym. Sci.*, A1, **22**, 623 (1984).
3476. Sudnik, M. V., Romantsev, M. F., Shapiro, A. B., and Rozantsev, E. G., *Izv. Akad. Nauk SSSR, Ser. Khim.*, **1975**, 2813.
3477. Suga, K., Tsunoda, T., Yamaoka, T., Watanabe, S., and Koyama, H., *Chem. Ind. (London)*, **1967**, 1362.
3478. Sugita, K., Hase, M., and Suzuki, S., *Bull. Chem. Soc. Japan*, **45**, 1921 (1972).
3479. Sugita, K., Ide, H., Tamura, K., and Suzuki, S., *Bull. Soc. Photogr. Sci. Tech. Japan*, **1973/74**, No. 23, 6.
3480. Sugita, K., Kilp, T., and Guillet, J. E., *J. Polym. Sci.*, **14**, 1901 (1976).
3481. Sugita, K., Muroga, H., and Suzuki, S., *Polym. J.*, **4**, 351 (1973).
3482. Sugita, K., and Suzuki, S., *Polym. J.*, **2**, 283 (1971).
3483. Sukegawa, S., Masaki, K., and Ozawa, T., *Kobunshi Kagaku*, **18**, 218 (1956).
3484. Sukigara, M., and Kikuchi, S., *Bull. Chem. Soc. Japan*, **40**, 1077 (1967).
3485. Sulliwan, W. F., *Pure Org. Chem.*, **1**, 157 (1972).
3486. Sumi, K., Furue, M., Nozakura, S. T., *J. Polym. Sci.*, A1, **22**, 3779 (1984).
3487. Sumida, Y., and Vogl, O., *Polym. J.*, **13**, 521 (1981).
3488. Sumida, Y., Yoshida, S., and Vogl, O., *Polym. Prepr.* **21**(1), 201 (1980).

738

3489. Sumimoto, H., Nobutoki, K., and Susaki, K., *J. Polym. Sci.*, A1, **9**, 809 (1971).
3490. Sumiyoshi, T., Henne, A., Lechtken, P., and Schnabel, W., *Z. Naturforsch.*, **39a**, 434 (1984).
3491. Sung, C. S. P., Gould, I. R., and Turro, N. J., *Macromolecules*, **17**, 1447 (1984).
3492. Sung, C. S. P., Lamarre, L., and Chung, K. H., *Macromolecules*, **14**, 1839 (1981).
3493. Sung, C. S. P., Lamarre, L., and Tse, M. K., *Macromolecules*, **12**, 666 (1979).
3494. Suppan, P., *J. Chem. Soc. Faraday Trans.* I, **71**, 539 (1975).
3495. Suppan, P., *Ber. Bunsengs. Phys. Chem.*, **72**, 321 (1968).
3496. Sutin, N., *J. Photochem.*, **10**, 19 (1979).
3497. Suzuki, F., Suzuki, Y., Nakanishi, H., and Hasegawa, M., *J. Polym. Sci.*, A1, **7**, 2319 (1969).
3498. Suzuki, M., Yamamoto, Y., Irie, M., and Haysahi, K., *J. Macromol. Sci. Chem.*, **10**, 1607 (1976).
3499. Suzuki, S., and Sugita, K., *Photogr. Sci. Eng.*, **15**, 464 (1971).
3500. Suzuki, S., and Sugita, K., *Bull. Chem. Soc. Japan*, **44**, 641 (1971).
3501. Suzuki, S., Sugita, K., and Negishi, K., *Bull. Soc. Photogr. Sci. Tech. Japan*, **1971**, No. 21, 35.
3502. Suzuki, T., and Tazuke, S., *Macromolecules*, **13**, 25 (1980).
3503. Suzuki, Y., Hasegawa, M., and Kita, N., *J. Polym. Sci.*, A1,**10**, 2473 (1972).
3504. Suzuki, Y., Tamaki, T., and Hasegawa, M., *Bull. Chem. Soc. Japan*, **47**, 210 (1974).
3505. Svegliado, G., and Zilio-Grandi, F., *J. Appl. Polym. Sci.*, **13**, 1113 (1969).
3506. Svirskaya, P., Danhelka, J., Redpath, A. E. C., and Winnik, M. A., *Polymer*, **24**, 319 (1983).
3507. Swartz, B. A., Cole, T., and Zewail, A. H., *Opt. Lett.*, **1**, 73 (1977).
3508. Tabata, M., and Sohma, J., *Polym. Degrad. Stabil.*, **1**, 139 (1979).
3509. Tabb, D. L., Sevcik, J. J., and Koenig, J. L., *J. Polym. Sci.*, A2, **13**, 815 (1975).
3510. Tada, K., Shirota, Y., and Mikawa, H., *J. Polym. Sci.*, B, **10**, 691 (1972).
3511. Tada, K., Shirota, Y., and Mikawa, H., *Macromolecules*, **6**, 9 (1973).
3512. Tada, K., Shirota, Y., and Mikawa, H., *J. Polym. Sci.*, A1, **11**, 2961 (1973).
3513. Tada, K., Shirota, Y., and Mikawa, H., *Micromolecules*, **7**, 549 (1974).
3514. Tagawa, S., and Schnabel, W., *Macromolecules*, **12**, 663 (1979).
3515. Tagawa, S., and Schnabel, W., *Makromol. Chem. Rapid Commun.*, **1**, 345 (1980).
3516. Tagawa, S., and Schnabel, W., *Chem. Phys. Lett.*, **75**, 120 (1980).
3517. Tagawa, S., Washio, M., and Tabata, Y., *Chem. Phys. Lett.*, **68**, 276 (1979).
3518. Taimr, L., and Pospisil, J., *Angew. Makromol. Chem.*, **39**, 189 (1974).
3519. Taimr, L., and Pospisil, J., *Angew. Makromol. Chem.*, **52**, 31 (1976).
3520. Takahashi, H., Sakuragi, M., and Hasegawa, M., *J. Polym. Sci.*, B, **9**, 685 (1971).
3521. Takai, Y., Mizutani, T., and Ieda, M., *Japan J. Appl. Phys.*, **17**, 651 (1978).
3522. Takakura, K., Hayashi, K., and Okamura, S., *J. Polym. Sci.*, B, **2**, 861 (1964).
3523. Takakura, K., Hayashi, K., and Okamura, S., *J. Polym. Sci.*, B, **3**, 565 (1965).
3524. Takakura, K., and Takayama, G., *Bull. Chem. Soc. Japan*, **38**, 328 (1965).
3525. Takakura, K., Takayama, G., and Ukida, J., *J. Appl. Polym. Sci.*, **9**, 3217 (1965).
3526. Takeishi, M., and Okawara, M., *J. Polym. Sci.*, B, **7**, 201 (1969).
3527. Takemoto, K., *J. Polym. Sci.*, Polym. Symp., No. 55, 105 (1976).
3528. Takeshita, T., Tsuji, K., and Seiki, T., *J. Polym. Sci.*, A, **10**, 2315 (1972).
3529. Takeuchi, Y., Sekimoto, A., and Abe, M., in *New Industrial Polymers* (Dequin, R. D., ed.), ACS Symp. Ser. No. 4, ACS Washington, D.C., 1972, p. 26.
3530. Takuma, K., Kajiwara, M., and Matsuo, T., *Chem. Lett.*, **1977**, 1199.
3531. Takuma, K., Sakamoto, T., and Matsuo, T., *Chem. Lett.*, **1981**, 815.
3532. Tamagaki, S., Liesner, C. E., and Neckers, D. C., *J. Org. Chem.*, **45**, 1573 (1980).
3533. Tamaki, T., Suzuki, Y., and Hasegawa, M., *Bull. Chem. Soc. Japan*, **45**, 1988 (1972).
3534. Tanaka, M., *J. Soc. Photogr. Technol., Japan*, **35**, 288 (1972).
3535. Tanaka, H., *J. Nat. Chem. Lab. Ind.*, **68**, 90 (1973).
3536. Tanaka, H., Azuma, C., Sanui, K., and Ogata, N., *Polym. J.*, **12**, 63 (1980).
3537. Tanaka, H., and Otomegawa, E., *J. Polym. Sci.*, A1, **12**, 1125 (1974).
3538. Tanaka, H., and Otsu, T., *J. Polym. Sci.*, A1, **15**, 2613 (1977).
3539. Tanaka, H., and Sato, Y., *J. Polym. Sci.*, A1, **10**, 3279 (1972).
3540. Tanaka, H., Takamuku, S., and Sakurai, H., *Bull. Chem. Soc. Japan*, **52**, 801 (1979).
3541. Tanaka, H., Tsuda, M., and Nakanishi, H., *J. Polym. Sci.*, A1, **10**, 1729 (1972).
3542. Tanaka, H., Tsuda, M., and Nakanishi, H., *J. Polym. Sci.*, A1, **10**, 2379 (1972).
3543. Tanaka, T., and Tokumaru, K., *Bull. Chem. Soc. Japan*, **43**, 3315 (1970).
3544. Tanford, C., *The Hydrophobic Effect: Formation of Micelles and Biological Membranes*, Wiley, New York, 1973.
3545. Tang, D. K., and Ho, S. Y., *J. Polym. Sci.*, A1, **22**, 1357 (1984).

3546. Tang, L., Lemaire, J., Sallet, D., and Mery, J. M., *Makromol. Chem.*, **182,** 3467 (1981).
3547. Tang, L., Sallet, D., and Lemaire, J., *Macromolecules*, **15,** 1432 (1982).
3548. Tang, L., Sallet, D., and Lemaire, J., *Macromolecules*, **15,** 1437 (1982).
3549. Tanielian, C., and Chaineaux, J., *J. Photochem.*, **9,** 19 (1978).
3550. Tanielian, C., and Chaineaux, J., *Photochem. Photobiol.*, **28,** 487 (1978).
3551. Tanielian, C., and Chaineaux, J., *J. Polym. Sci.*, A1, **17,** 715 (1979).
3552. Tanielian, C., and Chaineaux, J., *Europ. Polym. J.*, **16,** 619 (1980).
3553. Tarr, A. M., and Wiles, D. M., *Can. J. Chem.*, **46,** 2725 (1968).
3554. Tassin, J. F., and Monnerie, L., *J. Polym. Sci.*, A2, **21,** 1981 (1983).
3555. Taylor, H. A., Tincher, W. C., and Hamner, W. F., *J. Appl. Polym. Sci.*, **14,** 141 (1970).
3556. Taylor, L., *Chemtech*, **1979,** Sept., 542.
3557. Taylor, L. J., and Tobias, J. W., *J. Appl. Polym. Sci.*, **26,** 2917 (1981).
3558. Tazuke, S., *J. Phys. Chem.*, **74,** 2390 (1970).
3559. Tazuke, S., *High Polym. Japan*, **29,** 681 (1977).
3560. Tazuke, S., *Polym. Plast. Techn. Eng.*, **14,** 107 (1980).
3561. Tazuke, S., in *Developments in Polymer Photochemistry*, (Allen, S. S., ed.), Vol. 3, Applied Science Publishers, Ltd, London, 1982, p. 53.
3562. Tazuke, S., Asai, M., Ikeda, M., and Okamura, S., *J. Polym. Sci.*, **35,** 453 (1967).
3563. Tazuke, S., Asai, M., and Okamura, S., *J. Polym. Sci.*, A1, **6,** 1809 (1968).
3564. Tazuke, S., and Banba, F., *Macromolecules*, **9,** 451 (1976).
3565. Tazuke, S., and Banba, F., *J. Polym. Sci.*, A1, **14,** 2643 (1976).
3566. Tazuke, S., and Hayashi, N., *J. Polym. Sci.*, A1, **16,** 2729 (1978).
3567. Tazuke, S., and Hayashi, N., *Polym. J.*, **10,** 443 (1978).
3568. Tazuke, S., and Iwaya, Y., in *Contemporary Topics in Polymer Sciences*, Plenum Press, New York, 1980.
3569. Tazuke, S., Iwaya, Y., and Hayashi, R., *Photochem. Photobiol.*, **35,** 621 (1982).
3570. Tazuke, S., and Kimura, H., *J. Polym. Sci.*, A1, **15,** 2707 (1977).
3571. Tazuke, S., and Kimura, H., *J. Polym. Sci., B*, **16,** 497 (1978).
3572. Tazuke, S., and Kimura, H., *Makromol. Chem.*, **179,** 2603 (1978).
3573. Tazuke, S., Matoba, T., Kimura, H., and Okada, T., in *Modifications of Polymers* (Carraher, C. E., Jr., and Tsuda, M., eds) ACS Symp. Ser., No. 121, ACS Washington, D.C., 1980, p. 217.
3574. Tazuke, S., and Matsuyama, Y., *Macromolecules*, **8,** 280 (1975).
3575. Tazuke, S., and Matsuyama, Y., *Polym. J.*, **8,** 481 (1976).
3576. Tazuke, S., and Matsuyama, Y., *Macromolecules*, **10,** 215 (1977).
3577. Tazuke, S., and Okamura, S., *J. Polym. Sci.*, A1, **6,** 2907 (1968).
3578. Tazuke, S., and Okamura, S., *J. Polym. Sci.*, A1, **7,** 851 (1969).
3579. Tazuke, S., Ooki, H., and Sato, K., *Macromolecules*, **15,** 400 (1982).
3580. Tazuke, S., Sato, K., and Banaba, F., *Chem. Lett. Japan*, **1975,** 1321.
3581. Tazuke, S., Sato, K., and Banaba, F., *Macromolecules*, **10,** 1224 (1977).
3582. Tazuke, S., Sato, K., and Hayasahi, N., *J. Polym. Sci.*, A1, **15,** 671 (1977).
3583. Tazuke, S., and Suzuki, Y., *J. Polym. Sci., B*, **16,** 223 (1978).
3584. Tazuke, S., and Takasaki, R., *J. Polym. Sci.*, A1, **21,** 1517 (1983).
3585. Tazuke, S., and Takasaki, R., *J. Polym. Sci.*, A1, **21,** 1529 (1983).
3586. Tazuke, S., and Tanabe, T., *Macromolecules*, **12,** 848 (1979).
3587. Tazuke, S., and Tanabe, T., *Macromolecules*, **12,** 853 (1979).
3588. Tazuke, S., and Yuan, H. L., *Polym. J.*, **14,** 215 (1982).
3589. Tazuke, S., and Yuan, H. L., *Polym. J.*, **14,** 695 (1982).
3590. Tazuke, S., and Yuan, H. L., *J. Phys. Chem.*, **86,** 1250 (1982).
3591. Tazuke, S., Yuan, H. L., Iwaya, Y., and Sato, K., *Macromolecules*, **14,** 267 (1981).
3592. Tazuke, S., Yuan, H. L., Matsumaru, T., and Yamaguchi, Y., *Chem. Phys. Lett.*, **92,** 81 (1982).
3593. Teramoto, A., Hiratsuka, H., and Nishijima, Y., *J. Polym. Sci.*, A2, **5,** 37 (1967).
3594. Teramoto, A., Morimoto, M., and Nishijima, Y., *J. Polym. Sci.*, A1, **5,** 1021 (1967).
3595. Terrell, D. R., *Polymer*, **23,** 1045 (1982).
3596. Tezuka, T., and Narita, N., *J. Amer. Chem. Soc.*, **101,** 7413 (1979).
3597. Thijs, L., Gupta, S. N., and Neckers, D. C., *J. Org. Chem.*, **44,** 4123 (1979).
3598. Thomas, D. D., *Biophys., J.*, **24,** 439 (1978).
3599. Thomas, J. W. Jr., Frank, C. W., Holden, D. A., and Guillet, J. E., *J. Polym. Sci.*, A2, **20,** 1749 (1982).
3600. Thomas, M. J., and Foote, C. S., *Photochem. Photobiol.*, **27,** 683 (1978).
3601. Tickle, K., and Wilkinson, F., *Trans. Faraday Soc.*, **61,** 1981 (1965).
3602. Tieke, B., and Bloor, D., *Makromol. Chem.*, **182,** 133 (1981).

740

3603. Tieke, B., Graf, H. J., Wegner, G., Naegele, B., Ringsdorf, H., Banerjee, A., Day, D., and Lando, J. B., *Colloid. Polym. Sci.*, **255**, 521 (1977).
3604. Tieke, B., and Lieser, G., *J. Colloid. Interface Sci.*, **83**, 230 (1981).
3605. Tieke, B., and Lieser, G., *J. Colloid Interface Sci.*, **88**, 471 (1982).
3606. Tieke, B., Lieser, G., and Weiss, K., *Thin Solid Films*, **99**, 95 (1983).
3607. Tieke, B., and Wegner, G., *Makromol. Chem.*, **179**, 1639 (1978).
3608. Tieke, B., and Wegner, G., *Makromol. Chem.*, **179**, 2573 (1978).
3609. Tieke, B., Wegner, G., Naegele, D., and Ringsdorf, H., *Angew. Chem. Int. Ed. Engl.*, **12**, 764 (1976).
3610. Tien, H. T., in *Topics of Photosynthesis* (Barber, J., ed.), Vol. 3, Elsevier–North Holland, Amsterdam, 1979, p. 116.
3611. Tirrell, D., in *Photodegradation and Photostabilization of Coatings* (Pappas, S. P., and Winslow, F. H., eds), ACS Symp. Ser. No. 151, ACS Washington, D.C., 1981, p. 43.
3612. Tirrell, D., Bailey, D., Pinazzi, C., and Vogl, O., *Macromolecules*, **11**, 312 (1978).
3613. Tirrell, D., Bailey, D., and Vogl, O., in *Polymeric Drugs* (Donaruma, G., and Vogl, O., eds), Academic Press, New York, 1978.
3614. Tirrell, D., and Vogl, O., *Makromol. Chem.*, **181**, 2097 (1980).
3615. Tocker, S., *Makromol. Chem.*, **101**, 23 (1967).
3616. Todesco, R., Gelan, J., Martens, H., Put, J., and DeSchryver, F. C., *J. Amer. Chem. Soc.*, **103**, 7304 (1981).
3617. Toman, L., and Marek, M., *Makromol. Chem.*, **177**, 3325 (1976).
3618. Toman, L., Pilar, J., and Marek, M., *J. Polym. Sci.*, A1, **16**, 371 (1978).
3619. Toman, L., Pilar, J., Spevacek, J., and Marek, M., *J. Polym. Sci.*, A1, **16**, 2759 (1978).
3620. Tomlinson, W. J., Chandros, E. A., Fork, R. L., Pryde, C. A., and Lamola, A. A., *Appl. Opt.*, **11**, 533 (1972).
3621. Topp, M. R., *Chem. Phys. Lett.*, **32**, 144 (1975).
3622. Torikai, A., and Fueki, K., *Polym. Photochem.*, **2**, 297 (1982).
3623. Torikai, A., and Fueki, K., *Polym. Degrad. Stabil.*, **6**, 81 (1984).
3624. Torikai, A., Murata, T., and Fueki, K., *Polym. Photochem.*, **4**, 255 (1981).
3625. Torikai, A., Nishiyama, M., and Fueki, K., *Polym. Degrad.*, **4**, 281 (1984).
3626. Torikai, A., Suzuki, K., and Fueki, K., *Polym. Photochem.*, **3**, 379 (1983).
3627. Torikai, A., Takeuchi, T., and Fueki, K., *Polym. Photochem.*, **3**, 307 (1983).
3628. Torikai, A., Tsuruta, H., and Fueki, K., *Polym. Photochem.*, **2**, 227 (1982).
3629. Torkelson, J. M., Lipsky, S., and Tirrell, M., *Macromolecules*, **14**, 1601 (1981).
3630. Torkelson, J. M., Lipsky, S., Tirrell, M., and Tirrell, D. A., *Macromolecules*, **16**, 326 (1983).
3631. Toshima, N., Kuriyama, M., Yamada, Y., and Hirai, H., *Chem. Lett.*, **1981**, 793.
3632. Toth, L. M., and Johnston, H. S., *J. Amer. Chem. Soc.*, **91**, 1276 (1969).
3633. Tozzi, A., Cantatore, G., and Masina, F., *Text. Res. J.*, **48**, 433 (1978).
3634. Trozzolo, A. M., and Winslow, F. H., *Macromolecules*, **1**, 98 (1968).
3635. Tryson, G. R., and Shultz, A. R., *J. Polym. Sci.*, B, **17**, 2059 (1979).
3636. Tsubakiyama, K., and Fujisaki, S., *J. Polym. Sci.*, B, **10**, 341 (1972).
3637. Tsubomura, H., and Mulliken, R. S., *J. Amer. Chem. Soc.*, **82**, 5996 (1960).
3638. Tsuchida, E., Nishide, H., Shimidizu, N., Yamada, A., Kaneko, M., and Kurimura, Y., *Makromol. Chem. Rapid Commun.*, **2**, 621 (1981).
3639. Tsuda, K., and Kosegaki, K., *Makromol. Chem.*, **161**, 267 (1972).
3640. Tsuda, K., and Otsu, T., *Bull. Chem. Soc. Japan*, **39**, 2206 (1966).
3641. Tsuda, M., *J. Polym. Sci.*, A1, **7**, 259 (1969).
3642. Tsuda, M., *Yukigosei Kagaku*, **30**, 589 (1972).
3643. Tsuda, M., and Oikawa, S., in *Ultraviolet Light Induced Reactions in Polymers* (Labana, S. S., ed.), ACS Symp. Ser., No. 25, ACS Washington, D.C., 1976, p. 423.
3644. Tsuda, M., and Oikawa, S., in *Ultraviolet Light Induced Reactions in Polymers* (Labana, S. S., ed.), ACS Symp. Ser., No. 25, ACS Washington, D.C., 1976, p. 446.
3645. Tsuda, M., and Oikawa, S., *Polym. Eng. Sci.*, **23**, 993 (1983).
3646. Tsuda, M., Oikawa, S., Kanai, W., Yokota, A., Hijikata, I., and Uehara, H., *J. Vac. Sci. Technol.*, **19**, 259 (1981).
3647. Tsuda, M., Oikawa, S., and Nagayama, K., *Photogr. Sci. Eng.*, **27**, 118 (1983).
3648. Tsuda, M., Oikawa, S., Nakamura, Y., Nagata, H., Yokota, A., Nakane, H., Tsumori, T., Nakane, Y., and Mifune, T., *Photogr. Sci. Eng.*, **23**, 290 (1979).
3649. Tsunoda, T., Yamaoka, T., Osabe, Y., and Hata, Y., *Photogr. Sci. Eng.*, **20**, 188 (1976).
3650. Tsunoda, T., Yamaoka, T., and Nagamatsu, G., *Photogr. Sci. Eng.*, **17**, 390 (1973).

3651. Tsuji, K., *J. Polym. Sci.*, B, **11**, 351 (1973).
3652. Tsuji, K., *Rep. Progr. Polym. Phys. Japan*, **17**, 553 (1974).
3653. Tsuji, K., *Polym. Plast. Techn. Eng.*, **9**, 1 (1977).
3654. Tsuji, K., *Adv. Polym. Sci.*, **12**, 131 (1973).
3655. Tsuji, K., and Seiki, T., *J. Polym. Sci.*, B, **8**, 817 (1970).
3656. Tsuji, K., and Seiki, T., *Polym. J.*, **4**, 589 (1973).
3657. Tsuji, K., Seiki, T., and Takeshita, T., *J. Polym. Sci.*, A1, **10**, 3119 (1972).
3658. Tsukui, K., *Kobunshi Kagaku*, **26**, 602 (1969).
3659. Tsunoda, M., *J. Polym. Sci.*, A1, **2**, 2907 (1964).
3660. Tsunoda, M., *J. Soc. Sci. Phot. Japan*, **28**, 7 (1965).
3661. Tsunoda, M., Tanaka, M., and Murata, N., *Kobunshi Kagaku*, **22**, 107 (1965).
3662. Tsunoda, T., Tanaka, J., and Kobyashi, M., *Bull. Tech. Assoc. Graphic Arts Japan*, **11**, 21 (1969).
3663. Tsunoda, T., and Yamaoka, T., *J. Appl. Polym. Sci.*, **8**, 1379 (1964).
3664. Tsunoda, T., and Yamaoka, T., *J. Polym. Sci.*, A1, **3**, 3691 (1965).
3665. Tsunoda, T., Yamaoka, T., and Ijitsu, T., *Bull. Techn. Assoc. Graphic Arts Japan*, **13**, 11 (1971).
3666. Tsunoda, T., Yamaoka, T., and Nagamatsu, G., *Bull. Techn. Assoc. Graphic Arts Japan*, **14**, 18 (1972).
3667. Tsunoda, T., Yamaoka, T., Osabe, Y., and Hata, Y., *Photog. Sci. Eng.*, **20**, 188 (1976).
3668. Tsunooka, M., Cha, Y. S., and Tanaka, M., *Makromol. Chem. Rapid Commun.*, **2**, 491 (1981).
3669. Tsunooka, M., Cha, Y. S., Tanaka, M., and Konishi, F., *J. Polym. Sci.*, A1, **22**, 739 (1984).
3670. Tsunooka, M., Kotera, K., and Tanaka, M., *J. Polym. Sci.*, A1, **15**, 107 (1977).
3671. Tsunooka, M., Kusube, M., Tanaka, M., and Murata, N., *Kagyo Kagaku Zasshi*, **72**, 287 (1969).
3672. Tsunooka, M., Sasaki, H., and Tanaka, M., *J. Polym. Sci.*, B, **18**, 407 (1980).
3673. Tsunooka, M., Ueda, T., Tanaka, S., Tanaka, M., and Egawa, H., *J. Polym. Sci.*, B, **20**, 589 (1982).
3674. Tsunooka, M., Ueda, T., Tanaka, M., and Egawa, H., *J. Polym. Sci.*, A1, **22**, 2217 (1984).
3675. Tsunooka, M., Tanaka, S., and Tanaka, M., *Makromol. Chem. Rapid Commun.*, **4**, 539 (1983).
3676. Tu, R. S., in *UV Curing, Science and Technology* (Pappas, S. P., ed.), Vol. 2, Technology Marketing Co., Norwalk, CT, 1984, p. 143.
3677. Tuesday, C. S., ed., *Chemical Reactions in Urban Atmospheres*, Elsevier, New York, 1971.
3678. Turro, N. J., *Molecular Photochemistry*, Benjamin, New York,
3679. Turro, N. J., *Pure Appl. Chem.*, **49**, 405 (1977).
3680. Turro, N. J., *Modern Molecular Photochemistry*, Benjamin/Cummings, Menlo Park, 1978.
3681. Turro, N. J., and Aikawa, M., *J. Amer. Chem. Soc.*, **102**, 4866 (1980).
3682. Turro, N. J., Baretz, B. H., and Kuo, P. L., *Macromolecules*, **17**, 1321 (1984).
3683. Turro, N. J., and Cherry, W. R., *J. Amer. Chem. Soc.*, **100**, 7432 (1978).
3684. Turro, N. J., Chow, M. F., Chung, C. J., Weed, G. C., and Kraeutler, B., *J. Amer. Chem. Soc.*, **102**, 4843 (1980).
3685. Turro, N. J., Chow, M. F., Chung, C. J., and Tung, C. H., *J. Amer. Chem. Soc.*, **102**, 7391 (1980).
3686. Turro, N. J., Chow, M. F., and Burkhart, R. D., *Chem. Phys. Lett.*, **80**, 146 (1981).
3687. Turro, N. J., Chow, M. F., Chung, C. J., and Tung, C. H., *J. Amer. Chem. Soc.*, **105**, 1572 (1983).
3688. Turro, N. J., and Chung, C. J., *Macromolecules*, **17**, 2123 (1984).
3689. Turro, N. J., and Engel, R., *Mol. Photochem.*, **1**, 143, 235 (1969).
3690. Turro, N. J., Grätzel, M., and Braun, A. M., *Angew. Chem.*, **92**, 712 (1980).
3691. Turro, N. J., and Kraeutler, B., *Acc. Chem. Res.*, **13**, 369 (1980).
3692. Turro, N. J., and Okubo, T., *J. Phys. Chem.*, **86**, 1535 (1982).
3693. Turro, N. J., and Okubo, T., *J. Amer. Chem. Soc.*, **104**, 2985 (1982).
3694. Turro, N. J., Okubo, T., Chung, C. J., Emert, J., and Catena, R., *J. Amer. Chem. Soc.*, **104**, 4799 (1982).
3695. Turro, N. J., and Pierola, I. F., *J. Phys. Chem.*, **87**, 2420 (1983).
3696. Turno, N. J., and Pierola, I. F., *Macromolecules*, **16**, 906 (1983).
3697. Uberreiter, K., and Bruns, W., *Makromol. Chem.*, **68**, 24 (1968).
3698. Ueno, A., Anzai, J., and Kadoma, Y., *J. Polym. Sci.*, B, **15**, 411 (1977).
3699. Ueno, A., Anzai, J., Osa, T., and Kadoma, Y., *J. Polym. Sci.*, B, **15**, 407 (1977).
3700. Ueno, A., Anzai, J., Osa, T., and Kadoma, Y., *Bull. Chem. Soc. Japan*, **50**, 2995 (1977).

742

3701. Ueno, A., Anzay, J., Osa, T., and Kadoma, Y., *Bull. Chem. Soc. Japan*, **52**, 549 (1979).
3702. Ueno, A., Nohara, M., Toda, F., Uno, K., and Iwakura, Y., *J. Polym. Sci.*, A1, **13**, 2751 (1975).
3703. Ueno, A., Osa, T., and Toda, F., *J. Polym. Sci.*, B, **14**, 521 (1976).
3704. Ueno, A., Osa, T., and Toda, F., *Macromolecules*, **10**, 130 (1977).
3705. Ueno, A., Takahashi, K., Anzay, J., and Osa, T., *Macromolecules*, **13**, 459 (1980).
3706. Ueno, A., Takahashi, K., Anzay, J., and Osa, T., *J. Amer. Chem. Soc.*, **103**, 6410 (1981).
3707. Ueno, A., Takahashi, K., Anzay, J., and Osa, T., *Makromol. Chem.*, **182**, 693 (1981).
3708. Ueno, A., Takahashi, K., Anzay, J., and Osa, T., *Chem. Lett.*, **1981**, 113.
3709. Ueno, A., Toda, F., Iwakura, Y., *J. Polym. Sci.*, B, **12**, 287 (1974).
3710. Ueno, A., Toda, F., and Iwakura, Y., *J. Polym. Sci.*, A1, **12**, 1841 (1984).
3711. Ulinska, A., Dzierza, W., *Polimery*, **15**, 73 (1970).
3712. Ulinska, A., Koscielewska, A., and Mankowski, Z., *Polimery*, **10**, 334 (1963).
3713. Ulinska, A., and Kaminska, A., *Polimery*, **11**, 522 (1966).
3714. Ulinska, A., Mankowski, Z., and Koscielewska, A., *Polimery*, **10**, 442 (1965).
3715. Unruh, C. C., *J. Appl. Polym. Sci.*, **2**, 358 (1959).
3716. Unruh, C. C., and Smith, A. C., *J. Appl. Polym. Sci.*, **3**, 310 (1960).
3717. Uri, N., *Chem. Rev.*, **50**, 375 (1952).
3718. Uri, N., *Israel J. Chem.*, **8**, 125 (1970).
3719. Urruti, E. H., and Kilp, T., *Macromolecules*, **17**, 50 (1984).
3720. Ushiki, H., Horie, K., Okamoto, A., Mita, I., *Polym. J.*, **11**, 691 (1979).
3721. Ushiki, H., Horie, K., Okamoto, A., and Mita, I., *Polym. J.*, **13**, 4 (1981).
3722. Ushiki, H., Horie, K., Okamoto, A., and Mita, I., *Polym. J.*, **13**, 191 (1981).
3723. Ushiki, H., Horie, K., Okamoto, A., and Mita, I., *Polym. Photochem.*, **1**, 303 (1981).
3724. Usilton, J. J., and Patel, A. R., *Amer. Chem. Soc. Polym. Prepr.*, **18**(1), 393 (1977).
3725. Uzelmeier, C., *SPE Journal*, **26**, 69 (1970).
3726. Vala, M. T. Jr., Haebig, J., and Rice, S. A., *J. Chem. Phys.*, **43**, 886 (1965).
3727. Valeur, B., and Monnerie, L., *J. Polym. Sci.*, A2, **14**, 29 (1976).
3728. Valk, G., Kehren, M. L., and Daamen, I., *Angew. Makromol. Chem.*, **13**, 97 (1970).
3729. Van der Akker, J. A., Lewis, J. A., Jones, G. W., and Buchman, M. A., *Tappi*, **32**, 189 (1949).
3730. Van den Broek, A. J. M., Havekorn van Rijsewijk, H. C., Leglerse, P. E. J., Lippits, G. J. M., and Thomas, G. E., *J. Radiat. Curing*, **11**, Jan., 2 (1984).
3731. Van der Veen, G., Hoguet, R., and Prins, W., *Photochem. Photobiol.*, **19**, 197 (1974).
3732. Van der Veen, G., and Prins, W., *Natl. Phys. Sci.*, **230**, 70 (1971).
3733. Van der Veen, G., and Prins, W., *Photochem. Photobiol.*, **19**, 191 (1974).
3734. Vandewyer, P. H., and Smets, G., *J. Polym. Sci.*, Polym. Symp., No. 22, 231 (1968).
3735. Vandewyer, P. H., and Smets, G., *J. Polym. Sci.*, A1, **8**, 2361 (1970).
3736. Van Mil, J., Addadi, L., Gati, E., and Lahan, M., *J. Amer. Chem. Soc.*, **104**, 3429 (1982).
3737. Van Oosterhout, A. C. J., and Van Neerbos, A., *J. Radiat. Curing*, **9**, Jan., 19 (1982).
3738. Vansant, J., Toppet, S., Smets, G., Declerq, J. P., Germain, G., and Van Meerssche, M., *J. Org. Chem.*, **45**, 1565 (1980).
3739. Varghese, A. J., *Photophysiology*, **7**, 208 (1972).
3740. Venikouas, G. W., and Powell, R. C., *Chem. Phys. Lett.*, **34**, 601 (1975).
3741. Venkatarao, K., and Santappa, M., *J. Polym. Sci.*, A1, **5**, 637 (1967).
3742. Venkatarao, K., and Santappa, M., *J. Polym. Sci.*, A1, **8**, 3429 (1970).
3743. Verborgt, J., and Smets, G., *J. Polym. Sci.*, A1, **12**, 2511 (1979).
3744. Verdu, J., *J. Macromol. Sci. Chem.*, A, **12**, 551 (1978).
3745. Verdu, J., Michel, A., and Sonderhof, D., *Europ. Polym. J.*, **16**, 689 (1980).
3746. Vermes, J. P., and Beugelmans, R., *Tetrahed. Lett.*, **1969**, 2091.
3747. Verstraete, J. A., Noonan, J. M., and Neubert, R. W., U.S. Pat. 3,832,176 (1974).
3748. Vincent, P. S., and Roberts, G. G., *Thin Solid Films*, **68**, 135 (1980).
3749. Vieth, W., and Wuerth, W. F., *J. Appl. Polym. Sci.*, **13**, 695 (1969).
3750. Vilanove, R., Hervet, H., Gruler, H., and Rondalez, F., *Macromolecules*, **16**, 825 (1983).
3751. Vink, H., *Makromol. Chem.*, **67**, 105 (1963).
3752. Vink, P., *J. Polym. Sci.*, Polym. Symp., No. 40, 169 (1973).
3753. Vink, P., *Develop. Polym. Stabil.*, **3**, 117 (1980).
3754. Vink, P., *J. Appl. Polym. Sci.*, **35**, 265 (1979).
3755. Vink, P., Rotteveel, R. T., and Wisse, J. D. M., *Polym. Degrad. Stabil.*, **9**, 131 (1984).
3756. Vink, P., and Van Veen, T. J., *Europ. Polym. J.*, **14**, 533 (1978).
3757. Vink, P., and Wisse, J. D. M., *Polym. Degrad. Stabil.*, **4**, 51 (1982).

3758. Virt, J., Rosik, L., Kovarova, J., and Pospisil, J., *Europ. Polym. J.*, **16**, 247 (1980).
3759. Vlasov, G. P., Rudkovskaya, G. D., and Ovsyannikova, L. A., *Makromol. Chem.*, **183**, 2635 (1982).
3760. Vogl, O., and Yoshida, S., *Polym. Preprints, Japan*, **29**(4), 183 (1980).
3761. Voigt, J., *Die Stabilizerung der Kunstoffe gegen Licht und Wärme*, Springer Verlag, Berlin, 1966.
3762. Vollmert, B., *Polymer Chemistry*, Springer Verlag, Berlin, 1973.
3763. Vollenbroek, F. A., Spiertz, E. J., and Kroon, H. J. J., *Polym. Eng. Sci.*, **23**, 995 (1983).
3764. Völz, H. G., Kämpf, G., and Fitzky, H. G., *Progr. Org. Coat.*, **1**, 1 (1973).
3765. Völz, H. G., Kämpf, G., Fitzky, H. G., and Klaeren, A., *Amer. Chem. Soc. Div. Org. Coat. Plast. Chem. Preprint*, **42**, 660 (1980).
3766. Völz, H. G., Kämpf, G., Fitzky, H. G., and Klaeren, A., in *Photodegradation and Photostabilization of Coatings*, (Pappas, S. P., and Winslow, F. H., eds.), ACS Symp. Ser., No. 163, ACS Washington, D.C., 1981, p. 163.
3767. Völz, H. G., Kämpf, G., and Klearen, A., *Farbe u. Lack*, **82**, 805 (1976).
3768. Volz, R., Laustriat, G., and Coche, A., *J. Chim. Phys., Phys. Chim. Biol.*, **63**, 1253 (1966).
3769. Vrancken, A., *Farbe u. Lack*, **83**, 171 (1977).
3770. Wagner, H. M., and Purbrick, M. D., *J. Photogr. Sci.*, **29**, 230 (1981).
3771. Wagner, P. J., in *Creation and Detection of the Excited States* (Lamola, A. A., ed.), Vol. 1, Part A., Dekker, New York, 1971, p. 173.
3772. Wagner, P. J., *Mol. Photochem.*, **3**, 23 (1971).
3773. Wagner, P. J., *Topics in Current Chem.*, No. 66, 1 (1976).
3774. Wagner, P. J., *J. Photochem.*, **10**, 387 (1979).
3775. Wagner, P. J., *Acc. Chem. Res.*, **16**, 461 (1983).
3776. Wagner, P. J., and Chen, C. P., *J. Amer. Chem. Soc.*, **98**, 239 (1976).
3777. Wagner, P. J., Jellinek, K., and Kemppainen, A. E., *J. Amer. Chem. Soc.*, **94**, 7512 (1972).
3778. Wagner, P. J., Kelso, P. A., and Zepp, R. G., *J. Amer. Chem. Soc.*, **94**, 7480 (1972).
3779. Wagner, P. J., and Kemppainen. A. E., *J. Amer. Chem. Soc.*, **91**, 3085 (1969).
3780. Wagner, P. J., and Hammond, G. S., *Adv. Photochem.*, **5**, 168 (1968).
3781. Wagner, P. J., and Levitt, R. A., *J. Amer. Chem. Soc.*, **92**, 5806 (1970).
3782. Wahl, P., *J. Polym. Sci.*, **29**, 375 (1958).
3783. Waka, Y., Hamamoto, K., and Mataga, N., *Chem. Phys. Lett.*, **53**, 242 (1978).
3784. Wald, G., *Nature*, **219**, 800 (1968).
3785. Wall, L. A., and Brown, D. W., *J. Phys. Chem.*, **61**, 129 (1956).
3786. Wall, L. A., Harvey, M. R., and Tyron, M., *J. Phys. Chem.*, **60**, 1306 (1956).
3787. Wall, L. A., and Tryon, M., *Nature*, **178**, 101 (1956).
3788. Wall, L. A., and Tryon, M., *J. Phys. Chem.*, **62**, 697 (1958).
3789. Wall, L. A., and Frank, G. C., *Text. Res. J.*, **41**, 32 (1971).
3790. Wallace, T. J., and Gritter, R. J., *J. Org. Chem.*, **26**, 5256 (1961).
3791. Walling, C., and Gibian, M. J., *J. Amer. Chem. Soc.*, **87**, 3361 (1965).
3792. Walling, G., and Gibian, M. J., *J. Amer. Chem. Soc.*, **87**, 3413 (1965).
3793. Waltz, W. L., Hackelberg, O., Dorfman, L. M., and Wojucki, A., *J. Amer. Chem. Soc.*, **100**, 7259 (1979).
3794. Wamser, C. C., Hammond, G. S., Chang, C. T., and Baylor, C., Jr., *J. Amer. Chem. Soc.*, **92**, 6362 (1970).
3795. Wandelt, B., *Polym. Bull.*, **4**, 199 (1981).
3796. Wang, F. W., and Howell, B. F., *Polymer*, **25**, 1626 (1984).
3797. Wang, F. W., Lowry, R. E., and Grant, W. H., *Polymer*, **25**, 690 (1984).
3798. Wang, S. Y., *Photochemistry and Photobiology of Nucleic Acids*, Vols 1–2, Academic Press, New York, 1976.
3799. Wang, S. Y., *Nature*, **198**, 844 (1960).
3800. Wang, U., and Morawetz, H., *Makromol. Chem.*, **1**, 283 (1975).
3801. Wang, Y. C., and Morawetz, H., *Macromol. Chem. Suppl.*, **1**, 283 (1975).
3802. Wang, Y. C., and Morawetz, H., *J. Amer. Chem. Soc.*, **98**, 3611 (1976).
3803. Ward, H. R., and Wishnok, J. S., *J. Amer. Chem. Soc.*, **90**, 1085 (1968).
3804. Ware, W. R., *J. Phys. Chem.*, **66**, 455 (1962).
3805. Ware, W. R., and Novros, J. S., *J. Phys. Chem.*, **70**, 3247 (1966).
3806. Wasserman, H. II., and Murray, R. W., eds, *Singlet Oxygen*, Academic Press, New York, 1979.
3807. Watanabe, A., and Koizumi, M., *Bull. Chem. Soc. Japan*, **34**, 1086 (1961).

744

3808. Watanabe, S., Kato, M., and Kosaki, S., *J. Polym. Sci.*, A1, **22**, 2801 (1984).
3809. Waters, P. J., and Evans, N. A., *Text. Res. J.*, **48**, 251 (1978).
3810. Waters, W. A., and Wickham-Jones, C., *J. Chem. Soc.*, **1952**, 1420.
3811. Watkins, A. R., *Chem. Phys. Lett.*, **29**, 526 (1974).
3812. Watson, W. F., *Trans. IRI*, **19**, 32 (1953).
3813. Watt, W. R., in *Curing: Science and Technology* (Pappas, S. P., ed.), Vol. 2, Technology Marketing Co., Norwalk, CT, 1984, p. 247.
3814. Watt, W. R., Hoffman, H. T., Pobiner, H., Schkolnick, L. J., and Yang, L. S., *J. Polym. Sci.*, A1, **22**, 1789 (1984).
3815. Wayne, R. P., *Photochemistry*, Butterworth, London, 1970.
3816. Wayne, R. P., *Adv. Photochem.*, **7**, 311 (1979).
3817. Weber, G., *Biochem. J.*, **51**, 155 (1952).
3818. Weber, S., *Trans. Faraday Soc.*, **50**, 552 (1954).
3819. Weber, G., *J. Chem. Phys.*, **55**, 2399 (1971).
3820. Webber, S. E., and Avots-Avotines, P. E., *Macromolecules*, **12**, 708 (1979).
3821. Webber, S. E., Avots-Avotines, P. E., and Deumie, M., *Macromolecules*, **14**, 105 (1981).
3822. Webber, S. E., and Swenberg, C. E., *Chem. Phys.*, **49**, 231 (1980).
2823. Weeks, J. L., and Matheson, M. S., *J. Amer. Chem. Soc.*, **78**, 1273 (1956).
3824. Wegner, G., *Pure Appl. Chem.*, **49**, 443 (1977).
3825. Wegner, G., *Z. Naturforsch.*, **24b**, 824 (1969).
3826. Wegner, G., *Makromol. Chem.*, **134**, 219 (1970).
3827. Wegner, G., *Makromol. Chem.*, **145**, 85 (1971).
3828. Wegner, G., *J. Polym. Sci.*, B, **9**, 133 (1971).
3829. Wegner, G., *Makromol. Chem.*, **154**, 35 (1972).
3830. Wegner, G., *Pure Appl. Chem.*, **49**, 443 (1977).
3831. Wegner, G., *Makromol. Chem.*, Suppl., **6**, 347 (1984).
3832. Wegner, G., Fischer, E. W., and Munoz-Escalona, A., *Makromol. Chem.*, Suppl. **1**, 521 (1975).
3833. Wei, K., and Adelman, A. H., *Tetrahed. Lett.*, **1969**, 3297.
3834. Wei, K., Mani, J. C., and Pitts, J. N. Jr., *J. Amer. Chem. Soc.*, **87**, 4225 (1967).
3835. Weill, G., *C.R. Acad. Sci., Ser.*, B, **272**, 116 (1971).
3836. Weiner, S. A., *J. Amer. Chem. Soc.*, **93**, 425 (1971).
3837. Weiner, S. A., *J. Amer. Chem. Soc.*, **93**, 6978 (1971).
3838. Weir, N. A., *J. Appl. Polym. Sci.*, **17**, 401 (1973).
3839. Weir, N. A., in *Developments in Polymer Degradation* (Grassie, N., ed.), Vol. 1, Applied Science Publishers, Ltd, 1977, p. 67.
3840. Weir, N. A., *J. Polym. Sci.*, A1, **16**, 13 (1978).
3841. Weir, N. A., *J. Polym. Sci.*, A1, **16**, 1123 (1978).
3842. Weir, N. A., *Europ. Polym. J.*, **14**, 9 (1978).
3843. Weir, N. A., *J. Polym. Sci.*, A1, **17**, 3723 (1979).
3844. Weir, N. A., in *Developments in Polymer Degradation* (Grassie, N., ed.), Vol. 4, Applied Science Publishers, Ltd, London, 1982, p. 143.
3845. Weir, N. A., and Milke, T. H., *Makromol. Chem.*, **179**, 1989 (1978).
3846. Weir, N. A., and Milke, T. H., *J. Polym. Sci.*, A1, **17**, 3723 (1979).
3847. Weir, N. A., and Milke, T. H., *J. Polym. Sci.*, A1, **17**, 3735 (1979).
3848. Weir, N. A., and Milke, T. H., *Polym. Degrad. Stabil.*, **1**, 105 (1979).
3849. Weir, N. A., and Milke, T. H., *Polym. Degrad. Stabil.*, **1**, 181 (1979).
3850. Weir, N. A., and Milke, T. H., *J. Appl. Polym. Sci.*, Appl. Polym. Symp., No. 35, 289 (1979).
3851. Weir, N. A., and Milke, T. H., *Makromol. Chem.*, **180**, 1729 (1979).
3852. Weir, N. A., and Milke, T. H., *Europ. Polym. J.*, **16**, 141 (1980).
3853. Weir, N. A., Milke, T. H., and Nicholas, D., *J. Appl. Polym. Sci.*, **23**, 609 (1979).
3854. Weir, N. A., *Polym. Degrad. Stabil.*, **2**, 225 (1980).
3855. Weir, N. A., and Rujimethabhas, M., *Europ. Polym J.*, **18**, 813 (1982).
3856. Weir, N. A., Rujimethabhas, M., and Clothier, P. Q., *Europ. Polym. J.*, **17**, 431 (1981).
3857. Weir, N. A., Rujimethabhas, M., and Milke, T., *Polym. Photochem.*, **1**, 205 (1981).
3858. Weiss, G. H., *Proc. Natl. Acad Sci., U.S.*, **77**, 4391 (1980).
3859. Weller, A., *Pure Appl. Chem.*, **16**, 115 (1968).
3860. Wells, C. F., *Discuss. Faraday Soc.*, **29**, 219 (1960).
3861. Wells, C. F., *Trans. Faraday Soc.*, **56**, 1796 (1960).
3862. Wells, C. F., *Trans. Faraday Soc.*, **57**, 1703 (1961).
3863. Wells, C. F., *Trans. Faraday Soc.*, **57**, 1719 (1961).

745

3864. Werner, T., *J. Phys. Chem.*, **83**, 320 (1979).
3865. Werner, T., and Kramer, H. E. Λ., *Europ. Polym. J.*, **13**, 501 (1977).
3866. Werner, T., Woessner, G., and Kramer, H. E. A., in *Photodegradation and Photostabilization of Coatings* (Pappas, S. P. and Winslow, F. H., eds), ACS Symp. Ser., No. 151, ACS Washington, D.C., 1981, p. 1.
3867. Werner, T. C., and Hercules, D. M., *J. Phys. Chem.*, **73**, 2005 (1969).
3868. Westfahl, J. C., Carman, J. C., and Layer, R. W., *Rubb. Chem. Technol.*, **45**, 402 (1972).
3869. Wight, F. R., *J. Polym. Sci.*, B, **16**, 121 (1978).
3870. White, D. M., and Sonnenberg, J., *J. Amer. Chem. Soc.*, **88**, 3825 (1966).
3871. Whitfield, R. H., and Davies, D. I., *Polym. Photochem.*, **1**, 261 (1981).
3872. Whiteway, S. G., and Masson, C. R., *J. Amer. Chem. Soc.*, **77**, 1508 (1955).
3873. Whitten, D. G., *Rev. Chem. Intermed.*, **2**, 107 (1978).
3874. Whitten, G., *Acc. Chem. Res.*, **13**, 83 (1980).
3875. Wicks, Z. W. Jr., and Hill, L. W., in *UV Curing: Science and Technology* (Pappas, S. P., ed.), Vol. 2, Stamford Technology Marketing Co, Norwalk, CT, 1984, p. 77.
3876. Wicks, Z. W. Jr., and Kühhirt, W., *J. Paint Technol.*, **47**, 49 (1975).
3877. Wicks, Z. W. Jr., and Pappas, S. P., in *UV Curing: Science and Technology* (Pappas, S. P., ed.), Vol. 1, Stanford Technology Marketing Co., Norwalk, CT, 1978, p. 79.
3878. Wijnen, M. H. J., *J. Chem. Phys.*, **27**, 710 (1957).
3879. Wijnen, M. H. J., *J. Chem. Phys.*, **28**, 271 (1958).
3880. Wijnen, M. H. J., *J. Amer. Chem. Soc.*, **82**, 1847 (1960).
3881. Wiles, D. M., *Pure Appl. Chem.*, **50**, 291 (1978).
3882. Wiles, D. M., in *Singlet Oxygen: Reactions with Organic Compounds and Polymers* (Rånby, B., and Rabek, J. F., eds), Wiley, Chichester, 1978, p. 320.
3883. Wiles, D. M., *J. Appl. Polym. Sci.*, Appl. Polym. Symp., No. 35, 235 (1979).
3884. Wiles, D. M., and Carlsson, D. J., *Polym. Degrad. Stabil.*, **3**, 61 (1980/81).
3885. Wiles, D. M., Tovborg, J. P., and Carlsson, D. J., *Pure Appl. Chem.*, **55**, 1651 (1983).
3886. Wilkins, C. W. Jr., Reichmans, J., Chandross, E. A., and Hartless, R. L., *Polym. Eng. Sci.*, **23**, 1025 (1983).
3887. Wilkinson, F., *J. Phys. Chem.*, **66**, 2569 (1962).
3888. Wilkinson, F., *Adv. Photochem.*, **3**, 241 (1964).
3889. Wilkinson, F., in *Luminescence in Chemistry* (Bowen, E. J., ed.), Van Nostrand, London, 1968, p. 155.
3890. Wilkus, E., and Wright, A. N., *J. Polym. Sci.*, A1, **9**, 2071 (1971).
3891. Wilkus, E., and Wright, A. N., *J. Polym. Sci.*, A1, **9**, 2097 (1971).
3892. Williams, F. C., *Macromol. Chem.*, **3**, 85 (1979).
3893. Williams, G. E., and Gerrard, D. L., *J. Polym. Sci.*, A1, **21**, 1491 (1983).
3894. Williams, J. L. R., in *Preprints Soc. Plast. Eng. Reg. Techn. Conf.*, Ellenville, New York, 6 Nov. 1967, p. 123.
3895. Williams, J. L. R., *Fortsch. Chem. Forsch.*, **13**, 227 (1969).
3896. Williams, J. L. R., in *Polyelectrolytes* (Sélégny, E., ed.), Reidel Publishing Co., Dortrecht Holland, 1974, p. 507.
3897. Williams, J. L. R., and Borden, D. G., *Makromol. Chem.*, **73**, 203 (1963).
3898. Williams, J. L. R., and Daly, R. C., *Progr. Polym. Sci.*, **5**, 61 (1977).
3899. Williams, J. L. R., Farid, S. Y., Daly, R. C., Specht, D. P., Searle, R., Bordon, D. G., Chang, H. J., and Martic, P. A., *Pure Appl. Chem.*, **49**, 523 (1977).
3900. Williams, J. L. R., and Molaire, M. F., in *Kirk-Othmer Encyclopedia of Chemical Technology*, Vol. 17, Wiley, New York, 1982, p. 680.
3901. Williams, J. L. R., Specht, D. P., and Farid, S., *Polym. Eng. Sci.*, **23**, 1022 (1983).
3902. Williams, R. S., *J. Appl. Polym. Sci.*, **28**, 2093 (1982).
3903. Williamson, M. A., Smith, J. D. B., Castle, P. M., and Kaufman, R. N., *J. Polym. Sci.*, A1, **20**, 1875 (1982).
3904. Willner, I., Ford, W. E., Otvos, J. W., and Calvin, M., *Nature*, **280**, 823 (1979).
3905. Willson, G., in *Introduction to Microlithography* (Thompson, L. F., Willson, C. G., and Bowden, M. J., eds), ACS Symp. Ser., No. 219, ACS Washington, D.C., 1983, p. 80.
3906. Willson, G., Miller, R., McKean, D., Cleack, N., Tompkins, T., Hofer, D., Michl, J., and Dowing, J., *Polym. Eng. Sci.*, **23**, 1004 (1983).
3907. Wilson, D. C., and Drickamer, H. G., *J. Chem. Phys.*, **63**, 3649 (1975).
3908. Wilson, R., *J. Chem. Soc.*, B, **12**, 1581 (1968).
3909. Wilson, T., *J. Amer. Chem. Soc.*, **88**, 2898 (1966).

746

3910. Wilzbach, K. E., Ritscher, T. S., and Kaplan, L., *J. Amer. Chem. Soc.*, **89**, 1031 (1967).
3911. Winkler, D. E., *J. Polym. Sci.*, **35**, 3 (1969).
3912. Winnik, M. A., *Acc. Chem. Res.*, **10**, 173 (1977).
3913. Winnik, M. A., *Polym. Eng. Sci.*, **24**, 87 (1984).
3914. Winnik, M. A., Hua, M. H., Hougham, B., Williamson, B., and Croucher, M. D., *Macromolecules*, **17**, 262 (1984).
3915. Winnik, M. A., Li, X. B., and Guillet, J. E., *Macromolecules*, **17**, 699 (1984).
3916. Winnik, M. A., and Maharaj, U., *Macromolecules*, **12**, 902 (1980).
3917. Winnik, M. A., and Pekcan, O., *Macromolecules*, **16**, 1021 (1983).
3918. Winnik, M. A., Pekcan, O., and Egan, L., *Polymer*, **28**, 1767 (1984).
3919. Winnik, M. A., Redpath, A. E. C., Paton, K., and Danhelka, J., *Polymer*, **25**, 91 (1984).
3920. Winnik, M. A., Redpath, T., and Richard, D. H., *Macromolecules*, **13**, 328 (1980).
3921. Winnik, M. A., Redpath, A. E. C., Svirskaya, P., and Mar, A., *Polymer*, **24**, 473 (1983).
3922. Winnik, M. A., and Shum, E., *Macromolecules*, **9**, 875 (1976).
3923. Winnik, M. A., Trueman, R. E., Jackowski, G., Saunders, D. S., and Whittington, S. G., *J. Amer. Chem. Soc.*, **96**, 4843 (1974).
3924. Winslow, F. H., *Pure Appl. Chem.*, **49**, 495 (1977).
3925. Winslow, F. H., Hellman, M. Y., Matreyek, W., and Stills, S. M., *Polym. Eng. Sci.*, **6**, 1 (1966).
3926. Winslow, F. H., Matreyek, W., and Trozzolo, A. M., *Amer. Chem. Soc. Polymer Prepr.*, **10**, 1271 (1969).
3927. Wippelder, E., and Heusinger, H., *J. Polym. Sci.*, A1, **16**, 1779 (1978).
3928. Wischmann, K. B., in *Polymers in Solar Energy Utilization* (Gebelein, C. G., Williams, D. J., and Deanin, R. D., eds), ACS Symp. Ser., No. 220, ACS Washington, D.C., 1983, p. 115.
3929. Wissbrun, K. F., *J. Amer. Chem. Soc.*, **81**, 58 (1963).
3930. Witzman, H., and Helmshauss, A., *Z. Phys. Chem.*, **213**, 1 (1960).
3931. Wolff, C., and Grätzel, M., *Chem. Phys. Lett.*, **52**, 542 (1977).
3932. Wolinski, L., *Makromol. Chem.*, **181**, 2335 (1980).
3933. Wolinski, L., Turzynski, Z., and Zaleska, W., *Makromol. Chem.*, **183**, 3089 (1982).
3934. Wolinski, L., Witkowski, K., and Turzynski, Z., *Makromol. Chem.*, **180**, 2399 (1979).
3935. Wolinski, L., Witkowski, K., and Turzynski, Z., *Makromol. Chem.*, **181**, 1717 (1980).
3936. Wolinski, L., Witkowski, K., and Turzynski, Z., *Makromol. Chem.*, **185**, 75 and 725 (1984).
3937. Wolters, E. T. M., Van Gunst, C. A., and Paulen, H. J. G., in *Singlet Oxygen: Reactions with Organic Compounds and Polymers* (Rånby, B., and Rabek, J. F., eds), Wiley, Chichester, 1978, p. 282.
3938. Wood, D. G. M., and Kollman, I. M., *Chem. Ind. (London)*, **1972**, 423.
3939. Wright, A. N., in *Polymer Surfaces* (Clark, D. T., and Feast, W. J., eds) Wiley, Chichester, 1978, p. 155.
3940. Wrighton, M., *Chem. Rev.*, **74**, 401 (1974).
3941. Wrighton, M., and Bredesen, D., *J. Organomet. Chem.*, **50**, C35 (1973).
3942. Wrighton, M., and Markham, J., *J. Phys. Chem.*, **77**, 3042 (1973).
3943. Wu, S. K., Jiang, Y. C., Li, F. M., and Feng, K. D., *Polym. Bull.*, **8**, 275 (1982).
3944. Wu, S. K., Jiang, Y. C., and Rabek, J. F., *Polym. Bull.*, **3**, 319 (1980).
3945. Wu, S. K., Lucki, J., Rabek, J. F., and Rånby, B., *Polym. Photochem.*, **2**, 73 (1982).
3946. Wu, S. K., Lucki, J., Rabek, J. F., and Rånby, B., *Polym. Photochem.*, **2**, 125 (1982).
3947. Yabe, A., and Tsuda, M., *J. Polym. Sci.*, B, **9**, 81 (1971).
3948. Yabe, A., Tsuda, M., Honda, K., and Tanaka, H., *J. Polym. Sci.*, A1, **10**, 2379 (1972).
3949. Yamada, Y., Kashima, K., and Okawara, M., *J. Polym. Sci.*, B, **14**, 65 (1976).
3950. Yamada, Y., and Okawara, M., *Makromol. Chem.*, **152**, 153 (1972).
3951. Yamaoka, H., Ikeda, T., and Okamura, S., *Macromolecules*, **10**, 717 (1977).
3952. Yamaoka, T., *J. Radiat. Curing*, **7**, Oct., 4 (1980).
3953. Yamaoka, T., Ueno, K., Tsunoda, T., and Torige, K., *Polymer*, **18**, 81 (1977).
3954. Yamaoka, T., Tsunoda, T., Koseki, K., Tabayashi, I., in *Modification of Polymers* (Carraher, C. E. Jr., and Tsuda, M., eds), ACS Symp. Ser., 121, ACS Washington, D. C., 1980, p. 185.
3955. Yamamoto, N., Akaishi, S., and Tsubomura, H., *Chem. Phys. Lett.*, **15**, 458 (1972).
3956. Yamamoto, N., Asanuma, T., and Nishijima, Y., *J. Chem. Soc. Chem. Commun.*, **1975**, 53.
3957. Yamamoto, Y., Tanaka, K., Ohmichi, T., Ooka, M., and Nishijima, Y., *Rep. Progr. Polym. Phys. Japan*, **12**, 457 (1969).
3958. Yanari, S. S., Bovey, F. A., and Lumry, L., *Nature*, **200**, 242 (1963).
3959. Yang, N. C., Cohen, J. I., and Shani, A., *J. Amer. Chem. Soc.*, **90**, 3264 (1968).
3960. Yang, N. C., and Yang, D. D. H., *J. Amer. Chem. Soc.*, **80**, 2913 (1958).

747

3961. Yang, N. L., and Oster, G., *J. Phys. Chem.*, **74**, 856 (1970).
3962. Yang, N. L., Snow, A., Haubenstock, H., and Bramwell, F., *J. Polym. Sci.*, A1, **16**, 1909 (1978).
3963. Yang, Y. Y., Lucki, J., Rabek, J. F., and Rånby, B., *Polym. Photochem.*, **3**, 47 (1983).
3964. Yang, Y. Y., Lucki, J., Rabek, J. F., and Rånby, B., *Polym. Photochem.*, **3**, 97 (1983).
3965. Yano, S., *Rubb. Chem. Technol.*, **54**, 1 (1981).
3966. Yano, S., and Murayama, M., *J. Appl. Polym. Sci.*, **25**, 433 (1980).
3967. Yano, S., and Murayama, M., *Polym. Photochem.*, **1**, 177 (1981).
3968. Yanus, J. F., and Pearson, J. M., *Macromolecules*, **7**, 716 (1974).
3969. Yanus, J. F., and Pearson, J. M., *Macromolecules*, **7**, 951 (1974).
3970. Yassin, A. A., and Sabaa, M. W., *Polym. Degrad. Stab.*, **4**, 313 (1982).
3971. Yasunaga, T., Kubota, H., and Ogiwara, Y., *J. Polym. Sci.*, A1, **14**, 1617 (1974).
3972. Yeung, C. K., and Jasse, B., *Macromol. Chem.*, **185**, 541 (1984).
3973. Ygurabide, J., *J. Chem. Phys.*, **49**, 1018, 1026 (1968).
3974. Ygurabide, J., *Methods Enzymol.*, C, **26**, 498 (1972).
3975. Ygurabide, J., Dillon, M. A., and Burton, M., *J. Amer. Chem. Soc.*, **40**, 3040 (1964).
3976. Yip, R. W., Szabo, A. G., and Tolg, P. K., *J. Amer. Chem. Soc.*, **95**, 4471 (1973).
3977. Yohe, G. R., Hill, D. R., Dunbar, J. E., and Schneidt, F. M., *J. Amer. Chem. Soc.*, **95**, 2688 (1953).
3978. Yokawa, M., and Ogo, Y., *Makromol. Chem.*, **177**, 429 (1976).
3979. Yokota, A., Yabuta, M., Kanai, W., Kashiwagi, K., and Hijikata, I., *Polym. Eng. Sci.*, **23**, 993 (1983).
3980. Yokota, K., Tomioka, H., and Adachi, K., *Polymer*, **14**, 561 (1973).
3981. Yokota, K., Tomioka, H., Ono, T., and Kuno, F., *J. Polym. Sci.*, A1, **10**, 1335 (1972).
3982. Yokoyama, M., Funaki, M., and Mikawa, H., *Chem. Commun.*, **1974**, 372.
3983. Yokoyama, M., Tamamura, T., Atsumi, M., Yoshimura, M., Shirota, Y., and Mikawa, H., *Macromolecules*, **8**, 101 (1975).
3984. Yokoyama, M., Tamamura, T., Nakano, T., and Mikawa, H., *Chem. Lett.*, **1972**, 499.
3985. Yokoyama, M., Tamamura, T., Nakano, T., and Mikawa, H., *J. Chem. Phys.*, **65**, 272 (1976).
3986. Yoshida, H., and Rånby, B., *J. Polym. Sci.*, B, **2**, 1155 (1964).
3987. Yoshida, S., and Vogl, O., *Amer. Chem. Soc., Polym. Preprints*, **21**(1), 201 (1980).
3988. Yoshida, S., and Vogl, O., *Makromol. Chem.*, **183**, 259 (1982).
3989. Yoshitiara, K., and Kearns, D. R., *J. Chem. Phys.*, **45**, 1991 (1965).
3990. Yuan, H. L., and Tazuke, S., *Polym. J.*, **15**, 111 (1983).
3991. Yuan, H. L., and Tazuke, S., *Polym. J.*, **15**, 125 (1983).
3992. Young, A. H., Verbanac, F., and Protzman, T. F., in *Chemistry and Properties of Crosslinked Polymers* (Labana, S. S., ed.), Academic Press, New York, 1977, p. 191.
3993. Young, R. H., Brewer, D., and Keller, R. A., *J. Amer. Chem. Soc.*, **95**, 375 (1973).
3994. Zachariasse, K. A., and Kühnle, W., *Z. Phys. Chem. (Frankfurt A.M.)*, **101**, 267 (1976).
3995. Zachariasse, K. A., Kühlne, W., and Weller, A., *Chem. Phys. Lett.*, **73**, 6 (1980).
3996. Zahradnickova, A., Petruj, J., and Sedlar, J., *Polym. Photochem.*, **3**, 295 (1983).
3997. Zahradnickova, A., Sedlar, J., Kovarova, J., and Pospišil, J., *Polym. Photochem.*, **2**, 349 (1981).
3998. Zaleski, A., and Rabek, J. F., *Polimery*, **10**(1), 52 (1971).
3999. Zannucci, J. S., and Lappin, G. R., *Macromolecules*, **7**, 393 (1974).
4000. Zapolskii, D. B., *Vysokomol. Soedin.*, **7**, 615 (1965).
4001. Zepp, R. G., and Wagner, P. J., *Chem. Commun.*, **1972**, 167.
4002. Zeronian, S. H., *Text. Res. J.*, **41**, 184 (1971).
4003. Zewail, A. H., Batchelder, J. S., in *Polymers in Solar Energy Utilization* (Gebelein, C. G., Williams, D. J., and Deanin, R. J., eds), ACS Symp. Ser. No. 220, ACS Washington, D.C., 1983, p. 331.
4004. Zimmerman, G., Chow, L., and Paik, U., *J. Amer. Chem. Soc.*, **80**, 3528 (1958).
4005. Zimmermann, H. E., *Adv. Photochem.*, **1**, 183 (1963).
4006. Zinkovskaya, A. V., Fuki, V. K., Melnikov, M. Ya, and Foch, V. N., *J. Polym. Sci.*, B, **16**, 41 (1978).
4007. Zolotoj, N. B., Kuznecov, M. N., Ivanov, V. B., Karpov, G. V., Skurat, V. E., and Shlyapintokh, V. Yu., *Vysokomol. Soedin.*, A, **18**, 658 (1976).
4008. Zwarich, R., and Bree, A., *J. Mol. Spectr.*, **52**, 329 (1974).
4009. Zweifel, H., *Photogr. Sci. Eng.*, **27**, 114 (1983).
4010. Zweig, A., and Henderson, W. A., *J. Polym. Sci.*, A1, **13**, 717 (1975).

748

4011. Zweig, A., and Henderson, W. A., *J. Polym. Sci.*, A1, **13,** 993 (1975).
4012. Zwolenik, J. J., *J. Phys. Chem.*, **71,** 2464 (1967).
4013. *Chlorophyll Organization and Energy Transfer*, CIBA Foundation Symp., Excerpta Medica, Amsterdam, 1979.

Index